Biodesulfurization in Petroleum Refining

Scrivener Publishing
100 Cummings Center, Suite 541J
Beverly, MA 01915-6106

Publishers at Scrivener
Martin Scrivener (martin@scrivenerpublishing.com)
Phillip Carmical (pcarmical@scrivenerpublishing.com)

Biodesulfurization in Petroleum Refining

Nour Shafik El-Gendy
Hussein Nabil Nassar

Scrivener Publishing

WILEY

This edition first published 2018 by John Wiley & Sons, Inc., 111 River Street, Hoboken, NJ 07030, USA and Scrivener Publishing LLC, 100 Cummings Center, Suite 541J, Beverly, MA 01915, USA

For more information about Scrivener publications please visit www.scrivenerpublishing.com.

Wiley Global Headquarters
111 River Street, Hoboken, NJ 07030, USA

For details of our global editorial offices, customer services, and more information about Wiley products visit us at www.wiley.com.

Library of Congress Cataloging-in-Publication Data

ISBN 978-1-119-22358-0

Cover image: Oil Refinery at Dusk, Oleg Yermolov | Dreamstime.com
Cover design by Kris Hackerott

Set in size of 11pt and Minion Pro by Exeter Premedia Services Private Ltd., Chennai, India

10 9 8 7 6 5 4 3 2 1

Contents

Preface

Biotechnology is now accepted as an attractive means of improving the efficiency of many industrial processes and resolving serious environmental problems. One of the reasons for this is the extraordinary metabolic capability that exists within the bacterial world. Microbial enzymes are capable of biotransforming a wide range of compounds and the increasing worldwide attention paid to this concept can be attributed to several factors, including the presence of a wide variety of catabolic enzymes and the ability of many microbial enzymes to transform a broad range of unnatural compounds (xenobiotics), as well as natural compounds. Biotransformation processes have several advantages compared to chemical processes, such as: (i) Microbial enzyme reactions often being more selective; (ii) Biotransformation processes often being more energy-efficient; (iii) Microbial enzymes being active under mild conditions; and (iv) Microbial enzymes being environmentally friendly biocatalysts. Although many biotransformation processes have been described, only a few of these have been used as part of the industrial process. Many opportunities remain in this area.

Biotechnology has been successfully applied at the industrial level in the medical, fine chemical, agricultural, and food sectors. Petroleum biotechnology is based on biotransformation processes. Petroleum microbiology research is advancing on many fronts, spurred on most recently by new knowledge of cellular structure and function gained through molecular and protein engineering techniques, combined with more conventional microbial methods. Several applications of biotechnology in the oil and energy industry are becoming foreseen. Current applied research on petroleum microbiology encompasses oil spill remediation, fermenter- and wetland-based hydrocarbon treatment, bio-filtration of volatile hydrocarbons, enhanced oil recovery, oil and fuel biorefining, fine-chemical production, and microbial community based site assessment. The production of biofuels in large volumes is now a reality, although there are some concerns about the use of land, water, and crops to produce fuels. These come from the biofuels produced by agroindustrial wastes, lignocellulosic wastes, waste oils, and micro- and macro-algae. In the oil industry, biotechnology has found its place in bioremediation and microbial enhanced oil recovery (MEOR). There are other opportunities in the processing (biorefining) and upgrading (bio-upgrading) of problematic oil fractions and heavy crude oils. In the context of increasing energy demand, conventional oil depletion, climate change, and increased environmental regulations on atmospheric emissions, biorefining is a possible alternative to some of the current oil-refining processes. The major potential applications of biorefining are biodesulfurization, biodenitrogenation, biodemetallization, and biotransformation of heavy crude oils into lighter crude oils, i.e., upgrading heavy oils (degradation of asphaltenes and removal of metals). The most advanced area is biodesulfurization, for which pilot plants exist and the results obtained for biodesulfurization may be generally applicable to other areas of biorefining.

This book reviews the worldwide status of current regulations regarding fuel properties that have environmental impacts, such as sulfur and nitrogen content, cetane number, and aromatic content, summarizes the cumulative, and highlights the recent scientific and technological advances in different desulfurization techniques, including: physical (for example, adsorptive desulfurization ADS), chemical (for example, hydrodesulfurization HDS, oxidative desulfurization ODS), and biological (for example, bio-adsorptive desulfurization BADS, aerobic and anaerobic biodesulfurization BDS, and biocatalytic oxidation as alternative to BODS) techniques. It will also cover denitrogenation processes (physical, chemical and biological ones). Since basic nitrogen compounds inactivate HDS catalysts and non-basic compounds can be converted to basic ones during the refining/catalytic cracking process, they are also potential inhibitors of the HDS process. So, denitrogenation is advantageous both from an environmental point of view (reduction of NOx emissions) and from an operational point of view (to avoid catalyst deactivation, corrosion of refinery equipment, and chemical instability of refined petroleum).

The advantages and limitations of each technique are discussed. The application of molecular biology and the possibility of integration of bio-nano-technology in oil production plants, future oil refineries and bioprocessing of oil, for the production of ultra-low sulfur fuels are also summarized in this book. Challenges and future perspectives of BDS in the petroleum industry and their applications for detoxification of chemical warfare agents, or the production of other valuable products, such as:

surfactants, antibiotics, polythioesters, and various specialty chemicals are also covered in this book.

Dr. Nour Sh. El-Gendy
Professor of Petroleum and
Environmental Biotechnology

Dr. Hussein N. Nassar
Researcher of Petroleum and
Environmental Biotechnology

October 2017

1

Background

List of Abbreviations and Nomenclature

4,6-DMDBT	4,6-Dimethydibenzothiophene
4-MDBT	4-Methyldibenzothiophene
API	American Petroleum Institute
BT	Benzothiophene
BTEX	Benzene, Toluene, Ethylbenzene and Xylene
Cu	Cupper
DBT	Dibenzothiophene
EEB	European Environmental Burean
FCC	Fluid Catalytic Cracking
FSU	Former Soviet Union
HCR	Hydrocracking
HDS	Hydrodesulfurization
ICCT	International Council on Clean Transportation
IEA	International Energy Agency
LPG	Liquid Petroleum Gas
Ni	Nickel
NO_x	Nitrogen Oxides

NSO	Nitrogen, Sulfur and Oxygen
OPEC	Organization of the Petroleum Exporting Countries
PAHs	Polyaromatic Hydrocarbons
PASH	Polycyclic Aromatic Sulfur Heterocycles
Ph	Phytane
PM	Particulate Matters
ppm	Parts Per Million
Pr	Pristane
SO_x	Sulfur Oxides
Th	Thiophene
TLV	Threshold Limit Value
UE	Union European
ULS	Ultra-Low Sulfur
USA	United States of America
US-EPA	United States Environmental Protection Agency
V	Vanadium
WTI	West Texas Intermediate

1.1 Petroleum

Nowadays, although the percentage of energy obtained from fossil fuels has decreased, over 83% of the world's energy is still from fossil fuels, approximately half of which comes from crude oil (OPEC, 2013). Crude oil or petroleum (Black Gold) was formed under the surface of the earth millions of years ago. It is the most important renewable energy source. The largest growth in demand is from developing countries, but the largest consumers of oil are industrial nations. The OPEC has forecasted the demand for crude oil for a long-term period from 2010 to 2035, with an increasing capacity of 20 Mb/d, reaching 107.3 Mb/d by 2035 (Duissenov, 2013). It is estimated that the world consumes about 95 million barrels/per day (i.e. 5.54 trillion barrels/day) in many applications: industry, heating, transportation, generating electricity, and production of chemical reagents that can be used in making synthetics, polymers, plastics, pharmaceuticals, solvents, dyes, synthetic detergents and fabrics, fertilizers, pesticides, lubricants, waxes, tires, tars and asphalts, and many other products (Varjani, 2017). In a typical barrel, approximately 84% of the hydrocarbons present in petroleum are converted into energy-rich fuels (i.e. petroleum-based fuels); including gasoline, diesel, jet, heating, and other fuel oils, and liquefied petroleum gas. Constituents of crude oil are resulted from aerobic and anaerobic enzymatic degradation of organic matter under suitable conditions of temperature and pressure. Crude oils vary widely in appearance and viscosity from field to field. They range in color, odor, and in the properties they contain according to their origin and geographical place. Although all crude oils are essentially hydrocarbons that occur in the sedimentary rock in the forms of natural gas, liquid, semisolid (i.e. bitumen) or solid (i.e. wax or asphaltene), they differ in properties and in molecular structure (Berger and Anderson, 1981). It has been reported that the largest estimated crude oil reserves are in Canada, Iran, and Kazakhstan and approximately 56% of the world's oil reserves are in the Middle East. Thus, according to the regional basis, the Middle East accounts for nearly 48% of the world's reserves. Central and South America are the second with approximately 20%, with Brazil and Venezuela leading, and North America is the third with approximately 13%. Table 1.1 summarizes the world wide petroleum reserves as reported by Duissenov in 2013. However, there is a depletion of the high quality low sulfur content light crude oil coming with the increment of the production and use of high sulfur content heavy crude oil (Montiel *et al.*, 2009; Srivastava, 2012; Alves *et al.*, 2015). In the near future, with a harsh worldwide increase in energy demand, the petroleum industry will have to face the fact that sour crude oil and natural gas with high sulfur content is the only energy source of choice. For example, the sulfur content of crude oil input to refineries in USA was 0.88% in 1985, while it reached to 1.44% by 2013 (EIA, 2013).

The word of petroleum is derived from the Latin words "*petra*" and "*elaion*" (*petraoleum*) which mean rock and oil, respectively (Varjani, 2017). It is formed when large quantities of dead organisms, usually zooplankton and algae, are buried underneath sedimentary rock and subjected to both intense heat and pressure. It is a sticky, thick, flammable, yellow to black viscous mixture of gas, liquid, and solid hydrocarbons (Vieira *et al.*, 2007). It is also believed to be formed from the decomposition of animal and plants, where heat and geological pressure transform this organic matter into oil and gas during the geologic periods. Crude oil can exist either deep down in the earth's surface (onshore) or deep below the ocean beds (offshore).

Crude oils are roughly classified into three groups according to the nature of the hydrocarbons they contain: Paraffin–Base Crude Oils, Asphaltic–Base Crude Oils, and Mixed–Base Crude Oils (Varjani, 2014).

Crude oils are liquid, but may contain gaseous or solid compounds or both in solution. Crude oil varies considerably in its physical properties; the majority of crude oils are dark in color, but there are exceptions. There are also differences in odor. Many oils, such as those of Iran, Iraq, and Arabia have a strong odor of hydrogen sulfide and other sulfur compounds. There are, however, several kinds of crude oil which contain little sulfur and have unpleasant odor. This variation in properties is due to the differences in

Table 1.1 The Estimated Proven Reserve Holders as of January 2013 (Duissenov, 2013).

Country	Proven reserves (billions of barrels)	Share of total
Venezuela	297.6	18.2%
Saudi Arabia	265.4	16.2%
Canada	173.1	10.6%
Iran	154.6	9.4%
Iraq	141.4	8.6%
Kuwait	101.5	6.2%
UAE	97.8	6.0%
Russia	80.0	5.0%
Libya	48.0	2.9%
Nigeria	37.2	2.3%
Kazakhstan	30.0	1.8%
China	25.6	1.6%
Qatar	25.4	1.6%
United States	20.7	1.3%
Brazil	13.2	0.8%
Algeria	12.2	0.7%
Angola	10.5	0.6%
Mexico	10.3	0.6%
Ecuador	8.2	0.5%
Azerbaijan	7.0	0.4%
Oman	5.5	0.3%
India	5.48	0.3%
Norway	5.37	0.3%
World total	1,637.9	100
Total OPEC	1,204.7	73.6

composition and these differences greatly affect the methods of refining and the products obtained from it (El-Gendy and Speight, 2016).

A petroleum reservoir is an underground reservoir that contains hydrocarbons which can be recovered through a producing well as a reservoir fluid. In the reservoir, these fluids are usually found in contact with water in a porous media such as sandstone and sometimes limestone. Under natural conditions, the fluids are lighter than water; they always stay above the water level and migrate upward through the porous rocks until they are blocked by nonporous rock such as shale or dense limestone. Where petroleum deposits came to be trapped can be caused by geologic features such as folding, faulting, and erosion of the Earth's crust. In most oil fields, oil and natural gas occur together, gas being the top layer on top of crude oil under which water is found.

The oil industry classifies "crude" by the location of its origin and by its relative weight or viscosity ("light", "intermediate", or "heavy"). Light oils can contain up to 97% hydrocarbons, while heavier oils and bitumens might contain only 50% hydrocarbons and larger quantities of other elements. The sulfur content and the American Petroleum Institute (API) gravity are the two properties that determine the quality and value of the crude oil. The API is a trade association for businesses in the oil and natural gas industries. API gravity is a measure of the density of petroleum liquid compared to water. The petroleum can be classified as light (API > 31.1), medium (API 22.3–31.1), heavy (API < 22.3), and extra heavy (API < 10.0). But, in general, if the API gravity is greater than 10, it is considered "light," and floats on top of water. While if the API gravity is less than 10, it can be considered "heavy," and sinks in water (El-Gendy and Speight, 2016).

In most of the international standards, the sulfur content is expressed in ppm*w* S or mgS/ kg (Al-Degs *et al.*, 2016). The relative content of sulfur in natural oil deposits also results in referring to oil as "sweet", which means it contains relatively little sulfur (<0.5% S), or as "sour", which means it contains substantial amounts of sulfur (>0.5% S) (El-Gendy and Speight, 2016). Sweet oil is usually much more valuable than sour because it does not require as much refining and is less harmful to the environment. Light oils are preferred because they have a higher yield of hydrocarbons. Heavier oils have greater concentrations of metals and sulfur and require more refining. Petroleum geochemists are using aromatic sulfur compounds as a maturity parameter. The immature oils are characterized by a relatively high abundance of thermally unstable non-thiophenic sulfur compounds, while mature oils are marked by a relatively high concentration of the more stable benzo- and dibenzo-thiophenes (Wang and Fingas, 1995). Table 1.2 summarizes the classification of crude oil according to the sulfur content and API gravity (Duissenov, 2013). The major locations for sweet crude are the Appalachian Basin in Eastern North America, Western Texas, the Bakken Formation of North Dakota and Saskatchewan, the North Sea of Europe, North Africa, Australia, and the Far East including Indonesia.

Table 1.2 Crude oil classifications according to S-content and API-gravity (Duissenov, 2013).

Crude oil class	API-gravity	Sulfur content (wt.%)
Ultra-light	>50	<0.1
Light and sweet	35–50	<0.5
Light and medium sour	35–50	0.5–1
Light and sour	35–50	>1
Medium and sweet	26–35	<0.5
Medium and medium sour	26–35	0.5–1
Medium and sour	26–35	>1
Heavy and sweet	10–26	<0.5
Heavy and medium sour	10–26	0.5–1
Heavy and sour	10–26	>1

While those for sour ones are North America (Alberta, Canada, the United States' portion of the Gulf of Mexico, and Mexico), South America (Venezuela, Colombia, and Ecuador), and the Middle East (Saudi Arabia, Iraq, Kuwait, Iran, Syria, and Egypt) (Duissenov, 2013).

Oil is drilled all over the world. However, there are three primary sources of crude oil that set reference points for ranking and pricing other oil supplies:

Brent Crude is a mixture that comes from 15 different oil fields between Scotland and Norway in the North Sea. These fields supply oil to most of Europe.

West Texas Intermediate (WTI) is a lighter oil that is produced mostly in the U.S. state of Texas. It is "sweet" and "light" and considered very high quality. WTI supplies much of North America with oil.

Dubai crude, also known as Fateh or Dubai-Oman crude, is a light, sour oil that is produced in Dubai, part of the United Arab Emirates. The nearby country of Oman has recently begun producing oil. Dubai and Oman crudes are used as a reference point for pricing Persian Gulf oils that are mostly exported to Asia.

The OPEC Reference Basket is another important oil source. OPEC is the Organization of Petroleum Exporting Countries. The OPEC Reference Basket is the average price of petroleum from OPEC's twelve member countries: Algeria, Angola, Ecuador, Iran, Iraq, Kuwait, Libya, Nigeria, Qatar, Saudi Arabia, the United Arab Emirates, and Venezuela.

1.2 Petroleum Composition

The chemical composition of crude oils from different producing regions, and even from within a particular formation, can vary tremendously. That depends on the location of the oil field, its age, and the depth of the oil well (Varjani, 2017). Crude oil is a complex mixture of several hundred chemical compounds, mainly hydrocarbons (mostly alkanes) of various lengths. The approximate length range is C_5H_{12} to $C_{18}H_{38}$. Any shorter hydrocarbons are considered natural gas or natural gas liquids, while long-chain hydrocarbons are more viscous, and the longest chains are paraffin wax. Generally, petroleum consists of four main fractions; saturates, aromatics, resins and asphaltenes (SARA) (Varjani, 2014). The composition percentage within the petroleum is dependent on its type. However, they are typically present in petroleum at the following percentages: paraffins (15% to 60%), napthenes (30% to 60%), aromatics (3% to 30%), with asphaltics making up the remainder. But, in light oils, asphaltenes and resins constitute 1 to 5%. While in heavy oils, they may constitute up to 20% (Radwan, 2008). Light and less dense petroleum is more profitable as a fuel source due to its higher percentage of hydrocarbons. Since heavy and denser petroleum with high sulfur content is expensive to refine, it increases the price of gasoline (petro), diesel oil, and other important petroleum distillates. Heavy oil is characterized with high proportions of carbon and NSO and lower hydrogen and overall low quality. The world's reserves of light petroleum (light crude oil) are severely depleted and refineries are forced to refine and process more and more heavy crude oil and bitumen (El-Gendy and Speight, 2016).

Riazi (2005) has reported that crude oil composition can be summarized as follows:

- PONA (paraffins, olefins, naphthenes, and aromatics)
- PINA (paraffins, iso-paraffins, naphthenes, and aromatics)
- PNA (paraffins, naphthenes, and aromatics)
- PIONA (paraffins, iso-paraffins, olefins, naphthenes and aromatics)
- SARA (saturates, aromatics, resins, and asphaltenes)
- Based on elemental analysis (C, H, S, N, and O)

Most of the petroleum fractions are free of olefins, thus the petroleum composition can be expressed in terms of its PINA composition only. For light oils, the paraffin and iso-paraffin contents can be combined and the petroleum composition can be expressed as PNA. But, for heavy oils which are characterized by high concentrations of aromatics, resins, and asphaltenes, the petroleum composition is expressed as SARA. Elemental analysis is very important, as it gives an indication of the hydrogen and sulfur contents as well as the C/H ratio, which is diagnostic for the quality of petroleum and its products.

Hydrocarbons are the most abundant compounds in crude oils accounting for 50–98% of the total composition (Clarck and Brown, 1977). While carbon (80–87%) and hydrogen (10–15%) are the main elements in petroleum, sulfur (0.05–6%), nitrogen (0.1–2%), and oxygen (0.05–1.5%) are important minor constituents present as elemental sulfur or as heterocyclic constituents and functional groups (Chandra *et al.*, 2013). Compounds containing N, S, and/or O as constituents are often collectively referred to as NSO compounds. Crude oils contain widely varying concentrations of trace metals such as V, Ni, Fe, Al, Na, Ca, Cu, and U (Ni plus V, <1000 ppm w/w and Fe plus Cu, <200 ppm w/w) (El-Gendy and Speight, 2016).

There are known contaminates in petroleum, which are sulfur, vanadium, iron, and zinc. Sour crude oil needs a more expensive and longer refining process, thus higher price fuel distillates and petroleum products. The foul-smelling gas, or sewer gas, mainly comes from hydrogen sulfide, which results from the decay of organic matter. All vanadium compounds are considered toxic. The V is an oxidant; it is one of the main components of diesel fuel. It causes high temperature corrosion and contributes to the corrosion of oil transport pipelines, ships, and tanker trucks. Such corrosion would cause the petroleum to be contaminated with iron which can lead to sludge build-up in pumps, refinery exchangers, and other fuel delivery systems. Moreover, V can react with other contaminates, such as sodium and sulfur, producing vandates salts which increase the corrosion of steel. If the V concentration is >2 ppm in fuels, it would lead to severe corrosion in turbine blades and deteriorate the refractories in furnaces. Moreover, heavy metals like Ni, V, and Cu severely affect the catalytic activities of refining processes and decrease the production of valuable products. Zinc is another type of contaminant which never occurs as a natural component of oil. It generally comes from the recycling of lubricating oils and interferes with the removal of salts from petroleum. This increases the salt levels which would, consequently, increase the corrosion of refinery systems, engine parts, etc.

1.2.1 Petroleum Hydrocarbons

Petroleum is a complex mixture of different identifiable hydrocarbons. Hydrocarbons are organic compounds that contain only carbon and hydrogen (C_xH_y). Petroleum hydrocarbons consist of aliphatic (paraffins, olefins, and naphthenes) and aromatic compounds containing at least one benzene ring.

The alkanes, or aliphatic hydrocarbons, are a part of saturate fractions which represent the main constituent in crude oil. They consists of fully saturated normal alkanes (*n*-alkanes, also called paraffins) and branched alkanes of the general formula (C_nH_{2n+2}), with n ranging from 1 to usually around 40, although compounds with 60 carbons have been reported. Above C_{13}, the most important group of branched compounds is the isoprenoid hydrocarbons consisting of isoprene building blocks. Pristane (Pr, isomer of C_{19}) and phytane (Ph, isomer of C_{20}) are usually the most abundant isoprenoids and, while the C_{10}–C_{20} isoprenoids are often major petroleum constituents, extended series of isoprenoids (C_{20}–C_{40}) have been reported (Albaiges and Albrecht, 1979). The ratio of Pr to Ph is usually used to give information about the redox conditions at the time of sedimentation of the biogenic material. Moreover, this ratio can help on identification of the origin of oil in a given area (Peters *et al.*, 2005).

Pristane
2,6,10,14-Tetramethylpentadecane

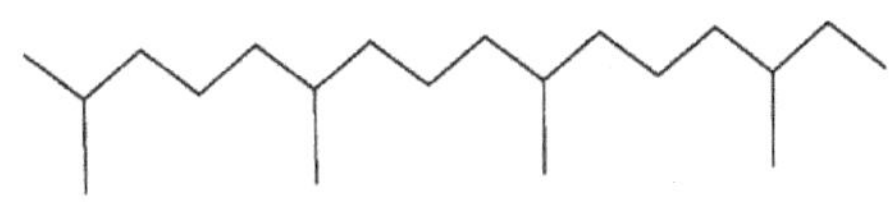

Phytane
2,6,10,14-Tetramethylhexadecane

For fuel purposes only, the alkanes from the following groups will be used: pentane and octane will be refined into gasoline, hexadecane and nonane will be refined into kerosene or diesel or used as a component in the production of jet fuel, and hexadecane will be refined into fuel oil or heating oil. Alkanes with less than five carbon atoms form natural petroleum gas and will either be burned away or harvested and sold under pressure as liquid petroleum gas (LPG). Hydrocarbons longer than 10 carbon atoms in length are generally broken down through the process known as "cracking" to yield molecules with lengths of 10 atoms or less. Alkanes with a carbon number >17 are considered paraffinic waxes and are the main cause for the increase in cloud and pour points (Maldonado *et al.*, 2006). In waxy crudes, alkanes can reach up to 60%, while in low-paraffinic oils the content is recorded at about 1.2% (Duissenov, 2013).

The olefins are the unsaturated non-cyclic hydrocarbons that have at least one double bond between the carbon-carbon atoms. The monoolefins have the general formula of C_nH_{2n}. When there are two double bonds, it is called diolefin or diene. Those hydrocarbons that have one or more double bonds between carbon atoms are called alkenes. Those with one or more triple bonds between carbon atoms are called alkynes. There are two types of isomerization in olefins: the structural isomers, which is related to the location of the double bond, and the geometric isomerism, which is related to the way the atoms are oriented in the space. The configurations are differentiated by the prefixes; *cis*- and *trans*-. Olefins and compounds with triple bonds (e.g. *cis*- and *trans*-2-butene, butadiene CH_2=CH–CH=CH_2, acetylene CH–CH, etc.) are not commonly found in petroleum because of their tendency to become saturated with hydrogen, but they are formed during the cracking reactions in the refining process. Olefins are valuable products in refineries as they are the precursors of polymers such as polyethylene.

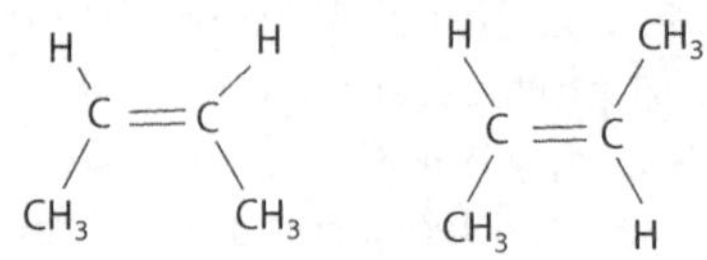

cis-and trans-2-butene

Many of the cycloalkanes or saturated ring structures, also called cycloparaffins or naphthenes, have the general formula C_nH_{2n}. These hydrocarbons display almost identical properties to paraffins, but have a much higher point of combustion. The content of naphthenes in crude oil can reach to 60%. However, in naphthenic crude oils it can reach to 80% (Duissenov, 2013). Saturated multi-rings attached to each other are called polycycloparaffins or polynaphthenes and are mainly found in heavy oils. Naphthenes that consist of important minor constituents like that of the isoprenoids which have specific animal or plant precursors (e.g., steranes, diterpanes, triterpanes, hopanes) serve as important molecular markers in oil spill and geochemical studies, as they are resistant to weathering. Moreover, they are used as internal preserved standard compounds in investigation of oil weathering (Albaiges and Albrecht, 1979; Daling *et al.*, 2002; El-Gendy *et al.*, 2014).

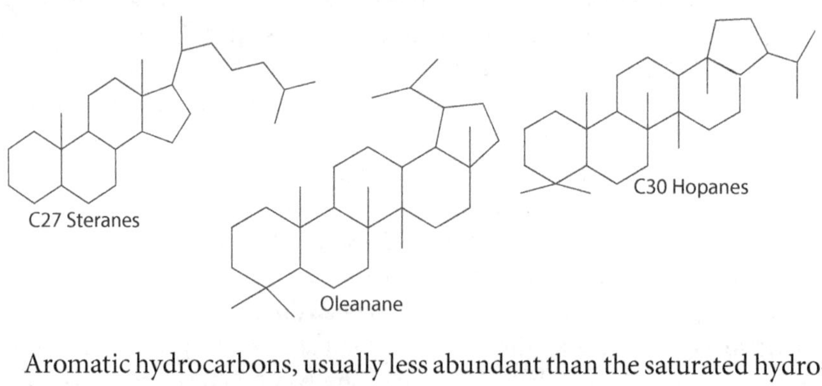

Aromatic hydrocarbons, usually less abundant than the saturated hydrocarbons, contain one or more aromatic (benzene) rings connected as fused rings (e.g., naphthalene) or lined rings (e.g., biphenyl). Some of the common aromatics that can be found in petroleum are benzene and its derivatives and methyl-, ethyl-, propyl-, or higher alkyl groups (i.e. alkylbenzens, with general formula of C_nH_{2n-6}, where $n \geq 6$). Monoaromatics consist of BTEX (the collective name for the benzene, toluene, ethylbenzene

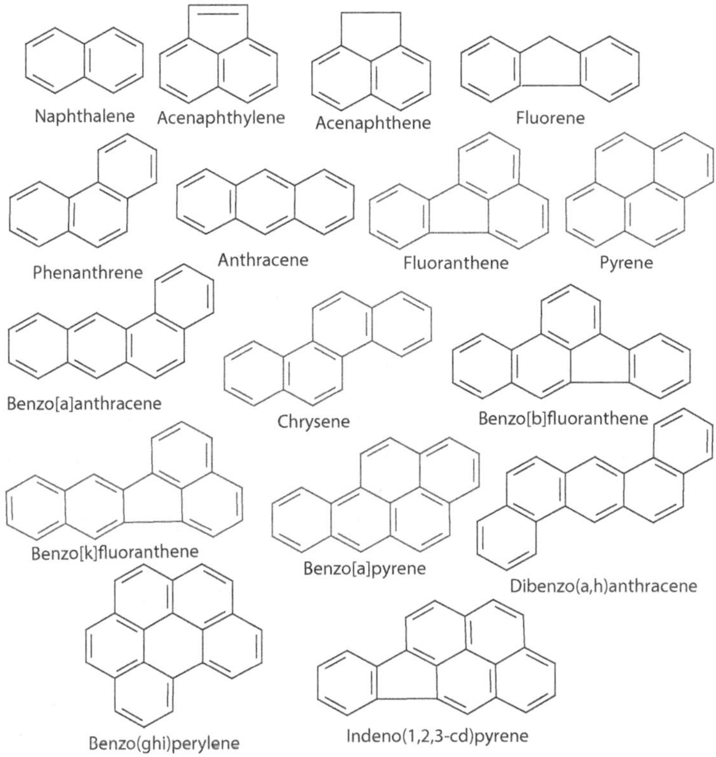

Figure 1.1 Molecular Structure of the 16-PAHs Selected by the US-EPA as Priority Pollutants.

and xylene) and mainly contribute to the octane number in gasoline. Polyaromatic hydrocarbons (PAHs) are compounds that consist of two or more aromatic rings. The United States Environmental Protection Agency (US-EPA) has selected 16 PAHs (Figure 1.1) as representative models for the toxic, mutagenic, and carcinogenic PAHs (Arun *et al.*, 2011). The PAHs up to three aromatic rings are considered low in molecular weight or light PAHs, while PAHs made up of four aromatic rings or higher are considered high in molecular weight or heavy PAHs (Wilkes *et al.*, 2016). The content of aromatics normally ranges from 15 to 20%. Usually, heavy crude oils contain more aromatics than the light ones. While in aromatic-base crude oil, aromatics content can reach to approximately 35% (Duissenov, 2013). Petroleum contains many homologous series of aromatic hydrocarbons consisting of unsubstituted or parent aromatic structures (e.g., phenanthrene) and like structures with alkyl side chains that replace hydrogen atoms. Alkyl substitution is most prevalent in 1-, 2-, and 3-ringed aromatics, although the higher polynuclear aromatic compounds (>3 rings) do contain alkylated (1–3 carbons) side groups. The polycyclic aromatics with more than 3 rings consist mainly of pyrene, chrysene, benzanthracene, benzopyrene, benzofluorene, benzofluoranthene and perylene structures. The naphthenoaromatic compounds consist of mixed structures of aromatic and saturated cyclic rings. This series increases in importance in the higher boiling fractions along with the saturated naphthenic series. The naphthenoaromatics appear related to resins, kerogen, and sterols.

Petroleum generation usually involves the formation of some naphthenoaromatic structures. The resins and asphaltenes fractions differ in the proportion of aromatic carbon (Speight, 2004).

1.2.2 Petroleum Non-Hydrocarbons

Petroleum non-hydrocarbons occur in crude oils and petroleum products in small quantity, but some of them have considerable influence on product quality. They can be grouped into six classes: sulfur compounds, nitrogen compounds, oxygen compounds, porphyrins, asphalthenes, and trace metals. Nitrogen is present in all crude oils in compounds as pyridines, quinolines, benzoquinolines, acridines, pyrroles, indoles, carbazoles, and benzocarbazoles (Clarck and Brown, 1977; Hunt, 1979; Tissot and Welte, 1984).

The porphyrins are nitrogen-containing compounds derived from chlorophyll and consist of four linked pyrroles rings. Porphyrins occur as organometallic complexes of vanadium (V) and nickel (Ni); V and Ni are the most abundant metallic constituents of crude oil, sometimes reaching thousands of part per million. They are present in porphyrins complexes and other organic compounds (Yen, 1975). Oxygen compounds in crude oils (0 to 2%) are found primarily in distillation fractions above 400°C and consist of phenols, carboxylic acids, ketones, esters, lactones, and ethers. Generally, organometallic compounds are precipitated with asphaltene and resin. Sulfur compounds comprise the most important group of nonhydrocarbon constituents. In many cases, they have harmful effects and must be removed or converted to less harmful compounds during refining process.

The heteroatom contaminants must be removed before the distillate fraction is further upgraded because they poison the hydrocracking (HCR), catalytic reforming, and fluid catalytic cracking (FCC) catalysts used in subsequent downstream refining processes (Burns *et al.*, 2008).

The resins and asphaltenes contain non-hydrocarbon polar compounds, with very complex carbon structure, but the resins have lower molecular weight and are soluble in n-alkanes. The resins are amorphous solids that are completely dissolved in petroleum (Speight, 2004). However, asphaltenes are colloidally dispersed in saturates and aromatic fractions (Speight, 2007). The resins act as peptizing agents, acting as surface-active molecules, keeping the asphaltenes in suspension and promoting the stability of the crude oil (Speight, 2004). Petroleum asphaltenes are considered the highest molecular weight compounds in petroleum (ranging from 600 to 3×10^5 and from 1000 to 2×10^6) (Speight and Moschopedis, 1981; Kawanaka *et al.*, 1989). They belong to the group of heterocyclic compounds composed of hydrogen, nitrogen, sulfur, and oxygen and minor proportions of nickel and vanadium in prophyrines are included in its general structure (Murgich *et al.*, 1999; Tavassoli *et al.*, 2012). Asphaltenes are considered the most polar fractions in petroleum, which are soluble in benzene or toluene but not in *n*-alkanes such as pentane, hexane, or heptane (El-Gendy and Speight, 2016). Based on the effect of *n*-alkanes, they appear as solid, amorphous, or friable particles (Strausaz *et al.*, 1999). Asphaltenes have a great influence upon the quality of crude oil, in such a way that they can lower its price and the price of its products if their concentration is too elevated, as many of the problems associated with recovery, separation, or processing of crude oil are related to the presence of high concentration of asphaltenes (Pineda-Flores and Mesta-Howard, 2001). Asphaltene fraction is thought to be largely responsible for adverse properties of oil; its high viscosity and its emulsion-, polymer-, and coke- forming propensities (Jacobs and Filby, 1983; Damste and De Leeuw, 1990).

Asphaltenes are dark brown to black, friable, solid components of crude oils. Their specific gravity is close to one and is highly aromatic. Asphaltene molecules carry a core of stacked, flat sheets of condensed fused aromatic rings linked at their edges by chains of aliphatic and/or naphthenic-aromatic ring systems. The condensed sheets contain NSO atoms and probably vanadium and nickel complexes (Strausz *et al.*, 1999).

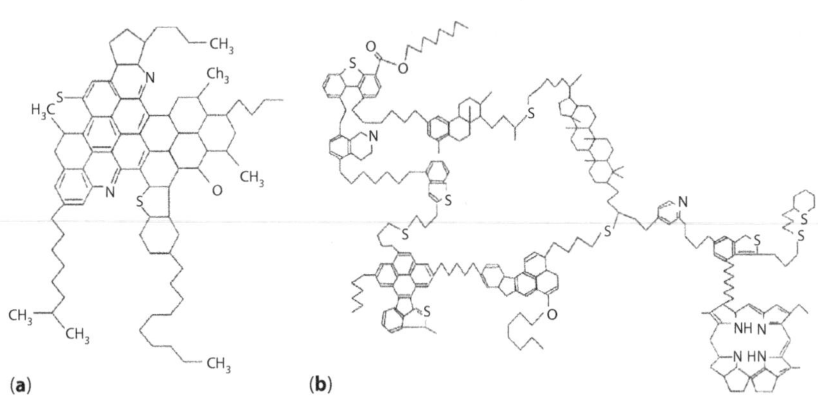

Figure 1.2 Examples of Proposed Molecular Structures of Asphaltenes. (A: Speight and Moschopedis, (1981) and B: Ancheyta *et al.*, 2009).

There are different models of asphaltene structure that coincide in showing them as a system of condensed aromatic hydrocarbons with side chains up to C_{30}, with a high proportion of heterocyclics such as benzothiophenes, dibenzothiophenes, pyrrol, pyridine, carbazole, benzocarabzole, thiophenes and sulfides, and porphyrins (Speight and Moschopedis, 1981; Murgich *et al.*, 1999; Strausz *et al.*, 1999; Pineda-Flores *et al.*, 2004; Sergun *et al.*, 2016). Figure 1.2 illustrates examples of the molecular structure of asphaltenes (Speight and Moschopedis, 1981; Ancheyta *et al.*, 2009). They are considered to be the byproduct of complex hetero-atomic aromatic macro-cyclic structures polymerized through sulfide linkages (Speight, 1970). Strausz (1988) suggested that the asphaltene molecular structure is composed of a number of polyaromatic clusters (up to 13 rings) with a few porphyrin groups bonded by heterocyclic groups and n-polymethylene bridges.

1.2.2.1 Problems Generated by Asphaltenes

It is possible to classify the problems associated to asphaltenes in five general groups: extraction, transport, processing, crude economical profit, and leaking (Pineda-Flores and Mesta-Howard, 2001).

With respect to extraction, asphaltenes have a large capability of blocking the porous spaces of the deposit, provoking a reduction of the permeability, and remarkably diminishing the crude's exit flux (Wu *et al.*, 2000).

When transporting petroleum through pipelines and metallic equipment in general, these compounds might be precipitated by the presence of ferric ions combined with acidic conditions, thus provoking the formation of a solid known as "asphaltenic mud" which deposits in conducts, blocking them and obstructing the free flow of crude. When this kind of mud develops, solvents, such as toluene and xylene, are applied in order to dissolve them. This process increases production costs and generates residues of a high degree of toxicity (Kaminski *et al.*, 2000).

As far as processing implies, asphaltenes affect petroleum refining, causing sulfur elimination to be decreased by developing a catalytic deactivation of the process through the formation of asphaltenic mud, which causes a general limitation in the maximal conversion of less-sulfured petroleum. Asphaltenes have no definite melting point and, therefore, remain in solid form, thus contributing to carbon residue (Wu *et al.*, 2000).

Regarding crude exploitation and economical profits, this is dependent on chemical composition in such a way that crude with a high content of asphaltenes (18–22%) is considered "heavy" and a low quality product. Since this represents major difficulties in its extraction and refining, economical profit notably diminishes.

Environmental petroleum leaking is the most evident way by which asphaltenes and microorganisms get in touch. If we refer specifically to microorganisms, these compounds present an influence upon their distribution and activity, as they might either have a toxic effect upon microorganisms or serve them as a source of carbon and energy.

One of the gravest problems related to these compounds in the environment resides in their resistance to biodegradation by microbial metabolic activity. Due to this fact, metabolic routes involved in this process are the less known these days, although there is some evidence suggesting that some microorganisms have the potential capability of transforming asphaltenes and, in the best case, eliminating them (Atlas, 1981; Guiliano *et al.*, 2000).

Nevertheless, asphaltenes are useful and contribute to some of the important end used products such as paving asphalts, road oils, polymer modified asphalts, and roofing asphalts.

1.3 Sulfur Compounds

Sulfur in crude oil comes, generally, from the decomposition of organic matter and with the passage of time and gradual settling into strata, the sulfur segregates from crude oil in the form of hydrogen sulfide (H_2S) that appears in the associated gas, while some portion of sulfur stays with the liquid.

Another theory behind the origin of sulfur compounds is the reduction of sulfates by hydrogen by bacterial action of the type *Desulforibrio desulfuricans.* Hydrogen comes from the reservoir fluid and the sulfate ions are kept in the reservoir rock. As a result, H_2S is generated which can react with the sulfates or rock to form sulfur that remains in the composition of crude. Moreover, under the conditions of pressure, temperature, and period of formation of the reservoir, H_2S can react with hydrocarbons to yield sulfur compounds (Wauquier, 1995).

$$4H_2 + SO_4^{2-} \xrightarrow{\textit{Desulforibrio desulfuricans}} H_2S + 2OH^- + 2H_2O$$

$$3H_2S + SO_2^{2-} \longrightarrow 4S + 2OH^- + 2H_2O$$

Sulfur is typically the third most abundant element in petroleum (Chauhan *et al.*, 2015; Peng and Zhou, 2016). It can vary from 0.04% to 14% (wt.%) in crude oil according to its type, geographical source, and origin (Voloshchuk *et al.*, 2009; ICCT, 2011). For example, the Middle East's crude oil is richer in sulfur content than those of others, such as Indonesia (Fedorak and Kropp, 1998). However, generally, the sulfur content of petroleum falls in the range 1–4% wt. Petroleum having <1% wt.

Table 1.3 Petroleum-Sulfur Content in Different Countries All Over the World (El-Gendy and Speight, 2016).

Location	Sulfur content (wt.%)
Argentina	0.06–0.42
Australia	0–0.1
Canada	0.12–4.29
Cuba	7.03
Denmark	0.2–0.25
Egypt	0.04–4.19
Indonesia	0.01–0.66
Iran	0.25–3.23
Iraq	2.26–3.3
Italy	1.98–6.36
Kuwait	0.01–3.48
Libya	0.01–1.79
Mexico	0.9–3.48
Nigeria	0.04–0.26
Norway	0.03–0.67
Russia	0.08–1.93
Saudi Arabia	0.04–2.92
United Kingdom	0.05–1.24
USA	0.05–5
Venezuela	0.44–4.99

sulfur is referred to as low-sulfur petroleum and petroleum with >1% wt. is referred to as high-sulfur petroleum (El-Gendy and Speight, 2016). The major locations where sweet crude oil is found include the Appalachian Basin in Eastern North America, Western Texas, the Bakken Formation of North Dakota and Saskatchewan, the North Sea of Europe, North Africa, Australia, and the Far East including Indonesia, while sour crude oil is more common in the Gulf of Mexico, Mexico, South America, and Canada. Moreover, crude oil produced by OPEC Member Nations also tends to be relatively sour, with an average sulfur content of 1.77%. Table 1.3 summarizes the sulfur content in crude oils from different countries all over the world (El-Gendy and Speight, 2016). The S-content in crude oils refined in the US is known to be higher than those in the Western Europe (Song, 2003). Moreover, the S-content in Egyptian Eastern-Desert crude oils is

Table 1.4 Average Regional and Global Crude Oil Quality for 2008 (Actual) and 2030 (Projected) (The International Council of Clean Transportation ICCT, 2011)

Region	2008 (actual)		2030 (projected)	
	API gravity	Sulfur wt.%	API gravity	Sulfur wt.%
North America	31.2	1.21	28.7	1.66
Latin America	25.1	1.59	23.5	1.57
Europe	37.1	0.37	37.4	0.38
Commonwealth of Independent States	32.5	1.09	35.1	0.97
Asia-Pacific	35.4	0.16	35.7	0.16
Middle East	34.0	1.75	33.9	1.84
Africa	36.5	0.31	37.1	0.26
World Average	33.0	1.1	32.9	1.3

higher than those of the Western-Desert (El-Gendy and Speight, 2016). Table 1.4 shows the estimated crude quality, in terms of API gravity and sulfur content, in various regions of the world for 2008 (actual) and 2030 (projected) (The International Council of Clean Transportation ICCT, 2011). Most of the sulfur present in crude oil is organically bound sulfur, where more than 200 organosulfur compounds have been identified in crude oils and can be summarized in four groups: as cyclic and noncyclic compounds, such as mercaptans (R-S-H), sulfides (R-S-R′), disulfides (R-S-S-R′), sulfoxides (R-SO-R′) where R and R′ are aliphatic, or aromatic groups and thiophenes (thiophene, benzothiophene, dibenzothiophene and their derivatives) with hydrogen sulfide and elemental sulfur (Fedorak and Kropp, 1998; Ma, 2010; Bahuguna *et al.*, 2011). Thiophenes compromise from 50% to 95% of the sulfur compounds in crude oils and their fractions (Mohebali and Ball, 2016). Moreover, one, two, and three sulfur atom containing compounds reach up to 74, 11, and 11% in heavy crude oils, respectively. Mercaptans content in crude varies from 0.1 to 15% mass from total content of sulfur compounds (Ryabov, 2009).

As a general rule, the higher the density of a crude oil, the lower its API and the higher its sulfur content are. Thus, the distribution of sulfur increases along with the boiling point of the distillate fraction. The sulfur content in crude oil fractions generally increases in the sequence: saturates < aromatics < resins < asphaltenes (Monticello and Finnerty, 1985). The types of organosulfur compounds that are predominant varies with

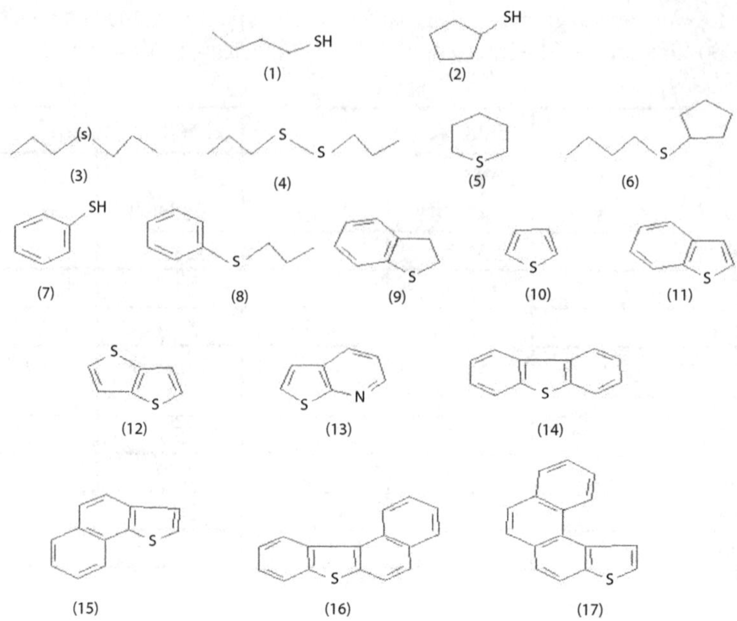

Figure 1.3 Representative Structures of the Various Types of Organosulfur Compounds Found in Petroleum. 1, alkane thiols; 2, cycloalkanes thiols; 3, dialkylsulfides; 4, polysulfides; 5, cyclic sulfides 6, alkylcycloalkylsulfides; 7, arene thiols; 8, alkyl aryl sulfides; 9, thiaindanes; 10, thiophenes 11, benzothiophenes; 12, thienothiophenes; 13, thienopyridines; 14, dibenzothiophenes 15, naphtha-thiophenes; 16, benzonaphthothiophenes; 17, phenanthrothiophenes.

the boiling range (Fedorak and Kropp, 1998). Figure 1.3 shows representative models of the various types of organosulfur compounds found in petroleum. Crude oils with higher viscosity and density usually contain high amounts of more complex sulfur compounds, although it is generally true that the sulfur content increases with the boiling point during distillation and sulfur concentration tends to increase progressively with increasing carbon number (Figure 1.4). Thus, crude fractions in the fuel oil and asphalt boiling range have higher sulfur content than those in the jet and diesel boiling range, which in turn are characterized by higher sulfur content than those in the gasoline boiling range. However, the middle distillates may actually contain more sulfur than those of the higher boiling fractions as a result of decomposition of the higher molecular weight compounds during distillation (Speight, 2000, 2006).

In fractions boiling below 150 °C, sulfur is present primarily as alkane (1) and cycloalkane (2) thiols, dialkyl (3) and alkyl cycloalkyl (6) sulfides, disulfides (4), 5- and 6- member monocyclic sulfides (5), thiophene (10), and thiophenes with one or two methyl groups. In fractions boiling

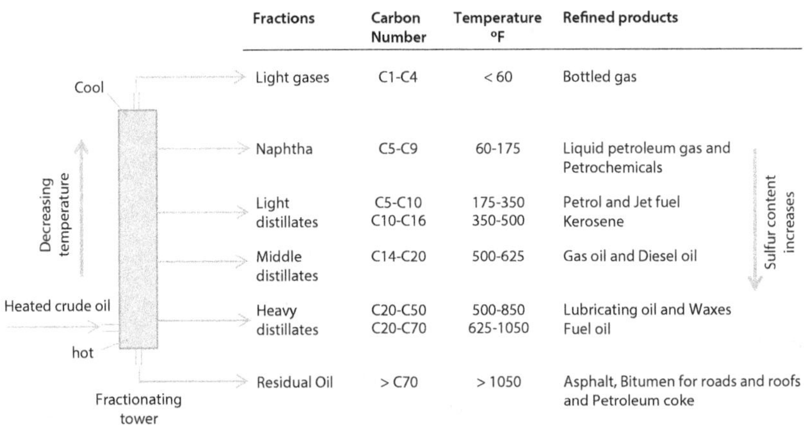

Figure 1.4 Distribution of Sulfur Content in Different Petroleum Distillates.

between 150 and 250 °C, many of these same compound types with arene thiols (7), alkyl aryl sulfides (8), polysulfides (4), and mono-, bi-, and tricyclic sulfides (5). As well as, thiaindanes (9), thiophenes (10) with up to four short side chains, and thiophenes condensed with one aromatic ring to form benzothiophenes (11), thienothiophenes (12), and thienopyridines (13) are typically major sulfur-containing constituents in this boiling range. The sulfur present in fractions boiling between 250 and 540 °C exists primarily in substituted thiophenes (10) and thiophene rings that are condensed to form substituted benzothiophene (11), dibenzothiophene (14), naphthothiophenes (15), benzonaphthothiophenes (16), and phenanthrothiophenes (17) (Figure 1.3). Many other complex compounds containing thiophene rings anellated with aromatic and naphthenoaromatic structures have been identified in this fraction. It was reported that >60% of the sulfur in higher boiling fractions of various crudes is present as substituted bezothiophenes. In some Middle East oils, the alkyl-substituted benzothiophenes and DBTs contribute up to 40% of the organic sulfur present (El-Gendy and Speight, 2016). Thus, benzothiophene and DBT and their alkylated derivatives are among the many condensed thiophenes that are an important form of organic sulfur in the heavier fractions of many crude oils, such as high boiling asphaltene or residue fractions. Condensed thiophenes are also the predominant form of sulfur identified in synthetic fuels derived from coal, oil shale, and tar sands. Sulfur compounds in the extremely high boiling fractions of petroleum (>540 °C) typically constitute approximately half of the total sulfur content of crude oils. While these compounds are the most difficult to analyze and identify, approximately 80% of the sulfur is estimated to be thiophenic in nature as part of larger complex molecules (Fedorak and Kropp, 1998).

1.4 Sulfur in Petroleum Major Refinery Products

Crude oil is rarely used in its raw form (with the notable exception when it is used for power generation), but must instead be processed into various products as liquefied petroleum gas (LPG), such as gasoline, diesel, solvents, kerosene, middle distillates, residual fuel oil, and asphalt. Refining is the processing of one complex mixture of hydrocarbons into a number of other complex mixtures of hydrocarbons. The safe and orderly processing of crude oil into flammable gases and liquids at high temperatures and pressures using vessels, equipment, and piping are subjected to stress and corrosion and requires considerable knowledge, control, and expertise. Refining process can be divided into three steps: separation (i.e. distillation), conversion, and treatment. It is carried out for two reasons: to separate hydrocarbons of different sizes to create different fuels and oils and to remove contaminants. Generally, a barrel of crude oil can produce approximately 42% gasoline, 22% diesel, 9% jet fuel, 5% fuel oil, 4% liquefied petroleum gases, and 18% other products (Riazi, 2005). In the distillation towers, petroleum is separated into fractions (i.e. distillates) according to weight and boiling point to light products (gas and gasolines), middle distillates (kerosene, automotive gas oil and heating gas oil), and heavy products (heavy fuel oil and bitumen). The lightest fractions, which include gasoline, rise to the top of the tower before they condense back to liquids. The heaviest fractions will settle at the bottom because they condense early. Light/sweet crude oil is generally more expensive and has great yields of higher value low boiling products such naphtha, gasoline, jet fuel, kerosene, and diesel fuel. Heavy, sour crude oil is generally less expensive and produces greater yields of lower value higher boiling products that must be converted into lower boiling products (El-Gendy and Speight, 2016). The conversion process is the the cahging of one hydrocarbon into another one. For example, the cracking process under high temperature and pressure of heavier, less valuable fractions of crude to lighter and more valuable products. Also, alkylation is another important conversion process, which is basically opposite to the cracking process. Then, finally, comes the highly technical, expensive, and most time consuming step of refining which is the treatment process. For example, optimizing the octane level in gasoline, desulfurization of diesel oil, ... etc. Figure 1.5 and Table 1.5 illustrate the type of common products obtained from crude oil distillation and cracking.

1.4.1 Gasoline

The most important refinery product is motor gasoline, a blend of hydrocarbons with boiling ranges from ambient temperatures to about 400 °F.

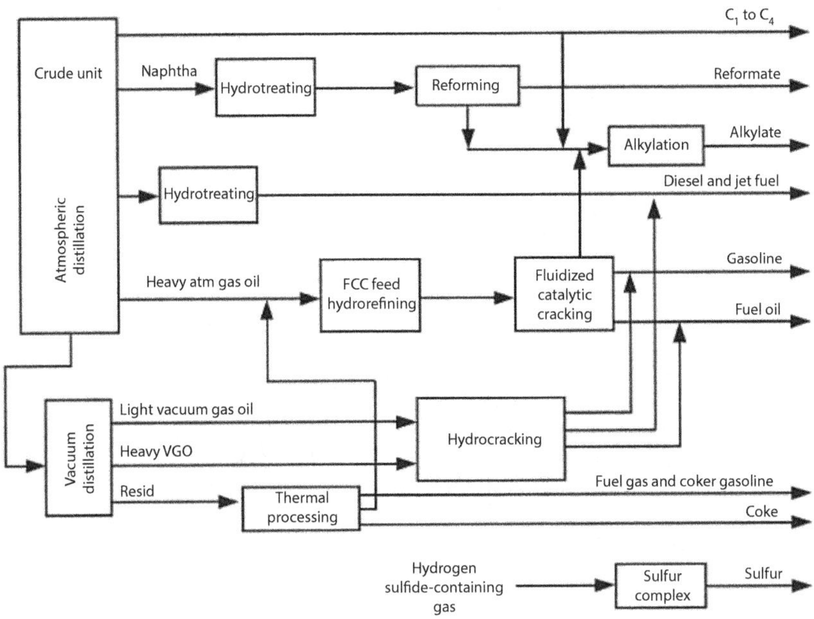

Figure 1.5 A Simplified Schematic Diagram for the Refining Process of Crude Oil (El-Gendy and Speight, 2016).

It is used as a fuel in spark-ignited internal combustion engines. The important qualities of gasoline are octane number (antiknock), volatility (starting and vapor lock), and vapor pressure (environmental control). Additives are often used to enhance performance and provide protection against oxidation and rust formation. Thiophene, benzothiophene, and their alkylated derivatives are the most common organosulfur compounds found in gasoline (Mc Farland *et al.*, 1998; Yin and Xia, 2001). The presence of sulfur in gasoline would cause corrosion in engine parts. Current gasoline desulfurization problems are dominated by the issues of sulfur removal from FCC naphtha, which contributes about 35% of gasoline pool, but over 90% of sulfur in gasoline. Deep reduction of gasoline sulfur (from 330 to 30 ppm) must be made without decreasing octane number or losing gasoline yield. The problem is complicated by high olefins contents of FCC naphtha which contribute to octane number enhancement, but can be saturated under HDS conditions (Song and Ma, 2003). Sulfur is a "poison" to FCC catalysts and many refineries have desulfurization units in front of the FCC that remove much of the sulfur from the FCC feed. However, untreated FCC products (FCC naphtha and light cycle oil) are the primary sources of sulfur in gasoline and diesel fuel. Hydrocracking has a notable advantage over FCC; the hydrogen input to the hydrocracker not only leads to

Table 1.5 Crude Oil Distillate Products (Riazi, 2005).

Distillate	Boiling range (°F)	Composition	Manufactured by	Principle uses
Gasoline	30–390	Complex mixture of materials. Contains additives to improve performance, but no polymerizable components, with very low sulfur content	Primary distillation cracking and reforming processes	Spark ignition and internal combustion engines
Kerosene	400–490	Paraffin hydrocarbons with substantial proportion of aromatics low sulfur content	Distillation, cracking	Agricultural tractors, lighting, heating, and aviation gas turbines
Diesel Oil	400–550	Saturated hydrocarbons often with high sulfur	Distillation, cracking	Diesel engines and furnace heating
Gas Oil	490–800	Saturated hydrocarbons	Distillation hydrodesulfurization	Diesel fuel, heating and furnaces. Fed to cracking units

cracking reactions but also to other reactions that remove hetero-atoms, especially sulfur, from the hydrocracked streams. These "hydrotreating" reactions yield hydrocracked streams (gasoline, jet fuel, and diesel fuel) with very low sulfur content and other improved properties. Hydrocracking is more effective in converting heavy gas oils and producing low-sulfur products than either FCC or coking, but hydrocrackers are more expensive to build and operate, in large part because of their very high hydrogen consumption. Thiophene and dimethylthiphene are the main thiophenic compounds in gasoline (Wauquier, 1995).

1.4.2 Kerosene

Kerosene is a light petroleum distillate, composed of a mixture of paraffins and naphthenes with usually less than 20% aromatics. It has a flashpoint above 38 °C and a boiling range of 160 °C to 288 °C and is used for lighting, heating, solvents, and blending into diesel fuel. It has an intermediate volatility between gasoline and heavier gas oils and can be used as fuels in some diesel engines and is often used as home heating fuel. Sulfides create the bulk of sulfur containing hydrocarbons in the middle distillates (kerosene and gas oil), where their content is equal to 50–80% of total sulfur compounds (Ryabov, 2009), while the disulfides are found in small quantities in petroleum fractions with a boiling point up to 300 °C. They account for 7–15% of the total sulfur (Ryabov, 2009). Dimethylthiphene and benzothiophes are the main thiophenic compounds in kerosene (Wauquier, 1995).

1.4.3 Jet Fuel

Jet fuel is a middle distillate kerosene product whose critical qualities are freeze point, flashpoint, and smoke point. Commercial jet fuel has a boiling range of about 191 °C to 274 °C and military jet fuel from 55 °C to 288 °C. The main sulfur compounds in jet fuel are thiols, sulfides, and traces of thiophenes.

1.4.4 Diesel Fuel

Diesel fuel is preferred in military vehicles. Diesel has higher energy content per volume than gasoline. Because the hydrocarbons in diesel are larger, it is less volatile and, therefore, less prone to explosion. Unlike gasoline engines, diesel engines do not rely upon electrically generated sparks to ignite the fuel. Diesel is compressed to high degree, along with air, creating high temperatures within the cylinder that lead to combustion. This process makes diesel engines highly efficient, achieving up to 40% better fuel economy than gasoline powered vehicles. Diesel fuel can be categorized according to its cetane number. No.1 diesel (super-diesel) has a cetane number of 45 and is used in high speed engines, trucks, and buses. No. 2 diesel has a 40 cetane number. Railroad diesel fuels are similar to the heavier automotive diesel fuels, but have higher boiling ranges up to 400 °C (750 °F) and a lower cetane number of 30.

The desirable qualities required for diesel fuel include controlled flash and pour points, clean burning, no deposit formation in storage tanks, and a proper diesel fuel cetane rating for good starting and combustion. In fractions used to produce diesel oil, most of the sulfur is found in BT, DBT, and their alkylated derivatives (Monticello and Finnerty, 1985). Deep reduction of diesel sulfur (from 500 to <15ppm sulfur) is dictated largely by 4-methyldibenzothiophene (4-MDBT) and 4,6-dimethydibenzothiophene (4,6-DMDBT) in which a sulfur atom is sterically hindered by substitutions in positions 4 and 6 is the most difficult to remove by HDS; 3,6-DMDBT has been shown to be the particularly recalcitrant to HDS (Kabe *et al.*, 1997). The deep HDS problem of diesel streams is exacerbated by inhibiting effects of the co-existing problem of polyaromatics and nitrogen compounds in the feed as well as H_2S in the product (Song and Ma, 2003).

Because of the similar distillation points of diesel and sulfur contaminants, they are produced together during the refining process. Due to the strict international regulations, additional steps are taken to remove the sulfur. That makes diesel fuel cost more than gasoline.

1.4.5 Heating/Fuel Oils

Domestic heating oils have boiling ranges of bout 400–700 °F. They consist of hydrocarbons of a chain length between eight and 21 carbon atoms. It is a general meaning that intermediate hydrocarbon liquid mixtures have lower volatility than that of kerosene. They contain a relatively high content of sulfur.

Heating/fuel oils are one of the "left-over" products of crude refining; they are less pure than other refined products, with a broader range of hydrocarbons. They are characterized with high flash points and are more prone to auto-ignition, producing more pollutants on combustion. However, No. 1 fuel oil is similar to kerosene and No. 2 fuel oil is very similar to No. 2 diesel fuel. Heavier grades of No. 3 and 4 are also available.

1.4.6 Bunker Oil

Bunker oil is one of the bottom products of petroleum refining. It is mainly used as marine fuel for shipping. It is a complex mixture of alkanes, alkenes with high content of aromatic hydrocarbons, and asphaltenes which leads to its high viscosity. It is characterized by high sulfur content, ranged between 1.5 to 4 wt.%. upon its combustion and it produces elevated concentrations of SO_x

emissions. It has been reported that the SO_2 from shipping activities reached 16.2 million tonnes by 2006 and it is expected to be raised to reach about 22.7 million tonnes by 2020 (The European Environmental Burean EEB, 2008).

1.5 Sulfur Problem

The depletion of continental crude oil deposits has forced the exploitation of deeper reservoirs containing petroleum rich in polynuclear aromatic sulfur heterocyclic (PASH) compounds and other unconventional oil reserves including heavy oil, extra heavy oil, and oil sands and bitumen, which comprise 70% of the world's total oil resources (US EIA, 2010). Also, with the rising of oil prices, the production of synthetic crude oil from unconventional resources, such as oil sands, is becoming increasingly economically viable (US EIA, 2010). Canada, Venezuela, and the US have extremely large deposits of heavy oil and oil shale. Unconventional oil reserves are characterized by large quantities of sulfur, nitrogen, nickel, and vanadium, for example, Athabasca oil sand bitumen has a sulfur content of 4.86% (Bunger *et al.*, 1979). The sulfur content in heavy crude oil varies from 0.1% to 15% (w/w) (Shang *et al.*, 2013).

Sulfur compounds are very undesirable because of their disagreeable odor, deleterious effect on color or color stability, and unfavorable influence on antiknock and oxidation characteristics. Organosulfur compounds in crude oil increase the viscosity of a large fraction making it non-amenable to the refinery process (Chauhan *et al.*, 2015). They have actual or potential corrosive nature, which would cause the corrosion of pipelines and pumping and refining equipment. The annual cost of corrosion worldwide is reported to be approximately \$3.3 trillion (Duissenov, 2013). Sulfur compounds contaminants, such as polysulfides, mercaptans, and aliphatic sulfides, can react with metal surfaces at high temperatures forming metal sulfides, organic molecules, and H_2Shydrogen sulfide. The corrosiveness of sulfur compounds in the form of uniform thinning, localized attack, or erosion-corrosion, increases with accumulating temperature, which is very dangerous, especially at high temperature and pressure since fire would occur upon sudden rupturing due to corrosion. Sulfur compounds are undesirable in refining processes as they tend to deactivate, i.e. poison the catalysts used in downstream processing and upgrading of hydrocarbons (El-Gendy and Speight, 2016). In liquid petroleum products, sulfur compounds would contribute to the formation of gummy deposits which would plug the filters of the fuel-handling systems of automobiles and other engines. Sulfur compounds in fuel oils would cause corrosion to the parts of the internal combustion engines and, consequently, their breakdown and premature failure (Collins *et al.*, 1997). Sulfur levels in automotive fuels have an unfavorable poisoning effect on the catalytic converters in automotive engines, thus increaing the particulate, CO, CO_2, SO_x, NO_x, and other combustion related emissions, that cause smog, global warming, and water pollution (Srivastava, 2012). The presence of sulfur in lubricating oils lowers the resistance to oxidation and increases the solid deposition on engine parts (Riazi *et al.*, 1999).

Consequently, refineries must have the capability to remove sulfur from crude oil and refinery streams to the extent needed to mitigate these unwanted effects, where corrosion by different sulfur compounds at temperatures between 260 and 540 °C is a general issue in petroleum refining and petrochemical processes. However, the higher the sulfur content of the crude, the greater the required degree of sulfur control and the higher the associated cost. Sulfur removal is also important for the new generation of engines, which are equipped with a nitrogen oxide (NO_x) storage catalyst, since sulfur in the fuel has poisoning effects on the catalyst (König *et al.*, 2001). The significant reduction of sulfur-induced corrosion and the slower acidification of engine lubricating oil would lead to lower maintenance costs and long maintenance intervals. These are also additional benefits of using ultralow sulfur diesel in diesel powered vehicles (Stanislaus *et al.*, 2010). It is believed that refining industries will spend about $37 billion on new desulfurization equipment and an additional $10 billion on annual operating expenses during the next decade to meet the new sulfur regulations. In addition to this opportunity in the refinery, there is also a large potential in the desulfurization of crude oil itself (Mohebali and Ball, 2008).

Emissions of sulfur compounds (e.g. SO_2 and fine particulate matter of metal sulfates) formed during the combustion of petroleum products are the subject of environmental monitoring in all developed countries. It has been reported that approximately 73% of the produced SO_2 is from anthropogenic origin and is due to the combustion of petroleum and its derivatives (*Aparicio et al., 2013).* As SO_2 is transported by air streams, it can be produced in one area and show its adverse impact in another remote place thousands of kilometers away (Soleimani *et al.*, 2007). These harmful emissions affect the stratospheric ozone, increasing the hole in the Earth's protective ozone layer (Denis, 2010). High levels of SO_2 cause bronchial irritation, trigger asthma, and prolonged exposure can cause cardio-pulmonary and lung cancer mortality (Mohebali and Ball, 2008). Furthermore, sulfur compounds have unfavorable influences on antiknock and oxidation characteristics. High concentration of sulfur in fuels dramatically decreases the efficiency and lifetime of emission gas treatment systems in cars (Mužic and Sertić-Bionda, 2103).

Incomplete combustion of fossil fuels causes the emission of aromatic sulfur and nitrogen compounds and oxidation of these compounds in the atmosphere would lead to the aerosol of sulfuric and nitric acids. For example, NO_x emission would be significantly increased by 66%, corresponding to an increase in sulfur content of gasoline from 40 to 150 ppm (US EPA, 1998). NO_x and CO_2 are thought by many to be the primary causes of "chemical smog", as well as "greenhouse gas" accumulation (Nasser, 1999). It has been reported that sulfur is the main cause of emissions of PM (Mohebali and Ball, 2016). Stanislaus *et al.* (2010), reported that approximately 2% of sulfur in diesel fuel can be directly converted to PM emissions. PM is carcinogenic and, moreover, PM and SO_x cause lung cancer and cardiopulmonary mortality. Soot and smoke both refer to particulate matter (PM) that gets trapped in gases during combustion. The visible, dark black component of smoke is carbon that has incompletely burned and, rather than forming CO_2, has formed solid carbon compounds known as amorphous carbon. These carbon compounds are collectively referred to as soot, which resulted from the use of the lower quality of fuel (i.e. a less refined one). Soot, in particular, contains large amounts of the mutagenic and carcinogenic PAHs. Diesel exhaust is considered the most carcinogenic of transportation fuels and accounts for approximately 25% of all smoke and soot in the atmosphere.

The health effects of sulfur oxides So_x include irritation in the eyes, nose, and skin and can lead to acute bronchitis and severe respiratory disorders when inhaled by humans (Grossman, 2001; Bordoloi *et al.*, 2014; Martinez *et al.*, 2015; Agarwal *et al.*, 2016; Martinez *et al.*, 2017).

Moreover, upon the emissions of SO_2 and NO_2 in the atmosphere, they react with hydrogen producing weak sulfurous acid, strong sulfuric acid, and nitric acid which are the main causes of acid rain and haziness that reduce the average temperature of an affected area (Derikvand *et al.*, 2014).

$$CH_4 + 2O_2 + N_2 \longrightarrow 2H_2O + CO_2 + NO + NO_2 + \text{Energy}$$

$$H_2O + N_2 + NO_2 \longrightarrow HNO_3 + HNO_2$$

$$3HNO_2 \longrightarrow HNO_2 + 2NO + H_2O$$

$$4NO + 3O_2 + 2H_2O \longrightarrow HNO_3$$

$$OH^{\bullet} + SO_2 \longrightarrow HSO^{\bullet}_3 + O_2$$

$$HSO^{\bullet}_3 + O_2 \longrightarrow HO^{\bullet}_2 + SO_3$$

$$SO_2 + O_2 \longrightarrow SO_3$$

$$SO_3 + H_2O \longrightarrow H_2SO_4$$

Acid rain has a deleterious effect as it causes soil pollution, destroys green area, kills forests, and damages crops, leather, cars, and buildings. It also poisons lakes and rivers leading to a devastating effect on their fauna and flora and a falling in fish population. Also, the presence of high levels of sulfate in water affects human health, as it causes diarrhea and dehydration (Grossman, 2001; Ardakani *et al.*, 2010; Mohebali and Ball, 2016; Paixao *et al.*, 2016).

H_2S present in crude oil and its low boiling distillates is one of the most corrosive sulfur compounds and would destroy paintings and buildings. Moreover, it is the main component of refinery sour waters, thus, it would cause corrosion problems in overhead systems of fractionation towers, hydrocracker and hydrotreater effluent streams, catalytic cracking units, sour water stripping units, and sulfur recovery units.

H_2S is a highly toxic gas that affects the nervous and respiratory systems causing symptoms that include headache, dizziness, and excitement. It is like cyanide in that it inhibits the cytochrome oxidase system essential for respiration (Manahan, 2003). It is reported that it may even have contributed to several mass extinctions in Earth's past (Duissenov, 2013). The personal risk arises when people are exposed to contaminated air with concentrations above the threshold limit value, T.L.V. (10 ppm). Deadly H_2S (1000–2000 ppm) can kill an operator in 10 seconds (Table 1.6). Fortunately, sulfides have a highly obnoxious smell which gives some warning of their danger.

H_2S is actually very flammable, so it could be used as a fuel if it were not for the fact that it is also deadly in relatively low concentrations. This deadly gas must be removed from petroleum in order to make it safer for use. The hydrogen sulfide can then be used to produce pure sulfur, a highly valued industrial element used in the production of fertilizer.

Light alkyl thiols, such as methanethiol, are fairly common air pollutants with odors that may be described as ultragarlic. Inhalation of even very low concentrations of the alkyl thiols in air can be very nauseating and result in headaches. Exposure to higher levels can cause increased pulse rate, cold hands and feet, and cyanosis. With extreme cases, unconsciousness, coma, and death may occur (Manahan, 2003).

Table 1.6 Toxicity of H_2S.

Low	Medium	High
Coughing	Fainting	Strangling
Burning sensation	Pallor	Rapid loss of conscious
Pain in eyes	Cramps	Loss of heart beats
Sleepiness	Breathing may stop, but may return	Loss of breathing Pupils dilated

The large-ring thiophenes and polycyclic aromatic sulfur heterocycles (PASHs) are among the most potent environmental pollutants and exhibit some mutagenic and carcinogenic activity (Murphy *et al.*, 1992; Jensen *et al.*, 2003). Observations and analyses associated with oil spills and chronic pollution have indicated that the PASHs are more resistant to microbial degradation, accumulated to a greater degree and departed more slowly than their polyaromatic hydrocarbon (PAHs) analogs (Durate *et al.*, 2001). For example, organic sulfur compounds found in crude oil usually serve as makers of oil pollution in fish and shell fish. GC/MS analysis of the extracts showed the presence of organic sulfur compounds of alkyl- (from mono- to penta-methyl) benzothiophenes, dibenzothiophene and alkyl-(from mono- to tri-methyl) dibenzothiophenes, and other organic sulfur compounds of alkyl-(from mono- to penta-methyl) naphthalenes (Ogata and Mujake, 1978 and Ogata and Fujisawa, 1983). It has been reported that DBT and its derivatives can persist for up to three years in the environment (Gundlach *et al.*, 1983).

1.6 Legislative Regulations of Sulfur Levels in Fuels

Worldwide environmental legislation has put limits for the sulfur levels in transportation fuels to decrease SO_x emissions. Consequently, refineries must have the capability to remove sulfur from crude oil and refinery streams to the extent needed to mitigate its negative impact and reach the restricted regulatory sulfur limits. Moreover, international cooperation is required to limit and control the SO_x emissions. The American Lung Association reported that low-sulfur fuel would reduce the soot amount from large trucks by approximately 90% (Riazi, 2005). Ultra-low sulfur (ULS) fuels significantly reduce total fuel cost by increasing fuel economy, since the considerable potential for reduction in greenhouse gas emissions adds further payback to health, environmental, and social benefits (Blumberg *et al.*, 2003).

Since 1979, Canada, Japan, the United States, and the European nations in particular have signed several agreements to reduce and monitor emissions of SO_2. Most of these agreements targeted transport fuel because it was one of the important sources of SO_2. For example, the US-EPA and other regulatory agencies worldwide have limited the amount of SO_2 released into the air and put strict regulations and limits for the sulfur content in transportation fuels; 15 and 30 ppm for diesel oil and gasoline, respectively (Agarwal *et al.*, 2016; Al-Degs *et al.*, 2016). Moreover, the sulfur content limits in the EU, set by Euro V norms, has reached 10 ppm since 2009 and 15 ppm for the US since 2006 to ultra-clean fuels <1 ppm in the near future, approaching zero-sulfur emissions upon fuel combustion (Calzada *et al.*, 2011; Zhang *et al.*, 2011; Martinez *et al.*, 2015; Al-Degs *et al.*, 2016). In China, the sulfur content in gasoline and diesel oil is targeted to be 10 ppm by 2016 (Yang *et al.*, 2014), while in most of the developed countries the acceptable sulfur level in gasoline has been restricted to be <10 ppm (Shang *et al.*, 2013). In India, the sulfur content in gasoline has lowered to 50 ppm since 2010 in 13 selected cities, whereas the nationwide sulfur gasoline level is 150 ppm. Several other countries around the world are moving forward with lowering the maximum fuel sulfur content. This is particularly true in the Middle East, Russia, South Africa, and some countries in Latin America. Saudi Arabia and Russia have put the limitation of 10 ppm of gasoline by 2013 and 2016, respectively. South Africa has agreed to enforce 10 ppm gasoline by 2017 (OPEC, 2011). However, a much higher level (350 ppm) is permitted for transportation fuels in Jordan and some other countries (Al-Ghouti and Al-Degs, 2014).

The European Fuel Quality Directive has required the on-road and off-road diesel fuel sulfur content to be set at 10 ppm since 2011. Same maximum level of 10 ppm was legislated in Japan, Hong Kong, Australia, New Zealand, South Korea, and Taiwan. In Canada and the US, the maximum level is 15 ppm since 2010 and 2012, respectively. China planned to reduce its on-road diesel sulfur to 350 ppm in July 2012. However, at the more regional level, Beijing has a diesel sulfur limit of 10 ppm, while the cities of Shanghai, Guangzhou, Shenzhen, Dongguan, and Nanjing have required a 50 ppm maximum since May 2012. India has also set a maximum level of 50 ppm for 20 urban centers, while its national specification is 350 ppm. Other Asian countries including Indonesia, Malaysia, the Philippines, and Thailand are putting plans together to catch the wave. The majority of Latin America has sulfur limits for diesel oil above 500 ppm. However, premium diesel in Argentina was set to 10 ppm since 2011 and Chile has a set limit of 50 ppm of diesel since 2006. In Brazil, diesel specification limits the amount of sulfur down to 10 ppm by 2013, Ecuador and Mexico are in their way to follow the sulfur level limitations. But, in Africa, the average sulfur content is in the range of 2,000 to 3,000 ppm for on-road diesel, and is much higher for off-road, while South Africa planned to reach to 10 ppm fuels by 2017. Briefly, with the exception of Africa, all regions worldwide, are projected to reach an average on-road sulfur content of <20 ppm by 2035 (OPEC, 2011). The projected sulfur content in gasoline and diesel oil for 2012–2035 are shown in Tables 1.7 and 1.8 respectively.

In developed countries, regulations have been set for heating oil, jet kerosene, and fuel oil. The EU has reduced the sulfur content distillate-based heating oil from 2,000 ppm to 1,000 ppm since 2008. Germany provides

Table 1.7 Expected Regional Gasoline Sulfur Content (OPEC, 2013).

	S-content in gasoline ppm					
Location	**2012**	**2015**	**2020**	**2025**	**2030**	**2035**
US and Canada	30	30	10	10	10	10
Latin America	520	255	130	45	30	20
Europe	13	10	10	10	10	10
Middle East	605	235	75	25	16	10
FSU	315	115	35	20	12	10
Africa	795	493	245	165	95	65
Asia-Pacific	205	130	65	35	20	15

Table 1.8 Expected Regional On-Road Diesel Sulfur Content (OPEC, 2013).

	S-content in on-road diesel ppm					
Location	**2012**	**2015**	**2020**	**2025**	**2030**	**2035**
US and Canada	15	15	15	10	10	10
Latin America	1085	440	185	40	35	20
Europe	13	10	10	10	10	10
Middle East	1725	415	155	70	20	10
FSU	440	175	60	15	10	10
Africa	3810	2035	930	420	175	95
Asia-Pacific	400	200	100	45	25	15

tax incentives for 50 ppm heating oil to enable the use of cleaner and more efficient fuel burners. Parts of North America plan to reduce sulfur levels in heating oil to 15 ppm before 2020 (Duissenov, 2013).

In industrialized regions in Europe, reductions in the sulfur content of jet fuel is aimed to reach 350 ppm by 2020, followed by other countries by 2025. Industrialized regions are also assumed to reach 50 ppm by 2025. However, the current jet fuel specifications still allow for sulfur content as high as 3,000 ppm, but there is jet fuel with 1,000 ppm sulfur content (OPEC, 2011). Marine bunker fuels are also subject to regulation. As of January 2012, the global sulfur cap was lowered from 4.5 wt.% to 3.5 wt.% and will be further lowered to 0.5 wt.% (5,000 ppm) as of January 2020. In September 2012, the European Parliament approved final legislation requiring all ships in EU waters to switch to 0.5% wt sulfur fuel or use corresponding technology allowing ships to reach the required emissions reduction by 2020 (OPEC, 2011; Tang, 2013).

References

Agarwal, M., Dikshit, P.K., Bhasarkar, J.B., Borah, A.J., Moholkar V.S. (2016) Physical insight into ultrasound-assisted biodesulfurization using free and immobilized cells of *Rhodococcus rhodochorous* MTTC3552. *Biochemical Engineering Journal*. 112: 105–111.

Albaiges, J., Albrecht, P. (1979) Fingerprinting marine pollutant hydrocarbons by omputerized gas chromatography-mass spectrometry. *International Journal of Environmental Analytical Chemistry*. 6: 171–190.

Al-Degs, Y.S., El-Sheikh, A.H., Al-Bakain, R.Z., Newman, A.P., Al-Ghouti, M.A. (2016) Conventional and upcoming sulfur-cleaning technologies for petroleum fuel: A review. *Energy Technology*. 6: 679–699.

Al-Ghouti, M.A., Al-Degs, Y.S. (2014) Manganese-loaded activated carbon for the removal of organosulfur compounds from high-sulfur diesel fuels. *Energy Technology*. 2: 802–810.

Alves, L., Paixao, S.M., Pacheco, R., Ferreira, A.F., Silva, C.M. (2015) Biodesulfurization of fossil fuels: energy, emissions and cost analysis. *RSC Advances*. 5: 34047–34057.

Ancheyta, J., Trejo, F., Rana, M.S. (2009) Asphaltenes chemical transformation during hydroprocessing of heavy oils. Founding Editor Heinemann H., Series Editor Speight J.G. CRC Press, Taylor and Francis Group, LLC, 6000 Broken Sound Parkway NW, Suite 300, Boca Raton, FL 33487–2742, USA.

Aparicio, F., Camú, E., Villarroel, M., Escalona, N., Baeza, P. (2013) Deep desulfurization by adsorption of 4,6-dimethyldibenzothiophene, study of adsorption on different transition metal oxides and supports. 58(4): 2057–2060.

Ardakani, M.R., aminsefat, A., Rasekh, B., Yazdiyan, F., Zargar, B., Zarei, M., Najafzadeh, H. (2010) Biodesulfurization of dibenzothiophene by a newly isolated, *Stenotrophomonas maltophilia* strain Kho1. *World Applied Sciences Journal*. 10(3): 272–278.

Arun, K., Ashok, M., Rajesh, S. (2011) Crude oil PAH constitution, degradation pathway and associated bioremediation microflora: An overview. *International Journal of Environmental Sciences* 1(7): 1420–1420.

Atlas, R.M. (1981) Microbial degradation of petroleum hydrocarbons: an environmental perspective. *Microbiological Reviews*. 45(1): 180–209.

Bahuguna, A., Lily, M.K., Munjal, A., Singh, R., Dangwal, K. (2011) Desulfurization of dibenzothiophene (DBT) by a novel strain *Lysinibacillus sphaericus* DMT-7 isolated from diesel contaminated soil. *Journal of Environmental Sciences*. 23: 975–982.

Berger, B.D., Anderson, K.E. (1981) Modern petroleum, A basic primer of the industry. 2nd edition, PennWell Books, Tulsa, OK, USA.

Blumberg, K.O., Walsh, M.P., Pera, Ch. (2003) Low sulfur gasoline and diesel: The key to lower vehicle emissions. Prepared for the May 2003 meeting in Napa, California, of The International Council On Clean Transportation (ICCT). http://www.theicct.org/sites/default/files/publications/Low-Sulfur_ICCT_2003.pdf

Bordoloi, N., Rai, S.K., Chaudhuri, M.K., Mukherjee, A.K. (204) Deep-desulfurization of dibenzothiophene and its derivatives present in diesel oil by a newly isolated bacterium *Achrobacter* sp. To reduce the environmental pollution from fossil fuel combustion. *Fuel Processing Technology*. 119: 236–244.

Bunger, J.W., Thomas, K.P., Dorrence, S.M. (1979) Compound types and properties of Utha and Athabasca tar sand bitumens. *Fuel*, 58:183–195.

Burns, A.W., Gaudette, A.F., Bussell, M.E. (2008) Hydrodesulfurization properties of cobalt-nickel phosphide catalysts: Ni-rich materials are highly active. *Journal of Catalysis*. 260 (2): 262–269.

Calzada, J., Alcon, A., Santos, V.E., Garcia-Ochoa, F. (2011) Mixtures of *Pseudomonas putida* CECT5279 cells of different ages: Optimization as biodesulfurization catalyst. *Process Biochemistry*. 46: 1323–1328.

Chandra, S., Sharma, R., Singh, K., Sharma, A. (2013) Application of bioremediation technology in the environment contaminated with petroleum hydrocarbons. *Annals of Microbiology*. 63:417–431.

Chauhan, A.K., Ahmad, A., Singh, S.P., Kumar, A. (2015) Biodesulfurization of benzonaphtothiophene by an isolated *Gordonia* sp. IITR100. *International Biodeterioration and Biodegradation*. 104: 105–111.

Clark, R.C., Jr., Brown, D.W. (1977) Petroleum properties and analyses in biotic and abiotic systems. In: Effects of petroleum on arctic and subarctic marine environments and organisms. Ed. by D. C. Malins. Acad. Press, New York I. pp. 1–89.

Collins, F.M., Lucy, A.R., Sharp, C.J. (1997) Oxidation desulfurization of oils via hydrogen peroxide and heteropolyanion catalysis. *Journal of Molecular Catalysis A: Chemical*. 117: 397–403.

Daling, P.S., Faksness, L.G., Hansen, A.B., Stout, S.A. (2002) Improved and standardized methodology for oil spill fingerprinting. *Environmental Forensics*. 3: 263–278.

Damste, J.S., De Leeuw, J.W. (1990) Analysis, structure and geochemical significance of organically-bound sulphur in the geosphere: state of the art and future research. *Organic Geochemistry*. 16: 1077–1101.

Denis, P.A. (2010) On the enthalpy of formation of thiophene. Theor. Chem. Acc. 127: 621–626.

Derikvand, P., Etemadifar, Z., Biria, D. (2014) Taguchi optimization of dibenzothiophene biodesulfurization by *Rhodococcus erythropolis* R1 immobilized cells in a biphasic system. *International Biodeterioration and Biodegradation*. 86: 343–348.

Duissenov, D. (2013) Production and processing of sour crude and natural gas – challenges due to increasing stringent regulations. Master's Thesis. Department of Petroleum Engineering and Applied Geophysics, Faculty of Engineering Science and Technology, Norwegian University of Science and Technology. NTNU, Trondheim, Norway.

Duarte, G.F., Rosado, A.S., Seldin, L., Araujo, W., Van Elsas, J.D. (2001) Analysis of bacterial community structure in sulfurous-oil-containing soils and detection of species carrying of dibenzothiophene desulfurization (dsz) genes. *Applied Environmental Microbiology*. 67:1052–1062.

Energy Information Administration EIA (2013). Petroleum & Other Liquids. Annual Energy Outlook 2013. Washington: US. www.eia.gov/outlooks/aeo/pdf/0383(2013).pdf

El-Gendy, N.Sh., Ali, H.R., El-Nady, M.M., Deriase, S.F., Moustafa, Y.M., Roushdy, M.I. (2014) Effect of different bioremediation techniques on petroleum biomarkers and asphaltene fraction in oil-polluted sea water. *Desalination and Water Treatment*. 52: 40–42.

El-Gendy, N.Sh., Speight J.G. (2016) Handbook of Refinery Desulfurization. CRC Press, Taylor and Francis Group, LLC, 6000 Broken Sound Parkway NW, Suite 300, Boca Raton, FL 33487–2742, USA.

Fedorak, P.M., Kropp, K.G. (1998) A review of the occurrence, toxicity and biodegradation of condensed thiophenes found in petroleum. *Canadian Journal of Microbiology*. 44: 605–622.

Grossman, M.J., Lee, M.K., Prince, R.C., Garrett, K.K., Minak-Bernero, V., Pickering, I. (2001) Deep desulfurization of extensively hydrodesulfurized middle distillate oil by *Rhodococcus* sp. Strain ECRD-1. A*pplied and Environmental Microbiology*. 67: 1949–1952.

Guiliano, M., Boukir, A., Doumenq, P., Mille G. (2000) Supercritical fluid extraction of BAL 150 crude oil asphaltenes. Energy and Fuels. 14(1): 89–94.

Gundlach, E.R., Boehm, P.D., Marchand, M., Atlas, R.M., Ward, D.M., Wolfe, D.A. (1983) The fate of Amoco Cadiz oil. Science. 221: 122–129.

Hunt, J. M. (1979) Petroleum geochemistry and geology. W. H. Freeman & Co., San Francisco, USA.

ICCT, International Council on Clean Transportation. US report, October 24 (2011).

Jacobs, F-S., Filby, R.H. (1983) Liquid chromatographic fractionation of oil-sand and crude oil asphaltenes. *Fuel*. 62(10): 1186–1192.

Jensen, A.-M. Finster, K.W., Karlson U. (2003) Degradation of carbazole, dibenzothiophene, and dibenzofuran at low temperature by *Pseudomonas* sp. strain C3211. Environmental Toxicology and Chemistry. 22: 730–735.

Kabe, T.A., Kamatsu, K., Ishihara, A., Otsuki, S., Godo, M., Zhang, Q., Qian, W. (1979) Deep Hydrodesulfurization of light gas oil. 1. kinetics and mechanisms of dibenzothiophene hydrodesulfurization. *Industrial and Engineering Chemistry Research*. 36: 5146–5152.

Kaminski, T., Fogler, H.S., Wolf, N., Wattana, P., Mairal, A. (2000) Classification of asphaltenes via fractionation and the effect of heteroatomic content on dissolution kinetics. *Energy and Fuels*. 14: 25–30.

Kawanaka, S., Leontaritis, K.J., Park, S.J., Mansoori, G.A. (1989) Thermodynamic and colloidal models of asphaltene flocculation, pp. 450–458. In ACS symposium series. Oil-field chemistry enhanced recovery and productions stimulation. ACS Washington D.C, USA.

König, A., Herding, G., Hupfeld, B., Richter, T., Weidmann, K. (2001) Current tasks and challenges for exhaust after treatment research, a viewpoint from the automotive industry. Topics in Catalysis. 16/17: 23–31.

Ma, T. (2010) The desulfurization pathway in *Rhodococcus.* Microbiology Monographs. 16: 207–230.

Maldonado, A.G., Doucet, J., Petitjean, M., Fan, B.T. (2006) Molecular similarity and diversity in chemoinformatics: from theory to applications. *Molecular Diversity.* 10(1):39–79.

Manahan, S.E. (2003) Biochemistry and Toxicological Chemistry. Third Edition, CRC Press, Taylor and Francis Group, LLC, 6000 Broken Sound Parkway NW, Suite 300, Boca Raton, FL 33487–2742, USA.

Martinez, I., Santos, V.E., Alcon, A., Garcia-Ochoa, F. (2015) Enhancement of the biodesulfurization capacity of *Pseudomonas putida* CECT5279 by co-substrate addition. *Process Biochemistry.* 50: 119–124.

Martinez, I., Santos, V.E., Garcia-Ochoa, F. (2017) Metabolic kinetic model for dibenzothiophene desulfurization through 4S pathway using intracellular compound concentrations. *Biochemical Engineering Journal.* 117: 89–96.

McFarland, B., Boron, D.J., Deever, W., Meyer, J.A., Johnson, A.R., Atlas, R.M. (1998) Biocatalytic sulfur removal from fuels: Applicability for producing low sulfur gasoline. *Critical Review of Microbiology.* 24:99–147.

Mohebali, G., a Ball A.S. (2016) Biodesulfurization of diesel fuels p past, present and future perspectives. *International Biodeterioration and Biodegradation.* 110: 163–180.

Mohebali, G., Ball, A.S. (2008) Biocatalytic desulfurisation (BDS) of petrodiesel fuels, Microbiology 154: 2169–2183.

Monticello, D.J., Finnerty, W.R. (1985) Microbial desulphurization of fossil fuels. *Annual Review of Microbiology.* 24:99–147.

Montiel C., Quintero R., Aburto J. (2009) Petroleum biotechnology: Technology trends for the future. *African Journal of Biotechnology* 8(12), 2653–2666.

Murgich, J., Abanero, A.J.Y., Strausz, P.O. (1999) Molecular recognition in aggregates formed by asphaltene and resin molecules from the Athabasca oil sand. *Energy and Fuels.* 13(2): 278–286.

Murphy, S.E., Amin, S., Coletta, K., Hoffmann, D. (1992) Rat liver metabolism of benzo[b]naphtho[2,1-d]thiophene. *Chemical Research in Toxicology* 5: 491–495.

Mužic, M., Sertić-Bionda, K. (2013) Alternative processes for removing organic sulfur compounds from petroleum fractions. *Chemical and Biochemical Engineering.* Q. 27(1): 101–108.

Nasser, W. (1999) CEO Energy biosystems corporation the woodlands; Texas before the senate environment and public works subcommittee on clean air, wetlands, private property and nuclear safety. https://www.epw.senate.gov/107th/nas_5-18.htm

Ogata, M., Fujisawa, K.J. (1983) Capillary GC/MS determination of organic sulfur compounds detected in oyster and mussel caught in the sea as an oil pollution index. *Journal of Chromatographic science.* 21(9): 420–424.

Ogata, M., Mujake, Y. (1978) Identification of organic sulfur compounds transferred to fish from petroleum suspension. *Acta Medica Okayama.* 32(6): 419–425.

Organization of Petroleum Exporting Countries OPEC (2011) World Oil Outlook. Helferstorferstrasse 17 A-1010, Vienna, Austria.

Organization of Petroleum Exporting Countries OPEC (2013) World Oil Outlook. Helferstorferstrasse 17 A-1010, Vienna, Austria.

Paixao, S.M., Arez, B.F., Roseiro, J.C., Alves, L. (2016) Simultaneously saccharification and fermentation approach as a tool for enhanced fossil fuels biodesulfurization. *Journal of Environmental Management*. 182: 397–405.

Peng, B., Zhou, Z. (2016) Study on growth characteristic and microbial desulfurization activity of the bacterial strain MP12. *Biochemical Engineering Journal*. 112: 202–207.

Peters, P.E., Walters, C.C., Moldowan, J.M. (2005) The biomarker guide, 2nd ed., Biomarkers in petroleum systems and earth history. Cambridge University Press Vol.1., New York, USA.

Pineda-Flores, G., Mesta-Howard, A.M. (2001) Petroleum asphaltenes: generated problematic and possible biodegradation mechanisms. *Revista Latinoamericana de Microbiologia*. 43(3): 143–150.

Pineda-Flores, G., Bollarguello, G., Mestahoward, A. (2004) A microbial mixed culture isolated from a crude oil sample that uses asphaltenes as a carbon and energy source. *Journal of Biodegradation*. 15: 145–151.

Radwan, S. (2008) Microbiology of oil-contaminated desert soils and coastal areas in the Arabian Gulf region. In: Dion P, Nautiyal CS (eds) Microbiology of extreme soils. Soil Biology 13. Springer, Berlin, pp 275–298.

Riazi M.R. (2005) Characterization and properties of properties of petroleum fractions. ASTM International Standards Worldwide, first edition, ASTM manual series: MNL50, 100 Barr Harbor PO Box C700 West Conshohocken, PA 19428–2959, USA.

Riazi, M.R., Nasimi, N., Roomi, Y. (1999) Estimating sulfur content of petroleum products and crude oils. *Industrial and Engineering Chemistry Research*. 38(11): 4507–4512.

Ryabov, V.D. (2009). Khimiya nefti i gaza (Oil and gas chemistry). Moscow: Forum.

Sergun, V.P., Cheshkova, T.V., Sagachenko, T.A., Min, R.S. (2016) Structural units with sulfur and ether/ester bonds in molecules of high- and low-molecular-weight asphaltenes of USA heavy oi. *Petroleum Chemistry*. 56(1): 10–15.

Shang, H., Du, W., Liu, Z., Zhang, H. (2013) Development of microwave induced hydrodesulfurization of petroleum streams: A review. *Journal of Industrial and Engineering Chemistry* 19: 1061–1068.

Sinninghe, Damste, J.S., De Leeuw, J.W. (1990) Analysis, structure and geochemical significance of of organically-bound sulphur in the geosphere: state of the art and future research. *Organic Geochemistry*. 16: 1077–1101.

Soleimani, M., Bassi, A., Margaritis, A. (2007) Biodesulfurization of refractory organic sulfur compounds in fossil fuels. *Biotechnology Advances*. 25: 570–596.

Song, C. (2003) An overview of new approaches to deep desulfurization for ultra-clean gasoline, diesel fuel and jet fuel. *Catalysis Today*. 86: 211–263.

Song, C., and Ma, X. (2003) New Design Approaches to ultra-clean diesel fuels by deep desulfurization and deep de-aromatization. *Applied Catalysis B Environmental.* 41(1–2): 207–238.

Speight, J.G. (1970) A structural investigation of the constituents of Athabasca bitumen by proton magnetic resonance spectroscopy. *Fuel* 49: 76–90.

Speight, J.G. (2004) Petroleum asphaltenes part 1: asphaltenes, resins and the structure of petroleum. *Oil and Gas Science and Technology-Rev. IFP*, 59(5): 467–477.

Speight, J.G. (2007) The chemistry and technology of petroleum. CRC Taylor and Francis. Boca Raton, Florida, USA, vol. 114.

Speight, J.G., Moschopedis, S.E. (1981) On the molecular structure of petroleum asphaltenes, in Chemistry of Asphaltenes, Bunger, J.W. and Li, N.C., Eds., Advances in Chemistry Series No. 195, American Chemical Society, Washington, D.C., 195, 1–15.

Speight, J.G. (2000) The desulfurization of heavy oils and residua. 2nd edition; Marcel Dekker; New York.

Speight, J.G. (2006) The chemistry and technology of petroleum. 4th edition; CRC Press; Taylor and Francis Group; Boca Raton; Florida.

Srivastava, C.C. (2012) An evaluation of desulfurization technologies for sulfur removal from liquid fuels. *RSC Advances.* 2: 759–783.

Stanislaus, A., Marafi, A., Rana, M.S. (2010) Recent advances in the science and technology of ultra-low sulfur diesel (ULSD) production. *Catalysis Today.* 153(1): 1–68.

Strausz, O.D. (1988) in Fourth UNITAR/UNDP International Conference on Heavy Crude and Tar Sands (Edmonton, Alberta, Canada, August 7–12, 1988) Alberta Oil Sand Technology and Research Authority (AOSTRA), vol. 2. pp. 608–628.

Strausz, O.P., Mojelsky, T.W., Faraji, F., Lown, E.M. (1999) Additional structural details on Athabasca asphaltene and their ramifications. *Energy and Fuels.* 13(2): 207–227.

Tang, Q., Lin, S., Cheng, Y., Liu, S.-J., Xiong, J.-R. (2013) Enhanced biodesulfurization of bunker oil by ultrasound pre-treatment with native microbial seeds. *Biochemical Engineering Journal.* 77: 58–65.

Tavassoli, T., Mousavi, S.M., Shojaosadati, S.A., Salehizadeh, H. (2012) Asphaltene biodegradation using microorganisms isolated from oil samples. *Fuel* 93: 142–148.

The European Environmental Burean EEB (2008) The European Federation for Transport and Environment (T&E), Seas at Risk (SAR) and The Swedish NGO Secretariat on Acid Rain 2004: Air Pollution from Ship. http://www.acidrain.org.pages/publications/reports/shipbrefing.July.Pdf.

The International Council of Clean Transportation (ICCT) (2011) The production of ultra-low sulfur gasoline and diesel fuel. Bethesda, Maryland 20827–0404, P.O. Box 34404, USA.

Tissot, B.P., Welte, D.H. (1978) Petroleum Formation and Occurrence. Springer-Verlag, New York.

Tissot, B.P., Welte, D.H. (1984) Petroleum formation and occurrence. Second Revised and Enlareed Edition. Springer-Verlag Berlin Heidelberg NewYork Tokyo.

U.S. Energy Information Administration (US EIA) (2010) International Energy Statistics. http://www.eia.gov/cfapps/ipdproject/IEDindex3.cfm?tid=5&pid=53&aid=1.

U.S. Environmental Protection Agency, EPA Staff Paper on Gasoline Sulfur Issues, EPA420-R-98-005 (Washington, DC, May 1998).

Varjani, S.J. (2014) Hydrocarbon degrading and biosurfactants (bioemulsifiers) producing bacteria from petroleum oil wells. Ph.D. Thesis, Kadi Sarva Vishwavidyalaya, Gandhinagar, India.

Varjani, S.J. (2017) Microbial degradation of petroleum hydrocarbons. *Bioresource Technology*. 223: 277–286.

Varjani, S.J., Rana, D.P., Jain, A.K. Bateja, S., Upasani, V.N. (2015) Synergetic ex-situ biodegradation of crude ol by halotolerant bacterial consortium of indigenous strains isolated from shore sites of Gujarat, India. *International Biodeterioration and Biodegradation*. 103: 116–124.

Vieira, P.A., Vieira, R.B., Franca, F.P., Cardoso, V.L. (2007) Biodegradation of effluent contaminated with diesel fuel and gasoline. *Journal of Hazardous Materials*. 140(1–2): 52–59.

Voloshchuk, A., Krasilnkova, O., Serbrykova, N., Artamonova, S., Aliev, A., Khozina, E., Kiselev, V. (2009) Selective adsorption of organic sulfur-containing compounds from diesel fuel using type Y-zeolite and γ-aluminum oxide. *Protection of Metals and Physical Chemistry of Surfaces*. 25: 512–517.

Wang, Z., Fingas, M. (1995): Use of Methyldibenzothiophenes as markers for differentiation and source identification of crude and weathered oils, Environmental Science and Technology. 29: 2842–2849.

Wauquier, J.P. (1995) Petroleum Refining V1: Crude Oil, Petroleum Products, Process Flowsheets. ED TECHNIP, Paris, France.

Wilkes, H., Buckel, W., Golding, R.I., Rabus, R. (2016) Metabolism of hydrocarbons in n-alkane utilizing anaerobic bacteria. *Journal of Molecular Microbiology and Biotechnology*. 26: 138–151.

Wu, J., Prausnitz, M.J., Firoozabadi, A. (2000). Molecular thermodynamics of asphaltene precipitation in reservoir fluids. *AIChE J*. 46: 197–206.

Yang, Y.-Z., Liu, X.-G., Xu, B.-S. (2014). Recent advances in molecular imprinting technology for the deep desulfurization of fuel oils. *New Carbon Material*. 29(1): 1–14.

Yen, T.F. (1975) The role of trace metals in petroleum. Ann Arbor Scientific Publishers, Ann Arbor, Mich.

Yin, C., Xia, D. (2001) A study of the distribution of sulfur compounds in gasoline produced in China. Part 1. A method for the determination of the distribution of sulfur compounds in light petroleum fractions and gasoline. *Fuel*. 80: 607–610.

Zhang, T.M., Li, W.L., Chen, X.X., Tang, H., Li, Q., Xing, J.M., Liu, H.Z. (2011) Enhanced biodesulfurization by magnetic immobilized Rhodococcus erythropolis LSSE8-1-vgb assembled with nano-gamma Al_2O_3. *World Journal of Microbiology and Biotechnology* 27(2): 299–305.

2

Desulfurization Technologies

List of Abbreviations and Nomenclature

+ve I	Positive Induction Effect
1,3,5-trimesic acid	TMA
2,4,6-TMDBT	2,4,6-Trimethyl Dibenzothiophene
2,5-DMT	2,5-Dimethyl Thiophene
2-MT	2-Methyl Thiophene
4,6-DMDBT	4,6-Dimethydibenzothiophene
4E6M-DBT	4-Ethyl, 6-Methyl-Dibenzothiphnene
4-MDBT	4-Methyldibenzothiophene
AAB	Acid Activated Bentonite
AC	Activated Carbon
ADS	Adsorptive Desulfurization
aEVM	Acid Treated Expanded Vermiculite
AIBN	Azoisobutyrontrile
ALF	Aliphatic Fuel
Alkyl-Ths	Alkyl-Thiophenes
ARF	Aromatic Fuel

ASA	Amorphous Silica-Alumina
AZ	SO_4^{2-}/Zro_2
bbl/day	Barrels Per Day
BDAC	Boron Doped Activated Carbon
BDS	Biodesulfurization
BP	British Petroleum
bpsd	Barrels Per Stream Day
BT	Benzothiophene
BTC	Breakthrough Curve
CCS	Carbon Catalyst Support
CEDS	Conversion and Extraction Desulfurization
CMSs	Carbon Microspheres
CNT/TiO_2	Carbon Nanotube/Tio_2
Co	Cobalt
CoMo	Cobalt–Molybdenum
Cu	Copper
Cx-DBTs	Alkylated-Dibenzothiophenes
CYHPO	Cyclohexanone Peroxide
DBT	Dibenzothiophene
DBTO	Dibenzothiophene Sulfoxide
$DBTO_2$	Dibenzothiophene Sulfone
DCA	9,10-Dicyanoanthracene
DMF	Dimethylformamide
DMSO	Dimethylsulfoxide
DOE	Design of Experiments
ECODS	Extractive Catalytic Oxidative Desulfurization
EDMA	Ethyleneglycol Dimethyl Acrylate
EDS	Extractive Desulfurization
ETO	Ethanol-to-Olefin
Eu-MOF	Europium Metal Organic Framework
EVM	Expanded Vermiculite
FA	Formaldehyde
FCC	Fluid Catalytic Cracking
Fe-TAML	Fe-Tetra Amido Macrocyclic Ligand
FOEB	Fuel Oil Equivalent Barrel
Ga	Gallium
H_2S	Hydrogen Sulfuide
HCN	Heavy Cut Naphtha
HCR	Hydrocracking
HDAr	Hydrodearomatization

HDN	Hydrodenitrogenation
HDS	Hydrodesulfurization
HDSCS	HDS of Concentrated Sulfur
HDT	Hydrotreating
HPA	Heterpolyacid
HPA-HTLcs	Heteropolyanion Substituted Hydrotalcite-Like Compounds
HPTM	Heteropoly Tungstate/Molybdate
HPW	$H_3PW_{12}O_{40}$
IBP	Initial Boiling Point
IFP	Institute Francais Du Petrole
INTEVEP SA	Instituto De Tecnología Venezolana Para El Petróleo Sociedad Anónima
Ir	Iridium
IRVAD	Irvine Robert Varraveto Adsorption Desulfurization
ISBL	In Side Battery Limits
LCN	Light Cut Naphtha
LCO	Light Cycle Oil
LHSV	Volume Hourly Space Velocity
LNB	Lukoil Neftochim Bourgas
LPG	Liquid Petroleum Gas
MAA	Methyl Acrylic Acid
MCN	Medium Cut Naphtha
MIP	Molecular Imprinting Polymer
MIT	Molecular Imprinting Technology
MNLB	Magnetite Nanoparticle Loaded Bentonite
MO	Molecular Orbital
MOFs	Metalorganic Frameworks
MoO_3	Molybdenum Oxide
MPS	Methyl Phenyl Sulfide
MS	Mesoporous Silica
MTBE	Methyl Tert-Butyl Ether
MWCNTs	Multiwall Carbon Nanotubes
MXF	Mixed Fuel
Ni	Nickel
NiMo	Nickel–Molybdenum
NIP	Non-Imprinted Polymer
NMP	N-Methyl Pyrrolidone
NPC	Nitrogen Polar Compounds
NSO	Nitrogen, Sulfur and Oxygen

NZSG	Near Zero Sulfur Gasoline
ODS	Oxidative Desulfurization
PA	Phsophotungstic Acid
PAHs	Polyaromatic Hydrocarbons
PASHs	Poly Aromatic Sulfur Heterocyclic
Pd	Palladium
PDT	Premium Distillates Technology
PMA	Phosphomolebdic Acid
POMs	Polyoxometalates
ppbw	Parts Per Billion Weight
ppmw	Parts Per Million Weight
Pt	Platinum
PTA	Phase Transfer Agent
PTFE	Polytetrafluoroethylene
R&D	Research and Development
RADS	Reactive Adsorptive Desulfurization
Rh	Rhodium
RON	Research Octane Number
Ru	Ruthenium
SARA	Saturates Aromatics Resins Asphaltenes
SARS	Selective Adsorption for Removing Sulfur
SCFB	Standard Cubic Feet/Barrel
SMIP	Surface Molecular Imprinting Polymer
SOx	Sulfur Oxides
SRB	Sulfate Reducing Bacteria
SRGO	Straight-Run Gas Oil
SRLGO	Straight-Run Light Gas Oil
SRT	Sulfur Removal Technology
STARS	Super Type II Active Reaction Sites
TAML	Tetra Amido Macrocyclic Ligand
TBHP	T-Butyl-Hydroperoxide
Th	Thiophene
UB	Untreated Bentonite
ULS	Ultra-Low-Sulfur
ULSD	Ultra-Low Sulfur Diesel
UOP	Universal Oil Products
US-EPA	United States Environmental Protection Agency
V	Vanadium
W	Tungsten

2.1 Introduction

Petroleum covers approximately 40% of the worldwide energy requirements (Al-Des *et al.*, 2016). The worldwide depletion of high quality low sulfur content crude oil reserves comes in parallel with the more stringent environmental regulations in every country around the world towards the reduction of all forms of greenhouse gas emissions, including the emissions of sulfur oxides (SO_x). Thus, refineries are facing many challenges, including the influx of heavy crude oils, increased fuel quality standards in terms of a severely diminished and regulated sulfur content to approximately 10 ppm for gasoline and diesel fuels (Bandosz *et al.*, 2006; Khamis and Palichev, 2012; Al-Degs *et al.*, 2016), and the need to reduce all forms of emissions to meet air pollution regulations. Babich and Moulijn (2003) reported that approximately \$10–15 billion and up to \$16 billion are estimated to be invested by European, US, and Canadian refineries, respectively, to meet the new restricted environmental clean-fuel legislation.

Ancheyta (2011) has reported that the range of sulfur-content in crude oils varies according to its classification: extra-light crude oil (0.02–0.2 wt.%), light crude oil (0.05–4 wt.%), heavy crude oil (0.1–5 wt.%), and, finally, extra-heavy crude oil (0.8–6 wt.%).

Based on the SARA compositions of crude oils (Figure 2.1), light crude oil is rich in light distillates, while heavy crude oil is rich in high molecular weight distillates and more polar residuum, where, the asphaltenes content with its high concentrations of polyromantic heterocyclic compounds and heavy metals increases with the decrease of the crude oil quality. According to Ancheyta (2011), the asphaltene content varies as follows: extra-light crude oil (0–<2 wt.%), light crude oil (<0.1–12 wt.%), heavy crude oil (11–25 wt.%), and, finally, extra-heavy crude oil (15–40 wt.%).

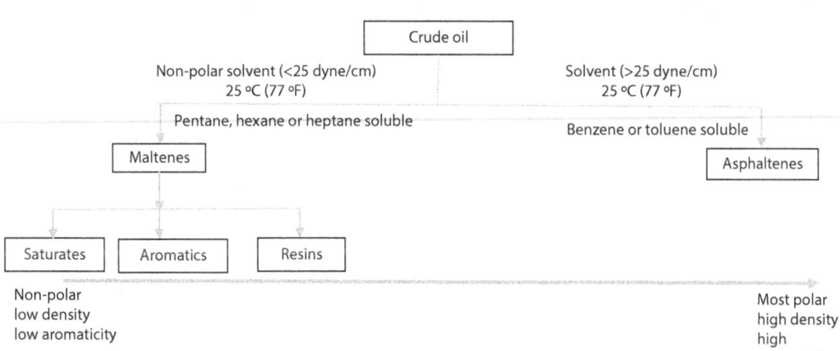

Figure 2.1 SARA Fractions in Crude Oil.

Moreover, the sulfur content in crude oils, increases with the increment of asphaltene content.

The higher the sulfur content in crude oil, the higher the sulfur content in its distillates. Generally, the sulfur concentration and types of sulfur compounds vary over the boiling range of the petroleum distillates (Table 2.1). Aitani *et al.* (2000) reported that sulfur is present mainly as thiophenic compounds (i.e. thiophene Th, benzothiopene BT, dibenzothiophene DBT, etc.) and their derivatives. They are mainly concentrated on resins fractions (approximately 70%) and the rest is found in the other fractions, but to higher extent in the asphaltenes. The main liquid fuels coming out from petroleum are gasoline range (naphtha and fluid catalytic cracking (FCC) –naphtha), jet fuel range (heavy cut naphtha (HCN) and middle distillate), diesel fuel range (middle distillate and light cycle oil (LCO)), and boiler fuel feeds (heavy oils and distillation resides) (Song, 2003). Mustafa *et al.* (2010) reported that alkyl sulfur compounds are concentrated in gasoline, while the high molecular weight aromatic organosulfur compounds are concentrated in diesel oil. Figure 2.2 illustrates some of the predominant sulfur compounds in gasoline and diesel oil. It has been reported that DBT and its derivatives constitute up to 50% of the sulfur content in diesel oil (Lee *et al.*, 2002). The recalcitrance of the sulfur compounds increases with the increase of the boiling points of the distillates. Thus, the desulfurization of thiols is much easier than the desulfurization of polyaromatic sulfur heterocyclic (PASH) compounds.

There are various reported desulfurization methods to remove sulfur from petroleum and its fractions (Figure 2.3). The most developed and commercialized technologies are those which catalytically convert organosulfur compounds with sulfur elimination. Such catalytic conversion technologies include conventional hydro-treating, hydro-treating with advanced catalysts and/or reactor design, and a combination of hydrotreating with some additional chemical processes to maintain fuel specifications. The main feature of the technologies of the second type is the application of physio-chemical processes different in nature from catalytic hydrodesulfurization (HDS) to separate and/or to transform organosulfur compounds from refinery streams. Such technologies are included as a key step in distillation, alkylation, oxidation, extraction, adsorption, or combination of these processes. Among these desulfurization techniques, HDS is currently considered as the most important one. However, HDS has several disadvantages in that it is energy intensive, costly to install and to operate, and does not work well on recalcitrant high molecular weight PASH compounds. Recent research has therefore focused on improving HDS catalysts and processes and also on the development of alternative

Table 2.1 Sulfur Content, Types, and Recalcitrance within Petroleum Distillates.

Distillation range °C	Distillate	S-content & recalcitrance towards desulfurization	S-compounds distribution
IBP-71	Light straight-run naphtha	Increases (↓)	Thiols, sulfides & traces of thiophenes
71–177	Medium straight-run naphtha		
177–204	Heavy straight-run naphtha		
204–274	Jet fuel		
274–316	Kerosene		Mainly; thiols, sulfuides & thiophenes. Low concentration of benzothiophenes, dibenzothiophenes & heavy sulfides.
316–343	Straight-run gas oil		Mainly; thiophenes, benzothiophenes, dibenzothiophenes & heavy sulfides. Low concentration of thiols & sulfides.
343–454	Light vacuum gas oil		Traces of thiols & sulfides. Lower concentration of thiophenes. Higher concentration of benzothiophenes, dibenzothiophenes & heavy sulfides.
454–538	Heavy vacuum gas oil		
>538	Vacuum residue		Mainly; benzothiophenes, dibenzothiophenes & heavy sulfides. Traces of thiols, sulfides & thiophenes.

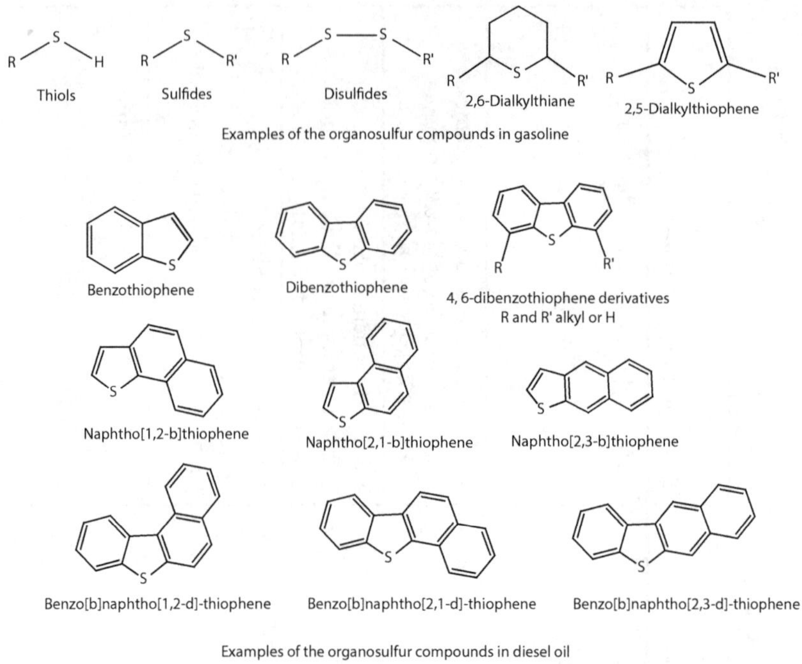

Figure 2.2 Examples of Organosulfur Compounds in Gasoline and Diesel Oil.

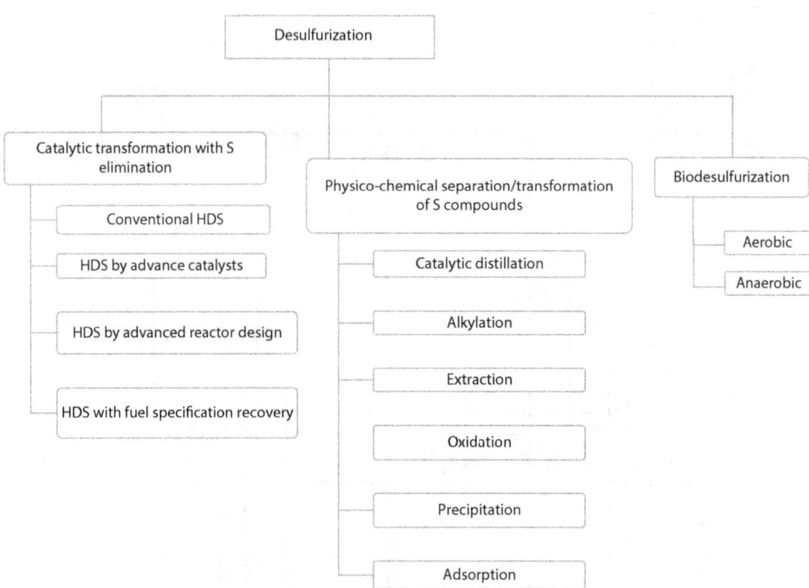

Figure 2.3 Desulfurization Technologies Classified According to the Nature of Sulfur Removal Key Processes.

technologies. Among these technologies are selective or reactive adsorptive desulfurization (ADS), oxidative desulfurization (ODS) combined with extractive desulfurization (EDS) or ADS, and biodesulfurization (BDS), which are advantageous because they can be operated in conditions that require less energy and hydrogen. They can also operate at ambient temperature and pressure with high selectivity, resulting in decreased energy costs, low emission, and no generation of undesirable side products. Moreover, the incorporation of non-HDS with HDS, will reduce the H_2 consumption required for the deep-HDS process (Rang *et al.*, 2006; Srivastava, 2012; Mužic and Sertić -Bionda, 2013).

2.2 Hydrodesulfurization

The conventional desulfurization process which is widely practiced for crude oil and its distillates (gasoline, kerosene and diesel oil) is hydrodesulfurization (HDS) (Heidarinasab *et al.*, 2016; Mohammed *et al.*, 2016). The terms hydrotreating, hydroprocessing, hydrocracking (HCR), and hydrodesulfurization (HDS) are industrially used referring to HDS and HCR. That is because the desulfurization and cracking processes occur simultaneously and are relative as to which is predominated. The main purpose of the HCR process is to reduce the boiling range of a feed to products with lower boiling ranges, while hydrotreating is a process to catalytically stabilize petroleum products by converting olefins to paraffins and/or removing contaminants such as nitrogen, oxygen, sulfur, halides, and trace metals by making them react with hydrogen (El-Gendy and Speight, 2016).

HDS is a technology which converts organic sulfur compounds (such as thiols, sulfides, and thiophenes) to hydrogen sulfuide (H_2S) and other inorganic sulfuides, under high temperature (200 to 455 °C) and high pressure (150 to 3000 psi) and uses hydrogen gas in the presence of sulfidized metal catalysts (e.g. CoMo/Al_2O_3 or NiMo/$Al_2O_{3)}$ (Lecrenary *et al.*, 1997; Rana *et al.*, 2007; Eswaramoorthi *et al.*, 2008; Al-Degs *et al.*, 2016). The produced H_2S is then catalytically air oxidized to elemental sulfur in Claus plants (El-Gendy and Speight, 2016).

$$\text{Organosulfur compunds} + H_2 \xrightarrow[\text{High temperature \& pressure}]{\text{Metal catalyst}} \text{Desulfurized organic compounds} + H_2S$$

$$2H_2S + 3O_2 \longrightarrow 2SO_2 + 2H_2O$$

$$SO_2 + 2H_2S \longrightarrow 3/2\, S_2 + 2H_2O$$

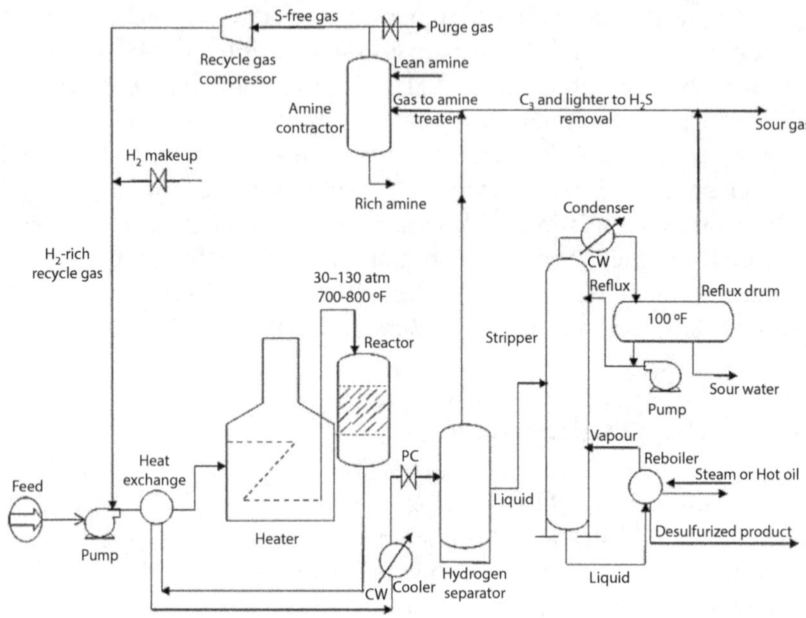

Figure 2.4 A Simple HDS-Process.

Bose (2015) reported that by 2005 approximately 64 million metric tons of sulfur were produced worldwide as byproducts form petroleum refineries and hydrocarbon processing plants. Figure 2.4 illustrates a simple HDS process. Sulfur is used for manufacturing sulfuric acid, medicine, cosmetics, fertilizers, and rubber products. Elemental sulfur is used as fertilizer and pesticide.

The HDS conditions depend on the required desulfurization degree and the type of sulfur compounds in the feed to be treated. The aliphatic S-compounds are known to be very reactive and can be completely removed during the conventional HDS process.

$$\text{Thiols: } R{-}SH + H_2 \rightarrow R{-}H + H_2S$$

$$\text{Sulfides: } R_1{-}S{-}R_2 + 2H_2 \rightarrow R_1{-}H + R_2{-}H + H_2S$$

$$\text{Disulfides: } R_1{-}S{-}S{-}R_2 + 3H_2 \rightarrow R_1{-}H + R_2{-}H + 2H_2S$$

However, S contained in thiophenes is more difficult to be removed since the lone pair electrons from S-atoms participates in the π-electron structure of the conjugated C=C system. Thus, resonance stabilization is around 120–130 kJ/mol, which is less than that of the benzene ring, which is 160–170 kJ/mol, but it is still sufficient to make the HDS energetically demanding (Gronowitz, 1985). The HDS of thiophenic compounds occurs through two pathways: the hydrogenation pathway (hydrogenation followed by hydrogenolysis) and the direct hydrogenolysis pathway (direct elimination of S atom via C–S bond cleavage) and they occur at different active sites. The least hydrogen utilization pathway is direct hydrogenolysis, but the resonance stabilization of sulfur in the thiophene ring makes this process difficult and obligates the HDS-pathway towards the hydrogenation-pathway, where aromatic ring saturation occurs before the occurrence of desulfurization. However, the PAHs are known to be the main inhibitors towards the hydrogenation pathway. Moreover, the equilibrium concentration of the hydrogenated product is low since there is a significant driving force for aromatization by dehydrogenation. Furthermore, the resonance stabilization of thiophene prevents cracking, which leads to the formation of coke during the FCC [Phillipson, 1971; Hatanaka *et al.*, 1998; Reddy *et al.*, 1998; IFP, 2000; Kaufmann *et al.*, 2000; Miller *et al.*, 2000; Nocca *et al.*, 2000; Corma *et al.*, 2001; ExxonMobi, 2001; Halbert *et al.*, 2001; Okamoto *et al.*, 2002; Song and Ma, 2003; Hatanaka, 2005; Shu and Wormsbecher, 2007; Javadli and de Klerk, 2012a).

Nickel–molybdenum (NiMo) catalysts remove contaminants by hydrogenation, while cobalt–molybdenum (CoMo) predominantly removes contaminants by direct hydrogenolysis (Ferreira *et al.*, 2013). Thus, CoMo-catalysts are preferred for the HDS of unsaturated hydrocarbon streams such as fluid catalytic cracking (FCC). It is also more efficient in batch reactors (Lecrenay *et al.*, 1997), but NiMo-catalysts are preferred for the HDS of recalcitrant PASHs, such as 4,6-dimethydibenzothiophene (4,6-DMDBT), thus it is applied more often in the HDS of fractions that require extreme conditions (Bataille *et al.*, 2000; Kim *et al.*, 2003). However, Wang *et al.* (2004) has reported that the relative significance of hydrogenolysis and hydrogenation routes depend mainly on Co/Mo ratios. Furthermore, the two routes could also be substantially affected by the presence of naphthalene or H_2S for the HDS of DBT or 4,6-DMDBT (Farag *et al.*, 1999a,b). Moreover, Ali and Siddiqui (1997) have reported that the catalysts' type has a critical influence on the composition and properties of the hydrotreated product. Furthermore, the interactions between the reactants and catalysts and effects of reaction conditions have a very significant effect on desulfurization efficiency (Al-Zeghayer and Jibril, 2006). One of the reasons that the large size of alkyl groups is difficult to desulfurize is that C-S bonds in the aromatic rings are more stable than the others (Tang and Shi, 2011).

Oil refiners depend on such a costly, extreme chemical process. Aitani *et al.* (2000) have reported that there are more than 35 HDS units all over the world, with a total installed capacity of 1.5 million bbl/day. Moreover,

Gupta and Roychoudhury (2005) reported worldwide petroleum refineries use the conventional HDS technology to desulfurize approximately 20 million bbl crude oil per day. Although HDS can easily remove inorganic sulfur or simple organic sulfur compounds (such as thiols, sulfides and disulfides, Figure 2.2), it is not effective for removing complicated PASHs such as DBT-derivatives and alkylated-DBT derivatives (Cx-DBTs) (Babich and Moulijn, 2003; Franchi *et al.*, 2003). Complete HDS of such recalcitrant compounds needs higher temperature, pressure, and more H_2 consumption, which has a negative impact on the octane number of gasoline and cetane index of diesel oil. That is due to the hydrogenation of olefins and PAHs (Scheme 2.1). Knowing that, the PAHs represent approximately 20% of diesel oil composition (Park *et al.*, 2011). As mentioned before, due to the resonance stabilization of sulfur in thiophene rings, a direct pathway is difficult to be achieved. Thus, consequently, HDS takes place through the hydrogenation (i.e. saturation) of the aromatic rings before the removal of sulfur (Scheme 2.1). However, Schulz *et al.* (1999) reported that, the presence of alkyl substituents on BTs and DBTs molecules might favour one of the possible HDS routes, which depends on the alkyl substituent position and, thus, to what extent the electron density is altered by the electron donating effect of alkyl groups. In addition, substituents in the vicinity of the sulfur atom cause steric hindrance and influence the HDS route (Kabe *et al.*, 1992). The DBT and alkyl derivatives substituted adjacent to the sulfur atom are refractory to HDS using conventional catalysts. The key sulfur compounds present in diesel oil fractions after conventional HDS are 4-MDBT and 4,6-DMDBT.

The quantum chemical calculation on the conformation and electron property of various sulfur compounds and their HDS intermediates shows that the hydrogenation pathway favours desulfurization of the refractory sulfur compounds by both decreasing the steric hindrance of methyl

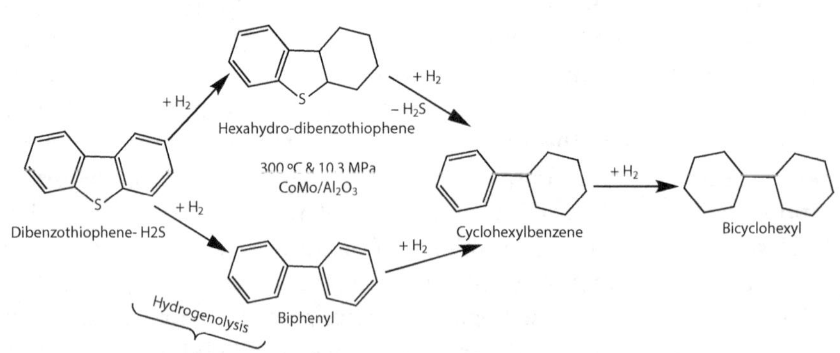

Scheme 2.1 Complete Hydrogenation of DBT in a HDS Process.

groups and increasing the electron density on the sulfur atom in the sulfur compounds. This steric hindrance increases with the increasing size of the alkyl groups (methyl < ethyl < propyl). The steric hindrance increases, as the alkylation is nearer to the S-atom, and, consequently, the recalcitrance moves towards the HDS process (Milenkovic *et al.*, 2000). The inhibition of the coexistent aromatics towards the HDS of the refractory sulfur compounds by competitive adsorption on the hydrogenation active sites becomes stronger in deep HDS (IFP, 2000; Westervelt, 2001; Golden *et al.*, 2002). Moreover, the produced H_2S from the reactive sulfur compounds in the early stage of the reaction is another inhibitor for HDS of the unreactive species (Vasudevan and Fierro, 1996). Thus, it should be removed from the recycle stream by scrubbing. Haldor–Topsoe indicates that decreasing the concentration of H_2S at the inlet to a cocurrent reactor by 3–6 vol.% can decrease the average temperature needed to achieve a specific sulfur reduction by 15–20 °C, and reduce final sulfur levels by more than two-thirds (Swain, 1991). More refractory sulfur compounds would require lower space velocity for achieving deeper HDS, as it would increase the reactant-catalyst contact time. Higher temperature facilitates more of the high activation–energy reactions, thus increasing the rate of HDS. But, the increase in temperature lowers the catalyst life time, which would consequently increase the overall cost as more catalyst will be needed. It will also affect the production stream while the unit is down for the catalyst change (the current catalyst life ranges from 6 to 60 months). High pressure also enhances the rate of HDS. Improvement of vapour–liquid contact increases the surface area of the catalyst. Akzo Nobel estimates that an improved vapour–liquid distributor can reduce the temperature necessary to meet a 50 ppmw sulfur level by 10 °C, which in turn would increase catalyst life and allow an increase in cycle length from 10 to 18 months. More volume of catalyst can enhance the HDS process and can be achieved through the expansion of catalyst bed volume or denser packing. Increase in reactor volume can enhance desulfurization. UOP assigns that doubling the reactor volume would reduce sulfur from 120 to 30 ppmw (Swain, 1991).

Generally, low hydrogenation efficiency is desirable during the HDS of FFC fractions for gasoline production, while high hydrogenation efficiency is desirable for HDS of refractory light cycle oil (LCO), which is rich in aromatics for diesel oil production (Leliveld and Eijsbouts, 2008). However, the main advantage of the HDS process is that it can desulfurize different oil feeds (such as gasoline, diesel oil, and jet fuel) with high sulfur content (10,000 ppm) to less than 100 ppm in one step (Al-Degs *et al.*, 2016).

The recalcitrance of sulfur compounds towards the HDS reaction can be ranked as follows (El-Gendy and Speight, 2016):

Thiophene < alkylated thiophene < benzothiophene < alkylated benzothiophene < dibenzothiophene and alkylated dibenzothiophene[1] < alkylated dibenzothiophene[2] < dibenzothiophene and alkylated dibenzothiophene[3]

[1]Without substituents at the 4 and 6 positions.
[2]With one substituent at either the 4 or 6 position.
[3]With alkyl substituents at the 4 and 6 positions.

One of the reasons that large size alkyl groups are difficult to desulfurize is that C-S bonds in the aromatic rings are more stable than the others (Tang *et al.*, 2013a). Thus, as mentioned above, the type of S-compounds affects the degree of HDS. For example, mercaptans and tetrahydrothiophene in FCC-naphtha are totally removed by using nickel and vanadium catalysts while Ths and BTs are more difficult to be removed. Thus, 4-MDBT and 4,6-DMDBT are known to be the most appropriate compounds for investigations of catalyst activity and proposed reaction mechanisms (Gates and Topsoe, 1997). These recalcitrant S-compounds should be promoted by hydrotreating and/or hydrocracking, where the unsaturated molecules are firstly saturated, then S, O, N, and metal atoms are taken away from hydrogenated oil by cracking (Myrstad, 2000). The commercial, binary-transition-metal catalysts are the most recommended and widely used catalysts in the worldwide refineries. For example, there is a synergetic effect between the active sites of molybdenum (Mo) or tungsten (W) and the Co or Ni as a promoter. In the Co/Mo/Al_2O_3 catalyst, the Co helps the Mo monolayer increase its stability and Co-Mo-S phase formation makes Co-Mo catalysts more active than Mo catalysts alone (Chianelli *et al.*, 2002). Hydrogen atoms generated from Co or Ni promoters under the working condition remove the sulfur atoms on Mo active sites, which would result in the unsaturated coordination of Mo atoms which are the active phase in catalytic desulfurization. These relations make the coordinative unsaturated site facilitate the adsorption of sulfur-containing reactants and the desorption of produced H_2S (Pimerzin *et al.*, 2013). However, Abrahamson (2015) reported that Co-promotor catalysts have higher HDS activity than Ni-promotor ones.

It has been reported that the application of the HDS process alone, to meet the new international standards for transportation fuels, is estimated to require a 3-fold increase in the catalyst volume/reactor size, which is non-economic and has a negative impact on the quality of the treated feed (Song and Ma, 2003). The desulfurization efficiency and selectivity depend on many factors: catalyst type, concentration of catalyst active species, support properties, catalyst synthesis route, the reaction conditions (i.e. the

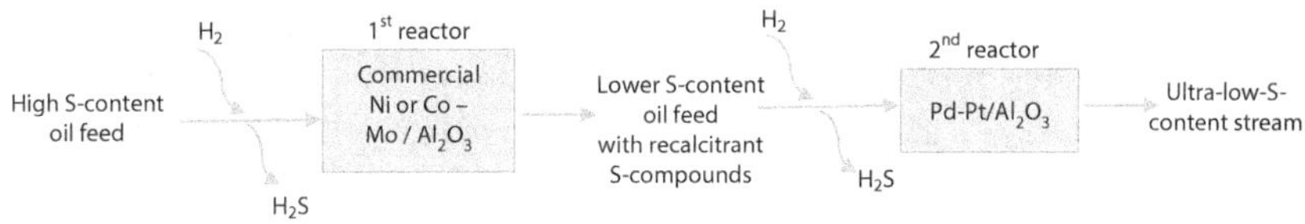

Figure 2.5 Schematic Diagram for a Simple Multi-Step HDS-Process.

sulfidizing protocol, temperature, and the partial pressure of H_2 and H_2S), the types and concentration of S-compounds in the feed stream, and finally the reactor and process design (Babich and Moulijn, 2003).

Noble metals and new carrier materials (such as amorphous silica-alumina ASA) with high desulfurization efficiency have been developed (Babich and Moulijn, 2003). However, the noble-metals-catalysts are recommended to be used after most of the organosulfur compounds and H_2S are removed, as they are susceptible for sulfur-poisoning. Thus, HDS with noble catalysts is recommended to be applied in a multi-step desulfurization process (Figure 2.5).

The commercial approaches applying two-stage or multi-stage hydrotreating processes are the Shell middle distillate hydrogenation process (Lucien *et al.*, 1994; Stork, 1996), the two-stage Haldor-Topose process (HDS/HDAr) in which the first stage uses their base metal HDS-catalyst TK-573 and the second stage uses their noble dearomatization TK-907/TK-908 or TK-915 catalysts (Cooper *et al.*, 1994), the IFP Prime D hydrotreating process reported to produce SRLGO with 10 ppmw S-content (Marchal *et al.*, 1994; EPA-Diesel RIA, 2000), United Catlysts and Sud-Chemie AG, which reported that the application of their ASAT catalyst in the treatment of feed distillate with 400 ppmw S, 127 ppmw N and 42.5 vol.% aromatics, managed to reduce the S to 8–9 ppm, eliminated N and reduced the aromatics to 2–5 vol.%, with hydrogen consumption of 800–971 standard ft^3/bbl (EPA-Diesel RIA, 2000) and the Criterion/ABB Lummus licensed their SynTechnology, which includes SynHDS for ultra-deep desulfurization and SynSat/SynShift for cetane improvement, aromatics saturation and density/T9 reduction (Suchanek, 1996; EPA-Diesel RIA, 2000; ABB Lummus Global, 2002).

The γ-Al_2O_3 is known to be the most widely used catalyst support because of its appropriate pore size distribution, large specific surface area, and its ability to provide high dispersion of active metal components (Salcedo, 2015; El-Gendy and Speight, 2016). The γ-Al_2O_3 supported molybdenum oxide catalysts promoted with cobalt or nickel have been widely used in conventional HDS processes (Segawa *et al.*, 2000). Active sites are formed when MoO_3 changes to MoS_2 by sulfurization (Arnoldy *et al.*, 1985). The hydrogenation route is the most important pathway in the HDS of DBT molecules with substituents on the 4- and 6-postition (Kabe *et al.*, 1993). The direct hydrogenolysis route is less favourable due to the steric hindrance (Robinson *et al.*, 1999). The molecule becomes more flexible upon hydrogenation of the aromatic rings and when the steric hindrance is relieved (Kabe *et al.*, 1993; Landau *et al.*, 1996). Consequently, catalysts with a relatively high hydrogenation activity must be considered. The Ni-promoted mixed sulfuide catalysts are known for their high hydrogenation activity (Van Veen *et al.*, 1993), but γ-Al_2O_3 has some disadvantages; the activity of some catalysts supported on alumina is lowered due to the occurrence of numerous chemical interactions between the alumina and the transition metal oxides. The formed species are very stable and completely resist sulfidizing. In addition, the coke formation is another disadvantage in the alumina-supported catalytic HDS-process, which deactivates and shortens the lifetime of the catalysts (Amini *et al.*, 2010). Moreover, nowadays, the application of nanocatalysts in the desulfurization process is very attractive because of their high stability and large specific surface area, thus, there are more active surface sites. Multiwall carbon nanotubes (MWCNTs) have been reported as a novel support material for desulfurizing catalysts (Mohammed *et al.*, 2016).

The concept of bi-functional catalysts has been proposed to increase the sulfur resistance of noble metal hydro-treating catalysts (Song, 1999). It combines catalyst supports with bimodal pore size distribution and two types of active sites. The first type of sites, placed in large pores, is accessible for organosulfur compounds and is sensitive to sulfur inhibition. The second type of active sites, placed in small pores, is not accessible for large S-containing molecules and is resistant to poisoning by H_2S. Since hydrogen can easily access the sites located in small pores, it can be adsorbed and transported within the pore system to regenerate the poisoned metal sites in the large pores. The practical applications of this concept have not been demonstrated yet. The Pd/Pt supported catalysts have been reported to possess high activity in the HDS of DBT and 4,6-DMDBT with a better selectivity to hydrogenation and higher thiotolerance than CoMo and NiMo catalysts (Kabe *et al.*, 2000; Barrio *et al.*, 2003; Niquille-Röthlisberger and Prins 2006; Niquille-Röthlisberger and Prins (2007); Baldovino-Medrano *et al.*, 2009). The Pt and Pt/Pd catalysts have also been reported as very active in the deep HDS of pre-hydrotreated straight-run gas oil (SRGO) under industrial conditions to 6 ppm-sulfur content, while simultaneously reducing the aromatics to 75% of their initial amount (Reinhoudt *et al.*, 1999). Bowker (2011) reported that silica-supported rhodium phosphide (Rh_2P/SiO_2) has a higher DBT-HDS activity than either silica-supported rhodium metal (Rh/SiO_2) or rhodium sulfide (sulf. Rh/SiO_2) and is also more active than the commercial Ni-Mo/Al_2O_3 catalyst, but the silica-supported palladium phosphide (Pd_3P/SiO_2, Pd_5P_2/SiO_2) catalysts are less active than commercial Ni-Mo/Al_2O_3 catalysts, due to their limited active site densities. However, they have higher DBT-HDS efficiency and more S-resistance than Pd/SiO_2. Kaluža and Gulková (2016) reported that the transition metal governed the ranking of the BT- HDS efficiency as follows:

$$\text{CoMo} \approx \text{NiMo} > \text{PtMo} \approx \text{RhMo} > \text{PdMo} > \text{IrMo} > \text{RuMo}.$$

Trakarnpruk *et al.* (2008) reported the application of an unsupported MoS_2 catalyst in the HDS of two types of diesel oils; the SRGO directly obtained from atmospheric distillation of crude oil and LCO produced in the FCC unit with a total sulfur content of 6100 and 310 ppm, respectively, in a high pressure batch reactor at 350 °C and 30 atm. Seventy percent desulfurization has been achieved in both oil feeds within 90 minutes. However, the reactivity of the refractory sulfur compound in the two feeds were ranked as follows: for LCO BT > DBT > 4,6-DMDBT, while for SRGO BT > 4,6-DMDBT > DBT. The SRGO used in that study contained a total nitrogen of 80 ppm and higher aromatics content than that of LCO (29.2 and 5 wt.%, respectively).

Akzo Nobel produced the trimetallic catalysts KF901 and KF902 for FCC-feed hydrotretament which maintains the hydrodenitrogenation (HDN) and aromatics saturation activity of the conventional Ni-Mo catalyst, while enhancing the HDS activity of the conventional Co-Mo ones at the same operating conditions and life cycle, but they are more expensive than conventional Ni-Mo catalysts since Co is more expensive (Reid, 2000).

However, various supports have been used to enhance the catalytic activity in HDS, including carbon (Farag *et al.*, 1999a,b), silica (Cattaneo *et al.*, 1999), zeolites (Breysse *et al.*, 2002), titania (Morales-Ortuño *et al.*, 2016), zirconia (Afanasiev *et al.*, 2002; Kaluža and Gulková, 2016), and silica-alumina (Qu *et al.*, 2003). The zeolite-based supported HDS catalysts are characterized by high hydrotreating (HDT) capacity, good distribution of products, and high catalytic activity for the removal of sterically hindered sulfur compounds, such as 4,6-DMDBT, due to the strong acidity, large surface area, and high hydrothermal/chemical stability of the zeolite-based supports (Wang *et al.*, 2014). However, it has been reported that the HY-zeolite is deactivated faster than CoMo/Al_2O_3 (Pawelec *et al.*, 1997). It has been reported that NiW/Al_2O_3 is less active than NiMo/Al_2O_3 in DBT HDS, but more active in 4-ethyl, 6-methyl-dibenzothiphnene (4E6MDBT) HDS because of its greater hydrogenation activity (Robinson *et al.*, 1999). Xu and Liu (2004) reported HDS of DBT and 4-MDBT by CoMo/γ-Al_2O_3. Moreover, at high sulfur levels, the ASA supported that noble metal catalysts are poisoned by sulfur and NiW/ASA catalysts become preferable for deep HDS and dearomatization (Robinson *et al.*, 1999). However, Reinhoudt *et al.* (1999) showed that ASA supported Pt/Pd catalysts are very promising to apply in deep desulfurization, provided that H_2S is removed efficiently. A major drawback of this though is the price of the noble metals. Moreover, the PtPd/ASA catalysts are excellent for streams with low or medium S-content and low aromatics, while the Pt/ASA catalysts are better for streams with higher aromatic contents. The TiO_2 supported Co or CoMo catalysts have higher activities than the Al_2O_3 supported ones (Abrahamson, 2015), but they are not commercialized because of their thermal instability, low surface area and poor mechanical properties (Dhar *et al.*, 2003). However, the TiO_2–Al_2O_3 supports have been extensively studied because of their commercial prospects. Yeetsorn and Tungkamani (2014) reported that the CoMo catalysts supported on titanium-rich (Al_2O_3–TiO_2) and pure titania carriers have higher HDS activity than those supported on pure alumina since the TiO_2 is acting as an electronic promoter in catalysts. But, to a certain limitation, the optimum concentration of tiania that expressed the highest HDS activity was estimated to be 1:0.75 (w:w) Al_2O_3–TiO_2. The ZrO_2–Al_2O_3 supported molybdenum catalysts promoted by Co and Ni received considerable attention for the HDS of Th, 4,6-DNDBT and real feed streams. The CoMo/ZrO_2–Al_2O_3 catalysts are reported to have higher activities than γ-Al_2O_3 supported catalysts (Flego *et al.*, 2001; Zhao *et al.*, 2001). Also, Lecrenay *et al.* (1998) reported that NiMo/ZrO2–Al_2O_3 has a higher HDS efficiency than NiMo/γ-Al_2O_3 for 4,6-DMDBT, gas, oil, and LCO. Deng *et al.* (2010) studied the effect of catalyst loading, hydrogen flow rate, and the operating temperature and pressure on the HDS of diesel oil in a slurry reactor using NiMoS/Al_2O_3 in a high pressure autoclave where the HDS efficiency increased proportionally with catalyst loading, increased temperature, pressure, and hydrogen flow rate. Xu *et al.* (2017) reported the preparation of the bimetallic catalyst, NiMo/SiO_2-$Al_2O_{3,}$ using gemini surfactant that has a high HDS efficiency towards 4,6-DMDBT and FCC-diesel oil. Krivtcova *et al.* (2015) reported that in the application of an aluminum-cobalt-molybdenum catalyst (GKD-202) for the HDS of diesel oil feed with total sulfur content of 1.4 wt.%, the rate of HDS of the sulfur compounds has been found to be ranked in the following decreasing order sulfides > BTs > DBTs. Moreover, the rate of

HDS decreases with the increase of the alkyl substituents concentrations. The zeolite-alumina supported catalysts have higher activity than alumina supported catalysts (Ali *et al.*, 2002; Marin *et al.*, 2004; Wan *et al.*, 2010; Sankaranarayanan *et al.*, 2011; Nakano *et al.*, 2013; Wang *et al.*, 2014). The noble-ASA/mixed zeolite supported catalysts, for example Pt–Pd/HY-MCM-41, exhibits high HDS efficiency with good S-tolerance (Zhang *et al.*, 2007). Moreover, the NiMo/HY-MCM-41 catalysts are known to have higher HDS activity than alumina supported catalysts (Ren *et al.*, 2008; Li *et al.*, 2009a). It has been reported that the addition of a small amount of HY zeolite to Al_2O_3 increases the HDS activity by 1.2 times compared with the Al_2O_3 supported CoMo catalyst (Ren *et al.*, 2008). Noble metals, especially platinum (Pt), supported on zeolites are reported to have high and stable activity in the HDS of thiophene (Kanda *et al.*, 2009) that is attributed to the enhanced adsorption of the reactant Th on these high specific surface area zeolites and the hydrocracking due to the high acidity of the zeolite substrate. Moreover, upon applying transition metals, an enhancement of hydrogenolysis reactivity over the metal occurs (Boukoberine and Hamada, 2016).

Recently, hierarchical zeolites have attracted much attention due to the presence of additional mesopores and/or macropores which increase the active phase and the access of large reactants as well as diffusion improvement by shortening the mass transfer pathway (Xiao and Meng, 2011; Chen *et al.*, 2012). It has been reported that noble metals supported on hierarchical zeolites are more efficient than those supported on microporous zeolites and Al_2O_3 (Sun and Prins, 2008).

Most of the carriers (i.e. the supports) are acidic in nature, however, MgO as an alkaline active support has been reported (Klicpera and Zdrazil, 2002; Heidarinasab *et al.*, 2016). Moreover, the higher resistance of MgO-support catalysts towards coking would facilitate its application at low or moderate pressure, but to overcome the limitation of its low specific surface area and low textural stability, the introduction of MgO on the formulation of Al_2O_3 has been suggested (Wu *et al.*, 2009). Not only this, but it will hinder the hydrogenation of the unsaturated compounds which have a great value in petroleum products (Zdrazil and Klicpera, 2001). The Mo/P_2O_5-promoted γ–Al_2O_3 favors the DBT-hydrogenation pathway, while the Mo/MgO favors the direct DBT-HDS pathway. Moreover, the Mo/MgO was reported to express higher DBT-desulfurization activity with an extremely low cooking than that of Mo/P_2O_5-promoted γ-Al_2O_3 in a fixed bed reactor at 320 °C under the atmospheric pressure (Heidarinasab *et al.*, 2016). In addition, the physical mixing of the two catalysts promoted Th reactivity. Although, the Th molecule is known to have a lower electron density over its sulfur heteroatom, thus it has an extremely low HDS reactivity under atmospheric pressure. The only possibility to activate Th is throughout its hydrogenation or, at least, its partial hydrogenation (Ma *et al.*, 1995) where the good synergism between MgO and $\gamma-Al_2O_3$ is the main reason for gaining the advantages of the dual-function catalyst (Trejo *et al.*, 2008; Heidarinasab *et al.*, 2016). The use of hydrotreating catalysts supported on transition metals containing (Ag^+, Cu^{2+} and Ni^{2+}) micro- and mesoporous materials are reported to have an excellent deep HDS activity (Yang, 2003; Hernández-Maldonado and Yang, 2004a; Gong *et al.*, 2009; Boukoberine and Hamada, 2016). Moreover, for the HDS of heavy and residual distillates containing recalcitrant S-compounds, the use of catalysts with an appropriate balance between cracking and hydrogenolysis functions and lengthy thermal stability are required (Boukoberine and Hamada, 2016).

A trickle-bed reactor is usually used in large refineries for the HDS process (Srivastava, 2012). Improved HDS efficiency can be achieved by applying the countercurrent process (Babich and Moulijn, 2003). ABB Lummus estimates that the counter-current design can reduce the catalyst volume needed to achieve 97% desulfurization by 16% relative the conventional cocurrent design with a cetane improvement (EPA-Diesel RIA, 2000). The SynAlliance (ABB Lummus, Criterion Catalyst Corp., and Shell Oil Co.) has patented a counter-current reactor design called SynTechnology where in a single reactor design, the initial portion of the reactor will follow a

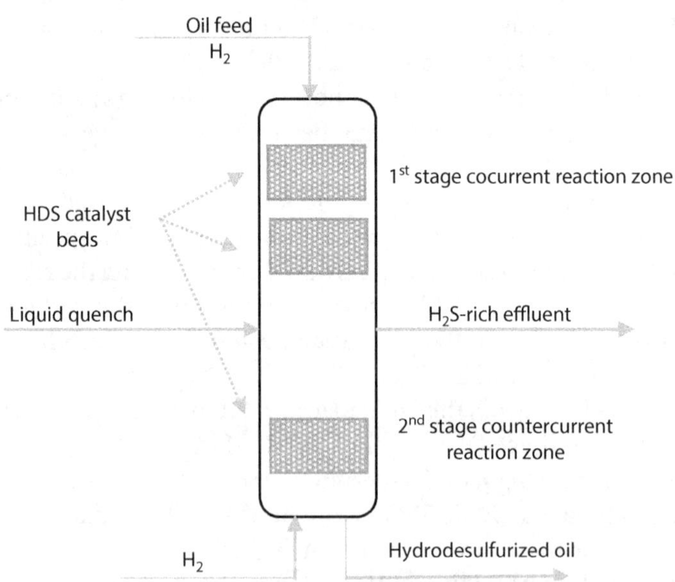

Figure 2.6 Schematic Diagram for the Cocurrent/Countercurrent HDS-Process.

co-current design, while the last portion of the reactor will be counter-current (Figure 2.6, Swain, 1991). This is very helpful in the application of very active, but S-sensitive noble-metal catalysts. The Scanraff's SynSat gas oil hydrotreating unit in Sweden uses noble-metal catalysts in the second countercurrent stage of the process. In the SynSat technologies, the countercurrent approach not only removes sulfur, but it also removes nitrogen and aromatics, where ULS-level (≈1 ppm) and aromatics level of 4 vol.% could be reached (van der Linde *et al.*, 1999).

The ebullated catalyst bed HDS process is a promising new advanced HDS process where the catalyst, the feed, and hydrogen are a good mix since they are in a fluidized state. Moreover, it is characterized by very good heat transfer efficiency, thus the overheating of the catalyst carrier and the formation of coke are minimized. The reactor operates in an isothermal condition with constant, rather small, pressure drops. The clogging and erosion of the catalyst are minimized. There is a good control on the catalyst activating in an ebullated bed HDS reactor throughout, with the flexibility of adding and removal of catalysts (Song, 2003). The T-Star IFP process uses the ebullated bed to desulfurize heavy feedstocks such as deep cut heavy vacuum gas oils, coker gas oils, and some residues. Integrating the T-Star process with hydrotreating process would produce diesel with an S-content of <50 ppm and gasoline with an S-content of 30–50 ppm (Billon *et al.*, 2000).

Reactors using monolithic catalyst supports are an attractive alternative to conventional multi-phase reactors (Kapteijn *et al.*, 2001), where instead of the catalyst trickle-bed, monolithic channels are present where bubble-train (or Taylor) flow occurs. Gas bubbles and liquid slugs move with constant velocity through the monolith channels approaching plug flow behaviour. Gas is separated from the catalyst by a very thin liquid film and during their travel through the channels the liquid slugs show internal recirculation. It has the advantages of optimal mass transfer properties and the achievement of sharp residence time distributions for gas and liquid compared to trickle flow (Kapteijn *et al.*, 2001; Nijhuis *et al.*, 2001). Larger channel geometries (internally finned monolith channels) might allow counter current flow at a relevant industrial scale. Monolithic catalysts can be prepared in different ways. They can be produced by direct extrusion of support material (often cordierite is used, but different types of clays or typical catalyst carrier materials such as alumina are also used), of a paste also containing catalyst particles (e.g. zeolites, V-based catalysts), or with a precursor of catalyst active species (e.g. polymers for carbon monoliths). The catalyst loading of the reactor, in this case, can be high (Kapteijn *et al.*, 2001). Alternatively, catalysts, supports, or their precursors can be coated into a monolith structure by a wash-coating (Beers *et al.*, 2000).

The polar nature of S- N- and O- containing compounds makes them have relatively high dielectric constants and dipole moments and more sensitive to microwave irradiation. The application of microwave irradiation in petroleum refinery processes (catalytic reforming, catalytic cracking, catalytic hydrocracking, hydro-dealkylation, and catalytic polymerization) was developed in the 1990s (Shang *et al.*, 2013). Microwave energy can preferentially heat these heterocyclic compounds and induce, for example, the desulfurization assisted by catalysts or solvents. The theory of use of microwaves in the desulfurization of petroleum streams at a relatively lower temperature is supported by the concept "hydrocarbon molecules are more transparent to microwaves than organo-sulfur or organo-sulfur-mettalic compounds" (Shang *et al.*, 2013). Thus, microwave energy would preferentially activate the sulfur compounds. The application of microwaves in HDS would achieve sustainable savings in capital and operating costs because catalytic reactions can be accelerated by the microwave energy, performed under less severe conditions (i.e. lower temperature and pressure), and shorter catalyst contact time. These make it possible to use smaller reaction vessels with reduced catalyst consumption (Loupy, 2006). Moreover, the microwave process is reported to be generally preferred to non-destructive desulfurization when no sensitizers or catalysts are used due to low microwave absorption. However, destructive desulfurization occurs upon their usage (Shang *et al.*, 2013). Furthermore, the pulsed mode microwave input is reported to be better than a continuous one for HDS reactions (Purta *et al.*, 2004).

Several catalysts are reported in microwave HDS processes, including powdered iron, charcoal on iron, palladium oxide-silica based material, calcium oxide, alkali metal oxide catalysts, and traditional hydro-treating catalysts (e.g. molybdenum sulfuide supported on porous γ-alumina promoted by cobalt or nickel (Co-Mo/Al_2O_3, Ni-Mo/Al_2O_3)). In addition, additives like boron, phosphorous, or silica can be used too. Unsupported catalysts like metal hydrides and metal powder are proved to be effective in the microwave HDS process, acting as hydrogen donor. Mutyala *et al.* (2010) reported the effectiveness of metal powders in the desulfurization of coal pitches and metal hydride catalysts in the desulfurization of hydrocracked petroleum pitch from Athabasca bitumen by using microwave heating. Wan and Kriz (1985) also reported the effectiveness of iron and copper catalysts in removing 70% of sulfur using microwave HDS in hydrocracked pitch. Activated carbon is reported to enhance the microwave absorption and, consequently, the HDS of high density crude oil (986.5 kg/m^3), where sulfur removal of approximately 65%, was recorded with a reduced coke yield, reduction in viscosity, and the treated crude met the pipeline regulations without dilution (Jackson and Soveran, 1995). Microwave sensitizers such as diethanolamine, silicon carbide, activated charcoal and serpentine are commonly used to improve microwave absorption (Shang *et al.*, 2013). However, when Miadonye *et al.* (2009) used activated charcoal and serpentine as microwave sensitizers, palladium oxide as catalyst, and ethanolamines as polar additives and hydrogen donors for HDS of Saudi Arabia heavy crude oil, the microwave absorption by crude oil was enhanced four-fold and doubled with charcoal and polar solvents, respectively, with no recorded significant changes upon using serpentine due to its poor heat transfer properties. Moreover, the HDS improved from 2.3 to 33.8%, upon using ethanolamine as a hydrogen donor within a short contact time (25 min). Furthermore, the HDS was higher for lighter fractions than heavier ones. Also, Khan and Al-Shafel Emad (2013) reported the increase of HDS efficiency of crude oil from 2.3 to 39% upon the application of microwaves.

However, the challenge is to apply microwave assisted HDS in the petroleum industry on a commercial scale since the reactor materials are either polytetrafluoroethylene (PTFE) or glass/quartz, which limit the maximum operating temperature and pressure required for HDS processes to achieve ultra-low sulfur (ULS) levels.

Generally, gasoline is made up of different fractions coming from reforming, isomerization, and FCC units (i.e., blending of straight run naphtha isomerate, reformate and alkylate products, FCC naphtha, and coker naphtha). Those coming from the reforming and isomerization units are produced from distillation cuts and, consequently, contain little or no sulfur since the S-containing compounds present in crude petroleum have generally high boiling points and the feedstocks used in the isomerization and reforming units are generally hydrotreated (Zhao *et al.*, 2010). On the other hand, the atmospheric residues or the vacuum distillates which constitute FCC feedstocks contain significant amounts of sulfur (approximately 0.5–2.5 wt.%). Since 2009, the EU, North American countries, and some Asian countries aimed successively to reach gasoline with a sulfur content <10 ppm. Depending on feed type, the typical sulfur content of FCC gasoline ranges from 150 ppm to 3,500 ppm. FCC gasoline sulfur is the biggest contributor to the gasoline pool (approximately 30–40%) and it is the major source of sulfur in gasoline (approximately 85–95%) (Song, 2003; Siddiqui and Aitani, 2007). The three main approaches for desulfurization of FCC-naphtha are the post-treatment of the FCC-naphtha products, the pre-treatment of the FCC-naphtha feed, and the in-situ HDS during the FCC-operation (Song, 2003). The sulfur-reduction in the FCC feed before its FCC-catalytic cracking reactor reduces the sulfur in the FCC-naphtha and LCO. However, not all refineries have the capability to carry out the FCC-feed deep hydrotreating as it needs more severe conditions including higher temperature and pressure (Shiflett, 2002; Reid, 2000). For achieving a 10 ppm ULS-gasoline target, most refineries install FCC-selective HDS-units as main countermeasures (Zhao *et al.*, 2010; Lee *et al.*, 2013). Shiflett and Krenzke (2001) reported that for the current commercial hydrotreater operations each 1 wt% sulfur removal results in about 18–20 Nm^3/m^3 feed (110–120 SCFB) of H_2 consumption, each 1000 ppm nitrogen removal results in about 5.9–6.1 Nm^3/m^3 (35–36 SCFB) of H_2 consumption, each 1 wt% aromatics removal yields about 5.0–8.4 (use half of these numbers if aromatics are reported as volume %), and each one unit increase in °API gravity requires about 17 Nm^3/m^3 feed (100 SCFB) of H_2 consumption as does each one unit increase in cetane number for diesel stocks. According to Imhof (2004), the investment required to desulfurize FCC feed ranges between $3,000–$6,000/bbl-stream-day, while Lesemann and Schult (2003) reported that the cost of the FCC additive option was estimated at $0.14/barrel of gasoline for 15% sulfur reduction compared to $1.3/bbl for the selective HDS at 90% sulfur removal, where the FCC additives are mainly supported by metal oxides having Lewis acid properties, such as Zn, Zr, Co, Ni, or Mn impregnated on alumina, hydrotalcite, titania, or Mg(Al)O, taking into consideration that the most successful additives are the ones combining high acidity and accessibility. Although there is a high cost for FCC additives, they are very recommendable because of their flexibility and ease of use. They can reduce sulfur in gasoline by the in-situ selective cracking of sulfur compounds into H_2S in the FCC riser (Bhore *et al.*, 2001; Roberie *et al.*, 2002; Chester *et al.* 2003; Zhao *et al.*, 2003). The activity of different additives for reduction of gasoline sulfur can be ranked in the following decreasing order: Zn/hydrotalcite > ZrO/alumina > Zn/titania > Mn/alumina. Andersson *et al.*, (1999) found that the Zn/hydrotalcite has a value of 80% reduction of sulfur in sulfur-spiked gasoline at the microactivity test level. Shan et al. (2002a,b) reported an additive comprising USY-zeolite/ZnO/alumina with excellent sulfur removal effects, where the optimum temperature for the desulfurization of Th and alky-Ths was found to be 400 °C. The Akzo Nobel's Resolve (Resolve 700, 750, 800 and 850) and the Grace Davison's (Saturn GSR6.1, SuRCA™, SATURN™, D-PriSM™ and RFG™) are examples of the commercial FCC catalyst additives converting organic sulfur into H_2S during the FCC-operation, thus reducing the S-content in liquid products (e.g. naphtha and LCO) (Skocpol, 2000; Purnell *et al.*, 2002; Zhao et l., 2002; Kuehler and Humphries, 2003; Krishnaiah and Balko, 2003).

On the other hand, Atlas *et al.* (2001) reported that the cost of lowering the sulfur content from 200 to 50 ppm has been estimated to be four times higher that reducing the sulfur content of a product from 500 to 200 ppm. Nevertheless, further reduction of sulfur concentration by HDS to <1 ppm will be necessary for future fuels and remains a challenging research and economic target.

As mentioned above, the main S-compounds in FCC-gasoline are thiols, sulfuides, tetrahydrothiophene, thiophenols, Th and its derivatives, and BT (Lappas *et al.*, 2002). Although most of these compounds are not present in FCC-feedstocks, they could either result from the direct transformation of the S-compounds present in the feedstock, throughout the cracking of long alkyl chain thiophenes, and the addition of H_2S to olefins or diolefins followed by cyclization, or they could come from the recombination of FCC-products (Leflaive *et al.*, 2002). However, the thiophene-sulfur represent approximately >60% of the total S-content in FCC-gasoline. Thus, alkyl-Ths are considered as representative models for S-compounds in FCC-gasoline (Yin *et al.*, 2002a,b; Brunet *et al.*, 2005). Scheme 2.2 summarizes the possible pathways for desulfurization of 2-methylthiophene over commercial sulfided CoMo catalysts. It is worthy to know that, relative to the diesel sulfur problem, it is not very difficult to remove sulfur from gasoline by HDS. However, the challenge is to reach to ULS-gasoline, while keeping the octane number.

The HDS of straight-run kerosene that is used for making jet fuels is reported to be more difficult than that from naphtha, but less difficult compared to that from gas oil.

Diesel oil is formed by the blending of straight run diesel, light cycle oil (LCO) from the FCC unit, hydrocracker diesel, and coker diesel. The LCO from FFC is one of the major diesel oil pools. It is characterized by high

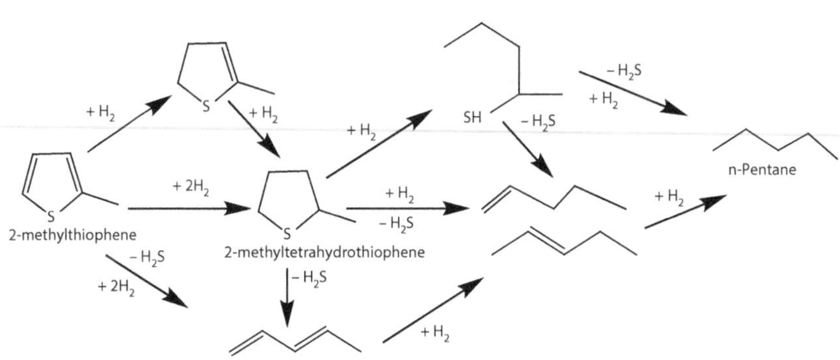

Scheme 2.2 Possible Pathways for Desulfurization of 2-Methylthiophene over Commercial Sulfided CoMo Catalysts.

sulfur and aromatic contents, especially the recalcitrant S-compounds, for example DBT, 4-MDBT, and 4,6-DMDBT. The ultra-deep desulfurization aims to reduce sulfur to <15 ppm in ULS-diesel oil. The problem comes from the low reactivity of 4,6-DMDBT due to the steric hindrance caused by the methyl groups. This prevents the interaction between the S-atom and the active sites of the catalysts. In addition to the inhibiting effects of PAHs and nitrogen compounds found in the SRGO and LCO diesel blends feedstock, PAHs compete with the PASHs on the surface of the hydrotreating catalysts. Moreover, H_2S competes with S-compounds which affects the direct C-S hydrogenolysis route. Much research has been performed on catalyst formylation to enhance hydrogenation of the aromatic ring in 4,6-DMDBT to induce isomerization of methyl-groups away from 4- and 6-positions by incorporating acidic features in the catalyst, removing the inhibiting H_2S, and controlling the reaction conditions for specific catalytic functions (Song, 2003). The general approach for pre-treating FFC feed and the in situ desulfurization during the FCC for reducing S-content in diesel feedstock are as in the case of naphtha desulfurization.

The known commercial processes for the HDS of FCC-gasoline are the selective HDS preserving octane number (Intitute Francais du Petrole (IFP's) Prime G+, Exxon Mobil's Selective Cat Naphtha hydrofining (SCANfining), catalytic distillation (CDHydro/CDHDS), RIPP's RSDS-I/II and OCT-M/OCT-MD, and deep desulfurization associated to octane recovery through alkane isomerization (ExxonMobil's Oct-Gain and UOP-INTEVEP's ISAL™). They are the main two strategies applied in the

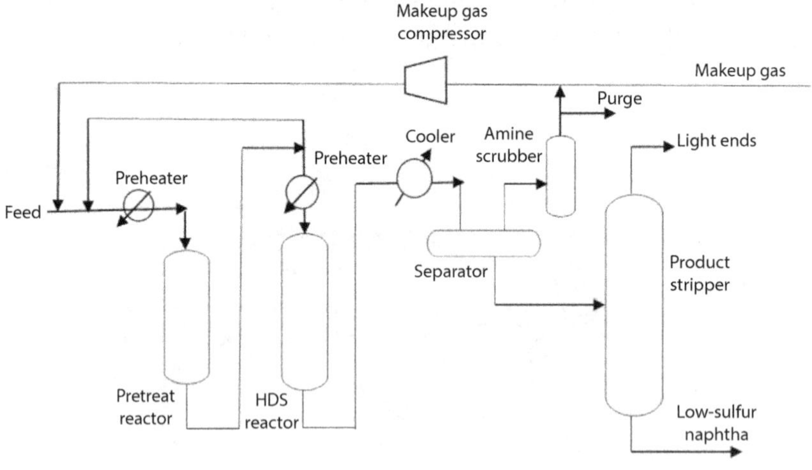

Figure 2.7 Schematic Diagram for ExxonMobil's SCANfining Process for Selective Naphtha HDS.

worldwide refineries for the desulfurization of FCC-gasoline. However, the selective HDS processes which are based on conventional catalytic fixed-bed technology and need low cost investments are, so far, the most attractive for the industry (Song, 2002, 2003; Brunet *et al.*, 2005; Lee *et al.*, 2013; El-Gendy and Speight, 2016).

For example, SCANfining (Figure 2.7) uses an RT-225 catalyst system which was jointly developed by ExxonMobil and Akzo Nobel specifically for the selective removal of sulfur from FCC naphtha by HDS with minimum hydrogenation of olefins and preservation of octane. It offers the ability to eliminate FCC naphtha product splitting towers and reduce hydrogen consumption to 30–50% less than in conventional hydro-finishing. This results in significant investment and operating cost savings (Kaufmann *et al.*, 2000; ExxonMobil, 2001; Halbert *et al.*, 2001; Song, 2003).

The Prime G+ (Figure 2.8) is less severe and has been commercially demonstrated for several years in many worldwide refineries. It is based on a combination of a selective hydrogenation unit which removes diolefins and light mercaptans, a splitter, and a selective HDS of mid and HCN cut through a dual catalytic system (HR-806 and HR-841) which were developed by IFP and commercialized by Axens. The HR-806 achieves the bulk of desulfurization, while HR-841 is a polishing catalyst which reduces sulfur and mercaptans with no activity for olefin hydrogenation (Nocca *et al.*, 2000; IFP 2000, 2001; Baco *et al.*, 2002).

The OCTGain process (Figure 2.9) uses a fixed-bed reactor to totally remove sulfur, saturate olefins, and then restore the octane to economically needed levels, producing products with <5 ppm sulfur and <1% olefins

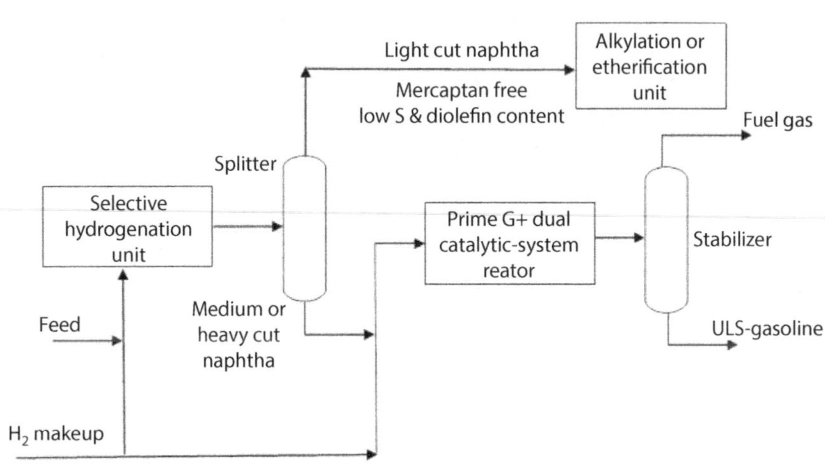

Figure 2.8 Schematic Diagram for IFP's Prime G+ Process.

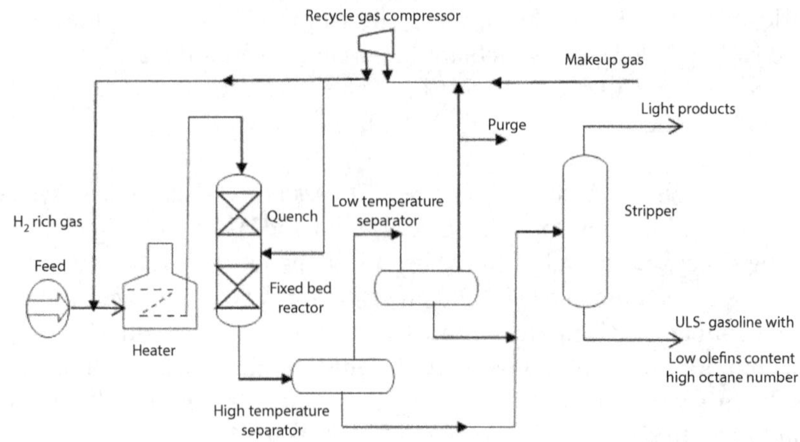

Figure 2.9 Schematic Diagram for ExxonMobil's OCTGain Process for Selective Naphtha HDS.

(Shih *et al.*, 1999; ExxonMobil, 2001). Qatar Petroleum OCTGain unit reported that to produce an ultra-low-sulfur (ULS <1 ppm) and low olefins gasoline, the process flexibility allows for the adjustment of an octane number between −2 and +2 RON of the feed (Chitnis *et al.*, 2003). The UOP-INTER-VEP's ISAl and ExxonMobil's OCTGain process are similar in the concept of process design and processing schemes, but the processing conditions and used catalysts are different. The ISAL™ process, which was jointly developed by INTEVEP SA and UOP, is designed as a low-pressure fixed-bed hydroprocessing technology for desulfurizing gasoline range feedstocks and selectively reconfigures lower octane components to restore product octane number. The catalyst used in this process is typically a combination of an HDS catalyst such as Co–Mo–P/Al_2O_3 and an octane-enhancing catalyst such as Ga–Cr/H-ZSM-5 catalysts in two beds (Salazar *et al.*, 1998; Babich and Moulijn, 2003).

The catalytic distillation desulfurization process developed by CDTech is significantly different from conventional hydrotreating (Rock, 2002; Rock and Shorey, 2003), where hydrotreating requires a distillation column after the fixed-bed hydrogenation unit while the CDTech eliminates the fixed-bed unit by the incorporation of the catalyst in the distillation column. The most important portion of the CDTech desulfurization process (Figure 2.10) is a set of two distillation columns loaded with a desulfurization catalyst in a packed structure. In this process, the LCN, MCN, and HCN are treated separately under optimal conditions for each. The first column, called CDHydro, treats the lighter compounds of FCC

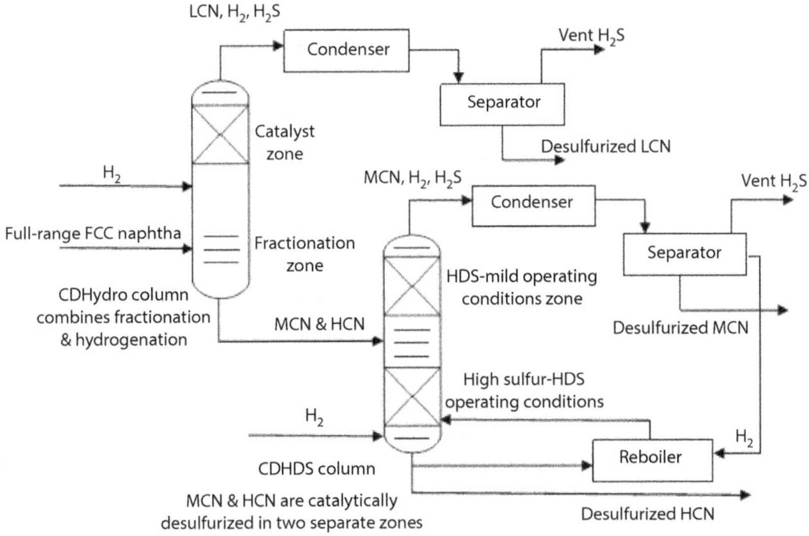

Figure 2.10 Schematic Diagram for Deep-Desulfurization of Full-Range FCC Naphtha.

gasoline and separates the heavier portion of the FCC gasoline for treatment in the second column. Catalytic distillation essentially eliminates catalyst fouling because the fractionation removes heavy coke precursors from the catalyst zone before coke can be formed and foul the catalyst bed. This prevents the catalyst from poisoning. Moreover, the clean hydrogenated reflux continuously washes the catalyst zone, leading to a longer catalyst life-time. In addition, mercaptans can react with diolefins at the bottom of the catalyst zone, forming heavy and thermally stable higher boiling point sufides (>C5 fraction), which can be easily fractionated to the bottom products. That, consequently, eliminates the separate mercaptans removal step (CDTech, 2000; Rock, 2002). The second column, called CDHDS, catalytically removes the sulfur from the heavier compounds of FCC gasoline (MCN and HCN) in two separate zones. The CDHDS is used in combination with CDHyrdro to selectively desulfurize gasoline with minimum octane loss. The HDS-conditions are optimized for each cut for maximum HDS efficiency with a minimum olefin saturation. Olefines go to the column-top where conditions are mild, while the concentrated sulfur moves to the column-bottom where severe conditions are present for higher HDS-efficiency (CDTech, 2000). The temperature and pressure of the CDTech process columns are lower than fixed-bed hydrotreating processes, particularly in the upper section of the distillation column, which is where most of the olefins are located. These operating conditions minimize yield and octane loss (Rock and Shorey, 2003). It has been estimated that the CDTech is approximately 25% less expensive than the conventional HDS process (Babich and Moulijn, 2003). Thus, it is very attractive for refineries and it has been demonstrated at Motiva's Port Arthur, Texas Refinery.

Moreover, Philips Petroleum Co., has proposed a combination between the pre-aromatization of FCC gasoline streams and conventional HDS where the aromatics content increases in the end product by approximately 68%, but the olefins are almost completely saturated and the S-content decreases from 300 to 10 ppm, while the octane number increases from 89 to 100 (Drake and Love, 2000), but due to the increased environmental limitations for the aromatics content in the gasoline, that technology is not widely applied.

The worldwide demand for diesel oil is expected to significantly increase. However, the worldwide reserve of high quality low sulfur content crude oil decreases. Thus, the problem of deep desulfurization of diesel fuel has become more serious due to the lower and lower limit of sulfur content in finished fuel products by a worldwide update in strict new regulatory specifications and the higher and higher sulfur contents in the available crude oils. The known approaches for deep desulfurization to produce ULS-diesel oil are ultra-deep HDS of middle distillates [MAKFining Premium Distillates Technology (PDT) by Akzo Noble, Exxon Mobil and Kellog Brown], ultra-deep HDS (SynHDS) and hydroaromatization (SynSat) and cetane improvement by ring opening (SynShift) of middle distillate [SynTehnology includes; new reactor design by SynAlliance including ABB Lummus, Criterion catalyst and Shell Global], two stage hydrotreating for ULS diesel fuel using industrially proven high-activity TK catalysts [Haldor Topose], ultra-deep HDS, HDN and hydrogenation of distillate fuels [Unionfining by UOP], hydrodearomatization (HDAr) of middle distillate in the second stage [PDT, by Akzo Noble, Exxon Mobil and Kellog Brown], FCC feed deep hydrotreating to remove sulfur before catalytic cracking which produces naphtha and LCO, and, finally, undercutting of S-rich narrow fraction of LCO from FCC [to remove a narrow boiling point range that is rich in refractory sulfur compounds and use it for off-road distillate fuels] (Turaga and Song, 2001; Song, 2002, 2003; Song and Ma, 2003).

There are many reported commercial catalysts for diesel oil-HDS that are marketed by Akzo Nobel (KF 752, KF 756, KF 757 and KF 848), Criterion (Century and Centinel), Haldor–Topsoe (TK-554 catalyst is analogous to Akzo Nobel's KF 756 catalyst and the more active TK-574 catalyst), United Catalyst/Sud-Chemie, and ExxonMobil. For example, Akzo Nobel reported the application of the Super Type II Active Reaction Sites (STARS) catalyst for desulfurizing diesel fuel at the BP Amoco refinery in Orangemouth,

UK. The original unit was designed to produce 35,000 bbl/day diesel fuel with an S-content of 500 ppmw from straight-run gas oil (SRGO) feed and LCO. However, the CoMo STARS catalysts are preferred for feed with high S-content (100–500 ppm) and works better at low pressures, while NiMo STARS catalysts are more suitable for low S-content feeds (<100 ppm) and work better at high pressures. They can work for a long-term run up to 400 days (Ma *et al.*, 1996; Topsoe *et al.*, 1999). The application of KF 757 catalysts in a reactor with a capacity 45,000 bbl/day day diesel fuel with an S-content of 10–20 ppmw has also been reported. KF 756 is reported to be widely used in Europe (≈20% of all distillate hydrotreaters operating by it), while KF 757 has been commercially used in at least three hydrotreaters (Swain, 1991; Turk *et al.*, 2002). Century and Centinel are reported to be 40–70 and 80% more active than the conventional HDS catalysts (Swain, 1991). They are claimed to combine superior hydrogenation activity and selectivity. However, the CoMo-Centinel catalysts are preferred for high S-content streams and work better at lower H_2 pressures, but the NiMo-Centinel catalysts are preferred for low S-content streams (<50 ppm) and work better at higher H_2 pressures. This occurred due to the better dispersion of the active metal on the catalyst substrate. In 2001, ExxonMobil, Akzo Nobel, and Nippon Ketjen introduced a new unsupported catalyst called NEBULA. It is a Mo-W-Ni trimetallic catalyst which is reported to be three-times better than any other hydrotreating catalysts available (Soled *et al.*, 2001). Haldor–Topsoe reported that the increase in temperature by 14 °C, using its advanced TK-574 CoMo catalyst in the processing of a mixture of SRGO and LCO, reduced the S-content from 120 to 40 ppmw (Swain, 1991).

During the FFC-process, the resonance stabilization of Th prevents cracking. Thus, most of the thiophenic sulfur compounds end up forming coke (Corma *et al.*, 2001). Aromatic hyrogenation during HCR would facilitate the desulfurization of heavy oils by cracking and/or hydrogenation. That would facilitate the selective ring opening and, consequently, improve the distillate quality (Santana *et al.*, 2006). Although HDS is used for the upgrading of heavy oil, its efficiency is limited by the properties of heavy oil and the high heavy metal content that deactivates the catalyst and causes deposit formation, coking, and fouling propensity which also causes catalyst deactivation, and high molecular size that causes mass transfer limitation and limits the access into the small catalyst pores. Moreover, the alkylated-thiophenic compounds increase the steric protection of the thiophenic sulfur and thus limits the adsorption to the active catalyst sites for HDS (Javadli and Klerk, 2012a).

Transportation fuels, such as gasoline, jet fuel, and diesel, are ideal fuels due to their high energy density, ease of storage and transportation, and established distribution network. One of the main advantages of HDS is versatility in treating different oil feed, such as gasoline, diesel, and jet fuel, with initial high S-content. For example, it can directly decrease the initial S-content of fuel oil from 21,900 ppmw by approximately 57% and the oil feed of 10,000 ppmw to less than 100 ppmw in one step (Al-Degs *et al.*, 2016). However, to meet the new international environmental regulations for producing ultra-clean transportation fuels with a low concentration of sulfur (<15 ppm) or near-zero sulfur products and to prevent the deactivation of catalysts in reforming process and electrodes in fuel cell system, higher temperatures and pressures with the developing of new expensive metal catalysts and reactors are required. This will consequently increase the energy consumption, the operations, and capital costs, as well as create more carbon dioxide emissions (Castorena *et al.*, 2002; Wild *et al.*, 2006; Yi *et al.*, 2013). In addition to desulfurization, this will also result in demetallization, carbon residue reduction, some denitrogenation, hydrocracking, and coking (Song, 2003).

Heavy oils, which are characterized by high S-content, can be treated through hydrocracking before the HDS process. Since hydrocracking facilitates aromatic hydrogenation, it enables desulfurization by both cracking and hydrogenation, but the main aim of hydrocracking is not the HDS of heavy oil, but the selective ring opening to improve the quality of its distillates (Santana *et al.*, 2006). There are less restrictions for S-levels in non-transportation fuels which are formed from vacuum gas oils and residual fractions from coking and FCC units, than that for transportation fuels since industrial fuels are used in stationary applications where sulfur emissions can be avoided by combustion gas cleaning processes. Deep desulfurization of the transportation fuels implies that more and more of the least reactive sulfur compounds must be converted. Keeping this in consideration, the increase of octane number and decrease of the aromatic and S- contents of the treated gasoline increase the cetane number and improve the density and polyaromatic content of diesel oil. The refining industry has made a great deal of progress towards developing more active catalysts and more economical processes to remove sulfur from gasoline and diesel fuel (oil) (Okada *et al.*, 2002a,b; Egorova and Prins, 2004; La Paz Zavala and Rodriguez, 2004). For example, California refineries are producing gasoline which contains 29 ppm sulfur on average (US EPA 2000) despite the high sulfur content of California and Alaska crude oils used as feed stocks up to 11,000 ppm. Due to incentives and regulations, 10 ppm sulfur diesel has been commercially available in Sweden for several years. But, although this process tends to improve diesel quality by raising cetane number, it decreases gasoline quality by lowering its octane number (Babich and Moulijn, 2003; Song and Ma, 2003). In Bulgaria, since 2009, the FCC based refinery, Lukoil Nentochim Bourgas (LNB), has been producing near zero sulfur gasoline (NZSG) (<10 ppm) by applying FCC feed hydrotreating. Later in 2010 by applying also Prime G FCC gasoline post treatment technology, the FCC gasoline (15 ppm S-content) is the cap for production of NZSG in the LNB refinery (Yankov *et al.*, 2011; Khamis and Palichev, 2012).

Briefly, worldwide, meeting the requirements of reaching to ultra-low sulfur fuels is believed to be the main challenge for the petroleum industry. As most of the gasoline in markets is coming from the cracked feeds with high concentrations of aromatics and olefins, that increases the difficulty of S-removal. Nowadays, two-stage HDS of diesel oil is used to reach to <10 ppm S-content. The first stage usually applies a conventional hydrogenating unit with moderate adjustments to the operating parameters to reach an S-content of approximately 50 ppm. However, the second stage requires higher hydrogen flow rate and purity, reducing space velocity and the choice of more expensive catalysts. Not only this, but it will add another problem, which is the increased production of H_2S, which would increase the stress on the Claus plant capacity, which produces hydrogen for the refineries. Accordingly, refineries have started to establish new complementary routes in addition to the HDS process, such as the incorporation of non-HDS techniques with HDS that will reduce hydrogen consumption and create the required high temperatures and pressures and expensive catalysts for the HDS of the recalcitrant S-compounds.

2.3 Oxidative Desulfurization

Oxidative desulfurization (ODS) is an innovative technology that can be used to reduce the cost of producing ultra-low sulfur diesel (ULSD). This technique is based on the polarity of sulfur compounds. It is well known that the electronegativity of sulfur is similar to that of carbon, consequently, the carbon-sulfur bond is relatively non-polar and the organo-sulfur compounds exhibit properties quite similar to their corresponding organic compounds, so their solubility in polar and non-polar solvents are nearly the same. But, upon oxidation of refractory organosulfur compounds, the polarity of the produced sulfoxides/sulfones increases and, consequently, their solubility in polar solvents increases and can be easily separated by extraction or adsorption distillation and thermal decomposition (Ito and van Veen, 2006; Nanoti *et al.*, 2009; Zongxuan *et al.*, 2011; Mužic and Sertić-Bionda, 2013; Al-Degs *et al.*, 2016). If the ODS is combined with distillation, the process scheme would be similar to the catalytic distillation desulfurization (CDTech) (Babich and Moulijn, 2003). The ODS occurs at relatively low temperatures and atmospheric pressures in presence of oxidizing reagents such as aqueous hydrogen peroxide, ozone, tert-butyl hydroperoxide, oxygen/aldehyde, potassium ferrate, etc. The ODS applying an oxidizing agent alone, such as H_2O_2, is slow. Thus, the reaction requires the presence of a homogenous catalyst such as organic acids (for example formic acid, acetic acid), cobalt-manganese-nickel acetates, polyoxometallic acids and their salts in an aqueous solution, or heterogeneous catalyst such as tungsten/zirconia, titanium/mesoporous silica, peroxy-carboxylic acid, functionalized hexagonal mesoporous silica, molybdenum-vanadium oxides supported on alumina, titania, ceria, niobia, silica, or cobalt-manganese-nickel oxides supported on alumina (Javadli and de Klerk, 2012b). The ODS efficiency of organic acids decreases approximately in the following order: formic acid > acetic acid > propanoic acid. Since the longer C-chain decreases the ODS-efficiency, due to its lower solubility in H_2O_2, it causes lower peroxyacid production and, consequently, decreases sulfur removal. Although formic acid can promote a higher removal in some cases, it is an unstable and corrosive reagent (de A. Mello *et al.*, 2009). Thus, in most cases, acetic acid is considered the best choice.

Ozone, as a highly reactive oxidant, was used for the ODS of LGO containing 150 ppm DBT and 145 ppm 4,6-DMDBT in a batch reactor where the DBTs were quantitatively converted to DBT sulfones and the 4,6-DMDBT was more readily oxidized than DBT (Otsuki *et al.*, 1999). Wang *et al.* (2009a) investigated ozone-mediated ODS of oil containing high thiophene concentrations using SO_4^{2-}/ZrO_2 (SZ) at ambient temperature and pressure. Kazakov *et al.* (2016) reported the ODS of straight-run residual fuel oil (with 2.86 wt. % total sulfur content) from Astrakhan gas condensate by ozonized air where the S-content reached 0.48 wt.% at optimum operating conditions of an ozonization temperature of 60 °C, rate of ozonized air circulation of 1222 nm^3/m^3, residual fuel oil-ozonized air contact time of 0.05 h, ozonized residual fuel oil thermolysis temperature of 300 °C, and thermolysis time of 2 h, whereupon hydrogen sulfide is removed from the residual fuel oil completely and the combustion heat increases. The increase of desulfurization degree with ozonization temperature was attributed to the decrease in viscosity of the reduced residual fuel oil which facilitates its mixing with ozonized air and increases the reaction of ozone with sulfur compounds. An increased rate of ozonized air circulation contributes to greater desulfurization due to an increased degree of oxidation reactions. With the increase of reaction time, the degree of desulfurization increases and is attributable to a higher degree of oxidation of sulfur compounds and their complete oxidation to sulfones.

Table.2.2 Electron densities on S-atoms and rate of ODS in some S-compounds using H_2O_2/organic acid oxidative system (Otsuki *et al.*, 2000).

S-compound	Electron density	Structure and reactivity
Methy phenyl sulfide	5.915	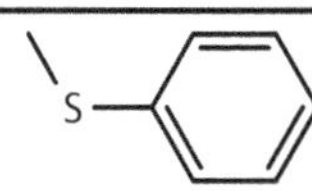
Thiophenol	5.902	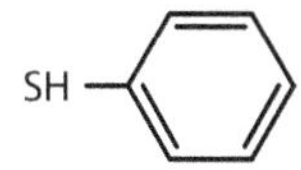
Diphenyl sulfide	5.860	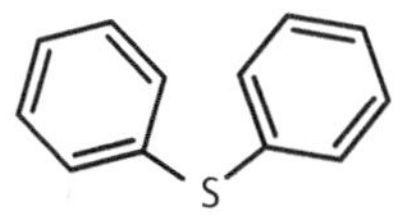
4,6-DMDBT	5.760	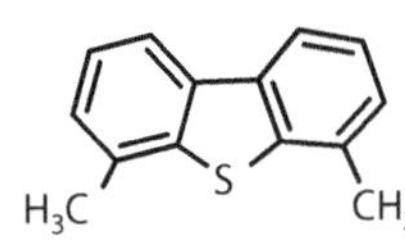
4-MDBT	5.759	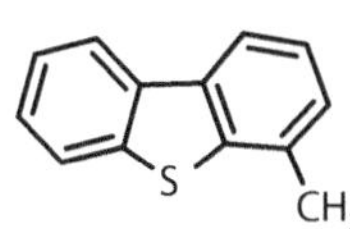 Decreases
DBT	5.758	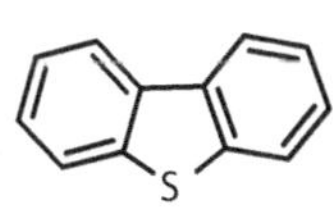
BT	5.739	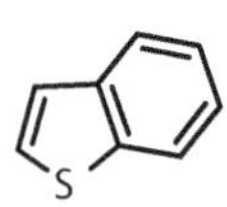
2,5-Dimethyl thiophene (2,5-DMT)	5.716	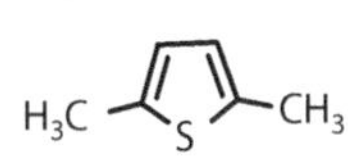
2-Methhyl thiophene (2-MT)	5.706	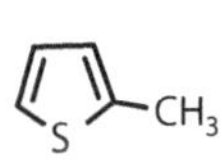
Thiophene	5.696	

Otsuki *et al.* (2000) reported that the higher the electron density on the S-atom, the higher the oxidative reactivity of the S-containing compound using H_2O_2 and formic acid. The electron densities on the S-atom of different S-compounds, as estimated by molecular orbital calculations, are summarized in Table 2.2. It is obvious, from Table 2.2, that the oxidative reactivity of S-compounds is opposite to that of hydrodesulfurization reactivity. Yao *et al.* (2004) confirmed the criteria where conversion of 4,6-DMDBT, DBT, and BT in *n*-heptane model oil was 100, 96, and 58%, respectively, using formic acid as a catalyst and H_2O_2 as an oxidant, since the electron donation of a methyl group increases the electron density on the S-atoms (Otsuki *et al.*, 2001). Ali *et al.* (2009) reported the ODS of a model system consisting of a toluene/hexane solvent containing 500 ppm DBT as a source of polyaromatic S-heterocyclic compounds in the presence of H_2O_2 and acetic acid/H_2SO_4 (2:1 molar ration) at 80 °C and under atmospheric pressure. The ODS of two real oil feeds was also performed, being FCC gasoline and hydro treated diesel, where the S-content decreased from 670 and 1045 ppm to 109 and 85 ppm, respectively. The oxidation process did not show any deleterious effects on the distillation profile and other characteristics of diesel fuel. Mamaghani *et al.* (2013) added sulfuric acid to increase the acidity of the sulfur compounds and catalytic activity of formic acid where a complete desulfurization of a model oil of BT, DBT, and 4,6-DMDBT was achieved by ODS to sulfone using H_2O_2 and formic acid, followed by liquid-liquid extraction using acetonitrile where the oxidation rates of benzo- and dibenzothiophene homologues depended on the amount and position of

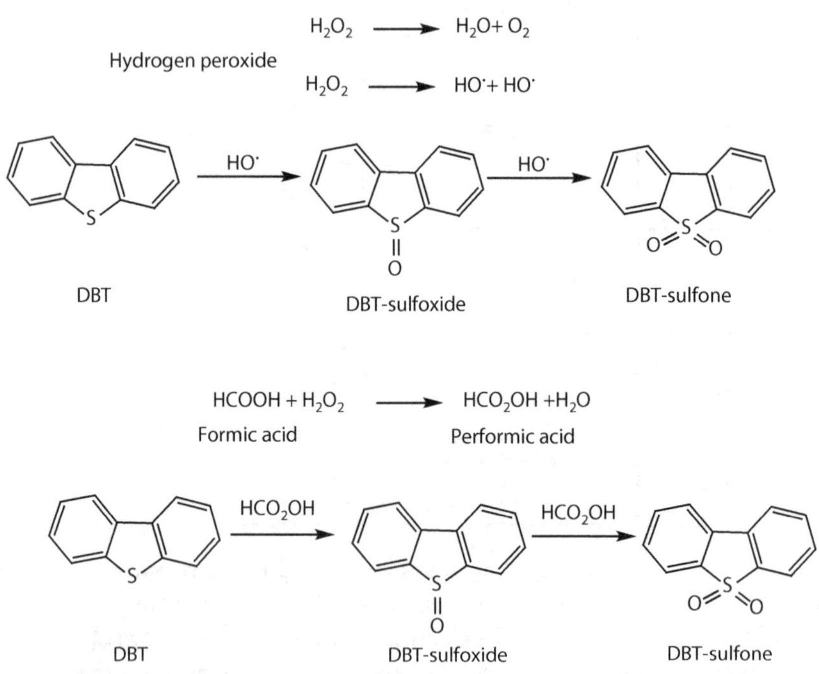

Scheme 2.3 Oxidative Desulfurization Reaction.

alkyl substituents. Farshi and Shiralizadeh (2015) reported that the ODS of heavy oil using an H_2O_2/acetic acid system is better than that performed using an H_2O_2/formic acid system where the S-content decreased from 2.75 wt.% to 1.14 wt.% at 60 °C under atmospheric pressure. The hydrogen peroxide/acetic acid system also proved to be efficient for the ODS of tire pyrolysis oil to be blended as heating fuel (Ahmad *et al.*, 2016).

The mechanism of the ODS using an H_2O_2/organic acid oxidative system can be summarized as in Scheme 2.3 (Haw *et al.*, 2010).

There are two competitive reactions for H_2O_2. The first reaction, which is undesirable because it increases the amount of water in the system and reduces the concentration of the oxidant, is undertaken through the thermal decomposition of H_2O_2. Thus, although the stoichiometric n_O/n_S ratio is 2, the oxidant is usually used in excess of the stoichiometric ratio because of the transport limitation in two-liquid-phase reaction systems and the unproductive decomposition of H_2O_2 to water and oxygen. Therefore, high temperatures are not recommended. The second reaction occurs through the production of strong oxidizing hydroxyl radicals ($HO^{\bullet}$). DBT is then oxidized to its corresponding sulfoxide or sulfones. In presence of acid (e.g., formic acid), H_2O_2 and acid interact and produce performic acid (CH_2O_3). Then, the performic acid gives its oxygen to DBT producing sulfoxide and sulfone.

The most widely used oxidants in the ODS process are reported to be the hydrogen peroxide and the t-butyl-hydroperoxide. However, H_2O_2 is more applied in the ODS as it is green and inexpensive. It is relatively cheap and readily available, with a worldwide annual production of approximately 4.3 million tonnes (100% H_2O_2), mainly for bleaching, pollution, and water treatment. Moreover, there is no production of undesirable by-products or liquid degradation products during the ODS process, other than water.

ODS using H_2O_2/organic acid oxidative system is characterized by an efficient and high desulfurization rate under mild operating conditions (60–65 °C and atmospheric pressure). However, oil-soluble liquid organic acids impact the treated fuel quality. Moreover, these liquid acids are unrenewable and have a high reclaiming cost. Not only this, but it is also a biphasic catalytic system where the S-containing compounds are present in the diesel phase, while the catalyst and H_2O_2 are present in the aqueous phase. It is well known that the reactions in a biphasic system are relatively slow due to the mass transfer limitations across the interface, where the reaction would take place at the interface or in the bulk of one of the phases. Thus, the process rate would be determined either by the reaction rate or the diffusion rate, so the use of surfactants (Krotz *et al.*, 2005), ultrasonic (Mei *et al.*, 2003; Mello *et al.*, 2009; Zhao and Sun, 2009), or increased mixing speed (Sharipov and Nigmatullin, 2005) were suggested as a solution, but this would increase the overall cost of the process, thus the application of solid catalysts is recommendable.

When a catalyst is employed, which may be homogeneous or heterogeneous, the relative oxidation reactivity is different: DBT > 4-MDBT > 4,6-DMDBT > BT. The same trend was reported applying H_2O_2/ionic liquids systems (Li *et al.*, 2009b). The relative reactivity of DBT and substituted DBT compounds appear to be related to the steric hindrance of the alkyl-substitutions at positions 4 and 6 in the DBT molecule (Campos-Martin *et al.*, 2004; 2010). Moreover, Zongxuan *et al.* (2011) reported that the aromaticity of the organic solvents also influences the oxidation rate.

However, upon a study performed by Krivtsov and Golovko (2014), the ODS of straight run diesel fraction (200–360 °C) with a high initial sulfur content (1.19%) included a sulfide sulfur of 0.26 wt.% and thiophenes (BTs and DBTs) 0.93 wt.%, and saturates, mono-, di-, and tri- aromatic hydrocarbons of 53.4, 28.7, 8.2 and 7.9 wt.%, respectively. A 96% S-removal was achieved at 35 °C within 8 h ODS using H_2O_2 and formic acid followed by the adsorption of oxidized products. Sulfides were found to be completely removed, while the rate of ODS of the BT and DBT homologue decreased with the increasing number and size of alkyl-substituents due to the enhancement of steric hindrances around the sulfur atom electron pairs with an increasing number of the alkyl substituents. But, compounds containing substituents in the 4- and 4,6-positions (dibenzothiophene homologues which are the most recalcitrant in HDS-process) were almost completely removed using the hydrogen peroxide–formic acid mixture since the combined positive inductive effect of the methyl groups in 4,6-DMDBT dominate over the steric hindrances to the electrophilic attack at a lone electron pair of the S-atom, as in the case of 4-MDBT. However, the rate of ODS of 4,6-DMDBT was lower than that of 4-MDBT because of the enhanced hindrance of the sulfur atom (by the second methyl group of 4,6-DMDBT). The rate of oxidation of the other identified DMDBT isomers was lower than that of DBT. Note that the isomers containing one of the methyl groups in the 4- or 6-position were characterized by higher values of the effective oxidation rate constants. That was attributed to a positive inductive effect (+I) which extends over the C–C bond chain and leads to an increase of electron density in the conjugated aromatic system, thereby facilitating electrophilic addition reactions. This is due to the significant predominance of the +I effect over the steric hindrance created by the methyl group, although 2-ethyl-DBT expressed the highest oxidation rate constant. The rate constant of 4-ethyl-DBT oxidation was almost two times below that of DBT because the steric effect of the ethyl group

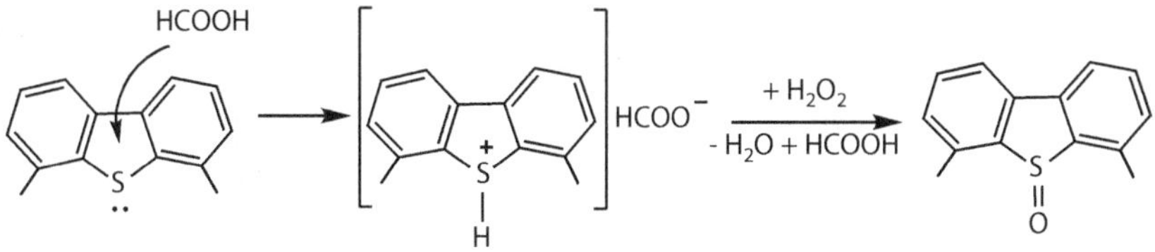

Scheme 2.4 ODS in Absence of Compounds that Facilitate Phase Transfer of Interacting Components.

in the 4-position dominates over its +*I* effect (in contrast to the methyl group). In conclusion, the relatively low effective rate constants for oxidation by the H_2O_2–HCOOH system are due to the fact that the reaction occurs predominantly through the protonation step in which the polar transition state is formed, not through the formation of performic acid (Scheme 2.4). In the absence of compounds that facilitate phase transfer of the interacting components, (Scheme 2.4) what probably occurs most is formic acid partially dissolves in the diesel fraction and an aromatic compound (hydrocarbon or sulfur compound) is protonated. The intermediate product (charge transfer complex), due to its polarity, is pushed to the aqueous solution/diesel fraction interface at which it is oxidized by a hydrogen peroxide molecule (Zhao *et al.*, 2007; di Giuseppe *et al.*, 2009; de Filippis *et al.*, 2010).

The methyl groups of 4-MDBT and 4,6-DMDBT significantly decrease the rate of HDS of these compounds. This is due to the quite close location of the alkyl groups, which sterically hinder the coordination of the DBT molecule through the lone electron pair of the sulfur atom to the catalyst active site. However, the +I effect plays a significant role in the ODS; the closer the alkyl substituent to the sulfur atom, the stronger the effect. Briefly, the interplay of the +I effect and steric hindrances due to the alkyl group size determines the reactivity of DBT homologues in their ODS. In many cases, the presence of alkyl substituents enhances the reactivity of a DBT homologue.

Te *et al.* (2001) reported the oxidative desulfurization (ODS) for recalcitrant sulfur compounds using a polyoxometalate/H_2O_2 decreases in the following order: DBT > 4-MDBT > 4,6-DMDBT. This is similar to HDS and is attributed to methylation and the steric hindrance effect where the activation energy of DBT, 4-MDBT, and 4,6-DMDBT oxidation were 53.8, 56, and 58.7 kJ/mole, respectively. Otsuki *et al.* (2001) reported that the γ-Al_2O_3 support itself can enhance the ODS where, in a flow reactor, more than 90% of 150 ppm DBT was oxidized in a decahydronaphthalene (decalin) solution using t-butyl hypochlorite (t-BuOCl) under ambient pressure at 50 °C. When Al_2O_3 was treated with KOH, the conversion of DBT decreased because the number of the acid sites on Al_2O_3 also decreased. Moreover, the ODS decreased with basic metal catalysts in the following order: γ-Al_2O_3> TiO_2 ≈ Diaion > SiO_2 > ZnO >> MgO (Otsuki *et al.*, 2001). The ODS of DBT is also reported to be inhibited by other compounds found in LGO and the extent of retardation is decreased in the order: nitrogen compounds > olefins > aromatics and paraffins (Otsuki *et al.*, 2001). The catalytic ODS of model oil DBT in decalin was performed using oil-soluble oxidant cyclohexanone peroxide (CYHPO) with a molybdenum oxide (MoO_3) catalyst supported on macroporous polyacrylic cationic exchange resin, D113, of weak acid series, where 100% ODS of DBT to $DBTO_2$ occurred at 100 °C in 40 min (Zhou *et al.*, 2007).

Wang *et al.* (2003) reported the complete desulfurization of hydrotreated oil feed with a total S-content of 39 ppm throughout the application of ODS using t-butyl hydroperoxide in the presence of a MoO_3/Al_2O_3 catalyst. However, the major drawback of this process is the high price of t-butyl-hydroperoxide and the waste treatment of t-butyl alcohol and sulfone. The t-butyl alcohols can be used as potential octane improving compounds for gasoline. Moreover, they can also be converted to MTBE, which can be used as fuel within the refinery while the sulfone stream could be transferred to a coker or bio-processor (Stanislaus *et al.*, 2010).

Chica *et al.* (2006) reported the complete conversion of S-compounds throughout the ODS of a partially hydrotreated LCO with a total S-content of 330 ppm in a continuous fixed bed reactor using a Ti-MCM-41S (sililated) catalyst in the presence of t-butyl hydroperoxide under atmospheric pressure at 373 K. Sampanthar *et al.* (2006) reported the ODS of diesel fuel of 430 ppm S-content using molecular oxygen in the air in the presence of Mn- and Co- containing oxide catalysts, supported on γ-Al_2O_3, at a temperature range of 130–200 °C and atmospheric pressure, followed by extraction using polar solvent, produced diesel fuel with a 40–60 ppm S-content. Binary mixed metal-oxide catalysts (for example 5%MnO_2/3%Co_3O_4/γ-Al_2O_3) showed better oxidation activity. To reach diesel oil with a lower S-content of ≈10–15 ppm, the obtained treated diesel (i.e. the oxidized and solvent extracted diesel) was passed into an activated basic γ-Al_2O_3 adsorbent-bed at room temperature. The ODS of model refractory S-compounds was found to decrease in the following order: trialkylsubstituted dibenzothiophene > dialkyl-substituted dibenzothiophene > monoalkyl-substituted dibenzothiophene > dibenzothiophene. This was attributed to the increased electron density of the sulfur atoms in di-substituted thiophenes which overcompensated for the steric hindrance of the C4 and C6 alkyl groups in the oxidative process. Moreover, the olefin content of the diesel was increased, while the aromatic content of the diesel was reduced substantially. The cetane index was also increased by approximately 20%. Density and other parameters were within the required limits. Furthermore, the lubricity of the treated diesel was increased by a substantial amount, which added advantageously to the applied catalyst and inexpensive molecular oxygen found in the air. In addition, the use of air as an oxidant also eliminates the need to carry out any oxidant recovery process that is usually required upon application of liquid oxidants. Furthermore, this allows for ease of integration into any existing refinery without major changes in the infrastructure.

The hydrothermally stable Ti-HMS, a hexagonal mesoporous molecular sieve with a pore size of about 3 nm, as a catalyst, was compared with Ti-MSU (long, worm-like pores) and TS-1 in the ODS of a model fuel (Th, BT or DBT in n-octane) in presence of H_2O_2 (Cui *et al.*, 2007). TS-1 could not oxidize BT or DBT since these molecules hardly penetrate the small pores of TS-1, but BT and DBT can penetrate the mesopores of Ti-MSU and Ti-HMS and oxidize. The Ti-HMS was the most effective catalyst for the oxidation of both BT and DBT with H_2O_2 where the ODS of DBT increased with the increase of the titanium content in the catalyst recording approximately 80% DBT conversion over Ti-HMS with a TiO_2/SiO_2 ratio of 0.056 and 60 °C, within 2 h reaction time.

The method of treatment by direct oxidation of sulfur-containing hydrocarbons into sulfur trioxide, on a heterogeneous catalyst with atmospheric oxygen, which can actively react with water forming sulfuric acid, as an environmentaly friendly process has been also reported (Boikov *et al.*, 2008).

García-Gutiérrez *et al.* (2008) prepared different Mo/γ-Al_2O_3 catalysts and investigated ODS activity for diesel fuel using H_2O_2 as the oxidizing reagent. The ODS was found to be highly affected by the presence of hepta- and octa- molybdate species on the supported catalyst with the use of a polar aprotic solvent. Moreover, the presence of phosphate markedly enhances ODS where the S-level decreased from 320 ppm to < 10 ppm at 60 °C under atmospheric pressure. Cedneño-Caero *et al.* (2008) reported the efficient ODS of commercial diesel with an initial S-content of 10005 ppm, using H_2O_2/V_2O_5, supported on niobia, with acetonitrile as extraction solvent.

In the view of the disadvantage of zeolites, Si *et al.* (2008) reported the preparation of different noble metal-loaded Ti-MWW (Au/Ti-MWW) catalysts, by impregnation method, then studied their ODS activity on model organic sulfur compounds in iso-octane (1,000 μg/mL of sulfur), 10 mL acetonitrile, and hydrogen peroxide (H_2O_2/sulfur molar ratio of 2:1–4:1) in an electrothermostatic glass flask equipped with a magnetic stirrer and a condenser. In this context, it was found that Au/Ti-MWW with different gold loading has higher ODS activity for DBT than Ti-MWW. Among these catalysts, Au/Ti-MWW with 0.05% gold gives the highest ODS activity, reaching 99% of 1000 ppm DBT, which was 26% higher than Ti-MWW at 323 K with an H_2O_2/S ratio of 3. Further, the activity of the 0.05% Au/Ti-MWW at 313 K was studied on different S-compounds which also expressed higher activity than Ti-MWW, where the conversion of TH, BT, DBT, and 4,6-DMDBT increased from 55% to 71%, 68% to 90%, 60% to 84% and 50% to 68%, respectively (Si *et al.*, 2008).

Rao *et al.* (2011) reported the ODS of hydrodesulfurized diesel using aldehyde and molecular oxygen in the presence of a cobalt phthalocyaninetetrasulfonamide catalyst. The reactivity of 4,6-DMDBT was higher than that of 4-MDBT. The S-content in diesel was reduced from 448 to 41 ppm, then to 4 ppm by oxidation, then extracted with acetonitrile, followed by adsorption by passing through a silica gel column.

Cui *et al.* (2012) reported the ODS of DBT using hydrogen peroxide (H_2O_2) by a recyclable amphiphilic catalyst with magnetic silica nanospheres covered with complexes between 3-(trimethoxysilyl)-propyldimethyloctadecyl ammonium chloride and phosphotungestic acid, under mild reaction conditions (600 rpm and 50 °C), where the amount of sulfur was decreased from 487 ppm to less than 0.8 ppm. The catalysts with oxidized products have been separated by external magnetic field and simply recycled by acetone eluent, then successfully reused three times with almost constant activity.

Li *et al.* (2009c) reported the preparation of three decatungstates with short carbon chains as the cations: tetrabutylammonium decatungstate ($[(C_4H_9)_4N]_4W_{10}O_{32}$), tetramethylammonium decatungstate ($[(CH_3)_4N]_4W_{10}O_{32}$), and benzyltriethylammonium decatungstate ($[(C_2H_5)_3NC_7H_7]_4W_{10}O_{32}$). These were then applied as catalysts in the [Bmim]PF_6 IL/H_2O_2 extractive catalytic oxidative desulfurization (ECODS) system. The sulfur level in the model oil (1000 ppm S) decreased to 8 ppm. The system could be recycled five times without losing its activity, then the sulfur removal decreased sharply. Moreover, the advantage of applying Ils is the elimination of the extra extraction step since the Ils act as extractants for all possible by-products (Al-degas *et al.*, 2016).

Yun and Lee (2013) reported the preparation of the amphiphilic phosphotungstic acid catalyst (A-PTA) as an example of polyoxometalate, with quaternary ammonium salt and its application in the ODS of light cycle oil (LCO). It expressed high activity, with an ODS conversion of 95% at an H_2O_2/S ratio of 10 and 353 K. That proved, the presence of polycyclic aromatic compounds in LCO feed plays a beneficial role in ODS by minimizing the deposits of oxidized products on the catalyst surface resulting in good ODS reactivity over the catalyst. The ODS of different refractory S-compounds in model feed oil was in the following decreasing order: DBT > 4-MDBT > 4,6-DMDBT > BT. These results were explained by the electron density on sulfur atoms and structural properties of the sulfur compounds where BT has the lowest electron density on its S-atom (Table 2.2), thus the lowest reactivity. Although 4,6-DMDBT and 4-MDBT have higher electron density on their S-atoms than that of DBT (Table 2.2), they expressed lower reactivity than DBT. That proved, the methyl groups become an obstacle for the approach of the sulfur atom to catalytic active sites. The nitrogen compounds reduced the overall ODS in the model oil feed. Sulfur and nitrogen compounds are competitive in oxidation and nitrogen compounds have higher reactivity than sulfur compounds in this reaction and the oxidized nitrogen compounds might reside on the catalyst surface and inhibit the following ODS, while the introduction of 1-methylnaphthalene fully enhanced the ODS activity and was attributed to the high solubility of the oxidized S or N compounds in the 2-ring aromatics. The poor solubility of sulfone in the aliphatic solvent thus proved lower activity of A-PTA in octane. However, the aromatic solvents improved the solubility of ODS products and increased ODS activity by minimizing the deposit of sulfones on the catalyst. Moreover, the type of aromatic solvents also affects ODS activity. For example, the ODS of BT was promoted in the presence of 1-ring aromatics of tetralin, while DBT-derivatives reacted faster upon the introduction of 2-ring aromatics of 1-methylnaphthalene, but, generally, the solubility of the oxidized compounds follows the order: 1-methylnaphthalene > tetralin > n-octane and, consequently, so does the ODS reactivity.

José da Silva and Faria dos Santos (2013) reported the ODS of model oil of DBT into iso-octane using keggin heterpolyacid (HPA) with an alumina catalyst where the model oil was oxidized to sulfones and extracted by acetonitrile with an S-decrease from 1000 ppm to < 1 ppm within 180 min at 60 °C. Tang *et al.* (2013b) applied $H_3PW_{12}O_{40}$, an HPA-catalyst with kegging structure incorporated into mesoporous TUD-1 materials, for the ODS of model oil DBT (500 ppm in n-octane) using H_2O_2 as an oxidant, where approximately 98% desulfurization occurred at 60 °C within 180 min using the catalyst 20HPW-TUD-1. It displayed excellent catalytic activity and recovering ability for ODS and has also been used three times without losing its activity.

Yu and Wang (2013) reported effective ODS by the heteropolyanion substituted hydrotalcite-like compound $Mg_9Al_3(OH)_{24}[PMo_{12}O_{40}]$

(MgAl-PMo_{12}) in a biphasic system using H_2O_2 as an oxidant and acetonitrile as extractant under atmospheric pressure at 60 °C. The S-content of diesel oil decreased from 493 ppm to 9.12 ppm, where the oxidative reactivity of different S-compounds decreased in the following order: DBT > 4,6-DMDBT > BT > Th because of the influence of electron density and spatial steric hindrance. The catalyst can be used for five successive cycles without losing its activity. Yu and Wang (2013) also reported the ODS activity of different heteropolyanion substituted hydrotalcite-like compounds to decrease in the following order: MgAl-PMo_{12} > MgAl-PW_{12} > MgAl-SiW_{12}. This was attributed to the central atom of phosphorous in the HPA-HTLcs being better than silicon. Moreover, coordinated molybdenum atoms are another important factor which express higher activity than that of tungsten.

Long *et al.* (2014) reported the ODS of DBT in model gasoline (DBT in n-octane, with an initial S-content of 400 ppm) using catalyst W/D152 which was prepared by depositing tungsten on resin D152, a macroporous polyacrylic cationic resin, in presence of oil-soluble cyclohexanone peroxide (CYHPO) as an oxidant, followed by washing with DMF to obtain the refined gasoline where the DMF was reused after regeneration. The conversion of DBT occurred and the sulfur content reached 99.1% and 3.52 ppm, respectively, at the optimal catalytic conditions of 100 °C, mass ratio of model gasoline to catalyst W/D152 of 100, molar ratio of CYHPO/DBT of 2.5, and reaction time of 40 min. The catalyst could be reused 7 times before the total sulfur content of the treated model gasoline was higher than 10 ppm. Increase in reaction temperature (>100 °C) decomposes the CYHPO and the catalyst is destroyed at 120 °C. The yield decreased at a longer reaction time (>40 min) due to the volatilization of gasoline in presence of CYHPO without the catalyst, where the ODS was recorded at 86.5%, while in presence of catalyst without CYHPO, the ODS recorded only 45.3%. Thus, the high ODS efficiency recorded for W/D152 is due to the good immobilization of tungsten on resin D152 and the role of the tungstate in the catalyst was to activate a CYHPO molecule so DBT could be rapidly oxidized to sulfones at the presence of catalyst W/D152. As a conclusion from that study, the W/D152-CYHPO reaction system overcomes some of the drawbacks and disadvantages of existing technologies and possesses good catalytic ODS potential to be applied for industrial applications.

Abu Bakar *et al.* (2015) reported the ODS of model diesel fuel by 4.35% WO_3/16.52% MoO_3/γ-Al_2O_3 at constant temperature of 60–65 °C under atmospheric pressure, followed by extraction using dimethylformamide. That removed 100% of DBT and 4,6-DMDBT and 92.5% of Th.

A solid superacid has advantages, such as acidity stronger than 100 % H_2SO_4 ($H_0 \leq -12$) (Gillespise and Peel, 1973), easy separation from the reaction mixture, non-corrosion to reactors, environmental friendliness, and high activity in many organic reactions including esterification (Li *et al.*, 2009d), isomerization of *n*-alkanes (Song *et al.*, 2010), acylation (Guo *et al.*, 2008) and alkylation (Zhang *et al.*, 2005).

Song *et al.* (2015) reported a novel procedure for ODS of a simulated light fuel oil (200 mg thiophene/L petroleum ether) and straight-run gasoline (169.5 mg/L initial S-content) using a K_2FeO_4 over SO_4^{2-}/ZrO_2 (SZ) superacid solid support which was prepared by impregnation. The presence of highly dispersed acidic sites over the support, a huge area of catalyst, and the formation of a tetragonal structure play an important role in ODS using K_2FeO_4, where the catalyst subjected to impregnation with 0.8 mol/L H_2SO_4, supported on ZrO_2, and calcined at 600 °C exhibited the best performance. At optimum conditions of 30 °C, 0.2 g SZ, 0.2 g K_2FeO_4, and 30 min reaction time, followed by methanol extraction at 15 °C for 10 min with the volume ratio of solvent/oil of 1, the straight-run gasoline desulfurization rate reached 89.2 % and the residual sulfur content was 18.3 mg/L. At a higher calcination temperature (>600 °C), a large amount of the zirconium oxide tetragonal phase was converted to the monoclinic phase, leading to poorer catalytic activity that may be due to the decomposition of a loaded sulfate at higher temperatures. Not only this, but sintering may also occur at excessive temperatures which would probably destroy the microporous structure of the catalyst and decrease its surface area. A lower sulfuric acid concentration (<0.8 mol/L) usually leads to lower SZ acid strength, resulting in poor catalytic performance, but a higher sulfuric acid concentration (>0.8 mol/L) would lead to oversaturation of the sulfate and

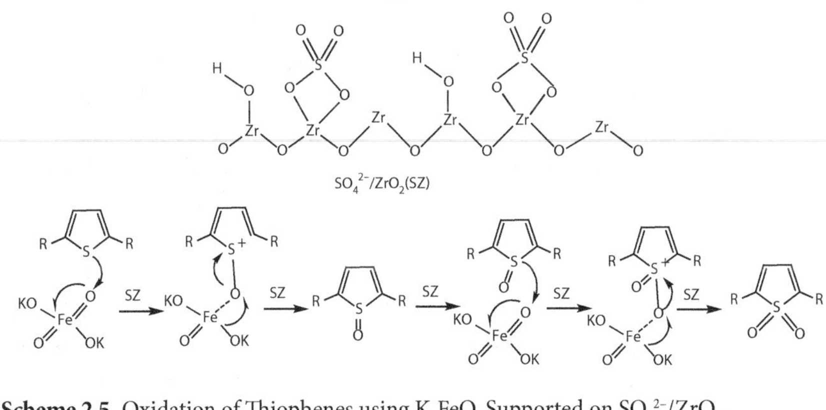

Scheme 2.5 Oxidation of Thiophenes using K_2FeO_4 Supported on SO_4^{2-}/ZrO_2.

would intensify sulfate agglomeration on the surface and cover some of the SZ active sites leading to lower catalytic activity. Furthermore, the huge surface area of SZ likely enhances contact between thiophene and K_2FeO_4 (Scheme 2.5), where only 37.9 % of the model fuel desulfurization rate was obtained in the presence of K_2FeO_4 without SZ, due to the low solubility of K_2FeO_4 in the organic phase, which led to poor contact between K_2FeO_4 and thiophene and, consequently, low ODS. The oxidation reaction takes place in two steps where oxidation of thiophene is considered to be a consecutive nucleophilic addition, during which thiophene first reacts with K_2FeO_4 to produce thiophene sulfoxide, which further reacts with another K_2FeO_4 molecule to form thiophene sulfone.

Rafiee and Rahpeyma (2015) reported the immobilization of 12-tunstophosphoric acid on silica-coated cobalt-ferrite ($CoFe_2O_4$) nanoparticles that can be magnetically separated. The ODS- efficiency of that catalyst, using H_2O_2 as an oxidant, decreased in the presence of N-compounds and the inhibition of indole was higher than that of quinoline. Due to the competitive inhibition, since N-compounds oxidizes faster than the S-compounds, and the higher electron density on the N-atom of indole, it is oxidized faster than quinoline under mild operating conditions. The prepared catalyst has also been used efficiently for four successive cycles.

Zhang *et al.* (2016) reported the preparation of a series of SO_4^{2-} promoted metal oxide solid superacids, including SO_4^{2-}/ZrO_2-WO_3, SO_4^{2-}/ZrO_2-MoO_2, and $SO4^{2-}/ZrO_2$-Cr_2O_3, by incrorporating different transition metals into a (SZ) superacid solid support which was prepared by impregnation. They were named as (SZW, SZM, and SZC, respectively) and then their ODS activity was tested on model oil (500 ppm S, DBT in n-octane) at 70 °C in the presence of hydrogen peroxide (nH_2O_2:Ns = 20) with catalyst dosage of 0.015 g/L in the oil, followed by extraction using DMF. Their acid strength is in the following decreasing order: SZC > SZM > SZW > SZ. Thus, the SZC expressed the highest ODS activity, recording approximately 91%, as the ODS is related to catalyst acid strength. Moreover, SZC can be reused successfully for five successive cycles.

Figueiredo dos Santos *et al.* (2016) reported the hydrothermal synthesis of ZSM-12 zeolite, which is impregnated onto TiO_2. The prepared TiO_2/ZSM-12 with 15% of titanium expressed 60% ODS for a model mixture of n-heptane with 5000 ppm thiophene in presence of H_2O_2, with acetonitrile as extractant, at 70 °C within 60 min and under atmospheric pressure.

Jiang *et al.* (2016) reported the preparation of a series of Brønsted acidic ionic liquids having a protonated amide- or lactam-based cation and investigated their activity as extractants and catalysts in extraction, combined with oxidative desulfurization (ECODS) of both model oil and diesel fuel, with H_2O_2 (30 wt%) as an oxidant. Each of them showed the obvious removal of BT and DBT in model oil. Among them, [HCPL][TFA] exhibited the best performance by completely removing BT and DBT in a short time, since the $HCPL^+$ cation was verified to exist in an enol form which was supposed to contribute to high desulfurization performance by improving the formation of peroxides. [HCPL][TFA] decreased the S-content of real hydrogenated diesel and straight-run diesel fuel from 659.7 and 11,034 ppm to 8.62 and 89.36 ppm, respectively.

Heteropolyacids, particularly the Keggin-type phosphomolybdic acids (HPMo), exhibit high acidic strengths and have good potential in ODS. However, HPMo suffers from some drawbacks such as low reactivity due to its small surface area and its difficulty to be recycled. To overcome these problems, it is usually loaded on an appropriate carrier such as silica, alumina, carbon, titania, zirconia, or mesoporous silica (MS), which facilitates increasing active sites on the surface and enhances catalytic activity. Mesoporous silica (MS) has attracted a great industrial interest for its low-cost and higher surface area. Yao *et al.* (2016) reported the preparation of MS support by hydrothermal method, then Eu-HPMo/MS catalyst with a specific surface area and pore volume of 309.7 m^2/g and 0.389 cm^3/g, respectively, that was prepared by impregnation method. The activity of

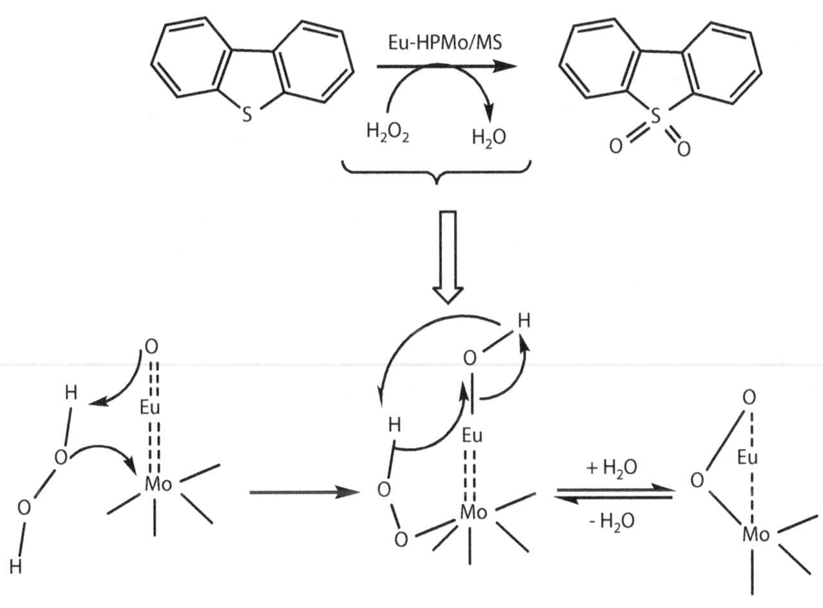

Scheme 2.6 ODS of DBT by Eu-HPMo/MS in Presence of H_2O_2.

the prepared catalyst was tested for the ODS of model oil (320 ppmw DBT in n-octane) in the presence of H_2O_2. The desulfurization rate reached approximately 100%, in the presence of 0.1 g catalyst, 3 n(H_2O_2)/n(S), within a 60 min reaction time at 323 K. The HPMo leads to the conversion of DBT for its high acidic strengths. The rare earth metal Eu is one of the most available promoters for facilitating the storage and release of oxygen. The high surface areas of the support also promote ODS activity. The catalyst can also be used five successive times. The elucidated mechanism of the ODS, using the Eu-HPMo/MS catalyst (Scheme 2.6), revealed that the active peroxo-species are generated from the reaction of Mo with H_2O_2. Then, the oxidation of the sulfur atom of DBT occurs by the nucleophilic attack of the active peroxo-species to form sulfoxide when DBT reacted with H_2O_2 at the interface. The Mo=O was lengthened by Eu, which facilitated the storage and release of oxygen.

Huang *et al.* (2017) reported the preparation of expanded vermiculite (EVM) that was treated with acid (aEVM) and then incorporated into $H_3PW_{12}O_{40}$ (HPW) by an ordinary impregnation method. The prepared HPW/aEVM catalyst expressed high ODS activity, reaching approximately 100% of 100 ppm S-content model oil (mixture of Th, BT, DBT, and 4,6-DMDBT in *n*-octane), at optimum operating conditions of 333 K, 6 O/S, a catalyst dosage of 0.04 g/10 mL, and reaction time of 0.5 h. The prepared catalyst can be reused for seven cycles with a slight decrease in its catalytic activity of approximately 1.89%. The ODS activity followed the order DBT > 4,6-DMDBT > Th > BT, which is related to the electron density on S-atom and molecular size of the S-compound.

Rafiee *et al.* (2017) reported the immobilization of different polyoxometalates (POMs), including $H_3PMo_{12}O_{40}$ (PMo_{12}), $H_5PMo_{10}V_2O_{40}$ ($PMo_{10}V_2$), $H_6PMo_9V_3O_{40}$ (PMo_9V_3), $H_7PMo_8V_4O_{40}$ (PMo_8V_4), $H_3PW_{12}O_{40}$ (PW), and $H_4SiW_{12}O_{40}$ (SiW) on carbon catalyst supports (CCS) prepared from naturally abundant potatoes which were used for the oxidative desulfurization ODS of model oils (DBT, BT and Th in 1/1 ethanol:n-heptane) in the presence of H_2O_2 at room temperature. The initial total S-concentration of the model oils varied between 250 and 1000 ppm. Moreover, the prepared catalysts were tested on real, light cycle oil with an initial S-content of 530 ppm. The natural potato, as a green and cheap source, provides a good support to design a composite nanorod structure via a simple hydrothermal method. Those green catalysts proved to be a unique, effective, and eco-friendly catalyst for selective oxidation sulfides using 30% aq. H_2O_2 as the co-oxidant in the ethanol:n-heptane solvent system, in model and real oil due to excellent yields, very short reaction time, room temperature reaction condition, and using air atmosphere, in addition to simple regeneration methods and reusability, and are all desirable advantages from the stand points of green sustainable chemistry. Moreover, the catalyst could be recovered and reused at least four times without a significant decrease in catalytic activity at an oxidation reaction with a negligible leaching of $PMo_{10}V_2$.

Guoxian *et al.* (2007) studied the ODS of model oil (DBT in n-octane) with hydrogen peroxide catalyzed by activated carbon-formic acid. The strong oxidizing agents and hydroxyl radicals generated from hydrogen peroxide can be produced on the surface of AC and the short life radicals produced during the surface reactions are likely to be resonance-stabilized on the carbon surfaces. This was attributed to the presence of Pt, Ag, and Pb metal ions such as Fe^{3+}, Cr^{3+}, Cu^{2+}, and Ag+ and metal oxides such as MnO_2 and Fe_2O_3 in AC, which can catalyse hydrogen peroxide to produce free radicals with very high oxidation potential ($OH^{\bullet}$ and HO_2). This results in the oxidation of DBT to form SO_4^{2-}, where activated carbon (AC) can catalyse hydrogen peroxide to produce hydroxyl radicals when pH is less than 7.0 which, consequently, oxidizes the adsorbed DBT on AC. Further, AC can disperse in the oil phase and increase the collision probability of DBT with the active oxygen species, accelerating the oxidation reaction. Thus, the carbon-formic acid systems have good catalytic activities in the oxidative removal of DBT with hydrogen peroxide. However, the catalytic performances of the carbons differ; the wood carbons have better catalytic performances than coal carbon, where the DBT conversion reaches 91.6%.

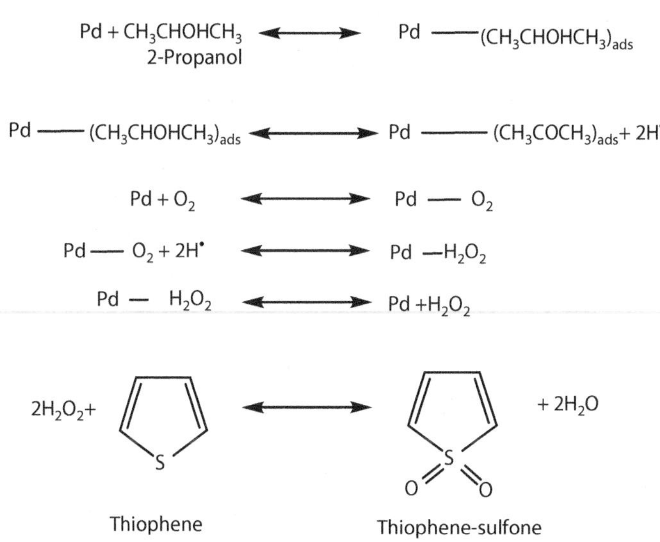

Scheme 2.7 Mechanism for In-Situ Oxidative Desulfurization.

In the presence of formic acid, the ODS was enhanced due to presence of both hydroxyl radicals and the performic acid in the reaction medium, thus the ODS of DBT reached 100% in presence of 12 mol/L formic acid within 30 min. Wang *et al.* (2017) also reported a single-stage reactor, in which the dehydrogenation of 2-propanol under alkaline conditions with in-situ production of H_2O_2, oxidation of thiophenes and adsorption of the produced sulfones from model fuel (250 ppm S) are demonstrated in one single step at mild temperature and under atmospheric pressure, using a palladium/activated carbon (Pd/C) catalyst (Scheme 2.7). It involves the dehydrogenation of 2-propanol, generation of an oxygen layer on the catalyst surface, stripping of the adsorbed acetone, releasing of H_2O_2 into the alkaline medium, and, finally, the ODS reaction of thiophenic compounds. The advantages of this process are the in-situ production of H_2O_2 and its rapid diffusion into the aqueous phase. The large specific surface area and pore volume of activated carbon guarantee the easy adsorption of the produced sulfones by simultaneous adsorption on activated carbon, eliminating the subsequent separation operation after the ODS-step. The accumulative sulfur on the activated carbon was also resolved by simple washing with anhydrous ethanol under mild operating conditions, without losing the catalytic adsorption efficiency of Pd/C. Thus, this system simplifies the ODS-process and decreases the overall operational cost.

One of the obstacles that faces the HDS-process, especially in catalytically cracked gasoline, is the high concentrations of olefins which could be hydrogenated, leading to a decrease in octane number (Song, 2003). Taking into consideration that olefin hydrocarbons can be produced not only by catalytic cracking, but also in the pyrolysis of shale oil, this will also cause difficulties during the desulfurization of oil fraction as well. One of the ways of coping with the problem of desulfurization of gasolines with increased olefin content can be preliminary reduction in the total sulfur content via hydrogen-free methods (Matsuzawa *et al.*, 2002; Wan and Yen, 2007), for example by ODS without the use of acid catalysts as pretreatment step before an HDS-unit. This will make it possible to conduct further hydrotreating under more mild conditions without a notable loss in octane number.

Akopyan *et al.* (2015) studied the effect of olefins (e.g. cyclohexene) and ether (e.g. dibutyl ether) on the ODS of model oil (Methylphenyl sulfide (MPS) or benzothiophene (BT) in *n*-octane) with an initial S-content of 1770 and 1880 ppmw, respectively, using different catalysts, including sodium molybdate, sodium tungstate and formic acid, H_2O_2, and Dacamid surfactant. The presence of the olefin decreases the oxidation of sulfides in all the cases, but the presence of the ether increases the ODS. The ODS of MPS is higher than that of BT, but the inhibition effect of olefins on the ODS of BT is higher than in case of MPS. Upon the ODS of 40–185 °C fraction of catalytically cracked gasoline that contains methylthiophene, alkylthiophenes, dimethylthiophene, benzothiophene, methyl-, dimethyl, and ethyl-benzothiophene, and dimethyl sulfuide, approximately 20% of olefins and an initial S-content of 2500 ppmw, ODS was very effective using sodium molybdate at a low temperature of 20 °C and a long reaction time of 12 h. Since the oxidation reaction oxidation is less selective at a higher temperature and the rates of side reactions increase, the oxidant is inefficiently consumed for the oxidation of olefins to epoxides and alcohols. In conclusion of this study, it is preferable to use neutral catalysts based on transition metal salts for ODS rather than catalysts of acidic nature.

Another obstacle is during the processing of high-boiling petroleum fractions by catalytic cracking, catalyst deactivation occurs due to high carbon residue value, which mainly occurs due to the presence of PASHs such as benzothiophene, dibenzothiophene, and naphthothiophene derivatives (Song, 2003). Applying ODS as a pretreatment step for high-sulfur heavy petroleum fractions (Sharipov and Nigmatullin, 2005; Zongxuan *et al.*, 2011; Javadli and de Klerk, 2012a,b) can promote the enhancement of the yield and quality of the products of catalytic cracking. Despite the low activity of peroxo complexes in the ODS of individual benzothiophenes, their use for ODS in biphasic technology is recommendable duc to the use of low percentage hydrogen peroxide solutions, the ease of isolation of oxidation products from the oxidation system, and the possibility of reusing the metal compound without its regeneration. Hydrogen peroxide, as an oxidant, presents itself in the aqueous phase, the substances to be oxidized present in the organic phase, and the peroxo-complex, which is formed due to the reaction of a metal salt with hydrogen peroxide, is transferred by a surfactant from the aqueous phase to the organic phase where oxidation can take place.

Guseinova *et al.* (2012) reported the ODS of catalytically cracked gasolines produced from a blend of Baku crude oils with hydrogen peroxide in formic acid medium under phase-transfer catalysis conditions, using pyridine and acetonitrile as ligands, in the presence of heterogeneous catalysts containing ions of Group VI–VIII transition metals. For example, Mo and Co ions decrease the S-content in gasolines from 200 to 33–40 ppm. The best results were obtained upon the usage of the Mo/Al_2O_3 catalyst used in an acidic medium. The data obtained in that study show that a unit volume of feedstock requires the oxidizing phase to contain 0.0025 g of Mo/Al_2O_3 catalyst, 0.0025 mL of 1:1 $H_2O_2/CO(NH_2)_2$ complex, and 0.2 mL of formic acid which facilitates lowering the total sulfur content of the catalytically cracked gasoline from 200 to 33 ppm within 6–7 h.

Rakhmanov *et al.* (2013) studied the ODS of a wide cut catalytically cracked gasoline fractions with an initial boiling point of < 205 °C, at 50 °C, containing 30% H_2O_2 as an oxidant with different surfactants, including lauryldimethylbenzylammonium chloride or N,N-diethanolamide of lauric acid $C_{11}H_{21}C(O)N(CH_2CH_2OH)_2$ and crown ethers as catalysts, where the following transition metal salts were used: Na_2MoO_4, Na_2WO_4, $VOSO_4$, or $[Cu(NH_3)_4]CO_3$, within a 2 h reaction time in presence of ethanol/water (1:9). After completion of the oxidation, the treated fuel stream was subjected to distillation to collect the fraction with an initial boiling point (IBP) of 70 °C. The best desulfurization results were achieved with sodium molybdate and an ammonia complex of copper in the presence of lauryldimethylbenzylammonium chloride, whereas crown ethers had a low activity. The main gain of this study was that relatively small amounts of the used oxidant hydrogen peroxide are best (0.123–0.38 mL or 0.001–0.003 mmol) because the economic performance of the process, along with the desulfurization characteristics, will play an important role in the industrial implementation of the oxidation method. The addition of organic compounds, in particular nitrogen bases, during the preparation of peroxo-complexes enhances their activity in oxidation reactions by virtue of ligand insertion in the inner sphere of the complex. Lauryldimethylbenzylammonium chloride can act as a ligand of this kind for the enhancement of the process of ODS. The total sulfur content of the IBP–70 °C fraction before oxidation was 355 ppm. The ODS and subsequent adsorption treatment of the fraction on alumina reduced the total sulfur content to 16 ppm upon the application of two successive ODS steps: in the first step, 1.86 mg $[Cu(NH_3)_4]CO_{3,}$ 0.123 mL 30% H_2O_2, 25% 6 μL NaOH were used and in the second step, 0.62 mg $Na_2MoO_4{\cdot}2H_2O$, 0.38 mL 30% H_2O_2 were used. Both steps were conducted for 2 h at 50 °C in presence of and lauryldimethylbenzylammonium chloride as a surfactant. The desulfurization percentage reached 96% after the two successive ODS steps followed by subsequent adsorption treatment on the alumina.

Rakhmanov *et al.* (2014a) also studied the ODS of straight run, nonhydrotreated diesel fractions (boiling range 178–342 °C, with initial S-content of 7260 ppm), that contain benzothiophene, dibenzothiophene, their alkyl-substituted derivatives, and thioxanthene, using hydrogen peroxide in the presence of different transition metal compounds (Na_2MoO_4, Na_2WO_4, $NaVO_3$, WO_3, tungstic acid, and heteropoly tungstate/molybdate $H_3PMo_6W_6O_{40}$ (HPTM)) in a biphasic system followed by the extraction of the oxidation products with dimethylformamide (DMF). The ligand in the inner coordination sphere of the peroxo complexes can be water or a specially added organic compound. The addition of an organic compound, in particular a nitrogen base, during the preparation of a peroxo complex enhances the activity of the complex in oxidative reactions, owing the incorporation of the ligand into its inner sphere. It was found that the addition of pyridine as a ligand intensifies the ODS with hydrogen peroxide in the presence of three different salts of transition metals and benzyllauryldimethylammonium chloride. However, unlike the case of Mo compounds, the addition of pyridine to the reaction mixture containing a W or V compound reduced the ODS. This was attributed to the change associated with the higher stability of peroxo-complexes of these metals with pyridine ligands compared to the other peroxo-complexes in which water serves as a ligand. The tested tungsten compounds (tungstic acid in the form of W(VI) oxide monohydrate and mixed HPTM) exhibited the highest activity, reducing the total sulfur content of the diesel fraction almost four- and six- fold, respectively. This was attributed to the fact that the formed peroxo-complexes are more reactive than in the case of tungstic acid salts. The high temperature and long reaction time were not favourable due to the decomposition of H_2O_2. The highest ODS, of approximately 82%, was performed using heteropoly tungstate/molybdate at 50 °C, for 6 h, with a sulfur:hydrogen peroxide:metal:pyridine ratio = 1:5:0.07:0.02 (mol), with surfactant benzyllauryldimethylammonium chloride. Rakhmanov *et al.* (2014b) reported that complexes of azacrown ethers with $NbCl_5$ lead to a decrease in the total sulfur content in the model mixtures to 13% of the initial amount. The structure of the azacrown ether used has little effect on the extent of desulfurization of the model mixture. The diesel fuel was simulated using mixtures containing methyl phenyl sulfide, benzothiophene, dibenzothiophene, dibenzyl sulfide, and thianthrene (1.4% S-content) dissolved in a mixture of tri-, hexa-, and hepta-decanes or individual hydrocarbons under the action of hydrogen peroxide in the presence of compounds of $NbCl_5$ and various crown ethers, where these compounds are oxidized to respective sulfoxides and sulfones. The most efficient oxidation occurs at 80 °C within 8 h. After the reaction, the sulfur content in the model mixture decreased to 40%. This was attributed to the possible complexation between the azacrown ether and $NbCl_2$, but further increase in temperature was inappropriate due to the decomposition of hydrogen peroxide. In the function of the complexing ligand, azacrown ethers play the role of a surfactant which facilitates the transfer of the resulting peroxo-complex from the aqueous phase to the organic phase, which form a two-phase system that consequently enhances the ODS. The data of that study suggested that an increase in the number of nitrogen atoms in the macrocycle, in general, contributes to an increase in the degree of oxidation of the sulfide components of the model mixture and has little effect on the oxidation of benzothiophene and dibenzothiophene.

In an another study, Rakhmanov *et al.* (2016) reported the ODS of a model oil mixture (a non-hydrotreated vacuum gas oil from TAIF-NK and a commercial diesel fuel of the Euro-5 standard from a Lukoil filling station with a sulfur content below10 ppm) with an initial S-content of 6300 ppm. The technology of a two-phase system with a phase-transfer catalyst has been employed for the desulfurization, using H_2O_2 and formic acid. The use of DMF for extraction makes it possible to extract up to 40% sulfur compounds, while the extraction with water is inefficient because of the low solubility of the oxidation products (dibenzothiophene sulfones and naphthothiophene sulfones); the degree of desulfurization did not exceed 3% in the case of its use. The optimum reaction time is 6 h and the hydrogen peroxide:sulfur molar ratio is 4:1, which led to approximately 74% desulfurization efficiency, while upon three successive ODS steps, 90% of total sulfur was removed.

A reduction in reserves of conventional crude oil necessitates a search for new hydrocarbon sources, of which oil shales are of particular interest because their reserves are comparable with those of conventional oil, therefore the development of a technology for shale processing into synthetic crude oil is a demanding task. An important step of this processing is the purification of the resulting synthetic crude oil for the removal of sulfur compounds, as it adversely affects many processes of oil refining, as discussed in chapter one. Akopyan *et al.* (2016) studied the ODS of shale oil, obtained by thermal extraction of organic matter from shale rock, using an oxidative catalytic system composed of 50% hydrogen peroxide, a molybdenum salt, and acids of different natures. The model mixture used in that study was prepared by dissolving shale oil in hydrofined straight-run gasoline (fraction 40–160 °C) in a shale oil; the gasoline had a ratio mass of 1:10 under vigorous stirring for a day with an initial S-content of 1375 ppm. Then, after the ODS reaction, the produced stream was applied to extractive desulfurization by DMF. It has been shown that the application of this method, in combination with extraction of the oxidation products of organic sulfur compounds, makes it possible to remove up to 94% of total sulfur from synthetic oil. The presence of H_2SO_4 is better that H_3PO_4 due to the propensity of sulfuric acid to form a strong peracid, which is an efficient oxidant of sulfur compounds. The presence of molybdenum salt is important for the in-situ generated molybdenum peroxo-complex, which accelerates the direct oxidation of sulfur compounds and acts as a catalyst for the formation of the peracid. The application of formic, sulfuric, and trifluoroacetic acids showed the best results, where trifluoroacetic showed a reduction of the sulfur content in desulfurized oil by almost six times. The maximal degree of removal of sulfur compounds was achieved by the application of the oxidizing catalytic system consisting of sodium molybdate, trifluoroacetic acid, and hydrogen peroxide in a Mo : S : H_2O_2 : CF_3COOH molar ratio of 1 : 100 : 600 : 100 in a reaction carried out for two hours at a temperature of 60 °C. In this case, the residual content of sulfur in the oil was 89 ppm with a degree of sulfur removal at 94%.

Compared to hydrodesulfurization (HDS), oxidative desulfurization (ODS) has several advantages: (i) refractory sulfur compounds, such as alkylated dibenzothiophene derivatives, are easily oxidized under low operating temperature and pressure; (ii) there is no use of expensive hydrogen so the process is safer and can be applied in small and medium size refineries, isolated refineries, and those located away from hydrogen pipelines; (iii) ODS avoids aromatic and olefin saturation and, thus, a low octane number; (iv) the overall capital cost and requirements for an ODS unit is significantly less than the capital cost for deep hydrodesulfurization-units; (v) like BDS, some of the sulfones can be converted to surfactants which could be sold to the soap industry, offsetting some of the overall cost of the ODS process (Gore, 2001; Guo *et al.*, 2011; Srivastava, 2012; Mužic and Setrić-Bionda, 2013).

However, the efficiency and economics of an ODS process is strongly dependent on the methods used for oxidizing sulfur compounds and, successively, the methods used for separating the sulfone and/or the sulfoxide derivatives from the oxidized fuels. Furthermore, the ODS process has some technological and economical problems. Solvent extraction of the sulfone-sulfoxide derivatives using polar solvents, such as γ-butyrolactone, n-methyl pyrrolidone (NMP), methanol, dimethylformamide (DMF), acetonitrile, dimethylsulfoxide (DMSO), sulfolane, and furfural are commonly used, which can be recovered and reused through a distillation process (Liu *et al.*, 2008). However, one of the major drawbacks of the solvent extraction method is the appreciable solubility of hydrocarbon fuels in polar solvents which leads to significant losses of usable hydrocarbon fuel. Such a loss is completely unacceptable on a commercial basis. In addition, sulfone derivatives are polar compounds and form strong bonding with polar solvents and it is difficult to remove them from the solvents to a level below 10 ppmw. This causes a build-up of sulfone derivatives in the solvent during solvent recovery (Nanoti *et al.*, 2009). Thus, ODS is a source of sulfonic waste which requires special treatment. Moreover, the homogeneous catalysts are difficult to separate from the reaction products and this limits their recycling. Furthermore, the preparation of new supported catalysts is the most desirable improvement of the ODS process.

On the other hand, the removal of sulfoxide- sulfone derivatives could be achieved using an adsorption technique using adsorbents such as silica gel, activated carbon, bauxite, clay, coke, alumina, silicalite (polymorph of silica), ZSM-5, zeolite β, zeolite x, and zeolite y, in addition to the mesoporous oxide-based materials. These have attracted much attention in recent years due to their large pore sizes and controlled pore size distribution which may be beneficial in allowing accessibility of large molecular size sulfone derivatives to surface-active sites. Stanislaus *et al.* (2010) reported that silica-alumina and silica gel are more effective for the adsorption of sulfones after the ODS of LGO than activated carbon, molecular sieves, γ-$Al_2O_{3,}$ zeolites, and ZSM-5, where the S-content in LGO was decreased from 730 to < 10 ppm applying ODS, followed by ADS using SiO_2-γ-Al_2O_3 or SiO_2. Ma et. (2007) reported the ODS of a model jet fuel (412 ppmw S) and a real jet fuel (JP-8, 717 ppmw S) in a batch system in the presence of molecular oxygen with free Fe(III) nitrate, Fe(III) bromide (denoted Fe-Fe), and carbon supported catalysts (Fe-Fe/ACMB) where the thiophenic compounds in the fuel were converted to their corresponding sulfone and/or sulfoxide compounds at 25 °C under atmospheric pressure. The Fe–Fe/ACMB catalysts expressed higher ODS activity and higher S-adsorptive capacity, which was 4 times higher than that for ACMB and about 1.4 times higher than that for Fe–Fe alone. The alkyl benzothiophenes with more alkyl substituents recorded higher oxidation reactivity and the ODS reactivity of the S-compounds decreased in the order of 2-methylbenzothiophene > 5-methylbenzothiophene > benzothiophene > dibenzothiophene. But, in a real JP-8 fuel, alkylated thiophenic compound, 2,3-dimethylbenzothiophene, was found to be the most refractory sulfur compound oxidized, while among the trialkylated–BTs, 2,3,7-trimethyl benzothiopene was found to be the most refractory. This was attributed to both the electronic and steric effects of the methyl groups which play an important role in determining oxidation reactivity. The ODS enhanced the adsorption onto commercial activated carbons (ACC and ACMB). Thus, upon the ODS on real jet fuel (JP-8, 717), using Fe–Fe/ACMB, the S-content decreased from 717 ppm to 279 ppm. Then, in the second ADS step, the S-content decreased from 279 ppm to 126 ppm. The main advantage of that method is its occurrence in presence of O_2 at ambient conditions without using peroxides and aqueous solvents and without involving the biphasic oil–aqueous-solution system.

However, one of the major drawbacks of the adsorption technique is the amount of oil treated per unit weight of adsorbent is low (Babich and Moulijn, 2003).

Nowadays, different catalysts (Wang and Yu, 2013; Kadijani *et al.*, 2014; Mjalli *et al.*, 2014), photochemical, and ultrasound oxidative desulfurization (Wu and Ondruschka, 2010; Liu *et al.*, 2014a; Wittanadecha *et al.*, 2014) are under research and development (R&D) to improve reaction efficiency.

The introduction of regulations stipulating ultra-low sulfur (ULS) content in fuels caused peroxides, such as hydrogen peroxide (H_2O_2), to become the most used oxidizing agents (Joskić *et al.*, 2014) and molecular oxygen (Murata *et al.*, 2003). The use of molecular oxygen may be appealing to refineries that already have the infrastructure for an oxidation facility to prepare blown asphalt. Other oxidizing agents include in-situ formed per-acids, organic acids, phosphate acid and hetero-polyphosphate acids, Fe-tetra amido macrocyclic ligand (Fe-TAML), Fenton and Fenton-like compounds, as well as solid catalysts such as those based on titanium-silica (tungsten-vanadium-titania, W-V-TiO_2), solid bases such as magnesium and lanthanum metal oxides or hydro-talcite compounds, iron oxides and oxidizing catalysts based on monoliths, and tert-butyl-hydroperoxide with catalyst supports that can effectively oxidize organic sulfur compounds into sulfones with less residue formation (Abdul Jalil and Falah Hassan, 2012; Mužic and Sertić-Bionda, K. 2013). The presence of a sulfone extracting agents such as acetonitrile in the oxidation phase has been also reported to enhance the ODS process (Hulea *et al.*, 2001; Yazu *et al.*, 2003; Ramírez-Verduzco *et al.*, 2004).

There are five processes based on the ODS technique used to reach ULS-fuels that have been reported to reach the commercialization stage.

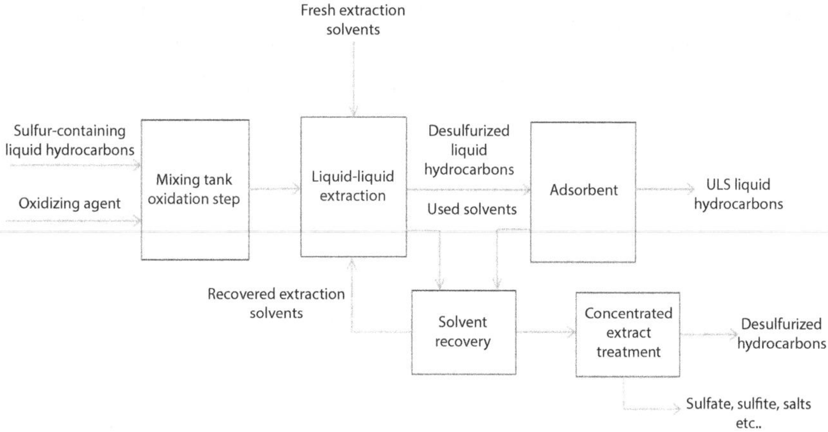

Figure 2.11 Simplified Flow Diagram for Conversion and Extractive Desulfurization (CEDS).

In 1996, PetroStar Inc. developed a technique combining conversion and extractive desulfurization (Figure 2.11 CEDS) to remove sulfur from diesel fuel (Chapados *et al.*, 2000). Briefly, the diesel fuel is oxidized by mixing with peroxyacetic acid (i.e. H_2O_2/acetic acid) at a low temperature, <100 °C, under atmospheric pressure. Then, liquid-liquid extraction takes place, producing low sulfur content diesel oil, followed by adsorption treatment to yield ultra-low sulfur diesel fuel. Recycling of extract solvent for re-use takes place and the concentrated extract is further processed to remove sulfur. In a bench scale pilot plant unit, the desulfurization of high S-content diesel oil (3500 ppm), applying the PetroStar-CEDS process, produced a product with <20 ppm S-content. Stanislaus *et al.* (2010) reported that further development is still required before the licensing and commercialization of the PetroStar-CEDS process.

Unipure Inc. and Texaco have jointly developed an ODS process (called ASR-2) based on peroxide (H_2O_2) oxidation in the presence of formic acid (HCO_2H) within a short residence time (5 minutes), mild temperature (120 °C), and at approximately atmospheric pressure where the oxidized fuel is first separated by a phase separator, washed and dried, and then passed over an alumina bed for the separation of sulfones by adsorption. The sulfone- loaded alumina bed can be regenerated by washing with methanol (Levy *et al.*, 2002; Ito and van Veen, 2006). ASR-2 plants, solely, are estimated to produce ULS-fuels from feeds with an S-conent of 50–1500 ppm. This process can desulfurize diesel oil or gasoline with an S-content of 1500 ppm to ≤5 ppm sulfur, at a cost approximately 50% less than the cost of a hydrotreater with an improvement in cetane number and API gravity (Levy, 2003). Unipure is undergoing research to develop an extension to its ASR-2 process that can cost-effectively desulfurize oil feeds with a high S-content (>1%). A demonstration plant, based on the Unipure ASR-2 process which can oxidatively desulfurize diesel oil (500 ppm S-content) to <8 ppm S-content, is in operation at Valero Energy Corp's Krotz Spring, LA refinery.

Unlike hydrogen peroxide, t-butyl-hydroperoxide (TBHP) is completely fuel soluble and would enhance the ODS process. Lyondell Chemical Technology, L.P. has developed ODS using *t*-butyl-hydroperoxide (TBHP) followed by extraction for sulfone separation (Karas *et al.*, 2008). This single liquid system is beneficial for simple reactor engineering, enabling the application of a fixed bed column, under mild temperature and pressure. The ODS takes place in less than 10 min, with a near quantitative conversion of thiophenes to sulfones, with the production of t-butyl alcohol as a co-product that can be easily removed from the treated fuel. This process was reported to be in a continuous operation for over 5 months, producing diesel with <10 ppm S (Liotta and Han, 2003). However, the process is still not available for licensing (Stanislaus *et al.*, 2010).

Uop Llc, in cooperation with EniChem S.P.A. (Gosling *et al.*, 2007), has developed a process for the autocatalytic (i.e. without a catalyst) production of organic hydroperoxides and ultra-low sulfur (ULS) diesel boiling range hydrocarbons. The autocatalytic production of hydroperoxides is believed to be initiated in the presence of organic compounds and oxygen by the disassociation of oxygen to produce free radicals which then proceed to react with the organic compounds to produce sulfones. Therefore, no solid catalyst is used in the production of organic hydroperoxides. The ENI-UOP ODS process takes place through three steps: an oxidant supply section (i.e. organic peroxide), sulfur reaction section, and a sulfone separation section (Figure 2.12). ***The first step*** is a circulating reaction loop in which the diesel feed is mixed with air to produce a hydroperoxide-containing stream, in the presence of an organic initiator, at 130 °C and 70 atm, in the absence of a catalyst. An induction period may be required during the initial preparation of the organic hydroperoxide in order to achieve the desired concentration of the organic hydroperoxide in the reactor and the recycle stream. In this regard, induction periods may also be essentially eliminated by the addition of a small amount of hydroperoxide other than the hydroperoxide product expected. In this context, the

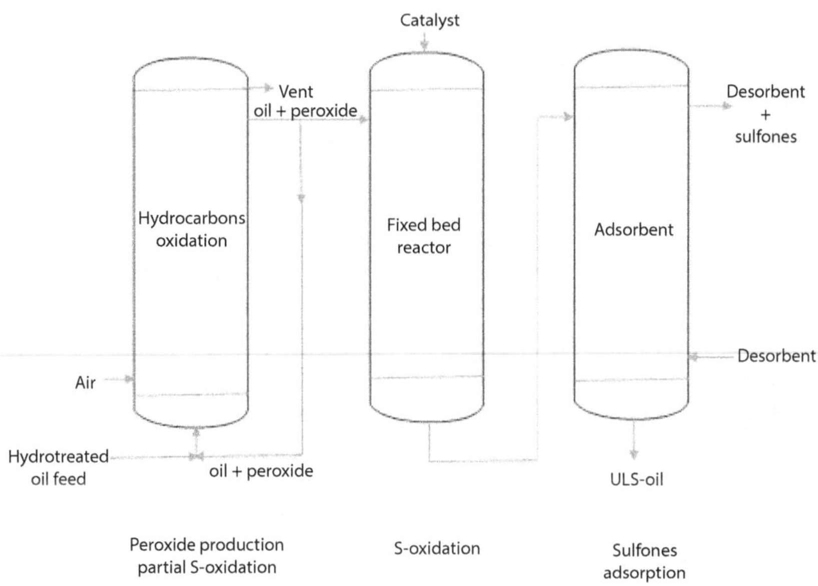

Figure 2.12 Simplified Flow-Chart of the Eni-UOP ODS Process.

added hydroperoxide is called an initiator. Hydroperoxides that may be suitable initiators are those which decompose under the reaction conditions quickly enough to reduce the induction period.

Examples of suitable initiators include cumene hydroperoxide and cyclohexybenzene hydroperoxide. Generally, hydroperoxide initiators are effective in amounts within a range of 0.2 - 1.5 wt.% of the fresh feedstock. The main advantage of this step is the in-situ production of the alkylhydroperoxide that would decrease the costs of storage, handling, and the use of conventional oxidants. ***In the second step***, the partially oxidized stream, containing about 2000 ppm oxygen as peroxide, is further oxidized to sulfone at a low temperature (<200 °F) and pressure (<100 psig). This occurs in a fixed bed reactor in the presence of a heterogeneous catalyst. The main advantages of this step are a more than 98% conversion of organic S-compounds occurring and the usage of organic peroxide omitting the need to recycle the conventional corrosive organic acid catalysts upon the application of H_2O_2. Finally, in ***the third step***, the separation of the polar sulfones by extraction or adsorption occurs. The adsorption technique is recommended more often as it is more cost-effective and removes any trace of undesirable by-products that might be formed during the oxidation and reaction steps.

The activation of H_2O_2 with polyoxometallates with a Keggin structure, such as $H_3PM_{12}O_{40}$ [where, M is Mo(VI) or W(VI)], produces more effective and selective oxidants as polyoxoperoxo complexes, such as $PO_4[MO(\mu\text{-}O_2)(O_2)_2]_4^{3-}$, for the oxidation of nucleophiles, such as sulfides and sulfur compounds with less nucleophilicity, such as DBT, under mild conditions to sulfoxides or sulfones in high yields, but the slow rate and excessive decomposition of H_2O_2 withdraw its application on an industrial scale. The use of ultrasound can significantly improve the ODS efficiency under phase transfer conditions (Javadli and de Klerk, 2012b; Srivastava, 2012; Al-Degas *et al.*, 2016). That would occur, by emulsification, throughout the improvement of the liquid-liquid interfacial area for viscous films containing gas-filled bubbles and cavitation bubbles. Thus, increases the interfacial area available, for the reaction, by increasing the effective local concentration of reactive species and enhancing the mass transfer in the interfacial region. Moreover, the ultrasonic irradiation can significantly improve the reaction efficiency, mainly due to cavitation when mechanical vibrations are produced and transmitted into the liquid as ultrasonic waves. This phenomenon involves the formation, growth, and implosive collapse of bubbles in liquids irradiated with high intensity ultrasound, creating shock waves, providing a unique set of conditions to promote chemical reactions and, thus, increasing the chemical reactivity in such systems. When the

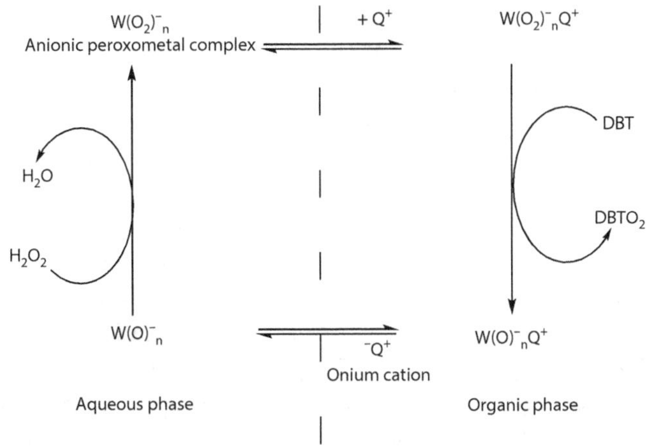

$W(O_2)^-_n$ is kegging heteropolytungstate anion, Q^+ is quaternary ammonium salts such as $Oc_4N^+Br^-$ with large lipophilic cation function as phase transfer agent (PTA)

Scheme 2.8 A Conceptual Model of Catalytic ODS in Ultrasound Assisted ODS.

compression of bubbles occurs during cavitation, short-lived localized hot spots can be generated. Cavitation causes bubble collapses rapidly and violently, providing temperatures of about 5000 K and pressures of about 1000 atm, with heating and cooling rates above 1010 K/s. This microenvironment, with extreme local conditions, is favourable to create active intermediates allowing the reaction to proceed instantaneously. The ultrasound assisted ODS would take place in four successive steps (Scheme 2.8). In the first, the metal precursor is peroxidized and disaggregated to form an anionic peroxometal complex in excess of H_2O_2. In the second step, quaternary ammonium salts, such as $Oc_4N^+Br^-$ with large lipophilic cation function, act as phase transfer agents (PTA) transferring the peroxometal anion into the organic phase. The phase transfer agents are surface-active species that lower surface tension and permit easy formation of microbubbles under ultrasound. In the third step, the S-compounds, such as DBT, are efficiently and selectively oxidized by the peroxometal complex. Finally, in the fourth step, the reduced oxo-species, that would have been dissociated with PTA, returns to the aqueous phase to restore the catalytic cycle (Mei *et al.*, 2003; Wan and Yen,2007).

Ultrasonic-assisted ODS of fuels has many advantages compared to HDS; it can be operated under atmospheric pressure and at relatively low temperatures. Furthermore, the advantage of ultrasonic-assisted ODS, rather than the conventional ODS, is the higher rate of the former technique that can lead to complete S-removal within a few minutes. It can also be performed without the addition of metallic catalysts (de A. Mello *et al.*, 2009; Duarte at al., 2011; Hosseini, 2012). Ultrasonic-assisted catalytic ozonation combined with the extraction process exhibits high catalytic efficiency for the removal of dibenzothiophene from simulated diesel oil (Zhao and Wang, 2013). Palaić *et al.* (2015) reported that the ultrasonic-assisted ODS of real diesel fuel revealed a significant reduction in S-content in a shorter reaction time relative to the ODS in a mechanically stirred system where ultrasonic-assisted ODS of diesel fuel with 4000 mg/kg S-content, in a batch reactor, using H_2O_2 as an oxidant and acetic acid as a catalyst, and *N,N*-dimethylformamide as extractant at solvent/oil ratio of 1.0, produced diesel fuel with a total S-content of 3 mg/kg within 30 min of oxidation at 70 °C and under atmospheric conditions. Kadijani *et al.* (2016) reported the application of response surface methodology based on central composite face-centered designs of experiments to optimize and study the effect of ultrasonic-assisted ODS of gasoil using tungstophosphoric acid catalyst and tetraoctylammonium bromide as a phase transfer agent in the presence of hydrogen peroxide as an oxidant. The results revealed that a mass of catalyst, mass of PTA, and ultrasonic wave amplitude affected the sulfur conversion positively, while the volume of oxidant could raise the sulfur conversion until a special point. Afterwards, every rise in the oxidant volume led to a decline in sulfur conversion where a maximum sulfur conversion of 95.92% has been achieved in the presence of 21.96 mL of oxidant, 1 g of catalyst, and 0.1 g of phase transfer agent, followed by liquid-liquid extraction using a polar acetonitrile solvent. Khodaei *et al.* (2016a) also reported that the application of RSM based on the Box–Behnken design has been employed to optimize the ultrasound-assisted oxidative desulfurization of non-hydrotreated kerosene (2490 ppmw S-content) in the presence of formic acid and H_2O_2, where 95.46 % S-removal from kerosene has been achieved in a sonication time of 10.5 min under optimal oxidation conditions (15.02 oxidant-to-sulfur molar ratio (no/ns), 107.8 formic acid-to-sulfur molar ratio nacid/ns, and 7.6 W/mL ultrasound power/fuel oil volume) and 90% of kerosene recovery occurred after the liquid–liquid extraction of the oxidized stream with acetonitrile that occurred with a decrease in the aromatic content of the treated kerosene from 15.9 to 10.9 vol%. Thus, the naphthene and olefin contents of the treated kerosene are slightly increased. There was a negligible increase in the water content of the kerosene after ultrasonic-assisted ODS treatment. In another study, Khodaei *et al.* (2016b) applied RSM based on the Box–Behnken design to optimize the ultrasound-assisted-ODS of a model light fuel oil 500 ppmw S (that is, 2.50125 g DBT/L toluene) using H_2O_2–HCOOH as the oxidation

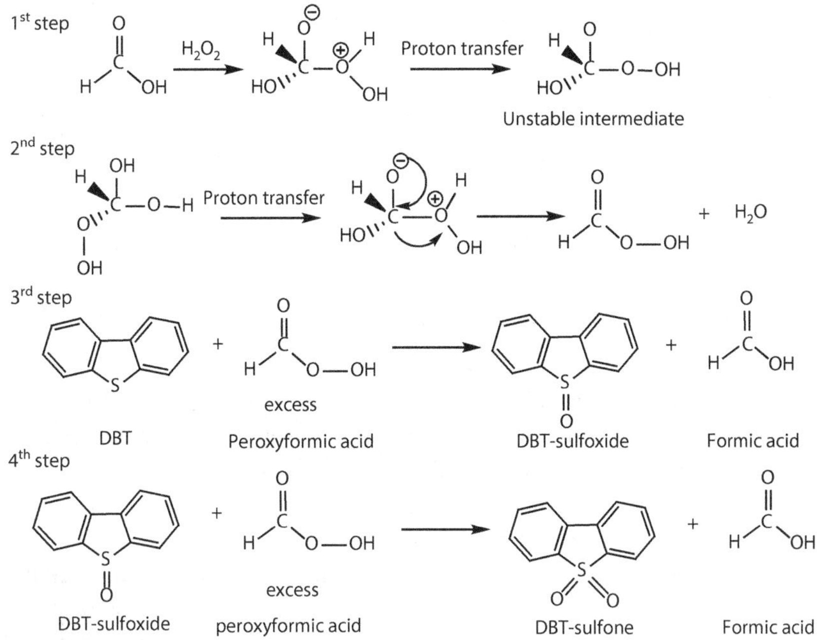

Scheme 2.9 Ultrasound Assisted-ODS of DBT in Presence of Hydrogen Peroxide and Formic Acid.

system. More than 97% sulfur conversion was achieved under the optimum conditions (oxidant to sulfur molar ratio of 26.7, acid to sulfur molar ratio of 74.6, and ultrasound power/fuel oil volume of 7 W/cm^3, at 50 °C within 80 s of sonication). The high conversion was attributed to the ultrasound irradiation which can enhance the overall oxidation rate due to an increase in the interphase mass transfer rate and cavitation, which creates a very fine emulsion between the aqueous and organic phases. The remarkably low reaction time emphasizes the application of this oxidation system in industrial applications. The ODS of DBT takes place in four successive steps (Scheme 2.9). Hydrogen peroxide (H_2O_2) reacts with the carbonyl group of formic acid. Then, the π bond between carbon and oxygen in the carbonyl group is broken and an unstable intermediate is formed (step 1), which is further decomposed to peroxyformic acid and water (step 2). DBT is oxidized to dibenzothiophene sulfoxide (DBTO) (step 3) and dibenzothiophene sulfone ($DBTO_2$) (step 4) and the formic acid is formed again in the aqueous phase.

SulphCo Inc. has developed ODS (Gunnerman, 2003) applying ultrasound power during sulfur oxidation by hydrogen peroxide, using

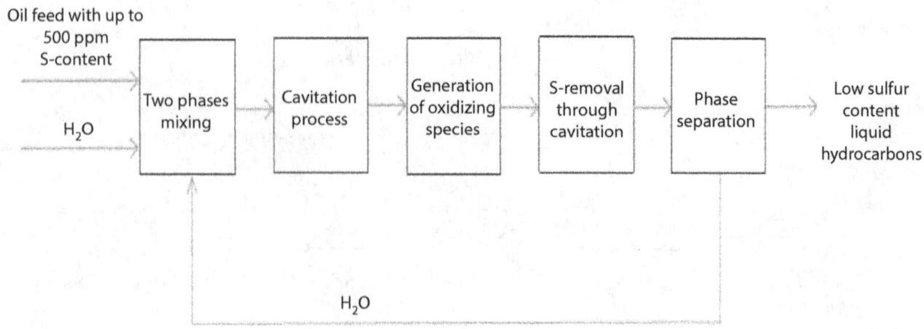

Figure 2.13 Simplified Flow Chart for Hydrodynamic Cavitative ODS.

tungsten phosphoric acid at 70 to 80 °C and atmospheric pressure, within a residence time of 1 min, where a desulfurization efficiency on the order of 80 and 90% occurred for crude oil and diesel oil, respectively (Dai *et al.*, 2008). Upon applying the SulphCo technology for upgrading of crude oil, a 50% decrease in S-content occurred with an increase in the API gravity up to 3 points and a reduction in viscosity by approximately 15%, has been reported (SulphCo, 2009). SulphCo has successfully used a 5000 barrel/day mobile "sonocracking" unit to duplicate, on a commercial scale, its proprietary process that applies ultrasonics at relatively low temperatures and pressures (Wu and Ondruschka, 2010). Stanislaus *et al.* (2010) reported that by the preliminary estimation of Betchel Corp., the SulphCo unit would cost 50% of what an equivalent hydrotreater would cost.

The first ultrasonic desulfurization unit has been installed at the IPLOM petroleum refinery near Genoa in Italy where it showed continuous desulfurization of diesel fuel to <10 ppm at a rate of up to 350 bbl/day (Stanislaus *et al.*, 2010).

Based on the same concept of acceleration of ODS reaction throughout the formation of cavitation, Suryawanshi *et al.* (2016) proposed a novel ODS process without the use of a catalyst (Figure 2.13). Briefly, the oil feed and aqueous phases (water) are mixed, under ambient temperature and pressure, then passed through a cavitating device (for example a vortex diode). The formed vapor cavities are then collapsed generating very high localized pressure (~ 1000 atm) and temperature (~ 10,000 K), as well as in-situ oxidizing species (hydroxyl radicals) which react with the S-moiety of thiophenic compounds in the organic phase. In the developed method, the removal of organic sulfur can be occurr by both mineralization, as well as oxidation, producing sulfones. But, due to the absence of acid catalysts, the latter mechanism may not be significant. Hydrodynamic cavitation generates hydroxyl radicals through the cleavage of water molecules

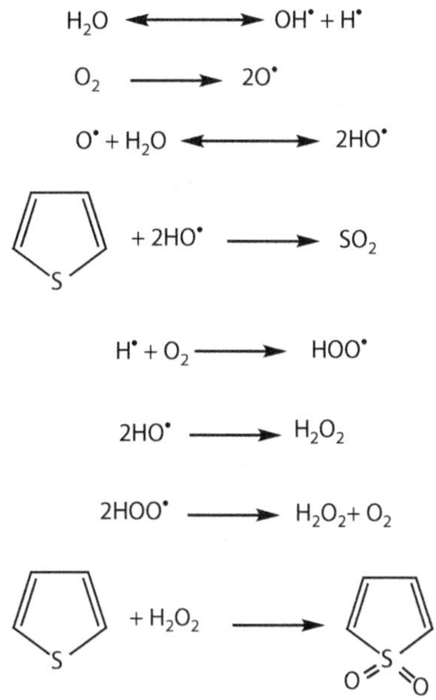

Scheme 2.10 Hydrodynamic Cavitative ODS.

to active oxidants, which would oxidize the S-compound to sulfones that would go to the aqueous phase. Alternatively, attacking the S-bond releases SO_2, HSO_3, H_2SO_4, and further mineralization of the organic skeleton to CO_2 and H_2O would also occur (Scheme 2.10). When applying this technique on commercial diesel oil with 30 ppm S-content, mainly of refractory S-compounds, a complete S-removal occurred.

The main advantages of this process are lower operational cost, significant ease of operation, and its effective deep desulfurization without using any catalyst, under ambient temperature and atmospheric pressure, at a low pressure drop of just 0.5 bar across the cavitating device. The aqueous phase can be recycled and reused after removing a purge stream (with corresponding make-up water). Hydrodynamic cavitation usually improves performance with scale-up. Thus, the proposed technique can be implemented effectively in large scale S-removal.

Recently, photo-oxidative desulfurization has attracted researchers. This occurs in a specially designed photoreactor, under mild reaction conditions (Javadli and de Klerk, 2012b) where the S-containing hydrocarbons are suspended in an aqueous-soluble solvent and then subjected to UV or visible irradiation, in presence of photo-catalysts such as TiO_2 (Tao *et al.*, 2009), ferric oxide (Fe_2O_3) (Zaki *et al.*, 2013), etc. The photochemical reaction can be assisted by a photosensitizer such as 9,10-dicyanoanthracene

(DCA) (Yazu *et al.*, 2001). The polar oxidized S-compounds are extracted directly into aqueous-soluble solvents such as acetonitrile (Ibrahim *et al.*, 2003). Finally, the treated hydrocarbon feed and the solvent phases are separated as in the extractive desulfurization.

Thus, to increase the yield and enhance the economic efficiency, the aromatics should be extracted from the solvent by liquid-liquid extraction using paraffinic solvents to be re-blended into the desulfurized fuel stream. The recovery of the photosensitizer from the solvent should be also done, applying the adsorption technique by using, for example, silica gel as an adsorbent, then recovered by desorption using the aqueous solvent (i.e, the acetonitrile). However, Tao *et al.* (2009) reported the rate of photo-oxidative desulfurization of sulfuides in kerosene is 100 times greater than that of DBTs.

The photo-oxidative desulfurization pathway for BT, DBT, and their derivatives was elucidated by Shiraishi *et al.* (1999a) as follows: BT is converted, first, to benzothiophene-2,3-dione, followed by hydrolysis, loss of carbon monoxide, and oxidation of the sulfur atom to be converted, finally, to 2-sulfobenzoic acid. 3-Methyl BT and 2,3-dimethyl BT produce 2-sulfobenzoic acid, benzenesulfonic acid, and 2-acetylbenzenesulfonic acid. DBT is first photooxidized to DBT sulfoxide, which is then further oxidized to form DBT sulfone or dibenz[c,e][1,2]oxathiin-6-oxide. The latter is then oxidized and converted to 2-sulfobenzoic acid via dibenz[c,e][1,2]oxathiin-6,6-dioxide. Benzothiophene-2,3-dicarboxylic acid is likely to be formed directly by the oxidation of DBT by a singlet oxygen, generated by photosensitization with DBT sulfone, while 4-MDBT and 4,6-DMDBT were found to produce the corresponding sulfobenzoic acid, dicarboxylic acid, and sulfone.

The order of reactivity of recalcitrant sulfur compounds in photo-oxidative desulfurization is reported to be the opposite of HDS:

dibenzothiophene < 4-methyl dibenzothiophene
< 4,6-dimethyl dibenzothiophene.

This represents a beneficial advantage for the application of photo-oxidative desulfurization (Hirai *et al.*, 1996), but there are some problems that should be solved to make it technically and economically feasible for its application on an industrial scale, for instance the need for better solvents to increase S-compound solubility and aromatic rejection and appropriate commercial techniques for solvent recovery. Not only this, but the order of photo-oxidative desulfurization was found to be dependent on the catalyst used, thus, it is not universal. For example, Zhu *et al.* (2014) reported the photocatalytic desulfurization of different sulfur compounds in model oil to be decreased in the following order:

$$\text{DBT} > \text{BT} > \text{dodecanethiol (RSH)} > \text{4,6-DMDBT}.$$

This is attributed to the steric hindrance and electron density around the S-atom since the S-conversion increases with the increase of the aromatic electron density. Although, the electron density on the S-atom records 5.758, 5.739, and 5.76 for DBT, BT, and 4,6-DMDBT, respectively. But, the photo-oxidation of 4,6-DMDBT was the lowest due to the predominance of the steric hindrance effect. That was also obvious in the case of RSH, where its long alkyl chain was an obstacle for the approach of its S-atom to the catalystic active sites.

Moreover, to increase the photo-transformation of S-compounds, a combination of a solvent and a photosensitizer should be optimized. Also, improvement of the separation processes and photosensitizer recovery is required. Babich and Moulijn (2003) reported that the stabilization of the photosensitizer on solid carriers would simplify the process and eliminate the process of photo-oxidant recovery from the fuels and the solvent.

Hirai *et al.* (1996) reported the desulfurization of DBT, 4-MDBT, and 4,6-DMDBT in one step under UV-irradiation ($\lambda > 280$ nm, to avoid the cleavage of C-C bond), using a high pressure mercury lamp in the presence of H_2O_2 where the sulfur was removed as sulfate in the water phase, under room temperature and atmospheric pressure. Moreover, the addition of 4-phenylbenzophenone is reported to enhance the photo-oxidation of S-compounds, in a light fuel, under UV-irradiation in the presence of 30% of H_2O_2 solution where the sulfur content of a light fuel was reduced from 0.2 wt.% to < 0.05 wt.% with a yield product of approximately 75% after 24 h of photo-irradiation (Hirai *et al.*, 1997). The total sulfur content in a FCC gasoline was reported to decrease from 309 to 68 ppm applying liquid-liquid extraction with acetonitrile and photo-oxidation with ultraviolet light from a high-pressure mercury lamp, with a total yield 90–96%. FCC gas oil with an S-content of 1800 ppm was desulfurized by liquid-liquid extraction, using acetonitrile, followed by a photochemical reaction in the presence of an oxidizing agent, to reach 508 ppm-sulfur with a final yield of approximately 90–95% (Ibrahim *et al.*, 2004).

To overcome the drawbacks of titania as a photo-catalyst and increase its photo-oxidative efficiency, titania coating of a multi wall carbon nano tube (MWCNT) by sol-gel method has been used to improve the photo-catalytic removal efficiency of DBT from an *n*-hexane solution using a 9 W UV lamp (Barmala *et al.*, 2015). Since carbon nanotubes (CNT) have chemical stability, high specific surface areas, and high electrical conductivity, they have electrical conductivity similar to that of copper, which makes them suitable electron acceptors when combined with semiconductors like titatnia. Moreover, the multi wall carbon nano tubes (MWCNTs) decrease the crystallite size of composite recombination chance. However, CNTs are black and a large amount in a CNT/semiconductor composite would disperse light and, consequently, prevent it from reaching the bulk solution, preventing light absorption by the solution and decreasing the efficiency of the process. Thus, it is mandatory to reach for an optimal amount of CNTs in a structure. By contrast, a low dosage of MWCNT causes recombination of the electron holes which also decreases the DBT removal rate. Furthermore, large amounts of catalyst in a solution also decreases the photo-catalytic efficiency for the same reason. It has been noticed, from that study, that by decreasing the MWCNT content in a composite, the ability of a composite to adsorb DBT decreases due to the reduction of specific surface area. However, increasing the MWCNT content in the composite decreases the time required to reach equilibrium. This, again, is due to increased surface area and adsorption ability. In conclusion, the DBT removal rate versus MWCNT content is found to follow a bimodal pattern related to the simultaneous influence of DBT adsorption on the catalyst surface and oxidation by the electron hole. The DBT molecules are easily adsorbed onto the composite, come into contact with the TiO_2 molecules, and are oxidized. As the MWCNT content increases, molecules adsorbed onto the surface are no longer in good contact with TiO_2 and photo-catalytic ability decreases. Thus, the two factors, adsorption and oxidation, influence the photo-catalytic process so the optimum MWCNT contents in the composite are found to be 0.25 g and 0.75 g MWCNT per 80 mL of sol. The rate of DBT removal was doubled in the presence of titania/MWCNT composite, relative to that in presence of titania alone.

Desulfurization of oil feeds, using liquid-liquid extraction by ionic liquids, cannot achieve the required ULS-fuels (Al-Degas *et al.*, 2016). However, ionic liquid ([Bmim]PF6) was reported to be used as the extractant and photochemical reaction medium to promote the oxidation of dibenzothiophene (DBT) in the presence of H_2O_2 where 99.5% removal of DBT from *n*-octane was achieved under mild conditions of room temperature and atmospheric pressure and the ionic liquid $[Bmim]PF_6$ was recycled eight times with a slight decrease in desulfurization efficiency (Zhao *et al.*, 2008). Zhu *et al.* (2014) reported that the desulfurization of DBT-model oil with amorphous TiO_2 was superior to that with anatase

TiO_2 and anatase - rutile TiO_2, recording 96.6, 23.6, and 18.2%, respectively, in presence of H_2O_2 and the ionic liquid [Bmim]BF_4. Abid (2015) reported the photo-oxidative desulfurization of DBT in a Y-shape catalytic microreactor using solar incident energy, TiO_2, and an H_2O_2 solution. It was also reported that the Cu-Fe co-catalyst was able to enhance the photo-activity of anatase TiO_2 under visible light illumination, where the photo-desulfurization of 100 ppm DBT, in model oil, in the presence of H_2O_2 reached 82.36 %.

Although, several studies proved the high selectivity to remove sulfur from different oil feeds, such as light oils, catalytic-cracked gasoline, and vacuum gas oils (Shiraishi *et al.*, 1998; 1999a,b; Ibrahim *et al.*, 2003; Zhao *et al.*, 2008). However, this technique is rather far from being applied on an industrial scale.

In conclusion, the efficiency and economics of the ODS process is strongly dependent on the methods used for oxidizing the sulfur compounds and, successively, the methods used for separating the sulfoxide and sulfone derivatives from the oxidized fuels. In addition, the key to successful implementation of ODS technology in most refinery applications is to effectively integrate the ODS unit with the existing diesel hydrotreating unit in a revamp situation.

In recent years, ODS has gained importance and is regarded as an excellent option after the HDS process since the Cx-DBTs derivatives are easily oxidized under low temperature and pressure conditions to form the

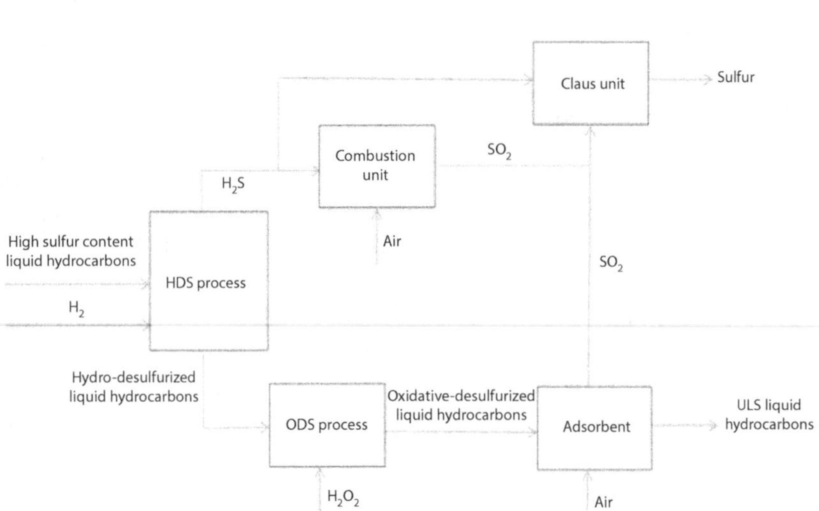

Figure 2.14 Simplified Flow Diagram for Integrated Hydrodesulfurization/Oxidative Desulfurization Process.

corresponding sulfoxide derivatives and sulfone derivatives. While sulfur compounds, such as disulfuide derivatives, are easy to be hydrodesulfurized, they are oxidized slowly. For this reason, ODS can be utilized as a second stage after existing HDS units, taking a low sulfur diesel (500 ppm) down to ultra-low sulfur diesel ULSD (<10 ppm) levels. Thus, this would lower the consumption of expensive hydrogen.

British Petroleum used hydrogen peroxide and phosphotungstic acid as a catalyst and tetraoctylammoniumbromide as the phase transfer agent in a mixture of water and toluene and applied this oxidative desulfurization process after the hydrodesulfurization process (Figure 2.14) (Collins *et al.*, 1997).

2.4 Selective Adsorption

Desulfurization by adsorption is a green technology wherein sulfur compounds are selectively removed through adsorption on the solid adsorbent leaving behind sulfur free fuel.

Adsorptive desulfurization (ADS) is considered as a promising approach to produce fuel cell grade gasoline and diesel at both a relatively low temperature and pressure, without using hydrogen gas which is advantageous compared to the conventional HDS method that uses both high temperature and hydrogen pressure. Moreover, it is characterized by easy operation, low cost, no pollution, and leads to deep desulfurization. The sulfur compounds can be removed from commercial fuels either via reactive adsorption by chemisorption, such as π-complexation, i.e. activated adsorption through interaction between adsorbent and adsorbate, or physical adsorption, such as van der Waals and electrostatic interactions, i.e. adsorption through intermolecular forces of attraction between molecules of the adsorbent and the adsorbate (Yang *et al.*, 2014). Physisorbed sulfur compounds can be easily removed from adsorbents by heating or decreasing pressure (Hernández-Maldonado and Yang, 2004a), so it is easy to regenerate the adsorbents. The zeolite-based adsorbents have been used as promising materials for selectively removing sulfur derivatives from diesel. Metal ion exchanged Y zeolites showed high selectivity and capacity for sulfur compounds using π-complexation between metal ion and sulfur compounds (Hernández-Maldonado and Yang, 2004b). The selectivity of the zeolite-based adsorbents varies according to fuel composition, such as aromatic and moisture concentrations (Bhandari *et al.*, 2006). The π–complexation adsorption is the most promising ADS, as it does not suffer from the steric hindrance that would occur by high molecular weight PASHs, such as 4,6-DMDBT (Hernández-Maldonado and Yang, 2004c). Molecular orbital (MO) calculations have shown that the π-complexation bonds between Cu or Ag and thiophene are stronger than those with benzene. Thus, π-complexation sorbents are selective for sulfur removal from transportation fuels (Yang *et al.*, 2001; Yang, 2003; Hernández-Maldonado and Yang, 2003a,b). The MO-calculations and experiments have shown that the refractory compounds (MDBT and DMDBT) bind strongly through π-complexation due to a better electron donation/back-donation ability. Within the π–complexation mechanism, the cations can form the usual σ bonds with their s-orbitals while their d-orbitals can back-donate electron density to the antibonding π–orbitals of the sulfur rings. Methyl groups should not have much effect on π-complexation because they would enhance adsorption through increased polarizabilities (hence, van der Waals interactions). Although the π-complexation bonds are stronger than those formed by van der Waals interactions, they are also weak enough to be broken by traditional engineering means, such as increasing temperature and/or decreasing pressure (King, 1987). This increases the opportunities for tailoring and developing new adsorbents for selective fuel desulfurization processes. The metals that can form strong π–complexation bonding are those that possess empty s-orbitals and available electron densities at the d-orbitals necessary for back donation, for example, the copper Cu(I) electronic configuration $1s^2 2s^2 2p^6 3s^2 3p^6 3d^{10} 4s^0$ (Hernández-Maldonado and Yang, 2004a,b,c; Baeza *et al.*, 2008).

Adsorbents, used industrially, are generally synthetic micro porous solids such as: (i) activated carbon, (ii) molecular sieve carbon, (iii) activated alumina, (iv) silica gel, (v) zeolite derivatives, (v) metal oxides, and (vi) clay. They are usually agglomerated with binders in the form of beads, extrudates, and pellets of a size consistent with the application that is considered (Song and Ma, 2003). The efficiency of adsorptive desulfurization (ADS) is mainly influenced by the adsorbents' properties, such as capacity, selectivity, stability, and ability to regenerate (Babich and Moulijn, 2003). Moreover, it is highly affected by its specific surface area which preferentially ranges between 200 to 2000 m^2/g (Alavi and Hashemi, 2014). The good potential of activated manganese zinc oxides in the ADS of crude oil was reported (Adekanmi and Folorunsho, 2012). Activated carbon, carbon fibers, and carbon nanotube can be used in ADS processes where the adsorption would be physical and depended on the size and volume of the pores or chemical that depends on the chemisorption surface properties (Bagreev *et al.*, 2004; Aparicio *et al.*, 2013; Gawande and Kaware, 2016).

Mužic *et al.* (2010) reported a comparative study for ADS diesel fuel with activated carbon (AC) and activated aluminium oxide as adsorbents in a LAM A1 batch adsorption apparatus at 50°C under atmospheric pressure. A 2^3 factorial design was applied to optimize and study the effect of three experimental parameters: time (t, min), initial sulfur concentration (C_o, mg/kg), and mass of activated carbon (m_{AC}, g). The ADS of AC was higher than that of activated aluminium oxide, where the maximum ADS efficiency of approximately 52.5% was obtained within 100 min, 16 mg/kg initial sulfur concentration, and 4 g AC/50 cm^3 diesel fuel. Mužic et al. (2010) reported in another study the ADS of diesel fuel, applying two Design of Experiments (DOE) methods. The experiments were carried out in a batch adsorption system using Chemviron Carbon SOLCARB™C3 activated carbon as the adsorbent. The first DOE method employed was a full factorial with three factors on two levels and five center points and the second was a Box-Behneken design with the same three factors, but on three levels. Response surface methodology (RSM) was applied to estimate the effects of the individual factors of time (t, min), initial sulfur concentration (C_o, mg/kg), and mass of activated carbon (m_{AC}, g), their interactions on sulfur concentration and sorption capacity were determined, and statistical models of the process were developed. The effects of time and adsorbent mass on output sulfur concentration recorded have very little effect at a lower level of initial S-concentrations, while at high initial S-concentrations they recorded a greater effect.

Desulfurization of a commercial diesel fuel by different adsorbents was studied in a fixed-bed adsorber operated at ambient temperature and pressure (Hernández-Maldonado and Yang, 2004c). Copper (auto-reduced) type-Y zeolites are found to be superior adsorbents for the removal of all sulfur compounds from commercial diesel fuels where the ADS capacity of the tested adsorbents decreased in the following order: AC/Cu(I)-Y > Cu(I)-Y > Selexsorb CDX (alumina) > CuCl/γ-Al_2O_3 > activated carbon > Cu(I)-ZSM-5. The GC/FPD analysis proved that the π-complexation sorbents selectively adsorbed highly substituted thiophenes, benzothiophenes, and dibenzothiophenes from diesel, which is not possible with conventional HDS reactors. The AC/Cu(I)-Y can produce 30 cm^3/g of diesel fuel with a total sulfur content of 150 ppbw and 20 cm^3/g of diesel fuel with a sulfur content of equal or less than 60 ppbw. This can be applied to a fuel cell which needs ULS-diesel oil. Furthermore, although the AC/Cu(I)-Y has selectivity towards the refractory S-compounds, the Cu(I)-ZSM-5 zeolite is promising for applications when selectivity toward small thiophene molecules is preferred. The Cu supported on zirconia is also reported to be a prominent adsorbent in the desulfurization of sulfur organic compounds. The adsorption capacity of Cu/ZrO_2 system would increase by increasing the surface area of the zirconia through the use of a lower calcination temperature (Baeza *et al.*, 2012).

Commercial diesel fuel liquid desulfurization, via adsorption under mild conditions, using commercial high specific surface area activated carbon (AC) was performed on both laboratory- and pilot-scale experiments in fixed-bed setups. Under laboratory-scale conditions, maximum sulfur removal exceeded 90%, while according to breakthrough curves the total sulfur content remained below 2 ppmw for up to 20–22 mL of processed diesel/g sorbent. Process scaling-up by a factor of 15 showed a moderately negative effect, with the respective breakthrough fuel amount (total sulfur ≤2 ppmw) being ~15–17 mL processed fuel/g sorbent. Several sorbent regeneration strategies were studied under laboratory-scale conditions. The one with the highest restoration of initial (i.e., fresh state) AC performance involved heating under a vacuum (200 mbarA) up to 200 °C and subsequent washing of the material with a binary organic solvent. The amount of solvent required was 50–55 mL/g sorbent where, from the second and up to the seventh cycle, desulfurization efficiency of the material was essentially stable, but from the eighth cycle and on, further performance depletion in desulfurization efficiency was recorded. Based on fresh/regenerated sorbent post-analysis, it was found that cycle-to-cycle degradation was due to a gradual decrease of the sorbent's surface area, mainly attributed to residual amounts of diesel-derived species remaining in its structure, thereby partially blocking its porosity. The most important advantage of this process is that the main properties of processed fuel remained essentially unaffected, although the removal of di- and polyaromatic compounds was notable (Baltzopoulou *et al.*, 2015). The ADS on activated carbons, natural clays, and other adsorbents prepared from natural sources will be discussed in detail in Chapter 5.

In an attempt to remove the recalcitrant and bulky PASHs from diesel fuel, Ganiyu *et al.* (2017) reported the modification of commercial activated carbon with different concentrations (0.5–10 wt.%) of boric acid to produce adsorbents with good physico-chemical properties and tested its activity for the ADS of model oil (4,6-DMDBT in iso-ocatne) at room temperature. Boron-doped adsorbents exhibited better adsorption capacity at low boron loadings of 0.5–2.5 wt.%, largely due to their preserved surface area compared to those with higher boron loadings (5–10 wt.%) and enhanced surface acidity compared to the unmodified AC. Since doping AC with metal improved the surface acidity, but at the same time reduced the surface area, which is as also an important property, it would have to be large enough to address the mass transport of a bulky adsorbate like 4,6-DMDBT across the pores of the adsorbent. The maximum ADS capacity of (8.50 mg/g), that was approximately 85% removal from 100 ppmw-S containing 4,6-DMDBT, was achieved using an adsorbent with 1 wt.% of boron loading (1 BDAC), within 5 min contact time. The prepared adsorbent 1BDAC expressed a great selectivity for 4,6-DMDBT in the presence of naphthalene. This was attributed to the ability of 4,6-DMDBT to be preferentially adsorbed on the active site of modified adsorbent through acid-base interaction resulting from the contribution of an oxygen containing functional group of AC and sulfur in 4,6-DMDBT, as well the boron-sulfur interaction, relevant to naphthalene (Yang *et al.*, 2007; Zhou *et al.*, 2009). A remarkable regeneration performance was exhibited by the 1BDAC adsorbent with only 7% loss in ADS capacity after five successive regenerations cycles.

For the use of wastes and a decreased cost of the activated carbon, waste rubber tires can be recycled to activated carbon via heating at 300 °C to isolate the produced oil, distilled diesel oil, and black tire crude oil, followed by carbonization in a muffle furnace at 500 °C for five hours to remove the ash as well as the carbon black. The produced char is then treated by an H_2O_2 solution to oxidize adhering organic impurities and, finally, calcined at 900 °C for five hours yielding a 473 m^2/g surface area AC, with a pore volume of 0.77 cm^3/g, and pore size of about 6 nm (Saleh and Danmaliki, 2016a,b). Upon its application for the ADS of model oil (150 ppm DBT in 85:15 *n*-hexane:toluene) in a fixed bed column, the best results were obtained at higher column length and dosage because the longer the column, the better the interaction between the PASHs and the adsorbents, while the higher the dosage the more active sites will be present for ADS. On the contrary, the lower the concentration and flow rates, the better the ADS% since the adsorbent can only take the maximum S-compounds at any given time before reaching saturation. Moreover, the lower the flow rate, the better the interaction of the model fuel with the adsorbent. Contact time did not show any significant effect and the ADS of the S-compounds on the AC was fast, reaching equilibrium within 5–10 min. Danmaliki and Saleh (2017) studied the effect of metal loading on AC for the ADS of recalcitrant S-compounds in both batch and fixed bed models. A model diesel fuel was formed with a mixture of 50 ppm thiophene, 50 ppm BT, and 50 ppm DBT in a solvent of 85% hexane and 15% toluene where the produced AC from waste tire was acid treated with HNO_3 and then loaded with10% cerium, iron, or cerium and iron to produce bimetallic composites by thermal co-precipitation method. The three composites named prepared namely AC/Ce, AC/Fe, and AC/Ce/Fe. Although, the AC had the highest surface area and pore volume of 460.27 m^2/g and 0.71 cm^3/g, respectively, and the highest surface oxygen-containing groups, it performed the least ADS efficacy. Moreover, although the AC/Ce/Fe recorded the least surface area and pore volume of 430.44 m^2/g and 0.64 cm^3/g, it performed the best in the adsorption of thiophene (31%), BT (30%), and DBT (75%). This was attributed to the changes in the chemical composition of the adsorbent, the acidic nature of cerium, and the crystal nature of iron. The ADS efficiency was in the following decreasing order: AC/Ce/Fe > AC/Ce > AC/Fe > AC. This confirmed the hypothesis of bimetallic modification of carbon's surface enhancing its selectivity and adsorptive capacity to refractory sulfur compounds. The adsorption of sulfur compounds and the breakthrough in all the adsorbents followed the order: DBT > BT > Th. Batch and column experiments carried out using AC/Ce/Fe revealed a high absorptive capacity and breakthrough for a DBT of 16 mg/g. The kinetic and isothermic analyses showed a synergistic effect of surface adsorption and intraparticle diffusion occurring concurrently and chemisorption and heterogeneous adsorption occurring on the surface. Thermal regeneration experiments carried out on AC/Ce/Fe showed stable efficiency in the adsorptive desulfurization after three regeneration cycles. In another study by Saleh *et al.* (2017), activated carbon prepared from waste rubber tires was successfully treated with manganese oxide to further improve their surface properties for ADS. The activated carbon-manganese oxide nanocomposite containing 10% optimum metal loading, of 160.98 m^2/g specific surface area, total pore volume of 0.141 cm^3/g, and average pore diameter of 9.27 nm, showed significant ADS efficiency within short time process for model oil (mixture of TH, BT and DBT in 85:15 *n*-hexane:toluene, with initial S-concentration of approximately 150 ppm). Adsorption capacities of 4.5 mg/g, 5.7 mg/g, and 11.4 mg/g were obtained for simultaneous adsorption of Th, BT, and DBT on the as-synthesized adsorbent in a batch process and the same trend was retained in the fixed bed model. The adsorbents performed excellently in the removal of DBT probably because of their structure containing more π-systems that provide more possible interactions, such as π-interaction with the active sites on the surfaces. Thus, dispersive interactions between the delocalized π-electrons, within the benzene rings of DBT and the electron-rich region on the nano-porous carbon aromatic ring, play a significant role in adsorption. In addition to the above mechanism, a feature common for the three S-compounds is the possible coordination with the metal species through S–M interactions. Several coordination configurations can be formed between the sulfur compounds and the metal including the n-type donor donating lone pairs of electrons that lie in the plane of the ring to the nanoparticles and the interaction between S and the oxygen atom on metal oxide nanoparticles.

Ni phosphides, supported on high surface area silica, have been reported for the ADS of diesel fuel to ultra-low-S-levels (Landau *et al.*, 2009). The zeolites are crystalline aluminosilicate materials, composed of tetrahedral units of SiO_4 and AlO_4. They have a stable and regular three-dimensional crystalline framework. Natural zeolites are reported to express high ADS capacity for OSCs from diesel oil, that reached 7.15 mg/g, with the liberation of H_2S (Scheme 2.11) (Mustafa *et al.*, 2010). Cheng *et al.* (2009) reported the deep desulfurization of FCC gasoline by selective absorption on various ion-exchanged nano-sized Y zeolites where the metal ion, as well as calcination conditions, have been found to significantly influence the ADS performance where the highest S absorption capacity was obtained at 100 °C on a HCuCeY sample calcined at 450 °C for 6 h in N_2 atmosphere. This was attributed to the interaction between Ce and Cu. Zeolite derivatives are efficiently used as adsorbents for selective adsorption of polar compounds and heterogenic organic compounds; they are hydrophilic and have large voids inside their structures (Mužic *et al.*, 2010). The ADS of gasoline with Ni-loaded Y type zeolites was found to decrease with an increase in temperature from 25 to 60 °C, where the adsorption capacity of the adsorbent decreased from 0.55 to 0.65 mg S/g adsorbent that was attributed to the exothermic nature of the process (Majid and Seyedeyn-Azad, 2010). The modified Y-type zeolite was popularly used as an adsorbent to remove sulfur from fuels via π-complexation, and copper (Cu^+) and silver-exchanged Y-type zeolites were effective to remove sulfur compounds from naphtha (Yang *et al.*, 2001; Takahashi *et al.*, 2002; Hernández-Maldonado and Yang, 2004a,b). Gallium-modified Y-zeolites are also reported for selective ADS of 4,6-DMDBT (Tang *et al.*, 2008).

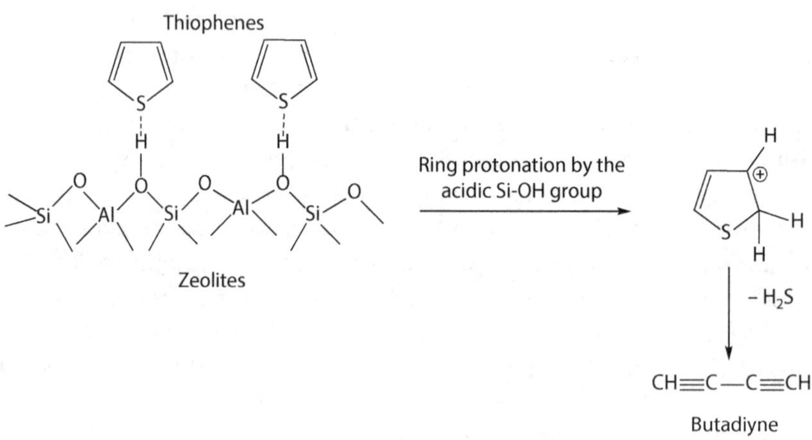

Scheme 2.11 Reactive ADS with the Liberation of H_2S.

Numerous oxides and hydroxides of Pb, Hg(II), and Ba have been found to efficiently perform reactive ADS with thiols producing solid metal thiolates (Nehlsen *et al.*, 2003). Shakirullah *et al.* (2009) studied the ADS-activity of different metal oxides, such as MnO_2, PbO_2, arsenic trioxide (ArO_3), Al_2O_3, MgO_2, ZnO_2 and silica on Jhal Magsi crude oil (collected from the Oil and Gas Development Corporation, Limited (OGDCL), Pakistan) and its distillate fractions, kerosene and diesel oil. Lead oxide (PbO_2) expressed the highest efficiency at 60.28%, 54.54%, and 52.57%, in the case of crude oil, diesel oil, and kerosene within 1 h, respectively. While the manganese oxide (MnO_2) performed maximum desulfurization activity of 58.55%, 49.38%, and 53.30% in the case of crude oil, diesel oil, and kerosene occurred within a longer time of 3 h. The reactivity of the metals with thiols depends on the metals' heat reaction, where a metal with small exothermic heat formation is the most reactive (Salem and Hamid, 1997). Since lead has low heat formation (ΔHrexn −121.4 kJ/mol), it expressed the fastest and maximum reactive-ADS. But, when the heat formation of metal thiolates is high (≥ −60 kJ/mol), the reaction rate is slow and, consequently, the reactive-ADS needs more time, taking into consideration that metal thiolates have relatively low melting points. Therefore, a descending desulfurization activity may occur with longer reaction time due to the decomposition of the formed metal thiolates formed. Kerosene has a high concentration of low molecular weight thiols and expressed higher desulfurization than that of diesel oil, throughout the faster reaction of thiols with the studied metals that was characterized by the high heat formation. Although PbO_2 expressed better performance than MnO_2, the use of MnO_2 is preferred over lead oxide since lead can be retained in the hydrocarbon stream which may contribute to engine fouling, in addition to environmental pollution in the form of lead particulates. These drawbacks will not occur with the application of MnO_2. This process is simple, employing inexpensive material and completes with minimum energy requirements compared to conventional HDS. Thiols react with metal ions, producing forming metal thiolates and metal oxides can be recovered by dilute-acid extraction. The choice of acid depends upon the desired metal salt (Scheme 2.12).

$$MnO_2 + 4\,RSH \longrightarrow Mn(SR)_4 + 2H_2O$$

$$Mn(SR)_4 + 4HNO_3 \longrightarrow Mn(NO)_3 + 4\,RSH$$

Scheme 2.12 Reactive-ADS of Thiols.

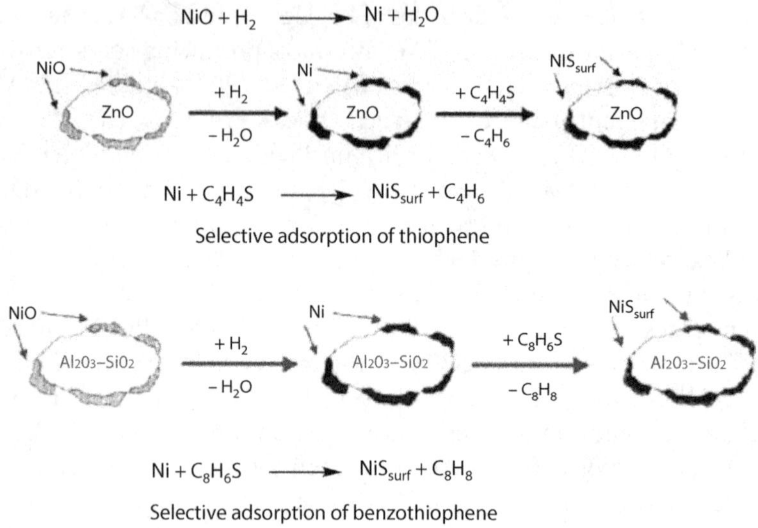

Scheme 2.13 Chemistry of SARS Process.

Adeyi *et al.* (2015) reported a 53% reduction from the initial S-content (0.175 wt.%) of diesel oil in an ADS batch process using activated manganese dioxide as the adsorbent.

McKinley and Angelici (2003) reported the selective removal of DBT and 4,6-DMDBT from simulated feedstock with Ag^+/SBA-15 and Ag^+/SiO_2 as adsorbents. The selective adsorption process for removing sulfur (SARS) at ambient temperature to achieve ultra-clean diesel and gasoline with an Ni-exchanged Y zeolite was reported (Velu *et al.*, 2005). Adsorbents are usually comprised of transition metals supported on base oxides, such as Ni/ZnO and $Ni/Al_2O_3–SiO_2$, where the Ni acts as the ADS-site (Scheme 2.13). The selective adsorption of various compounds on the nickel-based sorbent at room temperature is reported to increase in the following order: Naphthalene ≈ 1-methylnaphthalene < 4,6-DMDBT < DBT < quinoline < indole, while that of activated alumina follows: naphthalene ≈ 1-methylnaphthalene < 4,6-DMDBT ≈ DBT < indole < quinoline, while that of activated carbon follows: Naphthalene < 1-methylnaphthalene < DBT < 4,6-DMDBT < quinoline < indole (Kim *et al.*, 2006).

Song (2003) reported a process where the selective adsorption for the removal of sulfur compounds (SARS) is carried out and then followed by the HDS of concentrated sulfur (HDSCS) compounds (Figure 2.15) using high-activity catalysts, such as CoMo/MCM-41. Such a technique is much easier than the conventional HDS of diesel streams for two reasons: (1) it is more concentrated and reactor utilization is more efficient and (2)

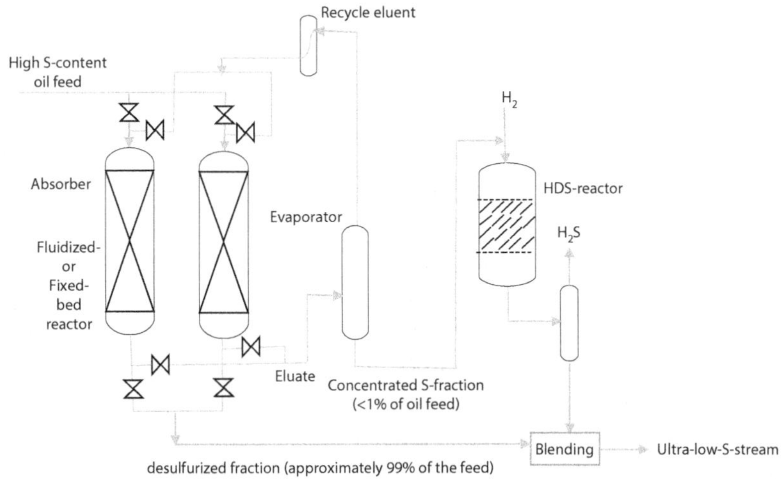

Figure 2.15 Simplified Flow Chart for Integrated SARS and HDSCS for Deep Desulfurization Process.

the rate of the HDS reaction is faster related to the removal of aromatics that inhibit the HDS by competitive adsorption in the hydrogenation sites (Ma *et al.*, 2002). Spontaneous monolayer dispersion was used to prepare MoO_3-MCM 41 and its ADS capacity has been tested at room temperature on gasoline with an initial S-content of 300 mg/kg (Shao *et al.*, 2012).

Moreover, acid activated alumina has more selective adsorption towards BT, DBT, and 4,6-DMDBT than zirconia and magnesia (Larrubia *et al.*, 2002; Srivastava and Srivastava, 2009). However, the steric hindrance expressed by the large molecular weight 4,6-DMDBT limited its ADS.

Removal of sulfur compounds can take place at elevated temperatures under a hydrogen atmosphere without the hydrogenation of aromatics, where the reactive ADS takes place under the presence of hydrogen in order to accelerate the reaction between the sulfur compounds and the adsorbing agent, where the S converts to H_2S and is subsequently adsorbed by the adsorbent, but less hydrogen is consumed than is the case of HDS processes. Research and development work has focused on different methods including fluidized bed, moving bed, and slurry for conveying the adsorbing agent through the reaction column and regeneration column. Materials that contain a combination of transition metal catalysts and sorbents are usually used for the reactive adsorption of S-compounds, for example, Cu-ZnO, Ni-Al_2O_3, Ni-ZnO, Ni/Al-SiO_2, NiP/SiO_2, Ni-SiO_2, Ni-SBA-15 (Babich and Moulijn, 2003; Stanislaus *et al.*, 2010).

Park *et al.* (2008) reported the reactive ADS of diesel oil by NiNPs supported on mesoporous silica materials (SBA-15 and KIT-6). The S-content decreased from 240 ppm to <10 ppm using 30% Ni/SBA-15 with an adsorption value of 1.7 mg/g.

In the USA, Conoco-Phillips Petroleum Company has developed the S-Zorb SRT process for the reactive-ADS of diesel oil. This process is based on fluidized bed technology, where it can decrease sulfur in feedstock containing greater than 2 mg/mL of sulfur to a very low level less than 0.005 mg/mL in the presence of hydrogen within an operating temperature and pressure of 300 to 400 °C and 1.9 to 3.45 MPa, respectively, where the sulfur compounds adsorbed on the adsorbent reacted with reduced metals, producing metal sulfides and newly formed hydrocarbon that would be released into the main stream. The spent adsorbent is continuously removed from the reactor and transported into a regeneration chamber. The sulfur is removed from the surface of the adsorbent by burning and the formed sulfur dioxide is sent to the sulfur plant. The adsorbent is then reduced with hydrogen and recycled back to the reactor (Hernández-Maldonado and Yang, 2004a; Mužic and Sertić-Bionda, 2013). H_2S is not released and the critical diesel fuel properties are unaffected. The process consumes relatively low H_2 and produces relatively lower CO_2 and NO_x. This process is estimated to decrease the overall operating and capital costs upon reaching the required product quality. A 6000 Barrels Per Stream Day (BPSD) gasoline unit has been in operation at the ConocoPhillips Borger refinery since April 2001.

In a study performed by Hagiwara and Echizen (Hagiwara and Echizen, 2001) for evaluation of different applied FCC gasoline desulfurization processes, including hydrodesulfurization processes Octgain 125, Octgain 220, and Scanfining from Exxon-Mobil, Prime G from IFP, CDHydro, and CD-HDS from CDTech, and S-Zorb from Conoco Phillips (Table 2.3), it was found that the S-Zorb ADS-process from ConocoPhillips and the CDTech HDS- process are better than other candidates for commercial implementation.

The Irvine Robert Varraveto Adsorption Desulfurization (IRVAD) process (Irvine, 1998) uses slurry and consumes very little (if any) hydrogen. The adsorption mechanism is based on the polarity of sulfur compounds. An alumina-based selective adsorbent counter-current contact liquid hydrocarbon stream up to 240 °C in a multistage adsorber was used. The adsorbent is regenerated in a continuous cross-flow reactivator using heated reactivation gas. The process operates at lower pressure, does not consume hydrogen or saturate olefins, and can effectively remove sulfur in liquid fuels and satisfy the demand for ultra-low sulfur fuels. However,

Table 2.3 A Comparison of FCC Gasoline Desulfurization Processes.

Parameters	Octgain 125	Octgain 220	Exxon scanfining	Prome G	CDTech	S-Zorb
Processing Capacity (bbl/day)	15,000	31,000	25,000	24,000	30,000	25,000
Investment Million ($US)	14.9	23.8	16.8	21.7	18.5	13.8
Hydrogen Consumption (m^3/m^3)	66	23	14	22	18	12
Power Consumption (kWh/m^3)	12.6	9.4	3.8	8.2	2.8	4.4
Steam Use (kg/m^3)	-	214	128	180	70	30
Home-Use Fuel (dm^3/m^3)	13.6	5.8	2.4	1.5	5.3	6.3
Catalyst Cost ($US/bbl)	0.43	0.22	0.22	0.01	0.25	0.27
Cooling Water (t/m^3)	6	5.4	3.2	3.1	1.3	3.1
Yield Loss (%)	5	0.7	0	0.8	0	0
Octane Loss	0	0.1	1	1.3	1	0.75
Required Capacity (10^3 m^3/year)	22,558	23,175	23,093	23,049	23,093	23,148
Construction Index	877	696	607	815	557	500
Depreciation Index	175	139	121	163	111	100
Variable Index	207	119	134	119	102	100
Evaluation Index	382	259	256	282	214	200

the adsorbent needs to be regenerated and recycled frequently (Sano *et al.*, 2005).

Xu *et al.* (2008) reported the RADS of FCC gasoline over Ni/ZnO of approximately 25.4 mg S/g adsorbent, with a 1.1 loss in octane number. Metal halides, such as $PdCl_2$ supported on SBA-15 and MCM-41, have also been found to be effective for the RADS desulfurization of jet fuel and Cu_2O supported on MCM-41 has been also reported to be more effective for ADS of jet fuel than that supported on SBA-15. All these adsorbents could be regenerated by heating in air and reused (Wang *et al.*, 2008a,b). Huang *et al.* (2010) studied the RADS of diesel oil over Ni/ZnO adsorbent under nitrogen and hydrogen, which followed two different mechanisms. The results indicated that ADS, in the presence of N_2, is achieved through physical and chemical adsorption and a severe decrease in the desulfurization activity of Ni/ZnO is observed with time on stream where the desulfurization capacity is very low. While in presence of H_2, the organic sulfur compounds in the diesel oil are first decomposed on the surface Ni of Ni/ZnO to form Ni_3S_2, followed by the reduction of Ni_3S_2 to form H_2S and then H_2S is stored in the adsorbent accompanied by the conversion of ZnO into ZnS (Scheme 2.14). The S-content decreased from 560 ppm to 2.6 ppm with an S-adsorption capacity of 0.333 g-S/g-NiO/ZnO, but the S-content in the adsorbent increased steadily with the RADS time on stream. Thus, after 122 h, certain amounts of sulfur-containing compounds (mainly4,6-DMDBT and 2,4,6-TMDBT) are still present in diesel oil which is attributed to the decrease of RADS activity of Ni/ZnO, along with the increase of Ni_3S_2 and ZnS contents in the adsorbent increment. Meanwhile, the desulfurization activity, as well as the acceptance ability of the adsorbent, is also decreased. As a result, at the later stage, the RADS efficiency decreases gradually and certain amounts of H_2S are released to the effluent rather than stored in the adsorbent.

Zhang *et al.* (2012) reported the preparation of special morphology ZnO materials with smaller crystallite size by the hydrothermal homogeneous precipitation method and the corresponding Ni/ZnO adsorbents were prepared by the incipient impregnation method. Reactive adsorption desulfurization (RADS) of model gasoline, using thiophene as model sulfur-containing compounds over Ni/ZnO adsorbent, was carried out in

$$2C_xH_yS + 3Ni + H_2 \longrightarrow 2C_xH_{y+2} + Ni_3S_2$$

$$Ni_3S_2 + 2H_2 \longleftrightarrow 3Ni + 2H_2S$$

$$ZnO + H_2S \longleftrightarrow ZnS + H_2O$$

Scheme 2.14 RADS on Ni/ZnO in Presence of H_2.

a fixed bed reactor in the presence of hydrogen. The results showed that the Ni/ZnO adsorbent using ZnO with a larger surface area and smaller crystal grains as an active component shows higher desulfurization activity and stability. During the adsorption process of thiophene on Ni/ZnO adsorbent, sulfur is trapped by zinc oxide and converted to zinc sulfide. Yin *et al.* (2012) reported the preparation of high-dispersed CeO_2 on mesoporous silica (SBA-15), where an adsorbent with 1.5 mol% Ce was able to remove 0.200 mmol S/g adsorbent from a model oil of thiophene and it also showed good ADS for BT and DMDBT. Sarda *et al.* (2012) investigated the selective ADS of a diesel fuel containing 325 mg S/kg, using adsorbents with different contents of Ni and Cu, supported on alumina and ZSM-5 zeolite, prepared by wet impregnation and ion exchange, respectively. The Ni/Al_2O_3 adsorbent was found to be more effective than the Cu/Al_2O_3 adsorbent. Up to 93.47% ADS occurred for a jet-A fuel with an initial total S-content of 949.03 mg/kg with a new adsorbent, Ni–Ce/Al_2O_3–SiO_2., which was prepared by extrusion molding, calcined at 600 °C for 3 hours. the main advantage of that method is that it was carried out at ambient temperature and pressure, under static condition (Shen *et al.*, 2012). Kong *et al.* (2013) reported the RADS of a model gasoline (1000 ppmw thiophene in a mixture of 35 wt.% of isopentene as olefins, 15 wt.% toluene as aromatic compounds, 10 wt.% of cyclohexane to mimic the cycloalkane and 40 wt.% n-heptane) where the presence of olefins decreased the desulfurization capacity from 360 mg S/g sorbent in a thiophene/n-heptane model gasoline to 322 mg S/g sorbent in a thiophene/mixture model gasoline, with slight loss of octane number, in a fixed bed reactor, at optimum operating conditions of 400 °C, 60 h^{-1} LHSV (volume hourly space velocity) and a pressure of 1.0 MPa. The RADS occurred through the reaction of active Ni metal with the S-containing compound in the presence of $H_{2,}$ producing metal sulfuide which could react with ZnO forming ZnS. This would lead to deep desulfurization with a low loss in octane number.

Wang *et al.* (2012) reported the RADS of 4,6-DMDBT from three different model diesel oils (1000 mg/kg 4,6-DMDBT in 100% n-dodecane, as aliphatic fuel (ALF), 10% benzene and 80% n-dodecane plus 10% naphthalene, as mixed fuel (MXF), and 90% benzene plus 10% naphthalene as aromatic fuel (ARF)) in a fixed-bed reactor packed with a reactive adsorbent that is composed of formaldehyde, the environmentally-friendly heteropoly acid, phosphotungstic acid, and mesoporous silica gel (19.1, 33.3, and 47.6 wt.%, respectively), at 80 °C under atmospheric pressure. The mechanism of RADS was suggested to depend on the condensation reaction of 4,6-DMDBT with formaldehyde using the phosphotungstic acid as a catalyst in pore spaces (Scheme 2.15). The presence of formaldehyde

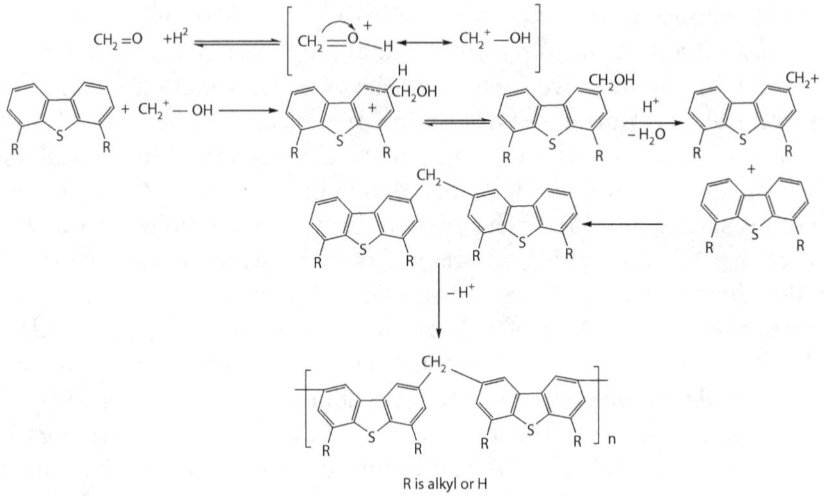

Scheme 2.15 The Suggested Mechanism for the Condensation Reaction Occurring in the Pore Spaces for Fuel Desulfurization.

led to 100% removal of 4,6-DMDBT within 90 min compared to 8.14% within three hours in its absence. Further, an optimum load of heteropoly acid is necessary, as it has a dramatic effect on the desulfurization rate. The rapidity of the reaction requires more heteropoly acid, which would result in a shrinkage of the specific surface area, pore diameter, and pore volume. The corresponding reduction of adsorbed formaldehyde brought about a decrease of the desulfurization rate. The sulfur breakthrough capacities decreased as follows: ALF > MXF > ARF due to the inhibitive effect of aromatics, which can competitively adsorb on active sites. Moreover, the sulfur breakthrough capacity per gram of substrate changed from 4.64 mg S to 4.59 mg S and 4.51 mg S when the aromatics went up from 0% to 20% to 100%. Thus, it is evident that the aromatics have less of an adverse effect on the reactive adsorbent than on an adsorbent based on π-complexation in the desulfurization of ARF. The capacity of the regenerated reactive adsorbent is almost completely recovered and the adsorbent sis reported to have almost the same high efficiency after being recycled twice.

It is known that an adsorbent is the carrier of the reaction characteristic for sulfur compounds, and that reagents and catalysts must be loaded on them before the desulfurization process. Thus, the adsorbent must match the catalyst, to guarantee the maximum desulfurization capacity, for example loading H_2SO_4 on carbonaceous material (Dai *et al.*, 2008) and HCl or phosphomolebdic acid (PMA) on silica materials (Wang *et al.*, 2008c,

2009b). As mentioned above, the reagent of the condensation reaction is formaldehyde (FA), which can be pre-loaded in the adsorbent via vapour adsorption. H_2SO_4 or HCl may be then dropped in the adsorbent, while PMA can be loaded by soaking method before loading FA. Silica gel is usually used to load peracetic acid for the oxidation reaction, which is used for commercial fuels following the condensation reaction. The weight ratio of peracetic acid to silica gel is reported to be usually 0.4/1.0.

The preparation of nano-ferrites (for example: Fe_3O_4, $NiFe_2O_4$, $CuFe_2O_4$ and $MnFe_2O_4$) by a reverse (water/oil) micro-emulsion method, containing cetyltrimethylammonium bromide, 1-butanol, cyclohexane, and a metal salt solution, using ammonium hydroxide as a co-precipitating agent, has also been reported (Zaki *et al.*, 2013). The complex nickel-iron oxide ($NiFe_2O_4$) expressed the highest adsorption capacity towards dibenzothiophene (166.3 μmol dibenzothiophene/g adsorbent) due to the high specific surface area of the oxide (233.4 m^2/g), large pore volume of approximately 0.3 cm^3/g, mesoporous framework, and strong acid sites.

Rare earth metal–organic framework materials composed of a metal ion center connect to a carboxyl group of the organic ligand (Lin *et al.* 2012). These metal organic frameworks (MOFs) are characterized by large pores, very high chemical and hydrothermal stability, and a Langmuir surface area which surpasses 500 m^2/g. Moreover, rare earth metals have a unique electronic structure where the 4f electron shell is not completely filled, with a different 4f electronic number that has a relatively high coordination number which provides a way to get new structure. This encourages the application of MOFs in ADS (Gustafsson *et al.*, 2010). Liu *et al.* (2014b) reported the preparation of the rare earth metal-organic frameworks (Ln-MOFs) materials, Ln(TMA)(H_2O)·(DMF), using the rare earth metal (Ln=Sm, Eu, Tb, Y) and 1,3,5-trimesic acid (TMA) as a metal ion center and ligand, respectively. That expressed good ADS capabilities for a thiophene/n-octane model oil with excellent reusability. The yttrium metal–organic framework material, Y(TMA)(H_2O)·(DMF), expressed a comparatively better activity for ADS with a desulfurization rate up to 80.7% and a sulfur adsorption capacity of 30.7 mgS/g(Y-MOFs). Habimana *et al.* (2016) reported the preparation of europium metal organic framework (Eu-MOF) using Europium as a metal ion center and 1,3,5-trimesic acid (TMA) as organic ligands under hydro-solvothermal conditions. Its ADS efficiency was tested using a thiophene/*n*-octane (1000 μg/g) model oil where the adsorption rate reached 64.70%, with an adsorption capacity of 24.59 mg S/g MO, under optimum operating conditions of 1:100 ($m_{adsorbent}$:$m_{model\ oil}$), 30 °C, and 4 h. The adsorption of thiophene over Eu-MOF absorbent occurs simultaneously through the monolayer physical adsorption and the multilayer chemical adsorption.

The adsorbent can be regenerated and reused for 5 successive cycles with a slight decrease in its activity, from 24.59 to 22.43 mg S/g, with a reduction in desulfurization rate from 64.70 to 59.03%.

It is reported that the film-shear reactor, as a flow process and in a batch reactor, was both very effective in the ADS of various recalcitrant thiophenes (80–90 % removal) in a model diesel fuel (thiophenes in mixture of 85:15 n-hexane/toluene) using a carbon nanotube/TiO_2

(CNT/TiO_2) as an adsorbent, but reactions in the film-shear reactor are faster and require far smaller TiO_2 concentrations to be effective (0.01–0.1 wt./vol% vs. 1–10% in the batch experiments). This is attributed to the intimate mix of the fuels and the CNT/TiO_2 adsorbent in the film-shear reactor. Moreover, the ADS increases as follows: Th < BT < DBT, that is attributed to the increased number of π–π dispersive interactions between the aromatic rings and the nanotubes as the number of aromatic rings increases. Further, the regeneration of CNT/TiO_2 can be achieved by heating at 400 °C without losing its activity (Siddiqui *et al.*, 2016).

Deep desulfurization of liquid fuel by molecular imprinting technology (MIT) is expected to find wide application since MIT is based on the application of biosensors, separation media, and affinity supports for the recognition of target molecules (Liu *et al.*, 2006). MIT has a unique predeterminative characteristic, specificity, and practicability since it is based on the creation of specific molecular recognition sites in polymers to identify template molecules. The prepared polymer is known as a molecularly imprinted polymer (MIP) because it is complimentary to the template in space structure and binding sites. Consequently, based on the selectivity mechanism of MIP towards the sulfur compounds in fuel oils, the prepared MIP using BT, DBT, and DBT-sulfone as templates would remove thiophene-like organosulfur compounds from fuel oils (Yang *et al.*, 2014).

Methacrylic acid based MIP with divinylbenzene as a cross-linker is reported to have a maximum retention capacity of 14.8 mgDBT/g at 25 °C (Castro *et al.*, 2001). Chang *et al.* (2003) reported preparation of MIP by bulk polymerization using DBT as the template, 4-vinyl pyridine as functional monomer, ethylene glycol dimethylacrylate EDMA as the cross-linker, and toluene as porogen with a maximum binding capacity of 48.3 mg/g at 20 °C and good removing capacity towards BTs and DBTs.

Surface molecular imprinting polymer (SMIP) technology is very promising in the deep desulfurization of liquid fuel oils since SMIP has high selectivity, fast adsorption, and good mechanical and thermal stability. SMIP can be prepared by grafting a thin imprinted polymer film with a large amount of binding sites onto inorganic supports, e.g. silica gel, TiO_2, $K_2Ti_4O_9$, and carbon microspheres. Silica gel is usually used for its high porosity, large surface area, good computability, mechanical property, and stability (Yang *et al.*, 2014). Hu *et al.* (2010) reported grafting polymerization using BT as a template, silica gel modified by KH-550 as support, methacrylic acid MAA as monomer, EDMA as cross-linker, azoisobutyrontrile AIBN as initiator, and toluene as solvent. This showed an adsorptive desulfurization capacity of 57.4 mg/g in gasoline at 25 °C and remained unchanged for 30 cycles. Liu *et al.* (2014c) reported the preparation of a double-template molecularly imprinted polymer (D-MIP) supported on carbon microspheres (CMSs), using benzothiophene (BT) and dibenzothiophene (DBT) as the template molecules for the removal of benzothiophene sulfides from fuel gasoline. The prepared SMIP was selectively able to bind BT and DBT at the same time. The D-MIP/CMSs had five times higher a binding capacity towards BT and DBT in simulated gasoline compared to non-imprinted polymers (D-NIP/CMSs). The adsorption equilibrium of D-MIP/CMSs was achieved within 90 min and the adsorption capacity reached 57.16 and 67.19 mg/ g for BT and (DBT), respectively. It can be also used for 10 times without significant loss in adsorption capacity. Moreover, its good efficiency in real oil (gasoline) promoted the application of D-MIP/CMSs as a new material for deep desulfurization of fuel oils.

As a brief of this review, ADS (Figure 2.16) has some problems to be solved. There are high proportions of aromatics compared to the amounts of sulfur compounds in middle distillates, such as kerosene and diesel and the aromatics can also be adsorbed on the desulfurization adsorbents. Thus, the adsorbents should be well designed to achieve suitable selectivity; when the selectivity is low, the adsorbents are easily regenerated. However, this can lead to heat loss because of the comparative adsorption. As the selectivity increases, the spent adsorbents become more and more difficult to be regenerated (Hernández-Maldonado and Yang, 2004a).

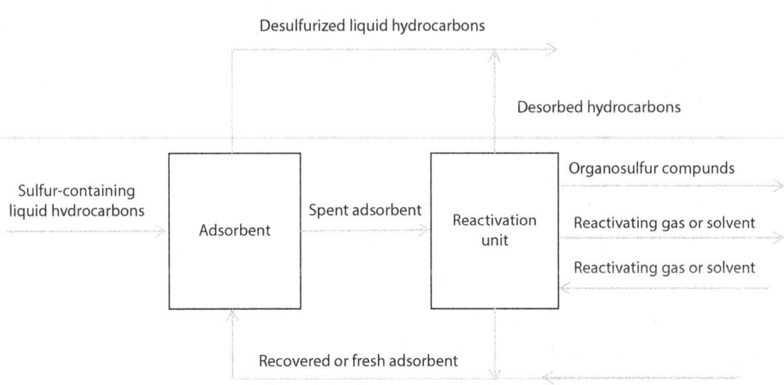

Figure 2.16 Desulfurization by Adsorption.

Consequently, the major challenge in ADS is to design an adsorbent material which can selectively adsorb the sulfur compounds from fuel without altering the aromatic content.

One of the new approaches for ADS applications is the ADS of methanthiol as one of S-impurities in propylene produced from bioethanol by ethanol-to-olefin (ETO) conversion. Depending on the fermentation route of a raw material of bioethanol, it contains dimethylsulfide and/or dimethylsulfoxide as sulfur impurities that are converted to methanethiol, ethanethiol, or hydrogen sulfuide during the ETO conversion. Poisoning of a metal catalyst, that is used for the polymerization of olefin, even by a trace amount of sulfur impurities can occur, thus desulfurization of olefin derived from bioethanol is mandatory (Bartholomew 2001). Among various porous materials, a hydrophilic zeolite with a low Si/Al ratio is one of the most suitable adsorbents for the desulfurization of olefin derived from bioethanol, since strong Lewis acid/base interaction between an acidic site (Al atom) and a sulfur compound is expected (Cui *et al.*, 2009). The breakthrough curve (BTC) for adsorption of methanethiol from bioethanol, in the presence of propylene, using a fixed-bed column packed with the LTA zeolite was reported, where the competitive adsorption of propylene on the LTA zeolite was found to be strongly dependent on a cation species exchanged in the micropores of the zeolite. The bivalent cation of zinc (Zn^{2+}) was proved to be the most effective one to increase the amount of methanethiol adsorbed on the LTA zeolite under the presence of propylene (Yamamoto *et al.*, 2011).

Solvent extraction and oxidation in the air are two methods to regenerate the desulfurization adsorbents, but there are disadvantages inherent in these two methods. The solvent extraction method suffers from the disadvantage that it is difficult to separate sulfur compounds from organic solvents and reuses these solvents. On the other hand, in the calcination method, sulfur compounds and aromatics are burned out which can cause a loss in the heat value of fuels. Recently, bioregeneration of the adsorbents is recommended (Li *et al.*, 2005, 2009e).

In conclusion, reactive adsorption desulfurization (RADS) is superior to ordinary physical adsorption as it involves the π-complexation between aromatic sulfur compounds and the adsorbent, which is stronger than van der Waals interaction. However, the π-complexation can be broken easily by heating or decreasing pressure, thereby it is easy to regenerate the adsorbent. So, briefly the S-Zorb SRT (Sulfur Removal Technology) is a representative of the hydrogenation-based ADS techniques and has advantages in less loss of octane number and relatively high desulfurization efficiency. However, it is not a cost efficient technology since the fuel has to be vaporized and reacted at 380–420 °C. Although, the ADS-techniques which are based on physical adsorption seem non-applicable due to the low selectivity for thiophenic compounds. Prominent success was reported in chemical adsorption and the IRVAD (Irvine Robert Varraveto Adsorption Desulfurization) method, the π-complexion method, and the SARS (selective adsorption for removing sulfur) method are well known representatives. The IRVAD method, as mentioned before, is reported to be suitable for many liquid hydrocarbons and more than 90% sulfur can be removed. However, the sulfur capacity of adsorbents is low and the adsorbents need to be frequently regenerated, but,the sulfur capacity of adsorbents can be considerably increased depending on the principle of π-complexion. However, the adsorption performance declines greatly where aromatic compounds, oxidants, and moisture are present in fuels. Furthermore, the SARS method depends on the preloading of transition metals on a supporting adsorbent, for example, $Ni/SiO_2–Al_2O_3$, and reached the same or a little higher sulfur capacity in comparison with the π-complexion method.

2.5 Biocatalytic Desulfurization

Microorganisms need sulfur to fulfil both growth and biological activity, as it forms around 1% of the dry weight of the bacterial cell. Some microorganisms have the ability to supply their needed sulfur from various sources due to the fact that sulfur exists in the structure of some enzymes cofactors, amino acids, and proteins. Some microorganisms have the ability to consume sulfur in thiophenic compounds and reduce the sulfur content in fuel (El-Gendy and Speight, 2016).

The concept of microbial desulfurization is not new. The first U.S. patents were issued in the 1950s (Strawinski, 1950; ZoBell, 1953). The two general approaches to microbial desulfurization are aerobic and anaerobic reactions (Monticello, 1994).

2.5.1 Anaerobic Process

Anaerobic reactions proceed more slowly than aerobic reactions, but generate the same products as conventional HDS technology, that is H_2S and desulfurized oil.

Mechanisms for the desulfurization of crude oil, using the hydrogenase enzyme activity of sulfate reducing bacteria (SRB), was first proposed in 1935. In 1953, a Texaco patent proposed the desulfurization of crude oil by hydrogenation in the presence of an SRB, such as *Desulfovibrio desulfuricans* (ZoBell, 1953). The first detailed study on anaerobic degradation of sulfur-containing hydrocarbons was published in 1971 (Kurita *et al.*, 1971). Anaerobic cultures isolated from oil well sludge were reported to catalyse the evolution of H_2S from thiophene and benzothiophene when methylviologen or hydrogen were supplied as substrates for hydrogenase. Later, a mixed culture containing *Desulfovibrio* strains, supplied with hydrogen and lactate, was shown to be capable of splitting C–S bonds in a range of sulfur-containing hydrocarbons by a reduction reaction mediated by hydrogenase. Later, several studies were performed on anaerobic-BDS (Kim *et al.*, 1990; Kohler *et al.*, 1984; Miller, 1992; Tilstra *et al.*, 1992; Finnerty, 1993; Yamada *et al.*, 2001; Kareem, 2016a,b).

The desulfurization of petroleum under anaerobic conditions would be attractive because it avoids costs associated with aeration, it has the advantage of liberating sulfur as a gas, and it does not liberate sulfate as a byproduct that must be disposed by some appropriate treatment (Ohshiro and Izumi, 1999). Due to low reaction rates, safety and cost concerns and the lack of identification of specific enzymes and genes responsible for anaerobic desulfurization by anaerobic microorganisms effective enough for practical petroleum desulfurization have not been found yet and an anaerobic biodesulfurization (BDS) process has not been developed. Consequently, aerobic BDS has been the focus of the majority of research in BDS (Le Borgne and Quintero, 2003).

2.5.2 Aerobic Process

Biocatalytic desulfurization (BDS) is often considered as a potential alternative and/or complementary to the conventional deep hydrodesulfurization (HDS) processes used in refineries. In this process, bacteria remove organosulfur from petroleum fractions without degrading the carbon skeleton of the organosulfur compounds. During the process, alkylated dibenzothiophene derivatives are converted to non-sulfur compounds. For example, dibenzothiophene (DBT) is converted to 2-hydroxybiphenyl (2-HBP) and sulfate under aerobic conditions. While under anaerobic conditions, DBT can be converted to biphenyl and hydrogen sulfide (H_2S). BDS offers mild processing conditions and reduces the need for hydrogen. Both these features would lead to high energy savings in the refinery (Dinamarca *et al.*, 2010). Caro *et al.* (2007) reported that BDS requires approximately two times less capital cost and 15% less operating cost in comparison with the HDS process (Caro *et al.*, 2007). Further, significant reductions in greenhouse gas emissions have also been predicted if BDS is used (Calzada *et al.*, 2011). However, the main drawbacks that should be overcome in the BDS area are the slow desulfurization rate and the challenge of the reuse and the life-time of the microorganisms without losing BDS-activity (Thaligari *et al.*, 2016).

As already mentioned, HDS is not equally effective in desulfurizing all classes of sulfur compounds present in fossil fuels. On the other hand, the BDS process is effective regardless of the position of alkyl substituents (Pacheco, 1999). However, the HDS process conditions are sufficient not only to desulfurize sensitive (labile) organosulfur compounds, but also to (1) remove nitrogen and metals from organic compounds, (2) induce saturation of at least some carbon–carbon double bonds, (3) remove substances having an unpleasant smell or color, (4) clarify the product by drying it, and (5) improve the cracking characteristics of the material (Swaty, 2005). Therefore, in consideration of such advantages, placing a BDS unit downstream of an HDS unit as a complementary technology (rather than as a replacement technology) to achieve ultra-deep desulfurization is a realistic consideration, giving a suggestion for multistage processes for desulfurization of fossil fuels (Monticello, 1996; Pacheco, 1999; Mohebali and Ball, 2008).

There have been few reports published about the BDS of diesel oil and whole crude oil desulfurization. However, most reports of crude BDS investigate only the total sulfur content of the crude before and after BDS and do not reveal sulfur specification data, which may guide further biocatalyst development. That is why this technology has not been commercialized, as the biological system has lacked reaction kinetics and specificity required for a commercial process, yet the investigations using model sulfur compounds such as DBT and the manipulation of the enzymatic pathways responsible for the BDS reactions have led to processes which are approaching commercial application, particularly in the BDS of diesel oil, whose primary sulfur compound is DBT (Monticello, 1994). In most of the studies, bacteria function in a "bioremediation mode", degrading sulfur-containing molecules to water-soluble products that can be extracted from oil. As a result, there have been several reports over the last 20 years of successful "microbial desulfurization" that was of limited commercial interest. The problem was that although they desulfurized the oil, they also consumed most of it.

The goal of biocatalytic desulfurization is to develop systems in which the bacteria or their enzymes catalyse very specific reactions to liberate sulfur and leave the hydrocarbon behind. This is analogous to the conventional technology where inorganic catalysts are used to facilitate the reaction of hydrogen gas and petroleum fractions to yield H_2S and desulfurized oil. The drawback with biological systems is that they are slow. Moroever, it is often difficult to separate desirable biochemical reactions from other reactions catalysed by the cells. New rDNA technologies enable researchers to isolate the genes that code for specific enzymes and manipulate them to increase specificity and reaction kinetics.

The Forthcoming chapters will discuss all the performed and aimed approaches and aspects in the BDS research field to make its way towards a commercialized process.

References

ABB Lummus Global, Inc. (2002) Hydrotreating-aromatic saturation. Hydrocarbon Processing. 81: 127.

Abdul Jalil, T., Falah Hasan, L. (2012) Oxidative desulfurization of gas oil using improving selectivity for active carbon from rice husk. *Diyala Journal for Pure Sciences*. 8(3): 68–81.

Abid, M.F. (2015) Desulfurization of gas oil using a solar photocatalytic microreactor. *Energy Procedia*. 74: 663–678.

Abrahamson J.P. (2015) Effects of catalyst properties on hydrodesulfurization of fluid catalytic cracking decant oils as feedstock for needle coke. MSc. Thesis, The Graduate School, Department of Energy and Mineral Engineering, The Pennsylvania State University, Old Main, State College, PA 16801, USA.

Abu Bakar, W.A.W., Ali, R., Abdul Kadir, A., Mokhtar, W.N.A.W. (2015) The role of molybdenum oxide based catalysts on oxidative desulfurization of diesel fuel. *Modern Chemistry and Applications*. 3(2): 1000150. http://dx.doi.org/10.4172/2329-6798.1000150.

Aburto, J., Borgne, S.L. (2004) Selective adsorption for dibenzothiophene sulfone by an imprinted and stimuli-responsive chitosan hydrogel. *Macromolecules* 37: 2938–2943.

Adekanmi, A.A., Folorunsho, A. (2012) Comparative analysis of adsorptive desulphurization of crude oil by manganese dioxide and zinc oxide. *Research Journal of Chemical Sciences*. 2(8): 14–20.

Adeyi, A., Giwa, A., Adeyi, V.A. (2015) Kinetics analysis of manganese dioxide adsorbent on desulphurization of diesel oil. *International Journal of Scientific and Engineering Research*. 6(7): 650–658.

Afanasiev P., Cattenot M., Geantet C., Matsubayashi N., Sato K., and Shimada S. (2002) (Ni)W/ZrO2 hydrotreating catalysts prepared in molten salts. *Applied Catalysis A: General*. 237(1/2): 227–237.

Ahmad, S., Ahmad, M.I., Naeem, K., Humayun, M., Zaeem, S.E., Faheem, F. (2016) Oxidative desulfurization of tire pyrolysis oil. *Chemical Industry and Chemical Engineering Quaternary*. 22(3): 249–254.

Aitani, A.M., Ali, M.F., Al-Ali, H.H. (2000) A review of non-conventional methods for the desulfurization of residual fuel oil. *Petroleum Science and Technology*. 18(5/6): 537–553.

Akopyan, A.V., Ivanov, E.V., Polikarpova, P.D., Tarakanova, A.V., Rakhmanov, E.V., Polyakova, O.V., A.V. Anisimov, Vinokurov, V.A. (2015) Oxidative desulfurization of hydrocarbon fuel with high olefin content. *Petroleum Chemistry.* 55(7): 571–574.

Akopyan, A.V., Kardasheva, Y.S., Eseeva, E.A., Plotnikov, D.A., Vutolkina, A.V., Kardashev, S.V., Rakhmanov, E.V., Anisimov, A.V., Karakhanov, E.A. (2016) Reduction of sulfur content in shale oil by oxidative desulfurization. *Petroleum Chemistry.* 56(8): 771–773.

Al Zubaidy I.A.H., Tarsh F.B., Darwish N.N., Majeed B.S.S.A., Sharafi A.A., Chacra L.A. (2013) Adsorption process of sulfur removal from diesel oil using sorbent materials. *Journal of Clean Energy Technologies (JOCET)* 1(1): 66–68.

Alavi, S.A., Hashemi, S.R. (2014) A review on diesel fuel desulfurization by adsorption process. Proceedings of International Conference on Chemical, Agricultural, and Biological Sciences (ICCABS'2014) Oct. 9–10, 2014 Antalya (Turkey). Pp. 33–36.

Al-Degs, Y.S., El-Sheikh, A.H., Al-Bakain, R.Z., Newman, A.P., Al-Ghouti, M.A. (2016) Conventional and upcoming sulfur-cleaning technologies for petroleum fuel: A review. *Energy Technology.* 6: 679–699.

Al-Ghouti M.A., and Al-Degs Y.S. (2014) Manganese-loaded activated carbon for the removal of organosulfur compounds from high-sulfur diesel fuels. *Energy Technology.* 2: 802–810.

Ali, M., Tatsumi T., Masuda T. (2002) Development of heavy oil hydrocracking catalysts using amorphous silica-alumina and zeolites as catalyst supports. *Applied Catalysis A: General.* 233(1/2): 77–90.

Ali, M.F., Al-Malki, A., and Ahmed, S. (2009) Chemical desulfurization of petroleum fractions for ultra-low sulfur fuels. *Fuel Processing Technology.* 90: 536–544.

Ali, S.A., Siddiqui, M.A.B. (1997) Dearomatization, cetane improvement and deep desulfurization of diesel feedstock in a single-stage reactor. *Reaction Kinetics and Catalysis Letters.* 61(2): 363–368.

Al-Zeghayer, Y.S., Jibril, B.Y. (2006) Kinetics of hydrodesulfurization of dibenzothiophene on sulfided commercial Co-Mo/γ-Al_2O_3 catalyst. *The Journal of Engineering Research.* 3(1): 38–42.

Amini, B., Jafari, J.K., Kalbasi, M., Khorami, P., Mohajeri, A., Parviz, D., Rashidi, A. (2010) Hydrodesulphurization nanocatalyst, its use and a process for its production. European Patent EP-2196260 A1.

Ancheyta, J. (2011). Modeling and Simulation of Catalytic Reactors for Petroleum Refining. 1st Ed., John Wiley & Sons, Inc., Publication, Hoboken, New Jersey, USA.

Andersson, P., Pirjamali, M., Jaras, S., Kizling, M. (1999). Cracking catalyst additives for sulfur removal from FCC gasoline. *Catalysis Today.* 53: 565–573.

Aparicio, F., Camú, E., Villarroel, M., Escalona, N., Baeza, P. (2013) Deep desulfurization by adsorption of 4,6-dimethyldibenzothiophene, study of adsorption on different transition metal oxides and supports. 58(4): 2057–2060.

Arnoldy, P., Van den Heikant, J.A.M., De Bok, G.D., Moulijn, J.A. (1985) Temperature-programmed sulfiding of $MoO_3Al_2O_3$ catalysts. *Journal of Catalysis* 92: 35–55.

Atlas, R.M., Boron, D.J., Deever, W.R., Johnson, A.R., McFarland, B.L., Meyer, J.A. (2001) Method for removing organic sulfur from heterocyclic sulfur-containing organic compounds. US patent number USH1986 H1.

Babich, I.V., Moulijn, J.A. (2003) Science and technology of novel processes for deep desulfurization of oil refinery streams: A review. *Fuel.* 82(6): 607–631.

Baco, F., Debuisschert, Q., Marchal, N., Nocca, J.-L., Picard, F., Uzio, D. (2002). Prime G+ process, desulfurization of FCC gasoline with minimized octane loss, in: Proceedings of the Fifth International Conference on Refinery Processing, AIChE2002 Spring National Meeting, New Orleans, LA, March11–14, 2002, pp. 180–188.

Baeza, P., Aguila, G., Gracia, F., Araya, P. (2008) Desulfurization by adsorption with copper supported on zirconia. *Catalysis Communications.* 9(5): 751–755.

Baeza, P., Aguila, G., Vargas, G., Ojeda, J., Araya, P. (2012) Adsorption of thiophene and dibenzothiophene on highly dispersed Cu/ZrO_2 adsorbents. *Applied Catalysis B: Environmental.* 111–112: 133–140.

Bagreev, A., Menendez, J.A., Dukhno, I., Tarasenko, Y., Bandosz, T.J. (2004) Bituminous coal-based activated carbons modified with nitrogen as adsorbents of hydrogen sulfide. *Carbon.* 42(3): 469–476.

Baldovino-Medrano, V.G., Giraldo, S.A., Centeno A. (2009) Highly HYD selective Pd–Pt/support hydrotreating catalysts for the high pressure desulfurization of DBT type molecules. *Catalysis Letters.* 130: 291–295.

Baltzopoulou, P., Kallis, K.X., Karagiannakis, G., Konstandopoulos, A.G. (2015) Diesel fuel desulfurization via adsorption with the aid of activated carbon: laboratory- and pilot-scale studies. *Energy and Fuels.* 29 (9): 5640–5648.

Bandosz, T.J. (2006) Desulfurization on activated carbons. In: Interface Science and Technology, Bandosz T.J. (Ed.) Elsevier, UK. 7: 231 –292.

Barmala, M., Moghadam, A.Z., Omidkhah, M.R. (2015) Increased photo-catalytic removal of sulfur using titania/MWCNT composite. *Journal of Central South University of Technology.* 22: 1066–1070.

Barrio, V.L., Arias, P.L., Cambra, J.F., Güemez, M.B., Pawelec, B., Fierro, J.L.G. (2003) Hydrodesulfurization and hydrogenation of model compounds on silica–alumina supported bimetallic systems. *Fuel* 82:501–509.

Bartholomew, C.H. Mechanisms of catalyst deactivation (2001). *Applied Catalysis A: General.* 212: 17–60.

Bataille, F., Lemberton, J.L., Michaud, P., Perot, G., Vrinat, M., Lemaire, M., Schulz, E., Breysse, M., Kasztelan, S. (2000) Alkyldibenzothiophenes hydrodesulfurization: Promotor effect, reactivity, and reaction mechanism. *Journal of Catalysis.* 191: 409–422.

Beers, A.E.W., Hoek, I., Nijhuis, T.A., Downing, R.S., Kapteijn, F., Moulijn, J.A., (2000) Structured catalysts for the acylation of aromatics. *Topics in Catalysis.* 13(3): 275–280.

Bhandari, V.M., Ko, C.H., Park, J.G., Han, S.S., Cho, S.H., Kim, J.N. (2006) Desulfurization of diesel using ion-exchanged zeolite. *Chemical Engineering Science*. 61: 2599–2608.

Bhore N.A., Chester A.W., Liu K., Timkin H.K.C. (2001). Reducing gasoline sulfur in FCC. World Patent WO 0121733 A1. Assigned to Mobil Oil Corporation.

Billon, A., Colyar, J.J., Kressmann, S., Morel, F. (2000) Deep conversion of heavy oils via combination of catalytic hydrocracking and deasphalting. World Petroleum Congress 2000, Calgary, Canada, June 2000.

Boikov, E.V., Vakhrushin, P.A., Vishnetskaya, M.V. (2008) Oxidative desulfurization of hydrocarbon feedstock. *Chemistry and Technology of Fuels and Oils*. 44(4): 271–274.

Bose D. (2015) Design parameters for hydrodesulfurization (HDS) unit for petroleum naphtha at 3500 barrels per day. World Science News. 9: 99–111.

Boukoberine, Y., Hamada, B. (2016) Thiophene hydrodesulfurization over CoMo/Al_2O_3-CuY catalysts: Temperature effect study. *Arabian Journal of Chemistry*. 9: S522 –S527.

Bowker, R.H. (2011) Hydrodesulfurization and hydrodeoxygenation over noble metal phosphide catalysts. MSc. Thesis, Western Washington University, 516 High Street, Bellingham, WA 98225, USA.

Breysse, M., Furimsky, E., Kasztelan, S., Lacroix, M., Perot, G. (2002) Hydrogen activation by transition metal sulfides. *Catalysis Reviews-Science and Engineering*. 44: 651–735.

Brunet, S., Mey, D., Pérot, G., Bouchy, C., Diehl, F. (2005). On the hydrodesulfurization of FCC gasoline: a review. *Applied Catalysis A: General*. 278: 143–172.

Calzada, J., Alcon, A., Santos, V.E., Garcia-Ochoa, F. (2011) Mixtures of Pseudomonas putida CECT 5279 cells of different ages: Optimization as biodesulfurization catalyst. *Process Biochemistry*. 46: 1323–1328.

Campos-Martin, J.M., Capel-Sanchez, M.C., and Fierro, J.L.G. (2004) Oxidative processes of desulfurization of liquid fuels. *Journal of Chemical Technology and Biotechnology*. 85: 879–890.

Campos-Martin, J.M., Capel-Sanchez, M.C., Perez-Presas, P. and Fierro, J.L.G. (2010) Highly efficient deep desulfurization of fuels by chemical oxidation. *Green Chemistry*. 6: 557–562.

Caro, A., Boltes, K., Letón, P., García-Calvo, E. (2007) Dibenzothiophene biodesulfurization in resting cell conditions by aerobic bacteria. *Biochemical Engineering Journal*. 35(2): 191 -197.

Castorena, G., Suarez, C., Valdez, I., Amador, G., Fernandez, L., Le Borgne, S. (2002) Sulfur-selective desulfurization of dibenzothiophene and diesel oil by newly isolated *Rhodococcus* sp. Strains. *FEMS Microbiolgy Letters*. 215(1): 157–161.

Castro, B., Whitcombe, M., Vulfson, E., Vazquez-Duhalt, R., Bárzana, E. (2001) Molecular imprinting for the selective adsorption of organosulphur compounds present in fuels. *Analytica Chimica Acta*. 435(1): 83–90

Cattaneo, R., Shido, T., Prins, R. (1999) Hydrotreatment and Hydrocracking of Oil Fractions. Elsevier, Amsterdam.

CDTech (2000) Hydrogenation. Hydrocarbon Processing. 79: 122–123.

Cedneňo-Caero, L., Gomez-Bernal, H., Fraustro-Cuevas, A., Guerra-Gomez, D.H., Cuevas-Garcia, R. (2008) Oxidative desulfurization of synthetic diesel using supported catalysts: Part III. Support effect on vanadium-based catalysts. *Catalysis Today*. (133/135): 244–254.

Chang, Y.H., Liu, B., Ying, H.J., He, M.F. (2003) Solid-phase extraction sorbent for organosulfur compounds present in fuels made by molecular imprinting. *Ion Exchange and Adsorption* 19(5): 450–456.

Chang, Y.H., Zhang, L., Ying, H.J., Li, Z.J., Lv, H., Ouyang, P.K. (2010) Desulfurization of Gasoline using Molecularly Imprinted Chitosan as selective adsorbents. *Applied Biochemistry and Biotechnology*. 160: 593–603.

Chapados, D., Bonde, S.E., Chapados, D., Gore, W.L., Dolbear, G., Skov, E. (2000) Desulfurization by selective oxidation and extract of sulfur-containing compounds to economically achieve ultra-low proposed diesel fuel sulfur requirements. Proceedings. NPRA Annual Meeting AM-00–25, San Antonio, Texas. March 26–28.

Chen, L., Li, X., Rooke, J.C., Zhang, Y., Yang, X., Tang, Y., Xiao, F., Su, B. (2012) Hierarchically structured zeolites: synthesis, mass transport properties and applications. *Journal of Materials Chemistry*. 22: 17381–17403.

Cheng, Z., Liu, X., Lu, J., Luo, M. (2009) Deep desulfurization of FCC gasoline by selective adsorption over nanosized zeolite-based adsorbents. *Reaction Kinetics and Catalysis Letters*. 97: 1–6.

Chester, A., Timken, H., Roberie T., Ziebarth, M. (2003). Gasoline sulfur reduction in FCC. U.S. Patent 20030089639 A1.

Chianelli, R.R., Berhault, G., Raybaud, P.., Kasztela, S., Hafnere, J., Toulhoatf, H. (2002) Periodic trends in hydrodesulfurization: in support of the Sabatier principle. *Applied Catalysis A: General*. 227: 83–96.

Chica, A., Corma, A., and Domine, M.E. (2006) Catalytic oxidative desulfurization (ODS) of diesel fuel on a continuous fixed0bed reactor. *Journal of Catalysis*. 242: 2999–308.

Chitnis, G.K., Richter, J.A., Hilbert, T.L., Dabkowski, M.J., Al Kuwari, I., Al Kuwari, N., Sherif, M. (2003) Commercial OCTGain's unit provides 'zero' sulfur gasoline with higher octane from a heavy cracked naphtha feed, AM-03–125, in: Proceedings of the NPRA 2003 Annual Meeting, San Antonio, TX, March 23–25, 2003.

Collins, F.M., Lucy, A.R., Sharp, C.J. (1997) Oxidation desulfurization of oils via hydrogen peroxide and heteropolyanion catalysis. *Journal of Molecular Catalysis A: Chemical*. 117: 397–403.

Cooper, B.H., Sogaard-Anderson, P., Nielsen-Hannerup, P. (1994) Production of Swedish class I diesel using dual-stage process, in: M.C. Oballa, S.S. Shih (Eds.), Catalytic Hydroprocessing of Petroleum and Distillates, Marcel Dekker, New York, 1994, pp. 279–290.

Corma, A., Martinez, C., Ketley, G., Blair, G. (2001) On the mechanism of sulfur removal during catalytic cracking. *Applied Catalysis A: General*. 208(1/2): 135–152.

Cui, H., Turn, S.Q., Reese, M.A. (2009) Removal of sulfur compounds from utility pipelined synthetic natural gas using modified activated carbons. *Catalysis Today*. 139: 274–279.

Cui, S., Ma, F., Wang, Y. (2007) oxidative desulfurization of model diesel oil over Ti-containing molecular sieves using hydrogen peroxide. *Reaction Kinetics and Catalysis Letters*. 92(1): 155–163.

Cui, X., Yao, D., Li, H., Yang, J., and Hu, D. (2012) Non-magnetic particles as multifunctional microreactor for deep desulfurization. *Journal of Hazardous Materials*. (205/206): 17–23.

Dai, W., Zhou, Y., Wang, S., Su, W., Sun, Y., Zhou, L. (2008) Desulfurization of transportation fuels targeting at removal of thiophene/benzothiophene. *Fuel Processing Technology*. 89(8): 749–755.

Danmaliki, G.I., Saleh, T.A. (2017) Effects of bimetallic Ce/Fe nanoparticles on the desulfurization of thiophenes using activated carbon. *Chemical Engineering Journal*. 307: 914–927.

de A. Mello, P., Duarte, F.A., Nunes, M.A.G., Alencar, M.S., Moreira, E.M., Korn, M., Dressler, V.L., and Flores, E.M.M. (2009) Ultrasound-assisted oxidative process for sulfur removal from petroleum product feedstock. *Ultrasonics Sonochemistry*. 16: 732–736.

de Filippis, P., Scarsella, M., Verdone, N. (2010) Oxidative desulfurization I: peroxyformic acid oxidation of benzothiophene and dibenzothiophene. *Industrial and Engineering Chemistry Research*. 49(10): 4594–4600.

Deng, Z., Wang, T., Wang, Z. (2010) Hydrodesulfurization of diesel in a slurry reactor. *Chemical Engineering Science* 65: 480–486.

Dhar, G.M., Srinivas, B.N., Rana, M.S., Kumar, M., Maity, S.K. (2003) Mixed oxide supported hydrodesulfurization catalysts—a review. Catalysis Today 86: 45–60.

di Giuseppe, A., Crucianelli, M., de Angelis, F., Crestini, C., Saladino, R. (2009) Efficient oxidation of thiophene derivatives with homogeneous and heterogeneous MTO/H_2O_2 systems: A novel approach for, oxidative desulfurization (ODS) of diesel fuel. *Applied Catalysis B: Environmental*. 89(1/2): 239–245.

Dinamarca, M.A., Ibacache-Quiroga, C., Baeza, P., Galvez, S., Villarroel, M., Olivero, P., Ojeda, J. (2010) Biodesulfurization of gas oil using inorganic supports biomodified with metabolically active cells immobilized by adsorption. *Bioresource Technology*. 101: 2375–2378.

Doyle, A., Tristao, M.L.B., Felcman, J. (2006) Study of fuel insolubles: Formation conditions and characterization of copper compounds. *Fuel*. 85(14/15): 2195–2201.

Drake, C.A., Love, S.D. (2000) Process for desulfurizing and aromatizing hydrocarbons. US Patent 6,083,379. Assigned for Phillips Petroleum Company.

Duarte, F.A., de A. Mello, P., Bizzi, C.A., Nunes, M.A.G., Moreira, E.M., Alencar, M.S., Motta, H.N., Dressler, V.L., and Flores, E.M.M. (2011) Sulfur removal from hydrotreated petroleum fractions using ultrasound-assisted oxidative desulfurization process. *Fuel*. 90: 2158–2164.

Egorova, M., Prins, R. (2004) Mutual influence of the HDS of dibenzothiophene and HDN of 2-methylpyridine. *Journal of Catalysis*. 221: 11–19.

El-Gendy, N.Sh., Speight, J.G. (2016) Handbook of Refinery Desulfurization. CRC Press, Taylor and Francis Group, LLC, 6000 Broken Sound Parkway NW, Suite 300, Boca Raton, FL 33487–2742, USA.

EPA-Diesel RIA, Regulatory Impact Analysis (2000): Heavy-duty Engine and Vehicle Standards and Highway Diesel Fuel Sulfur Control Requirements, United States Environmental Protection Agency, Air and Radiation, EPA420-R-00–026, December 2000.

Eswaramoorthi I., Sundaramurthy V., Das N., Dalai A.K., and Adjaye J. (2008) Application of multi-walled carbon nanotubes as efficient support to NiMo hydrotreating catalyst. *Applied Catalysis A: General.* 339(2): 187–195.

ExxonMobil, December 2001. http://www.exxonmobil.com/refiningtechnologies/fuels/mn scanfining.html.

Farag, H., Sakanishi, K., Mochida, I., Whitehurst, D.D. (1999a) Kinetic Analyses and inhibition by naphthalene and H_2S in hydrodesulfurization of 4,6-dimethyldibenzothiophene (4,6-DMDBT) over CoMo-based carbon catalyst. *Energy and Fuel.* 13(2): 449–453.

Farag, H., Whitehurst, D.D., Sakanishi, K., Mochida, I. (1999b) Carbon versus alumina as a support for Co–Mo catalysts reactivity towards HDS of dibenzothiophenes and diesel fuel. *Catalysis Today.* 50: 9–17.

Farshi, A., Shiralizadeh, P. (2015) Sulfur reduction of heavy oil by oxidative desulfurization (ODS) method. *Petroleum and Coal.* 57(3): 295–302.

Ferreira, A.S., Nicoletti, M.C., Bertini, Jr. J.R., Giordano, R.C. (2013) Methodology for inferring kinetic parameters of diesel oil HDS reactions based on scarce experimental data. *Computers and Chemical Engineering.* 48: 58–73.

Figueiredo dos Santos, M.R., Pedrosa, A.M.G., and Barros de Souza, M.J. (2016) Oxidative desulfurization of thiophene on TiO_2/ZSM-12 zeolite. Materials Research. 1991): 24–30.

Finnerty, W.R. (1993) Symposium on Bioremediation and bioprocessing presented before the division of petroleum chemistry, Inc. 205th National Meeting, American Chemical Society, Denver, Co. Pp. 282–285.

Flego, C., Arrigoni, V., Ferrari, M., Riva, R., Zanibelli, L. (2001) Mixed oxides as a support for new CoMo catalysts. *Catalysis Today*. 65: 265–270.

Franchi, E., Rodriguez, F., Serbolisca, L., Ferra, F. (2003) Vector development isolation of new promoters enhancement of catalytic activity of DSZ enzyme complex in Rhodococcus. sp. strain. *Oil & Gas Science and Technology – Rev. IFP.* 58(58): 515–520.

Ganiyu, S.A., Ajumobi, O.O., Lateef, S.A., Sulaiman, K.O., Bakare, I.A., Qamaruddin, M., Alhooshani, K. (2017) Boron-doped activated carbon as efficient and selective adsorbent for ultra-deep desulfurization of 4,6-dimethyldibenzothiophene. *Chemical Engineering Journal.* 321: 651–661.

García-Gutiérrez, J.L., Fuentes, G.A., Hernandez-Teran, M.E., Garcıa, P., Murrieta-Guevara, F., and Jimenez-Cruz F. (2008) Ultra-deep oxidative desulfurization of diesel fuel by the Mo/Al2O3-H_2O_2 system: The effect of system parameters on catalytic activity. *Applied Catalysis A: General* 334: 366- 373.

Gates, B.C., Topsoe, H. (1997) Reactivities in deep catalytic hydrodesulfurization: challenges, opportunities, and the importance of 4-methyldibenzothiophene and 4,6-dimethyldibenzothiophene. *Polyhedron.* 16(18): 3213–3217.

Gawande, P.R., Kaware, J. (2014) A review on desulphurization of liquid fuel by adsorption. *International Journal of Science and Research.* 3(7): 2255–2259.

Gawande, P.R., Kaware, J.P. (2016) Removal of sulphur from liquid fuels using low cost activated carbon -A review. *International Journal of Scientific Engineering and Applied Science.* 2(10): 64–69.

Gillespise, R.J., Peel, T.E. (1973) Hammett acidity function for some superacid systems. II. Systems sulfuric acid-[fsa], potassium fluorosulfate-[fsa], [fsa]-sulfur trioxide, [fsa]-arsenic pentafluoride, [sfa]-antimony pentafluoride and [fsa]-antimony pentafluoride-sulfur trioxide. *Journal of the American Chemical Society* 95(16): 5173–5178.

Golden, S.W., Hanson, D.W., Fulton, S.A. (2002) Use better fractionation to manage gasoline sulfur concentration. *Hydrocarbon Processing.* 81(2): 67–72.

Gong, Y., Dou, T., Kang, S., Li, Q., Hu, Y. (2009) Deep desulfurization of gasoline using ion-exchange zeolites: Cu(I)- and Ag(I)-beta. *Fuel Processing Technology.* 90(1): 122–129.

Gore, W. (2001) Method of desulfurization of hydrocarbons. US Patent 6274785.

Gosling, C.D., Gatan, R.M., and Barger, P.T. (2007) Process for producing hydrogen peroxides. US Patent 7297253.

Gosling, C.D., Gatan, R.M., Cavanna, A., Molinari, D. (2004) NPRA, Annual meeting, San Antonio, Technical paper AM04–48.

Gronowitz, S. (1985) Thiophene and its derivatives. Part one, Weissberger, A. and Taylor, E.C. Volume 44. In: The Series "The Chemistry of Heterocyclic Compounds" John Wiley and Sons, Inc., New York, USA.

Gunnerman, R.W. (2003) Continuous process for oxidative desulfurization of fossil fuels with ultrasound and products thereof. US patent 20030014911.

Guo, H.F., Yan, P., Hao, X.Y., Wang, Z.Z. (2008) Influences of introducing Al on the solid super acid SO_4^{2-}/SnO_2. *Materials Chemistry and Physics.* 112(3): 1065- 1068.

Guo, W., Wang, C., Lin, P., and Lu, X. (2011) Oxidative desulfurization of diesel with TBHP/Isobutyl aldehyde/air oxidation system. *Applied Energy.* 88: 175–179.

Guoxian, Y.U., Hui, C., Shanxiang, L.U., Zhongnan, Z. (2007) Deep desulfurization of diesel fuels by catalytic oxidation. *Frontiers of Chemical Engineering in China.* 1(2): 162–166.

Gupta, N., Roychoudhury, P. K. (2005) Biotechnology of desulfurization of diesel: Prospects and challenges. *Applied Microbiology and Biotechnology.* 66: 356–66.

Guseinova, A.D., Mirzoeva, L.M., Yunusov, S.G., Abbasov, M.F., Guseinova, I.S. (2012) Oxidative desulfurization of catalytically cracked gasolines produced from blended Baku oils. *Petroleum Chemistry.* 52(2): 126–129.

Gustafsson, M., Bartoszewicz, A., Martı´n-Matute, B., Sun, J., Grins, J., Zhao, T., Zou, X. (2010) A family of highly stable lanthanide metal–organic frameworks: structural evolution and catalytic activity. *Chemistry of Materials.* 22: 3316–3322.

Habimana, F., Huo, Y., Jiang, S., Ji, S. (2016) Synthesis of europium metal–organic framework (Eu-MOF) and its performance in adsorptive desulfurization. Adsorption. 22: 1147–1155.

Halbert, T.R., Stuntz, G.F., Brignac, G.B., Greeley, J.P., Ellis, E.S., Davis, T.J., Kamienski, P., Mayo, S. (2001). Akzo nobel catalyst symposium on SCANfining: A commercially proven technology for low sulfur gasoline. Noordwijkaan Zee, The Netherlands, June 10–13, 2001. http://www.prod.exxonmobil.com/refiningtechnologies/pdf/AkzoSCANfining2001.pdf.

Hatanaka, S. (2005) Hydrodesulfurization of selective catalytic cracked gasoline. *Catalysis Surveys from Asia*. 9(2): 87–93.

Hatanaka, S., Yamada, M., Sadakane, O. (1998) Hydrodesulfurization of catalytic cracked gasoline. 3. Selective catalytic cracked gasoline hydrodesulfurization on the Co–Mo/-Al_2O_3 catalyst modified by coking pretreatment. *Industrial and Engineering Chemistry Research*. 37(5): 1748–1754.

Haw, K.-G., Abu Bakar, W.A.W., Ali, R., Chong, J.-F., and Abdul Kadir, A.A. (2010) Catalytic oxidative desulfurization of diesel utilizing hydrogen peroxide and functionalized-activated carbon in a biphasic diesel-acetonitrile system. *Fuel Processing Technology*. 91(9): 1105–1112.

Heidarinasab, A., Soltanieh, M., Ardjmand, M., Ahmadpanahi, H., Bahmani, M. (2016) Comparison of Mo/MgO and Mo/γ-Al_2O_3 catalysts: impact of support on the structure and dibenzothiophene hydrodesulfurization reaction pathways. *International Journal of Environmental Science and Technology*. 13: 1065–1076.

Hernández-Maldonado, A.J., Yang, R.T. (2003a) Desulfurization of liquid fuels by adsorption via π-complexation with Cu(I)-Y and Ag-Y zeolites. *Industrial and Engineering Chemistry Research*. 42(1): 123- 129.

Hernández-Maldonado, A.J., Yang, R.T. (2003b) Desulfurization of commercial liquid fuels by selective adsorption via π-complexation with Cu(I)-Y Zeolite. *Industrial and Engineering Chemistry Research*. 42(13): 3103–3110.

Hernández-Maldonado, A.J., Yang, R.T. (2004a) Desulfurization of transportation fuels by adsorption. *Catalysis Reviews - Science and Engineering*. 46(2): 111–150.

Hernández-Maldonado, A.J., Yang, R.T. (2004b). Desulfurization of Diesel Fuels by Adsorption via π-Complexation with Vapor-Phase Exchanged Cu(I)–Y *Zeolites Journal of the American Chemical Society*. 126(4): 992–993.

Hernández-Maldonado, A.J., Yang, R.T. (2004c) New sorbents for desulfurization of diesel fuels via π-complexation. *AIChE Journal*. 50(4): 791–801.

Hirai, T., Ogawa, K., and Komasawa, I. (1996) Desulfurization process for dibenzothiophenes from light oil by photochemical reaction and liquid–liquid extraction. *Industrial and Engineering Chemistry Research*. 35(2): 586–589.

Hirai, T., Shiraishi, Y., Ogawa, K., and Komasawa, I. (1997) Effect of photosensitizer and hydrogen peroxide on desulfurization of light oil by photochemical reaction and liquid–liquid extraction. *Industrial and Engineering Chemistry Research*. 36(3): 530–533.

Hosseini, H. (2012) Novel methods for desulfurization of fuel oils. *International Journal of Chemical, Molecular, Nuclear, Materials and Metallurgical Engineering.* 6(11): 1072–1074.

Houalla M., Nag N.K., Sapre A.V., Broderick D.H., and Gates B.C. (1978) Hydrodesulfurization of dibenzothiophene catalyzed by sulfided CoO-$MoO_3\gamma$-Al_2O_3: The reaction network. *AIChE Journal.* 24: 1015–1021.

Hu, T.p., Zhang, Y.m., Zheng, L.H., Fan, G.z. (2010) Molecular recognition and adsorption performance of benzothiophene imprinted polymer on silica gel surface. *Journal of Fuel Chemistry and Technology.* 38(6): 722–729.

Huang, L., Wang, G., Qin, Z., Du, M., Dong, M., Ge, H., Wu, Z., Zhao, Y., Ma, C., Hu, T., Wang, J. (2010) A sulfur K-edge XANES study on the transfer of sulfur species in the reactive adsorption desulfurization of diesel oil over Ni/ZnO. *Catalysis Communications.* 11(7): 592–596

Huang, P., Luo, G., Kang, L., Zhu, M., and Dai, B. (2017) Preparation, characterization and catalytic performance of HPW/aEVM catalyst on oxidative desulfurization. *RSC Advances.* 7: 4681–4687.

Hulea, V., Fajula, F., and Bousquet, J. (2001) Mild oxidation with H2O2 over Ticontaing molecular sieves-a very efficient method for removing aromatic sulfur compounds from fuels. *Journal of Catalysis.* 198: 179–186.

Ibrahim, A., Xian S.B., and Wei, Z. (2003) Desulfurization of FCC gasoline by solvent extraction and photooxidation. *Petroleum Science and Technology.* 21: 1555–1573.

Ibrahim, A., Xian S.B., and Wei, Z. (2004) Desulfurization of FCC gas oil by solvent extraction, photooxidation, and oxidizing agents. *Petroleum Science and Technology.* 22(3/4): 287–301.

IFP, December 2001. http://www.ifp.fr/INTF/BI000GF2.html.

Imhof P. (2004). Economics of FCC additives for sulfur removal and emission control. Presented at Akzo Nobel Scope 2004 Symposium, Florence, Italy.

Institute Francais du Petrole IFP (2000) Gasoline desulfurization, ultra-deep. Hydrocarbon Processing. 79(9): 114

Ito, E., van Veen, J.A. (2006) On novel processes for removing sulphur from refinery streams. *Catalysis Today.* 116(4): 446–460.

Jackson, C., Soveran, D. (1995) Desulfurization of crude oils using microwave energy. P-110–300-C-95.

Javadli R., de Klerk A. (2012a) Desulfurization of heavy oil. Applied Petrochemical Research. 1: 3–19.

Javadli, N., de Klerk, A. (2012b) desulfurization of heavy oil-oxidative desulfurization (ODS) as potential upgrading pathway for oil sands derived bitumen. *Energy and Fuels.* 26: 594–602.

Jiang, B., Yang, H., Zhang, L., Zhang, R., Sun, Y., Huang, Y. (2016) Efficient oxidative desulfurization of diesel fuel using amide-based ionic liquids. *Chemical Engineering Journal.* 283(1): 89–96.

José da Silva, M., and Faria dos Santos, L. (2013) Novel oxidative desulfurization of a model fuel with H_2O_2 catalyzed by $AlPMo_{12}O_{40}$ under phase transfer catalyst-free conditions. *Journal of Applied Chemistry*. Volume 2013, ArticleID147945, http://dx.doi.org/10.1155/2013/147945.

Joskić, R., Dunja, M., and Sertić-Bionda, K. (2014) Oxidative desulfurization of model diesel fuel with hydrogen peroxide. Goriva i Maziva. 53(1): 11–18.

Kabe, T., Ishihara, A., Tajima, H. (1992) Hydrodesulfurization of sulfur-containing polyaromatic compounds in light oil. *Industrial & Engineering Chemistry Research*. 31: 1577–1580.

Kabe, T., Ishihara, A., Zhang, Q. (1993) Deep desulfurization of light oil. Part 2: hydrodesulfurization of dibenzothiophene, 4-methyldibenzothiophene and 4,6-dimethyldibenzothiophene. *Applied Catalysis A: General*. 97: L1-L9.

Kabe, T., Qian, W., Hirai, Y., Li, L., Ishihara, A. (2000) Hydrodesulfurization and hydrogenation reactions on noble metal catalysts: I. Elucidation of the behavior of sulfur on alumina-supported platinum and palladium using the 35S radioisotope tracer method. *Journal of Catalysis*. 190:191–180.

Kadijani, J.A., Narimani, E., and Kadijani, H.A. (2014) Oxidative desulfurization of organic sulfur compounds in the presence of molybdenum complex and acetone as catalysts. *Petroleum and Coal* 56(1): 116–123.

Kadijani, J.A., Narimani, E., Kadijani, H.A. (2016) Using response surface methodology to optimize ultrasound-assisted oxidative desulfurization. *Korean Journal of Chemical Engineering*. 33(4): 1286–1295

Kaluža, I., Gulková, D. (2016) Effect of promotion metals on the activity of MoS_2/ZrO_2 catalyst in the parallel hydrodesulfurization of 1-benzothiophene and hydrogenation of 1-methyl-cyclohex-1-ene. Reaction Kinetics, Mechanisms and Catalysis. 118: 313–324.

Kanda, Y., Iwamoto, H., Kobayashi, T., Uemichi, Y., Sugioka, M., (2009) Preparation of highly active alumina-pillared clay montmorillonite-supported platinum catalyst for hydrodesulfurization. *Topics in Catalysis*. 52(6/7): 765–771.

Kapteijn, F., Nijhuis, T.A., Heiszwolf, J.J., Moulijn, J.A. (2001) New non-traditional multiphase catalytic reactors based on monolithic structures. *Catalysis Today*. 66: 133–144.

Karas, L.J., Grey, R.A., and Lynch, M.W. (2008) Desulfurization process. US Patent 20080047875.

Kareem, S.A. (2016a) Anaerobic biodesulfurization of kerosene part I: Identifying a capable microorganism. IOSR *Journal of Biotechnology and Biochemistry*. 2(5): 46–53.

Kareem, S.A. (2016b) Anaerobic biodesulfurization of kerosene part II: Investigating its kinetics. *IOSR Journal of Biotechnology and Biochemistry*. 2(5): 37–45.

Kaufmann, T.G., Kaldor, A., Stuntz, G.F., Kerby, M.C., Ansell, L.L. (2000) Catalysis science and technology for cleaner transportation fuels. *Catalysis Today*. 62(1): 77–90.

Kazakov, A.A., Tarakanov, G.V., Ionov. N.G. (2016) Mechanisms of oxidative desulfurization of straight-run residual fuel oil using ozonized air. *Chemistry and Technology of Fuels and Oils*. 52(1): 33–37.

Khalili, F.I., Sultan, M., Robl, C., Al-Ghouti, M. (2015) Insights into the remediation characterization of modified bentonite in minimizing organosulphur compounds from diesel fuel. *Journal of Industrial and Engineering Chemistry.* 28: 282–293.

Khamis, F.S., Palichev, T.V. (2012) Production of ultra-low sulfur gasoline and assessment of the efficiency of ferrocene antiknock additive. *International Journal of Engineering and Applied Sciences.* 1(1): 11–16.

Khan, M.R., Al-Shafel, E.N. (2013) Microwave-promoted desulfurization of crude oil. US Patent 20130341247 A1.

Khodaei, B., Sobati, M.A., Shahhosseini, Sh. (2016a) Optimization of ultrasound-assisted oxidative desulfurization of high sulfur kerosene using response surface methodology (RSM). Clean Technologies and Environmental Policy. 18: 2677–2689.

Khodaei, B., Sobati, M.A., Shahhosseini, Sh. (2016b) Rapid oxidation of dibenzothiophene in model fuel under ultrasound irradiation. Monatshefte für Chemie - Chemical Monthly. DOI 10.1007/s00706–016-1801-z.

Kim, J.H., Ma, X., Song, C., Oyama, S.T., Lee, Y.-K. (2003) Kinetic Study of 4,6-Dimethyldibenzothiophene hydrodesulfurization over Ni phosphide, NiMo and CoMo sulfide catalysts. Fuel Chemistry Division Preprints. 48(1): 40–41.

Kim, J.H., Ma, X., Zhou, A., Song, C. (2006) Ultra-deep desulfur- ization and denitrogenation of diesel fuel by selective adsorption over three different adsorbents: a study on adsorptive selectivity and mechanism. *Catalysis Today,* 111: 74–83.

Kim, T.S., Kim, H.Y., Kim, B.H. (1990) Petroleum desulfurization by *Desulfovibrio desulfuricans* M6 using electrochemically supplied reducing equivalentBiotechnology Letters. 12(10): 757–760.

King, C.J. (1987) Separation Processes Based on Reversible Chemical Complexation. In: Handbook of Separation Process Technology. Rousseau, R.W., ed., Wiley, New York, USA.

Klicpera, T., Zdrazil, M. (2002) Preparation of High-activity MgOsupported Co–Mo and Ni–Mo sulfide hydrodesulfurization catalysts. *Journal of Catalysis.* 206: 314–320.

Kohler, M., Genz, I.L., Schicht, B., Eckart, V. (1984) Microbial desulphurization of petroleum and heavy petroleum fractions: anaerobic desulphurization of organic sulphur compounds of petroleum. Zentralblatt für Mikrobiologie. 139: 239–247.

Kong, A., Wei, Y. and Li, Y. (2013) Reactive adsorption desulfurization over a Ni/ZnO adsorbent prepared by homogeneous precipitation. *Frontiers of Chemical Science and Engineering.* 7: 170–176.

Krishnaiah, G., Balko, J. (2003) Reduce ultra-low sulfur gasoline compliance costs with Davison clean fuels technologies, S-BraneTM membrane process for sulfur removal and FCC catalyst technologies for clean fuels, AM-03–121, in: Proceedings of the NPRA 2003 Annual Meeting, San Antonio, TX, March 23–25, 2003.

Krivtcova, N.I., Tataurshikov, A.A., Ivanchina, E.D., Krivtsov, E.B. (2015) Mathematical modelling of diesel fuel hydrodesulfurization kinetics. Procedia Chemistry. 15: 180–186.

Krivtsov, E.B., and Golovko, A.K. (2014) The kinetics of oxidative desulfurization of diesel fraction with a hydrogen peroxide-formic acid mixture. *Petroleum Chemistry*. 54(1): 51–57.

Krotz, A.H., Mehta, R.C., and Hardee, G.E. (2005) Peroxide-mediated desulfurization of phosphorothioate oligonucleotides and its prevention. *Journal of Pharmaceutical Sciences*. 94(2): 341–352.

Kuehler, C.W., Humphries, A. (2003) Meeting clean fuels objectives with the FCC, AM-03-57, in: Proceedings of the NPRA 2003. Annual Meeting, San Antonio, TX, March 23–25, 2003.

Kurita, S., Endo, T., Nakamura, H., Yagi, T., Tamuja, N.J. (1971) Decomposition of some organic sulfur compounds in petroleum by anaerobic bacteria. *The Journal of General and Applied Microbiology*. 17: 185–198.

La Paz Zavala, C., Rodriguez, J.E. (2004) Practical Applications of a Process Simulator of Middle Distillates Hydrodesulfurization. *Petroleum Science and Technology*. 22(1/2): 61–71.

Landau, M.V., Berger, D., Herskowitz, M. (1996) Hydrodesulfurization of Methyl-Substituted Dibenzothiophenes: Fundamental Study of Routes to Deep Desulfurization. *Journal of Catalysis* 159: 236–245.

Landau, M.V., Herskowitz, M., Hoffman, T., Fuks, D., Liverts, E., Vingurt, D., Froumin, N. (2009) Ultradeep hydrodesulfurization and adsorptive desulfurization of diesel fuel on metal-rich nickel phosphides. *Industrial and Engineering Chemistry Research*. 48(11): 5239–5249.

Lappas, A., Valla, J., Vasalos, I., Kuehler, C., Francis, J., O'Connor, P., Gudde, N. (2002). Sulfur reduction in FCC gasoline. American Chemical Society Preprints-Division Petroleum Chemistry. 47:50–52.

Larrubia, M.A., Gutierrez-Alejandre, A., Ramirez, J., Busca, G. (2002) A FT-IR Study of the Adsorption of indole, carbazole, benzothiophene, dibenzothiophene, and 4,6-dibenzothiophene over solid adsorbents and catalysts. *Applied Catalysis A: General*. 224: 167–178.

Le Borgne, S., Quintero, R. (2003) Biotechnological processes for the refining of petroleum. *Fuel Processing Technology*. 81(2): 155–169.

Lecrenay, E., Sakanishi, K., Mochida, I. (1997) Catalytic hydrodesulfurization of gas oil and model sulfur compounds over commercial and laboratory made CoMo and NiMo catalysts: Activity and reaction scheme. *Catalysis Today*. 39(1/2): 13–20.

Lecrenay, E., Sakanishi, K., Mochida, I., Suzuka, T. (1998) Hydrodesulfurization activity of CoMo and NiMo catalysts supported on some acidic binary oxides. *Applied Catalysis A: General*. 175: 237–243.

Lee, M.-C., Cheng, S.-H., Hong, C.-T. (2013) Review and comparison of FCC gasoline selective hydrodesulfurization process. *Journal of Petroleum*. 49(4): 45–54.

Lee, S.H.D., Kumar, R., Krumpelt, M. (2002). Sulphur removal from diesel fuel contaminated methanol. *Separation and Purification Technology*. 26: 247 -258.

Leflaive, P., Lemberton, J.L., Pérot, G., Mirgain, C., Carriat, J.Y., Colin, J.M. (2002) On the origin of sulfur impurities in fluid catalytic cracking gasoline–Reactivity of thiophene derivatives and of their possible precursors under FCC conditions. *Applied Catalysis A: General*. 227: 201–215.

Leliveld, R.G., Eijsbouts, S.E. (2008) Howa70-year-oldcatalyticrefineryprocess is still ever dependent on innovation. *Catalysis Today*. 130:183–189.

Lesemann, M., Schult, C. (2003). Non-capital intensive technologies reduce FCC sulfur content. Hydrocarbon Processing. 2:69–76.

Levy, R.E. (2003) Refining: the oxidative route to ULS production. Petroleum Technology Quaternary. 8: 53–58.

Levy, R.E., Rappas, A.S., Nero, V.P., DeCanio, S.J. (2002) UniPure's breakthrough technologies for cost effective desulfurization of crudes and refined products. *World Petroleum Congress Proceeding*. 3: 339–340.

Li, H., He, L., Li, J., Zhu, W., Jiang, X., Wang, Y., Yan, Y. (2009d) Fatty acid methyl ester synthesis catalyzed by solid superacid catalyst $SO_4^{2-}/ZrO_2-TiO_2/La^{3+}$. *Applied Energy*. 87(1): 156–159.

Li, H., He, L., Li, J., Zhu, W., Jiang, X., Wang, Y., Yan, Y. (2009b) Deep oxidative desulfurization of fuels catalyzed by phosphotungstic acid in ionic liquids at room temperature. *Energy and Fuels*. 23(3): 1354–1357.

Li, H.M., Jiang, X., Zhu, W.S., Lu, J.D., Shu, H.M., and Yan, Y.S. (2009c) Deep oxidative desulfurization of fuel oils catalyzed by decatungstates in the ionic liquid of [Bmim]PF_6. *Industrial and Engineering Chemistry Research*. 48(19): 9034–9039.

Li, W., Tang, H., Liu, Q., Xing, J., Li, Q., Wang, D., Yang, M., Li, X., Liu, H. (2009e). Deep Desulfurization of Diesel by Integrating Adsorption and Microbial Method. *Biochemical Engineering Journal*. 44: 297–301.

Li, W.L., Xing, J.M., Xiong, X.C., Liu, H.Z. (2005) Bio-regeneration of π-complexation desulfurization adsorbents. *Science in China Series B: Chemistry* 48(6): 538–544.

Li, X., Zhou, F., Wang, A., Wang, L., Hu, Y. (2009a) Influence of templates on the overgrowth of mcm-41 over HY and the hydrodesulfurization performances of the supported Ni–Mo catalysts. *Industrial and Engineering Chemistry Research*. 48(6): 2870–2877.

Lin, Z.J., Yang, Z., Liu, T.F., Huang, Y.B., Cao, R. (2012) Microwave assisted synthesis of a series of lanthanide metal–organic frameworks and gas sorption properties. *Inorganic Chemistry*. 51: 1813–1820.

Liotta, F.J., Han, Y. (2003) Production of ultra-low sulfur fuels by selective hydroperoxide oxidation. Lyondell Chemical Compant. NPRA Annual Meeting, Newtown Square, PA, USA.

Liu, L., Zhang, Y., Tan, W. (2014a) Ultrasound-Assisted oxidation of dibenzothiophene with phosphotungstic acid supported on activated carbon. *Ultrasonics Sonochemistry*. 21(3): 970–974.

Liu, S., Wang, B., Cui, B., Sun, L. (2008) Deep desulfurization of diesel oil oxidized by Fe(vi) systems. *Fuel.* 87: 422–428.

Liu, W., Liu, X., Yang, Y., Zhang, Y., Xu, B. (2014c) Selective removal of benzothiophene and dibenzothiophene from gasoline using double-template molecularly imprinted polymers on the surface of carbon microspheres. *Fuel.* 117: 184–190.

Liu, X., Wang, J.Y., Li, Q.Y., Jiang, S., Zhang, T., Ji, S.F. (2014b) Synthesis of rare earth metal-organic frameworks (Ln-MOFs) and their properties of adsorption desulfurization. *Journal of Rare Earths.* 32:189–194.

Liu, X.J., Ouyang, C.B., Zhao, R., Shangguan, D.H., Chen, Y., Liu, G.Q. (2006) Monolithic molecularly imprinted polymer for sulamethoxazole and molecular recognition properties in aqueous mobile phase. *Analytica Chimica Acta.* 571(2): 235–241.

Long, Z., Yang, C., Zeng, G., Peng, L., Dai, C., He, H. (2014) Catalytic oxidative desulfurization of dibenzothiophene using catalyst of tungsten supported on resin D152. Fuel. 130: 19–24.

Loupy A. (2006) "Microwaves in organic synthesis", second ed., WILEY-VCH Verlag GmbH & Co kGaA, Weinheim.

Lucien, J.P., van den Berg, J.P., Germaine, G., van Hooijdonk, H.M.J.H., Gjers, M., Thielemans, G.L.B. (1994) Shell middle distillate hydrogenation process, in: M.C. Oballa, S.S. Shih (Eds.), Catalytic Hydroprocessing of Petroleum and Distillates, Marcel Dekker, New York, 1994, pp. 291–313.

Ma, X., Sakanishi, K., Isoda, T., Mochida, I. (1995) Quantum chemical calculation on the desulfurization reactivities of heterocyclic sulfur compounds. *Energy & Fuels.* 9: 33–37.

Ma, X., Sakanishi, K., Isoda, T., Mochida, I. (1996) Comparison of sulfided CoMo/Al_2O_3 and NiMo/Al_2O_3 catalysts, in: M.L. Occelli, R. Chianelli (Eds.), Hydrodesulfurization of Gas Oil Fractions and Model Compounds, in Hydrotreating Technology for Pollution Control, Marcel Dekker, New York. Pp. 183.

Ma, X., Sun, L., Song, C. (2002) A new approach to deep desulfurization of gasoline, diesel fuel and jet fuel by selective adsorption for ultra-clean fuels and for fuel cell applications. *Catalysis Today.* 77: 107–116.

Ma, X., Zhou, A., Song, C. (2007) A novel method for oxidative desulfurization of liquid hydrocarbon fuels based on catalytic oxidation using molecular oxygen coupled with selective adsorption. *Catalysis Today.* 123: 276–284.

Majid, D., Seyedeyn-Azad, F., (2010) Desulfurization of gasoline over nanoporous nickel-loaded Y-type zeolite at ambient conditions. *Industrial and Engineering Chemistry Research.* 49: 11254–11259.

Mamaghani, A.H., Fatemi, S., Asgari, M. (2013) Investigation of influential parameters in deep oxidative desulfurization of dibenzothiophene with hydrogen peroxide and formic acid. *International Journal of Chemical Engineering.* Volume 2013, Article ID 951045, http://dx.doi.org/10.1155/2013/951045.

Marchal, N., Kasztelan, S., Mignard, S. (1994) A comparative study of catalysts for the deep aromatic reduction in hydrotreated gas oil, in: M.C. Oballa, S.S. Shih

(Eds.), Catalytic Hydroprocessing of Petroleum and Distillates, Marcel Dekker, New York, 1994, pp. 315-327.

Marin, C., Escobar, J., Galvan, E., Murrieta, F., Zarate, R., Vaca, H. (2004) Light straight-run gas oil hydrotreatment over sulfided CoMoP/Al_2O_3-USY zeolite catalysts. *Fuel processing Technology*. 86(4): 391–405.

Matsuzawa, S., Tanaka, J., Sato, S., Takashi, I. (2002) Photocatalytic oxidation of dibenzothiophenes in acetonitrile using TiO_2: effect of hydrogen peroxide and ultrasound irradiation. *Journal of Photochemistry and Photobiology A: Chemistry*. 149(1/3): 183–189.

McKinley, S.G., Angelici, R.J. (2003) Deep desulfurization by selective adsorption of dibenzothiophenes on Ag^+/SBA-15 and $Ag^{+/}SiO_2$. *Chemical Communications*. 20: 2620–2621.

Mei, H., Mei, B.W., Yen, T.F. (2003) A new method for obtaining ultra-low sulfur diesel fuel via ultrasound assisted oxidative desulfurization. *Fuel*. 82(4): 405–414.

Mello, P.D.A., Duarte, F.A., Nunes, M.A.G., Alencar, M.S., Moreira, E.M., Korn, M., Gressler, V.L., Flores, E.M.M. (2009) Ultrasound-assisted oxidative process for sulfur removal from petroleum product feedstock. *Ultrasonics Sonochemistry*. 16(6): 732–736.

Miadonye, A., Snow, S., Irwin, D.J.G., Rashid Khan, M. (2009) Desulfurization of heavy crude oil by microwave irradiation. WIT Transactions on Engineering Sciences. 63: 455–465.

Mikhail, S., Zaki, T., Khalil, L. (2002) Desulfurization by an economically adsorption technique. *Applied Catalysis A: General*. 227(1/2): 265–278.

Milenkovic, A., Macaud, M., Schulz, E., Koltai, T., Loffreda, D., Vrinat, M., Lemaire, M. (2000) How could organic synthesis help the understanding of the problems of deep hydrodesulfurization of gas oils? Comptes Rendus de l'Académie des Sciences - Series IIC - Chemistry. 3 (6): 459–463.

Miller, J.T., Reagan, W.J., Kaduk, J.A., Marshall, C.L., Kropf, A.J. (2000) Selective hydrodesulfurization of FCC naphtha with supported MoS_2 catalyst. The role of cobalt. *Journal Catalysis*. 193(1): 123–131.

Miller, K.W. (1992) Reductive desulfurization of dibenzyldisulfide. *Applied and Environmental Microbiology*. 58(7): 2176–2179.

Mjalli, F.S., Ahmed, A.U., Al-Wahaibi, T., Al-Wahaibi, Y., AlNashef, I.M. (2014) Deep oxidative desulfurization of liquid fuels. *Reviews in Chemical Engineering*. 30(4): 337–378.

Mohammed, M.I., Abdul Razak, A.A., Shehab, M.A. (2016) Synthesis of nanocatalyst for hydrodesulfurization of gasoil using laboratory hydrothermal rig. *Arabian Journal of Science and Engineering*. doi:10.1007/s13369-016-2249-5.

Mohebali, G., Ball, A.S. (2008) Biocatalytic desulfurization (BDS) of petrodiesel fuels. *Microbiology*. 154, 2169–2183.

Monticello, D.J. (1994) Biocatalytic desulfurization. Hydrocarbon Processing. 39: 39–44.

Monticello, D.J. (1996) Multistage process for deep desulfurization of a fossil fuel. US Patent 5,510,265.

Monticello, D.J. (2000) Biodesulfurization and the upgrading of petroleum distillates. *Current Opinion in Biotechnology*. 11: 540–546.

Morales-Ortuño, J.C., Ortega-Domínguez, R.A., Hernández-Hipólito, P., Bokhimi, X., Klimova, T.E. (2016) HDS performance of NiMo catalysts supported on nanostructured materials containing titania. *Catalysis Today*.271: 127–139

Murata, S., Murata, K., Kidena, K., and Nomura, M. (2003) Oxidative desulfurization of diesel fuels by molecular oxygen. Preprints of Papers- American Chemical Society, Division of Fuel Chemistry. 48(2): 531.

Mustafa, F., Al-Ghouti, M.A., Khalili, F.I., Al-Degas, Y.S. (2010) Characteristics of organosulphur compounds adsorption onto Jordanian zeolitic tuff from diesel fuel. *Journal of Hazardous Materials*. 182(1/3): 97–107.

Mutyala, S., Fairbridge, C., Pare, J.R.J., Belanger, J.M.R., Hawkins, S.Ng.R. (2010) Microwave applications to oil sands and petroleum: A review. *Fuel Processing Technology*. 91: 127–135.

Mužic, M., Sertić-Bionda, K. (2013) Alternative processes for removing organic sulfur compounds from petroleum fractions. *Chemical and Biochemical Engineering*. Q. 27(1): 101–108.

Mužic, M., Sertić-Bionda, K., Adžamić, T., Gomiz, Z. (2010) A design of experiments investigation of adsorptive desulfurization of diesel fuel. *Chemical and Biochemical Engineering Quarterly*. 24(3): 253–264.

Myrstad, T. (2000). Effect of nickel and vanadium on sulphur reduction of FCC naphtha. *Applied Catalysis A: General*. 192 (2): 299–305.

Nakano, K., Ali, S.A., Kim, H., Kim, T., Alhooshani, K., Park, J., Mochida, I. (2013) Deep desulfurization of gas oil over NiMoS catalysts supported on alumina coated USY-zeolite. *Fuel Processing Technology*. 116: 44–51.

Nanoti, A., Dasgupta, S., Goswami, A.N., Nautiyal, B.R., Rao, T.V., Sain, B., Sharma, Y.K., Nanoti, S.M., Garg, M.O., and Gupta, P. (2009) Mesoporous silica as selective sorbents for removal of sulfones from oxidized diesel fuel. *Microporous and Mesoporous Materials*. 124: 94–99.

Nehlsen, J.P.; Benziger, J.B.; Kevrekidis, I.G. (2003) Removal of alkanethiols from a hydrocarbon mixture by a heterogeneous reaction with metal oxides. *Industrial and Engineering Chemistry Research*. 42(26): 6919–6923.

Nijhuis, T.A., Kreutzer, M.T., Romijn, A.C.J., Kapteijn, F., Moulijn, J.A. (2001) Monolithic catalysts as more efficient three-phase reactors. *Catalysis Today*. 66: 157–165.

Niquille-Röthlisberger A., and Prins R. (2007) Hydrodesulfurization of 4,6-dimethyldibenzothiophene over Pt, Pd, and Pt–Pd catalysts supported on amorphous silica–alumina. *Catalysis Today*. 123:198–207.

Niquille-Röthlisberger, A, Prins, R (2006) Hydrodesulfurization of 4,6-dimethyldibenzothiophene and dibenzothiophene over alumina-supported Pt, Pd, and Pt-Pd catalysts. *Journal of Catalysis*. 242:207–216.

Nocca, J.-L., Cosyns, J., Debuisschert, Q., Didillon, B. (2000) The domino interaction of refinery processes for gasoline quality attainment, AM-00-61, in: Proceedings of the NPRA Annual Meeting, March 2000, San Antonio, TX.

Ohshiro, T., Izumi, Y. (1999) Microbial desulfurization of organic sulfur compounds in petroleum. *Bioscience, Biotechnology and Biochemistry*. 63(1): 1–9.

Okada, H., Nomura N., Nakahara T. and Maruhashi K. (2002a) Analyses of substrate specificity of the desulfurizing bacterium *Mycobacterium* sp. G3. *Journal of Bioscience and Bioengineering*. 93(2): 228–233.

Okada, H., Nomura N., Nakahara T. and Maruhashi K. (2002b) Analysis of dibenzothiophene metabolic pathway in *Mycobacterium* strain G3. *Journal of Bioscience and Bioengineering*. 93(5): 491–497.

Okamoto, Y., Ochiai, K., Kawano, M., Kobayashi, M., Kubota, T. (2002) Effects of support on the activity of Co–Mo sulfide model catalysts. *Applied Catalysis A: General*. 226(1/2): 115–161.

Otsuki, S., Nonaka, T., Qian, W., Ishihara, A., Kabe, T. (1999) Oxidative desulfurization of middle distillate using ozone. Sekiyu Gakkaishi (*Journal of the Japan Petroleum Institute*). 42(5): 315–320.

Otsuki, S., Nonaka, T., Qian, W.H., Ishihara, A., Kabe, T. (2001) Oxidative desulfurization of middle distillate-oxidation of dibenzothiophene using t-butyl hypochlorite. Sekiyu Gakkaishi. (*Journal of the Japan Petroleum Institute*). 44(1): 18–24.

Otsuki, S., Nonaka, T., Takashima, N. (2000) Oxidative desulfurization of light gas oil and vacuum gas oil by oxidation and solvent extraction. *Energy and Fuels*. 14(6): 1232–1239.

Pacheco, M.A. (1999) Recent advances in biodesulfurization (biodesulfurization) of diesel fuel. Paper presented at the NPRA Annual Meeting, San Antonio, TX, 21–23 March 1999.

Palaić, N., Sertić-Bionda, K., Margeta, D., and Podolski, Š. (2015) Oxidative desulfurization of diesel fuels. *Chemical And Biochemical Engineering Quarterly*. 29(3): 323–327.

Park T.H., Cychosz K.A., Wong-Foy A.G., Dailly A., and Matzger A.J. (2011) Gas and liquid phase adsorption in isostructural Cu_3[biaryltricarboxylate]$_2$ microporous coordination polymers. *Chemical Communications*. 47: 1452–1454.

Park, J.G., Ko, C.H., Yi, K.B., Park, J.H., Han, S.S., Cho, S.H., Kim, J.N. (2008) Reactive adsorption of sulfur compounds in diesel on nickel supported on mesoporous silica. *Applied Catalysis B: Environmental*. 81: 244–250.

Pawelec, B., Navarro, R., Fierro, J. L.G., Cambra, J. F., Zugazaga, F., Guemez, M.B., Arias, P.L. (1997) Hydrodesulfurization over PdMo/HY zeolite catalysts. *Fuel*. 76(1): 61–71.

Phillipson, J.J. (1971) Kinetics of hydrodesulfurization of light and middle distillates: Paper Presented at the American Institute of Chemical Engineers Meeting, Houston, TX. USA.

Pimerzin, Al.A., Nikul'shin, P.A., Mozhaev, A.V., Pimerzin, A.A. (2013) Effect of surface modification of the support of hydrotreating catalysts with transition metal oxides (sulfides) on their catalytic properties. Petroleum Chemistry. 53(4): 245–254.

Purnell, S.K., Hunt, D.A., Leach, D. (2002) Catalytic reduction of sulfur and olefins in the FCCU: commercial performance of Davison catalysts and additives for clean fuels, AM-02–37, in: Proceedings of the NPRA 2003 Annual Meeting, SanAntonio, TX, March 17–19, 2002.

Purta, D.A., Portnoff, M.A., Pourarian, F., Nasta, M.A., Zhang, J. (2004) Catalyst for the treatment of organic compounds. US 2004077485.

Qu, L., Zhang, W., Kooyman, P.J., Prins, R. (2003) MAS NMR, TPR, and TEM studies of the interaction of NiMo with alumina and silica–alumina supports. *Journal of. Catalysis*. 215: 7–13.

Rafiee, E., Joshaghani, M., Abadi, P.G.-S. (2017) Oxidative desulfurization of diesel by potato based - carbon as green support for $H_5PMo_{10}V_2O_{40}$: Efficient composite nanorod catalyst. *Journal of Saudi Chemical Society*. 21: 599–609.

Rafiee, E., Rahpeyma, N. (2015) Selective oxidation of sulfurs and oxidation desulfurization of model oil by 12-tunstophosphoric acid on cobalt-ferrite nanoparticles as magnetically recoverable catalyst. *Chinese Journal of Catalysis*. 36: 1342–1349.

Rakhmanov, E.V., Anisimov, A.V., Tarakanova, A.V., Baleeva, N.S., Guluzade, D. (2013) Oxidative desulfurization of catalytically cracked gasoline with hydrogen peroxide. *Petroleum Chemistry.* 53(3): 201–204.

Rakhmanov, E.V., Baranova, S.V., Zixiao, W., Tarakanova, A.V., Kardashev, S.V., Akopyan, A.V., Naranov, E.R., Oshchepkov, M.S., Anisimov, A.V. (2014b) Hydrogen peroxide oxidative desulfurization of model diesel mixtures using azacrown ethers. *Petroleum Chemistry*. 54(4): 316–322.

Rakhmanov, E.V., Domashkin, A.A., Myltykbaeva, Zh.K., Kairbekov, Zh., Shigapova, A.A., Akopyan, A.V., Anisimov, A.V. (2016) Peroxide oxidative desulfurization of a mixture of nonhydrotreated vacuum gas oil and diesel fraction. *Petroleum Chemistry*. 56(8): 742–744.

Rakhmanov, E.V., Tarakanova, A.V., Valieva, T., Akopyan, A.V., Litvinovaa, V.V., Maksimov, A.L., Anisimov, A.V., Vakarin, S.V., Semerikova, O.L., Zaikov, Y.P. (2014a) Oxidative desulfurization of diesel fraction with hydrogen peroxide in the presence of catalysts based on transition metals. *Petroleum Chemistry*. 54(1): 48–50.

Ramírez-Verduzco, L.F., Torres-García, E., Gómez-Quintana, R., Gonzálezpeña, V., Murrieta-Guevara, F. (2004) Desulfurization of diesel by oxidation/extraction scheme: influence of the extraction solvent. *Catalysis Today*. 98: 289–294.

Rana, M.S., Samano, V., Ancheyta, J., Diaz, J. (2007) A review of recent advances on process technologies for upgrading of heavy oils and residua. *Fuel*. 86(9): 1216–1231.

Rang, H., Kann, J., Oja, V. (2006) Advances in desulfurization research of liquid fuel. *Oil Shale*. 23(2): 164–176.

Rao, T.V., Krishna, P.M., Paul, D., Nautiyal, B.R., Kumar, J., Sharma, Y.K., Nanoti, S.M., Sain, B., Garg, M.O. (2011). The oxidative desulfurization of HDS diesel: using aldehyde and molecular oxygen in the presence of cobalt catalysts. *Petroleum Science and Technology*. 29(6): 626–632.

Reddy, K.M., Wei, B., Song, C. (1998) Mesoporous molecular sieve MCM-41 supported Co–Mo catalyst for hydrodesulfurization of petroleum resids. *Catalysis Today*. 43(3/4): 261–272.

Reid, T.A. (2000) Hydrotreating Approaches to Meet Tier 2 Gasoline Regulations, Catalyst Courier No. 42, Akzo Nobel. Pp. 1–6.

Reinhoudt, H.R., Troost, R., Van Langeveld, A.D., Sie, S.T., Van Veen, J.A.R., Moulijn, J.A. (1999) Catalysts for second-stage deep hydrodesulfurisation of gas oils. *Fuel Processing Technology*. 61: 133–147.

Ren, J., Wang, A., Li, X., Chen, Y., Liu, H., Hu, Y. (2008) Hydrodesulfurization of dibenzothiophene catalyzed by Ni-Mo sulfides supported on a mixture of MCM-41 and HY zeolite. *Applied Catalysis A: General*. 344: 175–182.

Roberie, T.G., Kumar, R., Ziebarth, M.S., Cheng, W.-C., Zhao, X., Bhore, N. (2002). Gasoline sulfur reduction in FCC. U.S. Patent 6,482,315 B1. Assigned to W.R. Grace & Co.-Conn.

Robinson, W.R.A.M., van Veen, J.A.R., De Beer, V.H.J., van Santen, R.A. (1999) Development of deep hydrodesulfurization catalysts: II. NiW, Pt and Pd catalysts tested with (substituted) dibenzothiophene. *Fuel Processing Technology*. 61: 103–116.

Rock, K.L. (2002) Ultra-low sulfur gasoline via catalytic distillation, in: Proceedings of the Fifth International Conference onRefinery Processing, AIChE 2002 Spring National Meeting, New Orleans, LA, March 11–14, 2002, pp. 200–205.

Rock, R., Shorey, S. (2003) Producing low sulfur gasoline reliably, AM-03-122, in: Proceedings of the NPRA 2003 Annual Meeting, San Antonio, TX, March 23–25, 2003.

Ryan, G.H., Xie, X.J., Sofia, R.P., Isabel, M., Edward, I.S. (2012) Analysis of resonance Raman data on the blue copper site in pseudoazurin: Excited state π and σ charge transfer distortions and their relation to ground state reorganization energy *Journal of Inorganic Biochemistry*. 115: 155–162.

Salazar, J.A., Cabrera, L.M., Palmisano, E., Garcia, W.J., Solari, R.B. (1998) US Patent 5,770,047.

Salcedo, S.T.G. (2015) Hydrotreating of synthetic fuels with bulk and supported Ni-Mo(W) sulfide catalysts. MSc. Thesis, Technische Universität München, Arcisstraße 21, 80333 München, Germany.

Saleh, T.A., Danmaliki, G.I. (2016a) Influence of acidic and basic treatments of activated carbon derived from waste rubber tires on adsorptive desulfurization of thiophenes. *Journal of the Taiwan Institute of Chemical Engineering*. 60: 460–468.

Saleh, T.A., Danmaliki, G.I. (2016b) Adsorptive desulfurization of dibenzothiophene from fuels by rubber tyres-derived carbons: Kinetics and isotherms evaluation. *Process Safety and Environmental Protection*. 102: 9–19.

Saleh, T.A., Sulaiman, K.O., AL-Hammadi, S.A., Dafalla, H., Danmaliki, G.I. (2017) Adsorptive desulfurization of thiophene, benzothiophene and dibenzothiophene over activated carbon manganese oxide nanocomposite: with column system evaluation. *Journal of Cleaner Production*. 154: 401–412.

Salem, A.S., Hamid, H.S. (1997) Removal of Sulfur Compounds from Naphtha solutions using solid adsorbents. *Chemical Engineering Technology*. 20(5): 342–347.

Sampanthar, J.T., Xiao, H., Dou, J., Yin Nah, T., Rong, X., and Kwan, W.P. (2006) A novel oxidative desulfurization process to remove refractory sulfur compounds from diesel fuel. *Applied Catalysis B: Environment*. 63(1/2): 85–93.

Sankaranarayanan, T.M., Banu, M., Pandurangan, A., Sivasanker, S. (2011) Hydroprocessing of sun flower oil gas oil blends over sulfide Ni-Mo-AL-zeolite beta composites. *Bioresource Technology*. 102: 10717–10723.

Sano, Y., Kazomi, S., Ki-Hyonk, C., Yozo, K., Isao, M. (2005) Two-step adsorption process for deep desulfurization of diesel oil. *Fuel*. 84: 903–910.

Santana, R.C., Do P.T., Santikunaporn, M., Alvarez, W.E., Taylor, J.D., Surghrue, E.L., Resasco, D.E. (2006) Evaluation of different reaction strategies for the improvement of cetane number in diesel fuels. *Fuel*. 85(5/6): 643–656.

Sarda, K.K., Bhandari, A., Pant, K.K. Jain, S. (2012) Deep desulfurization of diesel fuel by selective adsorption over Ni/Al2O3 and Ni/ZSM-5 extrudates. *Fuel*. 93, 86–91.

Schulz, H., Böhringer, W., Ousmanov, F., Waller, P. (1999) Refractory sulfur compounds in gas oils. *Fuel Processing Technology*. 61: 5–41.

Segawa, K., Takahashi, K., Satoh, S. (2000) Development of new catalysts for deep hydrodesulfurization of gas oil. *Catalysis Today*. 63: 123–131.

Shakirullah, M., Ahmad, I. Ishaq, M., Ahmad, W. (2009) Study on the role of metal oxides in desulphurization of some petroleum fractions. *Journal of the Chinese Chemical Society*. 56: 107–114.

Shan, H., Li, C., Yuan, M., Yang, H. (2002a). Evaluation of a sulfur reduction additive for FCC gasoline. Preprints American Chemical Society- Division of Petroleum Chemistry. 47:55–57.

Shan, H., Yang, C., Zhao, H., Zhang, J. (2002b). Mechanistic studies on thiophene cracking over USY zeolite. *Catalysis Today*. 77:117–126.

Shang, H., Du, W., Liu, Z., Zhang, H. (2013) Development of microwave induced hydrodesulfurization of petroleum streams: A review. *Journal of Industrial and Engineering Chemistry*. 19: 1061–1068.

Shao, X., Zhang, X., Yu, W., Wu, Y., Qin, Y., Sun, Z., Song, L. (2012) Effects of surface acidities of MCM-41 modified with MoO3 on adsorptive desulfurization of gasoline. *Applied Surface Science*. 263: 1–7.

Sharipov, A.K., Nigmatullin, V.R. (2005) Oxidative desulfurization of diesel fuel (a review). *Petroleum Chemistry*. 45(6): 371–377.

Shen, Y., Xu, X. and Li, P. (2012) A novel potential adsorbent for ultra deep desulfurization of jet fuels at room temperature. RSC Advances. 2: 6155–6160.

Shiflett, W.K.A. (2002) User's guide to the chemistry, kinetics and basic reactor engineering of hydroprocessing, Proceedings of the Fifth International Conference on Refinery Processing,

AIChE 2002 Spring National Meeting, New Orleans, March 11–14, 2002. Pp. 101–122.

Shiflett W.K. and Krenzke, L.D. (2001) consider improved catalyst technologies to remove sulfur, Hydrocarbon Processing. 81:41–43.

Shih, S.S., Owens, P.J., Palit, S., Tryjankowski, D.A. (1999) Mobil's OCTGain process: FCC gasoline desulfurization reaches a new performance level, AM-99-30, in: Proceedings of the NPRA 1999 Annual Meeting, March 21–23, 1999.

Shiraishi, Y., Hirai, T., and Komasawa, I. (1998) A deep desulfurization process of light oil by photochemical reaction in an organic two-phase liquid-liquid extraction system. *Industrial and Engineering Chemistry Research*. 37(1): 203–211.

Shiraishi, Y., Hirai, T., Komasawa, I. (1999a) Identification of desulfurization products in the photochemical desulfurization process for benzothiophenes and dibenzothiophenes from light oil using an organic two-phase extraction system. *Industrial and Engineering Chemistry Research*. 38(9): 3300–3309.

Shiraishi, Y., Taki, Y., Hirai, T., and Komasawa, I. (1999b) Photochemical desulfurization of light oils using oil hydrogen peroxide aqueous solution extraction system: application for high sulfur content straight-run light gas oil and aromatic light cycle oil. *Journal of Chemical Engineering of Japan*. 32(1): 158–161.

Shu, Y., Wormsbecher, R. (2007) Catalytic FCC Gasoline: Sulfur Reduction Mechanism Excerpted from *The Journal of Catalysis*. Catalagram 102: 31–36. www.e-catalysts.com.

Si, X., Cheng, S., Lu, Y., Gao, G., He, M.-Y. (2008) Oxidative desulfurization of model oil over Au/Ti-MWW. Catalysis Letters. 122: 321–324.

Siddiqui, M.A.B., Aitani, A.M. (2007) FCC gasoline sulfur reduction by additives:
A review. *Petroleum Science and Technology*. 25: 299–313.

Siddiqui, M.N., Saleh, T.A., Anee, M.H., Basheer, M.C., Al-Arfaj, A.A., Tyler, D.R. (2016) Desulfurization of model fuels with carbon nanotube/TiO_2 nanomaterial adsorbents: comparison of batch and film-shear reactor processes. *Journal of Inorganic and Organometallic Polymers and Materials*. 26: 572–578.

Skocpol, R.C. (2000) RESOLVE prove its worth. http://www.akzonobel-catalysts.com/html/news/News link old/2000/RESOLVE.htm.

Soled, S.L, Miseo, S., Krycak, R., Vroman, H., Ho, T.C., and Riley, K.L. (2001) Nickel molybodtungstate hydrotreating catalysts (law444). U.S. Patent No. 6299760 B1.

Song, C. (1999) Designing sulfur-resistant, noble-metal hydrotreating catalysts. Chemtech. 3: 26–30.

Song, C. (2002) New approaches to deep desulfurization for ultra-clean gasoline and diesel fuels: an overview. American Chemical Society-Fuel Chemistry Division Preprints. 47(2): 438–444.

Song, C., Ma, X. (2003) New design approaches to ultra-clean diesel fuels by deep desulfurization and deep dearomatization. *Applied Catalysis B: Environmental.* 41(1–2): 207–238.

Song, C.S. (2003) An overview of new approaches to deep desulfurization for ultra-clean gasoline, diesel fuel and jet fuel. *Catalysis Today.* 86: 211–263.

Song, H., Cui, X., Jiang, B., Song, H., Wang, D., Zhang, Y. (2015) Preparation of SO_4^{2-}/ZrO_2 solid superacid and oxidative desulfurization using K_2FeO_4. Research on Chemical Intermediates. 41: 365–382.

Song, H., Dong, P.F., Zhang, X. (2010) Effect of Al Content on the n-Pentane Isomerization of the Solid Superacid Pt-SO_4^{2-}/ZrO_2-Al_2O_3. Acta Physico-Chimica Sinica. 26(8): 2229–2234.

Srivastava, A., Srivastava, V.C. (2009) Adsorptive desulfurization by activated alumina. *Journal of Hazardous Materials.* 170(2/3): 1133–1140.

Srivastava, C.C. (2012) An evaluation of desulfurization technologies for sulfur removal from liquid fuels. *RSC Advances.* 2: 759–783.

Stanislaus, A., Marafi, A., Rana, M.S. (2010) Recent advances in the science and technology of ultra low sulfur diesel (ULSD) production. *Catalysis Today.* 153: 1–68.

Stork, W.H.J. (1996) Performance testing of hydroconversion catalysts. Chapter 28. American Chemical Society Symposium Series. 634: 379–400.

Strawinski, R.J. (1950) Method of desulfurizing crude oil. US Patent 2521761 A.

Suchanek, A. (1996) How to make low-sulfur, low aromatics, high cetane diesel fuel? American Chemical Society Preprints-Division Petroleum Chemistry. 41: 583.

SulphCo Presentation (2009) "Oxidative Desulfurization". IAEE Houston Chapter, June 11, 2009.

Sun, Y., Prins, R. (2008) Hydrodesulfurization of 4,6-dimethyldibenzothiophene over noble metals supported on mesoporous zeolites. Angewandte Chemie International Edition. 47: 8478–8481.

Suryawanshi, N.B., Bhandari, V.M., Sorokhaibam, L.G., and Ranade, V.V. (2016) A non-catalytic deep desulfurization process using hydrodynamic cavitation. Scientific Reports. 6: 33021. DOI: 10.1038/srep33021.

Swain, E.J. (1991) US crude slate gets heavier, higher in sulfur. *Oil and Gas Journal.* 89(36): 59–61.

Swaty, T.E. (2005) Global refining industry trends: The present and future. Hydrocarbon Processing, September: 35–46.

Takahashi, A., Yang, F.H., Yang, R.T. (2002) New sorbents for desulfurization by -complexation: thiophene/benzene adsorption. *Industrial and Engineering Chemistry Research.* 41: 2487–2496.

Tang, K., Song, K.L., Duan, L., Li, X., Gui, J., Sun, Z. (2008) Deep desulfurization by selective adsorption on a heteroatoms zeolites prepared by secondary synthesis. *Fuel Processing Technology.* 89(1): 1–6.

Tang, L., Luo, G., Zhu, M., Kang, L., Dai, B. (2013b) Preparation, characterization and catalytic performance of HPW-TUD-1 catalyst on oxidative desulfurization. *Journal of Industrial and Engineering Chemistry.* 19(2): 620–626.

Tang, T., Zhang, L., Fu, W., Ma, Y., Xu, J., Jiang, J., Fang, G., Xiao, F.-S. (2013a) Design and synthesis of metal sulfide catalysts supported on zeolite nanofiber bundles with unprecedented hydrodesulfurization activities. *Journal of the American Chemical Society*. 135(31): 11437–11440.

Tang, X.-L., Meng, X., Shi, L. (2011) Desulfurization of model gasoline on modified bentonite. *Industrial and Engineering Chemistry Research*. 50: 7527–7533.

Tao, H., Nakazato, T., Sato, S. (2009) Energy-Efficient Ultra-Deep Desulfurization of Kerosene Based On Selective Photooxidation and Adsorption. *Fuel*. 88: 1961–1969.

Te, M., Fairbridge, C., Ring, Z. (2001) Oxidation reactivities of dibenzothiophenes in polyoxometalate/H_2O_2 and acid/H_2O_2 systems. *Applied Catalysis A: General*, 2001, 219(1/2): 267–280.

Thaligari, S.K., Srivastava, V.C., Prasad, B. (2016) Adsorptive desulfurization by zinc-impregnated activated carbon: characterization, kinetics, isotherms, and thermodynamic modeling. *Clean Technologies and Environmental Policy*.18(4): 1021–1030.

Tilstra, L., Eng, G., Olsonn, G.J., Wang, F.W. (1992) Reduction of sulphur from polysulphidic model compounds by the hyperthermophilic archaebacterium *Pyrococcus furiosus*. *Fuel*. 71(7): 779–783.

Topsoe, H., Knudsen, K.G., Byskov, L.S., Norskov, J.K., Clausen, B.S. (1999) Advances in deep desulfurization. Studies in Surface Science and Catalysis. 121: 13–22.

Trakarnpruk, W., Seentrakoon, B., Porntangjitlikit, S. (2008) Hydrodesulfurization of diesel oils by MoS_2 catalyst prepared by in situ decomposition of ammonium thiomolybdate. *Silpakorn University Science and Technology Journal* 2(1): 7–13.

Trejo, F., Rana, M.S., Ancheyta, J. (2008) CoMo/MgO–Al_2O_3 supported catalysts: an alternative approach to prepare HDS catalysts. *Catalysis Today*. 130: 327–336.

Turaga, U.T., Song, C. (2001) Deep hydrodesulfurization of diesel and jet fuels using mesoporous molecular sieve-supported Co-M0/MCM-41 catalysts. *American Chemical Society Preprints-Division Petroleum Chemistry*. 46: 275–279.

Turk, B.S., Gupta, R.P., Arena, B.A. (2002) Continuous catalytic process for diesel desulfurization. vols. 79–83. PTQ Summer, 2002.

United States Environmental Protection Agency US-EPA (2000) "Heavy-duty engine and vehicle standards and highway diesel fuel sulfur control requirements". EPA 420-F-00–057.

van der Linde, B., Menon, R., Davé, D., Gustas, S. (1999) Asian refining technology conference, DD-ARTC-99. Singapore, 21st–23rd April 1999. Pp. 24.

van Veen, J.A.R., Colijn, H.A., Hendriks, P.A.J.M., van Welsenes, A.J. (1993) On the formation of type I and type II NiMoS phases in NiMo/Al_2O_3 hydrotreating catalysts and its catalytic implications. *Fuel Processing Technology*. 35: 137–157.

Vasudevan, P.T., Fierro, J.L.G. (1996) A review of deep hydrodesulfurization catalysis. *Catalysis Reviews-Science and Engineering*. 38(2): 161–188.

Velu, S., Ma, X., Song, C. (2005) Desulfurization of JP-8 jet fuel by selective adsorption over a Ni-based adsorbent for micro solid oxide fuel cells. *Energy and Fuels* 19: 1116–1125.

Wan, G., Duan, A., Zhan, Y., Zhao, Z., Jiang, G., Zhang, D., Gao, Z (2010) Zeolite beta synthesized with acid-treated metakaolin and its application in diesel hydrodesulfurization. *Catalysis Today*. 149(1/2): 69–75.

Wan, J.K.S., Kriz, J.F. (1985) Hydrodesulphurization of hydrocracked pitch. US Patent 4545879.

Wan, M.W., Yen, T.F. (2007) Enhanced efficiency of tetraoctylammonium fluoride applied to ultrasound-assisted oxidative desulfurization (UAOD) process. *Applied Catalysis A: General*. 319: 237–245.

Wang, B., Zhu, P.J., Ma, H.Z. (2009a) Desulfurization from thiophene by SO_4^{2-}/ZrO_2 catalytic oxidation at room temperature and atmospheric pressure*Journal of Hazardous Materials*. 164(1): 256–264.

Wang, D., Qian, E.W., Amano, H., Okata, K., Ishihara, A., Kabe, T. (2003) Oxidation of dibenzothiophenes using tert-butyl hydroperoxide. *Applied Catalysis A*. 253: 91–99.

Wang, Q., Wang, S., and Yu, H. (2017) Oxidative desulfphurization of model fuel by in situ produced hydrogen peroxide on palladium/active carbon. *The Canadian Journal of Chemical Engineering*. 95: 136–141.

Wang, R., Yu, F (2013) Deep oxidative desulfurization of dibenzothiophene in simulated oil and real diesel using heteropolyanion-substituted hydrotalcite-like compounds as catalysts. *Molecules*. 18(11): 13691–13704.

Wang, S., Wang, R., Yu, H. (2012) Deep removal of 4,6- dimethyldibenzothiophene from model transportation diesel fuels over reactive adsorbent. Brazilian *Journal of Chemical Engineering*. 29(2): 421–428.

Wang, S.Q., Zhou, L., Long, L., Dai, W., Zhou, Y.P. (2008c) Thiophene capture with silica gel loading formaldehyde and hydrochloric acid. *Industrial and Engineering Chemistry Research*. 47: 2356–2360.

Wang, S.Q., Zhou, L., Su, W., Sun, Y., Zhou, Y.P. (2009b) Deep desulfurization of transportation fuels by characteristic reaction resided in adsorbents. *AICHE Journal*. 55(7): 1872–1881.

Wang, Y., Sun, Z., Wang, A., Ruan, A.L., Lifeng, L., Ren, J., Li, X., Li, C., Hu, Y., Yao, P. (2004) Kinetics of hydrodesulfurization of dibenzothiophene catalyzed by sulfided Co-Mo/MCM-41. *Industrial and Engineering Chemistry Research*. 43(10): 2324- 2329.

Wang, Y., Wang, B., Rives, A., Sun, Y. (2014) Hydrodesulfurization of transportation fuels over zeolite-based supported catalysts. *Energy and Environment Focus*. 3: 1–8.

Wang, Y., Yang, R.T., Heinzel, J.M. (2008a) Desulfurization of jet fuel by-complexation adsorption with metal halides supported on MCM-41 and SBA-15 mesoporous materials. *Chemical Engineering Science*. 63: 356–365.

Wang, Y., Yang, R.T., Heinzel, J.M. (2008b) Desulfurization of jet fuel JP-5 light fraction by MCM-41 and SBA-15 supported cuprous oxide for fuel cell applications. *Industrial and Engineering Chemistry Research*. 48: 142–147.

Westervelt, R. (2001) Suppliers benefits from an oil refining boom. Chemical Week. 163(11): 30–32.

Wild, P.J., Nyqvist, R.G., Bruijn, F.A., Stobbe, E.R. (2006) Removal of sulphur-containing odorants from fuel gases for fuel cell-based combined heat and power applications. *Journal of Power Sources*. 159: 995–1002.

Wittanadecha, W., Laosiripojana, N., Ketcong, A., Ningnuek, N., Praserthdam, P., and Assabumrungrat, S. (2014) Synthesis of Au/C catalysts by ultrasonic-assisted technique for vinyl chloride monomer production. *Engineering Journal*. 8(3): 65–71.

Wu, L., Jiao, D., Wang, J., Chen, L., Cao, F (2009) The role of MgO in the formation of surface active phases of CoMo/Al_2O_3–MgO catalysts for hydrodesulfurization of dibenzothiophene. *Catalysis Communications*. 11: 302–305.

Wu, Z., and Ondruschka, B. (2010) Ultrasound-assisted oxidative desulfurization of liquid fuels and its industrial application. *Ultrasonics Sonochemistry*. 17(6):1027–1032.

Xiao, F., Meng, X. (2011) Hierarchically structured porous materials: from nanoscience to catalysis, separation, optics, energy, and life science. edited by B.-L. Su, C. Sanchez, and X.-Y. Yang, Wiley-VCH Verlag GmbH and Co. KGaA, Weinheim, Germany.

Xu, J., Huang, T., Fan, Y. (2017) Highly efficient NiMo/SiO_2-Al_2O_3 hydrodesulfurization catalyst prepared from gemini surfactant-dispersed Mo precursor. *Applied Catalysis B: Environmental*. 203: 839–850.

Xu, W.Q, Xiong, C.Q, Zhou, G.L, Zhou, H.J. (2008) Removal of sulfur from FCC gasoline by using Ni/ZnO as adsorbent. *Acta Petrolei Sinica*. 24(6): 739–743. (Petroleum Processing Section).

Xu, Y., Liu, C. (2004) The hydrodesulfurization of 4-methyldibenzothiophene and dibenzothiophene over sulfided Mo/γ-Al_2O_3. Preprints of Papers-American Chemical Society, Division of Fuel Chemistry. 49(2): 990–993.

Yamada, K.O., Morimoto, M., Tani, Y. (2001) Degradation of dibenzothiophene by sulfate-reducing bacteria cultured in the presence of only nitrogen gas. *Journal of Bioscience and Bioengineering*. 91(1): 91–93.

Yamamoto, T., Chaichanawong, J., Thongprachan, N., Ohmori, T., Endo, A. (2011) Adsorptive desulfurization of propylene derived from bio-ethanol. Adsorption. 17: 17–20.

Yang, R.T. (2003). Adsorbents: Fundamentals and Applications. John Wiley & Sons, Inc., Hoboken, New Jersey, USA.

Yang, R.T., Takahashi, A., Yang, F.H. (2001) New sorbents for desulfurization of liquid fuels by π-complexation. *Industrial and Engineering Chemistry Research*. 40(26): 6236- 6239.

Yang, Y., Lu, H., Ying, P., Jiang, Z., Li, C. (2007) Selective dibenzothiophene adsorption on modified activated carbons. *Carbon*. 45(15): 3042–3044.

Yang, Y-Z., Liu, X-G., Xu B-S. (2014) Recent advances in molecular imprinting technology for the deep desulfurization of fuel oils. *New Carbon Materials*. 29(1): 1–13.

Yankov, V., Stratiev, D., Yalamov, A. (2011) Integration of The Processes FCC Feed Hydrotreatment and FCC Gasoline Posttreatment Through the Prime

G Process – Opportunity for Production of Euro V Gasolines at Increased Profitability. 45th International Petroleum Conference, June 13, 2011, Bratislava, Slovak Republic.

Yao, X.-Q., Wang, S.-J., Ling, F.-X., Li, F.-F., Zhang, J., Ma, B. (2004) Oxidative desulfurization of simulated light oil. *Journal of Fuel Chemistry and Technology*. 32(3): 318–322.

Yao, Y., Zhang, H., Lu, Y., Zhi, Y., Lu, S. (2016) Catalytic performance of supported Eu/phosphomolybdic acid modified mesoporous silica in the oxidative desulfurization of dibenzothiophene. *Reaction Kinetics, Mechanisms and Catalysis*. 118: 621–632.

Yazu, K., Yamamoto, Y., Furuya, T., Miki, K., Ukegawa, K (2001) Oxidation of dibenzothiophenes in an organic biphasic system and its application to oxidative desulfurization of light oil. *Energy Fuels*. 15(6): 1535–1536.

Yazu, K., Yamamoto, Y., Furuya, T., Mild, K., Ukegawa, K. (2003) Oxidation of dibenzothiophenes in an organic biphasic system and its application to oxidative desulfurization of light oil. *Energy anf Fuels*. 15: 1535–1536.

Yeetsorn R., Tungkamani S. (2014) Potential activity evaluation of CoMo/Al_2O_3-TiO_2 catalysts for hydrodesulfurization of coprocessing bio-oil. KMUTNB: *International Journal of Applied Science and Technology*. 7(4): 35–45.

Yi D., Huang H., Shi L. (2013) Desulfurization of model oil via adsorption by copper (II) modified bentonite. *Korean Chemical Society*. 34(3): 777–782.

Yin C., Zhu G., Xia D. (2002a) Determination of organic sulfur compounds in naphtha. Part I. Identification and quantitative analysis of sulfides in FCC and RFCC naphthas. American Chemical Society Preprints-Division Petroleum Chemistry. 47: 391–395.

Yin C., Zhu G., Xia D. (2002b) Determination of organic sulfur compounds in naphtha. Part II. Identification and quantitative analysis of thiophenes in FCC and RFCC naphthas. American Chemical Society Preprints-Division Petroleum Chemistry. 47: 398–401

Yin, Y., Xue, D.-M., Liu, X.-Q., Xu, G., Ye, P., Wu, M.-Y., Su, L.-B. (2012) Unusual ceria dispersion formed in confined space: a stable and reusable adsorbent for aromatic sulfur capture. *Electronic Supplementary Material (ESI) for Chemical Communications*. S1-S18.

Yu, F., and Wang, R. (2013) Deep oxidative desulfurization of dibenzothiophene in simulated oil and real diesel using heteropolyanion-substituted hydrotalcite-like compounds as catalysts. *Molecules*. 18: 13691–13704.

Yun, G.-N., Lee, Y.-K. (2013) Beneficial effects of polycyclic aromatics on oxidative desulfurization of light cycle oil over phosphotungstic acid (PTA) catalyst. *Fuel Processing Technology*. 114: 1–5.

Zaid, H.F.M., Kait, C.F., Abdul Mutalib, M.I. (2016) Preparation and characterization of Cu-Fe/TiO_2 photocatalyst for visible light deep desulfurization. *Malaysian Journal of Analytical Sciences*. 20(4): 713 – 725.

Zaki T., Saed D., Aman D., Younis S.A., Moustafa Y.M. (2013) Synthesis and characterization of MFe_3O_4 sulfur nanoadsorbants. *Journal of Sol-Gel Science and Technology*. 65: 269–276.

Zdrazil, M., Klicpera, T. (2001) High surface area MoO_3/MgO: preparation by reaction of MoO_3 and MgO in methanol or ethanol slurry and activity in hydrodesulfurization of benzothiophene. *Applied Catalysis A: General.* 216: 41–50.

Zhang, C., Zhang, J., Zhao, Y., Yonggeng, J., Sun, J., Wu, G. (2016) Study on the preparation and catalytic activities of SO_4^{2-} promoted metal oxide solid superacid catalysts for model oil desulfurization. *Catalysis Letter.* 146(7): 1256–1263.

Zhang, H., Meng, X., Li, Y., Lin, Y.S. (2007) MCM-41 overgrown on Y composite zeolite as support of Pd-Pt catalyst for polyaromatic compounds hydrogenation Industrial and Engineering Chemistry Research. 46(12): 4186–4192.

Zhang, Q., Jiao, Q.Z. Min, E. (2005) Iso-butane/1-butene alkylation reaction and deactivation of superacid (SO_4^{2-}/ZnO_2) catalysts. *Chemical Journal of Chinese Universities.* 26: 1130–1132.

Zhang, Y., Yang, Y., Han, H., Yang, M., Wang, L., Zhang, Y., Jiang, Z., Li, C. (2012) Ultra-deep desulfurization via reactive adsorption on Ni/ZnO: The effect of ZnO particle size on the adsorption performance. *Applied Catalysis B: Environmental.* 119–120: 13–19.

Zhao L., Chen Y., Gao J., and Chen Y. (2010) Desulfurization mechanism of FCC gasoline: A review. *Frontiers of Chemical Engineering in China.* 4(3): 314–321.

Zhao, D., Liu, R., Wang, J., Liu, B. (2008) Photochemical oxidation–ionic liquid extraction coupling technique in deep desulphurization of light oil. *Energy Fuels.* 22(2): 1100–1103.

Zhao, D., Ren, H., Wang, J., Yang, Y., Zhao, Y. (2007) Kinetics and mechanism of quaternary ammonium salts as phase-transfer catalysts in the liquid–liquid phase for oxidation of thiophene. *Energy and Fuels.* 21(5): 2543–2547.

Zhao, D.Z., and Sun, M.Z. (2009) Oxidative desulfurization of diesel fuel using ultrasound. *Petroleum Science and Technology.* 27 (17): 1943–1950.

Zhao, R., Liu, C., Yin, C., Liang, W. (2001) Support and CoPcTS effects on the catalytic activity and properties of molybdenum sulfide catalysts. *Petroleum Science and Technology.* 19(5/6): 495–502.

Zhao, X., Cheng, W.-C, Rudesill, J.A. (2002) FCC bottoms cracking mechanisms and implications for catalyst design for resid applications, AM-02–53, in: Proceedings of the NPRA 2003 Annual Meeting, San Antonio, TX, March 17–19, 2002.

Zhao, X., Cheng, W.-C., Rudesill, J.A., Wormsbecher, R.F., Deitz, P.S. (2003). Gasoline sulfur reduction catalyst for fluid catalytic cracking process. U.S. Patent 6,635,168. Assigned to W.R. Grace & Co.-Conn.

Zhao, Y., Wang, R. (2013) Deep desulfurization of diesel oil by ultrasound-assisted catalytic ozonation combined with extraction process. *Petroleum and Coal.* 55(1): 62–67.

Zhou, A., Ma, X., Song, C. (2009) Effects of oxidative modification of carbon surface on the adsorption of sulfur compounds in diesel fuel. *Applied Catalysis B: Environmental.* 87: 190–199.

Zhou, X., Zhao, C., Yang, J., Zhang, S. (2007) Catalytic oxidation of dibenzothiophene using cyclohexanone peroxide. *Energy and Fuels.* 21(1): 7–10.

Zhu, W., Xu, Y., Li, H., Dai, B., Xu, H., Wang, C., Chao, Y., Liu, H. (2014) Photocatalytic oxidative desulfurization of dibenzothiophene catalyzed by amorphous TiO_2 in ionic liquid. *Korean Journal of Chemical Engineering*. 31(2): 211–217.

ZoBell, C. E. (1953) Process of removing sulfur from petroleum hydrocarbons and apparatus. US Patent 2641564.

Zongxuan, J., Hongying, L., Yonga, Z., Can, L. (2011) Oxidative desulfurization of fuel oils. *Chinese Journal of Catalysis*. 32(5): 707–715.

3

Biodesulfurization of Natural Gas

List of Abbreviations and Nomenclature

AC	Activated Carbon
ADS	Adsorptive Desulfurization
AHS	Alkali Hydrogen Sulfide
APG	Associated Petroleum Gas
BAC	Biological Activated Carbon
BADS	Bioadsorptive Desulfurization
BBC	Bio-Bubble Column
BET	Brunauer-Emmet-Teller
BFW	Bioler Feed Water
BP	British Petroleum
BTF	Bio-Trickling Filter
BTU	British Thermal Unit
CFSTR	Continuous Flow Stirred Tank Reactor
CHP	Combined Heat And Power
CMC	Carbonoxylmethyl Cellulose
COS	Carbonyl Sulfide
CS_2	Carbon Disulfide

CSTR	Continuous Stirred Tank Reactor
DEA	Diethanolamine
DGA	Diglycolamine
DIPA	Diisopropylamine
DMDS	Dimethyl Disulfide
DMS	Dimethyl Sulfide
DO	Dissolved Oxygen
DR	Dubinin-Radushkevich
EC	Elimination Capacity
EDTA	Ethylenediaminetetraacetic Acid
ETS-2	Engelhard Titanosilicate-2
FBTB	Fixed-Bed Trickling Bioreactor
FGD	Flue Gas Desulfurization
FIC	Flow Influent Controller
GAC	Granular Activated Carbon
gal/ft^3	US Gallon/Cubic Feet
GHSV	Gas Hourly Space Velocity
GSB	Green Sulfur Bacteria
HPS	High Pressure Steam
HRT	Hydraulic Retention Time
HTN	*Halothiobacillus Neapolitanus* NTV01
IEA	International Energy Agency
IFP	Institut Français Du Petrole
IRB	Iron Reducing Bacteria
LEDs	Light Emitting Diodes
L_{high}	$Level_{high}$
LIC	Level Influent Controller
L_{low}	$Level_{low}$
LNG	Liquefied Natural Gas
LOR	Limited Oxygen Route
LPC	Low Pressure Steam
MDEA	Methyldiethanolamine
MDU	Micro-Aerobic Desulfurization Unit
MEA	Monoethanolamine
MIP	Molecular Imprinting
MM	Methyl Mercaptan
MnOx	Manganese Oxides
MT	Methanethiol
NGL	Natural Gas Liquification
NTA	Nitrilotriacetic Acid
OECD	Organisation For Economic Co-Operation And Development
OFA	Oil Fly Ash
OPUF	Open-Pore Polyurethane Foam
ORP	Oxidation Reduction Potential

p.a.	Per Annum
PC	Personal Computer
PID	Proportional + Integral + Derivative
ppb	Parts Per Billion
ppm	Parts Per Million
PVB	Polyvinylbutyral
Py-IR	Pyridine Adsorption Infrared Spectroscopy
redox	Reduction-Oxidation
RH	Rice Husk
RSM	Response Surface Methodology
S/N	Sulfide/Nitrate
S-BTF	Single-Stage Bio-trickling Filters
SCOT	Shell Claus Off-Gas Treating
SEM	Scanning Electron Microscope
SMIP	Surface Molecular Imprinting
SO_2	Sulfur Dioxide
SOB	Sulfide-Oxidizing Bacteria
SRB	Sulfate Reducing Bacteria
SRU	Sulfur Recovery Unit
SWS	Sour Water Stripper
T-BTF	Triple-Stage Bio-trickling Filters
tcf	Trillion Cubic Feet
TGA	Thermogravimetric Analyzer
TG-MS	Thermogravimetric-Mass Spectroscopy
TGTU	Tail-Gas Treatment Unit
UASB	Up-Flow Anaerobic Sludge Bed
USCM	Unreacted Shrinking Core Model
VOSCs	Volatile Organic Sulfur Compounds
XRD	X-Ray Diffractometer

3.1 Introduction

Hydrocarbon gases in a reservoir are called a natural gas or simply a gas. Natural gas is found in several different types of rocks including sandstone, coal seams, and shales. Thus, the generic term ***natural gas*** applies to gases commonly associated with petroliferous (petroleum-producing, petroleum-containing) geologic formations. Natural gas occurs under pressure in underground cavities. It can be found in petroleum reservoirs or in other reservoirs as the sole occupant. The term natural gas refers to hydrocarbon-rich gas; it is a gaseous fuel that is found in oil fields, gas fields, and coal beds (Speight, 2007; Carroll, 2010).

Natural gas generally contains high proportions of methane, is the lightest hydrocarbon (a single-carbon hydrocarbon compound), and is from the group of paraffins (CH_4); its content may range from 50 to 90%. Some of the higher molecular weight paraffins (C_nH_{2n+2}) generally contain up to six carbon atoms, such as ethane, propane, butane, pentane, and, all the isomers of butane and pentane may also be present in small quantities (Table 3.1). The hydrocarbon constituents of natural gas are combustible, but non-flammable, non-hydrocarbon components can be also found in small amounts and are regarded as contaminants. There is no single composition of components which might be termed typical natural gas. Before its refining, carbon dioxide (CO_2), hydrogen sulfide (H_2S), and mercaptans (thiols; R–SH), as well as trace amounts of other constituents may be present. Sulfur content is among the most important characteristics of natural gas, as well as of crude oil. Sulfur in natural gas is usually found in the form of hydrogen sulfide (H_2S), which can reach up to 30% by volume. However, it should be noted that corrosive hydrogen sulfide can make natural gas extremely toxic (Chapter 1) and must be fastidiously removed throughout natural gas treatment. Moreover, H_2S causes an irritating, rotten egg smell in concentrations above 1 ppm and at concentrations above 10 ppm the toxicological exposure limits are exceeded (Chapter 1).

Methane and ethane constitute the bulk of the combustible components, while carbon dioxide (CO_2) and nitrogen (N_2) are the major noncombustible (inert) components. However, in its purest form, such as the natural gas that is delivered to the consumer, it is almost pure methane. Trace amounts of rare gases, such as helium, may also occur and certain

Table 3.1 Natural Gas Composition (Heinz, 2008).

Components	Concentration
Methane CH_4	80–95%
Ethane C_2H_6	2–5%
Propane C_3H_8	1–3%
Butane C_4H_{10}	0–1%
C5 Alkanes and Higher Hydrocarbon	0–1%
Carbon Dioxide CO_2	1–5%
Nitrogen N_2	1–5%
Hydrogen Sulfide H_2S	0–6%
Oxygen O_2	0–0.2%
Helium	0–1%

natural gas reservoirs are a source of these rare gases. Just as petroleum can vary in composition, so can natural gas. Differences in natural gas composition occur between different reservoirs and two wells in the same field may yield gaseous products that are different in composition. Natural gas can be classified according to its constituents (Table 3.2) (Mokhatab *et al.*, 2006; Speight, 2007, 2014; El-Gendy and Speight, 2016).

There are several general definitions that have been applied to natural gas. Thus, lean gas is gas in which methane is the major constituent. Dry natural gas contains <0.1 gallon (1 gallon, US, = 264.2 m^3) of gasoline vapor (higher molecular weight paraffins) per 1000 ft^3 (1 ft^3 = 0.028 m^3). Wet gas contains considerable amounts of the higher molecular weight hydrocarbons (paraffins), in fact, more than 0.1 gal/1000 ft^3. Sour gas contains hydrogen sulfide, whereas sweet gas contains very little, if any, hydrogen sulfide. Residue gas is natural gas from which the higher molecular weight hydrocarbons have been extracted and casing head gas is derived from petroleum, but is separated at the separation facility at the wellhead (El-Gendy and Speight, 2016).

Associated or dissolved natural gas occurs either as free gas or as gas in the solution in the petroleum. Gas that occurs as a solution in the petroleum is dissolved gas, whereas the gas that exists in contact with the petroleum (gas cap) is associated petroleum gas (APG), while the non-associated gas is never linked to another product (El-Gendy and Speight, 2016).

In addition, natural gas condensate is a low-density mixture of hydrocarbon liquids that are present as gaseous components in the raw natural gas produced from many natural gas fields (Mokhatab *et al.*, 2006; Speight, 2014). As the name implies, the higher molecular weight constituents (typically hydrocarbon such as pentane, hexane, heptane, octane, and even nonane,

Table 3.2 Classification of Natural Gas According to its Composition (vol.%) (Rojey, 1997).

	Sweet dry gas (non-associated)	**Sour dry gas (non-associated)**	**Sweet wet gas (associated)**	**Sour wet gas (associated or condensate gas)**
Category	1	2	3	4
Ethane and Higher Hydrocrbons	<10	<10	>10	>10
H_2S	<1	>1	<1	>1
CO_2	<2	>2	<2	>2

and decane) condense out of the raw gas if the temperature is reduced to below the hydrocarbon dew point temperature of the raw gas. The natural gas condensate is also referred to as simply *condensate*, or *gas condensate*, or sometimes natural gasoline because the hydrocarbon constituents fall within the gasoline boiling range. Additionally, condensate may contain additional impurities such as hydrogen sulfide (H_2S), thiols (RSH, also called mercaptans), and carbon dioxide (CO_2) (El-Gendy and Speight, 2016).

There are three sources for raw natural gas: crude oil, gas, and condensation wells. Natural gas that comes with crude oil is generally referred to as associated gas or wet natural gas. Natural gas created from gas and condensation wells, during which there is very little or no oils termed, is non-associated or free gas. Gas wells differ from condensation wells, since the former type produces raw natural gas only, while the latter produces natural gas along with very light liquid hydrocarbon, known as natural gasoline as a results of its high octane number.

The first drilling of a gas well was done by William Hart, the "Father of Natural Gas", in 1821 in Fredonia, United States. Natural gas was discovered as a result of prospecting for crude oil drilling. Throughout the 19th century, natural gas was used locally as a source of light due to the lack of a safe structure for long-distance gas transport. After World War II, natural gas was extensively used because of the advances in engineering that allowed the construction of safe, reliable, long-distance pipelines for gas transportation (Speight, 1993, 2007; Kidnay *et al.*, 2011).

In its pure state, natural gas is colorless, shapeless, and odorless. It is a combustible gas and it gives off a significant amount of energy when burned (Speight, 2015). It is considered to be an environmentally friendly, clean fuel when compared with other fossil fuels such as coal and crude oil. The combustion of fossil fuels, other than natural gas, leads to the emission of enormous amounts of compounds and particulates that have negative impacts on human health (EIA, 1999).

Since natural gas is a reliable, cleaner-burning fuel, flexible, and plentiful, it underpins growing domestic and export production sectors. Moreover, natural gas is a very safe source of energy once transported, stored, and used. Consequently, now days it is used to generate electricity and to power appliances such as heaters and stoves. Not only for residential, commercial, and industrial heating, but also in many industrial processes including making fertilizers, glass, steel, plastics, paint, fabrics, ammonia production, and many other products. In the petrochemical industry, it is used as a feedstock or raw material, e.g. in the production of ethylene. Hydrogen, sulfur, and carbon black can be also produced using natural gas (Schoell, 1983; Mokhatab *et al.*, 2006; Tabak, 2009; Faramawy *et al.*, 2016).

Globally, natural gas accounts for 23.7% of primary energy consumption (BP Statistical Review of World Energy, 2015). The expected growth of the worldwide natural gas demands is 1.9% each year over the BP Energy Outlook (2015). Demand for natural gas will rise nearly 50% to 190 tcf in 2035, compared to 130 tcf for now. Gas demand in the forecasting period will be mainly driven by non-OECD countries, with growth averaging 3% p.a. to 2030, with the greatest demand coming from Asia (4.6% p.a.) and the Middle East (3.9% p.a.). Of the major sectors globally, growth is fastest in power (2.6% p.a.) and industry (2% p.a.) which matches with historic patterns. The usage by the power and industrial sectors accounts for over 80% of the expansion within the world demand for natural gas. Furthermore, compressed natural gas used in transport is confined to 2% of global transport fuel demand in 2030, with a three-fold increase from today's level (BP, 2016).

Natural gas proved reserves in the Middle East and the Europe & Eurasian regions account for 75% of whole world's reserves (Figure 3.1). However, 40% of the world's natural or associated gas reserves, currently identified as remaining to be produced, representing over approximately 2600 trillion cubic feet (tcf), are sour with both H_2S and CO_2 present most of the time. Among these sour reserves, more than 350 tcf contain H_2S in excess of 10% and almost 700 tcf contain over 10% CO_2 (Lallemand *et al.*, 2012; International Energy Agency, IEA, 2013; BP, 2016).

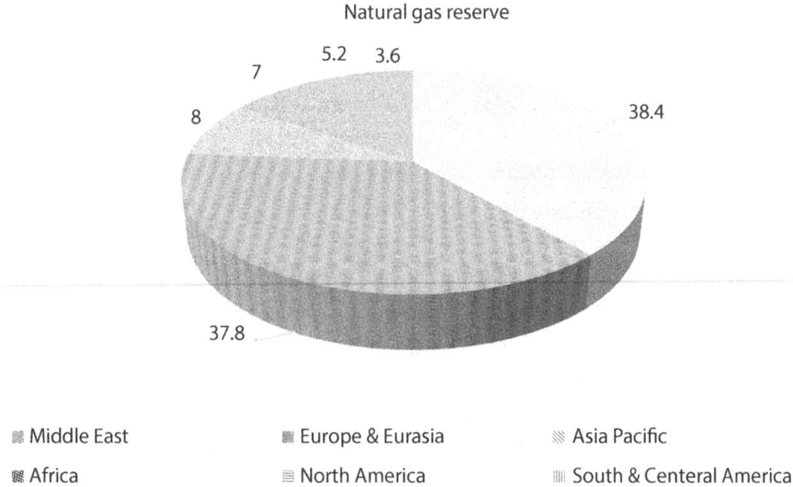

Figure 3.1 Distribution of Proved Reserves of Natural Gas in 2011, as Reported by (Duissenov, 2013).

It is worthy to know that natural gas can also be formed through the degradation of organic matters by microorganisms, such as methanogens, that biologically break down organic matters to produce methane (***bio-gas***). They are found in areas near the surface of the earth that are devoid of oxygen. The produced methane, therefore, may sometimes escape into the atmosphere. However, sometimes, it may be trapped and probably recovered. Notably, methanogens also live in the intestines of most animals, as well as humans. In this regard, the production of methane (biogas) by livestock is a relevant contributor to the release of greenhouse gases into the atmosphere (Heinz, 2008). Thus, biogas is considered a renewable energy that can be used directly as a fuel (its heating value is between 15 and 30 MJ/Nm^3) or as a raw material for the production of synthesis gas and/or hydrogen. Its main constituents are methane (CH_4) and carbon dioxide (CO_2), but biogas, like natural gas, also contains significant quantities of undesirable compounds (i.e. contaminants), such as hydrogen sulfide (H_2S), ammonia (NH_3), and siloxanes (Table 3.3). The percentage of those contaminants depends on the biogas source (Kwaśny *et al.*, 2015; Barbusiński and Kalemba, 2016). Table 3.4 summarizes the sulfur (S) content of some typical biogas substrates in percent of fresh matter (Naegele *et al.*, 2013).

Typically, methane content in biogas was reported to range between 35 and 75% vol. (Kwaśny *et al.*, 2015). However, it is also reported that the biogas mainly comprises of methane (60–70 %) and carbon dioxide (30–40 %) (Wu *et al.*, 2016). The H_2S content can vary between 100 to about 10,000 ppmv (0.0001–1 %vol) (Abatzoglou and Boivin, 2009). In other studies it is reported to range between 1–12 g/m^3 (Wu *et al.*, 2016) and 0.1–3 % (Li *et al.*, 2016). Naegele *et al.* (2013) reported that critical H_2S can be formed in biogas under anaerobic conditions by sulfur and sulfate reducing bacteria from animal manure and renewable energy crops, while Yang *et al.* (2015) attributed the formation of H_2S during the anaerobic digestion process by specific microorganisms, such as sulfate-reducing bacteria (SRB), due to the existence of sulfur-containing compounds in substrates. Sometimes, mercaptans (such as CH_3SH) are detected in biogas, as they result from the anaerobic fermentation of S-bearing organic molecules (i.e., proteins). Thus, one of the main barriers that slow down the application and commercialization of biogas as an energy source is the existence of H_2S gas during its production process. Same as natural gas, these contaminates should be removed, as they are toxic, corrosive, have bad smell, and generate harmful environmental emissions, for example SO_2 and H_2SO_4. In a hydrocarbon-based fuel cell system, as one of the most potential and suitable energy conversion devices for generating electricity

Table 3.3 Biogas Composition Produced by the Anaerobic Digestion of Different Substrates. http://www.biogas-renewable-energy.info/biogas_composition.html.

Component	Unit	Domestic waste	Sludge from wastewater treatment plants	Sewage*	Agricultural wastes	Wastes from agro-food industry
CH_4	% vol	50–60	60–75	55–65	60–75	68
CO_2		34–38	19–33	35–45	19–33	26
N_2		0–5	0–1	<1	0–1	ns
O_2		0–1	<0.5		<0.5	ns
H_2O	% vol (at 40 °C)	6	6		6	6
H_2S	mg/m^3	100–900	1000–4000	1000–4000	3000–10,000	100
NH_3		ns	ns		50–100	400
Aromatic Compounds		0–200	ns		ns	ns
Organic Halogenated or Fluoroorganic Compounds		100–800	ns		ns	ns

ns: not studied; *Barbusiński and Kalemba (2016).

for both mobile vehicles and stationary power plants including residential applications, the catalysts used, such as reforming catalysts and water–gas-shift catalysts are poisoned by the produced H_2S (Song, 2002). Biogas can be used for power generation, as fuel for vehicles, or for incorporation into the natural gas network, but H_2S concentration should be no more than 300, 15, and 20 mg/m^3, respectively (Wu *et al.*, 2016).

Biogas, natural gas, synthesis gas, and Claus process tail gas may contain H_2S. For large-scale (>15 tons S/ day) gas treatment, amine absorbers and Claus plants can be used to remove H_2S. For smaller quantities, liquid redox systems, based on reaction with iron chelates, are used. In recent years, biological techniques have been applied more frequently in waste gas treatment systems because they eliminate many of the drawbacks of the classical physico-chemical techniques (Madox, 1985). Some of the disadvantages of these classical methods for gas treatment are that they require relatively large investments and operational costs (e.g. special chemicals, equipment corrosion, high pressures, and temperatures) and they require special operational safety and health procedures (Gadre, 1989). In addition, they may produce waste products (e.g. spent chemical solutions or spent activated carbon). On the other hand, biodesulfurization (BDS) processes can proceed at around ambient temperatures and atmospheric pressure, thus eliminating the need for heat and pressure, reducing energy costs to a minimum. Moreover, different kinds of bacteria proved high efficiency in different bioreactor setups (Sublette and Sylvester, 1987a,b; Cline *et al*, 2003; Amirfakhri *et al*, 2006; Ma, *et al*, 2006; Syed *et al*, 2006; Aroca, *et al*, 2007). H_2S can be converted into elemental sulfur (S^0) via microbial processes. Various groups of organisms can oxidize sulfur compounds under aerobic or anaerobic conditions and reduce them to S^0.

Table 3.4 Sulfur Content in % of Fresh Matter and in g/kg for Typical Biogas Substrates. Naegele *et al.* (2013)

Source	**S-content (% fresh matter)**
Maize Silage	0.05–0.07
Grassland	0.05–0.08
Winter Oilseed Rape	0.061–1.14
Source	**S-content (g/kg)**
Cattle Manure	0.7–0.8
Poultry Manure	2–8–3.2
Cattle Slurry	4–6
Pig Slurry	6–7

Energy has become a crucial factor for humanity to continue economic growth and maintain a high standard of living, especially after the inauguration of the 55 industrial revolutions in the late 18th and 19th centuries (International Energy Agency EIA, 2011). According to the IEA report, the world will need 50% more energy in 2030 than today of which 45% will be accounted by China and 57% by India (Shahid and Jama, 2011; Atabani *et al.*, 2011). Consequently, there is a worldwide increase and demand for renewable energy sources. That comes with the increase in awareness about environmental protection. As a complementary or alternative for natural gas, biogas has attracted considerable attention within the scientific community (Mao *et al.*, 2015; Horváth *et al.*, 2016). For that reason, this chapter will cover, briefly, the research efforts on biodesulfurization of both natural gas and biogas.

3.2 Natural Gas Processing

The term gas processing is generally used to cover CO_2 removal, H_2S removal, water removal, hydrocarbon dew pointing, and gas sweetening. Gas sweetening may be a generic term for sulfur removal. Sometimes the term 'gas conditioning' is employed rather than gas processing. The actual practice of natural gas processing can be quite complex, but usually involves four main processes to remove the associated impurities oil and condensate removal, water removal, separation of natural gas liquids, and sulfur and carbon dioxide removal, as shown in Figure 3.2.

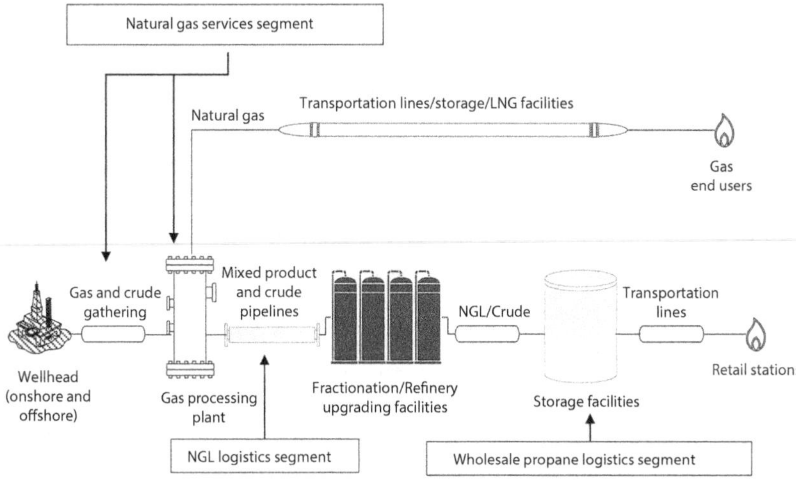

Figure 3.2 General Scheme of the Natural Gas Industry.

Carbon dioxide, hydrogen sulfide, and other contaminants are usually found in natural gas streams. CO_2, when combined with water, creates carbonic acid which is corrosive. CO_2 additionally reduces the British thermal unit (BTU) value of gas and, in concentrations of more than 2% or 3%, the gas is unmarketable. H_2S is an extremely toxic gas, causing a lot of environmental and health problems and it is also hugely corrosive to equipment (Chapter 1). The recovered hydrogen sulfide gas stream may be vented to atmosphere, flared in waste gas flares or modern smokeless flares, incinerated for sulfur removal and utilized for the production of elemental sulfur or sulfuric acid. If the recovered H_2S gas stream is not to be utilized as a feedstock for commercial applications, the gas is sometimes passed to a tail gas furnace during which the H_2S is oxidized to SO_2 and is then passed to the atmosphere out a stack.

Currently, the primary method of SO_2 disposal from flue gas is a scrubbing process with calcium-based sorbents (Bo *et al.*, 2007), which results in huge amounts of waste. Currently, other adsorbents are used, for example CuO supported on alumina, titania, and zirconia showed a direct relationship between the adsorption capacity of the sorbents and the surface area of the support. Also, the capacity and efficiency of the materials were directly related to the CuO content. Finally, increasing the temperature enhanced the adsorption capacity and the efficiency of the sorbents (Flores *et al.*, 2008). The Cu supported on alumina ($CuO/\gamma\text{–}Al_2O_3$) is reported to express higher ADS of simulated flue gas with high amounts of SO_2, relative to individual CuO and $\gamma\text{–}Al_2O_3$ (Qing-chun *et al.*, 2015).

$$Al_2O_3 + 3SO_2 + 3/2\,O_2 \longrightarrow Al_2(SO_4)_3$$

$$CuO + SO_2 + 1/2\,O_2 \longrightarrow CuSO_4$$

Moreover, the Ni impregnation on activated carbon and acid treated activated carbons was reported to enhance the ADS of SO_2 because nickel oxides can easily change their valence state, promote the redox cycle, and have also good oxidation performance because of their variable *d* electronic structure. Apart from the supported phase, carriers strongly affect the surface morphology and electronic structure of the metal particle, which can directly affect the stability and catalytic activity of catalysts (Efremenko and Sheintuch, 2003). Upon the ADS of SO_2 from a simulated gaseous mixture in a fixed bed flow microreactor, Ni- catalysts supported on untreated and acid treated coal-activated carbons were prepared by excessive impregnation method. The ADS activity of the prepared catalysts decreased in the following order: $Ni/AC\text{–}HNO_3 > Ni/AC\text{–}H_2SO_4 > Ni/AC > AC$ (Guo *et al.*, 2012). This was related to the specific surface area of those catalysts, where the specific surface area of the original activated carbons was 723 m^2/g, which was less than that of the activated carbons treated with nitric (831 m^2/g) and sulfuric acid (803 m^2/g). The increment of specific surface area with acid treatment is due to the elution of some impurities on the AC by acid, especially with HNO_3 because of its high oxidative ability. Although, the specific surface area of Ni/AC was lower than that of AC, recording 697 m^2/g, due to the presence of Ni, the ADS increased. The decrease in specific surface area is due to the presence of Ni species on activated carbons, which blocks some of the microporous surfaces. Moreover, acid treatment is known to affect the surface of functional groups of AC. Guo *et al.* (2012) noticed that the AC treated with nitric acid and sulfuric acid had significant increments in the less acidic C=O and C-O functional groups, which contributed in the enhancement of the ADS performance of the catalysts. Ni and NiO species coexist as microcrystals on the activated carbons before desulfurization, but after desulfurization Ni and NiO species disappear and Ni_2O_3 is observed, indicating that Ni species could be involved in the desulfurization reaction.

$$Ni + H_2O + SO_2 + O_2 \longrightarrow NiO + H_2SO_4$$

$$2NiO + H_2O + SO_2 + O_2 \longrightarrow Ni_2O_3 + H_2SO_4$$

$$Ni_2O_3 \xrightarrow{H_2SO_4} NiO + Ni + O_2$$

H_2SO_4 can be washed with excess condensed H_2O to recover the active sites and Ni_2O_3 can be retransformed into Ni and NiO and release O_2 in the H_2SO_4 medium. Thus, a new cycle of adsorption and oxidation of SO_2 and production and elution of H_2SO_4 begins. This continuous operation makes the removal of SO_2 possible.

The simultaneous desulfurization and denitrification are significant trends in the field of flue gas purification to reduce the cost of flue gas purification. Yi *et al.* (2007) reported the preparation of highly active absorbent fly ash, lime, and a few oxidizing $NaClO_2$ additives in which 93.7% and 65.5% removal of SO_2 and NO were achieved, respectively, at the optimal reaction temperature, additive quantity, humidity, and Ca/(S + N) molar ratio with this process shown to be approximately 60, 1.6%, 4.46%, and 1.2, respectively. The orthogonal experiments indicate that Ca/(S + N) is the main factor that influences the effectiveness of desulfurization and denitrification, which is followed by temperature and content of the absorbents.

Davini (2001) reported the good adsorption capacity of vanadium-impregnated palm shell activated carbon for SO_2 relative to Ni-impregnated and Fe-impregnated palm shell AC. Sumathi *et al.* (2010) also reported the preparation of metal modified palm shell activated carbon by impregnation with vanadium and cerium metal oxides (V_2O_5 and CeO_2), which expressed good removal of SO_2 and NO in gas streams. This was mainly attributed to the powerful oxidizing activity of cerium and vanadium (that is, the redox conversion between Ce^{3+} and Ce^{4+} and V^{5+} and V^{3+}) and their oxygen storage properties which could bring new oxygen surface groups, such as C–O, C=O, and COOH on the surface of the palm shell AC (Kaspar *et al.*, 1999; Tian *et al.*, 2009) where the maximum recoded adsorption capacity of 121.7 mg SO_2/g cerium modified AC and 3.5 mg NO/g cerium modified AC was at optimum conditions of 10% loading of cerium and at 150 °C. It was also observed that the cerium modified AC expressed higher adsorption capcity towards SO_2 and NO at temperatures ranging 100–300 °C than the vanadium modifed AC. However, it was also noticed that lower temperatures are more favorable for the removal of SO_2, while higher temperatures are more favorable for NO removal. The good removal of NO and SO_2 has been described by adsorption and reduction of cerium and vanadium metal in presence of oxygen, throughout the chemisorption of SO_2 and NO on the catalyst surface, then the transfer of oxygen from the catalytic sites to the carbon active sites and, finally, the desorption of oxygen from the carbon surface (Illan-Gomez *et al.*, 1999; Zhu *et al.*, 2000). Moreover, the capability of cerium and vanadium to develop the reduction of NO to N_2 would have indirectly contributed to the highest breakthrough time of NO in this system.

SO_2 sorption by cerium

$$xSO_2 + CeO_2 \longleftrightarrow xSO_3 + CeO_{2-x}$$

NO reduction by cerium

$$xNO + CeO_{2-x} \longleftrightarrow 1/2\, xN_2 + CeO_2$$

SO_2 sorption by vanadium

$$xSO_2 + V_2O_5 \longleftrightarrow xSO_3 + V_2O_{5-x}$$

While V_2O_5 becomes V_6O_{13} via redox reaction

$$V_6O_{13} + 2\,NO \longleftrightarrow 3\,V_2O_5 + N_2$$

At relatively low temperatures (150–250 °C), metal doped carbons express high catalytic activity due to the disassociation of NO chemisorption, which is accompanied with N_2O and N_2 evolution and oxygen accumulation on the catalyst surface (Mehandjiev *et al.*, 1997). However, CeO_2 capability for the storage and release of O_2, through the redox shift between Ce^{4+} and Ce^{3+} under oxidation and reduction conditions, is low at this relatively low temperature range (Qi and Yang, 2003). This would explain the moderate removal of NO from simulated flue gas at relatively low temperatures (<150 °C). However, at higher temperatures the Ce-impregnated palm shell AC will be in a more active state to oxidize and reduce NO. Not only this, but it is also known that at high temperatures, AC itself could decompose NO and reduce it to N_2 (Mehandjiev *et al.*, 1996). Further, higher temperature (>150 °C) means less water accumulation on the sorbent surface, allowing the metals to be more active. But, at higher temperatures (>250 °C) SO_2 molecules lose their kinetic energies, making the adsorption an exothermic process, which would indirectly lessen the adsorption of SO_2 onto the pores of metal impregnated AC (Guo and Lua, 2002).

Pyrolusite is an ordinary and economical mineral resource, with main metal oxide components of MnO_2, Fe_2O_3, and a few other transition metals. It was used to modify walnut shell-derived column activated carbon by blending method. Then, the activity of the prepared adsorbents was compared with that of MnO_2- and F_2O_3- modified activated carbons to study and investigate the advantages of pyrolusite addition to AC in the development of proper physiochemical properties and desulfurization activity compared to single metal oxide addition (Fan *et al.*, 2013). The desulfurization experiments showed that pyrolusite loaded carbons performed the best toward the removal of SO_2. Upon the optimal dosage of additives, the maximum sulfur capacity of activated carbon loaded by pyrolusite, MnO_2, and Fe_2O_3 were 227.8, 157.8, and 140.6 mg/g, which were 84.0, 27.5, and 13.6% higher than that of blank activated carbon, respectively. Thus, the desulfurization activity of prepared samples with the optimal additive dosage was in the order: pyrolusite-AC5 > MnO_2–AC10 > Fe_2O_3–AC2 > AC (the number after AC, represents the dosage of the loaded metal oxide). The decrease of sulfur capacity with the increase of the additive loading ratio might be related to the pore blockage caused by excess additive loading amounts. However, the remarkable performance of pyrolusite loaded activated carbon is related not only with its better development of texture property, higher surface area, larger pore volume, and relatively larger content of surface functional groups, but also the synergistic catalytic oxidation of SO_2 by manganese, iron, and other transition metals in pyrolusite. It was concluded from that study that the modification of AC from agricultural wastes by pyrolusite using a simple blending method is a very promising technique for improving the sulfur capacity of AC. This is due to its remarkable desulfurization activity, low cost, and relatively easy preparation process.

The proper ratio for metal loading on AC promotes the formation of specific porosity due to the reaction between the added metal oxides and the carbon substrate during carbon gasification process. However, excess addition of metal particles may promote aggregation and the formation of large crystals during preparation which would result in pore blockage and surface area decrease (Lee *et al.* 2002; Nguyen-Thanh and Bandosz 2003). Fan *et al.* (2014) reported the preparation of metal oxide modified activated carbon from walnut shell. The transition metal oxides powders (Co_2O_3, Ni_2O_3, CuO and V_2O_5) were blended with walnut shell carbon chars in presence of hot coal tar, which served as the major binder, and carboxymethyl cellulose (CMC) and/or polyvinyl butyral resin (PVB), which served as accessorial binders, then were activated at 900°C in the presence of CO_2 for 2 h. The prepared column activated carbons were named AC, AC-Co, AC-Ni, AC-Cu, and AC-V based on the metal oxides that were used for modification. A weight percentage attached refers to the metal oxide dosage. The pure AC prepared from walnut shells proved a good ADS capacity of 123.8 mg SO_2/g. The best metal oxide loading percentage was 2% and recorded an ADS capacity of 133.3, 147.5, 172.4, and 181.2 mg SO_2/g of AC-Co2, AC-Ni2, AC-Cu2, and AC-V2, respectively (where 2 refers to the percentage of metal oxide loading, i.e. 2%), which were 7.7, 19.1, 39.3, and 46.4 % higher than the AC ADS sulfur capacity, respectively. It is worth noting that for the 2 wt.% metal loading, the change of surface area and pore volume of AC was not evident. Thus, the development of texture properties was not the main reason for their better ADS. Thus, the increase in ADS capacity was related to the improved generation of oxygenate functional groups on the carbon surface with metal oxide additions. Also, the increment of basic functional groups is beneficial for the SO_2 adsorption process on AC due to the binding affinity. Thus, since AC-Cu2 possessed the highest basic functional group content, it exhibited good desulfurization activity. Moreover, the catalytic cycle depends on the change between chemical valences of the metal species. The complex metal species component in AC-V2 could partially explain its higher sulfur capacity compared to the other metal modified activated carbons. Thus, the improvement of sulfur capacity mainly resulted from the additional basic functional groups that were generated after modification and the high catalytic activities of the active metal oxides that were formed in the presence of O_2 during the desulfurization process. According to Doornkamp and Ponec (2000), the oxidation state of the added metal oxides is partly reduced during carbon activation. In the oxidative catalytic reaction, the gaseous oxygen is first absorbed onto these reduced and intermediate state metal species and converted into lattice oxygen. Active metal oxides with much higher catalytic activity are then formed by the reaction between the reduced and intermediate state metal species and the lattice oxygen atoms. For example, for the V_2O_5 modified AC V^{3+}, V^{2+}, and V^{+} species coexist on the carbon matrix after activation (Li *et al.*, 2009). Next, these active metal oxides reacted with SO_2 in the desulfurization system to form SO_3. This process was followed by H_2SO_4 generation with H_2O. However, the resulting metal species in their low oxidation state would be oxidized by O_2 in the desulfurization system to form active metal oxides again. Briefly, this catalytic cycle depends on the change between chemical valences of the metal species.

$$MO_{x-n} + O_2 \longrightarrow MO_{x-n} + O^{2-}_{(lattice)} \longrightarrow MOx \longrightarrow MO_{x-n} + O_{(ads)}$$

$$SO_{2(ads)} + O_{(ads)} \longrightarrow SO_{3(ads)} \xrightarrow{H_2O} H_2SO_4$$

Song *et al.* (2014) examined the feasibility of the application of activated carbon from waste tea in desulfurization and denitrification of gas stream. The results showed that less graphitization, lower microspore size, and a more nitrogenous basic group of the adsorbent enhanced its desulfurization ability. When well-developed mesopores were present in the adsorbent, the NO removal efficiency was decreased while more nitrogenous basic groups promoted the removal rate of NO. However, upon the simultaneous removal of SO_2 and NO, competing adsorption occurred. The removal efficiencies of SO_2 and NO increased in the presence of oxygen and steam in the flue gas. The adsorption of SO_2 and NO onto waste tea activated carbon, in absence of oxygen and steam, was physosoprtion, while the steam and oxygen promoted the chemical adsorption of SO_2.

Xiao-Li *et al.* (2014) reported the preparation of Fe catalysts supported on activated carbon, untreated and treated by HNO_3, using an ultrasound-assisted incipient wetness method. The desulfurization activity of the prepared catalysts was evaluated at a fixed bed reactor under a simulated flue gas. The results revealed that the adsorption capacity of the AC to SO_2 was 74 mg/g, which increased upon acid treatment to reach 84 mg/g. That was attributed to the increase of surface oxygen functional groups on activated carbon after HNO_3 treatment. The Fe-loading further increased the activity of the AC to reach 244 mg SO_2/g. This was related to iron species (active site) which promote the transformation of the adsorbed SO_2 into other stable forms. The ADS capacity was further increased for Fe-loaded-acid treated AC (Fe/AC-HNO_3) reaching 322 mg/g, owing to the different functional groups. The increase of mesopores volume is in the order AC < AC-HNO_3 < Fe/AC < Fe/AC-HNO_3. The increase of the mesopores may enhance adsorption abilities of samples, especially for large molecules of adsorbents, as SO_2 molecules. Furthermore, the average pore sizes of Fe/AC (or Fe/AC-HNO_3) slightly increased compared with that of AC (or AC-HNO_3), which was attributed to the aggregation of Fe species. The larger pore size provides a fast transporting channel for gas molecules (Li *et al.*, 2013) which is consequently conducive to the SO_2 adsorption. The Fe_3O_4 species were found to be the main active phases in the prepared samples and expressed excellent desulfurization activity at 353 K, but $Fe_2(SO_4)_3$ was detected after desulfurization (Xiao-Li *et al.*, 2014).

The adsorption of SO_2 on the surface of AC, in the presence of O_2 and H_2O, occurs as follows (Bagreev *et al.*, 2002):

$$SO_{2(gas)} \longrightarrow SO_{2(ads)}$$

$$O_{2(gas)} \longrightarrow 2O_{(ads)}$$

$$SO_{2(ads)} + O_{(ads)} \longrightarrow SO_{3(ads)}$$

$$SO_{3(ads)} + H_2O_{(ads)} \longrightarrow H_2SO_{4(ads)}$$

While on the basis of free-radical mechanism (Fu *et al.*, 2007), the adsorption of SO_2 on the surface of Fe_3O_4 can occur as follows:

$$Fe^{3+} + SO_3^{\bullet -} + H_2O \longrightarrow Fe^{2+} + HSO_4^{-} + H^{+}$$

$$SO_3^{\bullet -} + O_2 \longrightarrow SO_5^{\bullet -}$$

$$Fe^{2+} + SO_5^{\bullet -} + H^{+} \longrightarrow Fe^{3+} + HSO_5^{-}$$

$$Fe^{2+} + HSO_5^{-} \longrightarrow Fe^{3+} + SO_4^{\bullet -} + OH^{-}$$

$$Fe^{2+} + SO_4^{\bullet -} + H^{+} \longrightarrow Fe^{3+} + HSO_4^{-}$$

As detected from the above equations, upon the establishment of the Fe^{2+} and Fe^{3+} redox cycle, Fe_3O_4 reacts with generated H_2SO_4 on the carbon producing $Fe_2(SO_4)_3$. The generated $Fe_2(SO_4)_3$ gathered around the micropore mouth and surface and partially or totally blocked the micropores, leading to the increase in mass transfer resistance, which was the main reason for the deactivation of iron supported on activated carbon. This was proven by the SEM-analysis. The Fe_3O_4 on the carbon surface was considered the active center for SO_2 oxidation, whereas micropores are related with the diffusion of reactive molecules. When the pores are blocked by the generated $Fe_2(SO_4)_3$, which have larger molecular volume than Fe_3O_4, active sites are covered and isolated from SO_2 and O_2. Therefore, the blockage of $Fe_2(SO_4)_3$ may be the main reason for the deactivation of iron supported on activated carbon (Xiao-Li *et al.*, 2014).

Dube *et al.* (2015) investigated the use of bagasse ash as a siliceous material and ammonium acetate as a hydrating agent to enhance the removal of sulfur dioxide (SO_2) in a low temperature dry flue gas desulfurization (FGD) process using magnesium-based sorbent, where response surface methodology (RSM) was used to determine the effects of hydration temperature and time, amount of ammonium acetate, and bagasse ash on the surface area of the sorbent. A polynomial model was developed to relate the preparation variables to the sorbent surface area. The Brunauer-Emmet-Teller (BET) surface area results showed that the sur face area increased from 56 m^2/g to 219 m^2/g when the bagasse ash and ammonium acetate were used. This enhanced the ADS of SO_2 in the dry FGD process. The desulfurization experiments performed using a thermogravimetric analyzer (TGA) and (SO_2) removal efficiency reached approximately 99.9% within 2 h. The desulfurization reaction kinetics between SO_2 and the magnesium hydrated sorbent was described by an unreacted shrinking core model (USCM). The model showed the formation of a nonporous shielding layer that stopped further occurrence of gas (SO_2)/solid (magnesium sorbent) reaction and that product layer was determined to be the rate limit. Moreover, the scanning electron microscope (SEM) results illustrated the plugging of pores with the reaction products (for example magnesium hydrated sulfate silicate) during the sulfation process that led to a non-porous surface. The amount of bagasse ash was found to display the greatest influence on the surface area, followed by the hydration temperature and hydration period, respectively, while the amount of ammonium acetate showed a slight influence. The surface area increased as the amount of bagasse ash, hydration temperature, and hydration time increased, but a minimal surface increment occurred when more ammonium acetate was used. The USCM model elucidated that the reaction would have taken place throughout three sequential steps: (1) diffusion of SO_2 onto the sorbent surface, (2) reaction SO_2 and the sorbent on the sorbent surface, and (3) diffusion of SO_2 through the product (ash) layer towards the surface of the unreacted core.

The following sequential mechanism could explain why the hydrating agents enhance the sorbent surface area (Fillipou *et al.*, 1999; Birchalm *et al.*, 2001; Rocha *et al.*, 2004):

$$NH_4CH_3COO_{(aq)} \longrightarrow NH_4^+{}_{(aq)} + CH_3COO^-{}_{(aq)}$$

$$MgO_{(s)} + CH_3COO^-{}_{(aq)} + H_2O \longrightarrow Mg(CH_3COO)_{2(aq)} + 2OH^-{}_{(aq)}$$

$$NH_4^+{}_{(aq)} + H_2O \longrightarrow NH_{3(g)} + H_3O+_{(aq)}$$

$$MgO_{(S)} + H^+{}_{(aq)} \longrightarrow Mg^{2+} + OH^-{}_{(aq)}$$

$$CH_3COO^-{}_{(aq)} + H_2O \longrightarrow CH_3COOH_{(aq)} + OH^-{}_{(aq)}$$

$$MgO_{(aq)} + 2CH_3COOH_{(aq)} \longrightarrow Mg(CH_2COO)_{2(aq)} + 2OH^-{}_{(aq)}$$

Finally, the dissociation of magnesium complexes occurs and the magnesium hydroxide precipitation in the bulk of the solution formed due to the super-saturation. This is the main reason why hydrating agents improve the surface area of sorbents to be used in FGD.

$$Mg(CH_3COO)_{2(aq)} \longrightarrow 2CH_3COO^-{}_{(aq)} + Mg^{2+}{}_{(aq)}$$

$$Mg^{2+}{}_{(aq)} + 2OH^-{}_{(aq)} \longrightarrow Mg(OH)_{2(S)}$$

While the following sequential reactions could explain why the siliceous additive, that is the bagasse ash, enhances the sorbent surface area (Dube *et al.*, 2015):

$$Mg(OH)_{2(aq)} \longrightarrow Mg^{2+}{}_{(aq)} + 2OH^-{}_{(aq)}$$

$$Mg^{2+}{}_{(aq)} + H_4SiO_{4(aq)} \longrightarrow \underset{\text{magnesium hydrated silicate}}{MgH_2SiO_4.2H_2O_{(S)}}$$

Coconut-shell-based activated carbon was reported to exhibit good desulfurization activity for the simultaneous removal of H_2S and SO_2 from simulated Claus tail gas performance under a feed gas of H_2S (20,000 ppmv), SO_2 (10,000 ppmv), and N_2 (balance). The concentrations of H_2S and SO_2 in the simulated Claus tail gas reduced to less than 10 mg/m^3. The breakthrough sulfur capacity of coconut-shell-AC was 64.27 mg of S/g of sorbent at an adsorption temperature of 30 °C and a gas hourly space velocity of 237.7 h^{-1}. Micropores with sizes of around 0.5 nm in the AC were found to be the main active centers for adsorption of H_2S and SO_2, whereas mesopores showed little desulfurization activity for deep removal of H_2S and SO_2. Both physical and chemical adsorption coexisted in the process of desulfurization. The majority of sulfides were removed by physical adsorption and 11% of the sulfur compounds existing in the form of elemental sulfur (ca. 20 atom %) and sulfate (ca. 80 atom %) were derived from chemical adsorption. The mechanism of H_2S and SO_2 adsorption on the AC can be briefly summarized as follows: H_2S and SO_2 are first adsorbed on AC by physical adsorption and then partially oxidized to elemental sulfur and sulfate, respectively, by the oxygen adsorbed on AC. At the same time, the Claus reaction between H_2S and SO_2 occurs. The prepared AC was able to completely regenerate using water vapor at 450 °C with a stable breakthrough sulfur capacity during five adsorption–regeneration cycles (Shi *et al.*, 2015).

Isik-Gulsac (2016) studied the effects of relative humidity, carbon dioxide (CO_2), methane (CH_4), oxygen (O_2) presence, and gas hourly space velocity (GHSV) on H_2S adsorption dynamics of KOH/CaO impregnated activated carbon. The presence of water, O_2, and lower GHSV has beneficial effects on the activated carbon performance. However, CO_2 decreases the adsorption capacity due to its acidic characteristics and competition in adsorption between CO_2 and H_2S on the basic sites of activated carbon. The best adsorption capacity is obtained as 13 wt % in KOH/CaO impregnated activated carbon, in a CH_4 (60%)/CO_2 (38%)/O_2 (2%) gas atmosphere at ambient temperature, a relative humidity of 90, and 5000 h^{-1} GHSV. The specific surface area, micropore area, average pore diameter, total pore volume, and external surface area were found to be significantly decreased after the ADS. That was attributed to the deposition of the sulfur or sulfurous products of catalytic oxidation in the pores, which consequently decreases the surface area.

If H_2S and CO_2 are present in wet conditions, the following occurs:

$$CO_2 + H_2O \longleftrightarrow HCO_3^- + H^+$$

$$CO_2 + OH^- \longleftrightarrow HCO_3^-$$

$$HCO_3^- \longleftrightarrow CO_3^{2-} + H^+$$

$$H_2S \longleftrightarrow HS^- + H^+$$

$$H_2O \longleftrightarrow OH^- + H^+$$

As observed in the reactions, CO_2 transport is facilitated by bicarbonate ions and H_2S is similarly transported in the form of bisulfide ions. In sufficiently thin films, diffusion times are so short that negligible amounts of CO_2 are converted via the slow first two reactions that are responsible for the production of bicarbonate. However, steady-state carbonate, bicarbonate, and hydrogen ion concentrations are affected by the presence of carbon dioxide and the extent of reaction responsible for dissociation of H_2S is a function of pH. Thus, although the first two reactions have negligible effects upon CO_2 transport, their occurrence affects the permeation of H_2S because of their influence on the pH (Meldon and Dutta, 1994). Additionally, Bandosz (2002) and Bouzaza *et al.* (2004) stated that high-valent sulfur compound formation is promoted and H_2S dissociation is limited in an acidic environment, thus the removal of sulfur decreases.

Isik-Gulsac (2016) illustrated the mechanism for ADS of H_2S on KOH/CaO impregnated activated carbons as follows:

Physical or ionic adsorption of H_2S on a porous carbon surface in dry or wet conditions

$$H_2S_{gas} \longrightarrow H_2S_{ads}$$

$$H_2S_{liq} \longrightarrow HS^-_{ads} + H^+$$

$$HS^-_{ads} + H^+ \longrightarrow 2H^+ + S^{2-}$$

should be taken into consideration and the basic environment favors the formation of S^{2-}, which may promote the oxidation process and formation of elemental sulfur and partial conversion into sulfuric acid (Adib *et al.*, 1999).

Chemical adsorption of H_2S on metal oxides with sulfur formation:

$$KOH + H_2S \longrightarrow KHS + H_2O$$

$$2KOH + H_2S \longrightarrow K_2S + 2H_2O$$

$$KHS + 1/2\ O_2 \longrightarrow S + KOH$$

$$K_2S + 1/2O_2 + H_2O \longrightarrow S + 2KOH$$

In the presence of CO_2, reactions between KOH and CO_2 also occur which weaken the catalytic activity of the impregnated activated carbon for ADS.

$$2KOH + CO_2 \longrightarrow K_2CO_3 + H_2O$$

$$K_2CO_3 + H_2S \longrightarrow KHS + KHCO_3$$

The presence of oxygen enhancing the rate of ADS should also be taken into consideration because oxygen and H_2S adsorb on the free active sites of the carbon surface (C_f) and then react catalytically with the formation of elemental sulfur, sulfur dioxide, or sulfuric acid (Puri 1970; Choi *et al.* 2008):

$$C_f + 1/2\,O_2 \longrightarrow C(O)$$

$$C(O) + H_2S_{ads} \longrightarrow C_f + S_{ads} + H_2O$$

$$S_{ads} + O_2 \longrightarrow SO_{2ads}$$

$$SO_{2ads} + 1/2\,O_2 + H_2O \longrightarrow H_2SO_{4ads}$$

Moreover, in the presence of oxygen and humidity, $CaSO_4$ can be formed on the surface of activated carbon (Seredych *et al.*, 2008):

$$CaO + H_2O \longrightarrow Ca(OH)_2$$

$$Ca(OH)_2 + H_2S + 2O_2 \longrightarrow CaSO_4 + 2H_2O$$

Villegas (2016) reported the application of activated carbon prepared from different agricultural wastes including tropical bamboo, central American mahogany seed shells, Honduran mahogany seed shells, mamey zapote seeds, and corncobs in cleaning the combustion gases of diesel engines with CO, SO_2, NO_2, and H_2S, throughout a physical adsorption process.

Liu *et al.* (2017) reported the preparation of nitrogen-doped coconut shell activated carbon catalysts were prepared by urea or melamine impregnation followed by heat treatment and used for the removal of methyl mercaptan (CH_3SH) in a fixed bed quartz reactor. The AC sample has a specific surface area of 1508 m^2/g, a micropore volume of 0.571 cm^3/g, and a total pore volume of 0.807 cm^3/g. Nitric acid oxidation increased the total pore volume and decreased the micropore volume and specific surface area. This was attributed to the erosion and pore-widening effects of nitric acid treatment. Also, there was a small recorded decrease in the micropore volume with a small recorded increase in the total pore volume in the AC modified with urea, while the melamine-modified AC recorded a decrease in the micropore and total pore volumes. This was attributed to the introduced nitrogen-containing species, which create steric hindrances and partially prevent the access of nitrogen molecules into the micropores. The SEM analysis proved the porous structure of the prepared catalysts. Moreover, it is worth noting that the acid treatment with HNO_3 increased the acidic groups on the surface of AC, while the nitrogen doping increased the number of basic groups. The nitrogen in acidified AC by HNO_3 (ACO) is mainly in the form of nitrate and nitric oxides, which were introduced by the nitric acid treatment. The melamine-modified activated carbons (ACM or ACOM) samples have more nitrogen than those modified with urea (ACU or ACOU). This was attributed to the high content of nitrogen in the melamine molecule and its conversion to melamine resins at high temperatures. The dominant nitrogen species on the surface of nitrogen-doped activated carbons (ACU or ACM) are pyridinic-nitrogens. However, nitric acid oxidation leads to higher relative contents of pyridine-N-oxide and less pyrrolic-nitrogen. This was attributed to the nitric acid oxidation which increased the acidic groups and created more active sites on the surface of activated carbon, which is beneficial for nitrogen incorporation into the carbon matrix in a pyridinic-like nitrogen configuration. In addition, a small amount of pyridinic-nitrogen can be converted into pyridine-N-oxide. The ACM or ACOM contained lower relative amounts of pyridinic-nitrogen and quaternary-nitrogen than the corresponding ACU or ACOU. However, the melamine modification provided more nitrogen. Thus, the overall contents of pyridinic-nitrogen and quaternary-nitrogen in the activated carbons increased when melamine was used as a precursor. The CH_3SH capacity of the original activated carbon (AC) was only 161.8 mg/g. Nitrogen-doping increased the ADS capacity and the highest breakthrough CH_3SH capacity of approximately 602.1 mg/g was achieved by the ACOM sample with a surface nitrogen content of 4.41%. This was attributed to the pyridinic-nitrogen and quaternary-nitrogen which are known to be beneficial to the oxidation of S- and N- containing compounds because of their strong electron transfer abilities (Bashkova *et al.*, 2003; Bashkova and Bandosz, 2009; Sun *et al.*, 2013). Pyridinic- nitrogen can act as a Lewis basic site with a lone electron pair, promoting the dissociation of CH_3SH to thiolate ions. The breakthrough CH_3SH capacity of AC in the absence of moisture is at least three times smaller than that in the presence of moisture, while the breakthrough CH_3SH capacity of ACOM, under dry air conditions, is almost twice that measured without oxygen. Thus, water and oxygen are beneficial to remove CH_3SH. A thin water film can be formed on the carbon surface in the presence of water (Sun *et al.*, 2013) where CH_3SH dissolves into the water film and dissociates to the thiolate ion, which can be further oxidized to CH_3SSCH_3. In the absence of moisture, CH_3SH is first adsorbed onto the carbon surface and then the adsorbed CH_3SH can be oxidized to CH_3SSCH_3, which can be easily removed from the carbon surface. Thus, the regeneration of the adsorbent was easy and can be reused for four successive cycles retaining a CH_3SH capacity of 88.33%. Pyridinic-nitrogen and quaternary-nitrogen occupy the high-energy states (Bashkova *et al.*, 2005). The electrons could be transferred from the thiolate ion to the adsorbed oxygen, forming thiolate radicals and superoxide ions because the two types of nitrogen groups enhance the ion-exchange properties of the carbons. These superoxide ions can react with water forming the active hydroxyl radicals. All of these species could also facilitate oxidation. Finally, water and CH_3SSCH_3 are formed and stored in the pores. Considering the presence of water and active sites, such as hydroxyl radicals and oxygen radicals, CH_3SSCH_3 can be further oxidized into methyl methane thiosulfonate and methanesulfonic acid.

The aforementioned examples briefly promote the efficiency of using agricultural wastes as feasible alternatives for granular activated carbons preparation for the bioadsorptive desulfurization (BADS) of gas.

3.3 Desulfurization Processes

As already explained, there are several reasons why sulfur components need to be removed from natural gas before the gas may be 'shipped'. Even when further processing is to be done at site, sometimes sulfur removal up front will be required. This is very much the case in downstream plants where catalysts are used for processing. How sulfur components are removed will be subjected to local conditions and requirements. Engineers should always attempt to make the processing as cost efficient as possible under the constraints imposed.

Natural gas can be classified according to the amount of sulfur content (generally H_2S) within the produced gas (Rojey *et al.*, 1997; Carroll, 2010; Kidnay *et al.*, 2011, Speight, 2015 and Faramawy *et al.*, 2016). In this classification, the natural gas could be sweet or sour. Sweet gas contains no or a negligible quantity of H_2S, whereas sour gas contains unacceptable quantities of H_2S (more than 5 mg/Nm^3) (Maddox, 1974). There are several technologies available for removing H_2S from natural gas. They may be classified in six main classes, as illustrated in Figure 3.3.

3.3.1 Scavengers

There is no accurate definition of scavengers, however, the term is employed to explain chemicals injected into the gas stream to react with

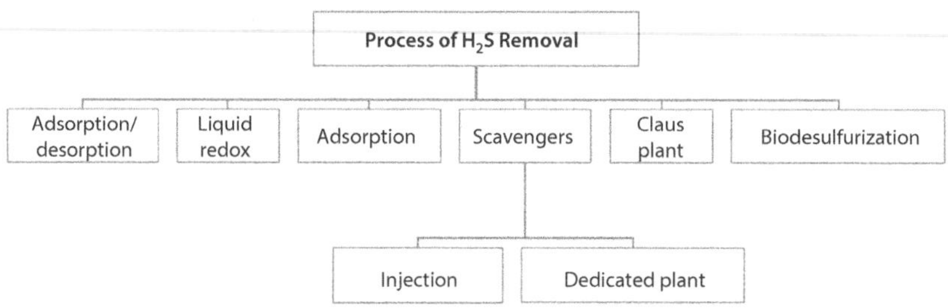

Figure 3.3 H_2S Removal Technologies.

the H_2S, mainly without side reactions. The removal or scavenge of acid gas impurities, like CO_2 and H_2S, from industrial gas streams is a significant operation in natural gas processing, hydrogen purifying, refinery off-gas treating, and synthesis gas for ammonia and methanol manufacturing. The industrial gas streams containing acid gas impurities must be purified in order to meet the requirements of the gas mixtures sequential processing (e.g., avoiding catalyst damage) or environmental regulation (exhaust of the gas mixtures as off-gas) (Lu *et al.*, 2006). The available methods to scavenge the gases can be largely divided into physical, chemical, and biochemical (Satoh *et al.*, 1988). These chemicals may be injected into pipelines or applied in special scavenging towers. They are generally used once the H_2S level is below 200 ppm (Dalrymple *et al.*, 1994). According to Dalrymple *et al.* (1994), H_2S scavengers can be categorized into three categories (Figure 3.4).

A scavenger ought to meet a number of criteria to be effective, such as the product should be able to efficiently remove all sulfide species under field condition, these being pH, pressure, and temperature and the reaction should be reliable, speedy, and irreversible. Moreover, the reaction products should be simple to dispose to the environment, the chemical should not exhibit incompatibilities with different components of the fluids, the chemical overdosing should not give rise to system difficulties, the chemical should not be corrosive, the chemical should be safe to handle and non-polluting to the environment, and the chemical should be readily available and cost-efficient (Nassar *et al.*, 2016).

Different scavengers are employed where the water-soluble scavengers are among the foremost common scavengers and are often the product of choice for applications at temperatures below 200 °F (Nagi, 1997). Oil-soluble scavengers are employed in high-temperature applications or when the water tolerance of the hydrocarbon is a problem (Garrett *et al.*, 1988). Metal-based scavengers answer the particular wants of very high temperature and high-H_2S concentration applications. These additives are used at temperatures above 177 °C to create thermally stable products and are able to achieve H_2S reduction levels other than H_2S scavengers (Speight, 1993).

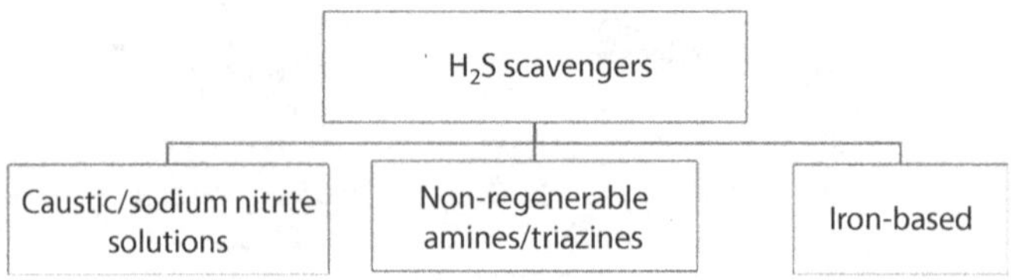

Figure 3.4 Types of H_2S Scavengers.

The H_2S scavenge from a gas stream can be carried out by adsorption onto a solid surface, catalytic oxidation, and absorption by a liquid solution (amine/alkaloamine) (Polychronopoulou *et al.*, 2005). However, many problems exist with the catalysts employed in these processes including the cost of renewing the inactivated catalysts, as well as the generation of secondary substances causing pollution and high energy requirement (Wubs *et al.*, 1991).

McKinsey (2003) published results on cow-manure compost as H_2S removal media. The data was rather inconclusive. The removal efficiency reported was around 80% for a gaseous stream containing 1500 ppmv H_2S. The removal rate was estimated to be 16–118 g H_2S/m^3 solids/h for residence times (empty bed) of 100 s. He did not have sufficient data to distinguish between the physical, chemical, and biological mechanisms of H_2S retention. The main utility of this work is the fairly comprehensive presentation of available technologies, including those essentially employed in scavenging H_2S in the oil industry on a large scale (solvent-based absorption and solid oxide scavengers). In addition, this review fairly well covers the technical and market survey in the field.

Kandile *et al.* (2014) prepared three H_2S scavengers by reacting monoethanolamine with formaldehyde in different ratios (1:1, 2:1 and 2:3) to give MF1, MF2, and MF3, respectively. The efficiency of scavengers increased with an increasing reaction time up to 50 min. Also, as concentration of scavengers and temperature increased, the removal efficiency of the scavengers increased. By comparing the efficiency of the prepared products with the commercial products EPRI-710 and EPRI 730, it was found that MF3 exhibited a similar efficiency comparing with the commercial scavenger EPRI 730 (currently used in the field) at different concentrations and temperatures.

One method that is usually used to overcome the issues related to the chemical treatment of H_2S, is the oxidation into elemental sulfur employing a metal chelating agent in the form of a liquid catalyst; this methodology uses metal ions, like Fe^{2+}, and different chelating agents, such as ethylenediaminetetraacetic acid (EDTA) and nitrilotriacetic acid (NTA), that are non-toxic, and thus, have no environmental pollution during the removal of H_2S (Husein *et al.*, 2010). Another methodology is sorbent injection into the gasifier. In situ desulfurization can be accomplished through the employment of (regenerable) transition metal oxides (Yumura and Furimsky, 1985). However, many of such processes are of limited efficiency and have high-energy costs. The employment of enormous amounts of solvents and catalysts in traditional removal processes increase the cost of treatment (Fuda *et al.*, 1991). The standard methods for H_2S removal include amine aqueous solution absorption-Claus process (Stirling, 2000). The Claus process has been most typically used to remove H_2S from natural gas or refinery plants. Claus plants typically convert 94–98% of sulfur compounds within the feed gas into elemental sulfur. Also, one of the more frequently used biological treatment methods of removing H_2S involves a combination of biological and chemical process, using the bacteria that oxidize iron from Fe (II) to Fe (III), *Thiobacillus ferrooxidans*. When this bacterium produces oxidized iron, this oxidized iron can oxidize H_2S to elemental sulfur and then *T. ferrooxidans* turns the Fe (II), again, into Fe (III) (Sasaoka *et al.*, 1992).

In another process, Bio-SR process bacteria can treat sour gas within chemical plants and refiners (Satoh *et al.*, 1988), the basis of the process being explained in Figure 3.5. A solution of ferric sulfate contacts sour gas in an absorber. The solution absorbs hydrogen sulfide and oxidizes it to elemental sulfur. At the same time, the ferric sulfate is reduced to ferrous sulfate.

$$H_2S + Fe_2(SO_4)_3 \longrightarrow S^o + 2FeSO_4 + H_2SO_4$$

Elemental sulfur is removed from the solution by a separator. Then, the solution goes to a bio-reactor. This is where the bacteria do their work. Upon contact with air, the bacteria oxidize the ferrous sulfate back to ferric sulfate (regeneration step).

$$2\,FeSO_4 + H_2SO_4 + 1/2\,O_2 \longrightarrow Fe_2(SO_4)_3 + H_2O$$

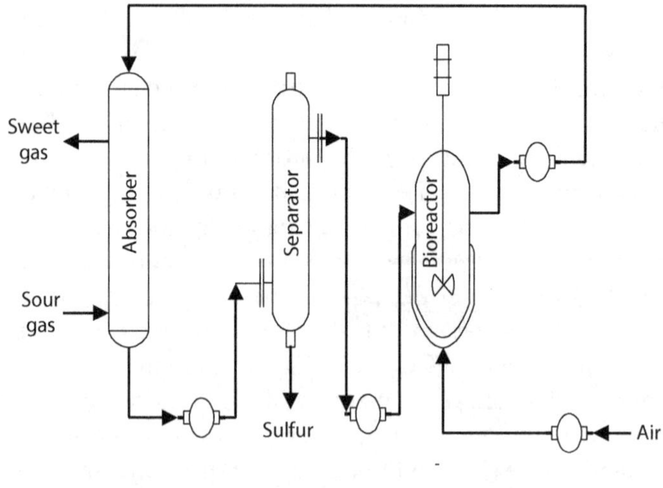

Figure 3.5 The Bio-SR Process Flow Scheme.

The oxidized solution is then recycled to the absorber to repeat the cycle. In this closed cycle, there is no waste and no special chemical.

Based on a similar principle, using chelated ferric ion for oxidation of H_2S to elemental sulfur, followed by a microbial process of regeneration of ferric ions using a mixed culture of iron oxidizing bacteria has been reported (Rai and Taylor, 1996). *T. ferroxidans*, for example, can be used to desulfurize petroleum and natural gas with the reaction being carried out in a closed vessel containing substrate mixed with a bacterial suspension (Das *et al.*, 1993).

$$S^{2-} + 2Fe^{3+} \longrightarrow S^{o} + 2Fe^{2+}$$

$$2Fe^{2+} + 1/2\,O_2 + H_2O \longrightarrow 2Fe^{3+} + 2OH^-$$

To decide which method to use, many factors should be considered, including the required extent of H_2S removal, the gas composition, temperature, volume, and pressure, and, further, the impact of sulfur recovery on the process economics and/or the environment (Wang, 2008).

3.3.2 Adsorption

It is doable to get rid of H_2S from natural gas by adsorption on a variety of materials, for instance zeolite molecular sieves. These are generally used to remove the last traces of water, $CO_{2,}$ and H_2S upstream of liquefied natural gas (LNG) condensation trains. A general method conception is shown in Figure 3.6.

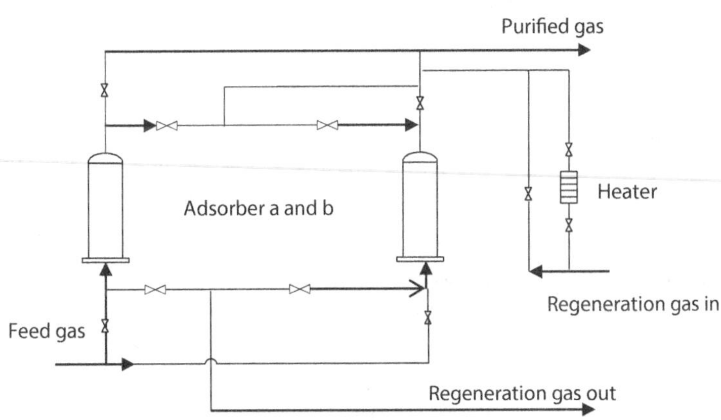

Figure 3.6 A Simple Diagram for H_2S Adsorption Plant.

Adsorption/oxidation by carbonaceous surfaces was widely studied as a technique to manage H_2S pollution. Boki and Tanada (1980) studied the adsorption of H_2S onto activated carbon surfaces at three different temperatures using a vacuum system. The authors discovered that a little quantity of H_2S that was chemisorbed attenuated with the decrease in temperature. Aranovich and Donohue (1995) and Bagreev *et al.* (2000) compared experimental adsorption isotherms for various adsorbates onto microporous adsorbents with those predicted using the Dubinin-Radushkevich (DR) equation. They found that the adsorption of H_2S follows the pore filling mechanism. These results indicate that the adsorption of H_2S is mainly physical when processed under dry and anaerobic conditions, where surface functionalities exhibit no impact on the adsorption process. Similar studies, with fifteen different activated carbons, showed that lower average pore size leads to stronger interaction between H_2S and, also, the carbon surface, which is due to the increase in adsorption potential (Bagreev *et al.*, 2000). All of those studies agree that physical adsorption of H_2S dominates under ideal conditions (i.e. vacuum, dry, anaerobic, and low temperatures).

Upon the application of activated carbon (AC) for the removal of H_2S, it is usually dosed with KI or sulfuric acid (H_2SO_4) to increase the reaction rate. Before entering the carbon bed, 4–6 % air is added to the biogas and H_2S is catalytically converted to elemental sulfur and water in biological filters as follows:

$$2H_2S + O_2 \longrightarrow 2S^0 + 2H_2O$$

The elementary sulfur is then adsorbed by the AC. Usually, the best efficiency is obtained at pressures of 700–800 kPa and temperatures of 50–70 °C that can be achieved through heat generation during compression. In a continuous process, the system consists of two vessels (Hagen *et al.*, 2001; Wellinger and Lindberg, 2005; Ryckebosch *et al.*, 2011): one vessel for adsorption and the other for regeneration. Regeneration can be performed with hot nitrogen (inert gas) or steam. The sulfur is vaporized and, after cooling, liquefied at approximately 130 °C. However, typically, the AC is replaced rather than regenerated (Hagen *et al.*, 2001; Wellinger and Lindberg, 2005; Ryckebosch *et al.*, 2011).

Bandosz (2002) summarized that activated carbons as hydrogen sulfide adsorbents under field conditions needed the correct combination of surface chemistry and carbon porosity; a more acidic environment encourages the formation of high-valent sulfur compounds and reduces the H_2S removal capacities. A basic environment favors the formation of elemental sulfur and increases the capability of sulfur removal. Also, pre-adsorbed water can enhance H_2S dissociation and removal from the gas stream.

Cal *et al.* (2000) carried out systematic experimental studies with unmodified and modified activated carbons at 550 °C and showed that both HNO_3 oxidation and Zn impregnation improved H_2S adsorption capability.

Since modification of surface chemistry is probably going to be effective for improving H_2S removal capability by activated carbons, different impregnates were studied. Bagreev and Bandosz (2002) evaluated the impact of NaOH on adsorption of H_2S and mentioned that the H_2S removal capacities were dominated by the presence of NaOH and were not sensitive to surface area and pore structure. Other reports on NaOH (Chiang *et al.*, 2000; Tsai *et al.*, 2001) or K_2CO_3 (Przepiorski *et al.*, 1998, 1999) impregnation also showed vital improvement within the H_2S removal capability. It is believed that the presence of alkaline chemicals facilitates the dissociation of H_2S on carbon surfaces.

The ADS-activity of CuO-AgO sorbents calcined at 700°C has been studied where CuO was used as a main active material, AgO was used as an additive material, and 25 wt% SiO_2 was used as a support material. The ADS activity and the activation energy of sorbent were found to be decreased as the content of AgO increased. The maximum ADS of a simulated gas recorded 14.95 g sulfur/100 g sorbent at 550°C using sorbent containing 1 wt% AgO (Lee *et al.*, 2005).

Yazdanbakhsh *et al.* (2014) reported the H_2S breakthrough capacity of copper exchanged Engelhard Titanosilicate-2 (ETS-2). The adsorbent efficiency remained unchanged up to 950 °C, but at a lower temperature, <750 °C, the adsorption capacity at breakthrough was 0.7 mol H_2S/mol copper, while at >750 °C the capacity of the adsorbent is halved. The change in H_2S capacity was attributed to the reduction of Cu^{2+} by the produced H_2 from the thermal dissociation of H_2S.

Liu et al. (2015a) reported an efficient hybrid adsorbent/photocatalytic composite (TiO_2/zeolite) for the H_2S removal and SO_2 captured by coating TiO_2 on the surface of cheap natural zeolite with an ultrasonic-calcination way. The TiO_2/zeolite expressed the highest H_2S removal capacity and lowest SO_2 emission, compared with the single zeolite adsorption and TiO_2 photo-catalysis. Moreover, the H_2S removal capacity and SO_2 capture capacity of TiO_2/zeolite were found to be enhanced in the presence of moisture in the biogas.

Saimura *et al.* (2017) reported the preparation of dual-layered paper-structured catalysts separately containing manganese oxides (MnO_x) and nickel/magnesium oxides (Ni/MgO) for sequential desulfurization and methane-steam reforming, respectively. The porous paper-structured MnOx adsorbents could remove H_2S down to 7 ppm or lower at 400 °C for 81 min from simulated gas, resulting in higher H_2S adsorption capacity, as compared to commercial powdered manganese-based adsorbents. That system was designed by stacking MnO_x papers upstream and Ni/MgO papers downstream in one reactor as a dual-layered form that enabled the continuous direct hydrogen production of 2174 ppm H_2S-containing simulated biogas at 400 °C, whereas single-layered Ni/MgO papers immediately lost their catalytic activity due to the H_2S poisoning. The Mn-based adsorbents are promising for capturing H_2S impurities in crude biogas, since manganese oxides (MnO_x) can react with H_2S and are stable even at high temperatures (400–1000 °C) (Westmoreland *et al.*, 1977; Bakker *et al.*, 2003).

Several studies on zeolites (Hernández *et al.*, 2011; Kwaśny *et al.*, 2015; Kwaśny and Balcerzak, 2016) show that zeolites have low efficacy of adsorption of H_2S as compared to activated carbons.

Li *et al.* (2012) reported the preparation of nano-ZnO by a homogeneous precipitation method and activated carbon from coconut shell, which was pretreated by two different acids: HNO_s and HCl. Then, AC-loading nano-ZnO was prepared by mechanical hybrid method. Then, the ADS capacity of the prepared AC-loading nano-ZnO was tested against H_2S gas. The load of activated carbon should be proper and the activated carbon loaded nano-ZnO reported to achieve the most effective desulfurization activity when the mass ratio of ZnO and AC is 0.3 (Li *et al.*, 2012) since excessive nano-ZnO powders could accumulate in the material surface and block the pores of the AC so that specific surface area and pore volume of material are reduced, hence the desulfurization properties. The desulfurization property of AC loading nano-ZnO desulfurizer modified by hydrochloric acid is better than that modified by nitric acid and the breakthrough time is up to 110 min. This is due to the strong oxydic ability of HNO_3, which reduces the specific surface area of the AC and, thus, its ADS activity. The most important thing is that the modification of nano-ZnO, with AC, did not alter its structure.

Phooratsamee *et al.* (2014) reported the preparation of AC from palm oil shells by carbonization of raw material in an inert atmosphere at 600 °C, followed by activation of char product by $ZnCl_2$, then, finally, alkali impregnation of activated carbon by NaOH, KI, and K_2CO_3 solutions. Then, the activity of the prepared AC was tested for H_2S absorption from biogas. The K_2CO_3-AC showed the highest surface area and the pore volume of 741.71 m^2/g and 0.4210 cm^3/g, respectively. Thus, it showed the best performance for biosorption of H_2S. The results of this research showed that

H_2S adsorption on alkaline impregnated activated carbon (K_2CO_3-AC and KI-AC) was higher than commercial activated carbon. This positive effect is mainly due to the chemical reaction in which H_2S reacts with alkali at carbon surface to produce alkali hydrogen sulfilde (AHS) and alkali sulfide. However, NaOH-AC showed lower adsorption activity than that of commercial AC, due to the less residual water vapor content which blocks the carbon surface.

$$H_2S + K_2CO_3\text{-AC} \longrightarrow KHS\text{-AC} + KHCO_3$$

$$H_2S + K_2CO_3\text{-AC} \longrightarrow K_2S\text{-AC} + H_2CO_3$$

$$H_2S + KI\text{-AC} \longrightarrow KHS\text{-AC} + HI$$

$$H_2S + 2(KI)\text{-AC} \longrightarrow K_2S\text{-AC} + 2HI$$

$$H_2S + NaOH\text{-AC} \longrightarrow NaHS\text{-AC} + H_2O$$

$$H_2S + 2NaOH\text{-AC} \longrightarrow Na_2S\text{-AC} + 2H_2O$$

In another study, activated carbon treated with Na_2CO_3 was reported to be the most effective sorbent, showing a breakthrough time of 1222 min at 0.5 ppm that is twice the time of the untreated-AC (Micoli *et al.*, 2014). Furthermore, Monteleone *et al.* (2011) reported that activated carbons impregnated with metal salts express high adsorption capacity due to the combination of micro-porosity and oxidative properties. Osorio and Torres (2009) reported a study dealing with biogas purification coming from the anaerobic digestion of sludges in a wastewater treatment plant. The purification apparatus contains scrubbing towers and filters of activated carbon at the end of the line. The H_2S in flow concentrations were quite high (~ 2000 ppm), while the effluent biogas from the scrubbing towers presented a H_2S concentration <1 ppm, zero, or undetectable values after adsorption of active carbons filters.

Aslam *et al.* (2015) reported the preparation of AC from waste oil fly ash (OFA) which is produced in large quantities from power generation plants through combustion of heavy fuel oil, applying physicochemical treatments using a mixture of HNO_3, H_2SO_4, and H_3PO_4 acids to remove non-carbonaceous impurities. The acid treated OFA was then activated by CO_2 at 990 °C. The physicochemical treatments of OFA increased the surface area from 4 to 375 m^2/g. The AC is further treated with HNO_3 and NH_4OH solutions in order to attach the carboxylic and amine groups on the surface, respectively. Then, its activity was tested for the removal of H_2S from a synthetic natural gas by carrying out breakthrough curves. The

NH_4OH-treated AC was found to be more effective for H_2S removal than acid-treated AC, recording a maximum adsorption capacity of 0.3001 mg/g for NH_4OH functionalized AC with 86.43% regeneration efficiency. The obtained results indicated that the presence of more acidic functionalities on the surface reduces the H_2S adsorption efficiency from the gas mixture.

As listed before, biogas desulfurization through the adsorption process is an attractive method due to its simple process and low starting cost. Currently available adsorbents, such as activated carbon, metal/metal oxide, and zeolite, are the most applied adsorbents for effective biogas desulfurization. Kaolinite is a naturally occurring material that has been used as an adsorbent material for several organic compounds. However, the presence of metal oxides in kaolin, such as aluminium (Al_2O_3), silica (SiO_2), iron (Fe_2O_3), potassium (K_2O), magnesium (MgO), and TiO, makes it possible to be used as H_2S adsorbent. Abdullah *et al.* (2017) studied the activity of a calcined kaolinite in ADS of H_2S using a fixed bed reactor in comparison with commercial activated carbon, commercial zeolite A, and pure zinc oxide. That recorded 0.087, 51.68, 2.1, and 58.30 mg H_2S/g sorbent, respectively. Moreover, as many other adsorbents, it was observed that the contact time between the gas and adsorbent is the key factor in determining the adsorption capacity. Dhage *et al.* (2008) reported that the increase in the gas flow rate reduces the contact time and, consequently, reduces the adsorption capacity. Furthermore, the increase in temperature increased the ADS capacity. That recorded 26 mg H_2S/100 g of adsorbent at 80 °C. It was also noted that at 30°C, the physical adsorption was dominant, meanwhile at a higher temperature, the mechanism of H_2S removal was governed by chemical adsorption. Thus, the presence of metal oxides on the kaolinite surface is suggested to play an important role in the ADS of H_2S.

The modification of a copper based metal-organic framework, (MOF-199) by incorporating activated carbon (AC) during its synthesis under hydrothermal conditions, has been reported (Shi *et al.*, 2017) where the XRD, SEM, and nitrogen adsorption results showed that the synthesized composites, with an amount of AC of less than 2% (MAC-2), had a more ordered crystallinity structure, as well as higher surface area, than the parent MOF-199. MAC-2 exhibited a maximum uptake of 8.46 and 8.53% for H_2S and CH_3SCH_3, respectively, which increased by 51 and 41% compared to that of MOF-199. This was attributed to the increased micropores and the number of copper metal sites resulted from AC incorporation. The pyridine adsorption infrared spectroscopy (Py-IR) confirmed the chemisorption process of H_2S, which led to the formation of CuS as well as the destruction of the MOF structure, whereas reversible chemisorption was found for CH_3SCH_3 adsorption, as testified by thermogravimetric- mass spectroscopy (TG-MS) and fixed-bed regeneration. However, there was a recorded obviously different increasing slope of the CH_3SCH_3 concentration curve, along with the corresponding color change during temperature programmed desorption which indicated that both physisorption and weak chemisorption were involved. Moreover, the used MAC-2 was almost totally regenerated by nitrogen purge at a temperature of 180 °C, with 94% of breakthrough capacity recovered after 3 cycles. This good regeneration behavior promoted MAC as a promising sorbent for CH_3SCH_3 removal.

3.3.3 Liquid Redox Processes

Liquid redox processes employ aqueous-based solutions containing metal ions, usually iron, which are capable of transferring electrons in reduction-oxidation (redox) reactions. The best familiar of these processes is the Stretford process that uses vanadium, valence state five, as an oxidant (Andrews, 1989).

The use of vanadium possesses different environmental problems; in addition, this process needs a large absorption tower and a large oxidation tank to regenerate the solution and the precipitated sulfur which is produced from H_2S. Thus, it is not an immediate choice for platform use. To overcome this problem using the redox principle many researchers make use of iron as an oxidant. The most publicized is the LO-CAT (presently owned by Merichem company), which is licensed by Gas Technology Products LLC. In this process, a non-toxic, chelated iron catalyst is employed to accelerate the reaction between H_2S and oxygen to form elemental sulfur (Hardison, 1991).

$$H_2S + O_2 \xrightarrow{Fe} S^o + H_2O \quad (1)$$

As implied by its generic name, liquid redox, all of the reactions in the LO-CAT process occurs in the liquid phase in spite of the fact that equation (1) is a vapor phase reaction. In the process, the sour gas is contacted in an absorber with the aqueous, chelated iron solution where the H_2S is absorbed and ionized into sulfide and hydrogen ions (Figure 3.7)

$$H_2S + H_2O \longrightarrow 2H^+ + S^- \quad (2)$$

Spent absorbent is pumped to the oxidizer for regeneration where oxygen in the air is employed as an oxidizing agent. There is a more recent version where the absorbent is enclosed within the oxidant unit. Additionally,

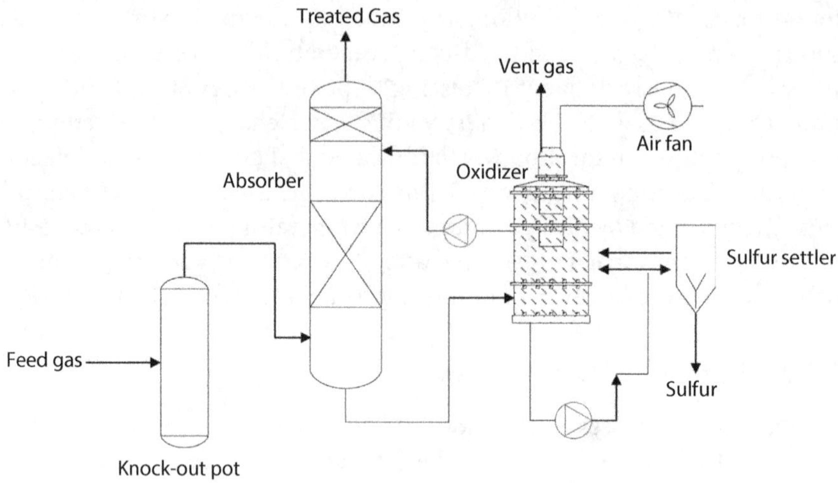

Figure 3.7 A Simple Diagram for LO-CAT Process.

a more recent conception is the Japanese Process Bio-SR that uses an unchelated form of iron which is regenerated microbially and it will be mentioned later. Shell also offers a microbiological process known as Paques (Benschop *et al.*, 2002; O'Brien *et al.*, 2007).

Successful operation of the liquid redox processes depends on a proper understanding of the fundamentals. The present development efforts may lead to improved processes in the future. Once their increased competitiveness opens a larger market, more development money is likely to be spent giving, hopefully, even better processes. The application range for these redox processes is generally 1–20 tons of sulfur per day and they are applicable to a vast range of gases.

In dry oxidation processes, the sulfur is removed from hydrogen sulfide with the employment of iron oxide (eq. 3). Iron oxide is used in the form of iron fillings, iron pellets, iron sponge, or steel wool (Shannon, 2000; Hansen, 2006).

$$Fe_2O_3 + 3H_2S \longrightarrow Fe_2S_3 + 3H_2O \tag{3}$$

The sulfur removal efficiency by iron oxide ranges from 0.20–0.716 kg of H_2S for every 1 kg of iron oxide (Wellinger and Linberg, 2000; Magomnang and Villanueva, 2014).

As shown in Figure 3.8, the sour gas from the gas holder/tank passes through the inlet section on which mass flow, temperature, humidity, and initial concentration of the H_2S from the sour gas is being measured.

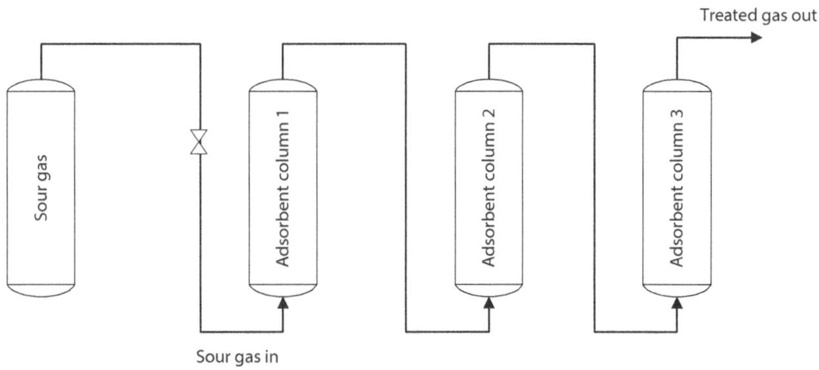

Figure 3.8 A Simple Diagram for the Dry Oxidation Processes.

It passes through the first stage purifying chamber and is passed through another until it reaches the last chamber. Upon exiting the respective chambers, the concentration of the H_2S is measured until it reaches the third stage chamber.

LO-CAT® (US Filter/Merichem) and SulFerox® (Shell/Dow) processes are currently the available chelated-Fe H_2S removal technologies. LO-CAT® can effectively treat any stream containing sulfur and, consequently, any biogas (Rouleau and Watson, 2014). Its typical economic niche is the removal needs of more than 200 kg of S/day. As an example for the real field application of Lo-CAT®, US Filter/Merichem company installed a LO-CAT®II H_2S oxidation system in Broward County, Floridato to treat up to 5 Nm^3/s of landfill gas containing up to 5000 ppmv H_2S. That led to capturing about 300 kg S/day (Abatzoglou and Boivin, 2009). While the SulFerox® license is jointly handled by Dow and Shell, Dow licenses the technology externally and Shell markets the process among its own company divisions. Institut Français du Petrole (IFP) has one such license and Gaz Integral Enterprise of France, a company like IFP, markets the SulFerox® process (Le Gaz Intégral, 2000) for S removal rates between 100 and 20,000 kg/day with high CO_2/H_2S ratios. Although CO_2 is not significantly removed, approximately 50–90% of mercaptans can be removed in either low- or high- pressure applications. The S-removal with SulFerox® costs around \$0.24–\$0.3/kg.

3.3.4 Claus Plants

The sulfur species in natural gas after its removal is usually in the form of H_2S. The most common means of recovering the sulfur contained in hydrogen sulfide is the Claus process. This process can recover 93–99% of the sulfur contained in its feed. Recovery depends upon feed composition, age of catalyst, and number of reactor stages.

The gas leaving the Claus plant is referred to as tail gas and is burnt to convert the remaining H_2S, that is fatal at low levels, to sulfur dioxide which has an abundant toxic limit. The off-gas stream is ventilated to the atmosphere or sent to a tail gas recovery plant.

3.3.4.1 *Classic Claus Plant*

The Claus process (Figure 3.9) is the most important gas desulfurizing process, recovering elemental sulfur from gaseous H_2S. It was developed by Carl Friedrich Claus in 1883. The H_2S containing feed gas is fed to the Claus furnace along with air. In this furnace, about one-third of the H_2S is converted to SO_2. The Claus technology can be divided into two process steps.

Thermal step: Hydrogen sulfide-laden gas reacts in a sub-stoichiometric combustion at temperatures above 850 °C, such that elemental sulfur precipitates within the downstream process gas cooler. The H_2S content and the concentration of other combustible components (hydrocarbons or ammonia) determine the location where the feed gas is burned. Claus gases (acid gas) with no further combustible contents, apart from H_2S, are burned in lances surrounding a central muffle by the following chemical reaction:

$$H_2S + 3/2\,O_2 \longrightarrow H_2O + SO_2$$

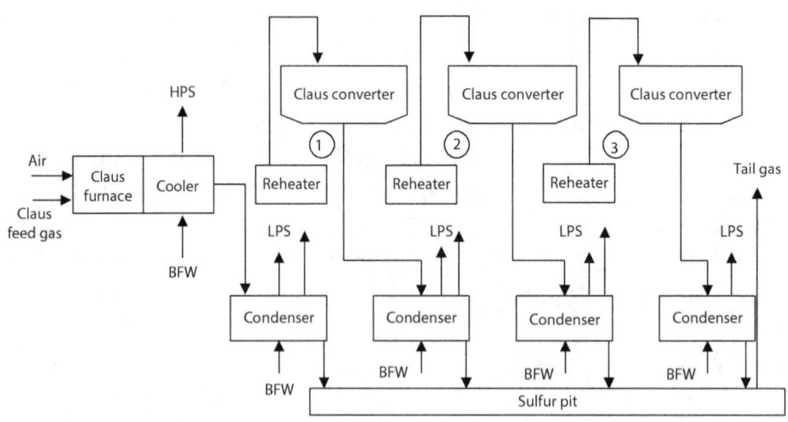

Figure 3.9 Classic Claus Plant.

*BFW= Boiler Feed Water, HPS = High Pressure Steam and LPS = Low Pressure Steam

Gases containing ammonia, like the gas from the refinery's sour water stripper (SWS), or hydrocarbons are converted in the burner muffle. Enough air is injected into the muffle for the complete combustion of all hydrocarbons and ammonia. The air to acid gas ratio is controlled so that, in total, 1/3 of all H_2S is transformed to SO_2. This ensures a stoichiometric reaction for the Claus reaction. The separation of the combustion processes ensures a correct dosage of the desired air volume required as a function of the feed gas composition. To minimize the process gas volume or get higher combustion temperatures, the air demand may also be covered by injecting pure oxygen. Many technologies utilizing high-level and low-level oxygen enrichment are available in the industry, which needs the employment of a special burner within the reaction furnace for this process option. Usually, 60 to 70% of the whole quantity of elemental sulfur produced in the process is obtained within the thermal process step.

The main portion of the hot gas from the combustion chamber flows through the tube of the process gas cooler and is cooled down so that the sulfur formed in the reaction step condenses. The heat given off by the process gas and the condensation heat evolved are utilized to produce medium- or low-pressure steam. The condensed sulfur is removed at the gas outlet section of the process gas cooler. A little portion of the process gas may be routed through a bypass inside of the process gas cooler. This hot bypass stream is added to the cold process gas through a three-way valve to regulate the inlet temperature needed for the first reactor.

Catalytic step: The Claus reaction continues in the catalytic step with activated aluminum (III) or titanium (IV) oxide and serves to boost the sulfur yield. H_2S reacts with the SO_2 formed during combustion in the reaction furnace and results in gaseous, elemental sulfur. This is often the known as a Claus reaction:

$$2\,H_2S + SO_2 \longrightarrow 2\,H_2O + 3/8\,S_8$$

The catalytic recovery of sulfur consists of three sub-steps: heating, catalytic reaction, and cooling plus condensation. These three steps are normally repeated a maximum of three times where an incineration or tail-gas treatment unit (TGTU) is added downstream of the Claus plant and only two catalytic stages are usually installed.

The first process step in the catalytic stage is that the gas heating process. It is necessary to prevent sulfur condensation in the catalyst bed, which might result in catalyst fouling. The desired bed operating temperature within the individual catalytic stages is achieved by heating the process gas in a re-heater until the desired operating bed temperature is reached. Many strategies of reheating are used in industry:

- Hot-gas bypass involves mixing the two process gas streams from the process gas cooler (cold gas) and the bypass (hot gas) from the first pass of the waste-heat boiler.
- Indirect steam reheaters, where gas also can be heated with high-pressure steam in a heat exchanger.
- Gas/gas exchangers, where the cooled gas from the process gas cooler is indirectly heated from the hot gas coming out of an upstream catalytic reactor in a gas-to-gas exchanger.
- Direct-fired heaters wre fired reheaters utilizing acid gas or fuel gas, which is burned sub-stoichiometrically to avoid oxygen breakthrough which can damage Claus catalysts.

The recommended operating temperature of the first catalyst stage is typically 315 °C to 330 °C (bottom bed temperature). The high temperature in the first stage also helps to hydrolyze carbonyl sulfide (COS) and carbon disulfide $(CS)_2$, which is formed in the furnace and would not otherwise be converted in the modified Claus process.

The catalytic conversion is maximized at lower temperatures, but care must be taken to make sure that each bed is operated above the dew point of sulfur. The operating temperatures of the subsequent catalytic stages are usually 240 °C for the second stage and 200 °C for the third stage (bottom bed temperatures).

In the sulfur condenser, the process gas coming from the catalytic reactor is cooled to between 150 and 130 °C. The condensation heat is used to generate steam at the shell side of the condenser. Before storage, liquid sulfur streams from the process gas cooler, the sulfur condensers, and from the final sulfur separator are routed to the degassing unit where the gases (primarily H_2S) dissolved in the sulfur are removed.

The tail gas from the Claus process, still containing combustible components and sulfur compounds (H_2S, H_2 and CO), is either burned in an incineration unit or further desulfurized in a downstream tail gas treatment unit. The 'Tail Gas' is a key point here. There is unconverted H_2S in this tail gas and that must be taken care of. The tail gas treatment is a process field of its own and is discussed further next.

3.3.4.2 Split-Flow Claus Plant

The classic Claus process must have a really high concentration of H_2S to make the plant work. For H_2S concentrations in the range of 25 to 40%, the split flow configuration can be utilized. During this process, the feed is split and one third or more of the feed goes to the furnace and the remainder joins the furnace exit gas before entering the first catalytic converter. Once two thirds of the feed are bypassed, the combustion air is adjusted to oxidize all the H_2S to SO_2 and, consequently, the necessary flame temperature can be maintained. The split-flow process has two constraints:

- Sufficient gas should be bypassed so the flame temperature is more than approximately 927 °C.
- The maximum bypass is only two thirds, as a result of one third of the H_2S must be reacted to form SO_2.

Thus, if air preheating is employed with the split-flow configuration, gases with as little as 7% H_2S can be processed.

3.3.4.3 *Oxygen Enrichment Claus Plant*

In this system, oxygen is provided to the Claus plant through a sparger into the air line to the reaction furnace. The oxygen can also be injected directly into the furnace through a lance or burner or feed air can be fully replaced by oxygen through a burner in the furnace. In some cases, complete feed air replacement can double Claus plant capacity. The resulting higher oxygen concentration increases the operating temperature of the furnace, which leads to steadier operation and greater ammonia destruction. As a result, the Claus reaction furnace maximizes its ability to combust the oxygen and air mixture and the H_2S produced throughout processing works with system converters and condensers to recover the sulfur. Increased oxygen also means lower nitrogen volumes which reduces gas volume in tail gas treatment.

3.3.4.4 *Claus Plant Tail Gas*

The tail gas treating unit converts the tiny quantity of sulfur compounds (<5%) that were not converted in the sulfur recovery unit (SRU) into H_2S and recycles them back to the SRU for additional processing. The foremost well-known tail-gas treatment technology is at the SCOT (Shell Claus Off-gas Treatment) plant (Kohl and Nielsen, 1997). The SRU tail gas is heated and sent to the catalytic reactor where, essentially, all of the sulfur compounds are converted into H_2S. The gas from the catalytic reactor is cooled in the waste heat exchanger and the quench tower. Excess water is removed in the cooling process and is sent to the sour water stripper. The cooled gas is then sent to the absorber column, where amine removes the H_2S and some of the CO_2 in the gas stream. The remaining gas (vent gas) is sent to the thermal oxidizing unit. The rich amine from the absorber is heated in the lean/rich exchanger and fed to the regenerator column. Steam, generated in the re-boiler, heats the amine and removes the H_2S and CO_2 from the amine. The lean amine from the stripper is cooled in the lean/rich exchanger and the lean solvent cooler and returned to the absorber. Amine has a natural affinity for both CO_2 and H_2S allowing this to be a very efficient and effective removal process. There are many various amines employed in gas treating:

- Monoethanolamine (MEA): Used in low pressure natural gas treatment applications requiring stringent outlet gas specifications
- Diethanolamine (DEA): Used in medium to high pressure treating and does not require reclaiming as MEA and DGA systems do
- Methyldiethanolamine (MDEA): Has a higher affinity for H_2S than CO_2, which allows some CO_2 "slip" while retaining H_2S removal capabilities
- Diisopropylamine (DIPA)
- Aminoethoxyethanol /diglycolamine (DGA)
- Formulated special solvents

MEA is the strongest base of the group and, as such, is the most reactive with acid gases. Many chemical reactions are involved and the reversibility of these reactions is the basis for the cyclic MEA scrubbing process. The SCOT absorption process is designed to absorb as little CO_2 as possible. The H_2S and CO_2 removed from the amine is cooled (and water removed) within the overhead condenser and recycled to the sulfur recovery unit for additional processing into sulfur.

The Shell Claus Off-Gas Treating (SCOT) process, which was announced to the industry in late 1972, was developed by Royal Dutch Shell laboratories in the Netherlands and is licensed in the U.S. by Shell Development and can be viewed as the industry standard classical process for small scale gas treatment. The SCOT process takes place in essentially three stages: heating and reducing all S-compounds to H_2S, cooling and quenching, and H_2S absorption, stripping, and recycling (Figure 3.10).

The reactions take place at 165–175 °C and the hydrogenated tail gas is cooled in the waste heat boiler, generating medium pressure steam, and then further cooled in a quench tower. The cool gas enters the absorber where it is contacted with a 25–50 wt.% methyldiethanolamine (MDEA) aqueous solution to absorb H_2S selectively. The rich MDEA solution flows down the stripper where the absorbed H_2S and CO_2 are stripped by upward

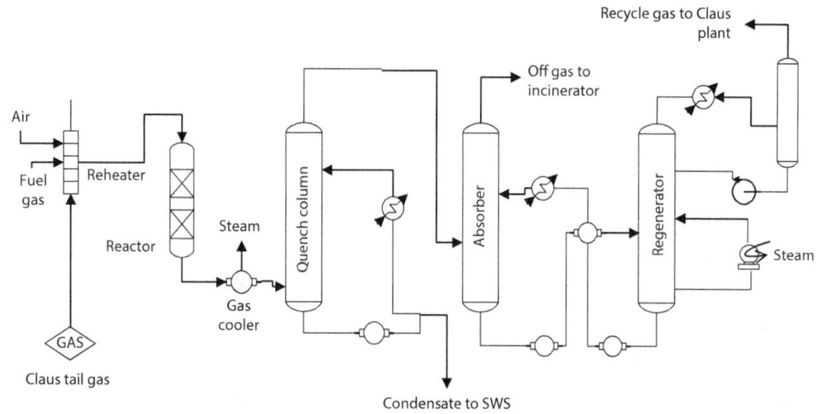

Figure 3.10 Simplified Block Flow Diagram of the SCOT Process.

rising steam. The regenerated solvent from the bottom of the stripper is recycled back to the absorber after cooling to 30–48 °C. H_2S and CO_2 leave the stripper at the top and recycle to the front-end Claus plant.

The SCOT reactor contains a bed of cobalt-nickel or cobalt-molybdenum catalyst that allows the reducing atmosphere to hydrogenate or hydrolyze most of the S-compounds contained in the heated tail gas feed to H_2S according to the following reactions:

$$SO_2 + 3H_2 \longrightarrow H_2S + 2H_2O$$

$$COS + H_2O \longrightarrow H_2S + CO_2$$

$$CS_2 + 2H_2O \longrightarrow 2H_2S + CO_2$$

$$1/x\,S_x + H_2 \longrightarrow H_2S$$

3.3.5 Absorption/Desorption Process

Chemical absorption is mainly based on the chemical affinity of H_2S for metallic cations. This process can be categorized to those involving oxidation of S^{2-} to S^0 and those based on the capture of S^{2-} through precipitation of its metallic salts, owing to their very low K_{sp} (water solubility product). Another option, which belongs to the second category, is the capture by aqueous alkaline solutions which rapidly react with diffused H_2S, but it is of low interest in industrial application due to the high reactivity of CO_2 with alkaline solutions and less selectivity towards H_2S. Not only this, but its main drawback comes from the high consumption of relatively expensive alkaline reactants (i.e. NaOH or CaO) throughout the capturing of CO_2.

The absorption/desorption processes supported with amines are those most widely used even for treatment of relatively low H_2S levels. There are many available processes and they may be divided into three categories, according to the absorbent: (1) physical absorption processes, (2) absorption combined with reversible reaction by carbonate solutions, and (3) amine solutions. A typical process based on an amine solution is shown in Figure 3.11. In this process, the acidic components react with an alkanolamine absorption liquid via an exothermic, reversible reaction in a gas/liquid contactor. These processes have the ability to remove a large quantity of H_2S economically and CO_2 may also be controlled if necessary. When CO_2 and H_2S are present, the following chemical reactions occur in an aqueous MDEA solution (Zare and Mirzaei, 2009):

$$CO_2 + H_2O \longleftrightarrow HCO^{3-} + H^+$$

$$HCO^{3-} \longleftrightarrow CO_3^{2-} + H^+$$

$$H_2O \longleftrightarrow OH^- + H^+$$

$$R.NH^+ \longleftrightarrow R.N + H^+$$

$$H_2S + R.N \longleftrightarrow HS^- + H^+$$

$$HS \longleftrightarrow S^{2-} + H^+$$

The absorber operates from ambient pressure up to 70 bar and from 25 to 70 °C. The energy consuming desorption of the acid gases is carried out at around 130 °C and at pressures from ambient up to 3 bar. Desorption

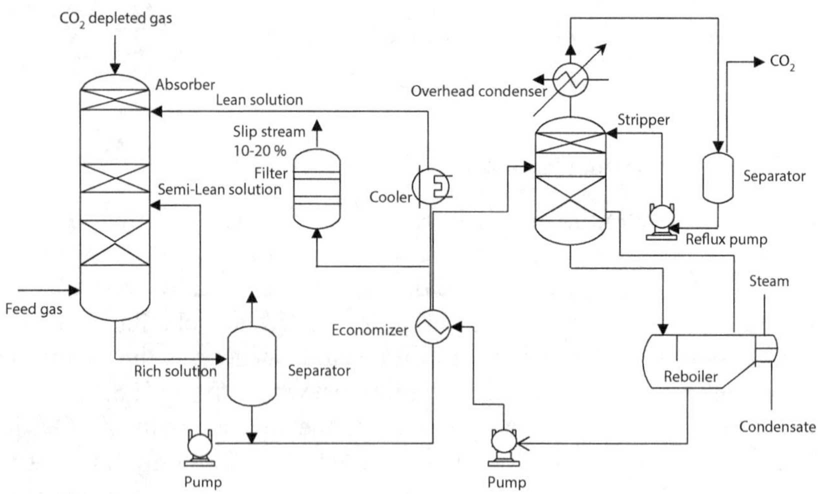

Figure 3.11 A High Pressure Absorber with a Split.

pressure may not necessarily be lower than the absorption pressure (e.g. tail gas treatment); this depends on the requirements of the connected Claus-plant. MDEA-plants process up to 40,0000 Nm^3/h feed gas in a single train (Bolhàr-Nordenkampf *et al.*, 2004).

For accurate plant design, it is of great importance to be able to predict the mass transfer behavior in the absorption and desorption column. Reactions which take place in the liquid phase can be divided, in principle, into two groups: reactions where equilibrium is controlled and reactions that are kinetically determined. The chemical reactions determine the composition of the different ion species in the liquid phase and, therefore, the enhancement of mass transfer. Equilibrium reactions are fast enough to assume chemical equilibrium throughout the entire liquid phase (Markus *et al.*, 2004). An equilibrium approach for the absorption is not suitable, if predictive capabilities of the model are necessary, as it is the case for selective H_2S and/or CO_2 removal in alkanolamine-systems. This can only be achieved using a rate-based non-equilibrium model. It is based on the mass and heat transfer between the liquid and vapor phase occurring on a height-increment of the structured and random packing, respectively. Mass and energy balances are connected by rate-equations across the interface using the two-film theory to calculate the transfer rates (Bolhàr-Nordenkampf *et al.*, 2004).

3.3.6 Biodesulfurization

Major problems in oil and gas operations result from the biogenic formation of H_2S in the reservoir. The presence of H_2S results in increased corrosion, iron sulfide formation, higher operating costs, reduced revenue, and constitutes a serious environmental and health hazard. Linkous (1993) reported a technology that removes and prevents the formation of biogenic H_2S which is based on the principle of 'Biocompetitive Exclusion' where the addition of low concentrations of a water soluble nutrient solution selectively stimulates the growth of an indigenous microbial population, thereby inhibiting the detrimental Sulfate Reducing Bacteria (SRB) population, which causes the generation of the H_2S. The versatility and low cost of this novel technology offers the petroleum industry a practical and cost effective methodology for the control of H_2S in oil and gas wells.

As mentioned before, H_2S contamination may be treated by biochemical, chemical, and physical methods (Burgess, 2001). Many physicochemical processes, like the dry gas reduction-oxidation (redox) process, liquid redox processes, and the liquid adsorption process, are usually used for desulfurization of gases containing H_2S. However, they have high capital costs, demand large energy inputs, and lead to the generation of secondary hazardous wastes (Pandey and Malhotra, 1999). Several physical means of controlling the formation of acid-mine drainage have been improved, but they have not been very successful. Therefore, efforts have been directed towards biological processes for the removal of H_2S that are characterized by small capital costs and low energy requirements.

BDS, via the injection of ambient air into the gas headspace in the digester, is the most widely used process for internal desulfurization applied in 90% of all biogas plants and followed by the addition of iron salts to the fermenting substrate. Depending on the

proportion of oxygen injected into the gas phase, H_2S is oxidized into elementary sulfur, sulfate-S, or sulfite-S. A reformation of S into H_2S by an unintended return of accumulated degradation products from the colonization surfaces into the fermentation substrate is one of the crucial drawbacks of this method. Recently, external-BDS methods have become much more popular and find their way into the market as it solves a lot of the drawbacks of the internal-BDS. Several methods for external-BDS have been reported, such as two-stage bio-scrubbers and one stage-bio-trickling filters.

In two-stage bio-scrubbers, the H_2S in the raw gas is absorbed, first, by contacting with water in a scrubber and, afterwards, biologically degraded in a bioreactor (van Groenestijn and Hesselink, 1993). A further option is a one-stage bio-trickling filter where the raw biogas passes through a fixed bed reactor filled with plastic packing materials. Washing water is trickled periodically over the fixed bed to promote a biofilm generation on the packing material enabling the immobilization of microorganisms. Dissolved H_2S is oxidized by certain microbial groups to elementary sulfur and sulfate (Naegele *et al.*, 2013; Barbusiński and Kalemba, 2016). Moreover, treatment of gaseous emissions contaminated with H_2S by bio-trickling filters inoculated with single cultures of sulfur oxidizer bacteria exhibit several advantages over physicochemical methods, such as shorter adaptation times, meaning a shortening and even absence of the bacterial lag phase and a higher BDS efficiency (Syed *et al.*, 2006; Aroca *et al.*, 2007).

One of the most common technologies for the biological treatment of sour gas is the THIOPAQ™ process (Figure 3.12) (Cline *et al*, 2003). It removes H_2S from gaseous streams by absorption into a mild alkaline solution, under a pressure of up to 75 barg, in a trayed or packed column (which is preferred) gas/liquid contactor, which is followed by the oxidation of the absorbed sulfide to elemental sulfur by a consortium of naturally occurring, colorless bacteria, *Thiobacilli,* in the THIOPAQ™ reactor. The volume of the bioreactor usually ranges from 5–2500 m^3 and operates

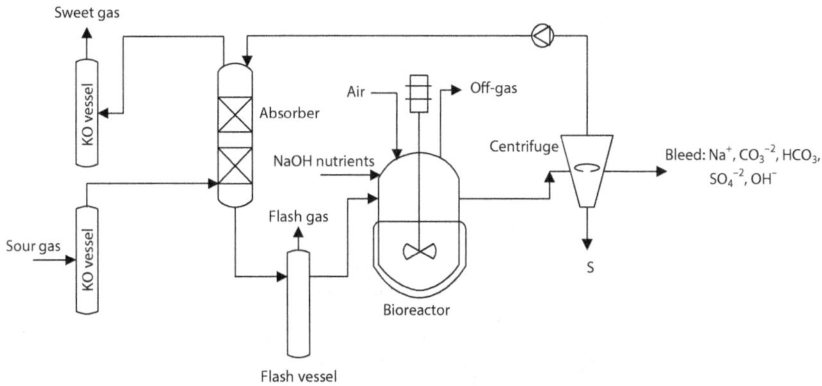

Figure 3.12 Simplified Block Flow Diagram of the THIOPAQ™ Process.

at atmospheric pressure at a temperature of around 30 °C. Beasley and Abry (2003) reported that virtually complete conversion of H_2S to elemental sulfur (>99.999%) can be achieved.

The following reactions may occur in the absorber at feed gas pressure using caustic soda as a solvent:

$$H_2S + NaOH \longrightarrow NaHS + H_2O$$

$$CO_2 + NaOH \longrightarrow NaHCO_3$$

$$NaHCO_3 + NaOH \longrightarrow Na_2CO_3 + H_2O$$

$$H_2S + Na_2CO_3 \longrightarrow NaHS + NaHCO_3$$

The sulfide formed in the absorber is removed in the bioreactor, where the following reactions may occur at atmospheric pressure and under aerobic conditions:

$$NaHS + 1/2\ O_2 \longrightarrow 1/8\ S_8 + NaOH$$

$$NaHS + 2O_2 + NaOH \longrightarrow Na_2SO_4 + H_2O$$

$$Na_2CO_3 + H_2O \longrightarrow NaHCO_3 + NaOH$$

$$NaHCO_3 \longrightarrow CO_2 + NaOH$$

Thiobacillus bacteria in the bioreactor can be considered equivalent to small catalyst particles that gain energy required for their growth from the conversion of the sulfide to elemental sulfur. Moreover, the caustic used to absorb H_2S gas is regenerated in the bioreactor and recycled back to the absorber and the produced elemental sulfur is gravimetrically separated in a decanter. In order to avoid accumulation of sulfate ions, a continuous bleed stream from the bioreactor is required.

The biodesulfurization rate of H_2S depends on several parameters which could be better controlled with a recirculated water phase (pH, temperature, addition of nutrients, removal of accumulated salts, etc.) (Smet *et al.*, 1998). A wide range of pH was reported especially for H_2S removal in biofilters. Gabriel and Deshuesses (2003) reported that the decrease in pH to pH2 resulted in an increased hydrogen sulfide degradation rate up to 99%. Jin *et al.* (2005) recorded 95% BDS at pH2. However, an optimum pH of 6 was reported for autotrophic bacteria. Smet *et al.* (1998) reported that within the genus of the sulfur compound bacteria *Thiobacillus* some species are found to favor pH-neutral environments and others favor low pH values. An insufficient availability of oxygen in the bioreactor would lead to an incomplete removal of H_2S (Gadre, 1989). An optimal stoichiometric demand of 4–6% (v/v) air in biogas is required for complete oxidation. However, in another study, the optimal desulfurization performance was obtained at 8–12% air fed into the biogas flow, but higher air application rates resulted in lower CH_4 content in the biogas and a lower calorific value (Naegele *et al.*, 2013). Furthermore, low and higher temperature are not desirable and approximately 30–40 °C is the optimum. The desirable bacteria to be used in a bioprocess to convert H_2S to S^0 should possess the subsequent basic features: reliable capability of converting H_2S to S^0, minimum nutrient inputs, and simple separation of S^0 from the biomass (Rattanapan and Ounsaneha, 2011). Moreover, a variety of bacteria are capable of H_2S oxidation and, hence, serve as potential candidates for gas desulfurization technology (Gadre, 1989).

3.3.6.1 *Photoautotrophic Bacteria*

The photosynthetic van Niel reaction is a well-studied anaerobic, inorganic acid gas bioconversion (van Niel, 1931):

$$2nH_2S + nCO_2 \longrightarrow 2nS + n(CH_2O) + nH_2O$$

Genera of the family *Chlorobiaciae* and *Chromatiaciae* catalyze this reaction and the reaction has been optimized for desulfurization of acid gas effluent containing H_2S, H_2, and CO_2. Fed-batch and chemostat cultures have shown that simultaneous control of molar flow rates of incoming gases and bioreactor photon flux is important to the optimization policy of the van Niel reaction (Maka and Cork, 1990). However, the requirement for light as an energy source in this process is a severe economic disadvantage.

Studies on microbial ecology related to phototrophic bacteria have shown that a species of green sulfur bacteria (GSB), *Cholorobium limicola,* (Larsen, 1952) is the best suited for sulfide removal and satisfies the criteria for a desirable bacterium (Syed and Henshaw 2003). GSB are non-motile and deposit elemental sulfur extracellularly. This feature makes GSB suitable where the recovery of elemental sulfur from sulfide-containing wastewater is desired (Syed *et al.*, 2006). A schematic diagram of the anaerobic acid gas bioconversion process is shown in Figure 3.13. In contrast of the Claus process, this system does not need oxygen because sulfide is anaerobically oxidized to elemental sulfur in the presence of CO_2 according to the van Niel reaction. The accumulation of sulfur compounds in the system reduced the total conversion of H_2S to elemental sulfur. However, the oxidation of H_2S to sulfate was suppressed when the reactor was operated with high H_2S concentrations and high space velocity (defined as the number of reactor volumes of feed gas that can be treated per hour).

A gas-fed batch reactor is a stirred tank type reactor (Figure 3.14) continuously or intermittently operated for the gas phase and cyclically operated for the liquid phase (nutritive solution). The microorganisms can be suspended in the solution or immobilized on different media (i.c. strontium alginate beads (Kim and Chang 1991)).

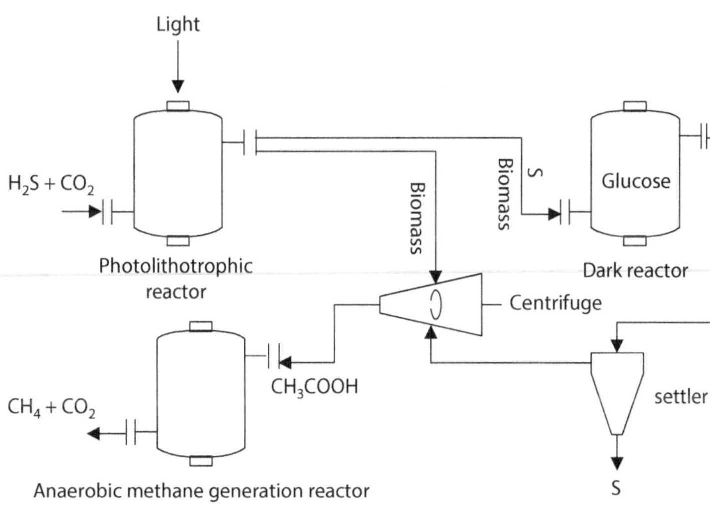

Figure 3.13 Phototrophic Microbial Acid Gas Bioconversion Plant.

Kim and Chang (1991) immobilized cells of *Chlorobium limicola* in a strontium-alginate matrix for transforming H_2S into elemental sulfur. They also addressed the problems of sulfate accumulation due to the undesirable side reaction:

$$H_2S + 2CO_2 + 2H_2O \longrightarrow 2(CH_2O) + H_2SO_4$$

The accumulation of sulfur and sulfate was found to be dependent on the light energy and feed rate of H_2S. The intensity of the light source was varied, keeping the exposed area of the reactor constant at loading rates of 32 or 64 mg S^{2-}/h/ L. A high irradiance resulted in sulfide removal, over half of which was converted to sulfate. At low irradiance, sulfide was not fully oxidized and accumulated in the reactor. Between these limits, a state was found in which neither sulfide nor sulfate was found in the reactor, while elemental sulfur and thiosulfate ($S_2O_3^{2-}$) were the only products. For white light (380 to 900 nm), this optimum state resulted in 60% of the influent H_2S becoming S^0 and 40% becoming $S_2O_3^{2-}$. For infrared light (700 to 900 nm), the optimum favored sulfur production was 97:3 (S^0:$S_2O_3^{2-}$). The optimum for infrared light also occurred at a lower irradiance level (219 W/m^2) than white light (406 W/m^2). These findings demonstrated the superiority of infrared light as a light source to produce elemental sulfur using GSB. It was suggested that light uptake by the transparent strontium alginate was easier because of light scattering and easy light permeation into the

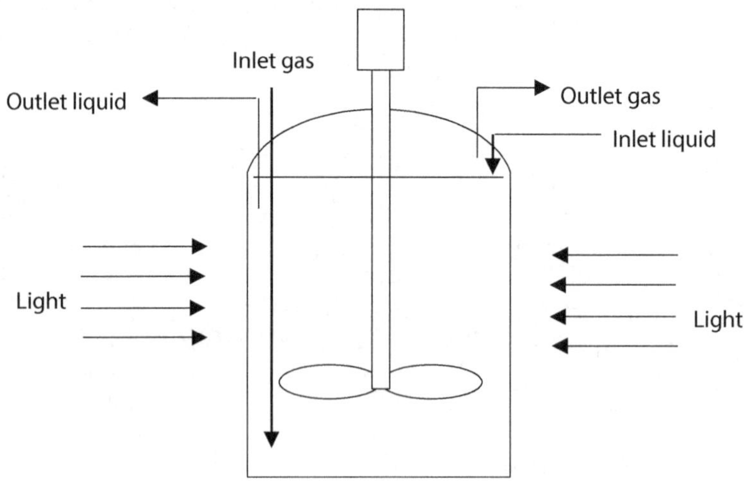

Figure 3.14 Fed-Batch or Continuous Flow Reactor.

interior space of the beads. Thus, the maximum oxidation rate of 3.8 mmol (H_2S)/L h, with negligible SO_4^{2-} accumulation, was achieved with immobilized cells. This experiment also tried to optimize light energy input by using light emitting diodes (LEDs) with a peak wavelength at 710 nm in a batch-fed stirred tank reactor. Experiments were performed separately using an incandescent bulb, LED710, and a combination of LED710 and a fluorescent bulb as the light sources and their individual performances during H_2S removal per unit luminous flux, (mmol/min) per g protein/L, and per W/m^2 were evaluated. They found that the maximum performances using LED710 and LED710 with a fluorescent bulb were 18.7 and 14.1 times the performance of a reactor with an incandescent bulb, respectively. Kim *et al.* (1996) investigated the performance of LEDs in a plate type photo-bioreactor. They observed that the maximum performance per unit luminous flux while using LEDs was 31 times that of an incandescent bulb. This efficiency was achieved by only supplying light within the wavelength range where absorption by bacteria was at a maximum.

In addition to the above-described system, a two-step bioprocess for the removal of sulfate from waste streams has been proposed (Jensen and Webb, 1995) (Figure 3.15). The first step of the process is bio-catalyzed by a strictly anaerobic acetate-degrading and sulfate reducing microorganism, for example *Desulfobacter postgateii*, while the second step is catalyzed by the strictly anaerobic photoautotroph *Chlorobium limicola*.

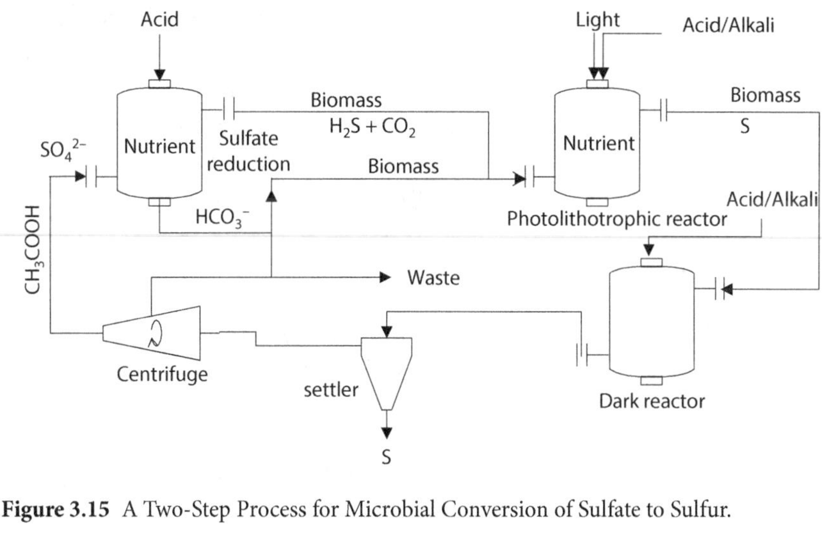

Figure 3.15 A Two-Step Process for Microbial Conversion of Sulfate to Sulfur.

First step

$$\text{Acetate} + SO_4^{2-} \xrightarrow{\textit{Desulfobacter postgateii}} H_2S + 2HCO_3^-$$

$$2HCO_3^- + 2H^+ \xrightarrow{\text{Acid}} 2H_2O + 2CO_2$$

Second step

$$H_2S + 1/2\,CO_2 \xrightarrow{\textit{Chlorobium limicola}} 1/4\ \text{Acetate} + S + 1/2\,H_2O$$

Theoretically, only 25% of the total acetate required in the first step can be produced by *Chlorobium.* The *Chlorobium* reactor does not require steam sterilization, as constant flushing with toxic concentrations of H_2S inhibits the growth of potential contaminating microorganisms. Studies on the optimization of *Desulfobacter* for sulfate reduction, utilizing a whole-cell immobilized *Desulfobacter chemostat,* have been reported by Grimm *et al.* (1984). The major disadvantages in the use of photosynthetic bacteria on a large scale are mainly attributed to their anaerobic nature and their requirement for radiant energy and, hence, extremely large transparent surface area.

Basu *et al.* (1996) used a continuous stirred tank reactor equipped with a sulfur separator to remove hydrogen sulfide from a gas stream containing 2.5% H_2S at 1 atmospheric pressure. When a sulfide loading rate was 94.4 mg/h/L, H_2S conversion by *Chlorobium thiosulfatophilum* ranged from 53% to 100% at a gas retention time of 12.2 and 23.7 min, respectively. The sulfur recovered from the process by gravity separation was 99.2% of the theoretical yield. The separation of elemental sulfur from the bioreactor contents is essential to attain its value as a chemical industry feedstock.

Henshaw *et al.* (1997) tested the separation of the sulfur by settling, filtration, and centrifugation. Centrifugation achieved the best separation results with 90% of the elemental sulfur and 29% of the bacteria were removed from the suspension. They noted that a continuous-flow suspended growth bioreactor system for sulfide removal/sulfur recovery requires two separation stages: one to separate S^0 from the bioreactor effluent and one to separate biomass from the liquid product of the primary separator.

A fixed-film reactor can reduce the requirement for two separators since the biomass remains in the reactor. Two types of phototube reactors are shown in Figures 3.16a and b. These are tubular type reactors that are operated continuously. The reactor can be horizontally oriented (Figure 3.16a), having several passes or spirals to improve the residence time in the reactor (Kobayashi *et al.*, 1983) or can be vertically oriented, as presented in Figure 3.16b (Henshaw *et al.*, 1999 and Syed and Henshaw, 2003). The material of the tube is transparent to light and impermeable to oxygen

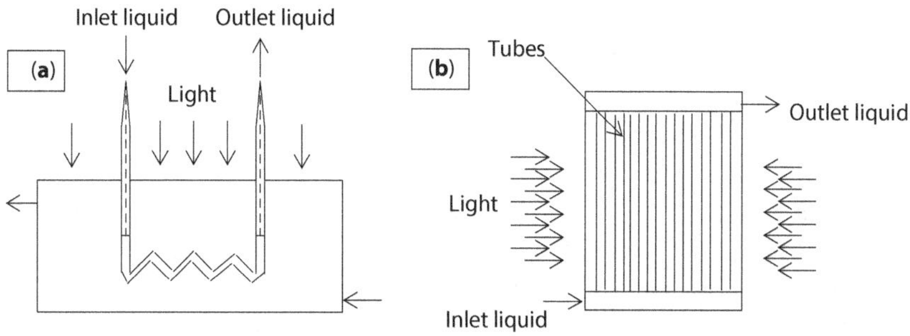

Figure 3.16 Phototube reactors (a) Horizontal (b) Vertical.

(Syed and Henshaw, 2003) and the bacteria develops on the inner wall of the tube reactor (fixed-film reactor).

Kobayashi *et al.* (1983) used a phototube reactor in which a sulfide containing reactor was passed through a 12.8 m long, 3.2 mm ID Tygon tube which was immersed in an illuminated water bath. The tube was able to achieve 95% sulfide removal in about 24.6 min while operated at a sulfide loading rate of 67 mg/h/L, postulating that a vertical attached growth configuration could eliminate or significantly reduce the problem of sulfur accumulation encountered by Kim and Chang (1991).

3.3.6.2 *Heterotrophic Bacteria*

The first extensive report on H_2S removal by a heterotrophic bacterium was revealed by Cho *et al.* (1992a), where *Xanthomonas* sp. strain DY44 isolated from dimethyl disulfide acclimated peat was employed in a batch system to remove H_2S from a gas stream. The maximum specific removal rate was 3.92 mmol (H_2S)/g DCW/h. The H_2S removal was not a consequence of chemolithotrophic activity, but was speculated to be a physiological H_2S detoxification process. The product of H_2S removal by *Xanthomonas* sp. strain DY44 was identified as a polysulfide. Therefore, a pH around neutral was sustained. H_2S could be removed as a single gas or in the presence of the sulfur-containing compounds methanethiol (MT), dimethyl sulfide (DMS), and dimethyl disulfide (DMDS). DMS and DMDS hardly affect H_2S removal, whereas the H_2O removal rate in the presence of MT was 40% of that in the presence of either DMS or DMDS. Process advantages that are claimed in the cell mass can be easily harvested for inoculation because of their rapid growth in nutrient mediums. Polysulfide, as a final product, is preferable to sulfate since a deterioration of microbial activity due to a decline in pH will not occur. H_2S removal efficiency can be enhanced by applying organic compounds to a peat biofilter inoculated with *Xanthomonas* sp. Strain DY44.

Xanthomonas sp. strain DY44 can also be used to enhance the removability of MT, DMS, and DMDS in mixed cultures with MT-, DMS-, and DMDS-degrading microorganisms where the degradability of these compounds is inhibited by the presence of H_2S. However, the specific H_2S removal rate of *Xanthomonas* sp. strain DY44 is lower than that of purified *Thiobacillus* Sp. Furthermore, application of a heterotrophic organism is not favorable if organic compounds are not readily available.

3.3.6.3 *Chemotrophic Bacteria*

Natural gas and biogas can be biologically desulfurized in additional units represented mainly by bio-filters, bio-trickling filters, and bio-scrubbers or directly into the anaerobic reactor by applying micro-aerobic conditions. All these processes are based on the S-cycle, and more specifically, in H_2S oxidation. In the aforementioned extra units H_2S is solubilized in a humid packed bed where aerobic species of sulfide-oxidizing bacteria (SOB) are immobilized and grown as a biofilm in the presence of O_2 (Noyola *et al.*, 2006). Elemental sulfur (S^0) and SO_4^{2-} are the thermodynamically stable by-products of biological H_2S oxidation, which have been proposed to proceed through several intermediates. Duan *et al.* (2005) suggested the following pathway of chemoautotrophs:

$$H_2S \longrightarrow S^0 \longrightarrow S_2O_3^{2-} \longrightarrow S_4O_6^{2-} \longrightarrow S_3O_6^{2-} \longrightarrow SO_3^{2-} \longrightarrow SO_4^{2-}$$

Tang *et al.* (2009) reported the main reactions that can be carried out by SOB:

$$H_2S + 1/2\,O_2 \longrightarrow S^0 + H_2O \quad \Delta G^0 = -209.4 \text{ kJ/reaction}$$

$$S^0 + 1/2\,O_2 + H_2O \longrightarrow SO_4^{2-} + 2H \quad \Delta G^0 = -587.41 \text{ kJ/reaction}$$

$$H_2S + 2\,O_2 \longrightarrow SO_4^{2-} + 2H \quad \Delta G^0 = -798.2 \text{ kJ/reaction}$$

$$S_2O_3^{2-} + H_2O + 2\,O_2 \longrightarrow SO_4^{2-} + 2H \quad \Delta G^0 = -818.3 \text{ kJ/reaction}$$

According to the above equations, Gibbs free energy (ΔG) is negative which means the reactions can happen spontaneously. At this point, it should also be noted that H_2S oxidation in biological systems occurs concurrently with chemical reactions (van der Zee *et al.*, 2007), where $S_2O_3^{2-}$ is the main by-product (Janssen *et al.*, 1995).

3.3.6.3.1 Thiobacilli

A number of chemotrophs are suitable for the biodesulfurization (BDS) of gases containing H_2S. These bacteria grow by using inorganic carbon (CO_2) as a carbon source and chemical energy from the oxidation of reduced inorganic compounds such as H_2S. In the presence of reduced organic carbon sources (glucose, amino acids, etc.), some of these bacteria (so-called mixotrophic microorganisms) can grow heterotrophically using the organic carbon as a carbon source and an inorganic compound as an energy source (Prescott *et al.*, 2003).

BDS of gases containing H_2S by chemotrophs occurs under aerobic conditions with O_2 as an electron acceptor or in anaerobic conditions with alternative electron acceptors (e.g. nitrate), depending on the type of bacteria (Prescott *et al.*, 2003; Syed *et al.*, 2006).

Thiobacillus sp. is widely used in studies of the degradation of H_2S and other sulfur compounds by biological processes (Sublette and Sylvester 1987a,b; Chung *et al.* 1996; Cha *et al.* 1999 and Oyarzún *et al.* 2003). These bacteria have the ability to grow under various environmental stress conditions, such as oxygen deficiency, acid conditions, etc. Many *Thiobacillus* sp. have acidophilic characteristics and are able to develop in conditions of low pH (1–6).

Thiobacillus ferroxidans is reported to oxidize ferrous ions at 30 °C and pH 2.2 at a rate 500,000 times as fast as the rate at which oxidation would occur in the absence of the bacteria (i.e in chemical oxidation) (Satoh *et al.*, 1988). In another microbial desulfurization process developed up to laboratory scale, where the direct conversion of H_2S to water soluble sulfate ions has been reported, the culture used was facultative anaerobic and autotrophic bacteria, *Thiobacillus denitrificans* (Sublette and Sylvester, 1987a,b).

Gadre (1989) reported a fixed-film bioreactor for H_2S removal from biogas (2% H_2S) using chemoautotrophic bacteria likely to be *Thiobacilli*. At a maximum volumetric removal rate of 3.2 mmol (H_2S)/L/h, the removal efficiency of 69.5% was not satisfactory. Insufficient oxygen availability in the bioreactor was thought to be the reason. The employing of fixed-film bioreactors, according to Gadre (1989), would be simple and effective since they have many advantages over other advanced bioreactors, such as rapid initial start-up, ability to withstand shock loadings, rapid restart after long shutdowns, elimination of mechanical mixing and biomass recycling, and better stability and efficiency. Removal of carbon disulfide and hydrogen sulfide from the exhaust gases of a cellulose filament manufacturing plant has been carried out on a pilot scale. Large volumes of exhaust gas, with a comparatively low concentration of H_2S and CS, were treated in a fixed bed bioreactor equipped with polypropylene or polyethylene packing material carrying immobilized microorganisms identified mainly as *Thiobacillus*. Figure 3.17 shows a schematic diagram of the process. The crude gas flows counter currently to a water-activated sludge suspension through the reactor. The oxidation products (SO_4^{2-}, H^+) are removed continuously from the reactor and neutralized by the addition of NaOH. The sulfate formed is removed from the settler by continuous addition of water and the surplus of sludge is also continuously harvested from the settler.

Cadenhead and Sublette (1990) described a study of H_2S oxidation by *Thiobacillus thioparus, Thiobacillus versutus, Thiobacillus neopolitanus,* and *Thiobacillus thiooxidans* the purpose of which was to determine whether any of these *Thiobacilli* offered clear advantages over *T. denitrificans* in the aerobic oxidation of H_2S. None of the organisms utilized were found to offer a clear advantage over *T. denitrificans;* all *Thiobacilli* showed lower biomass yields and lower NH_4^+ utilization than those reported by Sublette (1987) for aerobic oxidation of H_2S by *T. denitrificans* in similar batch conditions. No comparisons were made on the basis of oxidation rates.

Nishimura and Yoda (1997) used a multiple bubble-tray airtight contact tower (bio-scrubber) to scrub H_2S from the biogas produced by an anaerobic wastewater treatment process, as shown in Figure 3.18a. A two-reactor system (a gas-liquid contact tower and an aeration tank) was employed to separate the oxidation process from the absorption process to prevent air from mixing with the biogas. In the contact tower, H_2S from the biogas was absorbed into the mixed liquor and subsequently oxidized to sulfate by

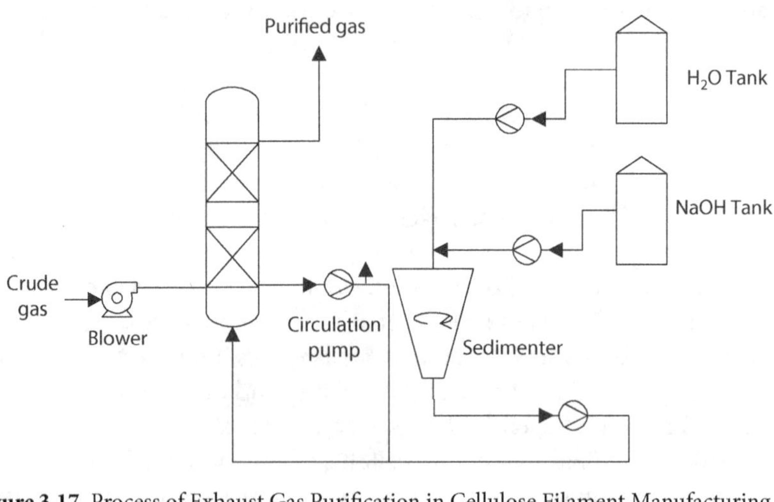

Figure 3.17 Process of Exhaust Gas Purification in Cellulose Filament Manufacturing Plant.

sulfur oxidizing bacteria after returning to the aeration tank. When treating 2000 ppm of H_2S in 40 m^3/h of biogas, more than 99% removal efficiency was achieved.

Kraakman *et al.* (2011) divided the bioreactors for gas treatment into two groups: turbulent and laminar contactors, The power consumption in laminar contactors systems (BTF) can be 1 or 2 orders of magnitude lower than in turbulent contactors. For the BTF, the nozzle at the top can disperse water evenly and form a thin liquid film on the surface of the carriers, which is good for H_2S transferring from gas phase to liquid phase. For the BBC however, the biogas enters from the bottom of the reactor to form bubbles and the diameter of the bubbles is about 1 mm between microbubbles and small bubbles. Smaller bubbles are good for mass transfer because they increase the gas-liquid contact area, while the turbulence of gas-liquid is weakened when the bubble is smaller. Thus, it is necessary for the BBC with smaller bubbles to adopt extra measures to improve the mixing of gaseous pollutants and carriers with biofilm.

Soreanu *et al.* (2008a; 2009) reported a maximum EC of 10.6–14.5 g $H_2S/m^3/h$ with 74–100 % of removal efficiency in an anoxic BTF.

Shell International Oil Products developed a process that uses naturally occurring *Thiobacillus* bacteria to remove hydrogen sulfide from natural gas by converting it to sulfur (Van Grinsven, 1999). *Thiobacillus thiooxidans* has a great tolerance for acidic conditions and can grow at a $pH < 1$ (Takano *et al.*, 1997; Devinny *et al.*, 1999).

A full scale plant located northeast of Brooks, Alberta, Canada uses the Shell-Paques process for natural gas desulfurization (Benschop *et al.*, 2002). H_2S is removed from a gaseous stream by absorption into a sodium carbonate/bicarbonate solution. The sulfide containing scrubbing liquid is treated in the bioreactor where it is mostly converted biologically to elemental sulfur. The bioreactor is supplied with a nutrient stream, air, makeup water, and sodium hydroxide. It is reported that normally less than 3.5% of the sulfide is converted to sulfate and a continuous bleed stream is required to avoid accumulation of sulfate. A compost filter is used to treat the trace H_2S present in the spent air from the bioreactor. Less than 4 ppmv effluent H_2S concentration is achieved when treating natural gas containing 2000 ppmv H_2S.

A bio-filter is a three phase bioreactor (gas, liquid, solid) made with a filter bed that has a high porosity, high buffer capacity, high nutrient availability, and high moisture retention capacity to ensure that the target microorganisms can grow on it, as shown in Figure 3.18b (Jorio and Heitz, 1999; Elias *et al.*, 2002; Daustos *et al.*, 2005). The contaminated gas is continuously fed in the bio-filter, while a nutrient solution is discontinuously

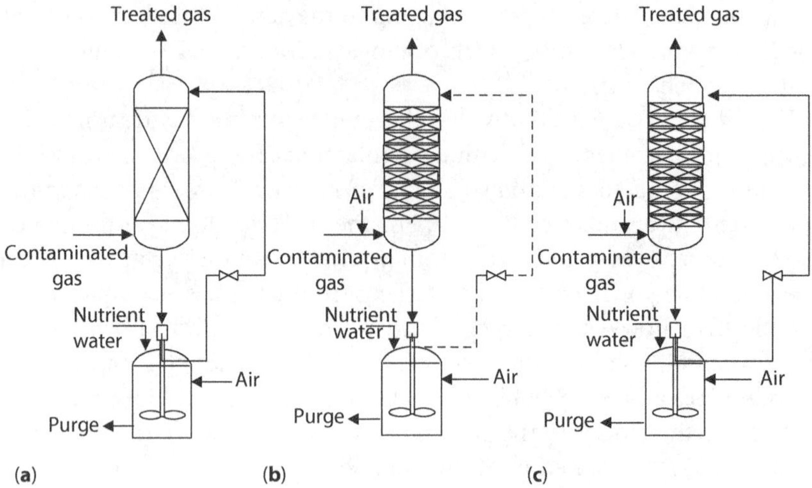

Bio-scrubber (**a**); Bio-filter (**b**); Bio-trickling filter (**c**)

Figure 3.18 Different Systems Used for H2S Removal.

added. Various types of bio-filter media have been used by researchers. Representative cases are discussed below.

Chung *et al.* (1997) used *Thiobacillus novellus* in a bio-filter for H_2S oxidation under mixotrophic conditions. A removal efficiency of 99.6% was achieved and the products were sulfate (83.6%) and sulfite (12.6%). Little conversion of sulfide to elemental sulfur was achieved.

Shareefdeen *et al.* (2002) reported the operation of a commercial bio-filter for the treatment of an air stream containing hydrogen sulfide, ammonia, dimethyl sulfide, methanethiol, and ethylamine. This proprietary wood-based (BIOMIX™) bio-filter achieved 96.6% removal of H_2S at an inlet concentration of 1.07 mg/m^3.

Elias *et al.* (2002) used packing material made up of pig manure and sawdust for bio-filtration purposes. More than 90% H_2S removal efficiency was attained at a loading rate of 45 g/m^3/h. No nutrient was added to the system and the porosity of the packing material decreased from 23.1 to 12.9%. However, this change in porosity did not affect the removal efficiency significantly and it was claimed that the bio-filter could be easily cleaned by flushing water through the inlet. The main by-product of the biodegradation process was sulfur (82% of total sulfur accumulation) accompanied by sulfates and thiosulfates (<18%).

Schieder *et al.* (2003) described the "BIO-Sulfex" bio-filter to remove H_2S from biogas, which uses thiobacteria attached on fixed bed material.

The biomass was aerated and the filter was flushed with nutrient containing liquid to remove sulfur from the system. Six BIO-Sulfex modules to treat biogas containing up to 5000 ppm H_2S were operated at flowrates of 10 to 350 m^3/h with 90% or more H_2S removal achieved.

Clark *et al.* (2004) operated a pilot-scale agricultural bio-filter to reduce odors from a swine manure treatment plant's exhaust air. Biofilters were packed with polystyrene particles and peat moss (3:1 ratio by volume). The packing volume was 1.89 m^3. The gas flowrate was 100 L/s. The inlet load contained 2–60 ppm hydrogen sulfide and 2–30 ppm ammonia. The addition of nutrients did not play an important role in the overall system performance. Average odor reduction was approximately 38% without addition of nutrients and 45% when nutrients were added. Increasing the temperature had a favorable effect during the acclimatization phase only.

The working principle of a bio-trickling filter is the same as for a biofilter except that the packed bed is continuously trickled over by an aqueous phase nutritive solution as shown in Figure 3.18c (Cox and Deshusses, 2001).

Cox and Deshusses (2001) used two laboratory scale bio-trickling filters (BTF) made of polypropylene inoculated with biomass from a toluene biodegrading filter operating at pHs of 4.5 and 7.0 to treat H_2S and toluene in a gas stream. There was no significant difference between the performances of the two reactors in terms of H_2S removal. At an inlet concentration of approximately 50 ppmv, complete consumption of H_2S was observed. However, the removal efficiency decreased to 70–80% when the inlet concentrations were raised to 170 ppmv. High removal efficiency for H_2S, in comparison to other reduced sulfur compounds, was obtained by Gabriel and Deshusses (2003) using *Thiobacillus* sp. in a bio-trickling filter (BTF). For inlet H_2S concentrations as high as 30 ppmv, the typical removal efficiency was 98%. Methyl mercaptan, carbonyl sulfide, and carbon disulfide removal efficiencies were 67, 44, and 35% at inlet concentrations of 67, 193, and 70 ppbv, respectively.

Soreanu *et al.* (2005) developed a laboratory-scale bio-trickling system, as shown in Figure 3.18c, in order to remove H_2S from digester biogas under anaerobic conditions. In these experiments, polypropylene balls inoculated with anaerobically digested sludge were used as packing material in the bioreactor (packing volume of 0.0062 m^3, 90% volume free). Sodium sulfite was added in the nutritive solution as an oxygen scavenging agent. Nitrate was used as an electron acceptor in the absence of oxygen. Removal efficiency greater than 85 % was achieved for an H_2S inlet concentration of 500 ppm and a gas flowrate of 0.05 m^3/h. Of particular interest, inhibition of the biological process by trace amounts of O_2 was noticed when a nitrate solution was used as the sole nitrogen/nutrient source.

Li *et al.* (2016) investigated the influence of the molar ratio of sulfide/nitrate (S/N) on biogas desulfurization performance in a bio-trickling filter (BTF) and a bio-bubble column (BBC). The results showed that with the decrease of the S/N ratios from 3.6 to 0.7, the removal efficiencies of H_2S increased from about 66 to 100%, while the removal of nitrate decreased from 100 to 70% in the two bioreactors. Thus, the BTF has a better and more stable desulfurization performance than the BBC does which could be attributed to their different gas-liquid contacting modes. After the start-up period, the removal efficiencies of H_2S were above about 95% in the BTF at different S/N ratios. However, in the BBC, the removal efficiencies of H_2S were about 78, 85, and 100% when the S/N ratios were about 2.4, 1.2, and 0.7, respectively. With the increase of the S/N ratios from 1.0 to 2.5 in the BTFs, the removal of H_2S in biogas was affected slightly, while the percentages of the produced sulfate decreased evidently. In addition, different supplying methods of nitrate wastewater, i.e. intermittent and continuous, did not affect the removal of H_2S significantly, while the intermittent addition of nitrate wastewater increased the percentages of sulfate and denitrification performance. The maximum elimination capacity (EC) of H_2S recorded 54.5 g $H_2S/m^3/h$ in the BTF.

3.3.6.3.1.a. Thiobacillus Denitrificans

Sublette and Sylvester (1987a) employed the anaerobic growth of *T. denitrificans* on H_2S in a continuous stirred-tank reactor. Although the removal of H_2S recorded no more than 3%, a maximum volumetric productivity of 2.3 mmol (H_2S)/L/h was recorded. They concluded that the efficiency of the process was low and the reactor volumes required for field applications would be impractical. Biomass concentration and the quality of the environment of the cells were identified as the two most important variables in maximizing volumetric productivity while maintaining reactor stability.

Sublette (1990) reported the application of the above process on a pilot scale for the treatment of a biogas from an anaerobic digester. The bioreactor consisted of a bubble column that received a gas feed of biogas plus compressed air. The off gas from the bubble column contained >9% oxygen. The contamination of biogas or methane by O_2 can be hazardous due to the explosive nature of such mixtures. The volumetric productivity was just 1.1 mmol (H-S oxidized)/L/h. However, a concentration of only 0.15% H_2S in the biogas made the *T. denitrifcans* reactor volumetrically feasible.

T. denitrificans have also been used for the oxidation of sulfide (H_2S, HS^-, S^{2-}) in sour gas to sulfate. A sulfide-tolerant strain of *T. denitrifcans* was co-immobilized with $CaCO_3$ in calcium alginate beads and contacted with sour gas under anaerobic conditions in a packed-bed column. The co-immobilized $CaCO_3$ has three functions; $CaCO_3$ acts as a buffer neutralizing the acid by-product of sulfide oxidation. Acting as a buffer, $CaCO_3$ converts to HCO_3^- and CO_2, which serve as carbon sources to support the growth of *T. denitrijcans.* Ca^{2+}-generated internally into the beads maintains the mechanical stability of the beads.

Buisman *et al.* (1990) tested three different continuous-flow reactor configurations: fixed-film CSTR (stirred-tank reactor), bio-rotor (a rotating cage containing reticulated polyurethane biomass support particles partly immersed in the reactor liquid), and a fixed-film up-flow reactor. For the up-flow and bio-rotor reactors, 95 to 100% sulfide removal efficiencies were achieved for loading rates up to 500 mg H_2S/h/L. The removal efficiency decreased rapidly above this loading rate. At 938 mg/h/L (bio-rotor) and 1040 mg/h/L (up-flow) loadings, sulfide removal efficiencies were 69 and 73%, respectively. At a 500 mg/h/L sulfide loading rate, the stirred-tank reactor's removal efficiency was approximately 62%.

Thiobacillus denitrificans is able to grow facultatively on reduced sulfur compounds by reducing nitrate (NO_3^-) to nitrogen gas (N_2) (Lampe and Zhang 1996; Kleerebezem and Mendez 2002). The oxidation of H_2S by *T. denitrificans* has also been applied in a two-stage microbial process for the removal of sulfur dioxide from a gas with net oxidation to sulfate. In reactors in a series, SO_2 was reduced to H_2S in the first stage by *Desulfovibrio desulfuricans.* The H_2S was then stripped with nitrogen and sent to a second stage where it was oxidized to SO_4^{-2} by *T. denitrifcans.* A sulfur balance demonstrated complete reduction to H_2S in the first stage and complete oxidation of H_2S to SO_4^{-2} in the second stage.

Malhautier *et al.* (2003) used two laboratory scale bio-filters packed with granulated digested sludge. One unit was fed mainly with H_2S and the other unit with NH_3. Complete H_2S removal (100%) and 80% NH_3 removal efficiency occurred. However, the authors concluded that the oxidation of high levels of H_2S might have a negative effect on the growth and activity of nitrifying bacteria.

Ma *et al.* (2006) described a biological removal of high concentrations of H_2S using the immobilized *Thiobacillus denitrificans* with peat moss, wood chip, ceramic, and granular activated carbon (GAC), separately, as shown in Figure 3.19. A GAC bioreactor had significant potential to treat H_2S odor gas as GAC provides a more uniform surface area and good resistance to crushing, allowing better operational control in areas such as gas adsorption capacity, gas distribution, and pressure drop. In addition, GAC provided higher bacterial adsorption capacity than other inert carriers

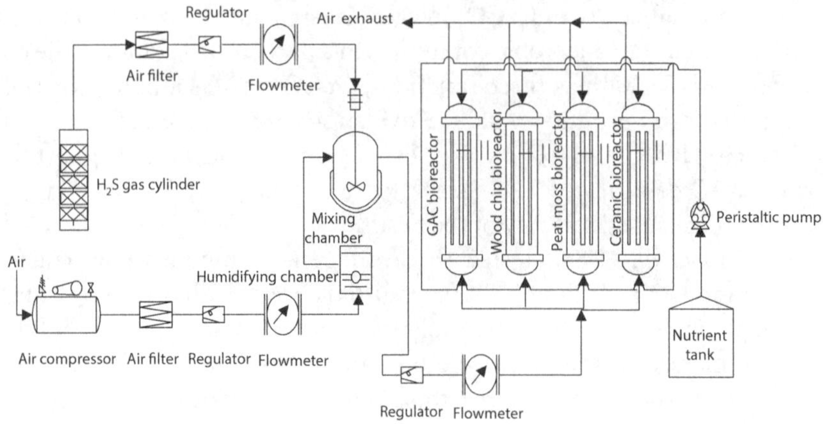

Figure 3.19 Immobilized-Cells Bioreactor System.

and could be regenerated. The GAC bioreactor achieved an average 96.8% removal efficiency of H_2S at the inlet concentrations of 110–120 mg/L of H_2S during a 160-day operating period. No significant acidification phenomenon occurred in this system during H_2S treatment because its main product was determined to be elemental sulfur.

3.3.6.3.1.b. Thiobacillus Thioparus

Tanji *et al.* (1989) described a system for the simultaneous removal of low concentrations of DMS, methyl mercaptan (MM), and H_2S from malodorous gases. *Thiobacillus thioparus* TK-m cells were immobilized on cylindrical porous polypropylene pellets and contacted with a sulfur-containing gas in a packed tower. Up to 95% removal of H_2S was achieved at a rate of 0.73 mmol/L/h. DMS was relatively less decomposable both by immobilized and freely dispersed *T. thioparus* TK-m. The removal rate of MM was intermediate. It was stressed that when treating large quantities of gas at low concentrations, a reduction in pressure drop was important to economize the operating costs.

Cho *et al.* (1992b) presented the first report of the application of an isolated microorganism to a practical deodorizing system. The capacity of *Thiobacillus thioparus* DW44 to remove hydrogen sulfide, MT, DMS, and DMDS from exhaust gas was displayed in a pilot-scale peat bio-filter. A schematic diagram of the peat bio-filter is shown in Figure 3.20. The moisture content of the peat was controlled at 6–70% by spraying with effluent water from a wastewater treatment plant. The temperature inside the bio-filter was kept above 8 °C at all times by preheating the inlet gas. No acclimation period was needed to reach such a high efficiency in the

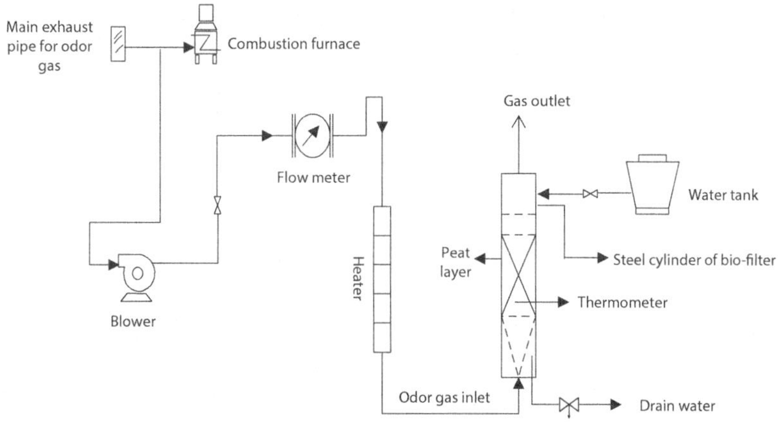

Figure 3.20 Pilot-Scale Peat Bio-Filter.

removal of the gases. The presence of heterotrophic bacteria and fungi utilizing organic substances extracted from the peat by the supplied wastewater did not seem to adversely affect the activity of *T. thiopurus* DW44.

Chung *et al.* (1996) immobilized *Thiobacillus thioparus* CH11 with Ca-alginate producing pellet packing material for the bio-filter. At a 28 s optimal retention time, the H_2S removal efficiency was more than 98%. Elemental sulfur or sulfate was produced depending on the inlet H_2S concentration.

Kim *et al.* (2002) investigated the simultaneous removal of H_2S and NH_3 using two bio-filters, one packed with wood chips and the other with granular activated carbon (GAC). A mixture of activated sludge (as a source of nitrifying bacteria) and *Thiobacillus thioparus* (for sulfur oxidation) was sprayed on the packing materials and the drain solution of the bio-filter was recirculated to increase the inoculation of microorganisms. Initially, both of the filters showed high (99.9%) removal efficiency. However, due to the accumulation of elemental sulfur and ammonium sulfate on the packing materials, removal efficiency decreased over time to 75 and 30% for H_2S and NH_3, respectively.

Oyarzún *et al.* (2003) used peat for the filter bed of a bio-filtration system inoculated with *Thiobacillus thioparus*. Supplemental nutrients were added and the initial moisture content was adjusted to 92%. The pH was also adjusted to 6.0. Full removal was achieved when fed with 355 ppm H_2S at 0.03 m^3/h. The removal efficiency decreased with increasing inlet H_2S concentrations and a maximum removal capacity of 55 $g/m^3/h$ was obtained.

Aroca *et al.* (2007) performed a comparative study on the removal of hydrogen sulfide in bio-trickling filters inoculated with *Thiobacillus thioparus* (ATCC23645) and *Acidithiobacillus thiooxidans* (ATCC19377). That proved the acid bio-trickling filter inoculated with *A. thiooxidans* perfoms better H_2S removal with the advantage being that the system does not require an exhaustive pH control of the liquid media, recording 370 g S/m^3/h at 45 sec of residence time and 405 g S/m^3/h of inlet load (91% of efficiency), while a maximum recorded removal capacity of the bio-trickling filter inoculated with *T. thioparus* was 14 g S/m^3/h at 30 g S/m^3/h of inlet load and 47% removal efficiency at a residence time of 26 sec within pH range of 5.5–7.0.

Tóth *et al.* (2015) employed a sulfur-oxidizing bacteria, *Thiobacillus thioparus* (immobilized on Mavicell B support), to develop a micro-aerobic bio-trickling filter reactor for the efficient elimination of H_2S from synthetic biogas as shown in Figure 3.21. To test the capability of this particular

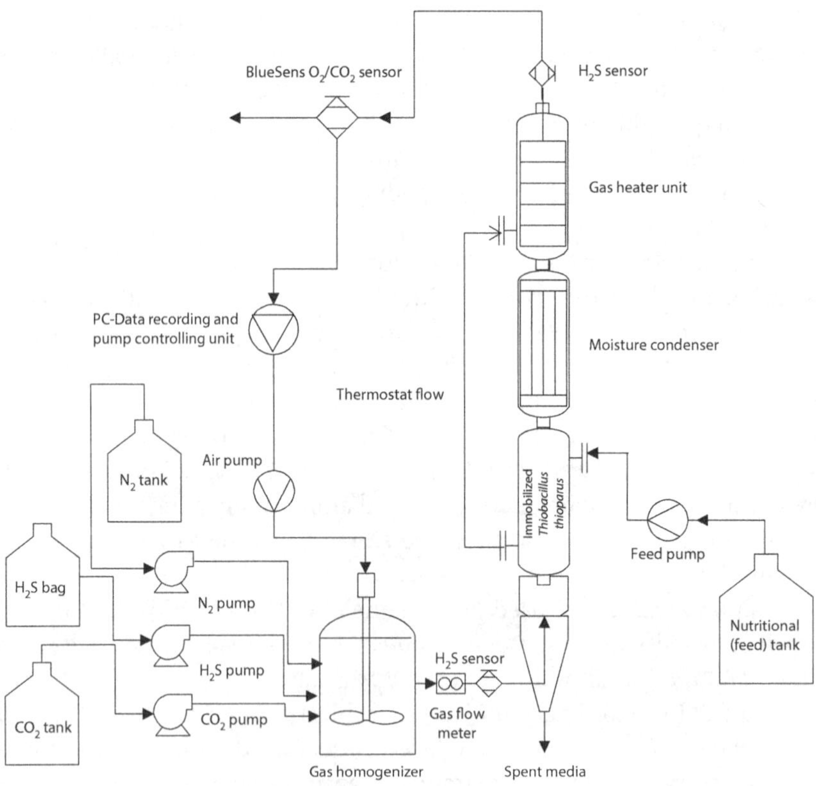

Figure 3.21 Continuous Micro-Aerobic Bioreactor.

strain in an oxygen-limited atmosphere, a fixed bed reactor was operated under 0.25–5.0 vol.% O_2 concentrations and its H_2S decomposing ability was statistically evaluated. It was found that the system achieved 100% H_2S elimination efficiency when at least 2.5 vol.% oxygen was provided.

The biological oxidation process of the absorbed sulfides to elemental sulfur by *Thiobacillus thioparus* TYY-1 was studied in an airlift-loop reactor (effective volume of 20 L) (Liu *et al.*, 2015b). The air aeration quantity was found to be the key influence factor of the desulfurization rate and elemental sulfur production rate. The best treatment effect of aeration was obtained at 120–160 L/h. The hydraulic remain time was found to be also important and the optimum hydraulic remain time was found to be 4–6 h under the influent concentration of S^{2-} for 200 mg/L. With these conditions after 20 days of operation, the results showed superior performance of the bioreactor for the desulfurization rate and elemental sulfur production rate, where the conversion products were mainly sulfur and the production rate of SO_4^{2-} was low. The removal efficiency of sulfide was more than 99.5% while the maximum yield of sulfur was 88% approximately.

In the raw natural gas or biogas, there are also some volatile organic sulfur compounds (VOSCs) such as methanethiol, dimethyl disulfide, dimethyl sulfide, carbon disulfide, and carbonyl sulfide (Mata-Alvarez and Llabrés, 2000; Böresson, 2001; Sheng *et al.*, 2008). These VOSCs could also be absorbed by the alkaline adsorbents and make a significant effect on the activity of sulfide-oxidizing bacteria (SOB) (Lobo *et al.*, 1999). Carbon disulfide was a common ingredient in sour natural gas, biogas, and some tail gases of plants. Currently, carbon disulfide is found to have a negative effect on desulfurization, but little research was done to investigate the inhibitory effect of carbon disulfide on the BDS process. As an example, Kim *et al.* (2005) reported the concentrations of carbon disulfide in landfill gas from four landfill sites to be in the range from 25 to 5352 ppb, where the highest ratio of H_2S to CS_2 was about 6:1, which was highly toxic to the strain *Thiobacillus thioparus*. Ziyu *et al.* (2014) have investigated the effect of different carbon disulfide concentrations (0.01, 0.05, 0.1, 0.15, and 0.2%) on the growing and resting cells of *Thiobacillus thioparus* CGMCC 4826, which was isolated from the effluent of sulfate reducing the bioreactor, and found that this strain could oxidize thiosulfate to elemental sulfur and sulfate.

$$\text{S-SO}_3^{2-} + 1/2\text{O}_2 \xrightarrow{\textit{Thiobacillus thioparus}} \text{S} + \text{SO}_4^{2-}$$

Although the carbon disulfide slightly dissolved in the water, little carbon disulfide could obviously inhibit the growth of cells. Carbon disulfide at a concentration of 0.01% has significantly inhibited the growth of cells, but has hardly affected the BDS efficiency of resting cells. Although carbon disulfide at concentration of 0.05% had a negative effect on the BDS efficiency of resting cells, the effect of inhibition could be relieved by the increased density of resting cells. Therefore, 0.05% was chosen to be the critical value of carbon disulfide for BDS. For the resting cells, the parameters of the Michaelis-Menten equation were calculated by the method of Lineweaver-Burk. The V_{max} of BDS was decreased from 27.93 to 14.0 $S_2O_3^{2-}$ mg/L/h and the K_m was increased from 0.264 to 0.884 mM with the concentration of carbon disulfide rising up from 0.0 to 0.1%, so, under the optimized BDS-process, it was necessary to adjust the concentration of carbon disulfide in the absorbent below 0.05% by renewing the absorbent. Finally, these results showed that the growth of cells was sensitive to carbon disulfide and the resting cells had a resistance to the low level of carbon disulfide (0.05%).

3.3.6.3.1.c. Thiobacillus Thiooxidans

Berzaczy *et al.* (1990) patented a microbiological conversion process for the removal of sulfur-containing pollutants such as H_2S, CS_2, COS, thioalcohols, thioethers, and thiophenes in a waste gas, especially from cellulose fiber manufacture. In this process, *Thiobacillus thiooxidans* cells immobilized on commercially available packing material are contacted with the gas in a packed-bed reactor. The cells are kept moist at all times by spraying with a nutrient solution. The metabolic products (mainly H_2SO_4) draining from the reactor are neutralized by the addition of lye and lime water in two stages. $CaSO_4$ precipitates and is removed as a waste product. The supernatant is returned for internal circulation to minimize loss of chemicals. To prevent salt concentration in the cycle liquid, a corresponding quantity of salt solution is drawn from the neutralization vessel and replaced with fresh water.

Duan *et al.* (2006) used biological activated carbon as a novel packing material to achieve a performance enhanced bio-filtration process in treating H_2S through an optimum balance and combination of adsorption capacity with the biodegradation of H_2S by Thiobacillus thiooxidans, immobilized on the material, as shown in Figure 3.22. The biofilm was mostly developed through culturing the bacteria in the presence of carbon pellets in mineral media. In two identical laboratory scale bio-filters, one was operated with biological activated carbon (BAC) and another with virgin carbon without bacteria immobilization. Various concentrations of H_2S (up to 125 ppmv) were used to determine the optimum column performance. A rapid startup (a few days) was observed for H_2S removal

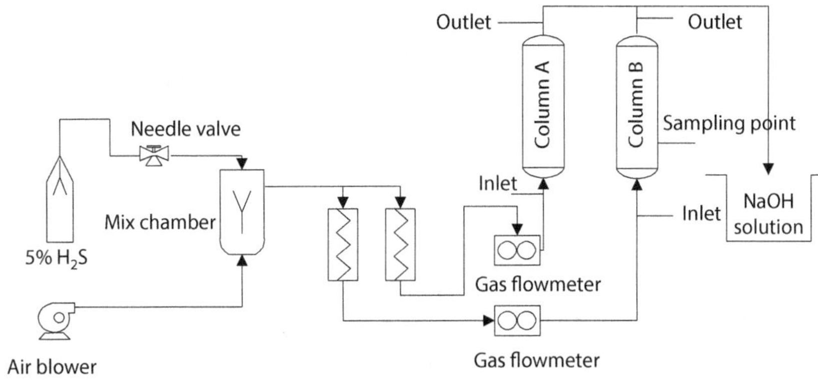

Figure 3.22 Bench-Scale Bio-Filter System.

in the bio-filter. At a volumetric loading of 1600 m^3/h (at 87 ppmv H_2S inlet concentration), the elimination capacity of the BAC (181 g H_2S m^3/h) at removal efficiency of 94% was achieved. If the inlet concentration was kept at below 30 ppmv, high H_2S removal (over 99%) was achieved at a gas retention time as low as 2 s, a value which is shorter than most previously reported for bio-filter operations.

3.3.6.3.1.d. Thiobacillus Ferrooxidans

Neumann *et al.* (1990) described a method for the removal of H_2S from biogas utilizing *Thiobacillus ferrooxiduns* cells in a packed bed. The micro-organisms are immobilized on peat, brushwood or pruning, refuse compost, or rubber. Air is fed continuously with gas to the reactor for oxidation of H_2S to S or SO_4^{2-}. Part of the clean gas/air mixture is recycled to the reactor inlet. To maintain a residual O_2 concentration in the purified gas below 3.0%, the feeding of air is adjusted by a controller. The controller receives a continuous signal from an oxygen probe on the outlet gas line, calculates a response compensating for the dynamics of the bio-filter system, and regulates the position of a control valve on the air line. The bed is kept moist at all times by a recirculating nutrient solution which also removes S/SO_4^{2-} and keeps the pH in its optimal range of 1.7–3.2. There is, however, no mention of flow rates or oxidation rates. This process applies to a situation, often encountered in practice, where the amount and composition of biogas are not constant. Only by adjusting the air flowrate to the amount and composition of biogas can the generation of an explosive mixture be avoided.

A promising alternative for the microbial treatment of H_2S containing gases is a Japanese method utilizing *T. ferrooxidans* as a means of reducing the costs of H_2S removal (Magota *et al.*, 1988). A layout of the BIO-SR process scheme is shown in Figure 3.5.

The process is basically dependent on the reaction of H_2S gas injected into a ferric sulfate solution in an absorber, producing a precipitate of elemental sulfur:

$$H_2S + Fe_2(SO_4)_3 \longrightarrow S + 2FeSO_4 + H_2SO_4$$

Depending on the gas flowrate and efficiency required, several types of absorbers are suitable, such as jet scrubbers, bubble-cap towers, or packed towers. Elemental sulfur is separated and recovered from the reduced solution of ferrous sulfate in a separator.

The sulfur separators can include settlers, filter presses, and sulfur melters, depending on the quality of elemental sulfur required.

The reactant, ferric sulfate, is regenerated from the ferrous sulfate solution by biological oxidation in an aerated bioreactor using *T. ferrooxidans* cells:

$$2FeSO_4 + H_2SO_4 + 1/2\,O_2 \xrightarrow{T.\ ferrooxidans} Fe_2(SO_4)_3 + H_2O$$

The oxidized solution is then recycled to the absorber to repeat the cycle. A distinct advantage of the process is that the reaction of H_2S with ferric sulfate is so rapid and complete that there remains no danger of discharging toxic waste gas. An H_2S removal efficiency of more than 99.99% has been attained in an existing commercial plant.

In general, the operating cost of the BIO-SR process is one-third that of conventional processes though the capital cost is only slightly lower. This gives an overall cost for the BIO-SR process, which is about 50% of the cost of conventional processes. Serious disadvantages of other microbial processes for H_2S removal are avoided in the BIO-SR process. H_2S does not inhibit the bacteria and SO_4^{2-} does not accumulate in the system. Also, contamination of the purified gas with O_2 is prevented.

Son and Lee (2005) improved a hybrid reactor by combining a chemical reduction reactor and a biological oxidation reactor to remove the toxic effect of H_2S on the cells and to enhance the H_2S removal rate. The microbial cells were immobilized on the surface of curdlan particles in order to enhance the Fe(II) oxidation rate through repeated fed-batch operation. As a result, the iron oxidation rate was four times faster than that obtained with the free cells. Iron solution, oxidized in an oxidation reactor by *Thiobacillus ferrooxidans*, was fed into the iron reduction reactor and the reduced iron solution was recycled into the iron oxidation reactor. The

X-ray diffractometer (XRD) data indicated that iron was precipitated along with elemental sulfur at the high concentration of H_2S, resulting in the iron oxidation rate being decreased with increasing reaction time.

Lin *et al.* (2013) used a pilot-scale chemical–biological process to remove H_2S from biogas, as shown in Figure 3.23. The inlet H_2S concentration was 3542 ppm and a removal efficiency of 95% was achieved with a gas retention time of 288 s. Purified biogas with an average of 59% CH_4 was collected for power production.

3.3.6.3.2 Other Examples of Chemotrophic Bacteria

Mesa *et al.* (2002) described a bio-scrubber system which can be integrated into a system to remove H_2S from biogas by a combination of chemical and biological processes. H_2S removal can be achieved by absorption in a ferric sulfate solution producing ferrous sulfate and elemental sulfur. Ferric sulfate can be regenerated by biological oxidation using *Acidithiobacillus ferrooxidans*. This study investigated the oxidation of ferrous iron by *ferrooxidans* which was immobilized on a polyurethane foam support where the support particles placed in an aerated column. Low cost polyurethane was chosen for being macroporous, having a large surface for microbial growth, and offering lower diffusion resistance to substrate transfer. However, ferric precipitates were accumulated on the support and on the air diffusers, which necessitated periodic interruptions of the process for cleaning. Precipitation, air supply, and chemical cost are the potential constraints for this process.

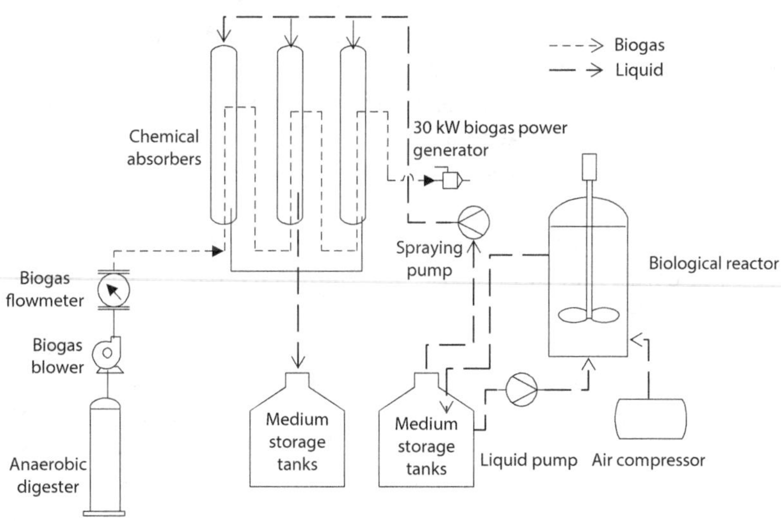

Figure 3.23 Pilot-Scale for Chemical–Biological Process.

Beggiatoa sp. and *Thiothrix* sp., both microaerophilic σ-proteobacteria, have mixotrophic nutritional functions because they are able to degrade H_2S using inorganic and organic (e.g. acetate) energy sources (Howarth *et al.*, 1999; Prescott *et al.*, 2003). *Pseudomonas acidovorans* and *Pseudomonas putida* are other mixotrophs which degrade both H_2S and organosulfur compounds (Chung *et al.*, 2001; Oyarzún *et al.*, 2003).

Chung *et al.* (2001) used bio-filters packed with co-immobilized cells *Pseudomonas putida* CH11 and *Arthobacter oxydans* CH8 for the removal of H_2S and NH_3, respectively, which are often present in off-gases of a livestock farm. In the 5–65 ppm range, H_2S and NH_3 removal efficiencies were greater than 96%. However, at higher concentrations, H_2S and NH_3 showed inhibitory effects on H_2S removal. They also assessed the environmental risk associated with the release of bacteria when treating large volumes of waste gases; the exhaust gas contained small amounts of bacteria (<19 CFU/m^3 in all cases) and was considered safe.

Sercu *et al.* (2005) studied the aerobic removal of hydrogen sulfide using a bio-trickling filter (BTF) packed with 1 L-polyethylene rings inoculated with *Acidithiobacillus thiooxidans* ATCC-19377. The inlet H_2S concentration was varied between 400 and 2000 ppm and the airflow rate was varied between 0.03 and 0.12 m^3/h. However, the system performance was not affected by changing the operational conditions and a maximal removal efficiency of 100% was obtained. During the experiment, the pH of the nutritive solution decreased to 2–3, but this did not affect the process performance.

Other studies that combined chemical and biological processes for both H_2S elimination and ferric iron regeneration by *Acidithiobacillus ferrooxidans* have been reported (Giro *et al.*, 2006; Alemzadeh *et al.*, 2009; Ho *et al.*, 2013). These processes are based on two reactions, as follows: the inlet H_2S is first oxidized with a ferric iron solution and yields elemental sulfur and the reduced ferrous iron is then re-oxidized by *A.* ferrooxidans in the biological process.

Cheng *et al.* (2013) investigated the integration of chemical and biochemical processes for desulfurization of simulated natural gas containing hydrogen sulfide (H_2S) using polyurethane foam as a support for the immobilization of *Acidithiobacillus ferrooxidans.* A good biological oxidation performance with a maximum oxidation rate of ferrous iron of 4.12 kg/m^3/h was recorded. That proved the effectiveness of the immobilizing matrix. The chemo-biochemical process is illustrated in Figure 3.24. A bubble column absorber was used as a chemical oxidation unit. Inlet and outlet were provided at the bottom and the top for untreated and treated simulated natural gas, respectively. The immobilized cell bioreactor was a glass column

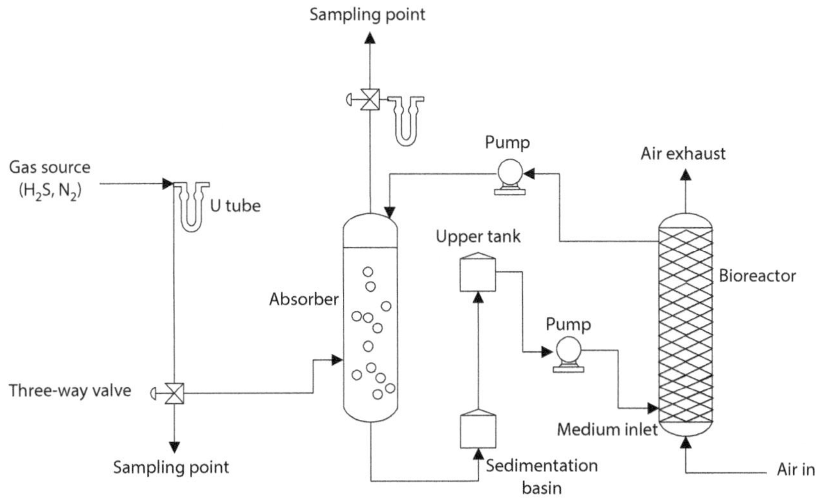

Figure 3.24 A Simple Diagram of Chemo-Biochemical Process for Desulfurization of Gas.

with a working volume of 0.56 L, where fresh modified 9K medium (Fe^{2+} = 8.5 kg/m^3) or reduced iron solution from the absorber was influx at the bottom of the bioreactor. Air was supplied with an air compressor through a filter and fed in at the bottom of the bioreactor. H_2S was produced by an equimolar reaction of Na_2S and H_2SO_4. After being introduced by passing nitrogen (99.9% v/v) over an H_2SO_4 solution where a solution of Na_2S was dripped, the simulated natural gas was introduced into the absorber and the gas flow rate was adjusted by a gas flowmeter. Elemental sulfur was retained through a sedimentation basin. Such chemo-biochemical processes (Figure 3.24) achieved removal efficiencies of about 80% when treating high concentrations of H_2S (15,000 ± 100 ppmv). This capacity was higher than those reported in literature, compared to 77 g H_2S/m^3/h (Pagella and De Faveri,2000), 250 g H_2S/m^3/h (Pandey *et al.*,2003), 22 g H_2S/m^3/h (Son and Lee, 2005), and 100 g H_2S/m^3/h (Li *et al.*, 2008).

Ho *et al.* (2013) used the chemical–biological process, as shown in Figure 3.25, using iron-tolerant *A. ferrooxidans* CP9 that maintained the balance of the Fe^{2+}/Fe^{3+} ratio and reached an H_2S removal efficiency of 98%. In the pilot-scale operations, the addition of glucose improved the biogas purification efficiency by increasing the cell density and ferrous oxidation efficiency. The H_2S loading reached 65.1 g S/m^3/h (3.3-fold higher than the laboratory-scale condition) with a removal efficiency of 96%. In addition, the factors of high tolerance for iron ions at 20 g/L and the rapid ferrous iron oxidation ability of *A. ferrooxidans* CP9 were important in

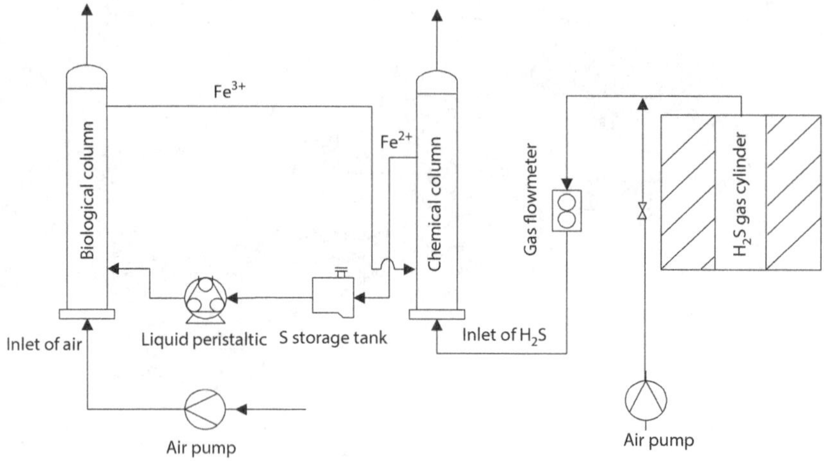

Figure 3.25 H_2S Treatment by Combined Chemical–Biological Reactor.

maintaining the balance of Fe^{2+}/Fe^{3+} concentrations. Although the exotic microbes appeared during the 311 d operation, the *A. ferrooxidans* CP9 cell density remained more than 10^8 CFU/g. These results clearly demonstrate that the chemical–biological process is a feasible method for removing a high H_2S concentration from biogas.

Vikromvarasiri *et al.* (2017) evaluated the efficiency of a bio-trickling filter inoculated with *Halothiobacillus neapolitanus* NTV01 (HTN) on H_2S removal from synthetic biogas. HTN is an obligatory chemolithoautotrophic bacteria able to tolerate and metabolize high sulfide concentrations. The HTN was isolated and purified from activated sludge and is a sulfur oxidizing bacteria able to degrade H_2S and thiosulfate to elemental sulfur and sulfate, respectively. Operational parameters in a short term operation were varied as follows: gas flow rate (0.5–0.75 LPM), inlet H_2S concentrations (0–1500 ppmv), and liquid recirculation rate (3.6–4.8 L/h). The maximum elimination capacity was found as 78.57 g H_2S/m^3 h, which had a greater performance than the previous studies.

Nazari *et al.* (2017) reported the isolation of Gram negative, motile, aerobic, and obligately chemolithoautotrophic and non-spore forming *Halothiobacillus* sp. ISOB 14 from contaminated soil with sulfur compounds for its ability to oxidize thiosulfate and used it as an electron donor. It showed a unique ability and high potential for usage in the removal of hydrogen sulfide. This bacterium also grows faster in a Postgate medium with 0.7% thiosulfate than in sulfur and sodium sulfide. Under optimum conditions, ISOB 14 showed the capability to remove thiosulfate at 100% after 18–24 h and produce sulfate up to 89.14% and 93.14% after 24 h and 72 h, respectively.

3.3.7 Other Approaches Concerning the Biodesulfurization of Natural Gas

Considerable efforts are still required concerning the packed media based biotechnologies; though effective, they have strict requirements in terms of both monitoring and maintenance due to the bacteria's high sensitivity to fluctuations in operational conditions, which translates into costs.

Recently, biogas biodesulfurization (BDS) with high concentrations of H_2S (>1000 ppmv) is mainly done by bio-filtration systems (Fortuny *et al.*, 2008; Montebello *et al.*, 2012). However, since biogas is initially oxygen-free, this means that the oxidants need to be supplied into a biogas BDS system for the oxidization of H_2S. Based on different electron acceptors, i.e. O_2 or NO_3^-/NO_2^-, BDS can be classified into two categories: aerobic and anaerobic processes (Li *et al.*, 2016). Currently, biogas aerobic desulfurization has mainly been applied, which requires a stoichiometric oxygen level depending on the inlet concentrations of H_2S in biogas (Tang *et al.*, 2004; Vanderzee *et al.*, 2007; Fdz-Polanco *et al.*, 2009; Fortuny *et al.*, 2011), but it is important to control the oxygen dosage in order to avoid reaching high concentrations of oxygen in biogas for safety reasons and because the residual oxygen or air in biogas after desulfurization can lead to a dilution of methane concentration, which will affect the further use of biogas (Fortuny *et al.*, 2008; Soreanu *et al.*, 2009). This can be solved upon the usage of NO_3^-/NO_2^- as electron acceptors during the BDS process. Moreover, biogas desulfurization could be coupled with wastewater denitrification when wastewater containing NO_3^-/NO_2^- is supplied (Soreanu *et al.*, 2008a; Deng *et al.*, 2009; Jing *et al.*, 2009; Turker *et al.*, 2011).

During anaerobic processes, the S/N molar ratio is the key parameter to control the level of H_2S oxidation, i.e. the end desulfurization products. At low concentrations of NO_3^-, H_2S is oxidized into S^0, while in the presence of enough concentrations of NO_3^-, H_2S is completely transformed to SO_4^{2-}. The anaerobic biological utilization of H_2S as an energy source for lithoautotrophic organisms can be described with the following reactions (Soreanu *et al.*, 2008b):

$$H_2S + NO_3^- \longrightarrow S^0 + NO_2^- + H_2O$$

$$5H_2S + 2NO_3^- \longrightarrow 5S^0 + N_2 + 4H_2O + 2OH^-$$

$$H_2S + 4NO_3^- \longrightarrow SO_4^{2-} + 4NO_2^- + 2H^+$$

$$5H_2S + 8NO_3^- \longrightarrow 5SO_4^{2-} + 4N_2 + 4H_2O + 2H^+$$

Zhou *et al.* (2015) also found that the S^0 is the dominant desulfurization product at high inlet loading rates of H_2S, while the sulfate is dominant at low loading rates in a micro-aerobic BTF.

Soft sulfur bacteria are typical anaerobic autotrophic microorganisms which can use CO_2 to produce microbial biomass and transfer H_2S to elemental sulfur or sulfate in the presence of inorganic nutrients and illumination (Wu *et al.*, 2016).

$$H_2S + CO_2 + O_2 + \text{Nutrients} \xrightarrow{\text{microbial chemoautotroph}} \text{Biomass} + S^0 \text{ or } SO_4^{2-} + H_2O$$

As previously mentioned, the evolved H_2S during the biogas fermentation process is removed after the anaerobic digestion process through different techniques, for example installing a set of additional desulfurization systems, such as iron oxide adsorption/oxidation, activated carbon adsorption, bio-filtration, etc. Although these methods can effectively remove H_2S from biogas, they have lots of disadvantages, such as large area required, high operation costs, and complex technological processes. Not only this, but the production of H_2S during the biogas fermentation process is reported to inhibit the anaerobic digestion process, thus reducing biogas production and leadig to poor biogas quality (Lar and Li, 2009); some novel desulfurization methods need to be developed to overcome these problems. One of these methods is in-situ H_2S removal during the biogas fermentation, which has some advantages, such as not requiring a set of additional desulfurization systems, ease of operation, simple process, and high efficiency (Jiang *et al.*, 2014).

This process can be performed through two techniques. The first is a micro-aerobic desulfurization process, in which a small amount of oxygen/air is supplied into anaerobic digesters so that the growth of SRB can be inhibited seriously and, at the same time, the introduced oxygen could react with the generated H_2S producing elemental sulfur. The other method is in-situ desulfurization throughout the addition of desulfurizers (such as iron-based oxidants) into the anaerobic digesters (Ripl and Fechter, 1991; Zhong *et al.*, 2004; Su *et al.*, 2012).

Diaz *et al.* (2011) reported reduction in evolved H_2S when the oxygen or air was supplied into the bioreactor. Kobayashi and Li (2011) also developed a self-agitated anaerobic reactor with the addition of air, in which about 99 % of H_2S in biogas could be removed.

As mentioned before, H_2S present in biogas can be oxidized to S^0 or SO_4^{2-} using nitrate and nitrite. Both nitrate and nitrite are normally available in most wastewater treatment plants and could be used to oxidize H_2S depending on the molar loading ratio of wastewater and biogas. However, a control approach is required in order to minimize the fluctuations in inlet and outlet H_2S concentrations in biogas and the oxidation potential of the wastewater used.

Turker *et al.* (2011) proved that combining sulfide removal with nitrate or nitrite removal not only allows the control of H_2S, but also improves nitrogen removal via autotrophic denitrification without using a carbon source. Biogas desulfurization was integrated with nitrogen removal in an industrial wastewater treatment plant (Turker *et al.*, 2011) and a control scheme (Figure 3.26) to monitor and control the concentration of hydrogen sulfide, combined with autotrophic denitrification process was developed. The mixture of nitrate and nitrite from a nitrification plant was used as a source of electron acceptors to oxidize sulfide in biogas. The performance of the process was monitored by an oxidation–reduction potential (ORP) sensor and the control scheme was developed to improve fluctuations in sulfide load to the bio-scrubber for stable operation, where the control scheme has been developed for biogas desulfurization using ORP under industrial conditions. The redox potential was maintained at about +50 to +100 mV in the activated sludge plant to monitor the performance of the nitrification process. The redox potential in the bio-scrubber was related to sulfide removal from biogas. More than 90% of H_2S was removed from the biogas with simultaneous nitrogen removal at wastewater/biogas ratios between 2 and 3. The process control algorithm was developed based

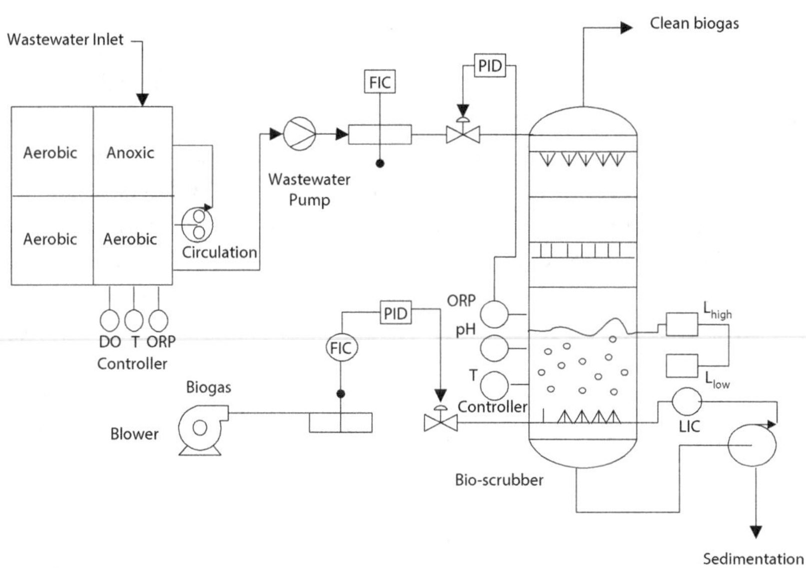

Figure 3.26 Integrated Process for Biogas BDS and Wastewater Biodenitrogenation.

on the measurement of redox potential in the effluent of the bio-scrubber. The effluent redox potential was related to sulfide removal and greater than 90% sulfide removals were correlated with approximately −100 mV effluent redox potentials. This value was used as a controlled variable and an influent wastewater flowrate containing a mixture of nitrate and nitrite was the manipulated variable. Therefore, the combination of anaerobic treatment, biogas production, and biogas cleaning with aerobic treatment could improve biogas desulfurization. This allows the integration of sulfur and nitrogen cycles to alleviate sulfur emissions. Thus, the advantage of the method developed by Turker *et al.* (2011) is that it combines sulfide removal with nitrogen removal which is normally required in most industrial wastewater treatment systems.

In another study concerning the simultaneous removal of sulfide and nitrite, Doğan *et al.* (2012) declared that bio-oxidation of sulfide under denitrifying conditions is a key process in the treatment of gas and liquids that are contaminated with sulfide and nitrite. A lab-scale continuous flow stirred tank reactor (CFSTR) was operated with nitrite as the electron acceptor for the evaluation of the effects of loading rates, hydraulic retention time (HRT), and substrate concentrations on the performance of the autotrophic denitrification process. The influent sulfide concentration was maintained at 0.16 kg/m^3, the HRT was decreased from 8.4 to 2 h and, for the entire study period, the sulfide removal efficiency was above 80% for the loading rates that ranged from 0.47 to 2.16 kg S^{2-}/m^{3}day. However, lower influent loading of NO_2^--N that corresponded to the stoichiometric ratios was used and the nitrite removal efficiency was close to 100%.

Klok *et al.* (2012) reported the application of haloalkaliphilic sulfide-oxidizing bacteria in gas lift bioreactors inoculated at oxygen-limiting levels, that is below an O_2/H_2S mole ratio of 1, where sulfide was oxidized to elemental sulfur and sulfate. This suggested that the bacteria reduced NAD^+ without the direct transfer of electrons to oxygen and that this is most likely the main route for oxidizing sulfide to elemental sulfur, which is subsequently oxidized to sulfate in oxygen-limited bioreactors. This pathway is called the limited oxygen route (LOR). Biomass growth under these conditions is significantly less abundant at higher oxygen levels.

Ramos *et al.* (2012) developed a new biotechnological process for the removal of H_2S from biogas. The desulfurization conditions present in micro-aerobic digesters were reproduced inside an external chamber called a micro-aerobic desulfurization unit (MDU) as shown in Figure 3.27. A 10 L-unit was inoculated with 1 L of digested sludge in order to treat the biogas produced in a pilot digester. After 128 d of incubation under optimum conditions, the average removal efficiency was 94%. Microbiological

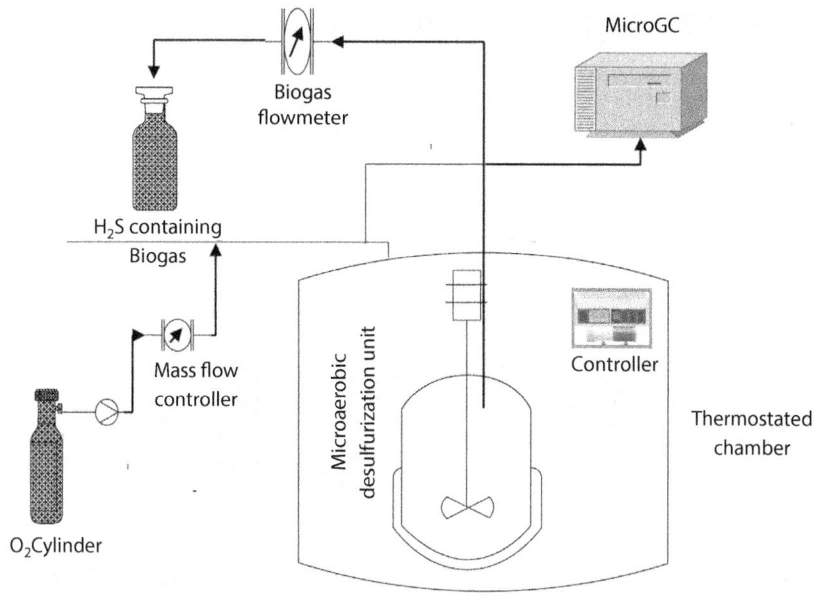

Figure 3.27 Micro-Aerobic Desulfurization Unit.

analysis confirmed the presence of at least three genera of sulfide-oxidizing bacteria. Approximately 60% of all thc H_2S oxidized was recovered from the bottom of the system in the form of large solid S^0 sheets with 98% w/w of purity. Therefore, this system could become a cost-effective alternative to the conventional bio-techniques for biogas desulfurization.

Wu *et al.* (2016) reported the in-situ removal of H_2S by the micro-supply of oxygen during anaerobic batch fermentation with rice straw as a raw material under mesophilic and thermophilic conditions at 35 and 55 °C, respectively. There were no obvious changes in biogas production and methane concentration with the addition of limited oxygen in both mesophilic and thermophilic fermentation. However, the suitable oxygen supply was found to be 2.0 ~ 4.0 times of theoretical demand under which the average desulfurization efficiency could be over 92% and the oxygen residues content complied with the requirements of biogas used as car fuel or incorporated into the gas network (no more than 0.5%). Moreover, the total biogas production increased slightly compared with the control experiment without oxygen. This was explained as follows: the hydrogen sulfide concentration is greatly reduced if more oxygen is being supplied, which reduces toxicity and increases methane production. It may also be attributed to the limited oxygen supply, causing facultative anaerobic bacteria in the fermentation system to use oxygen for hydrolysis which produces carbon dioxide or other gases. Upon the usage of air instead of oxygen, there was no obvious difference in biogas production, but the nitrogen content was higher compared to the case of pure oxygen. The desulfurization efficiency reached 93.9% and the nitrogen content could be controlled at about 1.69% when the air supply was 2.0 times that of theoretical air demand. Thus, air can be used as an oxygen source in the desulfurization of biogas for its utilization (such as combined heat and power (CHP)).

The practical suitability of the fixed-bed trickling bioreactor (FBTB) system could be proven while avoiding the disadvantages of internal biological desulfurization methods. Naegele *et al.* (2013) investigated the removal of hydrogen sulfide from biogas by external BDS in a full scale fixed-bed trickling bioreactor (FBTB) at a research biogas plant with a given output of 96 m^3 biogas/h and an H_2S concentration ranging between 500 ppm and 600 ppm (1 ppm = 1 cm^3/m^3) on average. The FBTB column has been designed to hold a packing volume of 2.21 m^3 at a gas retention time of 84 seconds being loaded at an average of 32.88 g $H_2S/m^3/h$. It was found that the temperature has a detectable effect, while the change in pH and air ratios have a little effect on BDS efficiency. The highest H_2S removal efficiency of 98% was recorded at a temperature range of 30–40 °C with a great decline to 21–45% at low temperatures ranging from 5–25 °C. Although different oxygen contents did not have a significant effect on desulfurization efficiency, 0.5% oxygen is recommended for general plant safety. It should be noted that, operation at pH7 caused a high consumption of operation resources, whereas pH2 imposed particularly high requirements for anti-corrosion.

In an attempt to enhance the biogas-BDS efficiency in bio-trickling filters, Chaiprapat *et al.* (2015) investigated the effects of triple stage and single stage bio-trickling filters with oxygenated liquid recirculation under the acidic condition at a pH down to 0.5. The packed bed of triple stage bio-trickling filters was divided into 3 stages (lower, middle and upper); each stage was 10 cm high with an attempt to improve the distribution of the dissolved O_2 within the bed. Single stage bio-trickling filters (S-BTF) with an equal size of total packed bed volume was operated for comparison with deigned triple stage bio-trickling filters (T-BTF) (Figure 3.28).

The coconut husk, which has been used as bed media, was inoculated with native wastewater microorganisms in order to establish an initial active biofilm. Coconut husk was used due to its rough surface, moisture storage capacity, inexpensiveness, and abundant availability. Triple stage bio-trickling filters (T-BTF) achieved a higher H_2S elimination capacity and removal efficiency of 175.6 g $H_2S/m^3/h$ and 89.0 %, than that in the case of applying the single stage bio-trickling filters which recorded

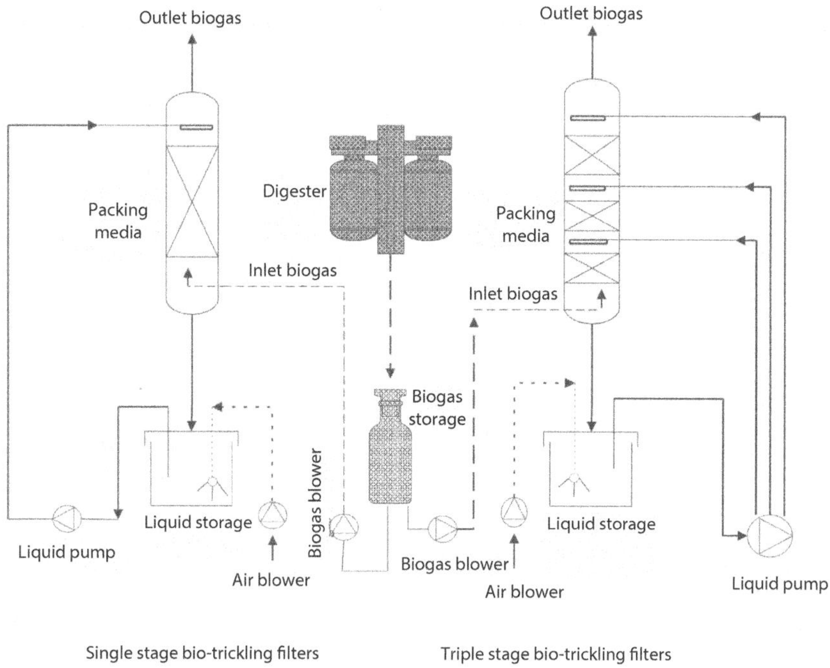

Figure 3.28 A Schematic Diagram for Acidic Bio-Trickling Filter Systems.

159.9 g $H_2S/m^3/h$ and 80.1%, respectively. This study clearly concluded that the O_2/H_2S ratio has a great impact on performance and can be manipulated by the distribution of the recirculating liquid flow. The oxidation of H_2S to elemental sulfur (S^0) requires only one-fourth of the O_2 needed for the conversion to sulfate. Thus, the production of sulfuric acid is enhanced with the increment of oxygen availability.

$$2HS^- + O_2 \longrightarrow 2S^0 + 2OH^- \qquad \Delta G^o = -129.5 \text{ kJ/mol}$$

$$2HS^- + 4O_2 \longrightarrow 2SO_4^{2-} + 2H^+ \qquad \Delta G^o = -732 \text{ kJ/mol}$$

The T-BTF has superiority in H_2S removal and sulfuric acid recovery relevant to the S-BTF because of the achieved more uniform distribution of dissolved oxygen in a step feed mode that could improve the performance of T-BTF over S-BTF. Moreover, the T-BTF has a potential for effective biogas clean up under highly acidic conditions (Chaiprapat *et al.*, 2015).

Pirolli *et al.* (2016) designed a simple and low maintenance bio-trickling filter (BTF) for desulfurization of the swine wastewater-derived biogas stream (Figure 2.29). The newly designed bio-trickling filter was filled

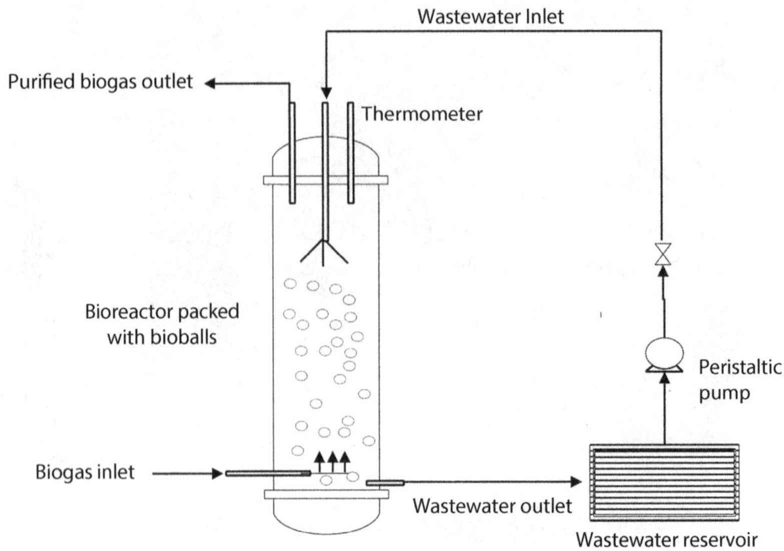

Figure 3.29 A Schematic for a New Design Bio-Trickling Filter Reactor (Pirolli *et al.*, 2016).

with polypropylene bio-balls as supporting material for the microorganism's biofilm fixation. BTF was continuously fed with wastewater effluent from an air sparged nitrification-denitrification bioreactor installed down-gradient from an up-flow anaerobic sludge bed reactor UASB-type digester. Thus, the swine wastewater was used as a source of electron acceptors, nutrients, and H_2S oxidizing bacteria inoculum to overcome the system's complexity and maintenance. Based on the measured consumption of NO_2^--N and NO_3^--N and the accumulation of SO_4^{2-} and S^0 over time, the biological H_2S removal in the BTF system was assumed to be carried out by chemolitotrophic denitrification.

$$S^{2-} + 1.6\ NO_3^- + 1.6\ H^+ \longrightarrow SO_4^{2-} + 0.8\ N_2 + 0.8\ H_2O$$

$$S^{2-} + 0.4\ NO_3^- + 2.4\ H^+ \longrightarrow S^0 + 0.2\ N_2 + 1.2\ H_2O$$

The predominant microbial population in the biofilm was found to be the hydrogenotrophic methanogens *Methanosarcinales*, *Methanobacteriales*, and *Methanomicrobiales*, and the latter represented the highest concentration. This was corroborated with the observed strong correlation between CO_2 removal and high CH_4 production as a source of renewable energy.

$$CH_3COO^- + H^+ \longrightarrow CH_4 + CO_2 \quad \Delta G^\circ = -75.7 \text{ kJ/mol}$$

$$4H_2 + CO_2 \longrightarrow CH_4 + 2H_2O \quad \Delta G^\circ = -135 \text{ kJ/mol}$$

The use of wastewater effluent from denitrification bioreactors as feeding solution for BTF, which contains sufficiently high concentrations of NO_2^- and NO_3^-, are known to exert toxic or inhibitory effects on methanogens. However, in that study, the presence of NOx at concentrations of 783 mg/L did not affect the methanogenic activity. Methanogens are also known to be highly sensitive and inhibited by low concentrations of oxygen. However, the presence of dissolved oxygen, as high as 2.2 mg/L, also did not affect the methanogenesis process. All of these added to the advantages of using this new design. Moreover, the desulfurization efficiency obtained with the new designed bio-trickling filter was 99.8% with a maximum elimination capacity of 1,509 g/m^3/h. The SO_4^{2-} and S^0 were the major metabolites produced from biological conversion of H_2S, with an average increase in methane production of $\approx 3.8 \pm 1.68$ g/m^3 in the filtered gas stream throughout 200 days of BTF operation.

Biodesulfurization of biogas has been largely studied under aerobic conditions (Fortuny *et al.*, 2008, 2011; Ramírez-Sáenz *et al.*, 2009; Chaiprapat *et al.*, 2011) with very few studies carried out under anoxic conditions (Soreanu *et al.*, 2009; Baspinar *et al.*, 2011). Although, the advantage of using anoxic BTFs over aerobic BTFs is that the biogas is not diluted with air and, therefore, the methane (CH_4) concentration is not reduced (Chaiprapat *et al.*, 2011; Montebello *et al.*, 2012). Furthermore, the electron acceptor mass transfer limitation is negligible because the nitrate solubility is very high [91.2 g/100 g at 25 °C] (Haynes, 2012). Consequently, for the production of pipeline grade methane, anoxic biofiltration is a more feasible technology as a pretreatment for H_2S removal than the more commonly used aerobic BTFs (Fernández *et al.*, 2013).

Fernández *et al.* (2014) studied the anoxic biofiltration process for H_2S removal from biogas using a BTF packed with open-pore polyurethane foam (OPUF) (Figure 3.30) to increase the elimination capacity (EC) and obtain a deeper understanding of the influence of the operating variables.

The BDS process was carried out throughout 620 days with a continuous supply of biogas. A BTF with a volume of 2.4 L (working volume bed) was packed with OPUF cubes. The biogas was produced by two up-flow anaerobic sludge bed reactors (UASB) of 200 L. In order to increase and set different H_2S concentrations, the biogas was passed through an H_2S generating column. Three nitrate mineral media were used to test three types of nitrate sources: $Ca(NO_3)_2{\cdot}4H_2O$, $NaNO_3$, and KNO_3. Biofilm formation was performed in-situ in the BTF (day 1–35). The results revealed that OPUF has great properties as a carrier for anoxic biofiltration and it can reach a critical EC of 130 g S/m^3/h (%BD ≈ 99 %) under the following conditions: inlet loads below 130 g S^{2-}/m^3/h, temperature of 30 °C, sulfate concentration below 33 g/L, a pH between 7.3 and 7.5, and a trickling liquid velocity higher than 4.6 m/h. In regard to nitrite concentration, high concentrations were reached without showing an inhibitory effect on the process, so nitrite could be used instead of nitrate as an electron acceptor source.

Ripl and Fechter (1991) employed iron oxides to control H_2S in anaerobic sludge through the conversion of H_2S into insoluble FeS without the inhibition of the anaerobic activity. Charles *et al.* (2006) reported the addition of an $FeCl_3$ solution (Fe^{3+}/S = 5:1) into a digester as an effective controller for H_2S emission throughout the precipitation of FeS. Su *et al.* (2012) reported the good efficiency of $Fe(OH)_3$ as a deslfurizer in an anaerobic digester. The supply of O_2 or air will need an external power and the amount of O_2 or air must be strictly controlled in order to avoid its inhibition on methanogens, decrease in methane content in biogas, and even the safety problem (Khanal *et al.*, 2003; Ramos and Fdz-Polanco, 2014). Moreover, the ferric agents (i.e. the iron-based oxidants) are relatively expensive when they are applied in large amounts in the practical in-situ desulfurization process.

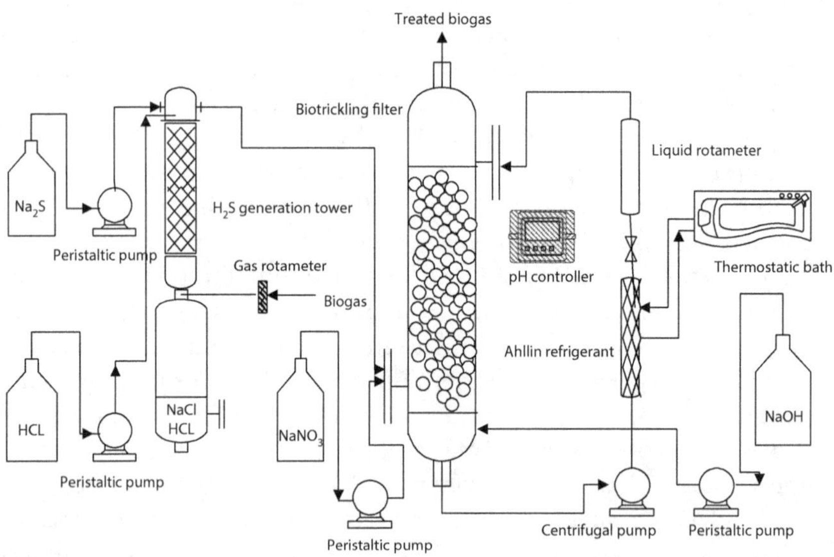

Figure 3.30 Bio-Trickling Filter Packed with Open Pore Polyurethane Foam.

Therefore, it is very necessary to explore a novel in situ desulfurizer for biogas desulfurization with low cost and high efficiency.

Natural iron ores, such as hematite and limonite, contain a high quantity of iron oxides and are widely available with low cost. Some natural iron ores have been adopted as the adsorbents in the H_2S abatement process and good desulfurization results were obtained (Sasaoka *et al.*, 1993). Five kinds of iron ores, limonite, hematite, manganese ore, magnetite, and lava rock were investigated as being suitable for in-situ desulfurization during anaerobic digestion in bioreactors for biogas production (Zhou *et al.*, 2016). These five ores showed good desulfurization for the evolved H_2S in biogas and ranked in the following increasing order: magnetite < lava rock < manganese ore < hematite < limonite. Thus, limonite expressed the best performance for H_2S removal. As limonite dosages increased (10–60 g/L), the contents of H_2S in biogas evidently decreased in the digesters with different initial sulfate concentrations (0–1000 mg/L). Not only this, but the amount of biogas from the digesters with limonite and hematite was slightly higher than that from the digesters without iron ore, which proved the promotion of biogas production due to hydrogen sulfide removal which leads, consequently, to the decrease of the H_2S inhibition on methanogenesis. Thus, limonite in the digesters expressed three functions: adsorption, oxidation, and precipitation, resulting in multiple control on H_2S in biogas (Figure 3.31). However, the manganese ore had a negative effect on biogas

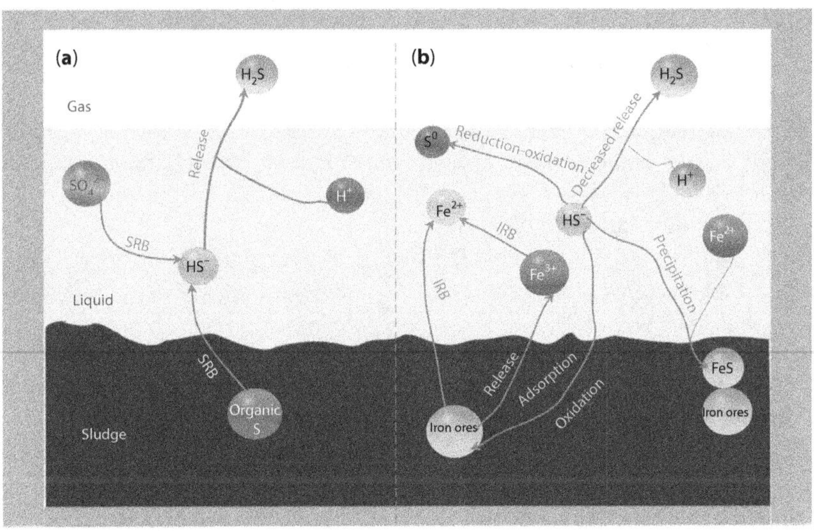

Figure 3.31 The Possible In-Situ Desulfurization Mechanisms by Iron Ores During Anaerobic Digestion System The release of H_2S in digesters without iron ores (b) the control of H_2S by iron ores.

yield, although it effectively controlled the H_2S content in biogas. This was attributed to the production of $MnCO_3$. As manganese captures the produced CO_2, it produced the insoluble $MnCO_3$ precipitates which would lead to a flocculation with the enzyme and, thus inhibit microbial activity (Zhong *et al.*, 2004).

References

Abatzoglou, N., Boivin, S. (2009) A review of biogas purification processes. Biofuels, Bioproducts and Biorefining. 3: 42–71.

Abdullaha, A.H., Mat, R., Abd Aziz, A.S., Roslanc, F. (2017) Use of kaolin as adsorbent for removal of hydrogen sulphide from biogas. Chemical Engineering Transactions. 56: 763–768.

Adib, F., Bagreev, A., Bandosz, T.J. (1999) Effect of pH and surface chemistry on the mechanism of H2S removal by activated carbons. *Journal of Colloid and Interface Science*. 216(2): 360–369.

Alemzadeh, I., Kahrizi, E., Vossoughi, M. (2009) Bio-oxidation of ferrous ions by *Acidithioobacillus ferrooxidans* in a monolithic bioreactor. *Journal of Chemical Technology and Biotechnology*. 84: 504–510.

Amirfakhri, J., Vossoughi, M., Soltanieh, M. (2006) Assessment of desulfurization of natural gas by chemoautotrophic bacteria in an anaerobic baffled reactor (ABR). *Chemical Engineering and Processing*. 45: 232–237.

Andrews E.M. (1989) Stretford Chemistry and Analytical Methods Handbook. GRI Report. Project 5086-253-1275. PB89-173595. GRI/GTI (Gas Technology Institute). Des Plaines. IL.

Aranovich G.L., Donohue, M.D. (1995) Adsorption-isotherms for microporous adsorbents. *Carbon*. 33:1369–1375.

Aroca, G., Urrutia H., Núñez, D., Oyarzún, P., Arancibia, A., Guerrero, K. (2007) Comparison of the removal of hydrogen sulfide in biotrickling filters inoculated with *Thiobacillus thiorapus* and *Acidithiobacillus thiooxidans*. *Electronic Journal of Biotechnology*. 10(4): 514–520.

Asai S., Konishi Y., Yabu T. (1990) Kinetics of absorption of hydrogen sulfide into aqueous ferric sulfate solutions. *American Institute of Chemical Engineers Journal*. 36: 1331–1338

Aslam, Z., Shawabkeh, R.A., Hussein, I.A., Al-Baghli, N., Eic, M. (2015) Synthesis of activated carbon from oil fly ash for removal of H_2S from gas stream. *Applied Surface Science*. 327: 107–115.

Atabani, A.E., Silitonga, A.S., Irfan, A.B., Mahlia, T.M.I., Masjuki, H.H., Mekhilef, S. (2012) A comprehensive review on biodiesel as an alternative energy resource and its characteristics. *Renewable and Sustainable Energy Reviews*. 16(4): 2070–2093.

Atabani, A.E., Badruddin, I.A., Mekhilef, S., Silitonga, A.S. (2011) A review on global fuel economy standards, labels and technologies in the transportation sector. *Renewable and Sustainable Energy Reviews*. 15: 4586–4610.

Bagreev A, Bashkova S, Bandosz TJ. (2002) Adsorption of SO_2 on activated carbons: the effect of nitrogen functionality and pore seizes. Langmuir.18: 1257–1264.

Bagreev A., Bandosz T.J. (2000) Study of hydrogen sulfide adsorption on activated carbons using inverse gas chromatography at infinite dilution. *Journal of Physical Chemistry B*. 104: 8841–8847.

Bagreev, A., Bandosz, T.J. (2002) A role of sodium hydroxide in the process of hydrogen sulfide adsorption/oxidation on caustic impregnated activated carbons. *Industrial and Engineering Chemistry Research*. 41: 672–679.

Bakker, W.J.W., Kapteijn, F., Moulijn, J.A. (2003) A high capacity manganese-based sorbent for regenerative high temperature desulfurization with direct sulfur production: conceptual process application to coal gas cleaning. *Chemical Engineering Journal*. 96: 223–235.

Bandosz, T.J. (2002) On the adsorption/oxidation of hydrogen sulfide on activated carbons at ambient temperatures. *Journal of Colloid and Interface Science*. 246(1): 1–20.

Barbusiński, K., Kalemba, K. (2016) Use of biological methods for removal of H_2S from biogas in wastewater treatment plants – A review. Architecture Civil Engineering Environment. 1: 103–111.

Bashkova, S., Bagreev, A., Bandosz, T.J. (2003) Adsorption/Oxidation of CH_3SH on Activated Carbons Containing Nitrogen Langmuir. 19(15): 6115–6121.

Bashkova, S., Bagreev, A., Bandosz, T.J. (2005) Catalytic properties of activated carbon surface in the process of adsorption/oxidation of methyl mercaptan. *Catalysis Today*. 99(3/4): 323–328.

Baspinar, A.B., Turker, M., Hocalar, A., Ozturk, I. (2011) Biogas desulphurization at technical scale by lithotrophic denitrification: Integration of sulphide and nitrogen removal. *Process Biochemistry* 46(4): 916–922.

Basu R., Clausen E.C., Gaddy J.L. (1996) Biological conversion of hydrogen sulfide into elemental sulfur. Environmental Progress & Sustainable Energy. 15: 234–238.

Beasley, T., Abry, R.G.F. (2003) Managing hydrogen sulfide the natural way. In: Papers of a Canadian Institute conference on the latest techniques for managing sour gas operations, Calgary, AB (Canada), 26–27 May 2003, Canadian Institute Conferences, [400] pages.

Bedell S.A., Kirby L.H., Buenger C.W. McGaugh M.C. (1988) Chelates role in gas treating. Hydrocarbon Process. 2: 63–66.

Benschop A., Janssen A., Hoksberg A., Seriwala M., Abry R., Ngai C. (2002) The shell-Paques/THIOPAQ gas desulphurization process: Successful startup first commercial unit. http://www.paques.nl.

Berzaczy L., Niedermayer E., Kloimstein, L., Windsperger. A. (1990) Biological exhaust gas purification in the rayon fiber manufacture (The Waagner-Biro/Glanzstoff Austria Process). *Chemical and Biochemical Engineering Quarterly*. 2: 201–203

Biogas composition, Biogas renewable energy – Information website on biogas, online access: http://www.biogas-renewable-energy.info/biogas_composition.html.

Birchalm, V.S.S., Rocha, S.D.F., Ciminelli, V.S.T. (2001) The effect of magnesite calcination conditions on magnesia hydration. Minerals Engineering. 13: 1629–1633.

Bo, Z., Yan, L., Yu-Qun, Z., Hui-Ling, T., Xiao-Wen, Z., Chang-He, C. (2007) Mass transfer and kinetics study on the sulfite forced oxidation with manganese ion catalyst. *Korean Journal of Chemical Engineering.* 24(3): 471–476.

Boki K., Tanada S. (1980) Adsorption of hydrogen sulfide on activated carbon. Chemical & Pharmaceutical Bulletin. 28: 1270–1275.

Bolhàr-Nordenkampf M., Friedl A., Koss U., Tork T. (2004) Modelling selective H_2S absorption and desorptionin an aqueous MDEA-solution using a rate-based non-equilibrium approach. *Chemical Engineering and Processing.* 43: 701–715.

Böresson, G. (2001). Inhibition of methane oxidation by volatile sulfur compounds (CH3SH and CS2) in landfill cover soils. Waste Management & Research. 19:314–319.

Bouzaza, A., Laplanche, A., Marsteau, S. (2004) Adsorption–oxidation of hydrogen sulfide of activated carbon fibers: Effect of the composition and the relative humidity of the gas phase. *Chemosphere.* 54(4): 481–488.

BP (2016) statistical review of world energy. 65th edition. https://www.bp.com/content/dam/bp/pdf/energy-economics/statistical-review-2016/bp-statistical-review-of-world-energy-2016-full-report.pdf.

BP Energy Outlook 2035. (2015) BP p.l.C. London, United Kingdom, 64th edition (February). https://www.bp.com/content/dam/bp/pdf/energy-economics/energy-outlook-2015/bp-energy-outlook-2035-booklet.pdf.

BP Statistical Review of World Energy. (2015) BP p.l.C. London, United Kingdom, 64th edition (June). https://www.bp.com/content/dam/bp/pdf/energy-economics/statistical-review-2015/bp-statistical-review-of-world-energy-2015-full-report.pdf.

Buenger C.W., Kushner D.S. (1988) The SulFerox process – Plant design considerations. Proceeding of the 67th Annual Convention GPA. Dallas. TX. March 14–16. p. 56–62.

Buisman C.J., Wit B., Lettinga G. (1990) Biotechnological sulphide removal in three polyurethane carrier reactors: stirred reactor, biorotor reactor and upflow reactor. *Water Research.* 24: 245–251.

Burgess J.E., Parsons S.A., Stuetz M. (2001) Developments in odor control and waste gas treatment biotechnology: a review. *Biotechnology Advances.* 19: 35–63.

Cadenhead P., Sublette K.L. (1990) Oxidation of hydrogen sulfide by Thiobacilli. *Biotechnology and Bioengineering.* 35: 1150–1154.

Cal M.P., Strickler B.W., Lizzio A.A. (2000) High-temperature hydrogen sulfide adsorption on activated carbon I. Effects of gas composition and metal addition. Carbon. 38: 1757–1765.

Carroll J.J. (2010) Acid gas injection and carbon dioxide sequestration. John Wiley & Sons Inc., Scrivener publishing LLC. Massachusetts. USA.

Cha J.M., Cha W.S., Lee J.H. (1999) Removal of organosulphur odour compounds by *Thiobacillus novellas* SRM, sulphur-oxidizing microorganisms. *Process Biochemistry*. 34: 659–665.

Chaiprapat S., Charnnok B., Kantachote D., Sung S. (2015) Bio-desulfurization of biogas using acidic biotrickling filter with dissolved oxygen in step feed recirculation. *Bioresource Technology*. 179: 429–435.

Chaiprapat, S., Mardthing, R., Kantachote, D., Karnchanawong, S. (2011) Removal of hydrogen sulfide by complete aerobic oxidation in acidic biofiltration. Process Biochemistry. 46(1): 344–352.

Charles, W., Cord-Ruwisch, R., Ho, G., Costa, M., Spencer, P. (2006) Solutions to a combined problem of excessive hydrogen sulfide in biogas and struvite scaling. *Water Science and Technology*. 53(6): 203–210.

Cheng, R., Zhou, W., Lin, L., Xie, T., Zhang, Y. (2013) Removal of high concentrations of H_2S from simulated natural gas by *Acidithiobacillus ferrooxidans* immobilized on polyurethane foam. *Journal of Chemical Technology and Biotechnology*. 88(5): 975–978.

Chiang H.L., Tsai J.H., Tsai C.L. Hsu Y.C. (2000) Adsorption characteristics of alkaline activated carbon exemplified by water vapor, H2S, and CH3SH gas. Separation Science and Technology. 35: 903- 918.

Cho K.S., Hirai M., Shoda M. (1992a) Degradation of hydrogen sulfide by *Xanthomonas* sp. strain DY44 isolated from peat. *Applied and Environmental Microbiology*. 158: 1183–1189.

Cho K.S., Hirai M., Shoda. M. (1992b) Enhanced removal efficiency of malodorous gases in a pilot-scale peat biofilter inoculated with *Thiobacillus thioparus* DW44. *Journal of Fermentation and Bioengineering*. 73: 46–50.

Choi, Y., Lee, J.W., Jang, S.C., Ahn, B.S., Choi, D.K. (2008) Adsorption dynamics of hydrogen sulfide in impregnated activated carbon bed. Adsorption. 1494/5): 533–538.

Chung Y.C., Huang C., Li C.F. (1997) Removal characteristics of H_2S by *Thiobacillus novellous* CH_3 biofilter in autotrophic and mixotrophic environments. *Journal of environmental science and health*. 5: 1435–1450.

Chung Y.C., Huang C., Tseng C.P. (1996) Operation optimization of *Thiobacillus thioparus* CH11 biofilter for hydrogen sulfide removal. *Journal of Biotechnology*. 52: 31–38.

Chung Y.C., Huang C., Tseng C.P. (2001) Biological elimination of H_2S and NH_3 from waste gases by biofilter packed with immobilized heterotrophic bacteria. *Chemosphere* 43: 1043–1050.

Clark O.G., Edeogu I., Feddes J., Coleman R.N., Abolghasemi A. (2004) Effects of operating temperature and supplemental nutrients in a pilot-scale agricultural biofilter. *Canadian Biosystems Engineering*. 46: 6.7–6.16.

Claus C.F. (1883) Invention concerning the production of sulfur resp. sulfurous acid from hydrogen sulfide. German Patent 28758.

Cline, C., Hoksberg, A., Abry, R., Janssen, A. (2003) Biological process for H_2S removal from gas streams: The Shell-Paques/THIOPAQTM gas desulfurization process, LRGCC 2003 Conference Proceedings, 1–17.

Cox H.H., Deshusses A.M. (2001) Co-treatment of H_2S and toluene in a biotrickling filter. *Chemical Engineering Journal*. 3901: 1–10.

Dalrymple D.A., Trofe T.W., Leppin D. (1994) Gas industry assesses new ways to remove small amounts of "H_2S" and "H_2S" scavenger selection index updated. *Oil & Gas Journal*. 92: 54–60.

Das, A., Roy, P., Mishra, A.K., (1993) Desulphuring of natural gas and petroleum oil by autotrophic *Thiobacillus ferrooxidans*. Letters in Applied Microbiology. 16(3): 164–166.

Dastous P.A., Soreanu G., Nikiema J., Heitz M. (2005) Biofiltration of three alcohols on a mature bed compost. In 2005 A&WMA Annual Conference Proceedings CD-ROM, Paper #1038. Pittsburgh, PA: Air & Water Management Association. Pittsburgh.

Davini, P. (2001) SO_2 and NO_x adsorption properties of activated carbons obtained from a pitch containing iron derivatives. Carbon. 39: 2173–2179.

Deng, L., Chen, H., Chen, Z., Liu, Y., Pu, X., Song, L. (2009) Process of simultaneous hydrogen sulfide removal from biogas and nitrogen removal from swine wastewater. *Bioresource Technology*. 100: 5600–5608.

Devinny S.J., Deshusses A.M., Webster S.T. (1999) Biofiltration for Air Pollution Control. Boca Raton, FL: CRC/Lewis Publishers.

Dhage P., Gaur V., Tatarchuk B.J. (2008) Promoted ZnO sorbents for low temperature H_2S removal: Influence of temperature, support surface area, pore size and pore. volume, 2008 AIChE Spring National Meeting, 6–10 April 2008, New Orleans, LA, United States, paper 231c.

Díaz I., Fdz-Polanco M. (2011) Robustness of the microaerobic removal of H2S from biogas. *Water Science & Technology*. 65: 1368–1374.

Díaz I., Pérez S.I., Ferrero E.M. Fdz-Polanco M. (2010) Effect of oxygen dosing point and mixing on the microaerobic removal of hydrogen sulphide in sludge digesters. *Bioresource Technology*. 102: 3768–3775.

Diaz, I., Perez, S.I., Ferrero, E.M., Fdz-Polanco, M. (2011) Effect of oxygen dosing point and mixing on the microaerobic removal of hydrogen sulphide in sludge digesters. *Bioresource Technology*. 102(4): 3768–3775.

Doğan, E.C., Türker, M., Dağaşan, Arslan, A. (2012) Simultaneous sulfide and nitrite removal from industrial wastewaters under denitrifying conditions. *Biotechnology and Bioprocess Engineering*. 17(3): 661–668

Doornkamp, C., Ponec, V. (2000) The universal character of the Mars and Van Krevelen mechanism. *Journal of Molecular Catalysis. A: Chemical*. 162: 19–32.

Duan H., Yan, R., Koe, L.C. (2005) Investigation on the mechanism of H_2S removal by biological activated carbon in a horizontal biotrickling filter. *Applied Microbiology and Biotechnology*. 69: 350–357.

Duan H., Koea L.C., Yan R., Chen, X. (2006) Biological treatment of H_2S using pellet activated carbon as a carrier of microorganisms in a biofilter. *Water Research*. 40: 2629–2636.

Dube, G., Osifo, P., Rutto, H. (2015) Preparation of bagasse ash/MgO/ammonium acetate sorbent properties and modeling their desulphurization reaction. *Environmental Progress and Sustainable Energy*. 34(1): 23–31.

Duissenov, D. (2013). Production and processing of sour crude and natural gas – challenges due to increasing stringent regulations. Master Thesis. Department of Petroleum Engineering and Applied Geophysics. Norwegian University of Science and Technology.

Efremenko, I. Sheintuch, M. (2003) Carbon-supported palladium catalysts. Molecular orbital study. *Journal of Catalysis*. 214: 53–67.

EIA. Energy Information Administration (1999) Natural Gas 1998: Issues and Trends. Office of Oil and Gas. U.S. Department of Energy. Washington. p. 49–71.

El-Gendy N.Sh., Speight J.G. (2016) Handbook of Refinery Desulfurization. CRC Press, Taylor and Francis Group, LLC, 6000 Broken Sound Parkway NW, Suite 300, Boca Raton, FL 33487–2742, USA.

Elias A., Barona A., Arreguy A., Rios J., Aranguiz I., Peñas J. (2002) Evaluation of a packing material for the biodegradation of H_2S and product analysis. *Process Biochemistry*. 37: 813–820.

Fan, L., Chen, J., Guo, J., Jiang, X., Jiang, W. (2013) Influence of manganese, iron and pyrolusite blending on the physiochemical properties and desulfurization activities of activated carbons from walnut shell. *Journal of Analytical and Applied Pyrolysis*. 104: 353–360.

Fan, L., Jiang, X., Jiang, W., Guo, J., Chen, J. (2014) Physicochemical properties and desulfurization activities of metal oxide/biomass-based activated carbons prepared by blending method. Adsorption. 20: 747–756.

Faramawy S., Zaki T., Sakr A. (2016) Natural gas origin, composition, and processing: A review. *Journal of Natural Gas Science and Engineering*. 34: 34–54

Fdz-Polanco, M., Diaz, I., Perez, S.I., Lopes, A.C., Fdz-Polanco, F. (2009) Hydrogen sulphide removal in the anaerobic digestion of sludge by micro-aerobic processes: pilot plant experience. *Water Science and Technology*. 60: 3045–3050.

Fernández, M., Ramírez, M., Gómez, J. M., and Cantero, D. 2014. Biogas biodesulfurization in an anoxic biotrickling filter packed with open-pore polyurethane foam. *Journal of Hazardous Materials*. 264: 529–535.

Fernández, M., Ramírez, M., Pérez, R. M., Gómez, J. M., Cantero, D. (2013) Hydrogen sulphide removal from biogas by an anoxic biotrickling filter packed with Pall rings. *Chemical Engineering Journal*. 225: 456–463.

Fillipou, D., Katiforis, N., Papassiopi, N., Adams, K. (1999). On the kinetics of magnesia hydration in magnesium acetate solutions. *Journal of Chemical Technology and Biotechnology*. 74: 322–328.

Flores, R., Rodas, A., Palma, C., Melendez, Y.I., Melendez, X.S., Gonzalez, L.L. (2008) Thermogravimetric studies on the effect of support in sorbent-materials applied to flue gas desulfurization. *Fuel*. 87(7): 1115–1126.

Fortuny, M., Baeza, J. A., Gamisans, X., Casas, C., Lafuente, J., Deshusses, M. A., Gabriel, D. (2008) Biological sweetening of energy gases mimics in biotrickling filters. *Chemosphere*. 71(1): 10–17.

Fortuny, M., Gamisans, X., Deshusses, M. A., Lafuente, J., Casas, C., Gabriel, D. (2011) Operational aspects of the desulfurization process of energy gases mimics in biotrickling filters. *Water Research*. 45(17): 5665–5674.

Fu, H.B., Wang, X., Wu, H.B., Yin, Y., Chen, J.M. (2007) Heterogeneous uptake and oxidation of SO_2 on iron oxides. *The Journal of Physical Chemistry C*. 111: 6077–6085.

Fuda K., Palmer A.D., Sears P., Blais D., Furimsky E. (1991) Chemical changes occurring during sulphidation and regeneration of iron-containing sorbents. *Fuel*. 70:100–106.

Gabriel, D., Deshusses, M.A. (2003) Retrofitting existing chemical scrubbers to biotrickling filters for H_2S emission control. Proceedings of the National Academy of Sciences of the United States of America. 100(11): 6308–6312.

Gadre, R.V. (1989) Removal of hydrogen sulfide from biogas by chemoautotrophic fixed-film bioreactor. *Biotechnology and Bioengineering*. 34(3): 410–414.

Garrett R.L., Carlton L.A., Denekas M.O. (1988) Methods for field monitoring of oil-based drilling fluids for hydrogen sulphide and water intrusions. SPE Drilling Engineering. 3: 296–302.

Giro, M.E.A., Garcia, O. Jr., Zaiat, M. (2006) Immobilized cells of *Acidithiobacillus ferrooxidans* in PVC strands and sulfide removal in a pilot-scale bioreactor. *Chemical Engineering Journal*. 28: 201–207.

Grimm D.T., Cork D.J., Uphaus R.A. (1984) Optimization of Desulfobacter for sulfate reduction. Development in Industrial Microbiology. 24: 709–716.

Guo, J.; Lua, A.C. (2002) Microporous activated carbons prepared rom palm shell by thermal activation and their application to Sulphur dioxide adsorption. *Journal of Colloid Interface Science*. 251: 242–247.

Guo, J.-X., Liang, J., Chu, Y.-H., Sun, M.-C., Yin, H.-Q., Li, J.-J. (2012) Desulfurization activity of nickel supported on acid-treated activated carbons. *Applied Catalysis A: General* (421/422): 142–147.

Hagen, M., Polman, E., Jensen, J., Myken, A., Jonsson, O., Dahl, A. (2001) Adding gas from biomass to the gas grid. Swedish Gas Center, Malmo Sweden, July 2001.Pp. 144. Report SCG118. http://www.dgc.eu/sites/default/files/filarkiv/documents/R0105_biomass_gas_grid.pdf.

Hansen C. (2006) Biogas Scrubbing System (Eliminating Corrosive Hydrogen Sulfide Gas). The Local Energy Farms Network – Reliable Renewable Energy for a Post Carbon World. Post Carbon Institute. http://www.energyfarms.net/node/1034.

Hardison L.C. (1991) Recent developments in acid gas treatment using the auto circulation LO-CAT H2S approach. AIChE Spring National Meeting, Houston, TX, April 7–11, Session 16c.

Haynes, W.M. (2012). Physcial constants of inorganic compounds, in: W.M. Haynes (Ed.), CRC Handbook of Chemistry and Physics, CRC Press/Taylor and Francis, Boca Raton, FL.

Heinz-Wolfgang H. (2008) Industrial Gases Processing. WILEY-VCH Verlag GmbH & Co. KGaA. Weinheim. ISBN: 978-3-527-31685-4.

Henshaw P.F., Bewtra J.K., Biswas N. (1997) Biological conversion of sulfide to elemental sulfur. *Indian Journal of Engineering & Material Sciences*. 5: 202–210.

Henshaw P.F., Medlar D., McEwen J. (1999) Selection of a support medium for a fixed-film green sulfur bacteria reactor. *Water Research*. 33: 3107–3110.

Henshaw P.F., Zhu W. (2001) Biological conversion of hydrogen sulfide to elemental sulfur in a fixed-film continuous flow photo-reactor. *Water Research*. 35: 3605–3610.

Hernández S.P., Scarpa, F., Fino, D., Conti, R. (2011) Biogas purification for MCFC application. *International Journal of Hydrogen Energy*. 36(13): 8112–8118.

Ho, K.L., Lin, W.C., Chung, Y.C., Chen, Y.P., Tseng, C.P. (2013) Elimination of high concentration hydrogen sulfide and biogas purification by chemical-biological process. *Chemosphere*. 92: 1396–1401.

Horváth, I.S., Tabatabaei, M., Karimi, K., Kumar, R. (2016) Recent updates on biogas production - a review. *Biofuel Research Journal* 10: 394–402.

Howarth R., Unz F.R., Seviour M.E., Seviour J.R., Blackall L., Pickup, W.R., Jones, G.J., Yaguchi J., Head I. (1999) Phylogenetic relationships of filamentous sulfur bacteria (*Thiothrix* spp. and *Eikelboom* type 021N bacteria) isolated from wastewater treatment plants and description of *Thiothrix eikelboomii* sp. nov., *Thiothrix unzii* sp. nov., *Thiothrix fructosivorans* sp. nov. and *Thiothrix defluvii* sp. nov. *International Journal of Systematic Bacteriology*. 49: 1817–1827.

Husein M.M., Patruyo L., Pereira-Almao P., Nassar N.N. (2010) Scavenging H_2S (g) from oil phases by means of ultra-dispersed sorbents. *Journal of Colloid and Interface Science*. 342:253–260.

Illan-Gomez, M.J., Raymund-Pinero, E., Garcia-Garcia, A., Linares-Solano, A., de Salinaz-Martinez-Salinaz, L. (1999) Catalytic NO_x reduction by carbon supporting metals. *Applied Catalysis B: Environmental*. 20: 267–275.

International Energy Agency, IEA (2011) Oil marketer reports, Total OECD12 month moving average demand Vs 160 year. 1year growth. Pans, France: Organization for economic Co-operation and development. https://www1.oecd.org/environment/cc/48073739.pdf.

International Energy Agency, IEA (2013) Resources to reserves. Oil, gas and coal technologies for the energy markets of the future. https://www.iea.org/publications/freepublications/publication/Resources2013.pdf

Isik-Gulsac, I. (2016) Investigation of impregnated activated carbon properties used in hydrogen sulfide fine removal. *Brazilian Journal of Chemical Engineering*. 33(4): 1021–1030.

Janssen A.J., Sleyster R., van der Kaa C., Jochemsen A., Bontsema J., Lettinga G. (1995) Biological sulphide oxidation in a fed-batch reactor. *Biotechnology and Bioengineering*. 47: 327–333.

Jensen A.B., Webb C. (1995) Treatment of H_2S-containing gases: A review of microbiological alternatives. Enzyme and Microbial Technology. 17:2–10.

Jiang, H., Li, T., Zhou, H. (2014) Method for determining desulfurizer addition in in-situ desulfurization process of methane. CN Patent CN104195183-A.

Jin, Y., Veiga, M.C., Kennes, C. (2005) Effects of pH, CO_2, and flow pattern on the autotrophic degradation of hydrogen sulfide in a biotrickling filter. *Biotechnology and Bioengineering*. 92(4): 462–471.

Jing, C., Ping, Z., Mahmood, Q. (2009) Simultaneous sulfide and nitrate removal in anaerobic reactor under shock loading. *Bioresource Technology*. 100: 3010–3014.

Johansen S.S., Licht D., Arvin E., Mosbaek H., Hansen A.B. (1997) Metabolic pathways of quinoline, indole and their methylated analogs by *Desulfobacterium indolicum* (DSM 383). Applied and Environmental Microbiology. 47: 292–300.

Jorio H., Heitz M. (1999) Treatment de l'air par biofiltration. *Canadian Journal of Civil Engineering*. 26: 402–424.

Kandile N., Razek T., Al-Sabagh A., Khattab M. (2014): Synthesis and evaluation of some amine compounds having surface active properties as H2S scavenger; *Egyptian Journal of Petroleum*; 23: 323–329.

Kaspar, J., Fornasiero, P., Granziani, M. (1999) Use of CeO_2-based oxides in the three-way catalysis. *Catalysis Today*. 50: 285–298.

Khanal, S., Shang, C., Huang, J. (2003) Use of ORP (oxidation-reduction potential) to control oxygen dosing for online sulfide oxidation in anaerobic treatment of high sulfate wastewater. *Water Science and Technology*. 47(12): 183–189.

Kidnay A.J., Parrish W.R., McCartney D.G. (2011) Fundamentals of Natural Gas Processing, second ed. CRC Press. Taylor and Francis Group. LLC.

Kim B.W., Chang H.N. (1991) Removal of hydrogen sulfide by *Chlorobium thiosulfatophilum* in immobilized cell and sulfur settling free-cell recycle reactors. Biotechnology Progress. 7:495–500.

Kim H., Kim J.Y., Chung S.J., Xie. Q. (2002) Long-term operation of a biofilter for simultaneous removal of H_2S and NH_3. Air & Waste Management Association. 52: 1389–1398.

Kim Y.J., Kim B.W. Chang H.N. (1996) Desulfurization in a plate-type gas-lift photobioreactor using light emitting diodes. *Korean Journal of Chemical Engineeri ng*. 13: 606–611.

Kim, K.H., Choi, Y.J., Jeon, E.C., Sunwoo, Y. (2005). Characterization of malodorous sulfur compounds in landfill gas. Atmospheric Environment. 39: 1103–1112.

Kleerebezem R., Mendez R. (2002) Autotrophic denitrification for combined hydrogen sulfide removal from biogas and post-denitrification. *Water Science and Technology*.45: 349–356.

Kleinjan W. (2005) Biologically Produced Sulfur Particles and Polysulfide Ions: Effects on a Biotechnological Process for the Removal of Hydrogen Sulfide from Gas Streams. PhD Thesis. Wageningen University. Wageningen. The Netherlands.

Klok, J.B.M., Van Den Bosch, P.L.F., Buisman, C.J.N., Stams, A.J.M., Keesman, K.J., Janssen, A.J.H. (2012) Pathways of sulfide oxidation by haloalkaliphilic bacteria in limited-oxygen gas lift bioreactors. *Environmental Science and Technology*. 46(14): 7581–7586.

Kobayashi H.A., Stenstrom M., Mah R.A. (1983) Use of photosynthetic bacteria for hydrogen sulfide removal from anaerobic waste treatment effluent. *Water Research*. 17: 579–587.

Kobayashi T., Li Y.Y., Kubot A.K., Harada H., Maeda T., Yu H.Q. (2012) Characterization of sulfide-oxidizing microbial mats developed inside a fullscale anaerobic digester employing biological desulfurization. *Applied Microbiology and Biotechnology*. 93: 847–857.

Kobayashi, T., Li, Y.Y. (2011) Performance and characterization of a newly developed self-agitated anaerobic reactor with biological desulfurization. *Bioresource Technology*. 102(10): 5580–5588.

Kohl A., Nielsen R. (1997) Gas Purification. 5th Edition. Gulf Publishing Company. Houston. USA.

Kraakman, N.J.R., Rocha-Rios, J., van Loosdrecht, M.C.M. (2011) Review of mass transfer aspects for biological gas treatment. *Applied Microbiology and Biotechnology*. 91: 873–886.

Kwaśny, J., Balcerzak, W. (2016) Sorbents used for biogas desulfurization in the adsorption process. *Polish Journal of Environmental Studies*. 25(1): 37–43.

Kwaśny, J., Balcerzak, W., Rezka, P. (2015) Application of zeolites for the adsorptive biogas desulfurization. Technical Transactions Environmental Engineering. 3-Ś: 40–45.

Lallemand, F., Perdu, G., Maretto, C., Weiss, C., Magne-Drisch, J., Lucquin, A.-C. (2012) Solutions for the treatment of highly sour gases. In Gas Treatment/ Processing, Digital Refining, Gas 2012, April 2012. http://www.digitalrefining.com/article/1000356.

Lampe D.G., Zhang T.C. (1996) Evaluation of sulfur-based autotrophic denitrification. In Proceedings of the HSRC/ WERC Joint Conference on the Environment. Great Plains/ Rocky Mountain Hazardous Substance Research Center. http://www.engg.ksu.edu/HSRC/96Proceed/lampe.pdf.

Lar, J.S., Li, X. (2009) Removal of H_2S during anaerobic bioconversion of dairy manure. *Chinese Journal of Chemical Engineering*. 17(2):273–277.

Larsen H. (1952) On the culture and general physiology of the green sulfur bacteria. *Journal of Bacteriology*. 64: 187–196.

Le Gaz Intégral (2000) The SulFerox process. http://www.gazintegral.com/sulferox/sulferox.htm.

Lee, H.-S., Kim, J.-Y., Yu, J.-K., Kil, I.-S., Kim, D.-H., Lee T.-J., Rhee, Y.-W. (2005) A Study of desulfurization ability and activation energy for CuO-AgO sorbents. *Korean Journal of Chemical Engineering*. 22(6): 889–893.

Lee, Y.W., Park, J.W., Choung, J.H., Choi, D.K. (2002) Adsorption characteristics of SO_2 on activated carbon prepared from coconut shell with potassium hydroxide activation. *Environmental Science and Technology*. 36: 1086–1092.

Li Y, Zhu CL, Lu T, Guo ZP, Zhang D, Ma J, Zhu, S. (2013) Simple fabrication of a Fe_2O_3/carbon composite for use in a high-performance lithium ion battery. *Carbon*. 52: 565–573.

Li, F., Yan, B., Jiang, W., Gong, Z. (2012) Preparation of activated carbon loading nano-ZnO and desulfurization properties under room temperature. In: Preparation of activated carbon loading nano-ZnO and desulfurization properties under room temperature. Zhang, W., ed., AISC 162, Springer-Verlag, Berlin, Heidelberg. Pp. 177–183.

Li, J.J., Kobayashi, N., Hu, Y.Q. (2009) Performance of V_2O_5/AC activatedwith the flue gas on SO_2 removal. *Journal of Environmental Engineering*. 4: 176–187.

Li, X., Jiang, X., Zhou, Q., Jiang, W. (2016) Effect of S/N ratio on the removal of hydrogen sulfide from biogas in anoxic bioreactors. *Applied Biochemistry and Biotechnology*. 180: 930–944.

Li, Z., Sun, T., Zhu, N., Cao, X., Jia, J. (2008) Comparative study of using different materials as bacterial carriers to treat hydrogen sulfide. *Applied Microbiology and Biotechnology*. 81: 579–588.

Lin W.C., Chen Y.P., Tseng C.P. (2013) Pilot-scale chemical–biological system for efficient H_2S removal from biogas. *Bioresource Technology*. 135: 283–291.

Linkous, C.A. (1993) Sulfur recovery from oil and gas refinery waste streams using semiconductor particulates. *Waste Management*. 13(5/7): 527–528

Liu, C., Zhang, R., Wei, S, Wang, J., Liu, Y., Li, M., Liu, R. (2015a) Selective removal of H_2S from biogas using a regenerable hybrid TiO_2/zeolite composite. *Fuel*. 157: 183–190.

Liu, Q., Ke, M., Liu, F., Yu, P., Hu, H., Li, C. (2017) High-performance removal of methyl mercaptans by nitrogen-rich coconut shell activated carbon. *RSC Advances*. 7: 22892- 22899.

Liu, Q.S., Xu, B., He, J.L., Sun, R., Chang, H.G. (2015b) Optimal conditions for bio-oxidation of biodesulfurization in natural gas sweetening process using *Thiobacillus*. *Advanced Materials Research*. 1104: 57–60

Lobo, R., Revah, S., Viveros-Garcia, T. (1999) An analysis of a trickle-bed bioreactor: Carbon disulfide removal. *Biotechnology and Bioengineering*. 63: 98–109.

Lu J., Zheng Y., He D. (2006) Selective absorption of H_2S from gas mixtures into aqueous solutions of blended amines of methyldiethanolamine and 2-tertiary-butylamino-2-ethoxyethanol in a packed column. *Separation and Purification Technology*. 52:209–217.

Ma, Y.-L., Yang, B.-L., Zhao, J.-L. (2006) Removal of H_2S by *Thiobacillus denitrificans* immobilized on different matrices. *Bioresource Technology*. 97(16): 2041–2046.

Maddox R.N. (1974) Gas and Liquid Sweetening. Second ed. John M. Campbell Ltd. Petroleum Series. Norman. OK.

Magomnang M., Villanueva P. (2014) Removal of Hydrogen Sulfide from Biogas using Dry Desulfurization Systems; International Conference on Agricultural, Environmental and Biological Sciences. Phuket (Thailand). http://dx.doi.org/10.15242/IICBE.C414016 65.

Magota H., Shiratori Y. (1988) Treatment of sour natural gas containing hydrogen sulfide. Japanese Patent 63,205,124.

Maka A., Cork. D. (1990) Quantum efficiency requirements for an anaerobic photobioreator. *Journal of Industrial Microbiology and Biotechnology*. 5: 337–354.

Malhautier L., Gracian C., Roux C.J., Fanlo L.J., Le Cloirec. P. (2003) Biological treatment process of air loaded with an ammonia and hydrogen sulfide mixture. Chemosphere. 50: 145–153.

Mao, C., Feng, Y., Wang, X., Ren, G. (2015) Review on research achievements of biogas from anaerobic digestion. *Renewable and Sustainable Energy Reviews.* 45: 540–555.

Markus B.N., Anton F., Ulrich K., Thomas T. (2004) Modelling selective H_2S absorption and desorption in an aqueous MDEA-solution using a rate-based non-equilibrium approach. *Chemical Engineering and processing.* 43: 701–715.

Mata-Alvarez, J., Llabrés, S.M. (2000). Anaerobic digestion of organic solid wastes: An overview of research achievements and perspectives. *Bioresource Technology.* 74: 3–16.

McKinsey Z.S. (2003): Removal of hydrogen sulfide from biogas using cowmanure compost, Thesis presented to the Faculty of the Graduate School of Cornell University; 1–104.

Mehandjiev, D., Bekyarova, E. Khristova, M. (1997) Study of Ni-impregnated active carbon II. Catalytic behavior in NO conversion. *Journal of Colloid Interface Science.* 192: 440–446.

Mehandjiev, D., Khristova, M., Bekyarova, E. (1996) Conversion of NO on Co-impregnated active carbon catalysts. Carbon. 34: 757–762.

Meldon, J.H., Dutta, A. (1994) Analysis of ultimate permselectivity for H_2S over CO_2 in alkaline solutions. *Chemical Engineering Science.* 49(5): 689–697.

Mesa M.M., Macías M., Cantero D. (2002) Biological iron oxidation by *Acidithiobacillus ferrooxidans. Chemical and Biochemical Engineering Quarterly.* 16: 69–73.

Micoli, L., Bagnasco, G., Turco, M. (2014) H_2S removal from biogas for fueling MCFCs: new adsorbing materials. *International Journal of Hydrogen Energy.* 39:1783–1787.

Mokhatab S., Poe W.A., Speight J.G. (2006) Handbook of Natural Gas Transmission and Processing. Elsevier Inc. USA.

Montebello, A.M., Fernández, M., Almenglo, F., Ramírez, M., Cantero, D., Baeza, M., Gabriel, D. (2012). Simultaneous methylmercaptan and hydrogen sulfide removal in the desulfurization of biogas in aerobic and anoxic biotrickling filters. *Chemical Engineering Journal.* 200: 237–246.

Montebello, A.M., Fernández, M., Almenglo, F., Ramírez, M., Cantero, D., Baeza, M., Gabriel, D. (2012) Simultaneous methylmercaptan and hydrogen sulfide removal in the desulfurization of biogas in aerobic and anoxic biotrickling filters. *Chemical Engineering Journal.* 200: 237–246.

Monteleone, G., De Francesco, M., Galli, S., Marchetti, M., Naticchioni, V. (2011) Deep H_2S removal from biogas for molten carbonate fuel cell (MCFC) systems. *Chemical Engineering Journal.* 173: 407–414.

Naegele, H.-J., Lindner, J., Merkle, W., Lemmer, A., Jungbluth, T., Bogenrieder, C. (2013) Effects of temperature, pH and O2 on the removal of hydrogen sulfide from biogas by external biological desulfurization in a full scale fixed-bed trickling bioreactor (FBTB). *International Journal of Agricultural and Biological Engineering.* 6(1): 69–81.

Nagl G. (1997) Controlling H_2S emission. Chemical Engineering. 104: 125–128.

Nassar I.M., Noor M.R., Morsi R.E., Abd El-Azeim A., Hashem A.I. (2016) *Renewable and Sustainable Energy Reviews*. 65: 101–112.

Nazari J. (2017) Isolation, identification, and characterization of a novel chemolithoautotrophic bacterium with high potential in biodesulfurization of natural or industrial gasses and biogas. *Energy Sources, Part A: Recovery, Utilization, and Environmental Effects*. 39(10): 971–977.

Neumann W., Riickauf H., Volk N., Michael A., Forkmann, R. (1990) Holzemann Metallverarbeitung GmbH; Technische Hochschule Merseburg "Carl Schlorlemmer". (Ger) Method for removal of hydrogen sulphide from a combustible waste gas. European Patent EP 402,704.

Nguyen-Thanh, D., Bandosz, T.J. (2003) Effect of transition-metal cations on the adsorption of H_2S in modified pillared clays. *Journal of Physical Chemistry. B*. 107: 5812–5817.

Nishimura S., Yoda M. (1997) Removal of hydrogen sulfide from an anaerobic biogas using a bio-scrubber. *Water Science & Technology*. 36: 349–356.

Noyola A., Morgan-Sagastume J.M., López-Hernández J.E. (2006) Treatment of biogas produced in anaerobic reactors for domestic wastewater: odor control and energy/resource recovery. Reviews in Environmental Science and Bio/Technology. 5: 93– 114

O'Brien M., Wentworth C., Lanning, A., Engert T. (2007) Shell-Paques biodesulfurization process directly and selectively removes H2S from high pressure natural gas – start-up report. Proceeding of the Laurance Reid Gas Conditioning Conference. Norman. OK.

Osorio, F., Torres, J.C. (2009) Biogas purifiation from anaerobic digestion in a wastewater treatment plant for biofuel production. *Renewable Energy*. 34: 2164–2171.

Oyarzún P., Arancibia F., Canales C., Aroca E.G. (2003) Biofiltration of high concentration of hydrogen sulphide using *Thiobacillus thioparus*. *Process Biochemistry*. 39: 165- 170.

Pagella, C., De Faveri, D.M. (2000) H_2S gas treatment by iron bioprocess. *Chemical Engineering Science*. 55: 2185–2194.

Pandey R.A., Malhotra S. (1999) Desulfurization of gaseous fuels with recovery of elemental sulfur: an overview. *Critical reviews in environmental science and technology*. 29: 229–268.

Pandey, R.A., Malhotra, S., Rajvaidya, A.S., Sharma, S., Peshwe, S. (2003) Bench scale evaluation of a chemo-biochemical process for desulphurization of gaseous streams containing hydrogen sulphide. *Water Air Soil Pollution*. 146: 243–260.

Phooratsamee, W., Hussaro, K., Teekasap, S., Hirunlabh, J. (2014) Increasing adsorption of activated carbon from palm oil shell for adsorb H_2S from biogas production by impregnation. *American Journal of Environmental Sciences*. 10(5): 431–445.

Pirolli, M., Busi da Silva, M.L., Mezzari, M.P., Michelon, W., Prandini, J.M., Soares, H.M. (2016) Methane production from a field-scale biofilter designed for desulfurization of biogas stream. *Journal of Environmental Management*. 177: 161–168.

Polychronopoulou K., Cabello G.F., López G.M., Fierro L.G., Bakas T. Efstathiou A.M. (2005) Novel Fe–Mn–Zn–Ti–O mixed-metal oxides for the low-temperature removal of H_2S from gas streams in the presence of H_2, CO_2, and H_2O. *Journal of Catalysis*. 236:205–220.

Postgate J.R. (1968) The sulphur cycle. In Inorganic Sulphur Chemistry. ed. G. Nickless. Amsterdam. Netherlands: Elsevier Publishing Company. P. 259–279.

Prescott M.L., Harley P.J., Klein A.D. (2003) Microbiology, 5th edition. NY: McGraw-Hill Companies. New York. USA.

Przepiorski J., Oya A. (1998) K_2CO_3-loaded deodorizing activated carbon fiber against H_2S gas: Factors influencing the deodorizing efficiency and the regeneration method. *Journal of materials science letters*. 17: 679–682.

Przepiorski J., Yoshida S., Oya, A. (1999) Structure of K_2CO_3-loaded activated carbon fiber and its deodorization ability against H_2S gas. *Carbon*. 37: 1881–1890.

Puri, B.R. (1970) In Chemistry and Physics of Carbon. Vol. 6, (Ed., P.A. Thrower), Marcel Dekker, New York. Pp. 191–282.

Qi., G., Yang, R.T. (2003) Performance and kinetics study for low-temperature SCR of NO with NH_3 over MNO_x-CeO_2 catalysts. *Journal of Catalysis*. 217: 434–441.

Qing-chun, Y., Yong, D., Fei, W., Yue-bin, F., Bin, Y., Bao-qiang, X., Da-chun, L. (2015) Comparison of desulfurization kinetics of copper oxide sorbent. *Journal of Central South University*. 22: 2902–2908.

Rai, C., Taylor, M. (1996) Microbial sweetening of sour natural gas using mixed cultures. Environmental Progress. 15(1): 6–11.

Ramírez-Sáenz, D., Zarate-Segura, P.B., Guerrero-Barajas, C., García-Peña, E.I. (2009) H2S and volatile fatty acids elimination by biofiltration: Clean-up process for biogas potential use. *Journal of Hazardous Materials*. 163(2/3): 1272–1281.

Ramos I., Díaz I., Fdz-Polanco M. (2012) The role of the headspace in hydrogen sulfide removal during microaerobic digestion of sludge. *Water Science & Technology*. 66: 2258–2264.

Ramos, I., Fdz-Polanco, M. (2014) Microaerobic control of biogas sulphide content during sewage sludge digestion by using biogas production and hydrogen sulphide concentration. *Chemical Engineering Journal*. 250: 303–311.

Rattanapan C., Ounsaneha W. (2012) Removal of Hydrogen Sulfide Gas using Biofiltration – a Review. *Walailak Journal of Science and Technology*. 9: 9–18.

Ripl, W., Fechter, L. (1991) Treating settled anaerobic sludge at base of water bodies–with additives which eliminate or convert toxic pollutants, esp. iron oxide to convert hydrogen sulphide. DE Patent DE4005064 A1.

Rocha, S.D.F., Mansur, M.B., Ciminelli, V.S.T. (2004) Kinetics and mechanistic analysis of caustic magnesia hydration. *Journal of Chemical Technology and Biotechnology*. 79: 816–821.

Rojey A., Jaffret C., Cornot-Gandolphe S., Durand B., Julian S., Valais M. (1997) Natural Gas: Production, Processing. Transport. Editions Technip. Paris.

Rojey, A. (1997) Natural gas: production, processing, transport. Institut Franpcais Du Pbetrole Publications.

Rouleau, W., Watson, J. (2014) LO-CAT®: A flexible hydrogen sulfide removal process. http://www.merichem.com/LO-CAT-Flexible-H2S-Removal-Process

Ryckebosch, E., Drouillon, M., Vervaeren, H. (2011) Techniques for transformation of biogas to biomethane. *Biomass and Bioenergy*. 35: 1633–1645.

Saimura, A., Shiratori, Y., Kitaoka, T. (2017) Dual-layered paper-structured catalysts for sequential desulfurization and methane-steam reforming of simulated biogas containing hydrogen sulfide. *Journal of Materials Science*. 52: 314–325.

Sasaoka E., Ichio T, Kasaoka S. (1992) High-temperature H_2S removal from coal-derived gas by iron ore. *Energy and Fuels*. 6:603–608.

Sasaoka, E., Sakamoto, M., Ichio, T., Kasaoka, S., Sakata, Y. (1993) Reactivity and durability of iron oxide high temperature desulfurization sorbents. *Energy and Fuels*. 7(5): 632–638.

Satoh, H. Yoshizawa, J., Kametanl, S. (1988) Bacteria help desulfurize gas. Hydrocarbon Processing. 76-D-76-F.

Schaack J.P., Chan F. (1989) H_2S scavenging: Formaldehyde-methanol, metallic oxide agents head scavengers list processes. *Oil and Gas Journal*. 87, 51–55; 81–82; 45–48; 90–91. 4 parts.

Schieder D., Quicker P., Schneider R., Winter H., Prechtl S., Faulstich M. (2003). Microbiological removal of hydrogen sulfide from biogas by means of a separate biofilter system: Experience with technical operation. *Water Science & Technology*. 48: 209–212.

Schoell M. (1983) Isotope techniques for tracing migration of gases in sedimentary basins. *Journal of the Geological Society*. 140: 415–422.

Sercu B., Núñez D., Langenhove V.H., Aroca G., Verstraete W. (2005) Operational and microbiological aspects of a bioaugmented two-stage biotrickling filter removing hydrogen sulfide and dimethyl sulfide. *Biotechnology and Bioengineering*. 90: 259–269.

Seredych, M., Strydom, C., Bandosz, T.J. (2008) Effect of fly ash addition on the removal of hydrogen sulfide from biogas and air on sewage sludge based composite adsorbents. Waste Management. 28(10): 1983–1992.

Shahid, E.M., Jama, J. (2011) Production of biodiesel: a technical review. *Renewable and Sustainable Energy Reviews*. 25(9): 4732–4745.

Shannon R. (2000) Biogas. conference proceedings http://www.rosneath.com.all/ipc6/ch08/shannon2/.

Shareefdeen Z., Herner B., Wilson S. (2002) Biofiltration of nuisance sulfur gaseous odors from a meat rendering plant. *Journal of Chemical Technology & Biotechnology*. 77: 1296–1299.

Sheng, Y.Q., Chen, F.Z., Wang, X.M., Sheng, G.Y., Fu, J.M. (2008) Odorous volatile organic sulfide in wastewater treatment plants in Guangzhou, China. *Water Environment Research*. 80: 324–330.

Shi, L., Yang, K., Zhao, Q., Wang, H., Cui, Q. (2015) Characterization and mechanisms of H_2S and SO_2 adsorption by activated carbon. *Energy and Fuels*. 29(10): 6678–6685.

Shi, R.-H., Zhang, Z.-R., Fan, H.L., Zhen, T., Shangguan, J., Mi, J. (2017) Cu-based metal-organic framework/activated carbon composites for sulfur compounds removal. *Applied Surface Science*. 394: 394–402.

Sisani, E., Cinti, G., Discepoli, G., Penchini, D., Desideri, U., Marmottini, F. (2014) Adsorptive removal of H_2S in biogas conditions for high temperature fuel cell systems. *International Journal of Hydrogen Energy*. 39(36): 21753–21766.

Smet, E., Lens, P., Van Langenhove, H. (1998) Treatment of waste gases contaminated with odorous sulfur compounds. *Critical Reviews in Environmental Science and Technology*. 28(1): 89–117.

Son H.J., Lee J.H. (2005) H_2S removal with an immobilized cell hybrid reactor. Process Biochemistry. 40: 2197–2203.

Song, C. (2002) Fuel processing for low-temperature and high temperature fuel cells: challenges, and opportunities for sustainable development in the 21st century. *Catalysis Today*. 77: 17–49.

Song, L., Zhang, B., Deng, W. (2014) Application of activated carbon from waste tea in desulfurization and denitrification. Huan Jing Ke Xue. 35(10): 3674–82 (Article in Chinese).

Sonta H., Shiratori T. (1990) Method of treating H_2S containing gases. Dova Mining Co. Ltd. US Patent 4,931,262, 1990.

Soreanu G, Al-Jamal M, Béland M. (2005) Biogas treatment using an anaerobic biosystem. In Proceedings of the 3rd Canadian Organic Residuals and Biosolids Management Conference, June 1–4, 2005. Calgary, AB. Pp. 502–513.

Soreanu, G., Beland, M., Falletta, P., Edmonson, K., Seto, P. (2008a) Laboratory pilot scale study for H_2S removal from biogas in an anoxic biotrickling filter. *Water Science and Technology*. 57: 201–207.

Soreanu, G., Béland, M., Falletta, P., Ventresca, B., Seto, P. (2009) Evaluation of different packing media for anoxic H_2S control in biogas. *Environmental Technology*. 30(12): 1249–1259.

Soreanu, G., Seto, P., Béland, M., Edmonson, K., Falletta, P. (2008b). Investigation on the use of nitrified wastewater for the steady-state operation of a biotrickling filter for the removal of hydrogen sulphide in biogas. *Journal of Environmental Engineering and Science*. 7: 543–552.

Speight J.G. (1993) Gas Processing: Environmental Aspects and Methods. Butterworth-Heinemann Ltd., Oxford, England.

Speight J.G. (2015) Liquid fuels from natural gas. In: Lee, S., Speight, J.G., Loyalka, S.K. (Eds.), Handbook of Alternative Fuel Technologies. second ed. Taylor and Francis Group. LLC. CRC Press. p. 157–178.

Speight, J.G. (2007) Natural Gas: A Basic Handbook. GPC Books, Gulf Publishing Company, Houston, TX.

Speight, J.G. (2014). The Chemistry and Technology of Petroleum, 5th Edition. CRC Press, Taylor & Francis Group, Boca Raton, FL.

Stanier R.Y., Ingraham J.L., Wheelis M.L., Painter P.R. (1986) The Microbial World, 5th edition. Englewood Cliffs. NJ. Prentice-Hall. Inc.

Stirling D. (2000) The sulfur problem cleaning up industrial feedstocks. First ed. The Royal Society of Chemistry. Cambridge. UK.

Su, L., Zhang, M., Shi, X. (2012) In situ odor control of sewage of sludge by addition of ferric hydroxide. Environmental Pollution Control. 34(12): 1–2.

Sublette K.L. (1988) Production of microbial biomass protein from autotrophic fermentation of hydrogen sulfide. *Biotechnology and Bioengineering*. 32: 408–409.

Sublette, K.L., Sylvester, N.D. (1987a) Oxidation of hydrogen sulfide by *Thiobacillus denitrificans*: desulfurization of natural gas. *Biotechnology and Bioengineering.* 29(2): 249–257.

Sublette. K.L. (1990) Microbial treatment of sour gases for the removal and oxidation of hydrogen sulphide. Gas Separation & Purification. 4: 91- 96.

Subletteb, K.L., Sylvester, N.D. (1987b) Oxidation of hydrogen sulfide by mixed cultures of *Thiobacillus denitrificans* and heterotrophs. *Biotechnology and Bioengineering*. 29(6): 759–761.

Sumathi, S., Bhatia, S., Lee, K.T., Mohamed, A.R. (2010) Performance of palm shell activated carbon impregnated with CeO_2 and V_2O_5 catalyst in simultaneous removal of SO_2 and NO_2. *Journal of Applied Sciences*. 10(2): 1052–1059.

Sun, F., Liu, J., Chen, H., Zhang, Z., Qiao, W., Long, D., Bashkova, S., Bandosz, T.J. (2009) The effects of urea modification and heat treatment on the process of NO_2 removal by wood-based activated carbon. *Journal of Colloid and Interface Science*. 333(1): 97–103.

Sun, W., Xia, C., Xu, M., Guo, J., Wang, A., Sun, G. (2013) Distribution and abundance of archaeal and bacterial ammonia oxidizers in the sediments of the Dongjiang River, a drinking water supply for Hong Kong. *Microbes and Environments*. 28(4): 457–465.

Syed M.A., Henshaw P.F. (2003) Effect of tube size on performance of a fixed-film tubular bioreactor for conversion of hydrogen sulfide to elemental sulfur. *Water Research*. 37: 1932–38.

Syed M.A., Henshaw P.F. (2005) Light emitting diodes and an infrared bulb as light sources of a fixed-film tubular photobioreactor for conversion of hydrogen sulfide to elemental sulfur. *Journal of Chemical Technology & Biotechnology.* 80:119–123.

Syed, M., Soreanu, G., Falleta, P., Beland, M. (2006) removal of hydrogen sulfide from gas streams using biological processes – A review. *Canadian Biosystems Engineering*. 48: 2.2–2.14.

Tabak, J. (2009). Natural Gas and Hydrogen. Facts on File. New York. USA.

Takano B., Koshida M., Fujiwara Y., Sugimori K., Takayangi S. (1997) Influence of sulfur-oxidizing bacteria on the budget of sulphate in Yugama crater lake. KusatsuShirane volcano. Japan. Biogeochemistry. 38: 227–253.

Tang K., Baskaran V., Nemati M. (2009) Bacteria of the sulphur cycle: an overview of microbiology, biokinetics and their role in petroleum and mining industries. *Biochemical Engineering Journal. J*. 44: 73–94.

Tang, Y.Q., Shigematsu, T., Ikba, Morimura, S., Kida, K. (2004). The effects of micro-aeration on the phylogenetic diversity of microorganisms in a thermophilic anaerobic municipal solid-waste digester. Water Research. 38: 2537–2550.

Tanji Y., Kanagawa T., Mikami, E. (1989) Removal of dimethyl sulfide, methyl mercaptan. and hydrogen sulfide by immobilized *Thiobacillus thioparus* TK-m. *Journal of Fermentation and Bioengineering*. 67: 280–285.

Tian, L., Li, C., Li, Q., Zeng, G., Gao, Z., Li, S., Fan, X. (2009) Removal of elemental mercury by activated carbon impregnated with CeO_2. *Fuel*. 88: 1687–1691.

Tóth G., Nemestóthy N., Bélafi-Bakó K., Vozik D., Bakonyi, P. (2015) Degradation of hydrogen sulfide by immobilized Thiobacillus thioparus in continuous biotrickling reactor fed with synthetic gas mixture. *International Biodeterioration and Biodegradation*, 105: 185–191.

Tsai J.H., Jeng F.T., Chiang H.L. (2001) Removal of H_2S from exhaust gas by use of alkaline activated carbon. Adsorption. 7: 357–366.

Tseng, H.H., Wey, M.Y., Fu, C.H. (2003) Carbon materials as catalyst supports for SO_2 oxidation: catalytic activity of CuO–AC. Carbon. 41(1): 139–149

Turker M., Baspinarb A., Hocalara Akif (2011): Monitoring and control of biogas desulphurization using oxidation reduction potential under denitrifiying conditions. *Journal of Chemical Technology and Biotechnology*. 87: 682–688.

U.S. Department of Energy (1988) Handbook on Biogas Utilization, Georgia Institute of Technology. Atlanta. Georgia. https://rirdc.infoservices.com.au/downloads/12-056

van der Zee F.P., Villaverde S., García-Encina P.A., Fdz-Polanco F. (2007) Sulphide removal by moderate oxygenation of anaerobic sludge environments. *Bioresource Technology*. 98: 518–524.

Van Grinsven, P. (1999) Shell unveils gas biodesulfurization process. *Oil and Gas Journal*. Pp. 32.

van Groenestijn, J.W., Hesselink, P.G.M. (1993) Biotechniques for air pollution control. Biodegradation. 4(4): 283–301.

van Groenestijn, J.W., Kraaman, N.J.R. (2005) Recent developments in biological waste gas purification in Europe, Chemical Engineering Journal. 113: 85–91.

Van Niel C.B. (1931) On the morphology and physiology of the purple and green sulphur bacteria. Archives of Microbiology. 3: 1–112.

Vanderzee, F.P., Villaverde, S., Garcia, P.A., Fdz-Polanco, F. (2007) Sulfide removal by moderate oxygenation of anaerobic sludge environments. *Bioresource Technology*. 98: 518–524.

Vikromvarasiri N., Champreda V., Boonyawanich S., Pisutpaisal N. (2017) Hydrogen sulfide removal from biogas by biotrickling filter inoculated with *Halothiobacillus neapolitanus*. *International Journal of Hydrogen Energy*.1–9.

Villegas, P.G. (2016) Preliminary study of activated carbon filters for pollutants removal in diesel engines. *International Journal of Sustainable Energy and Environmental Research*. 5(2): 31–45.

Wang D. (2008) Breakthrough behavior of H_2S removal with an iron oxide based CG-4 adsorbent in a fixed-bed reactor. MSc. Saskatoon. SK: Department of Chemical Engineering, University of Saskatchewan.

Weiland P. (2010) Biogas production: current state and perspectives. *Applied Microbiology and Biotechnology*. 85: 849–860.

Wellinger A., Linberg A. (2000) Biogas Upgrading and Utilization-IEA Bioenergy Task 24. International Energy Association. Paris. France. p. 2000

Wellinger, A., Lindberg, A. (2005) Biogas upgrading and utilisation. IEA Bioenergy Task 24: Energy from Biological Conversion of Organic Waste (2005). http://www.iea-biogas.net/files/daten-redaktion/download/publi-task37/Biogas%20 upgrading.pdf.

Westmoreland, P.R., Gibson, J.B., Harrison, D.P. (1977) Comparative kinetics of high-temperature reaction between hydrogen sulfide and selected metal oxides. *Environmental Science and Technology*. 11: 488–491.

Wu, M., Zhang, Y., Ye, Y., Lin, C. (2016) In situ removal of hydrogen sulfide during biogas fermentation at microaerobic condition. *Applied Biochemistry and Biotechnology*. 180: 817–825.

Wubs H.J., Beenackers C.M., Krish R. (1991) Absorption of hydrogen sulfide in aqueous solutions of iodine – a critical review. *Chemical Engineering Science*. 46:703–706.

Xiao-Li, L., Jia-Xiu, G., Ying-Hao, C., De-Ming, L., Hua-Qiang, Y., Ming-Chao, S., Yavuz, R. (2014) Desulfurization performance of iron supported on activated carbon. *Fuel*. 123: 93–100.

Yang, S., Tang, Y., Gou, M., Jiang, X. (2015) Effect of sulfate addition on methane production and sulfate reduction in a mesophilic acetate fed anaerobic reactor. *Applied Microbiology and Biotechnology*. 99(7): 3269–3277.

Yazdanbakhsh, F., Blasing, M., Sawada, J.A., Rezaei, S., Muller, M., Baumann, S., Kuznicki, S.M. (2014) Copper exchanged nanotitanate for high temperature H_2S adsorption. *Industrial and Engineering Chemistry Research*. 53: 11734–11739.

Yi, Z., Xiaoying, M., Zhouyan, C., Shuang-chen, M., Lidong, W. (2007) "Experimental study on simultaneous desulfurization and denitrification by highly active absorbent" Proceedings of International Conference on Power Engineering-2007, October 23–27, 2007, Hangzhou, China. Pp. 771–721.

Yumura M., Furimsky E. (1985) Comparison of calcium oxide, zinc oxide, and iron (III) oxide hydrogen sulfide adsorbents at high temperatures. *Industrial & Engineering Chemistry Process Design & Development*. 24:1165–1168.

Zare A.H., Mirzaei, S. (2009) Removal of CO_2 and H_2S using Aqueous Alkanolamine Solusions; *International Journal of Chemical, Molecular, Nuclear, Materials and Metallurgical Engineering*. 3: 50–59.

Zhong, J., Xue, Y., Chen, J. (2004) Effect of cobalt and manganese on activity of anaerobic sludge in treatment of waste water from production of purified terephthalic acid. Industrial Water and Wastewater. 35(3): 2004–2006.

Zhou, Q., Jiang, X., Li, X., Jiang, W. (2016) The control of H_2S in biogas using iron ores as in situ desulfurizers during anaerobic digestion process. *Applied Microbiology and Biotechnology*. 100: 8179–8189.

Zhou, Q., Liang, H., Yang, S., Jiang, X. (2015) The removal of hydrogen sulfide from biogas in a microaerobic biotrickling filter using polypropylene carrier as packing material. *Applied Biochemistry and Biotechnology*. 175: 3763–3777.

Zhu, Z., Liu, Z., Liu, S., Niu, H., Hu, T., Liu, T., Xie, Y. (2000) NO reduction with NH_3 over an activated carbon-supported copper oxide catalysts at low temperatures. *Applied Catalysis B: Environmental.* 26: 25–35.

Ziyu, S., Qiang, L., Dan, W., Hang, T., Maohua, Y., Jianmin, X. (2014) Inhibition of carbon disulfide on bio-desulfurization in the process of gases purification. *African Journal of Biotechnology.* 13(16): 1739–1744.

4

Microbial Denitrogenation of Petroleum and its Fractions

List of Abbreviations and Nomenclature

2-OHQn	2-hydroxyquinoline
AA	Anthranilic acid
ADN	Adsorptive Denitrogenation
ADS	Adsorptive Desulfurization
AntABC	Anthranilate-1,2-dioxygenase
ATCC	American Type Culture Collection
BDN	Biodenitrogenation
bpsd	Barrels Per Stream Day
BTs	Benzothiophenes
$CaCO_3$	Calcium Carbonate
CAR	Carbazole
CARDO	CAR 1,9a-dioxygenase
CARs	Carbazoles
CAT	Catechol
CGO	Coker Gas Oil

CUSs	Coordinatively Unsaturated Sites
DBF	Dibenzofuran
DBTs	Dibenzothiophenes
DCW	Dry Cell Weight
DESs	Deep Eutectic Solvents
ECODS	Extraction and Catalytic Oxidative Desulfurization
EDN	Extractive Denitrogenation
EPA	Environmental Protection Agency
FOEB	Fuel Oil Equivalent Barrels
H_2SO_4	Sulfuric Acid
HDN	Hydrodenitrogenation
HDS	Hydrodesulfurization
HGO	Heavy Gas Oil
HNO_3	Nitric Acid
HOADA	2-hyroxy-6-oxo-6-(2'-aminophenyl)-hexa-2,4-dienoic acid
IBL	Inside Battery Limits
ILs	Ionic Liquids
LCO	Light Cycle Oil
LGO	Light Gas Oil
MNPs	Magnetic Nanoparticles
MOFs	Metal-Organic Frameworks
MQn	Methylquinoline
NCCs	Nitrogen Containing Compounds
NOx	Nitrogen Oxides
NPAHs	Nitrogen Polyaromatic Heterocyclic Compounds
NPC	Nitrogen Polar Compounds
NRR	Nitrogen Removal Ratio
ODS	Oxidative Desulfurization
PAHs	Polycyclic Aromatic Hydrocarbons
PPM	Part Per Million
Ps	*Pseudomonas*
PVA	Polyvinyl Alcohol
Qn	Quinoline
Qns	Quinolines
RHC	Reactive Hydrocarbons
SOx	Sulfur Oxides
SRGO	Straight Run Gas Oil
SCCs	Sulfur Containing Compounds
TCA	Tricarboxylic Acid
PASHs	Polyaromatic Sulfur Heterocyclic Compounds

UD-HDS	ultra-deep HDS
ULS	Ultra-Low Sulfur
ULSD	Ultra-Low Sulfur Diesel

4.1 Introduction

Nitrogen is like sulfur, as it is considered a petroleum contaminant. Éigenson and Ivchenko (1977) reported that the higher the sulfur content in petroleum fractions, the higher its nitrogen content. Moreover, nitrogen containing compounds (NCCs) coexist with sulfur containing compounds (SCCs) in fossil fuels (Yi *et al.*, 2014). It interferes with the refining process, causes equipment corrosion and catalyst poisoning, reduces the process yields, and adds to the refining costs (Kilbane *et al.*, 2000). Moreover, it contributes to acid rain and atmospheric contamination (Hansmeier *et al.*, 2011). The presence of nitrogen containing compounds promotes tank corrosion and oil degradation during storage (Singh *et al.*, 2010). The total nitrogen content of crude oils averages around 0.1 to 0.9% (Speight, 2014), of which the non-basic compounds (e.g. carbazole) comprise approximately 70–75%, and the basic ones (e.g. quinoline, Qn) comprise 25–30% (Benedik *et al.*, 1998; Laredo *et al.*, 2002). In Egyptian crude oils, CAR and its derivative concentrations range from 0.3 to 1% (Bakr, 2009). Oil produced from shale deposits has higher concentrations of N-compounds (0.5–2.1%) and the waste materials derived from the processing of shale oil are heavily contaminated by nitrogen polyaromatic heterocyclic compounds (NPAHs) (Richard and Junk, 1984; Tissot and Welte, 1984). The nitrogen content in light cycle oil (LCO) and coker gas oil (CGO) are higher than that in straight run gas oil (SRGO) (Adam *et al.*, 2009).

The NPAHs are compounds where one or more carbon atoms in the fused ring structure of the polycyclic aromatic hydrocarbons (PAHs) are replaced by nitrogen atom(s) (Figure 4.1). The NPAHs found in crude oils fall into two classes: the 'non-basic' molecules that include pyrroles and indoles, but are predominantly mixed alkyl derivatives of carbazole (CAR), and the 'basic' molecules that are largely derivatives of pyridine, acridine, and quinolone (Benedik *et al.*, 1998; Asumana *et al.*, 2011). CAR, as an example for the non-basic NPAHs, can directly impact the refining processes in two ways: (1) during the cracking process CAR can be converted into basic derivatives which can be adsorbed onto the active Lewis and Brönsted acidic sites of the cracking and HDS catalysts and (2) it directly inhibits the HDS catalysts (Laredo *et al.*, 2001; Williams and Chishti, 2001; Choi *et al.*, 2004; Castorena *et al.*, 2006; Laredo *et al.*, 2015). The

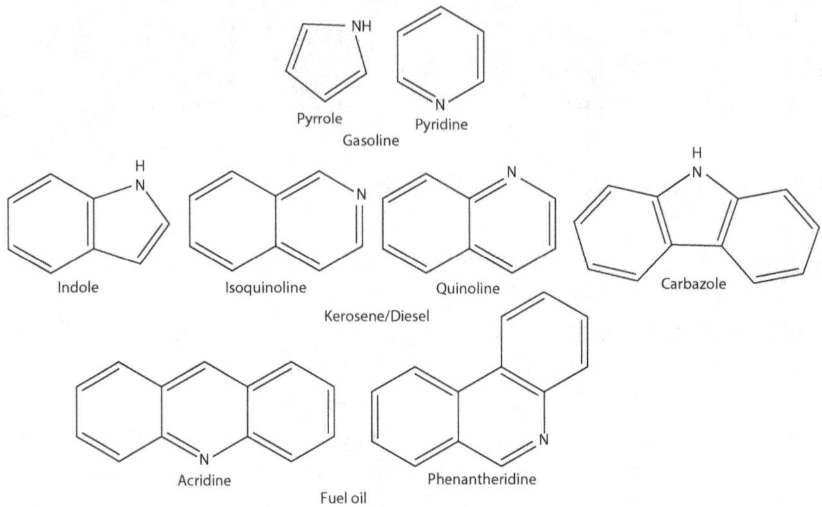

Figure 4.1 Different NPAHs Found in Crude Oil and its Fractions.

neutral NCCs polymerize forming gums that cause pores and intraparticular plugging of the catalysts (Dong *et al.*, 1997). It has been reported that increasing the nitrogen content in the feedstock from zero to 0.2 wt.% requires an increase of 100 °K to maintain the same hydrocracking conversion with Ni-Mo- or Pd-zeolite catalysts. Moreover, for a high hydrocracking efficiency, the nitrogen content in the feedstock has to be reduced to <0.01 wt.% (Weitkamp and Puppe, 1999). Thus, removal of CAR and other nitrogen-compounds would significantly increase the extent of catalytic cracking and, consequently, the gasoline yield. It has been reported that by 90% reduction in nitrogen-content, a 20% increase in gasoline yield occurs. This is a major economic improvement in low-margin, high volume refining processes (Benedik *et al.*, 1998). Moreover, since NCCs can be oxidized during the ODS process (Shiraishi *et al.*, 2002; Ishihara *et al.*, 2005; Souza *et al.*, 2009), they will lower the ODS efficiency due to the competitive parallel oxidation of NCCs. Also, extraction of NCCs can occur during the extractive desulfurization process (Xie *et al.*, 2008a; Huh *et al.*, 2009). The basicity of NCCs affects extraction and catalytic oxidative desulfurization (ECOD) (Caero *et al.*, 2006; Jia *et al.*, 2009).

The NPAHs tend to occur in a strong association with PAHs in the environment because they come from the same sources. They can be emitted due to the incomplete combustion of organic matter, including wood, waste, and fossil fuels (such as gasoline, diesel, and coal) (Furlong and Carpenter, 1982; Chuang *et al.*, 1991; Osborne *et al.*, 1997).

NPAHs are also released into the environment from spills, wastes, and effluents of several industrial activities, such as oil drilling, refining and storage, coal tar processing, chemical manufacturing, and wood preservation (Lopes and Furlong, 2001; Weigman, 2002; Wu *et al.*, 2014; Wang *et al.*, 2015). Both soil and groundwater are frequently contaminated by nitrogen, sulfur, and oxygen heterocycles at sites contaminated with petroleum and wood preservation wastes such as creosote. CAR is the major nitrogen heteroaromatic in coal-tar creosote, crude oil, tar sand sources, shale oil, and wood preserving wastes (Gieg *et al.*, 1996; Mushrush *et al.*, 1999). Although it is useful as an industrial raw material for dyes, medicines, insecticides, and plastics, CAR is also known to be an environmental pollutant. CAR is toxic, carcinogenic, and mutagenic (US EPA, 1986) and is considered an anthropogenic pollutant in air, soil, and water (Castorena *et al.*, 2006; Jong *et al.*, 2006; Guo *et al.*, 2008). It can readily undergo radical chemistry to generate the more genotoxic hydroxynitrocarbazoles (Jha and Bharti, 2002; Singh *et al.*, 2010). Despite its widespread use, little is known about the fate of CAR in the environment (Arcos and Argus, 1968; Singh *et al.*, 2011a,b). The NPAHs are hazardous to the environment since they are more polar, water soluble, and, consequently, more mobile and bioavailable in the environment than their corresponding PAHs analogs (Bleeker *et al.*, 2003; WHO, 2004). NPAHs, including CAR and its derivatives, have been detected in contaminated atmospheric samples, river sediments, and groundwater sites. This causes concern because CAR is known to be both mutagenic and toxic and even though it is not highly toxic itself, it readily undergoes radical chemistry to generate the genotoxic hydroxynitrocarbazole (Arcos and Argus, 1968; Zakaria *et al.*, 2015a,b, 2016). Moreover, the mutagenic potential of NPAHs increases with increasing the number of rings, thus Qn is known to be more mutagenic than pyridine (Mohammed and Hopfinger, 1983). The US-EPA considered Qns as toxic pollutants for their mutagenic and carcinogenic properties (Richard and Shieth, 1986; Bai *et al.*, 2010; Wang *et al.*, 2013). Qns are widely spread as they exist in oil shale, coal processing, creosote wood preservation, fossil fuel facilities, pharamceuticals', and Qn-dyes' industries (Qiao and Wang, 2010; Wang *et al.*, 2013). Growth impairment is also reported to increase with the increase in the number of rings per compound and with the increase in nitrogen content within the ring (Millemann *et al.*, 1984). NPAHs are more polar than their homocyclic analogues due to the incorporation of the free electrons on the N-atom in the ring system, which increases their water solubility thus lowering their absorption on the organic constituents of soil and aquifer materials. Upon combustion of fossil fuels, it produces SOx and NOx (Maass *et al.*, 2015) which contribute to acid rain, atmospheric contamination, and destruction of the ozone layer (Larentis *et al.*, 2011; Liu *et al.*, 2013).

The dominant precursors of acid deposition are anthropogenic sulfur oxides (SOx) and nitrogen oxides (NOx). Ozone formation involves NOx and reactive hydrocarbons (RHC). Anthropogenic hydrocarbon emissions, NOx, and SOx result primarily from petroleum refining and storage, other industrial process emissions, mobile sources, and burning of fossil fuels. Upon emission, SOx and NOx will react with atmospheric oxygen and water producing acid fogs and rains and cause the destruction of the ozone layer (Larentis *et al.*, 2011). About 25% of the acidity of rain is accounted for by the presence of nitric acid (HNO_3) and approximately 75% is related to the presence of sulfuric acid (H_2SO_4). Acid rains have many negative impacts on the ecosystem and environment, for example water streams become more acidic affecting the aquatic plants, microorganisms, and fish population, degradation of many soil minerals produces metal ions that are then washed away in runoff causing the release of toxic ions such as Al^{3+} into the water streams, and, moreover, loss of important minerals such as Ca^{2+} from the soil which would kill trees and damage crops. It also has a negative impact on buildings and monuments since it can dissolute marble and limestone which are mainly $CaCO_3$ (Scheme 4.1). Regions which are exposed to water and rains are eroded easily, while those which are more sheltered would be better preserved due to the formation of gypsum ($CaSO_4.2H_2O$). This attracts dust, carbon particles, dry-ash, and other dark pollutants, leading to the blackening of the surfaces. An even more serious situation arises when water containing calcium and sulfate ions penetrates the stone's pores. When the water dries, the ions form salt crystals within the pore system. These crystals can disrupt the crystalline arrangement of the atoms in the stone, causing the fundamental structure of the stone to be disturbed. Upon the disruption of the crystalline structure, cracking of the stone occurs.

$$2NO\,(g) + O_2\,(g) \longrightarrow 2NO_2\,(g)$$
$$3NO_2\,(g) + H_2O \longrightarrow 2HNO_3\,(aq) + NO\,(g)$$
$$2NO_2\,(g) + H_2O \longrightarrow HNO_3\,(aq) + HNO_2\,(aq)$$
$$2SO_2\,(g) + O_2\,(g) \longrightarrow 2SO_3\,(g)$$
$$SO_2\,(g) + H_2O \longrightarrow H_2SO_3\,(aq)$$
$$SO_3\,(g) + H_2O \longrightarrow H_2SO_4\,(aq)$$
$$H_2SO_4\,(aq) + CaCO_3\,(s) \longrightarrow Ca^{2+}\,(aq) + SO_4^{2-}\,(aq) + H_2O + CO_2$$

Scheme 4.1 Dissolution Mechanism of Marble and Limestone.

4.2 Denitrogenation of Petroleum and its Fractions

Emission standards for NOx from particular and light load transport vehicles was 0.07 g NOx/mile in 2009, but the EPA in March 2014 released new standards for vehicles and fuels to be fully implemented by 2025. Both the tailpipe and evaporative combined emissions of non-methane organic gas and NOx should be reduced from 109 mg/mile by 2017 to 30 mg/mile by 2025 (EPA, 2014). Moreover, there are stringent regulations that limited the N-content in diesel fuel from >70 ppm in 2003 to <0.1 ppm in 2010 (Jayaraman *et al.*, 2006; Anantharaj and Banerjee, 2010).

In order to meet the new stringent regulations for ultra-low sulfur (ULS) diesel, there are many studies being undertaken to overcome the obstacles of the low reactivity of recalcitrant high molecular weights of PASHs (such as 4,6-DMDBT) and the inhibition effect of nitrogen species in the HDS process. It has been also reported that high nitrogen contents in the oils leads to reduced oil stability and consequent storage problems (Williams and Chishti, 2001). Thus, nitrogen in middle distillates has to be removed to meet such standards (Bej *et al.*, 2001; Elise et a., 2013; Ayala *et al.*, 2016; Tao *et al.*, 2017).

4.2.1 Hydrodenitrogenation

Most of the N-containing compounds in petroleum are currently removed by the conventional chemical and physical refinery processes (Speight, 1980). NPAHs can be removed from petroleum by hydrotreatment under high temperature and pressure, but it is expensive and hazardous since HDN is more difficult than hydrodesulfurization (HDS) and needs more severe conditions, more hydrogen, and a higher energy is required to break the C-N bonds (Szymanska *et al.*, 2003). Besides the hydrogenation of aromatic rings, it is required prior to the attack of the C-N bonds for nitrogen removal (Figure 4.2).

Non-basic compounds can be converted to basic compounds during the refining/catalytic cracking process. Basic N-compounds are known to inactivate HDS catalysts, even at a few tens of ppm, though their competitive adsorption on the acidic active sites of the catalysts (Shin *et al.*, 2000 and 2001; Zeuthen *et al.*, 2001; Laredo *et al.*, 2003; Zepeda *et al.*, 2016). Thus, they are potential inhibitors for HDS process (Choi *et al.*, 2004; Laredo *et al.*, 2004). For example, CAR has an extremely low reaction rate in the HDN process and the position of the methyl group in methylated-CAR is determinant for their reactivity (Shin *et al.*, 2000), which poisons the HDS-catalysts. Egorova and Prins (2004a) reported that nitrogen

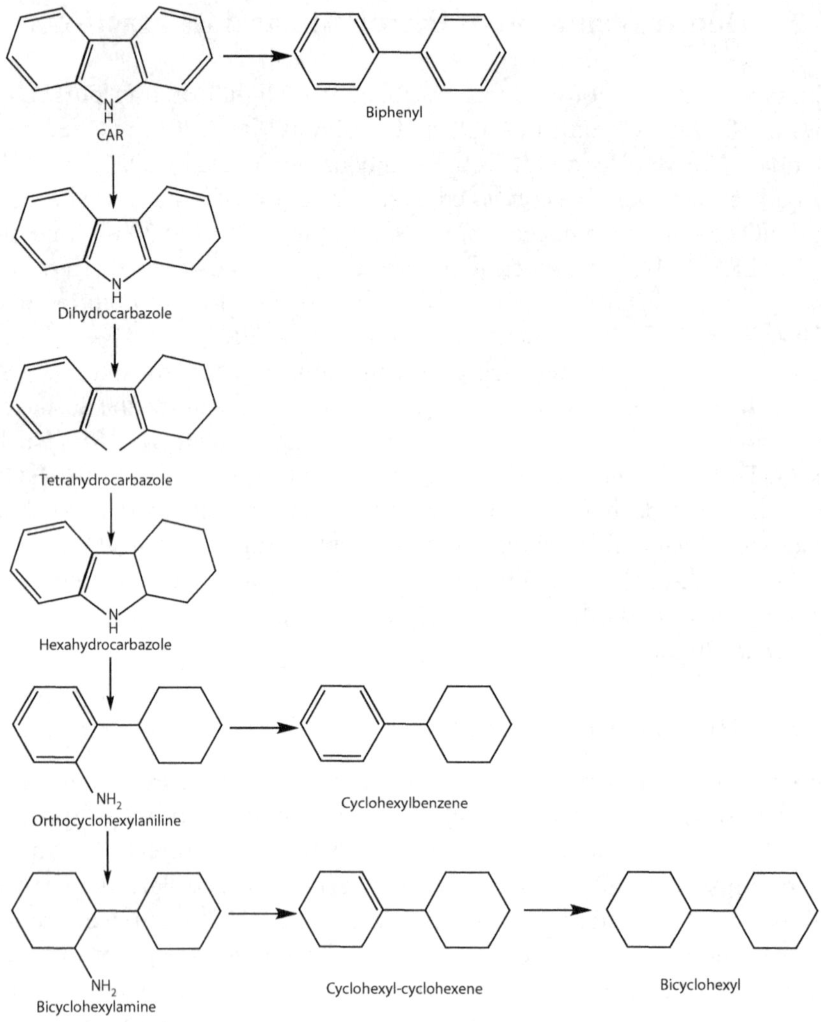

Figure 4.2 HDN-Pathway.

compounds are more rapidly denitrogenated on direct desulfurization sites than on hydrogenation sites. Thus, the hydrogenation pathway of HDS is strongly suppressed by nitrogen compounds rather than the direct desulfurization pathway (Zeuthen *et al.*, 2001; Ho *et al.*, 2003; Egorova and Prins, 2004b). Sano *et al.* (2004a) reported that the HDS of gas oil is affected by nitrogen species in feed oil with a nitrogen content higher than 60 ppm. Tao *et al.* (2017) reported that the HDS of straight-run gas oil (SRGO) is significantly inhibited by increasing N/S in feed oil. Moreover, the HDS of the recalcitrant PASHs (such as DBT, C1DBT, 4-MDBT and

4,6-DMDBT) are strongly inhibited with the increase of N/S rather than the simple sulfur compounds (such as Ths, BT, and C1BT). Consequently, denitrogenation before HDS is preferable. Although Qn displays a stronger inhibiting effects on HDS catalysts than indole, it is reported to be easily removed during the conventional hydrotreatment. The positive impact of its removal on HDS efficiency has not been clearly reported (Choudhary *et al.*, 2008). However, most of the denitrogenation studies focus on CAR removal. DBT and CAR are widely used as model compounds for the study of desulfurization and denitrogenation, respectively. It should be known that not only are nitrogen compounds less reactive than sulfur compounds in hydrotreatment, but the NH_3 produced during HDN can also inhibit the HDS catalysts. Thus, adsorptive denitrogenation is suggested to be performed before the HDS process (Murti *et al.*, 2003; Choi *et al.*, 2004). However, there are many ongoing studies to improve the cost and performance of the adsorptive pretreatment in terms of the type of the adsorbent, its cost, its efficiency for nitrogen and sulfur removal, its regeneration and reusability, and finally the recovery of the treated oil and its quality.

It has been reported that the most common catalysts used in the petroleum industry for nitrogen removal via catalytic hydrotreatment are those including nickel-molybdenum (Ni-Mo) supported on aluminium oxide, while Co-Mo catalysts are preferable for HDS (Katzer and Sivasubramanian, 1979; Williams and Douglas, 1985). However, molybdenum nitrides and carbides are reported to be more active than alumina-supported sulfided NiMo and CoMo ones for HDN and HDS (Choi *et al.*, 1992, 1995; Lee *et al.*, 1993). Landau *et al.* (1996; 1998ab) developed a novel catalyst system to reduce the sulfur and nitrogen contents in shale oil, in which the degree of HDS and HDN satisfied the requirements for furthering the hydrocracking process. This took place into three steps, where the first is HDS with wide-pore Ni-Mo-A1 catalysts to crack the C-S bonds in large molecules and then, deep HDN of lower-molecular-weight nitrogen products with zeolite-containing catalysts. The overall conversion of both sulfur and nitrogen in the two stages, under medium-severity conditions (380 °C and 15 MPa), was very high (>99.9%). Introduction of a zeolite into a Ni(Co)-Mo-AI catalyst had a significant effect on the hydrotreating performance. Then, finally, the third hydrocracking step of deeply denitrogenated shale oil at 350 °C and 14.8 MPa to lighter products within 4 h occured where 80% of the product was in the naphtha boiling range (Landau *et al.*, 1996). Hydrotreatment of shale oil in a trickle-bed reactor pilot plant was packed with two novel catalysts in series. The first catalyst was Ni-Mo supported on wide-pore alumina and the second catalyst was Co- Mo -Cr supported on a combined zeolite HY-alumina carrier was also reported (Landau *et al.*, 1998a). The desulfurization and denitrogenation conversions were higher than 99% at 380°C and 150 atm, while the hydrotreatment of shale oil in a fixed bed reactor pilot plant with two Ni-Mo-zeolite catalysts based on mono-(HY + Al_2O_3) and bizeolite (HY + H-ZSM-5 + Al_2O_3) supported at 50 atm of hydrogen pressure and 350 °C led to a desulfurization and denitrogenation conversion of higher than 99.7% and the nitrogen content in the liquid hydrocracking products at full conversion of atmospheric residue fraction of the shale oil was <1 ppm and the sulfur content <15 ppm (Landau *et al.*,.1998b). Williams and Chishti (2001) reported the hydrotreatment of shale oil at 400 °C for 12 h in a stirred reactor, first with a Ni-Mo catalyst and secondly with a Co-Mo catalyst. where both catalysts showed nearly the same HDN and HDS activities. The catalytic hydrotreating of the diesel fraction (200–360 °C) from Fushun shale oil was investigated in a fixed-bed reactor where a NiMoW catalyst showed good HDS and HDN efficiency producing oil with low sulfur, nitrogen, and alkene contents, reduced density, and increased cetane number, which can be used as a more valuable fuel (Hang *et al.*, 2010). Tang *et al.* (2013) reported the HDS and HDN of middle distillates of Chinese Huadian shale oil in a bench-scale trickle-bed reactor using a commercial NiMoW/Al_2O_3 catalyst. Tao *et al.* (2017) reported the HDS and HDN of straight-run gas oil over a NiW/Al_2O_3 catalyst.

One of the critical barriers for achieving ultra-deep HDS (UD-HDS) is the low reactivity of refractory high molecular weight PASHs and the strong inhibitory effect of H_2S, NH_3, nitrogen, and aromatic compounds, especially in the range of 0–100 ppmS (Zeuthen *et al.*, 2001; Shin *et al.*, 2003). Thus, for UD-HDS of oil feed, a pre-treatment adsorptive-desulfurization (ADS) and denitrogenation (ADN) is very recommendable (Sano *et al.*, 2004a).

4.2.2 Adsorptive Denitrogenation

Zeolite (Hermandez-Maldonado and Yang, 2004), activated carbon (Sano *et al.*, 2004b), activated alumina (Kim *et al.*, 2006), mesoporous silica (Kwon *et al.*, 2008), silica gel (Min, 2002), and silica-zirconia cogel (Bae *et al.*, 2006) are reported to have a good selectivity for N-species from gas oil at an ambient temperature and pressure. Metal-organic frameworks (MOFs) are very promising for their extraordinary functionality and porosity (Ahmed and Jhung, 2016).

Moreover, ADN on MOFs is more prominent than ADS due to the preferential adsorption of NCCs relative to the SCCs because of the exothermic and coordinative interaction of the NCCs with the coordinatively unsaturated sites (CUSs) of the MOFs (Maes *et al.*, 2011; Ahmed *et al.*, 2013; de Voorde *et al.*, 2013). However, the main drawbacks of MOFs are that they are not stable at elevated temperatures and they are structurally damaged in even moderately mild basic conditions. The first drawback is not of real concern, as adsorptive procedures are usually carried out at relatively low temperatures. Min *et al.* (2001) reported that adsorptive denitrogenation (ADN) of petroleum fractions improves HDS efficiency where the improvement of the degree of HDS is directly proportional to the degree of nitrogen removal in the pre-ADN of the oil feed (Min, 2002). Murti *et al.* (2003) reported that the pretreatment of gas oil for the removal of nitrogen species by silica or silica-alumina improves the HDS process. Choi *et al.* (2004) reported that HDS of 80 and 64% adsorptive denitrogenated light gas oil (LGO) produced 19 and 120 ppmS LGO, respectively, while the non-denitrogenated LGO produced 311 ppmS LGO. However, solvent extraction for non-basic nitrogen-compounds also exist, but approximately 30% of the oil goes into the extract phase, therefore it is less recommendable.

Streat *et al.* (1995) reported that a typical activated carbon (AC) has high specific surface area (600–2000 m^2/g) and a well-defined microporous structure (average pore size of about 1.5 nm). Furthermore, the oxygen containing functional groups on AC are reported to play an important role in the ADN and ADS of diesel fuels (Almarri, 2009; Almarri *et al.*, 2009). An activated carbon prepared from petroleum coke through KOH activation with an apparent surface area of 3000 m^2/g, MAXSORB-II, has been reported to be effectively applied for the ADN of gas oil at ambient temperature and pressure and was attributed to the presence of oxygen functional groups on the carbon surface (Sano *et al.*, 2004a,b). Many agro-industrial wastes, such as carbonized slash pine bark, palm tree cobs, oil palm shell, and chromium tanned leather, can be used for the preparation of cost effective activated carbon (Ali *et al.*, 2012). Household wastes, like olive stone, almond shells, apricot stones, peach stones, palm fruit bunch, and coconut shells, are also used for the production of low cost activated carbon (Anisuzzaman *et al.*, 2014). Sano *et al.* (2004b) reported that activated carbon and activated carbon fiber are considered to be versatile absorbents for gaseous and liquid substrates, which makes them good candidates for the pretreatment of oil feed at room temperature. Moreover, their surface structure and properties can be controlled for better adsorbent properties (Hayashi *et al.*, 2002; Shirahama *et al.*, 2002; Leboda *et al.*, 2003).

Most of the sulfur and nitrogen compounds in SRGO are BTs, DBTs, Qns, indoles, and CARs. Sano *et al.* (2004b) reported the adsorption of different S and N compounds from SRGO by activated carbon at room temperature. However, the selectivity of refractory S-compounds, such as 4,6-DMDBT, increases at high temperatures (50 °C). The ADN of SRGO over different adsorbents (i.e. activated carbon materials: MAXSORB-II, MGC-B, and OG-20A) was studied and recorded 77, 41, and 43, respectively, where the higher the oxygen content is, the higher the adsorption efficiency. Their surface area are not an important factor in and, however, their ADS capacity ranged between ~12–28% and is highly affected by their surface area. The HDS of pretreated SRGO with an N-content of 60 and 20 ppm, reduced the S-content from 193 ppm to 11 and 8 ppm, respectively, under conventional conditions. Moreover, HDN with almost total N-removal was also performed, along with HDS. The regeneration of the most efficient activated carbon, MAXSORB-II, by toluene at 80°C confirmed its regeneration and repeated removal of S and N species in SRGO without losing its activity. Alcohol or polar solvent is usually required to regenerate adsorbents (Min *et al.*, 2001), but applying a conventional hydrocarbon solvent like toluene, which is often found in refinery plant, adds to the advantages of MAXSORB-II. The high surface area, oxygen content, and lower solubility in SRGO of MAXSORB-II and its polar surface makes it good candidate for ADN, especially for basic N-species of high polarity and heavier S-species.

Zhang *et al.* (2010) reported the spontaneous physical adsorption of nitrogen containing compounds (NCCs) on the mesoporous molecular sieve adsorbent Ti-HMS. Koriakin *et al.* (2010) reported the selective ADN of NCCs from residue HDS-diesel by a lithium-modified mesoporous silica adsorbent. Generally, the adsorption selectivity on MOFs can be ranked as follows: NCCs > SCCs > aromatics > aliphatics (Maes *et al.*, 2011; Ahmed *et al.*, 2013; Ahmed and Jhung, 2016). Nuzhdin *et al.* (2010) reported the ADN of SRGO, LCO, and model mixtures by MOF through the coordination of N-atoms to Cr^{III} centers in the MOF. Li *et al.* (2013a) reported that the ADN is increased after the ammonium persulfate ($(NH_4)_2S_2O_8$) oxidation of the activated carbon.

Kim *et al* (2006) reported that the activated carbon (AC) shows higher adsorptive capacity and selectivity for both sulfur and nitrogen compounds in model diesel fuel than an activated alumina and nickel-based adsorbent, especially for the sulfur compounds with methyl substituents, such as 4,6-methyldibenzothiophene (4,6-DMDBT), where hydrogen bond interaction would play an important role in ADS and ADN over the activated carbon. Almarri *et al.* (2009) also reported that the activated carbon expresses higher ADN than the activated alumina. Wen *et al.* (2010) reported that the acidic functional oxygen-containing groups, carboxyl and carboxylic anhydride, support the selective adsorption of quinolone, while the basic groups of carbonyl and quinone support the adsorption of indole. Although the oxidation of a wood-based activated carbon (AC) by $(NH4)_2S_2O_8$ solutions decreases the specific surface area of AC samples, it increases the oxygen functional groups. Han *et al.* (2015) reported that upon the treatment of AC samples by saturated $(NH_4)_2S_2O_8$ solution, the total amount of carboxylic acid groups was significantly increased with the increase of oxidation time. After 0.75 h oxidation, the number of total carboxylic acid groups is more than tripled in contrast to that of AC-fresh. Moreover, the phenolic, ketone, and aldehyde groups also increase with the increase of oxidation time. On the application of the oxidized AC on light cycle oil (LCO), all the carbon samples showed higher adsorption capacities for N compounds than S compounds and the adsorption capacities for S compounds of oxidized AC samples were lower than that of AC-fresh. This was attributed to the decrease in the micro-porosity of oxidized AC samples and/or the weaker competitiveness than N compounds on oxygen functional groups. Therefore, the effect of textural properties on the adsorption capacity for S-compounds in LCO seems to be more remarkable than that of oxygen functional groups. The highest capacity for N compounds, 0.47 mmol N/g, was obtained when carbon was oxidized by a 15 wt.% $(NH4)_2S_2O_8$ solution for 1.5 h. Thus, the 1.5-h oxidation time is a critical point where the N capacity reaches a maximum. Afterwards, the stronger oxidation conditions cause more porosity loss. At that time, the positive effect from oxygen functional groups could not overcome the negative effects from micro-porosity, resulting in the decrease of adsorption capacity. The breakthrough of different oil compounds for AC-fresh increased in the following order: naphthalene ≈ 1-methylnaphthalene < indole < DBT < carbazole ≈ quinoline. The adsorption of quinoline, carbazole, and DBT is mainly dominated by physisorption, while that of indole is enhanced by a chemical effect. Carboxylic acid groups and lactone groups favor quinoline adsorption, while carboxylic anhydride groups and phenolic groups are likely to improve the adsorption of indole, carbazole, and DBT. Lopes *et al.* (2016) reported the ADS and ADN of diesel oil by activated carbon produced from coconut shell. Moreover, the ADN and ADS of the activated carbon impregnated with palladium chloride was over 85 and 60%, respectively, while the ADN of sulfuric acid-treated activated carbon was over 75%. The acidic extractants and adsorbents, such as the liquid acids, metal ion-loaded solvents, and MOFs, show better denitrogenation performance towards basic NCCs than the non-basic ones (Han *et al.*, 2013; Ahmed and Jhung, 2014, 2016), while the contrary occurs with aprotic solvents (Ali *et al.*, 2016). Han *et al.* (2015) reported that although the $(NH_4)_2S_2O_8$ oxidation of a commercial wood-based activated carbon WVB 1500 decreases its specific surface area, it increases its oxygen functional groups, thus facilitating the ADN and ADS of light cycle oil (LCO).

Santos *et al.* (2012) reported that the incorporation of molybdenum oxide on a commercial silica/alumina based adsorbent is promising for the removal of sulfur and nitrogen compounds, as it increases the amount of Brønsted acid sites. Shao *et al.* (2012) also reported the ADS activity of MoO_3–MCM 41 depends on the surface acidity, which is expressed by the ratio of the number of Lewis acid sites to the number of Brønsted acid sites measured by pyridine adsorption. Silveira *et al.* (2015) studied the effect of impregnation of nickel, cerium, molybdenum, and cobalt oxides in silica–alumina on the removal of nitrogen and sulfur compounds from a hydrotreated diesel where the impregnation of the metals oxides increased the density of acid sites and promoted the removal of nitrogen and sulfur compounds, where the adsorbent with impregnated molybdenum oxide expressed the best performance.

As has mentioned before, the removal of nitrogen polar compounds (NPC) is essential for the enhancement of the HDS process and the degree of improvement of HDS is directly proportional to the degree of nitrogen removal in a pretreatment step (Min, 2002). In an attempt for the denitrogenation process prior to HDS, the SK Corporation of Korea has developed a unique process, that is SK hydrodesulfurization (HDS) pretreatment technology, to produce ultra-low sulfur diesel (ULSD) with pretreatment of middle-distillate-range petroleum fractions. This technology has been successfully demonstrated through a 1000 bbl/day demonstration plant which has been running since May 2002. SK-HDS pretreatment technology is based on the adsorptive removal of a small amount of the NPC where a great improvement in deep HDS efficiency is observed with the removal of these NPC from the feedstock. The SK-HDS pretreatment process consists of multiple adsorbers, two stripping columns for desorbent recovery, associated pumps, and an overhead system. Briefly, the middle distillate feed is pretreated prior to the HDS process in the adsorbent vessels, followed by stripping to remove a small amount of desorbent. The pretreated fuel is then sent to the downstream HDS unit for sulfur removal. The rejected stream with high nitrogen content, which is approximately 2 vol%, can be either used as a blending stock for heavy petroleum products or processed in other refining units (Lee *et al.*, 2003). Min (2002) performed a pretreatment for high S-content LCO (1.34%S) and 200 ppm N with a pilot scale HDS unit operated under the conventional HDS conditions to determine the benefits of NCC removal prior to the HDS process. It was found that as the N-removal increased, the HDS performance increased. For example, at the same reaction temperature of around 354 °C, the product sulfur content decreased from 120 ppm to <20 ppm as the feed nitrogen content decreased from 200 to 50 ppm with a nearly complete removal of recalcitrant 4,6-DMDBT. Moreover, it was estimated that the reaction temperature gains increase significantly as the nitrogen removal ratio (NRR) increases. For example, with the target sulfur level of 50 ppm, the reaction temperature gain could be as much as 24 °C when approximately 80% of nitrogen is removed from the feed. Not only this, but if the demand becomes stronger to reduce the S-content in the fuel further, for example to 10 ppm, the reaction temperature gain is even higher at approximately 30 °C. Thus, as the regulation on sulfur content becomes more stringent the, SK HDS pretreatment process is expected to become a more effective tool for refiners to choose from to achieve the required regulation limit even in a low pressure HDS unit of as low as 30 bar. The other benefits of the SK process can be summarized as lower consumption of hydrogen, decreases in the aromatics-content by up to 1% point, acceptance of a heavier distillation range of feedstock than the conventional HDS approach, it requires no or minimum modification in the existing HDS units, causes virtually no downtime for the HDS unit, and finally, improves the color and storage stability of HDS products. Li *et al.* (2003) estimated that the adsorbent is expected to last over two years with little or no losses. The predicted adsorbent cost is $0.10 per barrel of charge regardless of unit capacity. Solvent losses are in the range of about $0.02 per barrel of feed. The inside battery limits (ISBL) investment cost for a stand-alone grass roots SK process is estimated to be US$ 400/bpsd for the first quarter of the year 2003 for plant capacities in the range of 25,000 to 30,000 bpsd at a US Gulf Coast location. Only minor modifications are required for the existing HDS unit to the couple SK pretreatment process upstream of the HDS unit. The investment cost is approximately US$ 300/bpsd for a 60,000-bpsd capacity unit. The total utility consumption of the process with fired heater design is estimated to be as follows:

- Fuel: 0.01 FOEB
- Cooling water: 1.0 t
- Electricity: 0.4 kW/barrel of feed

Pretreated feed consumes approximately 10 to 20% less hydrogen in the HDS unit at the same product outlet sulfur level due to the less severe operating conditions required and the significant reduction in denitrification requirements. This is also expected to result in a longer HDS catalyst life and a higher HDS product yield. The investment and operating cost can be lowered substantially if the pretreatment unit is heat-integrated directly with the downstream HDS unit.

Upon the integration of an SK-pretreatment unit with an HDS unit, it was estimated that the investment cost would be lowered by approximately 25% with significant operating cost savings where the total utility consumption of the integrated steam stripping case is as follows:

- Steam: 0.01 t
- Cooling water: 0.6 t
- Electricity: 0.33 kW/barrel of feed

The SK process is attractive for older, lower pressure units that require extensive revamping costs which are obviously very site specific and, therefore, must be developed on a case-by-case basis together with site-specific operating costs. The SK process also appears to be attractive for a new grass root unit due to the increased capital cost for a high-pressure unit versus a low-pressure HDS unit.

4.2.3 Extractive and Catalytic Oxidative Denitrogenation

The neutral NCCs can be selectively removed by ion-exchange resins which contain cationic and anionic sites. This would occur through the hydrogen-bonding of NCCs to the negative-sites, like that of chlorides, but the denitrogenation capacity of ion-exchange resins is lower than that of HDN (Xie *et al.*, 2010). The NNCs can be readily extracted by ionic liquids (ILs) (Zhang *et al.*, 2004). The denitrogenation by liquid-liquid extraction can be also applied, but in a multi-step process: first, the oxidation step, then a solvent extraction. This requires a large amount of oxidizing agents and solvents. Thus, it is a complex and cost ineffective method to be widely applied (Shiraishi *et al.*, 2002). NCCs could be removed simultaneously with SCCs in the ECOD of liquid fuels (Yi *et al.*, 2014) or ILs (Zhang *et al.*, 2004; Xie *et al.*, 2008a,b). Ali *et al.* (2016) reported the green and efficient denitrogenation of fuels using deep eutectic solvents (DESs). The 1:2 molar mixture of choline chloride and phenylacetic acid showed the best denitrogenation towards both basic and non-basic NCCs. Due to its high polarity and low solubility of oils in DES, the extraction efficiencies of pyridine and carbazole at 308 K with a 1:1 DES/oil mass ratio recorded 99.2% and 98.2%, respectively, the selectivity of NCCs to n-heptane were higher than 10,000 since the solubility of n-heptane recorded only 4.6 mg/g. The extraction efficiency of the DESs are not sensitive to the DES/oil mass ratio and temperature and can be used after four regeneration cycles without losing their activity. The DESs have interesting features similar to ILs, such as low volatility and tunable physicochemical characteristics, and since they are usually prepared from cheap and safe compounds, they are cheaper and less toxic than ILs. The most important factor is the acidity of the DES; too low an acidity leads to an insufficient affinity towards the basic NCCs, while too high an acidity, leads to a chemical reaction between the DES and the basic NCCs and, thus, a difficulty in DES regeneration. Moreover, the excess acidity of DES is not favorable of extractive denitrogenation of non-basic NCCs. Hansmeier *et al.* (2011) reported that the pyridinium-based ILs, [3-mebupy]$N(CN)_2$, [4-mebupy]$N(CN)_{2,}$ and [bmim]$C(CN)_{3,}$ can extract NCCs and SCCs from real oil feed, gasoline, and diesel better than conventional solvents with preference to the NCCs. The aromatic imidazolium ILs are better than those of cyclic thiophenium and tetrahedral trialkylsulfonium ILs for extractive denitrogenation (Asumana *et al.*, 2011). The dicyanamide-based ILs, [BMI][$N(CN_2)$], [EMI][$N(CN)_2$], [S_2][$N(CN_2)$], and [$EtMe_2S$][[$N(CN_2)$], are reported to be capable of effectively extracting NCCs from fuel oils with more preference to CAR than pyridine (Asumana *et al.*, 2011). The relatively inexpensive halogenated ILs are reported to be excellent in the selective extractive denitrogenation of liquid fuels that can be easily regenerated and recycled (Ceron *et al.*, 2012). Wang *et al.* (2014) reported that the extractive denitrogenation of Qn increases with the the increase of ILs acidity. Laredo *et al.* (2015) reported the synthesis of some tetraalkylammonium and choline ILs that expressed good extractive denitrogenation of CAR and indole, where the triethylmethylammonium butyrate and triethylmethylammonium acetate ILs presented good batch denitrogenation of SRGO. Hizaddin *et al.* (2015) reported the extraction of NCCs from diesel fuel using imidazolium-and pyridiium-based ILs with more preference towards the 5-membered ring compounds (pyrrole and indole) than the 6-membered ones (pyridine and Qn).

The biological process operates at ambient pressure and temperature (An *et al.*, 2010). Current research focuses on the microbial denitrogenation of non-basic nitrogen-compounds and their alkyl-derivatives since they represent the majority of the total nitrogen.

4.3 Microbial Attack of Nitrogen Polyaromatic Heterocyclic Compounds (NPAHs)

Several microorganisms capable of degrading NPAHs have been isolated from wastewater sludge, hydrocarbon-contaminated soil and water, industrial effluents, and coal- and shale-liquefaction sites. Several species of the genus *Pseudomonas* have been isolated that degrade CAR and its alkyl derivatives for its solvent tolerance. A variety of other microorganisms have also been reported to mineralize non-basic nitrogen compounds, including species of *Bacillus, Sphingomonas, Xanthomonas, Gordonia, Klebsiella, Burkholderia, Methylobacterium, Arthrobacter,* and *Novosphingobium, Pseudomonas, Cycloclasticus,* etc. (Singh *et al.*, 2011a,b; Zhao *et al.*, 2011; Nojiri, 2012; Morales and Le Borgne, 2010, 2017; Ayala *et al.*, 2016; Zakaria *et al.*, 2016).

The initial attack on NPAHs is normally by oxygenase enzymes that activate the dioxygen molecule and use it for regio- and stereo- selective oxidation by the insertion of molecular oxygen into the organic substrate. These enzymes function under mild conditions of pH and temperature, allowing high yields of hydroxylated products. Dioxygenases incorporate two atoms of molecular oxygen into one molecule of substrate, while monoxygenases add only one atom of oxygen (the other atom being reduced to water). Although there are different types of monooxygenases, most add a single hydroxyl group to an already hydroxylated substrate to generate a dihydroxy product. Dioxygenases that form cis-diols are composed of two or three components, forming cis-dihydrodiols or cis-diol carboxylic acids, respectively. The three-component dioxygenases are composed of a flavoprotein, a ferredoxin, and a terminal oxygenase. Dioxygenases often have broad substrate specificity and require only a minimal characteristic structure for substrate recognition (Marcelis *et al.*, 2003).

Among the non-basic nitrogen compounds (Benedik *et al.*, 1998), pyrrole and indole have been found to be readily degraded, while CAR is more recalcitrant to microbial degradation, although CAR is relatively resistant to microbial attacks. However, many laboratory investigations have been conducted to predict the fate of CAR in the environment (Fedorak and Westlake, 1984; Gieg *et al.*, 1996), but few reports about selective biodenitrogenation (BDN) of CAR are published (Kimura *et al.*, 1996; Sato *et al.*, 1997ab; Benedik *et al.*, 1998; Bressler *et al.*, 2002; Singh *et al.*, 2013; Zakaria *et al.*, 2016). CAR can be metabolized to give free amines, alcohols, phenols, ketones, aldehydes, and carboxylic acids (Ouchiyama *et al.*, 1993). Other pathways would liberate nitrogen from CAR in the form of ammonia (Rhee *et al.*, 1997). The functional groups are often combined in a single metabolite. Generally, there are two modes of aerobic attack on CAR: (1) The naphthalene-like attack which involves the dioxygenase attack at carbons number 3 and 4, followed by *meta*-cleavage (Grifoll *et al.*, 1995) and (2) the angular dioxygenase attack that involves the oxidation of carbons number 9a and 1.

Grifoll *et al.* (1995) reported that although the fluorine-degrading *Pseudomonas cepacia* strain F297 cannot grow on CAR, it can attack it through the naphthalene-like attack on carbons 3 and 4 producing 4-(3′-hydroxy-2′-indolyl)-2-oxo-3-butenoic acid Scheme 4.2).

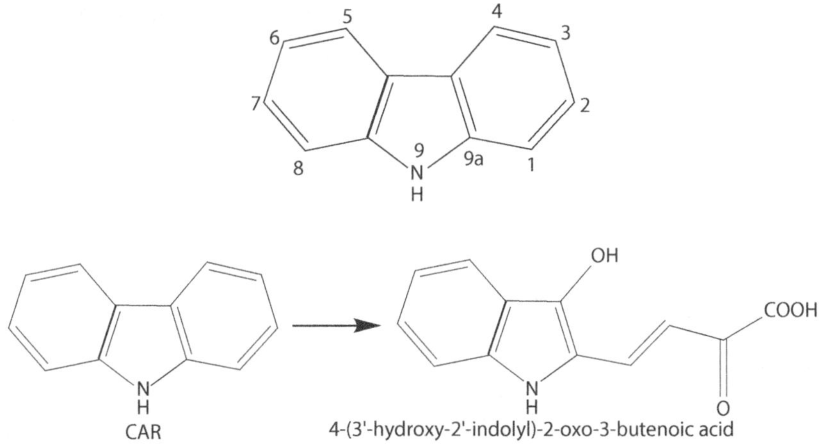

Scheme 4.2 Naphthalene-Like Attack of *Pseudomonas cepacia* F297 on Carbazole.

However, Yoon *et al.* (2002) reported that cells of *Pseudomonas rhodesiae* strain KK1 isolated from soil contaminated with coal tar pregrown on phenanthrene, show more rapid mineralization of CAR (2.8-times) than those pregrown on naphthalene. Thus, a possible close linkage between CAR and phenanthrene catabolism pathways is suggested.

In contrast to DBT-BDS, less is known about CAR-biodenitrogenating microorganisms. However, although it has a resistance to microbial attack, several studies have been performed concerning CAR-biodegradation. Gieg *et al.* (1996) reported several metabolites from the angular dioxygenase attack of *Pseudomonas* strain LD2 on CAR. The genes involved in CAR-BDN (car) have been identified and a car-gene cluster can be induced and constitutively expressed through the CAR addition to the microbial growth media (Shepherd and Lloyd-Jones, 1998; Kilbane *et al.*, 2002). The most studied strain capable of CAR metabolism is *Pseudomonas resinovorans* CA10 (Nojiri *et al.*, 1999; Nam *et al.*, 2002; Maeda *et al.*, 2003). Sato *et al.* (1997a, b) reported that the CAR 1,9a-dioxygenase (CARDO) enzyme which is responsible for angular dioxygenase attack of *Ps. resinovorans* CA10 on CAR is encoded by the carAa, carAc, and carAd genes. It produces an unstable intermediate that spontaneously forms 2′-aminobiphenyl-2,3-diol. This enzyme is reported to consist of four parts: two monomeric terminal oxygenases (CarAa, 43 kDa), ferredoxin (CarAc), and a ferredoxin reductase (CarAd, 36 kDa), while the two dioxygenases meta-cleavage enzymes are encoded by the carBa and carBb genes that produce 2-hyroxy-6-oxo-6-(2′-aminophenyl)-hexa-2,4-dienoic acid (HOADA) (Nojiri *et al.*, 1999, 2005; Nojiri and Omori, 2007; Diaz and

Garcia, 2010). The mechanism of CARDO has been clarified (Figure 4.3, Riddle *et al.*, 2003a; Nojiri *et al.*, 2005).

While the hydrolase enzyme 2-hydroxy-6-oxo-6-(2′-aminophenyl) hexa-2,4-dienoic is an acid (HOADA), hydrolase (CarC) is a protein of 290 amino acids encoded by the carC gene (870 nucleotides) (Riddle *et al.*, 2003ab). The CarBb is a unique extradiol dioxygenase member of the protocatechuate 4,5-dioxygenase family and CarC is a novel type of hydrolase having homology with hydrolases involved in the degradation of mono-aromatic compounds (Sato *et al.*, 1997a). The hydrolase is responsible for the production of 2-hydroxypenta-2,4-dienoic acid and anthranilic acid (AA) which then enters the TCA cycle after conversion to catechol (CAT) using the anthranilate 1,2-dioxygenase (AntABC). Then, the *ortho*-cleavage of catechol by a CatABC enzyme occurs and then, further metabolism occurs by *β*-ketoadipate pathway using the Pca enzyme. Formation of catechol is the step where nitrogen is removed, however, since it is degraded further, the whole molecule is lost and mineralized (Figure 4.4a). The 2-hydroxypenta-2,4-dienoate can also be metabolized using the *meta*-cleavage pathway enzymes, 2-hydroxypenta-2,4-dienoate hydratase (CarD), 4-hydroxy-2-oxo-valerate aldolase (CarE), and acetaldehyde dehydrogenase (CarF), produing TCA cycle intermediates (Figure 4.4a). Zakaria *et al.* (2016) reported the isolation of a Gram +ve *Bacillus clausii* BS1 with a higher CAR-degradation efficiency and tolerance to toxic metabolites relevant to the well-known biodenitrogenating

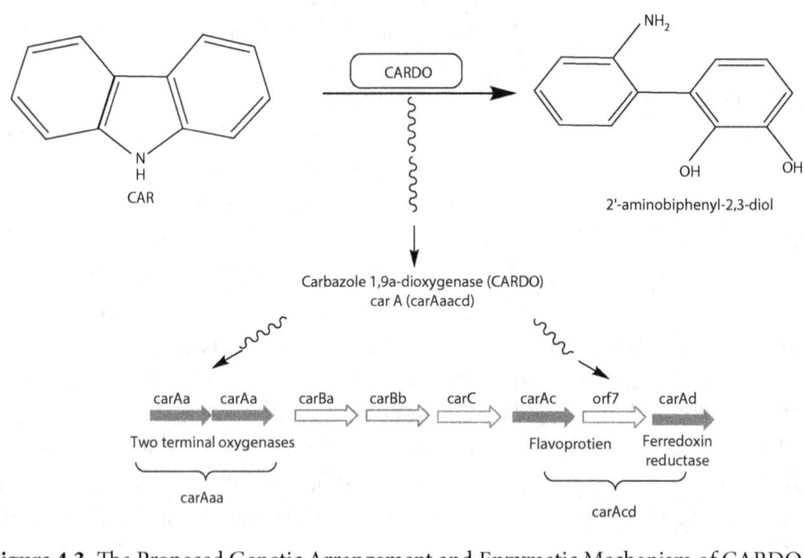

Figure 4.3 The Proposed Genetic Arrangement and Enzymatic Mechanism of CARDO (Riddle et al., 2003a).

bacterium strain *Pseudomonas resinovorans* CA10, recording 77.15 and 60.66% removal of 1000 ppm CAR with the production of 119.79 and 102.43 ppm anthranilic acid and 121.19 and 90.33 ppm catechol, as by-products, respectively. Gieg *et al.* (1996) reported the production of 3a,4-dihydro-2-hydroxypyrrolo(1,2-a)-quinoline-1,5-diene during CAR-metabolism

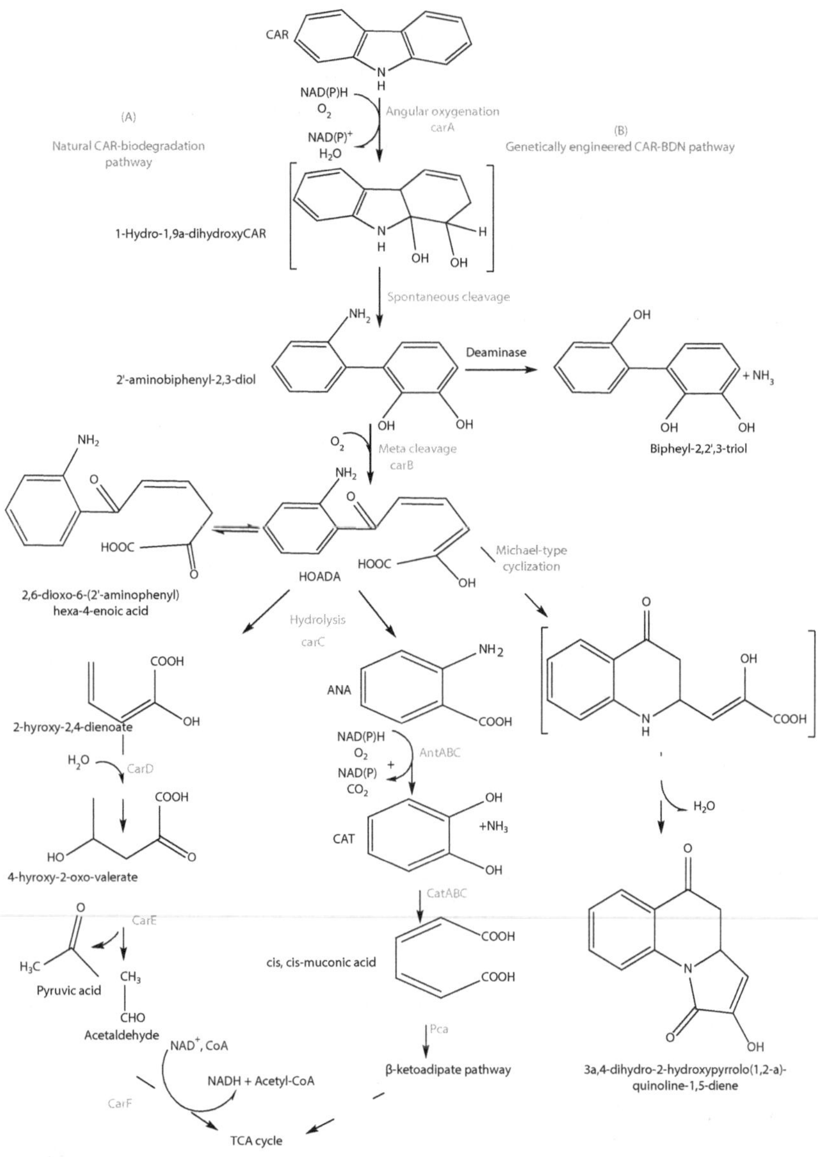

Figure 4.4 Pathways for Microbial Utilization of Carbazole.

by *Pseudomonas* sp. LD2. However, Shepherd and Lloyd-Jones (1998) reported that the angular dioxygenase from *Sphingomonas* sp. strain CB3 is a three-component enzyme, which is similar to biphenyl dioxygenase. Moreover, they deduced that a dehydrogenase, an extradiol dioxygenase, and a hydrolase enzymes are also involved in the CAR-biodegradation where 2-hydroxymuconic semi-aldehyde is produced from catechol. CARDO is reported to have the ability to oxidize a wide variety of PAHs, DBT, and biphenyl (Nojiri *et al.*, 1999), especially the polyaromatic heterocyclic compounds, by attacking the angular positions adjacent to the heteroatoms (Figure 4.5). Jong *et al.* (2006) isolated *Arthrobacter* sp. P1–1 from a PAHs-polluted soil from Hilo, HI, UA for its ability to degrade DBT (Figure 4.6) and CAR (Figure 4.7), producing different metabolites. DBT is degraded through three initial catabolic pathways: mono-oxygenation at an S-atom and 1,2- and 3,4- dioxygenations. The *ortho*- and *meta*-cleavage of DBT-diols are also reported. DBT is dioxygentaed by dioxygenase to cis-DBT-1,2-dihydrdiol and cis-DBT-3,4-dihydrodiol which produce 1,2- and 3,4-DBT-diol, respectively, by dihydrodiol dehydrogenase. This is attributed to the possession of P1–1 of either a dioxygenase system with a broad specificity or diverse dioxygenase systems, while the intermediates, trans-4-(3-hydroxy-benzo[b]thiophen-2yl)-2-oxobut-3-enoic acid (1a), 2-(2-carboxy-vinyl)-benzo[b]thiophene-3-carboxylic acid (1b), 3-(2-carboxy-vinyl)-benzo[b]thiophene-2-carboxylic acid (1c), and trans-4-(2-hydroxy-benzo[b]thiophen-3-yl)-2-oxo-but-3-enoic acid (1d) are metabolized by hydratase-aldolase (Eaton and Chapman, 1992). The *ortho*-cleavage is proved by the formation of benzo[b]thiophene-2,3-dicarboxylic acid and the *meta*-cleavage is proved by the formation of 2-hydroxy-benzo[b]thiophene-3-carboxylic acid, where both are converged to benzo[b]thiophene-2,3-diol that is further transformed into 2-mercaptobenzoic acid which is abiotically dimmerized to 2,2′-dithiosalicylic acid. Moreover, the CAR-metabolites prove lateral deoxygenation pathway, through 3,4-dioxygenation of CAR, followed by *ortho*- and *meta*-cleavage producing 2-aminobenzoic acid (i.e. anthranilic acid). Li *et al.* (2006) reported CAR-BDN by *Pseudomonas* sp. XLDN4–9 with a specific enzyme activity of 10.4 μmol/min/g dry cell weight. Gai *et al.* (2007) reported the cometabolic degradation of DBT and dibenzofuran (DBF), using CAR as a substrate through the angular dioxygenation pathway, by the Gram-negative bacterium, *Sphingomonas* sp. strain XLDN2–5, where CAR is the real inducer of the expression of degradation enzymes. Thus, the metabolic pathways are controlled by CAR-degrading enzymes. The marine bacterium *Neptuniibacter* sp. strain CAR-SF is reported to degrade some polyromantic heterocyclic compounds, including CAR and DBF

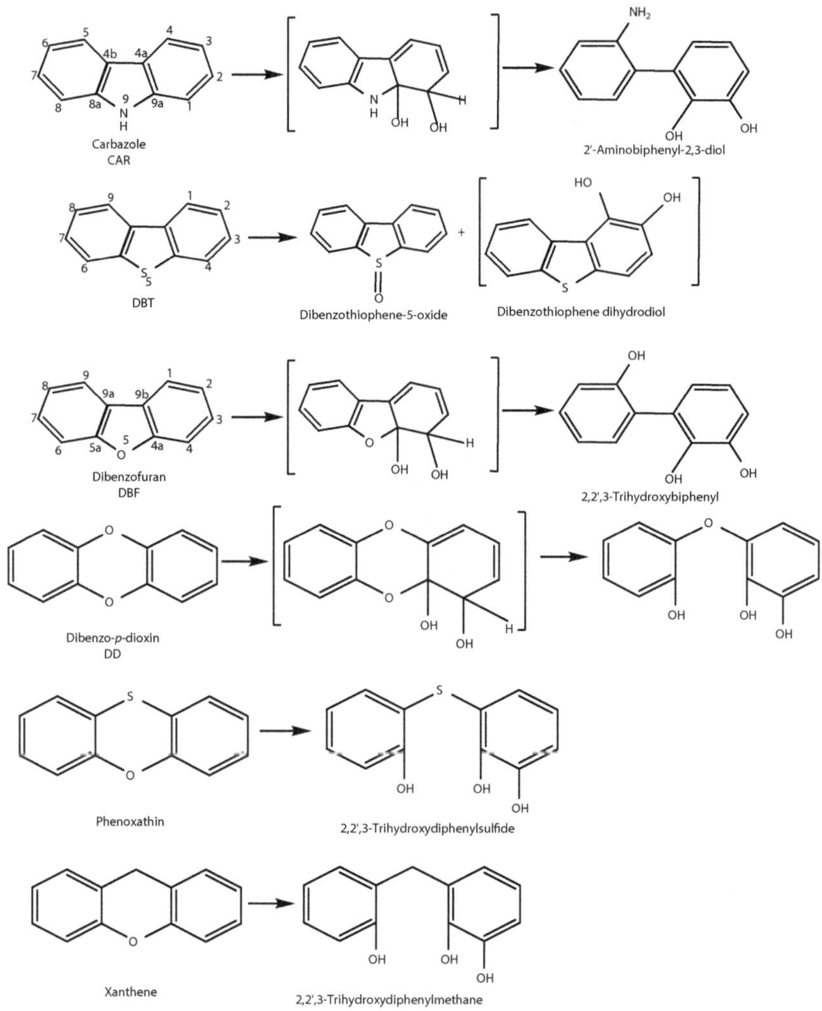

Figure 4.5 Oxidation Reaction Catalyzed by CARDO of *Pseudomonas* sp. Strain CA10 (Nojiri *et al.*, 1999).

(Fuse at al., 2003; Nagashima *et al.*, 2010). Maeda *et al.* (2009) reported the isolation of eleven marine bacteria that are capable of CAR-degradation to anthranilic acid where seven isolates were most closely (>97% similarity) to genera *Erythrobacter*, *Hyphomonas*, *Shingosinicella*, *Caulobacter*, and *Lysobacter*, while another four isolates were closely related (>97% similarity) to *Kordiimonas gwangyangensis* GW14-5. Larentis *et al.* (2011) reported the isolation of *Pseudomonas stutzeri* (ATCC 31258) for biodenitrogenation (BDN) of high concentration of CAR (1000 mg/L), while

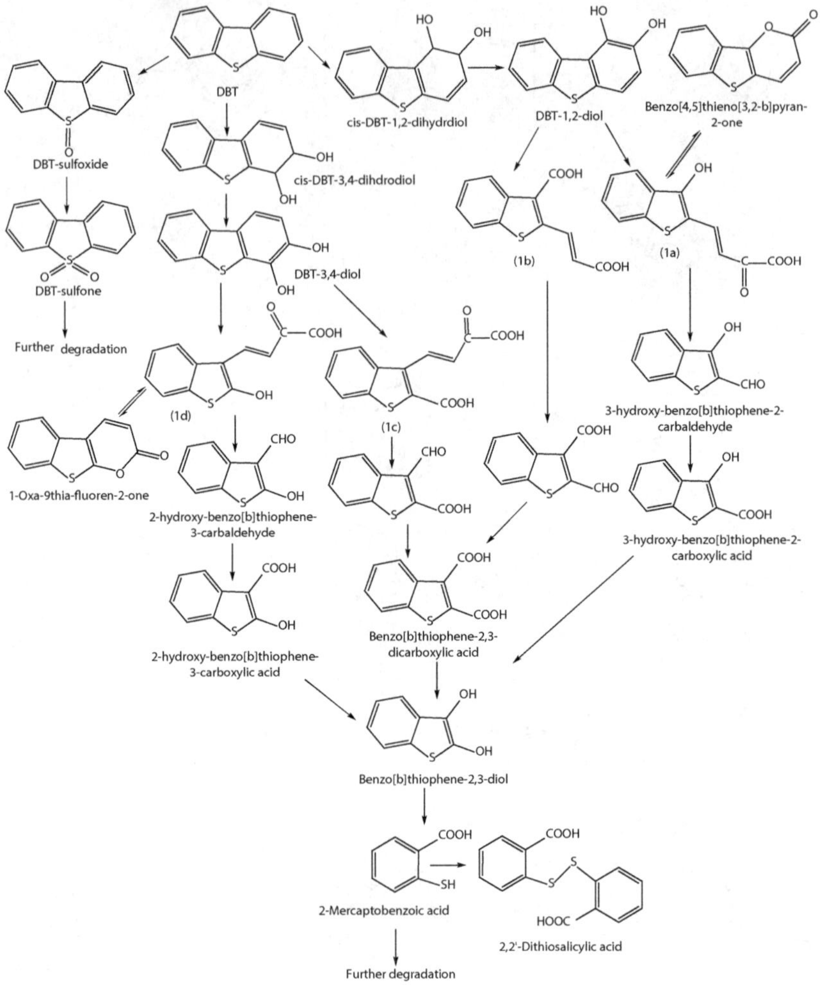

Figure 4.6 DBT-Biodegradation Pathway by *Arthrobacter* sp. P1–1 (Jong *et al.*, 2006).

1a. trans-4-(3-hrdroxy-benzo[b]thiophen-2yl)-2-oxobut-3-enoic acid; 1b. 2-(2-carboxy-vinyl)-benzo[b]thiophene-3-carboxylic acid; 1c. 3-(2-carboxy-vinyl)-benzo[b]thiophene-2-carboxylic acid; 1d. trans-4-(2-hydroxy-benzo[b]thiophen-3-yl)-2-oxo-but-3-enoic acid

Singh *et al.* (2013) reported a specific enzyme activity of 11.36 μmol/min/g DCW by a Gram negative *Pseudomonas* sp. strain GBS.5. Salam *et al.* (2014) reported the isolation of four CAR-degrading bacterial isolates from tropical African hydrocarbon-polluted soils: *Achromobacter* sp. strain SL1, *Pseudomonas* sp. strain SL4, *Microbacterium esteraromaticum* strain

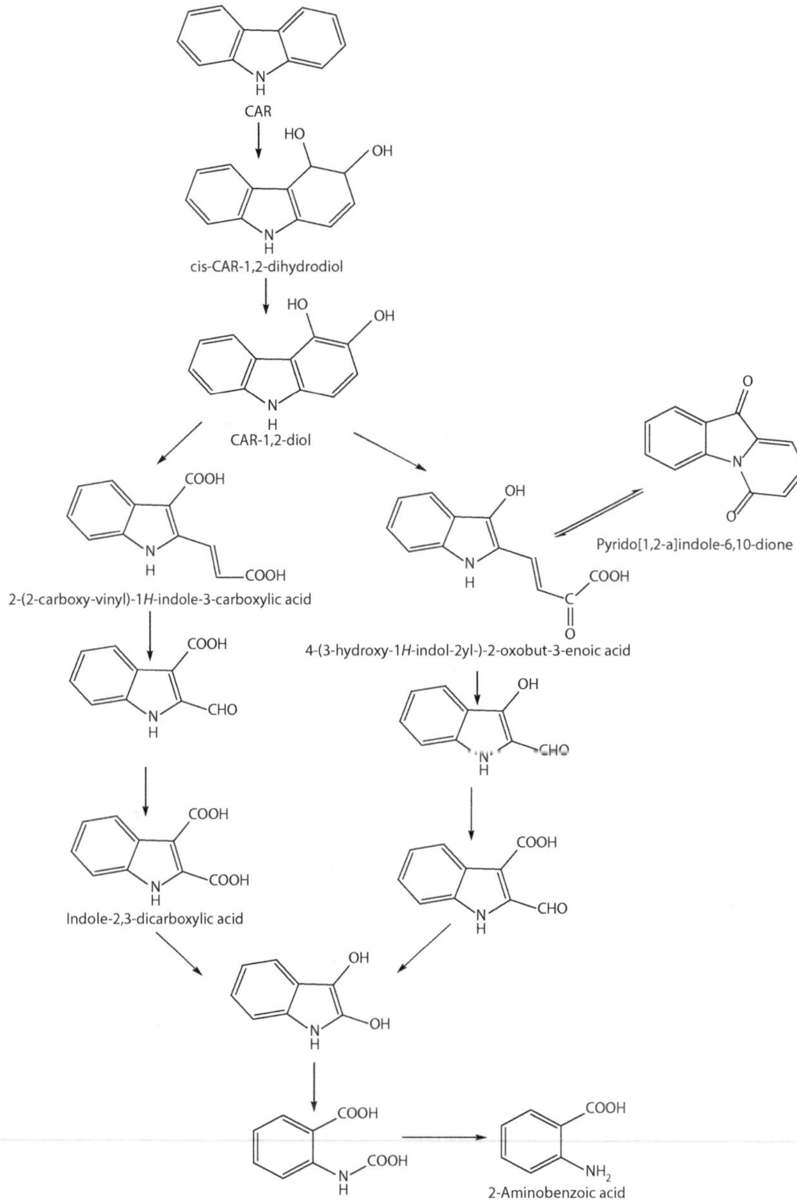

Figure 4.7 CAR-Biodegradation Pathway by *Arthrobacter* sp. P1–1 (Jong *et al.*, 2006).

SL6, and *Stenotrophomonas maltophilia* strain BA. These reported CAR-degradation rates of 0.057, 0.062, 0.036, and 0.050 mg/L/h, respectively. Salam *et al.* (2015) tested the ability of SL1, SL4, and SL6 to decontaminate sterile soil polluted with 100 mg CAR/kg and revealed CAR degradation rates of 0.093, 0.114, and 0.095 mg/kg/h, respectively, while their consortium recorded 0.121 mg/kg/h. Salam *et al.* (2015) also reported that the CAR-degradation rates increased when bioaugmenting those isolates in native soil contaminated with 100 mg CAR/kg, recording 0.127, 0.121, and 0.124 mg/kg/h, respectively, due to the positive interaction between them and the autochthonous microbial population.

Table 4.1 summarizes the specific enzyme activity for some CAR-degrading microorganisms. It has been reported that the biodegradation of CAR and its derivatives is affected by the position of the alkyl group and their number (Li *et al.*, 2006).

There are many reported bacteria capable for biodegradation of CAR. Most of them are Gram-negative bacteria, but there are also Gram-positive ones reported in the literature, as summarized in Table 4.2. However, fungi have been also reported for CAR-metabolism. *Cunnighamella elegans* is reported to introduce a hydroxyl group in position 3 of CAR (Holland *et al.*, 1986). *Cunninghamella echinulata* (ATCC 9244) is reported to produce four metabolites from N-methylcarbazole (NMCAR): CAR, N-hydroxymethylcarbazole (NHMCAR), 3-hydroxycarbazole, and 3-hydroxy-NHMCAR (Yang and Davis, 1992). Lobastova *et al.* (2004) reported the metabolism of CAR to 3-hydroxycarbazol as a major product and 1-hydroxy- and 2-hydroxy- carbazoles as minor products by *Aspergillus flavus* VKM F-1024.

Table 4.1 CAR-Degradation Ability of Microorganisms in One- or Two- Phase Systems.

Microorganisms	Specific enzyme activity mmol/ min g DCW	References
Sphingomonas sp. CDH-7	10	Kirimura *et al.* (1999)
Sphingomonas sp. GTIN11	8	Kilbane *et al.* (2002)
Pseudomonas sp.	1.2	Li *et al.* (2004)
A genetically engineered strain	28.8	Kayser and Kilbane (2004)
Burkholderia sp. strain IMP5GC	0.6	Castorena *et al.* (2006)
Pseudomonas sp. XLDN4–9	10.4	Li *et al.* (2006)
Novosphingobium sp. strain NIY3	1.7	Ishihara *et al.* (2008)
Acinetobacter sp. A1p6	7.96	Singh *et al.* (2011a)
Pseudomonas sp. strain GBS.5	11.36	Singh *et al.* (2013)

Table 4.2 CAR Biodegrading Microorganisms.

Microorganisms	Gram	References
Ralstonia sp. RJGII.123	negative	Grosser *et al.* (1991); Schneider *et al.* (2000)
Xanthamonas sp.	negative	Grosser *et al.* (1991)
Pseudomonas sp. CA06 and CA10	negative	Ouchiyama *et al.* (1993); Habe *et al.* (2001)
Pseudomonas stutzeri ATCC 31258	negative	Hisatsuka and Sato (1994)
Bacillus sp. KUKK-4,5	positive	Kobayashi *et al.* (1995)
Pseudomonas cepacia F297	negative	Grifoll *et al.* (1995)
Pseudomonas sp. LD2	negative	Gieg *et al.* (1996)
Sphingomonas CB3	negative	Shotbolt-Brown *et al.* (1996)
Pseudomonas stutzeri OM1	negative	Ouchiyama *et al.* (1998)
Sphingomonas sp. CDH-7	negative	Kirimura *et al.* (1999)
Pseudomonas putida ATCC 17484	negative	Loh and Yu (2000)
Ralstonia sp. RGII.123	negative	Schneider *et al.* (2000)
Sphingomonas sp. KA1	negative	Habe *et al.* (2002)
Pseudomonas rhodesiae KK1	negative	Yoon *et al.* (2002)
Sphingomonas spCDH-7	negative	Nakagawa *et al.* (2002)
Sphingomonas sp. GTIN11	negative	Kilbane *et al.* (2002)
Pseudomonas rhodesiae KK1	negative	Yoon *et al.* (2002)
Pseudomonas sp. C3211	negative	Jensen *et al.* (2003)
Neptuniibacter sp. CAR-SF	negative	Fuse *et al.* (2003); Nagashima *et al.* (2010)
Sphingomonas sp. CP19	negative	Bressler *et al.* (2003)
Janthinobacterium sp. strain J3	negative	Inoue *et al.* (2004)
Janibacter sp.	positive	Yamazoe *et al.* (2004)
Sphingomonas sp. strain KA1	negative	Inoue *et al.* (2004)
Burkholderia sp. strain IMP5GC	negative	Castorena *et al.* (2006)
Pseudomonas sp. XLDN4–9	negative	Li *et al.* (2006)

(*Continued*)

Table 4.2 Cont.

Microorganisms	Gram	References
Nocardioides aromaticivorans IC177	positive	Inoue *et al.* (2006)
Klebsiella sp. strain LSSE-H2 (CGMCC No. 1624)	negative	Li *et al.* (2008)
Novosphingobium sp. strain NIY3	negative	Ishihara *et al.* (2008)
Dietzia cinnamea P4	positive	Von der Weid *et al.* (2007)
Acinetobacter sp. A1p6	negative	Singh *et al.* (2011a)
Enterobacter sp.	negative	Singh *et al.* (2011b)
Pseudomonas sp. strain GBS.5	negative	Singh *et al.* (2013)
Achromobacter sp. strain CAR1389	negative	Farajzadeh and Karbalaei-Heidari (2012)
Achromobacter sp. strain SL1	negative	Salam *et al.* (2014)
Microbacterium esteraromaticum strain SL6	positive	Salam *et al.* (2014)
Bacillus clausii BS1	positive	Zakaria *et al.* (2016)

Indole is another example of NPAHs that can be released into the environment through cigarette smoke, coal-tar, and sewage. The observed microbial degradation pathways of indole differ significantly between microorganisms. Indole biodegradation pathways can be generally summarized into three pathways (Arora *et al.*, 2015). In the first pathway, the degradation of indole occurs via anthranilate, which can be further metabolized to denitrogenated products (Figure 4.8a and b). In the second pathway, indole has been found to be degraded via catechol (Figure 4.8c). The third pathway is a tryptophan independent pathway (Arora and Bae, 2014). Claus and Kutzner (1983) reported indole degradation by *Alcaligenes* sp. via a gentisate pathway (Figure 4.8a). Doukyu and Aono (1997) reported the mineralization of indole via isatin and isatoic acid in *Pseudomonas* sp. strain ST-200. Kamatht and Vaidyanathan (1990) reported indole degradation via catechol by *Aspergillus niger*. However, *Phomopsis liquidambari,* the endophytic fungus, initially oxidizes indole into oxindole and isatin which is then transformed to 2-dioxindole. This is further converted to 2-aminobenzoic acid via a pyridine ring cleavage (Chen *et al.*, 2013). The *Arthrobacter* sp. SPG reported to biotransform indole to indole-3-acetic acid via tryptophan pathway (Figure.4.9a., Arora and Bae, 2014). A Gram

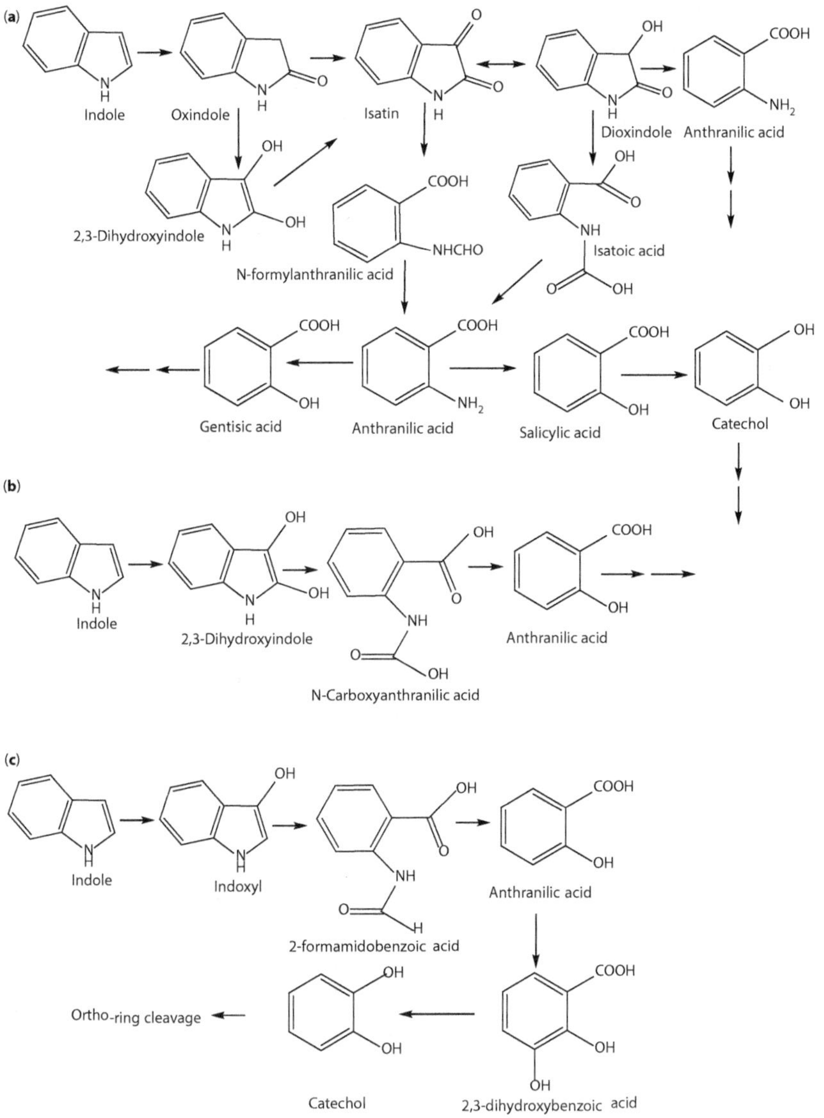

Figure 4.8 Different Biotransformation Pathways of Indole.

+ve phytopathogen *Rhodococcus fascians* is reported to metabolize indole to indole-3acetic acid via the indole-3-pyruvic acid pathway (Vandeputte *et al.*, 2005). *Bradyrhizobium japonicum* is reported to metabolize indole-3-acetic acid through two different pathways (Figure 4.9 b and c, Egebo *et al.*, 1991; Jensen *et al.*, 1995). *Ps. putida* strain 1290 is also reported to degrade indole-3-acetic acid via the catechol formation (Figure 4.9d,

Leveau and Lindow, 2005). Moreover, Proctor (1958) reported the biodegradation of indole-3-acetic to catechol via the production of 3-methylindole (Figure 4.9e). The anaerobic biodegradation of indole has been reported to wither under denitrifying, sulfate-reducing, or methanogenic conditions (Bak and Widdel, 1986; Madsen and Bollag, 1988; Madsen

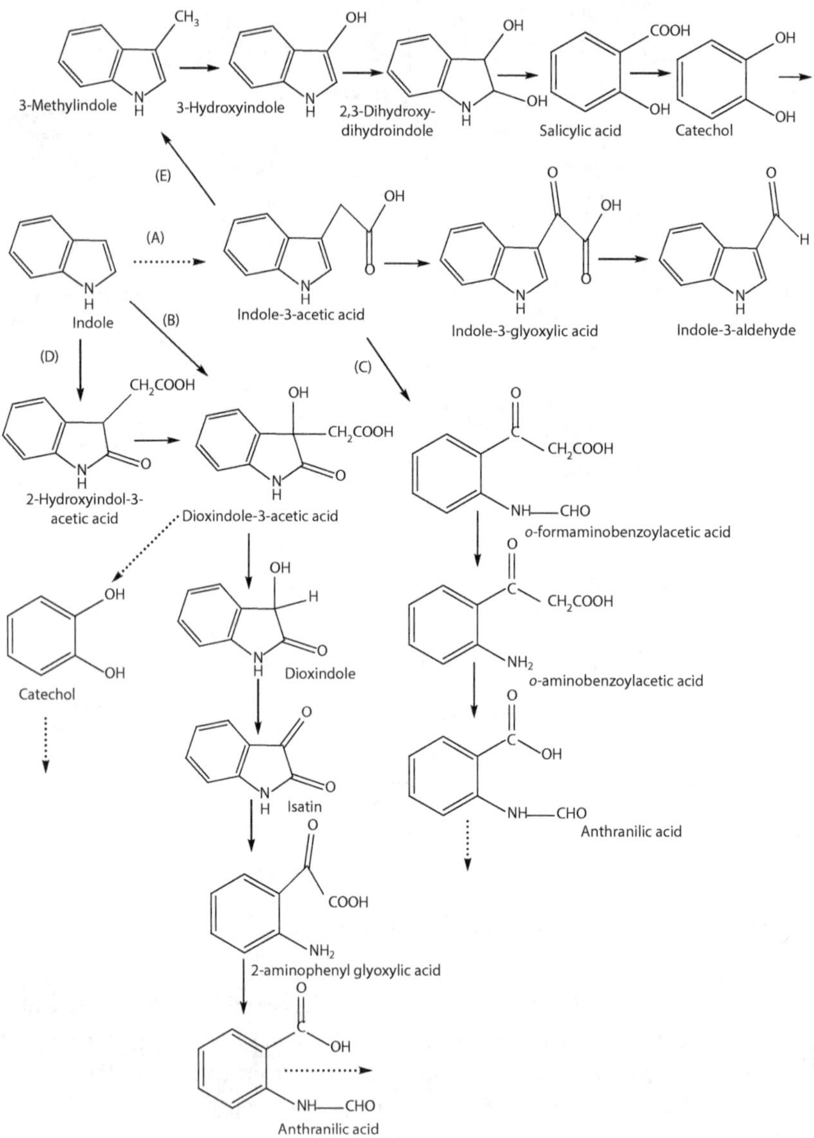

Figure 4.9 More Elucidated Biotransformation Pathways of Indole.

et al., 1988; Shanker and Bollag, 1990; Gu and Berry, 1991; Liu *et al.*, 1994; Gu *et al.*, 2002). Wang *et al.* (1984) reported mineralization of indole into carbon dioxide and methane by a consortium of methanogenic bacteria. Berry *et al.* (1987) reported the conversion of indole to oxindole under methanogenic conditions. The sulfate reducing bacteria *Desulfobacterium indolicum* DSM 3383 is reported to metabolize indole by indole oxygenase to give anthranilate via oxindole (Johansen *et al.*, 1997; Licht *et al.*, 1997).

Quinoline (Qn) is one of the major basic NPAHs compounds and has been reported as the most poisonous for catalysts (Le Borgne and Quintero, 2003). Many aerobic and anaerobic metabolisms of Qn and its derivatives have been reported (Atlas and Rothenburger, 1993; Solomon *et al.*, 1995; O'loughlin *et al.*, 1996). Aislabie *et al.* (1990) reported the selective transformation of Qn and methylquinoline (MQn) in shale oil by *Pseudomonas aeruginosa*, keeping the calorific value of the hydrocarbons within the fuel. Sutton *et al.* (1996) reported the utilization of 4-MQn by a Gram-negative bacterium Lep1 as a source of carbon and energy during growth in a liquid medium, under aerobic conditions, with the production of 2-hydroxy-4-methylquinoline and hydroxy-4-methylcoumarin. However, Qn has a generally suggested pathway by many researchers; the initial step is the hydroxylation at position 2 of the heterocyclic aromatic ring with the formation of 2-hydroxyquinoline (Shukla, 1986). *Pseudomonas stutzeri* is reported to degrade Qn by the pathway involving 2-hydroxyquinoline, 8-hydroxycoumarin, and 2,8-dihydroxyquinoline. The 8-hydroxycoumarin is further metabolized to 2,3-dihydroxyphenylpropionic acid, where the rate limiting step is the oxidation of the 2,8-dihydroxyquinoline (Shukla, 1989, Figure 4.10).

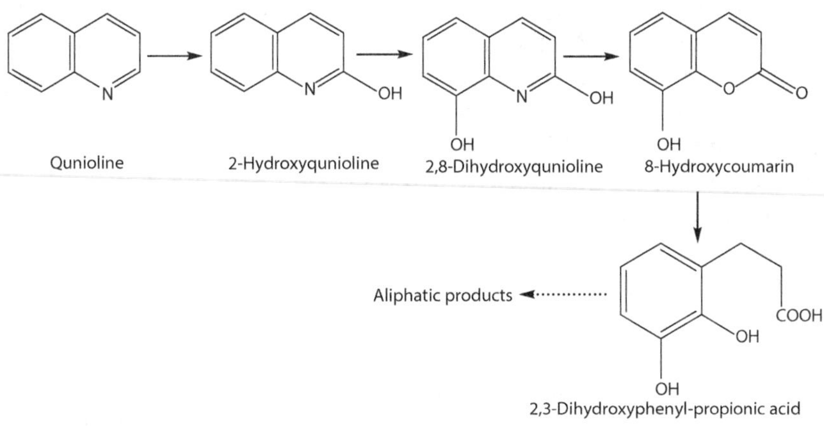

Figure 4.10 Metabolic Pathway for Quinoline by *Pseudomonas* stutzeri (Shukla, 1989).

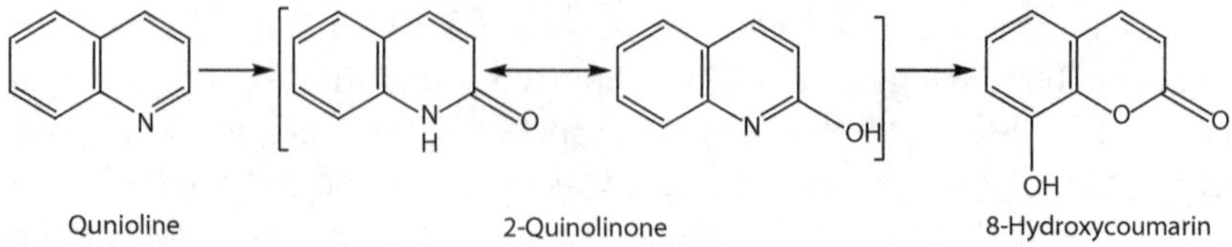

Figure 4.11 Quinoline BDN by *Ps. ayucida* IGTN9m (Kilbane *et al.*, 2001).

Pseudomonas putida is reported to transform both the homocyclic and the heterocyclic moieties of Qns (Boyd *et al.*, 1987). The attack on the homocyclic ring yielded the corresponding cis-hydrodiol derivatives and the monohydroxylated derivatives (e.g., 8-hydroxyquinoline or 5-hydroxyisoquinoline). Aislabie *et al.* (1989) reported the isolation of an *Acinetobacter* strain from oil and creosote contaminated soil for its ability to degrade isoquinoline. Although the methylquinolines (MQns) are reported to be resistant to microbial metabolism, Sutton *et al.* (1996) reported the isolation of *Pseudomonas putida* QP for its ability to degrade 4-MQn through the production of 2-hydroxy-4-methylquinoline and hydroxyl-4-methylcoumarin as intermediates. *Arthrobacter* sp. Rü61a (Dembek *et al.*, 1989) and *Pseudomonas putida* 33/1 (Bott *et al.*, 1990) reported to degrade 2-MQn.

Kilbane *et al.* (2000) reported the selective biodenitrogenation (BDN) of Qn without affecting its carbon-skeleton by *Pseudomonas ayucida* strain IGTN9m (ATCC N° PTA-806) (Figure 4.11), where approximately 5% nitrogen was removed by this isolate from petroleum without affecting its energetic value. Since IGTN9m converts Qn to 8-hydroxycoumarin without further degradation, no carbon is lost. Resting cells of IGTN9m selectively transform 68% of Qn from shale oil within 16 h. Kilbane *et al.* (2000) attributed the low BDN efficiency in oil to the low abundance of Qn relative to other NPAHs in oil and the narrow substrate range of Qn-degrading enzymes. Sona *et al.* (2010) reported the isolation of the Gram-positive *Bacillus licheniformis* strain CRC-75 from petroleum-contaminated soil for its capability for the selective metabolism of quinoline as an N-source but not as a C-source. Zhu *et al.* (2008) reported the utilization of Qn by *Rhodococcus* sp. QL2 via two simultaneous pathways: 5,6-dihydroxy-1H-2-oxoquinoline and 8-hydroxycoumarin.

Sugaya *et al.* (2001) reported the optimum QN-BDN conditions using *Comamonas* sp. TKV3–2-1 for its application in petroleum feed. The degradation rate reaches 1.6 mmol/g cell/h at 83% (v/v) petroleum/aqueous medium with a cell density of 28.5 g/L. Based on this, Sugaya *et al.* (2001) designed a BDN process (Figure 4.12).

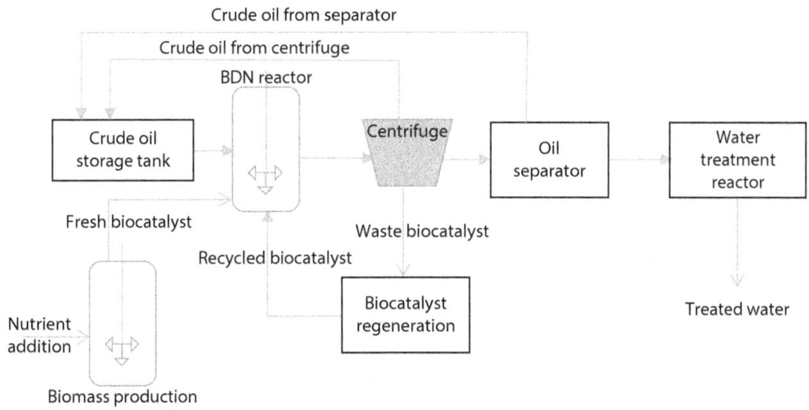

Figure 4.12 A BDN Process of Quinoline in Crude Oil Under Storage (as proposed by Sugaya *et al.*, 2010).

Desulfobacterium indolicum (DSM 3383) is reported to anaerobically metabolize Qn, isoQ, and 3-, 4-, 6-, and 8- MQ (Johansen *et al.*, 1997; Reineke *et al.*, 2008). However, it cannot degrade 2-MQn. Wang *et al.* (2013) reported the isolation of *Enterobacter aerogenes* TJ-D from acclimated activated sludge for its ability to utilize MQn as a sole carbon and nitrogen source under denitrifying conditions (Figure 4.13).

There are two general pathways reported for the biodegradation of pyridine. One involves a complete metabolic pathway for the degradation of pyridine by a *Bacillus* strain through hydroxylation reactions, followed by reduction (Figure 4.14a, Watson and Cain, 1975). The other reported for *Nocardia* strain Z1 throughout a reductive pathway not initiated by hydroxylations (Figure 4.14b, Rhee *et al.*, 1997).

4.4 Enhancing Biodegradation of NPAHs by Magnetic Nanoparticles

Nanotechnology represents a new generation of environmental-remediation technologies that could provide cost-effective solutions to some of the most challenging environmental clean-up problems. Magnetic nanoparticles (MNPs) are of special interest for their unique magnetic properties due to their reduced size and have potential use in many technological applications (Hyeon, 2003). Magnetite is used in a wide range of applications, including data storage (Hyeon, 2003), magnetic fluids (Chikazumi *et al.*, 1987), biotechnology (Gupta and Gupta, 2004), catalysis (Lu *et al.*,

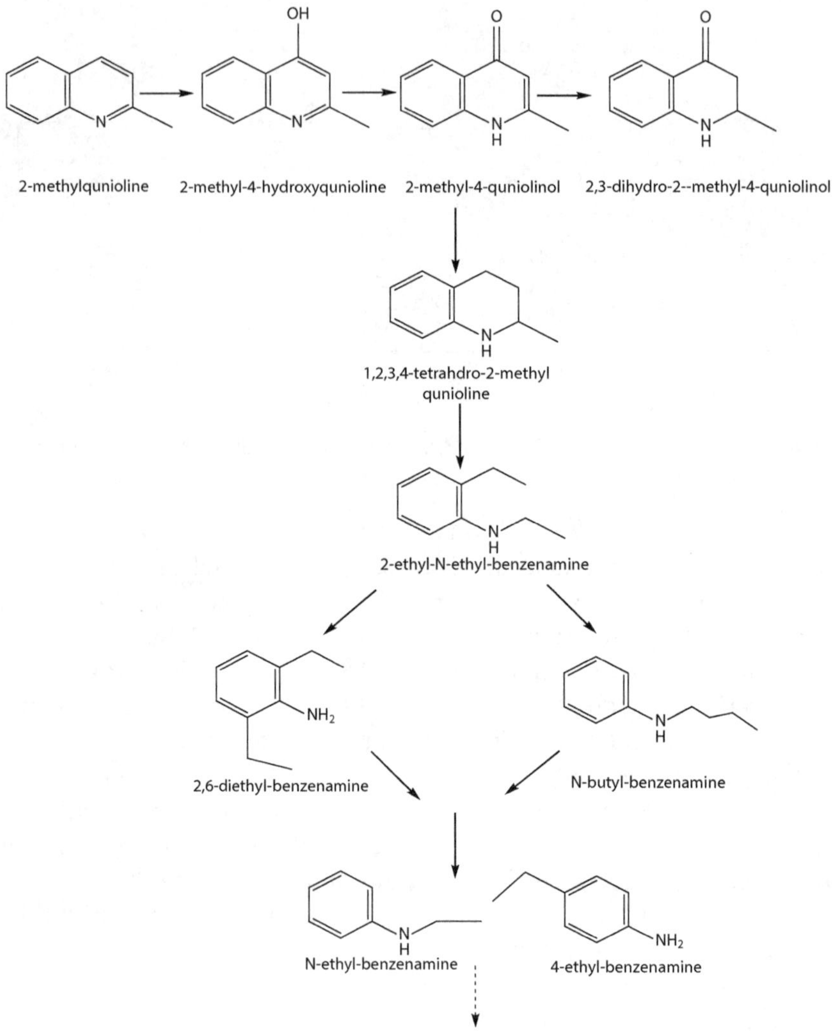

Figure 4.13 The Proposed Pathway for 2-MQn Metabolism Under Denitrifying Conditions (Wang *et al.*, 2013).

2004), magnetic resonance imaging (MRI) (Mornet *et al.*, 2006), and environmental remediation (Elliot and Zhang, 2001; Takafuji *et al.*, 2004). The magnetic Fe_3O_4 is one of the common iron oxides which has many important technological applications. The importance of MNPs are their applications in the separation of biomolecules for characterization or purification and they are a well-established alternative to centrifugal separation of biological solutions (Pankhurst *et al.*, 2003). They can be easily manipulated by permanent magnets or electromagnets, independent of normal

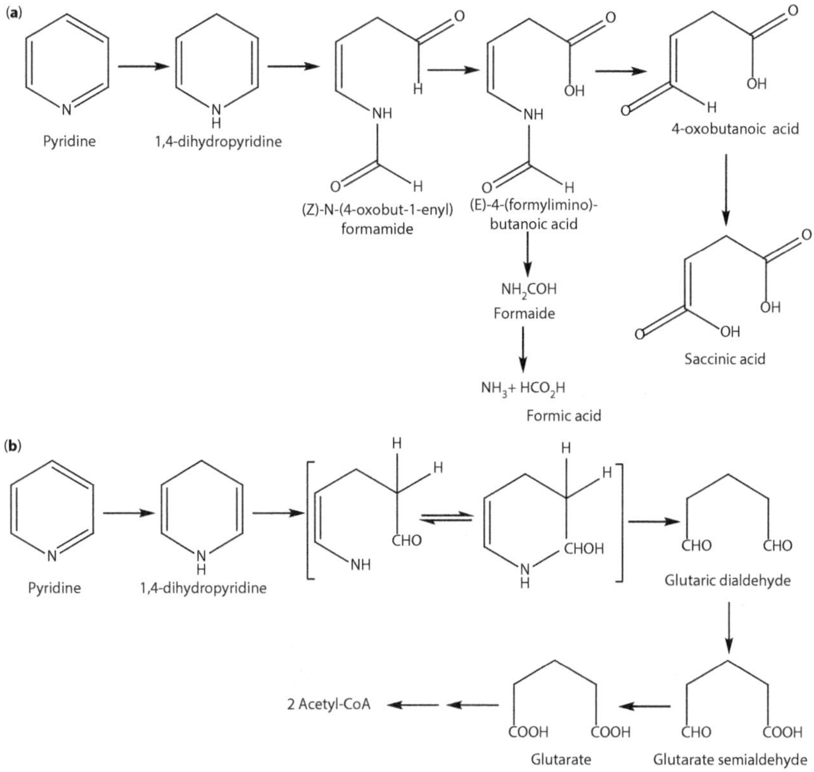

Figure 4.14 Metabolic Pathways for Pyridine Biodegradation.

microfluidic or biological processes. Therefore, the biggest advantage of the magnetic separation in biotechnology is ease of manipulation of biomolecules that are coated by magnetic particles. Upon placing the MNPs in solution, any target cells can be captured by the functionalized surfaces. By the use of a magnet at the side of the solution, a magnetic moment is induced in each of the freely floating particles and sets up a field gradient across the solution. The magnetized particles will move along the field lines and aggregate towards the permanent magnet, separating their bound target from the solution. MNPs can resolve many separation problems in industry and are being investigated for a number of different chemical separation applications. The catalysts are easily separated by utilizing the magnetic interaction between the magnetic nanoparticles (MNPs) and an external applied magnetic field that can be easily conjugated with biomolecules (Figure 4.15). Biocompatible MNPs have a wide range of applications in bioscience and they are also able to solve many separation problems in industry; small size enables the NPs to penetrate or diffuse in

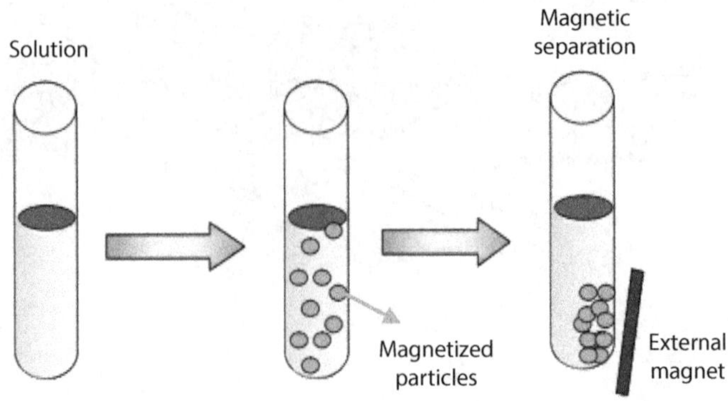

Figure 4.15 Magnetic Separation of Magnetized Particles.

contaminated areas where micro-sized particles fail to reach and nano-sized particles have higher reactivity to redox-amenable contaminants. The nanoscale particles (1–100 nm) afford very high surface areas without the use of porous absorbents and can be recovered for reuse. There have been wide studies of magnetic separation techniques of cells, proteins, viruses, bacteria, and other biomolecules which achieved enormous success (Olsvik *et al.*, 1994; Prestvik *et al.*, 1997; Arshady, 2001; Neuberger *et al.*, 2005).

Wang *et al.* (2007) stated that *Sphingomonas* sp. XLDN2–5, as a CAR-degrading strain, was entrapped in the mixture of Fe_3O_4 nanoparticles and gellan gum using a modified traditional entrapment method. The magnetically immobilized XLDN2–5 expresses higher CAR-biodegradation (3479 μg CAR/g wet weight cell/h) than the non-magnetically immobilized (1761 μg CAR/g wet weight cell/h) and free (3092 μg CAR/g wet weight cell/h) XLDN2–5 and can be used for eight successive cycles, where the specific degradation rate increased from 3479 to 4638 μg CAR/g wet weight cell/h in the 8th cycle due to the good growth of cells in the magnetic gellan gel beads. The observed decrease of the CAR-biodegradation rate in the non-magnetically immobilized matrix is attributed to mass transfer limitation and steric hindrance, while the increase in the magnetically immobilized matrix is due to the presence of MNPs which loose the binding of the sheets of the gellan gum matrix and the existence of many pores between the sheets of the gellan gum matrix. The highest CAR-biodegradation activity occurred at a concentration of 9 mg Fe_2O_3/mL and saturation magnetization of the magnetically immobilized cells of 11.08 μ/g.

Li *et al.* (2013b) reported the direct assembling of *Sphingomonas* sp. XLDN2–5 by Fe_2O_3 NPs, where the ration of cells/MNPs was 1:1. The average diameter of Fe_2O_3 was 20 nm with super-paramagnetic properties and a saturation magnetization of 45.5 μ/g. The resulting microbial cell/Fe_3O_4 biocomposite and free cells exhibited the same CAR-biodegradation efficiency. Thus, the MNPs did not exhibit any negative impact on XLDN2–5. This was attributed to the biocompatibility of MNPs, i.e. the coating layer itself, as it does not change the hydrophilicity of the cell surface. Moreover, this coating layer has a negligible effect on mass transfer, as its structure is looser than that of the cell wall. Thus, the microbial cell/Fe_3O_4 biocomposite produces a system that is not limited by diffusional limitations. The activity of the microbial cell/Fe_3O_4 biocomposite increased gradually during the recycling process, where complete removal of 3500 μg CAR occurred within 9 h for sixth successive cycles, but the same amount was completely removed in only 2 h within the 7th to the tenth cycles.

Zakaria *et al.* (2015b) reported the good assembling of magnetic Fe_3O_4 NPs (8–10 nm) on the Gram +ve long bacilli bacterial isolate *Bacillus claussi* BS1 (0.495 × 2.06 μm). The rate of CAR-biodegradation was doubled by coated cells, recording $t_{1/2}$ values of 31.36 and 64.78 h for coated and free-cells, respectively. The coated cells did not experience any mass transfer problem. MNPs express good adsorption capacity towards CAR (9.51 mmol/g), thus increasing the adsorption of CAR to the cells for biodegradation. The adsorption between Fe_3O_4 NPs and CAR is electrostatic, so the adsorption is reversible, i.e. it can be easily desorbed to the cells for BDN. The coated cells are characterized by high storage and operational stability and reusability. With the extension of storage time, coated cells hold a stable denitrogenation rate and could remove 97.80% of 1000 ppm CAR after being stored for 30 days at 4 °C, while the denitrogenation efficiency of the free cells decreased after being stored for seven days at 4 °C, recording 94.15% and lost nearly half of their activity, recording approximately 55.21% after their storage for 30 days at 4 °C. Coated cells can be used for four successive cycles without losing their activity, while the free cells lost 30% of their activity after the first cycle. Coated cells are also characterized by long-term BDN efficiency over 672 h, expressing approximately 96.4% BDN of 1000 ppm CAR. They have a lower sensitivity toward high concentrations of toxic byproducts and metabolites (anthranilic acid and catechol) than free cells. They also have the advantages of magnetic separation, which would resolve many operational problems in petroleum refinery. This would encourage the application of the MFe_3O_4NPs/*Bacillus claussi* BS1 biocomposite in BDN or biotransformation of NPAHs in the petroleum industry.

4.5 Challenges and Opportunities for BDN in Petroleum Industries

Benedik *et al.* (1998) reported that any fuel-upgrading process would be a low-margin (probably less than US$1/bbl or US$0.02/L, value added) and large volume (approximately 10^{10} L/year) commodity enterprise, where an efficient operation is essential for economic viability.

Bio-upgrading of petroleum and its fraction have not yet been applied on an industrial scale, but with the worldwide strict regulations for ultra-low nitrogen and sulfur fuels, which can be achieved by the high cost deep hydrotreating process, this also non-selectively modifies other components in treated petroleum fuels. There is an intensive demand for new technologies capable of reducing the hetero-atomic concentrations in fuels that raises the interest towards microbial approaches.

The microbial metabolism of petroleum-nitrogen leads to nitrogen removal and alleviation of the poisoning of refining catalysts. This will also eliminate the contribution of fuel nitrogen to NOx emissions. There are few reports about the microbial metabolism of CAR, as a model for nitrogenous compounds, directly related to petroleum distillates (Kirimura *et al.*, 1999; Kilbane *et al.*, 2002; Riddle *et al.*, 2003ab; Li *et al.*, 2004; Castorena *et al.*, 2006; Sona *et al.*, 2010). Moreover, processing of heavy oils is reported to require a temperature of 60–100°C for handling as a flowing fluid (Ayala *et al.*, 2016). Thus, thermophilic microbes and enzymes are required for industrial application. A Gram-positive thermophilic carbazole-degrading bacteria, *Anoxybacillus rupiensis* strain Ir3 (JQ912241), that can tolerate up to 80 °C with maximum activity at a temperature range of 55–65 °C has been isolated from hydrocarbon contaminated soil in Iraq (Fadhil *et al.*, 2014) and studies on the plasmid responsible for carbazole degradation have been performed (Al-Jailawi *et al.*, 2016).

One of the main limitations for the application of bioprocesses in petroleum industry, is the high consumption of water and high quantity of water required by cells. The toxicity of organic solvents for microorganisms is related to the logarithm of their partition coefficient in *n*-octanol and water (log P_{ow}) (Nielsen *et al.*, 2005). Organic solvents with a log P_{ow} range of 1.5 to 4 are known to be extremely toxic to living cells because they preferentially partition in the cytoplasmic membrane, disorganizing its structure and damaging vital functions, but solvents with a higher log P_{ow} (>4) are less toxic (Sikkema *et al.*, 1995). Thus, one of the important aspects for the application of BDN is the capability of the microorganism to tolerate high concentrations of nitrogenous compounds, a high oil to water ratio, and to be solvent tolerant. Li *et al.* (2004) reported the BDN of CAR in a cyclohexane/water system (1:1) by *Pseudomonas* sp. Castorena *et al.* (2006) managed to isolate a new thermotolerant bacterial isolate *Burkholderia* sp. strain IMP5GC which is capable of metabolizing high concentrations of CAR (up to 1200 mg/L). Moreover, the CAR metabolism increased in a biphasic-system with n-hexadecane and with water content as low as 30%, but the toxicity of hexadecane is not critical since log P_{ow} is 3.9. However, gas oil which is used as a feedstock for diesel fuel production is a complex mixture of hydrocarbons, PAHs, paraffins, olefins, naphthenes, PASHs, and NPAHs. Thus, due to its large diversity of organic compounds, it is toxic to microorganisms. Moreover, it produces inhibitory effects on CAR-metabolism, as cells could obtain nitrogen from other nitrogenous compounds. Not only this, but since gas oil is a complex mixture of organic molecules whose log P_{ow} depends on its nature, many low molecular weight molecules, including CAR, partition into microbial cells exerting their toxic effects. However, IMP5GC expresses a high BDN efficiency towards CAR and methyl-substituted CAR in gas oil, but, unfortunately, it is unable to metabolize di- or tri- methyl CARs in gas oil. Li *et al.* (2008) have isolated the Gram negative *Klebsiella* sp. LSSE-H2 from a mixture of dye-contaminated soil samples for its ability to biodenitrogenate high concentrations of CAR (19 mmol/L) with a specific activity toward CAR in a biphasic-system (3.3 μmol/min g wet cell) with n-dodecane/water. Sona *et al.* (2010) reported the BDN of crude oil (5% v/v oil/water ratio) using resting cells of *B. lichiformis* strain CRC-75. Riddle *et al.* (2003b) cloned and expressed car genes from the CAR-degrader *Pseudomonas* sp. LD2 in a solvent-tolerant *Ps. putida* Idho strain, where the recombinant strain efficiently degrades CAR in the organic phase.

Although the specific BDS of petroleum and its distillates has been reasonably investigated, there is a little information about the BDN of oil feed without affecting its calorific value. The economics of nitrogen-removal processes are affected by the amount of associated hydrocarbon lost from the fuel during the denitrogenation process. Generally, the currently well-established CAR-BDN pathway resembles that of the DBT-Kodama pathway, which results in the loss of fuel value (Jong *et al.*, 2006). This would, consequently, make the BDN of fuel streams economically unfeasible. Moreover, most of the CAR degrading microorganisms produce 2′-aminobiphenyl-2,3-diol as the first step in the CAR-BDN pathway. However, recovering the CAR-nitrogen as anthranilic acid or 2′-aminobiphenyl-2,3-diol, which is less inhibitory to refining catalysts, would solve part of that problem since the entire carbon-content of the fuel is preserved. This can be performed by mutant or recombinant strains. Moreover, a selective BDN of Qn by a mutant *Pseudomonas ayucida* strain ATCC No PTA-806 has been reported. This promotes its application in the BDN of nitrogen-containing coal, petroleum oil, lignite, and derived synthetic fuels while retaining their calorific value, but the mutant cannot remove all Qn-analogs, thus the total N-removal in the tested oil feed did not exceed 5 wt.% (Kilbane *et al.*, 2001, 2003). Microorganisms such as *Pseudomonas ayucida, Aneurinibacillus* sp., *Pseudomonas stutzeri*, *Yokenella* sp., and *Pseudomonas nitoreducens* are issued for the BDN of fossil fuels (Kilbane *et al.*, 2003).

Moreover, a dual microbial process for both selective BDS and BDN, with the overcome of the significant technical hurdles, would make the microbial refining processes and bio-upgrading of petroleum and its fractions feasible on a large scale (Monticello and Finnerty, 1985). Duarte *et al.* (2001), in PETROBRAS, the Brazilian oil company, have isolated *Gordonia* sp. strain F.5.25.8, which has the ability to utilize DBT through the 4S-pathway and CAR as a sole source of S and N, respectively. Santos *et al.* (2006) reported that F.5.25.8 can tolerate up to 42°C, which would add to its advantageous use in the industrial application of BDS/BDN complementary to hydrotreatment process. Moreover, it is reported to have a different genetic organization of the BDS (dsz) and BDN (car) gene clusters relative to *R. erythropolis* IGTS8 and *Pseudomonas* sp. IGTN9m, respectively. Kayser and Kilbane (2004) have combined the genes encoding carA in *Sphingomonas* GTIN11 (ATCC BAA-487) and *Ps. resinovorans* CA10 to create a truncated operon which encodes only the first step of CAR-BDN producing 2′-aminobiphenyl-2,3-diol. The genetically engineered bacteria can be applied for the transformation of CAR to the less toxic 2′-aminobiphenyl-2,3-diol in shale oil, petroleum products, and coal tar. The CAR content in a petroleum sample is reported to be reduced by 95% in a 2:10 petroleum/aqueous medium within 16 h using that genetically engineered bacteria. Moreover, Kayser and Kilbane (2004) inserted genes encoding amidases downstream of the artificial carA operon to accomplish the cleavage of the final C-N bond and produce biphenyl-2,2′,3-triol (Figure 4.4b), which is reintroduced to the fuel, keeping its fuel content. Yu *et al.* (2006) introduced the CAR dioxygenase gene (carAacd), which was amplified from *Pseudomonas* sp. strain XLDN4–9 into the excellent 4S-DBT-BDS bacteria, *Rhodococcus erythropolis* XP, and designated the recombinant as SN8. The recombinant *Rhodococcus erythropolis* SN8 expressed good BDS and BDN activities towards a wide range of recalcitrant alkyl CARs and DBTs in crude oil in just a one-step bioprocess. Moreover, upon the transformation of XP with a complete plasmid of a carABC gene cluster, which converts CAR to ANT, the resultant recombinant strain was able to metabolize DBT, CAR, and DBF to the non-toxic metabolites benzoate, anthranilate, and salicylate, respectively. However, Li *et al.* (2008) reported the simultaneous BDN/BDS of a model diesel oil/aqueous system (1:2) by a consortium of *Klebsiella* sp. LSSE-H2 (10 g DCW/L) and *Ps. delafieldii* R-8 (15 g DCW/L), at 92 and 94% of 10 mmol/L CAR and 3 mmol/L DBT, respectively, within 12 h at 30 °C. Tang and Hong (2014) reported diesel oil BDN using bacterial isolate HY9 in the presence of the surfactant Tween 80, where a maximum BDN efficiency of 16.7% is recorded when a 0.175 g surfactant was added to 25 mL diesel oil with an inoculum size of 3% (v/v) and 250 rpm mixing rate were applied. The repeated batch processes give the same BDN efficiency. The addition of Tween 20 enhances the CAR-metabolism by *Ps. stutzeri* (Larentis *et al.*, 2011). Moreover, the addition of the co-source of nitrogen would enhance BDN efficiency. Ishihara *et al.* (2008) reported that the addition of 0.2 g/L yeast extract enhances the growth and CAR utilization by *Novosphingobium* sp. strain NIY3, while Li *et al.* (2008) reported the CAR-BDN by *Klebsiella* sp. strain LSSE-H2 in the presence of 0.05 g/L yeast extract. Zakaria *et al.* (2015a) reported that the addition of surfactant Tween 80 (0.86% v:v) and yeast extract (0.868 g/L) increases the efficiency of microbial denitrogenation of CAR by *Bacillus clausii* BS1. Maass *et al.* (2015) reported the BDS/BDN of heavy gas oil (an intermediate fraction obtained from vacuum distillation used in the production of diesel and some lubricants) by *R. erythropolis* ATCC 4277 in a batch reactor that reached a maximum desulfurization and denitrogenation rate of 148 mg S/kg HGO/h and 162 mg N/kg HGO/h at 40% (v/v) HGO/water, respectively. The advantage of that BDS/BDN process is that N and S were removed directly from the petroleum fraction without the need for the addition of a chemical reagent or surfactant or the immobilization of the cells to increase the bioavailability. Moreover, ATCC 4277 tolerates a high concentration of oil feed. This would promote the applicability of the BDS/BDN process as an economic and environmentally/industrially viable technique.

Sugaya *et al.* (2001) reported that it is expected that if thte BDN of crude oil is applied, during its storage period it would effectively improve its quality and increases its products' yields (such as gasoline), since its nitrogen concentration would be decreased, which would overcome the problems of extra treatment period and cost.

The use of enzymes for BDN has been also reported. Laccase from *Coriolopsis gallica* has been reported for the BDN of CAR in a medium with 15% acetonitrile (Bressler, 2000). Moreover, the immobilization of bacteria is one of the recommendable methods for enhancing the BDN rate (Atlas and Rothenburger, 1993; Ulonska *et al.*, 2004; Balasubramaniyan and

Swaminathan, 2007). Anaerobic BDN has also been reported to modify the properties of heavy crude oil and enhance oil recovery (Fallon *et al.*, 2010).

Although the biochemistry and genetics of Qn and CAR as models for basic and non-basic NPAHs are available, the selective removal of nitrogen from crude oil is elucidated. However, the commercialization of the BDN process in petroleum industry is not practically considered. This might be due to the complexity of the NPAHs present in petroleum and its distillates, which are difficult to metabolize with one microbial strain. Moreover, the selective removal of nitrogen is still under investigation.

Moreover, the immobilization of bacteria is one of the recommendable methods for enhancing the BDN rate (Ulonska *et al.*, 2004; Balasubramaniyan and Swaminathan, 2007). Since immobilized microorganisms would overcome the problem of the cost of biocatalyst preparation and high volume ratio of water/oil, they enhance the stability of the biocatalyst, facilitate its recovery and reusability, and decrease the risk of contamination. Immobilization by the entrapment technique has some mass-transfer limitations, such as steric hindrance and limited diffusion. Thus, immobilization by adsorption technique reduces/eliminates these limitations. Moreover, applying nanoparticles (NPs) with large specific surface area and high surface energy is very recommendable.

Rothenburger and Atlas (1993) reported the biotransformation of Qns (Qn, isoQn and 6-MQn) in an immobilized *Pseudomonas putida* QP2 cells bioreactor (Figure 4.16). The immobilized-cell bioreactor showed a much higher efficiency for Qn removal and less sensitivity to higher Qn concentrations. Moreover, the metabolism of a 500 μl 6-MQn /L decane solution with a water-limited fixed-bed bioreactor, as a model for non-aqueous solution, recorded greater than 90% removal of 6-MQn within 24 h indicating the capability of immobilized cells to act the same as immobilized enzyme systems for the biotransformation of Qns in an organic solution.

The three phase airlift or fluidized bed bioreactor has been reported to overcome the problem of concentration gradients of toxic compounds (substrates or degradation products). The performance of the three-phase fluidized bed bioreactor depends on the stability and performance of the biofilm. The biofilm formation is a balanced process between accumulation and detachment of cells that is also affected by liquid shear stress, substrate conversion rate, substrate concentration, growth rate, biofilm density, biofilm thickness, type of microorganisms, and carrier.

Ulonska *et al.* (1995) reported the degradation of Qn through the coumarin pathway in a three-phase airlift reactor by an immobilized culture of *Comamonas acidovorans*. Buchtmann *et al.* (1997) reported that the immobilization of *Comamonas acidovorans* cells enables the three-phase fluidized

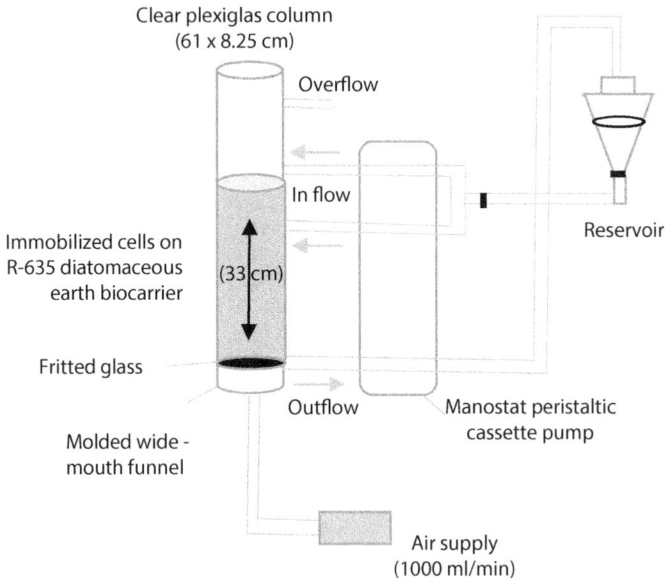

Figure 4.16 Immobilized-Cell Bioreactor (Heitkamp *et al.*, 1990).

bed reactor to operate at dilution rates 4–5 times higher than the critical one (μ_{max} = 0.42 h^{-1}), where a complete mineralization of Qn was achieved at a dilution rate of 2 h^{-1}. Jianlong *et al.* (2002) reported the isolation of the Gram-negative *Burkholderia pickettii* from the activated sludge of a coke-oven wastewater treatment plant for its ability to utilize Qn as the sole carbon and nitrogen source. *Burkholderia pickettii* cells were immobilized by a novel polyvinyl alcohol (PVA)-gauze hybrid carrier (1.5 × 1.5 cm small squares) with a final microbial cell density of about 0.15%. This showed higher biodegradation efficiency than PVA-gel immobilized cells, due to its multiple porous structure, with lower hindrance to the diffusion of oxygen, substrates, and metabolites and more space for bacterial growth than the PVA-gel beads (2–3 mm diameter). Castorena *et al.* (2008) reported the biofilm-immobilization of *Burkholderia* sp. IMP5GC and its application in a packed reactor (Figure 4.17) for a semi-continuous CAR-biodegradation process in a biphasic-system mixture of 85% gas oil and 15% light cycle oil/water of 0.1 (i.e. 90% O/W). The cells of IMP5GC were immobilized on the surface of porous glass cylinders 995 mm in diameter and 6 mm in height with a pore size of 100–160 μm. The reactor was packed with 170 cylinders covered with the IMP5GC-biofilm. The fuel mixture was replaced every 12 h, while the aqueous phase was recycled to the reactor. A small volume of the aqueous phase should be freshly added to restore its original volume

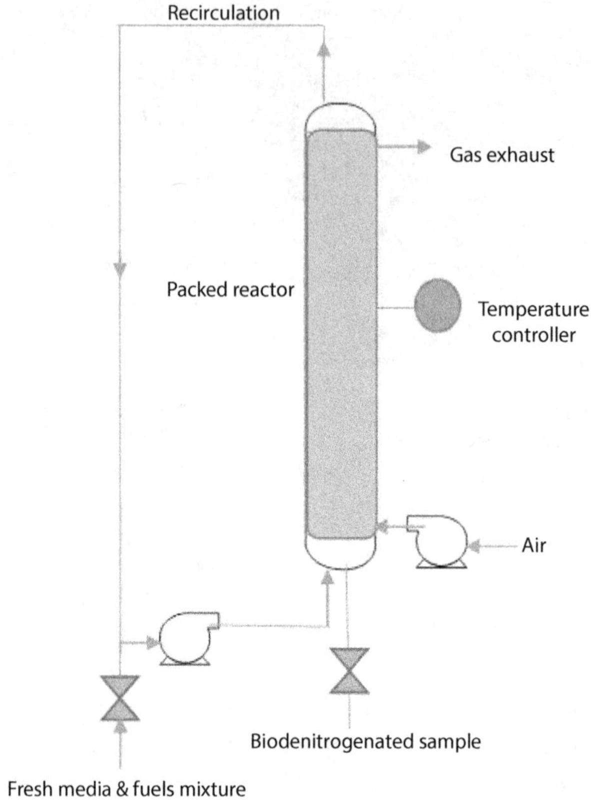

Figure 4.17 A Schematic Diagram for the Packed Reactor of a Semi-Continuous Fuel-BDN.

that is mainly lost by evaporation. The specific activity of the immobilized cells was lower than the free cells, recording 74 and 153.3 mg CAR/g protein/h, respectively. This might be attributed to the mass transfer limitation of CAR from medium to the cell cytoplasm, through the biofilm instead of just the aqueous phase, as well as other regulatory and metabolic changes that might have occurred due to the switching from planktonic to biofilm growth. However, the specific CAR-degrading activity of the immobilized cells was stable in the reactor for a period of 60 days and 120 batches of the fuel mixture. Most of the water was recycled back to the reactor for reuse and, thus, reduced the ratio of water/oil in the process (during the 60 days, barely 1 L of water was added, i.e. 10.81 fuel/ 1 L aqueous media). Complete removal of CAR occurred within 12 h, while 40 h were needed to remove 70% of 2-methyCAR and 3-methylCAR and 50% 4-methylCAR and 1-methyl-CAR from the oil mixture. This might be attributed to the depletion of required nutrient components in an aqueous medium and/or the accumulation of inhibitory metabolites during the long incubation period. The hydrocarbons profile of the treated oil did not change. Thus, for deeper BDN, it is suggested that the fuel mixtures be circulated through a sequence of successive semi-continuous packed reactors coupled with the addition of a fresh aqueous phase.

References

Adam, F., Bertoncini, F., Dartiguelongue, C., Marchand, K., Thiebaut, D., Hennion M. (2009) Comprehensive two-dimensional gas chromatography for basic and neutral nitrogen speciation in middle distillates. *Fuel*. 88: 938–946.

Ahmed, I., Jhung, S.H. (2014) Adsorptive denitrogenation of model fuel with CuCl-loaded metal–organic frameworks (MOFs). *Chemical Engineering Journal*. 251: 35–42.

Ahmed, I., Jhung, S.H. (2016) Adsorptive desulfurization and denitrogenation using metal-organic frameworks. *Journal of Hazardous Materials*. 301: 259–276.

Ahmed, I., Hasan, Z., Khan, N.A., Jhung, S.H. (2013) Adsorptive denitrogenation of model fuels with porous metal-organic frameworks (MOFs): effect of acidity and basicity of MOFs. *Applied Catalysis B: Environmental*. 129: 123–129.

Aislabie, J., Bej, A.K., Hurst, H., Rothenburger, S., Atlas, R.M. (1990) Microbial degradation of quinoline and methylquinolines. *Applied Environmental Microbiology*. 56(2):345–351.

Aislabie, J., Rothenburger, S., Atlas, R.M. (1989). Isolation of microorganisms capable of degrading isoquinoline under aerobic conditions. *Applied and Environmental Microbiology*. 55: 3247–3249.

Ali, I., Asim, M., Khan, T.A. (2012) Low cost adsorbents for the removal of organic pollutants from waste water. *Journal of Environmental Management*. 113: 170–183.

Ali, M.C., Yang, Q., Fine, A.A., Jin, W., Zhang, Z., Xing, H., Ren, Q. (2016) Efficient removal of both basic and non-basic nitrogen compounds from fuels by deep eutectic solvents. Green Chemistry. 18: 157–164.

Al-Jailawi, M.H., Mahdi, M.H., Fadhil, A.M.A. (2016) Effect of curing of *Anoxybacillus rupiensis* strain Ir3 (JQ912241) plasmid (s) in their ability for biodegradation of carbazole. *British Biotechnology Journal*. 10(1): 1–9.

Almarri, M.S. (2009) Selective adsorption for removal of nitrogen compounds from hydrocarbon streams over carbon-based adsorbents. Ph.D. Thesis. Pennsylvania State University.

Almarri, M., Ma, X.L., Song, C.S. (2009) Role of surface oxygen-containing functional groups in liquid-phase adsorption of nitrogen compounds on carbon-based adsorbents. *Energy and Fuels*. 23(8): 3940–3947.

An, S., Tang, K., Nemati, M. (2010) Simultaneous biodesulphurization and denitrification using an oil reservoir microbial culture: effects of sulphide loading rate and sulphide to nitrate loading ratio. *Water Research*. 44: 1531–1541.

Anantharaj, R., Banerjee T. (2010) COSMO-RS-based screening of ionic liquids as green solvents in denitrification studies. *Industrial and Engineering Chemistry Research*. 49: 8705–8725.

Anisuzzaman, S.M., Krishnaiah, D., Abang, S., Labadin, G.M. (2014) Adsorptive denitrogenation of fuel by oil palm shells a low cost adsorbent. *Journal of Applied Sciences*. 14(23): 3156–3161.

Arcos, J.C., Argus, M.F. (1968) Molecular geometry and carcinogenic activity of aromatic compounds: New perspectives. Advances in Cancer Research. 11: 305–471.

Arora, P.K., Bae, H. (2014) Identification of new metabolites of bacterial transformation of indole by gas chromatography-mass spectrometry and high performance liquid chromatography. *International Journal of Analytical Chemistry*. vol. 2014, Article ID 239641. http://dx.doi.org/10.1155/2014/239641

Arora, P.K., Sharma, A., Bae, H. (2015) Microbial degradation of indole and its derivatives. Journal of Chemistry. Volume 2015, Article ID 129159. http://dx.doi.org/10.1155/2015/129159

Arshady, R. (2001) Microspheres microcapsules and liposomes. London: Citus Books, ISBN: 9780953218769.

Asumana, C., Yu, G., Guan, Y., Yang, S., Zhou, S., Chen, X. (2011) Extractive denitrogenation of fuel oils with dicyanamide- based ionic liquids. Green Chemistry. 13: 3300–3305.

Atlas, M.R., Rothenburger, S. (1993) Hydroxylation and biodegradation of 6-methylquinoline by *Pseudomonas* in aqueous and no aqueous immobilized-cell bioreactors. *Applied and Environmental Microbiology*. 59(7): 2139–2144.

Ayala, M., Vazquez-Duhalt, R. Morales, M., Le Borgne, S. (2016) Application of microorganisms to the processing and upgrading of crude oil and fractions. Lee S.Y. (ed). Consequences of microbial interactions with hydrocarbons, oils, and lipids: Production of fuels and chemicals. Handbook of hydrocarbon and lipid microbiology. Springer International Publishing, AG.

Bachman, R.T., Johnson, A.C., Edyvean, R.G.J. (2014) Biotechnology in the petroleum industry: An overview. *International Biodeterioration and Biodegradation*. 86: 225–237.

Bae, Y.-S., Kim, M.-B., Lee, H.-J., Lee, C.-H. (2006) Adsorptive denitrogenation of light gas oil by silica–zirconia cogel. *AIChE Journal* 52: 510–521.

Bai, Y.H., Sun, Q.H., Xing, R., Wen, D.H., Tang, X.Y. (2010) Removal of pyridine and quinoline by bio-zeolite composed of mixed degrading bacteria and modified zeolite. *Journal of Hazardous Materials*. 181: 916–922.

Bak, F., Widdel, F. (1986) Anaerobic degradation of indolic compounds by sulfate-reducing enrichment cultures, and description of *Desulfobacterium indolicum* gen. nov., sp. nov. Archives of Microbiology. 146(2): 170–176.

Bakr, M.M.Y. (2009) Occurrence and geochemical significance of carbazoles and xanthones in crude oil from the Western Desert, Egypt. *Journal of King Abdulaziz University Earth Sciences.* 20(2): 127–159.

Balasubramaniyan, S., Swaminathan, M. (2007) Enhanced degradation of quinoline by immobilized *Bacillus brevis*. *Journal of the Korean Chemical Society.* 51(2): 154–159

Bej, S.K., Dalai, A.K., Adjay, J. (2001) Comparison of hydrogenation of basic and nonbasic nitrogen compounds present in oil sand derived heavy gas oil. *Energy and Fuels.* 15: 377–383.

Benedik, M.J., Gibbs, P.R., Riddle, R.R., Willson, R.C. (1998) Microbial denitrogenation of fossil fuels. Trends in Biotechnology. 16:390–395.

Berry, D.F., Madsen, E.L., Bollag, J.-M. (1987) Conversion of indole to oxindole under methanogenic conditions. *Applied and Environmental Microbiology.* 53(1): 180–182.

Bleeker, E.A.J., Wiegman, A.S., Droge, S.T.J., Kraak, M.H.S., van Gestel, C.A.M. (2003) Towards an improvement of the risk assessment of polycyclic (hetero) aromatic hydrocarbon, Report 2003-01 of the Department of Aquatic Ecology and Ecotoxicology. Vrije Universiteit, Amsterdam, the Netherlands. 27–32.

Bott, G., Schmidt, M., Rommel, T.O., Lingens, F. (1990) Microbial metabolism of quinoline and related compounds. V. Degradation of 1 *H*-4-oxoquinoline by *Pseudomonas putida* 33/l. Biological Chemistry Hoppe-Seyler. 371: 999–1003.

Boyd, D.R., McMordie, R.A.S., Porter, H.P. (1987) Metabolism of bicyclic azaarenes by *Pseudomonas putida* to yield vicinal cis-dihydrodiols and phenols. *Journal of the Chemical Society, Chemical Communications.* 22: 1722–1724.

Bressler, D.C., Gray, M.R. (2002) Hydrotreating chemistry of model products from bioprocessing of CARs. *Energy and Fuels.* 16(5): 1076–1086.

Bressler, D.C., Fedorak, P.M., Pickard, M.A. (2000) Oxidation of carbazole, p-ethylcarbazole, fluorine and dibenzothiophene by laccase of *Coriolopsis gallica.* Biotechnology Letters. 22: 1119–1125.

Bressler, D.C., Kirkpatrick, L.A., Foght, J.M., Fedorak, P.M., Gray, M.R. (2003) Denitrogenation of carbazole by combined biological and catalytic treatment. American Chemical Society, Petroleum Chemistry Division Preprints. 48: 44–46.

Buchtmann, C., Kies, U., Deckwer, W.-D, Hecht, V. (1993) Performance of three phase fluidized bed reactor for quinoline degradation on various supports at steady state and dynamic conditions. *Biotechnology and Bioengineering* 56(3): 295–303.

Buchtmann, C., Kies, U., Deckwer, W.D., Hecht, V. (1997) Performance of three phase fluidized bed reactor for quinoline degradation on various supports at steady state and dynamic conditions. *Biotechnology and Bioengineering* 56: 295–303.

Caero, L.C., Jorge, F., Navarro, A., Gutiérrez-Alejandre, A. (2006) Oxidative desulfurization of synthetic diesel using supported catalysts. Part II. Effect of oxidant and nitrogen-compounds on extraction-oxidation process. *Catalysis Today.* 116: 562–568.

Castorena, G., Acuña, M.E., Aburto, J., Bustos-Jaimes, I. (2008) Semi-continous biodegradation of carbazole in fuels by biofilm-immobilized cells of *Burkholeria* sp. strain IMP5GC. *Process Biochemistry.* 43: 1318–1321.

Castorena, G., Mugica, V., Le Borgne, S., Acuña, M.E., Bustos-Jaimes, I., Aburto, J. (2006) Carbazole biodegradation in gas oil/water biphasic media by a new isolated bacterium *Burkholderia* sp. Strain IMP5GC. *Journal of Applied Microbiology.* 100: 739–745.

Ceron, M.A., Guzman-Lucero, D.J., Palomeque, J.F., Martinez-Palou, R. (2012) Parallel microwave assisted synthesis of ionic liquids and screening for denitrogenation of straight-run diesel feed by liquid–liquid extraction. Combinatorial Chemistry and High Throughput Screening. 15: 427–432.

Chen, Y., Xie, X.-G., Ren, C.-G., Dai, C.-C. (2013) Degradation of N-heterocyclic indole by a novel endophytic fungus *Phomopsis liquidambari. Bioresource Technology.* 129: 568–574.

Chikazumi, S., Taketomi, S., Ukita, S., Mizukami, M., Miyajima, M., Setogawa, H., Kurihara, Y. (1987) Physics of magnetic fluid. *Journal of Magnetism and Magnetic Materials.* 65: 245–251.

Choi, J.G., Brenner, J.R., Colling, C.W., Demczyk, B.G., Dunning, J.L., Thompson, L.T. (1992) Synthesis and characterization of molybdenum nitride hydrodenitrogenation catalysts. *Catalysis Today.* 15(2): 201–222.

Choi, J.G., Brenner, J.R., Thompson, L.T. (1995) pyridine hydrodenitrogenation over molybdenum carbide catalysts. *Journal of Catalysis.* 154(1): 33–40.

Choi, K., Korai, Y., Mochida, I., Ryu, J.-W., Min, W. (2004) Impact or removal extent of nitrogen species in gas oil on its HDS performance: an efficient approach to its ultra-deep desulfurization. *Applied Catalysis B; Environmental.* 50: 9–16.

Choi, K.-H., Korai, Y., Mochida, I., Ryu, J.-W., Min, W. (2004) Impact of removal extent of nitrogen species in gas oil on its HDS performance: an efficient approach to its ultra-deep desulfurization. *Applied Catalysis B: Environmental.* 50: 9–16.

Choudhary, T.V., Parrott, S., Johnson, B. (2008) Unraveling heavy oil desulfurization chemistry: targeting clean fuels. *Environmental Science and Technology.* 42: 1944–1947.

Chuang, J.C., Mack, G.A., Kuhlman, M.R., Wilson, N.K. (1991) Polycyclic aromatic hydrocarbons and their derivatives in indoor and outdoor air in an 8-home study. Atmospheric Environment Part B-Urban Atmosphere. 25:369–380.

Claus, G., Kutzner, H.J. (1983) Degradation of indole by *Alcaligenes* spec. Systematic and Applied Microbiology. 4(2): 169–180.

de Voorde, B.V., Boulhout, M., Vermoortele, F., Horcajada, P., Cunha, D., Lee, J.S., Chang, J.-S., Gibson, E., Daturi, M., Lavalley, J.-C., Vimont, A., Beurroies, I., De Vos, D.E. (2013) N/S-heterocyclic contaminant removal from fuels by the mesoporous metal-organic framework MIL-100: the role of the metal ion. *Journal of American Chemical Society.* 135: 9849–9856.

Dembek, G., Rommel, T., Lingens, F., Hartmut, H. (1989) Degradation of quinaldine by *Alcaligenes* sp. and by *Arthrobacter* sp. *FEBS Letter.*, 246: 113–116.

Diaz, E., Garcia, J.L. (2010) Genetics engineering for removal of sulfur and nitrogen from fuel heterocycles. In Timmis K.N., McGenity T.J., van der Meer J.R., and de Lorenzo V. (eds.), "Handbook of hydrocarbon and lipid microbiology". Springer-Verlag GmbH, Berlin Heidlberg, Germany. Pp. 2787–2801.

Divakar, N.G., Subramanian, V., Sugumaran, M., Vaidyanathan, C.S. (1979) Indole oxygenase from the leaves of *Jasminum grandiflorum*. Plant Science Letters. 15:177–181.

Dong, D., Jeong, S., Massoth, F.E. (1997) Effect of nitrogen compounds on deactivation of hydrotreating catalysts by coke. *Catalysis Today*. 37: 267–275.

Doukyu, N., Aono, R. (1997) Biodegradation of indole at high concentration by persolvent fermentation with *Pseudomonas* sp. ST-200. Extremophiles. 1(2): 100–105.

Duarte, G.F., Rosado, A.S., Seldin, L., de Araujo, W., van Elsas, J.D. (2001) Analysis of bacterial community structure in sulfurous-oil-containing soils and detection of species carrying dibenzothiophene desulfurization (dsz) genes. *Applied Environmental Microbiology*. 67: 1052–1062.

Eaton, R.W., Chapman, P.J. (1992) Bacterial metabolism of naphthalene: Construction and use of recombinant bacteria to study ring cleavage of 1,2-dihydroxynaphthalene and subsequent reactions. *Journal of Bacteriology*. 174: 7542–7554.

Egebo, L.A., Nielsen, S.V.S., Jochimsen, B.U. (1991) Oxygendependent catabolism of indole-3-acetic acid in *Bradyrhizobium japonicum*. *Journal of Bacteriology*. 173(15): 4897–4901.

Egorova, M., Prins, R. (2004a) Competitive hydrodesulfurization of 4,6-dimethyldibenzothiophene, hydrodenitrogenation of 2-methylpyridine, and hydrogenation of naphthalene over sulfided NiMo/γ-Al_2O_3. *Journal of Catalysis*. 224(2): 278–287.

Egorova, M., Prins, R. (2004b) Mutual influence of the HDS of dibenzothiophene and HDN of 2-methylpyridine. *Journal of Catalysis* 221(1): 11–19.

Éigenson, A.S., Ivchenko, E.G. (1977) Distribution of sulfur and nitrogen in fractions from crude oil and residues. Chemistry and Technology of Fuels and Oils 13: 542–544.

Elise, B.F., Zhong, W.L., Zhao, T.L. (2013) Ultraclean fuels production and utilization for the twenty-first century: advances toward sustainable transportation fuels. *Energy and Fuels*. 27: 6335–6338.

Elliott, D.W., Zhang, W.X. (2001) Field assessment of nanoscale biometallic particles for groundwater treatment. *Environmental Science and Technology*. 35: 4922–4926.

Farajzadeh Z., Karbalaei-Heidari H.R. (2012) Isolation and characterization of a new Achromobacter sp. strain CAR1389 as a carbazole-degrading bacterium. *World Journal of Microbiology and Biotechnology*. 28: 3075–3080.

Fadhil, A.M.A., Al-Jailawi, M.H., Mahdi, M.S (2014) Isolation and characterization of a new thermophilic, carbazole degrading bacterium (*Anoxybacillus rupiensis*) Strain Ir3 (JQ912241). *International Journal of Advanced Research.* 2:795–805

Fallon, R.D., Hnatow, L.L., Jackson, S.C., Keeler, S.J. (2010) Method for identification of novel anaerobic denitrifying bacteria utilizing petroleum components as sole carbon source. US patent 7740063 B2.

Fedorak, P.M., Westlake D.W.S. (1984) Microbial degradation of alkylcarbazols in Norm Wells crude oil. *Applied and Environmental Microbiology*. 47(4): 858–862.

Furlong, E.T., Carpenter R. (1982) Azaarenes in Puget sound sediments. Geochimica. et. Cosmochimica. Acta., 46:1385–1396.

Furukawa, K. (2003) 'Super bugs' for bioremediation. Trends in Biotechnology. 21: 187–190.

Fuse, H., Takimura, O., Murakami, K., Inoue, H., Yamaoka Y. (2003) Degradation of chlorinated biphenyl, dibenzofuran, and dibenzop-dioxin by marine bacteria that degrade biphenyl, carbazole, or dibenzofuran. *Bioscience Biotechnology and Biochemistry*. 67: 1121–1125.

Gai, Z., Yu, B., Li, L., Wang, Y., Ma, C., Feng, J., Deng, Z., Xu, P. (2007) Cometabolic degradation of dibenzofuran and dibenzothiophene by a newly isolated carbazole-degrading *Sphingomonas* sp. strain. *Applied and Environmental Microbiology.* 73(9): 2832–2838.

Gieg, L.M. (1996) Ph.D Thesis. University of Alberta, Department of Biological Sciences.

Gieg, L.M., Otter, A., Fedorak, P.M. (1996) Carbazole degradation by *Pseudomonas* sp. LD2: metabolic characteristics and the identification of some metabolites. *Environmental Science and Technology*. 30: 575–585.

Grifoll, M., Selifonov, S.A., Gatlin, C.V., Chapman, P.J. (1995) Actions of a versatile fluorine degrading bacterial isolate on polycyclic aromatic compounds. *Applied and Environmental Microbiology*. 61(10): 3711–3723.

Grosser, R.J., Warshawsky, D., Robie, Vestal J. (1991) Indigenous and enhanced mineralization of pyrene, benzo[a]pyrene, and carbazole in soil. *Applied and Environmental Microbiology.* 57: 3462–3469.

Gu, J.-D., Berry, D.F. (1991) Degradation of substituted indoles by an indole-degrading methanogenic consortium. *Applied and Environmental Microbiology*. 57(9): 2622–2627.

Gu, J.-D., Fan, Y., Shi, H. (2002) Relationship between structures of substituted indolic compounds and their degradation by marine anaerobic microorganisms. Marine Pollution Bulletin. 45: 379–384.

Guo, W., Li, D., Tao, Y., Gao, P., Hu, J. (2008) Isolation and description of a stable carbazole-degrading microbial consortium consisting of *Chryseobacterium* sp. NCY and *Achromobacter* sp. NCW. Current Microbiology. 57: 251–257.

Gupta, A.K., Gupta, M. (2004) Synthesis and surface engineering of iron oxide nanoparticles for biomedical applications. Biomaterials. 26: 3995–4021.

Habe, H., Ashikawa, Y., Saiki, Y., Yoshida, T., Nojiri, H., Omori, T. (2002) *Sphingomonas* sp. strain KA1, carrying a carbazole dioxygenase gene homologue, degrades chlorinated dibenzo-pdioxins in soil. *FEMS Microbiology Letters.* 211(1): 43–49.

Habe, H., Ide, K., Yotsumoto, M., Tsuji, H., Hirano, H., Widada, J., Yoshida, T, Nojiri, H., Omori, T. (2001) Preliminary examinations for applying a carbazole-degrader, *Pseudomonas* sp. strain CA10, to dioxin-contaminated soil remediation. *Applied Microbiology and Biotechnology.* 56(5): 788–795.

Han, D.Y., Li, G.X., Cao, Z.B., Zhai, X.Y., Yuan, M.M. (2013) A study on the denitrogenation of Fushun shale oil. Energy Sources, Part A: Recovery, Utilization, and Environmental Effects. 35: 622–628.

Han, X., Lin, H., Zheng, Y. (2015) Adsorptive denitrogenation and desulfurization of diesel using activated carbons oxidized by $(NH_4)_2S_2O_8$ under mild conditions. *The Canadian Journal of Chemical Engineering.* 93: 538–548.

Han, X., Lin, H., Zheng, Y. (2015) Adsorptive denitrogenation and desulfurization of diesel using activated carbons oxidized by $(NH_4)_2S_2O_8$ under mild conditions. *The Canadian Journal of Chemical Engineering.* 93: 538–548.

Hansmeier, A.R., Meindersma, G.W., de Haan, A.B. (2011) Desulfurization and denitrogenation of gasoline and diesel fuels by means of ionic liquids. 13: 1907–1913.

Hayashi, J., Horikawa, T., Takeda, I., Muroyama, K., Ani, F.N. (2002) Preparing activated carbon from various nutshells by chemical activation with K_2CO_3. Carbon 40(13): 2381–2386.

Hegedus, L.L., McCabe, R.W. (1981) Catalyst poisoning. Catalysis Review-Science and Engineering. 23: 377–476.

Heitkamp, M.A., Camel, V., Reuter, T.J., Adams, W.J. (1990) Biodegradation of p-nitrophenol in an aqueous waste stream by immobilized bacteria. Applied and Environmental Microbiology. 56(10): 2967–2973.

Hermandez-Maldonado, A., Yang, R. (2004) Denitrogenation of transportation fuels by zeolites at ambient temperature and pressure. Angewandte Chemie International Edition. 43: 1004–1006.

Hisatsuka K., Sato, M. (1994) Microbial transformation of carbazole to anthranilic acid by *Pseudomonas stutzeri. Bioscience, Biotechnology and Biochemistry.* 58: 213 -214

Hizaddin, H.F., Hadj-Kali, M.K., Ramalingam, A., Hashim, M.A. (2015) Extraction of nitrogen compounds from diesel fuel using imidazolium- and pyridinium-based ionic liquids: Experiments, COSMO-RS prediction and NRTL correlation. Fluid Phase Equilibria 405: 55–67.

Ho, T.C. (2003) Inhibiting effects in hydrodesulfurization of 4,6- diethyldibenzothiophene. *Journal of Catalysis.* 219(2): 442–451.

Holland, H.L., Khan, S.H., Richards, D., Riemland, E. (1986) Biotransformation of polycyclic aromatic compounds by fungi. Xenobiotica. 16: 733–741.

Huh, E.S., Zazybin, A., Palgunadi, J., Ahn, S., Hong, J., Kim, H.S., Cheong, M., and Ahn B.S. (2009) Zn-containing ionic liquids for the extractive denitrogenation of a model oil: a mechanistic consideration. *Energy and Fuels.* 23: 3032–3038.

Hyeon, T. (2003) Chemical synthesis of magnetic nanoparticles. *Chemical Communications*. 8: 927–935.

Inoue, K., Habe, H., Yamane, H., Nojiri, H. (2006) Characterization of novel carbazole catabolism genes from gram-positive carbazole degrader *Nocardioides aromaticivorans* IC177. *Applied and Environmental Microbiology*. 72: 3321–3329.

Inoue, K., Widada, J., Nakai, S., Endoh, T., Urata, M., Ashikawa, Y., Shintani, M., Saiki, Y., Yoshida, T., Habe, H., Omori, T., Nojiri, H. (2004) Divergent structures of carbazole degradative car operons isolated from gram-negative bacteria. *Bioscience Biotechnology and Biochemistry*. 70: 1467–1480.

Ishihara, A., Dumeignil, F., Aoyagi, T., Ishikawa, M., Hosomi, M., Qian, E.W., Kabe, Y. (2008) Degradation of carbazole by *Novosphingobium* sp. strain NIY3. *Journal of the Japan Petroleum Institute*. 51(3): 174.179.

Ishihara, A., Wang, D.H., Dumeignil, F., Amano, H., Qian, E.W., Kabe, T. (2005) Oxidative desulfurization and denitrogenation of a light gas oil using an oxidation/adsorption continuous flow process. *Applied Catalysis A: General*. 279: 279–287.

Jayaraman, A., Yang, F.H., Yang, R.T. (2006) Effects of nitrogen compounds and polyaromatic hydrocarbons on desulfurization of liquid fuels by adsorption via π-complexation with Cu(I)Y zeolite. *Energy and Fuels*. 20: 909–914.

Jensen A.M., Finster K.W., Karlson U. (2003) Degradation of carbazole, dibenzothiophene, and dibenzofuran at low temperature by *Pseudomonas* sp. strain C3211. Environmental Toxicology and Chemistry. 22: 730–735.

Jensen, J.B., Egsgaard, H., van Onckelen, H., Jochimsen, B.U. (1995) Catabolism of indole-3-acetic acid and 4- and 5-chloroindole-3-acetic acid in *Bradyrhizobium japonicum*. *Journal of Bacteriology*. 177(20): 5762–5766.

Jha, A.M., Bharti, M.K. (2002) Mutagenic profiles of carbazole in the male germ cells of Swiss albino mice. Mutation Research. 500: 97–101.

Jia, Y.H., Li, G., Ning, G.L., Jin, C.Z. (2009) The effect of N-containing compounds on oxidative desulphurization of liquid fuel. *Catalysis Today*. 140: 192–196.

Jian, I.W., Xiangchun, Q., Liping, H., Yi, Q., Hegemann, W. (2002) Microbial degradation of quinoline by immobilized cells of *Burkholeria pickettii*. *Water Research*. 36(9): 2288–2296.

Jianlong, W., Xiangchun, Q., Liping, H., Yi, Q., Hegemann, W. (2002) Microbial degradation of quinoline by immobilized cells of *Burkholderia pickettii*. *Water Research* 36: 2288–2296.

Johansen, S.S., Licht, D., Arvin, E., Mosbaek, H., Hansen, A.B. (1997) Metabolic pathways of quinoline, indole and their methylated analogs by *Desulfobacterium indolicum* (DSM 383). *Applied and Environmental Microbiology*. 47(3): 292–300.

Kamatht, A.V., Vaidyanathan, C.S. (1990) New pathway for the biodegradation of indole in *Aspergillus niger*. *Applied and Environmental Microbiology*. 56(1): 275–280.

Katzer, J.R., Sivasubramanian, R. (1979) Process and catalyst needs for hydrodenitrogenation. *Catalysis Reviews: Science and Engineering*. 20: 155–208.

Kayser, K.J., Kilbane, J.J. II (2004) Method for metabolizing carbazole in petroleum. US Patent 6943006.

Kilbane, J.J. (2006) Microbial biocatalyst developments to upgrade fossil fuels. Current Opinion in Biotechnology. 17: 305–314.

Kilbane, J.J. II, Daram, A., Abbasian, J., Kayser, K. (2002) Isolation and characterization of *Sphingomonas* sp. GTIN11 capable of carbazole metabolism in petroleum. Biochemical and Biophysical Research Communications. 297: 242–248.

Kilbane, J.J. II., Ribeiro, C.M.S., Linhares, M.M. (2001) *Pseudomonas ayucida* useful for cleavage of organic C-N bonds. US patent 6221651 B1.

Kilbane, J.J. II., Ribeiro, C.M.S., Linhares, M.M. (2003) Bacterial cleavage of only organic C-N bonds of carbonaceous materials to reduce nitrogen content. US patent 6541240 B1.

Kilbane, J.J., Ranganathan, R., Cleveland, L., Kayser, K.J., Ribiero, C., Linhares, M.M. (2000) Selective removal of nitrogen from quinoline and petroleum by *Pseudomonas ayucida* IGTN9M. *Applied Environmental Microbiology.* 66(2): 688–693.

Kim, J.H., Ma, X., Zhou, A., Song, C. (2006) Ultra-deep desulfurization and denitrogenation of diesel fuel by selective adsorption over three different adsorbents: A study on adsorptive selectivity and mechanism. *Catalysis Today.* 111(1/2): 74–83.

Kimura, T., Yan, Z., Kodama, T., Omori, T. (1996) Isolation and characterization of Tn 5-induced mutants deficient in CAR catabolism. *FEMS Microbiology Letters.* 135(1): 65–70.

Kirimura, K., Nakagawa, H., Tsuji, K., Matsuda, K., Kurane, R., Usami S. (1999) Selective and continuous degradation of carbazole contained in petroleum oil by resting cells of *Sphingomonas* sp. CDH-7. *Bioscience Biotechnolgy Biochemistry.* 63: 1563–1568.

Kobayashi, T., Kurane, R., Nakajima, K., Nakamura, Y., Kirimura, K., Usami, S. (1995) Isolation of bacteria degrading carbazole under microaerobic conditions, i.e. nitrogen gas substituted conditions. *Bioscience Biotechnology and Biochemistry.* 59(5): 932–933.

Koriakin, A., Ponvel, K.M., Lee, C.H. (2010) Denitrogenation of row diesel fuel by lithium-modified mesoporous silica. *Chemical Engineering Journal.* 162: 649–655.

Kwon, J.-M., Moon, J.-H., Bae, Y.-S., Lee, D.-G., Sohn, H.-C., Lee, C.-H. (2008) Adsorptive desulfurization and denitrogenation of refinery fuels using mesoporous silica adsorbents. ChemSusChem. 1: 307–309.

Landau, M.V., Herskowitz, M., Givoni, D., Laichter, S., Yitzhaki, D. (1996) Medium severity hydrotreating and hydrocracking of Israeli shale oil. 1. Novel catalyst systems. *Fuel.* 75(7): 858–866.

Landau, M.V., Herskowitz, M., Givoni, D., Laichter, S., Yitzhaki, D. (1998a) Medium severity hydrotreating and hydrocracking of Israeli shale oil – II. Testing of novel catalyst systems in a trickle bed reactor. *Fuel.* 77(1–2): 3–13.

Landau, M.V., Herskowitz, M., Givoni, D., Laichter, S., Yitzhaki, D. (1998b) Medium severity hydrotreating and hydrocracking of Israeli shale oil: III. Hydrocracking of hydrotreated shale oil and its atmospheric residue for full conversion to motor fuels. *Fuel.* 77(14): 1589–1597.

Laredo, G., De Los Reyes, A., Cano, J., Castillo, J. (2001) Inhibition effects of nitrogen compounds on the hydrodesulfurization of dibenzothiophene. *Applied Catalysis A: General.* 207: 103–112.

Laredo, G.C., Altamirano, E., De Los Reyes, J.A. (2003) Inhibition effects of nitrogen compounds on the hydrodesulfurization of dibenzothiophene: Part 2. *Applied Catalysis A: General.* 243(2): 207–214.

Laredo, G.C., Leva, S., Alvarez, R., Mares, M.T., Castillo, J.J., Cano, J.L. (2002) Nitrogen compounds characterization in atmospheric gas oil and light cycle oil from a blend of Mexico crudes. *Fuel.* 81: 1341–1350.

Laredo, G.C., Likhanova, N.V., Lijanova, I.V., Rodriguez-Heredia, B., Castillo, J.J., Perez-Romo, P. (2015) Synthesis of ionic liquids and their use for extracting nitrogen compounds from gas oil feeds towards diesel fuel production. *Fuel Processing Technology.* 130: 38–45.

Larentis, A.L., Sampaio, H.C.C., Carneiro, C.C., Martins, O.B., Alves, T.L.M. (2011) Evaluation of growth, carbazole biodegradation and anthranilic acid production by *Pseudomonas stutzeri. Brazilian Journal of Chemical Engineering.* 28(1): 37–44.

Le Borgne, S., Quintero, R. (2003) Biotechnological processes for the refining of petroleum. *Fuel Processing Technology.* 81: 155–169.

Leboda, R., Skubiszewska-Ziba, J., Tomaszewski, W., Gunko, V.M. (2003) Structural and adsorptive properties of activated carbons prepared by carbonization and activation of resins. *Journal of Colloid and Interface Science* 263(2): 533–541.

Lee, K.S., Abe, H., Reimer, J.A., Bell, A.T. (1993) hydrodenitrogenation of quinoline over high-surface-area Mo_2N. *Journal of Catalysis.* 139(1): 34–40.

Lee, S.-W., Ryu, J.W., Min, W. (2003) SK hydrodesulfurization (HDS) pretreatment technology for ultralow sulfur diesel (ULSD) production. Catalysis Surveys from Asia. 7(4): 271–279.

Leveau, J.H.J., Lindow S.E. (2005) Utilization of the plant hormone indole-3-acetic acid for growth by *Pseudomonas putida* strain 1290. *Applied and Environmental Microbiology.* 71(5): 2365–2371.

Li, L., Li, Q., Li, F., Shi, Q., Yu, B., Liu, F., Xu, P. (2006) Degradation of carbazole and its derivatives by a *Pseudomonas* sp. *Applied Microbiology and Biotechnology.* 73: 941–948.

Li, L., Xu, P., Blankespoor, H.D. (2004) Degradation of carbazole in the presence of non-aqueous phase liquids by *Pseudomonas* sp. Biotechnology Letters. 26: 581–584.

Li, N., Zhu, J., Ma, X.L., Zha, Q.F., Song, C.S. (2013a) Tailoring O-containing groups on activated carbon and its effect on adsorptive denitrogenation of liquid hydrocarbons. *AIChE Journal.* 59(4): 1236–1244.

Li, Y., Du X., Wu C., Liu X., Wang X., and Ping Xu. (2013b) An efficient magnetically modified microbial cell biocomposite for carbazole biodegradation. Nanoscale Research. Letters. 8: 522.

Li, Y.-G., Li, W.-L., Huang, J.X., Xiong, X.-C., Gao, H.-S., Xing, J.-M., Liu, H.-Z. (2008) Biodegradation of carbazole in oil/water biphasic system by a newly isolated bacterium *Klebsiella* sp. LSSE-H2. *Biochemical Engineering Journal.* 41: 166–170.

Licht, D., Johansen, S.S., Arvin, E. (1997) Transformation of indole and quinoline by *Desulfobacterium indolicum* (DSM 3383). *Applied Microbiology and Biotechnology* 47(2): 167–172.

Liu, S.M., Jones, W.J., Rogers, J.E. (1994) Influence of redox potential on the anaerobic biotransformation of nitrogen heterocyclic compounds in anoxic freshwater sediments. *Applied Microbiology and Biotechnology.* 41(6): 717–724.

Liu, X., Zhang, Y., Han, W., Tang, A., Shen, J., Cui, Z., Vitousek, P., Erisman, J.W, Goulding, K., Christie, P., Fangmeier, A., Zhang, F. (2013) Enhanced nitrogen deposition over China. Nature. 494: 459–462.

Lobastova, T.G., Sukhodolskaya, G.V., Nikolayeva, V.M., Baskunov, B.P., Turchin, K.F., Donova, M. (2004) Hydroxylation of carbazoles by *Aspergillus flavus* VKM F-1024. *FEMS Microbiology Letters.* 235(1): 51–56.

Loh K.C., Yu Y.G. (2000) Kinetics of carbazole degradation by Pseudomonas putida in presence of sodium salicylate. *Water Research Journal.* 34: 4131–4138.

Lopes, A.R., Scheer, A.de-P., Silva, G.V., Yamamoto, C.I. (2016) Pd-impregnated activated carbon and treatment acid to remove sulfur and nitrogen from diesel. Revistamateria.21(2): 407–415.

Lopes, T.J., Furlong, E.T. (2001) Occurrence and potential adverse effects of semi volatile organic compounds in streambed sediment. United States. 1992–1995. Environmental Toxicology and Chemistry. 20:727–737.

Lu, A., Schmidt, W., Matoussevitch, N., Pnnermann, H., Bspliethoff, B., Tesche, B., Bill, E., Kiefer, W., Schvth, F. (2004) Nanoengineering of a magnetically separable hydrogenation catalyst. Angewandte Chemie International Edition in English. 43: 4303–4305.

Maass, D., Todescato, D., Moritz, D.E., Vladimir Oliveira, J., Oliveira, D., Ulson de Souza, A.A., Guelli Souza, S.M.A. (2015) Desulfurization and denitrogenation of heavy gas oil by *Rhodococcus erythropolis* ATCC 4277. *Bioprocess and Biosystems Engineering.* 38(8): 1447–1453.

Madsen, E.L., Bollag, J.M. (1988) Pathway of indole metabolism by a denitrifying microbial community. *Archives of Microbiology.* 151(1): 71–76.

Madsen, E.L., Francis, A.J., Bollag, J.M. (1988) Environmental factors affecting indole metabolism under anaerobic conditions. *Applied and Environmental Microbiology.* 54(1): 74–78.

Maeda, K., Nojiri, H., Shintani, M., Yoshida, T., Habe, H., Omori, T. (2003) Complete nuclitide sequence of carbazole/dioxin-degrading plasmid pCAR1 in *Pseudomonas resinovorans* strain CA10 indicates its mosaicity and the presence of large catabolic transporn Tn4676. *Journal of Molecular Biology.* 326: 21–33.

Maeda, R., Nagashima, H., Widada, J., Iwata, K., Omori, T. (2009) Novel marine carbazole-degrading bacteria. FEMS Microbiology Letters. 292: 203–209.

Maes, M., Trekels, M., Boulhout, M., Schouteden, S., Vermoortele, F., Alaerts, L., Heurtaux, D., Seo, Y.-K., Hwang, Y.K., Chang, J.-S., Beurroies, I., Denoyel, R., Temst, K., Vantomme, A., Horcajada, P., Serre, C., De Vos, D.E. (2011) Selective removal of N heterocyclic aromatic contaminants from fuels by Lewis acidic metal-organic frameworks. Angewandte Chemie International Edition. 50: 4210–4214.

Marcelis, C.L.M., Ivanova, A.E., Janssen, A.J.H., Stams, A.J.M. (2003) Anaerobic desulfurization of thiophenes by mixed microbial communities from oilfields. Biodegradation. 14: 173–182.

Miethling, R., Hecht, V., Deckwer, W.D. (1993) Microbial degradation of quinoline: kinetic studies with *Comamonas acidovorans* DSM6426. *Biotechnology and Bioengineering*. 42: 589–595.

Millemann, R.E., Birge, W.J., Black, J.A., Cushman, R.M., Daniels, K.L., Franco, P.J., Giddings, J.M., McCarthy, J.F., Stewart, A.J. (1984) Comparative acute toxicity to aquatic organisms of components of coal-derived synthetic fuels. Transactions of the American Fisheries Society. 113(1): 74–85.

Min, W. (2002) A unique way to make ultra-low sulfur diesel. *Korean Journal of Chemical Engineering*. 19(4): 601–606.

Min, W.-S., Choi, K., Khang, S.-Y., Min, D.-S., Ryu, J.-W., Yoo, K.-S., Kim, J.-H. (2001) Method for manufacturing cleaner fuels. US Patent No. 6248230.

Min, W. (2002) a unique way to make ultra-low sulfur diesel. *Korean Journal of Chemical Engineering*. 19(4): 601–606.

Mogollon, L., Rodriguez, R., Larrota, W., Ortiz, C., Torres, R. (1998) Biocatalytic removal of nickel and vanadium from petroporphyrins and asphaltenes. *Apllied Biochemistry and Biotechnology*. 70/72; 765–777.

Mohammed, S.N., Hopfinger, A.J. (1983) Intrinsic mutagenicity of polycyclic aromatic hydrocarbons: a quantitative structure activity study based upon molecular shape analysis. *Journal of Theoretical Biology* 102: 323–331.

Monticello, D.J., Finnerty, W.R. (1985) Microbial desulfurization of fossil fuels. *Annual Review of Microbiology*. 39: 371–389.

Montil, C., AQUintero, R., Aburto, J. (2009) Petroleum biotechnology: Technology trends for the future. *African Journal of Biotechnology*. 8(2): 2653–2666.

Morales, M., Le Borgne, S. (2010) Microorganisms utilizing nitrogen-containing hydrocarbons. In: Timmis K.N. (ed) Handbook of hydrocarbon and lipid microbiology. Springer, Berlin.

Morales, M., Le Borgne, S. (2017) Protocols for the isolation and preliminary characterization of bacteria for biodesulfurization and biodenitrogenation of petroleum-derived fuels. In: McGenity T.J., Timmis K.N., Nogales B. (ed) Hydrocarbon and lipid microbiology protocols. Springer. Berlin. Pp.201–218.

Mornet, S., Vasseur, S., Grasset, F., Veverka, P., Goglo, G.,Demourgues A.,Portier, J., Pollert, E., Duguet, E. (2006) Magnetic nanoparticle design for medical applications. *Progress in Solid State Chemistry*. 4: 237–247.

Murti, S.D.S., Yang, H., Choi, K.-H., Korai, Y., Mochida I. (2003) Influences of nitrogen species on the hydrodesulfurization reactivity of a gas oil over sulfide catalysts of variable activity. *Applied Catalysis A: General*. 252(2): 331–346.

Mushrush, G.W., Beal, E.J., Hardy, D.R., Hughes, J.M. (1999) Nitrogen compound distribution in middle distillate fuels derived from petroleum, oil shale, and tar sand sources. *Fuel Processing Technology*. 61: 197–210.

Nagashima H., Zulkharnain A.B., Maeda R., Fuse H., Iwata K. and Omori T. (2010) Cloning and nucleotide sequences of carbazole degradation genes from marine bacterium *Neptuniibacter* sp. strain CAR-SF. Current Microbiology 61:50–56.

Nakagawa, H., Kirimura, K., Nitta, T., Kino, K., Kurane, R., Usami, S. (2002) Recycle use of *Sphingomonas* sp. CDH-7 cells for continuous degradation of carbazole in the presence of $MgCl_2$. Current Microbiology. 44: 251–256.

Nam, I., Kim, Y.-M., Schmidt, S., Chang, Y.-S. (2006) Biotransformation of 1,2,3-tri- and 1,2,3,4,7,8-hexa-chlorodibenzo-p-dioxin by *Sphingomonas wittichii* strain RW1. *Applied Environmental Microbiology*. 72(1): 112–116.

Nam, J., Nojiri, H., Noguchi, H., Uchimura, H., Yoshida, T., Habe, H., Yamane, H., Omori, T. (2002) Purification and characterization of carbazole 1,9a-dioxygenase, a three-component dioxygenase system of *Pseudomonas resinovorans* strain CA10. *Applied Environmental Microbiology*. 68: 5882–5890.

Neuberger, T., Schopf, B., Hofman, H., Hofman, M., Rechenberg, B. (2005) Superparamagnetic nanoparticles for biomedical applications: Possibilities and limitations of a new drug delivery system. *Journal of Magnetism and Magnetic Materials*. 293: 483–496.

Nielsen, L.E., Kaday, D.R., Rajagopal, S., Drijber, R., Nickerson, K.W. (2005) Survey of extreme solvent tolerance in gram-positive cocci: membrane fatty acid changes in *Staphylococcus haemolyticus* grown in toluene. *Applied and Environmental Microbiology*. 71: 5171–5176.

Nojiri, H. (2012) Structural and molecular genetic analyses of bacterial carbazole degradation system. *Bioscience Biotechnology and Biochemistry*. 76: 1–18.

Nojiri, H., Omori, T. (2007) Carbazole metabolism by *Pseudomonas*. In: Ramos J.-L., and Filloux A. (eds) *Pseudomonas*. Springer, NY, USA. Pp107–145.

Nojiri, H., Ashikawa, Y., Noguchi, H., Nam, J.-W., Urata, M., Fujimoto, Z., Uchimura, H., Terada, T., Nakamura, S., Shimizu, K., Yoshida, T., Habe, H., Omori, T. (2005) Structure of the terminal oxygenase component of angular dioxygenase, carbazole, 1,9a-dioxygenase. *Journal of Molecular Biology*. 351(2): 355–370.

Nojiri, H., Nam, J.-W., Kosaka, M., Morii, K.-I., Takemura, T., Furihata, K., Yamane, H., Omori, T. (1999) Diverse oxygenations catalyzed by carbazole 1,9a-dioxigenase from *Pseudomonas* sp. strain CA10. *Journal of bacteriology*. 181(10): 3105–3113.

Nuzhdin, A.L., Kovalenko, K.A., Dybtsev, D.N., Bukhtiyarova, G.A. (2010) Removal of nitrogen compounds from liquid hydrocarbon streams by selective sorption on metal-organic framework MIL-101. Mendeleev Communications. 20: 57–58.

O'loughlin, E.J., Kehrmeyer, S.R., Sims, G.K. (1996) Isolation, characterization and substrate utilization of a quinoline-degrading bacterium. *International Biodeterioration and Biodegradation*. 32(2): 107–118.

Olsvik, O., Popovic, T., Skjerve, E., Cudjoe, K.S., Horns, E., Ugelstad, J., Uhlen, M. (1994) Magnetic separation techniques in diagnostic microbiology. *Clinical Microbiology Reviews*. 7: 43–54.

Osborne, P.J., Preston, M.R., Chen, H.Y. (1997) Azaarenes in sediments, suspended particles and aerosol associated with the River Mersey estuary. Marine Chemistry. 58:73–83.

Ouchiyama N., Miyachi S., Omori T. (1998) Cloning and nucleotide sequence of carbazole catabolic genes from *Pseudomonas stutzeri* strain OM1, isolated from activated sludge. *Journal of General and Applied Microbiology*. 44: 57–63.

Ouchiyama, N., Zhang, Y., Omori, T., Kodama, T. (1993) Biodegradation of CAR by *Pseudomonas* spp. CA06 and CA10. *Bioscience Biotechnology and Biochemistry*. 57: 455–460.

Pankhurst, Q.A., Connolly, J., Jones, S.K., Dobson, J. (2003) Applications of magnetic nanoparticles in biomedicine. *Journal of Physics D: Applied Physics*. 36: 167–181.

Prestvik, W.S., Berge, Mork, P.C., Stenstad, P.M., Ugelstad, J. (1997) Preparation and application of monosized magnetic particles in selective cell separation. In: Hafeli U., Schütt W., Teller J., and Zborowsk, M. (Eds.), Scientific and Clinical Applications of Magnetic Carriers, New York, Plenum Press. Pp.11–36.

Proctor, M.H. (1958) Bacterial dissimilation of indoleacetic acid: a new route of breakdown of the indole nucleus. Nature. 181(4619), article 1345.

Qiao, L., Wang, J.L. (2010) Biodegradation characteristics of quinoline by *Pseudomonas putida*. *Bioresource Technology*. 101: 7683–7686.

Reineke, A.K., Preiss, A., Elend, M., Hollender, J. (2008) Detection of methylquinoline transformation products in microcosm experiments and in tar oil contaminated groundwater using LC-NMR. *Chemosphere*. 70: 2118–2126.

Resnick, S.M., Torok, D.S., Gibson, D.T. (1993) Oxidation of carbazole to 3-hydroxycarbazole by naphthalene 1,2-dioxygenase and biphenyl 2,3-dioxygenase. *FEMS Microbiology Letters*. 113: 297–302.

Rhee, S.K., Lee, K.S., Chung, J.C., Lee, S.T. (1997) Degradation of pyridine by *Nocardioides* sp. strain OS4 isolated from the oxic zone of a spent shale column. *Canadian Journal of Microbiology*. 43(2): 205–209.

Richard, J.J., Junk, G.A. (1984) Steam distillation, solvent extraction, and ion exchange for determining polar organics in shale process waters. *Analytical Chemistry*. 56(9): 1625–1628.

Richards, D.J., Shieth, W.K. (1986) Biological fate of organic priority pollutants in the aquatic environment. *Water Research*. 20: 1077–1090.

Riddle, R.R., Gibbs, P.R., Willson, R.C., Benedik, M.J. (2003a) Purification and properties of 2-hydroxy-6-oxo-6-(2'-aminophenyl)hexa-2,4-dienoic acid hydrolase involved in microbial degradation of carbazole. *Protein Expression and Purification*. 28(1): 182–189.

Riddle, R.R., Gibbs, P.R., Willson, R.C., Benedik, M.J. (2003b) Recombinant carbazole-degrading strains for enhanced petroleum processing. *Journal of Industrial Microbiology and Biotechnology*. 30: 6–12.

Rothenburger, S., Atlas, R.M. (1993) Hydroxylation and biodegradation of 6-methylquinoline by *Pseudomonads* in aqueous and nonaqueous immobilized-cell bioreactors. *Applied and Environmental Microbiology*. 59(1): 2139–2144.

Saiki, Y., Habe, H., Yuuki, T., Ikeda, M., Yoshida, T., Nojiri, H., Omori, T. (2003) Rhizoremediation of dioxin-like compounds by a recombinant *Rhizobium tropici* strain expressing carbazole 1,9a-dioxygenase constitutively. *Bioscience Biotechnology and Biochemistry*. 67: 1144–1148.

Salam, L.B., Ilori, M.O., Amund, O.O. (2014) Carbazole angular dioxygenation and mineralization by bacteria isolated from hydrocarbon-contaminated tropical African soil. *Environmental Science and Pollution Research*. 21(15): 9311–9324.

Salam, L.B., Ilori, M.O., Amund, O.O. (2015) Carbazole degradation in the soil microcosm by tropical bacterial strains. *Brazilian Journal of Microbiology* 46(4): 1037–1044.

Sano, Y, Choi, KH, Korai, Y, Mochida, I. (2004a) Effects of nitrogen and refractory sulfur species removal on the deep HDS of gas oil. *Applied Catalysis B: Environmental*. 53(3): 169–74.

Sano, Y., Choi, K.-H., Korai, Y., Mochida, I. (2004b) Adsorptive removal of sulfur and nitrogen species from a straight run gas oil over activated carbons for its deep hydrodesulfurization. *Applied Catalysis B: Environmental*. 49: 219–225.

Santos, A.L., Reis, R.A., Rossa, V., Reis, M.M., Costa, A.L.H., Veloso, C.O. Henriques, C.A., Zotin, F.M.Z., Paredes, M.L.L., Silveira, E.B., Chiaro, S.S.X. (2012) Silica-alumina impregnated with cerium, nickel, and molybdenum oxides for adsorption of sulfur and nitrogen compounds from diesel. Materials Letters. 83 (2012) 158–160.

Santos, S.C., Alviano, D.S., Alviano, C.S., Pádula M., Leitão, A.C., Martins, O.B., Ribeiro, C.M., Sassaki, M.Y., Matta, C.P., Bevilaqua, J., Sebastián G.V., Seldin L. (2006) Characterization of *Gordonia* sp. strain F.5.25.8 capable of dibenzothiophene desulfurization and carbazole utilization. *Applied Microbiology and Biotechnology*.71: 355–362

Santos, S.C.C., Alvinao, D.S., Alviano, C.S., Pádula, M., Leitão, A.C., Martins, O.B., Ribeiro, C.M.S., Sassaki, M.Y.M., Matta, C.P.S., Bevilaqua, J., Sebasian, G.V., Seldin, L. (2001) Charachterization of Gordonia sp. strain F.5.25.8 capable of dibenzothiophene desulfurization and carbazole utilization. *Applied Microbiology and Biotechnology*. 71: 355–362.

Sato, S.I., Nam, J.W., Kasuga, K., Nojiri, H., Yamane, H., Omori, T. (1997b) Identification and characterization of genes encoding CAR 1,9α-dioxygenase in *Pseudomonas* sp. strain CA10. Bacteriology. 179:4850–4858.

Sato, S.I., Ouchiyama, N., Kimura, T., Nojiri, H., Yamane, H., Omori, T. (1997a) Cloning of genes involved in CAR degradation of *Pseudomonas* sp. strain CA10: nucleotide sequences of genes and characterization of meta-cleavage enzymes and hydrolase. Bacteriology. 179: 4841–4849.

Schneider, J. Grosser, R.J., Jayasimhulu, K., Xue, W., Kinkle, B., Warshawsky, D. (2000) Biodegradation of carbazole by *Ralstonia* sp. RJGII.123 isolated from a hydrocarbon contaminated soil. *Canadian Journal of Microbiology*. 46: 269–277.

Shotbolt-Brown, J., Hunter, W.F., Aislabie J. (1996) Isolation and description of carbazole-degrading bacteria. *Canadian Journal of Microbiology*. 42: 79–82. Jong, S.S., Young, S.K., Kyu, C., Qing, X.L. (2006) Degradation of dibenzothiophene and carbazole by *Arthrobacter* sp. P1-1. *International Journal and Biodegradation*. 58: 36–43.

Shanker, R., Bollag, J.-M. (1990) Transformation of indole by methanogenic and sulfate-reducing microorganisms isolated from digested sludge. *Microbial Ecology*. 20(2): 171–183.

Shao, X., Zhang, X., Yu, W., Wu, Y., Qin, Y., Sun, Z., Song, L. (2012) Effects of surface acidities of MCM-41 modified with MoO3 on adsorptive desulfurization of gasoline. *Applied Surface Science*. 263: 1–7.

Shepherd, J.M., Lloyd-Jones, G. (1998) Novel carbazole degradation genes of *Shingomonas* CB3: sequence analysis, transcription, and molecular ecology. *Biochemical and Biophysical Research Communications*. 247: 129–135.

Shin, S., Sakanishi, K., Mochida, I. (2000) Identification and reactivity of nitrogen molecular species in gas oils. *Energy and Fuel*. 14: 539–544.

Shin, S., Yang, H., Sakanishi, K., Mochida, I., Grudoski, D.A., Shinn, J.H. (2001) Inhibition and deactivation in staged hydrodenitrogenation and hydrodesulfurization of medium cycle oil over NiMoS/Al_2O_3 catalyst. *Applied Catalysis A: General*. 205(1/2): 101–108.

Shin, S., Yang H., Sakanishi K., Mochida I., Grudoski D.A., and Shinn J.H. (2003) Inhibition effects of nitrogen compounds on the hydrodesulfurization of dibenzothiophene: Part 2. *Applied Catalysis A: General*. 243(2): 207–214.

Shirahama, N., Moon, S.H., Choi, K.-H., Enjoji, T., Kawano, S., Korai, Y., Tanoura, M., Mochida, I. (2002) Mechanistic study on adsorption and reduction of NO_2 over activated carbon fibers. *Carbon*. 40(14): 2605–2611.

Shiraishi, Y., Tachibana, K., Hirai, T., Komasawa, I. (2002) Desulfurization and denitrogenation process for light oils based on chemical oxidation followed by liquid–liquid extraction. *Industrial & Engineering Chemistry Research*. 41: 4362–4375.

Shukla, O. P. (1989) Microbiological degradation of quinoline by *Pseudomonas-stutzeri*–the coumarin pathway of quinoline catabolism. *Microbios*. 59(238): 47–63.

Shukla, O.P. (1986) Microbial transformation of quinoline by a *Pseudomonas* Sp. *Applied and Environmental Microbiology*. 51(6): 1332–1342.

Sikkema, J., de Bont, J.A., Polman, B. (1995) Mechanisms of membrane toxicity of hydrocarbons. *Microbiological Reviews*. 59: 201–202.

Silveira, E.B., Veloso, C.O., Costa, A.L.H., Henriques, C.A., Zotin, F.M.Z., Paredes, M.L.L., Reis, R.A., Chiaro, S.S.X. (2015) Influence of metal oxides impregnated on silica-alumina in the removal of sulphur and nitrogen compounds from a hydrotreated diesel fuel stream. *Adsorption Science and Technology*. 33(2): 105–116.

Singh, G.B., Gupta, S., Gupta, N. (2013) Carbazole degradation and biosurfactant production by newly isolated *Pseudomonas* sp. strain GBS.5. International *Journal of Biodeterioration and Biodegradation*. 84: 35–43.

Singh, G.B., Gupta, S., Srivastava, S., Gupta, N. (2011a) Biodegradation of carbazole by newly isolated *Acinetobacter* spp. Bulletin of Environmental Contamination and Toxicology. 87: 522–526.

Singh, G.B., Srivastava, S., Gupta, N. (2010) Biodegradation of carbazole by a promising gram-negative bacterium. World Academy of Science, Engineering and Technology. 4: 681–684.

Singh, G.B., Srivastava, S., Gupta, S., Gupta, N. (2011b) Evaluation of carbazole degradation by *Enterobacter* sp. isolated from hydrocarbon contaminated soil. *Recent Research in Science and Technology*. 3(11): 44–48.

Solomon, B.O., Hecht, V., Posten, C., Deckwer, W.D. (1995) Estimation of the energetic parameters associated with aerobic degradation of quinoline by *Comamonas acidovorance* in continuous culture. *Journal of Chemical Technology and Biotechnology*. 62: 94–97.

Sona, K., Maryam, M., Esmat, F. (2010) Quinoline biodegradation by *Bacillus licheniformis* strain CRC-75. Iranian *Journal of Chemistry and Chemical Engineering*. 29(2): 151–158.

Souza, W.F., Guimarães, I.R., Guerreiro, M.C., Oliveira, L.C.A. (2009) Catalytic oxidation of sulfur and nitrogen compounds from diesel fuel. *Applied Catalysis A: General*. 360: 205–209.

Speight, J.G. (1980) The chemistry and technology of petroleum. Marcel Dekker, New York.

Speight, J.G. (2014) The chemistry and technology of petroleum. CRC, Taylor & Francis, Boca Raton.

Streat, M., Patrick, J.W., Perez, M.J.C. (1995) Sorption of phenol and para-chlorophenol from water using conventional and novel activated carbons. *Water Research*. 29: 467–472.

Sugaya K., Nakayama O., Hinata N., Kamekura K., Ito A., Yamagiwa K., and Ohkawa A. (2001) Biodegradation of quinoline in crude oil. *Journal of Chemical Technology and Biotechnology*. 76: 603–611.

Sutton, S.D., Pfaller, S.L., Shann, J.R, Warshawskii, D., Kinkle, B.K., Vestal, J.R. (1996) Aerobic biodegradation of 4-methylquinoline by a soil bacterium. *Applied and Environmental Microbiology*. 62(8): 2910–2914.

Szymanska A., Lewandowski M., Sayag C., and Djéga-Mariadassou G. (2003) Kinetic study of the hydrodenitrogenation of carbazole over bulk molybdenum carbide. *Journal of Catalysis*. 218: 23–31.

Takafuji, M., Ide, S., Ihara, H., Xu, Z. (2004) Preparation of poly-(1-vinylimidazole)-grafted magnetic nanoparticles and their application for removal of metal ions. *Chemistry of Materials*. 16: 1977–1983.

Takagi, T., Nojiri, H., Yoshida, T., Habe, H., Omori, T. (2002) Detailed comparison between the substrate specificities of two angular dioxygenases, dibenzofuran 4,4a-dioxygenase from *Terrabacter* sp. and carbazole 1,9a-dioxygenase from *Pseudomonas resinovorans*. Biotechnology Letters. 24(8): 2099–2106.

Tang, X., Li, S., Yue, C., He, J., Hou J. (2013) Lumping kinetics of hydrodesulfurization and hydrodenitrogenation of the middle distillate from Chinese shale oil. Oil Shale. 30(4): 517–535.

Tao, X., Zhou, Y., Wei, Q., Ding, S., Zhou, W., Liu, T., Li, X. (2017) Inhibiting effects of nitrogen compounds on deep hydrodesulfurization of straight-run gas oil over a NiW/Al_2O_3 catalyst. *Fuel.* 188: 401–407.

Tissot, B.B., Welte, D.H. (1984) Petroleum Formation and Occurrence. 2nd revised and Enlarged Ed., Springer-Verlag, Berlin Heidelberg New York Tokyo.

Ulonska, A., Deckwer, W.-D., Hecht, V. (1995) Degradation of quinoline by immobilized *Comamonas acidovorans* in a three-phase airlift reactor. *Biotechnology and Bioengineering.* 46: 80–87.

Ulonska, A., Deckwer, W.-D., Hecht, V. (2004) Degradation of quinoline by immobilized *Comamonas acidovorans* in a three-phase airlift reactor. *Journal of Biotechnology and Bioengineering.* 46(1): 80–87.

US Environmental Protection Agency (EPA) (1986) Health and environmental effects profile for carbazole. US Environmental Protection Agency, Wahington, DC. EPA/600/X-86/334 (NTIS PB88218789).

Vandeputte, O., Oden S., Mol, A., Vereecke, D, Goethals, K, El Jaziri, M, Prinsen, E. (2005) Biosynthesis of auxin by the gram-positive phytopathogen *Rhodococcus fascians* is controlled by compounds specific to infected plant tissues. *Applied and Environmental Microbiology.* 71(3) 1169–1177.

Von der Weid, I., Marques, J.M., Cunha, C.D., Lippi, R.K., Santos, S.C.C., Rosado, A.S., Lins, U., Seldin, L. (2007) Identification and biodegradation potential of a novel strain of *Dietzia cinnamea* isolated from a petroleum contaminated tropical soil. *Systematic and Applied Microbiology.* 30: 331–339.

Wang, H., Xie, C., Yu, S., Liu, F. (2014) Denitrification of simulated oil by extraction with H_2PO_4-based ionic liquids. *Chemical Engineering Journal.* 237: 286–290.

Wang, L., Li, Y., Duan, J. (2013) Biodegradation of 2-methylquinoline by *Enterobacter aerogenes* TJ-D isolated from activated sludge. *Journal of Environmental Sciences.* 25(7):1310–1318.

Wang, L., Tian, J., Li, Y. (2015) Nitrite accumulation and nitrous oxide emission during denitrification processes with quinoline or indole as the sole carbon source. *Journal of Chemical Technology and Biotechnology.* 90: 1317–1328.

Wang, X., Gai, Z., Yu, B., Feng, J., Xu, C., Yuan, Y., Deng, Z., Xu, P. (2007) Degradation of CAR by microbial cells immobilized in magnetic gellan gum gel beads. *Applied and Environmental Microbiology.* 73: 6421–6428.

Wang, Y.-T., Suidan, M.T., Pfeffr, J.T. (1984) Anaerobic biodegradation of indole to methane" *Applied and Environmental Microbiology.* 48(5): 1058–1060.

Watson, G.K., Cain, R.B. (1975) Microbial metabolism of the pyridine ring. *Biochemical Journal.* 146: 157–172.

Weigman, S. (2002) Photoenhanced toxicity of azaarenes to marine phytoplancton. Ph.D. thesis. Proefschrift Universiteit van Amsterdam. FNWI. University of Amsterdam, Amsterdam.

Weitkamp, J., Puppe, L. (1999) Catalysis and zeolites: Fundamentals and applications. Springer-Verlag Berlin Heidelberg GmbH, Germany.

Wen, J., Han, X., Lin, H., Zheng, Y., Chu, W. (2010) A critical study on the adsorption of heterocyclic sulfur and nitrogen compounds by activated carbon: equilibrium, kinetics and thermodynamics. *Chemical Engineering Journal*. 164: 29–36.

WHO (2004) Coal tar creosote. Concise International Chemical Assessment Document 62. World Health Organisation, Geneve, Switzerland.

Widada, J., Nojiri, H, Yoshida, T., Habe, H., Omori, T. (2003) Enhanced degradation of carbazole and 2,3-dichlorodibenzo-p-dioxin in soils by *Pseudomonas resinovorans* strain CA10. *Chemosphere*. 49: 485–491.

Williams, P.F.V., Douglas, A.G. (1985) Organic geochemistry of the British Kimmeridge Clay: 1. Composition of shale oils produced from Kimmeridge sediments. *Fuel*. 64(8): 1062–1069.

Williams, P.T., Chishti, H.M. (2001) Reaction of nitrogen and sulphur compounds during catalytic hydrotreatment of shale oil. *Fuel*. 80(7): 957–963.

Wu, G., Chen, D., Tang, H., Ren, Y., Chen, Q., Lv, Y., Zhang, Z., Zhao, Y.-L., Yao, Y., Xu, P. (2014) Structural insights into the specific recognition of N-hetrocycle biodenitrogenation-derived substrates by microbial amide hydrolases. *Molecular Microbiology*. 9(15): 1009–1021.

Xie, L.-L., Favre-Reguillon, A., Lemaire, M. (2008b). Selective removal of nitrogen-containing compounds from straight-run diesel fuel using ILs, International Solvent Extraction Conference (ISEC) Tucson, Az, USA. Pp.1325–1330.

Xie, L.L., Favre-Reguillon, A., Pellet-Rostaing, S., Wang, X.X., Fu, X. Z., Estager, J., Vrinat, M., Lemaire, M. (2008a) Selective extraction and identification of neutral nitrogen compounds contained in straight-run diesel feed using chloride based ionic liquid. *Industrial and Engineering Chemistry Research*. 47(22): 8801–8807.

Xie, L.-L., Favre-Reguillon, A., Wang, X.-X., Fu, X., Lemaire, M. (2010) Selective adsorption of neutral nitrogen compounds from fuel using ion-exchange resins. *Journal of Chemical and Engineering Data*. 55: 4849–4853.

Xie, L.L., Favre-Reguillon, A., Wang, X.X., Fu, X., Pellet-Rostaing, E. Toussaint, G., Geantet, C., Vrinat, M., Lemaire, M. (2008) Selective extraction of neutral nitrogen compounds found in diesel feed by 1-butyl-3-methyl-imidazolium chloride. Green Chemistry. 10: 524–531.

Xu, P., Yu, Bo., Li, F.L., Cai, X.F., Ma, C.Q. (2006) Microbial degradation of sulfur, nitrogen and oxygen heterocycles. *Trends in Microbiology*. 14(9): 398–405.

Yamazoe, A., Yagi, O., Oyaizu, H. (2004) Biotransformation of fluorine, biphenyl ether, dibenzo-*p*-dioxin and carbazole by *Janibacter* sp. Biotechnology Letters. 26: 479–486.

Yang, W., Davis, P.J. (1992) Microbial models of mammalian metabolism. Biotransformations of N-methylcarbazole using the fungus *Cunninghamella echinulata*. Drug Metabolism and Disposition. 20: 38–46.

Yi, N., Xue, G., HogShuai, G., XiangPing, Z., SuoJiang, Z. (2014) Simultaneous desulfurization and denitrogenation of liquid fuels using two functionalized group ionic liquids. Science China. 57(12): 1766–1773.

Yoon, B.-J., Lee, D.-H., Kang, Y.-S., Oh, D.-C., Kim, A.-II., Oh, K.-H., Kahng, H.-Y. (2002) Evaluation of carbazole degradation by *Pseudomonas rhodesiae* KK1 isolated from soil contaminated with coal tar. *Journal of Basic Microbiology.* 42(6): 434–443.

Yu, B., Xu, P., Zhu, S., Cai, X., Wang, Y., Li, L., Li, F., Liu, X., Ma, C. (2006) Selective biodegradation of S and N heterocycles by recombinant *Rhodococcus erythropolis* strain containing carbazole dioxygenase. *Applied and Environmental Microbiology.* 72(3): 2235–2238.

Yu, H., Li, S., Jin, G. (2010) Catalytic hydrotreating of the diesel distillate from Fushun shale oil for the production of clean fuel. *Energy and Fuel.* 24(8): 4419–4424.

Zakaria, B.S., Nassar, H.N., Abu Amr, S.S., El-Gendy, N.Sh. (2015a) Applying factorial design and response surface methodology to enhance microbial denitrogenation by tween 80 and yeast extract. *Petroleum Science and Technology.* 33(8): 880–892.

Zakaria, B.S., Nassar, H.N., EL-Gendy, N.Sh., El-Temtamy, S.A., Sherif, S.M. (2016) Denitrogenation of carbazole by a novel strain *Bacillus clausii* BS1 isolated from Egyptian coke. *Energy Sources, Part A: Recovery, Utilization, and Environmental Effects.* 38(13): 1840–1851.

Zakaria, B.S., Nassar, H.N., Saed, D., El-Gendy, N.Sh. (2015b) Enhancement of carbazole denitrogenation rate using magnetically decorated *Bacillus clausii* BS1. *Petroleum Science and Technology.* 33(7): 802–811.

Zepeda, T.A., Pswelec, B., Obeso-Estrella, R. (2016) Competitive HDS and HDN reaction over NiMoS/HMS-Al catalysts: diminishing of the inhibition of HDS reaction by support modification with P. *Applied Catalysis B: Environmental.* 180: 569–579.

Zeuthen, P., Knudsen, K.G., Whitehurst, D.D. (2001) Organic nitrogen compounds in gas oil blends, their hydrotreated products and the importance to hydrotreatment. *Catalysis Today.* 65(2/4): 307–314.

Zhang, H., Li, G., Jia, Y., Liu, H. (2010) Adsorptive removal of nitrogen-containing compounds from fuel. *Journal of Chemical and Engineering Data.* 55: 173–177.

Zhang, P., Jia, R., Zhang, Y., Shi, P., Chai, T. (2016) Quinoline-degrading strain *Pseudomonas aeruginosa* KDQ4 isolated from coking activated sludge is capable of the simultaneous removal of phenol in a dual substrate system. *Journal of Environmental Science and Health, Part A. Toxic/Hazardous Substances and Environmental Engineering.* 51(13): 1139–1148.

Zhang, S., Zhang, Q., Zhang, Z.C. (2004) Extractive desulfurization and denitrogenation of fuels using ionic liquids. *Industrial and Engineering Chemistry* Research. 43(2): 614–622.

Zhao, C., Zhang, Y., Li, X., Wen, D., Tang, X. (2011) Biodegradation of carbazole by the seven *pseudomonas* sp. strains and their denitrification potential. *Journal of Hazardous Materials.* 190: 253–259.

Zhu, S.-n., Liu, D.-q., Fan, L., Ni, J.-r. (2008). Degradation of quinoline by *Rhodococcus sp.* QL2 isolated from activated sludge. *Journal of Hazardous Materials.* 160: 289–294.

5

Bioadsorptive Desulfurization of Liquid Fuels

List of Abbreviations and Nomenclature

4,6-DMDBT	4,6-Dimethydibenzothiophene
4E6M-DBT	4-Ethyl, 6-Methyl-Dibenzothiphnene
AAB	Acid Activated Bentonite
AC	Activated Carbon
ACF	Activated Carbon Fiber
ACFH	Hydrogen-Treated Activated Carbon Fiber
ACN	Acetonitrile
ADS	Adsorptive Desulfurization
AgNPs	Silver Nanoparticles
BT	Benzothiophene
CAC	Commercial Activated Carbon
CAs	Carbon Aerogels
Cat-ODS	Catalytic-Oxidative Desulfuirzation
CCD	Central Composite Design
CICCW	Chemically Impregnated Coconut Coir Waste
CSAC	Coconut Shell Activated Carbon

DBDS	Dibenzyldisulphide
DBT	Dibenzothiophene
DBTO	Dibenzothiophene Sulfoxide
$DBTO_2$	Dibenzothiophene Sulfone
DBTS	Dibenzothiophene Sulfone
DHN	Decahydronaphthalene
DMBT	Dimethylbenzothiophene
DMDBT	Dimethyldibenzothiophene
DMDS	Dimethyldisulfide
DMS	Dimethylsulfide
DMSO	Dimethylsulfoxide
EDX	Energy Dispersive X-Ray Analysis
FESEM	Field Emission Scanning Electron Microscope
FLE	Fluorene
FTIR	Fourier Transform Infrared
GAC	Granular Activated Carbon
GO	Graphene Oxide
HD	Heavy Distillate
HH	Heavy Hydrotreated
IUPAC	International Union of Pure and Applied Chemistry
LD	Light Distillate
LH	Light Hydrotreated
LPG	Liquid Petroleum Gas
MD	Middle Distillate
MeOH	Methanol
MH	Middle Hydrotreated
MIH	Molecular Imprinting Hydrogel
MIP	Molecular Imprinted Polymer
MIT	Molecular Imprinting technique
MMT	Montmorollonite
MNLB	Magnetite Nanoparticles Loaded Bentonite
MWCNT	Multiwall Carbon Nanotubes
OSCs	Organosulfur Compounds
PAHs	Polyaromatic Hydrocarbons
PBU	Primary Building Unit
PM	Propylmercaptan
PWAC	Pin Wood Activated Carbon
RADS	Reactive Adsorptive Desulfurization
RH	Rice Husk
RHAC	Rice Husk Activated Carbon

RSM	Response Surface Methodology
S-AC	Sewage sludge-derived Activated Carbon
SARS	Selective Adsorption for Removing Sulfur
SBU	Secondary Building Unit
SEM	Scanning Electron Microscope
S-M	Sulfur-Metal
SMIP	Surface Molecular Imprinting Polymer
SRGO	Straight Run Gas Oil
TC	Thiophenic Compounds
Th	Thiophene
UB	Untreated Bentonite
ULS	Ultra-Low Sulfur
V_m	Micropores
V_{mic}	Micrpore Volume
V_s	Super- Micropores
V_u	Ultra-Micropores
WI	Wet Impregnation
XRD	X-Ray Diffraction
ZT	Zeolitic Tuff

5.1 Introduction

Adsorptive desulfurization (ADS) is a very promising and economical method for reaching ultra-low sulfur (ULS) fuel, with regard to energy consumption, since the adsorption process can be achieved at ambient pressure and temperature without (or with little) the use of expensive hydrogen and without any expensive catalyst. As has been previously summarized, (Chapter 2) ADS can be used for selective desulfurization of petroleum fractions from refractory organic sulfur compounds. The main factors affecting the efficiency of ADS are the specific surface area of the adsorbent, its porosity, and surface chemistry (i.e. functional groups). Furthermore, it is most influenced by adsorbent capacity, selectivity, stability, and ability to regenerate. Materials used as adsorbents for desulfurization include different forms of activated carbon (AC), modified activated carbon, zeolite, metallic oxides, and porous metals (Al-Degas *et al.*, 2016). Mostly, activated carbons are utilized for the adsorption of compounds that have weaker polarity from a gas phase or polar fluid-phase, such as the adsorption of organics in wastewater. Thus, the main challenges facing ADS using activated carbon are the development of easy remunerable adsorbents with a high adsorption capacity and selectivity for sulfur compounds with low polarity from non-polar fuel streams, especially the refractory S-compounds, that cannot be removed by HDS over the other aromatic and olefinic compounds present in fuel streams, which would increase the difficulty and cost of adsorbents regeneration, taking into consideration that relatively low selectivity is attributed to the fact that both cyclic aromatic sulfur compounds and non-sulfur aromatic compounds interact with AC via π–π interactions. Moreover, due to the growing concern for waste minimization, recovery, and reuse, as well as industrial applications, the regeneration performance of sulfur loaded adsorbent is mandatory. Another challenge that faces the industry is the adsorption capacity of many natural adsorbents is low. For example, economical clay materials displayed capacities of 1–4 mg S compound/ g clay, requiring huge amounts of adsorbent. Further, pressure swing adsorption is not effective due to the strong interaction of sulfur with the adsorbent. Large adsorbent beds are, therefore, required to minimize the number of turnovers and multiple beds are needed to keep a refinery on-stream. Repeated calcinations can also lead to a loss of surface area due to sintering, reducing the amount of sulfur a bed can remove. Most research efforts, therefore, focus on creating less expensive and higher surface area materials. In developing the selective adsorbent, the key consideration is to design adsorbent materials which selectively interact with sulfur in the presence of a large excess of aromatic compounds, which exist in concentrations of >20% in comparison with less than 1 wt.% sulfur compounds.

Activated carbon (AC) and carbon fibers (ACF) are well-known multipurpose adsorbents used for treating different gases and liquids (Gawande and Kaware, 2016). The high adsorptive capacity of AC is reported to be attributed to the high density of surface functional groups, large porosity, and intense specific surface area. Further, their surface structure and other surface properties can be adjusted in order to improve adsorption efficiency (Mužic and Sertić-Bionda, 2013). The specific application of AC as an adsorbent depends on the molecules that are being targeted for adsorption. Adsorption on AC can be physical, which depends on the size and volume of the pores, or chemical, which depends on the chemical properties of the surface, i.e. surface properties that are conducive to chemisorption. Activated carbon can be obtained from different natural sources, such as wood, peat, coal, lignite, coke, lignin, seeds, fruits, and others (Carvalho *et al.*, 2003). Its chemical composition includes elemental carbon followed by oxygen, sulfur, nitrogen, and hydrogen; their percent content is shown in (Table 5.1) (Kwaśny and Balcerzak, 2016).

Table 5.1 Composition of Activated Carbon (as Element Content %).

Constituents	%
C	85–95
O	6–7
S	1
N	0.5
H	0.5

Activated carbon is widely used due to its good porosity, high surface area, and high efficiency for adsorption and can be used in a batch vessel reactor, continuous stirred reactor, and fixed or fluidized bed reactor.

The advantage of the fluidized bed column reactors is their ability to achieve high mixing efficiency. However, after some time, increased wearing and erosion of the adsorbent particles occurs. These drawbacks of the fluidized bed adsorption column, as well as high prices of powder activated carbon, mean that the adsorptive desulfurization could be competitively carried out in fixed bed columns with granulated activated carbon (GAC). The saturated activated carbon must be regenerated in order for it to be used again. The most commonly used method of regeneration is thermal treatment, but chemical and biological treatments, as well as ultrasound could be used (Mei *et al.*, 2003; Javadli and De Klerk, 2012; Mužic and Sertić-Bionda, 2013; Hosseini and Hamidi, 2014).

The main disadvantage that may retard the practical application of activated carbon in fuel industry is its hydrophobic surface, which can attract mono- and di-aromatic components of fuel streams in high amounts (Seredych *et al.*, 2009). To overcome removal of non-sulfur components and, hence, improve adsorption selectivity of organosulfur compounds (OSCs), many surface modification procedures were adopted. Surface modification of ACs could be accomplished by different procedures to end up with surfaces of different chemistries or variable porosities. Particularly for OSC removal from diesel, AC of acidic nature and higher micro/mesopore volumes are preferable. For example, acidic functional groups work as centers for attracting cyclic OSCs via oxygen–sulfur interactions with lower adsorption of non-sulfur cyclic compounds. According to the IUPAC classification of pore dimensions, pores are classified as micropore (≤2 nm), mesopore (2–50 nm), and macropore (≥50 nm). A high correlation between DBT removal with micro/mesoporosity was reported where the rate of OSC adsorption is accelerated at higher porosities (Seredych and Bandosz, 2007; Seredych *et al.*, 2009; Al-Ghouti *et al.*, 2010; Bu *et al.*, 2011).

Thus, if the pore diameter is such that the adsorbate molecules are larger, lesser adsorption would take place due to steric hindrance. Toida (2003) demonstrated that AC having a specific pore structure that is a composite micro-mesoporous structure with a large pore volume, attained an efficient and selective removal of DBTs. Moreover, metal loading by AC is considered an effective modification procedure for the ADS of OSC in fuel streams since the π-complexation between partially filled *d*-orbitals is metal loaded on the surface with an electron donating S atom in DBT is proposed to explain the favorable interaction between metal-loaded adsorbents and sulfur compounds in fuel streams (Seredych *et al.*, 2009).

Activated carbons (ACs) have been extensively studied because of their high surface area, cost effectiveness, receptivity to modification, and high affinity for sulfur compounds removal from different fuels (Ania and Bandosz, 2006; Yu *et al.*, 2008; Zhou *et al.*, 2009; Bu *et al.*, 2011; Shi *et al.*, 2015a). Recently, nanoparticle-modified adsorbents attracted many researchers due to high surface area to volume ratio and short diffusion rate (Saed *et al.*, 2014; Zakaria *et al.*, 2015).

This chapter summarizes the worldwide researchers' efforts to modify the surface functional groups of carbon materials and other natural adsorbents in order to improve or extend their practical applications in the ADS process.

5.2 ADS by Agroindustrial-Wastes Activated Carbon

The utilization of waste materials to produce activated carbon for environmental remediation is of significant economic potentiality and can substitute commercial activated carbon (CAC).

Various waste materials can be used to produce activated carbon and some of the most commonly used are agriculture wastes such as coconut shell, pistachio shell, saw dust, walnut shell, rice straw, rice husk, tropical wood, almond shell ... etc. (Kim *et al.*, 2010).

Salem and Hamid (1997) examined ADS from naphtha solution using charcoal, GAC, and zeolite 13X. The results indicate that zeolite 13X has a high capacity for sulfur in low concentration ranges (<50 ppm) at ambient temperature. When the concentration increases, the capacity of AC can be 3 times greater than that with zeolite, while AC is superior to zeolite 13X at high temperatures (80 °C). At higher temperatures, zeolite may adsorb some olefins and aromatics from the solution, thus, competition for the sites between these hydrocarbons and sulfur compounds may be the main reason for the lower capacity of zeolite at higher sulfur concentrations. The high temperature used may also affect the sorption and desorption rates of such materials on the solid surface. However, this study suggests that naphtha solutions should be treated in two beds in industrial applications. The first bed contains activated carbon and the second is a smaller bed of zeolite 13X. The solution is treated initially in the first larger bed at 80 °C, cooled to ambient temperature, and then fed to the zeolite 13X bed.

Lee *et al.* (2002) investigated the ADS of diesel oil on ten different activated carbons and reported that coconut shell-based carbon, activated by high temperature, was more effective for ADS than coal-based or wood-based carbons.

Toida (2003) reported that AC having a specific pore structure attains an efficient and selective removal of dibenzothiophenes (DBTs). Moreover, mesoporous AC is one of the most important characteristics in liquid phase adsorption, owing to high surface areas (ranging from 164 to 1260 m^2/g) and pore volumes (up to 1.62 cm^3/g) (Binti Amiri, 2008).

Kim *et al.* (2006) suggested one potential new option for refinery applications is to use a sulfur-selective adsorption unit for the ultra-deep removal of organic sulfur following a conventional HDS unit, where such a combination could remove all of the sulfur from the liquid fuel products, where the ADS-activity of different adsorbents can be ranked as follows activated alumina < Ni/SiO_2–Al_2O_3 < activated carbon. Thus, activated carbon shows higher adsorption capacity and selectivity for sulfur compounds with methyl substituents with breakthrough and saturation capacities of 7.15 mg S/g and 16.29, respectively, since the oxygen-containing functional groups on the surface of activated carbon play an important role in increasing sulfur-adsorption capacity. Moreover, the adsorption capacity of AC can be ranked in the following increasing order BT < naphthalene < 2-methylnaphthalene < DBT < 4-MDBT < 4,6-DMDBT, which is dramatically different and almost opposite to that of nickel-based adsorbents. Thus, AC is very promising for SARS as a new approach to the ultra-deep desulfurization of diesel fuels at room temperature (Zhou *et al.*, 2006). The use of rice husk activated carbon for the adsorptive removal of sulfur compounds, e.g. dibenzothiophenes from kerosene, was reported (Shimizu *et al.*, 2007a).

Shimizu *et al.* (2007b), reported a comparative study for the ADS of kerosene as a promising hydrogen source of fuel cells for stationary home-use in Japan, using AC prepared from rice husks and microporous commercial activated carbon fibers (ACF) (FR-25, Kuraray Chemical Co., Ltd., Japan). The total S-content in the untreated kerosene was evaluated to be 14.1 mass ppm, while the content of sulfur in DBTs therein was 3.5 mass ppm. It has been noted that AC with high-content carbon and low-content ash has a large specific surface area and pore volume, while AC with low-content carbon and high-content ash provides small specific surface area and pore volume. Moreover, large specific surface area and total pore volume are found to be responsible for a reduction of the pore width. The increase of ultra-micropores (Vu) and super-micropores (Vs) enlarges the DBTs adsorption capacity. Upon the application of the ACF with specific surface area of 2336 m^2/g and pore volume of 1.052 cm^3/g and the KO-TO activated at 850 °C for 1 h that was characterized with a specific surface area and pore volume of 473 m^2/g and 0.267 cm^3/g, respectively, the ACF recorded remarkably larger Vs (0.614 cm^3/g), which contributed to larger specific surface area and total pore volume and were 5 and 4-fold greater than those of the activated KO-TO. Moreover, the Vu and Vm of the ACF were about two times larger than those of the activated KO-TO. This led to higher adsorption capacity of DBTs on the ACF of about 0.090 mg-S/g-adsorbent, while that of the activated KO-TO was 0.069 mg-S/g-adsorbent (that is 77% performance of the ACF).

The above DBTs adsorption results imply that larger super-micropores (Vs), which provide larger specific surface area and pore volume, are not so useful for the DBTs adsorption. The roles of ultra-micropores (Vu) and mesopores (Vm) on the DBTs adsorption seem to be important. According to the study performed by Meille *et al.* (1999), the mechanism of DBTs adsorption on a catalyst mainly depends of the π electrons in the aromatic rings. Thus, the interaction of π electrons on graphite like sheets constituting slit-shaped pores in activated carbons between π electrons in DBTs is likely to further promote the entrance of DBTs parallel to the wall surface of slit-shaped ultra-micropores. This suggested that the predominant adsorption sites of DBTs in the RHACs are deemed to be ultra-micropores. It is also worthy to note that at high S-content, it is likely that ultra-micropore volume dominates the capacity of DBTs adsorption and mesopores volume has influence on the kinetics of DBTs adsorption. However, at low S-content, the DBTs adsorption sites in ultra-micropores might not be fully filled, hence, not only ultra-micropores, but also mesopores, have important roles in leading DBTs to the adsorption sites and determining the DBTs adsorption capacity (Shimizu *et al.* 2007b).

Kumagai *et al.* (2009) reported the production of rice husk adsorbents throughout the carbonization and then activation with CO_2. The produced material was then characterized chemically through the quantification of acid and basic sites at the surface. It was noted that the number of acid and basic sites increased as a function of the time and temperature of activation. The organic S- and N- compounds contain a heteroatom bound to an aromatic hydrocarbon structure and have a high nucleophilicity due to the electron cloud of aromatic compounds and the non-bound electron pair in S or N atoms. Thus, adsorbents with acidic (electrophilic) characteristics are expected to have a greater capacity to adsorb the S- and N-compounds present in fuel streams. An improvement in the adsorption associated with acid groups was observed by Ania and Bandosz (2009), Zhou *et al.* (2009), and Singha *et al.* (2013). These groups are probably centers for the adsorption of dibenzotiophene (DBT) via oxygen-sulfur or sulfur-sulfur interaction. The acid groups present on the AC- surface are reported to increase the capacity of the adsorbent for the removal of DBT and 4,6 DMDBT. However, the amount of sulfur adsorbed is governed by the volume of micropores and acid groups located in larger pores can attract molecules of DBT and 4,6 DMDBT via specific interactions. Thus, to achieve a good performance, the AC needs to have specific characteristics, that is, a high volume of pores <1 nm and the presence of acid groups in the larger pores which would lead to better results. An appropriate proportion of the adsorption must occur via acid groups in order to avoid blocking the accessibility of small pores (Seredych *et al.*, 2009).

Mužic *et al.* (2010) reported that commercial-activated carbon is better than zeolite for the ADS of dimethylbenzothiophene (DMBT) and dimethyldibenzothiophene (DMDBT) from fuel. Aparicio *et al.* (2103) also reported in a comparative study that activated carbon has a higher adsorption capacity of 4,6-DMDBT (47.6 mg/g of support) compared to other supports (such as alumina, silica, and commercial and natural zeolites) and this is attributed to its high specific surface area, the presence of functional groups that make the acidic surface, and its isoelectric point (~4), thus, more electronic back donation with the aromatic rings in the molecule of 4,6-DMDBT occurs. Also, the presence of –OH groups on alumina surface creates a considerable adsorption capacity towards 4,6-DMDBT (21.2 mg/g of support).

Haw *et al.* (2010) reported the synthesis of granular functionalized AC from oil palm shell by a phosphoric acid activation method and carbonized at 500 °C and 700 °C for 1 h. The AC was characterized by a microporous structure with some contribution of mesoporisc properties. The Fourier Transform Infrared (FTIR) Spectroscopy results showed that higher activation temperature leads to fewer surface functional groups due to thermal decomposition. Micrographs from a field emission scanning electron microscope (FESEM) showed that activation at 700 °C creates orderly and well developed pores. Furthermore, X-ray diffraction (XRD) patterns revealed that pyrolysis has converted a crystalline cellulose structure of oil palm shell to an amorphous carbon structure. The prepared AC at 500 °C showed higher ODS for commercial diesel fuel with an initial S-content of

2189 ppm, where, at optimum conditions of a temperature at 50 °C, atmospheric pressure, 0.5 g activated carbon, 3 mol ratio of hydrogen peroxide to sulfur, 2 mol ratio of acetic acid to sulfur, and 3 oxidation cycles with 1 h for each cycle using acetonitrile as extraction solvent, the sulfur content in diesel was reduced from 2189 ppm to 190 ppm with 91.3% of total sulfur removed. The prepared AC was also used for ADS of commercial diesel fuel at 50 °C. The prepared functionalized-activated carbon acts both as catalyst and adsorbent in this Cat-ODS process. AC provides the adsorption surface for S-compounds and oxidation reaction to take place. It catalyzes the decomposition of H_2O_2 to ·OH or ·OOH free radicals which are powerful oxidants for the oxidation of S-compounds to their corresponding sulfones (Scheme 5.1). The results showed that sulfur oxidation without a catalyst gave only 16.2% of sulfur removal in the presence of acetonitrile, glacial acetic acid, and hydrogen peroxide, including the extraction of sulfur compounds by acetonitrile. Thus, AC enhances the rate of ODS and decreases the required amount of acetic acid. The presence of various surface functional groups, especially the acidic groups such as the C–O, C=O, P-O, and POOH, enhances the catalytic activity of the AC. These oxygen-containing functional groups, especially the carbonyl group, are

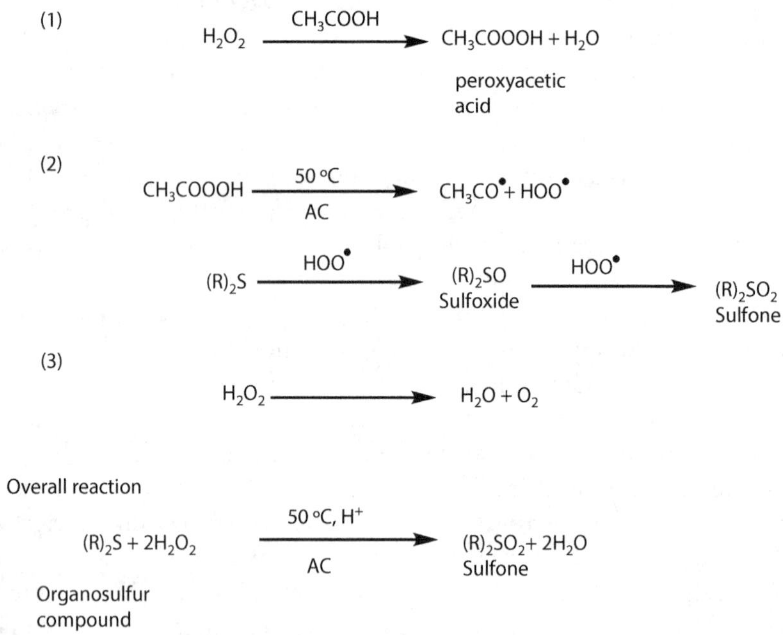

Scheme 5.1 Different ODS Reactions Between Oxidant and S-Compound in Presence of Activated Carbon (AC).

known to have a close relationship with the adsorptive and catalytic properties of the AC. As mentioned before, the acidic groups are acting as electron withdrawing groups or electron acceptors, while the sulfur atom in the S-compounds are electron donors. The interaction between them will lead to the formation of donor–acceptor complexes on the carbon surface. These acidic functional groups have a strong affinity towards the S-compounds, which causes the sulfur species to be initially physisorbed and then chemisorbed on the carbon surface. Once the S-compounds are chemisorbed, they form active species on the carbon surface, which is less stable and more readily oxidized by a nearby oxidant.

A comparative study for the adsorption of PAHs and PASHs from model and real diesels (with an initial S-content of 400 and 398 ppmw, respectively), on activated carbon (from different sources, including pitch, apricot, coconut, and wood), with surface areas that vary between 713 to 1403 m^2/g and particle sizes varied between 400–800 μm in both batch and fixed-bed adsorption systems, was performed by Bu *et al.* (2011), which illustrated that adsorption results showed that adsorptive affinities of molecules with polycyclic aromatic skeleton structures are primarily governed by the π–π dispersive interactions between the aromatic rings and the graphene layers of ACs. In addition, the electron donor–acceptor mechanism also plays an important role for S-containing molecules. Furthermore, for effective adsorption of large molecules, not only should the pore size of the adsorbent be at least larger than the critical diameter of the adsorbate, but the pore size should also be sufficiently large enough to reduce diffusional resistance during adsorption. Bu *et al.* (2011) concluded that the adsorption selectivity for different PAHs and PASHs increases as follows: naphthalene < fluorene < DBT < 4,6-DMDBT < anthracene < phenanthrene. It was also observed that the adsorption capacity of PASHs decreases significantly in the presence of PAHs as a result of the adsorption competition due to similar structure, molecular diameter, and adsorption mechanisms.

Patil *et al.* (2011) reported the preparation of activated carbon from black liquor, the waste of integrated pulp and the paper industry, uses eucalyptus and bamboo as raw materials and the Kraft pulping process. Phosphoric acid and nitrogen were used as intercalating agents. The prepared activated carbon showed good ADS activity on model oil (hexyl mercaptans/benzene).

Mužic *et al.* (2011) reported the desulfurization of diesel fuel in a fixed bed adsorption process using activated carbon, SOL-CARB C3 (Chemviron Carbon, Belgium), that yielded an output with a sulfur concentration <0.7 ppm.

Kumagai (2011) reported a comparative study for the ADS of BTs and DBTs from commercial kerosene by micro-and mesoporous granular rice husk activated carbon (RHAC) and microporous granular coconut shell activated carbon (CSAC). Fixed-bed flow desulfurization tests were conducted providing the breakthrough curves for BTs and DBTs. The adsorption isotherms for BTs and DBTs were also obtained in a batch-mode, correlating with the results of the fixed-bed flow test. RHAC and CSAC are found to be useful to remove the recalcitrant BTs and DBTs. However, RHAC showed a lower selectivity for DBTs and a higher selectivity for BTs than CSAC did. A larger adsorption capacity of DBTs in volume basis was observed on CSAC, which was attributed to its higher bed density, higher carbon content, and larger volume of ultra-micropores. Although RHAC had lower carbon content and a smaller volume of ultra-micropores, it showed a larger adsorption capacity of BTs and an acceptable adsorption capacity of DBTs in volume basis. This was explained by a larger volume of mesopores contributing to efficient transportation of BTs and DBTs towards their adsorption sites of ultra-micropores.

Hosseini (2012) reported the application of a packed bed of activated carbon as a solid adsorbent for removing the oxidized sulfur compounds after the ultrasound-assisted-ODS of different oil feeds in the presence of acetic acid and hydrogen peroxide, under ambient temperature and pressure, which showed that the use of solid absorbents has no negative effect on main fuel hydrocarbons and the unit scale is made more practical as the result of a very low solid volume used, compared with the liquid solution in a liquid-liquid extraction method. Also, upon applying a hydrogen peroxide-activated carbon-acetic acid system, the percentage of the remaining sulfur was much lower than that of a hydrogen peroxide-acetic acid system.

Kumar and Srivastva (2012) reported the application of response surface methodology (RSM) based on the central composite design (CCD) of experiments to optimize the ADS of model oil (DBT in iso-octane) using commercial activated carbon (CAC) from vegetable origin. The highest ADS, of approximately 53% of initial DBT at a concentration of 100 mg/L by CAC, was obtained with an adsorbent dosage of 20 g/L at 30 °C and within 6 h. The X-ray diffraction (XRD) analysis of the CAC revealed the presence of moganite (SiO_2), akdalaite ($(Al_2O_3)_4.H_2O$), tamarugite ($NaAl(SO_4)_2.6H_2O$), and fersilicate (FeSi) as major components, which would explain the high ADS capacity of the CAC.

Application of low temperature oxygen plasma was proposed to modify activated carbons (ACs) for enhancing their adsorption toward dibenzothiophene (DBT), in a model diesel fuel, by increasing the carbon surface oxygen-containing groups, thus significantly improving the ADS-capacities of AC towards DBT. More importantly, this novel modification method effectively keeps the mass losses of the ACs to fell within a very narrow range of ±2%, which overcomes a demerit of conventional thermal oxidation treatment, which resulted in a bad mass loss of 31.7–91.4%. The isotherms of DBT on the modified ACs were significantly higher than those on the original AC. Fixed-bed breakthrough experiments showed that the AC particles with a surface area of 1187 m^2/g were modified by oxygen plasma for 30, 60, and 120 min, expressed an increment in a working DBT-adsorption capacity with a treatment increment time of 35.1%, 44.7%, and 49.1% for AC30, AC60, and AC120, respectively, relevant to the original AC (Zhang *et al.*, 2012).

Al Zubaidy *et al.* (2013) reported the usage of carbonized date palm kernels without activation for the ADS process of diesel fuel. This was able to reduce the initial sulfur content (410 ppm) by approximately 34.15% with the addition of 6% adsorbent, which increased to approximately 55% S-reduction by using 10% of the natural origin AC.

AC with a specific surface area and pore volume of 788 m^2/g and 0.414 cm^3/g, respectively, was prepared from coconut shells and expressed good ADS of diesel oil with an initial S-content of 5200 mg/kg, resulting in diesel oil with a final S-content of 3390 mg/kg with approximately 35% desulfurization. However, the prepared activated carbon recorded approximately 82% desulfurization of hydro-desulfurized diesel oil with an S-content of 120 mg/kg, producing diesel oil with an S-content of 22 mg/kg. One of the important factors that would affect the absorptivity of the AC is its ash content. The lower its ash content is, the higher its adsorptive activity; the acceptable range is between 2–6%. On the other hand, the lower its fixed carbon content is, the lower its adsorption activity. Thus, the application of the ADS method by integrating an adsorption column system packed with low cost AC prepared from agricultural waste, in series with the hydrotreater, seems to be very promising as an upgrade to the existing HDS unit, would not be complicated, and would not raise the investment and operating costs as much as upgrading the HDS unit itself in order to achieve the desired ultra-low sulfur levels (Nkosi, 2014).

Daware *et al.* (2015) reported the ADS of diesel oil using neem leaves' powder, with approximately 65% of the initial S-content at optimum conditions of 3.5 h and 20% (w/v) with an adsorbent dosage 20 °C.

Ibrahim and Aljanabi (2015) reported the ADS of diesel oil samples with different initial S-contents (280, 380, and 480 ppm) using industrial grade granular and powder activated carbons in a batch laboratory scale setup, where the ADS increased with the increase of temperature from 30 to 50 °C.

This was attributed to the increase in the amount of sulfur compounds (mainly DBT) adsorbed with temperature, which was accompanied with an increase in the mobility of the adsorbate molecules in a solution and within the sorbent porous structure, consequently, overcoming the activation energy barrier. The ADS efficiency for constant diesel fuel volume increases in direct relationship with the increase in a sorbent dose up to 2 mL/g (diesel/AC ratio) due to the increase in surface area and, hence, more active sites are available for the adsorption of sulfur. Any additional decrease in the diesel/AC ratio above 2 mL/g is found to be impractical for batch systems, since beyond this ratio the system tends to be a slurry and more difficult to be treated. The ADS efficiency increased with the decrease of AC particle size due to the increase of the specific surface area, pore volume, and diameter, where the surface area and pore volume increased from 300 to 930 m^2/g and from about 0.19 to 0.5 cm^3/g, respectively, when the sorbent particle size decreased from 1.4 mm to 0.8 mm. However, further decrease in particle size was not favorable for the ADS process since the agglomerations of the small size AC particles impose this increase, reduce the actual contact area, and close the path of sulfur components to the active centers. Thus, 0.8 mm can be chosen as the best sorbent particle size. The overall sulfur removal rate is known to be limited by the mass transfer between phases. Thus, an increase in the mass transfer clearly improves the overall sulfur removal rate. Consequently, agitation affects the reaction rate by increasing the amount of collisions the atoms make with one another, therefore speeding up the reaction rate. However, the ADS decreased with the increment of the initial S-concentration for constant sorbent dose. This is due to the fact that for a fixed sorbent dose, the total available adsorption sites are limited, thereby, adsorbing almost the same amount of sulfur. Upon applying the selected optimum conditions, 2.5h, 50 °C, 2 mL/g diesel oil/AC, 0.8 mm AC particle size, and 1000 rpm for the ADS of diesel oil with an initial S-content of 500 ppm, S-content decreased to 240 ppm with recorded improvement in cetane number and a slight decrease in viscosity and specific gravity.

Approximately 20.94% desulfurization of diesel oil with an initial S-content of 398 ppm has been reported using 10 wt.% granular activated carbon (GAC) in a batch ADS process under ambient temperature and pressure within 1 h, with an improvement of the ignition quality of the treated diesel oil. The diesel index increased from 69.1 to 71.3, the cetane number increased from 59.7 to 61.3, keeping its calorific value, while the kinematic viscosity decreased from 9.03 to 8.92 cSt and its specific gravity slightly decreased from 0.819 to 0.817. This was attributed to the associated removal of some aromatic compounds and heavy metal from the diesel oil sample (Al Zubaidi *et al.*, 2015).

Anisuzzaman *et al.* (2017) also reported the effectiveness of mesoporous AC prepared from oil palm shells, which was characterized by a reasonable specific surface area of 15.41 m^2/g, pore volume 0.028 cm^3/g, and average pore diameter 4.2 nm for the ADS of model fuel (BT in a mixture of 75:25 v/v n-octane and *p*-xylene). The AC carbon was prepared by pyrolysis at 600 °C for 3 h, followed by activation at 600°C for 30 min with CO_2 at rate of 100 cm^3/min. The FTIR analysis revealed the presence of O=C=O, Si-O-Si group, and N-H group. The surface of the AC was neutral to slightly acidic. The SEM analysis proved the porosity of its surface. All of these enhanced its adsorptive desulfurization activity, which recorded 2.75 mg S/g adsorbent, and the chemisorption was suggested to be the rate limiting step that controlled the overall adsorption process.

Gawande and Kaware (2017) reported the preparation of activated carbon form coconut shell by thermal treatment in the presence of H_2SO_4 as an activating agent and its ADS activity was tested for commercial diesel fuel with an initial S-content of 334.5 ppm, where approximately 63.90% reduction in S-content occurred within 4 h using 20% w/v AC with a Mesh of 12 particles in size.

The S-content in heavy naphtha varies from 150 to 3000 mg/L depending upon the S-content of the crude oil (Nanoti *et al.*, 2011). Moreira *et al.* (2017) examined the activated carbon of porous and a high specific surface area of 606 m^2/g that was prepared from coconut shell for the ADS of synthetic sulfurized naphthenic oils composed of a mixture of different concentrations (5×10^{-4} and 5×10^{-3} w:w) of S-compounds: thiophene (Th), benzothiophene (BT), and dibenzothiophene (DBT) in decahydronaphthalene (DHN). The maximum adsorption capacity recorded 1.6×10^{-2}, 2.0×10^{-2}, and 1.9×10^{-2} kg/kg for Th, BT, and DBT, respectively. This was explained by the relation between the order of selectivity toward the S-containing compounds and the polarizability factor, where the stronger electronic characteristics are, the higher the adsorption capacity is. It is well known that the aromatic compounds possess considerable polarizability, which are more pronounced for polynuclear species, such as BT and DBT (Farag, 2007). Furthermore, the high adsorption capacity was attributed to the higher basicity of BT and DBT and there is greater interaction between thiophenes and acidic functional groups on the surface of the activated carbon, which increases the adsorption capacity (Fallah and Azizian, 2012; Fallah *et al.*, 2014). Moreover, the van der Waals and hydrogen bonding interactions could have played an important role in the adsorption of aromatics and S-containing compounds onto the AC. Thus, the carbon based adsorbent determines the physical nature of the adsorption of S-containing compounds, which can favor the generation of the adsorbent and enhance its feasibility for practical applications (Kim *et al.*, 2006; Farag, 2007). The effect of some classical adsorption inhibitors, such as water, naphthalene, phenol, and carbazole, representing polar, aromatic, oxygen, and nitrogen compounds, respectively, was also studied on the ADS of DBT (5×10^{-3} w:w). The inhibition effect was ranked in the following decreasing order: water > carbazole > naphthalene ≈ phenol. The strong inhibition effect of water due to the formation of stable emulsions, hinders the removal of S-compounds. Another important factor is that aromatic compounds with nitrogen heteroatoms are associated with higher selectivity compared to aromatic hydrocarbons, which consequently inhibits the adsorption of S-compounds (Song *et al.*, 2013; Zhang *et al.*, 2004). Moreover, the ADS of three types of naphthenic oils, obtained by vacuum petroleum distillation, were also investigated as a real case upon two treatment cases: hydrotreated oil feeds (LH, MH, and HH,) and non-hydrotreated feeds (LD, MD, and HD) at different temperatures (50–150 °C). The ADS increased with the increment of temperature and was attributed to the better accessibility of the active sites and greater mobility of the S-compounds within the porous sorbent structure, with on overcoming of the activation energy barrier at higher temperatures. The hydrotreated oil showed better ADS results than the untreated ones. The adsorption phenomenon was considered to proceed with an initially predominant resistance in the external fluid film and, after a transition time, the process was controlled by intraparticle mass transfer.

5.3 ADS on Modified Activated Carbon

There are several methods for the increment of the ADS activity of vegetable origin AC. Although the acidification of activated carbon, especially with nitric acid, can remove about 50% of inorganic components in the AC, it yields an increase in the micropore volume as well as the production of carboxyl functional groups on the surface of the AC, thus increasing its ADS activity (Yu *et al.*, 2013; Gokce and Aktas, 2014). The improved pore structures and surface acidity are significantly associated with the adsorbent's ability to remove bulky thiophenic compounds, such as dibenzothiophene and similar compounds (Seredych *et al.*, 2009). Similarly, oxidation of carbon materials offers oxygen-containing functional groups that are important for the adsorption of sulfur compounds (Qiu *et al.*, 2015; Shi *et al.*, 2015b). The impregnation of active metal ions can help increase the ADS efficiencies (Chapter 2) since the effect of a dopant on the textural and surface acidity properties of the modified AC leads to improved ADS of fuels (Ganiyu *et al.*, 2016; 2017).

Ania and Bandosz (2006) reported the improvement of the selective desulfurization capability of activated carbons (ACs), at room temperature, throughout the preparation of new metal-loaded carbon-based sorbents containing sodium, cobalt, copper, and silver highly dispersed within the carbon matrix, where the metals incorporated to the surface act not only as active sites for selective adsorption of sulfur-containing aromatic compounds, but also as structural stabilizers of the carbon materials and as catalyst initiators in RADS. Depending on the reactivity of the metal used, the adsorption capacity of the activated carbons significantly varied. Cobalt and copper loaded carbons showed the highest uptakes due to ill-defined catalytic synergetic effects. The presence of sulfur compounds in the structure of the carbon, as a result of the sulfonic moiety of the precursor, results in sulfur–sulfur specific interactions leading to an enhancement in the adsorption capacity for DBT removal. In another study, the ADS of DBT from model oil (DBT in n-hexane) reached to approximately 130 mg S/g upon applying adsorbent mesoporous carbons with highly dispersed copper, cobalt, and iron that were obtained from an organic polymer within amorphous silica powder, alumina, and zeolite 13X (Seredych and Bandosz, 2007). This is attributed to the high volume of carbon-mesopores and specific interactions of DBT with surface acidic groups and strong interactions of metals with DBT via S–M σ bonds or, in the case of copper, via interaction of metals (M) with disturbed π electrons of aromatic rings of DBT.

Alhamed and Bamufleh (2009) reported the preparation of granular activated carbon (GAC), of 1.71 mm particle size, from dates' stones through chemical activation by impregnation with $ZnCl_2$ as activator. Then, its ADS was tested on model oil (500 ppm DBT in n-decane) where approximately 86% of the DBT was adsorbed during the first three hours. The ADS increased gradually to reach equilibrium at around 92.6% within 48 h to reach 37.0 ppmw S. The ADS capacity of the prepared GAC was also tested on model oil (200 ppm 4,6-DMDBT and 500 ppm DBT in n-decane as a binary-system), which showed good adsorption capacity towards the 4,6-DMDBT. However, it was lower than that of DBT (Bamufleh,2009). The adsorption mechanism of DBT and 4,6-DMDBT on GAC samples is based on the assumption that the strong interaction takes place between sulfur atoms of these compounds and the active site on the adsorbent surface. The alkyl groups at 4 and/or 6 positions, which are adjacent to the sulfur atom of DBTs, decrease the access of the sulfur atom to the adsorbent active sites due to steric hindrance and, consequently, strongly affect the selective adsorption of the refractive sulfur compounds. Moreover, the higher molecular size of 4,6-DMBT decreases the ADS. The adsorption of DBT on the GAC sample is mainly physical. Linear regression of experimental data was able to predict the critical pore diameter for DBT adsorption (0.8 nm) and validated the reported impact of average pore diameter of activated carbon on the adsorption capacity, where DBT adsorption is hindered by the GAC with a critical pore diameter close to 0.8 nm. Thus, the adsorption is mainly affected by the size of the adsorbate and the pore size of the adsorbent. The critical size of DBT and 4,6-DMBDT are reported to be 0.8 and 0.9 nm, respectively (Jayne *et al.*, 2005). Briefly, the textural properties of the granular activated carbon play a key role. The faster adsorption rate can be attributed to the lower internal mass transfer resistance due to large pore size (Haj and Erkey, 2003).

In an attempt to decrease the cost of nano-ZnO ADS, an activated carbon prepared from coconut shell was used (Fen *et al.*, 2012). A nano-synthesized ZnO of 13 nm was prepared by a homogeneous precipitation method and then, AC/loaded nano-ZnO was prepared by a mechanical hybrid method where the ZnO powder was well adsorbed into the pores of the large and middle hole of activated carbon. The most important advantage of nano-ZnO/AC is that the structure of the synthesized ZnO was not affected after its loading on AC and the active component is still Zn^+. The most effective ADS activity, with a longer breakthrough time of 110 min, was recorded when the mass ratio of ZnO and C was 0.3, had a 2 h loading time, and acidified AC with HCl since, at higher ZnO loading and longer loading time, excess nano-ZnO powders would accumulate in the AC surface, block the holes, decrease the specific surface area and pore volume, and, thus, reduce the desulfurization properties. Moreover, the HNO_3-acidified AC has a lower specific surface area and porosity than that acidified with HCl, since nitric acid has strong oxydic ability. However, the HCl can remove some inorganic impurities in the hole of activated carbons, which, consequently, will improve the porosity (Biniak *et al.*, 1997). Adsorption of DBT as a major OSC in commercial diesel by AC is reported to be enhanced by the surface modification of commercial-AC with manganese dioxide (MnO_2) (Abu Safieh *et al.*, 2015), where the used diesel oil was mainly composed of 34.6 and 4.8 wt% for mono- and di-aromatic hydrocarbons, respectively, with an extremely high concentration of OSCs 7055 mg S/kg diesel with an initial DBT-concentration of 11890.0 mg/kg that is approximately represented by 30% of OSC in diesel. The results revealed that the ADS increased up to 26.8% (w:w) loading of Mn/AC, 43.8 mg DBT/g modified AC relevant to 18.4 mg DBT/g unmodified AC, within 30 min at a ratio of 11 cm^3/g (diesel:carbon). However, the ADS reduced at higher Mn-concentration (>26.8% (w:w) Mn/AC). This was attributed to several reasons; the average size of MnO_2 was <0.5 nm, thus, a large fraction of MnO_2 was deposited internally as the average pore diameter of AC is 2.0 nm. This decreased the specific surface area and porosity of the modified AC. Moreover, the deposition of large amounts of Mn (52.1% in Mn-AC) resulted in a high reduction in acidity and basicity of modified AC. The important achievement in that research is the remarkable reduction in mono- and di-aromatic hydrocarbons adsorption with modified-AC. This was attributed to the π-complexation between partially filled d-orbitals in Mn^{4+}(Mn^{4+} : [Ar]$3d^2$]) and an electron donating S-atom in DBT. Not only this, but the surface morphology of the AC is an another important factor, where adsorbents of pore diameter 0.7–1.0 nm are suitable for accessing DBT and other cyclic OSC, as the diameter of these OSCs are comparable to the dimension of these adsorbents. Further, the larger pore diameter is also enhancing DBT and other cyclic OSCs by reducing mass transfer resistance and increasing the rate of the ADS process (Ania and Bandosz, 2006). The mixing rate is also important to overcome the mass transfer limitation and accelerate the diffusion of DBT in the oil phase and attain equilibrium in a short time since oil streams are viscous and this would negatively affect DBT migration from the bulk of the oil phase to the surface. The modified adsorbent is satisfactorily regenerated using n-hexane at 65 °C due to the favorable hydrophobic–hydrophobic interactions with adsorbed DBT.

Nunthaprechachan *et al.* (2013) reported the application sewage sludge-derived AC for the ADS of DBT from a model oil (100 ppm DBT in n-octane) under ambient temperature and pressure, where the sludge was collected from a textile wastewater treatment plant in Thailand. The activated carbon was prepared first by pyrolysis under N_2 atmosphere, producing charcoal (donated S-AC), then followed by activation using KOH, $ZnCl_2$, or HNO_3. The S-AC prepared by KOH-activation at a KOH:char weight ratio of 6 with carbonization provided the highest adsorption capacity (14.25 mg/g or 71.4% DBT removal) compared with the other types of S-ACs, followed by that activated with $ZnCl_2$ and, finally, activated with HNO_3. That was 3.69-, 31.4-, and 1.28-fold greater than that prepared by $ZnCl_2$- and HNO_3-activation under the same conditions and the commercial-AC, respectively. The adsorption capacity of this S-AC was 1.3-fold higher than that of the commercial-AC (11 mg/g or 58.0% DBT removal). This was attributed to the oxygen-containing surface functional groups, especially the carbonyl groups of the S-ACs, which play a more important role in the adsorption of DBT than the textural properties of the AC. Since the carbonyl group is a good electron acceptor (Epiotis, 1972), it can combine with the two unpaired electrons in the DBT structure to form electron donor-acceptor complexes (Yu *et al.*, 2005). In addition, the presence of acidic functional groups, especially the carboxylic groups, on the carbon surface act as bifunctional groups that contain both a hydroxyl group (an electron donor) and a carbonyl group (an electron acceptor) and can form an electron donor–acceptor complex with DBT. Thus, the carbonization time and temperature are very critical of the ADS efficiency of the prepared AC. Raising the carbonization temperature from 600 °C to 800 °C or increment of carbonization time to be longer than 1 h resulted in a decrease in the DBT adsorption capacity and removal. This is caused by the fact that the generated carbonyl group is ready to decompose to CO in the temperature range from 700–900 °C (Figueiredo *et al.*, 1999).

Nazal *et al.* (2015) reported the improvement of surface properties of commercial coconut activated carbon (AC), multiwall carbon nanotubes (MWCNT), and synthesized graphene oxide (GO) by loading 5% and 10.9% aluminum (Al) in the form of aluminum oxide (Al_2O_3) particles. Then, their ADS activity was tested on model diesel fuel (DBT in *n*-hexane). Moreover, their selectivity towards DBT was also tested relative to thiophene as a model molecule for small aromatic heterocyclic compounds, as well as relative to naphthalene, which represents the availability of polyaromatic hydrocarbons (PAHs) with molecular structures close to that of DBT. The stock solution for the ternary mixture from these three compounds, thiophene/DBT/naphthalene, in n-hexane was prepared with concentrations of 250 mg/L for both thiophene and DBT and 1000 mg/L for naphthalene to simulate the actual availability of PAH in real diesel. For the pristine adsorbents, the decrease in surface area was AC > CNT > GO, while the trend in total pore volume was CNT > AC > GO. while the impregnated adsorbents showed lower surface areas and porosity relative to their corresponding pristine adsorbents. The prepared impregnated adsorbents showed good and stable activity and can be easily regenerated by heating and reused for at least 5 times without losing their activity. The adsorption capacities for DBT on the aluminum oxide modified adsorbents are improved by approximately two-fold, which is attributed to the introduction of Al_2O_3 as a Bronsted acid and Lewis acid in the environment of the base DBT and, also, as an additional adsorption site (Scheme 5.2). The highest ADS capacity was recorded for AC loaded with 5% Al (ACAL5), recording 85 ± 1 mg/g, with a high selectivity factor relative to naphthalene (54 mg/g) and thiophene (255 mg/g), while the selectivity of CNTAL5 recorded 127 and 7 for DBT/thiophene and DBT/naphthalene, respectively. The other prepared adsorbents followed the order: ACAL10 > AC > CNTAL5 > CNTAL10 > GOAL5 > CNT > GO > GOAL10. This was attributed to the decrease in surface area and porosity. The ADS for DBT by ACAL5 has also been tested for real diesel oil feed, which reduced by 30% from its initial concentration (43 mg/L) and at higher DBT concentration

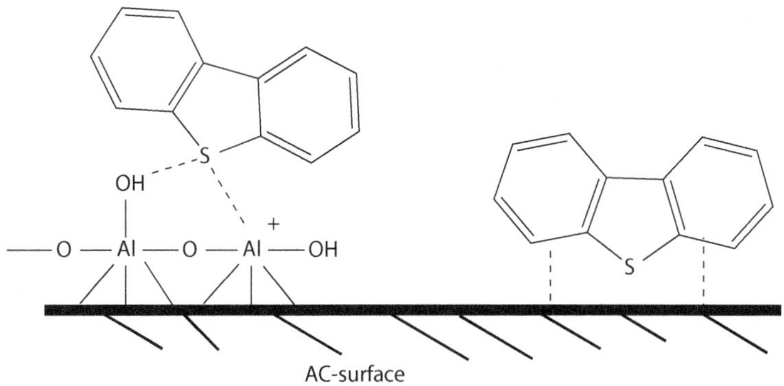

Scheme 5.2 Possible ADS Mechanism of DBT with Metal Impregnated Activated Carbon (AC).

(153 mg/L) the ADS recorded approximately 33%. The high selectivity factor towards DBT has been attributed to three main factors; first, the size of the DBT molecule which is closer to the size of the adsorbents' pores that allows them to be preferably trapped into the adsorbent. The second factor is the higher dipole moment, molar mass, and aromaticity of DBT, which leads to stronger van der Waals and π–π interactions with the adsorbent's surface. The third factor is the higher basicity of DBT relative to thiophene, giving it a stronger acid–base interaction with the Al_2O_3 (Lewis acid) on the adsorbent surface. However, the reduction in ADS efficiency in real diesel feed, relative to model one, was attributed to the severe medium of a real diesel which contains other competing aliphatic and aromatic sulfur compounds. In addition, the solubility of DBT in these components is much higher and its diffusion to the adsorbent surface is very low compared to its diffusion in the model diesel used in this study.

Lopes *et al.* (2016) studied the removal of S- and N- compounds from diesel oil throughout the adsorption onto activated carbon (AC) prepared from different sources, including babassu (*Attalea speciosa*) coconut shell and pine wood identified as CSAC and PWAC, respectively. CSAC2 and CSAC3 were selected for their large surface area and higher values for micropore area and volume, which were then pretreated by acid oxidation using HCl, HNO_3, and H_2SO_4 and alternatively by impregnation with palladium chloride. A reduction in the quantity of basic groups was verified for the treatment with all acids. Thus, increasing the adsorption capacity for sulfur compounds. This was attributed to the increase in the acid functional groups containing oxygen in the carbon structure, which provides an increased selectivity of the adsorbent for sulfur compounds, comparing the untreated carbon with oxidized carbon. Moreover, the oxidation with

HNO_3 led to an approximate 2-fold increase in the number of acid surface groups for the activated carbon sample CSAC2. Although, the oxidation with H_2SO_4 led to an almost 2-fold increase in the number of acid sites in relation to the oxidation with HNO_3. However, the samples treated with HNO_3 expressed higher ADS efficiency. After the impregnation process, there was a significant reduction in the quantity of basic surface groups of the samples in relation to the amount present on untreated samples. The reduction was 45% for carbon sample CSAC2 and 90% for carbon sample CSAC3. There was an increase of 76% in the quantity of acid sites for CSAC2 samples and 68% for CSAC3 samples, but due to the higher specific surface area, pore size, and volume of CSAC3, it showed higher susceptibility to oxidation with acids due to the increase in the available surface area. The adsorption capacity was conducted on commercial diesel oil with initial S and N contents of 305 and 133 ppmw, respectively, using 2% w/w adsorbent at 70 °C for 24 h. The adsorptive activity of the carbon samples increased proportionally to the increase in the acid groups, which was found to be Pd/AC > acid oxidized AC > AC, where the adsorption efficiency of the Pd/AC was over 85% and 60% for N- and S-compounds, respectively. Although, the number of acid surface groups observed for the samples of CAC2 and CAC3 impregnated with palladium is lower than that for the same samples oxidized with sulfuric acid, which indicated that the presence of palladium on the AC surface plays a significant role and acts together with the acid groups in the interaction with the sulfur compounds. Thus, this method can be an alternative pretreatment in the conventional hydrotreatment process. Palladium is reported to be physically adsorbed at the carbon surface and is present in the (2+) oxidation state (Pd^{2+}) and would act as an electrophile. Consequently, the palladium increases the surface acidity of the AC (Bedia *et al.*, 2010). It should be mentioned that ADS using CSAC2 was not highly affected by acid treatment, but its ADS efficiency highly increased by impregnation with pallidum. However, the CSAC3 sample showed improvement with both treatments. This was attributed to the better surface structure of CSAC3. For nitrogen removal, CASC3 expressed better ADN efficiency than that treated with H_2SO_4 or impregnated with palladium. Moreover, the treatments with HCl or HNO_3 did not improve the ADN capacity. The behaviors of CSAC2 and CSAC3 were the same for the treatment with HCl and HNO_3. This finding indicated that the S-compounds have a stronger attraction toward the acid sites than N-compounds. Another observation was that the presence of the element palladium on the carbon surface favors the adsorption of nitrogen compounds. These results show that it is possible to produce ACs with selective characteristics to enhance the adsorption of S- or N- compounds.

Thaligari *et al.* (2016) reported the ADS of model fuel (DBT/iso-ocatne) by zinc-impregnated granular activated carbon (GAC), that was prepared by wet impregnation method, was higher than that of GAC alone since the sulfur compounds coordinately interact with cations such as Zn^{2+} with the help of lone electron pairs of the sulfur atom (Zeng *et al.* 2008). Moreover, the presence of other functional groups on the surface of GAC, such as carbonyl, Si–O–Si, Si–H, and C–O–H would also help in the ADS. Consequently, the ADS of DBT would be enhanced onto Zn-GAC as compared to blank-GAC. It was elucidated that DBT uptake was found to be controlled by external mass transfer during the early stages of the adsorption process by particle diffusion within the pores of the adsorbent in the later stages.

Moradiganjeh and Aghajani (2016) reported the preparation of the magnetic adsorbent Fe_3O_4/C nanocomposite. The superparamagnetic iron oxide nanoparticles (Fe_3O_4 NPs) were prepared by co-precipitation method. The activated carbon (AC) was made throughout calcination almond shell powder at 700 °C for 60 min in a conventional furnace. Then, the surface of the Fe_3O_4 NPs was modified by the prepared AC through the co-precipitation method. The ADS- capacity of the magnetic adsorbent was tested on model oil (thiophene/n-hexane). The separation of adsorbents after each ADS-cycle was performed by a magnet. The regeneration of the adsorbent was performed by solvent extraction. The ADS was between 75 and 95% in the sulfur concentration range between 120–20 ppm within 60 min at room temperature. The amount of recyclability for methanol, toluene, and ethanol were 97.3, 91.5, and 87.6%, respectively.

Alzahrani *et al.* (2016) also reported the preparation of the magnetic adsorbent Fe_3O_4/C nanocomposite, so it would have the adsorption feature of AC and the magnetic properties of magnetic nanoparticles. The AC was first prepared from coconut shells through acid activation by phosphoric acid (70%), followed by carbonization at 400 °C for 30 min, then further activated at 800 °C for 10 min. Secondly, the acid activated AC carbon was immersed in a ferric nitrate solution in the presence of nitric acid (20%), followed by calcination of the mixture at 600 °C for 1h. The prepared Fe_3O_4/C was characterized by high porosity with a good specific surface area and a pore size of 533 and 0.54 m^2/g and cm^3/g, respectively. The ADS capability of the prepared Fe_3O_4/C was tested for different S-compounds, including sodium sulfide Na_2S, dimethyldisulfide $CH_3S_2CH_3$ (DMDS), and dimethylsulfoxide CH_3SOCH_3 (DMSO). The highest ADS rate was recorded for DMSO and was attributed to the oxygen atom, which reacted rapidly with the iron in the adsorbent followed by DMDS. The organic S-compounds adsorbed more rapidly than the inorganic sodium sulfide.

Nguyen and Lee (2015) also reported an amine-modified biochar derived from chicken manure with a high surface area of 334.6 m^2/g that can be utilized as a cheap and effective alternative adsorbent to commercial activated carbon for dimethyl sulfide (DMS) removal.

Wang *et al.* (2016) reported the preparation La(III)/AC adsorbents by ultrasonic impregnation. La(III) enhanced the ADS-capacity and selectivity, but the effect of some aromatics on the ADS capacity of La(III)/AC decreased in the following order: naphthalene > methylbenzene > benzene. The prepared La(III)/AC with a calcination temperature of 803 K and La-loaded amount of 2 wt.% possessed more effective ADS-performance. It was found that the adsorption mechanism was most likely to be π complex and the maximum adsorption capacities increased with the temperature increasing, thus, the ADS process on La(III)/AC is spontaneous and exothermic. Adsorption kinetics could be represented by a pseudo-second-order model, which suggested that chemical reaction along with electronic transfer and share seemed significant in the adsorption rate-controlling step.

Olajire *et al.* (2017) reported the green synthesis of silver nanoparticles (AgNPs) using *Cola nitida* seed shell extract and novel biomaterial (cobweb), that were named $AgNP^{skp}$ and $AgNP^{scw}$, respectively. Activated carbon was prepared from brewer's spent grains by carbonization at 300 °C followed by acid activation using H_3PO_4 and labeled as AC. Then, the AgNPs modified by AC were prepared using the wet impregnation (WI) method and labeled $AgNP^{skp}$/AC and $AgNP^{scw}$/AC. Their ADS-activity was tested on model oil (100–500 mg DBT in 50 mL n-heptane). The prepared AC was characterized by high specific surface area, pore volume, micropore volume (V_{mic}), and an average pore diameter of 411.7 m^2/g, 0.54 cm^3/g, 0.52 cm^3/g, and 13.2 nm, respectively. There was an obvious reduction in V_{mic} after impregnation of AgNPs due to their accumulation in micropores resulting in reductions in specific surface area and total pore volume. However, the average pore diameter increased to a range between 19–21 nm for $AgNP^{skp}$/AC and $AgNP^{scw}$/AC, respectively, in the mesoporous region that is recommended for DBT-ADS. As mentioned in Chapter 2, adsorbents of a pore diameter range of 0.7–1.0 nm have been reported to be suitable for accessing DBT and other cyclic organosulfur compounds (OSC) because the diameters of these compounds are comparable to the dimension of these adsorbents (Bu *et al*, 2011; Seredych *et al.*, 2012). The ADS capacity of modified AC was two-fold higher than unmodified. It is worthy to note that thiophenic compounds are known to interact with metal either via donation of a lone pair of electrons on sulfur forming a direct S-M σ-bond or via delocalization of π-electrons of aromatic ring forming π-type complexes with metals (Song, 2003; Yang *et al.*, 2006). Thus, the recorded high ADS capacity was attributed to attributed to the high activity of AgNPs throughout the release of Ag^0 (atomic) and Ag^+ (ionic) clusters on dissolution. Consequently, the strong acid–base interaction between the Ag(1) ion on the surface of the adsorbent and DBT occurs where the weak Lewis acid Ag^+ can form a complex with electron donor groups containing sulfur such as DBT (Capek, 2015). The oxidized silver metals possibly form complexes with DBT via S–Ag σ-bond and, thus, cause strong adsorption of DBT on the surface of activated carbon (Yang *et al.*, 2006), acting as an additional adsorption site (Scheme 5.3). The ADS-capacity was ranked as follows: $AgNP^{scw}$/AC > $AgNP^{skp}$/AC > ACB, recording maximums of 29.8, 25.7, and 13.9 mg DBT/g adsorbent, respectively, at room temperature, at 400 mg/L adsorbent, 500 mg/L DBT, and at a ratio of 4 DBT concentration/adsorbent dosage. The SEM analysis proved the increase in surface roughness after modification with AgNPs and the presence of pores on the adsorbent surface which would reduce the resistance to adsorbed molecules and, thus, facilitate their diffusion from model oil onto the adsorbent surface. The energy dispersive X-ray analysis (EDX) revealed that the AgNPs-modified ACs had a higher oxygen content compared to blank AC due to the availability of oxygenated functional groups on the AgNPs-modified ACs' surface. Moreover, the presence of a high load of active metals in all the adsorbents (e.g. Na, Al, Si, P, K, Ti, Cr, Fe, Co, Cu, Ag, and Au) indicated that the AC was not demineralized. Thus, all of these observations would explain the recorded high ADS efficiency of the prepared adsorbents due to the bond formation reaction between the sulfur atom and these metals and oxygen moieties. The recorded sharp decrease in specific surface area, pore volume, and micropore volume (V_{mic}), accompanied with a sharp increase in the average pore diameter of $AgNP^{scw}$/AC after ADS of DBT (from 213 to 136 m^2/g, from

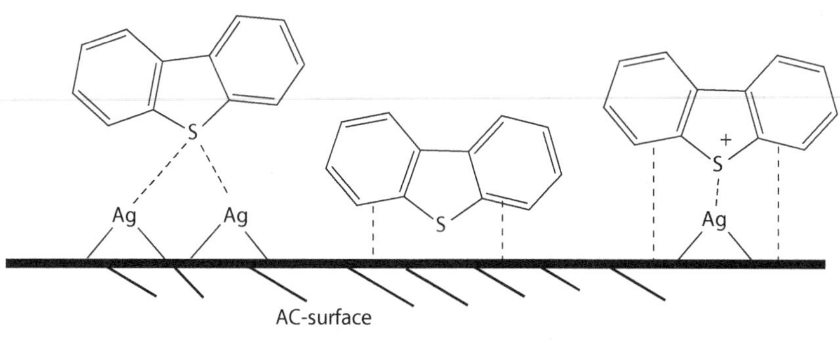

Scheme 5.3 Possible Interactions of DBT Molecules with AC and AgNPs/AC Adsorbents.

0.34 to 0.22 cm^3/g, from 0.33 to 0.21 $cm^3/9$, and from 21 to 127 nm, respectively) were attributed to the interaction that existed between DBT and AgNPs and the hydrophobic interaction of n-heptane on the adsorbent surface. The increase in the ADS with the increase in DBT concentration can be attributed to the increase in mass transfer driving force resulting from concentration gradient developed between the bulk solution and surface of the adsorbent (Anbia and Parvin, 2011), while the increase in DBT adsorption with increasing adsorbent dose can be attributed to increasing availability and accessibility of surface area and adsorption sites for the adsorption of DBT from model oil (Adeyi *et al.*, 2014). This is a chemical process involving adsorbate–adsorbent interactions, is endothermic in nature, feasible, and spontaneous, and the mechanism of adsorption was controlled by film and intra-particle diffusion. The adsorbents can be regenerated by simple thermal treatment and can be used for five successive cycles without losing its activity and for three further cycles with more than 80% of its original ADS capacity. Upon the application of $AgNP^{scw}$/AC on real oil feed, commercial gasoline that contained mercaptan, Th, BT, DBT, 4-MDBT, 4,6-DMDBT, and dibenzyldisulphide (DBDS) and spiked with DBT, the DBT concentration decreased from 115.6 mg/L to 79.2 mg/L with a removal percentage of about 31.5%. This relatively low ADS capacity, relative to that of model oil, was attributed to competitive inhibition with other S-compounds in gasoline and the increase of DBT-solubility due to the aromatic nature of the complex gasoline medium and its diffusion to the adsorbent surface which is very low compared with its diffusion in the model oil of a single component (n-heptane alone) used in this study.

5.4 ADS on Carbon Aerogels

Carbon aerogels (CAs) were studied as adsorbents for the desulfurization of liquid hydrocarbon fuels for fuel cell applications. Haji and Erkey (2003) studied the ADS of carbon aerogels (CAs) with two different average pore sizes (4 and 22 nm) on a model oil (DBT/n-hexadecane) at ambient temperature and atmospheric pressure. The CA with the larger average pore size expressed higher capacity towards DBT and, thus, a higher ADS rate. The ADS capacities recorded 11.2 and 15.1 mg S/g dry CA for the 4 nm CA and 22 nm CA, respectively. Although, the CAs were found to selectively adsorb DBT over naphthalene, which is a chemically similar but non-sulfur containing organic compound when both are present in a solution. However, the presence of naphthalene slightly reduced the amount of DBT adsorbed. Jayne *et al.* (2005) studied the ADS of carbon aerogels (CAs) in model oils (DBT or 4,6-DBT in n-hexadecane as a binary system), where they expressed a good adsorption capacity for both DBT and 4,6-DMDBT, but DBT displays higher adsorption intensity, capacity, and faster adsorption dynamics on CAs than 4,6-DMDBT does due to the steric effects caused by the two-methyl groups of 4,6-DMDBT. A pore volume diffusion model was developed to describe the dynamics of adsorption of both DBT and 4,6-DMDBT on CAs from model diesel and it was found to represent the experimental data fairly well. The model takes into account pore diffusion and assumes local equilibrium within the pores. The simulations indicated that the particle diameter and the pore size of the carbon aerogel adsorbent are critical parameters in the design of large-scale fixed-bed adsorbers for sulfur removal and the inner-particle diffusion mechanism for refractory sulfur compound adsorption using CAs' pore volume diffusion as the controlling factor.

5.5 ADS on Activated Carbon Fibers

Sano *et al.* (2005) reported an integrated process for the deep desulfurization of straight run gas oil (SRGO) (Figure 5.1), where conventionally hydrodesulfurized straight run gas oil (HDS-SRGO) with an S-content of 50 ppm was desulfurized to less than 10 ppm S over an activated carbon fiber (ACF), which is characterized by a low pressure drop and high performance among the AC materials examined. The adsorption bed used in the desulfurization of the HDS-SRGO was reused again to denitrogenate and desulfurize the refractory S-compounds in the SRGO as a pre-treatment step of the conventional HDS process to enhance the HDS reactivity over a conventional CoMo/SiO_2–Al_2O_3 catalyst at 340 °C under a pressure of 70 kg/cm^2, with a reaction time of 2 h. Such an integrated adsorption–reaction process makes it possible to utilize the maximum adsorption capacity of ACF and achieve ultra-deep desulfurization of SRGO. Regeneration of used ACF with a conventional solvent such as toluene, 1-methyl-naphthalene, or tetralin was proved very effective in restoring its adsorption capacity.

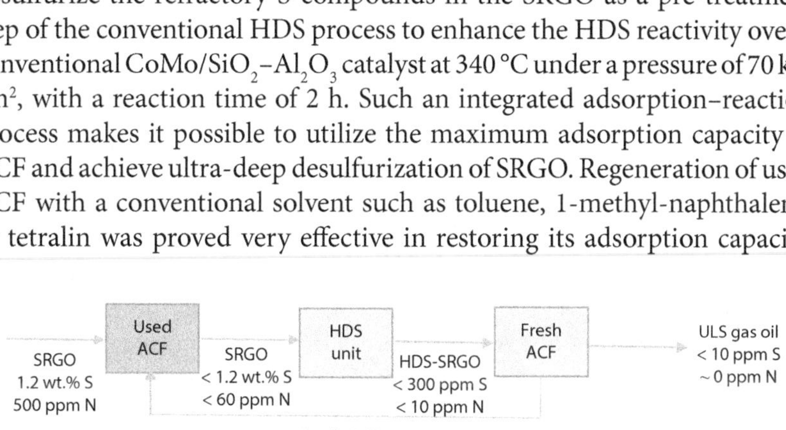

Figure 5.1 Two-step Adsorption Process Using Activated Carbon Fiber for Deep Desulfurization of Diesel Oil.

The life time of adsorbents, which is a very important factor in practical operation, is believed to be long enough to minimize the operation cost. The extracted nitrogen and sulfur species from ACF with solvent can be recovered by distillation. It should be noted that the addition of hydrocarbons having more than 10 carbons decreased the adsorption capacity of ACF. 1-methylnaphthalene of even 10% decreased the adsorption amount of sulfur to half. Thus, there is a limitation in the process to treat the highly aromatic gas oil, such as light cycle oil. This can be overcome by the use of anti-solvent, such as pressurized propane, which can relieve the inhibition by aromatics, but it would be an expensive process because of the applied high pressure for liquefying the propane. The overall process leads to the reduction of the initial investment. The used ACF, which treated HDS-SRGO to keep the elutant less than 10 ppm S, showed almost the same breakthrough profiles of nitrogen species in SRGO as the virgin one. Furthermore, the repeated use of ACF reduces not only oil loss, but also the frequency of regeneration which indicates the good economy of the present process.

A hydrogen-treated activated carbon fiber (ACFH) was selectively loaded with Ni, NiO, Cu, Cu_2O, and CuO and their ADS activity was studied for refractory thiophenic sulfur compounds, such as benzothiophene (BT), dibenzothiophene (DBT), and 4,6-dimethyldibenzothiophene (4,6-DMDBT) in single-solute systems from *n*-hexane solutions. All the metal-loaded adsorbents showed a high adsorption of tested S-compounds, where the Cu_2O- and NiO-loaded adsorbents exhibited the highest uptakes due to the more specific interactions between Cu^+ or Ni^{2+} and the studied thiophenic compounds. Further, the affinity of Cu-species towards the thiophenic compounds ranked as follows: Cu_2O (Cu^+) > CuO (Cu^{2+}) > Cu (Cu^0), while that for Ni-species; NiO (Ni^{2+}) > Ni. Although the three thiophenic compounds have

planar geometries and can orient themselves to enter the carbon slit-shaped ultra-micropores, the ADS ranked in the following decreasing order: 4,6-DMDBT > DBT > BT due to the higher intensity of specific and non-specific interactions of larger thiophenic compounds with the adsorbents. Moreover, 4,6-DMDBT has a thickness of 4.2 Å, compared to the 3.6 Å thickness of DBT and BT, that is closer to the width of carbon ultra-micropores (<7 Å). Therefore, there is an expected higher overlapping adsorption potential effect for 4,6-DMDBT than DBT and BT. Furthermore, 4,6-DMDBT and DBT also have larger molecular cross-sectional areas (than BT), that would, consequently, enhance their dispersion and π-π interactions with graphene sheets. Not only this, but the presence of methyl groups, the electron donating groups, increase the electron density of aromatic and sulfur rings in the 4,6-DMDBT molecule, thus enhancing the π-complexation interaction with Cu^+ (or Ni^{2+}). About 70% of the thiophenic compound uptake by metal loaded carbons is due to the thiophenic adsorption on the carbon surface and the remaining 30% is due to the adsorption of TC on metal sites, where the ADS of thiophenic compounds (TC) is primarily governed by dispersion interactions in carbon micropores, but specific interactions between the loaded metal species and TC molecules further increase TC uptake (Moosavi et al., 2015). Copper and nickel species interact with π-electrons of thiophenic compounds. Sulfur and metal atoms also interact by η^1-S and S-η^3 bonding (Xiao *et al.*, 2008). It is also reported that π-complexation bonds between Cu^+ (or Ni^{2+}) and a sulfur ring is stronger than that with a benzene ring. It is further hypothesized that Cu^+ or Ni^{2+} form stronger complexes than other oxidation states of copper or nickel (Hernández-Maldonado, 2004). The observed higher ADS by the ACFH-Cu^{2+} is due to the enhanced acid-base interaction of thiophenic compounds as a soft base with Cu^+ as a soft acid.

5.6 ADS on Natural Clay and Zeolites

Although the slit shape geometry of the pores of the activated carbon is suitable for the aromatic compounds, adsorption in comparison with the cylindrical zeolite pores is suitable for non-planer molecules adsorption (Rodríguez-Reinoso, 1998). However, naturally Jordanian occurring zeolites are reported to be good candidates for ADS for its low cost and good adsorptive characteristics. The fundamental building units in natural zeolites are tetrahedral (TO_4) of four oxygen ions surrounding a central ion (T) of either Si^{4+} or Al^{3+} as a primary building unit (PBU) (Mccusker *et al.*, 2001). The linking together of PBU gives a secondary building unit (SBU), which is a three-dimensional network in which each oxygen of a given tetrahedron is shared between this tetrahedron and one of four others (Barrer, 1978). The most common minerals in the Jordanian zeolite are phillipsite, faujasite, and chabazite (Yousef *et al.*, 1999). The important features of zeolites are high internal and external surface areas (Jiang and Ng, 2006), chemical and mechanical stability (Radak *et al.*, 1978; Stepan, 1999), and layered structure (Wang, 2007). The zeolitic tuff (ZT) has a good ADS capacity that can reach to 7.15 mg S/g, which can occur by two adsorption mechanisms; the interaction of thiophene with the Brønsted site of the ZT through S atoms and via the C–S bond cleavage in thiophene-derived carbocations to form unsaturated fragments on the Brønsted acid sites (Scheme 5.4), as the favorable interaction between S-compounds and

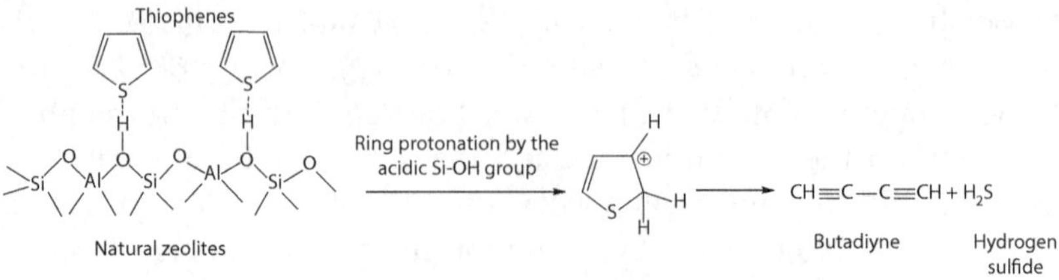

Scheme 5.4 Reactive ADS of Thiophenes on Natural Zeolites.

zeolite was mainly attributed to C–S bond cleavage in the solutes to form unsaturated fragments on the zeolite acidic sites. The spent ZT can be regenerated by simple washing with n-heptane with an efficiency of 81.5% (Mustafa *et al.*, 2010).

Shakirullah *et al.* (2012) reported ADS using natural clays or natural zeolites, including Kaolinite, Montmorollinte, Palygorskite, and Vermiculite, and compared its activity with that of charcoal and ion exchange resins, where the FT-IR analysis of the desulfurized fractions revealed that mostly high molecular weight thiols and thiophenic compounds were depleted during the adsorption process. The adsorption in clays occurs on the upper surface layer, lower basal layers, and then on the middle layers as time extends. Kaoilinte exhibited a maximum desulfurization yield of 60%, 76%, and 64% within 6 h at 40 °C in case of crude oil, kerosene, and diesel oil with initial S-contents of 1.174, 1.184, and 1.564, respectively, while the ion exchange resin recorded ADS of 47%, 49%, and 48% within 6 h, respectively. The charcoal expressed 45%, 40%, and 42% ADS in the case of crude oil, kerosene, and diesel oil, respectively within 3 h and decreased at a longer process time. This was attributed to its mineralogical composition and structure, which is mainly of simple alumino-silicate frame works. Moreover, its hydrophobic nature due to its heterogeneous surface, facilitates the interaction of nonpolar (oil) molecules with the clay particles and, hence, leads to selective adsorption of the sulfur bearing organic molecules. However, the adsorption of the sulfur containing molecules on the clay surface may be also attributed to the interaction with aluminosilcate framework, not only because of the attraction between the hydrophobic centers (Hezil *et al.*, 2008; Yin and Miller, 2012). The high activity of Kaoilinte is also attributed to its high specific surface area (30.08 m^2/g) and porous structure, which is enough to retain sulfur bearing molecules (Shakirullah *et al.*, 2012).

Clinoptilolite is an abundant, naturally occurring zeolite formed by diversification of volcanic ash in lake and marine waters millions of years

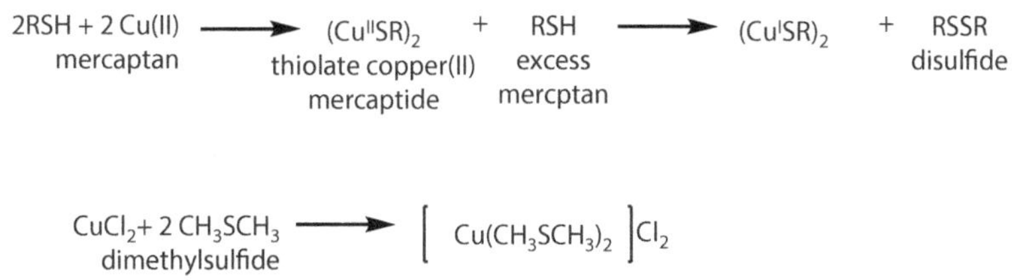

Scheme 5.5 The ADS of Mercaptans on Cu-Impregnated AC.

ago. Faghihian and Naeimi (2012) reported the application of natural zeolite clinoptilolite as a carrier for titanium dioxide for photodegradation of benzothiophene in n-hexane and as an adsorbent for the produced products to reach for ultra-desulfurization efficiency, where approximately 92% of 200 mg/L initial BT concentration was removed in the presence of TiO_2/zeolite catalyst that contained 15% TiO_2. The catalyst was easily regenerated by washing with n-hexane and calcination at 450 °C for 2 h. The catalyst can be used for 4 successive cycles, keeping 76% of its initial activity.

The structurally induced adsorptive characteristics and low cost of bentonite make it a good candidate for RADS. The acid activation of bentonite is known to increase its surface area and pore volume (Al-Degas *et al.*, 2016). Mikhail *et al.* (2002) reported the adsorption capacity on activated Egyptian bentonite to be 2.89 mg S/g. Tang *et al.* (2011) investigated the removal of mercaptans from model oil using bentonite modified with Cu^{2+}, Cu^{1+}, Fe^{3+}, and MnO_4^{1-}, where the high desulfurization capacity of bentonite modified, with Fe^{3+} and MnO_4^-, was attributed to the oxidation of mercaptans and that of Cu impregnated bentonite was attributed to the π-complexation. Bentonite adsorbent modified by $CuCl_2$ has been also investigated for the ADS of model oil (2000 ppm dimethyl sulfide (DMS) and propylmercaptan (PM) in n-hexane). The maximum ADS capacity (235 mg S/g adsorbent) was obtained at a Cu(II) loading of 15 wt.% that was calcinated at 150 °C, as a calcination temperature of 150 °C copper is present mostly in its own form of $CuCl_2$. However, as the calcination temperature increases, oxide compound (CuO) would replace the chloride compound which is unfavorable for the complex reaction and sulfur removability. The ADS was predicted to be via multilayer intermolecular forces and S-M (σ) bonds (Yi *et al.*, 2013) since Thiols are active organic sulfides and thiolate conjugate bases are easily formed (Scheme 5.5) (Doyle *et al.*, 2006). Moreover, the electronegativity of sulfur atoms is not so high and it is easy to lose the lone electron pairs. Atoms of $Cu^{II}(1s^2 2s^2 2p^6 3s^2 3p^6 3d^9 4s^0)$ can form the usual σ bonds with their empty *s*-orbitals and *p*-orbitals. Mixing one *s*-orbital and three *p*-orbitals gives four sp^3 hybrid orbitals which are directed toward a tetrahedral structure. Thus, the sulfur atoms of DMS provide lone pair electrons to Cu(II) and form the usual S-M (σ) bonds (Ryan *et al.*, 2012).

In another study by Xiaolin *et al.* (2011) of the ADS of model gasoline by bentonite adsorbents modified by silver nitrate, the modified bentonite was more effective for the ADS of alkylated dibenzothiophenes than the unmodified bentonite, where the S-capacity of the modified bentonite increased up to 7% Ag^+ ions loading, recording 2.2%. This was attributed to the strong nucleophilic nature of the silver ions which can perform complex reactions with alkyl dibenzothiophenes and express a stronger ability to adsorb more alkyl dibenzothiophenes compared to the raw bentonite, but excess silver ions would block the pores of the bentonite and the adsorption rate would decrease. Moreover, the calcination temperature of the modified bentonite was found to be very critical and at a higher temperature (>423 K), Ag_2O would replace Ag^+ and it would be unfavorable for the adsorption of alkyl dibenzothiophenes. Further, the ADS by modified Ag/bentonite prefers the low temperatures since the complex adsorption reaction is an exothermic reaction

While Khalili *et al.* (2015) reported an adsorption capacity of 4.7920 mg S/g activated bentonite at room temperature with an adsorbent particle size of a range of 300–150 mm, the higher efficiency of the acid activated bentonite towards organosulfur compounds (OSC) is attributed to the silicate–silicate bentonite structure that possesses Brönsted acid sites, which resulted from the dissociation of the water molecule in between silicate sheets. In addition, the clay surface after acid treatment would possess positive hydrogen sites. The increase of acid sites disturbs the charge equilibrium in the bentonite-clay lattice and creates strain that causes new active sites for adsorption to arise, thus, increasing the interaction with basic sulfur compounds.

Ishaq *et al.* (2017) compared the activity of untreated, acid activated, and magnetite nanoparticle loaded bentonite (UB, AAB, and MNLB, respectively) as adsorbents for a batch ADS process of model oil (DBT/n-heptane). The efficiency of ADS decreased in the following order: MNLB > AAB > UB. Acid treatment increased the porosity of the bentonite, which has a positive effect on its adsorption capacity because pores present on the adsorbent surface reduce resistance to adsorbed molecules and also facilitate their diffusion from solutions to the adsorbent surface. The surface of bentonite loaded with magnetite nanoparticles was also found to be porous and rough, which increased its adsorption capacity, unlike that of AAB and UB. Moreover, the high sulfur adsorption capacity of MNLB has been attributed to the catalytic activity of magnetite for the co-conversion or destruction of sulfur species. As an extra advantage, magnetite can be easily recovered by the magnetic separation in the application.

Zhang *et al.* (2016) reported that the sulfur compounds are adsorbed over Cu(I)- and Cu(II)-modified bentonite by a direct sulfur–adsorbent interaction, where copper-modified bentonite adsorbents significantly enhanced the desulfurization of liquid petroleum gas (LPG) in a fixed-bed flow sorption system. Optimum desulfurization with Cu(II)-modified bentonite adsorbents was obtained at a loading of 15 wt% Cu^{2+} and a calcination temperature of 150 °C.

Ahmad *et al.* (2017) reported wet impregnation of metals, Fe, Cr, Ni, Co, Mn, Pb, Zn, and Ag, on montmorollonite (MMT) clay and charcoal, then studied the ADS activity of the prepared adsorbents on kerosene and diesel oil with initial S-contents of 0.0542 and 1.041 wt.%, respectively, in a batch operation for 1 h at room temperature (25°C). The MMT has a higher ADS capacity than that of charcoal and the metal impregnated clay has enhanced ADS capacity compared to the untreated one. The untreated MMT and charcoal expressed the ADS-activity of about 16% and 21.98% for kerosene and, approximately, 43.96% and 27.80% for diesel, respectively. The ADS capacity for kerosene has been ranked in the following decreasing order: Zn-MMT (60%) > MnMMT (45.33%) > Co-MMT (40%) > Ni-MMT (41%), while the ADS capacity of diesel oil decreased as follows: Zn-MMT (62.48%) > Pb-MMT (55.7%) Ni-MMT (55.9%), while MnMMT and Co-MMT did not express a remarkable ADS for diesel oil. Furthermore, under optimum operating conditions of a 1-h stirring period at room temperature (25 °C), using an oil to Zn-MMT ratio of 20:1.5, the ADS of model oil (1000 ppm DBT in cyclohexane) reached 81%. The higher ADS efficiency of Zn-MMT was attributed to the improvement of the surface morphology of the MMT by impregnation with Zn, where the surface area, pore size, and pore volume of the MMT has been found to be increased many fold with Zn impregnation.

Choi *et al.* (2017) investigated dibenzothiophene sulfone ($DBTO_2$) adsorption utilizing clay material adsorbents such as activated clay, bentonite, and kaolinite. The adsorption capacity was ranked in the following decreasing order: activated clay (3.07 mg/g) > bentonite (2.36 mg/g) > kaolinite (1.49 mg/g). The results implied chemisorption as the rate-limiting step and a heterogeneous and endothermic adsorption onto the adsorbents. Utilizing the adsorbent of activated clay was spontaneous, while kaolinite was non-spontaneous at 298–328 K. Bentonite was found to be only non-spontaneous at 298 K. The hydroxyl and amine were the main functional groups involved in $DBTO_2$ adsorption onto the clay material adsorbents. The presence of the oxygen atom in $DBTO_2$ forms a bond with the hydrogen molecules of the hydroxyl and amine groups. Thus, activated clay displayed a good potential in adsorbing sulfones to achieve low sulfur fuel oil in an ODS process.

5.7 ADS on New Adsorbents Prepared from Different Biowastes

As has been mentioned before (Chapter 2), deep desulfurization of liquid fuel by green methods, such as molecular imprinted polymers (MIP) and surface molecular imprinted polymers (SMIPs), is expected to find wide application for selective ADS. This is because of mild conditions, simple operation, low investment, low pollution, high selectivity, no effect on octane value, and the possible reuse of the as-obtained benzothiophene-like compounds as fine chemicals. However, SMIPs is reported to be more promising and outperform MIPs and other commercial adsorbents because of its selective desulfurization efficiency, promising retention capacity for DBT and its derivatives, good mechanical strength, and high thermal stability. MIPs are easily prepared by copolymerizing a functional monomer with a cross-linker in the presence of a target solute (the template or DBT in the current case). Initially, the monomers are bonded by covalent or noncovalent forces with the template. Then, a suitable cross-linker is added with the initiator to start copolymerization. After polymerization, the template is eluted and a stable polymeric matrix containing cavities is left behind; it can selectively rebind the template in a later stage. SMIPs are simply obtained by grafting a thin imprinted polymer film with an intense number of binding sites onto a support, such as SiO_2, TiO_2, $K_2Ti_4O_9$, and carbon microsphere). The support is often added after the initial interaction between the template and the functional monomer. Recently, naturally occurring supports are reported to be useful in SMIPs (Liu *et al.*, 2014; Yang *et al.*, 2014; Al-Degs *et al.*, 2016).

Aburto and Borgne (2004a) reported the synthesis of imprinted chitosan hydrogels by bulk polymerization using glutaric dialdehyde as a cross-linker, DBT as the template, and acetonitrile/water as the solvent, where the adsorption capacity of DBT was 16.54 mg/g at 25 °C and increased to reach its maximum of 27.5 mg/g at 50 °C. The chitosan hydrogel involves a linear poly[β-(1→4)-2-amino-2-deoxy-d-glucopyranose] cross-linked through the amine groups with glutaric dialdehyde (Scheme 5.6). It was estimated that 0.96 mole of amine/mole of glucosamine are susceptible to be cross-linked with 1 mole of glutaric dialdehyde (Yang *et al.*, 2014).

In another study performed by Aburto *et al.* (2004b), hydrogels of chitosan were synthesized using glutaric dialdehyde as the crosslinking agent

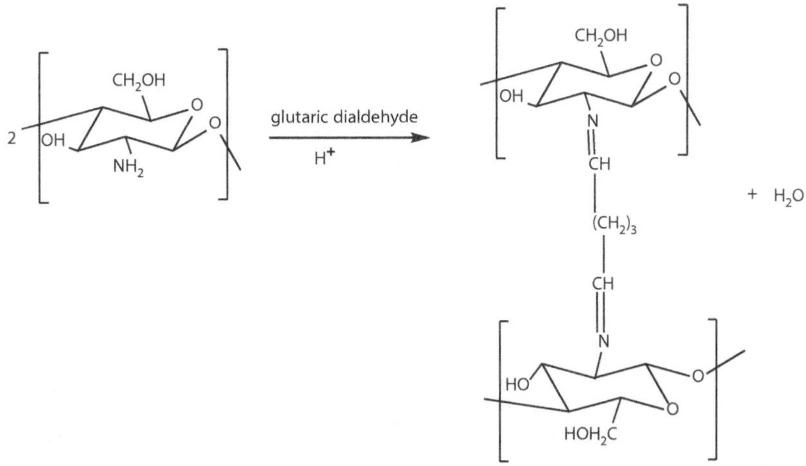

Scheme 5.6 Cross-Linking of Chitosan with Glutaric Dialdehyde.

in an acetonitrile/water solution (1:4, v/v). These hydrogels recognized and adsorbed the dibenzothiophene sulfone ($DBTO_2$) against other related compounds found in diesel, e.g. dibenzothiophene (DBT), fluorene (FLE), and 4,6-dimethyl DBT (4,6-DMDBT). In order to improve the recognition and adsorption of $DBTSO_2$, the latter compound served as a template for the building of recognition sites inside the hydrogel matrix through the use of the molecular imprinting technique (MIT). Despite this, the adsorption capacity of the molecular imprinted hydrogel (MIH) was not increase due to its heterogeneous solvent-rich macrostructure limiting mass transfer. However, the imprinting process of chitosan allowed both an enhancement of two orders of magnitude in ligand affinity constant and material homogeneity. Not only this, but the imprinting process improved ligand recognition and, consequently, hydrogel's specificity, as shown by a monolayer adsorption, while the adsorption isotherm of $DBTO_{2,}$ by the non-imprinted hydrogel, showed a multilayer adsorption due to non-specific interactions. The imprinted chitosan hydrogel is reported to express an S-adsorption capacity of 1.6 mg/g at 30 °C, while an ion-exchanged Y-zeolite reported an adsorption between 8 and 13 mg/g at 80 °C (Velu *et al.*, 2003). It was noticed that upon application of the prepared hydrogel on different S-compounds, the non-imprinted hydrogel showed a higher adsorption of every single compound compared with the imprinted hydrogel. This was explained from the basis of a higher number of unspecific sites in the non-imprinted hydrogel matrix, allowing a higher but non-selective adsorption of ligands. However, the adsorption ratio of $DBTO_2$ to other related compounds increased for the imprinted hydrogel. Moreover, the hydrogel can also be regenerated by solvent extraction with acetonitrile (ACN) and methanol (MeOH), i.e. a two-step extraction process using ACN followed by MeOH permitted DBTS elimination from the gel matrix of 66 and 90%, respectively. In contrast, the zeolitic and S-Zorb adsorbents require heat regeneration followed by reduction using an inert or hydrogen atmosphere with high-energy requirements (Hernández-Maldonado and Yang, 2003a,b).

Ying *et al.* (2006) used chitosan as the functional monomer and DBT as the template to prepare MIP by inverse suspension polymerization. The imprinted desulfurization adsorbents with homogeneous particle size were prepared using formaldehyde as the pre-cross-linker, epichlorohydrin as the cross-linker, and ethyl ether as the solvent. The adsorption capacity towards DBT was 17.53 mg/g at room temperature.

Chang *et al.* (2010) prepared chitosan MIP by the dispersion polymerization using DBT as the template, paraffin as the dispersed phase, Span 80 as the surfactant, and glutaric dialdehyde as the cross-linker. ADS of gasoline at 25 °C showed a capacity of 3.52 mg/g and remained unchanged after 10 adsorption-regeneration cycles.

The adsorption process on AC is, in most cases of physical nature, when a chemical bond is not formed. AC adsorbs thiophenic and aromatic compounds with no selectivity for thiophenic molecules throughout van der Waals and electrostatic interactions. Thus, larger molecules with higher polarizabilities are preferentially adsorbed by AC. However, adsorption on zeolite is supposed to be mainly chemisorption. Hadi *et al.* (2014) reported the preparation of Y-zeolite using Iraqi rice husk as a silica source from agricultural residues. The preparation of Cu(I)-Y-zeolite occurs firstly throughout the preparation of Na-Y–zeolite adsorbents, is then activated by impregnation with a copper using ion-exchange process, and is finally reduced by Cu^{2+} species to Cu^{+} which is desired for π -complexation. The desulfurization of a simulated diesel fuel (150 ppm benzothiophene and 148 ppm dibenzothiophene in 80:20 w:w n-dodecane/toluene) by three different adsorbents, commercial activated carbon (CAC), the prepared Cu-Y zeolite, and a layered bed of 15 wt.% activated carbon followed by Cu-Y zeolite, was studied in a fixed bed adsorption process operated at ambient temperature and pressure. The CAC breaks through almost immediately. This was attributed to the capability of CAC to adsorb both thiophenic and aromatic compounds by van der Waals and an electrostatic interaction with no selectivity for thiophenic molecules. Thus, larger molecules with higher polarizabilities are preferentially adsorbed by CAC. However, in the case of Cu-Y zeolite, the long delay of the breakthrough of these thiophenic compounds was a direct result of π -complication and followed that of molecular weights. It also indicated some degree of diffusion limitation in zeolite. The best performance was reported for the layered fixed bed reactor since the idea of using a guard bed of AC/Cu-Yzeolite is to pre-sorb not only aromatics, but also the largest sulfur-containing compounds, thus, extending the capacity of Cu Yzeolite for smaller sulfur compounds, which, consequently, delayed the breakthrough of the sulfur compounds and also sharpens their wave-fronts by eliminating pore blockage. Thus, the ADS capacity ranked in the following decreasing order: AC/Cu-Y zeolite > Cu-Y zeolite > AC when the fixed bed is packed with only Cu-Y zeolite, it produced about 20 cm^3 diesel fuel/g adsorbent with 2 ppmw S-content. But, when the fixed bed was designed as layers, the AC layer has a high selectivity towards large molecules, preventing them from entering the Cu-Yzeolite layer, protecting the pores of the zeolite from blocking. This led to the processing of more than 30 cm^3 of simulated diesel fuel per gram of Ac/Cu-Y zeolite adsorbent, with an S-content of <1 ppm.

It is worth noting that lignocellulosic wastes are preferred for chemical activation, especially with phosphoric acid H_3PO_4 (Jagtoyen and Derbyshire 1993). Ahmed and Ahmaruzzaman (2015) reported the preparation of a low-cost adsorbent obtained by H_3PO_4 impregnation of coconut coir waste for the removal of sulfur from feed diesel with an initial S-content of 2,050 mg/L. The chemically impregnated coconut coir waste (CICCW) has a high C-content (~55%) and the scanning electron microscopic analysis revealed that it had a rough, irregular, and highly heterogeneous surface, with randomly oriented pits of several sizes that promote a high surface area for adhering adsorbate molecules, favoring the adsorption of sulfur from feed diesel. The FTIR analysis revealed the presence of OH stretching of the hydroxyl group, skeletal vibration of C–C stretch, and the aromatic C–H and C=C–C stretching, would all also promote the sulfur adsorption onto a CICCW surface. It was also characterized with a high specific surface area of 1,254.67 m^2/g and as microporous, with a pore volume of 1.01 cm^3/g and average pore diameter of 3.24 nm. The adsorption process of S-compounds on the CICCW was found to be controlled by external mass transfer, followed by intra-particle diffusion and is spontaneous, physiosorpic, and of exothermic nature. The ADS capacity decreased with an increase in temperature. This was explained by the weakening of the adsorptive forces between the active sites of CICCW and sulfur. A maximum ADS of 65% occurred at optimum conditions of 5% (w/v) adsorbent concentration and 293 K within 3 h. A further increase in adsorbent concentration would lead to an overcrowding of particles that is the solid concentration effect phenomenon, which has an equilibrium of sulfur concentration at the adsorbent surface and bulk of the solution (Nowack *et al.* 1999). The regeneration of the adsorbents was performed by liquid-liquid extraction using methanol. The adsorbent was used for three cycles, retaining 98.04, 91.01, and 86.98% of its original adsorption capacity.

Yu *et al.* (2010) reported the preparation of bio-adsorbents from rice hull that was chemically activated by phosphoric acid. It showed good ADS for model oil (300 μg/g DBT in *n*-octane) and a commercial hydrotreated diesel fuel. It was proven that the high surface area and micro-pore volume of the adsorbent favored the adsorption of DBT and its derivatives. Richening of oxygen-containing compounds on the adsorbent surface was advantageous to the adsorption and removal of DBTs, at a phosphoric acid and rice hull weight ratio of 3:1, where using a two-step treatment (first heated in an oven at 170°C for 1 h and then activated at 450 °C for another 1 h) and a satisfactory adsorbent with an adsorption capacity of 28.89 mg S/g, was successfully achieved. If the silica in the adsorbent was further removed, the specific surface area of the adsorbent increased and, consequently, exhibited higher ADS performance, reaching 30.43 mg S/g for the model oil and 21.79 mg S/g for commercial diesel fuel. Both the textural structure and the surface chemical property, like acidic groups of a rice hull-based adsorbent, play important roles in its adsorption behaviors, where the formation of donor-acceptor complexes between surface acidic groups and DBT probably benefit DBT adsorption capacity. In a similar study performed by Zhang *et al.* (2010), the results showed that the DBT adsorption capacity from a model oil of an initial S-concentration of 300 μg/g increased with the increase of the specific surface area of the prepared adsorbent and the more medium, strong surface acid amount of the adsorbents.

Combined with the economic and environmental merits of the preparation procedure, activated carbons derived from hydrothermal carbonization of sucrose and subsequent KOH activation have been donated as promising candidates for potential practical applications as remarkable adsorbents for the ADS of refractory thiophenic compounds. The optimum carbon possessed a high adsorption capacity (41.5 mg S/g for 300 ppmw S model oil) and fast adsorption rate (97% saturated within 5 min), as well as relatively good selectivity for the adsorption of thiophenic compounds due to the abundance of small micropores, suitable mesopore fractions, and various oxygen functionalities present in the carbon (Shi *et al.*, 2015a).

The adsorption capacity of thiophene from model oil by pyrolysed rice husks without any pretreatment was found to be highly affected by the pyrolysis temperature, the initial S-concentration in the oil feed, the physical, textural, and chemical characteristics of the pyrolysed rice husk, the ADS-process temperature, and the oil/adsorbent ratio. Not only this, but the ADS was found to be higher under static rather than dynamic conditions, which adds to the economic-benefit of using low-cost pyrolysed rice husk adsorbent in the ADS-process (Uzunova *et al.*, 2016).

References

Abu Safieh, K.A., Al-Degs, Y.S., Sunjuk, M.S., Saleh, A.I., Al-Ghouti, M.A. (2015) Selective removal of dibenzothiophene from commercial diesel using manganese dioxide-modified activated carbon: a kinetic study. *Environmental Technology*. 36(1–4):98–105.

Aburto, J., Borgne, S.L. (2004a) Selective adsorption for dibenzothiophene sulfone by an imprinted and stimuli-responsive chitosan hydrogel. Macromolecules 37: 2938–2943.

Aburto, J., Mendez-Orozco, A., Borgne, S.L. (2004b) Hydrogels as adsorbents of organosulphur compounds currently found in diesel. *Chemical Engineering and Processing*. 43: 1587–1595.

Adeyi, A.A., Popoola, I.T., Yusuff, A.S., Olateju, A.S. (2014) Kinetics analysis and dosage effect of manganese dioxide adsorbent on desulphurization of crude oil. *Journal of Bioprocessing aAnd Chemical Engineering*. 2: 1–6.

Ahmad, W., Ahmad, I., Ishaq, M., Ihsan, K. (2017) Adsorptive desulfurization of kerosene and diesel oil by Zn impregnated montmorollonite clay. *Arabian Journal of Chemistry*. 10: S3263-S3269.

Ahmed, Md.J.K., Ahmaruzzaman, M. (2015) Adsorptive desulfurization of feed diesel using chemically impregnated coconut coir waste. *International Journal of Environmental Science and Technology*. (2015) 12:2847–2856

Al Zubaidi, I., Darwish, N.N., El Sayed, Y., Shareefdeen, Z., Sara, Z. (2015) Adsorptive desulfurization of commercial diesel oil using granular activated charcoal. *Journal of Advances in Chemical Engineering and Biological Sciences*. 2(1): 15–18.

Al Zubaidy, I.A.H., Tarsh, F.B., Darwish, N.N., Majeed, B.S.S.A., Sharafi, A.A., Chacra, L.A. (2013) Adsorption process of sulfur removal from diesel oil using sorbent materials. *Journal of Clean Energy Technologies (JOCET)* 1(1): 66–68.

Al-Degas Y.S., El-Sheikh A.H., Al-Bakain R.Z., Newman A.P., and Al-Ghouti M.A. (2016) Conventional and upcoming sulfur-cleaning technologies for petroleum fuel: A review. *Energy Technology*. 6: 679–699.

Al-Ghouti, M.A., Khalili, F.I., Al-Degs, Y.S. (2010) Minimization of organosulphur compounds by activated carbon from commercial diesels fuel: mechanistic study. *Chemical Engineering Journal* 162(10): 669–676.

Alhamed, Y.A., Bamufleh, H.S. (2009) Sulfur removal from model diesel fuel using granular activated carbon from dates' stones activated by $ZnCl_2$. *Fuel*. 88(1): 87–94.

Alzahrani, E., Abo-Dief, H.M., Mohamed, A.T. (2016) An investigation into the adsorption of sulphur compounds using iron oxide/activated carbon magnetic composite. *Advances in Environmental Biology*. 10(4): 204–219.

Anbia, M., Parvin, Z. (2011) Desulphurization of fuel by means of nanoporous carbon

adsorbent. Chemical Engineering Research and Design 89: 641–647.

Ania, C.O, Bandosz, T.J. (2009) Adsorption of dibenzothiophene from liquid phase by activated carbon. American Chemical Society Proceedings, Fuel Division. 88: 1975–1982.

Ania, C.O., Bandosz, T.J. (2006) Metal-loaded polystyrene-based activated carbons as dibenzothiophene removal media via reactive adsorption. *Carbon*. 44(12): 2404–2412.

Anisuzzaman, S.M., Abang, S., Krishnaiah, D., Razlan, M.A.R. (2017) Adsorptive desulfurization of model fuel by activated oil palm shell. *Indian Journal of Chemical Technology*. 24: 206–212.

Aparicio, F., Camú, E., Villarroel, M., Escalona, N., Baeza, P. (2013) Deep desulfurization by adsorption of 4,6-dimethyldibenzothiophene, study of adsorption on different transition metal oxides and supports. 58(4): 2057–2060.

Bamufleh, H.S. (2009) Single and binary sulfur removal components from model diesel fuel using granular activated carbon from dates' stones activated by $ZnCl_2$. *Applied Catalysis A: General*. 365: 153–158.

Barrer, R. (1978) Zeolites and Clay Minerals as Sorbents and Molecular Sieves, 1st ed., Academic press Inc., London & New York.

Bedia, J., Rosas, J.M., Rodrígues-Mirasol, J., Cordero, T. (2010) Pd supported on mesoporous activated carbons with high oxidation resistance as catalysts for toluene oxidation. *Applied Catalysis B: Environmental*. 94(1/2): 8–18.

Biniak, S., Szymanski, G., Siedlewski, J., Świątkowski, A. (1997) The characterization of activated carbons with oxygen and nitrogen surface groups. *Carbon* 35(12): 1799–1810.

Binti Amiri N. (2008) Preparation of activated carbons from waste tyres char impregnated with potassium hydroxide and carbon dioxide gasification. M.Sc. Thesis, University Sains, Malaysia.

Bu, J., Loh, G., Gwie, C.G., Dewiyanti, S., Tasrif, M., Borgna, A. (2011) Desulphurization of diesel fuels by selective adsorption on activated carbons: competitive adsorption of polycyclic aromatic sulphur heterocycles and polycyclic aromatic hydrocarbons. *Chemical Engineering Journal*. 166(1): 207–217.

Capek, I. (2015) Plasmonic nanoparticles and their conjugates: preparation, optical properties and antimicrobial activity. *Journal of Nanotechnology and Materials Science*. 2 (1): 1–18.

Carvalho, A.P., Cardoso, B., Pires, J., Brotas de Carvalho, M. (2003) Preparation of activated carbons from cork waste by chemical activation with KOH k. *Carbon*. 41: 2873–2884.

Chang, Y.H., Zhang, L., Ying, H.J., Li, Z.J., Lv, H., Ouyang, P.K. (2010) Desulfurization of Gasoline using Molecularly Imprinted Chitosan as selective adsorbents. *Applied Biochemistry and Biotechnology*. 160: 593–603.

Choi, A.E.S., Roces, S., Dugos, N., Arcega, A., Wan, M.-W. (2017) Adsorptive removal of dibenzothiophene sulfone from fuel oil using clay material adsorbents. *Journal of Cleaner Production* 161: 267–276.

Daware, G.B., Kulkarni, A.B., Rajput, A.A. (2015) Desulphurization of diesel by using low cost adsorbent. I*nternational Journal of Innovative and Emerging Research in Engineering*. 2(6): 69–73.

Epiotis, N.D. (1972) The importance of donor–acceptor interactions in photocycloadditions involving the carbonyl group. *Journal of American Chemical Society*. 94: 1946–1950.

Faghihian, H., Naeimi, Sh. (2012) Removal of benzothiophene from organic solution by a combined phtodegradtion-adsorption method. *International Journal of Chemical and Petrochemical Technology*. 2(1): 16–25.

Fallah, R.N., Azizian, S. (2012) Rapid and facile desulphurization of liquid fuel by carbon nanoparticles dispersed in aqueous phase. *Fuel*. 95: 93–96.

Fallah, R.N., Azizian, S., Reggers, G., Carleer, R., Schreurs, S., Ahenach, J., Meynen, V., Yperman, J. (2014) Effect of aromatics on the adsorption of thiophenic sulfur compounds from model diesel fuel by activated carbon cloth. *Fuel Processing Technology*. 119: 278–285.

Farag, H. (2007) Selective adsorption of refractory sulfur species on active carbons and carbon based CoMo catalyst. *Journal of Colloid and Interface Science*. 307: 1–8.

Fen, L., Bo, Y., Wei, J., Zhenlong, G. (2012) Preparation of activated carbon loading nano-ZnO and desulfurization properties under room temperature. Software Engineering and Knowledge Engineering: Theory and Practice. Part of Advances in Intelligent and Soft Computing book series (AINSC, volume 162): Pp. 177–183.

Figueiredo, J.L., Pereira, M.F.R., Freitas, M.M.A., Orfao, J.J.M. (1999) Modification of the surface chemistry of activated carbons. *Carbon*. 37: 1379–1389.

Ganiyu, S.A., Ajumobi, O.O., Lateef, S.A., Sulaiman, K.O., Bakare, I.A., Qamaruddin, M., Alhooshani, K. (2017) Boron-doped activated carbon as efficient and selective adsorbent for ultra-deep desulfurization of 4,6-dimethyldibenzothiophene. *Chemical Engineering Journal*. 321: 651–661.

Ganiyu, S.A., Alhooshani, K., Sulaiman, K.O., Qamaruddin, M., I Bakare, I.A., Tanimu, A., Saleh, T.A. (2016) Influence of aluminium impregnation on activated carbon for enhanced desulfurization of DBT at ambient temperature: role of surface acidity and textural properties, *Chemical Engineering Journal*. 303: 489–500.

Gawande, P.R., Kaware, J.P. (2016) Removal of sulphur from liquid fuels using low cost activated carbon -A review. *International Journal of Scientific Engineering and Applied Science*. 2(10): 64–69.

Gawande, P.R., Kaware, J.P. (2017) Desulphurization of diesel by coconut shell activated carbon. *International Journal of Informative and Futuristic Research*. 4(7) 6892–6897.

Gokce, Y., Aktas Z. (2014) Nitric acid modification of activated carbon produced from waste tea and adsorption of methylene blue and phenol. *Applied Surface Science*. 313: 352–359.

Hadi, N.M., Rashid, S.A., Abdalreda, S. (2014) Deep desulfurization of diesel fuel by guard bed adsorption of activated carbon and locally prepared Cu-Y zeolite. *Journal of Engineering*. 20(5): 146–159.

Hagiwara, T., Echizen, H. (2001) Gasoline production technology and methods, and an evaluation of their economic viability, survey. 4–2-1, Part 1- High-Quality (Low Sulfur Content) Gasoline. Petroleum Energy Center. Pp. 1–11.

Haji, S., Erkey, C. (2003) Removal of dibenzothiophene from model diesel by adsorption on carbon aerogels for fuel cell applications. *Industrial and Engineering Chemistry Research*. 42(26): 6933–6937.

Haw, K.-G., Abu Bakar, W.A.W., Ali, R., Chong, J.-F., Abdul Kadir, A.A. (2010) Catalytic oxidative desulfurization of diesel utilizing hydrogen peroxide and functionalized-activated carbon in a biphasic diesel-acetonitrile system. *Fuel Processing Technology*. 91(9): 1105–1112.

Hernández-Maldonado, A.J., Yang, R.T. (2003a) Desulfurization of commercial liquid fuels by selective adsorption via π-complexation with Cu(I)-Y Zeolite. *Industrial and Engineering Chemistry Research*. 42(13): 3103–3110.

Hernández-Maldonado, A.J., Yang, R.T. (2003b) Desulfurization of liquid fuels by adsorption via π-complexation with Cu(I)-Y and Ag-Y zeolites. *Industrial and Engineering Chemistry Research*. 42(1): 123- 129.

Hernández-Maldonado, A.J., Yang, R.T. (2004) Desulfurization of transportation fuels by adsorption. *Catalysis Reviews - Science and Engineering*. 46(2): 111–150.

Hezil, N., Guerfi, K., Hazourli, S., Hammadi, A., Zeroual, S. (2008) Study of the surface heterogeneity of hydrothermal kaolinite for low cost solar cells. *International Scientific Journal for Alternative Energy and Ecology* 6: 169–172.

Hosseini, H. (2012) Novel methods for desulfurization of fuel oils. *International Journal of Chemical, Molecular, Nuclear, Materials and Metallurgical Engineering*. 6(11): 1072–1074.

Hosseini. H., Hamidi, A. (2014) Sulfur removal of crude oil by ultrasound- assisted oxidative method. In proceedings: International Conference on Biological, Civil and Environmental Engineering (BCEE-2014). Dubai, United Arab Emirates (UAE), March 17–18. http://iicbe.org/siteadmin/upload/9150C0314090.pdf; accessed October 12, 2014.

Ibrahim, N.K., Aljanabi, S.K. (2015) Desulfurization and kinetic study of diesel fuel by batch adsorption on activated carbon. *Journal of Engineering and Technology*. 33(8): 1901 -1916.

Ishaq, M., Sultan, S., Ahmad, I., Ullah, H., Yaseen, M., Amir, A. (2017) Adsorptive desulfurization of model oil using untreated, acid activated and magnetite nanoparticle loaded bentonite as adsorbent. *Journal of Saudi Chemical Society*. 21(2): 143–151.

Jagtoyen, M., Derbyshire, F. (1993) Some considerations of the origins of porosity in carbons from chemically activated wood. *Carbon*. 32: 1185–1192.

Javadli R., de Klerk A. (2012) Desulfurization of heavy oil. Applied Petrochemical Research. 1: 3–19.

Jayne, D., Zhang, Y., Haji, S., Erky, C. (2005) Dynamics of removal of organosulfur compounds from diesel by adsorption on carbon aerogels for fuel cell applications. *International Journal of Hydrogen and Energy*. 30: 1287–1293.

Jiang, M., Ng, F. (2006) Adsorption of benzothiophene on Y zeolites investigated by infrared spectroscopy and flow calorimetry. *Catalysis Today*. 116(4): 530–536.

Khalili, F.I., Sultan, M., Robl, C., Al-Ghouti, M. (2015) Insights into the remediation characterization of modified bentonite in minimizing organosulphur compounds from diesel fuel. *Journal of Industrial and Engineering Chemistry*. 28: 282–293.

Kim, J.H., Ma, X., Zhou, A, Song, C. (2006) Ultra-deep desulfurization and denitrogenation of diesel fuel by selective adsorption over three different adsorbents: A study on adsorptive selectivity and mechanism. *Catalysis Today*. 111(1): 74–83.

Kim, S.H., Shon, H.K., Ngo, H.H. (2010) Adsorption characteristics of antibiotics trimethoprim on powdered and granular activated carbon. *Journal of Industrial and Engineering Chemistry*. 16: 344–349.

Kumagai, S. (2011) Adsorptive removal of residual sulfur compounds in commercial fuel oil by means of biomass-derived activated carbons. In: activated carbon: classifications, properties and applications. Kwiatkowski, J.F., Nova Science Publishers, Inc., Hauppauge, NY, USA. Pp. 421–432.

Kumagai, S., Shimizu, Y., Toida, Y., Enda, Y. (2009) Removal of dibenzothiophenes in kerosene by adsorption on rice husk activated carbon. Fuel. 88(10): 1975–1982.

Kumar, D.R., Srivastva, V.C. (2012) Studies on adsorptive desulfurization by activated carbon. Clean – Soil, Air, Water. 40(5): 545–550.

Kwaśny, J., Balcerzak, W. (2016) Sorbents used for biogas desulfurization in the adsorption process. *Polish Journal of Environmental Studies*. 25(1): 37–43.

Lee, Y.W., Park, J.W., Choung, J.H., Choi, D.K. (2002) Adsorption characteristics of SO_2 on activated carbon prepared from coconut shell with potassium hydroxide activation. *Environmental Science and Technology*. 36(5): 1086–1092.

Liu, W., Liu, X., Yang, Y., Zhang, Y., Xu, B. (2014) Selective removal of benzothiophene and dibenzothiophene from gasoline using double-template molecularly imprinted polymers on the surface of carbon microspheres. *Fuel*. 117: 184–190.

Lopes, A.R., Scheer, A.P. Silva, G.V., Yamamoto, C.I. (2016) Pd-Impregnated activated carbon and treatment acid to remove sulfur and nitrogen from diesel. Revista Matéria 21(2): 407- 415.

Mccusker, L., Liebau, F., Engelhardt, G. (2001) Nomenclature of structural and compositional characteristics of ordered microporous and mesoporous materials with inorganic hosts. *Pure and Applied Chemistry*. 73(2): 381–394.

Mei, H., Mei, B.W., Yen, T.F. (2003) A new method for obtaining ultra-low sulfur diesel fuel via ultrasound assisted oxidative desulfurization. *Fuel*. 82(4): 405–414.

Meille, V., Schulz, E., Lemaire, M., Vrinat, M. (1999) Hydrodesulfurization of 4-methyl-dibenzothiophene: A detailed mechanistic study. *Applied Catalysis A: General* 187: 179–186.

Mikhail S., Zaki T., & Khalil L. (2002) Desulfurization by an economically adsorption technique. *Applied Catalysis A: General*, 227:265–278.

Moosavi, E.S., Dastgheib, S.A., Karimzadeh, R. (2012) Adsorption of thiophenic compounds from model diesel fuel using copper and nickel impregnated activated carbons. Energies. 5: 4233–4250.

Moradiganjeh, J., Aghajani, Z. (2016) Synthesis and characterization of novel magnetic Fe_3O_4/C through the co-precipitation method and investigation of its desulfurization application. *Journal of Materials Science: Materials in Electronics*. 27:5948–5953.

Moreira, A.M., Brandão, H.L., Hackbarth, F.V., Maass, D., Ulson de Souza, A.A., Guelli U. de Souza, S.M.A. (2017) Adsorptive desulfurization of heavy naphthenic oil: Equilibrium and kinetic studies. Chemical Engineering Science. 172: 23–31.

Mustafa, F., Al-Ghouti, M.A., Khalili, F.I., Al-Degs, Y.S. (2010) Characteristics of organosulphur compounds adsorption onto Jordanian zeolitic tuff from diesel fuel. *Journal of Hazardous Materials*. 182: 97–107.

Mužic, M. Sertić-Bionda, K. (2013) Alternative processes for removing organic sulfur compounds from petroleum fractions. *Chemical and Biochemical Engineering Quarterly*. 27(1): 101–108.

Mužic, M. Sertić-Bionda, K., Gomzi, Z. (2010) A design of experiments investigation of adsorptive desulfurization of diesel fuel. *Chemical and Biochemical Engineering Quarterly*. 24(3): 253–264.

Mužic, M., Sterić-Bionda, K., Adzamic, T. (2011) Desulfurization of diesel fuel in a fixed bed adsorption column: Experimental study and simulation. *Petroleum Science and Technology*. 29(22): 2361–2371.

Mužic, M.M., Sertić-Bionda, K., Adzamic, T. (2009) Kinetic equilibrium and statistical analysis of diesel fuel adsorptive desulphurization. Goriva i Maziva (Fuels and Lubricants). 48(3): 373–394.

Nanoti, A., Dasgupta, S., Agnihotri, V., Gupta, P., Goswami, A.N., Garg, M.O., Tangstad, E., Stöcker, M., Karlsson, A., Vistad, Ø.B. (2011) A zeolite based vapor phase adsorptive desulfurization process for naphtha. Microporous and Mesoporous Materials. 146: 158–165.

Nazal, M.K., Khaled, M., Atieh, M.A., Aljundi, I.H., Oweimreen, G.A., Abulkibash, A.M. (2015) The nature and kinetics of the adsorption of dibenzothiophene in model diesel fuel on carbonaceous materials loaded with aluminum oxide particles. *Arabian Journal of Chemistry*. (Accepted in press).

Nguyen, M.-V., Lee, B.-K. (2015) Removal of dimethyl sulfide from aqueous solution using cost-effective modified chicken manure biochar produced from slow pyrolysis. Sustainability. 7: 15057–15072

Nkosi, M. (2014) Desulphurisation of petroleum distillates using adsorption method. A Research Report submitted to the Faculty of Engineering and the Built Environment, University of the Witwatersrand, in partial fulfilment of the requirements for the degree of Master of Science in Engineering, Johannesburg, South Africa.

Nowack K.O., Cannon F.S. and Arora H. (1999) Ferric chloride plus GAC for removing TOC. *Journal American Water Works Association* 91:65–78.

Nunthaprechachan, T., Pengpanich, S., Hunsom, M. (2013) Adsorptive desulfurization of dibenzothiophene by sewage sludge-derived activated carbon. *Chemical Engineering Journal.* 228: 263–271.

Olajire, A.A., Abidemi, J.J., Lateef, A., Benson, N.U. (2017) Adsorptive desulphurization of model oil by Ag nanoparticles-modified activated carbon prepared from brewer's spent grains. *Journal of Environmental Chemical Engineering.* 5: 147–159.

Patil, M.S., Bhattacharyulu, Y.C., Kulkarni, S.R. (2011) Desulfurization of hydrocarbon liquid fuels by adsorption. *Journal of Engineering Research and Studies.* 2(1): 81–98.

Qiu, J., Wang, G., Bao, Y., Zeng, D., Chen, Y. (2015) Effect of oxidative modification of coal tar pitch-based mesoporous activated carbon on the adsorption o benzothiophene and dibenzothiophene. *Fuel Processing Technology.* 129: 85–90.

Radak, V., Gal, I., Radulovic-Hercigonja, R., Seidel, D. (1978) Some physical and chemical properties of Fe(II)X, Fe(II)Y and Fe(III)Y zeolites. J. Inorg. Nucl. Chem. 40(1): 75–77.

Rodríguez-Reinoso, F. (1998) The role of carbon materials in heterogeneous catalysis. *Carbon.* 36: 159–175.

Ryan, G. H., Xie, X. J., Sofia, R. P., Isabel, M., Edward, I.S. (2012) Analysis of resonance Raman data on the blue copper site in pseudoazurin: Excited state π and σ charge transfer distortions and their relation to ground state reorganization energy. *Journal of Inorganic Biochemistry.* 115: 155–162.

Saed, D., Nassar, H.N., El-Gendy, N.Sh., Zaki, T., Moustafa, Y.M., Badr, I.H.A. (2014) The enhancement of pyrene biodegradation by assembling MFe_3O_4 nano-sorbents on the surface of microbial cells. Energy Sources, Part A: Recovery, Utilization, and Environmental Effects. 36(17): 1931–1937.

Salem, A.B.S.H, Hamid, H.S. (1997) Removal of sulfur compounds from naphtha solutions using solid adsorbents. *Chemical Engineering and Technology.* 20(5): 342–347.

Sano, Y., Sahara, K., Choi, K.-H., Korai, Y., Mochida, I. (2005) Two-step adsorption process for deep desulfurization of diesel oil. *Fuel.* 84: 903–910.

Seredych, M., Bandosz, T.J. (2007) Template-derived mesoporous carbons with highly dispersed transition metals as media for reactive adsorption of dibenzothiophene. *Langmuir.* 23(11): 6033–6041.

Seredych, M., Lison, J., Jans, U., Bandosz, T.J. (2009) Textural and chemical factors affecting adsorption capacity of activated carbon in highly efficient desulfurization of diesel fuel. *Carbon.* 47(10): 2491–2500.

Seredych, M., Wu, C.T., Brender, P., Ania, C.O., Vix-Guterl C., Bandosz, T.J. (2012) Role of phosphorus in carbon matrix in desulphurization of diesel fuel using adsorption process. *Fuel.* 92: 318–326.

Shakirullah, M., Ahmad, W., Ishaq, M., Khan, M.I. (2012) Desulphurization of liquid fuels by selective adsorption through mineral clays as adsorbents. *Journal of the Chilean Chemical Society.* 57(4): 1375–1380.

Shi, Y., Liu, G., Wang, L., Zhang, X. (2015b) Efficient adsorptive removal of dibenzothiophene from model fuel over heteroatom-doped porous carbons by carbonization of an organic salt. *Chemical Engineering Journal.* 259: 771–778.

Shi, Y., Zhang, X., Liu, G. (2015a) Activated carbons derived from hydrothermally carbonized sucrose: remarkable adsorbents for adsorptive desulfurization. ACS Sustainable Chemistry and Engineering. 3(9): 2237–2246.

Shimizu, Y., Kumagai, S., Takeda, K., Enda, Y. (2007) This research was in part supported by Industrial Technology Research Grant Program in 2006 from New Energy and Industrial Technology Development Organization (NEDO) of Japan". (acs.omnibooksonline.com/data/papers/2007_P035.pdf)

Shimizu, Y., Kumagai, S., Takeda, K., Toida, Y., Enda, Y. (2007) Adsorptive removal of sulfur compounds in kerosene by using rice husk activated carbon. Carbon – American Carbon Society - Print Edition. 3: 1629–1634.

Singha, S., Sarkar, U., Luharuka, P. (2013) Functionalized granular activated carbon and surface complexation with chromates and bi-chromates in waste water. Science of The Total Environment. 447: 472–487.

Song, C. (2003) An overview of new approaches to deep desulphurization for ultraclean gasoline, diesel fuel and jet fuel. *Catalysis Today*. 11: 211–263.

Song, H., Wan, X., Dai, M., Zhang, J., Li, F., Song, H. (2013) Deep desulfurization of model gasoline by selective adsorption over Cu–Ce bimetal ion-exchanged Y zeolite. Fuel Processing Technology. 116: 52–62.

Stepan, R. Selective adsorption of sulfur compounds from hydrocarbon mixtures, Unpublished Doctoral Dissertation, University of Western Ontario, London, Ontario, 1999.

Tang X.L., Meng X., Shi, L. (2011) Desulfurization of model gasoline on modified bentonite. *Industrial and Engineering Chemistry Research*. 50:7527–7533.

Thaligari, S.K., Srivastava, V.C., Prasad, B. (2016) Adsorptive desulfurization by zinc-impregnated activated carbon: characterization, kinetics, isotherms, and thermodynamic modeling. Clean Technologies and Environmental Policy. 18: 1021–1030.

Toida Y. (2003) adsorption desulfurization agent for desulfurizing petroleum fraction and desulfurization method using the same US 20050173297a1. International Patent WO 03 - 097771, Japan Energy Corporation.

Uzunova, S., Minchev, L., Uzunov, I., Toteva, V. (2016) Efficient adsorption of thiophene from model fuel by pyrolysed rice husks: factors of influence. Chemistry and Ecology. 32(10): 976–987

Velu, S., Ma, X., Song, C. (2003) Selective adsorption for removing sulfur from jet fuel over zeolite-based adsorbents. *Industrial and Engineering Chemistry Research*. 42(21): 5293–5304.

Wang, B. (2007) Zeolite deactivation during hydrocarbon reactions: characterization of coke precursors and acidity, product distribution, A Thesis submitted for the degree of Doctor of Philosophy of the University College London, United Kingdom.

Wang, J., Liu, H., Yang, H., Qiao, C.,Li, Q. (2016) Competition adsorption, equilibrium, kinetic, and thermodynamic studied over La(III)-loaded active carbons for dibenzothiophene removal. *Journal of Chemical and Engineering Data.* 61(10): 3533–3541.

Xiao, J., Li, Z., Liu, B., Xia, Q., Yu, M. (2008) Adsorption of benzothiophene and dibenzothiophene on ion-impregnated activated carbons and ion-exchanged Y zeolites. *Energy and Fuels.* 22: 3858–3863.

Xiaolin, T., Zheting, L., Li, S. (2011) Deep desulfurization via adsorption by silver modified bentonite. China Petroleum Processing and Petrochemical Technology. 13(3): 16–20.

Yang, X., Erickson, L.E., Hohn, K.L., Jeevanandam, P., Klabunde, K.L. (2006) Sol-gel $CuAl_2O_3$ adsorbents for selective adsorption of thiophene out of hydrocarbon. *Industrial and Engineering Chemistry Research.* 45: 6169–6174.

Yang, Y-Z., Liu, X-G., Xu B-S. (2014) Recent advances in molecular imprinting technology for the deep desulfurization of fuel oils. New Carbon Materials. 29(1): 1–13.

Yi, D., Huang, H., Shi, L. (2013) Desulfurization of model oil via adsorption by copper(II) modified bentonite. Korean Chemical Society. 34(3): 776–782.

Yin, X., Miller, J.D. (2012) Wettability of kaolinite basal planes based on surface force measurements using atomic force microscopy. *Minerals and Metallurgical Processing* 29(1): 13–19.

Ying H.-j., Chang Y-h., Hao, L.V. (2006) Solid-phase extraction sorbent for organosulphur com pounds present in fuels by molecular imprinting. *Journal of Nanjing University of Technology.* 28(1): 11–14.

Younis, M.K., Simo, S.M. (2015) Desulphurization of Tawke diesel fuel by adsorption on Na-Y type zeolite, local clay and active carbon. *International Research Journal of Pure and Applied Chemistry.* 1–7.

Yousef, R., Tutunji, M., Derwish, G., Musleh, S. (1999) Chemical and structural properties of Jordanian zeolitic tuffs and their admixtures with urea and thiourea: potential scavengers for phenolics in aqueous medium. *Journal of Colloid and Interface Science.* 216(2): 348–359.

Yu, C., Fan, X., Yu, L., Bandosz, T.J., Zhao, Z., Qiu, J. (2013) Adsorptive removal of thiophenic compounds from oils by activated carbon modified with concentrated nitric acid. *Energy and Fuels.* 27: 1499–1505.

Yu, C., Qiu, J.S., Sun, Y.F., Li, X.H., Chen, G., Bin, Z. (2008) Adsorption removal of thiophene and dibenzothiophene from oils with activated carbon as adsorbent: effect of surface chemistry. *Journal of Porous Materials.* 15: 151–157.

Yu, G.X., Lu, S.X., Chen, H., Zhu, Z.N. (2005) Diesel fuel desulfurization with hydrogen peroxide promoted by formic acid and catalyzed by AC. *Carbon.* 43: 2285–2294.

Yu, G.X., Sun, J., Hou, X.M., Zhou, X.L., Li, C.L., Chen, L.F., Wang, J.A. (2010) Adsorptive removal of dibenzothiophene in diesel fuel on an adsorbent from rice hull activated by phosphoric acid. *Advanced Materials Research.* 132: 133–140.

Zakaria, B.S., Nassar, H.N., Saed, D., El-Gendy, N.Sh. (2015) Enhancement of carbazole denitrogenation rate using magnetically decorated *Bacillus clausii* BS1. *Petroleum Science and Technology.* 33(7): 802–811.

Zeng Y-P, Ju S, Xing WH, Chen CL (2008) Adsorption of mercaptans from model gasoline on 13X loaded with Zn^{+2}. *The Canadian Journal of Chemical Engineering*. 86:186–191.

Zhang, Q., Shia, L., Meng, X. (2016) Deep adsorption desulfurization of liquid petroleum gas by copper-modified bentonite. RSC Advances. 6(12): 9589–9597.

Zhang, S., Zhang, Q., Zhang, Z.C. (2004) Extractive desulfurization and denitrogenation of fuels using ionic liquids. *Industrial and Engineering Chemistry Research*. 43: 614–622.

Zhang, W., Liu, H., Xia, Q., Li, Z. (2012) Enhancement of dibenzothiophene adsorption on activated carbons by surface modification using low temperature oxygen plasma. *Chemical Engineering Journal*. 209: 597–600.

Zhang, W.-L., Wan, K., Yu, G., Zhou X.-L., Li, C.-L. (2010) Preparation of adsorbent for desulfurization of diesel fuel through activating rice hull by phosphoric acid. *Acta Petrolei Sinica (Petroleum Processing Section)*. 26(4): 588–593.

Zhou, A., Ma, X., Song C. (2009) Effects of oxidative modification of carbon surface on the adsorption of sulfur compounds in diesel fuel. *Applied Catalysis B. Environmental*. 87(3/4): 190–199.

Zhou, A.N., Ma, X.L., Song, C.S. (2006) Liquid-phase adsorption of multi-ring thiophenic sulfur compounds on carbon materials with different surface properties. *The Journal of Physical Chemistry B*. 110(10): 4699–4707.

6

Microbial Attack of Organosulfur Compounds

List of Abbreviations and Nomenclature

1-MN	1-Methylnaphthalene
2CAT	2-Carboxythiophene
2FOT	2-Formylthiophene
2-HMBT	2-Hydroxymethyl benzothiophene
2-HMT	2-Hydroxymethyl thiophene
2-MT	2-Methylthiophene
3-MBT	3-Methylbenzothiophene
3-MT	3-Methylthiophene
4,6-DEDBT	4,6-Diethyldibenzothiophene
4-HMDBT	4-Hydroxymethyl dibenzothiophene
AaP	*A. aegerita* peroxygenase
ADS	Adsorptive Desulfurization
AED	Atomic Emission Detector
API	American Petroleum Institute
BDS	Biodesulfurization
BNT	Benzo-Naphtho-Thiophene

BSM	Benzyl methyl sulfide
BTO	Benzothiophene Sulfoxide
BTO_2	Benzothiophene Sulfone
BTs	Benzothiophenes
CrP	*C. radians* peroxygenase
CDW	Cell Dry Weight
CEES	2-Chloroethyl Ethyl Sulfide
DBDS	Dibenzyl disulfide
DBS	Dibenzyl sulfide
DBTO	Dibenzothiophene sulfoxide
$DBTO_2$	Dibenzothiophene sulfone
DBTs	Dibenzothiophenes
DCW	Dry Cell Weight
DMF	Dimethylformamide
DMS	Dimethyl sulfide
DMSO	Dimethyl sulfoxide
DPDS	Diphenyl disulfide
DTHT	2-n-dodecyltetrahydrothiophene
EBC	Energy Biosystems Corporation
EDS	Extractive Desulfurization
FTIR	Fourier-transform infrared spectroscopy
GC/AED	Gas Chromatography/Atomic Emission Detector
GC/FID	Gas Chromatography/Flame Ionization Detector
GC/MS	Gas Chromatography/Mass Spectrometry
GC/SCD	Gas Chromatography/Sulfur Chemiluminescence Detector
HBPSi	2′-hydroxybiphenyl 2-sulfinic acid
HBPSO	2′-hydroxybiphenyl-2-sulfonic acid
HDEBP	2-Hydroxy-3,3’-Diethylbiphenyl
HDS	Hydrodesulfurization
HFBT	3-Hydroxy-2-Formylbenzothiophene
HNESi	2-(2′-hydroxynaphthyl)ethen-1-sulfinate
HPEal	2-(2′-hydroxyphenyl)ethan-1-al
HPESi-	2-(2′-Hydroxyphenyl)ethen-1-Sulfinate
HTOP	4-[2-(3-hydroxy)thio-naphtenyl]-2-oxo-3-butenoic acid
LiP	Lignin Peroxidase
LLE	Liquid-Liquid Extraction
MnP	Manganese Peroxidase
MSM	Minimal Salts Medium
NTH	Naphtho[2,1-b]thiophene

O/W	Oil/Water
ODS	Oxidative Desulfurization
OSCs	Organosulfur Compounds
PASHs	Polyaromatic Sulfur Heterocyclic Compounds
PCR	Polymerase Chain Reaction
PFPS	bis-(3-penta furophenylpropyl)-sulfide
RE	Restriction Enzyme
T2C	Thiophene-2-Carboxylic Acid
TBL	Γ-Thiobutyrolactone
Ths	Thiophenes
THTA	2-tetrahydrothiophene-acetic acid
THTC	2-tetrahydrothiophene-carboxylic acid
TPO	2(5H)-Thiophenone
ULS	Ultra-Low Sulfur
ULSD	Ultra-Low Sulfur Diesel

6.1 Introduction

Desulfurization by microorganisms is potentially advantageous. Firstly, it is carried out in mild temperature and pressure conditions, therefore, it is considered an energy saving process (an advantage over HDS). Secondly, in biological activities, biocatalysts (enzymes) are involved, therefore, the desulfurization would be highly selective (another advantage). In this review, three different types of biodesulfurization (BDS) processes are described: destructive, anaerobic, and specific. Among these processes, specific biodesulfurization has gained more attention, therefore, a detailed review on this desulfurization type will be provided.

The American Petroleum Institute (API) identified 13 classes of sulfur compounds, including 176 individual structures in four crude oils. The distribution and characteristics of these different classes reflect the source, maturity, and alteration processes in the development of a particular crude (Coleman *et al.* 1971). However, the organosulfur compounds (OSCs) found in crude oil can be generally classified into two types: non-heterocyclics and heterocyclics (Chapter 1). The former are comprised of thiols, sulfides, and disulfides. Cyclic or condensed multicyclic organosulfur compounds are referred to as sulfur heterocyclics. Light fractions boiling below 200 °C contain mainly sulfides and thiols. These can easily be removed by chemical methods, for example HDS. Thiophenic compounds remaining in the heavier fractions are resistant to chemical processes. Among these, aromatic compounds, such as dibenzothiophene

(DBT) or its derivatives, are of significant importance because they have higher boiling points (BP > 200 °C) and it is difficult to remove them from atmospheric tower outlet streams (e.g. middle distillates) (Kawatra *et al.*, 2001). Derivatives of benzothiophene (BT) and dibenzothiophene (DBT) are the major sulfur compounds in certain types of crude oil. Thiophenic compounds are compromised from 50% to 95% of the sulfur compounds in crude oils and its fractions (Mohebali and Ball, 2016). Moreover, one, two, and three sulfur atom containing compounds reach up to 74, 11, and 11% in heavy crude oils, respectively (Adlakha *et al.*, 2016). Aromatic organosulfur compounds (OSCs) are reported to constitute up to 62% of the total sulfur content (Brons and Yu, 1995). Mercaptans content in crude varies from 0.1 to 15% mass from the total content of sulfur compounds (Ryabov, 2009). Mercaptans are a small portion of the gasoline sulfur compounds, whereas, thiophenes (e.g. thiophene and dimethylthiophene) are reported to be the major sulfur compounds found in gasoline produced by cracking heavy oil (Kim *et al.*, 1995; Shahaby and Essam El-din, 2017). Dimethylthiophene and benzothiophes are the main thiophenic compounds in kerosene (Kareem *et al.*, 2016). Benzothiophene (BT), non-β, single β, and di-β-substituted benzothiophenes (BP > 219 °C) are the typical thiophenic compounds that are found in diesel. Among organic sulfur compounds, some are considered recalcitrant. These compounds are chiefly stable aromatic sulfur-containing compounds which need a more invasive desulfurization procedure to remove their sulfur atom. The refractory portion of distillate/diesel fuels is attributed to thiophenic compounds such as DBT derivatives with 4 and/or 6 alkyl substituting groups (Soleimani *et al.*, 2007). It was reported that >60% of the sulfur in higher boiling fractions of various crudes is present as substituted bezothiophenes. In some Middle East oils, the alkyl-substituted benzothiophenes and DBTs contribute up to 40% of the organic sulfur present (El-Gendy and Speight, 2016). Thus, benzothiophene and DBT and their alkylated derivatives are among the many condensed thiophenes that are an important form of organic sulfur in the heavier fractions of many crude oils, e.g. high boiling asphaltene or residue fractions. Condensed thiophenes are also the predominant form of sulfur identified in synthetic fuels derived from coal, oil shale, and tar sands. Moreover, in fractions used to produce diesel oil, most of the sulfur is found in BT, DBT, and their alkylated derivatives (Monticello and Finnerty, 1985). It has been reported that DBT and its derivatives constitute up to 70% of the sulfur content in diesel oil (Mohebali *et al.*, 2007). Sulfur compounds in the extremely high boiling fractions of petroleum (>540 °C) typically constitute approximately half of the total sulfur content of crude oils. While these compounds are the most difficult to analyze and identify, approximately 80% of the sulfur is estimated to be thiophenic in nature as part of larger complex molecules (Fedorak and Kropp, 1998). These sulfur compounds end up in heavy petroleum products such as diesel and Bunker-C oil.

Although DBT is generally taken as the model compound of heterocyclic organosulfur present in diesel oil, alkylated DBTs represent a high proportion of these molecules. Crude oil contains a large number of molecules which contain sulfur, however, the major sulfur-containing molecules in the middle distillate fraction diesel fuel are alkylated dibenzothiophenes. Furthermore, a sterically hindered sulfur molecule, such as 4,6-dimethyl DBT, is quite representative of the compounds recalcitrant to HDS which should be removed to obtain ultra-low sulfur diesel oils (ULSD). Alkylated DBT compounds typically employed as a model in BDS processes are 1-methyl DBT, 2-methyl DBT, 3-methyl DBT, and 4- methyl DBT (Ishii *et al.*, 2005), 4,6-dimethyl DBT (Abbad Andaloussi *et al.*, 2003; Noda *et al.*, 2003; Ishii *et al.*, 2005), 4,6-diethyl DBT (Noda *et al.*, 2003; Ishii *et al.*, 2005), 4,6- dipropyl DBT (Okada, 2002a; Noda *et al.*, 2003; Ishii *et al.*, 2005), 4,6-dibutyl DBT (Okada, 2002a; Ishii *et al.*, 2005), and 4,6-dipentyl DBT (Noda *et al.*, 2003).

Based on the bond strengths summarized in Table 6.1, the C-S bonds in the sulfur heterocycles (thiophene, benzo-, and dibenzothiophene) will be broken preferentially (Bressler *et al.*, 1998). The C–C bond strengths are greater compared to C–S bond strengths. Aromaticity and addition of oxygen to a carbon atom adjacent to the sulfur atom weaken the bond strengths making the C–S bond more susceptible to cleavage. This is a common feature in the aerobic microbial conversion of sulfur compounds, where enzymes (dioxygenases) introduce oxygen molecules to facilitate C–S cleavage. According to the bond strengths (Table 6.1), the C–S bonds will also be attacked preferentially in the anaerobic reaction mechanism that is similar to the metal-catalyzed HDS reaction mechanism. However, the role of enzymes to enable an attack on the C–S bond in the anaerobic route is currently unknown.

Despite the obvious chemical similarity between DBT and BT, the two desulfurization pathways are mutually exclusive. Thus, BT cannot be desulfurized via the DBT-specific pathway and DBT cannot be desulfurized through the BT-specific pathway (Gilbert *et al.*, 1998). For example, cells of *R. erythropolis* IGTS8 pregrown in BT cannot desulfurize DBT. This suggests that the enzymes of BT desulfurization could be different from those of a DBT desulfurization system. However, it was reported that the first step of BTH desulfurization (sulfur oxidation) can be catalyzed by the same monooxygenase involved in DBT desulfurization (Kobayashi *et al.*,

Table 6.1 Bond Strength of Various C–S, C–C, and C–H Bonds (Bressler *et al.*, 1998).

C–S Bond	kJ/mol	C–C Bond	kJ/mol
Thiophene C–S	341	CH_3–CH_3	376
Benzothiophene C–S	339	CH_2=CH_2	733
Dibenzothiophene C–S	338	Benzene C–C	505
CH_3–SH	312	CH_3–CH_2–CH_3	330
CH_3-S-CH_3	308	CH_3-CO-CH_3	290
CH_3-SO_2-CH_3	280	**C–H Bond**	**kJ/mol**
CH_3-$SCH_2C_6H_5$	257	H-CH_3	438
CH_3-SO_2 $CH_2C_6H_5$	221	H-CH_2OH	410
		H-CHO	364

2000). Therefore, the organisms having these metabolic pathways are complementary in terms of their potential roles in development of a microbial fuel desulfurization technology (Oldfield *et al.*, 1998).

It is therefore reasonable to infer that the development of BDS processes is dependent on the establishment of a microbial system with the potential to desulfurize a broad range of OSCs present in crude oil fractions. However, there is no common model compound that can be used for all the various crude oil fractions. Although middle-distillate fractions contain a complex mixture of sulfur compounds, they can be represented by two general chemical types: thiophenes and aliphatic sulfides. DBT has been used as a model thiophene and benzylsulfide as a model aliphatic sulfide in the study of BDS (Mohebali and Ball, 2008).

This chapter summarizes the different microbial pathways involved in the desulfurization of different organosulfur compounds (OSCs), with a special emphasis on microorganisms capable of selective biodesulfurization of DBT.

6.2 Biodegradation of Sulfur Compounds in the Environment

A sulfur atom forms 0.5–1% of bacterial cell dry weight (CDW). Microorganisms require sulfur for their growth and biological activities. Sulfur generally occurs in the structure of some enzyme cofactors (such as Coenzyme A, thiamine, and biotin), amino acids, and proteins (cysteine, methionine, and disulfur bonds) (Soleimani *et al.*, 2007; Feng *et al.*, 2016). In natural systems, bacteria assimilate sulfur in very small amounts for their maintenance and growth. Microorganisms, depending on their enzymes and metabolic pathways, may have the ability to provide their required sulfur from different sources (Stoner *et al.*, 1990). Some microorganisms can consume the sulfur in thiophenic compounds, such as DBT, aerobically or anaerobically and reduce the sulfur content in fuel. Soil may provide a rich source of organic sulfur containing molecules. The sulfur-containing amino acids cysteine and methionine, sulfolipids, sulfonic acids, and sulfated polysaccharides have been found in almost every type of soil. It may be the result of the partly decomposed plant, animal, and microbial residues (Eriksen, 2008). Similarly, organic sulfur in coal and petroleum is present in varied forms including thiols, sulfies, disulfides, and thiophenes (Soleimani *et al*, 2007), so the samples from coal mine and coal heap sites could be a rich source of organosulfur-containing compounds which may likely promote the growth of organic sulfur-metabolizing microorganisms. The same is true for oil polluted soil samples; they would be rich in different forms of OSCs. The sulfur present in both agricultural and uncultivated soils is largely in the form of organic-bound sulfur either as sulfonates and sulfate esters and not as free as bioavailable inorganic sulfates (Singh and Schwan, 2011).

Sulfur is a component of the environment and there is a natural cycle of oxidation and reduction reactions which transforms sulfur into both organic and inorganic products. Sulfur in the form of sulfate constitutes about 0.1% of soils. Elemental sulfur leaches in soil as sulfate at a slow rate and it is slowly converted to sulfate in soil by the action of autotrophic bacteria (Raju *et al.*, 2013).

Bacteria which are able to transform sulfur-containing compounds for utilization of either the sulfur or the carbon skeleton are widespread in nature (Le Borgne and Ayala, 2010). The occurrence of desulfurizing bacteria in diverse environments and geographic locations suggests an important and fairly common survival strategy for some bacterial species (Duarte *et al.*, 2001; Kilbane, 2006). In aerobic soils, for instance, inorganic sulfate makes up less than 5% of total sulfur and most of the residual sulfur is present as peptides/amino acids, sulfate esters, and sulfonates (Kertesz, 2004).

Numerous organosulfur compounds (OSCs) are present in petroleum, making sulfur the third most abundant element in a typical crude oil after carbon and hydrogen. The OSCs contain thiol, sulfide, and thiophene moieties. However, in crude oils of higher density, where sulfur content is typically the highest, sulfur exists primarily in the form of condensed thiophenes. Thus, most studies of the biodegradation of OSCs have focused on compounds found in the aromatic fraction of petroleum, namely, the thiophenes, benzothiophenes, and dibenzothiophenes (Walker *et al.*, 1975; Fedorak and Westlake, 1983; Eastmond *et al.*, 1984; Shennan, 1996; Schulz *et al.*, 1999).

Dibenzothiophene (DBT) and its alkylated derivatives are present in most crude oils. Analyses of residual oil that has undergone biodegradation at contaminated sites have shown that these derivatives are among the most recalcitrant compounds in the aromatic fraction (Berthou *et al.*, 1981; Ogata and Fujisawa, 1983). Moreover, DBT and its alkylated derivatives are more recalcitrant and persistent than phenanthrene and its alkylated homologues in contaminated environments. This recalcitrant nature is also reflected by studies showing the accumulation of DBTs in sediments and in tissues of shellfish in contaminated marine environments (Friocourt *et al.*, 1982; Ogata and Fujisawa, 1983; Finkelstein *et al.*, 1997; Kropp and Fedorak, 1998). The persistence and accumulation of these compounds have led to the suggestion that they might serve as oil pollution markers and are of a great concern since these are potentially harmful environmental pollutants (Ogata and Mujake, 1978).

Despite the recalcitrance of the alkyl–DBTs, their biodegradation in the aromatic fraction of crude oil has been observed in laboratory cultures, in contaminated environments, and within natural petroleum reservoirs. Fedorak and Westlake (1983, 1984) showed that the susceptibility of DBTs in Prudhoe Bay Crude oil to biodegradation decreased with increasing alkyl substitutions: DBT > 2- and 3- MDBT > 4- and 1- MDBT > 4,6-DMDBT.

However, results from in–vitro experiments can conflict with the perceived persistence of dibenzothiophenic compounds in the marine environment. When testing the biodegradation rates of Arabian light crude oil in semi–continuous cultures, it was found that these aromatic sulfur compounds were fairly readily degraded (Oudot, 1984). In contrast, field observations by the same laboratory demonstrated persistence of dibenzothiophenes in marine sediments polluted by the Amoco Cadiz oil spill of March 1978 (Gundlach *et al.*, 1983).

It should be taken into consideration that inorganic sulfate is the preferred sulfur source for the growth of most micro-organisms, however, under sulfate-limited growth conditions, bacteria can utilize other sulfur-containing compounds like DBT as an alternative sulfur source (Kertesz 2000; Aggarwal *et al.*, 2012).

The development of BDS processes is dependent on the provision of a microbial system with the potential to desulfurize a broad range of organosulfur compounds present in crude oil fractions.

6.3 Microbial Attack on Non-Heterocyclic Sulfur-Containing Hydrocarbons

6.3.1 Alkyl and Aryl Sulfides

Compared to sulfur containing ring structures such as thiophenes (Ths), benzothiophenes (BTs) and dibenzothiophenes (DBTs), relatively little information is available on the microbial metabolism of compounds with sulfur moieties present within alkyl chains. These structures are important as bridges in the high-molecular-weight asphaltene components of petroleum (Chapter 1) (Speight and Moschopedis, 1981; Murgich *et al.*, 1999). Therefore, biological attacks on sulfides are of considerable interest for biological heavy oil viscosity reduction.

There are few reports that conclusively illustrate the bacterial cleavage of alkyl C–S bonds. Microbial oxidation (often stereospecific) of aromatic sulfides has been recorded. *Aspergillus niger* oxidized phenylbenzylsulfide to a mixture of the corresponding sulfoxide and sulfone (Dodson *et al.*, 1962). A coal-solubilizing strain of *Paecilomyces* sp. degraded various organic sulfur compounds by oxidative attack localized at the sulfur atom (Faison *et al.*, 1991) and the corresponding sulfones being generated as major products from ethylphenylsulfide and dibenzyldisulfide (DBDS) were cleaved to the corresponding thiols and other single-ring products. Desulfurization of dimethylsulfide (DMS) and methionine in *P. putida* has been reported (Figure 6.1) (Inoue *et al.*, 1997; Vermeij and Kertesz, 1999; Endoh *et al.*, 2003). A *Pseudomonas* isolate is able to utilize benzylmethylsulfide (BMS) as a sole source of sulfur, oxidized BMS at the sulfur atom followed by a C–S bond cleavage to yield benzoate and sulfate (Van Afferden *et al.*, 1993). Köhler *et al.* (1984) reported that mixed cultures containing sulfate reducing bacteria (SRB) desulfurized a variety of model compounds, including DBT, BT, DBS, and dibenzyldisulfide (DBDS). A mixed methanogenic culture which was derived from a methanogenic sewage digester reductively desulfurized DBDS to toluene with benzylmercaptan as an intermediate in a sulfur limited medium (Miller, 1992).

DBS was used as a model of aliphatic sulfides and DPDS was used as a model of disulfides. These structures are important as bridges in the high molecular weight asphaltene components of petroleum (Van Hamme *et al.*, 2003).

Corynebacterium sp. SY1 (Omori *et al.*, 1992) and *Rhodococcus erythropolis* D-1 (Izumi *et al.*, 1994) are reported to utilize DMS as a sole source of sulfur for growth, but the intermediates were not identified. *Gordona* strain CYKS1 is able to utilize methyl sulfide, benzyl sulfide, phenyl sulfide, and methyl disulfide as a sole sulfur sources (Rhee *et al.*, 1998).

Arobacterium MC501 and mixed culture XHCO both are able to utilize DBS as a sulfur source (Constanti *et al.*, 1996). *Rhodococcus* sp. strain SY1 is reported to convert dibenzylsulfoxide to benzylalcohol and toluene (Omori *et al.*, 1995). According to Tanaka *et al.* (2001), the first Gram negative isolate *Sinorhizobium* sp. KT55, which is capable of utilizing BT as a sole source of sulfur with the production of a phenolic product, could also utilize dimethyl sulfide (DMS) and dimethyl sulfoxide (DMSO) as a sole source of sulfur, but could not utilize DBT and Th.

A unique bacterial strain, *Rhodococcus* sp. strain JVH1 that used bis-(3-pentaflurophenylpropyl)-sulfide (PFPS) as a sole source of sulfur was isolated from an oil contaminated environment. GC/MS analysis revealed that JVH1 oxidized PFPS to sulfoxide and then a sulfone prior to cleaving the C–S bond to form an alcohol and, presumably, a sulfinate from which sulfur could be extracted for growth. Four known DBT-desulfurizing strains, including *Rhodococcus* sp. strain IGTS8, were all unable to cleave the C–S bond in PFPS but could oxidize PFPS to the sulfone via sulfoxide. Conversely, JVH1 was unable to oxidize DBT, but was able to use a variety of alkylsulfides, in addition to PFPS, as a sole source of sulfur. Overall, PFPS is an excellent tool for isolating bacteria capable of cleaving sub-terminal C–S bonds within alkyl chains (Van Hamme *et al.*, 2004).

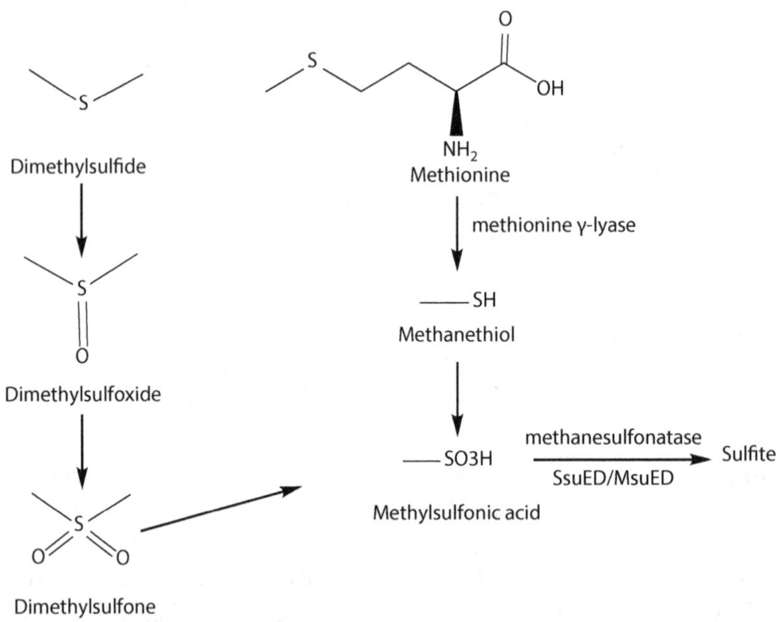

Figure 6.1 Desulfurization of Dimethylsulfide and Methionine by *P. putida* (Kertesz, 2004).

The extreme thermopile *Pyrococcus furiosus*, growing at 98°C, was shown to reduce organic polysulfides to H_2S using the reduction of sulfur as a source of energy (Tilstra *et al.*, 1992).

Rhodococcus rhodochrous IGTS8 (ATCC 53968), which was later identified as *R. erythropolis* IGTS8, was reported for its ability to utilize 2-chloroethyl ethyl sulfide (CEES) with the production of 2-chloroethanol and 2-chloroethanesulfinic acid (Kilbane and Jackowski, 1992). This indicated the cleavage of a C–S bond before the hydrolysis of the chlorine atom. Thus, IGTS8 is capable for the biodetoxification of the chemical warfare agent mustard (2,2′-dichlorodiethyl sulfide).

Itoh *et al.* (1997) reported that *Trametes versicolor* IFO30340 and *Tyromyces Palustris* IFO30339 metabolize DBS to benzylalcohol and benzylmercaptan. DBS metabolism in white rot fungi was studied by Van Hamme *et al.* (2003), where *Trametes trogii* UAMH 8156, *Trametes hirsute* UAMH 8165, *Phanerochaete chrysosporium* ATCC 24725, *Trametes versicolor* IFO30340 (formerly *Coriolus* sp.), and *Tyromyces Palustris* IFO30339 oxidized DBS to dibenzylsulfoxide prior to oxidation to dibenylsulfone.

Two basidiomycetes, white-rot *Coriolus versicolor* and brown-rot *Tyromyces palustris*, are reported for the complete degradation of Yperite (bis(2-chloroethyl) sulfide). Both fungi showed the same metabolic pathway (Wariishi et al., 2002). However, there were two distinct metabolic pathways detected during the fungal degradation of Yperite; the major path is a non-enzymatic hydrolytic dechlorination (path A in Figure 6.2) that generates thiodiglycol and the minor path is a direct sulfide cleavage reaction (path B in Figure 6.2) where the sulfide bond is cleaved prior to the hydrolytic dechlorination reaction, yielding chloroethanol and chloromercaptoethane, which were completely metabolized. The sulfoxide and sulfone of thiodiglycol were suggested to be formed only in the *C.*

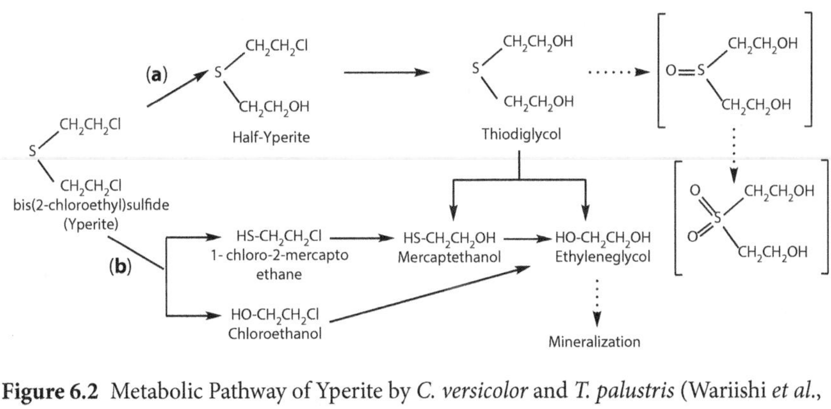

Figure 6.2 Metabolic Pathway of Yperite by *C. versicolor* and *T. palustris* (Wariishi *et al.*, 2002).

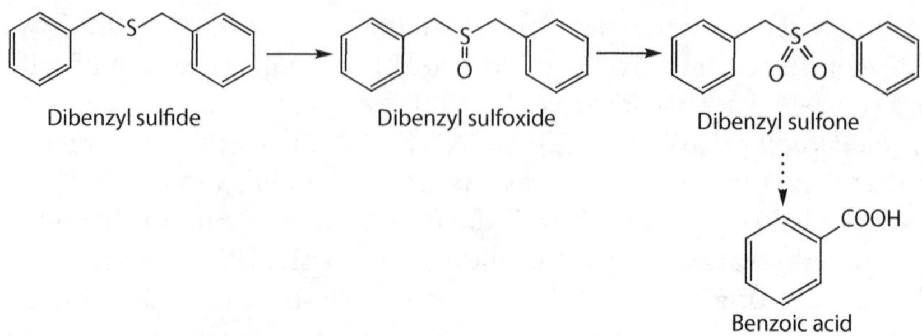

Figure 6.3 BDS of DBS by *Gordonia* sp. IITR100 (Ahmad *et al.*, 2015).

versicolor culture. This was attributed to the chemical stress responding to the gene expression of cytochrome P450 involved in a sulfoxidation reaction in *C. versicolor* (Ichinose *et al.* 2002a).

The BDS of DBS by *Gordonia* sp. IITR100 has been studied (Ahmad *et al.*, 2015) and the formation of these metabolites were detected: dibenzyl sulfoxide, dibenzyl sulfone, and benzoic acid (Figure 6.3). The same metabolites have been reported by the activity of *Rhodococcus* sp. strains JVH-1(Van Hamme *et al.*, 2004) and K1bD (Kirkwood *et al.*, 2005). Studies with recombinant *E. coli* revealed that enzyme DszC of IITR100 metabolizes DBS into dibenzyl sulfoxide and dibenzyl sulfone, but the reaction downstream to it is mediated by some enzyme other than its DszA, which suggests that this reaction in IITR100 might proceed by some enzyme that is distinct from DszABC. The DBS-metabolizing strain *Rhodococcus* sp. JVH1 also reported not carrying dszABC genes (Brooks and Van Hamme, 2012). In reactions where DBS and DBT were present together, both IITR100 and recombinant *E. coli* DszC exhibited preference for the desulfurization of DBS over DBT. The capability of IITR100 for the BDS of both thiophenic and sulfidic compounds promote its application in the BDS of petroleum fractions.

6.3.2 Non – Aromatic Cyclic Sulfur – Containing Hydrocarbons

Although n–alkyl–substituted tetrahydrothiophenes (Figure 6.4) are found in non-biodegraded petroleum, they are not found in petroleum that has undergone biodegradation in its reservoir. These observations suggested that this group of compounds with alkyl chain lengths from approximately C_{10} to at least C_{30} is biodegradable. Two of these sulfides, 2-*n*-dodecyltetrahydrothiophene (DTHT) and 2-*n*-undecyltetrahydrothiophene, were synthesized and their biodegradability was tested by using five Gram positive, *n*-alkyl-degrading bacterial isolates. The alkyl side chains of these

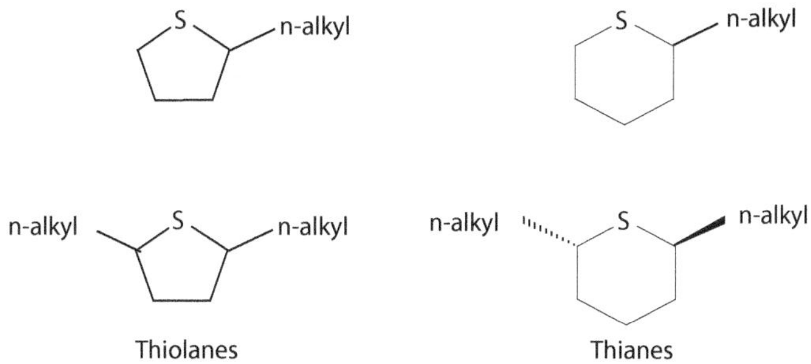

Figure 6.4 Examples of Thiolanes and Thianes Found in Non-Biodegraded Petroleum (the total number of carbon atoms in these compounds ranged from at least C_8 to C_{30}).

compounds were oxidized and the major intermediates found were 2-tetrahydrothiphenecarboxylic acid (THTC) and 2-tetrahydrothipheneacetic acid (THTA), respectively. Four *n*-alkane-degrading fungi were also shown to degrade DTHT, yielding both THTA and THTC. Although trace amounts of the corresponding sulfones and sulfoxides of 2-*n*-undecyl- and 2-*n*-dodecyl-tetrhydrothiophene were produced, the release of sulfur was not detected (Fedorak *et al.*, 1988).

Sulfolane (tetrahydrothiophene-1,1-dioxide) is used in the sulfinol process for natural gas sweeting. Many sour-gas processing plant spills, landfills, and leakage from unlined surface storage ponds have contaminated ground waters with sulfolane (Greene *et al.*, 1999). Aerobic sulfolane degradation has been demonstrated in laboratory and field-scale studies, however, anaerobic sulfolane biodegradation is unlikely to occur in the environment (Greene *et al*, 1998).

Chou and Swatloski (1983) studied the biodegradation of sulfolane in a laboratory-scale completely mixed activated sludge system. They found that sulfolane bio-oxidation generated acid requiring pH control and that there was a nearly stoichiometric release of the sulfur atom as sulfate. Thus, the heterocyclic ring of sulfolane can be broken by microorganisms (Klein, 1998).

Lee and Clark (1993) isolated a strain of *Pseudomonas maltophila*, that is able to degrade aromatic sulfonic acids, which could grow on minimal agar with sulfolane as a C-source. A bacterial strain WP1 determined to be most similar to *Variovorax paradoxus* which was isolated by Greene *et al.* (2000) and was reported for its ability to grow on sulfolane as a sole carbon and sulfur source and degrade it in a pathway similar to that of a 4S-pathway of DBT-BDS using *Rhodococcus* sp. strain IGTS8 (Gray *et al.*, 1996). Figure 6.5 illustrates the hypothesized release of sulfite from sulfolane by strain WP1.

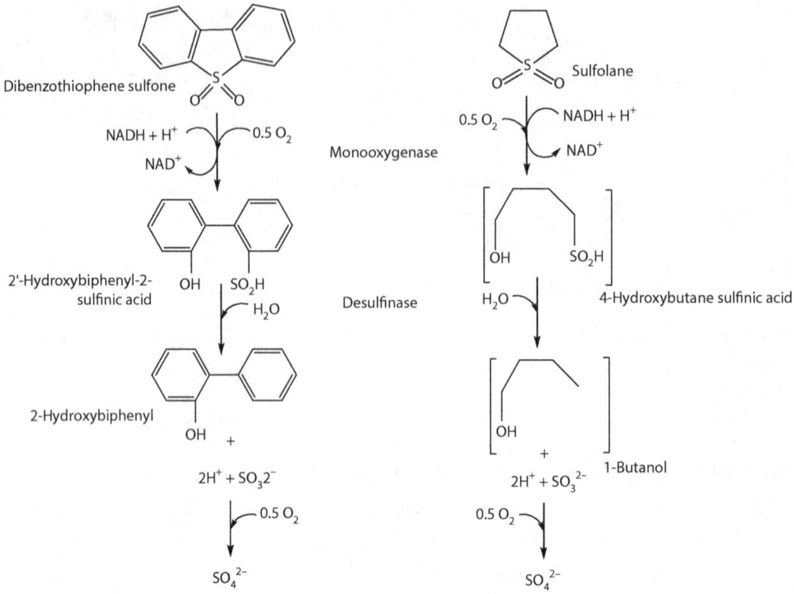

Figure 6.5 Release of Sulfite from DBT-Sulfone and Hypothesized Release of Sulfite from Sulfolane by Strain WP1 (Greene *et al.*, 2000) (the compounds in square brackets have not been detected in cultures of strain WP1).

6.4 Microbial Attack of Heterocyclic Sulfur – Hydrocarbons

Approximately 50% of the OSCs in crude oil are in the form of condensed thiophenic compounds (Frassinetti *et al.*, 1998). However, in another report, about 50–95% of sulfur in crude oil and derived fractions comes from sulfur heterocycles (Kilbane and Le Borgne, 2004). This occurs after refining at a high temperature leading to DBT concentrations higher than 70% of total sulfur (Kropp and Fedorak, 1998). The straight-run middle distillate feed stock contains about 95% of the OSCs as thiophenic compounds that include Ths, BT, and DBT. Each of these basic aromatic structures can have a variety of alkyl substituents (Folsom *et al.*, 1999). These compounds are recalcitrant and persistent in the biosphere and would be released into the environment through industrial processes, including gasification and liquefaction of coal, the refining of crude oil, and through oil spilling accidents (Berthou and Vignier, 1986). With the depletion of high quality, low sulfur content, light crude oil and the increase of reserves of low quality, high sulfur content, heavy oil, BT,

DBT, benzonapthothiophene (BNT), and their alkyl derivatives are found to constitute a major portion of the PASHs component in such heavy oil reserves (Liang *et al.*, 2006; Yang *et al.*, 2013; El-Gendy and Speight, 2016). Alkyl substituted DBTs seem to be the most difficult to remove among all organosulfur compounds, even for deep HDS treatment (Amorelli *et al.*, 1992). DBT is used as a model for the main organosulfur nucleus of the fuel matrix (Oldfield *et al.*, 1998). Although DBT is the preferential model compound, some reports also refer to benzothiophene in BDS studies (Matsui *et al.*, 2001a). The importance of BT as a model compound is relevant due to the fact that it presents a different metabolic pathway (Alves, 2007).

6.4.1 Thiophenes

Only a few reports of successful degradation or selective desulfurization of unsubstituted thiophene were published (Alam *et al.*, 1990; Grimalt *et al.*, 1991; Kim and Chon, 2002; Ardakani *et al.*, 2010; Nassar *et al.*, 2013). Kurita *et al.* (1971) reported anaerobic degradation of thiophene by bacterial culture isolated from oil–contaminated sludge. Moriya and Horikoshi (1993) reported a small amount of aerobic thiophene degradation by a *Bacillus* species isolated from deep-sea sediment. Neither study conclusively proved that thiophene could serve as a sole carbon source, nor did either study attempt to determine a pathway for thiophene degradation. A *Nocardioides* sp. mutant, PKSP12, utilized thiophene and thiophene 2-carboxylate aerobically as a carbon and energy source for growth (Sandhya *et al.*, 1995). A newly isolated *Gordona* strain, CYKS1, was grown aerobically in a minimal salts medium (MSM) with thiophene (Th), 2-methylthiophene (2-MT), and 3-methylthiophene (3-MT) as the sole sulfur sources, but the growth on 3-MT was lower than that on Th and 2-MT (Fedorak *et al.*, 1988). Many naturally occurring thiophenes are substituted at the 2 and 5 positions by aldehyde or carboxyl groups (Shennan, 1996). A yellow Gram-negative rod named R1, obtained by elective culture with thiophene-2-carboxylic acid (T2C) as substrate, was reported to use T2C for growth releasing the sulfur heteroatom as a sulfate (Cripps, 1973). Successive mutations of *Escherichia coli* yielded a strain that was able to degrade a variety of heterocyclic oxygen-and sulfur-containing ring compounds. In particular, this strain could use both furan-2-carboxylic acid and thiophene-2-carboxylic acid as sole carbon and energy sources (Abdul Rashid and Clark, 1987). Thiophene-2-carboxylic and thiophene-2-acetic acids served as growth substrates for *Rhodococcus* strains isolated from sewage sludge (Kanagawa and Kelly, 1987). T2C and thiophene-2-acetic acids also supported the growth of *Vibrio* YC1 isolated from oil-contaminated mud (Evans and Venables, 1990).

However, *Agrobacterium* MC501 and a mixed culture composed of *Agrobacterium* MC501, *Xanthomonas* MC701, *Corynebacterium* sp. MC401, and *Corynebactreium* sp. MC402, all isolated from a coal mine area by an enrichment culture with DBT, could both utilize thiophene-2-carboxylate as a sole source of sulfur (Constanti *et al.*, 1996).

Mixed cultures of petroleum-degrading bacteria can attack the side chains of several 2,5-dialkylthiophenes yielding various carboxylic acids, including 5-methyl-2-thiophenecarboxylic acid (Fedorak *et al.*, 1996) and 2,5-thiophenedicraboxylic acid (Finnerty, 1992). The latter compound appeared to be a key intermediate and some of the mixed cultures oxidized 5-methyl-2-thiophenecarboxylic acid to 2,5-thiophenedicarboxylic acid, which served as a growth substrate. Approximately 50% of the sulfur in this substrate was detected as a sulfate in the medium at the end of a 15-day incubation time. 2,5-thiophenedicarboxylic acid was detected in a mixed culture grown on 2,5-diundecylthiophene and 37% of the sulfur from this dialkylthiophene was detected as a sulfate in the medium after 35 days of incubation (Finnerty, 1992).

White basidiomycete *Coriolus versicolor* is reported to produce 2-formylthiophene (2FOT), 2-carboxythiophene (2CAT), 2(5H)-thiophenone (TPO), and γ-thiobutyrolactone (TBL), which were identified as metabolites from 2-hydroxymtheyl thiophene (2HMT) (Figure 6.6) (Ichinose *et al.*, 2002b). Although 2HMT and TBL act as nutrient sulfur sources, both compounds seemed to be toxic to the fungus at high concentrations. Therefore, gradual formation of TBL during the metabolism of 2HMT is advantageous for use as a sulfur source. Moreover, the pathway illustrated in Figure 6.6 occurs under non-ligninolytic conditions and none of these reactions were catalyzed by extracellular ligninolytic enzymes, such as lignin peroxidase (LiP) and manganese peroxidase (MnP). Thus, Ichinose *et al.* (2002b) concluded that although a thiophene ring is known to possess aromaticity, it is not degraded through the fungal ligninolytic system.

An unclassified aerobic, Gram +ve, soil bacterium, designated FE-9, was isolated and used hexadecane as its sole carbon and energy source and DBT or thianthrene as its sole sulfur source. This isolate was suspended in dimethylformamide (DMF) and used for BDS studies. When it was incubated with terthiophene (α-terthienyl, Figure 6.7) under a hydrogen atmosphere, the terthiophene was converted to H_2S and a highly unsaturated product tentatively identified as 1,3,5,7,9,11-dodecahexaene (Finnerty, 1993a).

6.4.2 Benzothiophenes and Alkyl-Substituted Benzothiophenes

Using Prudehoe Bay Crude oil as a substrate for enrichment cultures, Fedorak and Westlake (1983) studied the relative susceptibilities of

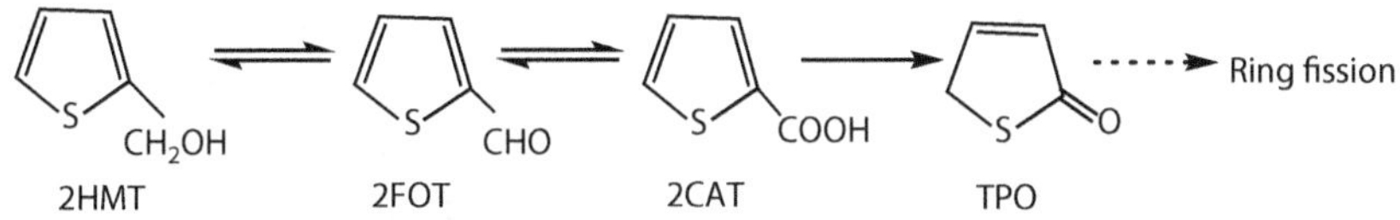

Figure 6.6 Metabolic Pathway of 2HMT by *C. versicolor* (Ichinose *et al.*, 2002b).

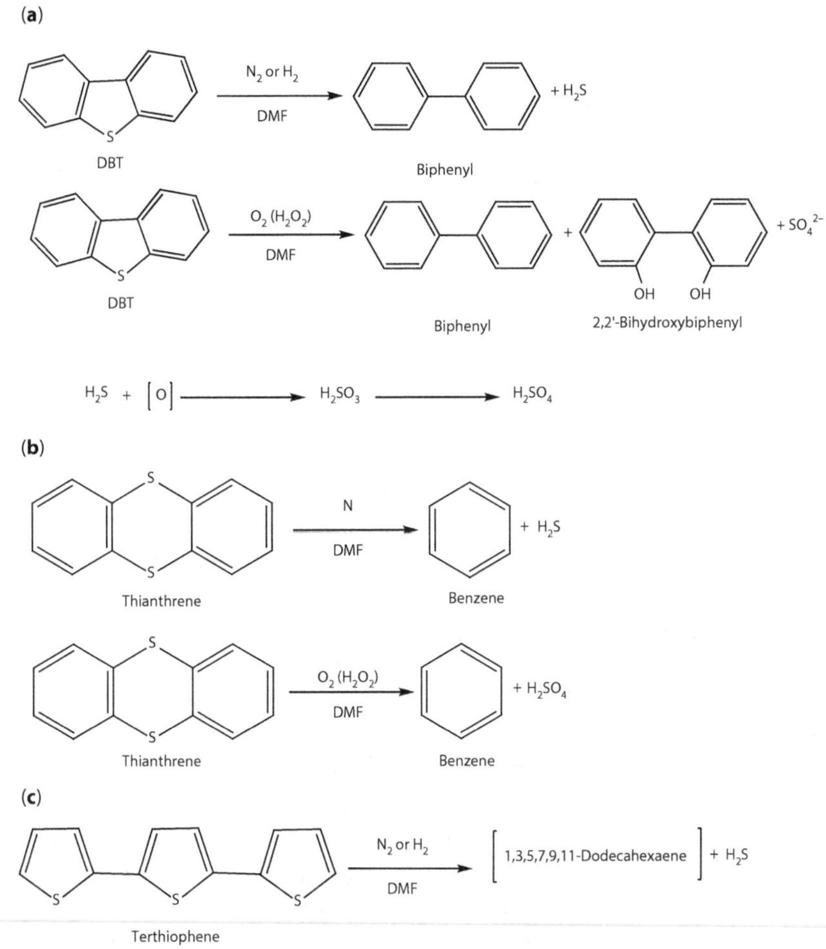

Figure 6.7 Non-Aqueous Biocatalytic Conversion of Sulfur-Containing Heterocycles by FE-9 (Finnerty, 1993a).

benzothiophene and alkyl-benzothiophenes to microbial attacks; in general, the larger the number of the alkyl-carbons, the more recalcitrant the compound. For example, the C_2-benzothiophenes were removed from the oil more readily than the C_3-benzothiophenes. There are several reports published on the aerobic co-metabolism of benzothiophene (BT). Sagardia *et al.* (1975) described the biodegradation of BT by *Pseudomonas aeruginosa* PRG-1 isolated from oil contaminated soil. BT did not support the growth of the isolate, but it was co-metabolized in cultures grown on 0.05% yeast extract. Growth of the culture resulted in emulsification of the oil phase and loss of 40% of the BT within 6 days. Bohonos *et al.* (1977) observed the biodegradation of BT by mixed populations of microorganisms in water samples from eutrophic and oligotrophic freshwater environments and aeration effluents of wastewater treatment plants. The microorganisms present in the water samples were capable of bio-transforming BT only when naphthalene was included in the incubations. The metabolites 2,3-dihydroxy-2,3-dihydrobenzothiophene (both the cis (Figure 6.8, structure 2) and trans (Figure 6.8, structure 4) isomers), benzothiophene-2,3-dione, and benzothiophene sulfoxide were tentatively identified.

Fedorak and Grbić-Galić (1991) studied the aerobic microbial co-metabolism of BT and 3-methylbenzothiophene (3-MBT) in a 1-methylnaphthalene (1-MN) degrading mixed enrichment culture and in pure cultures of *Pseudomonas* sp. strain BT1. Neither of the heterocyclic compounds would support the growth of the mixed culture, but were bio-transformed by the culture when it was grown on 1-MN, glucose, or peptone. Co-metabolism of BT yielded benzothiophene-2,3-dione (Figure 6.8, structure 8), whereas co-metabolism of 3-MBT (Figure 6.9, structure 1) yielded the corresponding sulfoxide (Figure 6.9, structure 2) and sulfone (Figure 6.9, structure 3). Strain BT1 was reported to mineralize ^{14}C-labeled naphthalene, biphenyl, and phenanthrene, but not saturated hydrocarbons. Isolate BT1 was found to be unable to further metabolize 3-methylbenzothiophene sulfone (Figure 6.9, structure3) in the presence of 1-MN. When 3-MBT was added to Prudhoe Bay crude oil, it was oxidized to the sulfoxide and sulfone by strain BT1 as it grew on the aromatic hydrocarbons in the crude oil. Benzothiophene-2,3-dione was found to be chemically unstable when incubated with Prudehoe Bay crude oil. The results of these studies with isolate BT1 led to the prediction that of the five other possible isomers of methyl-substituted benzothiophene, those that have a methyl group on the benzene ring would be co-metabolized to give methylbenzothiophene-2,3-diones, whereas those having a methyl group on the thiophene ring would give the corresponding sulfoxides and sulfones.

The metabolism of BT and all isomers of methyl-benzothiophene were further studied with *Pseudomonas* strains W1, F, and SB(G) (Kropp *et al.*, 1994a). In general, the types of abundant metabolites detected in extracts from these three new strains were similar to those previously reported for isolate BT1 (Table 6.2). The occurrence of sulfoxides and sulfones was more prevalent with the former three isolates. The oxidation of the methyl

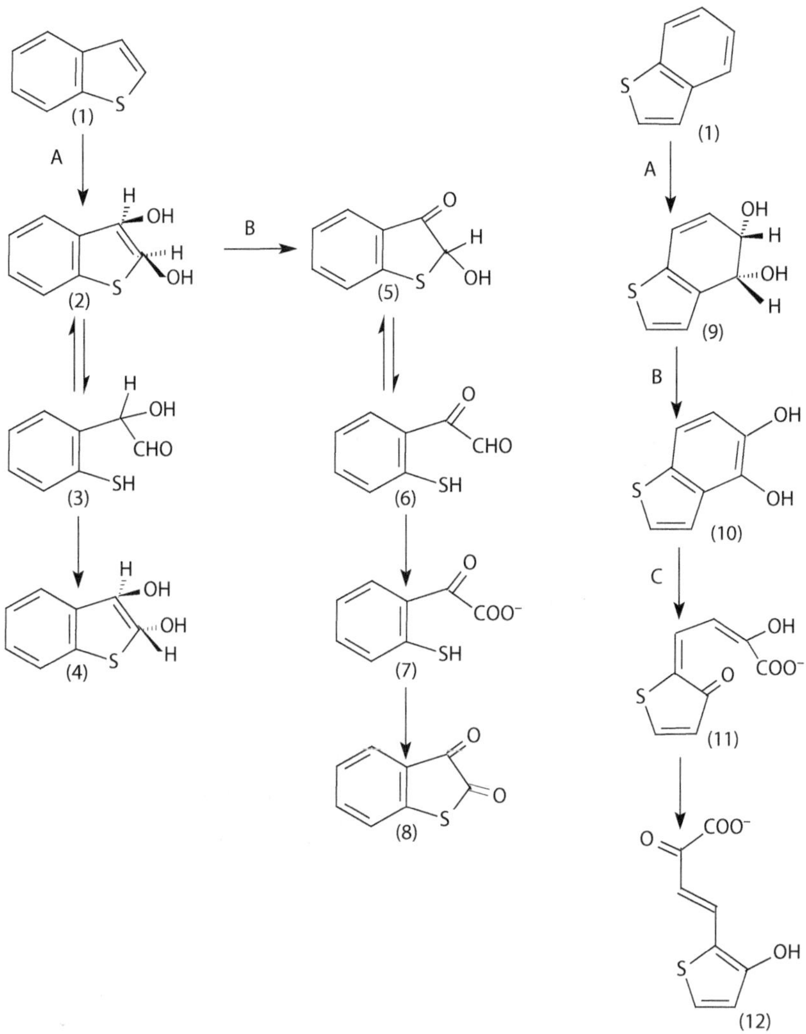

Figure 6.8 Pathways for the Biotransformation of Benzothiophene by *Pseudomonas putida* RE204.

(1) Benzothiophene; (2) cis-2,3-Dihydroxy-2,3-dihydrobenzothiophene; (3) 2'-Mercaptomandelaldehyde; (4) trans-2,3-Dihydroxy-2,3-dihydrobenzothiophene; (5) 2-Hydroxy-3-oxo-2,3-dihydrobenzothiophne; (6) 2-Mercaptophenylglyoxaldehyde; (7) 2-Mercaptophenylglyoxaldehyde; (8) Benzothiophene-2,3-dione; (9) cis-4,5-Dihydroxy-4,5-dihydrobenzothiophene; (10) 4,5-Dihydroxybenothiophene; (11) cis-4-(3-Oxo-2,3-dihydrothienyl)-2-hydroxybuta-2,4- dienoate; (12) trans-4-(3-hydroxy-2-thienyl)-2-oxobut-3-enoate.

Enzymes: (A) Isopropylbenzene-2,3-dioxygenase; (B) 2,3-Dihydroxy-2,3-dihydroisopropylbenzene dehydrogenase; (C) 3-Isopropylcatechechol-2,3-dioxygenase

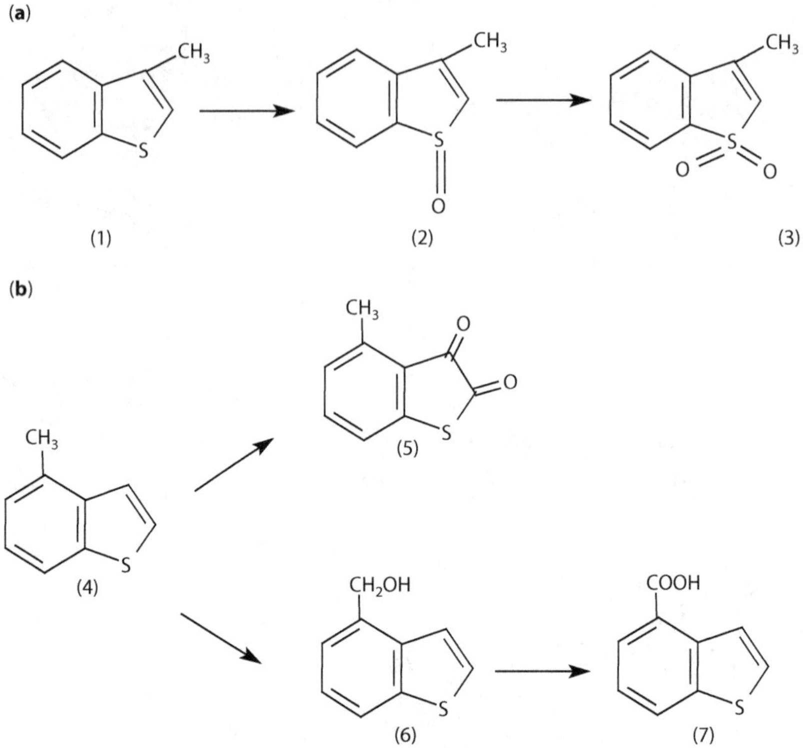

Figure 6.9 Examples for Bacterial Oxidation of Methyl-benzothiophenes.

A: Sulfoxidation of 3-methylbenzothiophene (1) to the corresponding sulfoxide (2), and sulfone (3) B: Oxidation of 4-methylbenzothiophene (4) to the corresponding: 2,3-dione (5), methanol (6) and carboxylic acid (7)

groups to carboxylic acids (Figure 6.9, structure 7) and, presumably, methanol (Figure 6.9, structure 6) was restricted to isolates W1 and F. Moreover, the 2,3-diones were produced from benzothiophenes and those methyl-benzothiophenes with a methyl group on the benzene ring.

Previous studies showed that BT, 3-, and 5- methyl-benzothiophenes are oxidized to their corresponding sulfones, which are not subsequently degraded (Fedorak and Grbić-Galić, 1991; Selifonov *et al.*, 1996). A filamentous bacterium belongs to the genus *Pseudonocardia* and the assigned strain designation DB1 was isolated, meaning it could not grow on benzothiophene, 3-, or 5- methyl benzothiophenes. It did not oxidize them to their sulfones (Bressler *et al.*, 1999). However, it could degrade the three benzothiophene sulfones. Benzothiophene sulfone (BTO_2) and 3-methyl-benzothiophene sulfone were more readily biodegraded than 5- methyl-benzothiophene sulfone and growth on the three compounds resulted in

Table 6.2 Summary of Sulfur-Containing Metabolites Produced from Benzothiophene (BT) and All of the Methyl-Benzothiophenes (MBT) by Four Bacterial Isolates (Kropp *et al.*, 1994a).

	Products found in cultures of each bacterial strain			
Substrate	**BT1**	**SB(G)**	**W1**	**F**
BT	2,3-Dione	Sulfoxide Sulfone $C_{16}H_{10}S$[a]	Sulfoxide Sulfone 2,3-Dione $C_{16}H_{10}S$	Sulfoxide Sulfone 2,3-Dione $C_{16}H_{10}S$
2-MBT	Sulfoxide Sulfone	Sulfoxide Sulfone	Sulfoxide Sulfone Carboxylic acid	Sulfoxide Sulfone Methanol Carboxylic Acid
3-MBT	Sulfoxide Sulfone	Sulfoxide Sulfone	Sulfoxide Sulfone Methanol Carboxylic acid	Sulfoxide Sulfone Methanol Carboxylic Acid
4-MBT	2,3-Dione	NT[b]	2,3-Dione Methanol Carboxylic acid	2,3-Dione $C_{18}H_{14}S$
5-MBT	2,3-Dione	Sulfone $C_{18}H_{14}S$	Carboxylic acid	Carboxylic acid 2,3-dione $C_{18}H_{14}S$
4-MBT and 6-MBT	*m*-tolyl methyl sulfoxide 2,3-Diones	*m*-tolyl methyl sulfoxide Sulfones $C_{18}H_{14}S$	*m*-tolyl methyl Sulfoxide 2,3-Diones Carboxylic acid $C_{18}H_{14}S$	*m*-tolyl methyl sulfoxide 4-MBT-2,3-diones Carboxylic acid $C_{18}H_{14}S$
7-MBT	Sulfoxide Sulfone 2,3-dione Several unidentified compounds	NT	Sulfoxide Sulfone 2,3-Dione *o*-tolyl methyl sulfoxide Methanol Carboxylic acid $C_{18}H_{14}S$	Sulfoxide Sulfone 2,3-dione *o*-tolyl methyl sulfoxide Methanol Carboxylic acid $C_{18}H_{14}S$

[a] Formula of high molecular weight product, [b] NT= not tested.

the release of 57, 62, and 28% of the substrate carbon as CO_2, respectively. The thiophene ring was also cleaved and between 44-88% of the sulfur from the consumed substrate was found as sulfate and/or sulfite.

Dimethylbenzothiophenes are among the sulfur heterocycles in petroleum that are known to be degraded by microbial activity. The co-metabolism of 6 of the 15 possible isomers of dimethylbenzothiophene by strains

BT1, W1, and F was also studied (Kropp *et al.*, 1996). Each isolate was grown on 1-methylnaphthalene (1-MN) or glucose in the presence of one of the dimethylbenzothiophenes and culture extracts were analyzed to identify nearly 30 sulfur-containing metabolites in total (Table 6.3). The methyl groups of all the isomers, except 4,6-, were oxidized to give hydroxymethylbenzothiophenes and methylbenzothiophene-carboxylic acids.

In most reports, these sulfur heterocycles do not serve as the sole carbon source for microbial growth, however, Watanapokasin *et al.* (2002) isolated a thermophilic bacterium that grows on benzothiophene as its sole carbon

Table 6.3 Summary of Sulfur-Containing Products Found in Extracts of Three Bacterial Cultures After Incubation with Various Dimethylbenzothiophenes for Seven Days (the products were found in cultures grown on 1-methylnaphtahlene or glucose) (Kropp *et al.*, 1996).

	Products found in cultures of *pseudomonas strain*		
Substrate	**BT1**	**W1**	**F**
2,3-dimethyl-benzothio-phene	Sulfoxide Sulfone Methanols	Sulfoxide Sulfone Methanols Carboxylic acids	Sulfoxide Sulfone Methanols Carboxylic Acids
2,7-dimethyl-benzothio-phene	Sulfoxide Sulfone Carboxylic acid	Sulfoxide Methanol Carboxylic acid	Sulfoxide Methanols Carboxylic Acid
3,5-dimethyl-benzothio-phene	None	Methanols Carboxylic acids	Methanols Carboxylic Acids
3,7-dimethyl-benzothio-phene	Sulfoxide Sulfone Methanols	Sulfoxide Methanols Carboxylic acids	Sulfoxide Methanols Carboxylic Acids
4,6-dimethyl-benzothio-phene	2,3-dione[a] 3(2H)-one 2(3H)-one $C_{20}H_{18}S$[b]	2,3-dione[a] 3(2H)-one 2(3H)-one	2,3-dione[a] 3(2H)-one 2(3H)-one $C_{20}H_{18}S$[b]
4,7-dimethyl-benzothio-phene	2,3-dione[a] 3(2H)-one 2(3H)-one $C_{20}H_{18}S$[b]	2,3-dione[a] 3(2H)-one 2(3H)-one Carboxylic acids $C_{20}H_{18}S$[b]	2,3-dione[a] 3(2H)-one 2(3H)-one Carboxylic acids $C_{20}H_{18}S$[b]

[a] likely a dimethyl-substituted 2-mercaptophenylglyoxalate at neutral pH
[b] several other high-molecular-weight products

and energy source and Sandhya *et al.* (1995) studied a mutated strain of *Nocardioides* that grew on benzothiophene as its sole carbon, sulfur, and energy source.

Recent studies have provided insight into the oxidation of the sulfur atom of benzothiophenes to form sulfoxides and sulfones by aromatic hydrocarbon-degrading bacteria. Recombinant bacteria expressing toluene dioxygenase (Allen *et al.*, 1995) and naphthalene dioxygenase (Selifonov *et al.*, 1996) have been shown to catalyze monooxygenation reactions of numerous organosulfur compounds, including 3-MBT, to form sulfoxides and sulfones (Path A, Figure 6.9) as end products.

Gordona sp. strain 213E (NCIMB 40816) grew in a pure culture in a mineral salts medium (MSM) containing fructose as a source of carbon and energy and benzothiophene (BT) as the sole source of sulfur (Gilbert *et al.*, 1998; Oldfield *et al.*, 1998), proved a BDS-pathway which is analogous to the DBT desulfurization pathway of *Rhodococcus* sp. strain IGTS8, in which 2-hydroxybiphenyl accumulates during growth with DBT as the sole sulfur source. The deduced pathway of BT desulfurization is illustrated in Figure 6.10 (Gilbert *et al.*, 1998).

Sinorhizobium sp. KT55 was the first Gram -ve isolate capable of utilizing BT as the sole source of sulfur (Tanaka *et al.*, 2001). The GC/MS analysis of metabolites of BT by this strain revealed BT sulfone, benzo[e][1,2]oxathiin-S-oxide, and *o*-hydroxystyrene, suggesting the pathway:

$$\text{BT} \rightarrow \text{BTO} \rightarrow \text{BTO}_2 \rightarrow \text{benzo[e][1,2]oxathiin-S-oxide} \rightarrow o\text{-hydroxystyrene}$$

The BDS activity of that strain was significantly repressed by methionine, cystine, sulfate, dimethylsulfoxide, and casamino acids (Tanaka *et al.*, 2001).

A benzothiophene-desulfurizing bacterium, which has a novel desulfurization pathway, was isolated and identified as *Gordonia rubropertinctus* strain T08; it metabolizes BT to *o*-hydroxystyrene, 2-coumaranone, and an unknown product (Figure 6.11) (Matsui *et al.*, 2001a).

Several studies to examine the possible coexistence of BT and DBT desulfurization activities in the same bacterial cells, were done and are listed as follows.

Paenibacillus sp. strain A11-2, which has been primarily isolated as a bacterial strain capable of desulfurizing DBT to produce 2-hydroxybiphenyl (2-HBP) at high temperatures, is found to desulfurize BT more efficiently than DBT. The desulfurized product was identified as *o*-hydroxystyrene by GC/MS and 1H-NMR analysis. DBT was assumed to be degraded in a

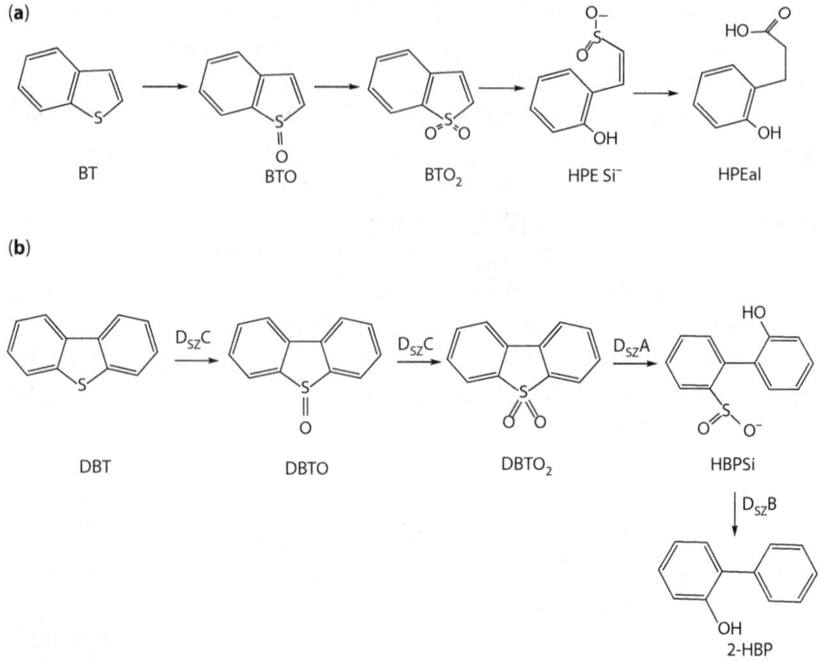

Figure 6.10 Comparison of Proposed BT Desulfurization Pathway with DBT Desulfurization Pathway (Gilbert *et al.*, 1998).

A: Proposed BT desulfurization pathway by *Gordona* sp. 213E
B: Proposed DBT desulfurization pathway by *Rhodococcus* sp. IGTS8

way analogous to the 4S pathway, which has been well known as a mode of DBT degradation. These results suggested that BT desulfurization may share, at least partially, the reaction mechanism with DBT desulfurization (Konishi *et al.*, 2000).

Thus, briefly, there are two reported divergent pathways for the microbial removal of sulfur from BT (Figure 6.12). In *Gordonia desulfuricans* strain 213E, the sulfinate group is removed with oxygenation of the molecule, yielding 2-(2′-hydroxyphenyl)ethan-1-al. This product is recovered as benzofuran due to dehydration under acidic extraction conditions (Gilbert *et al.*, 1998). In *Paenibacillus* sp. strain A11-2, the final product is *o*-hydroxystyrene produced through the desulfination of the molecule, which does not oxygenate the carbon atom (Konishi *et al.*, 2000). Only *Rhodococcus* sp. JVH1 and *Rhodococcus* sp. strain\ WUK2R have been reported to produce both end products from the desulfurization of BT (Kirimura *et al.*, 2002; Kirkwood *et al.*, 2007).

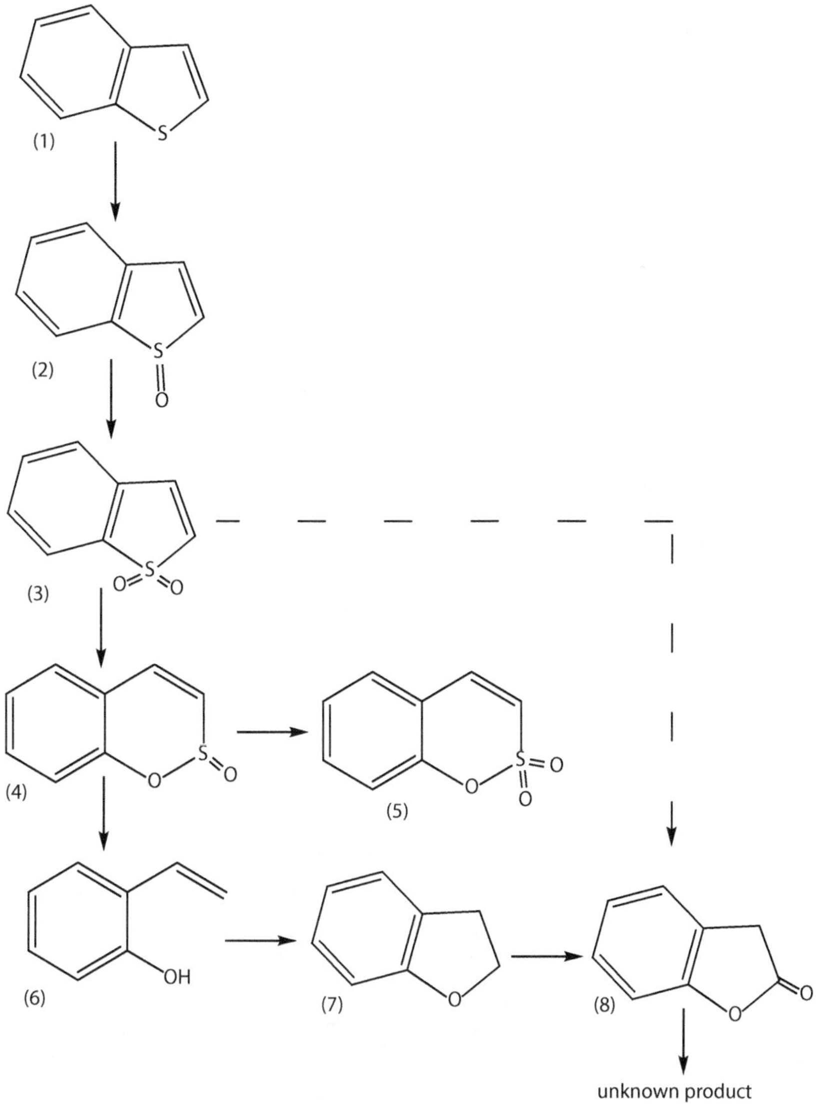

Figure 6.11 Postulated Pathway for BT Desulfurization by *G. rubropertinctus* T08 (Matsui *et al.*, 2001a). (1) BT; (2) BT-sulfoxide; (3) BT-sulfone; (4) BT-sultine; (5) BT-sultone; (6) *o*-Hydroxystyrene; (7) Coumarane; (8) 2-Coumaranone

The utilization of DBT, BT, and their various derivatives was tested by *Rhodococcus* sp. strain KA2-5-1. It has the ability to desulfurize a variety of alkyl DBTs through the 4S-pathway. In addition, it grew well in a medium containing 3-methyl, 2-ethyl, or 2,7-diethylbenzothiophene as the sole

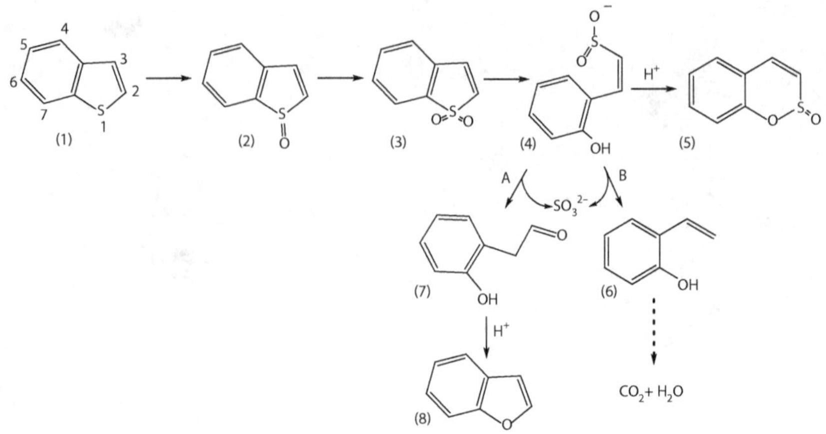

Figure 6.12 Divergent Pathways of Benzothiophene Metabolism by (A) *G. desulfuricans* strain 213E (Gilbert *et al.*, 1998), (B) *Paenibacillus* sp. strain A11-2 (Konishi *et al.*, 2000), (A and B) *Rhodococcus* sp. strain WU-K2R (Kirimura *et al.*, 2002) (1) Benzothiophene; (2) Benzothiophene-S-oxide (sulfoxide); (3) Benzothiophene-S,S-dioxide (sulfone); (4) 2-(2′-hydroxyphenyl)ethen-1-sulfinate; (5) Benzo[e][1,2]oxathiin-S-oxide (sultine); (6) *o*-Hydroxystyrene; (7) 2-(2′-hydroxyphenyl)ethan-1-al; (8) benzofuran. Sulfite has not been definitively shown as the final sulfur species. Products (5) and (8) formed by abiotic reactions under acidic extraction conditions.

sulfur source. On the other hand, no significant growth was observed when BT, 2-methyl, 5-methyl, 7-methyl, 7-ethyl, or 5,7-dimethylbenzothiophene was used as the sole sulfur source (Kobayashi *et al.*, 2000).

In order to enhance the BT degradation, a flavin-oxido reductase gene, dszD, from DBT desulfurizing *Rhodococcus erythropolis* strain KA2-5-1 (Kobayashi *et al.*, 2000) was expressed in BT desulfurizing *Rhodococcus* sp. strain T09 (Matsui *et al.*, 2000) using a *Rhodococcus* - *E. coli* shuttle vector. The BT degradation rate was increased about three-fold with BT grown recombinant cells (Matsui *et al.*, 2001b). Onaka *et al.* (2002) extracted the hydrophilic benzo[b]thiophene metabolites produced by *Rhodococcus* sp. strain T09 by a solid-phase extraction and a derivatization method, which prevented the dehydration that often occurs during liquid-liquid extraction (LLE), and were identified by GC with atomic emission detection (AED) and GC-MS. As a result, *cis*- or *trans*-2-(2-hydroxyphenyl) ethen-1-ol, previously reported as the tautomer 2-(2-hydroxyphenyl) ethan-1-al and 2-(2-hydroxyphenyl) ethen-1-sulfenic acid and its isomer, 1-(1-hydroxyethenyl) benzene-2-sulfenic acid, were identified. Benzothiophene desulfurization in culture broth at a neutral pH by the strain T09 was modified as a branched metabolism from a benzothiophene S-oxide based on the above compounds (Figure 6.13).

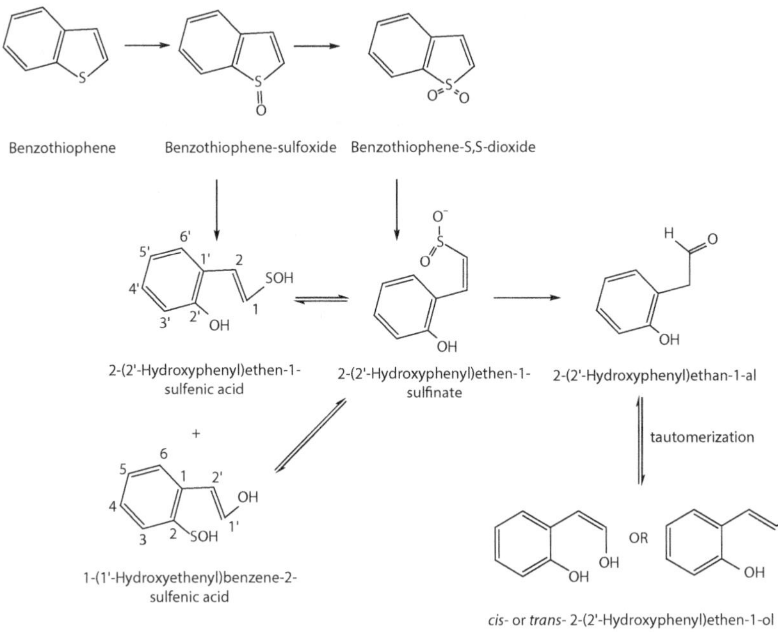

Figure 6.13 BT Desulfurization Pathway by *Rhodococcus* sp. Strain T09 (Onaka *et al.*, 2002).

Part of the BT-BDS-pathway is analogous to the '4S' pathway producing BT-sulfoxide and BT-sulfone (Gilbert *et al.*,1998; Tanaka *et al.*, 2001). Then, BT-sulfone is metabolized to benzo[c][l,2]oxathiin-S-oxide that undergoes transformation into either 2-(2′-hydroxypheny1) ethan-1-al (Kilbane, 2006) or *o*-hydroxystyrene (Tanaka *et al.*, 2001), as reported for *Gordonia* sp. 213E and *Sinorhizobium* sp. KT55, respectively.

Arensdorf *et al.* (2002) have shown that a mutant *R. erythropolis* BKO 53 is able to metabolize DBT, 2-MBT and 3-MBT although it cannot not transform 5-MBT.

There are some reports of anaerobic biodegradation of benzothiophene leading to cleavage of the thiophene ring. Using bacteria in oil sludge, Kurita *et al.* (1971) observed a hydrogen sulfide release from the benzothiophene ring, but no identification of the resulting organic compounds was given. Grbić-Galić (1989) described a study in which BT was degraded in methanogenic microcosms containing aquifer solids and water from a cresote-contaminated aquifer, *o*-hydroxybenzenesulfonic acid, phenylacetic acid, benzoic acid, and phenol were among the identified metabolites.

Anaerobic cometabolic conversion of BT was studied by Annweiler *et al.* (2001) with SRB enrichment culture growing with naphthalene as the sole source of carbon and energy. The SRB were not able to grow with BT as the primary substrate and metabolite analysis was performed with culture supernatants obtained by cometabolism. Experiments revealed the formation of three isomeric carboxybenzothiphenes. Two isomers were identified as 2-carboxybenzothiophene and 5-carboxybenzothiophene. In some experiments, further reduced dihydroxy-carboxy-benzothiophene was identified. No other products of BT degradation have been determined.

6.4.3 Naphthothiophenes

Naphtho[2,1-b]thiophene (NTH), which includes a BT structure, is an asymmetric structural isomer of DBT. In addition to DBT derivatives, NTH derivatives can also be detected in diesel oil following HDS treatment, although NTH derivatives are minor components in comparison to DBT derivatives. Recently, there is confirmation that alkyl-substituted derivatives of NTH and BT may be detected in diesel oil in addition to DBT derivatives (Chen *et al.*, 2010). Therefore, NTH may also be a model target compound for deeper desulfurization (Kirimura, 2002; Ishii, 2005). The desulfurization of three or more ringed-compounds, which are considered to inhibit biodesulfurization of crude oil, are rare and few applications have been published concerning the BDS of BT, NTH, and benzo-naphthothiophene (BNT), compared with DBT.

Kropp *et al* (1997a) reported the biodegradation of napththo[2,1-b] thiophene to 4-hydroxybenzothiphene-5-carboxylic acid and 5-hydroxybenzothiophene-4-carboxylic acid by different bacterial isolates denoted as BT1, W1, and F. Upon biodegradation of 1-methyl napththo[2,1-b]thiophene, 4-hydroxy-3-methylbenzothiphene-5-carboxylic acid was accumulated, but the degradation of napththo[2,3-b]thiophene was more difficult and persistent and yielded 5-hydroxybenzothiophene-6-carboxylic and 6-hydroxybenzothiophene-5-carboxylic acids.

Rhodococcus sp. strain WU-K2R was isolated from soil for its ability to grow in a medium with NTH as the sole sulfur source. Growing cells of WU-K2R degraded 0.27 mM NTH within 7 days. WU-K2R could also grow in the medium with NTH sulfone, BT, 3-MBT, or 5-MBT as the sole source of sulfur, but could not utilize DBT, $DBTO_2$, or 4,6-DMDBT. On the other hand, WU-K2R did not utilize NTH or BT as the sole carbon source. Using GC/MS analysis, desulfurized NTH metabolites were identified as NTH sulfone, 2,2′-hydroxynaphthylethene, and naphtha[2,1-b]furan. Moreover, since BT metabolites were identified as BT sulfone, benzo[c] [1,2]oxathiin-S-oxide, benzo[c][1,2]oxathiin-S,S-dioxide, *o*-hydroxystyrene, 2-(2′-phenyl)ethan-1-al, and benzofuran, it was concluded that

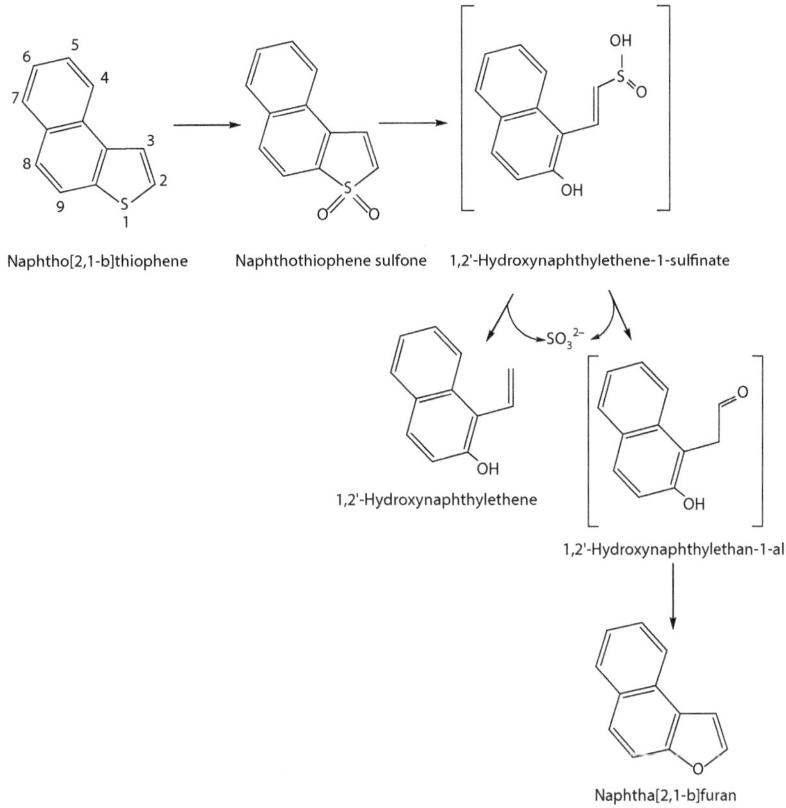

Figure 6.14 Proposed Pathway for NTH Desulfurization by *Rhodococcus* sp. Strain WU-K2R (Kirimura *et al.*, 2002).

WU-K2R desulfurized with asymmetric heterocyclic sulfur compounds, such as BT (Figure 6.12) and NTH (Figure 6.14), through sulfur specific degradation pathways with the selective cleavage of C–S bonds (Kirimura *et al.*, 2002).

Mycobacterium pheli WU-F1, which processes a high desulfurization ability toward DBT and its derivatives over a wide temperature range (20-50°C), can also grow at 50 °C in a medium with NTH or 2-ethylNTH, an alkylated derivative, as the sole source of sulfur. At 50 °C, the resting cells of WU-F1 degraded 67% and 83% of 0.81mM NTH and 2-ethylNTH, respectively, within 8 h. The GC/MS analysis identified 2-ethylNTH-desulfurized metabolites as 2-ethylNTH sulfoxide, 1-(2′-hydroxynaphthyl)-1-butene, and 1-naphthyl-2-hyroxy-1-butene. It was concluded that WU-F1 desulfurized 2-ethylNTH through a sulfur-specific degradation pathway with the selective cleavage of carbon-sulfur bonds (Figure 6.15). Therefore, *M.*

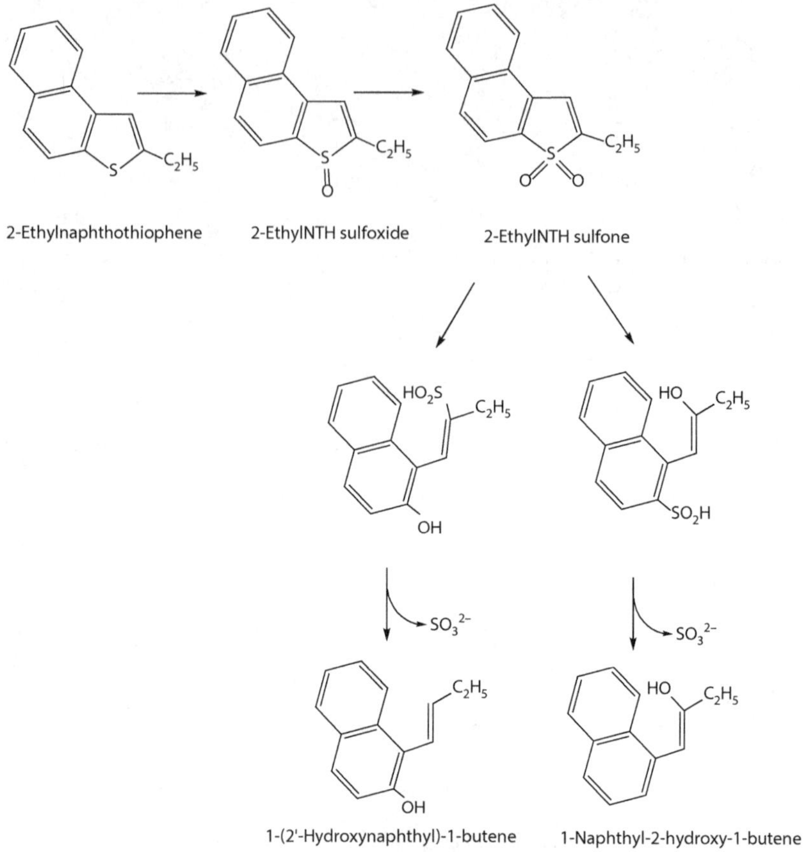

Figure 6.15 Proposed Pathway for 2-ethylNTH Desulfurization by *Mycobacterium pheli* WU-F1 (Furuya *et al.*, 2002).

pheli WU-F1 can effectively desulfurize asymmetric organosulfur compounds, NTH, and 2-ethylNTH, as well as symmetric DBT derivatives under high-temperature conditions, and it may be a useful desulfurizing biocatalyst possessing a broad substrate specificity toward organosulfur compounds (OSCs) (Furuya *et al.*, 2002).

Bacillus subtilis WU-S2B (Kirimura *et al.*, 2001) and *Rhodococcus erythropolis* XP (Yu *et al.*, 2006) have been reported to utilize BNT as a sulfur source and the reaction in both cases was accompanied with the formation of BNT-hydroxide, but the intermediates formed during the process and the enzymes involved in the activity were not studied, but the benzo [b] naphtho[2,1-d]thiophene (BNT) biodesulfurization pathway by *Gordonia* sp. IITR 100 has been elucidated (Chauhan *et al.*, 2015) (Figure 6.16).

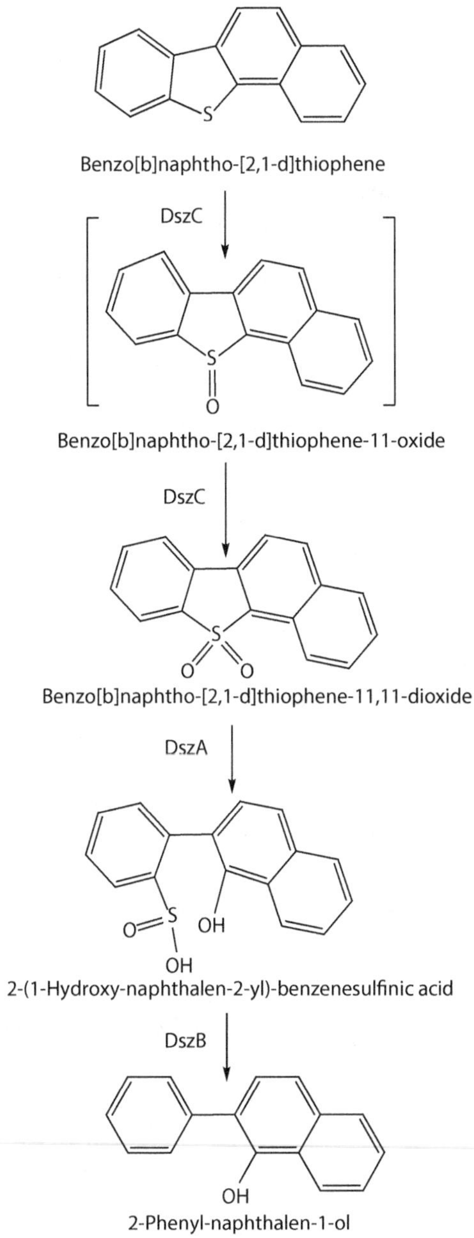

Figure 6.16 Proposed Pathway for BNT Desulfurization by *Gordonia* sp. IITR 100 (Chauhan *et al.*, 2015).

The *Mycobacterium* sp. G3 was reported to desulfurize benzo[b]naphtho[2,1-d]thiophene to the desulfurized hydroxy form and the methoxy form, which was converted from the hydroxy form (Okada *et al.*, 2002b).

6.4.4 Dibenzothiophene and Alkyl-Substituted Dibenzothiophenes

Dibenzothiophene (DBT) is considered to be particularly relevant as a model compound for the forms of thiophenic sulfur found in fossil fuels, such as crude oils, coals, or bitumen of a particular geographic origin and various refining intermediates and fuel manufactured from them (Rambosek *et al.*, 1999). This is not only because DBT and its derivatives have been reported to account for as much as 70 wt% of total sulfur content of West Texas crude oil and up to 40 wt% of the total sulfur content of some Middle East crude oils (El-Gendy and Speight, 2016), but DBT is also a commercially available compound and it models as a special recalcitrant form of organic sulfur in fossil fuels.

Three main pathways are known for the aerobic desulfurization of DBT (Klein *et al.*, 1994). The first is a completely destructive pathway in which DBT is mineralized to carbon dioxide, sulfate, and water. The second is a ring-destructive pathway in which the molecule is partially degraded with sulfur remaining in the organic skeleton. The third is a non-destructive pathway in which only sulfur is removed from DBT without breaking the hydrocarbon backbone of the molecule.

6.4.4.1 Aerobic Biodesulfurization of DBT

6.4.4.1.1 Dihydroxylation and DBT Ring Attack Without Sulfur Release

The ability to oxidize DBT by dihydroxylation without the release of sulfur appears to be wide spread among soil bacteria being reported in soil isolate strains of *Pseudomonas* (Monticello *et al.*, 1985; Fought and Westlake, 1988), *Beijerinckia* (Laborde and Gibson, 1977), and a mixture of *Acintobacter* and *Rhizobium* (Malik and Claus, 1976).

This pathway is a partial degradative pathway (Figure 6.17) called the "Kodama pathway" (Kodama *et al.*, 1970, 1973). It is analogous to that of naphthalene degradation (Kodama *et al.*, 1973). It involves oxidation of one of the benzene rings yielding a water-soluble product, 3-hydroxy-2-formylbenzothiophene (HFBT) being the most commonly identified product. Denome *et al.* (1993) cloned and sequenced a 9.8 kb DNA fragment from *Pseudomonas* strain C18 that encoded DBT-degrading enzymes. The isolated genes were referred to as the DOX (DBT oxidation) pathway.

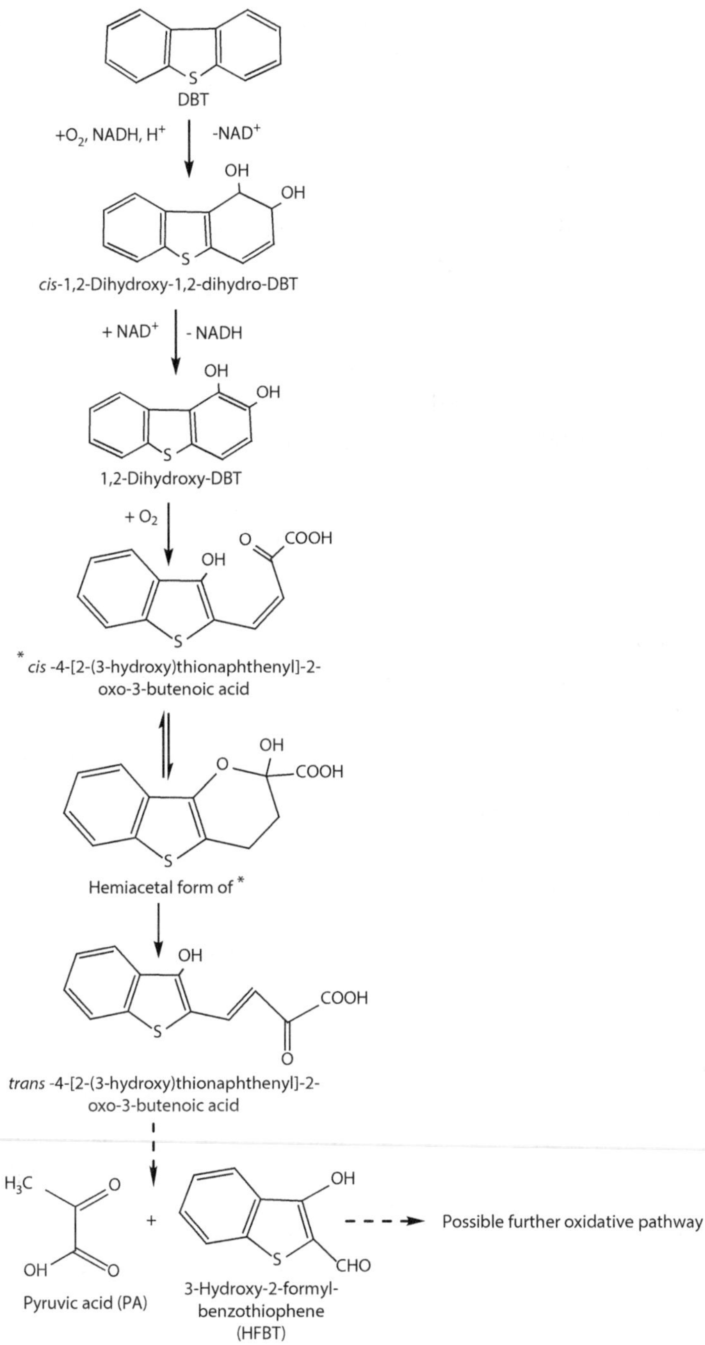

Figure 6.17 Kodama Pathway for the Biotransformation of DBT by Attacking the Benzene Ring.

The results indicated that a single genetic pathway controls the metabolism of DBT, naphthalene, and phenanthrene in strain C18.

The oxidation of DBT via the Kodama pathway was shown to be cometabolic because other substrates were required for growth and DBT oxidation to occur. Based on the large number of reports about bacteria producing HFBT (Kodama *et al.*, 1970; 1973; Laborde and Gibson, 1977; Fought and Westlake, 1990; FinkelStein *et al.*, 1997; Frassinetti *et al.*, 1998), it appears, that in an environment that is not sulfate limited, the Kodama-pathway is likely the most common method for microbial degradation of DBT (Bressler and Fedorak, 2001a,b). This process leads to the accumulation of the water soluble HFBT as an end product which is of lower carbon content than DBT. In this pathway, no real desulfurization of the organosulfur substrate occurs, but a partial degradation of DBT occurs (El-Gendy *et al.*, 2005).

Three 1–methylnaphthalene (1-MN) *Pseudomonas* strains were reported to oxidize the sulfur atom of 1,2,3,4-tetrahydroDBT to give the sulfoxide and sulfone (Kropp *et al.*, 1997b). They also degraded the benzene ring to yield 3-hydroxy-2-formyl-4,5,6,7-tetrahydrobenzothiophene.

Pseudomonas fluorescens 17 and 26 were found to be capable of a 96 to 99% degradation of DBT and, to a small extent, 4,6-dimethyldibenzothiophene (4,6-DMDBT). DBT could be utilized by enrichment cultures and *Pseudomonas* strains 17 and 26 as a sole source of carbon and energy, however, its transformation was enhanced in the presence of such co-substrates as glycerol and glucose. The products of DBT conversion (Figure 6.18) are 2,3-dihydroxybenzothiophene (4) and thiosalicyclic acid (5), as well as more complex products (FinkelStein *et al.*, 1997).

Saftić *et al.* (1993) studied the aerobic metabolism of DBT and in all four isomers of methyl DBT in pure cultures of three *Pseudomonas* sp. (strains BT1, W1, and F), the three isolates were shown to grow on 1-MN and cometabolize DBT to benzothiophene-2,3-dione, HFBT, DBTO, and $DBTO_2$. For all isomers of methyl DBT, the unsubstituted ring was preferably oxidized and cleaved by the Kodama pathway to form methyl-substituted HFBTs and benzothiophene-2,3-diones. In addition, other oxidation products were detected for some of the isomers, including methylDBT sulfones and DBT methanols.

Kropp *et al.* (1997c) described the bio-transformations of three isomers of dimethylDBT by pure cultures of *Pseudomonas* strains BT1, W1, and F and four petroleum-degrading mixed cultures. 1-MN and the aromatic fraction of Prudhoe Bay crude oil served as the growth substrates for the pure and mixed culture, respectively. Culture extracts were analyzed by GC with FPD, MS, and FTIR detectors to identify oxidation products.

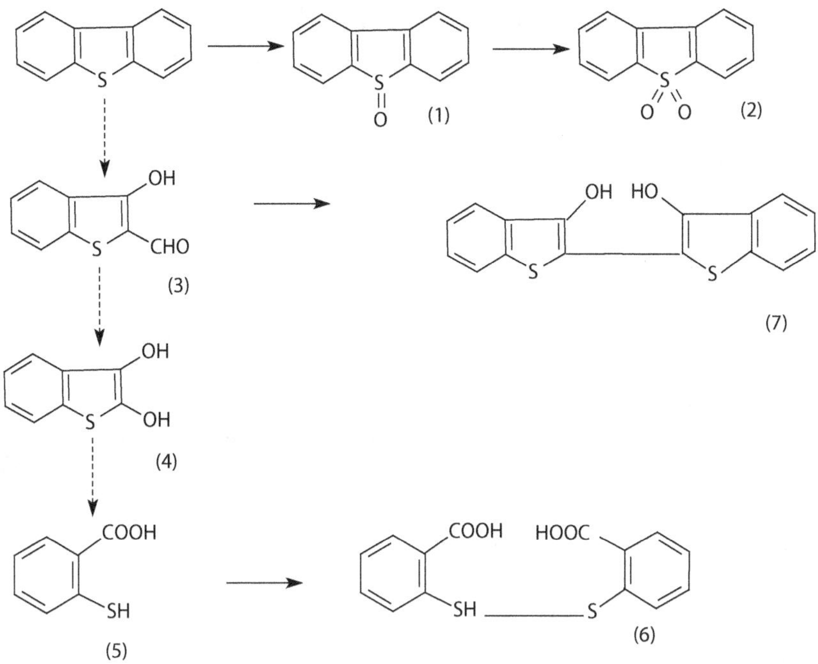

Figure 6.18 Putative Pathway for DBT Degradation by *P. fluorescens* 17 (FinkelStein *et al.*, 1997). (1) DBT-sulfoxide; (2) DBT-sulfone; (3) 3-Hydroxy-2-formylbenzothiophene; (4) 2,3-Dihydroxybenzothiophene, (5) Thiosalicyclic acid, (6) 2,2′-Dithiosalicyclic acid, (7) 2,2′-bibenzo[b]thiophene]-3,3′-diol.

One isomer of dimethylDBT studied (3,4-) had both methyl groups on the same homocylic ring, while the other two isomers (2,8- and 4,6-) both had a single methyl group on each homocyclic ring. The susceptibilities of the dimethylDBTs depended upon the positions of the methyl groups. All the cultures tested were able to degrade the unsubstituted ring of the 3,4-isomer to give 6,7-dimethyl-HFBT and 6,7-dimethylbenzothiophene-2,3-dione, among other products. None of the pure cultures were able to oxidize the 4,6-isomer, which has a methyl group on each of the homocyclic rings, and only slight losses were observed in the petroleum-degrading mixed cultures. However, isolate BT1 was able to oxidize and cleave a methyl-substituted ring of the 2,8-isomer, yielding 5-methylbenzothiophene-2,3-dione and two other products. This 2,3-dione was also observed in extracts of two of the four mixed cultures, which were capable of moderate degradation of the 2,8-isomer.

Many bacteria are able to oxidize mesophilically complex aromatic sulfur compounds with the formation of water soluble sulfur containing products via the Kodama pathway, such as *Bacillus* sp., *Micrococcus* sp., *pseudomonas* sp., *Rhizobium* sp., *Acinetobacter* sp., *Arthrobacter* sp., and the filamentous marine fungus *Cunninghamella elegans* (Kim *et al.*, 1996). On the other hand, *Sulfolobus acidocaldarius* was reported to degrade DBT via the Kodama pathway at high temperature of 70 °C (Kargi and Robinson, 1984).

6.4.4.1.2 Direct Oxidation of DBT at the Sulfur Heteroatom to the Sulfoxide and Sulfone

In some reports of benzene ring dihyroxylation of DBT, oxidation at the heteroatom to DBTO was also detected (Kodama *et al.*, 1970; Laborde and Gibbson, 1977; Frassinetti *et al.* 1998). Other workers, including Mormile and Atlas (1988) and Constanti *et al.* (1994), demonstrated that a *Pseudomonas ptudia* strain and a series of uncharacterized isolates from oil-contaminated soils (Stoner *et al.*, 1990) could degrade DBT by two pathways, one yielding the dihydroxylation products trans-HTOB and HFBT and the other forming DBTO and $DBTO_2$ (Figure 6.19). The sulfoxide and sulfone were completely degraded, but without a detectable release of the sulfur heteroatom (Mormile and Atlas, 1988).

Sphingomonas LB126 was isolated from PAH-contaminated soil and was able to use DBT co-metabolically using pyruvate as a carbon source. GC/MS analysis of the ethylacetate extract showed that LB126 produces DBTO and $DBTO_2$ as the dead end products (Van Herwijnen *et al.*, 2003).

Direct oxidation of the sulfur heteroatom was the predominant reaction found when the fungi *Cunninghamella elegans* (Holland *et al.*, 1986), *Rhizopus arrhizus* (Holland *et al.*, 1986), *Mortieralla isabellina* (Holland *et al.*, 1986), and white rot fungus *Pleurotus ostreatus* (Bezalel *et al.*, 1996) were incubated aerobically with DBT and similar aromatic sulfur heterocycles, such as thioxanthane and its monomethyl derivatives (Crawford and Gupta, 1990). These compounds were transformed to their corresponding sulfoxides and sulfones without further degradation. No evidence for ring hydroxylation or methyl-group oxidation was detected. *P. ostreatus* acts on PAHs like non-ligninolytic fungi, such as *C. elegans*, where cytochrome P-450 monooxygenase is responsible for the initial step in the attack and oxidation of DBT to DBT-sulfone (Crawford and Gupta, 1990; Schlenk *et al.*, 1994).

Horseradish peroxidase (Klyachko and Kilbanov, 1992; Wu *et al.*, 1994), purified bovine blood haemoglobin (Wu *et al.*, 1994), cytochrome c (Vazquez-Duhalt *et al.*, 1993; Wu *et al.*, 1994), and myoglobin (Wu *et al.*, 1994) converted DBT to DBTO. A more pragmatic approach for a biotechnological process used haemoglobin (a slaughterhourse waste product),

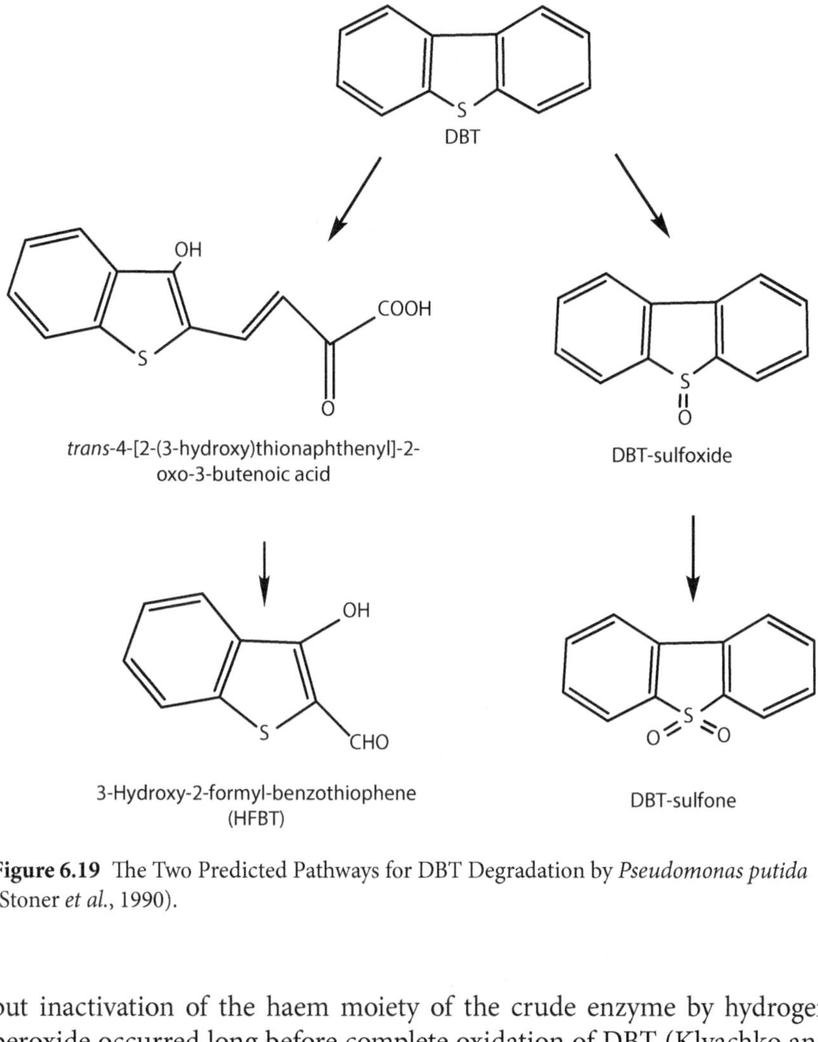

Figure 6.19 The Two Predicted Pathways for DBT Degradation by *Pseudomonas putida* (Stoner *et al.*, 1990).

but inactivation of the haem moiety of the crude enzyme by hydrogen peroxide occurred long before complete oxidation of DBT (Klyachko and Kilbanov, 1992).

6.4.4.1.3 DBT Degradation by Oxidation of Sulfur Heteroatom Followed by C–S Bond Cleavage

Complete degradation of DBT to biomass and sulfate was reported for a mixed culture of *Arthrobacter* and *Pseudomonas* strains isolated for growth on DBT as a sole source of carbon energy and sulfur (Shennan, 1996).

A *Brevibacterium* sp. (Van Afferden *et al.*, 1990) and an *Arthrobacter* sp. (Dahlberg, *et al.*, 1993), isolated following screening for the use of sulfur, were reported to degrade DBT completely through intermediates DBTO and $DBTO_2$ to benzoate and sulfate (Figure 6.20). The *Brevibacterium*

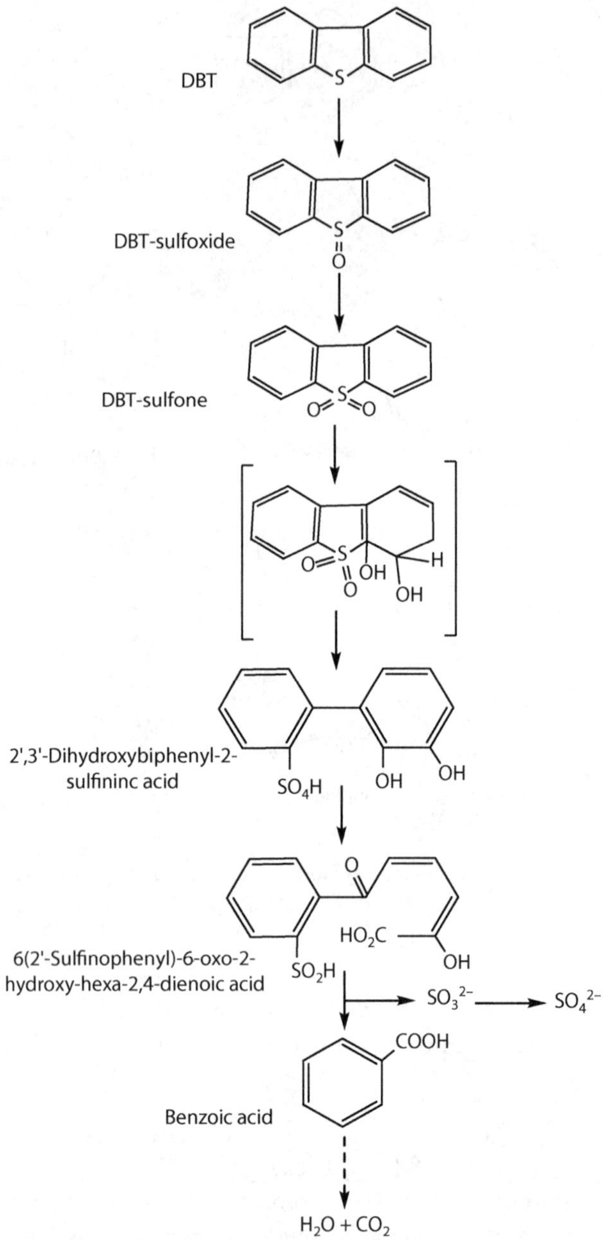

Figure 6.20 Proposed Metabolic Pathway for the Degradation of DBT by a *Brevibacterium* sp. (Van-Afferden *et al.*, 1993).

strain was also reported to desulfurize DBT-supplemented crude oil without attacking the non-sulfur hydrocarbons (Van-Afferden *et al.*, 1993). However, without added DBT, this same strain could not reduce the sulfur content of emulsified heavy oil or bitumen and gave a reduction of only 21 wt.% sulfur in light fuel oil (Klein *et al.*, 1994). *Arthrobacter* DBTS2 reported to mineralize DBTO and $DBTO_2$, but not DBT to sulfate and benzoate (Sato and Clarck, 1995). Nevertheless, Bressler and Fedorak (2000) reported that *Brevibacterium* sp. strain DO initially oxidizes the S-atom to sulfoxide then sulfone, where, only after the production of sulfone, the angular attack occurs (Figure 6.20). This pathway leads to complete mineralization of DBT with the release of S-atom as sulfite, which then oxides to sulfate.

6.4.4.1.4 Sulfur-Specific Cleavage of the C–S Bond Without Further Attack to the Hydrocarbon Skeleton

6.4.4.1.4.1 Biodesulfurization to Hydroxybiphenyl

The biocatalytic activity that is needed for a BDS process is one that proceeds with a non-destructive pathway, removing the sulfur atom from DBT without breaking the hydrocarbon backbone of the molecule because this would decrease the fuel value of the desulfurized product. Thus, microorganisms which carry out the sulfur assimilation necessary for the incorporation of sulfur into the biomass without degrading the carbon skeleton of DBT, are the main important factor for the successfulness of the BDS-process.

The first report of a microorganism capable of removing the sulfur heteroatom from DBT without altering the hydrocarbon structure was published by the Atlantic Research Corporation (1985). *Pseudomonas* strain CB1 was isolated from soil and coal mine samples enriched for cultures able to grow on benzoate as the sole source of carbon and energy in the presence of high concentrations of DBT. Sulfur-specific attacks on DBT by strain CB1 were followed using ^{35}S-DBT and ^{14}C-DBT (Constanti *et al.*, 1996). ^{35}S-sulfate was liberated from DBT, leaving 2,2′-dihydroxybiphenyl and no ^{14}C carbon dioxide was released. However, later reports indicated that the CB1 culture was unstable and subsequently lost.

In a review published in 1989, Kilbane emphasized the need for the selection of organisms capable of sulfur-specific attacks on sulfur-hydrocarbons if coal or crude oil are to be economically desulfurized leaving the aromatic hydrocarbon skeleton intact. Kilbane (1989) proposed a hypothetical thermodynamically favorable set of reactions that, if they ever existed in nature, could specifically remove sulfur from DBT. The pathway was named as a '4S-pathway' (Figure 6.21) and implied consecutive

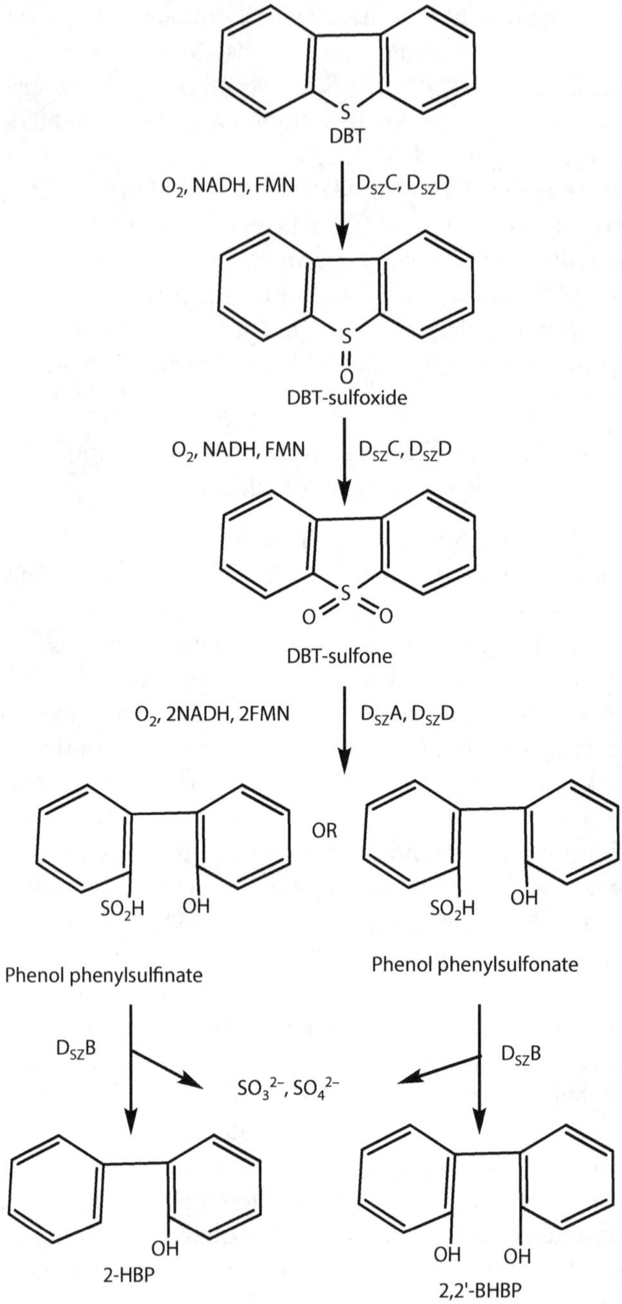

Figure 6.21 DBT-BDS via the 4S-Pathway.

oxidation of DBT sulfur to sulfoxide (DBTO), sulfone ($DBTO_2$), sulfinate (HPBSi), and/or sulfonate (HPBSo) and then, finally, to the oil soluble product 2-HBP or 2,2′-BHBP, respectively, which finds its way back into the petroleum fractions while retaining the fuel value (Krawiec, 1990).

To obtain experimental evidence for the 4S-pathway, a mixed culture, IGTS7, was developed using a bioreactor selection and mutagenesis procedure with DBT as the sole source of sulfur (Kilbane, 1990a). IGTS7 metabolized DBT, predominantly, to 2-HBP. From this mixed culture, a strain of *Rhodococcus erythropolis* IGTS8 (ATCC 53968) able to cleave the C–S bond of DBT was isolated (Kilbane, 1990b). DBT was metabolized to yield 2-HBP and sulfate, where the amount of DBT degraded was directly proportional to the concentration of bacterial cells present, corresponding to the minimum amount of sulfur required for bacterial growth (0.1 mM sulfur allows growth of 109 cell/cm^3). IGTS7 was also claimed to give substantial sulfur removal (60–70%) from crude oil after 7 d. The isolation and characterization of *Rhodococcus erythropolis* IGTS8 led to major advancements in the investigations of BDS (Kayser *et al.*, 1993).

From the mid-1990s, several groups followed the strategy of using DBT as the sole source of sulfur for isolation of cultures capable of sulfur-specific oxidation. Details of the selection strategies and techniques used by different research groups have been reviewed (Krawiec, 1990). This biocatalytic activity has been reported for numerous bacterial strains (Gallagher *et al.*, 1993; Olson *et al.*, 1993; Wang and Krawiec, 1994) and the use of this pathway has been proposed for the desulfurization of petroleum in production fields and refineries (Kim *et al.*, 1996; Soleimani *et al.*, 2007), since the 4S-pathway microorganisms desulfurize the OSCs leaving the hydrocarbon skeleton intact, thus keeping the calorific value of the treated feedstocks.

The "4S" pathway for sulfur removal was first reported for *Rhodococcus erythropolis* IGTS8 by Gallagher *et al.* (1993). Besides *R. erythropolis* IGTS8 (Kilbane and Jackowski 1992), other *Rhodococcus* that are also reported to follow this 4S-pathway are *Rhodococcus* strains UM3 and UM9 (Purdy *et al.*, 1993), *R. erythropolis* D1 (Izumi *et al.*, 1994; Ohshiro *et al.*, 1994; Ohshiro *et al.*, 1995; Mamie *et al.*, 1996; Ohshiro *et al.*, 1997), *R. erythropolis* N1-36 (Wang and Krawiec, 1994, 1996), *R. erythropolis* I-19 (Wang and Krawiec, 1996), *R. erythropolis* DS-3 (Li *et al.*, 2006a,b), *Rhodococcus* sp. ECRD1 (Grossman *et al.*, 1999), *Rhodococcus* sp. B1 (Denis-Larose *et al.*, 1997), *Rhodococcus* sp. SY1 (Omori *et al.*, 1992), *Rhodococcus* sp. KT462 (Tanaka *et al.*, 2002), and *R. erythropolis* KA2-5-1(Kobayashi *et al.*, 2000). Among these, IGTS8 has been studied most extensively. *R. erythropolis* IGTS8 was isolated by Kilbane and Bielaga (1990) and was used by Energy

Biosystems Corp. (EBC) for the development of their commercial microbial desulfurization plan.

However, other strains have been isolated for utilizing DBT via the 4S-pathway, including *Corynebacterium* sp. SY1 (Folsom *et al.*, 1999) and two other *Corynebacterium* isolates (Omori *et al.*, 1992), *Gordona* sp. CYKS1 (Rhee *et al.*, 1998), *Nocardia* sp. CYKS2 (Chang *et al.*, 2000), *Corynebacterium* sp. ZD-1 (Wang *et al.*, 2004), and *Rhodococcus* sp. strain SA11 (Mohamed *et al.*, 2015) removed sulfur from DBT without further degradation of the product 2-HBP. The sulfur heteroatom was either released as sulfate or it was assumed to be incorporated into microbial biomass.

Agrobacterium MC501 and a mixed culture composed of *Agrobacterium* MC501, *Xanthomonas* MC701, *Corynebacterium* sp. MC401, and *Corynebacterium* sp. MC402, all previously isolated from a coal mine area by an enrichment culture with DBT, were used to study DBT and $DBTO_2$ desulfurization. Both cultures were able to use DBT and $DBTO_2$ as a sole source of sulfur for growth. These compounds were metabolized to 2-hydroxybiphenyl (2-HBP) and sulfate (Constanti *et al.*, 1996).

A coal-metabolizing strain of the fungus *Paecilomyces* could also carry out sulfur-specific oxidation of DBT; 2,2′–dihydroxybiphenyl was produced from DBT and diphenylsulfide from thianthrene (only one of its two sulfur heteroatoms being attacked) (Faison *et al.*, 1991).

Two routs of sulfur-specific metabolism of DBT by IGTS8 were tentatively identified. Stationary phase cells produced 2-HBP as the desulfurized product and used 2′-hydroxybiphenyl-2-sulfinate as the key intermediate. Under growth conditions (with glycerol as a carbon and energy source), the intermediates were DBTO, $DBTO_2$, and 2′-hydroxybiphenyl-2-sulfonate, with 2,2′-dihydroxybiphenyl as the final product. The fate of the released sulfur was free sulfate (Gallagher *et al.*, 1993; Olson *et al.*, 1993; Oldfield *et al.*, 1997; Setti *et al.*, 1999).

Omori *et al.* (1992) reported that, the soil isolate *Corynebacterium* sp. strain SY1, utilizes DBT and a wide range of organic and inorganic sulfur compounds as a sole source of sulfur, such as DBT sulfone, dimethyl sulfide, dimethyl sulfoxide, dimethyl sulfone, CS_2, FeS_2, and even elemental sulfur. Strain SY1 metabolizes DBT to DBT sulfoxide and DBT sulfone and 2-HBP, which subsequently nitrated to produce at least two different hydroxynitobiphenyls during cultivation (Figure 6.22). This was attributed to the presence of nitrate in the medium composition and, upon replacement of nitrate by ammonium, the hydroxynitrobiphenyls were produced in trace amounts. Nitration to 2HBP might be a way for the detoxification of 2-HBP. However, introduction of a nitro group into 2HBP by SY1 is not desirable as nitrogen oxides are also regarded as a main cause of acid rain.

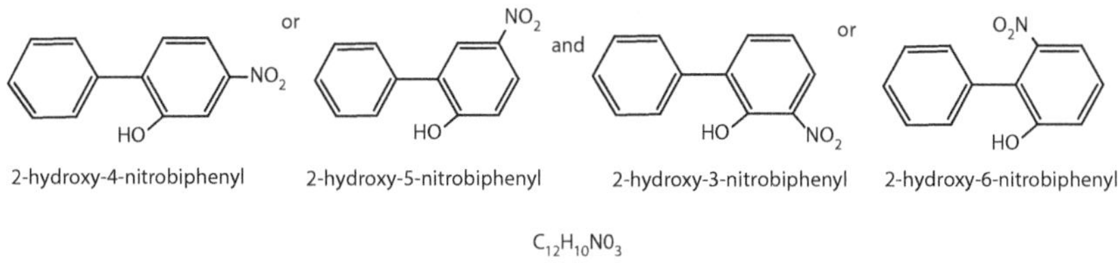

Figure 6.22 The Suggested Hydroxynitrobiphenyls Produced by *Corynebacterium* sp. strain SY1 Omori *et al.* (1992).

El-Gendy *et al.* (2006) reported the isolation of yeast strain *Candida parapsilosis* Nsh45 from oil polluted sea water for its ability to desulfurize DBT via the 4S-pathway producing 2,2′-BHBP.

Nassar *et al.* (2013) reported the isolation of a Gram +ve *Brevibacillus invocatus* C19 (NCBI Gene Bank Accession No. KC999852) from a mineral coke sample, with a sulfur content of 3.8 wt.%, for its ability to selectively desulfurize DBT via the 4S-pathway to 2-HBP as a dead end product.

Khedkar and Shanker (2015) reported an *Actinomycete* R3 isolated from a petroleum contaminated soil sample. This strain is able to desulfurize a DBT and DBT-sulfone to 2-HBP, BT to benzothiophene sulfoxide, then to benzothiophene-S,S-dioxide and 4-hydroxybenzothiophene. Furthermore, it has the ability to biotransform thianthrene to thianthrene-S-oxide. Observations from the polymerase chain reaction (PCR) amplifications, restriction enzyme (RE) digestions, and southern hybridization indicate an overall homology in the R3 and *R. erythropolis* IGTS8 dsz gene sequences, but with some minor variations.

In another study, Nassar *et al.* (2016) reported the isolation of the Gram +ve *Brevibacillus brevis* strain HN1 (accession no. KF018281) from an Egyptian coke sample with an S-content of 2.74 wt.% for its ability to selectively desulfurize DBT via the 4S-pathway to 2-HBP as a dead end product.

6.4.4.1.4.2 Biodesulfurization to Methoxybiphenyl

Ohshiro *et al.* (1996), Monticello (2000), and Kim *et al.* (2004) reported the methoxylation pathway from 2-HBP to 2-MBP as a way to overcome the toxicity of 2-HBP, as 2-MBP would make a less inhibitory effect on the microbe due to the loss of the hydroxy structure of phenolic compounds, thus expressing a lower inhibitory effect on growth and BDS efficiencies of microorganisms (Figure 6.23). This is considered as an extended 4S-pathway and called the 4SM-pathway (Xu *et al.*, 2007).

Later, a methoxylation pathway from 2-HBP to 2-MBP in DBT desulfurization was reported for *Mycobacterium* G3 (Okada *et al.*, 2002b), *Rhodococcus* sp. KT462 (Tanaka *et al.*, 2002), *Sphingomonas subarctica* T7b

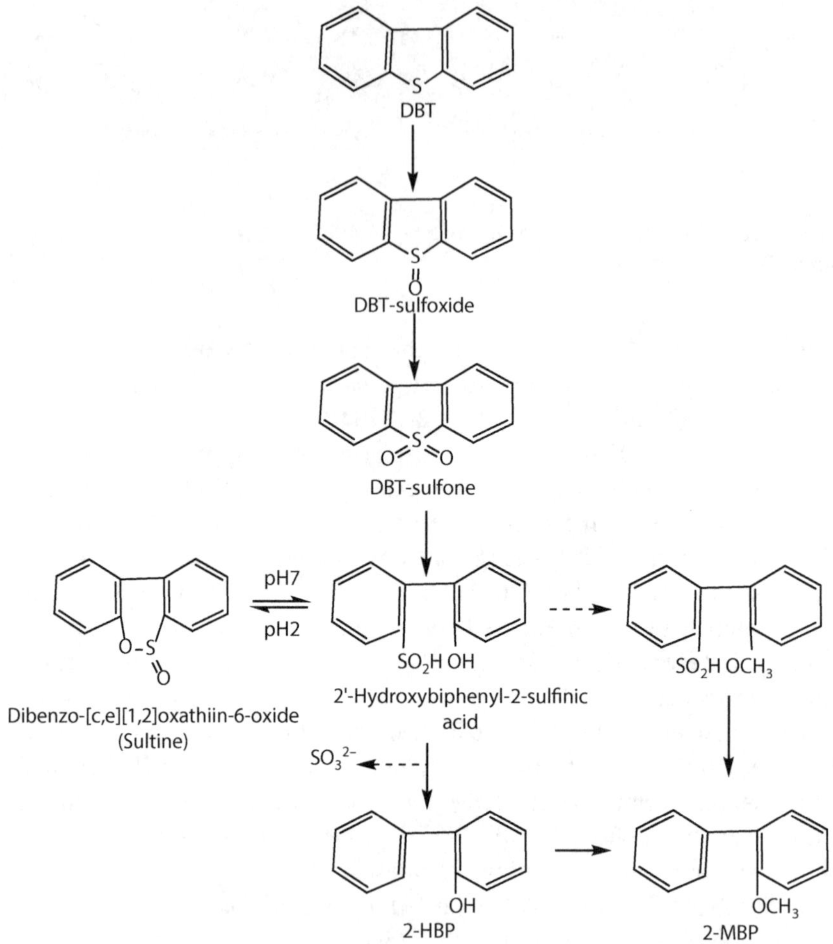

Figure 6.23 4S-Pathway Accompanied by Methoxylation (Okada *et al.*, 2002b).

(Gunam *et al.*, 2006), *Rhodococcus erythropolis* FSD2 (Zhang *et al.*, 2007), and *Mycobacterium* sp. (Chen *et al.*, 2009). The desulfurization activity of *Mycobacterium* G3 is reported to be repressed by sulfate ions at concentrations above 0.15 mM in the growth medium (Nekodzuka *et al.*, 1997). However, Okada *et al.* (2002b) observed the metabolism of DBT in cultures with sulfate ions at 0.1 mM, but not at 0.5 mM. The methoxylation from HBP to MBP was confirmed under both conditions, which revealed that the methoxylation activity was independent of the desulfurization activity and was not affected by sulfate ions in the growth medium.

Onaka *et al.* (2000) also reported that desulfurization of 3-M-DBT by *R. erythropolis* KA2-5-1 produced 4-methyl-2-hydroxybiphenyl as an end metabolite. The DBT metabolic pathway in *Mycobacterium* strain G3, which is classified as a desulfurizing bacterium with the 4S-pathway, was analyzed. 2-HBP, which is the end metabolite in the DBT desulfurization reaction, and 2-methoxybiphenyl (2-MBP) were found in the reaction mixture. 4,6-DMDBT, 4,6DEDBT, and benzo[b]naphtha[2,1-d]thiophene were also metabolized to their methoxy forms (Okada *et al.*, 2002 a,b). Li *et al.* (2003) reported that desulfurization of 4-M-DBT by *Mycobacterium* sp. X7B resulted in the production of 2-hydroxy-3′-methyl-biphenyl. *Achromobacter* sp. reported the flow of the 4S-pathwy upon the desulfurization of 4-MDBT, producing hydroxymethylbiphenyl. The GC–MS profile of the metabolites of 4-MDBT exhibited the molecular ion peak at m/z 184 showing that the ultimate product of metabolism is 2-methoxybiphenyl (2-MBP) (Bordoloi *et al.*, 2014).

Two bacterial strains, which have been identified as *Paenibacillus* strains, are capable of efficiently cleaving carbon-sulfur (C–S) bonds in DBT and its various methylated derivatives at temperatures up to 60 °C, where both growing and resting cells of these bacteria can release sulfur atoms as sulfate ions and leave the monohydroxylated hydrocarbon moieties intact. Moreover, when either of these *Paenibacilli* was incubated at 50 °C with light gas oil (LGO) previously processed through hydrodesulfurization, the total sulfur content in the oil phase was clearly decreased (Konishi *et al.*, 1997).

6.4.4.2 Aerobic Biodesulfurization of Alkylated DBT

Arthrobacter species (Lee *et al.*, 1995; Serbolisca *et al.*, 1999) and *Rhodococcus* sp. strain ECRD-1 (Macpherson *et al.*, 1998; Grossman *et al.*, 1999) could selectively remove sulfur from sterically hindered 4,6-diethyldibenzothiophene (4,6-DEDBT) and 2-hydroxy-3,3′-diethylbiphenyl (HDEBP) was identified as the sulfur-free end product (Figure 6.24).

Similarly, it was shown by Ohshiro *et al.* (1996a) that *R. erythropolis* H-2 removed the sulfur atom from 2,8-dimethyldibenzothiophene, 4,6-dimethyldibenzothiophene, and benzo[b]naphthi[2,1-d]thiophene. The product from the desulfurization of the latter compound was identified as α-hydroxy-β-phenylnaphthalene by GC/MS and NMR. The research by Ohshiro *et al.* (1996a) appears to be the first evidence of a microbial attack of a four-ring condensed thiophene.

Resting cells of *Mycobacterium* strain G3 were reported to desulfurize 4,6-dipropylDBT at 37 °C to 4,6-dipropyl DBT sulfone and 2-hydroxy-3,3′-dipropyl biphenyl in an oil/water biphasic system within 15 h. However, 4,6-dibutyl DBT-sulfone and 2-hydroxy-3,3′-dibutyl biphenyl

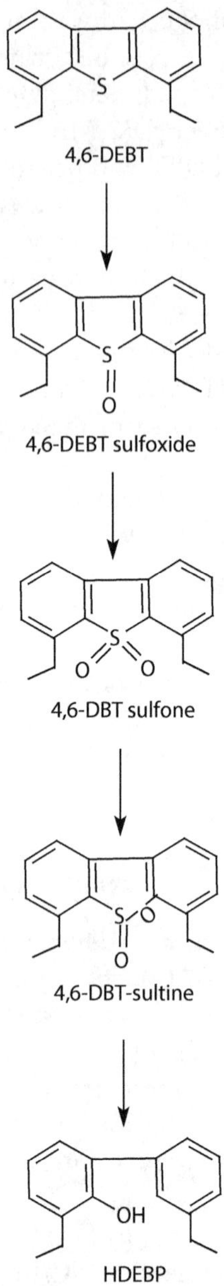

Figure 6.24 BDS of 4,6-Diethyl-DBT by *Rhodococcus* sp. Strain ECRD-1 (Grossman *et al.*, 1999).

were detected only in aqueous 4,6-dibutyl DBT cultures (i.e. one phase systems) with high initial inoculum biomasses and longer incubation periods (21 h). The same occurred with 4,6-dipentyl DBT, but only 2-hydroxy-3,3′-dipentyl biphenyl was produced (Okada *et al.*, 2002a). The difference in reactivity between the oil/water two-phase reaction system and the water phase reaction system was assumed to be the efficiency of contact between cells and substrates. In the case of the oil/water two-phase reaction system, alkyl DBT as a hydrophobic substrate was dissolved in the entire oil phase at the same concentration. However, in the water phase reaction system, the hydrophobic substrate dispersed in the water phase might have the tendency to gather on the cell surface, which is assumed to be a more hydrophobic field in the reaction mixture and to concentrate around the cells (Okada *et al.*, 2002a).

The thermophilic bacterium *Paenibacillus* sp. A11-2, which can utilize DBT as the sole sulfur source at a high temperature (45–55 °C), was investigated for its ability to cleave C–S bonds in the DBT ring with an asymmetrical alkyl substitution, such as methyl, dimethyl, trimethyl, ethyl, and propyl DBTs to their corresponding alkylated HBPs in a rate that differs according to the position of the numbers and the length of alkyl substituents (Okada *et al.*, 2001).

Feng *et al.* (2016) identified the metabolic pathway of DBT and 4,6 DMDBT in *Gordonia* sp. JDZX13 and its key pathway genes (Figure 6.25).

6.4.4.3 Anaerobic Biodesulfurization of DBT

A *Nocardioform actinomycete* FE-9 grown on n-hexadecane with DBT as the sole sulfur source, formed biphenyl, traces of 2-HBP, dihydroxybiphenyl,

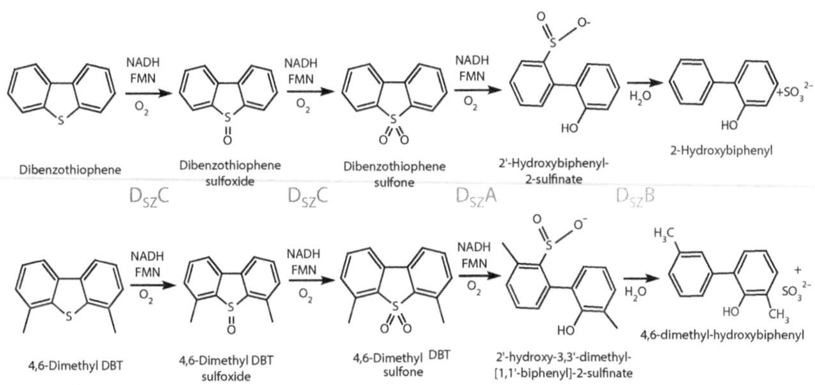

Figure 6.25 BDS of DBT and 4,6-Dimethyl-DBT by *Gordonia* sp. JDZX13. (Feng *et al.*, 2016).

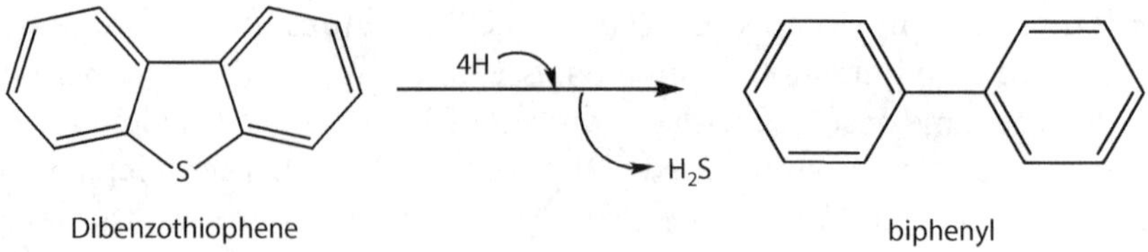

Figure 6.26 Anaerobic Biodesulfurization of DBT by *D. desulfuricans* M6.

and sulfate under air atmosphere and biphenyl and H_2S under nitrogen or hydrogen atmospheres, but the highest conversion rates were under a hydrogen atmosphere. This strain could remove both the sulfur heteroatoms in thianthrene, converting it to benzene and sulfate under an oxygen atmosphere. Terthiophene was transformed to H_2S and an unsaturated product tentatively identified as 1,3,5,7,9,11-dodecahexaene (Figure 6.7) (Finnerty, 1992, 1993b).

During the reductive conversion, DBT is used as the sole electron acceptor and sulfur is removed selectively (Kim *et al.*, 1990a). *Desulfovibrio desulfuricans* M6 is a sulfate reducing bacterium isolated from soil, characterized by its high hydrogenase activity, that produces biphenyl as the main product during BDS of DBT (Figure 6.26) (Kim *et al.*, 1995). To remove sulfur compounds in petroleum, sulfate reducing bacteria *Desulfomicrobium escambium* and *Desulfovibrio longreachii* were isolated so they could degrade DBT in the presence of only nitrogen gas in a pathway (Yamada *et al.*, 2001) different from the anaerobic one already reported, in which biphenyl is detected as the main product (Kim *et al.*, 1990b).

6.5 Recent Elucidated DBT-BDS Pathways

El-Gendy (2004) reported that the Gram +ve *Arthrobacter* sp. NSh39 isolated from oil polluted Egyptian soil possess two different pathways for the metabolism of DBT (Figure 6.27). The HPLC and GC/MS analysis of the produced metabolites, revealed that NSh39 utilizes DBT via an aromatic ring oxygenation, ring cleavage, and the release of pyruvic acid with the production of thiosalicylic acid and benzothiophene (BT) as an intermediate. Although BT has not been reported to be produced as a BDS intermediate, BT was produced as an end product by *Pseudomonas* sp. TG232 from the BDS of DBT as reported by Kilbane (1990a). *Pseudomonas fluorescens*17 are also reporteded to produce thiosalicylic acid from DBT through the production of HFBT via the Kodama pathway (FinkelStein *et al.*, 1997). The observed Kodama-pathway (El-Gendy *et al.*, 2004) was branched to

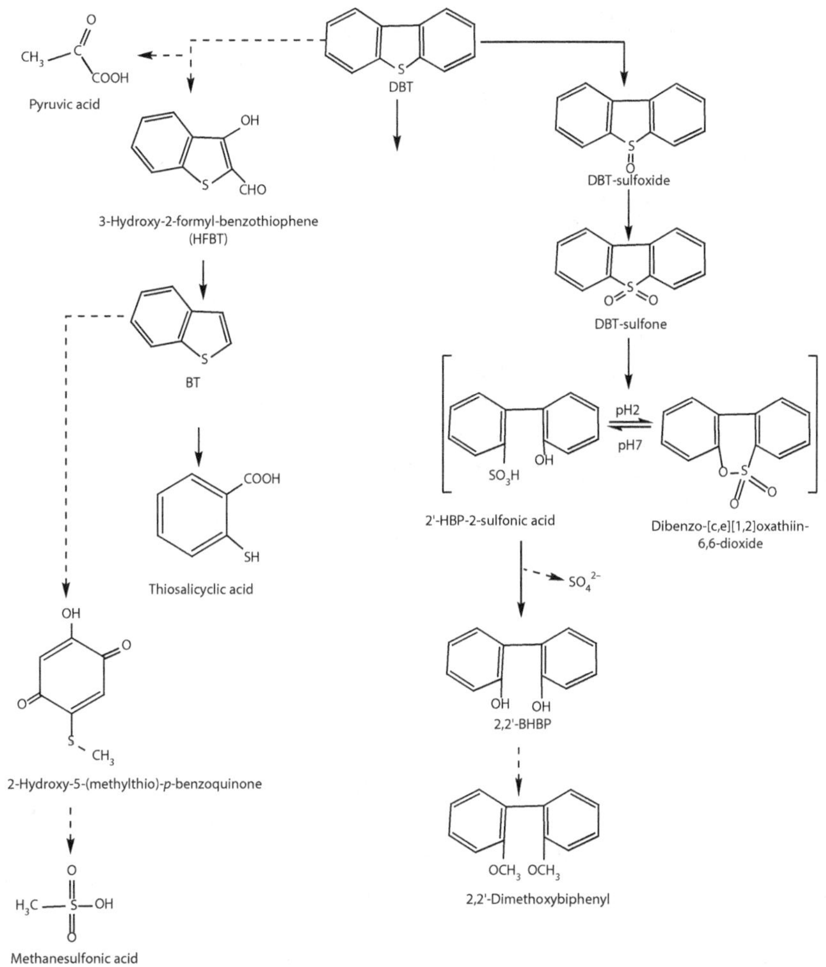

Figure 6.27 DBT-BDS Pathway by *Arthrobacter* sp. NSh39 (El-Gendy, 2004).

another pathway producing methane sulfonic acid as a dead end product via the production of 2-hydroxy-5-methylthio-*p*-benzoquinine. The other pathway is a 4S-pathway producing 2,2′-bihydroxybiphenyl (2,2′-BHBP), which was further metabolized to 2,2′-dimethoxybiphenyl (2,2′-DMBP). The presence of sultone (i.e. dibenzo-[c,e][1,2]oxathiin-6,6-dioxide) is attributed to the spontaneous cyclization of 2′-HBP-2-sulfonic acid (2-HBPSo) under the acidic conditions used in the work up and preparation of samples for analysis (Olson *et al.*, 1993; Oldifield et a;., 1997; Gibert *et al.*, 1998; Reichmuth *et al.*, 2000). Many phenolic compounds are harmful biologically (Omari *et al.*, 1992; Nekodzuka *et al.*, 1997; Arafa *et al.*,

2001; Kayser *et al.*, 2002) and 2,2′-BHBP is generally considered as a phenol derivative. 2,2′-BHBP is also reported to be an inhibitor for growth and desulfurization activity (Ohshiro *et al.*, 1996b; Monticello *et al.*, 2000; Kim *et al.*, 2004). Thus, NSh39 produced 2,2′-DMBP throughout the methoxylation of 2,2′-BHBP for the detoxification and protection of microorganisms from its toxic effect (Okada *et al.*, 2002a,b). Moreover, El-Gendy *et al.* (2004) observed that the accumulation of DBTO and $DBTO_2$ appeared to be a side reaction in the metabolism of DBT, as the accumulation of 2,2′-BHBP, 2′-HBP-2-sulfonic acid, and sultone were in minor amounts. Lee *et al.* (1995) reported that the removal of sulfur by *Arthrobacter* sp. is tied to the sulfur requirements of the cell. Serbolisca *et al.* (1999) found that the specific desulfurization of *Arthrobacter* sp. DS7 was completely inhibited when the strain was grown in the presence of low concentrations of sulfate. So, *Arthrobacter* sp. NSh39 might have been directed to utilize DBT through the Kodama pathway to overcome the inhibition effect caused by the presence of sulfate ions. It is possible that the enzyme that affects oxygenation of the aromatic nucleus also catalyzes sulfoxygenation (Laborde and Gibson, 1977; Frassinetti *et al.*, 1998). However, the further oxidation up to the production of 2,2′-BHBP might be an easier way for NSh39 to get the SO_4^{2-} ions necessary for cell growth and maintenance, which might have been assimilated by the cells and incorporated into sulfur-containing amino acids and vitamins necessary for growth (Monticello, 1998). According to Kilbane (1990b), sulfur can be liberated from DBT via the Kodama-pathway, but there is no specificity to the metabolism of sulfur, rather sulfur is metabolized only in the course of the overall degradation of the DBT molecule. The consequence of this is that to achieve 50% desulfurization, 50% of the substrate is consumed. El-Gendy *et al.* (2004) also observed that HFBT was not greatly accumulated, which was in agreement with results obtained by Bressler and Fedorak (2001 a,b) who explained this phenomenon by two possible reasons: the chemical instability of HFBT and its susceptibility to mineralization. However, HFBT is the main product formed from DBT by *Beijerickia* (Laborde and Gibson, 1977), *Pseudomonas* (Fought and Westlake, 1990), and *Rhizobium meliloti* (Frassinetti *et al.*, 1998), which are reported to be unstable intermediates in DBT biodegradation, but were fairly stable at pH11 and were also stable enough in neutral sulfate-free mediums. Other studies have shown that sulfur from thiophene rings can be used as the sole sulfur source for bacterial growth (Bressler *et al.*, 1999; Greene *et al.*, 2000), so it is likely that HFBT would serve as a sole sulfur source. However, it has yet to be determined whether this is true.

El-Gendy *et al.* (2005) reported that the Gram –ve *Agrobacterium* sp. NSh10 isolated from hydrocarbon polluted Egyptian soil desulfurizes

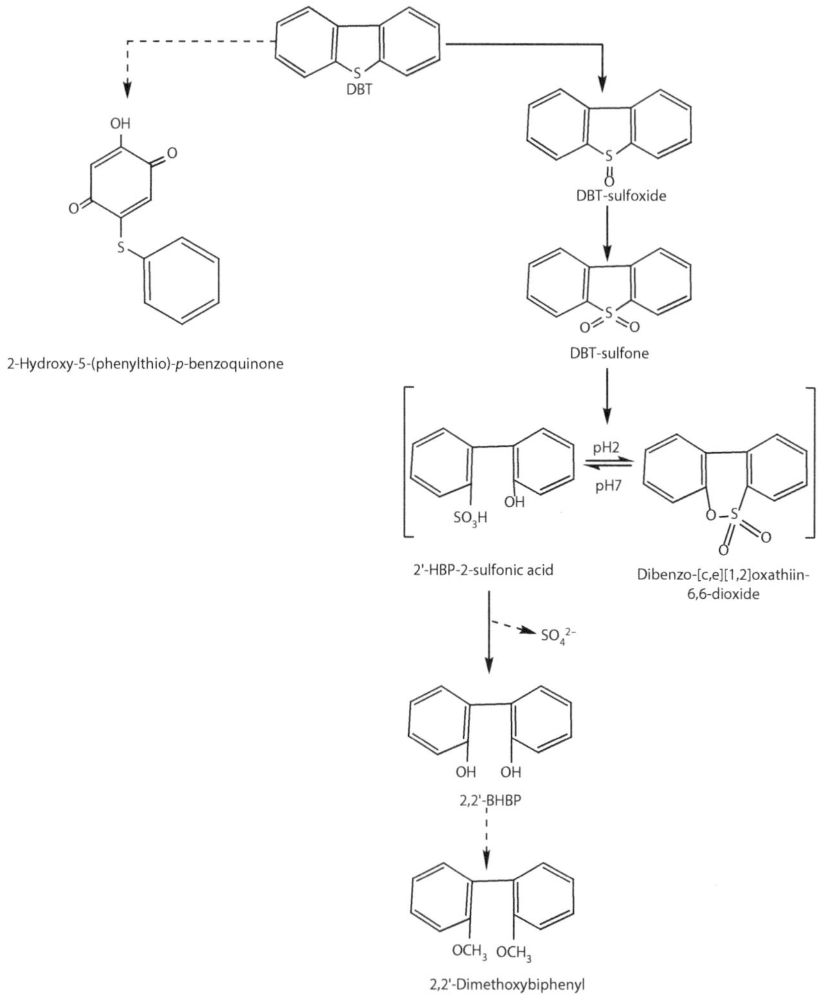

Figure 6.28 DBT-BDS Pathway by *Agrobacterium* sp. NSh10 (El-Gendy *et al.*, 2005).

DBT to water soluble benzoquinone as a major product and 2,2′-BHBP as a minor product (Figure 6.28). *Agrobacterium* MC501 is reported to desulfurize DBT and $DBTO_2$ through the 4S-pathway producing 2-HBP instead of 2,2′-BHBP (Constanti *et al.*, 1996).

El-Gendy (2006) reported the isolation of two bacterial strains from oil polluted soil samples, including Gram +ve *Aureobacterium* sp. NShB1 and Gram –ve *Enterbacter* sp. NShB2, for their abilities to degrade DBT to benzoate through the production of 2-HBP and 2,2′-BHBP, respectively (Figure 6.29). The presence of sultine, dibenz-[c,e][1,2]oxathiin-6-oxide, and sultone

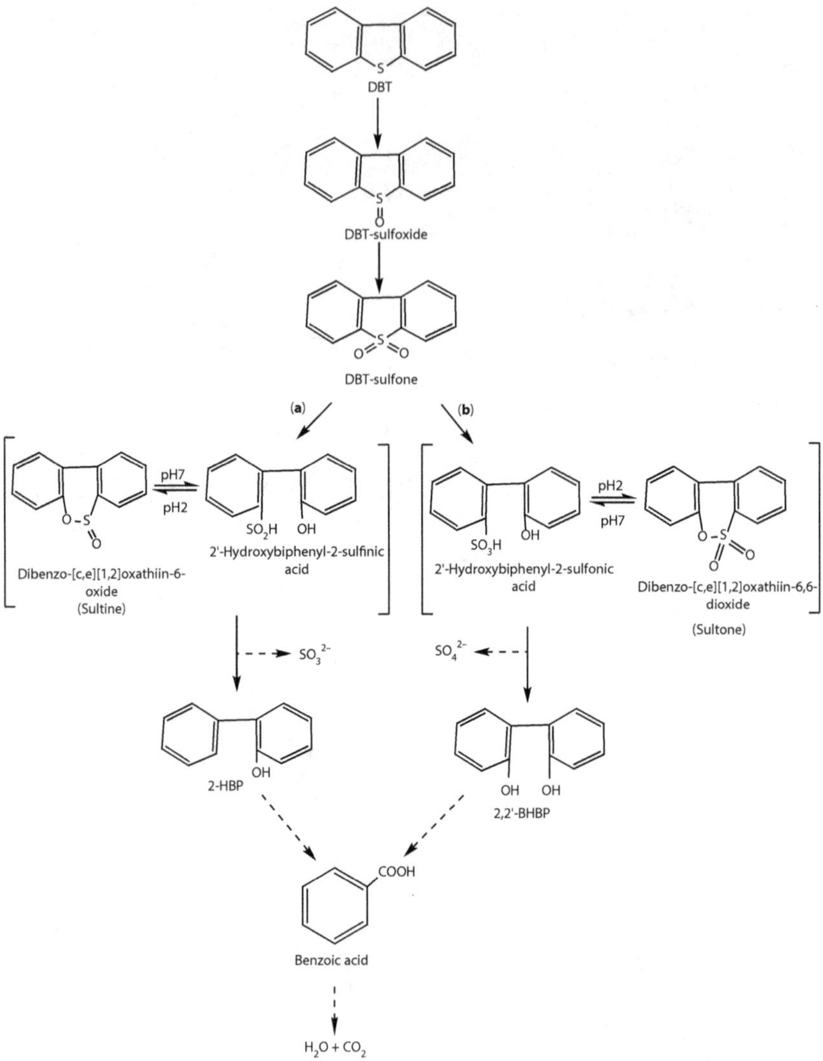

Figure 6.29 DBT-Biodegradation Pathway by (A) *Aureobacterium* sp. NShB1 and (B) *Enterobacter* sp. NShB2 (El-Gendy, 2006).

dibenz-[c,e][1,2]oxathiin-6,6-dioxide (Figure 6.29) is due the cyclization of 2′-HBP-2-sulfinic acid and 2′-HBP-2-sulfonic acid under the acidic conditions used in the work up and preparation of samples for analysis (Olson *et al.*, 1993; Oldfield *et al.*, 1997, Gilbert *et al.*, 1998; Reichmuth *et al.*, 2000).

El-Gendy and Abo State (2008) reported that the Gram +ve *Staphylococcus gallinarium* NK1 isolated from an Egyptian drain receiving oil refinery wastewater desulfurizes DBT through the Kodama pathway,

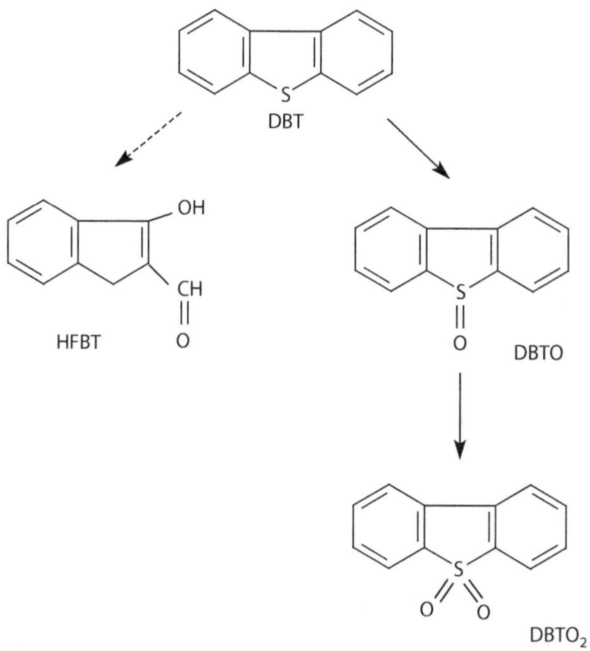

Figure 6.30 DBT-Biodegradation Pathway by *Staphylococcus gallinarum* NK1 (El-Gendy and Abo State, 2008).

producing the yellow water soluble HFBT as a major product with a minor accumulation of DBT-sulfoxie and DBT-sulfone (Figure 6.30). NK1 also has great biodegradation capabilities on DBT, phenanthrene, and naphthalene. These results were in agreement with results obtained by Frassinetti *et al.* (1998), where a nodulating isolate *Rhizobium meliloti* produced 6 degradation intermediates. Three of these products were identified as HFBT, benzothienopyran-2-one, and DBTO, but DBTO was obtained as a dead end product without further metabolism to 2,2′-BHBP. According to Laborde and Gibson, (1977) metabolism of DBT by *Beijerinckia* species was through the Kodama pathway and the formation of DBTO appeared to be a side reaction in the metabolism of DBT, but the produced oxide was not metabolized further. Their possible explanation for the side minor production of DBTO, concomitantly with the Kodama pathway, is that the enzyme which affected oxygenation of the aromatic nucleus of DBT also catalyzed sulfoxygenation. Mormile and Atlas (1988) demonstrated that a *Pseudomonas putida* could degrade DBT by two parallel pathways, one yielding HFBT and the other forming DBTO and $DBTO_2$. The sulfoxide and sulfone were completely degraded, but without detectable release of sulfur heteroatom.

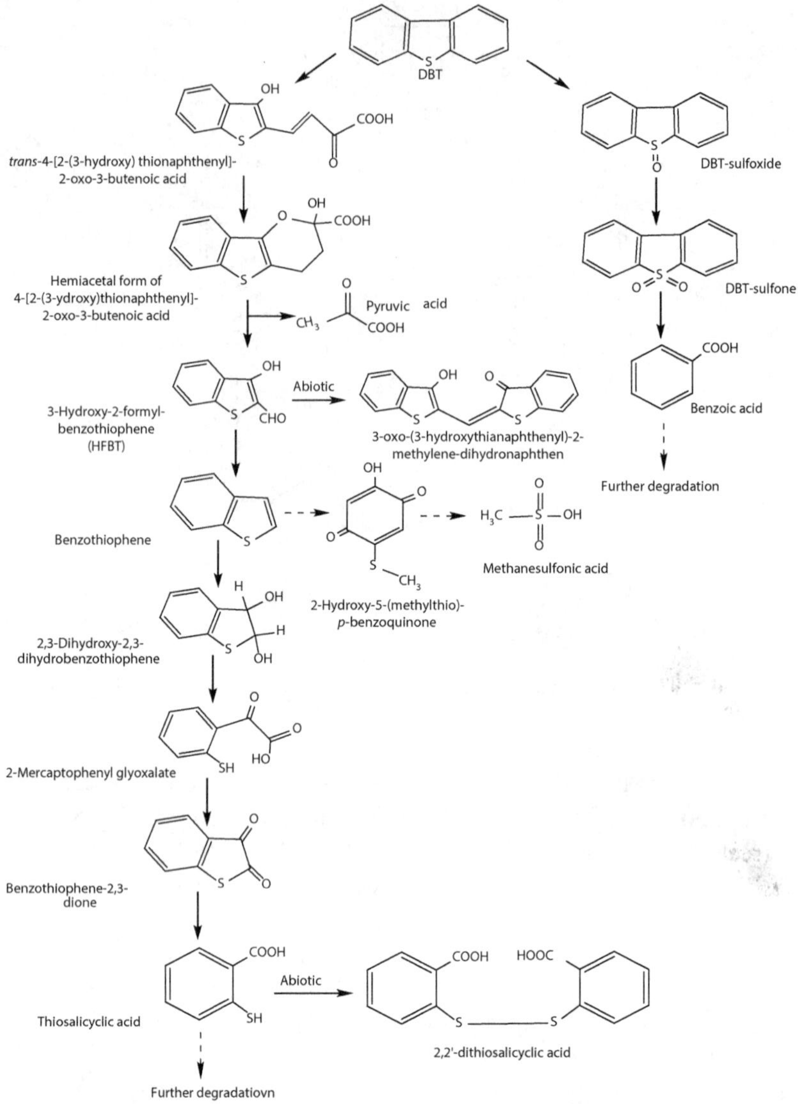

Figure 6.31 DBT-Biodegradation Pathway by *Bacillus sphaericus* HN1 Nassar *et al.* (2010).

Nassar (2010) reported the biodegradation of different components of diesel oil by *Bacillus spharicus* HN1 isolated for its ability to degrade DBT to benzoic acid and thiosalicyclic acid through the 4S and Kodama pathways, respectively (Figure 6.31). Briefly, the GC/MS revealed the production of 3-oxo-(3-hydroxythianaphthenyl)-2-methylene-dihydronaphthene as a major peak, which was attributed to the abiotic dehydration of 2 molecules of HFBT and BT which were also detected as minor products. BT was further degraded to produce 2,3-dihydroxy-2,3-dihydrobenzothiophene, leading to the production of another major product, 2-mercaptophenyl glyoxalate, which was furtherly degraded to produce thiosalicylic acid, which was abiotically dimerized to produce 2,2′-dithiosalicylic acid as one of the end products. A new identified byproduct, 2-hydroxy-5-(methylthio)-*p*-benzoquinone, which produced methanesulfonic acid was also detected.

Bohnos *et al.* (1977) reported 3-hydroxybenzothiophene and BT-2,3-dione that may originate from HFBT. Fedorak and Grbić-Galić, (1991) reported that *Pseudomonas alcaligenes* DBT2 oxidized BT to water soluble products identified as BT-sulfoxide, 2,3-dihydrobenzothiophene-2,3- diol, and BT-2,3-dione. Eaton and Nitterauer (1994) showed that when an acidic condition is used to extract a culture supernatant containing 2-mercaptophenyl glyoxalate, an acid-catalyzed dehydration occurs to form BT-2,3-dione. The same is true for Nassar (2010) conditions. Finkelstein *et al.* (1997) reported that *Pseudomonas fluorescens* 17 is able to utilize DBT as a sole source of carbon and energy via the Kodama pathway, producing HFBT, 2,3-dihydroxybenzothiophene and thiosalicylic acid, as well as 2,2′-dithiosalicylic acid. Seo *et al.* (2006) reported that *Arthrobacter* sp. P1-1 is able to degrade DBT to DBT-diols that are transformed from 1,2- and 3,4- dioxygenation of DBT to produce BT-2,3-dicarboxylic acid and hydroxyl-BT carboxylic acid, respectively. These products are converted into BT-2,3-diol which is transformed to 2,2′-dithiosalicylic acid via thiosalicylic acid.

The detection of DBTO, $DBTO_2$, and HFBT suggests that *Bacillus spharicus* HN1 possess a mono-oxygenase system in addition to a di-oxygenase system (Nassar *et al.*, 2010). Gai *et al.* (2007) reported that *Sphingomonas* sp. possess two different pathways for the metabolism of DBT. The first supports cell growth and produces HFBT as the end metabolite, while the second leads to the production of DBT-5-oxide and seems to be a dead end route. Kilbane and Jackowski (1992) reported that *Bacillus sphaericus* has a monooxygenase system, in addition to a dioxygenase system, due to the detection of DBT-sulfone and metabolites of the Kodama pathway. Nassar *et al.* (2010) explained the performance of the minor 4S-patway as an easier way for HN1 to get the SO_4^{2-} ions, which are necessary for cell growth and maintenance, that might be assimilated by the cells and incorporated into sulfur-containing amino acids and vitamins necessary for growth (Monticello, 1998b). Moreover, Nassar (2010) explained the further degradation of 2-HBP to benzoic acid, as a route *Bacillus spharicus* HN1 performs to overcome the toxicity of 2-HBP (Omori *et al.*, 1992; Nekodzuka *et al.*, 1997; Arafa *et al.*, 2001; Kayser *et al.*, 2002).

El-Gendy *et al.* (2010) reported the complete mineralization of DBT by the halo-tolerant Gram +ve bacterial isolate *Corynebacterium variabilis* sp.

Sh42 (Figure 6.32), which also showed great biodegradation capabilities on different polyaromatic compounds. The obtained suggested pathway in that study proposed that an aromatic ring of 2,2′-BHBP is degraded via a site-specific monooxygenase that hydroxylates aromatic compounds at the C-3 position, where there is a hydroxyl group at C-2 and alkyl or phenyl rest at C-1. Interestingly, 2,2′,3-trihydroxybiphenyl also serves as a substrate for the monooxygenase activity producing 2,2′,3,3′-tetrahydroxybiphenyl. Therefore, the monooxygenase also hydroxylates the C-3′position of

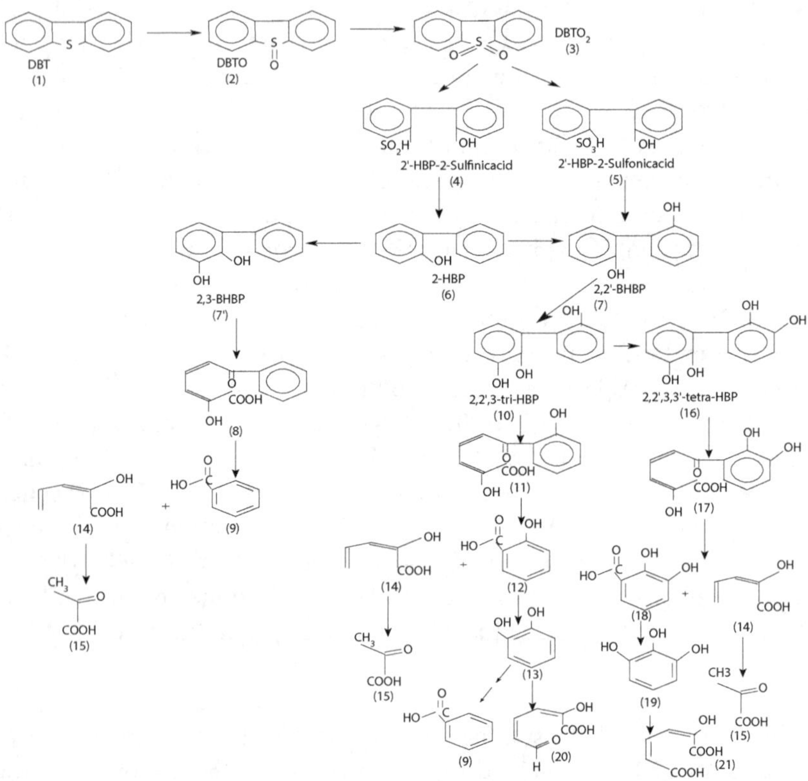

Figure 6.32 DBT-Biodegradation Pathway by *C. variabilis* sp. Sh42 (El-Gendy *et al.*, 2010). 1, dibenzothiophene (DBT); 2, DBT-sulfoxide; 3, DBT-sulfone; 4,2-hydroxybiphenyl sulfinic acid; 5, 2-hydroxybiphenyl sulforic acid; 6,2-hydroxybiphenyl; 7, 2,2′-bihydroxybiphenyl; 7′, 2,3-bihydroxybiphenyl; 8, 2-hydroxy-6-oxo-phenylhexa-2,4-dienoic acid;9, benzoic acid; 10, 2,2′,3-trihydroxybiphenyl; 11, 2-hydroxy-6-(2-hydroxyphenyl)-6-oxo-2,4-hexadienoic acid; 12, salicylic acid; 13, catechol; 14, 2-hydroxy-2,4-pentadienoic acid; 15, pyruvic acid; 16, 2,2′,3,3′-tetrahydroxybipheny; 17, 2-hydroxy-6-(2,3-dihydroxyphenyl)-6-oxo-2,4-hexadienoic acid; 18, 2,3-dihydroxybenzoic acid; 19, pyrogallol; 20, 2-hydroxymuconic semialdehyde; 21, 2-hydroxymuconic acid.

2,2′,3-trihydroxybiphenyl. This finding provides additional evidence for the previously suggested relaxed specificity of the monooxygenase with respect to the molecular rest at the C-1 position of the aromatic backbone structure reported by Kohler *et al.* (1988). There was a yellow coloration observed in phenolic cultures inoculated with Sh42, which might be due to the production of 2,2′,3,3′-tetrhydroxybiphenyl. Kohler *et al.* (1993) reported the formation of yellow meta-cleavage compounds from 2,2′,3-trihydroxybiphenyl (2-hydroxy-6-(2-hydroxyphenyl)-6-oxo-2,4-hexadienoic acid) and 2,2′,3,3′-tetrahydroxybiphenyl (2-hydroxy-6-(2,3-dihydroxyphenyl)-6-oxo-2,4-hexadienoic acid), which did not remain stable in an aqueous solution. They also reported the formation of yellow meta-cleavage metabolite 2-hydroxymuconic semi-aldehyde from catechol produced from salicylate monooxygenases of salicylic acid, produced through the biodegradation of 2,2′,3-trihydroxybiphenyl, which was produced from the monooxygenase of 2,2′-bihydroxybiphenyl. The proposed pathway for the metabolism of 2,2′-bihydroxybiphenyl presented in Figure 6.32 indicates that the first intermediate, 2,2′,3-trihydroxybiphenyl, may be metabolized via two different routes. On one hand, it serves as a substrate to the extradiol ring cleavage dioxygenase and, on the other hand, it can be turned over by the monooxygenase. The conversion of produced catechol and pyrogallol to 2-hydroxymuconic semi-aldehyde and 2- hydroxymuconic acid, respectively, indicates that the extra-diol ring cleavage dioxygenase activity from strain Sh42 is a broad-spectrum *meta*-cleavage dioxygenase because it is able to turn over various 2,2′,3-trihydrox- and 2,2′,3,3′-tetrahydroxybiphenyl, catechol, and pyrogallol. 2-HBP can also be degraded by dioxygenation of vicinal *ortho-meta* carbons of the unsubstituted ring producing 2,2′-BHBP and then 2,2′,3-trihydroxybiphenyl, which is furtherly degraded through a *meta*-cleavage pathway as discussed before. Sondossi *et al.* (2004) reported that *Comamonas testosterone* B-356 is able to metabolize monohydroxybiphenyls through the biphenyl catabolic pathway leading to the production of benzoic acid and 2-hydroxypentanoate.

Ismail *et al.* (2016) reported that the BDS of DBT by resting cells of mixed culture AK6 produced 2-hydroxybiphenyl (2-HBP) in addition to trace amounts of phenylacetate. AK6 is a mixed culture and 2-HBP is a toxic phenolic biocide that can damage cell membranes, thus it was speculated that phenylacetate might have originated from 2-HBP via an unknown detoxification mechanism by one or more of the AK6 bacterial components. However, the likelihood of 2-HBP conversion to phenylacetate by AK6 bacteria, lacking the BDS-activity, could not be ruled out. Ismail *et al.* (2016) postulated that phenylacetate might not be considered as a new intermediate in an extended 4S pathway, rather it might be a product of cometabolism of a sulfur-free substrate (2-HBP). The proposed transformation of 2-HBP to phenylacetate could probably enhance the BDS activity of AK6 by eliminating the toxic and inhibitory effect of 2-HBP. This was confirmed by the slight ability of AK6 to grow on 2-HBP as a carbon source.

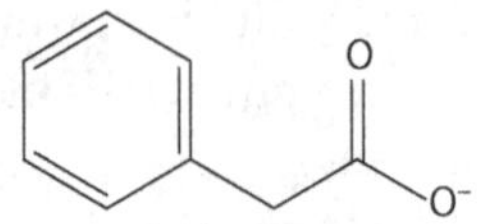

Phenylacetate

The above ring-destructive pathways are not commercially useful for the petroleum industry because the use of the carbon skeleton of sulfur compounds by the bacteria reduces the fuel's calorific value, but they are highly recommended for the bioremediation of oil polluted environments to degrade the recalcitrant organosulfur compounds.

However, there are new pathways performed without altering the hydrocarbon skeleton of the OSCs. Most of them are initially performed via the conventional 4S-pathway and then branched to new novel paths.

Biodesulfurization of DBT throughout the 4S-pathway was reported for *Serratia marcescens* (UCP 1549), producing 2-HBP. Nevertheless, to overcome the toxicity of 2-HBP, it bio-transforms 2-HBP to biphenyl (Figure 6.33) (Casullo de Araújo *et al.*, 2012). The further metabolism of 2-HBP to biphenyl has been also reported for *Rhodococcus* spp Eu-32 (Akhtar *et al.*, 2009).

El-Gendy *et al.* (2014) reported the isolation of Gram +ve *R. erythropolis* HN2 (accession no. KF018282) from an Egyptian coke sample with an S-content of 2.74 wt.% for its ability to desulfurize DBT via the 4S-pathway to 2-HBP. The produced 2-HBP was detoxified to 2-methoxybiphenyl (2-MBP), then further oxidized to form 2-methoxy [1,1′-biphenyl]-2-ol, then subsequently methoxylated to yield 2,2′-dimethoxybiphenyl (2,2′-DMBP) (Figure 6.34). Methoxylation of 2-HBP is a way for microorganisms to overcome the toxicity of 2-HBP. Also, one of the minor produced dead end products was elucidated to be 2,2′,3-trihydroxybiphenyl.

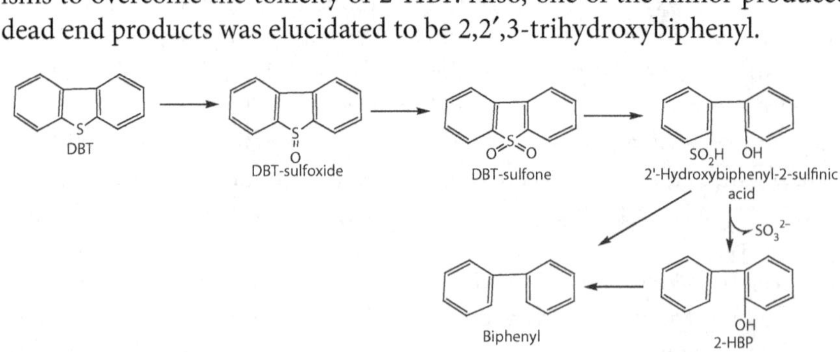

Figure 6.33 DBT-BDS Pathway via the Production of Biphenyl as a Dead End Product.

Pseudomonas aeruginosa S25 isolated from oil contaminated soil in Iraq was reported to overcome the toxicity of 2-HBP throughout the production of 2-methoxybiphenyl (2-MBP). However, the GC/MS analysis of the culture extract revealed two more metabolites: 1,2-naphthalenediol-2-ethyl-1,2,3,4-tetrahydro- and cis and 4-methoxybenzhydrol (Figure 6.35) (Al-Jailawi *et al.*, 2015).

Fungi are capable of metabolizing a wide range of aromatic hydrocarbons through cytochrome P450 and their extracellular enzymes (Van Hamme *et al.*, 2003; Etemadifar *et al.*, 2006; Husaini *et al.*, 2008). Crawford and Gupta (1990) reported that the fungus *C. elegans* converted 99% of DBT to DBT-5-oxide and DBT-sulfone after 7 days. This DBT metabolism occurred by the P-450 cytochrome, but not further because of the lack of appropriate enzymes. *Paecylomyces* sp. desulfurized DBT by producing 2,2′-di-ydroxybiphenyl using a sulfur-specific oxidation pathway (Faison *et al.*, 1991). However, the fungal conversion of DBT by white basidiomycete *Coriolus versicolor* to DBT-5-oxide and -5-dioxide, as the major products from DBT after a 2-day incubation, occurred only under ligninolytic conditions (i.e high carbon and low nitrogen), strongly suggesting that the S-oxidation reactions were catalyzed by ligninolytic enzyme(s). It also oxidizes 4-MDBT, 2-MBT, and 7-MBT to their corresponding sulfones,

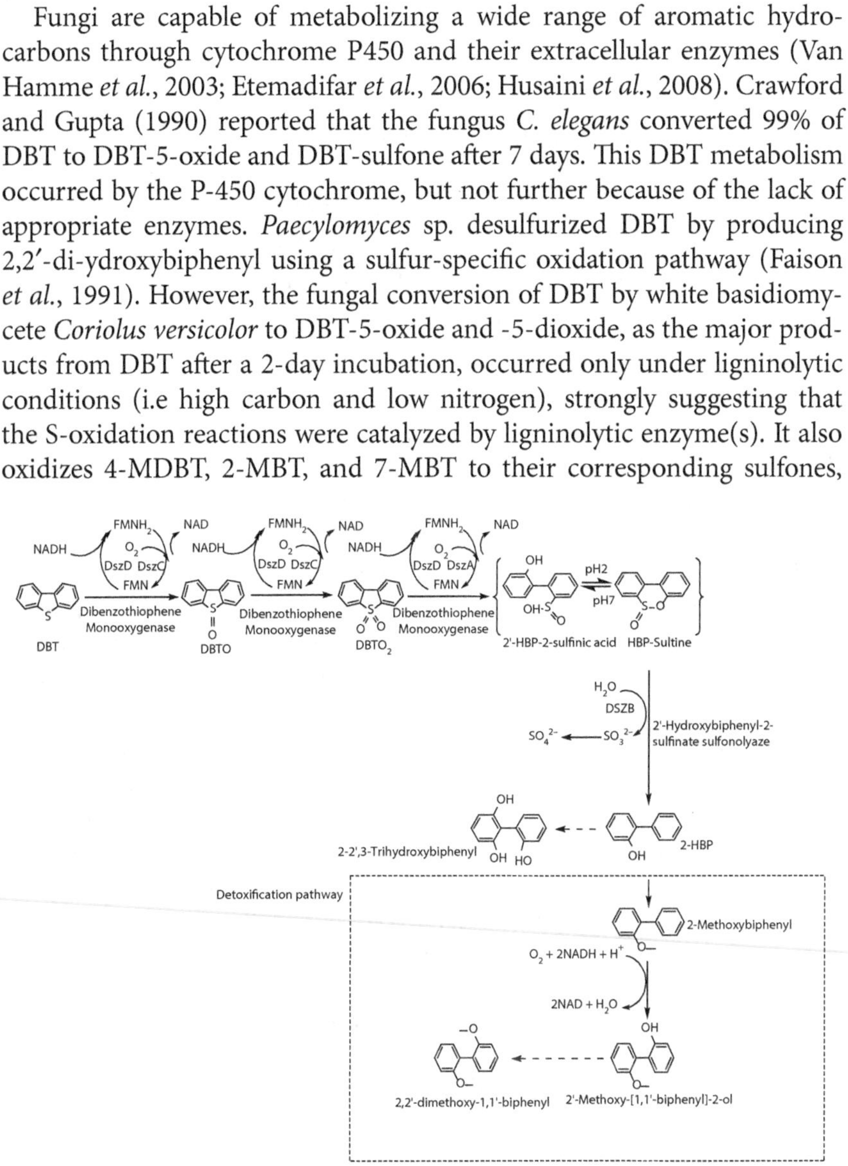

Figure 6.34 DBT-BDS Pathway by *Rhodococcus erythropolis* HN2 (El-Gendy *et al.*, 2014).

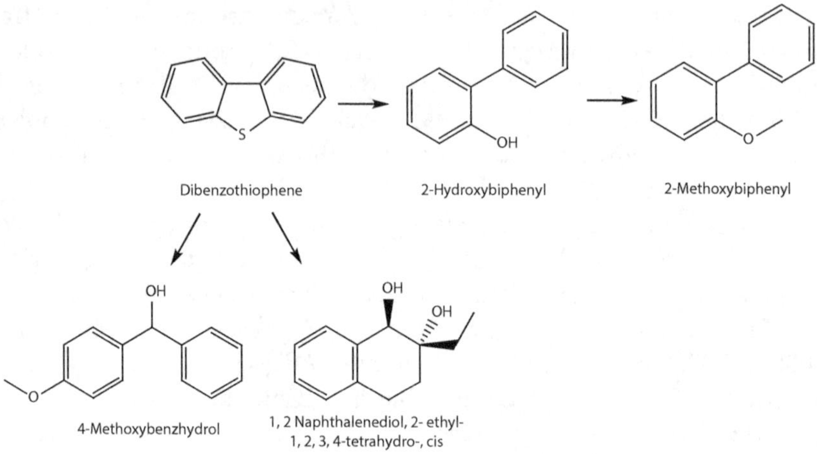

Figure 6.35 Different Metabolites of DBT-BDS by *P. aeruginosa*-S25 (Al-Jailawi *et al.*, 2015).

suggesting also the involvement of cytochrome P450 (Ichinose *et al.*, 2002b). *Rhodosporidium toruloides* strain DBVPG 6662 has been reported to metabolize DBT to DBT-5-oxide and DBT-5-dioxide, but does not produce biphenyl (Baldi *et al.* 2003). Conversion of DBT to DBT-sulfone by *Bjerkandera* sp. BOS55 with peroxidase-manganase enzyme, followed by ring cleavage and production of 4-methoxy benzonic acid has been reported by Eibes *et al.* (2006). Also, oxidation of biphenyl to the hydroxylated derivatives 4,4′-dihydroxybiphenyl, 3,4-dihydroxybiphenyl, 2-hydroxybiphenyl, and ring cleavage product 4-phenyl-2-pyrone-6-carboxylic acid by filamentous fungus *Talaromyces helices* was reported by Pinedo-Rivilla *et al.* (2009).

The conversion of dibenzothiophene by the whole cells of mushrooms *Agrocybe aegerita* TMA1 and *Coprinellus radian* DMSZ 888 (Aranda *et al.*, 2009) occurred; *A. aegerita* oxidized DBT (110 μM) by 100% within 16 days into eight different metabolites. Among the latter, were mainly S-oxidation products (DBT sulfoxide, DBT sulfone) and in lower amounts, ring-hydroxylation compounds (e.g., 2-hydroxyDBT), while *C. radians* converted about 60% of DBT into DBT sulfoxide and DBT sulfone as the sole metabolites. Metabolites detected during DBT conversion suggest that *A. aegerita* and *C. radians* oxidize DBT via two pathways: S-oxidation leading to the formation of DBT sulfoxide and sulfone, as well as ring hydroxylation resulting in differently hydroxylated benzene rings (Figure 6.36). Furthermore, both fungal species were described to produce a new type of extracellular heme-thiolate protein combining activities of classic peroxidases, haloperoxidases, and P450 monooxygenases (Ullrich and Hofrichter

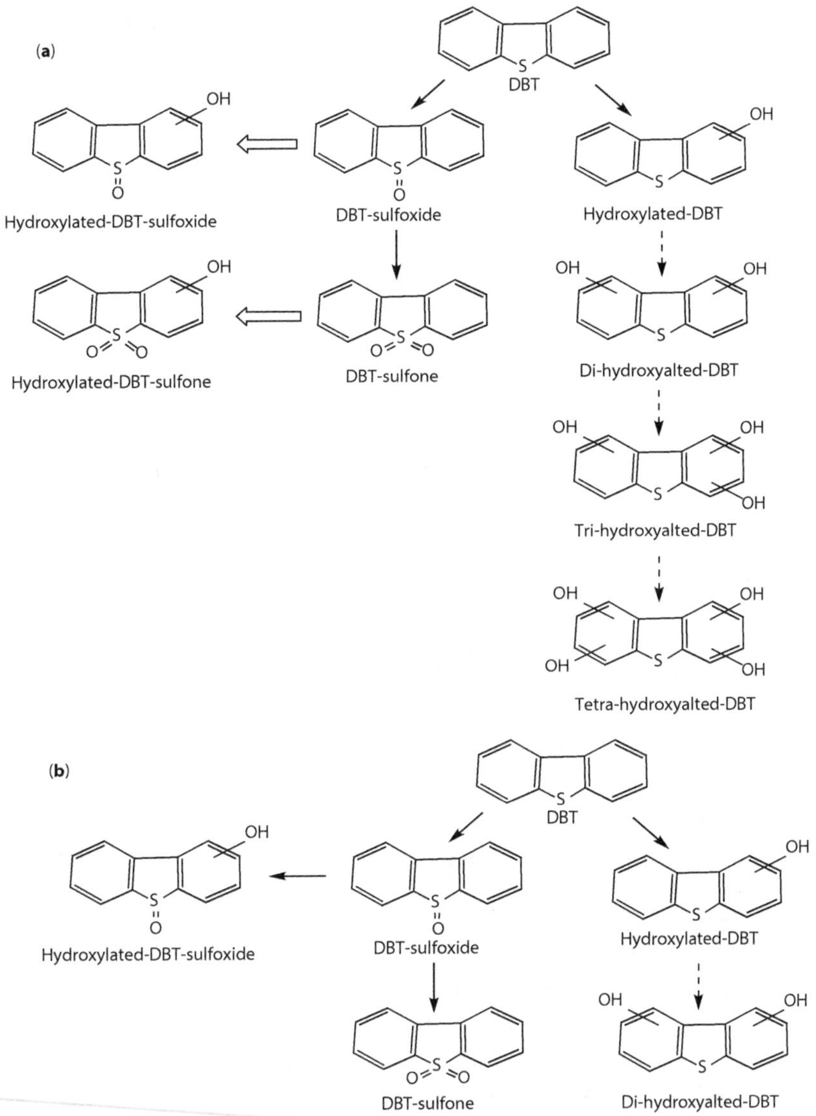

Figure 6.36 Proposed Pathways for the Conversion of DBT by *A. aegerita* (a) and *C. radians* (b). Dotted and hollow arrows indicate reactions solely observed in the in-vitro and in-vivo experiments, respectively (Aranda *et al.*, 2009).

2007). These enzymes are now referred to as aromatic peroxygenases (in earlier publications they were also named haloperoxidases or haloperoxidase-peroxygenases) (Anh *et al.*, 2007; Ullrich and Hofrichter 2005). *A. aegerita* peroxygenase (AaP) and *C. radians* peroxygenase (CrP) catalyze the oxidation of phenolic compounds, aryl alcohols, and bromide, as well as the oxygenation/hydroxylation of aromatic substrates (Anh *et al.*, 2007; Kinne *et al.*, 2008; Ullrich *et al.*, 2004; Ullrich and Hofrichter 2005). These enzymes were found to be able for bio-tranformation of DBT. In vitro tests with purified peroxygenases were performed to compare the product pattern with the metabolites formed in vivo. Using ascorbic acid as a radical scavenger, a total of 19 and seven oxygenation products were detected after DBT conversion by the peroxygenases of *A. aegerita* (AaP) and *C. radians* (CrP), respectively, whereas ring hydroxylation was favored over S-oxidation by AaP (again 2-hydroxy-DBT was identified) and CrP formed DBT sulfoxide as major product. This finding suggests that fungal peroxygenases can considerably differ in their catalytic properties. Using $H_2{}^{18}O_2$, the origin of oxygen was proven to be the peroxide. Based on these results, it was proposed that extracellular peroxygenases may be involved in the oxidation of heterocycles by fungi, also under natural conditions. The results of the in vitro tests indicated that there are considerable differences concerning the specificity of *A. aegerita* peroxygenase (AaP) and *C. radians* peroxygenase (CrP). Thus, in the absence of ascorbic acid, sulfoxidation by AaP was not observed at all and in its presence, only traces of DBT sulfoxide and sulfone were detected. In contrast, sulfoxidation of DBT was the favored reaction of CrP both in the presence and absence of ascorbic acid. The differences between both enzymes could be explained by structural differences in the active sites, which may favor either the transfer of an oxygen atom to the sulfur or to the adjacent benzene ring. Some authors suggested for lignin peroxidase, which oxidizes DBT directly at the sulfur, that less reactive compounds, such as DBT, are possibly sterically hindered and do not bind closely to the prosthetic heme group (Vazquez-Duhalt, 1999). Although the similarity in the catalytic properties of AaP and CrP is evident (peroxygenation), there are differences between both enzymes, not only concerning the oxidation of DBT, but differences were also observed in the hydroxylation of naphthalene and toluene (Anh *et al.*, 2007).

Exophiala spinifera FM strain (accession no. KC952672), isolated from oil contaminated soil, has been reported to desulfurize 99% of 0.3 mM DBT within 7 d of incubation, at 30 °C and 180 rpm shaking, in the presence of 1% (w/v) glucose, throughout the 4S-pathway, producing 2-hydroxy biphenyl (2-HBP) (Elmi *et al.*, 2015). To overcome the toxicity of 2-HBP, the FM strain bio-transformed 2-HBP to the less toxic 1,3-benzenediol, with 5-hexyl as a dead end product. It is hypothesized that 2-HBP converts to di-hydroxy biphenyl by 2-hydroxybiphenyl 3-monooxygenase enzyme that is an inducible flavoenzyme involved in the degradation of the toxic 2-hydroxybiphenyl produced by *Exophiala* and O_2 through an oxidative

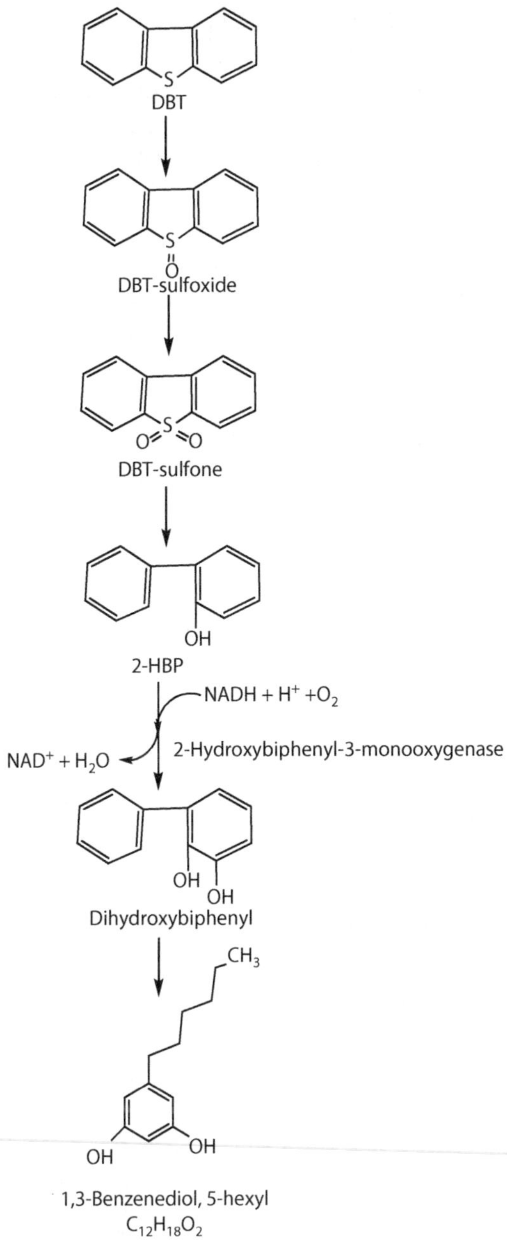

Figure 6.37 Proposed Modified 4S-Pathway by *Exophiala spinifera* Strain FM (Elmi *et al.*, 2015).

meta-cleavage pathway (Figure 6.37). In previous studies, the investigators found that dihydroxylate biphenyl production induces breakage on aromatic rings and reduces toxicity of biphenyl compounds and increases the fungal tolerance to these compounds (Romero *et al.*, 2005).

2-Hydroxymethyl benzothiophene (2HMBT) and 4-Hydroxymethyl dibenzothiophene (4HMDBT) are reported to be metabolized more quickly in a high nitrogen level culture medium of lignin-degrading white basidiomycete *Coriolus versicolor*. They are converted to their xyloside without the occurrence of sulfur oxidation (Figure 6.38). It also converts 2HMBT and 4HMDBT to their corresponding sulfones via a minor

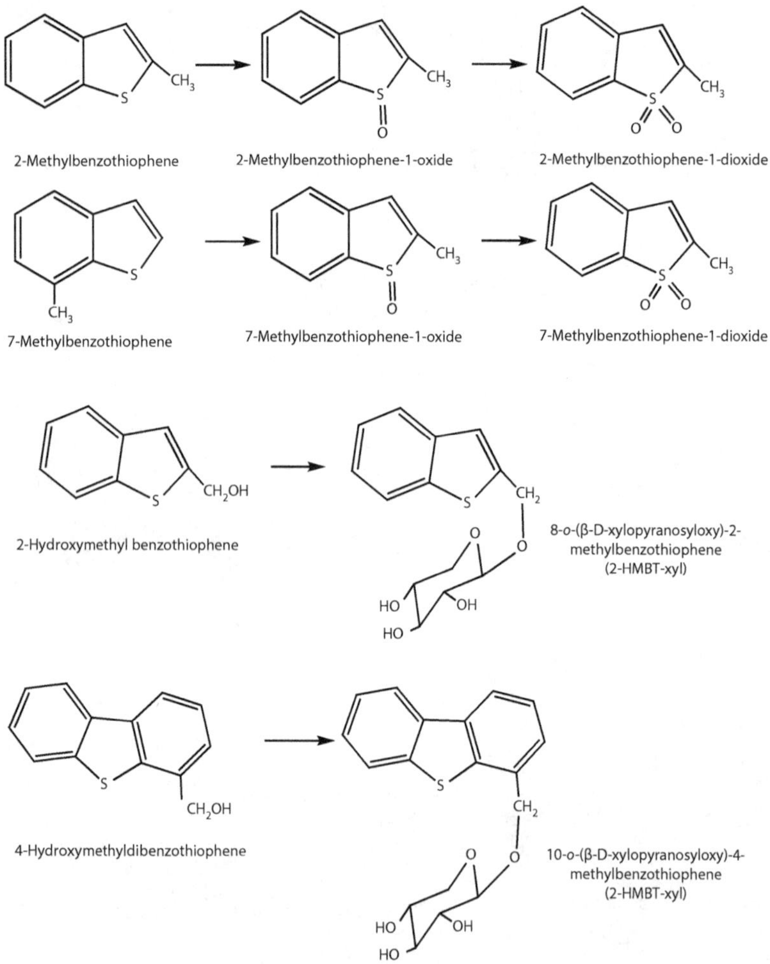

Figure 6.38 Proposed Metabolic Pathways for Aromatic Thiophene Derivatives by *Coriolus versicolor* (Ichinose *et al.* 2002b).

pathway (Ichinose *et al.*, 2002b). The carboxyl and formyl substituents, 2-carboxybenzothiophene (2CBT) and 4-carboxydibenzothiophene (4CDBT), were reduced to form a hydroxymethyl group, then xylosidated (Ichinose *et al.*, 2002b)

The glycosylation reaction has been reported for *P. chrysosporium*, *T. palustris*, and *C. elegans*. Interestingly, they have been reported to catalyze glucosylation (Cerniglia *et al.*, 1989; Kondo and Imamura 1989; Masaphy *et al.*, 1995). On the other hand, *C. versicolor*, *Dichomitus squalens* (basidiomycete), and *Rhizoctonia solani* (fungi imperfect) were shown to catalyze xylosylation (Ichinose *et al.*, 1999; Kondo *et al.*, 1993; Reddy *et al.*, 1997; Sutherland *et al.*, 1992). Glycosylation is a metabolic response to chemical stress in fungal cells. Thus, it appears to reduce the toxicity of aromatic compounds by increasing water-solubility.

References

Abbad-Andaloussi, S., Lagnel, C., Warzywoda, M., Monot, F. (2003) Multicriteria comparison of resting cell activities of bacterial strains selected for biodesulfurization of petroleum compounds. *Enzyme and Microbial Technology*. 32(3/4): 446–454.

Abdulrashid, N., Clark, D.P. (1987) Isolation and genetic analysis of mutations allowing the degradation of furans and thiophenes by *Escherichia coli. Journal of Bacteriology*. 169:1267–1271

Adlakha, J., Singh, P., Ram, S.K., Kumar, M., Singh, M.P., Singh, D., Sahai, V., Srivastava, P. (2016) Optimization of conditions for deep desulfurization of heavy crude oil and hydrodesulfurized diesel by *Gordonia* sp. IITR100. 184: 761–769.

Aggarwal, S., Karimi, I.A., Kilbane, J.J. II, Lee, D.Y. (2012) Roles of sulfite oxidoreductase and sulfite reductase in improving desulphurization by *Rhodococcus erythropolis*. Molecular BioSystems. 8: 2724–2732.

Ahmad, A., Chauhan, A.K., Kushwaha, H.N., Javed, S., Kuma, A. (2015) Preferential desulfurization of dibenzyl sulfide by an isolated *Gordonia* sp. IITR100. 3 Biotech. 5: 237–243.

Akhtar, N., Ghauri, M.A., Anwar, M.A., Akhtar, K. (2009) Analysis of the dibenzothiophene metabolic pathway in a newly isolated *Rhodococcus* spp. FEMS Microbiology Letters. 301: 95–102.

Alam, K.Y., Worland, M.J., Clark, D.P. (1990) Analysis and molecular cloning of genes involved in thiophene and furan oxidation by E. coli. Applied iochemistry and Biotechnology. 24:843- 855.

Al-Jailawi, M.H., Al-Faraas, A.F., Yahia, A.I. (2015) Isolation and identification of dibenzothiophene biodesulfurizing bacteria. *American Journal of Bioscience and Bioengineering*. 3(5): 40–46.

Allen, C.C.R., Boyd, D.R., Dalton, H., Sharma, N.D., Haughey, S.A., McMordie, S., McMurray, B.T., Sheldrake, G.N., Sproule, K. (1995) Sulfoxides of high enantiopurity from bacterial dioxygenase-catalyzed oxidation. *Journal of the Chemical Society, Chemical Communications*. 1: 19- 120.

Alves, L., Melo, M., Mendonca, D., Simoes, F, Matos., J, Tenreiro, R., Girio, F.M. (2007) Sequencing, cloning and expression of the dsz genes required for dibenzothiophene sulfone desulfurization from *Gordonia alkanivorans* strain 1B. *Enzyme and Microbial Technology*. 40:1598–1603.

Alves, L., Salgueiro, R., Rodrigues, C., Mesquita, E., Matos, J., Gírio, F.M. (2005) Desulfurization of dibenzothiophene, benzothiophene, and other thiophene analogs by a newly isolated bacterium, *Gordonia alkanivorans* Strain 1B. *Applied Biochemistry and Biotechnology*. 120: 199–208.

Amorelli, A., Amos, Y.D., Halsig, C.P., Kosman, J.J., Jonker, R.J., de Wind, M., Vrieling J. (1992) Estimate feedstock process ability. Hydrocarbon Processing. 71: 93–106.

Anh, D.H., Ullrich, R., Benndorf, D., Svatos, A., Muck, A., Hofrichter, M. (2007) The coprophilous mushroom Coprinus radians secretes a haloperoxidase that catalyzes aromatic peroxygenation. *Applied and Environmental Microbiology*. 73: 5477–5485.

Annweiler, E., Michaelis, W., Meckenstock, R.U. (2001) Anaerobic cometabolic conversion of benzothiophene by a sulfate-reducing enrichment culture and in a tar- oil-contaminated aquifer. *Applied and Environmental Microbiology*. 67:5077–5083.

Arabian, D., Najafi, H., Farhadi, F., Dehkordi, A.M. (2014) Biodesulfurization of simulated light fuel oil by a native isolated bacteria *Bacillus cereus* HN. *Journal of Petroleum Science and Technology*. 4(1): 31–40.

Arafa, M.A., Al-Fantoukh, S.A., Sano, T. (2001) Biodesulfurization activity of dibenzothiophene by mutant strain of *Rhodococcus erythropolis*; 4th Chemistry in Industry Conference and Exhibition. Petroleum Refining and Petrochemicals Session. pp.1–7.

Aranda, E., Kinne, M., Kluge, M., Ullrich, R., Hofrichter, M. (2009) Conversion of dibenzothiophene by the mushrooms *Agrocybeae gerita* and *Coprinellus radians* and their extracellular peroxygenases. *Applied Microbiology and Biotechnology*. 82: 1066–1414.

Ardakani, M.R., Aminsefat, A., Rasekh, B., Yazdiyan, F., Zargar, B., Zarei, M., Najafzadeh, H. (2010) Biodesulfurization of dibenzothiophene by a newly isolated *Stenotrophomonas maltophila* strain Kho1. *World Applied Sciences Journal*. 10(3): 272–278.

Arensdorf, J.J., Loomis, A.K., DiGrazia, P.M., Monticello, D.J., Pienkos, P.T. (2002) Chemostat approach for the direct evolution of biodesulfurization gain-of-function mutants. *Applied and Environmental Microbiology*. 68: 691–698.

Atlantic Research Corporation (1985). United States Patent No. 4562156.

Baldi, F., Pepi, M., Fava, F. (2003) Growth of *Rhodosporidium toruloides* strain DBVPG 6662 on dibenzothiophene crystals and orimulsion. *Applied and Environmental Microbiology*. 69(8): 4689–4696.

Berthou, F., Gourmelun, Y., Dreano, Y., Fricourt, M.P. (1981) Application of gas chromatography on glass capillary columns to the analysis of hydrocarbon pollutants from the Amoco Cadiz oil spill. *Journal of Chromatography A.* 203:279–292.

Berthou, F., Vignier, V. (1986) Analysis and fate of dibenzothiophene derivatives in the marine environment. *International Journal of Environmental Analytical Chemistry*. 27(1/2): 81–96.

Bezalel, L., Hadar, Y., Fu, P.P., Freeman, J.P., Cerniglia, C.E. (1996) Initial oxidation products in the metabolism of pyrene, anthracene, fluorene, and dibenzothiophene by the white rot fungus *Pleurotus ostreatus*. *Applied and Environmental Microbiology*. 62(7): 2554–2559.

Bohonos, N., Chou, T.W., Spanggord, R.J. (1977) Some observations on biodegradation of pollutants in aquatic systems. *Japanese Journal of Antibiotics*. 30: 275–285.

Bordoloi, N.K., Rai, S.K., Chaudhuri, M.K., Mukherjee, A.K. (2014) Deep-desulfurization of dibenzothiophene and its derivatives present in diesel oil by a newly isolated bacterium *Achromobacter* sp. to reduce the environmental pollution from fossil fuel combustion. Fuel Processing Technology. 119: 236–244.

Bressler, D.C., Fedorak, P.M. (2001a) Purification, stability and mineralization of 3-hydroxy-2-formylbenzothiophene, a metabolite of dibenzothiophene. *Applied and Environmental Microbiology*. 67:821–826.

Bressler, D.C., Fedorak, P.M. (2001b) Identification of disulfides from the biodegradation of dibenzothiophene. *Applied and Environmental Microbiology*. 67:5084– 5093.

Bressler, D.C., Leskiw, B.K., Fedorak, P.M. (1999) Biodegradation of benzothiophene sulfones by a filamentous bacterium. *Canadian Journal of Microbiology*. 45: 360–368.

Bressler, D.C., Norman, J.A., Fedorak, P.M. (1998) Ring cleavage of sulfur heterocycles: how does it happen? Biodegradation. 8: 297–311.

Brons, G., Yu, J.M. (1995) Solvent deasphalting effects on whole Cold Lake bitumen. *Energy Fuels*. 9: 641–647.

Brooks, S.L., Van Hamme, J.D. (2012) Whole-genome shotgun sequence of R*hodococcus* species strain JVH1. *Journal of Bacteriology*. 194(19): 5492–5493.

Casullo de Araújo, H.W., Siva, M.C.d.F Lins, C.I.M., Elesbão do Nascimento, A., Alves da Silva, C.A., Campos-Takak, G.M. (2012) Oxidation of dibenzothiophene (DBT) by *Serratia marcescens* UCP 1549 formed biphenyl as final product. Biotechnology for Biofuels. 5:33. http://www.biotechnologyforbiofuels.com/content/5/1/33

Cerniglia, C.E., Campbell, W.L., Freeman, J.P., Evans, F.E. (1989) Identification of a novel metabolite in phenanthrene metabolism by the fungus *Cunninghamella elegans*. *Applied and Environmental Microbiology*. 55: 2275–2279.

Chang, J.H., Chang, Y.K., Ryu, H.W., Chang, H.N. (2000) Desulfurization of light gas oil in immobilized-cell systems of *Gordona* sp. CYKS1 and *Nocardia* sp. CYKS2. FEMS Microbiology Letters. 182(2): 309–312.

Chauhan, A.K., Ahmad, A., Singh, S.P., Kumar, A. (2015) Biodesulfurization of benzonaphthothiophene by an isolated *Gordonia* sp. IITR 100. *International Biodeterioration and Biodegradation*. 104: 105–111.

Chen, H., Cai, Y.B., Zhang, W.J., Li, W. (2009) Methoxylation pathway in biodesulphurization of model organosulfur compounds with Mycobacterium sp. Bioresource Technology. 100: 2085–2087.

Chen, T.C., Shen, Y.H., Lee, W.J., Lin, C.C., Wan, M.W. (2010) The study of ultrasound-assisted oxidative desulfurization process applied to the utilization of pyrolysis oil from waste tires. *Journal of Cleaner Production*. 18(18): 1850–1858.

Chou, C.C., Swatloski, R.A. (1983) Biodegradation of sulfolane in refinery wastewater. Proceedings of the 37th Industrial Waste Conference. Ann Arbor Science, Ann Arbor, Mich. Pp. 559–566.

Coleman, H.J., Hopkins, R.L., Thompson, C.J. (1971) Highlights of some 50 man-years of petroleum sulphur studies. *International Journal of Sulphur Chemistry* B. 41–61.

Constanti, M., Giralt, J., Bordons, A. (1994) Desulfurization of dibenzothiophene by bacteria. *World Journal of Microbiology and Biotechnology*. 10: 510–516.

Constanti, M., Giralt, J., Bordons, A. (1996) Degradation and desulfurization of dibenzothiophene sulfone and other sulfur compounds by *Agrobacterium* MC501 and a mixed culture. *Enzyme and Microbial Technology*. 19(3): 214–219.

Crawford, D.L., Gupta, R.K. (1990) Oxidation of dibenzothiophene by *Cunninghamella elegans*. Current Microbiology. 21: 229–231.

Cripps, R.E. (1973) The microbial metabolism of thiophen-2- carboxylate. *Biochemical Journal*. 134:353–366.

Dahlberg, M.D., Rohrer, R.L., Fauth, D.J., Sprecher, R., Olson, G.J. (1993) Biodesulfurization of dibenzothiophene sulfone by *Arthrobacter* sp. and studies with Illinois no. 6 coal. *Fuel*. 72:1645–1649.

Davoodi-Dehaghani, F., Vosoughi, M., Ziaee, A.A. (2010) Biodesulfurization of dibenzothiophene by a newly isolated Rhodococcus erythropolis strain. Bioresource Technology 101: 1102–1105.

Denis-Larose, C., Labbe, D., Bergeron, H., Jones, A.M., Greer, al-Hawari, J., Grossman, M. J., Sankey, B.M., Lau, P.C. (1997) Conservation of plasmid-encoded dibenzothiophene desulfurization genes in several *Rhodococci*. *Applied and Environmental Microbiology*. 63: 2915–2919.

Denome, S.A., Olson, E.A., Young, K.D. (1993) Identification cloning of genes involved in specific desulfurization of dibenzothiophene by *Rhodococcus* sp. strain IGTS8. *Applied and Environmental Microbiology*. 59:2837–2843

Derikvand, P., Etemadifar, Z., Biria, D. (2015a) RSM optimization of dibenzothiophene biodesulfurization by newly isolated strain of *Rhodococcus erythropolis* PD1 in aqueous and biphasic systems. Microbiology. 84(1): 65–72.

Derikvand, P., Etemadifar, Z., Saber, H. (2015b) Sulfur removal from dibenzothiophene by newly isolated paenibacillus validus strain PD2 and process optimization in aqueous and biphasic (Model-Oil) systems. *Polish Journal of Microbiology*. 64(1): 47–54.

Dodson, R.M., Newman, N., Tsuchuja, H.M. (1962) Microbiological transformations. XI. The preparation of optically active sulfoxides. *Journal of Organic Chemistry*. 27: 2707–2708.

Duarte, G.F., Rosado, A.S., Seldin, L., Araujo, W., Van Elsa, J. (2001) Analysis of bacterial community structure in sulfurous-oil-containing soils and detection of species carrying dibenzothiophene desulfurization (dsz) Genes. *Applied and Environmental Microbiology*. 67(3): 1052–1062.

Eastmond, D.A., Booth, G.M., Lee, M.L. (1984) Toxicity, accumulation, and elimination of polycyclic aromatic sulfur heterocycles in *Daphnia magna. Archives of Environmental Contamination and Toxicology*. 13: 105–111.

Eaton, R.W., Nitterauer, J.D. (1994) Biotransformation of benzothiophene by isopropyl benzene- degrading bacteria. *Journal of Bacteriology*. 176: 3992–4002.

Eibes, G., Cajthaml, T., Moreira, M.T., Feijoo, G., Lema, J.M. (2006) Enzymatic degradation of anthracene, dibenzothiophene and pyrene by manganese peroxidase in media containing acetone. *Chemosphere*. 64: 408–414.

El-Gendy, N.S. (2006). Biodegradation potentials of dibenzothiophene by new bacteria isolated from hydrocarbon polluted soil in Egypt. Biosciences, Biotechnology Research Asia. 3(1a): 95–106.

El-Gendy, N.S., Abo State, M.A. (2008) Isolation, characterization and evaluation of *Staphylococcus gallinarum* NK1 as a degrader for dibenzothiophene, phenanthrene and naphthalene. *Egyptian Journal of Petroleum*. 17(2):75–91.

El-Gendy, N.Sh. (2004) Biodesulfurization potentials of crude oil by bacteria isolated from hydrocarbon polluted environments in Egypt. A PhD Thesis, Faculty of Science, Cairo University, Cairo, Egypt.

El-Gendy, N.Sh., Farahat, L.A., Moustafa Y.M., El-Temtamy, S.A. (2005) Biodesulfurization of crude and diesel oil by *Agrobacterium* Sp. NSh10. In conference proceedings Towards a Safe and Clean Environment TSCE'05. El-Menia and Hurghada, Egypt, 15–19 April, 2005.

El-Gendy, N.Sh., Farahat, L.A., Mustafa, Y.M., Shaker, N., El-Temtamy, S.A. (2006) Biodesulfurization of crude and diesel oil by *Candida parapsilosis* NSh45 isolated from Egyptian hydrocarbon polluted sea water. Biosciences, Biotechnology Research Asia. 3(1a): 5–16.

El-Gendy, N.Sh., Moustafa, Y.M., Habib, S.A., Ali, Sh. (2010) Evaluation of *Corynebacterium variabilis* Sh42 as a degrader for different poly aromatic compounds. *Journal of American Science*. 6(11): 343–356.

El-Gendy, N.Sh., Nassar, H.N., Abu Amr, S.S. (2014) Factorial design and response surface optimization for enhancing a biodesulfurization process. *Petroleum Science and Technology*. 32(14): 1669–1679.

El-Gendy, N.Sh., Speight J.G. (2016) Handbook of Refinery Desulfurization. CRC Press, Taylor and Francis Group, LLC, 6000 Broken Sound Parkway NW, Suite 300, Boca Raton, FL 33487–2742, USA.

Elmi, F., Etemadifar, Z., Emtiazi, G. (2015) A novel metabolite (1,3-benzenediol, 5-hexyl) production by Exophiala spinifera strain FM through dibenzothiophene desulfurization. *World Journal of Microbiology and Biotechnology*. 31:813–821.

Endoh, T., Kasuga, K., Horinouchi, M., Yoshida, T., Habe, H., Nojiri, H., Omori, T. (2003)

Characterization and identification of genes essential for dimethyl sulfide utilization in *Pseudomonas putida* strain DSI. *Applied Microbiology and Biotechnology*. 62: 83–91.

Eriksen, J. (2008) Soil sulfur cycling in temperate agricultural system. In: Sulfur: a missing link between soils, crops, and nutrition. Joseph Jez, Ed., ASA-CSSA-SSSA. 50: 25–44.

Etemadifar, Z., Emtiazi, G., Peimanfar, S. (2006) Removal of dibenzothiophene, biphenyl and phenol from waste by *Trichosporon* sp. Scientific Research and Essays. 1(3):072–076

Evans, J., Venables, W. (1990) Degradation of thiophene-2- carboxylate, furan-2-carboxylate, pyrrole-2-carboxylate and other thiophene derivatives by the bacterium Vibrio YC1. *Applied Microbiology and Biotechnology*. 32:715–720.

Faison, B.D., Clark, T.M., Lewis, S.N., Ma, C.Y., Sharkey, D.M., Woodward, C.A. (1991) Degradation of organic sulfur compounds by a coal-solubilizing fungus. *Applied Biochemistry and Biotechnology*. 28/29: 237–251.

Fedorak, P.M., Payzant, J.D., Montgomery, D.S., Westlake, D.W.S. (1988) Microbial degradation of n-alkyl tetrahydrothiophenes found in petroleum. *Applied and Environmental Microbiology*. 54(5):1243–1248.

Fedorak, P.M., Westlake, D.W.S. (1983) Microbial degradation of organic sulfur compounds in Prudhoe Bay crude oil as determined by glass capillary gas chromatography. *Canadian Journal of Microbiology*. 27: 432–443.

Fedorak, P.M., Westlake, D.W.S. (1984) degradation of sulfur heterocycles in Prudhoe Bay crude oil as determined by soil enrichments. Water Air Soil Pollution. 21: 225–230.

Fedorak, P.M.; Coy, D.L., Peakman, T.M. (1996) Microbial metabolism of some 2,5-substituted thiophenes. Biodegradation. 7:313–327.

Fedorak, P.N., Grbić-Galić, D. (1991) Aerobic microbial cometabolism of benzothiophene and 3-methylbenzothiophene. *Applied and Environmental Microbiology*. 57:932–940.

Finkelstein, Z.I., Baskunov, B.P., Vavilova, L.N., Golovleva, L.A. (1997) Microbial transformation of dibenzothiophene and 4,6-dimethyldibenzothiophene. Microbiology. 66: 402–407.

Finnerty W.R. (1993a) Organic sulfur biodesulfurization in non-aqueous media. *Fuel*. 72: 1631–1634.

Finnerty, W.R. (1992) Fossil resource biotechnology: Challenges and prospects. Current Opinion in Biotechnology. 3:277–282.

Finnerty, W.R. (1993b) Symposium on Bioremediation and Bioprocessing Presented before the Division of Petroleum Chemistry, Inc. 205th National Meeting, American Chemical Society, Denver, Co, March 28-April 2, Pp.282–285.

Folsom, B.R., Schieche, D.R., Digrazia, P.M., Werner, J., Palmer, S. (1999) Microbial desulfurization of alkylated dibenzothiophenes from a hydrodesulfurized middle distillate by *Rhodococcus erythropolis* I-19. *Applied and Environmental Microbiology*. 65: 4967–4972.

Fought, J., Westlake, D. (1990) Expression of dibenzothiophene degradative genes in two *Pseudomonas* species. *Canadian Journal of Microbiology*. 36:718–724.

Fought, J.M., Westlake, D.W.S. (1988) Degradation of polycyclic aromatic hydrocarbons and aromatic heterocycles by a *Pseudomonas* species. *Canadian Journal of Microbiology*. 34: 1135–1 141.

Frassinetti, S., Setti, L., Corti, A., Fairly, P., Montevechi, P., Vallini, G. (1998) Biodegradation of dibenzothiophene by a nodulating isolate of *Rhizobium meliloti*. Canadian Journal of Microbiology. 44: 289–297.

Friocourt, M. P., Berthou, F., Picart, D. (1982) Dibenzothiophene derivatives as organic markers of oil pollution. *Toxicological and Environmental Chemistry*. 5(3–4): 205–215.

Furuya, T., Kirimura, K., Kino, K., Usami, S. (2002) Thermophilic biodesulfurization of naphthothiophene and 2-ethylnaphthothiophene by a dibenzothiophene-desulfurizing bacterium, *Mycobacterium* phlei WU-F1. *Applied Microbiology and Biotechnology*. 58(2): 237–240.

Gai, Z., Yu, B., Li, L., Wang, Y., Ma, C., Feng, J., Deng, Z., Xu, P. (2007) Cometabolic degradation of dibenzofuran and dibenzothiophene by a newly isolated carbazole-degradaing *Sphingomonas* sp. strain. *Applied and Environmental Microbiology* 73(9):2832–2838.

Gallagher, J.R., Olson, E.S., Stanley, D.C. (1993) Microbial desulfurization of dibenzothiophene: a sulfur-specific pathway. FEMS Microbiology Letters. 107: 31–36.

Gilbert, S.C., Morton, J., Buchanan, S., Oldfield, C., McRoberts, A. (1998) Isolation of a unique benzothiophene-desulphurizing bacterium, *Gordona* sp. strain 213E (NCIMB 40816), and characterization of the desulphurization pathway. Microbiology. 144: 2545–2553.

Gray, K.A., Pogrebinsky, O.S., Mrachko, G.T., Lei, X., Monticello, D.J., Squires, C.H. (1996) Molecular mechanisms of biocatalytic desulphurization of fossil fuels. Nature Biotechnology. 14: 1705–1709.

Grbic-Galic, D. (1989) Microbial degradation of homocyclic and heterocyclic hydrocarbons under anaerobic conditions. Developments in industrial microbiology. 30:237–253.

Greene, E.A., Beatty, P.H., Fedorak, P.M. (2000) Sulfolane degradation by mixed cultures and a bacterial isolate identified as a *Variovorax* sp. Archives of Microbiology. 174 :111–119.

Greene, E.A., Coy, D.L., Fedorak, P.M. (1999) Laboratory evaluations of factors affecting biodegradation of sulfolane and diisopropanolamine. *Bioremediation Journal*. 3: 299–313.

Greene, E.A., Gieg, L.M., Coy, D.L., Fedorak, P.M. (1998) Sulfolane biodegradation potential at sour natural gas plant sites. *Water Resarch*. 18: 158–173.

Grimalt, J.O., Grifoll, M., Solanas, A.M., Albaigés, J. (1991) Microbial degradation of marine evaporitic crude oils. Geochimica et Cosmochimica Acta. 55: 1903–1913.

Grossman, M.J., Lee, M.K., Prince, R.C., Garrett, K.K., George, G.N., Pickering, I.J. (1999) Microbial desulfurization of a crude oil middle-distillate fraction: analysis of the extent of sulfur removal and the effect of removal on remaining sulfur. *Applied and Environment Microbiology*. 65:181–188.

Gunam, I.B.W., Yaku, Y., Hirano, M., Yamamura, K., Tomita, F., Sone, T., Asano, K. (2006) Biodesulfurization of alkylated forms of dibenzothiophene and benzothiophene by *Sphingomonas subarctica* T7b. *Journal of Bioscience and Bioengineering*. 101: 322–327.

Gundlach, E.R., Boehm, P.D., Marchand, M., Atlas, R.M., Ward, D.M., Wolfe, D.A. (1983) The fate of Amoco Cadiz oil. Science. 221: 122–129.

Holland, H.L., Khan, S.H., Richards, D., Riemland, E. (1986) Biotransformation of polycyclic aromatic compounds by fungi Xenobiotica.16: 733–741.

Husaini, A., Roslan, H.A., Hii, K.S.Y., Ang, C.H. (2008) Biodegradation of aliphatic hydrocarbon by indigenous fungi isolated from used motor oil contaminated sites. *World Journal of Microbiology and Biotechnology*. 24(12): 2789–2797.

Ichinose, H., Nakamizo M, Wariishi, H., Tanaka, H. (2002b) Metabolic response against sulfur-containing heterocyclic compounds by the lignin-degrading basidiomycete *Coriolus versicolor. Applied Microbiology and Biotechnology*. 58(4): 517–26.

Ichinose, H., Wariishi, H., Tanaka, H. (1999) Bioconversion of recalcitrant 4-methyldibenzothiophene to water-extractable products using lignin-degrading basidiomycete *Coriolus versicolor*. Biotechnology Progress. 15: 706–714.

Ichinose, H., Wariishi, H., Tanaka, H. (2002a) Identification and characterization of novel cytochrome P450 genes from white-rot basidiomycete *Coriolus versicolor. Applied Microbiology and Biotechnology*. 58: 97–105.

Inoue, H., Inagaki, K., Eriguchi, S.l., Tamura, T., Esaki, N., Soda, K., Tanaka, H. (1997)

Molecular characterization of the *mde* operon involved in L-methionine catabolism of *Pseudomonas putida. Journal of Bacteriology.* 179: 3956–3962.

Isbister, J.D., Doyle, R.C. (1985) Mutant microorganism and its use in removing organic sulfur compounds. United States Patent No. 4,562,156.

Ishii, Y., Kozaki, S., Furuya, T., Kino, K., Kirimura, K. (2005) Thermophilic biodesulfurization of various heterocyclic sulfur compounds and crude straight-run light gas oil fraction by a newly isolated strain *Mycobacterium phlei* WU-0103. Current Microbiology. 50: 63–70.

Ismail, W., El-Sayed, W.S., Abdul Raheem, A.S., Mohamed, M.E, El Nayal, A.M. (2016) Biocatalytic desulfurization capabilities of a mixed culture during non-destructive utilization of recalcitrant organosulfur compounds. Frontiers in Microbiology. Volume 7: Article 266. doi: 10.3389/fmicb.2016.00266

Itoh, N., Yoshikda, M., Miyamoto, T., Ichinose, H., Wariishi, H., Tanaka, H. (1997) Fungal cleavage of thioether bondfoundinY perite FEBS Letters. 412: 281–284.

Izumi, Y., Oshiro, T., Ogino, H., Hine, Y., Shimao, M. (1994) Selective desulfurization of dibenzothiophene by *Rhodococcus erythropolis* D-1. *Applied and Environmental Microbiology*. 60: 223–236.

Jiang, X., Yang, S., Li, W., 2014. Biodesulfurization of model compounds and deasphalted bunker oil by mixed culture. *Applied Biochemistry and Biotechnology.* 172: 62–72.

Kanagawa, T. and Kelly D.P. (1987) Degradation of substituted thiophenes by bacteria isolated from activated sludge. Microb. Ecol. 13:47–57.

Kareem, S.A., Aribike, D.S., Susu, A.A., Nwachukwu, S.C.U. (2016) Anaerobic biodesulfurization of kerosene part I: Identifying a capable microorganism. *Chemical Engineering and Process Techniques.* 2(2): 1028.

Kargi, F., Robinson, J.M. (1984) Microbial oxidation of dibenzothiophene by the thermophilic organism *Sulfolobus acidocaldarius. Biotechnology and Bioengineering.* 26(7): 687–690.

Kawatra, S.K., Eisele, T.C. (2001) Coal desulfurization, high-efficiency preparation methods. Taylor & Francis, New York.

Kayser, K.J., Cleveland, L., Park, H.S., Kwak, J.H., Kolhatkar, A., Kilbane, J. (2002) Isolation and characterization of moderate thermophile *Mycobacterium phlei* GTIS10, capable of dibenzothiophene desulfurization. *Applied Microbiology and Biotechnology.* 59: 737–746.

Kayser, Y., Bielagajones, B.A., Jackowski, K., Odusan, O., Kilbane, J.J. (1993) Utilization of organusulphur compounds by axenic and mixed cultures of *Rhodococcus rhodochrous* IGTS8. *Journal of General Microbiology.* 139: 3123–3129.

Kertesz, M.A. (2000) Riding the sulfur cycle - metabolism of sulfonates and sulfate esters in Gram-negative bacteria. *FEMS Microbiology Reviews.* 24: 135–175.

Kertesz, M.A. (2004) Metabolism of sulphur-containing organic compounds. In: *Pseudomonas.* Volume 3, edited by Juan-Luis Ramos. Kluwer Academic / Plenum Publishers, New York. Pp. 323–357.

Khedkar, S., Shanker, R. (2015) Isolation and classification of a soil actinomycete capable of sulphur-specific biotransformation of dibenzothiophene, benzothiophene and thianthrene. *Journal of Applied Microbiology.* 118: 62–74.

Kilbane 2nd, J.J., (2006) Microbial biocatalyst developments to upgrade fossil fuels. Current Opinion in Biotechnology. 17: 305–314.

Kilbane II, J.J., Jackowski, K. (1996) Biocatalytic detoxification of 2-chlroethyl ethyl sulfide. Journal of Chemical Technology and Biotechnology. 65: 370–374.

Kilbane, J.J. (1989) Desulfurization of coal: the microbial solution. Trends in Biotechnology. 7: 97–101

Kilbane, J.J. (1990a) Sulfur-specific microbial metabolism of organic compounds. Resources, Conservation and Recycling. 3: 69–79.

Kilbane, J.J. (1990b) Biodesulfurization: Future prospects in coal cleaning. Proceedings 7th Annual International Pittsburgh Coal Conference. Pp..373–381.

Kilbane, J.J. (2006) Microbial biocatalyst developments to upgrade fossil fuels. Current Opinion in Biotechnology. 17: 305–314.

Kilbane, J.J., Bielaga B.A. (1990b) Toward sulfur-free fuels. ChemTech. 20:747–51.

Kilbane, J.J., Jackowski, K. (1992) Biodesulfurization of water soluble coal derived material by *Rhodococcus rhodochrous* IGTS8. *Biotechnology and Bioengineering.* 40: 1107–1114.

Kilbane, J.J., Le Borgne, S. (2004) Petroleum biorefining: the selective removal of sulfur, nitrogen, and metals. In: Petroleum Biotechnology, Developments and Perspectives ed.

Vazquez-Duhalt, R. and Quintero-Ramirez, R., Amsterdam: Elsevier. Pp. 29–65.

Kim C.G., Chon, B. (2002) Oxygen uptake characteristics of soil inoculum amended with thiophene derivatives. *Korean Journal of Chemical Engineering.* 19:773–779.

Kim, B. H., Kim, H. Y., Kim, T. S., Park, D. H. (1995) Selectivity of desulfurization activity of *Desulfovibrio desulfuricans* M6 on different petroleum products. *Fuel Processing Technology*, 43(1): 87–94.

Kim, B.Y., Kim, H.Y., Kim, T.S., Park, D.H. (1996) Selectivity of desulfurization activity of *Desulfovibrio desulfuricans* M6 on different petroleum products. *Fuel Processing Technology*. 43: 87–94.

Kim, C.G., Chon, B. (2002) Oxygen uptake characteristics of soil inoculum amended with thiophene derivatives. *Korean Journal of Chemical Engineering.* 19:773–779.

Kim, H.Y., Kim, T.S., Kim, B.H. (1990b) Degradation of organic sulfur compounds and the reduction of dibenzothiophene to biphenyl and hydrogen sulfide by *Desulfovibrio desulfuricans* M6. Biotechnology Letters. 10: 761–764.

Kim, T.S., Kim, H.Y., Kim, B.H. (1990a) Petroleum desulfuriazation by *Desulfovibrio desulfuricans* M6 using electrochemically supplied reducing equivalent. Biotechnology Letters. 10: 757–760.

Kim, Y.J., Chang, J.H., Cho, K.S., Ryu, H.W., Chang, Y.K. (2004) A physiological study on growth and dibenzothiophene (dibenzothiophene) desulfurization characteristics of Gordonia sp. CYKS1. *Korean Journal of Chemical Engineering.* 21: 436–441.

Kinne, M., Ullrich, R., Hammel, K.E., Scheibner, K., Hofrichter, M. (2008) Regioselective preparation of (R)-2-(4-Hydroxyphenoxy)propionic acid with a fungal peroxygenase. Tetrahedron Letters. 49: 5950–5953.

Kirimura, K., Furuya, T., Nishii, Y., Ishii, Y., Kino, K., Usami, S. (2001) Biodesulfurization of dibenzothiophene and its derivatives through the selective cleavage of carbon-sulfur bonds by a moderately thermophilic bacterium *Bacillus subtilis* WU-S2B. *Journal of Bioscience and Bioengineering.* 91: 262–266.

Kirimura, K., Furuya, T., Sato, R., Ishii, Y., Kino, K., Usami, S. (2002) Biodesulfurization of naphthothiophene and benzothiophene through selective cleavage of carbon-sulfur bonds by *Rhodococcus* sp. strain WU-K2R. *Applied and Environmental Microbiology*. 68(8): 3867–3872.

Kirkwood, K.M., Ebert, S., Foght, J.M., Fedorak, P.M., Gray, M.R. (2005) Bacterial biodegradation of aliphatic sulfides under aerobic carbon- or sulfur-limited growth conditions. *Journal of Applied Microbiology*. 99: 1444–1454.

Kirkwood, K.M., Foght, J.M., Gray, M.R. (2007) Selectivity among organic sulfur compounds in one- and two-liquid-phase cultures of *Rhodococcus* sp. strain JVH1. Biodegradation. 18. 473–480.

Klein, J. (1998) Technological and economic aspects of coal biodesulfurization. Biodegradation. 9: 293–300.

Klein, J., Van afforded, M., Pfeiffer, F., Schacht, S. (1994) Microbial desulfurization of coal and oil. *Fuel Processing Technology*. 40: 297–310.

Klyachko, N.L., Klibanov, A.M. (1992) Oxidation of Dibenzothiophene Catalyzed by Hemoglobin and Other Hemoproteins in Various Aqueous-Organic Media. *Applied Biochemistry and Biotechnology*. 37(1): 53–68.

Kobayashi, M., Onaka, T., Ishii, Y., Konishi, J., Takaki, M., Okada, H., Ohta, Y., Koizumi, K., Suzuki, M. (2000) Desulfurization of alkylated forms of both dibenzothiophene and benzothiophene by single bacterial strain. FEMS Microbiology Letters. 187: 123–126.

Kodama Central Research Institute of Electric Power Industry, Abiko, Chiba Prefecture, K., Nakatani Research Laboratories, Fujisawa Pharmaceutical Co., Ltd., Higashiyodogawa-ku, Osaka, S., Umehara, K., Shimizu, K., Minoda, Y., Yamada, K. (1970) Microbial conversion of petro-sulfur compounds part III. Isolation and identification of products from dibenzothiophene. *Agricultural and Biological Chemistry*. 34: 1320–1324.

Kodama Central Research Institute of Electric Power Industry, Abiko, Chiba Prefecture, K., Umehara Central Research Institute of Electric Power Industry, Abiko, Chiba Prefecture, K., Shimizu Research Laboratories, Fujisawa Pharmaceutical Co., Ltd., Higashiyodogawa-ku, Osaka, K., Nakatani Research Laboratories, Fujisawa Pharmaceutical Co., Ltd., Higashiyodogawa-ku, Osaka, S., Minoda Department of Agricultural Chemistry, Faculty of Agriculture, The University of Tokyo, Y., Yamada Department of Agricultural Chemistry, Faculty of Agriculture, The University of Tokyo, K. (1973) Identification of Microbial Products from Dibenzothiophene and Its Proposed Oxidation Pathway. *Agricultural and Biological Chemistry*. 37(1): 45–50.

Kohler, H.P.E., Kohler-Staub, D., Focht, D.D. (1988) Degradation of 2-hydroxybiphenyl and 2,2′-dihydroxybiphenyl by *Pseudomonas* sp. strain HBP1. *Applied and Environmental Microbiology*. 54: 2683–2688.

Kohler, H.P.E., Schmid, A., Van der Maarel, M. (1993) Metabolism of 2,2′-dihydroxybiphenyl by *Pseudomonas* sp. strain HBP1: Production and consumption of 2,2′,3-trihydroxybiphenyl. *Journal of Bacteriology*. 175: 1621–1628.

Köhler, M., Genz, I.-L., Schicht, B., Eckart, V. (1984) Microbial desulfurization of petroleum and heavy petroleum fractions. 4. Anaerobic degradation of organic sulfur compounds of petroleum. Zentralblatt Fur Mikrobiologie. 139: 239–247.

Kondo, R., Imamura, H. (1989) Formation of lignin model xyloside in polysaccharides media by wood-rotting fungi. Mokuzai Gakkaishi 35: 1001–1007.

Kondo, R., Yamagami, H., Sakai, K. (1993) Xylosylation of phenolic hydroxyl groups of the monomeric lignin model compounds 4-methylguaiacol and vanillyl alcohol by *Coriolus versicolor. Applied and Environmental Microbiology.* 59: 438–441.

Konishi, J., Ishii, Y., Onaka, T., Okumura, K., Suzuki, M. (1997) Thermophilic carbon-sulfur-bond-targeted biodesulfurization. *Applied and Environmental Microbiology.* 63 (8): 3164–169.

Konishi, J., Onaka, T., Ishii, Y., Suzuki, M. (2000) Demonstration of the carbon–sulfur bond targeted desulfurization of benzothiophene by thermophilic *Paenibacillus* sp. strain A11-2 capable of desulfurizing dibenzothiophene. FEMS Microbiology Letters. 187: 151–154.

Krawiec, S. (1990) Bacterial desulfurization of thiophenes: screening techniques and some speculations regarding the biochemical and genetic basis. Developments in industrial microbiology. 31: 103–114

Krawiec, S. (1990a) Transformation of dibenzothiophene by axenic cultures of *Sulfolobus acidocaldarius* and other Bacteria: A Critiqu. In: Biotechnology Applied to Fossil Fuels, Donald L. Wise (ed.), Marcel Dekker, Inc. New York. Pp. 569–601.

Krawiec, S. (1990b) Bacterial desulfurization of thiophenes: screeing techniques and some speculations regarding the biochemical and genetic bases. In: Pierce GE (ed) Developments in Industrial Microbiology. Society for Industrial Microbiology. 31: 103–114.

Kropp K.G., Saftic S., Andersson J.T., Fedorak P.M. (1996) Transformations of six isomers of dimethylbenzothiophenes by three *Pseudomonas* strains. Biodegradation. 7: 203–221

Kropp, K.G., Andersson, J.T., Fedorak, P.M. (1997a) Bacterial degradation of naphtothiophenes. *Applied and Environmental Microbiology.* 63: 3463–3473.

Kropp, K.G., Andersson, J.T., Fedorak, P.M. (1997b) Bacterial transformations of 1,2,3,4-tetrahydrodibenzothiophene and dibenzothiophene. *Applied and Environmental Microbiology.* 63(8): 3032–3042.

Kropp, K.G., Andersson, J.T., Fedorak, P.M. (1997c) Biotransformations of three dimethyl dibenzothiophenes by pure and mixed bacterial cultures. *Environmental Science and Technology.* 31: 1547–1554.

Kropp, K.G., Fedorak, P.M. (1998) A review of the occurrence, toxicity, and biodegradation of condensed thiophenes found in petroleum. *Canadian Journal of Microbiology.* 44: 605–622.

Kropp, K.G., Goncalves, J.A., Andersson, J.T., Fedorak, P.M. (1994a) Bacterial transformations of benzothiophene and methylbenzothiophenes. *Environmental Science & Technology.* 28: 1348–1356.

Kropp, K.G., Goncralves, J.A., Andersson, J.T., Fedorak, P.M. (1994b) Microbially mediated formation of benzonaphthothiophenes from benzo[b]thiophenes. *Applied and Environmental Microbiology.* 60: 3624- 3631.

Kropp, K.G., Saftic, S., Andersson, J.T., Fedorak, P.M. (1996) Transformations of six isomers of dimethylbenzothiophenes by three *Pseudomonas* strains. Biodegradation. 7: 203–221.

Kurita, S.; Endo, T.; Nakamura, H.; Yagi, T., Tamuja, N. (1971) Decomposition of some organic sulfur compounds in petroleum by anaerobic bacteria. *Journal of General and Applied Microbiology*. 17: 185–198.

Laborde A.L., Gibson, D.T. (1977) Metabolism of dibenzothiophene by a *Beijerinckia* species. *Applied and Environmental Microbiology*. 34(6): 783–790.

Le Borgne, S., Ayala, M. (2010) Microorganisms utilizing sulfur-containing hydrocarbons. In: Timmis, K.N. (Ed.). Handbook of Hydrocarbon and Lipid Microbiology. Springer-Verlag. Berlin Heidelberg. Germany.

Lee M.K., Senius J.D., Grossman M.J. (1995) Sulfur-specific microbial desulfurization of sterically hindered analogs of dibenzothiophene. *Applied and Environmental Microbiology*. 61: 4362–4366.

Lee, M.Y., Dordick, J.S. (2002) Enzyme activation for non-aqueous media, Current Opinion in Biotechnology. 13(4):376–384.

Lee, N.A., Clark, D.P. (1993) A natural isolate of *Pseudomonas maltophila* which degrades aromatic sulfonic acids. FEMS Microbiol Lett 107:151–156.

Lee, S.H.D., Kumar, R., Krumpelt, M. (2002): Sulfur removal from diesel fuel contaminated methanol. *Separation and Purification Technology*. 26(2/3): 247–258.

Li, F.L., Xu, P., Ma, C.Q., Luo, L.L., Wang, X.S. (2003) Deep desulfurization of hydrodesulfurization-treated diesel oil by a facultative thermophilic bacterium *Mycobacterium* sp. X7B. FEMS Microbiology Letters. 223: 301–307.

Li, F.L., Zhang, Z.Z., Feng, J.H., Cai, X F., Xu. P. (2007a) Biodesulfurization of DBT in tetradecane and crude oil by a facultative thermophilic bacterium *Mycobacterium goodii* X7B. *Journal of Biotechnology*. 127: 222–228.

Li, W., Wang, M.D., Chen, H., Chen, J.M., Shi, Y. (2006b) Biodesulfurization of dibenzothiophene by growing cells of *Gordonia* sp. in batch cultures. Biotechnology Letters 28: 1175–1179.

Liang, F., Lu, M.M., Birch, M.E., Keener, M.C., Liu, Z.F. (2006a) Determination of polycyclic aromatic sulfur heterocycles in diesel particular matter and diesel fuel by gas chromatography with atomic emission detection. *Journal of Chromatography A*. 1114:145–153

MacPherson, T., Greer, C.W., Zhou, E., Jones, A.M., Wisse, G., Lau, C.K., Sankey, B.M., Grossman, J., Hawari, J. (1998) Application of SPME/GC-MS to characterize metabolites in the biodesulfurization of organosulfur model compounds in bitumen. *Environmental Science & Technology*. 32:421–426.

Malik, K.A, Claus, D. (1976) Microbial degradation of dibenzothiophene. In: "Abstracts of the Fifth International Fermentation Symposium. (Dellweg H. Ed.)". Berlin. Germany. 23(03): 421.

Mamie, Z.L., Charles, H.S., Daniel, J.M. (1996) Genetic analysis of the *dsz* promoter and associated regulatory regions of *Rhodoccus erythropolis* IGTS8. *Journal of Bacteriology*. 11(178): 6409–6418.

Masaphy, S., Levanon, D., Henis, Y., Venkateswarlu, K., Kelly, S.L. (1995) Microsomal and cytosolic cytochrome P450 mediated benzo(a)pyrene hydroxylation in *Pleurotus pulmonarius*. Biotechnology Letters 17: 967–974.

Matsui, T., Hirasawa, K., Konishi, J., Tanaka, Y., Maruhashi, K., Kurane, R. (2001a) Microbial desulfurization of alkylated dibenzothiophene and alkylated benzothiophene by recombinant *Rhodococcus* sp. strain T09. *Applied Microbiology and Biotechnology*. 56: 196–200.

Matsui, T., Onaka, T., Maruhash, K., Kurane (2001b) Benzo(b)thiophene desulfurization by *Gordonia rubropertinctus* strain T08. *Applied Microbiology and Biotechnology.* 57: 212–215.

Matsui, T., Onaka, T., Tanaka, Y., Tezuka, T., Suzuki, M., Kurane, R. (2000) Alkylated benzothiophene desulfurization by *Rhodococcus* sp. strain T09. *Bioscience, Biotechnology, and Biochemistry*. 64:596–599.

Miller, K.W. (1992) Reductive desulfurization of dibenzyldisulfide. *Applied and Environmental Microbiology*.58(7): 2176–2179.

Mohamed, M.E.S., Al-Yacoub, Z.H., Vedakumar, J.V. (2015) Biocatalytic desulfurization of thiophenic compounds and crude oil by newly isolated bacteria. Frontiers in Microbiology. 6:1–12.

Mohebali, G., Ball, A.S. (2008) Biocatalytic desulphurization (BDS) of petrodiesel fuels. Microbiology. 154: 2169–2183.

Mohebali, G., Ball, A.S. (2016) Biodesulfurization of diesel fuels – Past, present and future perspectives. *International Biodeterioration and Biodegradation*. 110:163–180.

Mohebali, G., Ball, A.S., Rasekh, B., Kaytash, A. (2007) Biodesulfurization potential of a newly isolated bacterium, *Gordonia alkanivorans* RIPI90A, *Enzyme and Microbial Technology*. 40: 578–584.

Monticello, D.J. (1998) Riding the fossil fuel biodesulfurization wave. Proceedings of the 5th World Petroleum Congress. Published by John Wiley & Sons. pp. 901–906.

Monticello, D.J. (2000) Biodesulfurization and the upgrading of petroleum distillates. *Current Opinion in Biotechnology*. 11: 540–546.

Monticello, D.J., Finnerty, W.R. (1985) Microbial desulfurization of fossil fuels. Annual Review in Microbiology. 39:371–389.

Moriya, K., Horikoshi, K. (1993) A benzene-tolerant bacterium utilizing sulfur compounds isolated from deep sea. *Journal of Fermentation and Bioengineering*. 76:397–399.

Mormile, M.R., Atlas, R.M. (1988) Mineralization of the dibenzothiophene biodegradation products 3-hydroxy-2-formyl benzothiophene and dibenzothiophene sulfone. *Applied and Environmental Microbiology*. 54:3183–3184.

Murgich, J. Abanero, A.J., Strausz, P.O. (1999) Molecular recognition in aggregates formed by asphaltene and resin molecules from the Athabasca oil sand. *Energy and Fuel* 13(2): 278–286.

Nassar, H. N. El-Gendy N. Sh., Farahat, L. A., Hawas A. E., and Mahdy, H. M. (2010) Study of diesel oil biodegradation by *Bacillus sphaericus* HN1. The 13th International Conference in Petroleum, Mineral Wealth and Development, Cairo, Egypt, 7–9 February, 2010.

Nassar, H.N. (2010) Potentials of microorganisms isolated from Egyptian hydrocarbon polluted sites on degradation of polycyclic aromatic sulfur heterocycles (PASHs) compounds. A M.Sc. Thesis. Al-Azhar University.

Nassar, H.N., Deriase, S.F., El-Gendy, N.Sh. (2016) Modeling the relationship between microbial biomass and total viable count for a new bacterial isolate used in biodesulfurization of petroleum and its fractions. *Petroleum Science and Technology*. 34(11/12): 980–985.

Nassar, H.N., El-Gendy, N.Sh., Abo State M., Moustafa, Y.M., Mahdy, H.M., El-Temtamy S.A. (2013) Desulfurization of dibenzothiophene by a novel strain *Brevibacillus invocatus* C19 isolated from Egyptian coke. *Biosciences, Biotechnology Research Asia*. 10(1): 29–46.

Nekodzuka, S., Nakajima-Kambe, T., Nomura, N., Lu, J., and Nakahara, T. (1997) Specific desulfurization of dibenzothiophene by *Mycobacterium* sp. strain G3. *Biocatalysis and Biotransformation*. 15: 17–27.

Noda, K., Watanabe, K., Maruhashi, K. (2003) Isolation of the Pseudomonas aeruginosa gene affecting uptake of dibenzothiophene in n-tetradecane. *Journal of Bioscience and Bioengineering*. 95: 504–511.

Ogata, M., Fujisawa, K. (1983) Capillary GC/MS Determination of Organic Sulfur Compounds Detected in Oyster and Mussel Caught in the Sea as an Oil Pollution Index. *Journal of Chromatographic Science*. 21(9): 420–424.

Ogata, M., Mujake, Y. (1978) Identification of organic sulfur compounds transferred to fish from petroleum suspension. *Acta Medica Okayama*. 32(6): 419–425.

Ohshiro T, Hirata T, Izumi Y (1995) Microbial desulfurization of dibenzothiophene in the presence of hydrocarbon. *Applied Microbiology and Biotechnology*. 44: 249–252.

Ohshiro, T., Hine, Y., Izumi, Y. (1994) Enzymatic desulfurization of dibenzothiophene by a cell-free system of *Rhodococcus erythropolis* D-1. FEMS Microbiology Letters. 118: 341–344.

Ohshiro, T., Hirata, T., Hashimoto, I., Izumi, Y. (1996a) Characterization of dibenzothiophene desulfurization reaction by whole cells of *Rhodococcus erythropolis* H-2 in the presence of hydrocarbon. *Journal of Fermentation and Bioengineering*. 82: 610–612.

Ohshiro, T., Suzuki, K., Izumi, Y. (1996b) Regulation of dibenzothiophene degrading enzyme of *Rhodococcus erythropolis* D-1. *Journal of Fermentation and Bioengineering*. 81: 121–124.

Ohshiro, T., Suzuki, K., Izumi, Y. (1997) Dibenzothiophene (DBT) degrading enzyme responsible for the first step of DBT desulfurization by *Rhodococcus erythropolis* D-1: purification and characterization. *Journal of Fermentation and Bioengineering*. 83(2): 233–237.

Okada, H., Nomura, N., Nakahara, T., Maruhashi, K. (2002a) Analyses of substrate specificity of the desulfurizing bacterium *Mycobacterium* sp. G3. *Journal of Bioscience and Bioengineering*. 93(2): 228–233.

Okada, H., Numura, N., Nakahara, T., Muruhashi, K. (2001) Cultivation of a desulfurizing bacterium, *Mycobacterium* sp. G3. Biotechnology Letters. 23: 2047–2050.

Okada, I., Nomura, N., Nakahara, T., Maruhashi K. (2002b) Analysis of dibenzothiophene metabolic pathway in *Mycobacterium* strain G3. *Journal of Bioscience and Bioengineering*. 93(5):491–497.

Oldfield, C., Pogrebinsky, O., Simmods, J., Olson, E.S., Kulpa, C.F. (1997) Elucidation of the metabolic pathway for dibenzothiophene desulfurization by *Rhodococcus* sp. strain IGTS8 (ATCC53968). Microbiology. 143: 2961–2973.

Oldfield, C., Wood, N.T., Gilbert, S.C., Murray, F.D., Faure F.R. (1998) Desulfurization of benzothiophene and dibenzothiophene by actinomycete organisms belonging to the genus *Rhodococcus*, and related taxa. *Antonie Van Leeuwenhoek International Journal of General and Molecular Microbiology*. 74: 119–132.

Olson, E.S., Stanley, D.C., Gallagher, J.R. (1993) Characterization of intermediates in the microbial desulfurization of dibenzothiophene. *Energy and Fuels*. 7: 59–163.

Omori, T., Monna, L., Saiki, Y., Kodama, T. (1992). Desulfurization of dibenzothiophene by *Corynebacterium* sp. strain SY1. *Applied and Environmental Microbiology*. 58(3): 911–915.

Omori, T., Saiki, Y., Kasuga, K., Kodama, T. (1995) Desulfurization of alkyl and aromatic sulfides and sulfonates by dibenzothiophene desulfurizing *Rhodococcussp*. strain SY1. *Bioscience, Biotechnology and Biochemistry*. 59(7): 1195–1198.

Onaka, T., Kobayashi, M., Ishii, Y., Okumura, K., Suzuki, M. (2000) Application of solid-phase extraction to the analysis of the isomers generated in biodesulfurization against methylated dibenzothiophenes. *Journal of Chromatography A*. 903: 193–202.

Onaka, T., Matsui, T., Maruhashi, K. (2002) Metabolite detection in *Rhodococcus* benzothiophene desulfurization using reversed-phase solid-phase extraction and a derivatization technique. Biotechnology Letters. 24: 1961–1966.

Oudot, J. (1984) Rates of microbial degradation of petroleum components as determined by computerized capillary gas chromatography and computerized mass spectrometry. *Marine Environmental Research*. 13(4): 277–302.

Pinedo-Rivilla, C., Aleu, J., Collado, I.G. (2009) Pollutants biodegradation by fungi. *Current Organic Chemistry*. 13(12): 1194–1214.

Praveen Reddy, P., Umamaheshwara Rao, V (2015) Dibenzothiophene desulfurizing Streptomyces species isolated from oil contaminated sites. *International Journal of Science and Research*. 5(12): 1671–1674.

Purdy, R.F., Lepo, J.E., Ward, B. (1993) Biodesulfurization of organic sulfur compounds. Current Microbiology. 27: 219–222.

Raju, K.S., Sridevi, V., Chandana Lakshmi, M.V.V., Thomas, K.K. (2013) Biodesulfurization of fossil fuels- An overview. *International Journal of Pharma and Bio Sciences*. 4(3B): 719 – 727.

Rambosek, J., Piddington, C.S., Kovacevich, B.R., Young, K.D., Denome, S. (1999) Recombinant DNA encoding a desulfurization biocatalyst. US Patent no. WO9401563.

Rath, K., Mishra, B., Vuppu, S. (2012) Biodegrading ability of organosulfur compound of a newly isolated microbe *Bacillus* sp. KS1 from the oil contaminated soil. *Archives of Applied Science Research*. 4(1): 465–471.

Reddy, G.V.B., Joshi, D.K., Gold, M.H. (1997) Degradation of chlorophenoxyacetic acids by the lignin-degrading fungus *Dichomitus squalens*. Microbiology. 143: 2353–2360.

Reichmuth, D.S., Hittle, J.L., Blanch, H.W., Keasling J.D. (2000) Biodesulfurization of dibenzothiophene in *Escherichia coli* is enhanced by expression of a *vibrio harveyi* oxidoreductase gene. *Biotechnology and Bioengineering*. 67(1): 72–79.

Rhee S.K., Chang, J.H., Chang, Y.K., and Chang, H.N. (1998) Desulfurization of dibenzothiophene and diesel oils by a newly isolated *Gordona* strain CYKS. *Applied an Environmental. Microbiol.ogy*. 64(6): 2327–2331.

Romero, M.C., Hammer, E., Hanschke, R., Arambarri, A.M., Schauer, F. (2005) Biotransformation of biphenyl by the filamentous fungus *Talaromyces helices*. Microbial Biotechnology. 21: 101–106.

Ryabov, V.D. (2009). Oil and gas chemistry. Moscow: Forum.

Saftić, S., Fedorak, P.M., Andersson, J.T. (1993) Transformations of methyldibenzothiophenes by three *Pseudomonas* isolates. *Environmental Science and Technology*. 27: 2577–2584

Sagardia, F., Rigau, J.J., Martinez, L.A., Fuentes, F. and Lopez C. (1975) Degradation of benzothiophene and related compounds by a soil Pseudomonas in an oil-aqueous environment. *Journal of Applied Microbiology*. 29: 722–725.

Sandhya, S.; Prabu, SK., and Sundari, R.B.T. (1995) Microbial degradation of dibenzothiophene by Nocardioides. *Journal of Environmental Science and Health, Part A*. 30:1995–2006.

Sato, H., Clark, D. P. (1995) Degradation of dibenzothiophene sulphoxide and sulphone by *Arthrobacter* strain dibenzothiophene S2. Microbios. 83: 145–159.

Schlenk, D., Bever, R.J., Vertino, A.M., Cerniglia, C.E. (1994) Cytochrome P-450 catalyzed S-oxidation of dibenzothiophene by *Cunninghamella elegans*. Xenobiotica .24: 1077–1083.

Schulz, H., Böhringer, W., Ousmanov, F., Waller, P. (1999) Refractory sulfur compounds in gas oils. *Fuel Processing Technology*. 61: 5–41.

Selifonov, S.A., Grifoll, M., Eaton, R.W., Chapman, P.J. (1996) Oxidation of naphthenoaromatic and methyl-substitued aromatic compounds by naphthalene 1,2-dioxygenase. *Applied and Environmental Microbiology*. 62: 507–514.

Seo, J.S., Keum, Y.S., Cho, I.K., Li, Q.X. (2006) Degradation of dibenzothiophene and carbazole by *Arthrobacter* sp. P1-1. *International Biodeterioration and Biodegradation*. 58: 36–43.

Serbolisca, L., de Ferra, F., Margrit, I. (1999) Manipulation of the DNA coding for the desulfurizing activity in a new isolate of *Arthrobacter* sp. *Applied Microbiology and Biotechnology*. 52: 122–126.

Setti, L., Farinelli, P., Di Martino, S., Frassinetti, S., Lanzarini, G., Pifferi, P.G. (1999) Developments in destructive and non-destructive pathways for selective desulfurizations in oil-biorefining processes. *Applied Microbiology and Biotechnology*. 52: 111–117.

Shahaby, A.F., Essam El-din, K.M. (2017) Desulfurization of crude oil and oil products by local isolated bacterial strains. *International Journal of Current Microbiology and Applied Sciences*. 6(4): 2695–2711.

Shennan, J.L. (1996) Microbial attack on sulfur-containing hydrocarbons, implications for the biodesulfurization of oils and coals. *Journal of Chemical Technology and Biotechnology*. 67: 109–123.

Singh S.P.; Schwan A.L. (2011) Sulfur metabolism in plants and related biotechnologies. In: Moo-Young, M. (Ed.), Comprehensive Biotechnology, second ed. Elsevier. 4: 257–271.

Singh, P., Srivastava, P. (2013) An improved protocol for electroporation in members of the genus *Gordonia*. *Journal of Microbiological Methods*. 95(2): 114–116.

Soleimani, M., Bassi, A. Margaritis, A. (2007) Biodesulfurization of refractory organic sulfur compounds in fossil fuels. Biotechnology Advances. 25(6):570–596.

Sondossi, M., Barriault, D., Sylvestre, M. (2004) Metabolism of 2,2- and 3,3-dihydroxybiphenyl by the biphenyl catabolic pathway of *Comamonas testosteroni* B-356. *Applied and Environmental Microbiology*. 279: 174–181.

Speight, J.G., Moschepodis, S.E. (1981) On the molecular nature of petroleum asphaltenes. In: Bunger J. W. & Li N. C. (Eds). Chemistry of asphaltenes American Chemical Society, Washington D.C. Pp. P.1–15.

Stoner, D.L., Wey, J.E., Barrett, K.B., Jolley, J.G., Wright, R.B., Dugan, P.R. (1990) Modification of water- soluble coal- derived products by dibenzothiophene-degrading microorganisms. *Applied and Environmental Microbiology*. 56(9): 2667–2676.

Sutherland, J.B., Selby, A.L., Freeman, J.P., Fu, P.P., Miller, D.W., Cerniglia, C.E. (1992) Identification of xyloside conjugates formed from anthracene by *Rhizoctonia solani*. Mycological Research. 96: 509–517.

Tanaka, Y., Matsui, T., Konishi, J., Maruhashi, K., Kurane, R. (2002) Biodesulfurization of benzothiophene and dibenzothiophene by a newly isolated *Rhodococcus* strain. *Applied Microbiology and Biotechnology*. 59: 325–328.

Tanaka, Y., Onaka, T., Matsui, T., Maruhashi, K., Kurane, R. (2001) Desulfurization of benzothiophene by the gram-negative bacterium, *Sinorhizobium* sp. KT55. Current Microbiology. 43: 187–191.

Tilstra, L., Eng, G., Olson, G.J. and Wang, F.W. (1992) Reduction of sulphur from polysulphidic model compounds by the hyperthermophilic *Archaebacterium Pyrococcus* furiosus. *Fuel*. 71: 779–783.

Ullrich, R., Hofrichter, M. (2005) The haloperoxidase of the agaric fungus Agrocybe aegerita hydroxylates toluene and naphthalene. FEBS Letters. 579: 6247–6250.

Ullrich, R., Hofrichter, M. (2007) Enzymatic hydroxylation of aromatic compounds. Cellular and Molecular Life Sciences. 64: 271–293.

Ullrich, R., Nüske, J., Scheibner, K., Spantzel, J., Hofrichter, M. (2004) Novel haloperoxidase from the agaric basidiomycete *Agrocybe aegerita* oxidizes aryl alcohols and aldehydes. *Journal of Applied and Environmental Microbiology*. 70: 4575–4581.

Van Afferden, M., Schacht, S., Klein, J., Trüper, H.G. (1990) Degradation of dibenzothiophene by *Brevibacterium* sp. DO. *Archives of Microbiology*. 153: 324–328.

Van Afferden, M., Trappe, D., Beyer, M., Tüpper, H.G., Klein, J. (1993) Biochemical mechanisms for the desulfurization of coal-relevant organicsulfur compounds. *Fuel* 72(12): 1635–1643.

Van Hamme, J.D., Fedorak, P.M., Foght, J.M., Gray, M.R., Dettman, H.D. (2004) Use of a novel fluorinated organosulfur compound to isolate bacteria capable of carbon-sulfur bond cleavage. *Applied and Environmental Microbiology*. 70(3): 1487–1493.

Van Hamme, J.D., Wong, E.T., Dettman, H., Gray, M.R., Pickard, M.A. (2003a) Dibenzyl sulfide metabolism by White Rot Fungi. *Applied and Environmental Microbiology*. 69(2): 1320–1324.

Van Herwijnen, R., Wattiau, P., Bastiaens, L., Daal, L., Jonker, L., Springael, D. (2003b) Elucidation of the metabolic pathway of fluorine and cometabolic pathways of phenanthrene, fluoranthene, anthracene and dibenzothiophene by *Sphingomonas* sp. LB126. Research in Microbiology. 154: 199–206.

Vazquez-Duhalt, R. (1999) Cytochrome c as a biocatalyst. *Journal of Molecular Catalysis B: Enzymatic*. 7: 241–249.

Vazquez-Duhalt, R., Semple, K.M., Westlake, D.W.S., Fedorak P.M. (1993) *Enzyme and Microbial Technology*. 15: 936.

Verma, N., Sharma, R., Kaur, R. (2016) Microbial desulphurization study of dibenzothiophene and crude oil by a soil isolate. *International Journal of Science and Research Methodology*. 4(4): 133–145.

Vermeij, P., Kertesz, M.A. (1999) Pathways of assimilative sulfur metabolism in *Pseudomonas putida*. *Journal of Bacteriology*. 181(18): 5833–5837.

Walker, J.D., Colwell, R.R., Petrakis, L. (1975) Microbial petroleum degradation: application of computerized mass spectrometry. *Canadian Journal of Microbiology*. 21: 1760–1767.

Wang, P., Krawiec, S. (1994) Desulfurization of dibenzothiophene to 2-hydroxybiphenyl by some newly isolated bacterial strains. *Archives of Microbiology*. 161(3): 266–271.

Wang, P., Krawiec, S. (1996) Kinetic analyses of desulfurization of dibenzothiophene by *Rhodococcus erythropolis* in batch and fed-batch cultures. *Applied and Environmental Microbiology*. 61: 1670–1675.

Wang, T., He, F., Li, M., Hou, Y., Guo, S. (2004) Alkyldibenzothiophenes: molecular tracers for filling pathway in oil reservoirs. Chinese Science Bulletin. 49: 2399–2404.

Wariishi, H., Itoh, N., Yoshida, M., Tanaka, H. (2002) Complete degradation of Y perite, a chemical warfare agent, by basidiomycetes. Biotechnology Letters. 24: 501–505.

Watanapokasin, Y., Nuchfoang, S., Nilwarangkoon, S., Sarangbin, S., Kakizono, T. (2002) Isolation and characterization of thermophilic benzothiophene degrading Mycobacterium Sp. *Applied Biochemistry and Biotechnology.* 98–100:301–309.

Wu, S., Lin, J., Chan, S.I. (1994) Oxidation of dibenzothiophene catalyzed by heme-containing enzymes encapsulated in sol-gel glass. *A new form of biocatalysts, Applied Biochemistry and Biotechnology.* 47: 11–20.

Xu, P., Yu, B., Li, F., Cai, X.F., Ma, C.Q. (2006) Microbial degradation of sulfur, nitrogen and oxygen heterocycles. *Trends in Microbiology.* 14(9): 398–405.

Yamada, K.O., Morimoto, M., Tani, Y. (2001) Degradation of dibenzothiophene by sulfate-reducing bacteria cultured in the presence of only nitrogen gas. *Journal of Bioscience and Bioengineering.* 91:91–93.

Yang, B., Hou, W., Zhang, K., Wang, X. (2013) Application of solid- phase micro extraction to the determination of polycyclic aromatic sulfur heterocycles in Bohai Sea crude oils. *Journal of Separation Science.* 16:2646–2655.

Yu, B., Xu, P., Shi, Q., Ma, C. (2006) Deep desulfurization of diesel oil and crude oils by a newly isolated *Rhodococcus erythropolis* strain. *Applied and Environmental Microbiology.* 72(1): 54–58.

Zhang, H., Shan, G., Liu, H., Xing, J. (2007) Surface Modification of γ-Al2O3 Nanoparticles with gum Arabic and its applications in adsorption and biodesulfurization. Surface and Coatings Technology. 201: 6917–6921.

7

Enzymology and Genetics of Biodesulfurization Process

List of Abbreviations and Nomenclature

2-HBP	2-hydroxybiphenyl
2-MBP	2-Methoxybiphenyl
Ala	Alanine
APS	Adenylsulfate
Asn	Asparagine
ATCC	American Type Culture Collection
ATP	Adenosine triphosphate
BDS	Biodesulfurization
BDSM	Biodesulfurizing Microorganism
BP	Biphenyl
BT	Benzothiophene
CoA	Coenzyme A
DBT	Dibenzothiophene
DBT-MO	Dibenzothiophene Monooxygenase
DBTO	Dibenzothiophene Sulfoxide

$DBTO_2$	Dibenzothiophene Sulfone
DCW	Dry cell weight
DHBP	2,2'-dihydroxybiphenyl
DMSO	Dimethylsulfoxide
DNA	Deoxyribonucleic acid
Dox	DBT oxidation
Dsz	Desulfurization
Dsz	Desulfurization genes
DszA	Desulfurization enzyme A
dszA	Desulfurization gene A
*dsz*AB	Desulfurization genes A & B
DszB	Desulfurization enzyme B
dszB	Desulfurization gene B
DszC	Desulfurization enzyme C
dszC	Desulfurization gene C
EDTA	Ethylenediaminetetraacetic acid
FMN	Flavin Mononucleotide
FMNH	Flavin Mononucleotide, reduced form
G	Guanine
GFP	Green Fluorescent Protein
HbpA	2-Hydroxybiphenyl 3-monooxygenase
HbpC	Extradiol Dioxygenase
HbpD	Hydrolase
HBPS	2'-hydroxybiphenyl-2-sulfinate
HDS	Hydrodesulfurization
He-Ne laser	Helium-Neon laser
$HpaC_{St}$	4-hydroxyphenylacetate hydroxylase from *Sulfolobus tokodaii*
$HpaC_{Tt}$	4-hydroxyphenylacetate hydroxylase from *Thermus thermophiles*
Kb	Kilo base
kDa	Kilo Dalton
K_m	Michaelis constant
LGO	Light gas oil
Log *P*	Solubility and partition coefficient
mRNA	Messenger RNA
NAD	Nicotinamide Adenine Dinucleotide
NADH	Nicotinamide Adenine Dinucleotide, reduced form
ORF	Open Reading Frame
OSCs	Organo sulfur compounds
PAPS	3-Phosphoadenyl Sulfate

PASHs	Polyaromatic Sulfur Heterocyclics
PCR	Polymerase Chain Reaction
pdb	protein data base
pET	Plasmid for Expression under control of T7 promoter
ppm	Parts per million
Rsgfp	Red-shifted green fluorescence protein gene
S1	Sulpeptide 1
SDS–PAGE	Sodium dodecyl sulfate-polyacrylamide gel electrophoresis
SOR	Sulfite Oxidoreductase
sox	Sulfur oxidation gene
SR	Sulfite Reductase
T	Thymine
VHb	Vitreoscilla hemoglobin

7.1 Introduction

Biodesulfurization (BDS) has attracted a lot of attention as a complementary system to hydrodesulfurization (HDS) to remove the organosulfur compounds from fuel before combustion to reduce sulfur emissions into the atmosphere (Ohshiro and Izumi, 2002 and Gray *et al.*, 2003).

Dibenzothiophene (DBT) and its alkylated derivatives are the most abundant form of organic sulfur in petroleum oils (Kilbane, 2006). These compounds are difficult to desulfurize via the conventional HDS process that is currently used in petroleum refinery, so scientists have focused on the desulfurization of organic sulfur using microorganisms to apply as commentary for HDS to decrease the sulfur content of petroleum products.

Numerous microorganisms have been isolated and are capable of utilizing DBT as a sole source of carbon, sulfur, and energy (Kodama *et al.*, 1973; Monticello *et al.*, 1985; Van Afferden *et al.*, 1990; Bressler *et al.*, 1998). However, the complete degradation of organosulfur compounds is not useful for upgrading crude oils and derived fuels. The selective cleavage of the carbon-sulfur bonds is preferred. In this manner, the sulfur is selectively removed and the calorific value of treated fuel remains intact. The first microorganism that showed the ability to break the carbon sulfur bonds of DBT selectively, removing the S-atom and leaving the carbon skeleton intact, was *Rhodococcus erythropolis* IGTS8 (ATCC 53968), isolated by Kilbane (1989). Subsequently, numerous other bacteria capable of selectively cleaving carbon-sulfur bonds of DBT were isolated and characterized (El-Gendy *et al.*, 2006; Mohebali *et al.*, 2007; Davoodi-Dehaghani *et al.*, 2010; Gunam *et al.*, 2013; Nassar *et al.*, 2013; El-Gendy *et al.*, 2014).

DBT is widely used as a model compound for polyaromatic sulfur heterocyclics (PASHs) (Xu *et al.*, 2006). Other PASHs have been rarely studied (Schreinier *et al.*, 1988; Kayser *et al.*, 1993; Kirkwood *et al.*, 2005, 2007; Ahmad *et al.*, 2014). The desulfurization of these compounds is done by microbial catabolic pathways, encoded either on plasmids or on the chromosome, that, generally, are not necessary for growth, but rather are allowed to overcome a specific environmental condition (e.g. temperature, pH, mixing rate…etc.). The key to improving the biocatalysts' mass production is the identification of the genes responsible for the desulfurization of PASHs and manipulation of the system through genetic engineering techniques (Chen *et al.*, 1999; Alves *et al.*, 2005).

Factors, such as the cell-reducing power, cytoplasmic oxygen levels, transmembrane transferring of substrates and products, solvent tolerance, and the ability of the cells to access and uptake the PASHs compounds, may strongly influence BDS efficiency. A large number of recombinant microorganisms have been engineered to overcome the major problems of the BDS process, so this chapter aims to illustrate the BDS process from the biochemical, genetic, and molecular enzyme views and discuss the extensive research on this subject to improve the applicability of BDS process in the industrial sector.

7.2 Genetics of PASHs BDS Pathway

BDS processes have been described by three different types of reactions: (1) sulfur oxidation, (2) carbon-carbon cleavage, and, finally, (3) sulfur-specific cleavage (Chapter 6) (Gupta *et al.*, 2005; Xu *et al.*, 2006; Soleimani *et al.*, 2007; Mohebali and Ball, 2016). Some fungal laccases and bacterial ring-hydroxylating dioxygenases catalyze the sulfur oxidation of DBT to DBT-sulfone and other sulfur-containing hydroxylated derivatives as dead-end products (Gupta *et al.*, 2005; Xu *et al.*, 2006; Soleimani *et al.*, 2007; Mohebali and Ball, 2016).

7.2.1 Anaerobic BDS Pathway

Kim *et al.* (1995) isolated sulfate-reducing bacteria, *Desulfovibrio desulfuricans* M6, which anaerobically reduces DBT to biphenyl and H_2S as shown in Figure 7.1 (Kim *et al.*, 1995). Other anaerobic biodesulfurizing microorganisms (BDM), such as *Desulfomicrobium escambium* and *Desulfovibrio longreachii*, have been reported to desulfurize DBT following a pathway in which biphenyl was not identified. The BDS of oil anaerobically avoids

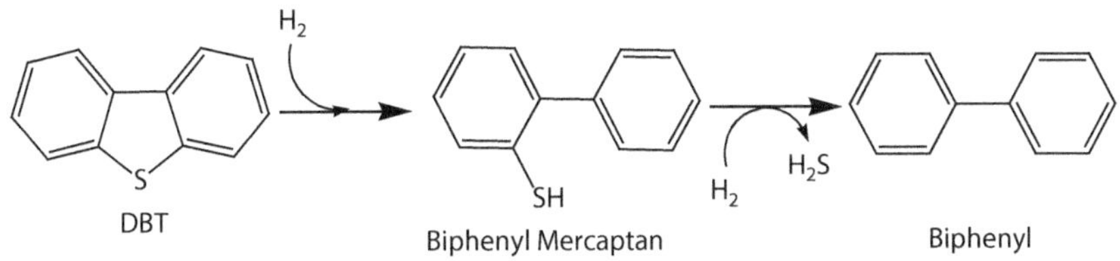

Figure 7.1 Proposed Metabolic Pathway for the Anaerobic BDS of DBT by *Desulfovibrio desulfuricans* M6 (Kim *et al.*, 1995).

costs related to aeration and has the advantage of releasing sulfur as a gas. However, the low rate and extent of petroleum desulfurization by currently available anaerobic cultures and the lack of knowledge on the biochemistry and genetics of such microorganisms, make the development of a commercial process unlikely in the near future (Gupta *et al.*, 2005).

7.2.2 Aerobic BDS Pathway

7.2.2.1 Kodama Pathway

The Kodama pathway is the oxidative and carbon-carbon cleavage of DBT, which consists of three main steps including 1) lateral dioxygenation of one of the homocyclic rings, 2) ring cleavage, and 3) hydrolysis yielding hydroxy-formyl-benzothiophene as the end product, as shown in Figure 7.2 (Gupta *et al.*, 2005). The degradation of DBT by the Kodama pathway is commonly plasmid encoded in different *Pseudomonas* strains (Ohshiro and Izumi, 1999). A 9.8-kb DNA fragment from a 75-kb plasmid of *Pseudomonas* sp. strain C18 was shown to contain nine genes, *doxABDEFGHIJ* (*dox* for DBT oxidation), responsible for the conversion of naphthalene to salicylate. DBT serves as an alternate substrate for these naphthalene-degrading enzymes, as shown in Figure 7.2 (Denome *et al.*, 1993; Martínez *et al.*, 2016). The genes and enzymes of this pathway have not yet been fully characterized (Bressler and Fedorak 2000; Martínez *et al.*, 2016).

Five major pathways are shown in Figure 7.2: the 4S pathway two extended 4S pathways, E1 and E2, the Kodama pathway, the complete degradation pathway and the BT degradation pathway. The metabolites shown are DBT dibenzothiophene; DBTO dibenzothiophene sulfoxide; $DBTO_2$ dibenzothiophene sulfone; HBPS 2'-hydroxybiphenyl-2-sulfinate; 2-HBP 2-hydroxybiphenyl; 2-MBP 2-methoxybiphenyl; BP biphenyl; **I** *cis*-1,2-dihydroxy-1,2-dihydro DBT; **II** 1,2-dihydroxy-DBT; **III** *cis*-4-[2-(3-hydroxy)-thionaphthenyl]-2-oxo-3-butenoate; **IV**

3-hydroxy-2- formyl-benzothiophene; BT benzothiophene; **A** benzothiophene S-oxide; **B** benzothiophene S,S- dioxide; **C** benzo[c][1,2]oxathiin S,S-dioxide; **D** *o*-hydroxystyrene.

7.2.2.2 Complete Degradation Pathway

Another type of carbon-carbon cleavage pathway for the degradation of DBT and DBT-sulfoxide (DBTO) has been described in *Brevibacterium* sp. DO and *Arthrobacter* sp. DBTS2, respectively. These two organisms utilize DBT and DBTO as sole sources of carbon, sulfur, and energy. Although their desulfurization pathway was reported to be partly similar to the 4S pathway, as shown in Figure 7.2, it was still carbon destructive because the PASHs are degraded finally to CO_2, sulfite, and water (Ohshiro and Izumi, 1999). The genes/enzymes of this pathway have not yet been characterized. Unfortunately, this pathway has not been studied in more detail

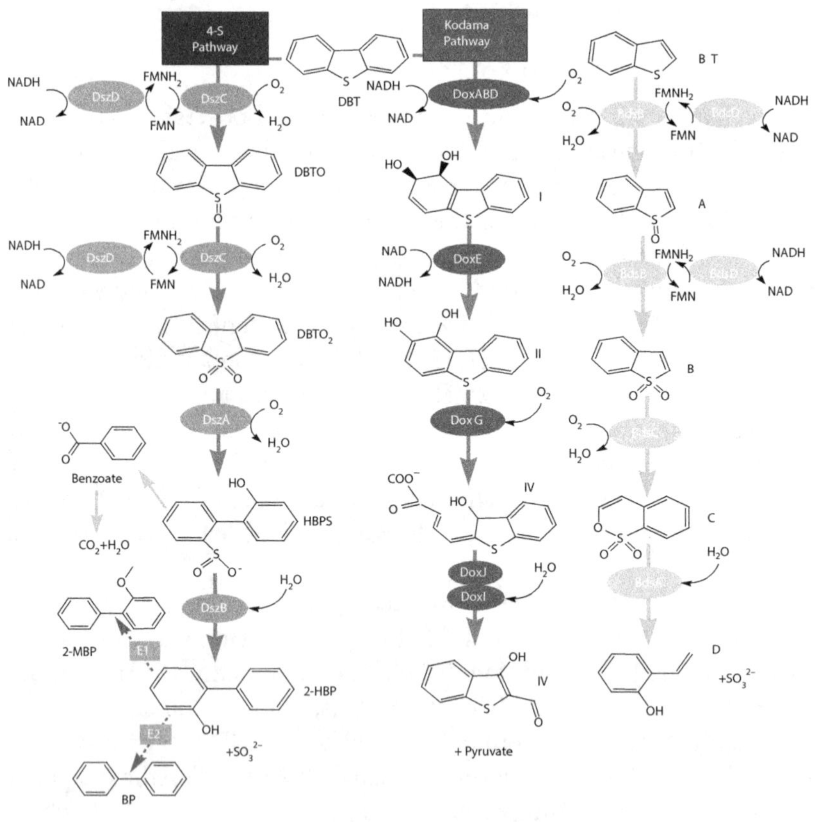

Figure 7.2 The Main Aerobic BDS Pathways of DBT and BT.

and there is no further information concerning its enzymology or genetics. Although it is less important in terms of BDS technology, the bacteria using this metabolic pathway are potentially useful in the formulation of mixed microbial inocula for the polyaromatic hydrocarbon bioremediation process.

Kohler *et al.* (1988) demonstrated that there are two major routes for the bacterial metabolism of hydroxybiphenyls HBPs. One route proceeds via hydroxylation of the already hydroxylated aromatic ring by NADH-dependent monooxygenase followed by meta cleavage. This route is employed by the strains isolated on 2-hydroxy- and 3-hydroxybiphenyl (Kohler *et al.*, 1993). The other route proceeds via dioxygenation of the non-hydroxylated aromatic ring, subsequent rearomatization, and meta-cleavage. This pathway is utilized by *pseudomonas* sp. strain FH23 (Higson and Focht, 1989).

For 2-HBP degradation by *Pseudomonas* sp. strain HBP1, Kohler *et al.* (1993) reported that the metabolic pathway was followed via a site specific monooxygenase that hydroxylates aromatic compounds at the C-3 position when there is a hydroxy group at C-2 and an alkyl or phenyl at C-1 and the product of this reaction is 2,3-dihydroxybiphenyl, which undergoes meta-cleavage by an extradiol dioxygenase enzyme which is further hydrolyzed by a hydrolase enzyme to benzoic acid, as shown in Figure 7.3.

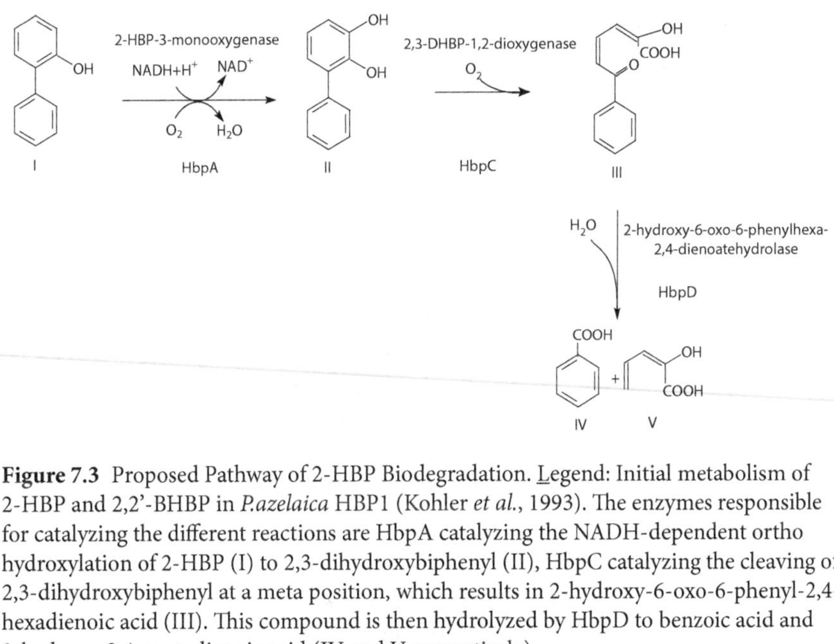

Figure 7.3 Proposed Pathway of 2-HBP Biodegradation. Legend: Initial metabolism of 2-HBP and 2,2'-BHBP in *P.azelaica* HBP1 (Kohler *et al.*, 1993). The enzymes responsible for catalyzing the different reactions are HbpA catalyzing the NADH-dependent ortho hydroxylation of 2-HBP (I) to 2,3-dihydroxybiphenyl (II), HbpC catalyzing the cleaving of 2,3-dihydroxybiphenyl at a meta position, which results in 2-hydroxy-6-oxo-6-phenyl-2,4-hexadienoic acid (III). This compound is then hydrolyzed by HbpD to benzoic acid and 2-hydroxy-2,4-pentadienoic acid (IV and V, respectively).

The same mechanism is operative in 2,2'-BHBP-grown cells and the product of the monooxygenase attack on 2,2'-BHBP was 2,2',3-trihydroxybiphenyl. 2,2',3-trihydroxybiphenyl also serves as a substrate for the monooxygenase activity to give 2,2',3,3'-tetrahydroxybiphenyl. Therefore, the monooxygenase also hydroxylates the C-3' position of 2,2',3-trihydroxybiphenyl. This provides evidence that NADH-dependent 2-HBP monooxygenase is a relatively broad-spectrum enzyme (Sondossi *et al.*, 2004).

The proposed pathway for the metabolism of 2,2'-BHBP is presented in Figure 7.4, where it was observed that the first intermediate, 2,2',3-trihydroxybiphenyl, can be metabolized via two different routes. On one hand, it serves as a substrate to the extradiol ring cleavage dioxygenase and, on the other hand, it can be turned over by the monooxygenase to form 2,2',3,3'-tetrahydroxybiphenyl which undergoes *meta*-cleavage by the dioxygenase enzyme.

Consequently, the extradiol ring cleavage dioxygenase acts on both 2,2',3-tri- and 2,2',3,3'-tetrahydroxybiphenyl, producing yellow meta-cleavage products that are unstable in the aqueous incubation environment. Both meta-cleavage compounds could undergo cyclization by non-enzymatic processes. The meta-cleavage product acts as a substrate for hydrolase enzymes, which produces salicylic acid and 2,3-dihydroxybenzoic acid, respectively.

7.2.2.3 4S-Pathway

Isbister and Koblynski (1985) isolated a strain of *Pseudomonas* sp., CB-1, that could desulfurize DBT through a sulfur-specific pathway (Gallagher *et al.*, 1993). The intermediates were DBTO and DBT-sulfone ($DBTO_2$) and the end product was 2-hydroxybiphenyl (2-HBP); unluckily, this strain was lost before the metabolic pathway could be fully characterized (Gallagher *et al.*, 1993). Isolation of a suitable bacterium was obtained after 40 years of research effort, when Kilbane (1990) at the Gas Technology Institute (formerly Institute of Gas Technology), USA isolated *Rhodococcus erythropolis* IGTS8. Since then, many researchers have succeeded to isolate different bacterial species capable of desulfurizing DBT and its alkylated forms via the sulfur-specific pathway (Figure 7.2) (Kilbane, 1992; Purdy *et al.*, 1993; Kayzer *et al.*, 1993; Izumi *et al.*, 1994; Lee *et al.*, 1995; Ohshiro *et al.*, 1995; Folsom *et al.*, 1999; Kobayashi *et al.*, 2000; Maghsoudi *et al.*, 2001; Castorena *et al.*, 2002; Akbarzadeh *et al.*, 2003; Labana *et al.*, 2005; Yu *et al.*, 2006; Li *et al.*, 2007; Xiong *et al.*, 2007; Ma, 2010; Derikvand and Etemadifar, 2014; El-Gendy *et al.*, 2014; Nassar, 2015).

The biochemical pathway of selective sulfur removal from DBT has been termed "*dsZ* or 4S pathway" due to the progressive oxidation of sulfur

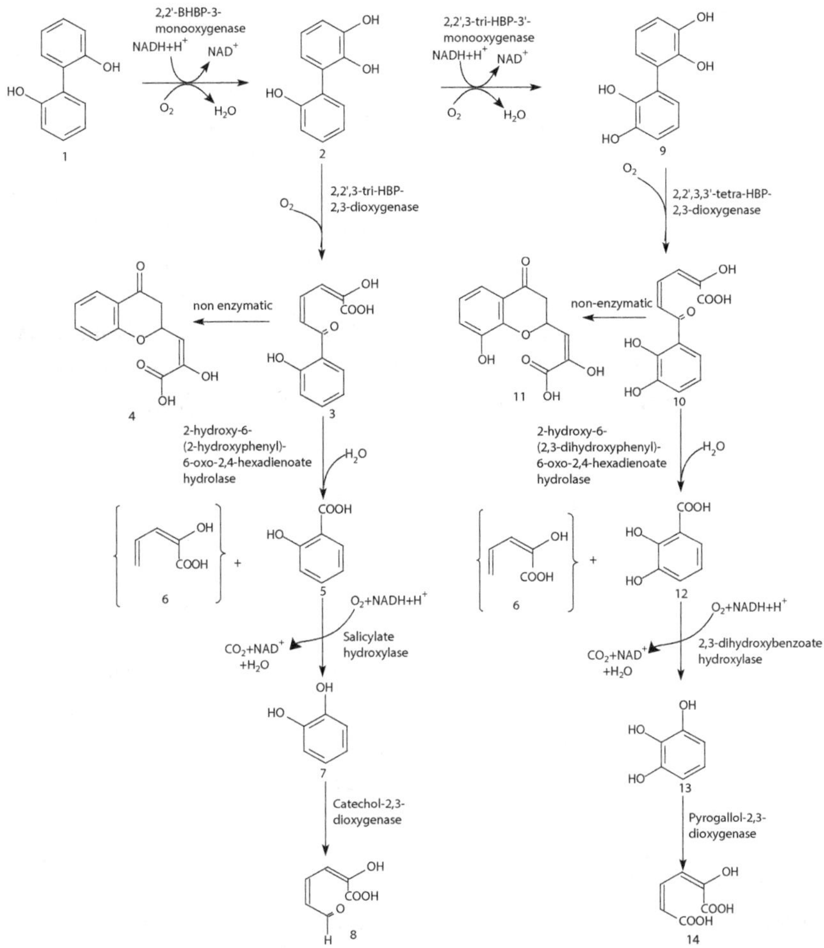

Figure 7.4 Proposed Pathway of 2,2'-BHBP Biodegradation. Legend: Pathway proposed for the metabolism of 2,2'-bihydroxybiphenyl by *Pseudomonas* sp. strain HBP1 (Sondossi *et al.*, 2004). (1) 2,2'-BHBP; (2) 2,2',3-trihydroxybiphenyl; (3) 2-hydroxy-6-(2-hydroxyphenyl)-6-oxo-2,4-hexadienoic acid; (4) 3-(chroman-4-on-2-yl) pyruvate; (5) salicylic acid; (6) 2-hydroxy-2,4-hexadienoic acid; (7) catechol; (8) 2-hydroxymuconic semialdehyde; (9) 2,2',3,3'-tetrahydroxybiphenyl; (10) 2-hydroxy-6-(2,3-dihydroxyphenyl)-6-oxo-2,4-hexadienoic acid; (11) 3-(8-hydroxychroman-4-on-2-yl) pyruvate; (12) 2,3-dihydroxybenzoic acid; (13) pyrogallol; (14) 2-hydroxymuconic acid.

that occurs through 4 steps, as shown in Figure 7.2 (Kayzer *et al.*, 1993; Gallagher *et al.*, 1993; Le Borgne and Quintero, 2003). Through this pathway, DBT molecules are converted to 2-HBP and sulfite by three enzymes DszA, DszB, and DszC (Figure 7.2), which are localized on plasmid-encoded *dsz* operon.

The genes involved in DBT metabolism have been called *bds* (Ohshiro *et al.*, 2005), *dsz* (Piddington *et al.*, 1995), *mds* (Nomura *et al.*, 2005), *sox* (Denome *et al.*, 1994), and *tds* (Ishii *et al.*, 2000) because several other unrelated genes have already been labeled *sox* and, to avoid confusion with these other genes, the *sox* designation has been generally rejected. Bds, Dsz, Tds and Mds have all been accepted as gene products and the abbreviation Dsz is the most popular one (Mohebali and Ball, 2008).

When the *dsz* genes and their proteins were compared with other enzymes and genes by sequences deposited in databases, such as the GenBank and the Swiss-Prot, the researchers showed no significant homology, suggesting that the desulfurization genes *dszC*, *dszA*, and *dszB* are encoded by specific enzymes (Piddington *et al.*, 1995).

7.3 The Desulfurization *dsz* Genes

The *dszABC* genes encoded the enzymes for the four-step conversion of DBT into 2-HBP and sulfate as shown in Figure 7.5 (Denome *et al.*, 1993, 1994; Piddington *et al.*, 1995). The length of *dszABC* genes is approximately 3750 bp and was transcribed into polycistronic mRNA to encode the DszA, DszB, and DszC enzymes, respectively. The *dszA* termination codon and *dszB* initiation codon are 4-bp overlaps (Figure 7.6) and the gap between *dszB* and *dszC* is only a 13-bp (Piddington *et al.*, 1995). Therefore, *dszB* appears to be expressed at a level lower than the gene on either side (Li *et al.*, 1996).

The *dszC* gene encodes the DBT monooxygenase (DszC) that catalyzes the oxidations of DBT to DBTO and $DBTO_2$. Subsequently, the *dszA* gene encodes the $DBTO_2$ monooxygenase that catalyzes the oxidation of $DBTO_2$ to HBPS and, finally, the *dszB* gene encodes the HBPS sulfinolyase (desulfinase) that catalyzes the conversion of HBPS into 2-HBP and sulfite as shown in Figure 7.4 (Gray *et al.*, 1996).

At the same time, the 4S pathway needs a chromosome-encoded additional gene, named *dszD*, which is responsible for the NADH-FMN

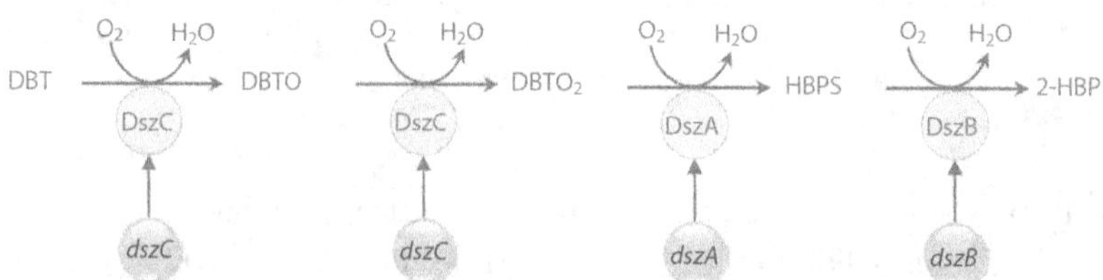

Figure 7.5 *dszABC* Genes Encode the Enzymes for the Four-Step Conversion of DBT to 2-HBP.

oxidoreductase (DszD) that allows the regeneration of the $FMNH_2$ cofactor necessary for the DszC and DszA catalyzed reactions as illustrated in Figure 7.4 (Gray *et al.*, 2003). The genome sequence of the desulfurizing bacteria *R. erythropolis* XP also localized genes *dszABC* on a plasmid, while *dszD* is located on a chromosome (Tao *et al.*, 2011a).

The phenotype of three *dsz* genes are a cluster and represent a 4 kb located on large plasmid, organized as one operon and transcribed in the same direction (Denome *et al.*, 1993, 1994; Piddington *et al.*, 1995). Denis-Larose *et al.* (1997) reported that the *dsz* genes clustered in IGTS8 were located on a 150 kb mega-plasmid.

Piddington *et al.* (1995) reported that the genomic DNA of *R. erythropolis* IGTS8 was partially digested by the restriction enzyme TaqI into fragments from 0.5 to 23 kb and the fragments more than 5 kb long were then ligated into plasmid *pRR-6* and transformed into a negative mutant of IGTS8. The cloned fragment 6.7-kb was found able to desulfurize the DBT. A 4-kb BstBI-SnaBI fragment was the minimal DNA sequence for DBT desulfurization which was responsible for desulfurization in IGTS8. This DNA fragment was found to comprise three ORFs (open reading frames) designated dszA, dszB, and dszC. Functional analyses for this cluster revealed that the product of dszC converts DBT directly to $DBTO_2$ and that the products of dszA and dszB act in concert to convert $DBTO_2$ to 2-HBP (Figure 7.4). The *dsz* pathway has been well studied using purified DszA, DszB, and DszC enzymes from several bacteria (Kilbane, 2006).

The characterization of *dszABC* genes showed a difference in the termini of each gene. Kilbane and Robbins (2007) failed to amplify the correct genes from *Gordonia amicalis* F.5.25.8 using PCR primers based on the 5′ and 3′ regions of *R. erythropolis* IGTS8 *dszABC* genes. The 5′ and 3′ termini of the *dsz*, *bds*, and *tds* genes are not conserved regions. Kilbane and Robbins

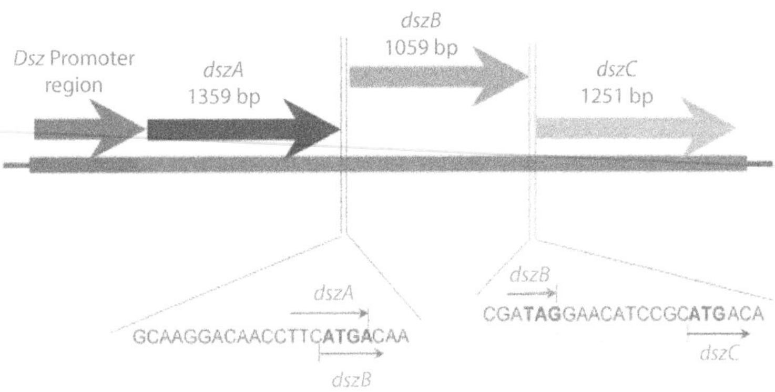

Figure 7.6 The *dszABC* Operon of *R. erythropolis* IGTS8.

(2007) identified conserved and unique regions inside the three *dsz* genes, which could be useful to design PCR primers used to amplify a partial sequence from the *dsz* genes. The use of PCR technique for amplification of the complete *dsz* genes from uncharacterized cultures is still challenging.

The transcription start site of the *dszABC* operon mapped 46 nucleotides upstream from the ATG initiation codon of the *dszA* gene. At least three regions that affect dsz expression have been identified: 1) the first region (-263 to -244) reduces dsz repression, but its deletion did not affect repression of gene expression, 2) the second region (-146 to -121) could bind an activator, and 3) the third region (-98 to -57) could be a repressor binding site, as shown in Figure 7.7 (Li *et al.*, 1996).

The characterization of the *Gordonia alkanivorans* RIPI90A dsz operon showed a high matching with that of *Rhodococcus erythropolis* IGTS8, except in case of *dsz* promoter sequences. The full promoter activity in strain RIPI90A was done by the promoter region from −156 to −50 (Shavandi *et al.*, 2010).

Recently, bacterial cultures that possess identical *dsz* gene sequences can have very different Dsz phenotypes. This was clearly illustrated by examining the desulfurization activity of *Mycobacterium phlei* GTIS10 having *dszABC* gene sequences identical to *R. erythropolis* IGTS8; the temperature at which maximum desulfurization activity was detected in the cultures was about 50 °C and 30 °C, respectively (Kayzer *et al.*, 2002; Kilbane, 2006). Characterization of four bacterial cultures capable of utilizing DBT as the sole source of sulfur revealed that these cultures had identical *dsz* genes, but the cultures differed significantly with regard to their substrate range, desulfurization activity, and yield of metabolites (Abbad-Andaloussi *et al.*, 2003). A comparative study of *M. phlei* SM120-1 and *M. phlei* GTIS10, aimed at increasing the understanding of the Dsz trait at intra-species levels, revealed

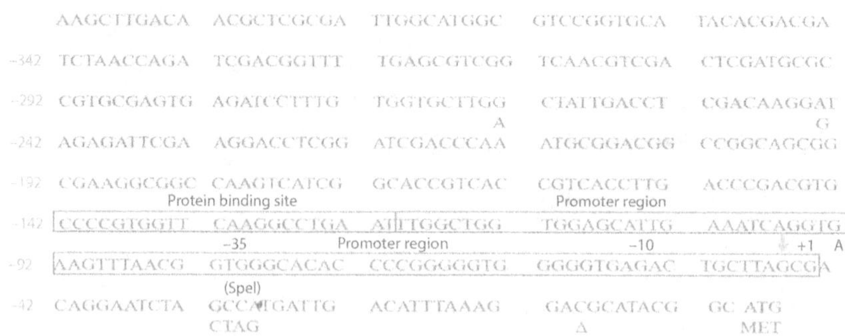

Figure 7.7 Protein Binding Site and Promoter Region of *dsz* Operon in *R. erythropolis* IGTS8.

considerable differences in the phenotypic and genotypic characteristics of these two desulfurizing strains (Srinivasaraghavan *et al.*, 2006). The range of Dsz phenotypes observed in different cultures may reflect the ability of each bacterial species or strain to provide cofactors and reaction substrates under the conditions tested. The transport of substrates and products might also contribute to desulfurization activity, as demonstrated by the fact that cell-free lysates of desulfurization cultures can exhibit a broader substrate range than the intact cell culture (Kilbane, 2006).

The distribution of the *dsz* genes in a wide variety of mesophilic bacteria strongly suggest that these genes are usually subjected to horizontal gene transfer in nature (Duarte *et al.*, 2001; Gray *et al.*, 2003; Kilbane, 2006). Additionally, desulfurization has also been shown to occur in thermophilic bacteria. The first identified thermophilic BDM was *Paenibacillus* sp. A11-2, whose desulfurization *tdsABC* genes match the *dszABC* genes (Ishii *et al.*, 2000). Other DszABC/TdsABC identities have been found in the thermophilic *Paenibacillus naphtalenovorans* 32O-Y, but with low similarity with Dsz and Tds proteins from *R. erythropolis* IGTS8 and *Paenibacillus* sp. A11–2 (26–48%, respectively) (Wang *et al.*, 2015).

Kirimura *et al.* (2004) reported that the desulfurization genes of thermophilic *Bacillus subtilis* WU-S2B (*bdsABC* genes) and *Mycobacterium phlei* WU-F1 are matching to each other and they exhibit about 61% and 58% identity with the *dszABC* and *tdsABC* genes, respectively. The flavin reductase from *M. phlei* WU-F1 (Frm) exhibited a high activity over a wide range of temperatures and the overexpression of the *frm* gene encoding this protein with the *bdsABC* genes was critical to achieve a high desulfurization efficiency (Furuya *et al.*, 2005). Moreover, *M. phlei* WU-0103 is able to degrade a wide range of DBT, benzothiophene (BT), and naphthothiophene derivatives (Ishii *et al.*, 2005).

Enhancement of the BDS activity was highly affected by several factors, such as the reducing power (NADH and $FMNH_2$), the cytoplasmic oxygen levels, and the ability of the cells to utilize and uptake the aromatic compounds (Kilbane, 2006). Therefore, *Rhodococcus* cells are able to desulfurize the very hydrophobic compounds directly from the oil, although no DBT transport genes have been identified in *Rhodococcus* or *Mycobacterium* strains (Kilbane, 2006; Monticello, 2000).

One of the limitations of the desulfurizing enzymes of fossil fuels is that they are relatively specific for DBT. Currently, significant progress has been made to overcome this problem using the technique of "gene shuffling" to develop versions of the *dsz* genes that encode new, hybrid enzymes with greater substrate tolerance (Monticello, 2000). A second problem is that expression of the *dsz* genes in their native hosts is repressed in the presence of inorganic sulfate, cysteine, or methionine (Li *et al.*, 1996). To solve this problem, analysis of the promoter region of the *dszABC* operon in *Rhodococcus erythropolis* IGTS8 indicated the presence of a possible regulatory hairpin structure and a bandshift analysis revealed that a putative repressor protein may be involved in the sulfate-mediated repression observed (Li *et al.*, 1996; Omori *et al.*, 1995). To avoid the sulfate repression, the *dsz* genes were placed under the control of the *tac* promoter in various *Pseudomonas* strains, either on a broad host range plasmid or on the chromosome. With these constructs, DBT could be efficiently desulfurized in growing *P. aeruginosa* cells [95% of 0.2 mM DBT transformed in 24 h, compared with 18% for *R. erythropolis IGTS8* (Gallardo *et al.*, 1997)].

BDS of DBT derivatives is limited not only by the metabolic activity of the BDSM, but also by the structure of the studied compounds. In this way, Bhatia and Sharma (2010) reported that it is difficult to desulfurize long-chain alkylated DBT because an increase in the carbon number of the alkyl substituent group results in an increase in size and hydrophobicity. Nevertheless, several bacterial isolates have been shown to desulfurize sterically hindered DBTs (Mohebali and Ball, 2016). Thus, some monooxygenases, such as MdsC from *Mycobacterium* sp. G3, have a broad substrate range and oxidize different alkyl DBTs to the corresponding sulfone form (Nomura *et al.*, 2005).

Despite the chemical similarity between DBT and BT, the two desulfurizing pathways are mutually exclusive in most bacteria (Gilbert *et al.*, 1998). Only a few bacteria that have *dszABC* genes are responsible for desulfurizing both DBT and BT, although the involvement of Dsz enzymes in BT desulfurization has not yet been established (El-Said Mohamed *et al.*, 2015a). Comparative transcriptomics studies in *Gordonia terrae* C-6, a strain capable of desulfurizing BT and its derivatives but not DBT and its derivatives, revealed a three-gene operon, named *bdsABC*, responsible for BT desulfurization through a pathway similar to the 4S pathway, as shown in Figure 7.4 Wang *et al.* (2013), reported that the expression of *bdsABC* genes in *E. coli* allowed the conversion of BT into o-hydroxystyrene as a final product.

7.4 Enzymes Involved in Specific Desulfurization of Thiophenic Compounds

7.4.1 The Dsz Enzymes

The 4S pathway proceeds via two cytoplasmic monooxygenases (DszC and DszA) and a desulfinase (DszB) which are encoded by an operon (*dszABC*).

The pathway requires molecular oxygen and NADH (Oldfield *et al.*, 1997) and the overall reaction (for neutral pH) is as follows:

$$DBT + 3O_2 + 4NADH + 2H^+ \longrightarrow 2\text{-}HBP + SO_3^{2-} + 3H_2O + 4NAD^+$$

The two monooxygenases, DszA and DszC, require NADH as cofactor for activity and need an NADH-FMN flavin reductase to regenerate the NADH (Ohshiro *et al.*, 1994; Gray *et al.*, 1996; Oldfield *et al.*, 1997; Matsubara *et al.*, 2001). DszC and DszA do not use NADH directly, but use $FMNH_2$ from the NADH-FMN oxidoreductase (Figure 7.8).

Lei and Tu (1996) reported that purified DszC requires reduced $FMNH_2$ as a co-substrate (the purified enzyme contains no bound FMN cofactor). An NADH-dependent FMN oxidoreductase has been identified in strain IGTS8 and named as DszD. The DszD is a 25 kDa protein which is not encoded by the *dsz* operon and presumably serves as a source of FMNH for DszC and DszA in vivo (Gray *et al.*, 1996).

Ohshiro *et al.* (1999) studied DBT desulfurization in vitro using a cell-free extract where flavin reductase was not required for the enzymes reaction in vitro because cell-free extracts have the oxidoreductase enzymes that are required in DBT desulfurization.

Several researchers have reported that the desulfurization reactions by *R. erythropolis* IGTS8 occur inside the cells, where the Dsz enzymes are soluble

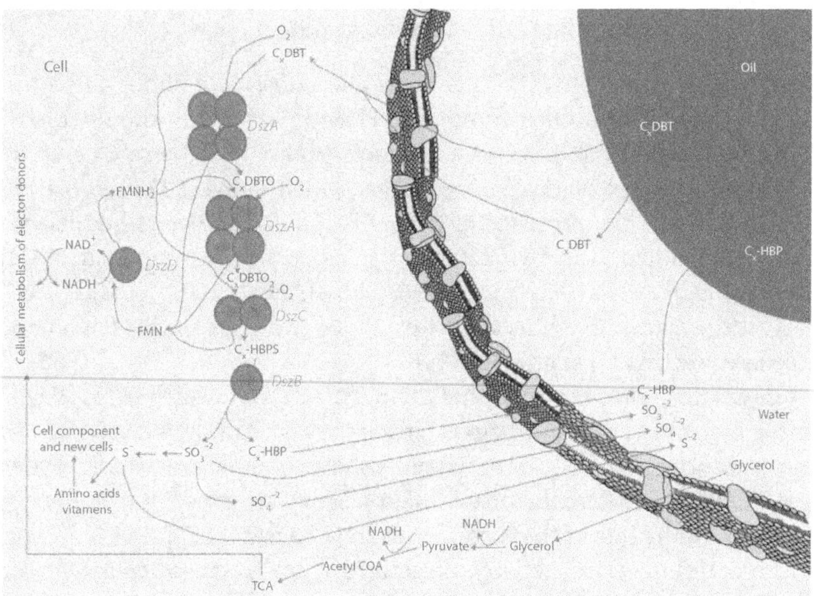

Figure 7.8 Biodesulfurizing Enzymes Involved in 4S Pathway.

and, presumably, found in the cytoplasm (Gray *et al.*, 1996; Oldfield *et al.*, 1997; Monticello, 2000). Although the intracellular metabolism of DBT by strain IGTS8 has been reported, there is no evidence that DBT is actively transported inside the cell (Monticello, 2000). Kayzer *et al.* (1993) and Patel *et al.* (1997) reported that the desulfurization activity of IGTS8 was associated with the external surface of the cells. Kilbane and Le Borgne (2004) confirmed that the BDS process does not occur intracellularly, but in association with the external surface of cells. Consequently, it is believed that the desulfurization pathway may function in association with the cell membrane, such that extracellular substrates and intracellular cofactors can both be accessed.

PASHs may employ a stress response on the desulfurizing bacteria, including a change in the structural-functional properties of the host cell membrane that may result in overexpression of different stress proteins, antioxidants, and enzymes associated with the energetic metabolism (Wang *et al.*, 2013). Thus, a proteomic study with *Chelatococcus* cells grown in DBT revealed the upregulation of putative transporters that may play an important role in transporting DBT and its metabolites across the cell membrane, as well as the upregulation of the ATP synthase enzyme to provide the required energy supply for the transporters. Several stress proteins and chaperons were also upregulated, confirming their role in DBT desulfurization and the adaptation of bacteria to PASHs (Bordoloi *et al.*, 2016).

7.4.1.1 DszC Enzyme (DBT-Monooxygenase)

DszC is approximately 45 kDa protein and consists of about 417 amino acids translated from ORF comprised 1251 bp located in the *dsz* operon (Denome *et al.*, 1994). DszC is a monooxygenase enzyme catalyzing two consecutive monooxygenation reactions which convert DBT to DBTO, then to $DBTO_2$. The presence of DBTO is difficult to detect because it is readily consumed. The two catabolic steps require oxygen and NADH (Denome *et al.*, 1994; Gray *et al.*, 1996; Oldfield *et al.*, 1997; Xi *et al.*, 1997). DszC seems exceptional because it catalyses two consecutive flavomonooxygenase reactions (Lei and Tu, 1996).

Gray *et al.* (1996) reported that the purified DszC, which is responsible for the first and second oxidation steps of the 4S pathway, is shown to have a peak absorption at λ_{281nm}. This enzyme shows homology to the acyl coenzyme A enzyme, where the first oxidation step (rate constant 0.06 min^{-1}) is one-tenth of the rate of the second step (rate constant 0.5 min^{-1}).

The optimum DszC activity, originated from *R. erythropolis* D-1, was found to be at 40 °C and pH 8 and the enzyme demonstrated heat stability up to 40 °C (Ohshiro *et al.*, 1997). No metal was detected to enhance the DszC activity, providing that this enzyme has no metal cofactor (Gray *et al.*, 1996; Li *et al.*, 1996 and Soleimani *et al.*, 2007).

DBT-MO can act on the derivatives of DBT, such as 4,6-dimethyl DBT, 2,8-dimethyl DBT, and 3,4-benzo-DBT, but it does not show any activity on carbazole, dibenzofuran, and fluorine. The DBT-MO from *Rhodococcus*, compared with other genera, has been shown to have a higher specific reaction rate for sparsely alkylated DBTs (Arensdorf *et al.* 2002).

Konishi *et al.* (2002) purified a BT and DBT monooxygenase (TdsC) which catalyzes the oxidation of the sulfur atoms in BT and DBT molecules from *Paenibacillus* sp. strain A11–2. The molecular mass of the purified enzyme and its subunit were determined to be 200 kDa and 43 kDa by gel filtration and sodium dodecyl sulfate polyacrylamide gel electrophoresis (SDS-PAGE), respectively, indicating a tetrameric structure. The N-terminal amino acid sequence of the purified TdsC completely matched the amino acid sequence deduced from the nucleotide sequence of the *tdsC* gene. The optimal temperature and pH for the TdsC reaction were 65 °C and pH 9, respectively. A comparison of the substrate specificity of TdsC and DszC, the homologous monooxygenase purified from *Rhodococcus erythropolis* strain KA2–5–1, demonstrated a contrasting pattern towards alkylated DBTs and BTs. The activity of TdsC toward BT was about eight times higher than that toward DBT. On the other hand, the activity of TdsC was ten times lower than that of DszC toward DBT. Alkylated BTs were found to be poor substrates for DszC except for 3-methyl BT (Kobayashi *et al.*, 2000). Among the alkylated DBTs examined, TdsC exhibited extremely low activity against 1-methyl DBT and 4-methyl DBT. In contrast to DszC, TdsC showed 2.5 times higher activity against 4,6-diethyl DBT rather than DBT and did not oxidize 4,6-dipropyl DBT. Neither enzyme oxidized 4,6-dibutyl DBT, which indicated a limited catalytic ability.

The crystal structure of DszC from *Rhodococcus* strains revealed that two distinct conformations, named "open" and "closed," occur in the flexible lid loops adjacent to the active site and might show the status of the free and ligand-bound DszC enzymes. Site-directed mutagenesis study shows that mutations in the residues involved either in catalysis or in flavin or substrate binding result in a complete loss of enzyme activity, suggesting that the accurate positions of flavin and substrate are crucial for enzyme activity. The DszC C-terminus, which is located in the interior of the protein, plays a crucial role in tetramerization and the formation of the substrate-binding pocket (Duan *et al.*, 2013; Liu *et al.*, 2014; Zhang *et al.*, 2014; Guan *et al.*, 2015). Different mutations at the active site of DszD were evaluated and enable the design of DszD mutants with increased activity (Sousa *et al.*, 2016).

7.4.1.2 DszA Enzyme ($DBTO_2$-Monooxygenase)

DszA is approximately 50 kDa protein, consists of about 453 amino acids, and, translated from ORF, comprises 1359 bp located in the *dsz* operon (Denome *et al.*, 1994). DszA is a second monooxygenase enzyme that catalyzes the conversion of $DBTO_2$ to HPBS, also utilizing $FMNH_2$ as a co-substrate (Denome *et al.*, 1994; Oldfield *et al.*, 1997; Xi *et al.*, 1997), with a reaction rate 5 to 10-fold higher than DszC (Gray *et al.*, 1996). The reaction requires oxygen and is NADH-dependent. The oxygen atoms are derived from molecular oxygen (Lei and Tu, 1996 and Oldfield *et al.*, 1997). Gray *et al.* (1996) showed that in *R. erythropolis* IGTS8, DszA, like DszC, utilizes $FMNH_2$ as a co-substrate and is apparently derived from the flavin reductase reaction. This molecule would attack a C–S bond of the substrate ($DBTO_2$), leading to its cleavage by the expulsion of sulfinate. DszA seems to proceed via a flavin-N5-oxide intermediate (Adak and Begley, 2016).

Ohshiro *et al.* (1999) isolated the DszA enzyme from a thermophilic *R. erythropolis* D-1 that was found to have a molecular mass of 97 kDa and to consist of two subunits with identical masses of 50 kDa. The N-terminal amino acid sequence of the purified DszA completely coincided with the deduced amino acid sequence for DszA from *R. erythropolis* IGTS8, except for a methionine residue at the latter N-terminal.

The optimum DszA activity, originating from *R. erythropolis* D-1, was found to be at 35 °C and pH 7.5. The optimum temperature of the flavin reductase required for the DszA activity was the same as the one for DszC. However, flavin reductase showed its highest activity at pH 6.0. The activity of flavin reductase, at DszA optimum conditions (35 °C, pH 7.5), was 80% of its activity at pH 6.0 (Ohshiro *et al.*, 1999). Addition of metal ions, Fe^{3+}, Fe^{2+}, and Cu^{+}, to the pure enzyme DszA did not increase the desulfurization rate (Gray *et al.*, 1996; Li *et al.*, 1996; Soleimani *et al.*, 2007).

Oldfield *et al.* (1997) reported that the purified DszA enzyme from *R. erythropolis* IGTS8 acted not only on DBT sulfone, but was also able to catalyze the conversion of dibenz[c,e][1,2]oxathiin 6,6-dioxide (sultone) to 2,2'-dihydroxybiphenyl (DHBP).

Gray *et al.* (1996) demonstrated that 10 mM EDTA did not inhibit the activity of DszA from *R. erythropolis* IGTS8. On the contrary, the activity of DszA from *R. erythropolis* D-1 was inhibited (≈ 50%) by 1 mM EDTA. Moreover, 2,2'- bipyridine, 8-quinolinol, and the other metal chelating reagents, such as Mn^{2+} and Ni^{2+}, also inhibited the activity of the enzyme, suggesting that a metal might be involved in its activity.

DszA recognizes the sulfone moiety within the structure of DBT sulfone and sultone. Ohshiro and Izumi (2000) demonstrated that by using the purified DszA enzyme from *R. erythropolis* D-1, sultone presented 54% activity as a substrate, compared with DBT sulfone, and 2-HBP was formed as a product. Moreover, dibenz[c,e][1,2] oxathiin 6-oxide (sultine) showed 23% activity and yielded DHBP as a product. However, DszA did not act on DBT and HBPS. Although sultine was non-enzymatically hydrolyzed to form HBPS, it was also oxidized to sulfonic acid during shaking. It was supposed that once sultone was non-enzymatically formed from sultine, it was immediately converted to 2-HBP by DszA.

7.4.1.3 DszB Enzyme (HBPS- Desulfinase)

DszB is approximately 40 kDa protein and consists of about 365 amino acids, which is translated from ORF and comprises 1095 bp located in the *dsz* operon (Denome *et al.*, 1994). HPBS desulfinase is a novel enzyme in that it can specifically cleave the carbon–sulfur bond of HBPS to give 2-HBP (the end product of the 4S pathway) (Oldfield *et al.*, 1997) and the sulfite ion without the aid of any other protein components or coenzymes (Denome *et al.*, 1994). On this basis, DszB is classified as an aromatic sulfinic acid hydrolase catalyzing the following reaction:

$$\text{HBPS} + H_2O \longrightarrow \text{2-HBP} + SO_3^{2-} + 2H^+$$

The three-dimensional structure of DszB revealed that this enzyme belongs to a new family of desulfinases that share a catalytic cysteine residue and whose overall fold is similar to that of the periplasmic substrate-binding components of sulfur-regulated ABC-type transporters (Lee *et al.* 2006).

It has been reported that DszB uses a nucleophilic attack of a water molecule on the sulfinate sulfur to form 2-HBP in a rate limiting reaction (McFarland, 1999; Ohshiro and Izumi, 1999). Although the 4S pathway genes are expressed as an operon, DszB is present at concentrations several-fold less than DszA and DszC in the cytoplasm (Gray *et al.* 1996; Li *et al.* 1996).

The optimum activity of DszB was determined to be at 35 °C and pH 7.5. The enzyme demonstrated heat stability up to 28 °C, however, temperatures above 30 °C drastically decreased the enzyme's relative activity (Nakayama *et al.* 2002). The DszB required no metal ions for its activity, however, Cu^{2+} and Zn^{2+} proved to be significant inhibitors (Watkins *et al.* 2003).

DszB desulfinase is present in the cytoplasm at concentrations lower than those of DszA and DszC and is the slowest of the three Dsz enzymes, becoming the rate-limiting step in the whole BDS process (Gray *et al.* 1996). Moreover, it was recently shown that HBPS, the substrate of DszB, can be secreted out of the cell to the medium and is unable to enter into the cell again, reinforcing the critical role of DszB in BDS efficiency (Martínez *et al.* 2016).

Watkins *et al.* (2003) reported that DszB is the rate-limiting enzyme of the 4S pathway because the activity of DszB is the lowest among enzymes of desulfurization metabolism. It is also the least-studied enzyme since only a very small quantity is produced. Lee *et al.* (2004, 2006) elucidated the 3D structure of DszB, which was the first X-ray crystallographic study of enzymes involved in DBT desulfurization.

Lee *et al.* (2006) compared the amino acid sequences between DszB and reported thermophilic and thermostable homologs (TdsB and BdsB). The Tyr63 and Gln65 amino acids were selected as targets for mutagenesis to increase DszB activity. The promising mutant enzymes (replaced with these two residues by other amino acids) were purified and their properties were examined. Between the wild-type and mutant enzymes, Y63F had higher catalytic activity, but similar thermostability and Q65H showed higher thermostability, but less catalytic activity and affinity for the substrate. Furthermore, the double mutant enzyme Y63F-Q65H was purified and overcame these drawbacks. Thus, this mutant enzyme had higher thermostability without loss of catalytic activity or affinity for the substrate.

Ohshiro *et al.* (2007) confirmed that the mutation at positions 63 and 65 of DszB developed the maximum activity and thermal stability, respectively, and that the double mutation increased thermostability without losses in maximal activity. Hence, to develop the BDS as an applied process, it is necessary to improve DszB further by structural analysis of the mutant enzymes in the near future.

7.4.1.4 DszD Enzyme (Flavin-Oxidoreductase Enzyme)

In BDS research, two monooxygenases, DszA and DszC, have been shown to be active in a desulfurization reaction with the aid of an oxidoreductase encoded by *dszD*. The *dszD* gene was cloned and sequenced from *R. erythropolis* IGTS8. The gene encoded a protein of ~20 kDa and was present on the chromosome of IGTS8, rather than on the plasmid that contains *dszABC* (Gray *et al.*, 2003).

The flavin reductases (DszD enzyme) are associated with monooxygenases since monooxygenases (DszC and DszA) cannot work in the absence of these reductases. Matsubara *et al.* (2001) purified and characterized the flavin reductase from *R. erythropolis* D-1, where the N-terminal amino acid sequence of the purified flavin reductase was found to be identical to that of the DszD of *R. erythropolis* IGTS8. The gene (*frm*) encoding flavin reductase from *Mycobacterium phlei* WU-F1 has been cloned and the cloned enzyme has been purified and characterized. The deduced amino acid sequence of the *frm* product shared only 33% identity with that of the flavin reductase gene (*dszD*) from *R. erythropolis* IGTS8 (Furuya *et al.*, 2005).

Ishii *et al.* (2000) isolated and cloned the gene *tdsD*, encoding the flavin reductase which coupled with the monooxygenases of *Paenibacillus* sp. A11-2. However, it was found that there was no detectable sequence similarity between TdsD and the *Rhodococcus* DszD (Gray *et al.*, 2003).

In studies on enzymology and the genetics of the 4S pathway, Dsz proteins were isolated from the soluble fractions of lysates of *Rhodococcus* IGTS8 grown with DMSO or DBT as the sole sulfur source. All enzymes were colorless and isolated, indicating no tightly associated chromophores.

The optimum activity of DszD was determined to be at a temperature of 35 °C and optimal pH 6 and the enzyme retained 30% of its activity after heat treatment at 80 °C for 30 min (Furuya *et al.*, 2005).

Purified Dsz enzymes from *Mycobacterium goodii* X7B showed that DszC can use both $FMNH_2$ and $FADH_2$ (Li *et al.*, 2009) and DszD is a homodimer with each subunit binding one FMN or FAD (Li *et al.*, 2012).

Xi *et al.* (1997) studied the enhanced desulfurization activity of DszC and DszA by increasing the concentrations of flavin reductase and suggested that the two are terminal oxygenases. The reaction rate with 1 unit/mL of flavin reductase was linear for 10–15 min, whereas it was linear for more than 20 min at a lower concentration.

Koike *et al.* (1998) demonstrated that the flavin reductase activity of *R. erythropolis* D-1 was inhibited by 7-hydroxycoumarin, but not by other coumarin derivatives, including dicoumarol which inhibited FRase I activity (a flavoprotein possessing FMN as a prosthetic group) and was used for the analysis of its crystal structure.

Ohshiro *et al.* (2002) searched for non-DBT-desulfurizing microorganisms producing a flavin reductase that coupled more efficiently with DszC than that produced by the DBT-desulfurizing bacterium *R. erythropolis* D-1 and found *Paenibacillus polymyxa* A-1 to be a promising strain. The coupling reactions of *P. polymyxa* A-1 flavin reductase with DszC and DszA proceeded more efficiently (3.5 and 5-fold, respectively) than those of *R. erythropolis* D-1 flavin reductase when identical enzyme activities of each flavin reductase were added to the reaction mixture. The result of the coupling reaction may be beneficial in developing an efficient enzymatic desulfurization system and in elucidating the mechanisms of interaction between the monooxygenases, reductases, and reduced flavin species.

Mingfang *et al.* (2003) studied the effects of NADH on the desulfurization activity of *Pseudomonas delafieldii* R-8 whole cell in the model oil system of dodecane. The results showed that the desulfurization activity was increased by the addition of NADH. The maximum specific desulfurization activity was 1.16 mg DBT/g/h when the concentration of $NADHNa_2$ was 400 mg/L. It was 55% higher than that without the addition of $NADHNa_2$. In the practical BDS of fuel, it is uneconomical to increase the microbial activity by means of supplementing NADH. In fact, the amount of NADH needed for fuel desulfurization could be acquired through the microbial metabolism of carbon sources, such as glucose, glycerol, and ethanol.

In contrast to what has been observed in strain IGTS8, the DszC ortholog (BdsC) appears to be the rate-limiting enzyme in the desulfurization pathway of *B. subtilis* WU-S2B (Ohshiro *et al.*, 2005). This BdsC monooxygenase is also able to oxidize indole to the blue pigment indigo in the presence of a heterologous flavin reductase in *E. coli* and this feature was used to clone a gene (*frb*) encoding a flavin reductase from *B. subtilis* WU-S2B (Furuya *et al.* 2004). The flavin reductase Frb shows no appreciable similarity to DszD, but is stable over wide temperature and pH ranges of 20–55 °C and 2–12, respectively. Frb contained FMN and exhibited both flavin reductase and nitroreductase activities (Takahashi *et al.*, 2009).

7.5 Repression of *dsz* Genes

The inhibition caused by the intermediates of the 4S pathway is also an important bottleneck in BDS and the two final products (sulfate and 2-HBP) are responsible for the major inhibitory effects on Dsz enzymes and cell growth (Chen *et al.*, 2008; Alves and Paixão, 2011; Abin-Fuentes *et al.*, 2013).

Through the BDS process, sulfate and 2-HBP are formed as end products of the 4S pathway. Sulfate is a preferred sulfur source for bacterial cells if it presents the bacteria stopping the synthesizing of enzymes involved in desulfurization of alternative organosulfur compounds (OSC) (Kertesz, 2000). Under sulfate limited conditions, several species of bacteria synthesize a set of extra proteins which are required for the metabolism of alternative sulfur sources. These so-called sulfate starvation induced proteins are synthesized only in the absence of 'preferred sulfur sources,' including sulfate. These proteins may be enzymes and transport systems involved in processing alternative sulfur sources from the environment (Kertesz, 2000). The Dsz enzymes can possibly be considered in this type.

Although there is no indication about the inducibility of *dsz* genes, many investigators reported that the expression of dsz genes in different isolates was fully repressed by sulfates, sulfides, and sulfur-containing amino acids, such as cysteine and methionine, which create an important bottleneck when designing efficient DBT desulfurizer biocatalysts (Li *et al.*, 1996). These compounds inhibit the promoters of *dsz* genes or enzyme synthesis at the transcriptional level, but are not inhibitors for the enzymes (Li *et al.*, 1996; McFarland, 1999). Due to sulfate repression, it is difficult to obtain in the commercial BDS process. Sulfate repression represents a major block to a massive production of desulfurizing cells (Le Borgne and Ayala, 2010). In order to make a fast BDS process for industrial application, large-scale production of active BDSMs is required (Mohebali and Ball, 2008). A lot of OSCs have been investigated for the induction of *dsz* activity, but no one induces activity in the presence of sulfates (Li *et al.*, 1996). Many researchers have reported that the desulfurization activity in different bacteria was completely repressed by sulfate or other readily bioavailable sulfur sources (Lee *et al.*, 1995; Piddington *et al.*, 1995; Li *et al.*, 1996; Ohshiro *et al.*, 1996; Wang and Krawiec, 1996; Rhee *et al.*, 1998; Serbolisca *et al.*, 1999; Kertesz, 2000; Matsui *et al.*, 2002; Noda *et al.*, 2002; Gunam *et al.*, 2006). However, it was reported by other researchers that there is no indication of feedback inhibition of the 4S pathway enzymes themselves; the presence of sulfate in growth media inhibits expression of BDS activity, but when adding sulfate to resting cells grown on DBT, does not inhibit desulfurization activity (Li *et al.*, 1996; Wang and Krawiec, 1996; Guobin *et al.*, 2006).

Li *et al.* (1996) reported that the transcription of three dsz genes were organized by a single promoter on the region from -121 to -44. The region from -146 and -121 could bind a protein (Figure 7.7), such as an activator, and removal of this region decreased the gene expression, but not the repression. However, the region between −98 and −57 could be a repressor binding site that overlaps the promoter −10 and −35 regions. Therefore, regulation of *dsz* operon is considered a negatively controlled repressible operon, meaning that during sulfate starvation the aporepressor is not able to bind to the repressor binding site. However, in the presence of sulfates or sulfur-containing amino acids, the corepressors molecules bind to the aporepressor to change its configuration to bind to the repressor binding site to prevent transcription.

To apply the BDS process on an industrial scale, the following two stages must be considered: (1) growing the biodesulfurizing microorganisms (BDSM) in a suitable medium to obtain the highest possible desulfurizing activity and (2) harvesting active cells and using them in the form of resting cells in a BDS process. The usage of the suitable growth medium is necessary to produce a high biomass of resting cells, which express the highest BDS efficiency. So, sulfate contamination of the growth medium is the main block to the expression because the Dsz phenotype is repressed by sulfate (Kilbane and Le Borgne, 2004; Gupta *et al.*, 2005).

Many researchers studied the expression of the *dsz* genes in a heterologous host as a way to overcome the problem of *dsz* gene repression. Gallardo *et al.* (1997) reported that the *dsz* cluster from *R. erythropolis* IGTS8 could be engineered as a DNA cassette under the control of heterologous regulatory signals to increase the ability of *Pseudomonas* strains to efficiently desulfurize DBT. Promoter replacement has also been examined by Serbolisca *et al.* (1999) and Matsui *et al.* (2002) as a possible alternative to the expression of the *dsz* genes in a heterologous host as a means to mitigate the repression problem.

To achieve the highest biomass production of desulfurizing resting cells, repression can also be avoided by several microbiological methods, including lowering the repressor (sulfate) content of the growth medium, substituting sulfate with an alternative sulfur source, and producing resting cell biomass on sulfate or another suitable sulfur sources, followed by inducing the Dsz phenotype in the resting cells using DBT (Honda *et al.*, 1998; Chang *et al.*, 2001; Ma *et al.*, 2006a; Mohebali *et al.*, 2008).

Many researchers have used synthetic media containing DBT as the sulfur compound for the growth of BDSM as a way to overcome the problem of dsz genes repression by inorganic sulfates. BDS activity has been improved by limiting the amount of DBT added to the medium (Yoshikawa *et al.*, 2002). However, biomass production using DBT is considered to be commercially impractical as a result of its high price and low water solubility and growth inhibition by 2-HBP. Therefore, researchers have tried to find a suitable substrate alternative to DBT as a sulfur source for growing cells, such as 2-aminoethanesulfonic acid (Yoshikawa *et al.*, 2002) and dimethyl sulfoxide (DMSO) (Mohebali *et al.*, 2008).

The produced sulfite from DBT desulfurization through the 4S pathway should be assimilated into bacterial biomass. Aggarwal *et al.* (2011a) developed a genome-scale in a silico metabolic model of *R. erythropolis* to try to determine the possible pathways of incorporation of sulfur from DBT into biomass. *R. erythropolis* was suggested to convert sulfite into sulfide that can be assimilated into biomass by sulfite reductase (reducing sulfite to sulfide that can be incorporated into sulfur-containing amino acids, CoA, and other metabolites) and/or sulfite oxidoreductase (oxidizing sulfite to sulfate) pathways (Aggarwal *et al.*, 2012).

The inhibitory effects of 2-HBP were noted early in many studies of BDS (Wang *et al.*, 1996; Nekodzuka *et al.*, 1997; Pan *et al.*, 2013; Nassar, 2015). Only about half of the desulfurized DBT was converted to 2-HBP in some

BDS studies, so that the 2-HBP accumulated intracellularly (Nakashima *et al.*, 2004; Ismail *et al.*, 2016).

The desulfurization enzymes DszA, DszB, and DszC were inhibited by 2-HBP at concentrations that are lower than the accumulate concentration of 2-HBP intracellularly (Abin-Fuentes *et al.*, 2013). Furthermore, 2-HBP can be toxic to desulfurizing cells, so it is important to differentiate between cell toxicity and enzyme inhibition in studies with bacterial cells (Abin-Fuentes *et al.*, 2013; Martinez, *et al.*, 2016).

DszB is the rate-limiting enzyme in the 4S pathway and this enzyme is reported to be more tolerable to 2-HBP than other DSZ enzymes (Reichmuth *et al.*, 2004; Abin-Fuentes *et al.*, 2013; Kilbane, 2017). Watkins *et al.* (2003) reported that a DszB enzyme purified from *R. erythropolis* IGTS8 was able to retain activity in the presence of 2-HBP. Abin-Fuentes *et al.* (2013) reported that the required concentrations of 2-HBP to decrease the enzymatic activity by 50% for DszA, DszB, and DszC were 60, 110, and 50 mM, respectively, while the accumulated 2-HBP, intracellularly, was 1100 mM concentration, so it was necessary to make the BDS process with low concentrations of DBT to define the maximum catalytic rate of desulfurization enzymes. Additionally, it was demonstrated that the intracellular concentration of 2-HBP cannot be reduced below inhibitory levels even by the presence of 50 g/L of a resin selected because of its superior ability to bind 2-HBP.

An examination of the DNA sequences responsible for the translation of DszC has not been reported, but is worthy of investigation. There is a 4-basepair overlap between the *dszA* and *dszB* genes of *R. erythropolis* IGTS8 (Piddington *et al.*, 1995) and the *bdsA* and *bdsB* genes of *Bacillus subtilis* WU-S2B and *Mycobacterium phlei* WU-F1 (Kirimura *et al.*, 2004), but a 64-bp overlap in the *dszA* and *dszB* genes of *Gordonia alkanivorans* strain 1B (Alves, *et al.*, 2007; Kilbane, 2017). This gene overlap in the desulfurization operon may occur in order to permit translational coupling to limit the expression of *dszB* and minimize the inhibitory effects of the accumulation of intracellular 2-HBP.

Extended 4S pathways have also been reported in some thermophilic *Mycobacterium* strains. In these pathways, 2-HBP becomes partially methoxylated to 2-methoxybiphenyl (2-MBP), likely by the action of an *o*-methyltransferase, as shown in Figure 7.2 (Xu *et al.*, 2006). 2-MBP has a lower inhibitory effect than 2-HBP on cell growth and desulfurization activity. This extended 4S pathway was also found in other *Mycobacterium* strains (Chen *et al.*, 2009) and was recently reported in Gram-negative bacteria, such as *Achromobacter* sp., *Chelatococcus* sp. (Bordoloi *et al.*, 2016), and *Rhodococcus erythropolis* HN2 (El-Gendy *et al.*, 2014). Although the *dszABC* genes were identified in this extended 4S pathway, no candidate genes responsible for the production of 2-MBP were reported (Yu, *et al.* 2015). Also, there has been no identification of the genes in *Rhodococcus* spp. Eu-32 that are responsible for the conversion of 2-HBP to biphenyl and the genome of this microorganism has not yet been determined (Akhtar *et al.*, 2015). To compare gene expression patterns when the *Mycobacterium goodie* X7B is grown in the presence and absence of 2-HBP, microarray experiments could be helpful in identifying the genes involved in the conversion of 2-HBP to 2-MBP and to identify the source of the methyl group in the formation of 2-MBP. The characterization of the genes involved in the formation of 2-MBP could not only enable the development of the BDS process by optimizing the production of 2-MBP (by reducing the accumulation of 2-HBP), but could also enable the improvement of other types of an extended 4S pathway that produce end-products with even lower inhibition than 2-MPB. (Akhtar *et al.*, 2009, 2015; Kilbane, 2017).

7.6 Recombinant Biocatalysts for BDS

A multi-enzymatic system of the 4S pathway and its cofactor requirements prevents the use of purified enzymes rather than whole cells for an applied BDS process, so resting cells are considered to be the best biocatalysts. Naturally isolated BDSM showed a very low BDS efficiency compared to the requirements of an industrial process and, consequently, genetic manipulation has been employed to overcome the major problems of the biocatalysts that prevent higher desulfurization efficiency as illustrated in Table 7.1.

The first attempts to construct a recombinant BDSM was done by cloning *dsz* genes from *R. erythropolis* IGTS8 into a plasmid of *R. erythropolis* UV1 (mutant IGTS8) and *Rhodococcus fascians* (Denome *et al.*, 1993). Since then, a high number of recombinant microorganisms were developed to overcome the major problems of BDS (Table 7.1).

Gou *et al.* (2003) reported that *Rhodococcus erythropolis* LSSE8–1 attacked and oxidized the C-S bond in DBT selectively and produced 2-HBP. By feeding $DBTO_2$ as a sole sulfur source, it produced both DBT and 2-HBP. It revealed that the reaction from DBT to $DBTO_2$, catalyzed by the cell, is reversible. $DBTO_2$ showed distinct poisonous effect to the cell, so there is no $DBTO_2$ accumulated in the culture. After treatment by lysozme, the plasmid DNA of the strain is isolated by alkaline method to be used as the template of PCR reaction. Three *dsz* gene fragments of 1.3, 1.0, and 1.2 kb, respectively, were amplified. Each fragment was ligated with a PGEM-T

Table 7.1 Genetic Engineering Approaches to Overcome the Major Problems of the BDS Process (Martínez *et al.*, 2016)

Problems	Genetic engineering approaches
Repression of the *Pdsz* promoter by sulfate	- *dszABC* gene expression under heterologous promoters (*Ptrc, Phsp*......)
	- Disruption of the *dszB* gene for HBPS accumulation
	- Transposome mutagenesis of a host-encoded *cbs* gene
	- Mutagenesis of sufur metabolism-related genes (e.g. *cbs*)
Low DszABCD activity and thermostability	- Cloning and expression of the *dsz* genes in multicopy plasmids under the control of strong heterologous promoters
	- Cloning and expression of the dsz genes in thermophiles (e.g., *Thermus thermophilus*)
Low concentrations of NADH	- Optimizing metabolic fluxes using systems biology tools (e.g. Genome-Scale Metabolic Models)
Low concentrations of flavin reductase	- Cloning and expression of heterologous flavin reductases in *dszABC*-harboring recombinant strains
Inhibitory effect of 2-HBP on Dsz enzymes	- Further biotransformation of 2-HBP by expressing heterologous enzymes (e.g., *carABC* genes)
	- Disruption of the *dszB* gene
	- Engineering synthetic microbial consortia that express individualized Dsz enzymes
DszB-dependent rate-limiting step	- Optimizing *dszB* expression by replacing its translation signals
	- Re-arranging the *dsz* operon (e.g., *dszBCA*)
Production of HBPS outside the cells	- Increase of DszB activity/production
	- Membrane permeabilization (e.g. expression of HBPS uptake systems)
Uptake of DBT from the oil phase	- Expression of hydrophobic compounds transport systems (e.g., *hcuABC* genes)
	- Heterologous expression of *dsz* genes in appropriate recipient cells (e.g., biosurfactant producers)
Narrow range of DBT-derived (e.g. alkyl-DBTs) substrates	- Gain-of-function mutations of *dszA* and *dszC* genes
	- Molecular protein evolution (e.g., RACHITT on DszC)
Low viability of the biocatalyst in petroleum	- Expressing the *dsz* genes in solvent-resistant strains (e.g., *Pseudomonas putida* Idaho)

vector and cloned into *E. coli* DH5α. The cloned DNA is sequenced and the result shows that *dsz* related genes are highly conservative. The identities of dszA and dszB, with respect to IGTS8, were 100% and the identity of dszC with that of IGTS8 was 99%.

The *dszABC* genes from a DBT desulfurizing microorganism, *Gordonia alkanivorans* RIPI90A, were cloned and sequenced by Shavandi *et al.* (2009). The overall nucleotide sequence similarity between the *dszABC* genes of *G. alkanivorans* RIPI90A and those of *Rhodococcus erythropolis* IGTS8 and *Gordonia nitida* were 83.1% and 83.2%, respectively. A gene transfer system for *G. alkanivorans* RIPI90A was established by employing the *Escherichia coli–Rhodococcus* shuttle vector pRSG43 as a suitable cloning vector. This stable vector was applied for both cloning and efficient expression of the *dsz* genes under the control of a *lac* promoter. In the presence of inorganic sulfate and sulfur-containing amino acids, the maximum desulfurization efficiency of recombinant resting cells was increased up to 2.6-fold compared to the highest desulfurization activity of native resting cells.

The toxicity and inhibitory effect of 2-HBP could be avoided by engineering a biocatalyst that can convert this phenolic compound into a less toxic metabolite. In this way, Yu *et al.* (2006) introduced the gene cassette, *carABC,* encoding enzymes responsible for degrading carbazole to anthranilic acid from *Pseudomonas* sp. strain XLDN4–9 into a DBT desulfurizing microorganism *Rhodococcus erythropolis* XPDN. The resultant strain, *Rhodococcus erythropolis* XPDN, could simultaneously transform 2-HBP to benzoate.

Classification and cloning of different Dsz enzymes into different hosts allows them to overcome process limitations, e.g. 2-HBP feedback inhibitions, that are difficult to achieved when using monocultures.

Martínez *et al.* (2016) demonstrated that the 4S pathway consists of three serial and irreversible modules, as shown in Figure 7.4, and it is possible to develop a collection of synthetic *dsz* cassettes formed by different combinations of the *dsz* genes that were functional both alone and by combination in different *P. putida*. Based on the sequence of the *R. erythropolis* IGTS8 *dsz* genes (Gray *et al.*, 1996; Oldfield *et al.*, 1997; Piddington *et al.*, 1995; Tao *et al.*, 2011a), different modified genes were synthesized and called *dszA1*, *dszB1*, *dszC1,* and *dszD1* to engineer a synthetic *dszB1A1C1-D1* cassette. This was achieved with the following:

1. the native *dszABC* operon was rearranged to avoid overlapping genes and to increase the expression of the *dszB* gene that encodes the rate-limiting step of the 4S pathway;

2. the native gene expression signals that are under sulfur-dependent control were substituted by optimized transcriptional (based on the *lacI*^{q} /P$_{tac}$ and lacI^q/P$_{lac}$ regulatory couples) and translational (consensus ribosome binding site and optimized codon usage) signals for *Pseudomonas* strains that are not subject of sulfate-dependent inhibition;
3. the Y63F amino acid substitution (previously reported to enhance the activity and stability of the *DszB* desulfinase) (Ohshiro *et al.*, 2007) was engineered in the synthetic *dszB1* gene; and
4. a synthetic *dszD1*gene (encoding the flavin reductase needed for the two monooxygenase reactions of the 4S pathway is not linked to the *dszABC* genes in strain IGTS8) was included in the *dsz* cassette.

To avoid the potential toxic effect of an increased expression of the *dszD* gene (Galán *et al.*, 2000), it was arranged as an independent operon under the control of the weak P$_{lac}$ promoter and separated from the *dszB1A1C1* operon. The modularity of the *dsz* synthetic cassette was facilitated by removing some restriction enzyme sites present in native *dsz* genes and adding other sites that allow the easy removal/insertion of synthetic genes in different vectors and under the control of different regulatory elements. The *dszB1A1C1-D1* gene cassette was cloned and expressed in a broad-host range vector, plasmid pIZ1016, and was able to replicate in a wide variety of Gram-negative bacteria (Moreno-Ruiz *et al.*, 2003), giving rise to plasmid *pIZdszB1A1C1*-D1.

In a resting cell process, *P. putida* KT2440 (*pIZdszB1A1C1- D1*) converted all DBT to nearly stoichiometric amounts of 2-HBP (3-fold increase in the production of 2-HBP) after a 180 min incubation, suggesting that the *dsz*B1A1C1-D1 synthetic module behaves as an attractive and efficient alternative to previous recombinant *dsz* cassettes. Thus, the generation of an artificial 4S pathway by combining the *dszC1-D1* and the *dszB1A1-D1* cassettes in an optimized synthetic bacterial consortium and the *dszB1* cassette, as a cell-free extract, is promising instead of the use of naturally existing or recombinant *dsz* pathways for enhanced conversion of DBT into 2-HBP.

This new strategy will facilitate further improvements of the BDS process to achieve target desulfurization efficiencies (1– 3 mmol 2-HBP/g DCW/h) that have been reported to be needed for industrial applications (Kilbane, 2006). The compartmentalization or segregation of functional units of the 4S pathway should also facilitate and support the development of faster fuel desulfurization hybrid technologies, which could involve a combination of biological and physicochemical methods, for example the *dszC1D1* harboring biocatalysts could be used to efficiently convert DBT into the more polar $DBTO_2$ which could be easily separated from the fuel and further desulfurized by physicochemical methods or the $DBTO_2$ delivered by oxidative desulfurization technologies (Campos-Martín *et al.*, 2004) could be efficiently desulfurized by using the *dszB1A1-D1* harboring biocatalysts.

Pseudomonas azelaica HBP1 is one of the few bacterial strains known to completely metabolize 2-HBP. Strain HBP1 was isolated from a wastewater treatment plant and can efficiently grow on a 2.7 mM 2-HBP concentration (Kohler *et al.*, 1988). The strain utilized 2-HBP with three specific enzymes, named HbpA (a hydroxylase), HbpC (an extradiol dioxygenase), and HbpD (a hydrolase), to form benzoate and 2-hydroxy-2,4-pentadienoic acid which are then subsequently utilized by regular metabolic pathways (Kohler *et al.*, 1988, 1993; Jaspers *et al.*, 2001). The metabolic pathway is regulated by an HbpR protei, which activates transcription of the *hbpCA* and *hbpD*-genes in the presence of 2-HBP from two promoters named P_C and P_D, respectively (Jaspers *et al.*, 2001). The *Pseudomonas azelaica* HBP1 and related strains express a MexAB-OprM efflux system that confer tolerance to the toxic 2-HBP, thus being suitable candidates for the expression of *dsz* genes (Czechowska *et al.*, 2013; El-Said Mohamed *et al.*, 2015b).

Mohebali and Ball (2016) reported that the genetically engineered BDSM that contain several copies of the dsz genes were able to desulfurize alkylated derivatives of DBT which are the main target molecules when extremely low sulfur levels are desired.

Matsui *et al.* (2001) examined the gene dosage effect of the reductase gene, *dszD*, and the *dsz* operon to enhance the DBT desulfurization activity of the recombinant biodesulfurizing *Rhodococcus* sp. strain T09. DBT degradation activity of recombinant *Rhodococcus* sp. T09/pRKPP was increased to about 3.5 fold by the introduction of the *dszD* gene, while DBT desulfurization activity remained the same with the production of dibenzo[1,2]oxathiin-6-oxide which was caused by insufficient activity of the last desulfurization step involving a desulfinase. Introduction of an additional *dsz* operon resulted in a 3.3 fold increase in DBT desulfurization activity (31 μmol/ g (DCW)/ h) compared with that of T09/pRKPP (9.5 μmol/ g (DCW) /h).

Hirasawa *et al.* (2001) constructed an *E. coli–Rhodococcus* shuttle vector with the desulfurization gene cluster, *dszABC*, and the related reductase gene, *dszD*, cloned from KA2–5-1 and reintroduced into KA2–5-1 which was efficiently expressed. The DBT-desulfurization ability of the recombinant strain carrying two *dszABC* and one *dszD* on the vector was about 4-fold higher than that of the parent strain.

Li *et al.* (2008b) used a different methodology to improve the DszB level within the host cell and the *dsz* operon was reconstructed by uncoupling the translation of *dszB* from that of the *dszA* gene. More improvement occurred by modifying both transcription and translation levels and a rearranged *dszBCA* operon was constructed to reach a 12-fold increase of desulfurization efficiency when expressed in *R. erythropolis* DS-3. It should be taken into account that when the *dszABC* operon does not contain the reductase-encoding *dszD* gene, constructing recombinant biocatalysts that produce the wanted amounts of NADH-FMN reductase for the activity of the DszC and DszA enzymes is critical to reach a maximum desulfurization efficiency.

Galán *et al.* (2000) used the purified *hpaC* gene (from *E. coli)* that encodes the NADH-FMN flavin reductase of the 4-hydroxyphenylacetate degradation pathway to construct a recombinant *P. putida* KTH2 (pESOX3), which carried the *dszABC* genes from *R. erythropolis* IGTS8 and the *hpaC* gene to be stably inserted into the chromosome. The resulting recombinant strain showed a higher BDS efficiency more than 10 fold compared to the parental strain lacking the *hpaC* reductase gene.

He–Ne laser (632.8 nm) irradiation was introduced to improve the DBT biodegradation ability of *Gordonia* sp. WQ- 01 with various energy doses by varying power density (5 to 25 mW) and exposure time (5 to 20 min). Results showed that the frequency of positive mutation depended on the He–Ne laser energy dose and output power, where the optimum values for energy dose and output power were 22.9 J cm−2 and 20 mW, respectively. The BDS efficiency of a stable positive mutant strain *Gordonia* sp. WQ-01A was about 4 fold higher than that of the wild strain under the optimum conditions of laser irradiation, indicating He–Ne laser irradiation to be a feasible way to improve desulfurization capacity (Jia *et al.*, 2006).

Kamali *et al.* (2010) performed amino acid sequence comparisons and structural predictions based on the crystal structure of available pdb (protein data base) files for three flavin reductases, including PheA2 (phenol 2-hydroxylase from *Bacillus thermoglucosidasius*), $HpaC_{Tt}$ (4-hydroxyphenylacetate hydroxylase from *Thermus thermophiles*), and $HpaC_{St}$ (4-hydroxyphenylacetate hydroxylase from *Sulfolobus tokodaii*), with the closest structural homology to IGTS8 DszD. Threonine, at position 62 in DszD, was substituted with Asn (Asparagine) and Ala (Alanine) by site-directed single amino acid mutagenesis. Variants HpaCSt-Asn53 (T62N mutant) and alanine (T62A mutant) showed a 5 and 7 fold increase in activities based on the recombinant wild type DszD, respectively. This study revealed the critical role of residue 62 in desulfurizing enzyme activity.

The ability of Dsz enzymes to act with the wild substrate range is one of the important factor for the improvement of BDS as a commercial process.

Chemostat enrichment is a classical microbiological method that is well suited for selection and isolation of desired mutants that is often disregarded as a chemostat selection, which can permit to indicate the rare mutants within a large population. Arensdorf *et al.* (2002) used a two-phase sulfur-limited chemostat to select for gain-of-function mutants with mutations in the *dsz* gene of *Rhodococcus erythropolis* IGTS8, enriching growth in the presence of PASHs that could not support growth of the wild-type strain. Mutations arose that allowed growth with octyl sulfide and 5-methylbenzothiophene as sole sulfur sources. An isolate from the evolved chemostat population was genetically characterized and found to contain mutations in two genes, *dszA* and *dszC*. A conversion of G (Guanine) to T (Thymine) in *dszC* at codon 261 resulted in a V261F mutation that was determined to be responsible for increasing the desulfurization of 5-methylbenzothiophene. By using a modified RACHITT (random chimeragenesis on transient templates) method, mutant DszC proteins containing all possible amino acids at that position were generated and this mutant set was assayed for the ability to metabolize 5-methylbenzothiophene, alkyl-thiophenes, and DBT. No mutant with further improvements in these catalytic activities was identified, but several clones lost all activity, confirming the importance of codon 261 for enzyme activity.

BDS of BT, one of the major sulfur compounds in gasoline, is poorly characterized biochemically and only a few strains are known to be able to use BT as a sole sulfur source.

Petrella *et al.* (2007) improved the desulfurization capability of DBT desulfurizing microorganism *Rhodococcus* sp. DS7 (which is not able to desulfurize BT) to desulfurize both DBT and BT. Heterologous chromosomal DNA, randomly constructed from two BT desulfurizing strains, *Rhodococcus* sp. ATCC 27778 and *Gordonia* sp. ATCC 19067, was electroporated into *Rhodococcus* DS7. After transformation of electrocompetent cells of *Rhodococcus* sp. DS7 with genomic DNA from two different BT desulfurizing strains, two recombinant microorganisms called SMV165 and SMV166 combined the ability of DS7 to completely desulfurize DBT with a higher BT desulfurization efficiency up to 90% after 88 h, but it is not yet known which gene or genes responsible for this new metabolic pathway have been transferred from the donor strains to DS7 or even how this pathway is structured.

Wang *et al.* (2013) cloned and overexpressed *bdsABC* operon in *E. coli* Rosetta (DE3) with the expression vector pET28a. As expected, a phenolic compound was accumulated (indicated by the positive result in the Gibbs′ assay) when BT was incubated with cell extracts of *E. coli* Rosetta (DE3), overexpressing pET28a-*bdsABC*. NADH, FMN, and *dszD*, a NAD(P)H-dependent flavin reductase involved in the DBT-BDS pathway, were essential for the activity of these cell extracts, indicating that some of these enzymes involved in the BT-BDS pathway are also flavin-dependent. The recombinant strain of *E. coli* Rosetta (DE3) desulfurized BT to yield *o*-hydroxystyrene by a mechanism analogous to the 4S-BDS pathway, as shown in Figure 7.2.

Another problem facing the application of the BDS process on the industrial scale is the isolation of thermophilic bacterial strains and enzymes due to the fact that the petroleum refining process needs high temperatures. Kirimura *et al.* (2004) cloned the desulfurizing gene cluster containing *bdsA*, *bdsB*, and *bdsC* from thermophilic bacteria *Bacillus subtilis* WU-S2B (able to desulfurize DBT and alkylated DBTs through specific cleavage of the carbon-sulfur bonds over a temperature range up to 52 °C) into *Escherichia coli* JM109. The activities of DBT degradation at 50 °C and DBT desulfurization (2-HBP production) at 40 °C in resting cells of the recombinant *E. coli* JM109 were approximately five and two times higher, respectively, than those in the parent strain *B. subtilis* WU-S2B.

Parravicini *et al.* (2016) investigated the structural features of enzymes cloned from a *Rhodococcus sp. AF21875* isolated from polluted environmental samples and their resistance to temperature (20–95 °C) and organic solvents (5, 10, and 20 % v/v methanol, acetonitrile, n-hexane, and toluene). Amplified gene sequences were cloned in the expression vector pET22 and used to transform *E. coli* BL21 cells. Changes in protein structures were assessed by circular dichroism and intrinsic fluorescence spectroscopy. All Dsz proteins are unfolded by temperatures in the range 45–60 °C and by all tested solvents. These results suggest that stabilization of the biocatalysts by protein engineering will be necessary for developing BDS technologies based on Dsz enzymes.

Kilbane (2006) explored the possibility of cloning and expression of the *dsz* genes from mesophilic bacteria in extreme thermophiles, such as *Thermus thermophilus*. The *dszC* gene from *R. erythropolis* IGTS8 was successfully expressed in *T. thermophilus*, but the thermostability of the other Dsz enzymes prevented the functional expression of full desulfurization pathway in this host.

Santos *et al.* (2007) studied the influence of operational conditions (pH, temperature, and oxygen transfer rate) on the initial reaction rate values of the 4S pathway, mainly in reactions, which are catalyzed by monooxygenases (DszC and DszA). The optimal values of the operational conditions obtained in this study carried out in vivo are not the same as those observed from in vitro studies. In this study, the optimum pH value was 9, whereas near 7.4 was the optimum when in vitro determinations with DszA and DszB enzymes were performed (Ohshiro *et al.*, 1999; Watkins *et al.*, 2003). The optimum temperature value given in the literature is 35 °C for the reactions catalyzed in vitro by DszB and DszA, but, in this work, 30 °C was the best temperature for all the 4S reactions.

Zhu *et al.* (2016) reported the draft genome sequence of strain *Geobacillus thermoglucosidasius* strain W-2 which may provide genetic information of a thermophilic mechanism of BDS. *Geobacillus thermoglucosidasius* strain W-2 is a thermophilic bacterium which is capable of desulfurizing organosulfur compounds. The genome sequencing of strain W-2 was carried out using the Illumina HiSeq 2500 and Illumina paired-end (PE) libraries were constructed. The final genome draft of strain W-2 contains 51 codons, with a total size of 3,894,555 bp and an average G+C content of 43.34%. The average codon length was 76,664 bp, with the largest codon being 560,848 bp. As a result, 3,935 protein-encoding genes, one rRNA operon, and 76 tRNA genes for all 20 amino acids were predicted. Genes encoding for three putative alkanesulfonate monooxygenases, seven putative sulfonate ABC transporters, and two putative sulfate permeases were identified in the draft genome. They were hypothesized to be responsible for organosulfur compound degradation in addition to some genes that encode nitronate monooxygenases, which catalyze oxidative denirification of nitroalkanes to carbonyl compounds and nitrites. It may be the first time that the two groups of enzymes responsible for BDS and biodenitrification pathways have been found in thermophilic bacteria. As thermophilic enzymes offer major biotechnological advantages over mesophilic enzymes, the draft genome of strain W-2 may provide an excellent platform for further improvement of this organism for BDS and other biotechnological applications at elevated temperatures.

Calzada *et al.* (2009) found that the activity of each enzyme of the 4S pathway depends on the cellular age of *Pseudomonas putida* CECT5279 cells. The DszB enzyme was produced along the growth curve of *P. putida* CECT5279 at a constant rate and its activity reached a sharp maximum at an early exponential phase and rapidly decreased at a middle exponential phase, causing an efficient transformation of HBPS into 2-HBP. To overcome this problem, a two-step resting-cell process was designed to increase the BDS efficiency of the *P. putida* CECT5279 recombinant cells (harboring the BDS genes *dszABC* from *R. erythropolis* IGTS8 and the gene *hpaC* from *E. coli* W). In the first step, *P. putida* CECT5279 was grown for 10 h, showing the maximum activities of DszA and DszC monooxygenases, and was able to transform all DBT into a 50% mixture of HBPS and 2-HBP after 75 min; this transformation rate did not increase further even after 150 min of reaction. In the second step, the cells were removed from the reaction medium and a *P. putida* CECT5279 was grown for 5 h and those presenting the maximum DszB activity were added. Directly after the addition, the remaining HBPS was efficiently converted into 2-HBP, showing a complete transformation after 400 min of the whole resting-cell process. It should be noted that the second step proceeds more slowly than the first step, likely because the activity of DszB is partially inhibited by the high concentration of 2-HBP.

Martínez *et al.* (2015) studied the enhancement of the BDS efficiency of recombinant *Pseudomonas putida* CECT5279 (harboring the genes *dszABC* from *R. erythropolis* IGTS8 and the gene *hpaC* from *E. coli* W) by using different co-substrates that related to the Krebs cycle, such as citric and succinic acids, in order to increase the global reducing power of the cell. Acetic acid was also selected due to the metabolic importance of its active derivative, acetyl CoA. The three tested compounds (acetic, citric, and succinic acids) provided positive results, but the best improvement in the BDS efficiency was achieved using acetic acid at a concentration of 1.5%. The initial intracellular concentrations of ATP and NADH also increased to 58 and 42%, respectively, and the initial rate of all of the reactions of the 4S route were increased by 42–117%.

Another important problem in BDS is the mass transfer or low substrate solubility from the oil to the cytoplasm of the biocatalysts. Noda *et al.* (2003) cloned *hcuABC* genes (hydrophobic compound uptake genes of *P. aeruginosa* NCIMB9571) in a recombinant *P. putida* IFO13696 strain carrying the *dszABCD* genes using a transposon vector. All of the recombinant strains have the ability to desulfurize up to 82% of 1 mM DBT in model oil (n-tetradecane) after 24 h due to the increase in the uptake of DBT from the solvent except one, named PARM1. PARM1 was unable to desulfurize DBT in n-tetradecane, but was able to desulfurize it in water. The transposon tagging analysis indicated that the tranposon was inserted into *hcuA* of the open reading frames *hcuABC*. The full-length *hcuABC* genes, when transformed into PARM1, achieved 87% recovery of the desulfurization activity of DBT in n-tetradecane, but partial *hcuABC* genes achieved only 0–12%. These results indicated that DBT desulfurization in the oil phase by recombinant *P. aeruginosa* NCIMB9571 required the full-length *hcuABC* gene cluster. The *hcuABC* gene cluster is related to DBT uptake from the oil phase into the cell.

Wang *et al.* (2011) introduced *hcuABC* genes from *P. delafieldii* R-8 into the desulfurizing bacteria *R. erythropolis* LSSE8–1 and the desulfurization efficiency of the resulting recombinant strain was increased by 19% and 13% in the case of DBT and 4,6-DMDBT, respectively.

Watanabe *et al.* (2003a) reported that the rate-limiting step in the desulfurization reaction of highly alkylated Cx-DBTs is the transfer process from the oil phase into the cell. When using a recombinant *Mycobacterium* sp. strain MR65 harboring *dszABCD* genes to desulfurize alkyl-DBTs in n-hexadecane, the desulfurizatioin rate of 2,4,6,8-tetraethyl DBT in hexadecane, by recombinant strain MR65, was only about 40% of that of 4,6-dipropyl DBT, while the specific desulfurization activity of the DszC enzyme for 2,4,6,8-tetraethyl DBT (C_8-DBT) was twice that for 4,6-dipropyl DBT (C6-DBT). The desulfurization ability for Cx-DBTs by a recombinant strain depends on the carbon number substituted at positions 4 and 6.

Watanabe *et al.* (2002) cloned the desulfurization gene cluster, *dszABC*, and the flavin reductase gene, *dszD*, from *Rhodococcus erythropolis* KA2–5-1 into the chromosomes of *Pseudomonas aeruginosa* NCIMB 9571 by using a transposon vector. Resting cells of the recombinant strain, PAR41, desulfurized 63 mg sulfur/L of light gas oil (LGO) containing 360 mg sulfur/L. The desulfurization activity for LGO by the resting cells of strain PAR41, grown with n-tetradecane (50% v/v), was much higher (10~18-fold) than in glucose-grown cells. *P. aeruginosa* NCIMB 9571 is able to take up water-insoluble compounds from an oil phase which is enhanced by n-tetradecane.

Watanabe *et al.* (2003b) used a transposon-transposase complex to transfer the *dsz* gene cluster from *Rhodococcus erythropolis* strain KA2–5-1 into *R. erythropolis* strain MC1109 (unable to desulfurize LGO). Resting cells of the recombinant strain decreased the sulfur concentration of LGO from 120 mg/L to 70 mg/L in 2 h. Also, it was shown that a transposon-induced mutation likely altered Δ9-unsaturated fatty acid composition, increased the fluidity of cell membranes, and enhanced the desulfurization activity in a *R. erythropolis* strain.

Feng *et al.* (2016) studied the inhibitory effects of major end-products (2-HBP and sulfate ions) on cell growth, transcriptional levels of *dszA/dszB/dszC,* the desulfurization efficiency of *Gordonia* sp. JDZX13, and the optimal oil/aqueous ratio used to weaken major feedback inhibition effects. Compared to 2-HBP, the accumulation of sulfate ions would generate more negative feedback inhibition effects on desulfurization enzyme activity. Compared to the equal 2-HBP concentration, the negative effect of sulfate ions on the desulfurization gene was more significant. Under a 0.2 mmol/L sulfate ion, the transcriptional levels of the desulfurization gene *dszA/dszB/*

dszC were repressed only 0.015, 0.004, and 0.011 times, respectively, than the original level, while under 0.2 mmol/L 2-HBP, the transcriptional levels of the desulfurization genes *dszA*/*dszB*/*dszC* were 0.77, 0.59, and 0.56 times, respectively. These results revealed that the expression of *dszB* and *dszC* down-regulated compared to *dszA*, which indicates that the third and fourth steps of the 4S pathway were inhibited by sulfate ions, while 2-HBP was a potential inhibitor of the second (*dszC*) and fourth steps (*dszB*) of the 4S pathway. These results also prove that the DBT removal performance of the resting cells was greatly inhibited under a higher sulfate concentration. However, the highest DBT-removal efficiency (44.14%) was achieved in the 1:3 oil/aqueous ratio system. Considering the amount of water used, the DBT-removal efficiency of the oil/aqueous ratio 1:2 system reached 0.188 mmol /L water, which was the highest among all the systems. Thus, the oil/aqueous ratio of 1:2 (improved by 100.7%) was determined as the best choice for better balancing the feedback inhibition effects of 2-HBP and sulfate ions. The overall mechanism of DBT BDS was summarized in Figure 7.8.

The oxygen transfer issues are quite important because BDSM utilizes oxygen not only as a co-substrate of monooxygenases, but also in their endogenous metabolism.

The Michaelis constant (K_m) of the oxygenase for oxygen is relatively high, so it might be necessary to maintain significant oxygen pressure during bioconversion to allow the oxygenase to compete for oxygen with endogenous respiration (Duetz *et al.*, 2001). Vitreoscilla hemoglobin (VHb) technology is considered a promising strategy for improving the supply, transfer, and store of oxygen in vivo. Consequently, Xiong *et al.* (2007) introduced the *vgb* gene that encodes the VHB into *R. erythropolis* LSSE8-1. The recombinant strain bearing *vgb* showed a higher biomass yield and desulfurizing activity compared to the wild type. The sulfur content of diesel oil was reduced from 261.3 mg/L to 70.1 mg/L by the recombinant strain.

Ayala *et al.* (1998) studied the integration of a biocatalytic oxidation of organosulfides and thiophenes with hemoproteins, such as fungal peroxidases, to form more soluble sulfoxides and sulfones with a further bacterial metabolism to remove the sulfur from the latter. Straight-run diesel fuel containing 1.6% sulfur was biocatalytically oxidized with chloroperoxidase from *Caldariomyces fumago* in the presence of 0.25 mM hydrogen peroxide. The reaction was carried out at room temperature and the organosulfur compounds were successfully transformed to their respective sulfoxides and sulfones which were then removed by distillation. The resulting fraction after distillation contained only 0.27% sulfur.

Microorganisms with a high tolerance to organic solvents are suitable and important in many biotechnological fields, such as BDS and biocatalysis (de Bont, 1998). *Pseudomonas* sp. was found to be an ideal candidate for BDS because it is organic solvent-tolerant and has a high growth rate. Several biotechnological properties for the design of biocatalysts targeted to industrial BDS processes are present in *Pseudomonas* species. For example, while the solvent tolerance of *Rhodococcus* is the lowest reported (log *P* values from 6.0 to 7.0), that of the genus *Pseudomonas* (log *P* values from 3.1 to 3.4) is the highest known (Inoue and Horikoshi, 1991).

Tao *et al.* (2006) constructed a solvent-tolerant desulfurizing bacterium, *Pseudomonas putida* A4, by introducing a desulfurizing gene, *dszABCD* from *Rhodococcus erythropolis* XP, into the solvent-tolerant strain *P. putida* Idaho, which can grow in the presence of more than 50% toluene, m-xylene, p-xylene, 1,2,4-trimethylbenzene, and 3-ethyltoluene when A4 is contacted with sulfur refractory compounds dissolved in hydrocarbon solvent and IT maintained the same substrate desulfurization traits as observed in *R. erythropolis* XP. Resting cells of *P. putida* A4 could desulfurize 86% of DBT at 10% (v/v) *p*-xylene in 6 h. In the first 2 h, the desulfurization occurred with a rate of 1.29 mM DBT /g DCW/ h. Tao *et al.* (2011b) reported that the combination of a solvent-resistant bacterial host, such as *P. putida* S12 strain, with an organic solvent-responsive expression system that pushes the transcription of the *dsz* genes, resulted in a high efficiency of DBT desulfurization in a biphasic reaction with n-hexane.

Kawaguchi *et al.* (2012) constructed a recombinant strain by introducing a desulfurizing gene *dszABCD* from *Rhodococcus erythropolis* IGTS8 into an organic solvent-tolerant bacterium, *Rhodococcus opacus* B-4, to avoid transcriptional inhibition by the sulfate end-product. DBT consumption rates increased by 80% in oil (n-hexadecane)/water biphasic reaction mixtures and resting cells were predominantly localized in the emulsion layer. Desulfurization efficiency in biphasic reaction mixtures increased with increasing concentrations of DBT and was not inhibited by 2-HBP accumulation. The low desulfurization efficiency under biphasic conditions by the combined effects of attenuated feedback inhibition of 2-HBP and reduced DBT uptake limitations was improved due to the high diffusion of 2-HBP from cells and accumulation of both DBT and biocatalysts in the emulsion layer, so that the solvent-tolerant and hydrophobic bacterium *Rhodococcus opacus* B-4 appears suitable for BDS reactions in solvents containing a minimum ratio of water.

Torktaz *et al.* (2012) mutated and changed the structure of the DszC enzyme to increase protein hydrophobicity, stability, and improve its performance in the oil phase in a biphasic system. The mutation in surface amino acids of DszC to hydrophobe residues changed hydrophobicity properties of this protein. After an acting mutation, the DszC enzyme had more tendencies to oil surface in a manner that its stably increased. Consequently, this mutated enzyme can work better in a biphasic system by increasing hydrophobicity and can maintain an active structure in longer times by increasing stability.

Pan *et al.* (2013) illustrated a new modern technology of synthetic biology and metabolic engineering to obtain improved desulfurization biocatalysts. A synthetic gene (Sulpeptide, S1) encoding high quantities of the sulfur containing amino acids methionine and cysteine was designed, constructed, and cloned between *dszA* and *dszB* in the *dsz* operon, in vector pRESX, under control of the *Rhodococcus kstD* promoter and transformed this construct into the *dsz*-negative *Rhodococcus opacus* in addition to subjecting the transformant to the repeated cultivation in a medium with DBT as a sole sulfur source. Because the S1 peptide gene was inserted inside the *dsz* operon, the increased production of desulfurization enzymes also resulted in the increased production of the S1 peptide, thereby increasing the nutritional demand for sulfur. *R. opacus* cultures containing either pRESXdszABC or pRESXdszAS1BC, after 40 sub-culturing events, showed 23-fold and 35-fold improvements, respectively.

Aggarwal *et al.* (2011b) reported the first attempt at reconstructing a genome-scale flux-based model to analyze sulfur utilization by R. *erythropolis*. This model provided a clear and comprehensive picture of the pathways that assimilated the sulfur from DBT into biomass and successfully elucidated that sulfate promotes higher cell growth than DBT and its existence in the medium reduces DBT desulfurization rates as shown in Figure 7.9. In addition, it closely predicts the growth rates reported in literature and suggests that ethanol and lactate yield higher cell growth and desulfurization rates than citrate, fructose, glucose, gluconate, glutamate, and glycerol. Furthermore, it was used to identify essential genes and reactions for desulfurization activity (Aggarwal *et al.*, 2011a). Also, the genome-scale flux-based model showed that the presence of sulfur containing amino acids, cysteine and methionine, in the medium can reduce the desulfurization activity of *Gordonia alkanivorans*. This model confirmed that the BT is favorably desulfurized over DBT by *G. alkanivorans,* which may be due to BT's required lower energy in terms of NADH, so that NADH plays an important role in desulfurization and reengineers *G. alkanivorans* for the improved supply/regeneration of NADH, which is promising for increasing desulfurization efficiency (Aggarwal *et al.*, 2013).

Aggarwal *et al.* (2012) proposed two alternate, reasonable hypotheses for explaining the formation of sulfate from sulfite through DBT desulfurization, as shown in Figure 7.9. The first is based on the presence and toxicity of excess sulfite and the second is based on enzyme classification. In the first hypothesis, the cell can convert the sulfite from the metabolism of DBT into sulfide by sulfite reductase (SR) and then assimilated biomass. However, it can convert the excess of sulfite into extracellular sulfate by using sulfite oxidoreductase (SOR) to avoid the toxicity in sulfite. In the second hypothesis, the cell cannot directly assimilate the sulfite into biomass by SR, so it must first use SOR to produce extracellular sulfates and then recapture that sulfate into biomass by SR. Therefore, the activity of SOR and SR, in addition to efficiently expressing the *dsz* genes and the requirement of reducing cofactors, play a critical role in the improvement of DBT desulfurization. Decreasing the expression of the *cysJ* gene to achieve lower SR activity would starve cells for sulfur and result in an increased rate of DBT utilization provided that elevated levels of SOR are maintained to avoid the toxic accumulation of sulfite.

A proteomic study in *Chelatococcus* cells grown in the presence of DBT suggests that synthesis of sulfur containing amino acid cysteine is a predominant mechanism of sulfur assimilation in this desulfurizing microorganism. The cysteine is further converted to methionine and CoA is also incorporated into proteins. Moreover, synthesis of thioredoxin by *Chelatococcus* sp. improved the synthesis of cell division proteins, transcription regulators, stress response proteins, and activation of several metabolic enzymes when grown in the presence of DBT by the interaction of thioredoxin with the target proteins. The proteomic analysis also revealed that *Chelatococcus* sp. expressed enolase and ATP synthase subunits alpha and

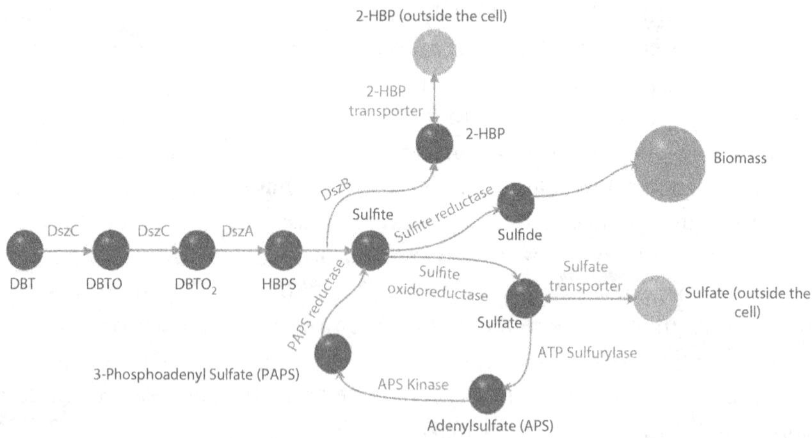

Figure 7.9 Formation and Consumption of Sulfite and Sulfate for Biomass Formation During BDS of DBT.

beta during its growing on DBT as the sole sulfur source. The enolase is a glycolytic enzyme and it is actively involved in the transportation of organic compounds across the cell membrane. Therefore, the role of enolase is to transport toxic metabolites 2-HBP / 2-MBP and/or excess sulfite to outside the cell membrane and, consequently, decrease the toxicity of accumulated metabolites inside the cell, so the involvement of this protein in sulfur assimilation is a target for improving desulfurization (Bordoloi *et al.* 2016).

The overexpression of flavin reductase in recombinant bacteria was necessary to improve the activities of DszC and DszA.

Lei *et al.* (1997) improved BDS activities of purified DszC and DszA enzymes from *R. erythropolis* IGTS8 in vitro when activated with flavin reductase from *Vibrio harveyi*. Increasing the amount of flavin reductase protein in the reaction mixture linearly increased the rate of reaction for both DszC and DszA. The amount of produced $DBTO_2$ or 2-HBP increased with time and the reaction rates were linear for 10 to 15 min with 1 unit/mL of flavin reductase and for 20 min or longer with lower concentrations of flavin reductase. These results suggest that both DszC and DszA are terminal oxygenases and that another protein, a flavin reductase, is required for the full activity of both.

Reichmuth *et al.* (1999) engineered the *E. coli* DH10B strain that contained two compatible plasmids: pDSR3 encoded all three of the enzymes that converted DBT to 2-HBP and pDSR2 contained a *Vibrio harvey* NADH-FMN oxidoreductase gene. The expression of oxidoreductase caused an increase in the rate of DBT removal, but a decrease in the rate of 2-HBP production. The maximum rate of DBT removal was 8 mg/g DCW/h, so that designing of an operon that expresses the proper amount of FMN: NADH reductase to existed *dszABC* enzymes is important to reach a maximum desulfurization efficiency. It should be kept in mind that insufficient FMN: NADH reductase would make NADH the limiting step in DBT oxidation, while a high concentration of $FMNH_2$ will increase H_2O_2 formation which would be lethal to cells (Gaudu *et al.* 1994; Galan *et al.* 2000).

In order to provide the required amount of reduced flavin to Dsz monooxygenases, Galan *et al.* (2000) inserted *hpaC*, oxidoreductase from *E. coli*, into a different recombinant *Pseudomonas putida* strain bearing the *dszABC* gene cluster. They also used catalase in the desulfurization medium to minimize the probability of an H_2O_2 formation, which might be produced by the non-enzymatic reoxidation of $FMNH_2$ under high oxygen concentrations. The addition of *hpaC* flavin reductase increased DBT desulfurization 7–10 times over 30 min. These results confirmed that the expression of an oxidoreductase with the *dsz* genes is able to improve the rate of DBT desulfurization.

The expression of dsz genes in most desulfurizing bacteria is repressed by sulfate through a repressor-binding site that may be in the promoter, so it is important to design a new promoter that cannot be repressed by sulfate as a way to enhance the desulfurization efficiency.

Gallardo *et al.* (1997) improved the BDS phenotype and mitigated sulfur repression by engineering the *dszABC* operon from IGTS8 under the control of heterologous broad-host-range regulatory signals (*$lacI^q/P_{trc}$*) and stably inserted it into the chromosomes of different *Pseudomonas* strains. These recombinant bacteria associate significant industrial and environmental behaviors through the production of biocatalysts able to desulfurize DBT more efficiently than the native host in addition to its ability to produce biosurfactants. The *dsz* cassette was inserted into plasmid pVLT31 under the control of a hybrid promoter, *Ptac*, that has been shown to be functional in a wide range of bacteria. The resulting plasmid was transferred into *P. putida and P. aeruginosa* PG201, which cannot use DBT as the sole carbon and/or sulfur source and produce rhamnolipid biosurfactants. The final recombinant bacterium was named *P. aeruginosa* EGSOX. After 48 h of incubation, cultures of strain IGTS8 still contained DBT. However, this compound was exhausted by the two engineered *Pseudomonas* strains. *P. aeruginosa* EGSOX had the fastest metabolism of DBT, transforming 95% of the DBT at 24 h of incubation. Only 18% of the DBT was transformed by *R. erythropolis* IGTS8 and 40% was transformed by *P. putida* EGSOX. The depletion of DBT was associated with 2-HBP accumulation in all three strains, indicating that 2-HBP is a dead-end metabolite that cannot be further catabolized or used as a carbon source. This data demonstrated that the IGTS8-derived *dsz* cassette was efficiently expressed, allowing the elimination of sulfur with no loss of DBT carbon atoms, both in *P. putida* EGSOX and *P. aeruginosa* EGSOX. Moreover, in comparison with wild-type *R. erythropolis* IGTS8, the two recombinant biocatalysts showed enhanced BDS ability and were able to keep their desulfurization phenotype even in sulfate-containing media. However, many *Pseudomonas* strains were unable to desulfurize DBT in the oil phase and this will restrict their application in industry. Similar approaches that mitigate sulfur repression and increased desulfurization efficiency by expression of the *dsz* genes under the control of heterologous promoters have been described in *Rhodococcus* strains (Gupta *et al.*, 2005).

Matsui *et al.* (2002) cloned the putative *Rhodococcus rrn* promoter region in the 16SrRNA region from benzothiophene desulfurizing bacteria *Rhodococcus* sp. strain T09 and the DBT desulfurizing gene, *dsz*, was expressed under the control of the putative *rrn* promoter in strain T09 using a *Rhodococcus–E. coli* shuttle vector pRHK1. Strain T09, harboring the expression vector pNT, could desulfurize DBT in the presence of inorganic sulfate, methionine, or cysteine, while the Dsz phenotype was completely repressed in recombinant cells carrying the gene with the control of the native *dsz* promoter under the same conditions.

At the same time, Noda *et al.* (2002) constructed a promoter–probe transposon using a promoterless red-shifted green fluorescence protein gene (*rsgfp*). A 340-bp putative promoter element, kap1, was isolated from a recombinant strain, *Rhodococcus erythropolis* strain KA2–5-1, that had been shown to have high fluorescence intensity. The promoter element of kap1 was not repressed by 1 mM of sulfate and it gave about 2 fold greater activity than the 16S ribosomal RNA promoter in *R. erythropolis*. Therefore, kap1 is a suitable tool for improving the BDS efficiency of *Rhodococcus* sp.

Tanaka *et al.* (2002) used a transposome technique to isolate the two mutants of *R. erythropolis* KA2–5-1, which express about 5 fold higher levels of desulfurization efficiency than that by cell-free extracts from the wild-type in the presence of sulfate. Gene analysis of the mutants revealed that the transposon was inserted inside the *cbs* gene encoding a cystathionine synthase. A putative promoter, $P_{kap1,}$ which is not affected by sulfate was isolated from *R. erythropolis* KA2–5-1 and successfully used for expressing desulfurization genes.

Franchi *et al.* (2003) used an efficient *Rhodococcus* gene expression system for the optimization of the entire reaction. The system is based on plasmids derived from a large *Rhodococcus* endogenous plasmid and the utilization of new regulatory sequences which promote an efficient expression of the genes controlled. Two different promoter sequences were identified and named P600 and P57, which promote high levels of transcription of exogenous genes. New recombinant strains were constructed in which the *dszD* gene was either inserted in the chromosome under the control of the P57 promoter in a single copy (strain D10) or cloned on a plasmid under the control of its own promoter (strain D9). The comparison between wild type and D9 cells grown on DBT as the only source of sulfur indicates that a 2–3-fold increase in reductase activity was obtained by cloning the dszD gene on a plasmid, thus increasing its copy number/cell, while D10 cells, which harbour a second copy of dszD on the chromosome under the control of the P57 promoter, showed an 11-fold increase in reductase activity compared to wild type cells. These results indicate that the expression of the *dszD* gene is increased several folds by changing the expression system and the gene expression regulated by the P57 promoter is not affected by the presence of sulfate during cell growth.

Radwan (2015) cloned a promoterless fragment of *dsz* operon *R. erythropolis* HN2 under control of a T7 promoter in plasmid pET 29a(+)

expressed in *E. coli* BL21 on a sulfate-containing medium. Desulfurization activity of the recombinant proteins was two times higher than the extracted proteins from the parent strain HN2. Furthermore, expression of *dsz* genes in fast-growing microorganisms *E. coli*, under a sulfate non-repressible promoter (T7 promoter), could solve the problem of biocatalysts' mass production. Moreover, expression of *dsz* genes in *E. coli* on a sulfate-containing medium could also help in decreasing the time and cost of producing the biocatalyst.

Several approaches have been implemented by genetic engineering to improve desulfurizing enzyme activities of DszC and DszB because the reactions catalyzed by DszC and DszB have been widely recognized as rate-limiting steps in the microbial desulfurization pathway.

Coco *et al.* (2001) improved rates of substrate conversion and specificity of DszC enzymes by a molecular protein evolution technique (RACHITT) that allowed in vitro recombination of *dszC* genes from *Rhodococcus* and *Nocardia* strains. The activity of *dszC* was about 20-fold higher than that of either parent and it must have increased the rate of the whole pathway.

Ting *et al.* (2006) cloned the *dszC* gene from a DBT-desulfurizing bacterium, *Rhodococcus* sp. DS-3, into plasmid pET28a and then transformed it into *E. coli* BL21 strain. The soluble DszC in the supernatant was purified by Ni^{2+} chelating a His–Tag resin column and sodium dodecyl sulfate-polyacrylamide gel electrophoresis (SDS–PAGE) to electronics purity. The activity of purified DszC was 0.36 U. DszC acted not only on DBT, but also on thioxanthen-9-one as well as DBT derivatives. In the reaction using 4-methyl DBT and 4,6-dimethyl DBT, the enzyme activities were 38.4% and 17.1% compared with using DBT. On the contrary, DszC showed no activity toward other heterocyclic compounds, dibenzofurane, and fluorene, where other O and N atoms were substituted for a sulfur atom of DBT, respectively. These results suggested that DszC specifically recognized the sulfur atom and catalyzed the mono-oxygenation reaction.

Borolle *et al.* (2003) investigated an improved desulfinase enzyme by using the point mutagenesis of the *dszB* gene. The mutations were introduced by an error-prone PCR (polymerase chain reaction). Plasmid DNA (pACKI), isolated from the transformed cells, was digested with the appropriate restriction enzymes to release the *dszB* containing DNA fragments generated from error prone PCR. This fragment was then ligated back into the vector portion of pESOX1. These mutagenized plasmids (pACKD) were then transformed into *E. coli* JM109. Transformant cells were then screened for desulfinase activity. The mutants 8B1 and 6A10 showed about 2.5 and 4.0 times higher activity than the native strain. Sequencing of the *dszB* isolated from the improved strains was conducted to find if this was due to a change in the gene sequence. No change was observed from the parent sequence, however, the possibility that a mutation in the regulatory portion of the DNA sequence being responsible for the improved activity cannot be ruled out. Another possible reason for the observed increase in activity may be the presence of a different copy number of the plasmid carrying the *dszB* gene in different strains.

The ratio of mRNA of *dszA, dszB,* and *dszC* in the native *dsz* operon was 11:3.3:1, indicating that the translation levels of the desulfurization enzymes decreased according to their positions in the operon due to polar effects on *dsz* gene transcription. However, western blot analysis indicated that the expression level of *dszB* was far lower than that of *dszC.* These results suggest that the translation of *dszB* mRNA was not as efficient as *dszA* mRNA. Gene analysis revealed that the termination codon of *dszA* and the initiation codon of *dszB* overlapped, whereas there was a 13-bp gap between *dszB* and *dszC.* To overcome this problem, Li *et al.* (2007b) removed the overlapping structure between *dszA* and *dszB* (as described before) by overlap polymerase chain reaction (PCR) and expressing the redesigned *dsz* operon *in R. erythropolis,* named *R. erythropolis* DR-2. Western blot analysis revealed that *R. erythropolis* DR-2 produced more DszB than *R. erythropolis* DR-1, which holds the original *dsz* operon. The desulfurization activity of resting cells prepared from *R. erythropolis* DR-2 was about 5 fold higher than that of *R. erythropolis* DR-1, so the enhanced expression level increased the metabolic rate of HBPS in the cells and contributed to the improved desulfurization rate of *R. erythropolis* DR-2.

Li *et al.* (2008) presented a genetic rearrangement strategy for optimizing the metabolic pathway of DBT. By using recombinant PCR, the *dsz* operon of *R. erythropolis* DS-3 was rearranged, as shown in Figure 7.10, according to the catalytic capabilities of the Dsz enzymes and their reaction orders in the 4S pathway. Fragments of the 400-bp 5' upstream segment (5'-U-S) and the 400-bp 3' downstream segment (3'-D-S) of *dszABC* and the *dszA* and *dszBC* segments, including the overlap regions, were amplified by PCR, then the ligated 5' upstream-*dszBC* segment (5'-U-S-*dszBC*) and the ligated dszA-3' downstream segment (dszA-3'-D-S) were produced by overlap PCR via their overlap regions, and finally, the 5' upstream-*dszBC* segment and the *dszA*-3' downstream segments were linked together by overlapping PCR via their overlap region to yield the reconstructed *dsz* operon, as shown in Figure 7.10. After rearrangement, the ratio of *dszA*, *dszB*, and *dszC* mRNAs in the cells was changed from 11:3.3:1 to 1:16:5. The desulfurization rate of the recombinant strain containing the rearranged *dsz* operon was 12 times faster than that of the native *dsz* operon. The maximum desulfurization rate was only about 26 mmol DBT/g

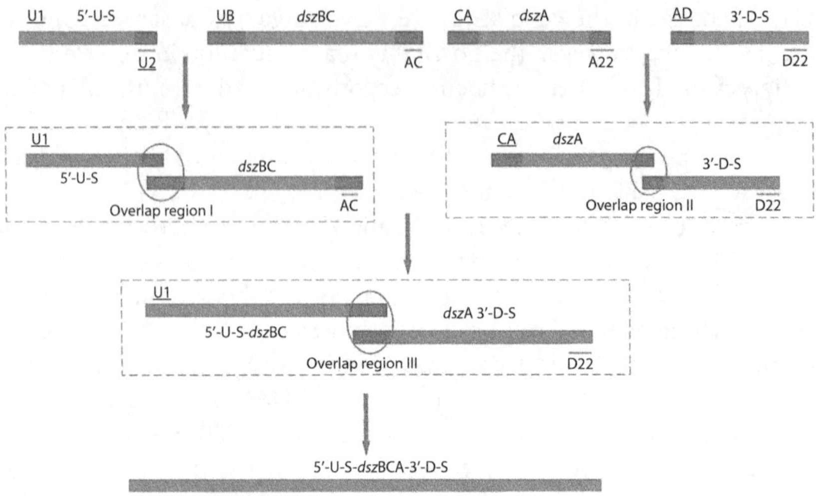

Figure 7.10 A Genetic Rearrangement Strategy for Improving the Metabolic Pathway of DBT (gray bars represent genes and dark gray bars represent overlap regions).

DCW/h for the strain containing the native *dsz* operon. After removing the overlapped structure before the initiation codon of *dszB*, the rate was 120 mmol DBT/g DCW/h. The recombinant strain containing the rearranged *dsz* operon had the highest desulfurization rate, about 320 mmol DBT/g DCW/h. Therefore, the enhanced expression levels of DszC and DszB increased the desulfurization rate of the recombinant strain.

Rehab *et al.* (2011) cloned the *dszABC* genes into the pVLT31 vector and the plasmid harboring *dszABC* genes were transferred into *E. coli* BL21 which already contained the *dszD* gene. The highest desulfurization efficiency, up to 90%, was achieved by recombinant *E. coli* BL21, while *Rhodococcus erythropolis* IGTS8 was desulfurized only 65% after 99 hours. In the next step, the pUC18 plasmid harboring *dszAB* genes of *Rhodococcous* FMF was transferred into *E. coli* DH5α in order to evaluate the effect of elimination of the *dszC* gene on the rate of the BDS process. This analysis demonstrated that 2-HBP production of recombinant *E. coli* DH5α, using a broken 4S pathway, was less than that of *Rhodococcous* FMF, so that the existence of the third gene (*dszC*) is necessary for the enhancement of desulfurization activity.

The overexpression of the *dszB* gene by increasing its copy number and optimizing its ribosome-binding site was used by Reichmuth *et al.* (2004) to overcome the rate-limiting step caused by low DszB activity. Reichmuth *et al.* (2004) mutated the untranslated 5' region of *dszB* using degenerate oligonucleotides. Because neither DszB activity nor the amount of DszB protein produced could be directly measured, it was difficult to determine the exact cause for the lack of 2-HBP production. To clarify the results of genetic manipulations, they chose to tag the production of the desulfurization transcripts and proteins by creating transcriptional and translational fusions with a fluorescent protein. This permitted a quick, straightforward, and direct determination of the amount of the desulfurization protein produced. This technique does not measure the activity of the proteins, but activity screens could be used after protein production was optimized using fluorescent fusion tags. The protein used for those fusions was GFP (Green Fluorescent Protein). GFP has been widely used for the quantitative measurement of protein production and is known to be stable for a period of several days, allowing an integrative and quantitative measure of protein production (Albano *et al.*, 1998; Cha *et al.*, 2000). After screening only 96 mutants, several of them showed increased green fluorescence and two showed increased DszB activity. When co-transformed with the full *dszABC* operon, the mutant *dszB* increased the rate of desulfurization 9 fold relative to the native *dszB*.

Ma *et al.* (2006b) cloned the *dsz* genes from *Rhodococcus erythropolis* DS-3 into the chromosomes of *Bacillus subtilis* ATCC 21332 and UV1, which can secrete biosurfactant, using an integration vector, pDGSDN, yielding two recombinant strains, *B. subtilis* M29 and M28, in which the integrated dsz genes were expressed efficiently under the promoter, Pspac. The desulfurization efficiency of M29 was 16.2 mg DBT/L/h at 36 h, which is significantly higher than that of *R. erythropolis* DS-3, and *B. subtilis* M28 showed no product inhibition. The interfacial tension of the supernatant caused by M29 varied from 48 mN/m to 4.2 mN/m, lower than that of the recombinant strain M28, revealing that the biosurfactant secreted by M29 may have an important function in the DBT desulfurization process.

Wang *et al.* (2017) cloned *dszAS1BC* (containing a peptide "Sulpeptide 1" or "S1" with a high proportion of methionine and cysteine residues to act as a sulfur sink that was inserted into the *dsz* operon of *Rhodococcus erythropolis* IGTS8) and the intact *dsz* operon (*dszABC*) into vector pRESX under control of the *Rhodococcus* kstD promoter and then transformed it into the desulfurization-negative strain CW25 of *Rhodococcus qingshengii*. The resulting strains, CW25[pRESX-*dszABC*] and CW25[pRESX-*dszAS1BC*], were subjected to adaptive selection by repeated passages at a log phase (up to 100 times) in a minimal medium with DBT as sole sulfur source. Growth rates increased by 7 fold and 13 fold in CW25[pRESX-*dszABC*] and CW25[pRESX-*dszAS1BC*], respectively, and these increases were stable. For both transformants, the BDS efficiency of CW25[pRESX-dszABC] and CW25[pRESX-dszAS1BC] was 36-fold and 33-fold compared to the parent strain in each case. The adaptations of CW25[pRESX-dszAS1BC] were explained by a 3–5X increase in plasmid copy numbers from those of the initial transformed cells. The increases in growth rate and DBT metabolism were not correlated to any mutations in genes, but they were closely correlated to an increase in the copy number of the genes encoding DBT metabolism. It may be that the important changes that occurred during the adaptation include epigenetic and stable physiological changes. These results suggested that future research should first optimize growth on sulfate, under ideal conditions, and only then should the *dsz* genes be introduced and adaptive evolution experiments be performed.

References

Abbad-Andaloussi, S., Lagnel, C., Warzywoda, M., Monot F. (2003) Multi-criteria comparison of resting cell activities of bacterial strains selected for biodesulfurization of petroleum compounds. Enzyme and Microbial Technology. 32: 446–454.

Abin-Fuentes, A., Mohamed, M.E., Wang, D., Prather K. (2013) Exploring the mechanism of biocatalyst inhibition in microbial desulfurization. *Applied and Environmental Microbiology*. 79:7807–7817.

Adak, S., Begley T.P. (2016) Dibenzothiophene catabolism proceeds via a flavin-N5-oxide intermediate. *Journal of the American Chemical Society*. 138: 6424–6426

Aggarwal, S., Karimi, I.A., Ivan, G.R. (2013) In silico modeling and evaluation of *Gordonia alkanivorans* for biodesulfurization. Molecular BioSystems. 9: 2530–2540.

Aggarwal, S., Karimi, I.A., Lee, D.Y. (2011b) Flux-based analysis of sulfur metabolism in desulfurizing strains of *Rhodococcus erythropolis*. FEMS Microbiology Letters. 315:115–121.

Aggarwal, S., Karimi, I.A., Kilbane, J.J., Lee, D.Y. (2012) Roles of sulfite oxidoreductase and sulfite reductase in improving desulfurization by *Rhodococcus erythropolis*. Molecular BioSystems. 8 :2724–2732.

Aggarwal, S., Karimi, I.A., Lee, D.Y. (2011[a]) Reconstruction of a genome-scale metabolic network of *Rhodococcus erythropolis* for desulfurization studies. Molecular BioSystems. 7:3122–3131

Ahmad, A., Chauhan, A.K., Javed, S., Kumar A. (2014) Desulfurization of thianthrene by a *Gordonia* sp. IITR100. Biotechnology Letters. 36:2209–2214

Akbarzadeh, S., Raheb, J., Aghaei, A., Karkhane A.A. (2003) Study of desulfurization rate in *Rhodococcus* FMF native bacterium. *Indian Journal of Biotechnology*. 1: 36–40.

Akhtar, N., Ghauri, M.A., Anwar, M.A., Akhtar, K. (2009) Analysis of the dibenzothiophene metabolic pathway in a newly isolated *Rhodoccocus* spp. FEMS Microbiology Letters. 301: 95–102.

Akhtar, N., Ghauri, M.A., Anwar, M.A., Heaphy, S. (2015) Phylogenetic characterization and novelty of organic sulphur metabolizing genes of *Rhodococcus* spp. Eu-32. Biotechnology Letters. 37: 837–847.

Albano, C., Randers-Eichhorn, L., Bentley, W. and Rao, G. (1998) Green fluorescent protein as a real time quantitative reporter of heterologous protein production. Biotechnology Progress. 14:351–354.

Alves L. and Paixão S.M. (2011) Toxicity evaluation of 2-hydroxybiphenyl and other compounds involved in studies of fossil fuels biodesulphurisation. *Bioresource Technology*. 102:9162–9166.

Alves, L., Salgueiro, R., Rodrigues, C., Mesquita, E., Matos, J., Gírio F.M. (2005) Desulfurization of dibenzothiophene, benzothiophene, and other thiophene analogs by a newly isolated bacterium, *Godonia alkanivoransstrain* 1B. *Applied Biochemistry and Biotechnology*. 120:199–208.

Arensdorf, J.J., Loomis, A.K., DiGrazia, P.M., Monticello, D.J., Pienkos P.T. (2002) Chemostat approach for the directed evolution of biodesulfurization gain-of-function mutants. *Applied and Environmental Microbiology.* 68:691–698

Ayala, M., Tinoco, R., Hernandez, V., Bremauntz, P., Vazquez-Duhalt R. (1998) Biocatalytic oxidation of fuel as an alternative to biodesulfurization. *Fuel Processing Technology*. 57:101–111.

Bhatia, S., Sharma D.K. (2010) Biodesulfurization of dibenzothiophene, its alkylated derivatives and crude oil by a newly isolated strain *Pantoea agglomerans* D23W3. *Biochemical Engineering Journal*. 50: 104–109.

Bordoloi, N.K., Bhagowati, P., Chaudhuri, M.K., Mukherjee A.K. (2016) Proteomics and metabolomics analyses to elucidate the desulfurization pathway of *Chelatococcus* sp. PLoS One. 11: e0153547.

Borole, A.P., Hamilton, C.Y., Miller, K., Brian, D., Matthew, G., Shong R. (2003) Improving Enzyme Activity and Broadening Selectivity for Biological Desulfurization and Upgrading of Petroleum Feedstocks. Technical Report. University of Tennessee. United States. http://www.osti.gov/scitech/servlets/purl/812037.

Bressler, D.C., Fedorak P.M. (2000) Bacterial metabolism of fluorene, dibenzofuran, dibenzothiophene, and carbazole. *Canadian Journal of Microbiology*. 46:397–409

Bressler, D.C., Norman, J.A., Fedorak P.M. (1998) Ring cleavage of sulfur heterocycles: how does it happen?. Biodegradation. 8: 297–311.

Calzada, J., Zamarro, M.T., Alcón, A., Santos, V.E., Díaz, E., García, H.L., García-Ochoa F. (2009) Analysis of dibenzothiophene desulfurization in a recombinant *Pseudomonas putida* strain. *Applied and Environmental Microbiology*. 75:875–877

Campos-Martín, J.M., Capel-Sánchez, M.C., Fierro J.L.G. (2004) Highly efficient deep desulfurization of fuels by chemical oxidation. Green Chemistry. 6: 557–562.

Castorena, G., Suarez, C., Valdez, I., Amador, G., Fernandez, L., Le Borgne S. (2002) Sulfur-selective desulfurization of DBT and diesel oil by newly isolated *Rhodococcus* sp. Strains. FEMS Microbiology Letters. 215:157–161.

Cha, H., Wu, C., Valdes, J.., Rao, G., Bentley, W. (2000) Observations of green fluorescent protein as a fusion partner in genetically engineered *Escherichia coli*: monitoring protein protein expression and solubility. *Biotechnology and B ioengineering*. 67:565–574.

Chang, J.H., Kim, Y.J., Lee, B.H., Cho, K.S., Rye, H.W., Chang, Y.K., Chang H.N. (2001) Production of a desulfurization biocatalyst by two-stage fermentation and its application for the treatment of model and diesel oils. Biotechnology Progress. 17: 876–880.

Chen, H., Cai, Y.B., Zhang, W.J., Li, W. (2009) Methoxylation pathway in biodesulfurization of model organosulfur compounds with *Mycobacterium* sp. *Bioresource Technology*. 100:2085–2087.

Chen, H., Zhang, W.J., Cai, Y.B., Zhang, Y., Li W. (2008) Elucidation of 2-hydroxybiphenyl effect on dibenzothiophene desulfurization by *Microbacterium* sp. strain ZD-M2. *Bioresource Technology*. 99:6928–6933.

Chen, W., Brühlmann, F., Richins, R.D., Mulchandani A. (1999) Engineering of improved microbes and enzymes for bioremediation. Current Opinion in Biotechnology. 10: 137–141.

Coco, W.M., Levinson, W.E., Crist, M.J., Hektor, H.J., Darzins, A., Pienkos, P.T., Squires, C.H., Monticello D.J. (2001) DNA shuffling method for generating highly recombined genes and evolved enzymes. Nature Biotechnology. 19:354–359.

Czechowska, K., Reimmann, C., van der Meer, J.R. (2013) Characterization of a MexAB-OprM efflux system necessary for productive metabolism of Pseudomonas azelaica HBP1 on 2-hydroxybiphenyl. Front Microbiol. 4:203

Davoodi-Dehaghani, F., Vosoughi, M., Ziaee A.A. (2010) Biodesulfurization of dibenzothiophene by a newly isolated *Rhodococcus erythropolis* strain. *Bioresource Technology*. 101: 1102- 1105.

de Bont, J.A.M. (1998) Solvent-tolerant bacteria in biocatalysis. Trends in biotechnology. 16:493–499.

Denis-Larose, C., Labbe, D., Nergron, H., Jones, A.M., Greer, C.W., Al-Hawari, J., Grossman, M.J., Sankey, B.M., Lau P.C.K. (1997) Conservation of plasmid-encoded dibenzothiophene desulfurization genes in several *rhodococci*. *Applied and Environmental Microbiology*. 63: 2915–2919.

Denome, S.A., Oldfield, C., Nash, L.J., Young K.D. (1994) Characterization of the desulfurization genes from *Rhodococcus* sp. strain IGTS8. *Journal of Bacteriology*. 176: 6707–6716

Denome, S.A., Olson, E.S., Young K.D. (1993) Identification and cloning of genes involved in specific desulfurization of dibenzothiophene by *Rhodococcus* sp. strain IGTS8. *Applied and Environmental Microbiology*. 59:2837–2843.

Derikvand, P., Etemadifar Z. (2014) Improvement of biodesulfurization rate of alginate immobilized *Rhodococcus erythropolis* R1. Jundishapur. *Journal of Microbiology*. 7: e9123.

Duan, X., Zhang, L., Zhou, D., Ji, K., Ma, T., Shui, W., Li, G., Li X. (2013) Crystallization and preliminary structural analysis of dibenzothiophene monooxygenase (DszC) from *Rhodococcus erythropolis*. Acta Crystallographica Section F, Structural biology and crystallization communications. 69:597–601

Duarte, G.F., Rosado, A.S., Seldin, L., De Araujo, W., Van Elsas J.D. (2001) Analysis of bacterial community structure in sulfurous-oil-containing soils and detection of species carrying dibenzothiophene desulfurization (*dsz)* genes. *Applied and Environmental Microbiology*. 67: 1052–1062.

Duetz, W.A., van Beilen, J.B., Witholt B. (2001) Using proteins in their natural environment: potential and limitations of microbial whole-cell hydroxylations in applied biocatalysis. Current Opinion in Biotechnology. 12:419–425.

El-Gendy, N.Sh., Nassar, H.N., Abu-Amr, S.S. (2014) Factorial design and response surface optimization for enhancing a biodesulfurization process. *Petroleum Science and Technology*. 32: 1669–1679.

El-Said Mohamed, M., Al-Yacoub, Z.H., Vedakumar, J.V. (2015b) Biocatalytic desulfurization of thiophenic compounds and crude oil by newly isolated bacteria. Front Microbiol. 6:112.

El-Said Mohamed, M., García, J.L., Martínez, I., Del Cerro, C., Nogales, J., Díaz E. (2015a) Genome sequence of *Pseudomonas azelaica* strain Aramco. Genome Announcements. 3: e00037–15.

Feng, S., Yang, H., Zhan, X., Wang W. (2016) Enhancement of dibenzothiophene biodesulfurization by weakening the feedback inhibition effects based on a systematic understanding of the biodesulfurization mechanism by *Gordonia* sp. through the potential. *Royal Society of Chemistry Advances*. 6(86): 82872–82881.

Folsom, B.R., Schieche, D.R., Digrazia, P.M., Werner, J., Palmer S. (1999) Microbial desulfurization of alkylated dibenzothiophenes from a hydrodesulfurized middle distillate by *Rhodococcus erythropolis* I-19. *Applied Environmental Microbiology*. 65: 4967–4972.

Franchi, E., Rodriguez, F., Serbolisca, L., deFerra F. (2003) Vector development isolation of new promoters and enhancement of the catalytic activity of the Dsz enzyme complex in *Rhodococcus* sp. strains. Oil & Gas Science and Technology. 58:515–520.

Furuya, T., Takahashi, S., Ishii, Y., Kino, K., Kirimura K. (2004) Cloning of a gene encoding flavin reductase coupling with dibenzothiophene monooxygenase through coexpression screening using indigo production as selective indication. *Biochemical and Biophysical Research Communications*. 313:570–575.

Furuya, T., Takahashi, S., Iwasaki, Y., Ishii, Y., Kino, K., Kirimura K. (2005) Gene cloning and characterization of *Mycobacterium phlei* flavin reductase involved in dibenzothiophene desulfurization. *Journal of Bioscience and Bioengineering*. 99:577–585.

Galan, B., Dı´az, E., Garcı´a J.L. (2000) Enhancing desulfurization by engineering a flavin reductase- encoding gene cassette in recombinant biocatalyst. Environmental Microbiology. 2: 687–694.

Gallagher, J.R., Olson, E.S., Stanley D.C. (1993) Microbial desulfurization of dibenzothiophene: a sulfur-specific pathway. FEMS Microbiology Letters. 107: 31–36.

Gallardo, M.E., Ferrandex, M.E.A., De Lorenzo, V., Garcia, J.L., Diaz, E. (1997) Designing Recombinant Pseudomonas Strains to Enhance Biodesulfurization. *Journal of Bacteriology*. 179, 7156–7160.

Gaudu, P., Touati, D., Niviere, V., Fontecave M. (1994) The NAD(P)H: flavin oxidoreductase from *Escherichia coli* as a source of superoxide radicals. *Journal of Biological Chemistry*. 269:8182–8185.

Gilbert, S.C., Morton, J., Buchanan, S., Oldfield, C., McRoberts A. (1998) Isolation of a unique benzothiophene-desulphurizing bacterium *Gordonia* sp. strain 213E, and characterization of the desulphurization pathway. *Microbiology*. 144: 2545–2553.

Gou, Z., Luo, M., Li, X., Xing, J., Liu H. (2003) Desulfurization metabolite of *Rhodococcus erythropolis* LSSE8-1 and its related desulfurizational gene fragments. Chinese Science Bulletin. 48(24):2703–2709.

Gray, K.A., Mrachko, G.T., Squires C.H. (2003) Biodesulfurization of fossil fuels. *Current Opinion in Microbiology*. 6: 229- 235.

Gray, K.A., Pogrebinsky, O.S., Mrachko, G.T., Lei, X., Monticello, D.J., Squires C.H. (1996) Molecular mechnisms of biocatalytic desulfurization of fossil fuels. Nature Biotechnology. 14: 1705–1709.

Guan, L.J., Lee, W.C., Wang, S., Ohshiro, T, Izumi, Y., Ohtsuka, J., Tanokura M. (2015) Crystal structures of apo-DszC and FMN-bound DszC from *Rhodococcus erythropolis* D-1. *FEBS Journal*. 282:3126–3135

Gunam, I., Yamamura, K., Sujaya, I., Antara, N., Aryanta, W., Tanaka, M., Tomita, F., Sone T. and Asano K. (2013) Biodesulfurization of dibenzothiophene and its derivatives using resting and immobilized cells of *Sphingomonas subarctica* T7b. *Journal of Microbiology and Biotechnology*. 23(4): 473–82.

Gunam, I.B.W., Yaku, Y., Hirano, M., Yamamura, K., Tomita, F., Sone, T., Asano K. (2006) Biodesulfurization of alkylated forms of dibenzothiophene and benzothiophene by *Sphingomonas subarctica* T7b. *Journal of Bioscience and Bioengin eering*. 101(4): 322–327.

Guobin, S., Huaiying, Z., Jianmin, X., Guo, C., Wangliang, L., Huizhou L. (2006) Biodesulfurization of hydrodesulfurized diesel oil with *Pseudomonas delafieldii* R-8 from high density culture. *Biochemical Engineering Journal*. 27: 305–309.

Gupta, N., Roychoudhury, P.K., Deb J.K. (2005) Biotechnology of desulfurization of diesel: prospects and challenges. *Applied Microbiology and Biotechnology*. 66:356–366

Higson, F.K., Focht. D.D. (1989) Bacterial metabolism of hydroxylated biphenyls. *Applied and Environmental Microbiology*. 55: 946–952.

Hirasawa, K., Ishii, Y., Kobayashi, M., Koizumi, K., Maruhashi K. (2001) Improvement of desulfurization activity in *Rhodococcus erythropolis* KA2-5-1 by genetic engineering. *Bioscience, Biotechnology, and Biochemistry*. 65: 239–246.

Honda, H., Sugiyama, H., Saito, I., Kobayashi T. (1998) High cell density culture of *Rhodococcus rhodochrous* by pH-stat feeding and dibenzothiophene degradation. *Journal of Fermentation and Bioengineering.* 85: 334–338.

Inoue, A., Horikoshi K. (1991) Estimation of solvent-tolerance of bacteria by the solvent parameter log P. *Journal of Fermentation and Bioengineering.* 71:194–196

Isbister, J.D., Kobylnski E.A. (1985) In: Processing and Utilization of High Sulfur Coals (Attia, Y.A., Ed.), pp. 627–641, Elsevier, Amsterdam.

Ishii, Y., Konishi J., Suzuki, M., Maruhashi K. (2000) Cloning and expression of the gene encoding the thermophilic NAD(P)H-FMN oxidoreductase coupling with the desulfurization enzymes from *Paenibacillus* sp. A11-2. *Journal of Bioscience and Bioengineering.* 90:591–599.

Ishii, Y., Kozaki, S., Furuya, T., Kino, K., Kirimura K. (2005) Thermophilic biodesulfurization of various heterocyclic sulfur compounds and crude straight-run light gas oil fraction by a newly isolated strain *Mycobacterium phlei* WU-0103. Current Microbiology. 50:63–70.

Ismail, W., El-Said, W., Raheem, A.S.A., Mohamed, M.E., El Nayal, A.M. (2016) Biocatalytic desulfurization capabilities of a mixed culture during non-destructive utilization of recalcitrant organosulfur compounds. Frontiers in Microbiology. 7: 1–14.

Izumi, Y., Ohshiro, T., Ogino, H., Hine, Y., Shimo M. (1994) Selective desulfurization of dibenzothiophene by *Rhodococcus erythropolis* D-1. *Applied Environmental Microbiology.* 60:223–226.

Jaspers, M.C.M., Schmid, A., Sturme, M.H.J., Goslings, D.A.M., Kohler, H.P.E., van der Meer, J.R. (2001) Transcription organization and dynamic expression of the *hbpCAD* genes, which encode the first three enzymes for 2-hydroxybiphenyl degradation in *Pseudomonas azelaica. Journal of Bacteriology.* 183: 270–279.

Jia, X., Wen, J., Sun, Z., Caiyin, Q., Xie S. (2006) Modeling of DBT biodegradation behaviors by resting cells of *Gordonia* sp. WQ-01 and its mutant in oil–water dispersions. *Chemical Engineering Science.* 61:1987–2000.

Kamali, N., Tavallaie, M., Bambai, B., Karkhane, A.A., Miri M. (2010) Site-directed mutagenesis enhances the activity of NADH- FMN oxidoreductase (DszD) activity of *Rhodococcus erythropolis.* *Biotechnology Letters.* 32(7): 921–927.

Kawaguchi, H., Kobayashi, H., Sato K. (2012) Metabolic engineering of hydrophobic *Rhodococcus opacus* for biodesulfurization in oil-water biphasic reaction mixtures. *Journal of Bioscience and Bioengineering.* 113:360–366.

Kayser, K.J., Bielaga-Jones, B.A., Jackowski, K., Odusan, O., Kilbane J.J. (1993) Utilization of organosulfur compounds by axenic and mixed cell cultures of *Rhodococcus rhodochrous* IGTS8. *Journal of General Microbiology.*139: 3123–3129.

Kayser, K.J., Cleveland, L., Park, H.S., Kwak, J.H., Kolhatkar, A., Kilbane J. (2002) Isolation and characterization of moderate thermophile *Mycobacterium phlei* GTIS10, capable of dibenzothiophene desulfurization. *Applied Microbiology and Biotechnology*. 59:737–746.

Kertesz, M.A. (2000) Riding the sulfur cycle – metabolism of sulfonates and sulfate esters in gram-negative bacteria. *FEMS Microbiology Reviews*. 24: 135–175.

Kilbane, J.J. (1989) Desulfurization of Coal: the microbial solution. *Trends in Biotechnology*. 7: 97–101.

Kilbane, J.J. (1990) Sulfur specific microbial metabolism of organic compounds. *Resources, Conservation and Recycling*. 3:69–79.

Kilbane, J.J. (1992) Mutant microorganisms useful for cleavage of organic C–S bonds. US patent number 5,104,801.

Kilbane, J.J. (2017) Biodesulfurization: How to Make It Work? *Arabian Journal for Science and Engineering*. 42(1):1–9.

Kilbane, J.J., Le Borgne S. (2004) Petroleum biorefining: the selective removal of sulfur, nitrogen, and metals. In Petroleum Biotechnology – Studies in Surface Science and Catalysis, vol. 151 ed. Vazquez-Duhalt, R. and Quintero- Ramirez, R. Elsevier, New York, pp. 29–66.

Kilbane, J.J., Robbins J. (2007) Characterization of the *dszABC* genes of *Gordonia amicalis* F.5.25.8 and identification of conserved protein and DNA sequences. *Applied Microbiology and Biotechnology*. 75: 843–851.

Kilbane, J.J. (2006) Microbial biocatalyst developments to upgrade fossil fuels. *Current Opinion in Microbiology*. 17:1–10.

Kim, B.H., Kim, H.Y., Kim, T.S., Park D.H. (1995) Selectivity of desulfurization activity of *Desulfovibrio desulfuricans* M6 on different petroleum products. *Fuel Processing Technology*. 43:87–94.

Kirimura, K., Harada, K., Iwasawa, H., Tanaka, T., Iwasaki, Y., Furuya, T., Ishii, Y., Kino K. (2004) Identification and functional analysis of the genes encoding dibenzothiophene-desulfurizing enzymes from thermophilic bacteria. *Applied Microbiology and Biotechnology*. 65:703–713

Kirkwood, K.M., Ebert, S., Foght, J.M., Fedorak, P.M., Gray M.R. (2005) Bacterial biodegradation of aliphatic sulfides under aerobic carbon-or sulfur-limited growth conditions. *Journal of Applied Microbiology*. 99:1444–1454

Kirkwood, K.M., Foght, J.M., Gray M.R. (2007) Selectivity among organic sulfur compounds in one-and two-liquid-phase cultures of *Rhodococcus* sp. strain JVH1. *Biodegradation* 18:473–480.

Kobayashi, M., Onaka, T., Ishii, Y., Konishi, J., Takaki, M., Okada, H., Ohta, Y., Koizumi, K., Suzuki, M. (2000) Desulfurization of alkylated forms of both dibenzothiophene and benzothiophene by a single bacterial strain. *FEMS Microbiology Letters*. 187:123–126.

Kodama, K., Nakatani, S., Umehara, K., Shimizu, K., Minoda, Y., Yamada K. (1973) Identification of microbial products from dibenzothiophene and its proposed oxidation pathway. *Agricultural and Biological Chemistry*. 37: 45–50.

Kohler, H.P.E., Kohler-Staub, D., Focht, D.D. (1988) Degradation of 2-hydroxybiphenyl and 2,2-dihydroxybiphenyl by *Pseudomonas* sp. strain HBP1. *Applied and Environmental Microbiology*. 54: 2683–2688.

Kohler, H.P.E., Schmid, A., van der Maarel M. (1993) Metabolism of 2,2-dihydroxybiphenyl by *Pseudomonassp.* strain HBP1: production and consumption of 2,2,3-trihydroxybiphenyl. *Journal of Bacteriology.* 175: 1621–1628.

Koike, H., Sasaki, H., Kobori, T., Zenno, S., Saigo, K., Murphy, M.E.P., Adman, E.T., Tanokura M. (1998) 1.8 A crystal structure of the major NAD-(P)H:FMN oxidoreductase of a bioluminescent bacterium, Vibrio fischeri: overall structure, cofactor and substrate- analog binding, and comparison with related flavoproteins. *Journal of Molecular Biology.* 280:259–273.

Konishi, J., Okada, H., Hirasawa, K., Ishii, Y., Maruhashi K. (2002) Comparison of the substrate specificity of the two bacterial desulfurization systems. Biotechnology Letters. 24: 1863–1867.

Labana, S., Pandey, G., Jain R.K. (2005) Desulfurization of dibenzothiophene and diesel oils by bacteria. *Letters in Applied Microbiology.* 40:159–163.

Le Borgne, S., Ayala M. (2010). Microorganisms utilizing sulfur- containing hydrocarbons. In: Handbook of Hydrocarbon and Lipid Microbiology. Timmis K.N. (Ed.), Heidelberg Springer. 3: 2129- 2141.

Le Borgne, S., Quintero R. (2003) Biotechnological processes for the refining of petroleum. *Fuel Processing Technology.* 81: 155–169.

Lee, M.K., Senius, J.D., Grossman M.J. (1995) Sulfur-specific microbial desulfurization of sterically hindered analogos of dibenzothiophene. *Applied Environmental Microbiology.* 61: 4362–4366.

Lee, W.C., Ohshiro, T., Matsubara, T., Izumi, Y., Tanokura M. (2006) Crystal structure and desulfurization mechanism of 20-hydroxybiphenyl-2-sulfinic acid desulfinase. *Journal of Biological Chemistry.* 281:32534–32539.

Lee, W.C., Ohshiro, T., Matsubara, T., Izumi, Y., Tanokura M. (2004) Crystallization and preliminary X-ray analyses of desulfurization enzyme DszB and its C27S mutant complexed with biphenyl-2-sulfinic acid. Acta Crystallographica Section D. 60:1636–1638

Lei, X., Squires, C.H., Monticello, D.J., Child D. (1997) A Flavin Reductase Stimulates DszA and DszC Proteins of *Rhodococcus erythropolis* IGTS8 in vitro. *Biochemical and Biophysical Research Communications.* 230:73–75

Lei, B., Tu, S. (1996) Gene Overexpression, purification and identification of a desulfurization enzyme from *Rhodococcus* sp. *strain IGTS8 as a sulfide/sulfoxide monooxygenase Journal of Bacteriology.* 178: 5699–5705.

Li, G., Ma, T., Zhang, Q., Wang, C., Chen Q. (2007[a]) Sulfur-selective desulfurization of dibenzothiophene and diesel oil by newly isolated *Rhodococcus erythropolis* NCC1. *Chinese Journal of Organic Chemistry.* 25: 400–405.

Li G.Q., Li S.S., Qu S.W., Liu Q.K., Ma T., Zhu L., Liang F.L., Liu R.L. (2008[b]) Improved biodesulfurization of hydrodesulfurized diesel oil using *Rhodococcus erythropolis* and *Gordonia* sp. *Biotechnology Letters.* 30:1759–1764.

Li, G.Q., Li, S.S., Zhang, M.L., Wang, J., Zhu, L., Liang, F.L., Liu, R.L., Ma T. (2008a) Genetic rearrangement strategy for optimizing the dibenzothiophene biodesulfurization pathway in *Rhodococcus erythropolis. Applied and Environmental Microbiology.* 74:971–976.

Li, G.Q., Ma, T., Li, S.S., Li, H., Liang, F.L., Liu RL (2007b) Improvement of dibenzothiophene desulfurization activity by removing the gene overlap in the *dsz* operon. *Bioscience, Biotechnology and Biochemistry*. 71:849–854.

Li, J., Feng, J., Li, Q., Ma, C., Yu, B., Gao, C., Wu, G., Xu P. (2009) Both FMNH2 and FADH2 can be utilized by the dibenzothiophene monooxygenase from a desulfurizing bacterium *Mycobacterium goodii* X7B. *Bioresource Technology*. 100:2594–2599.

Li, M., Squires, C., Monticello, D., Childs J. (1996) Genetic analysis of the *dsz* promoter and associated regulatory regions of *Rhodococcus erythropolis* IGTS8. *Journal of Bacteriology*. 178: 6409–6418.

Li, Q., Feng, J., Gao, C., Li, F., Yu, C., Meng, L., Zhang, Z., Ma, C., Gu, L., Wu, G., Xu P. (2012) Purification and characterization of a flavin reductase from the biodesulfurizing bacterium *Mycobacterium goodii* X7B. *Process Biochemistry*. 47:1144–1149

Liu, S., Zhang, C., Su, T., Wei, T., Zhu, D., Wang, K., Huang, Y., Dong, Y., Yin, K., Xu, S., Xu, P. and Gu, L. (2014) Crystal structure of DszC from *Rhodococcus* sp. XP at 1.79 Å. Proteins. 82:1708–1720.

Ma, C.Q., Feng, J.H., Zeng, Y.Y., Cai, X.F., Sun B.P., Zhang Z.B., Blankespoor, H.D., Xu P. (2006a). Methods for the preparation of a biodesulfurization biocatalyst using *Rhodococcus* sp. *Chemosphere*. 65: 165–169.

Ma, T. (2010) The desulfurization pathway in *Rhodococcus*: In Biology of *Rhodococcus*, Microbiology Monographs Series. Edited by Alvarez HM, Berlin Heidelberg: Springer, 16: 207–230.

Ma, T., Li, G.Q., Li, J., Liang, F.L., Liu R.L. (2006b) Desulfurization of dibenzothiophene by *Bacillus subtilis* recombinants carrying *dszABC* and *dszD* genes. *Biotechnology Letters*. 28:1095–1100.

Maghsoudi, S., Vossoughi, M., Kheirolomoom, A., Tanaka, E., Katoh, S. (2001) Biodesulfurization of hydrocarbons and diesel fuels by *Rhodococcus* sp. strain P32C1. Biochemical Engineering. 8:151–156.

Martínez, I., Santos, V.E., Alcón, A., García-Ochoa F. (2015) Enhancement of the biodesulfurization capacity of *Pseudomonas putida* CECT5279 by co-substrate addition. *Process Biochemistry*. 50:119–124.

Martínez, J., García, L., Díaz E. (2016) Genetic Engineering for Removal of Sulfur from Fuel Aromatic Heterocycles. Springer International Publishing AG 2016 S.Y. Lee (ed.), Consequences of Microbial Interactions with Hydrocarbons, Oils, and Lipids: Production of Fuels and Chemicals, Handbook of Hydrocarbon and Lipid Microbiology.

Matsubara, T., Ohshiro, T., Nishina, Y., Izumi Y. (2001) Purification, characterization, and over- expression of flavin reduced involved in dibenzothiophene desulfurization by *Rhodococcus erythropolis* D-1. *Applied and Environmental Microbiology*. 67:1179–1184

Matsui, T., Hirasawa, K., Koizumi, K.I., Maruhashi, K., Kurane R. (2001) Optimization of the copy number of dibenzothiophene desulfurizing genes to increase the desulfurization activity of recombinant *Rhodococcus* sp. *Biotechnology Letters*. 23:1715–1718.

Matsui, T., Noda, K., Tanaka, Y., Maruhashi, K., Kurane R. (2002) Recombinant *Rhodococcus* sp. strain T09 can desulfurize DBT in the presence of inorganic sulfate. *Current Microbiology*. 45: 240–244.

McFarland, L.B. (1999) Boidesulfurization. *Current Opinion in Microbiology*. 2: 257–264.

Mingfang, L., Zhongxuan, G., Jianmin, X., Huizhou, L., Jiayong C. (2003) Microbial desulfurization of model and straight-run oils. Journal of Chemical Technology and Biotechnology. 78: 873.

Mohebali, G., Ball A. S. (2008) Biocatalytic desulfurization (BDS) of petrodiesel fuels. *Microbiology*. 154: 2169–2183.

Mohebali, G., Ball A.S. (2016) Biodesulfurization of diesel fuels – past, present and future perspectives. *International Biodeterioration and Biodegradation*. 110:163–18.

Mohebali, G., Ball, A.S., Kaytash, A., Rasekh B. (2008) Dimethyl sulfoxide (DMSO) as an alternative sulfur source to DBT for the production of desulfurizing resting cells of *Gordonia alkanivorans* RIPI90A. *Microbiology*. 154: 878–885.

Mohebali, G., Ball, A.S., Rasekh, B., Kaytash A. (2007) Biodesulfurization potential of a newly isolated bacterium *Gordonia alkanivorans* RIPI90A. *Enzyme and Microbial Technology*. 40: 578- 584.

Monticello, D.J. (2000) Biodesulfurization and the upgrading of petroleum distillates. *Current Opinion in Biotechnology*. 11:540–546.

Monticello, D.J., Bakker, D.T., Finnerty, W.R. (1985) Plasmid- mediated degradation of dibenzothiophene by *Pseudomonas* species. *Applied and Environmental Microbiology*. 49:756–760.

Moreno-Ruiz, E., Hernáez, M.J., Martínez-Pérez, O., Santero, E. (2003) Identification and functional characterization of *Sphingomonas macrogolitabida* strain TFA genes involved in the first two steps of the tetralin catabolic pathway. *Journal of Bacteriology*. 185: 2026–2030.

Nakashima, N., Tamur, T. (2004) A novel system for expressing recombinant proteins over a wide temperature range from 4 to 35°C. *Biotechnology and Bioengi neering* . 86: 136–148.

Nakayama, N., Matsubara, T., Ohshiro, T., Moroto, Y., Kawata, Y., Koizumi, K., Hirakawa, Y., Suzuki, M., Maruhashi, K., Izumi, Y. and Kurane, R. (2002) A novel enzyme, 2′-hydroxybiphenyl- 2-sulfinate desulfinase (DszB), from a dibenzothiophene- desulfurizing bacterium *Rhodococcus erythropolis* KA2–5-1: gene overexpression and enzyme characterization. Biochimica et Biophysica Acta.1598:122–30.

Nassar, H.N. (2015) Development of biodesulfurization process for petroleum fractions using nano-immobilized catalyst. Ph.D. Degree. Al-Azhar University.

Nassar, H.N., El-Gendy, N.Sh., Abo-State, M.A., Moustafa, Y.M., Mahdy, H.M., El-Temtamy, S.A. (2013) Desulfurization of Dibenzothiophene by a Novel Strain *Brevibacillus invocatus* C19 Isolated from Egyptian Coke. *Biosciences, Biotechnology Research Asia*. 10(1): 29- 46.

Nekodzuka, S., Toshiaki, N., Nakajima-Kambe, T., Nobura, N., Lu, J., Nakahara, Y. (1997) Specific desulfurization of dibenzothiophene by *Mycobacterium* strain. *Biocatalysis and Biotransformation*.15:17–27.

Noda, K., Watanabe, K., Maruhashi K. (2002) Cloning of *Rhodococcus* promoter using a transposon for dibenzothiophene biodesulfurization. *Biotechnology Letters*. 25: 1875–1888.

Noda, K., Watanabe, K., Maruhashi K. (2003) Recombinant *Pseudomonas putida* carrying both the *dsz* and *hcu* genes can desulfurize dibenzothiophene in n-tetradecane. *Biotechnology Letters*. 25:1147–1150.

Nomura, N., Takada, M., Okada, H.., Shinohara, Y., Nakajima-Kambe, T., Nakahara, T., Uchiyama H. (2005) Identification and functional analysis of genes required for desulfurization of alkyl dibenzothiophenes of *Mycobacterium* sp. G3. *Journal of Bioscience and Bioengineering*. 100:398–402.

Ohshiro, T., Izumi Y. (1999) Microbial desulfurization of organic sulfur compounds in petroleum. *Bioscience, Biotechnology, and Biochemistry*. 63:1–9.

Ohshiro, T., Izumi Y. (2000) Purification, characterization and crystallization of enzymes for dibenzothiophene desulfurization. *Bioseparation*. 9: 185–188.

Ohshiro, T., Izumi Y. (2002) Desulfurization of fossil fuels, In Bitton, G. (ed.), Encyclopedia of environmental microbiology. Wiley and Sons, New York, p. 1041–1051.

Ohshiro, T., Aoi, Y., Torii, K., Izumi, Y. (2002) Flavin reductase coupling with two monooxygenases involved in dibenzothiophene desulfurization: purification and characterization from a non- desulfurizing bacterium, *Paenibacillus polymyxa* A-1. *Applied Microbiology and Biotechnology*.59: 649–657.

Ohshiro, T., Hine, Y., Izumi Y. (1994) Enzymatic desulfurization of dibenzothiophene by a cell-free system of *Rhodococcus erythropolis* D-1. FEMS Microbiology Letters.118: 341–344.

Ohshiro, T., Hirata, T., Izumi Y. (1995) Microbial desulfurization of dibenzothiophene in the presence of hydrocarbon. *Applied Microbiology and Biotechnology*. 44:249–252.

Ohshiro, T., Ishii, Y., Matsubara, T., Ueda, K., Izumi, Y., Kino, K., Kirimura K. (2005) Dibenzothiophene desulfurizing enzymes from moderately thermophilic bacterium *Bacillus subtilis* WU-S2B: purification, characterization and overexpression. *Journal of Bioscience and Bioengineering*. 100: 266–273.

Ohshiro, T., Kojima, T., Torii, K., Kawasoe, H., Izumi Y. (1999) Purification and characterization of dibenzothiophene (DBT) sulfone monooxygenase, an enzyme involved in DBT desulfurization, from *Rhodococcus erythropolis* D-1. *Journal of Bioscience and Bioengineering*. 88(6): 610–616.

Ohshiro, T., Ohkita, R., Takikawa, T., Manabe, M., Lee, W.C., Tanokura, M., Izumi, Y. (2007) Improvement of 2-Hydroxybiphenyl-2-sulfinate Desulfinase, an Enzyme Involved in the Dibenzothiophene Desulfurization Pathway, from *Rhodococcus erythropolis* KA2–5-1 by Site-Directed Mutagenesis. *Bioscience, Biotechnology, and Biochemistry*. 71:2815–2821

Ohshiro, T., Suzuki, K., Izumi Y. (1996) Regulation of dibenzothiophene degrading enzyme activity of *Rhodococcus erythropolis* D-1. *Journal of Fermentation a nd Bioengineering.* 81, 121–124.

Ohshiro, T., Suzuki, K., Izumi Y. (1997) Dibenzothiophene (DBT) degrading enzyme responsible for the first step of DBT desulfurization by *Rhodococcus erythropolis* D-1: purification and characterization. *Journal of Fermentation an d Bioengineering.* 83: 233–237.

Oldfield, C., Pogrebinsky, O., Simmonds, J., Olson, E.S., Kulpa, C.F. (1997) Elucidation of the metabolic pathway for dibenzothiophene desulfurization by *Rhodococcus* sp. IGTS8 (ATCC 53968). *Microbiology.* 143:2961–2973.

Omori, T., Monna, L.S., Kodama, T. (1995) Desulfurization of Alkyl and Aromatic Sulfides and Sulfonates by Dibenzothiophene Desulfurizing *Rhodococcus* sp. strain SY-1, *Bioscience, Biotechnology, and Biochemistry.* 59: 1195–1198.

Pan, J., Wu, F., Wang, J., Yu, L., Hassanzadeh, K., Stark, B.C., Kilbane, J.J. (2013) Enhancement of desulfurization activity by enzymes of the *Rhodococcus dsz* operon through coexpression of a high sulfur peptide and directed evolution. *Fuel.* 112:385–390.

Parravicini, F., Brocca, S., Lotti M. (2016) Evaluation of the Conformational Stability of Recombinant Desulfurizing Enzymes from a Newly Isolated *Rhodococcus* sp. Molecular Biotechnology. 58(1):1–11.

Patel, S.B., Kilbane, J.J., Webster D.A. (1997) Biodesulfurization of dibenzothiophene in hydrophobic media by *Rhodococcus* sp. strain IGTS8. *Journal of Chemical Technology and Biotechnology.* 69: 100 –106.

Petrella, P, De Ferra, F., Rodriguez, F., Luca, S., Elisabetta F. (2007) In vivo evolution of the *Rhodococcus* sp. strain DS7: selection of recombinants able to desulfurize both dibenzothiophene and benzothiophene. *Biocatalysis and Biotransformation.* 25:318–327.

Piddington, C.S., Kovacevich, B.R., Rambosek J. (1995) Sequence and molecular characterization of a DNA region encoding the dibenzothiophene desulfurization operon of *Rhodococcus* sp. strain IGTS8. *Applied and Environmental Microbiology.* 61:468–475.

Purdy, R.F., Lepo, J.E., Ward B. (1993) Biodesulfurization of organic-sulfur compounds, Current Microbiology. 27:219–222.

Radwan A. (2015) Genetic Improvement of Desulfurizing Bacteria. MSc Degree, Al-Azhar University.

Raheb, J., Rasekh, B., Irani, S., Hajipour, M.J., Mozaffari Tabatabaei, M., Kefayati, M.E. and Memari B. (2012) The study of biodesulfurization activity in recombinant *E. coli* strain by cloning the *dsz* genes involve in 4S pathway. *Journal of Sciences, Islamic Republic of Iran.* 22(3): 213–219.

Reichmuth, D.S., Blanch, H.W., Keasling T.D. (2004) Dibenzothiophene biodesulfurization pathway improvement using diagnostic GFP fusions. *Biotechnology and Bioengineering.* 88: 94–99.

Reichmuth, D.S., Hittle, J.L., Blanch, H.W., Keasling J.D. (1999) Biodesulfurization of dibenzothiophene in *Escherichia coli* is enhanced by expression of *Vibrio harveyi* oxidoreductase gene. *Biotechnology and Bioengineering.* 67: 72–79.

Rhee, S.K., Chang, J.H., Chang, Y.K., Chang, H.N. (1998) Desulfurization of dibenzothiophene and diesel oils by a newly isolated *Gordonia* strain CYKS1. *Applied and Environmental Microbiology.* 64: 2327–2331.

Santos, S.C., Alviano, D.S., Alviano, C.S., Goulart, F.R., de Pa′dula, M., Leitao, A.C., Martins, O.B., Ribeiro, C.M., Sassaki, M.Y., Matta, C.P., Bevilaqua, J., Sebastian, G.V., Seldin L. (2007) Comparative studies of phenotypic and genetic characteristics between two desulfurizing isolates of *Rhodococcus erythropolis* and the well- characterized *R. erythropolis* strain IGTS8. *Journal of Industrial Microbiology and Biotechnology.* 34: 423–431.

Schreinier, B.P., Stevens, S.E., Sing T. (1988) Oxidation of thianthrene by the ligninase of *Phanerochaete chysosporium. Applied Environmental Microbiology.* 54:1858–1860

Serbolisca, L., de Ferra, F., Margarit, I. (1999) Manipulation of the DNA coding for the desulfurizing activity in a new isolate of *Arthrobacter* sp. *Applied Microbiology and Biotechnology.* 52: 122–126

Shavandi, M., Sadeghizadeh, M., Khajeh, K., Mohebali, G., Zomorodipour, A. (2010) Genomic structure and promoter analysis of the *dsz* operon for dibenzothiophene biodesulfurization from *Gordonia alkanivorans* RIPI90A. *Applied Microbiology and Biotechnology.* 87:1455–1461

Shavandi, M., Sadeghizadeh, M., Zomorodipour, A., Khajeh K. (2009) Biodesulfurization of dibenzothiophene by recombinant *Gordonia alkanivorans* RIPI90A. *Bioresource Technology.* 100:475–479.

Soleimani, M., Bassi, A., Margaritis A. (2007) Biodesulfurization of refractory organic sulphur compounds in fossil fuels. *Biotechnology Advances.* 25: 570–596.

Sondossi, M., Barriault, D., Sylvestre, M. (2004) Metabolism of 2,2- and 3,3- dihydroxybiphenyl by the biphenyl catabolic pathway of *Comamonas testosteroni* B-356. *Applied and Environmental Microbiology.* 279: 174–181.

Sousa, S.F., Sousa, J.F., Barbosa, A.C., Ferreira, C.E., Neves, R.P., Rubeiro, A.J., Fernandes, P.A., Ramos M.J. (2016) Improving the biodesulfurization of crude oil and derivatives: a QM/MM investigation of the catalytic mechanism of NADH-FMN oxidoreductase (DszD). *The Journal of Physical Chemistry.* 120:5300–5306

Srinivasaraghavan, K., Sarma, P.M., La B. (2006) Comparative analysis of phenotypic and genotypic characteristics of two desulfurizing bacterial strains, *Mycobacterium phlei* SM120–1 and *Mycobacterium phlei* GTIS10. *Letters in Applied Microbiology.* 42(5): 483- 489.

Takahashi, S., Furuya, T., Ishii, Y., Kino, K., Kirimura K. (2009) Characterization of a flavin reductase from a thermophilic dibenzothiophene-desulfurizing bacterium, *Bacillus subtilis* WU-S2B. *Journal of Bioscience and Bioengineering.* 107:38–41.

Tanaka, Y., Matsui, T., Konishi, J., Maruhashi, K., Kurane R. (2002) Biodesulfurization of benzothiophene and dibenzothiophene by a newly isolated *Rhodococcus* strain. *Applied Microbiology and Biotechnology* 59:325–328.

Tao, F., Tang, H., Gai, Z., Su, F., Wang, X., He, X., Xu P. (2011b) Genome sequence of *Pseudomonas putida* Idaho, a unique organic-solvent-tolerant bacterium. *Journal of Bacteriology*. 193:7011–7012.

Tao, F., Yu, B., Xu, P., Ma C.Q. (2006) Biodesulfurization inbiophasic systems containing organic solvents. *Applied and Environmental Microbiology*. 72:4604–4609.

Tao, F., Zhao, P., Li, Q., Su, F., Yu, B., Ma, C., Tang, H., Tai, C., Wu, G., Xu P. (2011a) Genome sequence of *Rhodococcus erythropolis* XP, a biodesulfurizing bacterium with industrial potential. *Journal of Bacteriology*. 193:6422–6423.

Ting, M., Shanshan, L., Guoqiang, L., Renjing, W., Fenglai, L., Rulin L. (2006) Cloning and expressing DBT (dibenzothiophene) monooxygenase gene (dsz C) from *Rhodococcus* sp. DS-3 in *Escherichia coli*. *Frontiers of Biology in China*. 1(4): 375–380.

Torktaz, I., Etemadifar, Z., Derikvand P. (2012) Comparative modeling of DszC, an enzyme in biodesulfurization, and performing in silico point mutations for increasing tendency to oil. *Bioinformation*. 8:246–250.

Van Afferden, M., Schacht, S., Klein, J., Tüpper H.G. (1990) Degradation of dibenzothiophene by *Brevibacterium* sp. DO. *Archives of Microbiology*. 153: 324–328.

Wang, J., Butler, R.R., Wu, F., Pombert, J.F., Kilbane, J.J., Stark B.C. (2017) Enhancement of microbial biodesulfurization via genetic engineering and adaptive evolution. *PLoS ONE*. 12(1): 1–20.

Wang, J., Davaadelger, B., Salazar, J.K., Butler, R.R., Pombert J.F. (2015) Isolation and characterization of an interactive culture of two *Paenibacillus* species with moderately thermophilic desulfurization ability. *Biotechnology Letters*. 37:2201–2211.

Wang, P., Krawiec S. (1996) Kinetic analysis of desulfurization of dibenzothiophene by *Rhodococcus erythropolis* in batch and fed-batch cultures. *Applied and Environmental Microbiology*. 62: 1670–1675.

Wang, P., Humphrey, A.E., Krawiec S. (1996) Kinetic analysis of desulfurization of dibenzothiophene by *Rhodococcus erythropolis* in continuous cultures. *Applied and Environmental Microbiology*. 62: 3066–3068.

Wang, W., Ma, T., Lian, K., Zhang, Y., Tian, H., Ji, K., Li G. (2013) Genetic analysis of benzothiophene biodesulfurization of *Gordonia* terrae strain C-6. PLoS One. 8: e84386

Wang, Z., Wang, D., Li, Q., Li, W., Tang, H., Xing J. (2011) Enhanced biodesulfurization by expression of dibenzothiophene uptake genes in *Rhodococcus erythropolis*. *World Journal of Microbiology and Biotechnology*. 27:1965–1970.

Watanabe, K., Noda, K., Maruhashi, K. (2003b) Enhanced desulfurization in a transposon-mutant strain of Rhodococcus erythropolis. *Biotechnology Letters*25:1299–1304

Watanabe, K., Noda, K., Konishi, J., Marushashi K. (2003a) Desulfurization of 2,4,6,8-tetraethyl dibenzothiophene by recombinant *Mycobacterium* sp. strain MR65. *Biotechnology Letters*. 25:1451–1456.

Watanabe, K., Noda, K.I., Ohta, Y., Maruhashi, K. (2002) Desulfurization of light gas oil by a novel recombinant strain from *Pseudomonas aeruginosa. Biotechnology Letters*. 24: 897–903.

Watkins, L.M., Rodriguez, R., Schneider, D., Broderick, R., Cruz, M., Chambers, R., Ruckman, E., Cody, M., Mrachko G.T. (2003) Purification and characterization of the aromatic desulfinase, 2-(2′-hydroxyphenyl) benzenesulfinate desulfinase. *Archives of Biochemistry and Biophysics*. 415:14–23.

Xi, L., Squires, C.H., Monticell, D.J., Childs, J.D. (1997) A flavin reductase stimulates DszA and DszC proteins of *Rhodococcus erythropolis* IGTS8 in-vitro. *Bioc hemical and Biophysical Research Communications*. 230: 73–75.

Xiong, X., Xing, J., Li, X., Bai, X., Li, W., Li, Y., Liu H. (2007) Enhancement of biodesulfurization in two-liquid systems by heterogeneous expression of *Vitreoscilla hemoglobin*. Applied Environmental Microbiology. 73: 2394–2397.

Xu, P., Yu, B., Li, F., Cai, X.F., Ma C.Q. (2006) Microbial degradation of sulfur, nitrogen and oxygen heterocycles. *Trends in Microbiology*. 14:398–405.

Yoshikawa, O., Ishii, Y., Koizumi, K., Ohshiro, T., Izumi, Y., Maruhashi K. (2002) Enhancement and stabilization of desulfurization activity of *Rhodococcus erythropolis* KA2-5-1 by feeding ethanol and sulfur components. *Journal of Bioscience and Bioengineering*. 94: 447–452.

Yu, B., Tao, F., Li, F., Hou, J., Tang, H., Ma, C., Xu, P. (2015) Complete genome sequence of *Mycobacterium goodie* X7B, a facultative thermophilic biodesulfurization bacterium with industrial potential. *Journal of Biotechnology*. 212: 56–57.

Yu, B., Xu, P., Quan, S., Cuiqing M. (2006) Deep desulphurization of diesel oil and crude oils by a newly isolated *Rhodococcus erythropolis* srain. *Applied and Environmental Microbiology*. 72:54–58.

Zhang, L., Duan, X., Zhou, D., Dong, Z., Ji, K., Meng, W., Li, G., Li, X., Yang, H., Ma, T., Rao Z. (2014) Structural insights into the stabilization of active, tetrameric DszC by its C-terminus. *Proteins*. 82: 2733–2743.

Zhu, L., Li, M., Guo, S., Wang W. (2016) Draft Genome Sequence of a Thermophilic Desulfurization Bacterium, *Geobacillus thermoglucosidasius* Strain W-2. *Genome Announcements*. 4(4): 1–2.

8

Factors Affecting the Biodesulfurization Process

List of Abbreviations and Nomenclature

C_x^{max}	Maximum Biomass Concentration (g/L)
R^2_{adj}	Adjusted Correlation Coefficient
q_{O_2}	Specific Oxygen Uptake Rate (h^{-1})
q_p	Specific HBP production rate (h^{-1})
q_s	Specific DBT Consumption rate (h^{-1})
μ	Specific Growth Rate (h^{-1})
μ_{max}	Maximum Specific Growth Rate (h^{-1})
2,2'-BHBP	2,2'-Bihydroxybiphenyl
2-HBP	2-Hydroxybiphenyl
2HMBT	2-hydroxymethyl benzothiophene
2HMT	2-hydroxymethyl thiophene
2-MBP	2-Methoxybiphenyl
3-MBT	3-Methylbenzothiophene
4,6-DBDBT	4,6-Dibutyldibenzothiophene
4,6-DEDBT	4,6-Diethyldibenzothiophene

4,6-DMDBT	4,6-Dimethyldibenzothiophene
4,6-DPDBT	4,6-Dipropyldibenzothiophene
4HMBT	4-hydroxymethyl benzothiophene
4-MDBT	4-Methyldibezothiophene
7-EBT	7-Ethylbenzothiophene
7-HBT	7-Hexylbezothiophene
7-MBT	7-Methylbenzothiophene
7-PBT	7-Propylbenzothiophene
AH	Acid hydrolysis
ATP	Adenosine triphosphate
BBD	Box-Behnken Design
BDS	Biodesulfurization
BNT	Benzonaphthothiophene
BP	Biphenyl
BT	Benzothiophene
C*	Saturation Concentration of Dissolved Oxygen,
CCD	Central Composite Design
C_o	Initial Concentration of the Dissolved Oxygen
DBS	Dibenzylsulfide
DBT	Dibenzothiophene
DBTO	Dibenzothiophene Sulfoxide
$DBTO_2$	Dibenzothiophene Sulfone
DCW	Dry Cell Weight
D_{DBT}	The desulfurizing capability index ($\%g_{cell}$/L/h)
DGGE	Denaturing Gradient Gel Electrophoresis
DMF	Dimethylformamide
DMSO	Dimethylsulfoxide
DO	Dissolved Oxygen
DOE	Design of Experiment
DPS	Diphenylsulphide
E	Fractional Approach to Equilibrium
ED	Entner–Doudoroff
FAD	Flavin Adenine Dinucleotide
FFD	Fractionation Factorial Design
FWHM	Full Width at Half Maximum
Gr	Cell Growth (mg/L)
H/A	Hydrocarbon to Aqueous Phase Ratio
HB	Higher is Better
HBPS	2′-Hydroxybiphenyl-2-sulfinic acid
HCLN	High Carbon and Low Nitrogen

HGO	Heavy Gas Oil
HN	High nitrogen
JA	Jerusalem Artichoke
JAJ	Jerusalem Artichoke Juice
JAJ_t	Treated Jerusalem Artichoke Juice
k_{cat}	Turnover Number (min^{-1})
k_La	Gas Liquid Mass Transfer Coefficient (s^{-1})
K_m	Saturation Constant or Michaelis constant (g/L)
K_s	Monod Saturation Constant (g/L)
LB	Lower is Better
LN	Low nitrogen
MBC	Minimum bacteriocidal concentration
MIC	Minimum Inhibitory Concentration
NAD	Nicotinamide Adenine Dinucleotide
NADH	Nicotinamide Adenine Dinucleotide, reduced form
NADP	Flavin Adenine Dinucleotide Phosphate
NADPH	Flavin Adenine Dinucleotide Phosphate, reduced form
NB	Nominal is Better
O/W	Oil/Water
OA	orthogonal arrays
OFP	Organic Fraction Phase
OSCs	Organosulfur compounds
OTR	Oxygen Transfer Rate
OUR	Oxygen Uptake Rate
PASHs	Polyaromatic Sulfur Heterocyclic Compounds
$P_{C/O}$	Partition Coefficients between the Biocatalyst and the Oil Phases
$P_{C/W}$	Partition Coefficients between the Biocatalyst and the Aqueous Phases
$P_{M/W}$	Partition coefficient between membrane and water
$P_{O/W}$	Partition coefficient of a solvent
Q_{DBT}	Maximum Value of the Specific Desulfurization Rate (mmol/g_{cell}/h)
q_{HBP}	2-HBP specific production rate
R^2	Correlation Coefficient
RMSE	Root Mean Square Errors
RPS	Recycled Paper Sludge
RSM	Response Surface Methodology
S/N	Signal to Noise Ratio
SBM	Sugar Beet Molasses

SFM	Sulfur Free Medium
SSE	Sum of Squares Errors
SSF	Simultaneous Saccharification and Fermentation
STBR	Stirred Tank Bioreactor
TCA	Tricarboxylic Acid Cycle
T_G	Generation Time
Th	Thiophene
U_g	Superficial Gas Velocity (L/min)
ULS	Ultra-low sulfur
ULSD	Ultra-low sulfur diesel
X_{BDS}	Biodesulfurization yield
Y_{DBT}	2-HBP Yield as a Measure for BDS Efficency
YE	Yeast Extract
v	Working Volume (L)

8.1 Introduction

There are two pathways involved in the biodesulfurization (BDS) pathways: a ring destructive one that is in the C-C cleavage via the Kodama pathway and complete mineralization pathway and the non-destructive pathway that is in the C-S cleavage, which is the 4S-route (Chapter 6). The recommendable pathway for the petroleum industry is the 4S-pathway, retaining the calorific value of fuel.

It has been frequently reported in the literature that there are several factors affecting BDS rate and efficiency: substrate diffusion problems, the presence of inhibition effects, and the necessity of cofactors re-generation, but the cell membrane transport rate, the reduced cofactors concentration, and HpaC Flavin reductase activity are not the only factors limiting the BDS rate (Alcon *et al.*, 2005; 2008). Some points other than the main physicochemical factors are also important to achieve a successful BDS process with a high desulfurization rate. For example, resting cells are preferable over growing cells (Chang *et al.*, 2000). Further, the immobilization of biocatalysts helps in easy separation and reusability of the biocatalysts (Ansari *et al.*, 2007). The biphasic reaction systems are the ones to be applied in the real industry, enhancing the solubility and availability of hydrophobic substrates to the biocatalyst and limiting biocatalyst inhibition by hampering the accumulation of 2-HBP. However, the viability of bacteria would be affected by the toxicity of solvent, mass, and oxygen transfer (Marcelis *et al.*, 2003; Kawaguchi *et al.*, 2011). The pH, process temperature, biomass concentration, oil/water O/W ratio, and initial S-concentration are other important factors that should be considered for a successful BDS process.

It should also be noted that the physicochemical parameters affecting the BDS capacity interact with each other and most of the published studies investigated different factors and discussed their mutual effect, not only their individual effects.

This chapter illustrates the different physicochemical parameters affecting the BDS-process and summarizes the worldwide studies for enhancing its rate.

8.2 Effect of Incubation Period

It is very important to know the time at which the biocatalyst would give the highest BDS efficiency. The shorter the time, the higher the BDS rate, and the more beneficial it will be for the industrial application and commercialization of BDS-process.

The time dependent desulfurization of DBT into 2HBP has been studied for *Rhodococcus erythropolis* IGTS8, *Rhodococcus erythropolis* D-1, *Rhodococcus erythropolis* KA2–5-1, *Mycobacterium* sp. G3, *Rhodococcus* sp. P32C1, *Bacillus subtilis* WU-S2B, *Mycobacterium phlei* WU-F1, and *Paenibacillus* sp. All-2 (Kilbane and Bielage, 1990; Li *et al.*, 1996; Ohshiro *et al.*, 1996a; Konishi *et al.*, 1997; Oldfield *et al.*, 1997; Ishii *et al.*, 2000; Maghsoudi *et al.*, 2001; Kirimura *et al.*, 2001; Furuya *et al.*, 2001a).

Yoshikawa *et al.* (2002), Okada *et al.* (2002), and del Olmo *et al.* (2005a) reported that the capacity of *R. erythropolis* IGTS8 expressed its maximum DBT-BDS capacity in shaken flask cultures at the end of the exponential phase of growth of the cells (between 15 and 25 h of growth time). In another study, high values of BDS capacity were maintained for cells cultured during longer periods when dissolved oxygen (DO) concentration was constant and equal to 10% of saturation. The maximum BDS capacity was similar for stirrer speeds of 250 and 400 rpm, reaching approximately 80% for 30 h of growth. Afterwards, this capacity slightly decreased. However, upon working at 150 rpm, this decrease took place at a faster interval as a consequence of the oxygen transport limitation during growth (Gomez *et al.*, 2006a). Other researchers reported that it is not a must that the time for maximum DBT removal be the same for obtaining maximum growth and 2-HBP. This phenomenon is mainly related to the feedback inhibition of the end products of BDS. Thus, it is important during the optimization of a BDS to know the incubation period that would give appropriate microbial growth with maximum enzymatic activity and BDS-efficiency.

Okada *et al.* (2002) reported the methoxylation of 2-HBP to 2-methoxybiphenyl (2-MBP) in the DBT-BDS using *Mycobacterium* strain G3. The time course of methoxylation and desulfurization in a growing cell culture has been investigated. The desulfurization activity was induced at the end of the exponential phase and the highest level of desulfurization activity was observed in the middle of the stationary phase. On the other hand, the methoxylation activity was observed in the early growth phase, up to 4 d after inoculation, and decreased in the middle of the stationary phase. This indicated that both activities were expressed by the independent regulation systems. In a 1/1 (O/W) biphasic batch BDS process of model oil (100 mM 4,6-dipropylDBT in n-tetradecane) using resting cells of *Mycobacterium* strain G3, the desulfurization recorded 4% at 37 °C within 2 h and, by increasing the incubation period to 15 h, the desulfurization rate of 4,6-dipropyl DBT increased by approximately three-fold.

Wang *et al.* (2004) reported the maximum desulfurization activity for a batch DBT-BDS process occurred at a late exponential growth phase of *Corynebacterium* sp. ZD-1 and recommended the preparation of active resting cells of ZD-1 in that period. Verma *et al.* (2016) reported maximum DBT removal using growing cells of *Bacillus* sp. E1 within 72 h and remained sustained thereafter, while maximum 2-HBP occurred within 48 h recording 1.103 mM and decreased at a longer incubation period. This was attributed to the transformation of 2-HBP to biphenyl and 2-methoxybiphenyl. Zakharyant *et al.* (2004) reported that the enrichment of *Rhodococcus erythropolis* Ac-1514D and *Rhodococcus ruber* Ac-1513D, in a selective liquid medium containing DBT before their inoculation into a batch DBT-BDS process, decreased the lag phases from 120–122 h to 20–22. That is approximately a six-fold decrease. *R. ruber* Ac-1513D recorded about 63–65% DBT removal from the initial concentration of 0.43 mM within 80 h, while *R. erythropolis* recorded about 78–80%.

In a batch DBT-BDS using growing cells of *Lysinibacillus sphaericus* DMT-7, Bahuguna *et al.* (2011) reported 97% BDS of 0.2 mM DBT within 30 d with the production of only 121 μmol/L 2-HBP after 15 d of incubation which remained nearly sustained recording 122.5 μmol/L after 30 d. The onset of 2-HBP production occurred within 24 h, recording 5.1 μmol/L. A sharp decrease in DBT occurred within 15 d, recording 77% out of which 60% desulfurized to 2-HBP. However, an exponential increase in the growth of DMT-7 occurred up to 12 d. Bahuguna *et al.* (2011) also reported that the induction period prior the application of the microorganisms in BDS process is recommended. Upon the performance of a comparative study for a batch BDS of DBT (0.2 mmol/L) with *Lysinibacillus sphaericus* DMT-7 previously induced by pre-growing on 0.2 mmol/L DBT for 7 d and a non-induced one by pre-growing DMT-7 on 0.5 mmol/L $MgSO_4$ for 7 d, a difference in the onset of 2-HBP production was observed. The DBT-BDS to 2-HBP started within 6 h, producing 5.2 μmol/L. Then, the concentration of 2-HBP in the cultures increased linearly up to 107 μmol/L up to 5 d, attaining maximum production of 129.2 μmol/L of 2HBP (65% desulfurization) within 10 d of growth and thereafter it became stationary. In contrast to the induced culture, the non-induced DMT-7 cultures, pre-grown on $MgSO_4$, required a prolonged lag phase of 24 h and 15 d for the accomplishment of a maximum DBT desulfurization of 60% accompanied by production of 119.5 μmol/L 2-HBP. The inducibility of desulfurization pathways has been also reported by Kayser *et al.* (1993), Ohshiro *et al.* (1996a), and Ohshiro and Izumi (1999).

Maass *et al.* (2015) reported that modeling of the phenomena associated with logarithmic phase appears to be the most important in the BDS experiments since the highest rates of DBT removal and 2-HBP production occur within the growth phase. The same was reported in this study, where a high BDS rate occurred during the logarithmic phase regardless of the initial DBT concentration. Nassar *et al.* (2017a) also reported the sharp depletion of DBT within 72 h of incubation by *Rhodococcus erythropolis* HN2 during the logarithmic growth phase regardless of the initial DBT concertation. The absence of a lag phase in all cultures of the studied initial DBT concentrations (100–1000 ppm) was attributed to the adaptation of NH2 to high concentrations of DBT as it was isolated from an Egyptian coke sample of high S-content (2.74%) and previously enriched on 1000 ppm DBT (El-Gendy *et al.*, 2014). Nevertheless, it is not clear up till now, whether the decrease in activity at longer incubation periods is due to the exhaustion of a carbon source or due to accumulation of 2-HBP and other by-products which express some toxicity to the microorganism.

8.3 Effect of Temperature and pH

Several reports have been published concerning the effects of operational temperature and pH.

Enzymes are reported to be affected by the pH variations in the system because of the dependence of their 3D shape on the pH. The impact of pH variation is not only the on shape of enzymes, but it also affects the electrical charge properties of the substrate in such a way that the substrate cannot bind to the enzyme active site. Thus, the enzyme cannot undergo its catalytic activity (Berg *et al.*, 2002). Barrios (2011) reported that the change in pH has a great effect on the microbial metabolism, as well as the solubilization and adsorption/desorption of ions and hydrocarbons to be metabolized.

Changes in pH would occur due to the production of metabolites. Thus, it should be controlled sometimes with diluted NaOH solution or tris-HCl buffer. There would be a drop in pH if it is not controlled. Most reports apply 30 °C and pH values between 6–5 – 7.5 (Omori *et al.*, 1995; Patel *et al.*, 1997; Ohshiro *et al.*, 1998; Folsom *et al.*, 1999; Kaufman *et al.* 1999; Setti *et al.*, 1999; Abbad-Andaloussi *et al.*, 2003; Maghsoudi *et al.*, 2001; Matsui *et al.*, 2001; Ardakani *et al.*, 2010; Derikvand *et al.*, 2014). Lu *et al.* (2003) reported that the BDS efficiency of the lyophilized cells of *Pseudomonas delafieldii* R-8 is not affected by the change in pH (4.6–8.5) which would be favorable for the fuel desulfurization. However, upon the application of growing cells of R-8, the growth and, consequently, the BDS activity were inhibited with an initial pH below 5. Kim *et al.* (2004) reported the highest growth and BDS efficiency of *Gordonia* sp. CYKS1 is at pH 7–8, while at a pH < 6 neither a significant desulfurization of DBT nor an observable cell growth occurred.

There are various reported DBT-desulfurizing microorganisms that work via the 4S-pathway that can be categorized according to the optimal working temperature as mesophilic (Patel *et al.*, 1997; Ohshiro *et al.*, 1996b; Kirimura *et al.*, 2001; Wang *et al.*, 2004; Nassar *et al.*, 2017b), moderately thermophilic (Furuya *et al.*, 2003), and hyper-thromophilic (Kargi and Robinson, 1984; Gün *et al.*, 2015). However, most of the reported BDS processes are working at 30 °C. Furuya *et al.* (2006) reported that the first and third enzymes in the 4S-pathway ($D_{SZ}C$ and $D_{SZ}B$) are more sensitive to temperature changes, compared to other enzymes, and are BDS-rate limiting. Most of the isolated *Rhodococcus* strains that are capable for DBT-BDS throughout the 4S-pathway are reported to be mesophilic, including *R. erythropolis* IGTS8 (Li *et al.*, 1996), *R. erythropolis* D-1 (Ohshiro *et al.*, 1996a), and *Rhodococcus* sp. strain P32CI (Maghsoudi *et al.*, 2001).

Furuya *et al.* (2001b) reported that during the BDS of DBT by the thermophilic *Mycobacterium phlei* WU-F1, conversion of DBT to $DBTO_2$ would be stopped at high temperatures, while the conversion of $DBTO_2$ and other intermediate compounds would continue within the same high temperature. Furuya *et al.* (2001b) concluded that the activity of the first enzyme in the 4S-pathway, which oxidizes DBT to $DBTO_{2,}$ is more sensitive to temperature variations and is considered as the BDS rate-limiting enzyme.

Jiang *et al.* (2002) studied the effect of initial pH (2.9–10.29) and incubation temperature (20–40 °C) on the BDS-activity of *Pseudomonas delafieldii* R-8 and the optimum pH and temperature were 7.2 and 32 °C. Xu *et al.* (2002) studied the effects of different initial pH on batch BDS of 0.5

mmol/L DBT by *Rhodococcus*. sp. 1awq strain within 48 h and found 30 °C and pH7 to be the best.

Martin *et al.* (2005) reported the great influence of incubation temperature on the growth and BDS activity of *Pseudomonas putida* CECT5279. The highest values of the growth parameters, such as specific growth rate μ (h^{-1}) and maximum biomass concentration C_x^{max} (g/L), are obtained when the cells were grown at 26 °C and these parameters reached their lowest values at 32 °C. Nevertheless, the best BDS-capability was obtained when the bacterium growth occurred at 30 °C.

The effect of incubation temperature (25, 30, 45 °C) and initial pH (4–10) on batch DBT-BDS (0.2 mM in ethanol) using growing cells of *Rhodococcus* spp Eu-32 has been studied (Akhtar *et al et al.*, 2009) and showed that isolate Eu-32 is a neutrophilic mesophile and the maximum BDS-efficiency was recorded within 72h at a temperature of 30 °C and pH 7.0.

Arabian *et al.* (2014) proved that *Bacillus cereus* HN performs good BDS capacity over a wide range of temperature (30–40 °C), but the optimum is 40 °C, while Praveen Reddy and Umamaheshwara Rao (2015) reported the isolation of two DBT-biodesulfurizing *Streptomyces* species, *Streptomyces* sp. VUR PPR 101 and *Streptomyces* sp. VUR PPR 102, from oil contaminated sites with a maximum growth at 30 °C and pH 7.

In a study performed for the optimization of bacterial growth and BDS efficiency of *R. erythropolis* HN2, it was observed that the optimum pH and temperature for microbial growth (pH 7 and 30 °C) were slightly different from that obtaining maximum BDS efficiency (pH 6.69 and 27.47 °C). This was explained by the ability of HN2 to survive and propagate on the available added co-substrates, 0.09 M glycerol and 0.35 g yeast extract (Nassar *et al.*, 2017). Similar observation was reported by Konishi *et al.* (1997) and ascribed this phenomenon in part to the weak desulfurizing activities of cell enzymes at such low temperatures. Nassar *et al.* (2017b) attributed the recorded large decrease in cell growth and, consequently, the decrease in the BDS activity of HN2 at higher incubation temperatures (> 30 °C) to the solubility of DBT. Moreover, at a low temperature (<25 °C), the insoluble DBT was not highly available for the microbial cells to induce its enzymatic activity. Although DBT is known to be an intrinsically xenobiotic water insoluble compound (Marzona *et al.*, 1997), the water solubility of DBT is reported to increase with the increase in temperature (Kayser *et al.*, 2002) and, according to Kim *et al.* (2004), the water-soluble compounds play an inhibitory role on cell growth and enzyme activities.

A study performed by Gunam *et al.* (2016), in biphasic batch biodesulfurization (1:5 O/W) of model oil 200 ppm DBT, in n-tetradecane using resting cells of *Agrobacterium tumefaciens* LSU20 of age 4 d at 150 rpm, showed the highest BDS rate at 37 °C and pH7 in the presence of glucose as a C-source, recording a DBT-removal percentage of 76.9% within 96 h. Similarly, Gunam *et al.* (2012) reported a DBT removal percentage of approximately 75.21%, using *Pseudomonas* sp. strain KWN5, under the same aforementioned conditions of LSU20, but KWN5 expressed nearly the same BDS efficiency within a pH range of 6.5–7.5, recording ≈ 72–75%, but KWN5 lost about 40% of its activity at pH < 5.5.

8.4 Effect of Dissolved Oxygen Concentration

Biodesulfurization (BDS) is the generic term which defines all processes where microorganisms catalyze the desulfurization reaction and remove the recalcitrant S-compounds under mild pressures and temperatures (Monticello and Finnerty, 1985). It has been described that aerobic, anaerobic, and facultative anaerobic microorganisms effectively degrade DBT and its analogs (McFarland, 1999), but the anaerobic BDS is characterized by a slow rate (Ohshiro and Izumi, 1999). The BDS capacity and intracellular enzyme activities are highly affected by the dissolved oxygen concentration in the broth. The 4S-BDS pathway is highly sensitive to oxygen availability even under oxygen limiting conditions (Martinez *et al.*, 2016). Del Olmo *et al.* (2005b) reported that the concentration of the dissolved oxygen can be controlled by the agitation speed. However, sparging of gas increases turbulence and bubble release at the liquid surface produces dramatic changes in local velocity driven by surface tension. Consequently, it would have a negative influence on the performance of suspended cells. Thus, the mixing and oxygen transfer rates in a bioreactor depend on the power dissipated by agitation and aeration. If the stirrer speed was restricted to avoid shear effects, mixing or mass transfer would limit the performance of the culture. On the other hand, if agitation in the bioreactor exceeds a certain level, the hydrodynamic forces can affect the cells. Consequently, the optimal situation will take place when the mixing and mass transfer rates satisfy the oxygen and nutrient uptake rates by the cells and the overall process rate is determined by the cell metabolism (Gomez *et al.*, 2015; Escobar *et al.*, 2016).

It is worthy to know that a correct understanding of the role of oxygen transfer and uptake rates (OTR and OUR, respectively) in aerobic BDS processes is a crucial aspect to the design, operation, and scale-up of bioreactors (Garcia-Ochoa and Gomez., 2009).

In a study performed by del Olmo *et al.* (2005b) on *Rhodococcus erythropolis* IGTS8, the highest growth rate and maximum biomass concentration were obtained by culturing at 30 °C and pH 6.5 controlled by a diluted NaOH addition and a dissolved oxygen concentration constant value of 20% saturation. Under these conditions, optimal BDS capability was also developed by the cells under the same operating conditions although similar BDS capability was also obtained when pH was maintained at 6.5 using either diluted NaOH or Tris HCl buffer. On the other hand, high values of BDS were maintained during more growth time when dissolved oxygen was maintained at a constant value of 10% of saturation.

It has been reported that the dissolved oxygen concentration is a key factor in the flow of source carbon for cell growth and biodesulfurization capacity of *R. erythropolis* IGTS8 and that both may be modified by power input. This not only affects the growth rate, but also the enzymes developed by the cells which are afterwards used in BDS, being measured by resting cell assays (Gomez *et al.*, 2006a). In a study performed by Gomez *et al.* (2015) on *R. erythropolis* IGTS8, it was observed during the growth curve on DBT at a different agitation speed, the dissolved oxygen (DO) concentration decreased to reach a minimum value corresponding to a point halfway through the exponential growth phase where the demand for oxygen by the cells is the highest. However, the change in oxygen concentration with time was lower with the increase in agitation speed. Moreover, under very low agitation conditions (100–150 rpm), the DO concentration reached nearly 0% only after a few hours of growth (between 12 and 15 h) remaining in that position until the end of the growth stage. The culture is oxygen limited under these operating conditions and the oxygen transfer rate controls the overall process rate and, consequently, the growth rate increases when the stirrer speed is increased. The specific growth rate and maximum cell growth were low at agitation speeds < 250 rpm. This was attributed to the fluid dynamic conditions in the bioreactor under that condition, resulting in the oxygen transport rate being insufficient (lower than the maximum uptake rate) and the DO becoming the limiting nutrient. It increased and remained nearly constant within an agitation speed of 250- 450 rpm, recording approximately 0.26 h^{-1} and 1.8 / L, respectively. Then, both decreased at a higher agitation speed recording 0.146 h^{-1} and 1.018 g/L at 700 rpm. Since the cells were adversely affected by the intense mechanical agitation, this proved that either hydrodynamic forces or oxygen level are affecting the cell metabolism, which can be detected by a decrease in the specific oxygen uptake rate, q_{O_2}.

The growth rates of *P. putida* CECT5279 (Gomez *et al.*, 2006b) and KTH2 (Escobar *et al.*, 2016) are reported to be strongly dependent on the availability of dissolved oxygen concentration. This, consequently, affects the activities of the intracellular enzymes involved in the 4S-pathway. Escobar *et al.* (2016) proved the inactivation by oxidation of DszB desulfinase when the OTR increases, that is increasing the dissolved oxygen concentration available for the cells.

As the aeration in a bioreactor is somehow linked with the hydrodynamic damage to suspended cells (Chalmers, 1994; Garcia *et al.*, 2001), Gomez *et al.* (2015) studied the effect of the gas flow rate on *R. erythropolis* IGTS8 biomass growth in the STBR at a constant stirrer rate of 250 rpm and varied gas flow rate of 1–10 L/min. This revealed that a constant specific growth rate of μ and maximum growth C_x^{max} of 0.26 h^{-1} and 1.73 g/L, respectively, and the evolution of dissolved oxygen (DO) concentrations were similar over the different studied air flow rates. Therefore, the change of the aeration rate between 1 and 10 L/min does not affect the specific growth rate and no hydrodynamic stress effect is detected under these conditions.

8.5 Effect of Agitation Speed

The effect of the mixing rate (that is the agitation speed or shaking speed) is a very important factor in bioprocesses. One of the challenges facing the scale-up of the BDS process is the turbulence effect. The increase of the power dissipation per volume unit, usually caused by an increase of the agitation, is often beneficial in most bioprocesses due to the improvement of mass transfer and mixing rates. However, excessive agitation causes a concomitant increase in hydrodynamic forces, which may become a negative factor due to the interaction of turbulence with living cells, also called hydrodynamic stress. These restrictions have great influence on the bioreactor operating conditions, the scale-up criteria, and the selection of the type of bioreactor to be used.

The shear effects of flow turbulence inside the bioreactor have a great influence on the cellular response, their morphology (Märkl *et al.*, 1991; Sahoo *et al.*, 2003; Chisty, 2010), their metabolic and enzymatic activities (Hewitt *et al.*, 1998; Sahoo *et al.*, 2003), and, consequently, affects the growth (Sahoo *et al.*, 2003; Hodaifa *et al.*, 2010), the production (Calik *et al.*, 2004; Olmos *et al.*, 2013), and the nutrient consumption rates (Prokop and Bajpai, 1992; Olmos *et al.*, 2013). The effects of agitation can be evaluated in terms of the threshold values of either shear rate, shear stress, or stirrer speed, above which the growth rate and cell viability (Meijer *et al.*, 1994; Yepez and Maugeri, 2005; Kao *et al.*, 2007), product yield (Kao *et al.*, 2007; Olmos *et al.*, 2013), or some other critical culture parameters are significantly affected (Bronnenmeier and Markl, 1982; Arnaud *et al.*, 1993).

It is important to reach the optimal agitation level which yields the optimal growth rate or biomass concentration with the optimal enzymatic activities. For cell suspension cultures, the independent effect of agitation is difficult to quantify as it is coupled with various other phenomena (mixing and oxygen transfer rates and hydrodynamic stress, mainly). In many cases, microbial growth rate is claimed to increase by improving the oxygen transfer rate, therefore, the positive effect of agitation is asserted, which is also able to compensate for the possible cell damage by shear stress. When understanding the influence of hydrodynamic conditions in the bioreactor on the biocatalyst production of biodesulfurizing strains, it is important to separate the positive influence of the transport rate from the possible damage due to shear effects. Thus, for an adequate description of the system, it is necessary to know the dependence of the oxygen transfer coefficient on fluid dynamics and the evaluation of the associated physical and biological parameters.

Wang *et al.* (2006) reported the maximum BDS efficiency of model oil (1 mmol/L DBT in n-hexadecane) by resting cells of *Corynebacterium* sp. ZD-1 was 1:2 (O/W) and 250 r/min. This was discussed at lower rotation rate, where increasing the shake rate improves the mixing of the two phases, thus accelerating the transfer of DBT to the cells in the water phase and enhancing the transport of oxygen as well. Thus, the concentration of 2-HBP increased with the increasing rotation rate before it reached 250 r/min, but at a higher rotation rate, the mass-transfer limited the BDS process and turned it to a controlled reaction. Moreover, too high a rotation rate would be harmful to the bacteria.

Abin-Fuentes *et al.* (2013) applied a mixing speed of 500 rpm to minimize the mass transport limitations in a batch BDS using *R. erythroplos* IGTS8 in a bioreactor with a bi-phasic system (aqueous/model oil of DBTin n-hexadecane).

Gomez *et al.* (2015) studied the effect of agitation speed on growing cells of *R. erythropolis* in a stirred tank and shaken flasks. The growth rate was found to be increased when the stirrer speed increased between 100 and 450 rpm. However, at higher stirrer speeds between 500 and 700 rpm, the cell growth rate was sharply decreased. This decrease, with a concomitant decrease in μ and C_x^{max}, indicated that the hydrodynamic forces or the oxygen level affected the cell metabolism and can be detected by a decrease of the specific oxygen uptake rate, q_{O_2} (Garcia-Ochoa *et al.*, 2015). Not only this, but Calzada *et al.* (2009a,b) proved that this behavior of metabolic changes also affects the BDS-capacity, which is directly related to the 4S-route enzyme (DszA, DszB, DszC and DszD) activity. Gomez *et al.* (2015) concluded that the values of the growth parameters obtained for the runs of *R. erythropolis* IGTS8, when the cells are cultured in shaken flasks (in orbital shaker at 210 rpm) and in the bioreactor (STBR) between 250 and 450 rpm, were very similar. Those recorded were 0.26 h^{-1} for the specific growth rate and 1.8 g/L for the maximum biomass concentration. Thus, the specific growth rate in shaken flasks at 210 rpm in orbital shaker is the same as that found in STBR between 250 and 450 rpm, with the air flow rate used (2 L/min) and the tank, stirrer, and gas distributor geometry employed in that study. Therefore, fluid dynamic conditions must be similar. Unless ptherwise therwise stated, the relationship between OTR, OUR, and the shear stress must be similar in both types of operations.

Upon studying the effect of agitation speed on the growth and BDS capacity of *P. putida* KTH2 (Escobar *et al.*, 2016) in a stirred tank bioreactor (STBR), the obtained experimental results indicated that cultures were influenced by the fluid dynamic conditions in the bioreactor. An increase in the stirrer speed, from 400 to 700 rpm, expressed a positive influence on the cell growth rate. This was attributed to the consequence of the combined effect of an improvement of the mass transfer rate and a decrease in the resistance to the transport of nutrients to cells, mainly because the culture is oxygen limited (that is under aerobic conditions). However, the increase of agitation from 700 to 2000 rpm hardly expressed any influence on the growth rate. Thus, it was concluded that no hydrodynamic stress was observed for *P. putida* KTH2 until 2000 rpm, where the kinetic growth parameters, increased with the stirring speed from 400–600 rpm and remained constant between 700–2000 rpm. Thus, the μ_{max} and C_x^{max} increased from 0.439 to 0.520 h^{-1} and from 2.983 to 4.916 g/L when the speed increased from 400 to 700 rpm, respectively. Although, other microorganisms, such as *R. erythropolis* IGTS8, are reported to suffer this stress on growth at much lower stirrer speeds (Gomez *et al.*, 2015). The slightly lower values of growth parameters at relatively lower speeds (< 700 rpm) was attributed to the amount of dissolved oxygen, as the oxygen transfer rate is the limiting step of the overall growth process. The consumption rate of nutrients, such as glycerol and glutamic acid, also increased with a speed of 400–700 rpm. The effect of fluid dynamics on the cell development of the BDS capacity of the cells during the cell growth was different and, consequently, the activities of the intracellular enzymes involved in the 4S-pathway changed with the dissolved oxygen concentration. The enzyme activities have been evaluated in cells at several growth times and different hydrodynamic conditions. An increase of the agitation from 100 to 300 rpm had a positive effect on the overall BDS capacity of the cells during growth. However, the BDS capacity decreased at higher stirrer speeds, the activity of the enzyme monooxygenases DszC and DszA was dramatically decreased, and the highest value of the activity of the DszB enzyme was obtained with cells cultured at 100 rpm and decreased at a higher speed, becoming the step that controls the overall rate of the process (that is to say, the rate limiting step). Escobar *et al.* (2016) proved that the desulfinaze enzyme $D_{SZ}B$ is adversely affected at two levels: by the growth time of the cells and by the increase of the stirring speed during growth. Thus, at higher growth agitation speed, with higher improvement in the oxygen transferred in the culture, although *P. putida* KTH2 showed an improved specific growth rate, it expressed a dramatic decrease in the BDS capacity.

The BDS efficiency of heavy crude oil was reported to increase with the increase of rotation speed, recording 48% and 76% at 180 and 250 rpm, respectively, in shaken flasks in a batch BDS with a 1/3 O/W ratio using growing cells of *Gordonia* sp. IITR100 (Adlakha *et al.*, 2016). However, upon the application in a pilot study, commercial diesel oil with an initial S-content of 50 ppm in 5 L bioreactor of 2 L working capacity using a 1/3 O/W phase ratio and 15 g/L sucrose was used as a carbon source. The impeller speed was optimized to ascertain uniform dispersion of oil droplets in an aqueous medium and obtain a representative sample using different agitation speeds of 300–800 rpm. Though the dispersion as measured by uniformity of the oil in the outlet samples was found to be satisfactory at the stirrer speed of 400 rpm, it was not considered adequate for proper oxygen and other nutrient transport. Therefore, the range was narrowed down with a higher starting agitation (500–625 rpm). Higher agitation rates (> 625 rpm) were not considered as they might lead to cell damage by shear stress without giving any significant increase in the transport of nutrients. Agitation at 600 rpm was found to be optimal and the S-content reduced to 15 ppm within 3 days and no further removal occurred at longer incubation periods. However, the growth continued after 3 days and that was explained by the presence of sulfur released in an aqueous phase and with a residual carbon source which could have still supported the growth. Reduction in sulfur was also accompanied by increased bisphenol levels. Sucrose concentrations decreased from about 15 g/L to 5.6 g/L by the end of the 4th day and no further decrease was observed thereafter.

Peng and Zhou (2016) studied the effect of an initial biomass concertation (8.4 and 25.6 g DCW/L) and agitation speed (300 and 600 rpm) on a batch BDS of crude oil (0.3 O/W) using resting cells of *Rhodococcus* sp. MP12 that were pregrown on DMSO. The maximum desulfurization activity, which was measured as sulfate production rate (μmole sulfate/g DCW/h), occurred within the first 48 h and then decreased with further increments of the incubation period. The low concentration of initial biomass expressed higher BDS efficiency than the higher biomass concentration. The best was 8.4 g DCW/L, however, the maximum desulfurization activity of high density cells (25.6 g DCW/L) increased sharply by approximately 90% when the agitation speed was raised from 300 to 600 rpm. This was explained by a higher agitation speed of 600 rpm better breaking up high density cell aggregates and, thus, improving the mass and oxygen transmission due to decreased aggregate size, lower fraction of cells in aggregates, and higher fractions of free and oil-adhered cells in a crude oil system, which leads to an increase in the total desulfurization rate, but upon the increment of agitation speed from 300 to 600 rpm with low cell density cultures, there was no statistical significant influence on the desulfurization activity ($p > 0.05$). This indicated that the agitation speed at 300 rpm might have been enough to break up cell aggregates of low density cells to overcome mass transport limitations and, when increasing the agitation speed to 600 rpm, it would have been hard to continually improve the biocatalyst's kinetic rate for desulfurization activity.

Nassar *et al.* (2017b) observed that the growth and BDS activity of *R. erythropolis* HN2 is highly affected by the shaking speed, where the optimum speed was found to be 150 rpm, recording a BDS of 80% from the initially added 1000 ppm DBT. The growth was sharply decreased at a higher shaking speed (> 150 rpm) and, consequently, the BDS decreased to 73%. This was explained because at a relatively lower range of shaking speed (< 150 rpm), increasing the shaking speed improved the mixing and contact of an insoluble substrate (DBT) and microbial cells (HN2) and also would have enhanced the transport of oxygen. However, Wang *et al.* (2006), Purwanto *et al.* (2009), and Adlakha *et al.* (2016) reported that a high shaking speed is harmful to the bacteria due to the turbulence increment which would cause cell rapture. Thus, at a higher shaking speed, the mass-transfer limiting the BDS-process would be reaction controlled.

8.6 Effect of Initial Biomass Concentration

Several studies concluded that the BDS yield is substantially higher at lower oil fraction phases and DBT concentrations, but higher cell concentrations (Maghsoudi *et al.*, 2001; Abbad-Andaloussi *et al.*, 2003; Luo *et al.*, 2003; Jia *et al.*, 1996; Rashtchi *et al.*, 2006). A high cell density can reduce operating costs (Arabian *et al.* 2014), however, Nassar *et al.* (2017b) observed, upon the optimization of initial inoculum size for a shaken flask batch DBT-BDS process using growing cells of *R. erythropolis* HN2, that the increase of an initial inoculum size expressed a negative impact on the microbial growth and, consequently, BDS-efficiency. According to Abusham *et al.* (2009), the high initial inoculum size would cause a lack of oxygen and depletion of nutrients in the culture medium.

The effect of the initial biomass concentration (20–120 g DCW/L) of resting cells of *Rhodococcus* sp strain P32C1 (which was formerly identified as *Corynebacterium* sp. P32C1 by the National Collection of Industrial and Marine Bacteria (NCIMB Japan)) in a batch BDS of model oil 24 mM DBT in n-hexadecane and 50 vol.% (O/W) bi-phasic system was studied (Maghsoudi *et al.*, 2001). The production rates of 2HBP at higher cell concentrations were lower, probably due to mass transfer limitation, especially for oxygen transfer needed for the oxidation reaction of DBT, but the maximum conversions of DBT were higher.

The effect of the initial biomass concentration (20–80 mg DCW/L) of resting cells of Gram-negative *P. delafieldii* R-8 in a biphasic BDS-system (1:1 O/W) of model oil 1 mM DBT in dodecane was that the DBT-BDS rate was decreased with increasing cell concentration due to the mass-transfer resistance, which is the controlling step in these cases. Also, at a higher initial biomass concertation the separation after the BDS process was difficult and the results showed that almost all cell pastes suspended at the organic–aqueous interface when the cell concentration was 20 mg/mL after the centrifugation to separate the oil from the aqueous phase after finishing the reaction. Not only this, but it was also observed, at high cell concentration, that part of the cells suspended at the organic–aqueous interface while others precipitated at the bottom of the centrifuge bottle, which might mean that these cells precipitated at the bottom did not contact well with the oil phase to react with DBT. From such phenomena, it was speculated that the desulfurization process might be limited by the rate of surface renewal as a new biocatalyst interacts with the organic–aqueous interface to acquire the DBT substrate as found by Kaufman *et al.* (1998).

The effect of the initial biomass concentration of resting cells of *Corynebacterium* sp. ZD-1 in a batch BDS process of 0.5 mmol/L DBT dissolved in ethanol was studied (Wang *et al.*, 2004), where the production rate of 2-HBP was higher at a lower initial biomass concentration and recorded its maximum (0.067 mmo/L) at 9.2 g DCW/L. This was attributed to the mass transfer limitation at higher cell concentrations, especially, oxygen transfer, which is needed for the oxidation of DBT.

Caro *et al.* (2007a) reported in aqueous phase study, with an initial DBT concentration of 54.27 μM, that the maximum specific HBP production rate of 6.3 mmol HBP produced/kg DCW/h with 70% DBT removal at *P. putida* CECT 5279 initial biomass concentration of 8 g/L which decreased at a higher biomass concentration probably due to oxygen mass transfer or the inhibitory effect of 2-HBP, while *R. erythropolis* recorded complete

DBT removal at an initial biomass concentration of 28 g/L, with a constant maximum specific 2-HBP production rate of 5 mmol HBP produced/kg DCW/h within a range of 2 to 16 g/L initial biomass concentration.

The effect of initial biomass concertation in batch shaken flasks of a DBT–BDS process was studied using *Rhococcus* spp Eu-32 and it was depicted that the rate of BDS decreased at a higher initial inoculum size (> 4 g/L) (Akhtar *et al.*, 2009). This indicated that the oxygen is a limiting factor for dibenzothiophene desulfurization by the growing cells. Mohebali *et al.* (2007) reported that increasing cell concentrations led to a decrease in the dibenzothiophene degradation by the formation of cellular flocs due to the hydrophobic nature of biodesulfurizing bacteria. Akhtar *et al.* (2009) assumed that the same phenomenon may have been taking place in the case of Eu-32 because the formation of cellular colloidal material at high cell mass concentrations could limit the BDS activity by lowering the mass transfer of DBT and dissolved oxygen available to the growing bacterial aggregates.

Arabian *et al.* (2014) reported that the BDS efficiency in a biphasic system (aqueous/500 ppm DBT in dodecane 1:2 O/W), using *Bacillus cereus* HN, increases with the increment of biomass concentration due to the increase of the available biocatalyst, but to a certain limit. At a higher cell density ($3x10^7$), the percentage of desulfurization would fall. This can be explained by the mass-transfer limitations and inability of some cells to be in contact with the organic phase. In fact, the rate at which DBT can be converted by cells is higher than the mass transfer rate of DBT from the organic phase into the cells. Therefore, the mass transfer of DBT from an organic phase into the cells controls the overall process.

8.7 Effect of Biocatalyst Age

One of the important factors that recently gained attention is the age of the biocatalyst. Table 8.1 summarizes the published BDS-efficiencies of some bacterial strains at the selected optimal resting cell age.

A maximum DBT conversion, using *P. putida* CECT5279, is reported to be achieved when cells collected at 9 h of growth time are employed for DBT desulfurization in resting cell conditions (Martin *et al.*, 2004; 2005). It was proved in another study of BDS using resting cells of CECT5279 that the transport of all 4S-route intermediates across the cell membrane is not a mass transport controlling resistance and that neither the intracellular concentrations of reducing cofactors or the NADH dependent reductase HpaC have influence in the desulfurization rate (Alcon *et al.*, 2005).

Table 8.1 The DBT-BDS Efficiency of Reported Resting Cell Strains.

Microorganism	C-source	Optimal cell age (h)	D_{DBT} (% g_{cell}/L/h)	Q_{DBT} (mmol/g_{cell}/h)	References
R. erythropolis IGTS8	111 mM glucose	24	5.16	0.008	del Olmo *et al.* (2005b)
RIPI-22	13.9 mM benzoate	30	–	0.002	Rashtchi *et al.* (2006)
Rhodococcus sp. P32C1	54 mM glycerol	35–50	–	0.03	Maghsoudi *et al.* (2001)
P. putida CECT5279	136 mM glutamic acid	9	21.65	0.008	Martin *et al.* (2004, 2005)
	104 mM citrate	12	9.81	0.004	
	111 mM glucose	8	1.14	0.006	
Ralstonia eutropha	111 mM glucose	5	23	0.017	Dejaloud *et al.* (2017)

Moreover, upon measuring the evolution of in vivo enzymatic activities of DszA, DszB, and DszC, along the growth curve of *P. putida* CECT5279 maximum activities of both flavin dependent monooxygenases, DszC and DszA, were found in cells of 23 h of growth time and a maximum activity for the desulfinase, DszB, was observed in cells of 5 h of growth time of *P. putida* CECT5279. These different patterns of expression of monooxygenases, DszA and DszC, and desulfinase, DszB, along the growth curve explained the behavior of cells collected at 9 h of growth time (Calzada *et al.* 2009a). Thus, upon using a biocatalyst mixture of resting cells collected after 5 and 23 h in a 1:1 ratio, the BDS efficiency was higher relative to simple biocatalysts and composed exclusively by 9 h cells (Calzada *et al.* 2009b). However, upon another study for the optimization of the biocatalyst mixture, a total biomass concentration of 2.1 g DCW/L of a cell mixture containing 66.7% of 23 h growth time cells that has a 1:2 ratio of cells collected at 5 and 23 h of growth time, expressed the best BDS-efficiency in resting cells among all the proposed biocatalyst formulations. Thus, this was the best DBT conversion and reduction time needed for BDS (Calzada *et al.*, 2011). Del Olmo *et al.* (2005a) reported that the maximum DBT-BDS activity of *R. erythropolis* IGTS8 is achieved at the end of the exponential phase of the growth of the cells (between 15 and 25 h of growth time).

Rashtchi *et al.* (2006) reported that the activity of strain RIPI-22 is sensitive to the growth phase. The cells harvested in the late-exponential growth phase displayed the maximum specific activity which was 2.03 mmol 2HBP/kg DCW/h. Gomez *et al.* (2015) studied the effects of *R. erythropolis* of different biocatalyst ages (16, 24, and 30 h) in shaken flasks at 210 rpm and STBR at different agitation intensities (from 100 to 700 rpm). Short age biocatalysts (16 h) yielded a lower HBP concentration, thus expressimg a lower BDS capacity than those of longer ages (24 and 30 h), which expressed the same BDS capacity.

However, Gün *et al.* (2015) reported the BDS of DBT by the inoculation of the batch process by *Sulfolobus solfataricus* P2 cells grown at the mid-log phase.

Escobar *et al.* (2016) studied the effect of *P. putida* KTH2 biocatalyst age on desulfurization efficiency. The results revealed that, under relatively low agitation conditions (from 100 to 300 rpm), the DBT consumption rate was faster on more advanced age cells; that is, higher HBP concentration is obtained with cells of 23 h of growth time. However, at relatively higher agitation speeds (about 500 rpm), there were increases in the DBT consumption rate and HBP production rate to 15 h of growth age. Nevertheless, for cells of 23 h age, a decrease has been observed in the consumption of DBT and a smaller HBP production, but with a higher agitation speed of 800 rpm, all cell ages performed lower DBT consumption and, consequently, lower production of HBP.

8.8 Effect of Mass Transfer

Another important parameter is the influence of mass transfer on the overall reaction rate and process yield as the reaction is carried out in a complex medium with two immiscible liquid phases. This parameter is essential for the implementation of the BDS-technology of fuels. It depends on the hydrophobicity of the compound and its alkylation degree, molecular structure, and weight.

Marcelis *et al.* (2003) have indicated that the greatest resistance to the mass transport of compounds from organic to an aqueous phase occurs in the water phase, but there has also been indication that it is the microbial process which is the global rate limiting step.

Caro *et al.* (2007a) reported the effect of mass transfer limitation in a biphasic system with an initial DBT concentration of 54.27 μM, using *R. erythropolis* IGTS8 and *P. putida* CECT 5279, which proved the results mainly depend on the hydrophobicity of the biocatalyst itself. The desulfurization yield increased with the increase of an initial biomass concentration recording 41.4% at 24 g/L of initial CECT 5279 biomass concentration with a low production rate of 2-HBP. Nevertheless, the BDS yield of IGTS8 was higher, recording 80%, with an initial biomass concentration of 8 g/L. Then, is kept stable at a higher initial biomass concentration due to the saturation of the interface surface by the biocatalyst cells. Thus, increasing the interface surface, the process would achieve better conversion levels, but the specific 2-HBP production rate was also low due to both the inhibition and mass transfer limitations. Caro *et al.* (2007b) reported that in spite of the lower HBP inhibition effects with biphasic media, other pathway compounds accumulated in the aqueous phase can be responsible for inhibition.

Upon combining the oil fraction phase, substrate concentration, and cellular density effects, mass transfer limitation is responsible for the BDS yield in resting cell system conditions in low oil fraction and low substrate concentrations, but where high cell densities were used (Jia *et al.*, 2006; Caro *et al.*, 2007b).

8.9 Effect of Surfactant

DBT as a model for recalcitrant organosulfur compounds (OSCs) is known to be a very hydrophobic compound exhibiting water solubility around

0.005 mM. This very low concentration can be increased by the action of biosurfactants produced endogenously by the applied microorganisms (i.e. the biocatalyst) used in the BDS process (Maghsoudi *et al.*, 2001). However, it has been reported that the capability of the bacteria to pick up the DBT from the organic phase and its subsequent insertion inside the cells is independent of the produced biosurfactants (Noda *et al.*, 2003). *R. erythropolis* IGTS8 is known to be characterized by a high hydrophobicity (Abbad-Andaloussi *et al.*, 2003) and the uptake of the DBT takes place on the interface surface (Monticello, 2000; Le Borgne and Quintero, 2003). However, high mass transfer limitations occur when this surface is saturated by adhered biocatalyst cells (Maghsoudi *et al.*, 2001). Marzona *et al.* (1997) reported that the addition of cyclodextrins improves the diffusion of DBT into the aqueous phase upon the formation of DBT-cyclodextrin complexes. Moreover, it takes up the HBP, avoiding the potential inhibition effects produced by the 4S-pathway's final product accumulation. Cyclodextrins cannot be metabolized by the microbial cells and have no toxic effects on the cells (Setti *et al.*, 2003). Upon the addition of 5% (v/v) of the non-ionic surfactant Tween-80 (polyethylene glycol sorbitan monooleate) and an oil-in-water emulsifier in a batch BDS process of 0.3 mM DBT using *Gordonia* sp. CYKS1, the growth rate and BDS-capacity increased from 0.06 h^{-1} and 4.70 μmol/L/h to 0.08 h^{-1} and 5.90 μmol/L/h (Kim *et al.* 2004). Feng *et al.* (2006) reported the enhancement effect of Tween 80 on the BDS of *Rhodococcus* species 1awq in an aqueous one-phase system where the addition of 0.4% Tween 80 enhanced the growth and BDS efficiency of 0.5 mM DBT. The specific desulfurization activity was 1 μmol/g/min. This activity was 35% higher than that seen without Tween 80. This was explained by the nonionic chemical surfactant's, Tween 80, ability to enhance BD activity in both aqueous and biphasic systems by reducing the concentrations of the products around the cells. Tween 80 can also reduce the concentrations of hydrophobic substrates associated with the cells. As long as the concentrations support adequate reaction rates, this reduction will not limit the overall conversion. If a substrate is also inhibitory at high concentrations, the addition of Tween 80 is, theoretically, stimulatory.

Wang *et al.* (2006) studied the effect of different surfactants, including Triton-100X, Tween-80, Brij 35, and cyclodextrin, on the biphasic batch of BDS (DBT in n-hexadecane) in 72 h at 30 °C, pH, and 150 r/min, using resting cells of *Corynebacterium* sp. ZD-1, that were harvested at the late exponential phase. The solubility of DBT was found to linearly increase with the increase of surfactant concentration. The solubility enhancement efficiencies of surfactants above the critical micelle concentration (CMC) followed the order Triton-100X > Tween-80 > Brij 35 > cyclodextrin, but both Brij and Triton-100X inhibited DBT desulfurization. β-cyclodextrin produced almost no effect. Nevertheless, Tween-80 increased the desulfurization efficiency. This was attributed to the toxicity of Brij and Triton-100X to the bacteria. Although β-cyclodextrin did not inhibit cells growth, its ability for DBT solubilization was much less than Tween-80, where Tween 80 increased the growth in DBT-cultures because Tween-80 increased the absorption and degradation of the bacteria to DBT and supplied the bacteria with enough sulfur-source necessary for growth. The optimum value of Tween-80 concentration for desulfurization was found to be 0.5 g/L. The amount of 2-HBP formed with 0.5 g/L Tween-80 present was about 50% more than that formed without surfactant. This was better than the effect of cyclodextrine on *R. erythropolis* IGTS8 (Setti *et al.*, 2003).

In studies of BDS on bunker oil (heavy oil), BDS efficiency was reported to be only 2–3% by different microorganisms (Jiang *et al.*, 2014). However, the activity is reported to increase up to 36–50% by use of either deasphalted oil or the use of surfactants, which decrease the viscosity of heavy oil (Li and Jiang, 2013; Jian *et al.*, 2014). Throughout the addition of 0.5 g/L Tween-80 in a batch BDS of heavy crude oil of 1/3 O/W using *Gordonia* sp. IITR100, the BDS percentage increased from 48% to 65%. The higher BDS efficiency was nearly equivalent to that obtained after two successive rounds of 7 days. Thus, addition of surfactants eliminated an extra BDS-step. This was attributed to the addition of surfactants, which resulted in overcoming mass transfer problems (Adlakha *et al.*, 2016).

Span-80 and Tween-80 are reported for the enhancement of BDS efficiency using *Pseudomonas delafieldii* (Li *et al.*, 2008). The addition of non-ionic surfactants, Tween-80 or Span-80, to a biphasic system, inoculated with immobilized *R. erythropolis* R1 alginate beads, also enhanced the BDS efficiency and increased the production of 2-HBP (Derikvand *et al.*, 2014) since both surfactants have an oleate-chain with 18 carbons and an unsaturated bond that could improve the stability of the O/W interface layer. The surfactants would enhance DBT dissolution in the aqueous phase and also facilitate its penetration into the immobilizing beads. Moreover, 2-HBP as a product of DBT-BDS, had a limited solubility in the aqueous phase and displayed an inhibitory role on the BDS efficiency by aggregation around the cells. Since, the surfactants also increase the solubility and dispersion of 2-HBP, they reduce its concentration around the cells and, consequently, improve the BDS efficiency (Li *et al.*, 2008a). However, Span-80 expresses a higher BDS efficiency than that obtained with Tween-80 (Derikvand *et al.*, 2014) since the average droplet sizes produced by Span 80 are smaller than those of Tween 80 (Schmidts *et al.*, 2009) and reported higher DBT removal efficiency, probably due to differences in the size of the droplets.

However, it should be noted that in oil desulfurization processes the production of a large amount biosurfactants by biocatalysts would enhance the bioavailability of organic sulfur compounds in fuel oils and at the same time increase the reaction rate by improving the contact between the oil and aqueous phases. However, it should be noted that the biosurfactants could cause the formation of excessively stable emulsion and, thus, may cause serious phase separation problems in real processes and the toxicity of OSCs would be increased in the presence of biosurfactants.

8.10 Effect of Initial Sulfur Concentration

Sulfur accounts for approximately 0.5–1% of the dry weight of the cell (Kertesz, 2000), thus, the required sulfur level for cell growth is limited. For example, *Rhodococcus* sp. 1awq was found to require only 0.5 mmol/L Na_2SO_4 for the achievement of maximum cell growth (Xu *et al.*, 2002). Although it is known that as water-solubility of compounds increases, its inhibitory effect on cell growth and enzymatic activity increases. Otherwise, with the extremely low solubility of DBT in water (around 0.005 mM), it precipitates on being added to the aqueous culture medium; DBT in solid form in the medium would be expected to have negligible effects, if any, on desulfurization activity and cell growth, but DBT is intrinsically a xenobiotic compound and there have been several reports addressing that a high concentration of DBT had inhibition effects on cell growth and desulfurization activity (Oshiro *et al.*, 1995; Setti *et al.*, 1996; Goindi *et al.*, 2002; Ansari *et al.*, 2007; Dejaloud *et al.*, 2017). This might be attributed to the secretion of biosurfactants, as most of the microbial strains with DBT desulfurization activity are known to secrete some biosurfactants to solubilize and thus enhance the bioavailability of DBT. Consequently, upon the secretion of the biosurfactant, the effects of the amount of DBT added would not be limited by its intrinsic solubility to water.

The different reported biodesulfurizing microorganisms have different optimum DBT concentrations and incubation periods. For example, *Gordonia alkanivorans* strain 1B reported 64.8% of 0.5 mM DBT within 120 h (Van der Ploeg and Leisinger, 2001), *Microbacterium* ZD-M2 reported 100% of 0.2 mM DBT within 58 h (Li *et al.*, 2005), *Rhodococcus erythropolis* R1 reported 100% of 0.3 mM DBT within 72 h (Etemadifar *et al.*, 2008), *Microbacterium* sp. NISOC-06 reported 90.6% of 1 mM DBT within 120 h (Papizadeh *et al.*, 2010), and *Stenotrophomonas* sp. NISOC-04 reported 82% of 0.8 mM DBT within 48 h (Papizadeh *et al.*, 2011).

The bacterial growth of *Rhodococcus* sp. strain P32C1 on different initial concentrations of DBT 0.05–0.5 mM was generally the same. However, the resting cells collected at the late exponential phase, from the 0.1 mM DBT cell culture, showed the highest BDS-activity on 0.5 mM DBT which was 30 mmol 2-HBP/kg DCW/h (Maghsoudi *et al.*, 2001). In a study performed on model oil, DBT in n-hexadecane and the BDS efficiency of resting cells of *Staphylococcus* sp. strain S3/C increased with the increase of initial DBT concentration, recording its maximum at 57% at 463 mg/L, with initial sulfur/biomass (0.16). Further increase in initial sulfur to 800 mg/L (S/B 0.31) declined the sulfur removal to 32% and attributed this to the limitation of O_2 in the biphasic reaction mixture (Goindi *et al.*, 2002). A similar observation was reported by Guerinik and Al-Mutawah (2003) for model oil with DBT in n-tetradecane, where the specific desulfurization activity increased as DBT concentration increased. However, there is an optimum DBT concentration beyond which there is no effect on the activity. This is the point where mass transfer ceases to be a problem and some other factor, like product inhibition, becomes operational. Moreover, the BDS was limited by the availability of DBT to the organism and the interfacial mass transfer through the aqueous-organic layer was confirmed to be a limiting factor. In another study performed by Kim *et al.* (2004), the BDS rate increased as the initial DBT concentation increased from 0.3 to 1.5 mM, recording approximately 4.7 and 12.5 μmol/L/h, respectively, and then sharply decreased at a higher initial DBT concentration recording 4.5 μmol/L/h at 4.0 mM DBT. Ansari *et al.* (2007) reported the inhibitory effects of DBT for *Shewanella putrefaciens* NCIMB 8768 at concentrations ≥ 0.6 mM.

Casullo de Araújo *et al.* (2012) reported that DBT expressed the same minimum inhibitory concentration (MIC) and minimum bactericidal concentration (MBC) (3.68 mM) on *Serratia marcescens* UCP 1549. Moreover, the specific growth rate in a batch BDS of different initial DBT concentrations at 28 °C, recorded 0.15, 0.11, and 0.05 h^{-1} with a generation time (T_G) of 4.6, 6.27, and 13.8 h, with 0.5, 1.0, and 2.0 mM of DBT, respectively. However, the specific desulfurization rate recorded 0.038, 0.022, and 0.038 h^{-1}, respectively. After 96 h, the percentage desulfurization was 50, 84, and 98%, respectively, with a complete removal of 0.5 and 2 mM DBT within 120 h, but complete removal of 1 mM occurred within a longer incubation period of 144 h.

The BDS of model oil (DBT in dodecane) using *Bacillus cereus* HN in a biphasic system 1:10 O/W showed an increase in the BDS efficiency as the DBT concentration increased and was attributed to the enhancement in the DBT diffusion to the oil-water interface (Arabian *et al.*, 2014).

The specific growth rate of the acidophilic and hyper-thermophilic *Sulfolobus solfataricus* P2 was reported to decrease with the increase of DBT concertation (Gün *et al.*, 2015) and recorded its highest BDS at 0.1 mM DBT, producing 1.23 μ mol 2-HBP/h/g DCW in a one phase BDS system.

Dejaloud *et al.* (2017) reported that the increase of initial DBT concentration form 0.03 mM to 0.05 mM increased the biomass concentration and BDS-efficiency of *Ralstonia eutropha* (PTCC1615). However, further increase in initial DBT-concentration decreased the desulfurization activity of the cell.

8.11 Effect of Type of S-Compounds

Not all microorganisms utilize the same pathway for the BDS of different organosulfur compounds (OSCs). Also, not all the microbial strains utilize all the OSCs. They also never use all the OSCs with the same rate or efficiency. This is related to the different molecule structures of substrates such as the extent of alkyl group substitution (Grossman *et al.*, 2001), as well as the enzymes involved in the BDS of each S-compound.

The induction of the dsz operon by DMSO in *Rhodococcus* sp. 1AWQ has shown similarities to DBT (Ma *et al.*, 2006) and Mohebali *et al.* (2008) showed that the 2-HBP production of resting cells grown on DMSO was higher than that of DBT. In another study, Jiang *et al.* (2002) reported that the shortest induction time and highest growth rate for *Pseudomonas delafieldii* R-8 occurred with DMSO, relative to DBT and $MgSO_4$. Not only this, but Jiang *et al.* (2002) reported the broad versatility of two bacterial isolates, *Pseudomonas delafieldii* R-8 and *Nocardia globerula* R-9, to utilize different S-compounds as a sole S-surfur source. The most important observation was their ability to desulfurize 4,6-DMDBT at a higher rate than DBT, recording a desulfurization rate of 2.46 and 2.25 mg DMDBT/g DCW/h and 2.26 and 1.98 mg DBT/g DCW/h for R-8 and R-9, respectively, with a higher desulfurization rate for DBT-sulfone relative to DBT at 2.35 and 2.45 mg $DBTO_2$/g DCW/h, respectively. They are also capable of desulfurizing BT and DBS, recording 1.34 and 1.04 mg BT/g DCW/h and 1.93 and 2.06 mg DBS/g DCW/h, respectively. The Gram +ve *Mycobacterium* sp. ZD-19 reported to desulfurize different S-compounds in the following order: Th>BT>DPS > BT>4,6-DMDBT (Chen *et al.*, 2008a). However, the rate of BDS of DBT was decreased in the presence of 4,6-DMDBT, although DBT appeared to be attacked preferentially by ZD-19 enzymes when DBT and 4,6-DMDBT coexisted. This is refered to as the substrate competitive inhibition. Resting cells of ZD-19 recorded a specific desulfurization rate of 2.984, 1.756 mM S/kg DCW/h for DBT and 4,6-DMDBT, respectively, in a single substrate system, with an overall desulfurization percentage of 100 and 57.3% from 0.5 mM DBT and 4,6-DMDBT, respectively, within 8 h. However, in a binary substrate system, the specific desulfurization rate recorded 1.78 and 0.624 mM S/kg DCW/h, with a BDS percentage of 91.6 and 53.2 of 0.3 mM DBT and 0.2 mM 4,6-DMDBT, respectively, with a total S-removal of 77% within 8 h.

In an attempt to study the substrate specificity, Konishi *et al.* (2002) used the cell-free extracts of a desulfurizing mesophile, *Rhodococcus erythropolis* KA2–5-1 (the Dsz system), and *Escherichia coli* JM109, which possesses the desulfurizing genes of thermophile *Paenibacillus* sp. A11–2 (the Tds system), to investigate the reactivity of desulfurizing enzymes toward 4,6-dialkyl dibenzothiophenes (4,6-dialkyl DBTs) and 7-alkyl benzothiophenes (7-alkyl BTs). Both systems desulfurized DBT and all the studied 4,6-dialkyl DBTs, including 4,6-DMDBT, 4,6-DEDBT, and 4,6-DPDBT, to their corresponding hydroxybiphenyls, but they did not desulfurize 4,6-DBDBT. Although some alkylated BTs were degraded by the Dsz system, no desulfurized compounds were detected. This was attributed to the limit of the catalyzing ability of the desulfurizing enzymes. The desulfurization activity of various alkylated DBTs has been examined by the recombinant KA2–5-1/pRKPBP cells using an oil-water resting cell system (Kobayashi *et al.*, 2001), where the bacterial cells were not able to incorporate highly alkylated DBTs, such as 4,6-dipropyl DBT, from the oil phase. The reactivity of the Tds system toward alkylated BTs was higher than that of DBT. In contrast to the Dsz system, the Tds system yielded desulfurized compounds from BT and all of the studied alkylated BTs: 7-MBT, 7-EBT, 7-PBT and 7-HBT. The 7-Hexyl BT was more efficiently degraded by Tds than the Dsz system (70% vs 30% of 0.5 mM in 1 h reaction), although the rate of degradation was the lowest. Konishi *et al.* (2002) also pointed out the strong effect of the position of the alkyl chain on the reactivity of Dsz enzymes. An alkyl chain near the sulfur atom of the BT molecule is evidently desirable for enzyme reactivity in contrast to chemical catalysts. The presence of alkyl groups attached to the thiophene ring of the BT molecule is important for complete desulfurization in the Dsz system. The reactivity in the Dsz system was also found to be heavily affected by the position of the alkyl group attached to the ring of the BT molecules. The substrates with the alkyl group on a non-thiophene ring of the BT molecules exhibited poor reactivity. The desulfurized phenolic products were observed for only 2-methyl, 2-ethyl, 3-methyl, and 2,7-diethyl BTs. The Dsz system exhibited a reduced desulfurization rate with the increase in the length of the alkyl chain attached to the DBT molecule. 4,6-dimethyl- and diethyl-DBTs were completely desulfurized to corresponding phenolic compounds, whereas a small amount of desulfurized compounds (less than 5% of the initial substrate concentration, 0.5mM) were detected by GC analysis from 4,6-dipropyl DBT in spite of the decrease in substrate. This was explained by the probability of the conversion of most of the 4,6-dipropyl DBT to water-soluble intermediates, such as sulfonate, which could not be extracted by the organic solvent. The reactivity of Tds toward DBT was weaker than that of the Dsz system, where the Tds enzymes exhibited much stronger activity (about 1.5–2 times) toward alkylated BTs than toward alkylated DBTs. All the studied substrates, as well as 7-alkyl BT, were desulfurized to corresponding phenolic compounds and the reactivity had almost no relation with the length or the position of the alkyl chain attached to the BT molecule. Interestingly, 4,6-dimethyl DBT and 4,6-diethyl DBT, which are bulkier substrates than DBT, were more rapidly desulfurized than DBT in contrast to the Dsz system. This study proved the efficiency of cell free extracts over free whole cells. For example, *R. erythropolis* KA2–5-1 did not grow on BT, 5- methyl BT, and was unable to utilize 7-alkyl BT as a sulfur source, although it can grow on 3-methyl BT (Kobayashi *et al.*, 2000).

Rhodococcus sp. WU-K2R preferentially degraded BT for a BT/naphthothiophene mixture (Kirimura *et al.*, 2002). *Rhodococcus* sp. strain K1bD can individually desulfurize DBT and 1,4-dithiane, but when they were supplemented together, only DBT-BDS occured (Kirkwood *et al.*, 2005). Similarly, the desulfurization of benzothiophene or 1,4-dithiane by *Rhodococcus* sp. strain JVH1 was delayed significantly in the presence of benzyl sulfide in the culture medium (Kirkwood *et al.*, 2007). In another study, the BDS of BNT by a mixed culture was decreased significantly in the presence of DBT (Jiang *et al.*, 2014). The same was reported for *Gordonia* sp. IITR 100, where complete removal of 0.3 mM BNT and DBT occurred within 6 d when incubated individually. However, in their presence together, the BDS of DBT was not affected, but the BNT-BDS was completely inhibited for up to four days. Then, the desulfurization of BNT was observed between the fourth and the sixth day of incubation when the level of DBT dropped to < 0.01 mM, but the reaction stopped again thereafter. Moreover, when the recombinant *E. coli*-DszC was applied, it expressed lower DBT-BDS efficiency ($\approx 60\%$) than that for BNT. Although the rate of DBT-BDS in binary system remained nearly the same as in single-substrate system, the BDS of BNT was completely inhibited in a binary substrate system of DBT and BNT. The observed inhibition during the metabolism of BNT by *E. coli*-DszC into BNT-sulfone suggested that the competition between the two substrates sets in the first step itself. The results are consistent with an earlier study (Jiang *et al.*, 2014) where the biodesulfurization of BNT by a mixed culture decreased significantly in the presence of >0.1 mM DBT.

G. alkanivorans strain 1B showed vigorous growth on sulfate, sulfide, sulfite, and elemental sulfur and on DBT, DBT sulfone, BT, 2-methylthiophene, and 2-mercaptoethanol, in the presence of glucose as a C-source, which clearly demonstrates that the *G. alkanivorans* strain 1B utilizes different sources of sulfur-containing aromatic hydrocarbons, including alkylated thiophene, BT, and DBT, suggesting its ability to decrease the sulfur content of fossil fuels. However, although 1B can utilize DBT and DBT-sulfone individually as sole S-sources, upon their presence together the bacterial growth stopped and, consequently, so did the BDS capability. Upon utilizing a binary substrate system of BT and DBT, the specific desulfurization rate recorded 0.954 and 0.813 μmol DBT/g DCW/h, respectively (Alves *et al.*, 2005). In another study, Alves *et al.* (2008) reported the specific desulfurization rates in the model oil for DBT, 4-MDBT, and 4,6-DMDBT to be 0.78, 0.68. and 0.41 μmol/g DCW/h, respectively, in a batch culture inoculated with *G. alkanivorans* strain 1B.

Pantoea agglomerans D23W3 showed good BDS potential on BT, DBT, and 4,6-DMDBT, but it could not desulfurize 4,6-DEDBT (Bhatia and Sharma, 2010). This was attributed to an increase in the C-number of the alkyl substituent group resulting in increased d size and hydrophobicity, which, consequently, increased the expulsion of the compound from the aqueous phase which rarely comes in contact with the bacteria present in the aqueous.

Ahmad *et al.* (2015) studied the BDS of dibenzylsulfide (DBS) and dibenzothiophene (DBT) as model compounds for sulfides and thiophenes in petroleum and its fractions in single and binary substrate systems using *Gordonia* sp. IITR100. In a single substrate system, desulfurization of DBS by IITR100 was marginally better than DBT and both the chemicals were utilized completely within 3 and 4 days of incubation, respectively, but when the two chemicals were present together, an enhanced preference for the desulfurization of DBS was observed. Here, while nearly all of DBS was metabolized after 4 days of incubation, approximately 50 % of DBT was still present in the medium after the same period.

Aggarwal *et al.* (2013) reported that *G. alkanivorans* prefers BT over DBT as a sulfur source. This preference may be driven by the lower NADH requirements for BT metabolism, rather than the higher affinity of the transport system for BT.

One mmol/L DBT in n-dodecane was reported to be completely desulfurized with an almost stoichiometric production of 2-HBP, within 10 h at pH7, 30 °C, and 170 rpm, in a batch bi-phasic BDS process (1:1 O/w) using resting cells of *Nocardia globerula* R-9 collected at the late exponential phase, while 1 mmol/L 4,6-DMDBT was desulfurized by about 70% using the same conditions. The initial desulfurization rate for 4,6-DMDBT was about 5.9 mmol S/kg DCW/h, which was about 40% of that for DBT (Mingfang *et al.*, 2003). While using a binary substrate system (0.5 mmol/L DBT and 0.5 mmol/L 4,6-DMDBT), both compounds were desulfurized simultaneously without preference for either one, which is of great importance in the practical desulfurization of petroleum products. The initial desulfurization rates for DBT and 4,6-DMDBT were 4.0 and 3.0 mmol S/kg DCW/h, respectively, but the overall desulfurization rates for both compounds, when present together, were lower than those found when treated separately (Mingfang *et al.*, 2003). This was attributed to competitive inhibition of substrates. Similar phenomenon was reported for *Rhodococcus erythropolis* KA2–5-1 (Morio *et al.*, 2001). In another study performed by Luo *et al.* (2003) using resting cells of the Gram negative *P. delafieldii* R-8 in biphasic system (1:1 O/W) of model oil 10 mM DBT or 4,6-DMDBT, BDS efficiencies of 97 and 86% were recorded, respectively, within 18 h of reaction time with mean specific desulfurization rates of 11.4 and 9.4 mmol S/kg DCW/h, respectively, while the volumetric desulfurization rate, with respect to the volume of dispersion, (mixture of the oil and water) was 0.29 and 0.24 mmol S/$L_{dispersion}$/h for DBT and 4,6-DMDBT, respectively.

The desulfurization rate of DBT-sulfone is reported to be twice that of DBT in a growing cell culture of *Corynebacterium* sp. ZD-1 (Wang *et al.*, 2004). The acidophilic and thermophilic *S. solfataricus* P2 was reported to utilize DBT, DBT-sulfone, and 4,6-DMDBT, but cannot utilize BT as sole S-source. Thus, it was concluded that *S. solfataricus* P2 has a metabolic pathway specific for DBT and its derivatives and a recorded growth rate μ of 0.0179 and 0.0172 h^{-1} for DBT-sulfone and 4,6-DMDBT, respectively (Gün *et al.*, 2015).

Derikvand *et al.* (2015a) mentioned that the presence of the dszC gene in *Paenibacillus validus* (strain PD2) confirmed that DBT desulfurization occurred through the 4S pathway. Upon studying its ability to desulfurize different S-compounds, including thiophene, DBT, DMSO, and $MgSO_4$, a higher cell density was obtained when DMSO was used as the sole sulfur source. This was attributed to the easy metabolization of DMSO over other sulfur sources (Bustos-Jaimes *et al.*, 2003). Moreover, the growth rate was inhibited by production of 2-HBP when DBT is used as the sulfur source. The growth on $MgSO_4$ was more than that on DBT or thiophene, which was attributed to the lower water solubility of DBT and thiophene and its lower bioavailability to the cells. Although the dsz operon is reported to be repressed by $MgSO_4$ as a sole sulfur source during the growth (Li *et al.*, 1996), which explained the low BDS activity of resting cells grown on $MgSO_4$, which is the re-expression of the dsz operon.

The maximum BDS activity of the PD2 strain resting cells was achieved when they were grown on DMSO, where the maximum growth and the highest induction in the dsz operon obtained by *Paenibacillus validus* (strain PD2) was in the presence of dimethyl sulfoxide (DMSO) as the sole sulfur source.

El-Gendy *et al.* (2015) studied the main and interactive effects of polyaromatic sulfur heterocyclic compounds (PASHs) on the growth and biodegradation efficiencies of *Bacillus sphaericus* HN1, which was isolated for its ability to utilize DBT as a sole sulfur and carbon source (Figure 8.1).

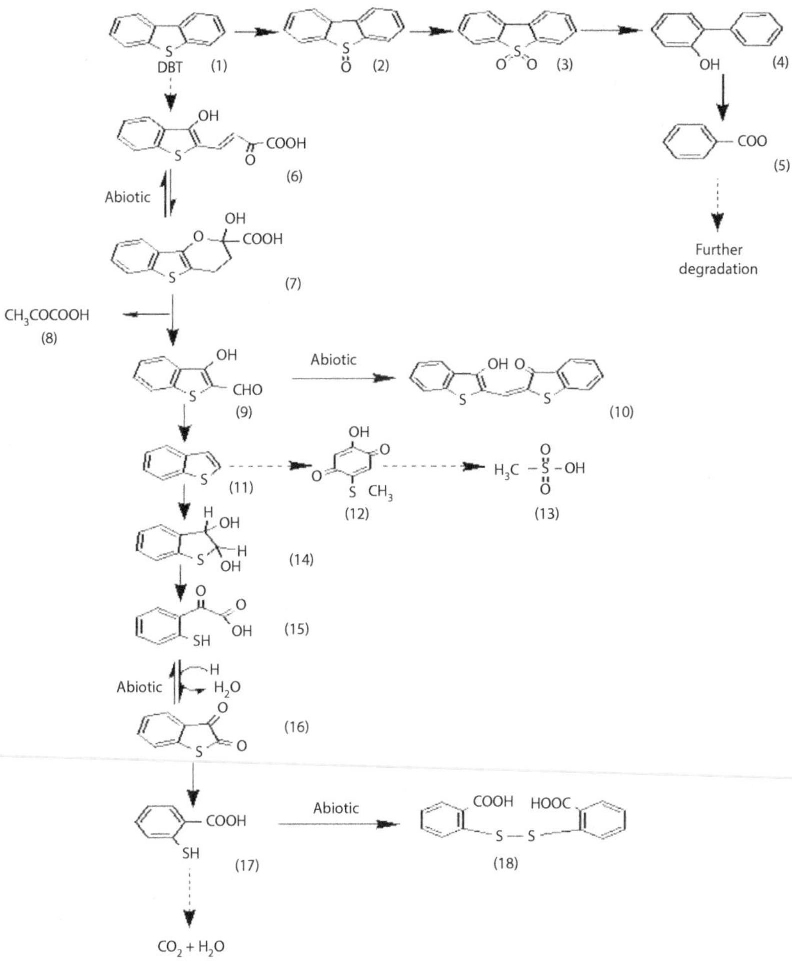

Figure 8.1 Proposed Pathway for Biodegradation of DBT by *Bacillus sphaericus* HN1 (Nassar, 2009).

The biodegradation efficiency of the studied PASHs, Th, BT, and DBT, has been found to be ranked in the following decreasing order: BT > DBT > Th, whether in a single- or tertiary-substrate system. The biodegradation of Th decreased in a multi-substrate system by ≈ 14%, 18%, and 20% in the presence of BT, DBT, and in tertiary mixtures systems, respectively. The biodegradation of BT was not affected by the presence of other PASHs, while the biodegradation of DBT was slightly enhanced by ≈ 4% in presence of BT, but decreased in a binary mixture with Th by ≈ 12% and tertiary mixture of the three studied PASHs by ≈ 10.4%.

The overall removal of total PASHs decreased in the following decreasing order: single-substrate batch system > binary-substrate batch system > tertiary-substrate batch system, recording average biodegradation percantages of ≈ 94%, 87%, and 75%. This was in agreement with the bacterial growth trend in the three studied systems and might be attributed to the increase of total sulfur concentration, which might have expressed higher toxicity to bacterial cells and, consequently, a decrease in their enzymatic activity.

Regardless of the type of PASHs and whether they are in single- or multi-substrate systems, bacterial growth was stopped or depleted before the complete removal of the PASHs. This was attributed to the biodegradation process itself, which might produce toxic metabolites to the cell, but from the time profile of each PASH, it was obvious that the degradation trait was expressed at increasing levels during the bacterial growth cycle even after reaching its maximum growth. This indicated that stationary phase cells have the ability to continue degrading PASHs. Sahinkaya and Dilek (2007) mentioned an important point; in the calculation of degradation rate values, total biomass concentration is considered. However, only a fraction of biomass is responsible for the degradation of a particular substrate.

Ismail *et al.* (2016) reported the changes in an AK6- mixed culture community structure according to the provided sulfur sources (DBT, 4-MDBT, 4,6-DMDBT, or BT). The major denaturing gradient gel electrophoresis (DGGE) bands represented members of the genera *Sphingobacterium*, *Klebsiella*, *Pseudomonas*, *Stenotrophomonas*, *Arthrobacter*, *Mycobacterium*, and *Rhodococcus*. However, *Sphingobacterium* sp. and *Pseudomonas* sp. were abundant across all cultures utilizing any of the tested thiophenic S-compounds, but *Mycobacterium*/*Rhodococcus* spp. were restricted to the 4-MDBT culture which had the highest of the species' richness and diversity.

Peng and Zhou (2016) reported that DMSO is a kind of sulfur source better than DBT or MgSO4 for the growth of *Rhodococcus* sp. MP12. Not only this, but it was found that resting cells previously grown on sodium sulfate did not express any BDS activity on DBT. This raised the possibility that sulfate can significantly repress the desulfurizing activity of the enzymes in the genus of *Rhodococcus*, but cells that harvested cells in mid- and late-exponential phases during growth with DBT or DMSO as sulfur sources seem to have better desulfurization activity in a DBT system. The maximum BDS activity happens in a mid-exponential phase when the OD is between 4 and 6 for DBT or DMSO cells. A maximum value of desulfurization activity (10.5 μmol/g DCW/h) occurred with cells pre-grown for 48 h with DMSO as a sulfur source. This was double that of cells pre-grown for 36 h with DBT, which recorded 4.3 μmol/g DCW/h. Thus, DMSO was also a better sulfur source than DBT or $MgSO_4$ for the expression of desulfurizing enzymes of *Rhodococcus* sp. MP12 (Peng and Zhou, 2016).

8.12 Effect of Organic Solvent and Oil to Water Phase Ratio

Various hydrocarbons are present in gasoline (C6 –C9) and diesel oil (C12 –C23) (Mingfang *et al.*, 2003). As hexadecane comprises about 400 g/kg of certain diesel oils, it is considered a representative aliphatic hydrocarbon in diesel oil. However, the effect of an organic solvent or, in another word, the influence of an organic fraction phase (OFP) is a very important factor for the success of any BDS process.

The antimicrobial action of a solvent is correlated to its hydrophobicity and can be measured as the logarithm of the octanol–water partition coefficient log $P_{O/W}$ (Laane *et al.* 1987; Osborne *et al.* 1990; Sikkema *et al.*, 1994). According to this scale, enzymes and microorganisms present a minimum of activity with solvents with log P values of 0–2 and 2–4, respectively, after which the use of solvents with increasing log P values will result in increased biocatalyst stability. However, the actual concentration of the solvent in the bacterial cell membrane depends both on the solvent concentration in the water phase and on the partitioning of the solvent from the water phase to the membrane. Sikkema *et al.* (1994) proposed an equation to correlate the log $P_{O/W}$ value of a solvent and its partitioning value between the membrane and water, log $P_{M/W.}$

$$\log P_{M/W} = 0.97 \times \log P_{O/W} - 0.64$$

Hydrophobic solvents, with a log $P_{O/W} > 4$, accumulate in the membrane, but will not reach a high membrane concentration and are not toxic because of their low water solubility. On the contrary, solvents with a log

$P_{O/W}$ between 1 and 4 present higher water solubility values, while also being able to partition to biological membranes, resulting in relatively high concentrations of these solvents in the membranes and high toxicity to the cells (de Bont, 1998).

Many organic solvents are reported in BDS studies, such as diethylether, ethanol, DMSO, dimethylformamide (DMF), n-hexane, xylene, dodecane, n-tetradecane, and n-hexadecane, where upon the choice of the solvent its toxicity on the cells activity should be considered. Solvents with high partition coefficient to the membrane, such as n-dodecane and n-hexadecane, are not toxic to bacterial cells and are applied in many BDS processes in model oil studies.

For example, Mingfang *et al.* (2003) examined the effect of n-hexane, cyclohexane, heptane, n-octane, dodecane, and hexadecane on the BDS of *Nocardia globerula* R-9. The results revealed that the desulfurization rate increased with the increase of the number of carbon atoms in the range of C6–C16. The resting cells of *Nocardia globerula* R-9 showed higher desulfurization activity in hexadecane and dodecane than in hexane and heptane and showed that it is more appropriate to apply *Nocardia globerula* R-9 in the BDS of diesel oil rather than gasoline.

Jiang *et al.* (2002) studied the effect of different solvents (hexadecane, ethanol, DMF, and adding DBT, directly in aqueous system) on the BDS activity of *Pseudomonas delafieldii* R-8. The concentration of DBT in a hexadecane broth decreased from 0.2 to 0.002 mmol/L within 80 h. It was about 1.4 and 1.2 times faster than in DBT-ethanol and DBT-DMF cultures, respectively, but the slowest was with the direct use of DBT powder. This was explained by the ability of hexadecane to increase the contact area between cells and DBT, which increases the growth and, consequently, the BDS-rate. The specific activity of the non-genetically engineered strain RIPI-22 in both biphasic and aqueous reaction systems were detected as 6.5 and 4.52 μmol 2-HBP/g DCW/h, respectively. It was concluded that the reason of the differences is the cell's affinity to the organic phase is because of its surface hydrophobicity in the condition; the cells could use the substrate directly from oil fractions (Rashtchi *et al.*, 2006). The solubility of DBT in water is in the order of 0.005 mM and can be increased a little by surfactant produced by cells. Since cells were suspended in the aqueous phase, the increased rates in higher hydrocarbon fractions might suggest the transfer of DBT through the interface between the aqueous and hydrocarbon phase or the adsorption of cells at the interface (Maghsoudi *et al.*, 2001).

Alves and Paixão (2011) reported that *Gordonia alkanivorans* strain 1B is more sensitive to DMF than *R. erythropolis* D1. The IC_{50} of DMF was found to be 6.56% and 9.91%, while the IC_{50} of ethanol was found to be 10.56% and 10.52% for 1B and D1, respectively. This highlights the preferential use of ethanol as a solvent in BDS studies (Maghsoudi *et al.*, 2001; Guerinik and Al-Mutawah, 2003; Mingfang *et al.*, 2003; Wang *et al.*, 2004; Labena *et al.*, 2005; Tangaromsuk *et al.*, 2008; Chen *et al.*, 2009; Akhtar *et al.*, 2009; Boniek *et al.*, 2010; Davoodi-Dehaghani *et al.*, 2010; Pikoli *et al.*, 2014; Ismail *et al.*, 2016; Papizadeh *et al.*, 2017). However, methanol is also reported as a solvent in the BDS of DBT by *Lysinibacillus sphaericus* DMT-7 (Bahuguna *et al.*, 2011). Neverthless, ethylether was used as a solvent in the BDS OF DBT by *Candida parapsilosis* Nsh45 (El-Gendy *et al.*, 2006), *R. erythropolis* HN2 (El-Gendy *et al.*, 2014), *Brevibacillus invocatus* C19 (Nassar *et al.*, 2013), and *Brevibacillus brevis* HN1 (Nassar *et al.*, 2016). Morever, n-hexane was also reported as a solvent for the BDS of DBT using *R. erythropolis* HN2 (Nassar *et al.*, 2017a,b). The oil/water phase ratio affects the bioavailability of OSCs at the interface (Mohamed *et al.*, 2015; Adlakha *et al.*, 2016). From a practical point of view, the biodesulfurization (BDS) process has to be performed with really high proportions of organic solvent. Thus, for more economic benefits, it is preferable to work with the lowest water content and as much high BDS efficacy as can be reached for a high initial S-content oil feed. In a bi-phasic system (i.e. oil/water), the transfer of OSCs from the oil to the aqueous phase is one of the most determinant parameters in the BDS process. Therefore, using hydrophobic biocatalysts, such as *Rhodococcus* and *Gordona* cells, are preferable because they are capable of being joined at the oil/water interface for the uptake of OSCs there. Their ability of stabilizing the oil/water emulsion is related to cell surface hydrophobicity that may be related to cell surface long mycolic acids (Mohebali *et al.*, 2007). Reaching for microbial isolates tolerable to high contraptions of oil fraction is recommended for the success of the BDS process. For example, the presence of heavy gas oil (HGO) in the culture medium of *Rhodococcus erythropolis* ATCC 4277 promotes a higher cell growth, especially for HGO concentrations from 10 to 50 % (v/v). Other researchers reported a good cell adaptation when using approximately 20 % (v/v) of the organic phase (Izumi and Ohshiru, 2001; Kirimura *et al.*, 2001; Naito *et al.*, 2001; Li *et al.*, 2008b). The presence of hydrocarbons in abundance, as well as nitrogen and sulfur compounds, contributes to a better nutrition of the microorganism acting as a substrate and co-substrate. However, the decrease in cell concentration for higher concentrations of the oil fraction phase is due to the resistance of the microorganism decreasing with the increase of the toxicity of the system, making it impossible to maintain cell viability under more severe conditions (Carvalho *et al.*, 2004, 2005).

Moreover, the 4S-pathway is reported to be energetically expensive due to a great consumption of reducing equivalents and oxygen before the production of the free sulfur product (Tao *et al.*, 2006). Thus, the success of BDS as an alternative or complementary conventional desulfurization technique, on industrial scale, to produce ultra-low sulfur (ULS) fuels depends on the design of the microbial strains which remove sulfur in the presence of great proportions of organic solvents, with a higher rate or longer stability of desulfurization activity even at high temperatures. These criteria come in parallel with the worldwide depletion in reserves of high quality, low S-content conventional crude oil and the growing increment of utilizing the low quality, high S-content crude oil. The organosulfur compounds (OSCs) contribute to viscosity and other problems, such as pipeline corrosion and environmental pollution (Chapter 1). One of the recently recommended ways for upgrading heavy crude oils is the application of biodesulfurization (BDS). Most of the studies on heavy crude oil BDS have been limited to microorganisms which can either desulfurize aromatic or aliphatic organosulfur compounds.

Several studies have been focused on the isolation of new extremophilic microorganisms or the development of genetic modifications on natural strains (Xu *et al.*, 2006), improving the design of solvent tolerant strains which metabolize a broad range of organic sulfur compounds (Gunam *et al.*, 2006; Kilbane, 2006; Yu *et al.*, 2006a,b) under high temperatures (Li *et al.*, 2007) and with two-layer and continuous bioreactors (Yang *et al.*, 2007).

Several other studies have been performed on the Gram –ve *Pseudomonas* species for being solvent tolerable with good biodesulfurizing activity (Setti *et al.*, 1994, 1997; Luo *et al.*, 2002; Martin *et al.*, 2004; Alcon *et al.*, 2005; Tao *et al.*, 2006). Moreover, a two-phase system has been tested in many BDS studies, using *Rhodococcus* sp., in which hexane, heptane, and xylene were mainly used as the oil phase (Maghsoudi *et al.*, 2001; Ma *et al.*, 2006). The specific production rates increased with the increase in the DBT concentration and the hydrocarbon fraction using resting cells of *Rhodococcus* sp. P32C1 (Maghsoudi *et al.*, 2001).

Effects of different O/W phase ratios have been studied for the batch biodesulfurization of model oil DBT in n-hexadecane using resting cells of *Staphylococcus* sp. strain S3/C (Goindi *et al.*, 2002). The maximum extent of desulfurization (57%) with an overall rate of sulfur removal (2.2 mg S/L/h) occurred at an optimum hydrocarbon to an aqueous phase ratio (H/A) of 2:1, whereas the maximum specific sulfur removal rate (0.8 mg S/h/g DCW) at H/A was 3:1. The recorded low specific sulfur removal rate (0.2 mg S/h/g DCW) at a low hydrocarbon phase (0.5) was attributed to the limitation of effective contact of the hydrocarbon phase with the cells suspended in the aqueous phase under experimental conditions, while the observed repression of the specific sulfur removal rate at H/A 10:1 was attributed to the mass transfer limitation of O_2 to the high concentration of cells suspended in the aqueous phase. A similar effect of a decrease in the sulfur removal rate at a high cell concentration in high H/A was reported on *Rhodococcus* sp. P32C1 (Maghsoudi *et al.* 2001). Tolerance of the resting cells of *Staphylococcus* sp. strain S3/C at an H/A as high as 5:1 was observed by its recorded specific sulfur removal rate of 0.4 mg S/h/g DCW. However, in a study performed by resting cells of *Pseudomonas delafieldii* R-8, when the phase ratio was lower than 1 (O/W), the reaction proceeded more efficiently in comparison with that without the addition of dodecane (i.e. the non-aqueous or oil phase). The optimum phase ratio was found to be 1:4 (O/W) with an initial desulfurization rate of 4.6 mmol S/kg DCW/h. The possible reason for this was the formation of oil-in-water emulsion, where the interfacial area was increased resulting in the increase of the specific desulfurization rate. However, the high dodecane concentration (1.5 and 2.0 O/W) did not increase appreciably with the specific desulfurization rate. DBT was hardly desulfurized in dodecane without the addition of an aqueous phase. This was attributed to the activity of enzymes which need water.

Wang *et al.* (2006) reported that the optimum BDS of model oil (0.5 mmol/L DBT in n-hexadecane) by resting cells of *Corynebacterium* sp. ZD-1 was 1:2 (O/W) and the specific production rate of 2-HBP was about 1.72 times that in the aqueous phase and under higher oil held-up with an O/W ratio of over 2:1 and far less BDS activity was detected. This was explained because under low concentration, an oil-in-water emulsion was formed with the oil/water interface increasing with the proportion of oil. Meanwhile, hexadecane decreased the feedback inhibition of the biocatalyst due to the by-products' accumulation in the water phase. However, because water was necessary for enzyme activity, too high a concentration of hexadecane resulted in a low desulfurization rate.

A comparison study has been performed by Caro *et al.* (2007b) on the aerobic, Gram +ve *Rhodococcus erythropolis* IGTS8 and the genetically modified strain *Pseudomonas putida* CECT 5279 for the BDS process in both aqueous and biphasic media of model oil (DBT/n-hexadecane). This was undertaken over different experimental conditions, including different oil fraction phase (OFP) percentages, substrate concentrations, and cellular densities, using resting cells as an operation mode. Generally, when cell densities were not too high, both biocatalysts achieved better DBT conversion with higher biomass concentrations and lower oil fractions and

DBT concentrations. It was proved that the *P. putida* CECT 5279 strain is more sensitive for DBT mass transfer limitation and the encouragement results for the recommendation of applying the hydrophobic IGTS8 in biphasic systems is practical in the BDS process. Because of the oil–water partition coefficient of 2-HBP, its inhibition effect is lowered in a biphasic system. The evaluation was based on measuring serval parameters, such as BDS yields and BDS percentages (X_{BDS}) defined as the ratio between HBP concentrations produced and the initial substrate concentration used. The specific DBT consumption (q_s) and HBP production (q_p) rates were achieved after analyzing the samples after a prescribed time interval. They were found to decrease with the increase of OFP and there was no recorded BDS at OFP higher than 50% (i.e. > 1:1 O/W v/v). The CECT 5279 expressed a longer lag phase for BDS than that of IGTS8. This was attributed to the disability of CECT 5279 to uptake DBT directly from the solvent and in the absence of a co-solvent, such as ethanol, in the aqueous phase where there is practically no DBT bioavailable in the aqueous phase for the bacteria to start BDS. IGTS8 recorded an increase in specific DBT consumption and HBP production rates up to 40% (v/v) of OFP and then a drop to zero with 75% (v/v). This was explained by the increment of the organic interface surface with the OFP increment (Luo *et al.*, 2003) and the high capability of *Rhodococci's* to uptake organic compounds from the oil interface (Monticello, 2000). Thus, rates were higher until an oil–water proportion near 1:1, after which emulsion conditions might change (when there were water drops instead of oil ones) by avoiding cell-oil drop contact or where toxic oil effects were excessively influenced. In an aqueous phase system, the decrease of BDS efficiency was more pronounced with IGTS8 when the initial DBT concentration increased from 25 to 271 M, since the studied concentration range overcame the aqueous DBT solubility and the bioavailability did not go up enough. Furthermore, the steady production rate values of 2-HBP were attributed to its inhibitory effect, as it is a strong conversion inhibitor in the 4S-pathway (Ohshiro *et al.*, 1995a; Patel *et al.*, 1997; Monticello, 2000). CECT 5279 recorded a higher q_s and q_p in a biphasic system than in an aqueous one. This was explained by the higher bioavailability of DBT due to better diffusion in the aqueous phase, where the highest HBP production activity recorded 3.32 mmol HBP/kg DCW/h for an initial DBT concentation of 271 M. However, IGTS8 activity was not changed in the biphasic system, recording a maximum HBP production of 5.7 mmol HBP/kg DCW/h, for an initial DBT concentation of 271 M. The consumption rate of CECT 5279 was lower than that of IGTS8 in a biphasic system, recording a maximum value of 43 mmol DBT removed/kg DCW/h.

The growth rate of *Pantoea agglomerans* D23W3 was 50% slower when more than 10% of different oil feed, including light crude oil, heavy crude oil, diesel oil, hydrodesulfurized diesel, and aviation turbine fuel, were applied (Bhatia and Sharma, 2010). The oil/water phase ratio of 1:9 (i.e. 10%) was used in different BDS studies (Rhee *et al.*, 1998; Li *et al.*, 2003, 2007). Arabian *et al.* (2014) reported the BDS of model oil (500 ppm DBT in dodecane) by *Bacillus cereus* HN, where the BDS efficiency decreased with the increase of OFP and the highest of ≈ 60% occurred at 1/10 O/W.

Gün *et al.* (2015) reported the BDS of model oil (0.1 mM DBT in xylene) by the acidophilic and hyper-thermophilic *S. solfataricus* P2 in an oil/water phase ratio of 40% (v/v) which recorded approximately 22% within 72 h.

The resting cells of *Bacillus* strain KS1 also expressed higher DBT-BDS activity in a biphasic system of model oil DBT in tetradecane (1:1 O/W) than in an aqueous phase (Rath *et al.*, 2012). The resting cells of *S. subarctica* T7b in a biphasic system were reported to exhibit the complete removal of DBT (1.36 mM) within 24 h (Gunam *et al.*, 2013). Verma *et al.* (2016) reported a higher BDS efficiency in a biphasic system (0.2 mM DBT in n-hexadecane) by resting cells of *Bacillus* sp. E1 than in free resting cells, which recorded 63.6 ± 1.93 % with a 2-HBP production of 1.291 ± 0.095 mM in a biphasic system, while the other recorded 54.1 ± 2.19 % BDS% with a 2-HBP production of 0.913 ± 0.014 mM. The higher BDS efficiency in a biphasic system is due to the fact that in a biphasic system, 2-HBP accumulates in the oil phase, expressing lower toxic effects to the cells (Guobin *et al.*, 2006). Moreover, the dissolution of DBT in the non-aqueous phase increases the bioavailability of DBT to the cells (Davoodi-Dehaghani *et al.*, 2010).

Usually, upon the BDS-studies performed on heavy crude oil, dilution with organic solvents are performed. For example, Grossman *et al.* (1999) reported ten-fold dilution of crude oil with decane and a 30% reduction in total sulfur was observed using *Rhodococcus* sp. ECRD-1. Similarly, Ishii *et al.* (2005) observed a 52% reduction in a 12 fold diluted crude straight run light gas oil fraction using *Mycobacterium phlei* WU-0103. Adlakha *et al.* (2016) reported five fold dilution of heavy crude oil using n-hexadecane. This dilution was found to be appropriate for accuracy in measurements in experiments. Adlakha *et al.* (2016) mentioned that further oil dilution (> 5 fold) would involve the cost of additional diluent. In that study, the effects of different oil/water (O/W) phase ratios were studied for the BDS of heavy oil with an S-content of 4.53 mass percentge, which was decreased to 1.083 ± 0.5% upon the 5 fold dilution using n-hexadecane, using growing cells of *Gordonia* sp. IITR100. The recorded maximum S-decrease was 48% at

180 rpm and 30 °C with either 1:3 or 1:5 (O/W) after a 7 d incubation period. The least S-removal was recorded for an O/W phase ratio of 1:9. This was attributed to the high water volume which leads to the dilution of the enzyme activity and a decrease in overall desulfurization activity. Also, the 1:1 O/W expressed a low BDS efficiency, since higher oil content causes mass transfer limitations and limits the oxygen supply (Caro *et al.*, 2008). The 1:20 O/W was reported for the BDS of crude oil by resting cells of *Rhodococcus erythropolis* XP (Yu *et al.*, 2006a). Torkamani *et al.* (2008) reported an oil/water ratio of 1:10 for the BDS of Kuhemond heavy crude oil. An oil/water ratio of 1:50 has been reported for the BDS of heavy oil (Bunker oil) by a mixed culture (Jiang *et al.*, 2014).

The effect of the O/W phase ratio was studied for the BDS of model oil 0.3 mM DBT in n-hexadecane using resting cells of *Gordonia* sp. JDZX13 in an attempt to decrease the feedback inhibition of the end products, 2-HBP and sulfate ions (Feng *et al.*, 2016). The DBT removal efficiencies of the 2:1, 1:1, 1:2, and 1:3 systems were 20.89%, 13.89%, 24.40%, 41.92%, and 44.14%, respectively. Compared to the aqueous phase, although an oil phase was introduced, the removal efficiency decreased with the increase of the OFP. The reason for this might be that a higher oil/aqueous ratio reduced the effective space of cell growth, resulting in a lower biomass for further BDS (Alves and Paixão, 2011; Srivastava, 2012). However, the highest DBT-removal efficiency was achieved in the 1:3 O/W ratio system. Considering the amount of water used, the DBT-removal efficiency of the oil/aqueous ratio 1:2 system reached 0.188 mmol/L water, which was the highest among all the systems. Thus, the oil/aqueous ratio of 1:2 (improved by 100.7%) was determined as the best choice for better balancing the feedback inhibition effects of 2-HBP and sulfate ions.

8.13 Effect of Medium Composition

The production of biomass with high BDS efficiency is important since the BDS-degree developed during microbial-growth is useful not only to compare the results achieved under different media and conditions, but also to compare different microorganisms with desulfurization capability. This mainly depends on the composition of the media applied for the BDS process. For example, the effect of an addition of co-substrate, either as a sulfur source (such as dimethylsulfoxide DMSO or sulfate) or carbon source (such as glucose, glycerol citrate, etc). Not only this, but one of the main limitations to the BDS industrial application is the high costs of the nutritional medium compounds.

Usually, some solvents are used to dissolve DBT (such as n-hexane, ethanol, diethylether, etc.) before its addition in a culture medium. Thus, the medium usually inevitably contains a certain amount of these solvents. Some reports pointed to the enhancement of the rate of BDS through the addition of these solvents. For example, the addition of DBT dissolved in ethanol provided more rapid growth and desulfurization than only DBT powder (Setti *et al.*, 1995).

Carbon and nitrogen concentration in a BDS-culture medium significantly affected the BDS process (del Olmo *et al.*, 2005a; Maass *et al.*, 2015; Porto *et al.*, 2017). The carbon source that is most commonly employed is glucose (Izumi *et al.*, 1994; Ohshiro *et al.*, 1994; Wang and Krawiec, 1996; Oldfield *et al.*, 1997; Folsom *et al.*, 1999; Kobayashi *et al.*, 2000; Yan *et al.*, 2000; Matsui *et al.*, 2001; Nakayama *et al.*, 2002), but glycerol is also used (Kilbane and Jackowski, 1992; Kayser *et al.*, 1993; Denome *et al.*, 1994; Kishimoto *et al.*, 2000; Maghsoudi *et al.*, 2001; Abbad-Andaloussi *et al.*, 2003) and other sources, such as ethanol (Wang and Krawiec, 1994; Onaka *et al.*, 2000; Yan *et al.*, 2000), sodium succinate (Omori *et al.*, 1995; Setti *et al.*, 1999; Abbad-Andaloussi *et al.*, 2003), and tetradecane (Ohshiro *et al.*, 1995a) have been reported, while the common nitrogen source that is usually employed is ammonium as NH_4Cl, but ammonium nitrate is also reported (Omori *et al.*, 1995; Setti *et al.*, 1997; Abbad-Andaloussi *et al.*, 2003). Different organic S-compounds are also reported as sulfur sources. Some of these are 4S route compounds (Kayser *et al.*, 1993; Denome *et al.*, 1994; Ohshiro *et al.*, 1996b; Wang and Krawiec, 1996; Setti *et al.*, 1999; Yan *et al.*, 2000; Maghsoudi *et al.*, 2001), including dimethylsulfoxide (DMSO) (Oldfield *et al.*, 1997; Folsom *et al.*, 1999; Kayser *et al.*, 1993; Omori *et al.*, 1995; Abbad-Andaloussi *et al.*, 2003) and others, such as thiophenes, sulfides, mercaptans, naphthalenes, and amminoacids (Kayser *et al.*, 1993; Izumi *et al.*, 1994; Omori *et al.*, 1995). Recently, new studies have been reported for using some natural agro-industrial wastes, such as molasses, cassava waste, trub, etc., in the BDS-culture media since they are rich in essential components required for microbial growth and high enzymatic activities.

The pulp and paper industries generate large amounts of waste throughout the year (Thomas, 2000). Concentrated sludge generated by wastewater treatment facilities of recycled paper plants is currently a major disposal problem concerning the paper industry that has to be urgently solved (Oral *et al.*, 2005). Recycled paper sludge (RPS) (after neutralization) is reported to be made up of approximately 35% cellulose, 10% xylan, and 20% lignin (on a dry-weight basis) and the remaining is mainly inorganic ash. Due to this high polysaccharide content, RPS appears as a promising feedstock for the formulation of inexpensive culture media (Van Wyk and Mohulatsi, 2003), providing their polymeric carbohydrates are broken down into fermentable monosaccharides. This hydrolysis can be carried out by chemical or enzymatic methods. The latter is advantageous since it is more specific and allows milder operation conditions leading to reduced production of biologically inhibitory compounds (such as sugar and lignin degradation products) and the biocatalysts are potentially reusable (Wen *et al.*, 2004).

Alves *et al.* (2008) investigate the possibility of using the RPS hydrolysates obtained either with dialyzed enzymes (dialyzed hydrolysate) or non-dialyzed enzymes (non-dialyzed hydrolysate) as a carbon source in a concentration of 10 g/L glucose for DBT-BDS using *Gordonia alkanivorans* strain 1B in a batch culture containing 0.25 mM DBT. The maximum specific growth rates, μ_{max}, recorded 0.051 and 0.035 h^{-1} in the presence of non-dialyzed hydrolysate and dialyzed hydrolysate, respectively, compared to 0.019 h^{-1} in the presence of commercial grade glucose as the only carbon source. However, 125 μM 2-HBP with the BDS of 250 μM DBT was only observed in the presence of dialyzed hydrolysate with a maximum specific productivity of 2-HBP at 1.1 μmol/g (DCW)/h. Dialysis was performed aiming to remove possible sulfur compounds present on the enzymatic formulation that could be more easily assimilated than DBT by bacterial cells, unless a recombinant strain (lacking dsz substrate repression) is used. Then, Alves *et al.* (2008) investigated the capability of using the dialyzed hydrolysate as a complete culture medium. The results revealed that RPS hydrolysate, without additional nutrients and with phosphates or with phosphates and magnesium, did not allow a significant growth or 2-HBP production. However, RPS hydrolysate with ammonia allowed high initial growth and 2-HBP, but both were lost after 72 and 96 h, respectively. This was attributed to the absence of some nutrients, but a complete consumption of the 250 μM DBT and glucose occurred in the cases of RPS hydrolysate with ammonia + phosphate, RPS hydrolysate + phosphates + ammonia + magnesium, RPS hydrolysate + phosphates + ammonia + magnesium + Zinc, and RPS hydrolysate + phosphates + ammonia + magnesium + trace elements solution (all the nutrients were added according to the components of sulfur-free medium (SFM) as described by Alves *et al.*, 2007) and $ZnCl_2$ in 10 mg/L. Any of the listed formulas can be used as complete mediums for DBT-BDS by *G. alkanivorans* strain 1B. Nevertheless, the highest specific maximum growth rate, μ_{max} 0.03 h^{-1}, and 2-HBP production rate, 5.7 μM/h, occurred with RPS hydrolyzate + phosphates + ammonia + magnesium + zinc due to the presence of Zn. Thus, upon the application of that formula for the batch BDS of model oil containing 2 mM each of DBT, 4-MDBT, and 4,6-DMDBT dissolved in n-heptane, a final ration of 1/10 O/W occured. Total consumption of glucose occurred within 96 h with a μ_{max} of 0.062 $h^{-1}$and decrease of total S-content occurred by 2.6 fold reaching 2.23 mM after 168 h of cultivation. The specific desulfurization rates after 24, 48, and 72 h were 22.2, 11.1, and 4.8 μmol/g DCW/h, respectively (Alves *et al.*, 2008). Realistically, for a potential industrial application, an inexpensive culture medium would have to be employed in a large-scale process. Therefore, the cost of commercial enzyme formulations might economically hamper the suggested process by Alves *et al.* (2008), unless the enzymes added in the hydrolysis step are recovered and reused. However, Alves *et al.* (2008) suggested that this might be achieved by developing a process based on an enzymatic membrane reactor in which a semi-permeable ultrafiltration membrane is used to retain the enzymes in the reactor while preserving their activity.

Carob kibbles are another agro-industrial waste characterized by a high content of soluble sugars, mainly sucrose, glucose, and fructose, which are easily extractable by water, producing sugar-rich syrups (Mansom *et al.*, 2010), but these wastes have high concentrations of sulfate, thus to use carob pulp liquors for BDS processes, it is necessary to reduce the sulfate concentration to minimum levels. Silva *et al.* (2013) studied the removal of such sulfate by precipitation, using $BaCl_{2,}$ producing the less toxic, low water soluble $BaSO_4$ (0.00285 g/L at 30 °C) (Guo *et al.*, 2009). During this treatment, an accumulation of barium sulfate occurs which can constitute an environmental problem in large-scale BDS application. However, Silva *et al.* (2013) mentioned that this drawback can become an economic advantage, suggesting that the recovered barium sulfate at the end of the process can be used as a component of oil well drilling (the main application of this material) or as a component of white pigment for paints. The treatment of carob pulp liquor by 0.5% of $BaCl_2$ for 21 h led to carob pulp liquor with a maximum sulfate concentration of 15–18 mg/L sulfate. Upon the application of treated carob pulp liquor in the ratio of 1:6 as the sole C-source to a batch BDS process, using *Gordonia alkanivorans* strain 1B, with an initial DBT concentration of 400 μmol/L dissolved in DMF, a maximum BDS efficiency was measured by the produced 2-HBP and recorded as 237 μmol/L (Silva *et al.*, 2013).

The commercially and readily available sugar beet molasses (SBM), which is rich in sucrose (~50 %, w/v) that can be hydrolyzed to glucose and fructose, has been used for the enhancement of the BDS efficiency of *G. alkanivorans* 1B (Alves and Paixão, 2014a). Untreated SBM revealed good microbial growth with μ_{max} of 0.067 h^{-1} and a biomass yield of 0.471 g/g within 3 d, but no DBT-BDS has occurred. Upon the pretreatment of SBM by 0.25% $BaCl_2$ overnight (16–18 h) to precipitate the unwanted sulfates (denoted SBM_t), the BDS efficiency as measured by the produced 2-HBP and recorded at 200 μM from the BDS of 250 μM DBT. Alves and Paixão (2014a) also investigated the effect of different SBM pretreatment methods on DBT-BDS: (i) DBT desulfurization using directly SBM_t, (ii) DBT desulfurization using SBM_t after acidic hydrolysis (SBM_t–AH) to convert sucrose to glucose and fructose, and (iii) DBT desulfurization using SBM_t in a SSF approach with *Zygosaccharomyces bailii* strain Talf1 crude enzymatic extract exhibiting invertase activity. This showed that the addition of SBM_t in a dilution of 1:50 i.e. 12 g/L sucrose in a batch culture of 250 μM DBT, within 96 h, 1B produced 122 μM 2-HBP and a μ_{max} of 0.04 h^{-1} with a biomass yield of 0.377 g/g sugar, maximum 2-HBP production rate of 2.56 μM/h, and a maximum 2-HBP specific production rate q_{HBP} 2.2 μmol/g DCW/h. This was attributed to the presence of some nutrients in this complex alternative carbon source, which can promote a faster bacterial metabolism. Upon the application of SBM_t–AH, the DBT-BDS was further enhanced recording μ_{max} of 0.0575 h^{-1}, with a biomass yield of 0.488 g/g sugar within 3 days of incubation, which was approximately 44 % higher than in the growth with SBM_t that was not hydrolyzed. Maximum production of 2-HBP, 168 μM, was achieved after 66 h with a maximum and specific production rate of q_{HBP} 5.76 μM/h and 2.61 μmol/g DCW/h, respectively. The rate of fructose utilization was higher than that of glucose; the maximum fructose consumption rate was 0.167 g/L/h and the maximum glucose consumption rate changed from 0.078 g/L/h (during the high availability of fructose) to 0.134 g/L/h (after the low availability of fructose). However, in order to make the process more eco-friendly and decrease the costs associated with the sucrose hydrolysis of SBM_t, the utilization of invertases in a simultaneous saccharification and fermentation approach was also studied. Alves and Paixão (2014a) performed another set of DBT desulfurization assays using SBMt and a crude enzymatic extract containing inulinase/invertase activity produced by the yeast *Z. bailii* strain Talf1 (1 % v/v) in the SSF approach. The SSF process allowed a faster growth of strain 1B with a μ_{max} of 0.08 h^{-1} and a higher biomass yield of 0.72 g/g sugar during 55 h. The 2-HBP recorded a maximum quantity of 249.5 μM at 47.5 h of culture, which corresponds to the total desulfurization of DBT present in the culture medium (250 μM). The maximum 2-HBP production rate achieved was 7.78 μM/h and the q_{HBP} was 3.12 μmol/g DCW/h. Alves and Paixão (2014a) estimated that the application of SBM in BDS as carbon source instead of sucrose would lead to a cost reduction of 3-fold. Moreover, SBM utilization in an SSF process, with invertase crude extract, contributes to an improved DBT desulfurization due to possible nutrients/inducers from the molasses or enzymatic extract.

Jerusalem artichoke (JA, *Helianthus tuberosus*), also known as topinambur, is a species of sunflower and is cultivated worldwide for animal consumption and/or sugar (Baldini *et al.*, 2004). Recently, it founds its way to use in biofuels, including bioethanol and biodiesel production and in the biochemical industry (Cheng *et al.*, 2009; Liang *et al.*, 2012; Guo *et al.*, 2013; Li *et al.*, 2013). The JA can grow on poor land, being resistant to frost, drought, pests, diseases, and saline soils and it is characterized by high fertility and demands low amounts of fertilizers (Zhao, 2011). As it is drought-resistant, it can be used to improve soil and water conservation in arid regions. Moreover, JA tubers accumulate inulin, a polydisperse fructan polymer composed by linear chains of β-2,1-linked D-fructofuranose molecules terminated by a glucose residue through a sucrose-type linkage at the reducing end. Inulin is soluble in water and, after hydrolytic breakdown by inulinases, it releases fructose and glucose producing sugar-rich juices, containing up to 95% fructose (Makino *et al.* 2009; Kango and Jain 2011). Thus, JA is considered as a natural source of fructose for the industry and an attractive alternative carbon source for microbial growth and further application to bioprocesses that use fructophilic microorganisms, such as BDS (Paixão *et al.*, 2013; Alves and Paixão 2014b). However, the sulfate concentration of JA reaches up to > 450 mg/L. Thus, sulfate removal is a must before its application in BDS process.

Silva *et al.* (2014) reported the acid hydrolysis of Jerusalem artichoke (JA) to produce fructose rich juice to be used as a carbon source for batch DBT-BDS using the fructophilic bacterium *G. alkanivorans* 1B. The amounts of sugars (sucrose, glucose, and frustose) in the acid hydrolyzed JA juice (JAJ) are about 85% higher than those in JA. Upon the application of JAJ in a batch BDS of DBT with an initial concentration of 400 μmol/L, where inulin was the main C-source, the μ_{max} recorded 0.0546 h^{-1} due to the low available fermentable sugars (4 g/L) and the high sulfate concentration, 77.8 mg/L, (as JAJ was added 1:6, so it was diluted) stopped the BDS and no 2-HBP was produced. Nevertheless, upon the application of acid hydrolyzed JAJ with approximately 25 g/L glucose and fructose, the μ_{max} increased to 0.0816 h^{-1} due to the presence of high C-nutrients (sugars and growth inducing compounds, such as vitamins and amino-acids), but still no 2-HBP was produced due to the high concentration of sulfate (~ 75 mg/L). Upon the treatment of JAJ and acid hydrolyzed JAJ, denoted JAJ_t with 0.5% $BaCl_2$ at pH8.73, the production of 2-HBP was observed with both juices; for JAJ_t with < 1 g/L of fermentable sugars, the bacterial growth achieved a μ_{max} 0.0428 h^{-1}. The 2-HBP production began at 18 h, reaching a maximum concentration of 43 μmol/L by the end of the growth. The maximum production rate was 1.74 μmol/L/h and the specific production rate

($q_{2\text{-HBP}}$) was 1.23 μmol/g DCW/h. These results showed the effectiveness of $BaCl_2$ treatment, where, despite the small growth, the production of 2-HBP was observed, meaning that only a residual sulfate level was introduced in a test medium and it was non-inhibitory for the BDS process. The limiting factor in that process was the very low level of available fermentable sugars for growth. In acid hydrolyzed JAJ_t, with residual sulfates (<2 g/L) and all sugar content as fermentable sugars bioavailable (25 g/L), the BDS results were remarkable: μ_{max} 0.06 h^{-1} with total consumption of the available carbon source within 96 h. Total consumption of DBT occurred with a nearly stoichiometric production of 2-HBP, which was initiated within 18 h and attained ~ 400 μmol/L within 73 and 96 h, that corresponded to an overall production rate of 4.48 μmol/L/h and to a maximum production rate of 28.2 μmol/L/h, with a related $q_{2\text{-HBP}}$ of 5.06 μmol/g DCW/h. Using the $BaCl_2$ treated hydrolyzed JAJ_t, the amount of 2-HBP produced (μmol) per gram of sugar consumed averaged 16 μmol/g (Silva *et al.*, 2014).

Cassava wastewater has a high pollution potential, primarily because of its high organic load. However, this residue also has essential nutritional components, such as the sugar combination of sucrose, fructose, glucose, and mainly soluble starch (Barros *et al.*, 2013), and ammonium ions (NH_4^+) and nitrate (NO_3^-) as nitrogen sources (Barros *et al.*, 2008). Trub is a precipitation by-product from the wort boiling process in brewing industries. Its composition is comprised of 40–70% protein, 10–20% hop bitter acid, 7–8% polyphenols, 7–10% carbohydrate, and 1–2% fatty acid (Kunze, 2014). In the beer manufacturing process, the complete removal of trub is essential to ensure the quality of the final product in terms of the yield of bitter substance (Kühbeck *et al.*, 2007). Thus, the application of these wastes in the BDS process would solve some waste management problems and would have a positive impact on the economy, environment, and energy.

Porto *et al.* (2017) reported a specific desulfurization rate of 453 mg S/kg HGO/h, with a BDS efficiency of 75% within 12 h using *R. erythropolis* ATCC 4277 in the presence of cassava waste water as a source of carbon co-substrate and nitrogen, while, upon the application of hot trub, the BDS percentage reached 70% within only 1 h of incubation with a specific desulfurization rate of 5073 mg S/kg HGO/h.

In a process for diesel oil BDS, ethanol was added for the enhancement of phase separation, as it acts as a de-emulsifier to three phase (diesel oil/biocatalyst/aqueous phase) emulsion. It promoted the phase separation, recovery of desulfurized diesel oil, and microbial cells from the very stable emulsion from the BDS-reactor (Choi *et al.*, 2003). After the phase separation, the major portion of the added ethanol remains in the aqueous phase containing microbial cells, which can serve as a good carbon source for the growth of the recovered cells to be recycled to the reactor, replacing, at least partially, the carbon source requirements.

However, Kim *et al.* (2004) reported that the growth rate and BDS efficiency performed by *Gordonia* sp. CYKS1, in presence of ethanol, was higher than that in the presence of glucose. The BDS-rate increased about 2.5 times (from 1.71 to 4.02 μmol/L/h) as the ethanol concentration increased from 3 to 20 g/L. The Monod-model equation was found to fit the growth of *Gordonia* sp. CYKS1on ethanol and glucose, with maximum specific growth rates (μ_{max}) of 0.2 and 0.08 and saturation constants (K_m) of 2.76 and 6.29 g/L, respectively.

BDS can be enhanced by promoting the oxidoreductase activity via the addition of a carbon source. To test whether biodesulfurization can be enhanced by oxidoreductase co-expression, three different sources of carbon were used; glucose, glycerol, and acetate were used to supplement the resting cells medium of *Rhodococcus* species supplemented with 1% DBT. The rate of HBP production was highest with glucose, followed by glycerol and acetate, in a decreasing order, recording 5.82, 4.77, and 2.48 μmol 2-HBP/g-cell/h, respectively. This corresponds with the theoretical yields of NADH from the carbon sources of 10, 6, and 0.5 for glucose, glycerol, and acetate, respectively (Guerinik and Al-Mutawah, 2003).

However, in a study performed by Luo *et al.* (2003) on resting cells of *Pseudomonas delafieldii* R-8 in biphasic batch BDS process (1:1 O/W) of model oil (1 mM DBT in dodecane), using different aqueous reaction media, such as saline (0.85 g NaCl/L), 0.1 M phosphate buffer (pH 7.0), and sulfur free medium (SFM), the results showed that the rates of DBT consumption and 2-HBP production were almost linear in the initial stage of reaction. The DBT was completely removed within 8 h with an almost stoichiometric production of 2-HBP in all the studied aqueous phases. The most valuable observation was the proceeding of DBT-BDS by the R-8 lyophilized cells in the non-growth media (i.e. saline and phosphate buffer) being as efficient as in the growth medium (i.e. SFM). Not only this, but the results also also that the DBT-BDS by whole cells of R-8 did not need the addition of any other cofactors or reducing agents.

In another study performed by Xu *et al.* (2002) on batch BDS of 0.5 mmol/L DBT by *Rhodococcus.* sp. 1awq strain within 48 h at 30 °C and pH7, glycerol showed the highest growth followed by citrate, glucose, and acetate in a decreasing order. The effect of glycerol concentration was also studied and higher concentrations (>0.5%) were not favorable. Not only this, but higher concentrations of NH_4Cl (>0.5%) were not favorable either in the presence of 0.4% glycerol or 1 mmol/L Na_2SO_4.

Alves *et al.* (2004) reported the BDS of 478 μM DBT by *Gordonia alkanivorans* strain 1B in the presence of glucose as a co-substrate to 168 μM, within 120 h of culture, with a specific desulfurization rate of 1.03 μmol DBT/g DCW/h. This rate is similar to that obtained with *Gordonia* sp. strain CYKS1 at 0.917 μmol DBT/g DCW/h, which reduced the concentration of DBT from 320 to 50 μM within 120 h (Rhee *et al.*, 1998). The maximum extracellular concentration of 2-HBP, however, was somewhat lower at about 120 μM, which is only 27% of the consumed DBT (450 μM).

Labana *et al.* (2005) reported the batch BDS of 27 ppm DBT dissolved in ethanol by growing cells of *Rhodococcus* sp. and *Arthrobacter sulfureus* isolated from oil-contaminated soil/sludge samples in the presence of 10 mM sodium succinate as a C-source, where the S-level decreased to 8 and 10 ppm, recording a BDS percentage of 70 and 63%, respectively, within a 30 d incubation period.

Yan *et al.* (2000) studied the relative efficacy of ethanol, glucose, and glycerol as sole carbon sources for the growth and desulfurizing activity of *R. erythropolis.* They reported ethanol to yield the highest growth and desulfurizing rates, followed by glucose, then glycerol. The recorded results were as follows: the highest growth rate of 1.39 h^{-1} and the highest desulfurizing rate of 0.18 mmol HBP/g DCW/h for ethanol. In contrast, the rates were 0.60 h^{-1} and 0.08 mmol HBP/g DCW/h for glucose and 0.59 h^{-1} and 0.07 mmol HBP/ g DCW/h for glycerol. Similar observations were obtained by Aggarwal *et al.* (2011) using *R. erythropolis*, where the desulfurization and growth rates, relative to those of ethanol, decrease in the following order: ethanol (0.18 mmol HBP/g DCW/h as 100% and 1.39 h^{-1} as 100%) > lactate (67%) > citrate (48%) > glutamate (44%) > glucose = fructose (43%) >glycerol (42%) > gluconate (40%). These results can be attributed to the NADH and carbon metabolism. Ethanol yields more NADH during this metabolism than glucose and glycerol. The more the NADH, the more the bacterial growth and it enables more of the cells to increase the flux (i.e. the desulfurizing rate) of the 4S pathway. Aggarwal *et al.* (2013) investigated the effect of 17 carbon sources (acetate, citrate, ethanol, formate, fructose, fumarate, gluconate, glucose, glutamate, glycerol, lactate, malate, oxaloacetate, oxoglutarate, pyruvate, and succinate) on growth and the BDS activity of *G. alkanivorans*. Ethanol was found to be the best for growth and BDS activity, recording μ_{max} 0.027 h^{-1} and a BDS-efficiency of 3.3 mmol HBP/g DCW/h. Relative to the maximum activity that occurred with ethanol (considered as 100%), those obtained with other studied sugars ranked as follows: fumarate (80%) > oxoglutarate (78.79%) > pyruvate (78.43%) > glutamate (78.24%) > succinate (78%) > acetate ≈ fructose ≈ glucose ≈ lactate (76.86%) > glycerol (75%) > citrate

(71.88%) > oxaloacetate (69.70%) > malate (69.11%) > formate (50%). This ranking remained unchanged even for BT as the sole sulfur source. This was attributed to the production of NADH. For the cell to consume 1 mol of DBT as a sulfur source via the 4S pathway, it requires 4 moles of NADH (Aggarwal *et al.*, 2011). Additionally, NADH is required for other growth related activities. The carbon nutrient is the main source of this energy. It affects the cofactor regeneration in cellular metabolism. Therefore, a carbon source that provides more NADH during its metabolism is likely to support higher desulfurization and growth. One mole of ethanol generates two additional moles of NADH (Aggarwal *et al.*, 2011), which is the highest among all the studied 17 substrates. So, it can be concluded that any changes in medium design or genetic manipulations that increase NADH regeneration and supply within the cellular metabolism are likely to enhance desulfurization activity.

Recently, Alves and Paixão (2014b) described *Gordonia alkanivorans* strain 1B as being fructophilic. This means it prefers fructose to glucose as a carbon source to grow, growing faster and, thus, desulfurizing at faster rates in the presence of fructose, where, upon studying the effect of different carbon sources in a batch BDS of 500 μM DBT, the lowest values for the growth rate (0.025 h^{-1}) and overall 2-HBP production rate (1.80 mM/h) by the strain 1B were obtained in glucose grown cultures. Strain 1B grew considerably in the presence of glucose and sucrose, but the bacterial growth was much higher when fructose was used as the only carbon source. In fact, after five days of growth, the optical density obtained in a fructose medium was about 5 and 3.3 fold relative to that of glucose and sucrose, respectively. Moreover, this higher growth in the presence of fructose led to a higher 2-HBP production (248.1 mM), thus, the desulfurization ability was 5 and 6 fold higher, relevant to that obtained in the presence of sucrose (53.9 mM) and glucose (41.0 mM), respectively. However, no significant bacterial growth was observed in culture media containing cellobiose or xylose as the only carbon source. Strain 1B grew about 3.6 fold faster when fructose was present as a carbon source instead of glucose. The amount of 2-HBP produced μmol/g of sugar consumed by *G. alkanivorans* in glucose, sucrose, fructose, and fructose + glucose recorded 19.6, 20.8, 30, and 24.2 μmol/g consumed sugar, respectively. When using sucrose, the increase in the growth rate exhibited by strain 1B led to a higher biomass productivity, which induced a slight increase in the 2-HBP production rate (1.91 μM/h). Conversely, in terms of a 2-HBP specific production rate (q_{HBP}), the value obtained was markedly lower (0.718 μmol/g/h in sucrose versus 1.22 μmol/g/h in glucose). When a mixture of glucose and fructose was used as a carbon source, strain 1B reached a value of q_{HBP} 1.90 μmol/g/h, close to that in fructose (q_{HBP} 2.12 μmol/g/h). The highest values for both cell growth (μ_{max} 0.091 h^{-1}) and 2-HPB production (9.29 μM/h) were obtained when strain 1B was desulfurizing DBT in the presence of fructose as the only carbon source, indicating a fructophilic behavior by this bacterium. The greater number of functional cells conducted a more effective BDS process by strain 1B, as they attained a q_{HBP} about 74% higher than in glucose grown cultures, increasing from 1.22 to 2.12 μmol/g/h. Furthermore, the significant BDS enhancement has been better observed in terms of the overall 2-HBP production rate, which increased over 5 fold from 1.80 μM/h (in presence of glucose) to 9.29 μM/h (in presence fructose). This fact is in agreement with the highest value of biomass productivity by strain 1B in fructose, which resulted in a higher number of cells fulfilling DBT-desulfurization. The greater number of functional cells conducted a more effectivene BDS process by strain 1B, as they attained a q_{HBP} about 74% higher than those in glucose grown cultures. Moreover, this significant BDS enhancement can be better observed in terms of the overall 2-HBP production rate, which increased over 5 fold from 1.80 mM/h (in glucose) to 9.29 mM/h (in fructose). Fructophilic pathways were usually associated with yeasts; in fact, until the discovery of this characteristic, lactic bacteria were the only exception (Sutterlin, 2010; Endo, 2012). Since the utilization of microorganisms in the industry requires cheap carbon sources to be productive, the fact that this strain is fructophilic opens new possibilities in terms of available carbon sources, allowing the exploration of some fructose rich residues and plants that might both make the process cheaper and more efficient.

Alves and Paixão (2014b) also studied the effect of fructose pulse by fructose addition to a batch DBT-BDS using *G. alkanivorans* 1B and glucose as a C-source in the middle of an exponential phase that was after 96 h of cultivation, which led to the enhancement of the BDS-process; the growth rate increased from 0.025 h^{-1} to 0.072 h^{-1}, the biomass production rate increased from 0.005 g/L/h to 0.147 g/L/h, biomass yield increased from 0.327 g/g to 0.455 g/g, the rate of 2-HBP production increased from 0.72 μM/h to 11.59 μM/h, and the specific production rate of 2-HBP q_{HBP} increased from 1.7 μmol/g/h to 4.96 μmol/g/h. Finally, upon the comparison of the efficiency of 1B resting cells pregrown on glucose or fructose, 150 μM DBT, and harvested at the late exponential phase on a batch BDS of 500 μM DBT, resting cells pre-grown in fructose, showed an overall 2-HBP specific production rate of 3.6 μmol/g DCW/h, which was about 67% higher than that obtained using resting cells pre-grown in glucose (2.16 μmol/g DCW/h) (Alves and Paixão, 2014b).

Dejaloud *et al.* (2017) investigated the importance of the maintenance energy during BDS of different concentrations of DBT, where glucose as the most common energy substrate in biotechnology industries was used at two different initial concentrations. The energy-limited growing cell culture of *Ralstonia eutropha* PTCC1615, using glucose at 55 mM, was used to describe cell behavior when the availability of the energy source became limited (denoted case-A). The energy-sufficient medium prepared using glucose at 111 mM, as the initial concentration, showed the amount of energy source to be in considerable excess over other cell requirements, such as nitrogen or sulfur sources (denoted case-B). PTCC1615 showed sufficient growth in both cases, without any lag phase, reaching the stationary phase within 10 h with complete consumption of glucose and 40–50 mM remaining in case-A and B, respectively. This was accompanied with a decrease in pH due to the production of acetic acid. It is worth noting that the yield of 2-HBP as a measure for BDS-efficiency (Y_{DBT}%) was higher in the energy-sufficient media when compared with energy-limited cultures, regardless of the initial DBT-concentration. The recorded maximum Y_{DBT} value of 90.7% was obtained in initial DBT and glucose concentrations of 0.05 and 111 mM, respectively. The specific growth rate (μ h^{-1}) was higher in energy-limited cultures of DBT (0.56–0.61 h^{-1}) than in energy-sufficient media (0.32–0.41 h^{-1}), since all the cells' efforts in the "B" cultures are directed toward the DBT-BDS process and the cells focused less on biomass formation. But, in a control culture with NH_4SO_4 instead of DBT, the recorded μ (0.63 h^{-1}) with 111 mM glucose was higher than that with 55 mM glucose (0.49 h^{-1}). This indicated that the cells' need for NAD(P)H and $FMNH_2$ is low in the absence of the DBT-BDS process. In that study, with DBT as a toxic substrate, showed that its presence in growing cell cultures of cases "A" and "B" cultures exert a tension on the cells and the formation of 2-HBP in the DBT-BDS process may be considered as the product formation case, which has been described before as the formation and excretion of only one predominant product (Tsai and Lee, 1990). Considering the dynamic behavior of the cells, this type of expression occurs under certain conditions (Tsai and Lee, 1990; Zeng and Deckwer, 1995). The activity of the enzymes responsible for the product's synthesis is one of the mentioned conditions. For instance, the responsible enzymes are expressed under the control of the dsz operon in the *Rhodococcus* sp. strain IGTS8 during the BDS process. The cellular pools of the produced NADH and $FMNH_2$ coenzymes should be directed toward the BDS process and the excess energy in the cells in the form of accumulated coenzymes can be relieved in this way (Figure 8.2). The response of chemoheterotrophic microbes to the energy substrate in a particular medium defines the cell behavior in regulating the metabolic activities by balancing between catabolic and anabolic reactions (Russell and Cook, 1995). However, in

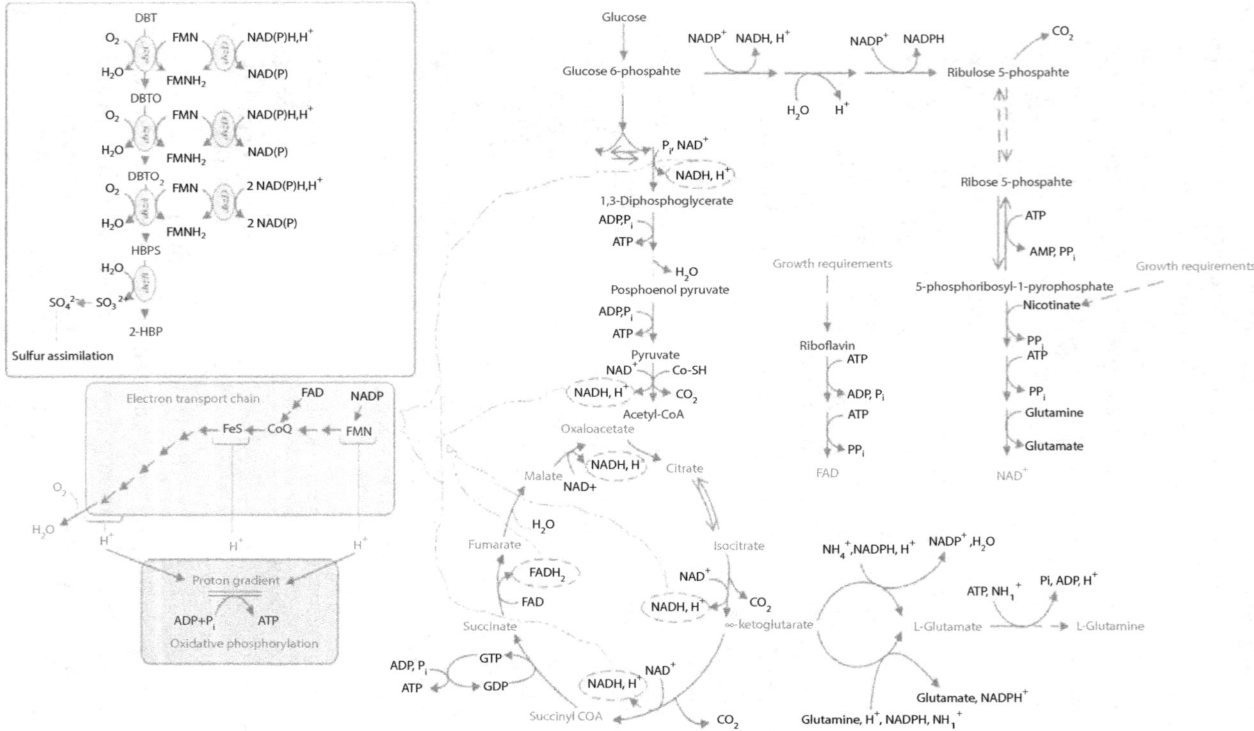

Figure 8.2 Certain Molecular Pathways Operating in a Hypothetical Aerobic Bacterium Involved in the DBT-BDS Process This bioenergetics scheme focuses on the cell's need for hydrogen and electron donor molecules, such as NAD^+/NADH and $NADP^+$/NADPH. The pentose phosphate pathway starts from glucose 6-phosphate and this pathway is actively involved in the formation of NAD^+ and FAD. Ribose-5-phosphate from this pathway is also involved in the synthesis of $NADP^+$. The determination of the roles of NAD^+ and FAD in ATP synthesis through the electron transport chain and electron phosphorylation is also shown. Involvement of the intermediates of the TCA cycle during amino-acid synthesis through glutamate–glutamine formation is also presented (Zubay, 1998; Kilbane and Robbins, 2007; Berg *et al.*, 2010).

many cases, the amount of energy (ATP) generated through exogenous oxidizable substrates, such as glucose, is not directly proportional to the cell growth yield. Although the logistic model is referred to as the substrate-independent model in some texts (Zhang *et al.*, 2013), the initial concentration of the substrate actually determines the highest achievable cell density in a particular environment. In other words, C_x^{max} reflects the capacity of the environment to support cell growth. The important point to realize is the aerobic cell requirements of the carrier molecules that are needed to transfer electrons of the energy substrate (i.e., glucose) to O_2 as the final electron acceptor molecule (Figure 8.2). Aerobic respiration (in terms of the microbe's bioenergetics) involves the formation of 3 and 2 moles of ATP, respectively, per 1 mol of NADH or $FADH_2$ being oxidized (Berg al., 2010). The NADH requirements for the DBT-BDS process have been reported and this electron rier was found to be more efficient than NADPH (Ohshiro *et al.*, 1994; Boniek etal., 2015). The findings using this approach showed that 4 moles of NADH are needed to convert 1 mol of DBT to 1 mol of 2-HBP through the 4 S pathway (Oldfeld *et al.*, 1997). Although, the inhibitory concentration of substrate or product are the main effective factors on cell growth inhibition and the increased requirement of maintenance energy which, adversely, cause a reduction in the specific growth rate and cell growth yield. However, in that study, no substrate or product inhibitory effects were observed.

The optimum production medium for *R. erythropolis* IGTS8 with high BDS-capability is reported to be composed of 20 g/L glucose, using 670 ppm NH_4^+ and 1300 M DMSO as carbon, nitrogen, and sulfur sources, respectively (del Olmo *et al.*, 2005a). It is proven that the addition of ammonium as a nitrogen source is necessary for the good development of the enzymatic pool during the growth of the microorganism used afterwards in the 4S-BDS route (del Olmo *et al.*, 2005a).

The effect of nitrogen in the form of ammonium tartrate on the BDS efficiency of the lignin-degrading white basidiomycete *Coriolus versicolor* has been investigated in the presence of glucose as a carbon source (Ichinose *et al.*, 2002b). The fungal conversion of 2-hydroxymtheyl thiophene (2HMT) was examined under either high nitrogen, HN (12 mM), or low nitrogen, LN (1.2 mM). The same metabolites were produced under the two applied conditions, but the rate of desulfurization was twice as fast in the HN culture as that in the LN culture. This was attributed to the higher mycelium dry weight, which was twice that obtained in LN culture. Moreover, 2-hydroxy-methyl benzothiophene (2HMBT) and 4-hydroxymethyl dibenzothiophene (4HMDBT) were metabolized more effectively in an HN culture medium than in an LN medium. However, upon the investigation on higher S-compounds, the conversion of DBT, 4-MBT, and 7-MBT to their corresponding sulfones, only observed under ligninolytic conditions, i.e. high carbon and low nitrogen (HCLN), suggesting that the S-oxidation reactions were catalyzed by ligninolytic enzyme(s) (Ichinose *et al.*, 2002b).

Martin *et al.* (2005) reported that the highest growth rate and maximum biomass concentration of *Pseudomonas putida* CECT5279 were obtained when glutamic acid was used as the carbon, while the lowest occurred upon the usage of citrate as the C-source. Furthermore, the highest percentage of DBT-BDS was obtained in a non-buffered medium with glutamic acid, but it remained constant for more time in buffered media, although, in this case, the value of the desulfurizing capability was smaller. Glucose showed lower growth and BDS for *Pseudomonas putida* CECT5279, which contradicts with *R. erythropolis* IGTS8 (del Olmo *et al.*, 2005a). The maximum values of the D_{DBT} reported by Martin *et al.* (2004) using glutamic acid as the carbon source in the *P. putida* resting cells assay were close to the values reported in that of Martin *et al.* (2005). Involvement of this amino acid in nitrogen metabolism in bacteria should be referenced (Figure 8.2). The excess or limitation of nitrogen in a growth medium is clearly described in the literature in terms of the ratio of alpha-ketoglutarate (an intermediate of TCA cycle) to glutamine, where the low ratio reflects the excess of nitrogen in a prepared medium, while the high level of this ratio is indicative the nitrogen limitation (Zubay, 1998). The BDS-enhancing role of some intermediates of the TCA cycle have also been reported recently (Martinez *et al.*, 2015).

Sagardia *et al.* (1975) reported that *Pseudomonas aeruginosa* PRG1 cannot grow on BT alone in BSM; 0.05% yeast extract was a suitable substrate for its growth and attack on BT. Premchaiporn and Akaracharanya (2000) reported that YE is a growth limiting factor of *Bacillus* K10 grown in a BSM-DBT medium. The concentration of YE higher than 2.0% (w/v) inhibited DBT desulfurization by *Bacillus* K10. Hirano *et al.* (2004) reported DBT-removal through the Kodama pathway by *Xanthobacter polyaromaticivorans* sp. in the presence of 0.05 g/L of YE in a carbon and sulfur-free medium. Kim *et al.* (2004) reported that the removal rate of the PASHs increased with increasing amounts of YE. However, Ohshiro *et al.* (1996a) and Xiaojuan *et al.* (2008) reported that upon adding YE, this could remarkably increase cell growth, but did not increase the removal rate of DBT and Th. Yu *et al.* (2006) reported that YE is an excellent nitrogen source for bacterial growth because it contains all the metal ions and required micro-nutrients. Zahra *et al.* (2006) reported that *Trichosporon* sp. gave better growth when glucose (5 g/l) and yeast extract (0.05 g/l) were added to BSM containing DBT. Wang *et al.* (2015) reported batch

DBT-BDS by a thermophilic mixed culture of *Paenibacillus* spp. "32O-W" and "32O-Y" in a minimal medium supplemented with glucose and yeast extract. However, Salmani *et al.* (2015) reported the best conditions for BDS activity of *R. erythropolis* IGTS8 were 2 g/L NH_4Cl and 10 g/L glycerol.

Alves *et al.* (2005) reported that glucose, sucrose, n-hexadecane, and mannitol, in a decreasing order, are the most suitable sources of carbon to support the bacterial growth of *G. alkanivorans* strain 1B in a batch BDS of 0.2 mM DBT. Strain 1B showed poor growth on tetrahydrofuran and anthrone and for the other sources of carbon tested, including DBT sulfone and 2-HBP, no bacterial growth was observed. Moreover, no bacterial growth was detected when DBT was used as sole S- and C- sources. This added to the privilege of strain 1B, as it does not affect the hydrocarbon skeleton. Li *et al.* (2007) studied the effect of different C-sources, including glycerol, glucose, sucrose, ethanol, trisodium citrate dehydrate, sodium succinate, potassium acetate, and paraffin, on the growth and BDS of 0.2 mmol/L DBT using growing cells of *R. erythropolis* LSSE8–1. Glycerol expressed the highest growth and BDS efficiency, followed by other C-sources, but Paraffin showed the lowest growth and BDS efficiency. Li *et al.* (2007) also studied the effect of different N-sources, including ammonium chloride, urea, ammonium nitrate, ammonium acetate, and magnesium nitrate. NH_4Cl was found to be the best N-source for obtaining high growth and BDS efficiency. Upon the growth on different S-sources to get efficient resting cells with high BDS capabilities, DBT, DMSO, sodium sulfate, and taurine were tested. Resting cells of 48 h age grown on DBT expressed the highest BDS capabilities. DBT and DMSO was used to obtain high cell density culture in a jar-fermenter. 1 g/L of yeast extract was supplemented to enhance cell growth and the function of the small quantity of DBT with the inducement of Dsz enzymes.

Bordoloi *et al.* (2014) reported that the addition of glycerol as a carbon source in the BDS process of hydrodesulfurized diesel oil by resting cells of *Achromobacter* sp. led to good BDS efficiency without compromising the quality of oil.

Abo-State *et al.* (2014) investigated the effect of different C-sources as co-substrates in shaken flask batch DBT-BDS processes using *Brevibacillus invocatus* C19 and *Rhodococcus erythropolis* IGTS8. Within the studied five carbon sources (glycerol, saccharides, glucose and sucrose, and organic acid salts, sodium lactate and sodium succinate), glucose and glycerol expressed the highest biodesulfurization efficiency of 1000 ppm DBT for C19 and IGTS8, respectively. This recorded good growth with a BDS percentage of 82.3 and 67.34%, with non-stoichiometric production of 2-HBP of 48.06 and 40.67 ppm, respectively.

Gün *et al.* (2015) studied the effect of different C-sources, including D-glucose, D-arabinose, D-mannitol, and ethanol, at an initial concentration of 2 g/L on the growth of the acidophilic and thermophilic biodesulfurizing *S. solfataricus* P2. This recorded a maximum growth rate of 0.0164 h^{-1} and a biomass concentration of 0.149 g DCW/L using glucose, while the other C-sources did not support the growth. This was attributed to the possession of *S. solfataricus* by a semi-phosporylative Entner–Doudoroff (ED) pathway for sugar metabolism (Sato and Atomi, 2011; Kouril *et al.*, 2013). Thus, since glucose is the first metabolite necessary to initiate glycolysis, better glucose utilization than the other sugars would occur. Further, Sato and Atomi (2011) proposed well-defined pentose mechanisms in *S. solfataricus* for both D- and L-arabinose that include intermediates which are not heat stable. Thus, these products may become degraded while enough ATP is accumulated to allow cells to survive. Kouril *et al.* (2013) also reported unstable intermediate metabolites in the semi-phosporylative ED pathway during glucose metabolism using hyperthermophiles that grow at extreme temperatures. Therefore, a similar type of unstable intermediate production in the pentose mechanism may prevent the growth of *S. solfataricus* cells under scarce sugar supplies. Upon studying the effect of different glucose concentrations on P1, the highest growth rate and biomass concentation of 0.0339 h^{-1} and 0.157 g DCW/L were obtained at 20 g/ L, respectively. Moreover, the time for achieving maximum growth decreased from 60.9 h using 2 g/L glucose to 29.5 h at 20 g/L glucose with an observed decrease in lag phase and time required for reaching the stationary phase.

Papizadeh *et al.* (2017) studied the effect of different C-sources, including benzoate, glucose, and glycerol, on the BDS of DBT using *Entreobacter* sp. strain NISOC-03, where the BDS efficiency decreased from 64% in the presence of benzoate to 19.6% in the presence of the other C-sources. In comparison, in the presence of benzoate and DBT as the carbon and sulfur sources, the rate of the DBT biodesulfurization was 3.26 times higher than that in the presence of glucose.

The addition of some nutrients and nitrogen sources, such as yeast extract (El-Gendy *et al.*, 2014; Gün *et al.*, 2015) and glutamic acid (del Olmo *et al.*, 2005; Martin *et al.*, 2005), have been reported to enhance the rate of BDS.

Yan *et al.* (2008) reported that the addition of nicotinamide and riboflavin as cofactor precursors to the 4S-pathway would enhance the BDS efficiency. The addition of nicotinamide and riboflavin would also increase the BDS efficiency in multiple uses of immobilized cells. However, their addition to batch DBT-BDS, using *G. alkanivorans*, expressed no effect on growth or BDS-activity (Aggarwal *et al.*, 2013). This was explained by the possibility of their action at the regulatory or transcriptional level rather than the metabolic level as hypothesized by Yan *et al.* (2008).

Derikvand *et al.* (2014) observed the loss of BDS efficiency (X_{BDS}%) of γ-Al_2O_3NPs/alginate immobilized *R. erythropolis* R1 after 4 successive cycles from 81% to 59%, in spite of the preservation of cell viability, which was attributed to the great loss in the cofactors ($NADH_2$ and $FMNH_2$). But, the addition of 10 mmol/L nicotinamide and 40 mmol/L riboflavin as precursors of $NADH_2$ and $FMNH_2$, respectively, to the BDS media in BDS activity, enhanced and maintained approximately 77% BDS efficiency after the 4th cycle.

Moreover, the addition of co-substrates can regenerate some of the key cofactors involved in microbial metabolism, such as NADH and NAD(P) H, by means of the coupled enzyme approach, which would, consequently, enhance the yield of a whole cell-based bioprocess (Wichmann and Vasic-Racki, 2003). The 4S-pathway is deeply influenced by the availability of ATP because of the crucial importance of this cofactor in the energetic metabolism and the 4S-pathway as an energy-intensive bioprocess. NADH regeneration also implies an enhancement of the availability of other key cofactors, such as ATP (Zhou *et al.*, 2009), which is the most important energy source for metabolic reactions. Currently, the study of its role on the BDS process is becoming more important. It has been confirmed that the uptake of DBT into cells is ATP-driven (Wang *et al.*, 2011) and ATP is also used in the formation and consumption of sulfite and sulfate for biomass formation during DBT metabolism (Aggarwal *et al.*, 2012). Martinez *et al.* (2015) reported that it is possible to improve the yield of the BDS process using resting cells of *Pseudomonas putida* CECT5279 by adding short chain organic acids, such as acetic, citric, and succinic acids. There was an optimal concentration of acetic and succinic acids that maximizes the BDS capacity (1.5% in both cases). The use of citric acid did not present the same trend; an improvement of the BDS capacity was observed, but there was a wider range of concentrations where the maximum improvement was reached. In all the situations of single age cells, a strong inhibition of the process was observed at higher concentrations (3%). However, upon the application of a consortium of mixed aged resting cells (67% of cells grown for 23 h and 33% of cells grown for 5 h), citric acid performed the best yield. The addition of this co-substrate also provided higher intracellular concentrations of some key metabolites, such as NADH and ATP. Upon using cells of a single age as a biocatalyst, it was possible to enhance the yield of the process up to 140% in a batch process and, when employing the optimal mixture of the cells' age, an improvement of 122% was achieved in a fed-batch process. The initial intracellular concentrations of

ATP and NADH also increased to 58 and 42%, respectively, and the initial rate of all of the 4S route was increased. The consumption of both cofactors was very similar with and without a co-substrate (0.50 $\mu mol_{NADH}/gX$ and 30 $nmol_{ATP}/gX$, respectively) and the differences in the concentrations of the cofactors at the end of the process were much higher when acetic acid was supplied (74% 6for NADH and 181% for ATP). Moreover, the addition of co-substrates decreased the loss of 2-HBP with time and improved the life of the biocatalyst.

The addition of DMSO in bi-phasic systems is reported to increase the BDS efficiency as it increases the cell wall permeability, thus increasing the uptake of organosulfur compounds by microorganisms. Adlakha *et al.* (2016) reported that the addition of 0.2% DMSO increased the BDS of heavy crude oil from 29.73% to 67.71% within 72 h using resting cells of *Gordonia* sp. IITR100 in a batch BDS-process of 1/3 O/W. At that DMSO concentration, the possible intracellular concentrations of cytoplasmic enzymes and the proper orientation of the bacterial membrane were maintained. However, as the concentration of the permeabilizing agent (i.e. DMSO) was increased, a decrease in the BDS efficiency was observed. This indicated the occurrence of irreversible damage done to the cells. Moreover, DMSO could not be used with growing cells because it could act as a sulfur source as well as preventing the uptake of organosulfur compounds present in crude oil.

Upon optimization of medium composition for obtaining good growth for two biodesulfurizing *Streptomyces* species, *Streptomyces* sp. VUR PPR 101 and *Streptomyces* sp. VUR PPR 102, it was found that *Streptomyces* sp. VUR PPR101 expressed relatively more growth than *Streptomyces* sp. VUR PPR102 in the presence of all the studied carbon sources and for all the studied N-sources except for casein. However, both the species showed maximum growth in the presence of glucose and the least in the presence of cellulose. The obtained maximum growth of both species was found to be decreasing according to the following order: Glucose > Maltose > Starch > Glycerol > Mannitol > Fructose > Sucrose > Cellulose. The optimal nitrogen sources for *Streptomyces* sp. VUR PPR 101 and *Streptomyces* sp. VUR PPR 102 were found to be yeast extract and potassium nitrate, respectively. The least growth was expressed by both bacteria with urea (Praveen Reddy and Umamaheshwara Rao, 2015). Moreover, methionine was found to be the best amino acid source for both the *Streptomyces* species. The other studied amino acids, pheylalanine, tyrosine, alanine, arginine, lysine, proline, glutamine, and cysteine, also expressed good growth and the optimum salt concentration (i.e. NaCl) was found to be 1% (Praveen Reddy and Umamaheshwara Rao, 2015).

Amino acids, such as arginine, histidine, isoleucine, leucine, lysine, phenylalanine, tryptophan, tyrosine, and valine, are reported not to affect the growth nor desulfurization activity of *G. alkanivorans*. In contrast, cysteine and methionine had strong effects on desulfurization. The sulfur-containing amino acids, such as cysteine and methionine, decreased desulfurization activity of *G. alkanivorans* (Aggarwal *et al.*, 2013). Similar to *R. erythropolis*, *G. alkanivorans* has the ability to utilize cysteine as a sole sulfur source, which is energetically less expensive than DBT (1 mole DBT requires additional 4 moles of NADH). Therefore, the cell prefers to consume cysteine rather than DBT and inhibit the DBT-BDS (Aggarwal *et al.*, 2011), while the reduced desulfurization in the presence of methionine may be due to the inability of *G. alkanivorans* to produce all the sulfur-containing metabolic precursors solely from methionine. For instance, they cannot produce cysteine, L-homocysteine, coenzyme A, etc. solely from methionine and, hence, need an additional sulfur source such as DBT or BT. However, Alanine, asparagine, aspartate, glutamine, glutamate, glycine, proline, serine, and threonine improved growth and desulfurization greatly. These, in contrast to cysteine and methionine, can serve as sole carbon sources as well. Thus, they supplement glucose and promote higher growth and cofactor regeneration. Since sulfur is essential for growth, higher growth leads to greater sulfur usage and higher desulfurization (Aggarwal *et al.*, 2013).

8.14 Effect of Growing and Resting Cells

Growing and resting cells can be used for the BDS process, but resting cells are preferable as they offer greater desulfurization yields (Ohshiro *et al.*, 1996b; Konishi *et al.*, 1997) and overcome the problem of cofactor (NADH) regenerations, which is an important factor and supports the first three oxygenation steps in the 4S-pathway (Luo *et al.*, 2003; Alcon *et al.*, 2005). Moreover, the aqueous phase can also be reduced using high densities of resting cells. They are also reported to give higher BDS efficiencies than the growing ones (Caro *et al.*, 2007b). Ohshiro *et al.* (1994) and Setti *et al.* (1997) reported that NADH is a fundamental limiting factor with growing cell conditions. However, in a study performed by Adlakha *et al.* (2016), the BDS of heavy crude oil applying resting cells was lower than that obtained using growing cells of *Gordonia* sp. IITR100. But, upon applying resting cells of *Ralstonia eutropha* (PTCC1615) collected at the mid-exponential growth phase in batch a BDS of DBT (0.05 mM), the specific BDS rate was much higher than that using growing cells, recording 0.01674 and 0.002 mmol/g_{cell}/h, respectively.

8.15 Inhibitory Effect of Byproducts

As it has been previously mentioned (Chapter 6), the 4S pathway is a four-step enzymatic pathway that converts DBT to 2-hydroxybiphenyl (HBP) and/or 2,2′-bihydrobiphenyl (2,2′-BHBP) and sulfate (Figure 8.3). The first two steps are the conversion of DBT to DBT sulfoxide (DBTO) and then to DBT sulfone ($DBTO_2$). These steps are catalyzed by the enzymes DszC monooxygenase and DszD oxidoreductase in synchrony. The third step is the conversion of $DBTO_2$ to 2-(2′-hydroxyphenyl) benzene sulfinate (HBPS), which is catalyzed by DszA monooxygenase and DszD oxidoreductase in synchrony. The final step is the conversion of HBPS to HBP and sulfite by DszB desulfinase (Gray *et al.*, 2003).

However, it has been reported by several researchers that the main obstacle making BDS not commercially viable are the process limitations caused not only by mass transfer problems, its slower rate than the conventional HDS process, and the short life time of the biocatalyst, but also the cellular deactivation caused by the inhibitory effects of products such

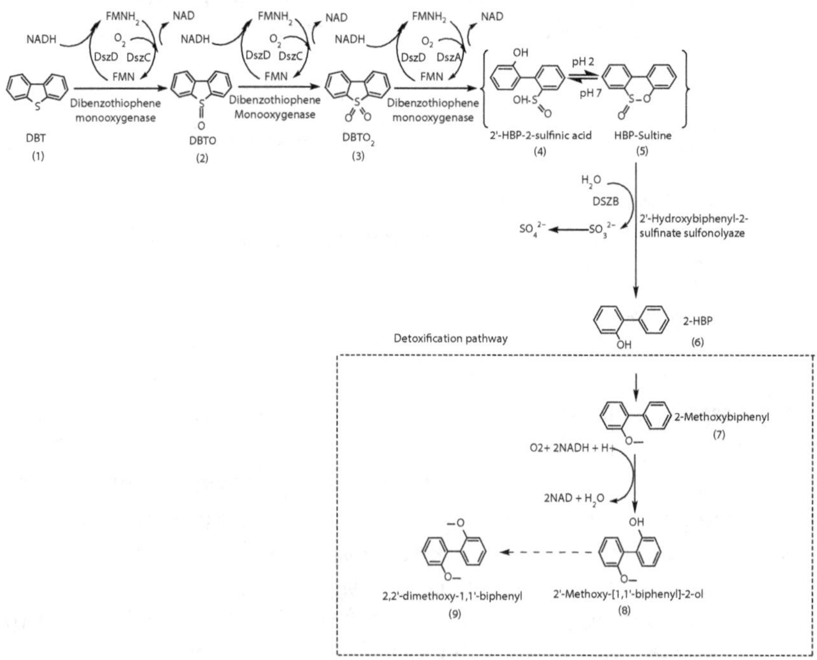

Figure 8.3 4S-Pathway by *Rhodococcus erythropolis* HN2 (El-Gendy et al., 2014)

as biphenyl (BP), 2-HBP, and/or 2,2′-BHBP (Oshiro *et al.*, 1995, 1997; Setti *et al.*, 1996; Maxwell and Yu, 2000; Abin-Fuentes *et al.*, 2013; Akhtar *et al.*, 2016), and also, sulfate accumulation (Li *et al.*, 1996). The inhibition effect of 2,2′-BHBP is known to be lower than that of 2-HBP (Kim *et al.*, 2004).

The amount of sulfate detected in BDS cultures are always non-stoichiometric with the amount of desulfurized DBT. According to Kilbane (1990), the sulfate released is consumed by the bacteria to satisfy their growth requirements and any remaining sulfur is stored by the organisms for future use. Consequently, very little sulfate sulfur would be left to be released in the media.

Although the sulfate enhances the growth rate, the produced sulfate in the 4S-pathway is also reported to cause repression of dsz genes (Kayser *et al.*, 1993; Ohshiro *et al*, 1995b; Wang and Krawiec, 1996; Oldfield *et al.*, 1997; Kim *et al.*, 2004; Wang *et al.*, 2004; del Olmo *et al.*, 2005a; Alves *et al.*, 2005; Mohebali *et al.*, 2008; Silva *et al.*, 2013). The inhibition of DBT desulfurization activity by sulfate is considered to be a gene-level regulation. The expression of dsz genes that are involved in the desulfurization of DBT is strongly repressed by sulfates (Li *et al.*, 1996). The organism needs 4 moles of NADH per 1 mol of DBT to use DBT to get the needed sulfur for growth. In contrast, the organism does not need this extra NADH for metabolizing sulfate. Thus, sulfate promotes higher growth at lower energy and so the organism prefers sulfate consumption over DBT conversion. Only when sulfate is limited, does it desulfurize DBT (Aggarwal *et al.*, 2011). Omori *et al.* (1995) also observed enhanced desulfurizing rates arising from the removal of byproduct sulfate from a succinate-based medium. Wang *et al.* (2004) reported the complete inhibition of DBT-BDS by growing cells of *Corynebacterium* sp. ZD-1 with almost no 2-HBP production when the concentration of SO_4^{2-} was above 0.2 mmol/L. This was related to sulfate ions and their strong inhibition effect on the production of DBT-desulfurization enzymes. However, it has no effect on the BDS activity of resting cells. The BDS of 100 ppm DBT in the presence and absence of sulfate was investigated using *Pantoea agglomerans* D23W3, which expressed 23 and 92% removal after 24 h, respectively, which proved the suppression of BDS in presence of sulfate (Bhatia and Sharma, 2010). An amount as low as 6 mg/L of sulfate is enough to cause more than 22% inhibition in DBT-BDS by *Gordonia alkanivorans* strain 1B (Silva *et al.*, 2013), while 60 mg/L sulfate was reported to completely inhibit the BDS activity of *G. alkanivorans* 1B and the IC_{50-72h} for desulfurization activity was estimated to be 13.6± 0.6 mg/L (Silva *et al.*, 2013). From the process point of view, the removal of such feedback regulation is very essential for the success of the BDS-process.

It should be known that most of the published studies prove the non-stoichiometric relation between the consumption of DBT and the production of 2-HBP and/or 2,2′-BHBP and attributed this to the accumulation of 2-HBP and other 4S-pathway intermediates (up to a certain threshold) inside and on the surface of the cells (Alves *et al.*, 2005; Derikvand *et al.*, 2015b). On the other hand, Wang and Krawiec (1994) suggested that the difference might be owed to the volatile characteristics of 2-HBP. Alves *et al.* (2005) reported the maximum extracellular concentration of 2-HBP was about 120 μM, which is only 27% of the consumed DBT (450 μM). This non-stoichiometric accumulation of 2-HBP had also been observed in *Nocardia* sp. strain CYKS2 (Chang *et al.*, 1998), *Bacillus subtilis* strain WU-S2B (Kirimura *et al.*, 2001), *Corynebacterium* sp. ZD-1 (Wang, 2004), *Rhodococcus* spp. Eu-32 (Akhtar *et al.*, 2009), *Lysinibacillus sphaericus* DMT-7 (Bahuguna *et al.*, 2011), *Bacillus pumillus*-related strain (Buzanello *et al.*, 2014), *Actinomycete* R3 (Khedkar and Shanker, 2015), 10 new isolates of genera *Gordonia*, *Amycolatopsis*, *Microbacterium,* and *Mycobacterium* (Akhtar *et al.*, 2016), and another three bacterial isolates, *P. putida* TU-S2, *B. pumilus* TU-S5, and *R. erythropolis* TU-S7 (Shahaby and Essam El-din, 2017). In a batch DBT-BDS by bacterial consortium AK6, approximately 90% DBT removal was recorded, but only 11% of the utilized DBT substrate were recovered as 2-HBP (Ismail *et al.*, 2016). This non-stoichiometric relation can be attributed to the 4S-pathway itself since the 4S pathway for the conversion of DBT to HBP is a multistep pathway catalyzed by a series of enzymes with different activities, controls, and regulations (Gray *et al.*, 1996). For example, the enzyme dibenzothiophene sulfone monooxygenase (catalyzing conversion of $DBTO_2$ to DBT- sulfinate) has five times the activity of the enzyme desulfinase (catalyzing conversion of DBT-sulfinate to HBP) (Gray *et al.*, 1996). A maximum consumption rate of DBT (11 μmol/g DCW/h) was observed at the early exponential growth phase of *Rhodococcus* sp. strain SA11 and the maximum 2-HBP formation rate (4 μmol/g DCW/h) was reached in the mid exponential growth phase (Mohamed *et al.*, 2015). This was also observed in another study by Bordoloi *et al.* (2014), where *Achromobacter* sp. desulfurized 0.5 mM DBT to 0.03 mM 96 h post incubation, while the concentration of 2-HBP produced from the DBT never exceeded 0.01 mM. However, some microbial strains reported the formation of 2-HBP to be almost equal to DBT removal (i.e. the stoichiometric production of 2-HBP with the depletion of DBT), for example, *Pseudomonas delafieldii* R-8 (Lu *et al.*, 2003), *Nocardia globerula* R-9 (Minfang *et al.*, 2003), and *Mycobacterium* sp.

ZD-19 (Chen *et al.*, 2008a). However, in a study performed by Alves and Paixão (2011) on *Gordonia alkanivorans* 1B and *Rhodococcus erythropolis* D1, it was proven that microorganisms exhibiting faster generation times could be more resistant to 2-HBP accumulation during a BDS process. In other words, strains presenting a lower growth rate are characterized by a higher sensitivity to 2-HBP and attributed this to lower cell numbers in contact with the toxicant (i.e. 2-HBP). The inhibitory concentration that causes an inhibition of 50% of the bacterial baseline respiration rate i.e., the IC_{50}, were recorded. The acute toxicity assay revealed that the 2-HBP IC_{50-3h} values were 0.52 mM for strain D1 and 0.62 mM for strain 1B, while the chronic toxicity assays revealed that the 2-HBP IC_{50-72h} and $IC_{50-120h}$ values were 0.27 mM and 0.21 mM, respectively, for *G. kanivorans* 1B and 0.49 mM and 0.38 mM for *R. erythropolis* D1, respectively. However, Dejaloud *et al.* (2017) reported that no inhibition of cell growth was observed at low concentrations of 2-HBP (<0.1 mM).

Usually, the production of 2-HBP occurs within 48 h (Monticello, 2000) and the presence of $DBTO_2$ till 120 h in cultures is unique since $DBTO_2$ is usually present for a short time, as it is a transition phase in the pathway during the conversion of DBT to 2-HBP (Gray *et al.*, 1996; Chen *et al.*, 1998b). However, in batch DBT-BDS culture using *Pantoea agglomerans* D23W3, $DBTO_2$ accumulated up to 120 h and 2-HBP appeared only after 120 h. Thus, the conversion of $DBTO_2$ to 2-HBP was considered as the rate limiting step in the 4S-pathway by *Pantoea agglomerans* D23W3 (Bhatia and Sharma, 2010). The delay in accumulation of 2-HBP is beneficial for the continuous desulfurization of DBT, as the accumulation of 2-HBP leads to the inhibition of the microbial activity to desulfurize DBT (Gray *et al.*, 1996; Chen *et al.*, 1998b).

Upon the exogenous addition of 2-HBP in a batch BDS of DBT using *Gordonia* sp. CYKS1 (Kim *et al.*, 2004), it was found to retain a 90% desulfurization rate upon the addition of 0.1 mM 2-HBP compared to that with an exogenous addition of HBP. However, upon the exogenous addition of ≥ 0.15 mM 2-HBP, no significant cell growth or DBT-BDS were observed, but upon the exogenous addition of 2,2′-BHBP, most of the added DBT (0.3 mM) was desulfurized in 72 hours in the presence of 0.2 mM of 2,2'-BHBP. About 0.25 mM of DBT was desulfurized when 0.4 mM of 2,2'-BHBP was added, but no significant DBT-BDS was observed in the presence of 0.8 mM 2,2'-BHBP. Upon the exogenous addition of Na_2SO_4, the BDS capacity decreased. When 0.07 mM of sulfate was initially added, only about 0.02 mM of DBT was desulfurized in 72 hours, while 0.27 mM of DBT was desulfurized in the case of the control in which only DBT was the sole sulfur source. The desulfurization rate at 0.07 mM of sulfate was 3.13 μmol/L/h being about 70% of that (4.70 μmol/L/h) obtained in the control, but no significant DBT-BDS occurred at a higher initial sulfate concentration $\geq$ 0.7 mM. The initial exogenous addition of 2-HBP was also reported to inhibit the growth and BDS activity of *Rhodococcus* sp. MP12, where the cells hardly grew when the initial 2-HBP concentration was above 0.2 mM in a growth medium with 300 μL/L DMSO as a sulfur source. In this case, the increase of 2-HBP from 0.0 to 0.8 mM decreased the relative cell growth activity from 100% to 0% (Peng and Zhou, 2016), but the desulfurization activity of MP12 cells decreased sharply by 60% when the initial 2-HBP concentration in the DBT system reached 150 M and, further, increased the initial 2-HBP concentration to 250 M, which obviously led to the significant inhibitory effect on desulfurization activity. Moreover, it was noted that there was no significant inhibitory effect on desulfurization activity when the initial 2-HBP concentration was less than 100 M. This indicated that the exogenous 2-HBP had a significant inhibition on the desulfurization activity of MP12 cells grown with DMSO as a sulfur source and the desulfurization activity decreased with an increase of initial 2-HBP concentration. Breifly, initial 2-HBP expresses an inhibitive effect on both growth characteristics and the desulfurization activity of strain MP12. However, Abin-Fuentes *et al.* (2013) concluded that both generated endogenously or supplied exogenously 2-HBP reduces the overall biocatalyst activity and 2-HBP added exogenously cannot be used to quantitatively predict the effect of HBP on a typical BDS process where DBT is added exogenously and HBP is generated endogenously within the cytoplasm of the cells.

In a study performed by Gray *et al.* (1996), where *R. erythropolis* IGTS8 lysates were initially supplied with 200 M DBT and the concentrations of DBT, $DBTO_2$, HBPS, and HBP were monitored over time, after 10 min all DBT was depleted, the HBPS concentration was approximately 130 M, and the HBP concentration was approximately 50 M. From 10 to 60 min, the HBPS concentration decreased steadily from 130 to 0 M and the HBP concentration increased from 50 to 200 M, but no $DBTO_2$ was detected at any time. The DszB of *Rhodococcus erythropolis* IGTS8, which catalyzes the slowest reaction, has a turnover number (k_{cat}) of about 2 min^{-1} while DszA and DszD have k_{cat} values of 60 and 300 min^{-1}, respectively (Gray *et al.*, 1996).

Watkins *et al.* (2003) appropriately modeled the DszB kinetics by the Michaelis-Menten model with a Michaelis constant (K_m) of 0.90 ± 0.15 μM and a k_{cat} of 1.3± 0.07 min^{-1}. The activity of DszC from *R. erythropolis* D-1 was reported to be 30.3 nmol $DBTO_2$/mg DszC/min (Ohshiro *et al.*, 1997).

It is worth noting that DBTO is usually not typically observed during the BDS process because its rate of consumption is much faster than its rate of production (Gray *et al.*, 1996). Abin-Fuentes *et al.* (2013) explained the data obtained by Gray *et al.* (1996) by another study performed by resting cells of *R. erythropolis* IGTS8 as follows: the k_{cat} of DszA was approximately 7 times that of DszC. The consumption rate of $DBTO_2$ was found to be significantly greater than its generation rate. This explained the non-detection of $DBTO_2$ in that study. Moreover, the buildup of HBPS within the first 10 min is consistent with the fact that its consumption rate (DszB k_{cat} 1.7 min^{-1}) is significantly slower than its generation rate (DszA k_{cat} 11.2 min^{-1}). The fact that the HBP concentration accumulates to over 130 M within the first few minutes indicated that DszC would have been severely inhibited by HBPS at that point in time. As HBPS is consumed, DszC inhibition by HBPS is relieved, but then HBP inhibition of DszC (and DszA and DszB) becomes more significant. Furthermore, HBPS inhibition of DszC is responsible for maintaining a low BDS rate at the beginning of the BDS process when HBP levels are still low. Once HBP levels rise, HBP inhibition of DszA, DszB, and DszC is mostly responsible for the inhibition of the BDS rate.

Several researchers reported that the decrease in BDS activity is attributed to the accumulation of 2-HBP (Naito *et al.*, 2001; Schilling *et al.*, 2002; Kilbane *et al.*, 2006). Honda *et al.* (1998) reported the growth inhibition of *R. erythropolis* IGTS8 by HBP at concentrations > 200 μM; the same was also reported by Ohshiro *et al.* (1996b) for *R. erythropolis* H-2. Nakayama *et al.* (2002) reported the inhibition of *R. erythropolis* KA2–5-1 DszB enzymatic activity and its reduction by 50% at HBP concentrations of ≈ 2,000 μM. It has also been reported in other published studies that 2-HBP is toxic to bacterial cells and that once the concentration of 2-HBP reached 0.2 mM, the BDS of DBT was inhibited by *Rhodococcus erythropolis* D-1 (Ohshiro *et al.*, 1996a), *Gordona* strain CYKS1 (Rhee *et al.*, 1998), and *Mycobacterium* sp. X7B (Li *et al.*, 2003). Akhtar *et al.* (2009) observed a complete growth inhibition of *Rhodococcus* Eu-32 when the concentration of 2-HBP was > 0.4 mM. However, the 2-HBP concentration above 47.6 μM expressed an inhibitory effect on growth and, consequently, the BDS activity of *S. solfataricus* P2 (Gün *et al.*, 2015).

Corynebacterium sp. ZD-1 showed no growth on DBT with an exogenous addition of 0.1 mmol/L 2-HBP, but it did not affect the growth on $NaSO_4$, although 1 mmol/L 2-HBP expressed toxicity on it. This was attributed to the inhibitory effect of 2-HBP on DBT-BDS, which would lead to a lack of S-supply necessary for microbial growth (Wang *et al.*,

2004). Caro *et al.* (2008) also studied the inhibition effect of 2-HBP on the BDS activity of *R. erythropolis* IGTS8 throughout the exogenous addition of 50 μM HBP in an aqueous system using resting cell suspension of 2 g DCW/L, supplied with only one of the 4S pathway compounds (either DBT, DBTO, $DBTO_2$, or HBPS). The disappearance rate of DBTO and HBPS were significantly reduced by the presence of 50 M HBP, suggesting the inhibitory effect of HBP on the enzymes responsible for the conversion of DBTO and HBS, DszC and DszB, respectively. In a similar study on resting cells of *Microbacterium* sp. strain ZD-M2, the DBT-desulfurization rate was found to decrease significantly when HBP was added exogenously at concentrations ranging from 0 to 2000 M (Chen *et al.*, 2008b). However, the exogenous addition of HBP does not give an accurate indication of the effect of HBP on the BDS efficiency, since, upon the exogenous addition of HBP, a significant fraction would be retained by the cell wall and may never reach the cytoplasm, where the inhibition of the desulfurization enzymes occurs. On the other hand, HBP generated endogenously within the cytoplasm is immediately at the location where it can be inhibitory to the desulfurization enzymes. Therefore, the specific HBP loading of the biocatalyst (mg HBP/g DCW) that leads to a certain level of reduction in BDS activity may be significantly larger when HBP is added exogenously.

Schilling *et al.* (2002) studied the kinetics of DBT-BDS by *R. erythropolis* IGTS8 of cell density 66 g DCW/L in a model oil (19,000 M DBT in n-hexane 1:1 v/v). This was found to follow the first order decay, with a decay constant of 0.072 h^{-1}, and the loss of biocatalyst activity was attributed to the exposure of increasing HBP concentrations. The cells were active for 24 h, and, although only 7,000 out of 19,000 M DBT in oil was consumed, the final concentration of HBP accumulated in the oil phase after 24 h was only 3,300 M. This discrepancy has been suggested to be due to the accumulation of pathway intermediates or the retention of DBT and/or HBP within the biocatalyst. The amount of DBT or HBP retained by the biocatalyst depends on the partition coefficients between the biocatalyst and the oil or aqueous phases, $P_{C/O}$ or $P_{C/W}$, respectively (where C, O, and W represent cells, oil, and water, respectively). Based on the finding reported by Lichtinger *et al.* (2000), the Gram +ve *R. erythropolis*, as a member of the *Actinomycetes* family, is characterized by a cytoplasmic membrane surrounded by a thick cell wall composed of a thick peptidoglycan structure to which fatty acid molecules are attached. Moreover, the cell wall is composed of mycolicacids of C-length ranges between 30–45 C-atoms that are perpendicular to the cell surface, which makes the cell wall of *R. erythropolis* highly hydrophobic.

Several researchers reported that the decrease in BDS activity is attributed to the accumulation of 2-HBP (Naito *et al.*, 2001; Schilling *et al.*, 2002; Kilbane, 2006). Honda *et al.* (1998) reported the growth inhibition of *R. erythropolis* IGTS8 by HBP at concentrations > 200 μM. Nakayama *et al.* (2002) reported the inhibition of *R. erythropolis* KA2–5-1 DszB enzymatic activity and its reduction by 50% at HBP concentrations of ≈ 2,000 μM. Mohamed *et al.* (2015) reported that DBT-cultures of *Rhodococcus* sp. strain SA11 exhibited around 30% growth retardation at 0.3 mM 2-HBP. Moreover, upon the exogenous addition of 0.5 mM 2-HBP during the log growth phase, a sharp drop in growth occurred within 1 h after addition.

The activity of a *R. erythroplis* IGTS8 resting-cell suspension of 15.5 g DCW/liter was decreased by 90% when the HBP concentration in the aqueous medium reached 40 M (Abin-Fuentes *et al.*, 2013). The dose-response experiments performed by Abin-Fuentes *et al.* (2013) to identify major inhibitory interactions in the most common 4S-BDS pathway revealed that HBP is responsible for three of the four major inhibitory interactions identified. The concentrations of HBP that led to a 50% reduction in the enzymes' activities (IC_{50}s) for DszA, DszB, and DszC were found to be 60±5 μM, 110±10 μM, and 50± 5 μM, respectively. The fact that the IC_{50}s for HBP are all significantly lower than the cytoplasmic HBP concentration suggests that the inhibition of the desulfurization enzymes by HBP is responsible for the observed reduction in biocatalyst activities concomitant with HBP generation.

Gün *et al.* (2015) reported the effects of using different inorganic S-sources (elemental sulfur, sodium sulfite, sodium sulfate, potassium persulfate, and potassium disulfite) on the growth of the acidophilic and hyper-thermophilic *S. solfataricus* P2 and revealed a high growth rate and maximum biomass concentration of 0.0220 h^{-1} and 0.651 g DCW/L in sulfate sources, respectively. This was followed by the disulfite source recording 0.0254 h^{-1} and 0.623 g DCW/L, then the sulfite, recording 0.0226 h^{-1} and 0.628 g DCW/L, respectively, while the elemental S recorded the lowest at 0.0165 h^{-1} and 0.586 g DCW/L, respectively. However, all the studied inorganic sulfur sources led to rapid cell death after a maximum biomass cell density was obtained, except for in the elemental sulfur case which showed a sustained stationary phase. The observed rapid microbial death with sulfate and sulfite was explained by the excess uptake of these anions by the cells, leading to a demand for counter ion balance, which can be maintained by excess accumulation of cations to cells, causing an osmotic imbalance.

Feng *et al.* (2016) studied the detailed effects of the major end products 2-HBP and sulfate ions (0, 0.05, 0.1, or 0.2 mmol/L) on cell growth and transcriptional levels of dszA/dszB/dszC and desulfurization efficiency of *Gordonia* sp. JDZX13 in a batch BDS of 0.3 mmol/L DBT at 35 °C and 170 rpm within 80 h, while the resting cells were used as the indicator of desulfurization enzyme activity at 30 °C, 170 rpm, and 12 h. 2-HBP was found to be highly toxic to cell growth and, consequently, desulfurization activity. The results revealed the decrease in the specific growth rate with the increase of exogenous 2-HBP concentration recording 0.147, 0.139, 0.131, and 0.11 h^{-1}. It was also depicted that the lag phase was lengthened with the increase of 2-HBP concentration, while DBT was completely removed in the control flasks (i.e. without 2-HBP) and at 0.05 mmol/L recorded 98.8% and 20.2% with 0.1 and 0.2 mmol/L, respectively. The effect on BDS efficiency, applying resting cells recorded 20.89%, 15.71%, 11.51%, and 3.31%, respectively. These results confirmed the strong inhibition effect of the desulfurization enzyme activity. The effects of different 2-HBP densities on the transcriptional level of key desulfurization related genes and fluorescent quantitative PCR was performed. With an increase in 2-HBP, the transcriptional levels of the desulfurization genes dszA/dszB/dszC gradually decreased. Compared to the blank system, the transcriptional levels of the desulfurization genes dszA/dszB/dszC were 0.89, 0.71, and 0.65 times, respectively, in 0.05 mmol/L 2-HBP. Compared to the blank system, the transcriptional levels of desulfurization genes dszA/dszB/dszC were 0.81, 0.66, and 0.59 times, respectively, in the 0.1 mmol/L 2-HBP system. Compared to the blank system, the transcriptional levels of the desulfurization genes dszA/dszB/dszC were 0.77, 0.59. and 0.56 times, respectively, in the 0.2 mmol/L 2-HBP system. The effect of SO_4^{2-} depicted a specific growth rate of 0.147 h^{-1}, where the values were 0.171 h^{-1}, 0.163 h^{-1}, and 0.157 h^{-1} with the addition of 0.05, 0.1, or 0.2 mmol/L SO_4^{2-}, respectively. At low sulfate ion concentration cell growth was accelerated. In contrast to the situation of 2-HBP, which was eliminated from the cell, part of the sulfate ion remains in the cell acted as a sulfur source for synthetic cysteine, methionine, vitamins, and other material necessary for life and the other part is excreted (Caro *et al.*, 2008). However, there was a slight inhibition with an increase in sulfate ion concentration compared to lower sulfate ion concentration. The reason for this may be that the sulfur source was not the limiting factor and feedback inhibition was gradually generated by excess sulfate ions (Soleimani, *et al.*, 2007; Abro *et al.*, 2014). The BDS effeminacy using growing cells was also affected by the intial SO_4^{2-} concentrations. In the blank and 0.05 mmol/L sulfate ion systems, DBT was completely eliminated, while 95.6% and 15.6% DBT removal were recorded with 0.1 mmol/L and 0.2 mmol/L sulfate ions systems, respectively. Thus, sulfate increases the growth rate, but inhibits the BDS efficiency. This was attributed to the fact that excessive sulfate ions inhibit the expression of desulfurization genes or even directly reduce enzyme activity, which, consequently, lower the desulfurization efficiency (Caro *et al.*, 2007b; McFarland *et al.*, 1998; Abin-Fuentes *et al.*, 2013). Moreover, the BDS efficiency of the resting cells, compared to the control flasks (i.e. without exogenous addition of sulfate), slightly decreased with sulfate ion concentrations, recording 23.70%, 33.12%, and 37.80% for 0.05, 0.1, and 0.2 mmol/L sulfate ions systems, respectively. This confirmed that the desulfurization enzyme activity was slightly inhibited under higher sulfate ion concentration. To confront the increase in sulfate ions, the transcription of desulfurization genes, dszA/dszB/dszC, first, upregulated and then decreased. Under 0.05 mmol/L sulfate ions, the transcriptional levels of the desulfurization genes dszA/dszB/dszC were 3.61, 2.00, and 2.62 times, respectively, its original level. This was attributed to the utilization of sulfate ions for the synthesis of cellular substances which sequentially stimulate the expression of the desulfurization gene (Li *et al.*, 2006; Kilbane, 2006; Chauhan *et al.*, 2015). Under a 1 mmol/L sulfate ion, the transcriptional levels of dszA/dszB/dszC were only 0.33, 0.14, and 0.53 times, respectively, the original level. Different from 2-HBP, the expression of dszB and dszC down-regulated compared to dszA, thus the third and fourth steps were inhibited by sulfate ions. Compared to the equal 2-HBP concentration, the negative effect of sulfate ions on the desulfurization gene was more significant. Under the 0.2 mmol/L sulfate ion, the transcriptional levels of the desulfurization gene dszA/dszB/dszC severely down-regulated and were only 0.015, 0.004, and 0.011 times, respectively, the original level. These results also prove that the DBT removal performance of the resting cells was greatly inhibited under a higher sulfate concentration.

Thus, it can be concluded that accumulation of 2-HBP and sulfate ions would form prominent feedback inhibition effects on cell growth and biodesulfurization efficiency. Generally, the feedback inhibition effects of sulfate ions can be weakened by eliminating excess sulfate ions into the extracellular aqueous phase. Concerning 2-HBP, since it is hydrophobic in water and prefers to return back into the oil phase, it is well known that in a biphasic system, bacterial cells primarily grew in the aqueous phase and then a large number of cells gathered at the oil–aqueous interface due to the hydrophobic interaction between the cells (Gün *et al.*, 2015), so Feng *et al.* (2016) suggested overcoming such bottleneck via using the biphasic system with an appropriate O/W phase ratio.

8.16 Statistical Optimization

Although there are many published studies about the BDS-process concerning isolation of new biodesulfurizing microorganisms, studies for the physiology of microorganisms and their gene modifications, as well as the metabolic pathways and kinetics of desulfurization in one phase and two phase systems of model and real oil feed, there are few reports on the significance of operational factors and their optimization through statistical methods.

It is difficult to identify the important factors and interactions between a large number of variables. Usually, one factor at a time technique is the conventional optimization procedure and involves altering one parameter at a time, keeping all other parameters constant, which enables one to assess the impact of those particular parameters on the process performance. However, it is time consuming.

The DOE or the experimental design is a powerful statistic-based approach to design experiments in order to achieve a predictive knowledge of a complex, multi-variable process with the fewest acceptable trials. It enables designers to simultaneously determine the individual and interactive effects of many factors that could affect the output results in any design. Thus, DOE provides a full insight of interaction between design elements; it helps to pin point the sensitive parts and areas in designs that cause problems in the yield, so designers would be then able to fix these problems to achieve robust performance and produce higher yield designs prior to going into production (Buasri *et al.*, 2014). The essence of DOE is to plan informative experiments, analyze the resulting data to get a good model, and from the model, create meaningful maps of the system.

Taguchi orthogonal array experimental design is based on mixed levels of highly fractional factorial designs and other orthogonal arrays (OA) to perform the fewest number of experiments in a timely manner and at lower costs (Taguchi, 1986). It distinguishes between control variables which are the factors that can be controlled (i.e. inner array), and noise variables (i.e. outer array), which are factors that cannot be controlled except during experiments in the lab. Taguchi design provides information about the interaction between the controllable and noise variables. Each run in the inner array would be performed for all of the combinations in the outer array. A signal to noise (S/N) ratio, which summarizes the mean and variance information, is defined and data analysis would be carried out for this ratio using ANOVA to determine the optimum conditions, as well as the contribution of each factor in the experimental results (Biria and Balouchi, 2013). There are three applicable types of S/N ratio, depending on the optimization criteria: (1) lower is better (LB), (2) nominal is better (NB), and (3) higher is better (HB). The DOE using the Taguchi approach can economically satisfy the needs of problem solving and process design optimization projects. By applying this technique, engineers, scientists, and researchers can significantly reduce the time required for experimental investigations. It helps in examining the effect of different process parameters on the mean and variance of performance characteristics, which determine the proper functioning of the process. It is very effective with a nominal number of parameters (3–50) with few interactions between them and very few significant contributing ones.

Montgomery (2013) has noted that by using experimental design, engineers can determine which subset of the process variables has the greatest influence on process performance. The results of such experiments can lead to improved process yield and reduced design and development time and operation cost. Also, Box and Wilson applied the idea of DOE to industrial experiments and developed the response surface methodology RSM (Cavazzuti, 2013).

Response Surface Methodology (RSM) is one of the well-known statistical methods which is utilized for evaluating the important factors and finding a relationship between effective variables and one or more responses in a system. In another word, RSM is a collection of mathematical and statistical techniques that are useful for modeling and analysis of a process in which a response of interest is influenced by several variables and the objective is to optimize this response in that complex process. Accordingly, RSM is a 3-D response surface plotted on the basis of the predicted model equation to investigate the interaction among the variables and to determine the optimum condition (range) of each factor. The response surface of the response variable is mapped out and the process is moved as close to the optimum as possible, taking into account all constraints, supposing that the outputs are defects or yields and the goal is to minimize defects and maximize the yield. If these optimal points are in the interior of the region in which the experiment is to be conducted, we need a mathematical model that can represent curvature so that it has a local optimum. Response surface models may have quadratic and possibly cubic terms to account for curvature. The RSM helps in understanding the pattern in which the dependent variables are affected by the corresponding changes in the independent variables, improving the product of predicted property values, and predicting the interactive effects of two or more factors and the effects caused by the collective contributions of the measured response. This method is often employed after a "vital few" controllable factors have been identified in order to find the factor settings that optimize the response (Deriase *et al.*, 2012; Alhassan *et al.*, 2014). A contour plot is a graph that can be used to explore the potential relationship between three variables (i.e. the two independent variables and the response variable), where it displays a 3-D relationship in two dimensions, with independent variables plotted on the x- and y- scales and response values represented by the contours. In another word, a contour plot is like a topographical map in which x-, y-, and z-values are plotted instead of longitude, latitude, and elevation, where the contour lines corresponding to different levels will not cross each other and the contour lines of the same level may appear to intersect (Anderson and Whitcomb, 2015).

In this approach, analysis of variance (ANOVA) and regression techniques are employed to estimate a low degree polynomial model for optimization of the levels of significant explanatory variables in a limited number of experiments (El-Gendy *et al.*, 2014). Central Composite Design (CCD) and Box-Behnken Design (BBD) of RSM are fractional factorial designs for the optimization of variables with a limited number of experiments.

The Box-Wilson or central composite design, which is commonly called the "central composite design (CCD)", is a design that contains an imbedded factorial or fractional factorial design with center points that are augmented with a group of "star points" which allows estimation of curvature. The addition of center points to the 2k design, based on the idea of replicating some of the runs in a factorial design, runs at the center to provide an estimate of error and allows the experimenter to distinguish between two possible models of the first and second order. The CCD is a very effective design for fitting a second-order response surface model.

Thus, the CCD has a different structure from that of BBD, where in the former a ball is used, in which all of the corner points lay on the surface, while in the BBD, the ball is located inside the box defined by a wire frame that is composed of the edges of the box. Also, the number of observations for BBD are lower than those of CCD (Rezaei *et al.*, 2013). Based on the CCD matrix, the studied independent factors would vary within a defined range to reach the optimum condition for the response (Koohikamali *et al.*, 2012). In order to reduce the effects of any uncontrolled factor on the response, the sequence of the experiments in the designed matrix can be randomized (Montgomery, 2013). For k factors, 2k star points and one central point are added to the 2^k full factorial, bringing the sample size for the central composite design to $2^k + 2k+1$, i.e. CCD is a 2^k full factorial to which the central point and the star points are added.

The Box-Behnken design (BBD) was devised by George E.P. Box and Donald Behnken in 1960 (Cavazzuti, 2013) to achieve a design where each factor is placed at one of three equally spaced points that are usually coded as -1, 0, and +1, where at least three levels for each factor are needed, taking into consideration that the design should be sufficient to fit a quadratic model and the ratio of the number of experimental points to the number of coefficients in the quadratic model should be reasonable.

One factor at a time technique was applied to select the optimal carbon, nitrogen, and sulfur sources for *Rhodococcus erythropolis* LSSE8–1, which is capable of desulfurizing DBT via the 4S-pathway to 2-HBP as an end product (Li *et al.*, 2007). Glycerol expressed the highest growth and BDS capacity of 3.850×10^{-3} mmol/L/h. NH_4Cl as an N-source showed the highest growth and BDS capacity of 3.96×10^{-3} mmol/L/h. Resting cells of age 48 h that were grown on DBT, expressed the highest BDS of 7.46×10^{-3} mmol/L/h. Upon applying Taguchi methodology, five factors, namely glycerol, ammonium chloride, DMSO, magnesium chloride, and a trace element solution at four levels with an orthogonal array layout of L16 (4^5), were selected. The analysis of variance techniques to determine which factors were statistically significant for growth revealed that the influence of these variables on the growth (48 h), on a fermenter scale, was found to decline in the order: DMSO > magnesium chloride > ammonium chloride > trace element solution > glycerol. The resulting optimum conditions were glycerol 10 g/L, ammonium chloride 3 g/L, DMSO 3 mmol/L, magnesium chloride 2 g/L, and a trace element solution 0.5%. This showed an enhanced cell production of 70% from 3.40 to 5.78 g dry cells/L at 48 h cultivation (Li *et al.*, 2007).

Upon experimental and numerical approaches for studying different factors affecting the growth and DBT removal efficiencies of *Bacillus sphaericus* HN1, Nassar (2009) found that the best equation model correlating both of the above mentioned response variables were incubation period (day), shaking speed (rpm), temperature (°C), and pH, represented as a simple k degree polynomial in the form:

$$y = b_0 + \sum_{k=1}^{n} a_k x^k$$

where y stands for response (dependent) variable (the remaining DBT concentration (ppm i.e. mg/L) in the cultures or HN1 cell growth expressed in dry weight (mg/L)) and x represents the independent variable (the above mentioned key process variables applying one factor at a time). The b_0 represents the free regression coefficient parameter, a_k's are the regression coefficients, and n is the degree of the polynomial. Table 8.2 represents the degree of the resulting polynomial.

Table 8.2 Degree of the Predicted Polynomial for Batch DBT-BDS using *Bacillus sphaericus* HN1, Nassar (2009).

	Dependent (response) variable	
Independent variable	**DBT (ppm)**	**Cell growth (mg/l)**
Incubation Period (day)	4*	4*
Shaking Speed (rpm)	2	2
Temperature (°C)	3*	4*
pH	3*	5

where, the astrike * indicates whether or not the polynomial includes a free parameter (b_0) (i.e. coefficient $b_0 = 0.0$).

The validity of the fitted models was evaluated using the analysis of variance (ANOVA) with the values of R^2 (unadjusted coefficient of multiple determination) and R^2_{adj} (adjusted coefficient of multiple determination). R^2_{adj} is computed using the formula:

$$\bar{y} = \frac{\left(\sum_{i=1}^{N} y_i\right)}{N}$$

$$R^2 = \frac{\sum_{i=1}^{N} (\hat{y}_i - \bar{y})^2}{\sum_{i=1}^{N} (y_i - \bar{y})^2}$$

$$R^2_{adj} = 1 - \frac{N-1}{N-P}(1 - R^2)$$

where N represents the number of observations (data points), p is the number of parameters (regression coefficients), $\hat{y}_i$ is the ith observation of the dependent variable y, y_i represents the experimental value of the variable, $\bar{y}$ is the average value of y, and R^2 and R^2_{adj} have a value closer to 1, indicating a better fit,

The Fischer's (F) value is the ratio of the mean regression sum of squares divided by the mean error sum of squares. The value of prob(F) is the probability that all of the regression coefficients are zero. In general, the (F) value with a low probability (prob(F)) value indicates high significance of the regression model. The student t-test (t) can be used as a tool to check the significance of the regression coefficient. The prob(t) value is the probability of obtaining the estimated value of the regression coefficient. The smaller the value of prob(t), the more significant the coefficient is.

Nassar (2009) observed that the values of the calculated $R^2 > 0.9406$, $R^2_{adj} > 0.901$, F ratio, prob(F), and prob(t) were statistically significant and proved the goodness of fit of the predicted model equations. The optimum conditions for maximum growth and DBT-removal using *Bacillus sphaericus* HN1 were found to be 10 days' incubation period, 200 rpm shaking speed, 30 °C, and pH7. These recoded an approximately complete removal of 250 ppm DBT with a maximum growth of approximately 8×10^9 cells/mL.

One factor at a time technique was also applied by Nassar *et al.* (2017b) to optimize the growth and BDS efficiency of *R. erythropolis* HN2 in shaken flasks of batch a BDS of 1000 ppm DBT. Several model equations (Gaussian, polynomial, and exponential functions) were investigated to assess the relationships between the variable operating conditions and both cell growth and the percentage of BDS.

The relationships between the change in microbial growth (CFU/mL), with respect to variations in initial pH, incubation temperature, and shaking speed, were found to be best described by the Gaussian model for fitting peaks (Giraud, 2008)

$$F(x) = \sum_{i=1}^{n} a_i \cdot \exp\left[-\left(\frac{x-b_i}{c_i}\right)^2\right]$$

where "**a**" is the height of the curve's peak or amplitude (i.e. the curve maximum) and the graph expands vertically as "**a**" increases, "**b**" is the position of the center of the peak, that is, as "**b**" increases the graph is shifted to the right and upon the decease of "**b**", the graph is shifted to the left, and "**c**" controls the width of the curve and is related to the full width at half maximum (FWHM) of the peak. As "**c**" is made small, the graph shrinks horizontally, and as "**c**" is made large, the graph expands horizontally. "**n**" is the number of peaks to fit.

The effects of initial pH, incubation temperature, and initial inoculum size on BDS-efficiency were found to be best described by a polynomial function of nth degree

$$f(x) = p_1 x^n + p_2 x^{n-1} + \cdots + p_n x + p_{n+1}$$

where P_1, ..., P_{n+1} are polynomial coefficients and x is the independent variable.

However, the exponential function which calculates the natural exponential of all the data was found to be the best at describing the effect of both the shaking speed on BDS-efficiency and initial inoculum size at maximum achieved microbial growth (CFU/mL). The general form of exponential function can be written as:

$$f(x) = a\ exp(bx) + c\ exp(hx)$$

where a, b, c, and h are exponential coefficients and x is the independent variable.

Accepted mathematical model equations have been characterized by three common validity tests: (1) high correlation coefficient (R^2) and adjusted correlation coefficient (R^2_{adj}), which indicate the applicability and reliability of a given model, (2) the accepted model must have the least values of sum of squares errors (SSE) and root mean square errors (RMSE) relative to those obtained for other tested mathematical models, and (3) a close agreement between the calculated and experimental values.

MATLAB software package 7.9.0 (R2009b) was used for numerical investigation, regression analysis of the data obtained, and estimation of the coefficients of the regression equation with 95% confidence bounds. The optimum values of the selected predictor (explanatory) variables were obtained by solving the corresponding predicted model equation using LINGO software package version 6.01 for mathematical optimization (http://www.LINDO.com). The predicted mathematical model equations that best represent the effects of different studied physicochemical operational parameters on microbial growth and DBT-BDS efficiency of growing cells of *R. erythropolis* HN2 in shaken flasks of batch BDS of 1000 ppm DBT, together with the corresponding goodness of fit parameters, are illustrated in Table 8.3. Thus, the optimum condition for maximum growth of HN2 was predicted to be a pH7, 29.48 °C, initial OD_{600nm} of 0.1 which is equivalent to 3.73×10^4, and 155.9 rpm, while those for maximum BDS efficiency were a pH6.69, 27.47 °C, initial OD_{600nm} of 0.1, and 154.19 rpm. These recorded a BDS efficiency of approximately 82% in shaken flasks of a batch BDS of 1000 ppm DBT using *R. erythropolis* HN2 with a maximum growth of 3.2×10^9 cells/mL within 8 d of incubation.

Taguchi design procedure was applied to study the effects of cell density, pH, and phase ratio at three levels on the desulfurization reaction of a biphasic system at 30 °C and 175 rpm using model oil (2.7 mM DBT in hexadecane) and resting cells of a newly isolated strain of RIPI-22,

Table 8.3 Predicted Model Equations Together with Corresponding Goodness of Fit Parameters for Batch DBT-BDS Using Growing Cells of *R. erythropolis* HN2 (Nassar *et al.*, 2017b).

Physicochemical parameters	Model equations	Optimum predicted value	R^2	R^2_{adj}	SSE	RMSE
Initial pH	$CFU/mL = (4.003e+9)\exp\left[-\left(\frac{(pH-7)}{0.4357}\right)^2\right]$	7	0.9716	0.9527	3.561e+17	3.445e+8
	$\%BDS = -8.84(pH)^2 + 118.3(pH) - 319.9$	6.69	0.988	0.980	37.9	3.554
Incubation Temperature	$\ln(CFU/mL) = (21.99)\exp\left[-\left(\frac{(T-29.48)}{18.2}\right)^2\right]$	29.48 °C	0.9751	0.9585	1.84	0.7842
	$\%BDS = 0.005411(T)^3 - 0.8483(T)^2 + 34.36(T) - 340.6$	27.47 °C	0.9938	0.9846	24.71	3.515
Initial Inoculum Size	$CFU/mL = (7.894e+9)\exp(-8.152 * (OD_{600}))$	0.1 OD_{600}	0.9964	0.9952	3.185e+16	1.03e+8
	$\%BDS = -168.1(OD_{600})^3 + 257.9(OD_{600})^2 - 154.9(OD_{600}) + 93.37$	0.1 OD_{600}	0.9931	0.9726	6.015	2.453
Shaking Speed	$CFU/mL = (3.248e+9)\exp\left[-\left(\frac{(V-155.9)}{48.07}\right)^2\right]$	155.9 rpm	1.00	1.00	3.031e+13	3.893e+6
	$\%BDS = -(2.606)\exp(0.0231 * V) + (18.66)\exp(0.01448 * V)$	154.19 rpm	0.9962	0.985	11.41	3.378

where the CFU/mL is the total viable count of HN2 (i.e. the microbial growth), T is the process temperature (°C), OD is the optical density of the microbial cells culture (i.e. the absorbance at λ_{600nm}), and V is the shaking speed (rpm).

harvested in the late exponential phase (Rashtchi *et al.*, 2006), where An $L_9(3^4)$ OA, which has 8 d.f., was used. The amount of 2HBP production as an indication for BDS-efficiency was measured in the hydrocarbon phase as a result of each trial condition. After performing the experiments, the analysis of data was accomplished using the standard approach determination of the main effects and formation of the ANOVA table and signal to noise (S/N) analysis by WinRobust™ and Qualitek-4™ software. The data analysis revealed that the extent of desulfurization by RIPI-22 was hardly dependent on the pH, but the volume ratio of the hydrocarbon-aqueous phase significantly affected the desulfurization activity. The extent of desulfurization increased with raising cell density from 7 to 11 g/L, where the higher cell concentration (> 11 g/L) did not significantly influence the desulfurization. In addition, the results showed that DBT-BDS was well accomplished in a low phase ratio and the maximum activity occurred at a 30% (O/W) phase ratio. Using two aforementioned softwares, the optimum desulfurizing conditions using the newly isolated strain RIPI-22 were determined to be pH 6, 11 g DCW/L, and a hydrocarbon fraction of 30% (O/W) (Rashtchi *et al.*, 2006).

Taguchi optimization was applied for preparation of an immobilized biocatalyst of *Rhodococcus erythropolis* R1 (NCBI GenBank Accession No. GU570564) resting cells in Ca-alginate beads (Derikvand *et al.*, 2014). This was throughout the consideration of some important factors for achieving a successful BDS using immobilized cells, such as the alginate concentration, size of the beads, the concentration of surfactants, and γ-Al_2O_3 nanoparticles. The age of the resting cells was 72 h cultivated on 0.3 mM DBT. The immobilized beads were applied on a bi-phasic system (aqueous/1 mM DBT in n-tetradecane 2:1 v/v). The aim was to maximize the biodesulfurization yield (X_{BDS}). Thus, the S/N ratio with HB characteristics was utilized in the applied Taguchi design and was calculated as follows:

$$\frac{S}{N} = -10\log_{10}\left[\frac{1}{n}\Sigma\left(\frac{1}{X_{BDS}}\right)^2\right]$$

where n is the number of repetitions and X_{BDS} are the experimental results of BDS capacity (i.e. BDS-yield) and was calculated as follows:

$$X_{BDS} = \frac{C_{2-HBO,20}}{C_{DBTo}} \times 100$$

where $C_{2\text{-HBP},\,20}$ is the experimental concentration of 2-HBP at the end of prescribed time, t = 20 h, and C_{DBTo} is the initial DBT concentration at time zero.

Taguchi L18 orthogonal array for the three parameters is in three levels (3^3) and one parameter in two levels (2^1), with a layout of $2^1 \times 3^3$ was used indicating 18 experimental runs in duplicate. Moreover, the effects of factor levels on the BDS efficiency were determined employing analysis of variance (ANOVA) and the statistically significant factors were distinguished for P value < 0.05. The contribution percentage of the factors in the final results was calculated as well. Derikvand *et al.* (2014) found that the influence of the studied variables on the 2-HBP production was found to decline in the order: nano-γ-Al_2O_3 > surfactant > bead size > alginate concentration. The results revealed that all factors under investigation were significant (P < 0.05). The BDS efficiency increased by decreasing the bead size and was explained as the lower size of the bead, the higher the surface to volume ratio would be and thus, more interactions would occur between the cells and DBT. Surfactants were highly statistically, significantly effective in BDS (P-value = 0.001). Nevertheless, the addition of the Span-80 had a higher efficiency than Tween-80. The nano-γ-Al_2O_3 was found to be absolutely critical and was considered as the most effective factor in the BDS of DBT (percentage of contribution = 55.2% and P-value = 0). The optimum conditions were found to be 20% (w/w) nano γ-Al_2O_3, 1.5 mm bead size, 1% (w/v) alginate concentration, and 0.5% (v/v) Span-80 instead of Tween-80, which recorded X_{BDS}% of 81%.

RSM was used to optimize the pH, temperature, and shaking speed of a BDS-process using *Stenotrophomonas maltophila* strain Kho1. The optimum conditions were found to be pH7.2, 29 °C, and 180 rpm, respectively, which recorded a maximum growth of 0.42 g DCW/L and 63.15 μM 2-HBP from 1 mM DBT within 96 h (Ardakani *et al.*, 2010).

RSM based on CCD was applied to optimize and study the interactive effects of the oil/water (O/W) phase ratio and initial sulfur concentration in shaken flasks of batch desulfurization of diesel oil using growing cells of *G. alkanivorans* RIPI90A (Irani *et al.*, 2011). A quadratic polynomial model was predicted to identify the relationship between the response BDS percentage (Y) and the studied variables, X_1: O/W and X_2: initial S-content:

$$Y = 8.83 + 0.17X_1 + 10.33X_2 + 0.5X_1^2 + 1.5X_2^2 - 0.25X_1X_2$$

ANOVA analysis revealed that the linear and square effect of the initial S-concentration were found to be more statistically significant (p < 0.1), while the contour plot representing the response, i.e. the S-removal (mg/L) and the combined effect of initial sulfur concentration (mg/L) and oil/water phase ratio (%), showed that the S-removal increased with the increase in the initial sulfur substrate concentration and decrease of oil/water phase ratio.

Response optimization helps to identify the factor settings that optimize a single response or set of responses. It is useful in determining the operating conditions that will result in a desirable response. Thus, upon optimization, the optimum conditions, which are defined as the best combination of factors set for achieving the optimum response were found to be 30% O/W, 28 mg/L initial S-concentration for a predicted response of 21.25 mg/L S-removal, and a desirability score of 1 (Irani *et al.*, 2011).

Irani *et al.* (2011) also used RSM based on a CCD to study the interactive effect and optimize the independent variables, such as superficial gas velocity (U_g, L/min) and working volume (v, L), which are related to the liquid level above the riser section, on the response gas liquid mass transfer coefficient (k_La, s^{-1}) for a loop sparger in an airlift bioreactor for an emulsion of 30:70 diesel/water for a batch BDS using growing cells of *G. alkanivorans* RIPI90A.

The k_La was calculated as the slope of the linear equation:

$$-\ln(1-E) = k_L a(t - t_0)$$

where the E is the fractional approach to equilibrium and can be estimated using following equation:

$$E = (C - C_0)/(C^* - C)$$

where C* is the saturation concentration of dissolved oxygen and C_o is the initial concentration of the dissolved oxygen at time, $t_{0,}$ when a hydrodynamic steady-state has been reestablished upon the beginning of aeration and C is the dissolved oxygen concentration at any time, t.

The relationship between the two variables, U_g (X_1) and v (X_2), and the response, k_La, was predicted to be:

$$Y = 0.02 - 0.001751X_1 + 0.003213X_2 - 0.005606X_1^2 \\ - 0.006956X_2^2 - 0.03325X_1X_2$$

ANOVA analysis revealed that all linear, square, and interaction terms of X_1 and X_2 ($P < 0.05$) were found to be significant on the k_La. The contour plot between the response, i.e. k_La and the combined effect of superficial gas velocity, Ug, and working volume, v, showed that the value of k_La increased in the beginning and decreased after increasing from middle values. The optimum condition which is defined as the best combination of factor setting for achieving the optimum response was found to be a superficial gas velocity (2.5 L/min) and working volume (6.6 L) for a predicted response of 0.0206 s^{-1} and desirability score of 1 (Irani *et al.*, 2011). Upon applying these conditions in a bioreactor with 30% diesel/water and an initial S-content of 28 mg/L, the recorded S-removal was 14 mg/L within 30 h (Irani *et al.*, 2011).

Based on four levels of full factorial design (4^2), a statistical design of experiments was used to investigate two cases of DBT-removal in batch processes using *Bacillus sphaericus* HN1, involving factors of yeast extract and dimethylsulfoxide (DMSO) or magnesium sulfate for first and second cases, respectively. Four quadratic polynomial model equations were predicted finding out how significant the effects of these variables (factors) and their interactions are in practice (Deriase *et al.*, 2012).

The first case exhibited the effect of yeast extract (X_1, g/L) and DMSO (X_2, mM) concentrations on cell growth (Y_1, DCW, mg/L) and DBT removal were expressed as the remaining DBT (Y_2, mg/L).

$$Y_1 = 6.02 + 0.96X_1 - 0.081X_2 + 0.059X_1X_2 - 0.27X_1^2 - 0.015X_2^2$$

$$Y_2 = 4.39 - 1.34X_1 - 0.4X_2 - 0.17X_1X_2 + 0.46X_1^2 + 0.14X_2^2$$

The second case exhibited the effect of yeast extract (X_1, g/L) and $MgSO_4$ (X_3, g/L) concentrations on cell growth (Y_2, DCW, mg/L) and DBT removal expressed as the remaining DBT (Y_4, mg/L).

$$Y_3 = 6.32 + 0.78X_1 + 1.07X_3 - 0.14X_1X_3 - 0.25X_1^2 - 1.7X_3^2$$

$$Y_4 = 4.47 - 1.8X_1 - 4.1X_3 - 1.15X_1X_3 + 0.75X_1^2 + 0.47X_3^2$$

It was depicted that yeast extract (YE) significantly influenced the growth pattern and DBT-removal of *B. sphaericus* HN1, in the two studied cases with an increase of YE concentration, an increase of cell growth occurred, and the remaining concentration of DBT decreased, i.e. high removal of DBT occurred. DMSO expressed negative influence on both growth and DBT-removal. $MgSO_4$ had a positive effect on growth, but negative influence on DBT removal. Moreover, the interaction effect of YE with $MgSO_4$ was more significant than that of YE with DMSO.

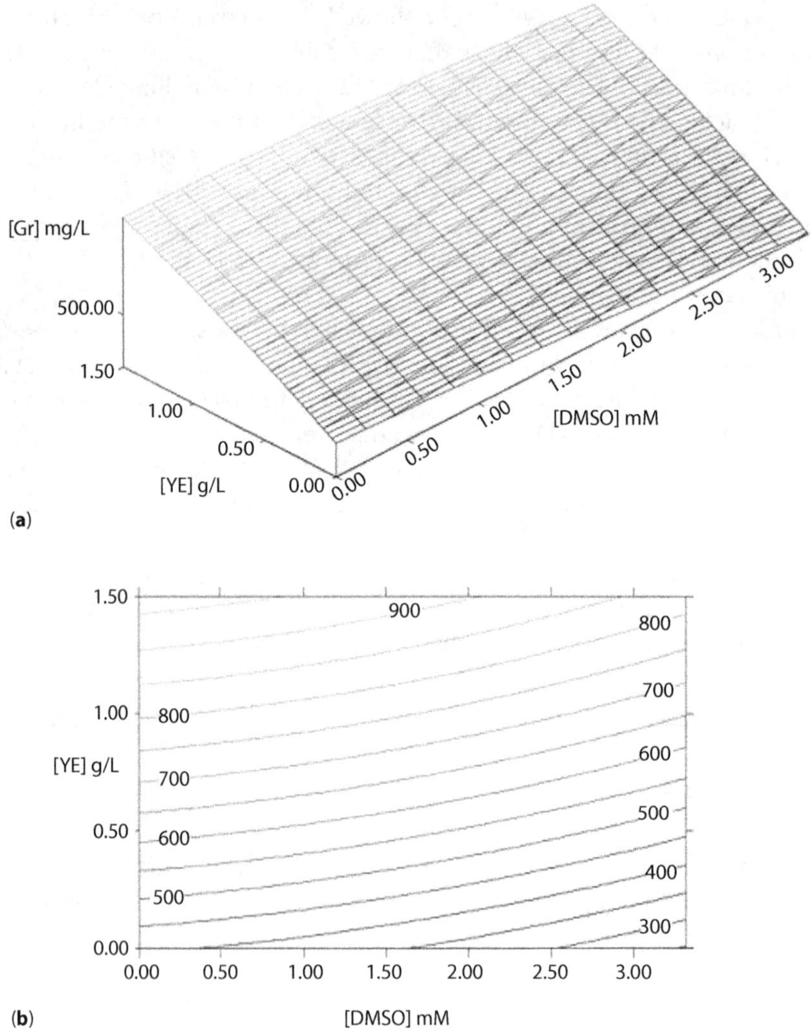

Figure 8.4 Response Surface 3D-Plot (a) and 2D Contour Plot (b) for the Effect of YE (g/L) and DMSO (mM) on Cell Growth (Deriase et al., 2012).

One of the greatest advantages of the response surface methodology (RSM) is to illustrate a correlation between variables in the form of three-dimensional (3D) surface and two-dimensional (2D) contour plots to visualize the effect of the studied factors.

Figures 8.4a and b show the surface and contour plots of the effect of YE (g/L) and DMSO mM on cell growth, Gr (mg/L). They illustrates the significant increase of cell growth with an increment of YE concentration

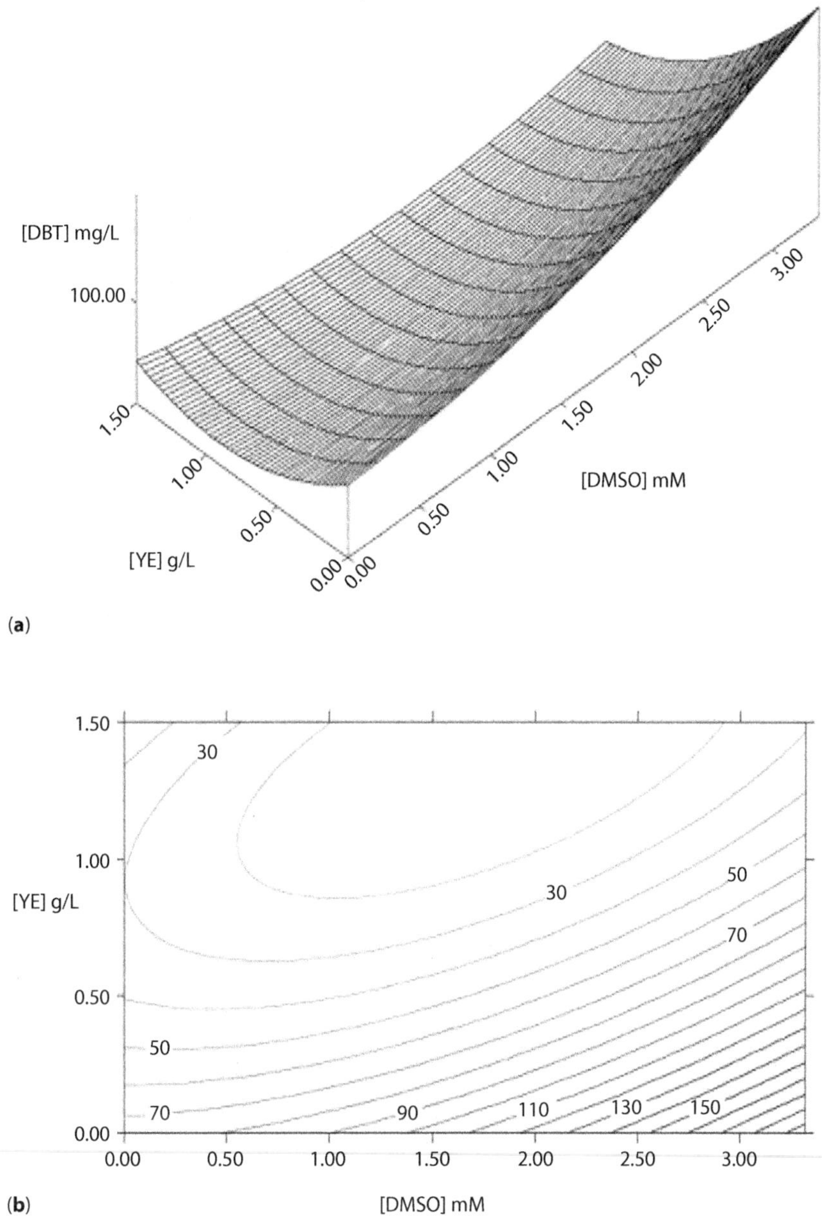

Figure 8.5 Response Surface 3D-Plot (a) and 2D Contour Plot (b) for the Effect of YE (g/L) and DMSO (mM) on the Remaining DBT (i.e. the removal efficiency) (Deriase et al., 2012).

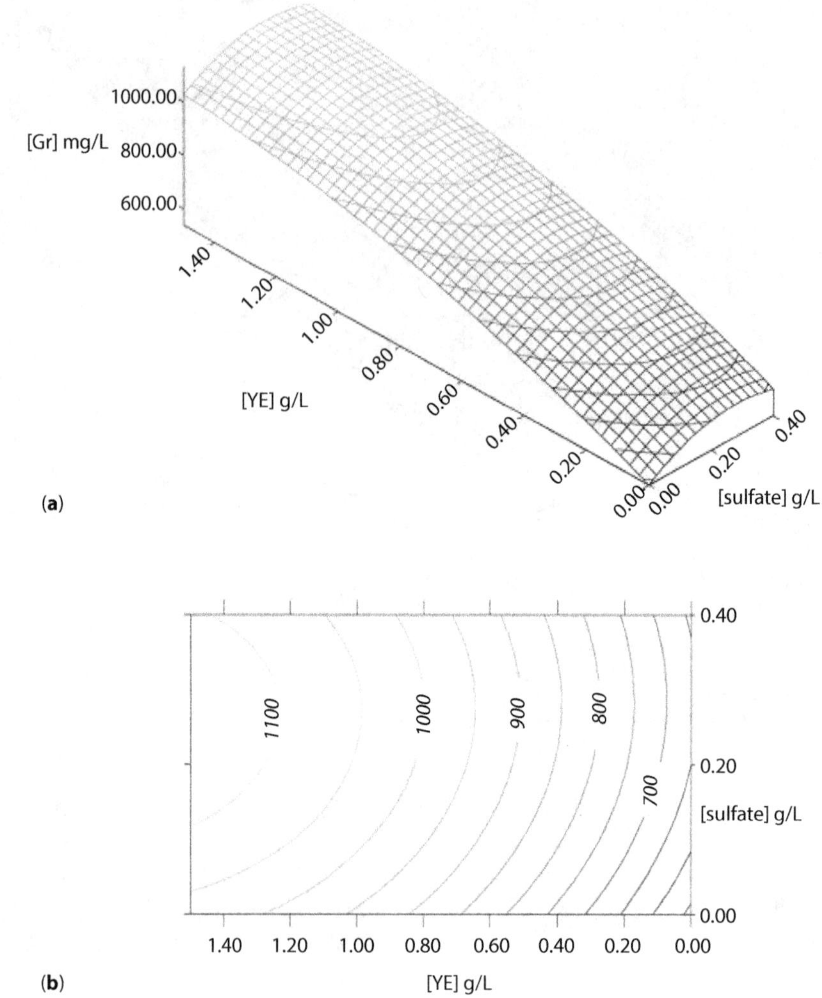

Figure 8.6 Response Surface 3D-Plot (a) and 2D Contour Plot (b) for the Effect of YE (g/L) and $MgSO_4$ (g/L) on Cell Growth (Deriase et al., 2012).

and decrement of DMSO concentration, i.e. higher values of Gr (mg/L) were noticed at a lower concentration of DMSO.

Figures 8.5a and b show the surface and contour plots of the interaction effects of YE g/L and DMSO mM on DBT concentration mg/L. The lowest remaining concentration of DBT, reflecting the highest removal, was noticed at a high level of YE concentration and a low level of DMSO concentration.

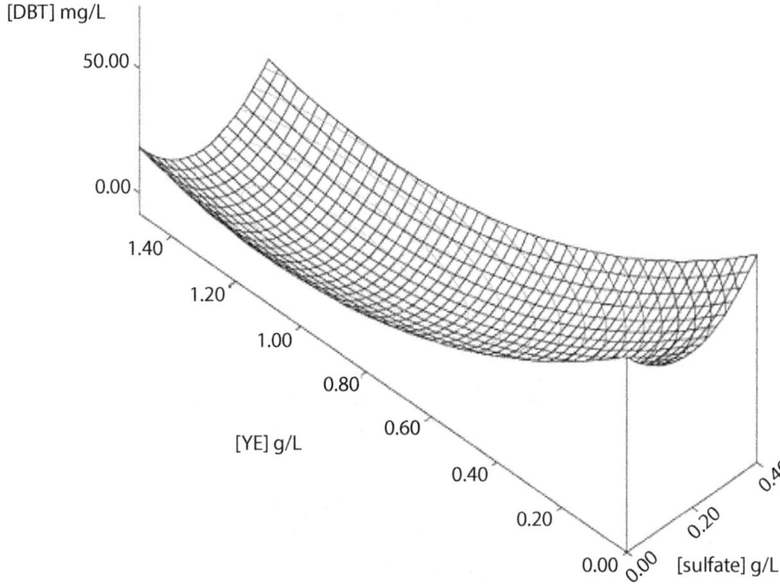

(a)

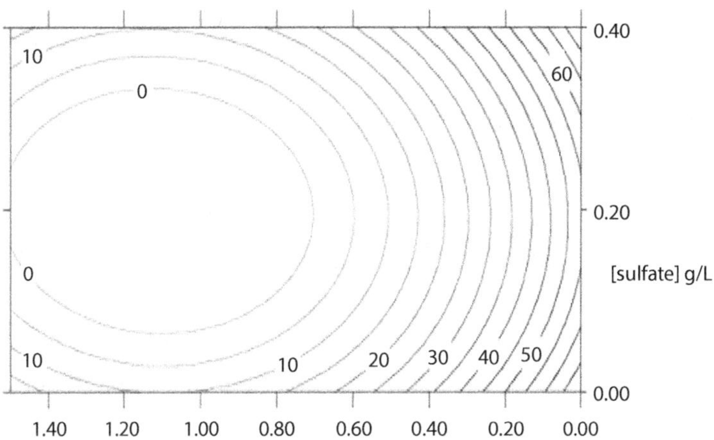

(b)

Figure 8.7 Response Surface 3D-Plot (a) and 2D Contour Plot (b) for the Effect of YE (g/L) and $MgSO_4$ (g/L) on the Remaining DBT (i.e. the removal efficiency) (Deriase et al., 2012).

It is clear from the surface and contour plots for the interaction effect of YE (g/L) and $MgSO_4$ (g/L) on Gr (mg/L) of HN1 (Figure 8.6a and b), that increasing concentrations of YE and $MgSO_4$ enhanced microbial growth.

It was evident from the surface and contour plots for the interaction effect of YE (g/L) and $MgSO_4$ (g/L) on removed DBT (mg/L) using HN1 (Figure 8.7a and b) that the remaining DBT concentration significantly decreased with increasing both YE and $MgSO_4$ concentrations.

The elliptical contour plots (Figures 8.6b and 8.7b) indicate that the interactions between the corresponding variables YE and $MgSO_4$ are more significant than those of YE and DMSO.

LINGO optimization revealed that the optimum values of the test variables were 1.5 g/L YE and 0.86 mM DMSO, which corresponded to a maximum cell growth of 954 mg DCW/L and a minimum DBT concentration of 21.47 mg/L that is ≈ 91.4% removal from the initial concentration of 250 mg/L, while a maximum cell growth of 1,136 mg/L and a minimum DBT concentration of 4.25 mg/L equivalent to ≈ 98.3% removal were achieved at 1.497 g/L YE and 0.25 g/L $MgSO_4$ (Deriase *et al.*, 2012).

Arabian *et al.* (2014) reported the application of RSM, based on CCD, to optimize the effect of three variables: biocatalyst cell density (A; cell/mL), oil phase fraction (B), and initial DBT concentration (C, ppmw), with three levels on BDS using Gram +ve *Bacillus cereus* HN in a biphasic system (aqueous/DBT in dodecane) at 30° C, 180 rpm, and 48 h. A quadratic model equation was found to be the best correlating to studied variables with the response, i.e. BDS percentage:

$$BDS\% = 56.48 + 1.57A - 4.74B + 4.99C - 6.34\,A^2 + 2.56\,B^2$$

The optimum conditions were predicted to be a biocatalyst cell density of 3.6×10^7 (cell/mL), oil phase fraction of 0.2, and initial DBT concentration of 1086 ppmw, that provided ≈ 79% biodesulfurization (Arabian *et al.*, 2014).

Bordoloi *et al.* (2014) reported the application of RSM based on a central composite design (CCD) of four independent variables: incubation time (h) (C_1), initial DBT concentration (% v/v) (C_2), initial inoculum size (% v/v) (C_3), and medium pH (C_4) with three levels to optimize the bacterial growth and BDS efficiency of *Achromobacter* sp. and to study their interactive effects on the BDS process. The overall second-order polynomial regression equation showing the empirical relationship between the bacterial growth (Y) of the four tested were predicted to be:

$$
\begin{aligned}
Y = {} & 0.185501 - 0.002603C_1 - 0.075348C_2 - 0.023098C_3 - 0.005338C_4 \\
& + 4\times10^{-6}C_1^2 + 9.335\times10^{-3}C_2^2 + 1.085\times10^{-3}C_3^2 - 1.479\times10^{-3}C_4^2 \\
& + 1.86\times10^{-4}C_1C_2 + 3.32\times10^{-4}C_1C_3 + 2.47\times10^{-4}C_1C_4 \\
& + 8.062\times10^{-3}C_2C_3 + 6.469\times10^{-3}C_2C_4 + 4.844\times10^{-3}C_3C_4
\end{aligned}
$$

The coefficients of the model, including the significance of each coefficient as determined by t-test and p-values, showed that the incubation time, DBT level, and inoculum level had significant effects ($p < 0.05$) on bacterial growth. The optimum conditions were predicted to be 3.5 mM of DBT and 3.5% (v/v) inoculum size at pH 10.0 of the medium for 132 h post incubation at 37 °C (Bordoloi *et al.*, 2014).

Based on five levels of full factorial design, a statistical design of experiments was used to investigate two cases of DBT-BDS in batch processes, involving yeast extract and glucose or glycerol as factors for *Brevibacillus invocatus* C19 and *Rhodococcus erythropolis* IGTS8, respectively. Cubic and quadratic predictive models, significantly describing the interactive relationships between dependent and independent variables, were established for C19 and IGTS8, respectively. The statistical analysis and optimization revealed the optimum values of glucose and yeast extract concentrations for maximum BDS efficiency and microbial growth in a batch DBT-BDS using *Brevibacillus invocatus* C19 were 0.05 M glucose and 0.3 g/L YE, which recorded 95% BDS of 1000 ppm DBT with the production of 65.20 ppm 2-HBP and the cell biomass was nearly doubled, while for *R. erythropolis* IGTS8, the optimum values for glycerol and YE were 0.07 M and 0.1 g/L, respectively, which recorded a production of 47.82 ppm 2-HBP with a percentage BDS of 76.81% where the growth was nearly doubled (Abo-State *et al.*, 2014).

Based on the five levels of full factorial design, RSM was used for modeling, optimization, and studying the interactive effects of two variables: nitrogen source, YE and carbon co-substrate, and glycerol in a batch process for DBT-BDS using a new Gram-positive bacterial isolate *R. erythropolis* HN2 (accession no. KF018282) (El-Gendy *et al.*, 2014). Four quadratic polynomial model equations were predicted to correlate the studied variables A: concentration of YE (g/L) and B: concentration of glycerol (M) and four responses, Y_1, Y_2, Y_3, Y_4, which represent cell growth ln(cell/mL), DBT-BDS percentage, the produced 2-HBP mg/L, and the end product 2,2′-DMBP mg/L, respectively.

$$Y_1 = 20.39 + 0.23A + 3.56B - 0.26A^2 - 1.04B^2 - 0.39AB$$

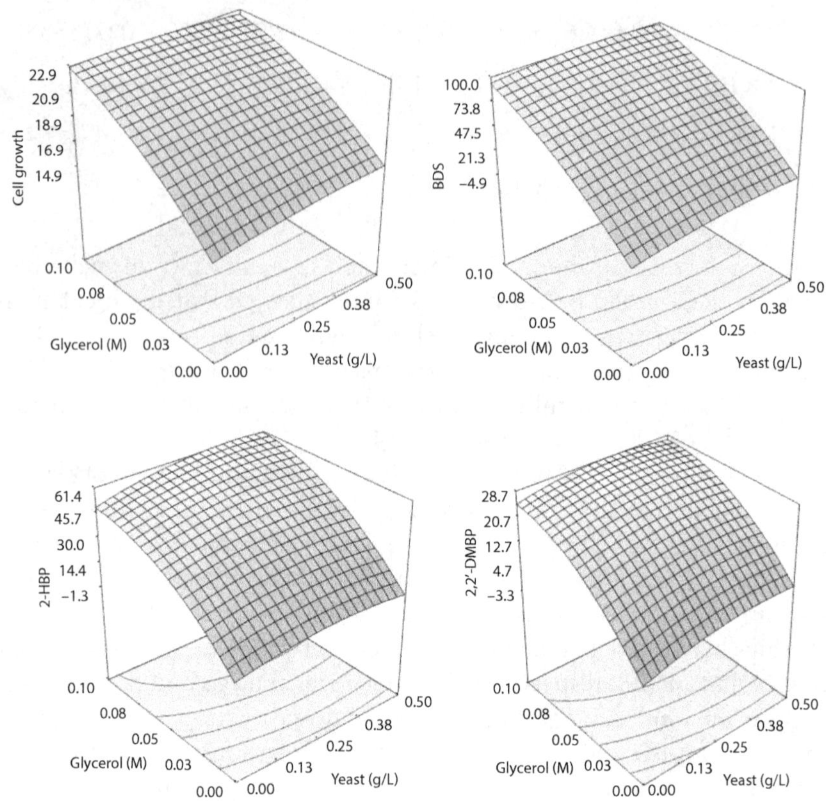

Figure 8.8 The Response Surface Plots of Studied Variables on Different Responses: Cell Growth, BDS Percentage, 2-HBP Production and 2,2′-DMBP in Batch DBT-BDS by *R. erythropolis* HN2 (El-Gendy *et al.*, 2014).

$$Y_2 = 64.1 + 4.01A + 46.89B - 4.93A^2 - 11.2B^2 - 1.99AB$$

$$Y_3 = 44.85 + 3.95A + 26.93B - 4.98A^2 - 12.02B^2 + 1.75AB$$

$$Y_4 = 22.03 + 2.16A + 13.07B - 2.67A^2 - 6.51B^2 - 0.97AB$$

The statistical analysis of the data obtained revealed that all the responses were very sensitive to the changes in glycerol concentration. However, they were less sensitive to changes in YE concentration. The YE expressed no significant effect on cell growth, but expressed a relatively significant effect on DBT-BDS and production of 2-HBP and 2,2′-DMBP, but the increase in glycerol concentration expressed a highly significant effect on cell growth,

DBT-BDS, and production of 2-HBP and 2,2′-DMBP, where it was obvious from the RSM 3-D plots (Figure 8.8) that within the studied YE concentrations, with the increase of glycerol concentration, cell growth, BDS activity, the production of 2-HBP and 2,2′-DMBP increased, reaching their maximum values at 0.1 M glycerol.

The optimum operating conditions for DBT-BDT using *R. erythropolis* HN2 were found to be 0.35 g/L yeast extract and 0.09 M glycerol in a batch shaken flasks BDS of 1000 ppm DBT, pH7, 30 °C, and 150 rpm, which recorded, 23, 97%, 65 mg/L, and 29 mg/L, cell growth (ln[cells/mL]), DBT-BDS percentage, and production of 2-HBP and 2,2-DMBP mg/L, respectively (El-Gendy *et al.*, 2014).

Quadratic and 2F1 models were established on the basis of the results obtained from the CCD of experiments to represent the relationship between the responses, i.e. surface tension and emulsification index of the biosurfactants and the multiple independent variables, namely dilution rate, recycle ratio, and inlet sulfur concentration in a continuous 2.5 dm^3 B.Braun chemostat with a working volume of 1.5 dm^3 (Bandyopadhyay and Chowdhury, 2014):

$$Y = 28.99 + 0.98A - 0.47B - 8.78C - 0.68AB - 0.78AC + 0.18BC + 0.31A^2 + 0.58B^2 + 7.8C^2$$

$$E = 45.93 - 1.01A + 0.56B + 13.26C + 0.46AB$$

where Y is surface tension (dynes/cm), E is the emulsification index (%), A is the dilution rate (D) (h^{-1}), B is the recycle ratio (R), and C is the inlet sulfur concentration (S_0 mg/L). The quadratic effects of A, B, and C are found to be significant ($p < 0.0001$). Process optimization has been done using the Design Expert software and it revealed that the optimum values of surface tension and the emulsification index were 28 dynes/cm and 60.2 at S_0 of 540 ppm, D of 0.01 h^{-1}, and R value of 0.4 with a desirability of 0.72.

Poto *et al.* (2017) applied CCD for the optimization of BDS-culture medium for the enhancement of heavy gas oil (HGO) BDS using *R. erythropolis* ATCC 4277 in a biphasic system at 20% HGO and 80% aqueous phase. In a CCD of three variables, different concentrations of YE, glucose, and process time when yeast extracted concentration, was fixed at 4.0 g/L to the highest BDS percentage occurred with a minimum glucose concentration (1.0 g/L) within 72 h and specific desulfurization rate of 73 mg/kg HGO/h was recorded. However, there was an interesting result in the test without glucose at a shorter processing time of 12 h obtaining a specific desulfurization rate of 418 mg sulfur/kg HGO/h and BDS percentage of 72.7%. From that preliminary study, Porto *et al.* (2017) concluded that sometimes the BDS percentages are similar, but the desulfurization rate and the desulfurization kinetics are different. Under some conditions desulfurization occurred quickly, while in other cases it gradually increased with time. The effects of estimated analysis for glucose and yeast extract variables were also carried out for a second-order model at a 95% confidence interval and a significant level of $\alpha = 0.05$ (Porto *et al.*, 2017). In the tests with a fixed glucose concentration, the linear term of the yeast extract concentration and the interaction between glucose and yeast extract concentrations had a significant effect on BDS capacity. The desirability profile indicated that the responses were maximized by using just 10 g/L of yeast extract and 12 h of BDS was enough. Moreover, for the tests with fixed YE concentration, only the linear term of time had a significant effect on sulfur removal. Thus, it seemed that glucose concentration can be minimized in the medium without affecting process efficiency. The desirability profile confirmed that the best process conditions were those that minimized glucose concentration and the process time. In another CCD of different concentrations of C-sources (glucose, malt extract) and an N-source (yeast extract) to achieve maximum BDS of HGO, the highest growth of 4.35 g/L occurred at optimum concentrations of 2 g/L glucose, 6.5 g/L YE, and 9.62 g/L malt extract, that were accompanied with a BDS percentage of 51.4% and specific desulfurization rate of 295 mg/kg HGO/h, while the maximum BDS percentage of 75% with the highest specific desulfurization rate of 454 mg/kg HGO/h occurred within 12 h at optimum concentrations of 2 g/L, 12.39 g/L, and 5 g/L of glucose, YE, and malt extract, respectively (Porto *et al.*, 2017). The observed lower biomass concentration was attributed to the fact that CCD was designed to achieve higher sulfur removal and not higher cellular concentrations. Other studies also noticed that the best biodesulfurization capability was obtained under different conditions to those for the best biomass growth (Del Olmo *et al.*, 2005ab). Porto *et al.* (2017) showed that the optimum nutritional composition of 1.55 g/L glucose, 5.15 g/L malt extract, and 12.39 g/L yeast extract revealed a specific desulfurization rate of 423 mg S/kg HGO/h and a constant BDS percentage of 70% from 8 to 12 h and a high BDS efficiency in a shorter time.

Fractional factorial design (FFD) of a 2^{6-1} was realized using $CaCO_3$ (g/L, D), glucose (g/L, B), malt extract (g/L, C), yeast extract (g/L, A), agitation (rpm, F), and temperature (°C, E) as independent variables to permit the identification of the culture conditions which present a significant influence on the cell growth of *R. erythropolis* ATCC 4277 (Todescato *et al.*, 2017). A total of 36 runs with different combinations of six factors were carried out. A multiple regression analysis was applied to experimental results in order to obtain an equation capable of predicting biomass production BC_{Re}:

$$BC_{Re} = 1.93 + 9.95A - 0.99D - 0.44E + 0.83F + 0.67AE$$

Thus, a higher concentration of yeast extract expressed a clear influence in the quantity of biomass produced. Moreover, the interaction effect of yeast extract and agitation speed presented a major positive effect in biomass concentration. However, the evolution in the biomass concentration was dimensioned by an increase in the calcium carbonate ($CaCO_3$), concentration, and temperature.

A central composite design (CCD, 2^4) was accomplished in order to optimize the variables by a response surface methodology (RSM) for *R. erythropolis* ATCC 4277. The factor levels were decided based on the range obtained with the previous fractional factorial design (FFD) (Todescato *et al.*, 2017). Thus, the factors that were not statistically significant were maintained at the lowest level, settling the malt extract and glucose concentrations in 5.0 and 2.0 g/L, respectively. The remaining variables were displaced considering the effects noted in the FFD: yeast extract concentration (g/L, A) and agitation speed (rpm, D) with positive effects and concentration of $CaCO_3$ (g/L, B) and the temperature (°C, C) with negative effects. Statistical analysis permitted the development of a model that describes the *R. erythropolis* ATCC 4277 cell growth BC_{Re}:

$$BC_{Re} = 5.04 - 0.56A^2 - 0.34B^2 - 0.57C^2 + 0.39D^2 - 0.38BC$$

The higher coefficients of yeast extract and $CaCO_3$ concentrations, temperature, and agitation indicate that cell growth of *R. erythropolis* ATCC 4277 was highly affected by these parameters.

The optimal conditions, yielding 6 g/L cell density, were found to be 2 g/L glucose, 5 g/L malt extract, 6.15 g/L yeast extract and 1.16 g/L $CaCO_3$, temperature of 23.7 °C, and agitation of 180 rpm. Moreover, the optimized conditions resulted in a maximum specific growth rate of 0.116 h^{-1}, growth yield coefficient of 0.675 g/g, and Monod saturation constant of 20 g/L (Todescato *et al.*, 2017).

Doehlert uniform design (Doehlert, 1997) was applied to study the effect of two factors, X_1 or $BaCl_2$ concentration and X_2 or exposure time for sulfate precipitation in carob pulp liquor; the response, Y, to be applied in the BDS process using *Gordonia alkanivorans* strain 1B (Silva *et al.*, 2013).

A second order polynomial model equation was predicted to correlate the variables with the response:

$$Y = 40.67 - 18.51X_1 - 3.73X_2 - 6.29X_1X_2 - 7.76X_1^2 - 5.87X_2^2$$

The results indicated that a $BaCl_2$ concentration of 0.125% is not enough to obtain significant sulfate precipitation, independent of the exposure time. However, at higher $BaCl_2$ concentrations, the longer the exposure time is, the greater the occurrence of sulfate precipitation. Furthermore, the exposure time was found to have five fold less influence on the final sulfate concentration than the amount of $BaCl_2$ used in the process. The response surface showed that a small variation of $BaCl_2$ concentration produces a response in the sulfate concentration. The optimum conditions to obtain a carob pulp liquor with minimal sulfate concentration (0–5 mg/L) were found to be a $BaCl_2$ concentration above 0.4% with an exposure time of 36 h. However, combination of 0.5% of $BaCl_2$ and a time of exposure of 21 h produced carob pulp liquor which led to a maximum BDS efficiency of 400 μmol/L DBT producing 237 μmol/L 2-HBP.

In a similar study, statistically validated experimental design followed the Doehlert distribution for two factors, $BaCl_2$ concentration (X_1) and pH (X_2), to study their effects on the sulfate precipitation (response, Y) in the acid hydrolysate of Jerusalem artichoke juice (JAJ) and a second order polynomial model was predicted to describe the process (Silva *et al.*, 2014).

$$Y = 29.41 - 99.97X_1 - 43.39X_2 + 66.23X_1X_2 + 72.13X_1^2 + 33.39X_2^2$$

Then, to evaluate the treatment effect on both microbial growth and BDS ability by *G. alkanivorans*, the different JA juices, treated and untreated, were added to the desulfurization medium (dilution in a 1:6 ratio to adjust the initial concentration to ~25 g/L of total sugars) and tested as carbon sources in a batch BDS with an initial DBT concentration of 400 μmol/L. The surface response methodology according to a Doehlert distribution for the studied two factors revealed that both factors, $BaCl_2$ concentration and initial pH, have relevant effects on the sulfate removal from hydrolyzed JAJ. When the concentration of $BaCl_2$ was doubled at pH 5.27 and pH 8.73, the average amount of final sulfate concentration decreased about 6.7 and 5.4 fold, respectively, indicating a less efficient sulfate removal at a high pH 8.73. This also confirmed that for lower concentrations of $BaCl_2$ (0.25%), increasing pH from 5.27 to 8.73 decreased the sulfate 3.2 fold, while for higher concentrations of $BaCl_2$ (0.5%), the same pH increase led to a 2.6 fold drop in sulfate. Thus, a more efficient sulfate removal occurs at a higher pH, but with low $BaCl_2$ concentration, which was also indicated by the negative coefficients of X_1^2 and X_2^2 and positive coefficients of the interactive effect X_1X_2 in the predicted polynomial equation. The response surface revealed that the relative effect of $BaCl_2$ concentration on the final sulfate concentration is greater than that of pH. Small concentrations of $BaCl_2$ seem to be less effective at an acidic pH. The maximum removal of sulfate occurred around pH7 and a $BaCl_2$ concentration of 0.5–0.55%. However, the optimum conditions were predicted to be 0.5% (w/v) $BaCl_2$ and pH 8.73. The most efficient treatments, with very similar values of DBT removal and 2-HBP production, were the treatment with 0.5% $BaCl_2$ and pH 8.73 achieved at a maximum of 353 μmol/L 2-HBP and the treatments with 0.375 and 0.63% at pH 7 reached 330 μmol/L and 328 μmol/L 2-HBP, respectively. However, upon the application of the predicted optimum conditions, μ_{max} reached 0.06 h^{-1} with a total consumption of the available carbon source (glucose and fructose ~ 25 g/L) by the end of the growth (96 h). Production of 2-HBP was also initiated at 18 h, but attained ~ 400 μmol/L between 73 and 96 h, corresponding to an overall production rate of 4.48 μmol/L/h and to a maximum production rate of 28.2 μmol/L/h with a related q_{2HBP} of 5.06 μmol/g DCW/h with a total consumption of the DBT. Using the $BaCl_2$ treated hydrolyzed JAJt, the amount of 2-HBP produced (μmol) per gram of sugar consumed was ≈ 16 μmol/g (Silva *et al.*, 2014).

Lin *et al.* (2014) reported 100% degradation of DBT by the new isolate *Pseudomonas* sp. LKY-5, via the application of RSM based on Box-Behnken design (BBD). A total of 29 experiments were conducted on four factors: initial DBT concentration (X_1), temperature (X_2), pH (X_3), and agitation rate (X_4) at three levels with BBD. A second order polynomial equation was found to best describe the interactive effects of the four studied variables on the DBT-removal percentage (Y):

$$
\begin{aligned}
Y = 188.47 - 1.23X_1 + 9.6X_2 + 52.29X_3 - 4.02X_4 + 0.01X_1X_2 \\
- 0.04X_1X_3 + 1.52\times10^{-3}X_1X_4 + 0.34X_2X_3 - 0.02X_2X_4 \\
- 0.14X_3X_4 + 1.37\times10^{-3}X_1^2 - 0.17X_2^2 - 2.2X_3^2 + 0.02X_4^2
\end{aligned}
$$

The value of determination coefficient R^2=0.9534 indicated a satisfactory agreement between the quadratic model and experimental data. The analysis of variance ANOVA proved that DBT removal was more significantly affected (P< 0.0001) by the initial DBT concentration compared with the other three studied parameters. The optimum operation conditions with 100% DBT-removal were found to be 100 mg/L initial DBT-concentration at a temperature range of 25–35 °C, an agitation rate of 140–180 r/min, and a pH value of 6.5–8.5 within 7 days, while, upon applying, the RSM optimized settings included an agitation rate of 140 r/min, a temperature of 33°C, and a pH value of 8.28 and 56 percent biodegradation of DBT at a high initial concentration of 200 mg/L was achieved after a cultivation period of 7 days (Lin *et al.*, 2014).

DBT concentration, temperature, and pH were optimized statistically for growing and resting cells of *Paenibacillus validus* (strain PD2) by using RSM (Derikvand *et al.*, 2015a). All the parameters in growing cells had a significant effect on 2-HBP production during the BDS of DBT by *P. validus* PD2. However, in resting cells in a temperature range of 20–40 °C was not a significant factor. Maximum BDS for growing cells was obtained at 0.41 mM DBT concentration, pH 6.92, and temperature of 31.23 °C, while for resting cells, optimum pH, temperature, and DBT concentration were 6.62, 27.73 °C, and 7.86 mM, respectively (Derikvand *et al.*, 2015a).

Box-Behnken RSM was applied to the study of interactive effects of initial DBT concentration, temperature, and pH on the BDS activity of *Rhodococcus erythropolis* PD1 and used to determine the optimum value of these factors for both growing and resting cell conditions in aqueous and biphasic (2:1 aqueous/model oil; DBT in n-tetradecane v/v) systems, respectively (Derikvand *et al.*, 2015b), where a 3-factor and 3-level Box-Behnken design (BBD) was used to determine the optimum level of the studied factors and to study their interactive relationship to the BDS efficiency. The used resting cells were the bacterial cells grown and harvested at the late exponential phase. The BDS efficiency was evaluated by the concentration of the produced 2-HBP.

A quadratic polynomial equation was predicted to best fit and describe the relationship between the BDS efficiency as expressed by the production of 2-HBP and the studied three factors in a one phase aqueous system:

$$Y = 2.06 - 0.19X_1 - 0.13X_2 - 0.13X_3 + 0.037X_1X_2 + 0.047X_2X_3 \\ - 0.24X_1^2 - 0.3X_2^2 - 0.042X_3^2$$

where Y is the response value (i.e. the produced 2-HBP mM), X_1 is the initial DBT concentration (mM), X_2 is temperature (°C), and X_3 is initial pH. Positive and negative signs before each term indicate synergistic and antagonistic effects, respectively (El-Gendy *et al.*, 2014). The ANOVA results for 2-HBP production by growing cells have shown that the initial DBT concentration, initial pH and incubation temperature, and the interactive effects of the initial DBT concentration with incubation temperature and that of incubation temperature with the initial pH have significant effects on the BDS efficiency in an aqueous system ($p<0.05$). The optimum conditions for maximum efficiency in a one-phase aqueous system, using *R. erythropolis* PD1, were found to have an initial DBT concentration of 0.38 mM and initial pH of 6.88 at 27.57 °C, which produced 0.21 mM 2-HBP within 48 h (Derikvand *et al.*, 2015b), while the quartic model equation describing the effect of three studied variables on the BDS efficiency in a bi-phasic system using resting cells of *R. erythropolis* PD1 was predicted to be:

$$Y = 0.78 + 0.053X_1 - 0.061X_2 - 0.055X_1X_2 - 0.1X_1X_3 \\ - 0.093X_1^2 - 0.095X_2^2$$

The ANOVA test proved the significant effects of initial DBT concentration, incubation temperature, and their interactive effect ($p< 0.05$). The optimum predicted conditions using the resting cells of PD1 in biphasic system were found to be 7.73 mM, 26.13 °C, and 6.29 initial DBT concertation, incubation temperature, and initial pH, respectively, which performed a maximum BDS activity of 0.46 μM 2-HBP/g DCW/min, within 20 h (Derikvand *et al.*, 2015b). The optimum higher initial DBT concentration (7.73 mM) in the bi-phasic system, relative to the lower value in an aqueous phase system (0.38 mM), was explained by the low solubility of DBT in an aqueous phase and its high solubility in *n*-tetradecane (organic phase) which reduces its toxic effect on bacteria. Moreover, the produced 2-HBP accumulation in the cells would be lower because of its high solubility in the solvent phase (Kawaguchi *et al.*, 2012).

A 3-factor and 3-level Box-Behnken design (BBD) based on RSM was applied to determine the optimum level of variables of DBT concentration (X_1: 2, 6, and 10 mM), incubation temperature (X_2: 20, 30, and 40 °C), and pH (X_3: 5, 7, and 9) and to study their relationship in a biphasic (1:2 O/W) batch BDS process of model oil (DBT in n-tetradecane) using magnetic nanoparticle (MNPs) immobilized cells of *R. erythropolis* R1 (Etemadifar *et al.*, 2014). A model of coded units, after removing non-significant parameters, was depicted to correlate the response, i.e. the BDS-efficiency, represented as the amount of produced 2-HBP:

$$Y = 1.97 - 0.19X_1 + 0.12X_2 - 0.063X_1X_2 - 0.4X_1^2 - 0.41X_2^2$$

where the equation indicates a quadratic linear relationship between variables and 2-HBP (Y).

According to the depicted model, DBT concentration, temperature, and interaction between them were significant, but pH and its interaction with other factors was not statistically significant. The interaction of each of the two independent factors was shown by a response surface with a contour plot, while another factor is fixed at the level of zero. The fitted surface and contour plots were between DBT concentration and temperature, DBT concentration and pH, and temperature and pH. The highest 2-HBP production was obtained when all factors were at the middle level: 6.76 mM, 29.63 °C and pH6.84.

References

Abbad-Andaloussi, S., Lagnel, C., Warzywoda, M., Monot, F. (2003) Multicriteria comparison of resting cell activities of bacterial strains selected for biodesulfurization of petroleum compounds. *Enzyme and Microbial Technology*. 32(3/4): 446–454.

Abin-Fuentes, A., Mohamed, M.E., Wang, D.I.C., Prather, K.L.J. (2013) Exploring the mechanism of biocatalyst inhibition in microbial desulfurization. *Applied and Environmental Microbiology*. 79(24): 7807–7817.

Abo-State, M.A., El-Gendy, N.Sh., El-Temtamy, S.A., Mahdy, H.M., Nassar, H.N. (2014) Modification of Basal Salts Medium for Enhancing Dibenzothiophene Biodesulfurization By *Brevibacillus invocatus* C19 and *Rhodococcus erythropolis* IGTS8. *World Applied Sciences Journal*. 30(2): 133–140.

Abro, R., Abdeltawab, A.A., Al-Deyab, S.S., Yu, G., Qazi, A.B., Gao, S.R., Chen, X.C. (2014) A Review of Extractive Desulfurization of fuel oils using ionic liquids. *RSC Advances*. 4: 35302–35317.

Abusham, R., Raja, N., Rahman, R., Abu, B., Mahiran, B. (2009) Optimization of physical factors affecting the production of thermo-stable organic solvent-tolerant protease from a newly isolated halo tolerant *Bacillus subtilis* strain Rand. Microbial Cell Factories. 8: 20–28.

Adlakha, J., Singh, P., Ram, S.K., Kumar, M., Singh, M.P., Singh, D., Sahai, V., Srivastava, P. (2016) Optimization of conditions for deep desulfurization of heavy crude oil and hydrodesulfurized diesel by *Gordonia* sp. IITR100. 184: 761–769.

Aggarwal, S., Karimi, I.A., Kilbane, J.J., Lee. D.Y. (2012) Roles of sulphite oxidoreductase and sulphite reductase in improving desulfurization by *Rhodococcus erythropolis. Molecular BioSystems*. 8: 2724–2732.

Aggarwal, S., Karimi, I.A., Lee, D.L. (2011) Flux-based analysis of sulfur metabolism in desulfurizing strains of *Rhodococcus erythropolis. FEMS Microbiology Letters*. 315: 115–121.

Aggarwal, S., Karimi, I.A., Lee, D.Y. (2011) Reconstruction of a genome-scale metabolic network of *Rhodococcus erythropolis* for desulfurization studies. *Molecular BioSystems*. 7: 3122–3131.

Aggarwall, S., Karimil I.A., Ivan, G.R. (2013) In silico modeling and evaluation of *Gordonia alkanivorans* for biodesulfurization. *Molecular BioSyststem*. 9: 2530–2540.

Ahmad, A., Chauhan, A.K., Kushwaha, H.N., Javed, S., Kuma, A. (2015) Preferential desulfurization of dibenzyl sulfide by an isolated *Gordonia* sp. IITR100. 3 Biotech. 5: 237–243.

Akhtar, N., Ghauri, M.A., Anwar, M.A., Akhtar, K. (2009) Analysis of the dibenzothiophene metabolic pathway in a newly isolated *Rhodococcus* spp. *FEMS Microbiology Letters*. 301: 95–102.

Alcon, A., Martin, A. B., Santos, V.E., Gomez, E., Garcia-Ochoa, F. (2008). Kinetic model for DBT desulphurization by resting whole cells of *Pseudomonas putida* CECT5279. *Biochemical Engineering Journal*. 39: 486–495.

Alcon, A., Santos, V.E., Martin, A.B., Bustos, P., Garcia-Ochoa, F. (2005) Biodesulfurization of DBT with *Pseudomonas putida* CECT5279 by resting cells: influence of cell growth time on reducing equivalent concentration and HpaC activity. *Biochemical Engineering Journal*. 26: 168–175.

Alhassan, Y., Kumar, N., Bugaje, I.M., Mishra, C. (2014) Optimization of Gossypium arboretum seed oil biodiesel production by central composite rotatable model of response surface methodology and evaluation of its fuel properties. *Journal of Petroleum Technology and Alternative Fuels*. 5(1): 1–12.

Alves, L., Marques, S., Matos, J., Tenreiro, R., Gírio FM. (2008) Dibenzothiophene desulfurization by *Gordonia alkanivorans* strain 1B using recycled paper sludge hydrolyzate. *Chemosphere*. 70: 967–973.

Alves, L., Melo, M., Mendonça, D., Simões, F., Matos, J., Tenreiro, R., Gírio, F.M. (2007) Sequencing, cloning and expression of the dsz genes required for dibenzothiophene sulfone desulfurization from *Gordonia alkanivorans* strain 1B. *Enzyme and Microbial Technology* 40: 1598–1603.

Alves, L., Paixão, S.M. (2011) Toxicity evaluation of 2-hydroxybiphenyl and other compounds involved in studies of fossil fuels biodesulphurisation. *Bioresource Technology*. 102: 9162–9166.

Alves, L., Paixão, S.M. (2014a) Enhancement of dibenzothiophene desulfurization by *Gordonia alkanivorans* Strain 1B using sugar beet molasses as alternative carbon source. *Applied Biochemistry and Biotechnology*. 172: 3297–3305.

Alves, L., Paixão, S.M. (2014b) Fructophilic behaviour of *Gordonia alkanivorans* strain 1B during dibenzothiophene desulfurization process. *New Biotechnology*. 31(1): 74–79.

Alves, L., Salgueiro, R., Rodrigues, C., Mesquita, E., Matos, J., Gírio, F.M. (2005) Desulfurization of dibenzothiophene, benzothiophene, and other thiophene analogs by a newly isolated bacterium, *Gordonia alkanivorans* Strain 1B. *Applied Biochemistry and Biotechnology*. 120: 199–208.

Alves, L., Salgueiro, R., Rodrigues, C., Mesquita, E., Matos, J., Gírio, F.M. (2005) Desulfurization of dibenzothiophene, benzothiophene, and other thiophene analogs by a newly isolated bacterium, *Gordonia alkanivorans* Strain 1B. *Applied Biochemistry and Biotechnology*. 120: 199–208.

Anderson, J.M., Whitcomb, J.P. (2015) DOE simplified: practical tools for effective experimentation. 3rd ed.; CRC Press, Taylor & Francis Group, New York, NY, USA.

Ansari, F., Prayuenyong, P., Tothill, I. (2007) Biodesulfurization of dibenzothiophene by *Shewanella putrefaciens* NCIMB 8768. *Journal of Biological Physics and Chemistry*. 7: 75–78.

Ansari, F., Prayuenyong, P., Tothill, I. (2007) Biodesulfurization of dibenzothiophene by *Shewanella putrefaciens* NCIMB 8768. *The Journal of Biological Physics and Chemistry*. 7: 75–78.

Ansari, F., Prayuenyong, P., Tothill, I.E. (2007) Biodesulfurization of dibenzothiophene by *Shewanella putrefaciens* NCIMB 8768. *The Journal of Biological Physics and Chemistry*. 7:75–77.

Arabian, D., Najafi, H., Farhadi, F., Dehkordi, A.M. (2014) Biodesulfurization of simulated light fuel oil by a native isolated bacteria *Bacillus cereus* HN. *Journal of Petroleum Science and Technology*. 4(1): 31–40.

Ardakani, M.R., Aminsefat, A., Rasekh, B., Yazdiyan, F., Zargar, B., Zarei, M., Najafzadeh, H. (2010) Biodesulfurization of dibenzothiophene by a newly isolated *Stenotrophomonas maltophila* strain Kho1. *World Applied Sciences Journal*. 10(3): 272–278.

Arnaud, J.P., Lacroix, C., Foussereau, C., Choplin, L. (1993) Shear stress effects on growth and activity of *Lactobacillus delbrueckii* subsp. Bulgaricus. *Journal of Biotechnology*. 29: 157–175.

Bahuguna, A., Lily, M.K., Munjal, A., Singh, R.N., Dangwal, K. (2011) Desulfurization of dibenzothiophene (DBT) by a novel strain *Lysinibacillus sphaericus* DMT-7 isolated from diesel contaminated soil. *Journal of Environmental Sciences*. 23(6): 975–982.

Baldini, M., Danuso, F., Turi, M., Vannozzi, G.P. (2004) Evaluation of new clones of Jerusalem artichoke (*Helianthus tuberosus*) for inulin and sugar yield from stalks and tubers. *Industrial Crops and Products*. 19: 25–40.

Bandyopadhyay, S., Chowdhury, R. (2014) parameter optimization of surface active properties and quantification of biosurfactant produced in continuous stirred tank bioreactor during biodesulfurization of diesel. *International Journal of Innovative Research in Advanced Engineering*. 1(9): 217–224.

Barrios, S.M.Y. (2011) Bioremediation: A tool for the management of oil pollution in marine ecosystems. *Biotecnología Aplicada*. 28: 69–76.

Barros, F.F.C., Ponezi, A.N., Pastore, G.M. (2008) Production of biosurfactant by *Bacillus subtilis* LB5a on a pilot scale using cassava wastewater as substrate. *Journal of Industrial Microbiology and Biotechnology*. 35:1071–1078.

Barros, F.F.C., Simiqueli, A.P.R., de Andrade, C.J., Pastore, G.M. (2013) Production of enzymes from agroindustrial wastes by biosurfactant-producing strains of *Bacillus subtilis. Biotechnology Research International*. volume (2013)103960. http://dx.doi.org/10.1155/2013/103960

Berg, J.M., Tymoczko, J.L., Stryer, L. (2002) Biochemistry. 5th edition. W. H. Freeman and Company. New York, USA.

Berg, J.M., Tymoczko, J.L., Stryer, L. (2010) Biochemistry, 7th edn. Freeman WH, New York.

Bhatia, S., Sharma, D.K. (2010) Mining of genomic databases to identify novel biodesulfurizing microorganisms. *Journal of Industrial Microbiology andBiotechnology*. 37(4): 425–429.

Biria, D., Balouchi, A. (2013) Investigation of the role of chemotaxis in bacterial transport through saturated porous media using Taguchi approach. Colloids and Surfaces A: Physicochemical and Engineering Aspects. 436: 542–548.

Boniek, D., Figueiredo, D., dos Santos, A.F.B., de Resende Stoianof, M.A. (2015) Biodesulfurization: a mini review about the immediate search for the future technology. Clean Technologies and Environmental Policy. 17: 29–37.

Boniek, D., Figueiredo, D., Pylro, V.S., Duarte, G.F. (2010) Characterization of bacterial strains capable of desulfurisation in soil and sediment samples from Antarctica. *Extremophiles*. 14: 475–481.

Bordoloi, N.K., Rai, S.K., Chaudhuri, M.K., Mukherjee, A.K. (2014) Deep-desulfurization of dibenzothiophene and its derivatives present in diesel oil by a newly isolated bacterium *Achromobacter* sp. to reduce the environmental pollution from fossil fuel combustion. *Fuel Processing Technology*. 119: 236–244.

Bressler, D.C., Fedorak, P.M. (2000) Bacterial metabolism of fluorene, dibenzofuran, dibenzothiophene, and carbazole. *Canadian Journal of Microbiology*. 46: 397–409.

Bronnenmeier R, Märkl H (1982) Hydrodynamic stress capacity of microorganisms. *Biotechnology and Bioengineering*. 24: 553–578.

Buasri, A., Worawanitchaphong, P., Trongyong, S., Loryuenyong, V. (2014) Utilization of scallop waste shell for biodiesel production from palm oil – optimization using Taguchi method. APCBEE Procedica. 8: 216–221.

Bustos-Jaimes I., Amador, G., Castorena, G., Le Borgne, S. (2003) Genotypic characterization of sulfur-oxidative desulfurizing bacterial strains isolated from Mexican refineries. Oil Gas Sciience and Technology. 58: 521–526.

Buzanello, E.B., Rezende, R.P., Sousa, F.M.O., Marques, E.L.S., Loguercio, L.L. (2014) A novel *Bacillus pumilus*-related strain from tropical landfarm soil is capable of rapid dibenzothiophene degradation and biodesulfurization. BMC Microbiology. 14:257. http://www.biomedcentral.com/1471-2180/14/257.

Calik, P., Yilgör, P., Ayhan, P., Demir, A. (2004) Oxygen transfer effects on recombinant benzaldehyde lyase production. *Chemical Engineering Science*. 59: 5075–5083.

Calzada, J., Heras, S., Alcon, A., Santos, V.E., Garcia-Ochoa, F. (2009a) Biodesulfurization of dibenzothiophene (DBT) using *Pseudomonas putida* CECT 5279: a biocatalyst formulation comparison. *Energy and Fuels*. 23: 5491–5495.

Calzada, J., Zamarro, M.T., Alcon, A., Santos, V.E., Diaz, E., Garcia, J.L., Garcia-Ochoa, F. (2009b) Analysis of dibenzothiophene desulfurization in a recombinant *Pseudomonas putida* strain. *Applied and Environmental Microbiology*. 75: 875–877.

Caro, A., Leton, P., Garcia-Calvo, E., Setti, L. (2007a) Enhancement of the dibenzothiophene biodesulfurization by using b-cyclodextrins in oil-to-water media. *Fuel.* 86: 2632–2636

Caro, A,, Boltes, K., Leton, P., Garcia-Calvo, E. (2008) Description of by-product inhibition effects on biodesulfurization of dibenzothiophene in biphasic media. Biodegradation. 19: 599–611.

Caro, A., Boltes, K., Letón, P., García-Calvo, E. (2007b) Dibenzothiophene biodesulfurization in resting cell conditions by aerobic bacteria. *Biochemical Engineering Journal.* 35:191–197.

Carvalho CCCR, Da Fonseca MMR (2005) The remarkable Rhodococcus erythropolis. Appl Environ Microb 67:715–726

Carvalho, C.C.C.R., Parreño-marchante, B., Neumann, G., Da Fonseca, M.M.R. (2005a) Adaptation of *Rhodococcus erythropolis* DCL14 to growth on n-alkanes, alcohols and terpenes. *Applied and Environmental Microbiology.* 67: 383–388.

Carvalho, C.C.C.R., Poretti, A., Da Fonseca, M.M.R. (2004) Solvent toxicity in organic-aqueous systems analyzed by multivariate analysis. *Applied and Environmental Microbiology.* 26: 361–375.

Casullo de Araújo, H.W., Siva, M.C.d.F Lins, C.I.M., Elesbão do Nascimento, A., Alves da Silva, C.A., Campos-Takak, G.M. (2012) Oxidation of dibenzothiophene (DBT) by *Serratia marcescens* UCP 1549 formed biphenyl as final product. Biotechnology for Biofuels. 5:33. http://www.biotechnologyforbiofuels.com/content/5/1/33

Cavazzuti, M. (2013) Optimization methods: From theory to design. Springer, Verlag Berlin Heidelberg, 2013. DOI:10.1007/978-3-642-31187-1_2.

Chalmers, J.J. (1994) Cells and bubbles in sparged bioreactors. *Cytotechnology.* 15: 311–320.

Chang, J.H., Chang, Y.K., Cho, K.S., Chang, H.N. (2000) Desulfurization of model and diesel oils by resting cells of *Gordona* sp. *Biotechnology Letters.* 22: 193–196.

Chang, J.H., Rhee, S.-K., Chang, Y.K., Chang, H.N. (1998) Desulfurization of diesel oils by a newly isolated dibenzothiophene-degrading *Nocardia* sp. strain CYKS2. *Biotechnology Progress.* 14(6): 851–855.

Chauhan, A.K., Ahmad, A., Singh, S.P., Kumar, A. (2015) Biodesulfurization of benzonaphthothiophene by an isolated *Gordonia* sp. IITR 100. *International Biodeterioration and Biodegradation.* 104: 105–111.

Chen, H., Cai, Y.B., Zhang, W.J., Li, W., (2009) Methoxylation pathway in biodesulfurization of model organosulfur compounds with Mycobacterium sp. *Bioresource Technology.* 100: 2085–2087.

Chen, H., Zhang, W.J., Cai, Y.B., Zhang, Y., Li, W. (2008b) Elucidation of 2-hydroxybiphenyl effect on dibenzothiophene desulfurization by *Microbacterium* sp. strain ZD-M2. *Bioresource Technology.* 99: 6928–6933.

Chen, H., Zhang, W.J., Chen, J.M., Cai, Y.B., Li. W. (2008a) Desulfurization of various organic sulfur compounds and the mixture of DBT + 4,6DMDBT by *Micobacterium* sp. ZD-19. *Bioresource Technology.* 99: 3630–3634.

Cheng, Y., Zhou, W., Gao, C., Lan, K., Gao, Y., Wu, Q. (2009) Biodiesel production from Jerusalem artichoke (*Helianthus tuberosus* L.) tuber by heterotrophic microalgae Chlorella protothecoides. *Journal of Chemical Technology and Biotechnology*. 84: 777–781

Chisty, Y. (2010) In: M.C. Flickinger (ed) Shear Sensitivity in Encyclopedia of Industrial Biotechnology: Bioprocess, Bioseparation, and Cell Technology. John Wiley & Sons, New York.

Choi, O.K., Cho, K.S., Ryu, H.W., Chang, Y.K. (2003) Enhancement of phase separation by the addition of de-emulsifiers to three phase (diesel oil/biocatalyst/aqueous phase) emulsion in diesel biodesulfurization. *Biotechnology Letters*. 25(1): 73–77.

Davoodi-Dehaghani, F., Vosoughi, M., Ziaee, A.A. (2010) Biodesulfurization of dibenzothiophene by a newly isolated *Rhodococcus erythropolis* strain. Bioresource Technology. 101(3): 1102.

de Bont, J.A.M. (1998) Solvent-tolerant bacteria in biocatalysis. *Trends in Biotechnology*. 16: 493–499.

Dejaloud, A., Vahabzadeh, F., Habibi, A. (2017) *Ralstonia eutropha* as a biocatalyst for desulfurization of dibenzothiophene. *Bioprocess and Biosystems Engineering*. 40: 969–980.

del Olmo, C.H., Alcon, A., Santos, V.E., Garcia-Ochoa, F. (2005a) Modeling the production of a *Rhodococcus erythropolis* IGTS8 biocatalyst for DBT biodesulfurization: influence of media composition. *Enzyme and Microbial Technology*. 37: 157–166.

del Olmo, C.H., Santos, V.E., Alcon, A., Garcia-Ochoa, F. (2005b) Production of a *Rhodococcus erythropolis* IGTS8 biocatalyst for DBT biodesulfurization: influence of operational conditions. *Biochemical Engineering Journal*. 22: 229–237.

Denome, S.A., Oldfield, C., Nash, L.J., Young, K.D. (1994) Characterization of the desulfurization genes from *Rhodococcus* sp. strain IGTS8. *Journal of Bacteriology*. 176: 6707–6716.

Deriase, S. F., El-Gendy, N.Sh., Nassar, H.N. (2012) Enhancing biodegradation of dibenzothiophene by *Bacillus sphaericus* Hn1 using factorial design and response surface optimization of medium components. *Energy Sources, Part A: Recovery, Utilization, and Environmental Effects*. 34(22): 2073–2083.

Derikvand, P., Etemadifar, Z., Biria, D. (2014) Taguchi optimization of dibenzothiophene biodesulfurization by *Rhodococcus erythropolis* R1 immobilized cells in a biphasic system. *International Biodeterioration and Biodegradation*. 86: 343–348.

Derikvand, P., Etemadifar, Z., Biria, D. (2015b) RSM optimization of dibenzothiophene biodesulfurization by newly isolated strain of *Rhodococcus erythropolis* PD1 in aqueous and biphasic systems. *Microbiology*. 84(1): 65–72.

Derikvand, P., Etemadifar, Z., Saber, H. (2015a) Sulfur removal from dibenzothiophene by newly isolated *Paenibacillus validus* strain PD2 and process optimization in aqueous and biphasic (Model-Oil) systems. *Polish Journal of Microbiology*. 64(1): 47–54.

Doehlert, D.H. (1970) Uniform shell designs. *Journal of Applied Statistics*. 19: 231–239.

El-Gendy, N.Sh., Nassar, H.N., Abu Amr, S.S. (2014) Factorial design and response surface optimization for enhancing a biodesulfurization process. *Petroleum Science and Technology*. 32(14): 1669–1679.

El-Gendy, N.Sh., Nassar, H.N., Younis, S.A. (2015) Main and interactive effects of polyaromatic sulfur heterocyclic compounds on growth and biodegradation efficiencies of *Bacillus sphaericus* HN1: modeling and statistical analysis. *Petroleum Science and Technology*. 33(11): 1167–1181.

Endo, A. (2012) Fructophilic lactic acid bacteria inhabit fructose-rich niches in nature. *Microbial Ecology in Health and Disease Journal*. 23: 18563. doi: 10.3402/mehd.v23i0.18563.

Escobar, S., Rodriguez, A., Gomez, E., Alcon, A., Santos, V.E., Garcia-Ocho, F. (2016) Influence of oxygen transfer on *Pseudomonas putida* effects on growth rate and biodesulfurization capacity. *Bioprocess and Biosystems Engineering*. 39: 545–554.

Etemadifar, Z., Derikvand, P., Emtiazi, G., Habibi, M.H. (2014) Response surface methodology optimization of dibenzothiophene biodesulfurization in model oil by nanomagnet immobilized *Rhodococcus erythropolis* R1. *Journal of Materials Science and Engineering* B 4(10): 322–330.

Etemadifar, Z., Emtiazi, G., Christofi, N. (2008). Enhanced desulfurization activity in protoplast transformed *Rhodococcus erythropolis*. *American-Eurasian Journal of Agricultural and Environmental Sciences*. 3: 285–29.

Feng, J., Zeng, Y., Ma, C., Cai, X., Zhang, Q., Tong, M., Yu, B., Xu, P. (2006) The surfactant Tween 80 enhances biodesulfurization. *Applied and Environmental Microbiology*.72: 7390–7393.

Feng, S., Yang, H., Zhan, X., Wang, W. (2016) Enhancement of dibenzothiophene biodesulfurization by weakening the feedback inhibition effects based on a systematic understanding of the biodesulfurization mechanism by *Gordonia* sp. through the potential "4S" pathway. RSC Advances. 6: 82872–82881.

Folsom, B.R., Schieche, D.R., Digrazia, P.M., Werner, J., Palmer, S. (1999) Microbial desulfurization of alkylated dibenzothiophenes from a hydrodesulfurized middle distillate by *Rhodococcus erythropolis* I-19. *Applied and Environmental Microbiology*. 65: 4967–4972.

Furuya, T., Ishii, Y., Nada, K., Kino, K., Kirimura, K. (2003) Thermophilic biodesulphurization of hydrodesulfurized light gas oils by *Mycobacterium phlei* WU-F1. *FEMS Microbiology Letters*. 221: 137–142.

Furuya, T., Kirimura, K., Kino, K., Usami, S. (2001a) Thermophilic biodesulfurization of naphthothiophene and 2- ethylnaphthothiophene by dibenzothiophene-desulfurizing bacterium, *Mycobacterium phlei* WU-F1. *Applied Microbiology and Biotechnology*. 58: 237–240.

Furuya, T., Kirimura, K., Kino, K., Usami, S. (2001b) Thermophilic biodesulfurization of dibenzothiophene and its derivatives by *Mycobacterium phlei* WU-F1. *FEMS Microbiology Letters*. 24: 129–133.

Furuya, T., Kirimura, K., Kino, K., Usami, S. (2006) Thermophilic biodesulfurization of dibenzothiophene and its derivatives by Mycobacterium phlei WU-F1, *FEMS Microbiology Letters*. 204: 129–133.

García Camacho, F., Molina Grima, E., Sánchez Mirón, A., González Pascual, V., Chisti, Y. (2001) Carboxymethyl cellulose protects algal cells against hydrodynamic stress. *Enzyme and Microbial Technology*. 29: 602–610.

Garcia-Ochoa, F., Escobar, S., Gomez, E. (2015) Specific oxygen uptake rate as indicator of cell response of *Rhodococcus erythropolis* cultures to shear effects. *Chemical Engineering Science*. 122: 491–499.

Garcia-Ochoa, F., Gomez, E. (2009) Bioreactor scale-up and oxygen transfer rate in microbial processes: An overview. *Biotechnology Advances*. 27: 153–176.

Giraud, C. (2008) Estimation of Gaussain graphs by model selection. *Electronic Journal of Statistics*. 2: 542–563.

Goindi, H.K., Saini, V.S., Verma, P.S., Adhikari, D.K. (2002) Dibenzothiophene desulfurization in hydrocarbon environment by *Staphylococcus* sp. resting cells. *Biotechnology Letters*. 24: 779–781.

Gomez, E., Alcon, A., Escobar, S., Santos, V.E., Garcia-Ochoa, F. (2015) Effect of fluid dynamic conditions on growth rate and biodesulfurization capacity of *Rhodococcus erythropolis* IGTS8. *Biochemical Engineering Journal*. 99: 138–146.

Gomez, E., Santos, V., Alcon, A., Garcia-Ochoa, F. (2006a) Oxygen transport rate on *Rhodococcus erythropolis* cultures: effect on growth and BDS capability. *Chemical Engineering Science*. 61: 4595–4604.

Gomez, E., Santos, V., Alcon, A., Martin, A.B., Garcia-Ochoa, F. (2006b) Oxygen-uptake and mass-transfer rates on the growth of *Pseudomonas putida* CECT5279: influence on biodesulfurization (BDS) capability. *Energy and Fuels*. 20: 1565–1571.

Gray, K.A., Mrachko, G.T., Squires, C.H. (2003) Biodesulfurization of fossil fuels. *Current Opinion in Microbiology*. 6: 229–235.

Gray, K.A., Pogrebinsky, O.S., Mrachko, G.T., Lei, X., Monticello, D.J., Squires, C.H. (1996) Molecular mechanisms of biocatalytic desulphurization of fossil fuels. *Nature Biotechnology*. 14: 1705–1709.

Gray, K.A., Pogrebinsky, O.S., Mrachko, G.T., Xi, L., Monticello, D.J., Squires, D.J. (1996) Molecular mechanisms of biocatalytic desulfurization of fossil fuels Nature Biotechnology. 14: 1705–1709.

Grossman, M.J., Lee, M.K., Prince, R.C., Garrett, K.K., George, G.N., Pickering, I.J. (1999) Microbial desulfurization of a crude oil middle-distillate fraction: analysis of the extent of sulfur removal and the effect of removal on remaining sulfur. *Applied and Environmental Microbiology*. 65(1): 181–188.

Grossman, M.J., Lee, M.K., Prince, R.C., Minak-Bernero, V., George, G.N., Pickering, I.J. (2001) Deep desulfurization of extensively hydrodesulfurized middle distillate oil by *Rhodococcus* sp. strain ECRD-1. *Applied and Environmental Microbiology*. 67: 1949–1952.

Guerinik, K., Al-Mutawah, Q. (2003) Isolation and characterization of oil-desulphurizing bacteria. *World Journal of Microbiology and Biotechnology*. 19: 941–945.

Gün, G., Yürüm, Y., Doğanay, G.D. (2015) Revisiting the bio desulfurization capability of hyperthermophilic archaeon *Sulfolobus solfataricus* P2 revealed DBT consumption by the organism in an oil/water two-phase liquid system at high temperatures. *Turkish Journal of Chemistry*. 39: 255 – 266.

Gunam, I.B.W., Iqbal, M., Arnata, I.W., Antara, N.S., Anggreni, A.A.M.D., Setiyo, Y., Gunadnya, I.B.P. (2016) Biodesulfurization of dibenzothiophene by a newly isolated *Agrobacterium tumefaciens* LSU20. *Applied Mechanics and Materials*. 855:143–149.

Gunam, I.B.W., Kenta, Y., Nengah, I.S., Nyoman, S.A., Wayan, R.A., Michiko, T., Fusao, T., Teruo, S., Kozo, A. (2013) Biodesulfurization of dibenzothiophene and its derivatives using resting and immobilized cells of *Sphingomonas subarctica* T7b. *Journal of Microbiology and Biotechnology*. 23: 473–482.

Gunam, I.B.W., Triani, I.G.A.L., Antara, N.S., Duniaji, A.S., Setiyo, Y., Supata, D.A. (2012) Biocatalytic desulfurization of dibenzothiophene by *Pseudomonas* sp. strain KWN5. In: Proceedings of 4th International Conference on Biociences and Biotechnology. Advancing Life Sciences for Health [Anti-Aging Development in Particular] and Food Security. 21st-22nd September, 2012, Udayana University, Denpasar, Bali, Indonesia. Pp. 245–248.

Gunam, I.B.W., Yaku, Y., Hirano, M., Yamamura, K., Tomita, F., Sone, T., Asano, K., (2006) Biodesulfurization of alkylated forms of dibenzothiophene and benzothiophenes by *Sphingomonas subarctica* T7b. *Journal of Bioscience and Bioengineering*. 101: 322–327.

Guo, L., Zhang, J., Hu, F., Ryu, D.D., Bao, J. (2013) Consolidated bioprocessing of highly concentrated Jerusalem artichoke tubers for simultaneous saccharification and ethanol fermentation. *Biotechnology and Bioengineering*. 110: 2606–2615.

Guo, S., Evans, D.G., Li, D., Duan, X. (2009) Experimental and numerical investigation of the precipitation of barium sulfate in a rotating liquid film reactor. *AIChE Journal*. 55: 2024–2034.

Guobin, S., Huaiying, Z., Jianmin, X., Guo, C., Wangliang, L., Huizhou, L. (2006) Biodesulfurization of hydrodesulfurized diesel oil with *Pseudomonas delafieldii* R-8 from high density culture. *Biochemical Engineering Journal*. 27(3): 305–309.

Hewitt, C.J., Boon, L.A., McFarlane, C.M., Nienow, A.W. (1998) The use of flow cytometry to study the impact of fluid mechanical stress on *Escherichia* coli W3110 during continuous cultivation in an agitated bioreactor. *Biotechnology and Bioengineering*. 59: 612–620.

Hirano, S.I., Haruki, M., Takano, K., Imanaka, T., Morikawa, M., Kanaya, S. (2004) Gene cloning and in vivo characterization of a dibenzothiophene dioxygenase from *Xanthobacter polyaromaticivorans*. *Applied Microbiology and Biotechnology*. 69: 672–681.

Hodaifa, G., Martinez, M.E., Orpez, R., Sanchez, S. (2010) Influence of hydrodynamic stress in the growth of *Scenedesmus obliquus* using a culture medium based on olive mill waste water. *Chemical Engineering Process*. 49: 1161–1168.

Honda, H., Sugiyama, H., Saito, I., Kobayashi, T. (1998) High cell density culture of *Rhodococcus rhodochrous* by pH-stat feeding and dibenzothiophene degradation. *Journal of Fermentation and Bioengineering*. 85: 334–338.

Ichinose, H., Nakamizo M, Wariishi, H., Tanaka, H. (2002b) Metabolic response against sulfur-containing heterocyclic compounds by the lignin-degrading basidiomycete *Coriolus versicolor*. *Applied Microbiology and Biotechnology*. 58(4): 517–26.

Ichinose, H., Wariishi, H., Tanaka, H. (2002a) Identification and characterization of novel cytochrome P450 genes from white-rot basidiomycete *Coriolus versicolor*. *Applied Microbiology and Biotechnology*. 58: 97–105.

Irani, Z.A., Mehrnia, M.R., Yazdian, F., Soheily, M., Mohebali, G., Rakesh, B. (2011) Analysis of petroleum biodesulfurization in an airlift bioreactor using response surface methodology. *Bioresource Technology*. 102: 10585–10591.

Ishii, Y., Konishi, J., Okada, H., Hirasawa, K., Onaka, T., Suzuki, M. (2000) Operon structure and functional analysis of the genes encoding thermophilic desulfurizing enzymes of *Paenibacillus* sp. A11-2. *Biochemical and Biophysical Research Communications*. 270: 81–88.

Ishii, Y., Kozaki, S., Furuya, T., Kino, K., Kirimura, K. (2005) Thermophilic biodesulfurization of various heterocyclic sulfur compounds and crude straight-run light gas oil fraction by a newly isolated strain *Mycobacterium phlei* WU-0103. *Current Microbiology*. 50: 63–70.

Ismail, W., El-Sayed, W.S., Abdul Raheem, A.S., Mohamed, M.E, El Nayal, A.M. (2016) Biocatalytic desulfurization capabilities of a mixed culture during non-destructive utilization of recalcitrant organosulfur compounds. Frontiers in Microbiology. Volume 7: Article 266. doi: 10.3389/fmicb.2016.00266

Izumi, Y., Ohshiru, T. (2001) Purification and characterization of enzymes involved in desulfurization of dibenzotiophene in fossil fuels. *Journal of Molecular Catalysis A: Chemical*. 11(1–4): 1061–1064.

Izumi, Y., Oshiro, T., Ogino, H., Hine, Y., Shimao, M. (1994) Selective desulfurization of dibenzothiophene by *Rhodococcus erythropolis* D-1. *Applied and Environmental Microbiology*. 60: 223–236.

Jia, X., Wen, J., Sun, Z., Caiyin, Q., Xie, S. (2006) Modeling of DBT biodegradation behaviors by resting cells of *Gordonia* sp WQ-01 and its mutant in oil–water dispersions. *Chemical Engineering Science*. 61: 1987–2000.

Jiang C., Liu, H., Xie, Y., Chen, J. (2002) Isolation of soil bacteria species for degrading dibenzothiophene. *Chinese Journal of Chemical Engineering*. 10(4): 420–426.

Jiang, X., Yang, S., Li, W. (2014) Biodesulfurization of model compounds and deasphalted bunker oil by mixed culture. *Applied Biochemistry and Biotechnology*. 172: 62–72.

Kango, N., Jain, S.C. (2011) Production and properties of microbial inulinases: recent advances. *Food Biotechnology*. 25: 165–212.

Kao, P.-M., Chen, C.-I., Huang, S.-C., Chang, Y.-C., Tsay, P.-Y., Liu, Y.-C. (2007) Effects of shear stress and mass transfer on chitinase production by *Paenibacillus* sp. CHE-N1. *Biochemical Engineering Journal*. 34: 172–178.

Kargi, F., Robinson, J.M. (1984) Microbial oxidation of dibenzothiophene by the thermophilic organism *Sulfolobus acidocaldarius*. *Biotechnology and Bioengineering*. 26: 687–690.

Kaufman, E.N., Borole, A.P., Shong, R., Sides, J.L., Juengst, C. (1999) Sulfur specificity in the bench-scale biological desulfurization of crude oil by *Rhodococcus* IGTS8. *Journal of Chemical Technology and Biotechnology*. 74: 1000–1004.

Kaufman, E.N., Harkins, J.B., Borole, A.P. (1998) Comparison of batch stirred and electrospray reactors for biodesulfurization of dibenzothiophene in crude oil and hydrocarbon feedstocks. *Applied Biochemistry and Biotechnology*. 73: 127–144.

Kawaguchi, H., Kobayashi, H., Sato, K. (2011) Metabolic engineering of hydrophobic *Rhodococcus opacus* for biodesulfurization in oil–water biphasic reaction mixtures. *Journal of Bioscience and Bioengineering*. 113: 360–366.

Kawaguchi, H., Kobayashi, H., Sato, K. (2012) Metabolic engineering of hydrophobic *Rhodococcus opacus* for biodesulfurization in oil–water biphasic reaction mixtures. *Journal of Bioscience and Bioengineering*. 113: 360–366.

Kayser, K.J., Cleveland, L., Park, H.S., Kwak, J.H., Kolhatkar, A., Kilbane, J. (2002) Isolation and characterization of moderate thermophile *Mycobacterium phlei* GTIS10, capable of dibenzothiophene desulfurization. *Applied Microbiology and Biotechnology*. 59: 737–746.

Kayser, Y., Bielagajones, B.A., Jackowski, K., Odusan, O., Kilbane, J.J. (1993) Utilization of organusulphur compounds by axenic and mixed cultures of *Rhodococcus rhodochrous* IGTS8. *Journal of General Microbiology*. 139: 3123–3129.

Kertesz, M.A. (2000) Riding the sulfur cycle - metabolism of sulfonates and sulfate esters in Gram-negative bacteria. *FEMS Microbiology Reviews*. 24: 135–175.

Khedkar, S., Shanker, R. (2015) Isolation and classification of a soil actinomycete capable of sulphur-specific biotransformation of dibenzothiophene, benzothiophene and thianthrene. *Journal of Applied Microbiology* 118(1), 62–74.

Kilbane, J.J. (2006) Microbial biocatalyst development to upgrade fossil fuels. *Current Opinion in Microbiology*. 17: 305–314

Kilbane, J.J. II (1990) Microbial removal of organic sulfur from coal: current status and research needs. Bioprocessing and Biotreatment of Coal (Wise DL, ed), Marcel Dekker Inc., New York. Pp. 487–506.

Kilbane, J.J., Bielage, B.A. (1990) Genetic study of biodesulfurization. In: Proceedings: First International Symposium of the Biological Processing of Coal (Yunker S, Rhee K, eds.). Electric Power Research Institute, Inc., Palo Alto, California. Pp.2/15–2/32.

Kilbane, J.J., Jackowski, K. (1992) Biodesulfurization of water soluble coal derived material by *Rhodococcus rhodochrous* IGTS8. *Biotechnology and Bioengineering.* 40: 1107–1114.

Kilbane, J.J., Robbins, J. (2007) Characterization of the dszABC genes of Gordonia amicalis F.5.25.8 and identifcation of conserved protein and DNA sequences. *Applied Microbiology and Biotechnology.* 75: 843–851.

Kim, Y.J., Chang, J.H., Cho, K.-S., Ryu, H.W., Chang, Y.K. (2004) A physiological study on growth and dibenzothiophene (DBT) desulfurization characteristics of *Gordonia* sp. CYKS1. *Korean Journal of Chemical Engineering.* 21(2): 436–441.

Kirimura, K., Furuya, T., Nishii, Y., Ishii, Y., Kino, K., Usami, S. (2001). Biodesulfurization of dibenzothiophene and its derivatives through the selective cleavage of carbon-sulfur bonds by a moderately thermophilic bacterim *Bacillus subtilis* WU-S2B. *Journal of Bioscience and Bioengineering.* 91(3): 262–266.

Kirimura, K., Furuya, T., Sato, R., Ishii, Y., Kino, K., Usami, S. (2002) Biodesulfurization of naphthothiophene and benzothiophene through selective cleavage of carbon-sulfur bonds by *Rhodococcus* sp. strain WU-K2R. *Applied and Environmental Microbiology.* 68(8): 3867–3872.

Kishimoto, M., Inui, M., Omasa, T., Katakura, Y., Suga, K., Okumura, K. (2000) Efficient production of desulfurizing cells with the aid of expert system. *Biochemical Engineering Journal.* 5: 143–147.

Knightes, C.D., Peters, C.A. (2006). Multisubstrate biodegradation kinetics for binary and complex mixtures of polycyclic aromatic hydrocarbons. *Environmental Toxicology and Chemistry.* 25: 1746–1756.

Kobayashi, M., Horiuchi, K., Yoshikawa, O., Hirasawa, K., Ishii, Y., Fujino, K., Sugiyama, H., Maruhashi, K. (2001) Kinetic analysis of microbial desulfurization of model and light gas oils containing multiple alkyl dibenzothiophenes. *Bioscience, Biotechnology and Biochemistry.* 65: 298–304.

Kobayashi, M., Onaka, T., Ishii, Y., Konishi, J., Takaki, M., Okada, H., Ohta, Y., Koizumi, K., Suzuki. M. (2000) Desulfurization of alkylated forms of both dibenzothiophene and benzothiophene by a single bacterial strain. *FEMS Microbiology Letters.* 187: 123–126.

Konishi, J., Ishii, Y., Osaka, T., Okumura, K., Suzuki, M. (1997) Thermophilic carbon-sulfur-bond-targeted biodesulfurization. *Applied and Environmental Microbiology.* 63(8): 3164–3169.

Konishi, J., Okada, H., Hirasawa, K., Ishii, Y., Maruhashi, K. (2002) Comparison of the substrate specificity of the two bacterial desulfurization systems. *Biotechnology Letters.* 24: 1863–1867.

Koohikamali, S., Tan, C.P., Ling, T.C. (2012) Optimization of sunflower oil transesterification process using sodium methoxide. *Scientific World Journal.* Volume 2012, Article ID 475027, 8 pages. doi:10.1100/2012/475027.

Kouril, T.; Esser, D.; Kort, J.; Westerhoff, H. V.; Siebers, B.; Snoep, J. L. Intermediate instability at high temperature leads to low pathway efficiency for an in vitro reconstituted system of gluconeogenesis in Sulfolobus solfataricus. FEBS J. 2013, 280, 4666–4680.

Kühbeck, F., Müller, M., Back, W., Kurz, T., Krottenthaler, M. (2007) Effect of hot trub and particle addition on fermentation performance of *Saccharomyces cerevisiae. Enzyme and Microbial Technology.* 41: 711–720.

Kunze, W. (2014) Technology Brewing and Malting. 5th revised English ed. Versuchs- und Lehranstalt für Brauerei (VLB), Berlin.

Laane, C., Boeren, S., Vos, K., Veeger, C. (1987) Rules for optimization of biocatalysis in organic solvents. *Biotechnology and Bioengineering.* 30: 81–87.

Labana, S., Pandey, G., Jain, R.K. (2005) Desulphurization of dibenzothiophene and diesel oils by bacteria. *Letters in Applied Microbiology.* 40: 159–163.

Le Borgne, S.; Quintero, R. Biotechnological processes for the refining of petroleum. Fuel Process Technol. 2003, 81, 155–169.

Li W, Jiang X. (2013) Enhancement of bunker oil biodesulfurization by adding surfactant. *World Journal of Microbiology and Biotechnology.* 29:103–8.

Li, F., Zhang, Z., Feng, J., Cai, X., Xu, P. (2007) Biodesulfurization of DBT in tetradecane and crude oil by a facultative thermophilic bacterium *Mycobacterium* goodi X7B. *Journal of Biotechnology.* 127, 222–228.

Li, F.L., Xu, P., Ma, C.Q., Luo, L.L., Wang, X.S. (2003) Deep desulfurization of hydrodesulfurization-treated diesel oil by a facultative thermophilic bacterium *Mycobacterium* sp. X7B. *FEMS Microbiology Letters.* 223: 301–307.

Li, G., Li, S., Qu, S., Liu, Q., Ma, T., Zhu, L., Liang, F., Liu, R. (2008) Improved biodesulfurization of hydrodesulfurized diesel oil using *Rhodococcus erythropolis* and *Gordonia* sp. *Biotechnology Letters.* 30(10): 1759–1764.

Li, L., Li, L., Wang, Y., Du, Y., Qin, S. (2013) Biorefinery products from the inulin-containing crop Jerusalem artichoke. *Biotechnology Letters.* 35: 471–477.

Li, M., Squires, C., Monticello, D., Childs, J. (1996) Genetic analysis of the *dsz* promoter and associated regulatory regions of *Rhodococcus erythropolis* IGTS8. Journal of Bacteriology. 178: 6409–6418.

Li, W., Wang, M.D., Chen, H., Chen, J.M., Shi, Y. (2006) Biodesulfurization of dibenzothiophene by growing cells of Gordonia sp. in batch cultures. *Biotechnology Letters.* 28: 1175–1179.

Li, W., Ying, Z., Miao, D.W., Yao, S. (2005) Biodesulfurization of dibenzothiophene and other organic sulfur compounds by a newly isolated *Microbacterium* strain ZD-M2. *FEMS Microbiology Letters.* 247: 45–50.

Li, Y., Xing, J., Li, W., Xiong, X., Li, X., Liu, H. (2007) Medium optimization of *Rhodococcus erythropolis* LSSE8-1 by Taguchi methodology for petroleum biodesulfurization. *Korean Journal of Chemical Engineering.* 24(5): 781–786.

Li, Y., Xing, J., Xiong, X., Li, W., Gao, H., Liu, H. (2008a) Improvement of biodesulfurization activity of alginate immobilized cells in biphasic systems. *Journal of Industrial Microbiology and Biotechnology.* 35: 145–150.

Liang, Z.X., Li, L., Li, S., Cai, Y.H., Yang, S.-T., Wang, J.F. (2012) Enhanced propionic acid production from Jerusalem artichoke hydrolysate by immobilized

Propionibacterium acidipropionici in a fibrous-bed bioreactor. *Bioprocess and Biosystems Engineering*. 35, 915–921.

Lichtinger, T., Reiss, G., Benz, R. (2000) Biochemical identification and biophysical characterization of a channel-forming protein from *Rhodococcus erythropolis*. *Journal of Bacteriology*. 182: 764–770.

Lin, L., Zhao, C., Liu, Q., Zhang, Y., Liu, C., Xue, J. (2014) Optimization for microbial degradation of dibenzothiophene by *Pseudomonas* sp. LKY-5 using response surface methodology. *China Petroleum Processing and Petrochemical Technology*. 16(1): 19–26.

Luo, M.F., Xing, J.M., Gou, Z.X., Li, S., Liu, H.Z., Chen, J.Y. (2003) Desulfurization of dibenzothiophene by lyophilized of *Pseudomonas delafieldii* R-8 in the presence of dodecane. *Biochemical Engineering Journal*. 13: 1–6.

Luo, M.F., Xing, J.M., Gou, Z.X., Li, S., Liu, H.Z., Chen, J.Y. (2002) Desulfurization of dibenzothiophene by lyophilized cells of *Pseudomonas delafieldii* R-8 in the presence of dodecane. *Biochemical Engineering Journal*. 13: 1–6.

Ma, C.Q., Feng, J.H., Zeng, Y.Y., Cai, X.F., Sun, B.P., Zhang, Z.B., Blankespoor, H.D., Xu, P. (2006) Methods for the preparation of a biodesulfurization biocatalyst using *Rhodococcus* sp. *Chemosphere*. 65: 165–169.

Maass, D., Todescato, D., Moritz, D.E., Oliveira, J.V., Oliveira, D., Ulson de Souza, A.A., Guelli Souza, S.M. (2015) Desulfurization and denitrogenation of heavy gas oil by *Rhodococcus erythropolis* ATCC 4277. *Bioprocess and Biosystems Engineering*. 38: 1447–1453.

Maghsoudi, S., Vossoughi, M., Kheirolomoom, A., Tanaka, E., Katoh, S. (2001) Biodesulfurization of hydrocarbons and diesel fuels by *Rhodococcus* sp. strain P32C1. *Biochemical Engineering Journal*. 8:151–156.

Makino, Y., Treichel, H., Mazutti, M.A., Maugeria, F., Rodrigues, M.I. (2009) Inulinase bio-production using agroindustrial residues: screening of microorganisms and process parameters optimization. *Journal of Chemical Technology and Biotechnology*. 84: 1056–1062.

Mansom, T., Nunes, C., Raposo, S., Lima-Costa, M.E. (2010) Carob pulp as raw material to produce the biocontrol agent *P. agglomerans* PBC-1. *Journal of Industrial Microbiology and Biotechnology*. 37: 1145–1155.

Marcelis, C.L.M., Leeuwen, M.V., Polderman, H.G., Janssen, A.J.H., Lettinga, G. (2003) Model description of dibenzothiophene mass transfer in oil/water dispersions with respect to biodesulfurization. *Biochemical Engineering Journal*. 16: 253–264.

Märkl, H., Bronnenmeier, R., Wittek, B. (1991) The resistance of microorganisms to hydrodynamic stress. *International Chemical Engineering*. 31: 185–197.

Martin, A.B., Alcon, A., Santos, V.E., Garcia-Ochoa, F. (2004) Production of a biocatalyst of *Pseudomonas putida* CECT5279 for dibenzothiophene (DBT) biodesulfurization for different media compositions. *Energy and Fuel*. 18: 851–857.

Martin, A.B., Alcon, A., Santos, V.E., Garcia-Ochoa, F. (2005) Production of a biocatalyst of *Pseudomonas putida* CECT5279 for DBT biodesulfurization: influence of the perational conditions. *Energy and Fuel*. 19: 775–782.

Martinez, I., Santos, V.E., Alcon, A., Garcia-Ochoa, F. (2015) Enhancement of the biodesulfurization capacity of Pseudomonas putida CECT5279 by co-substrate addition. *Process Biochemistry* 50: 119–124.

Martinez, I., Santos, V.E., Gomez, E., Garcia-Ochoa, F. (2016) Biodesulfurization of dibenzothiophene by resting cells of *Pseudomonas putida* CECT5279: influence of the oxygen transfer rate in the scale-up from shaken flask to stirred tank reactor. *Journal of Chemical Technology and Biotechnology*. 91: 184–189.

Marzona, M., Pessione, E., Di Martino, S., Giunta, C. (1997) Benzothiophene and dibenzothiophene as the sole sulfur source in *Acinetobacter*: Growth kinetics and oxidation products. *Fuel Processing Technology*. 52: 199–205.

Matsui, T., Hirasawa, K., Konishi, J., Tanaka, Y., Maruhashi, K., Kurane, K. (2001b) Microbial desulfurisation of alkylated dibenzothiophene and alkylated benzothiophene by recombinant *Rhodococcus* sp. Strain T09. *Applied Microbiology and Biotechnology*. 56: 196–200.

Maxwell, S., Yu, J., (2000) Selective desulphurisation of dibenzothiophene by a soil bacterium: microbial DBT desulphurisation. *Process Biochemistry*. 35: 551–556.

McFarland, B.L. (1999) Biodesulfurization. *Current Opinion in Microbiology*. 2,257–264.

McFarland, B.L., Boron, D.J., Deever, W., Meyer, J.A., Johnson, A.R., Atlas, RM. (1998) Biocatalytic sulfur removal from fuels: applicability for producing for producing low sulfur gasoline. *Critical Reviews in Microbiology*. 24: 99–147.

Meijer, J.J., ten Hoopen, H.J.G., van Gameren Y.M., Luyben, K.Ch.A.M., Libbenga, K.R. (1994) Effects of hydrodynamic stress on the growth of plant cells in batch and continuous culture. *Enzyme and Microbial Technology*. 16: 467–477.

Mingfang, L., Zhongxuan, G., Jianmin, X., Huizhou, L., Jiayong, C. (2003) Microbial desulfurization of model and straight-run diesel oils. *Journal of Chemical Technology and Biotechnology*. 78: 873–876.

Mohebali G., Ball, A.S., Kaytash, A., Rasekh. B. (2008) Dimethyl sulfoxide (DMSO) as the sulfur source for the production of desulfurizing resting cells of *Gordonia alkanivorans* RIPI90A. *Microbiology*. 154: 878–885.

Mohebali, G., Ball, A.S., Rasekh, B., Kaytash, A. (2007) Biodesulfurization potential of a newly isolated bacterium, *Gordonia alkanivorans* RIPI90A. *Enzyme and Microbial Technology*. 40: 578–584.

Montgomery, D.C. (2013) Design and analysis of experiments. 8th ed., John Wiley and Sons, Inc., New York, USA.

Monticello, D.J., Finnerty, W.R. (1985) Microbial desulfurization of fossil fuels. *Annual Review of Microbiology*. 39: 371–389.

Morio, K., Keizo, H., Osamu, Y., Kazuaki, H., Yoshitaka, I., Kazuhito F, Hiroshi, S., Kenji, M. (2001) Kinetic analysis of microbial desulfurization of model and light gas oils containing multiple alkyl dibenzothiophenes. *Bioscience, Biotechnology, and Biochemistry*. 65: 298–304.

Naito, M., Kawamoto, T., Fujino, K., Kobayashi, M., Maruhashi, K., Tanaka, A. (2001) Long-term repeated desulfurization by immobilized *Rhodococcus erythropolis* KA2–5-1 cells. *Applied Microbiology and Biotechnology*. 55: 374–378.

Nakayama, N., Matsubara, T., Ohshiro, T., Moroto, Y., Kawata, Y., Koizumi, K., Hirakawa, Y., Suzuki, M., Maruhashi, K., Izumi, Y., Kurane, R. (2002) A novel enzyme, 2-hydroxybiphenyl-2-sulfinate desulfinase (DszB), from a dibenzothiphene-desulfurizing bacterium *Rhodococcus erythropolis* KA2-5-1: gene over-expression and enzyme characterization. Biochimica et Biophysica Acta (BBA)1598(1/2): 122–130.

Nassar, H.N. (2009) Potentials of Microorganisms Isolated from Egyptian Hydrocarbon Polluted Sites on Degradation of Polycyclic Aromatic Sulfur Heterocycles (PASHs) Compounds. MSc Thesis, Botany and Microbiology Department, Faculty of Science, AL-Azhar University, Cairo- Egypt.

Nassar, H.N., Deriase, S.F., El-Gendy, N.Sh. (2017a) Kinetic modeling and simulation of a batch biodesulfurization process by *Rhodococcus erythropolis* HN2. Petroleum Science and Technology. 35(20): 1995–2002.

Nassar, H.N., Deriase, S.F., El-Gendy, N.Sh. (2017b) Statistical optimization of biomass production and biodesulfurization activity of *Rhodococcus erythropolis* HN2. Petroleum Science and Technology. 35(20): 1951–1959.

Ohshiro, T., Hine, Y., Izumi, Y. (1994) Enzymatic a cell-fee desulfurization of dibenzothiophene by system of *Rhodococcus erythropolis* D-1. *FEMS Microbiology Letters*. 118: 341–344.

Ohshiro, T., Hirata, T., Hashimoto, I., Izumi, Y. (1996b) Characterization of dibenzothiophene desulfurization reaction by whole cells of *Rhodococcus erythropolis* H-2 in the presence of hydrocarbon. *Journal of Fermentation and Bioengineering*. 82: 610–612.

Ohshiro, T., Hirata, T., Izumi, Y. (1995a) Microbial desulfurization of dibenzothiophene in the presence of hydrocarbon. *Applied Microbiology and Biotechnology*. 44: 249–252.

Ohshiro, T., Izumi, Y. (1999) Review microbial desulfurization of organic compounds in petroleum. *Bioscience, Biotechnology and Biochemistry*. 63: 1–9.

Ohshiro, T., Kobayashi, Y., Hine, Y., Izumi, Y. (1995b) Involvement of flavin coenzyme in dibenzothiophene degrading enzyme system from *Rhodococcus erythropolis* D-1. *Bioscience, Biotechnology and Biochemistry*. 59(7): 1349–1351.

Ohshiro, T., Suzuki, K., Izumi, Y. (1996a) Regulation of dibenzothiophene degrading enzyme of *Rhodococcus erythropolis* D-1. *Journal of Fermentation and Bioengineering*. 81: 121–124.

Ohshiro, T., Suzuki, K., Izumi, Y. (1997) Dibenzothiophene (DBT) degrading enzyme responsible for the first step of DBT desulfurization by *Rhodococcus erythropolis* D-1: purification and characterization. *Journal of Fermentation and Bioengineering*. 83(2): 233–237.

Ohshiro, T., Suzuki, K., Izumi, Y. (1998) Regulation of dibenzothiophene degrading enzyme activity of *Rhodococcus erythropolis* D-1. *Journal of Fermentation and Bioengineering*. 81: 121–124.

Okada, I., Nomura, N., Nakahara, T., Maruhashi K. (2002) Analysis of dibenzothiophene metabolic pathway in *Mycobacterium* strain G3. Journal of Bioscience and Bioengineering. 93(5):491–497.

Oldfeld, C., Pogrebinsky, O., Simmonds, J., Olson, E.S., Kulpa, C.F. (1997) Elucidation of the metabolic pathway for dibenzothiophene desulphurization by *Rhodococcus* sp. strain IGTS8 (ATCC 53968). *Microbiology*. 143: 2961–2973.

Oldfield, C., Pogrebinsky, O., Simmonds, J., Olson, E.S., Kulpa, C.H. (1997) Elucidation of the metabolic pathway for dibenzothiophene desulfurization by *Rhodococcus* sp. strain IGTS8 (ATCC 53968). *Microbiology* UK. 143: 2961–2973.

Olmos, E., Mehmood, N., Haj-Husein, L., Goergen, J.L., Fick, M., Delaunay, S. (2013) Effects of bioreactor hydrodynamics on the physiology of *Streptomyces*. *Bioprocess and Biosystems Engineering*. 36:259–272.

Omori, T., Saiki, Y., Kasuga, K., Kodama, T. (1995) Desulfurization of alkyl and aromatic sulfides and sulfonates by dibenzothiophene-desulfurizing *Rhodococcus* sp. strain SY1. *Bioscience, Biotechnology, and Biochemistry*. 59(7): 1195–1198.

Onaka, T., Kobayashi, M., Ishii, Y., Okumura, K., Suzuk,i M. (2000) Application of solid-phase extraction to the analysis of the isomers generated in biodesulfurization against methylated dibenzothiophenes. J*ournal of Chromatography A* 903: 193–202.

Oral, J., Sikula, J., Puchyr, R., Hajny, Z., Stehlik, P., Bebar, L. (2005) Processing of waste from pulp and paper plant. *Journal of Cleaner Production*. 13: 509–515.

Osborne, S.J., Leaver, J., Turner, M.K., Dunnill, P. (1990) Correlation of biocatalytic activity in an organic/aqueous two-liquid phase system with solvent concentration in the cell membrane. *Enzyme and Microbial Technology*. 12: 281–291.

Paixão, S.M., Teixeira, P.D., Silva, T.P., Teixeira, A.V., Alves, L. (2013) Screening of novel yeast inulinases and further application to bioprocesses. *New Biotechnology*. 30: 598–606.

Papizadeh, M., Ardakani, M.R., Motamedi, H., Rasouli, I., Zarei, M. (2011) C–S targeted biodegradation of dibenzothiophene by *Stenotrophomonas* sp. NISOC-04. *Applied Biochemistry and Biotechnology*. 165: 938–948.

Papizadeh, M., Roayaei ardakani, M., Ebrahimipour, G., Motamedi, H. (2010) Utilization of dibenzothiophene as sulfur source by *Microbacterium* sp. NISOC-06. *World Journal of Microbiology and Biotechnology*. 26: 1195–1200.

Papizadeh, M., Roayaei Ardakani, M., Motamedi, H. (2017) Growth-phase dependent biodesulfurization of dibenzothiophene by *Enterobacter* sp. strain NISOC-03. *Pollution*. 3(1): 101–111.

Patel, S.B., Kilbane, J.J., Webster, D.A. (1997) Biodesulphurisation of dibenzothiophene in hydrophobic media by *Rhodococcus* sp. strain IGTS8. *Journal of Chemical Technology and Biotechnology*. 69: 100–106.

Peng, B., Zhou, Z. (2016) Study on growth characteristic and microbial desulfurization activity of the bacterial stain MP12. *Biochemical Engineering Journal*. 112: 202–207.

Pikoli, M.R., Aditiawati, P., Akhmaloka, Indrianiastuti, D., Wijayanti, R. Growth of *Bacillus megaterium* CSK2, *Bacillus subtillus* CSK3 and *Bacillus subtillus* CSK4 from coal mixed soil in dibenzothiophene-containing medium. *Asian*

Journal of Microbiology, Biotechnology and Environmental Sciences. 16(2): 455–460.

Porto, B., Maass, D., Oliveira, J.V., de Oliveira, D., Yamamoto, C.I., Ulson de Souzaa, A.A., Ulson de Souza, S.M.A.G. (2017) Heavy gas oil biodesulfurization by *Rhodococcus erythropolis* ATCC 4277: optimized culture medium composition and evaluation of low-cost alternative media. *Journal of Chemical Technology and Biotechnology.* 92: 2376–2382.

Potumarthi, R., Subhakar, C., Pavani, A., Jetty, A. (2008) Evaluation of various parameters of calcium-alginate immobilization method for enhanced alkaline protease production by *Bacillus licheniformis* NCIM-2042 using statistical methods. Bioresource Technology. 99: 1776–1786.

Praveen Reddy, P., Umamaheshwara Rao, V (2015) Dibenzothiophene desulfurizing *Streptomyces* species isolated from oil contaminated sites. *International Journal of Science and Research.* 5(12): 1671–1674.

Premchaiporn, P., Akaracharanya, A. (2000) Desulfurization of dibenzothiophene by *Bacillus* K10. *Journal of scientific research, Chulalongkorn University.* 25: 177–185.

Prokop, A., Bajpai, R.K. (1992) The sensitivity of biocatalysts to hydrodynamic shear stress. *Advances in Applied Microbiology.* 37: 165–232.

Purwanto, L.A., Ibrahim, D., Sudrajat, H. (2009) Effects of agitation speed on morphological changes in *Aspergillus niger* hyphae during production of tannase. *World Journal of Chemistry.* 4: 34–38.

Rashtchi, M., Mohebali, G.H., Akbarnejad, M.M., Towfighi, J., Rasekh, B., Keytash, A. (2006) Analysis of biodesulfurization of model oil system by the bacterium, strain RIPI-22. *Biochemical Engineering Journal.* 29: 169–173.

Rath, K., Mishra, B., Vuppu, S. (2012) Biodegrading ability of organosulfur compound of a newly isolated microbe *Bacillus* sp. KS1 from the oil contaminated soil. *Archives of Applied Science Research.* 4(1): 465–471.

Rezaei, R., Mohadesi, M., Moradi, G.R. (2013) Optimization of biodiesel production using waste mussel shell catalyst. *Fuel.* 109: 534–541.

Rhee, S.-K., Chang, J.H., Chang, Y.K., Chang, H.N. (1998) Desulfurization of dibenzothiophene and diesel oils by a newly isolated *Gordona* strain, CYKS1. *Applied and Environmental Microbiology.* 64(6): 2327–2331.

Russell, J.B., Cook, G.M. (1995) Energetics of bacterial growth: balance of anabolic and catabolic reactions. *Microbiological Reviews.* 59: 48–62.

Ryu, H.W., Wee, Y.J. (2001) Characterization of bioconversion of fumarate to succinate by alginate immobilized *Enterococcus faecalis* RKY1. *Applied Biochemistry and Biotechnology.* 91: 525–535.

Sagardia, F., Rigau, J.J., Martinez, L.A., Fuentes, F., Lopez, C. (1975) Degradation of benzothiophene and related compounds by a soil *Pseudomonas* in an oil-aqueous environment. *Applied Microbiology.* 29: 722–725.

Sahinkaya, E., Dilek, F. B. (2007). Biodegradation kinetics of 2,4-dichlorophenol by acclimated mixed cultures. *Journal of Biotechnology* 127: 716–726.

Sahoo, S., Verma R.K., Suresh, A.K., Rao, K.K., Bellare, J., Suraishkumar, G.K. (2003) Macro-level and genetic-level responses of bacillus subtilis to shear stress. *Biotechnology Progress.* 19: 1689–1696.

Salmani, Z., Gheshlaghi, R., Mahdavi, M.A. (2015) Effect of carbon and nitrogen sources on biodesulfurization (BDS) activity by *Rhodococcus erythropolis* IGTS8. Proceeding of the 9th International Chemical Engineering Congress and Exhibition. (IChEC 2015). Shiraz, Iran, 26–28 December, 2015.

Sato, T.; Atomi, H. Novel metabolic pathways in archea. Curr. Opin. Microbiol. 2011, 14, 307–314.

Schilling, B.M., Alvarez, L.M., Wang, D.I.C., Cooney, C.L. (2002) Continuous desulfurization of dibenzothiophene with *R. rhodochrous* IGTS8 (ATCC 53968). *Biotechnology Progress.* 18: 1207–1213.

Schmidts, T., Dobler, D., Nissing, C., Runkel, F. (2009) Influence of hydrophilic surfactants on the properties of multiple W/O/W emulsions. *Journal of Colloid and Interface Science.* 338: 184–192.

Setti, L., Bonoli, S., Badiali, E., Giuliani, S. (2003) Inverse phase transfer biocatalysis for a biodesulfurization process of middle distillates. 44(1): 80–83.

Setti, L., Farinelli, P., Di Martino, S., Frassinetti, S., Lanzarini, G., Pifferi, P.G. (1999) Developments in destructive and non-destrucitve pathways for selective desulfurizations in oil biorefining processes. *Applied Microbiology and Biotechnology.* 52: 111–117.

Setti, L., Lanzarini, G., Pifferi, P.G. (1994) Diffusion as a rate controlling step in heavy oil biodesulfurization processes. *Fuel Processing Technology.* 40, 311–317.

Setti, L., Lanzarini, G., Pifferi, P.G. (1995) Dibenzothiophene biodegradation by a *Pseudomonas* sp. in model solutions. *Process Biochemistry.* 30(8): 721–728.

Setti, L., Lanzarini, G., Pifferi, P.G. (1996) Immobilized cells for applications in non-conventional systems. *Progress in Biotechnology.* 11: 777–784.

Setti, L., Lanzarini, G., Pifferi, P.G. (1997) Whole cell biocatalysis for an oil desulfurization process. *Fuel Processing Technology.* 52: 145–153.

Shahaby, A.F., Essam El-din, K.M. (2017) Desulfurization of crude oil and oil products by local isolated bacterial strains. *International Journal of Current Microbiology and Applied Sciences.* 6(4): 2695–2711.

Sikkema, J., Weber, F.J., Heipieper, H.J., de Bont, J.A. (1994) Cellular toxicity of lipophilic compounds: mechanisms, implications, and adaptations. *Biocatalysis.* 10: 113–122.

Silva, T.P., Paixão, S.M., Roseiro, J.C., Alves, L. (2014) Jerusalem artichoke as low-cost fructose-rich feedstock for fossil fuels desulphurization by a fructophilic bacterium. *Journal of Applied Microbiology.* 118: 609–618.

Silva, T.P., Paixão, S.M., Teixeira, A.V., Roseiro, J.C., Alves, L. (2013) Optimization of low sulfur carob pulp liquor as carbon source for fossil fuels biodesulfurization. *Journal of Chemical Technology and Biotechnology.* 88: 919–923.

Silva, T.P., Paixão, S.M., Teixeira, A.V., Roseiro, J.C., Alves, L. (2013) Optimization of low sulfur carob pulp liquor as carbon source for fossil fuels biodesulfurization. *Journal of Chemical Technology and Biotechnology.* 88: 919–923.

Soleimani, M., Bassi, A., Margaritis, A. (2007) Biotechnology Advances. 25: 570–596.

Srinivasulu, B., Adinarayana, K., Ellaiah, P. (2003) Investigations on neomycin production with immobilized cells of *Streptomyces marinensis* Nuv-5 in calcium alginate matrix. AAPS PharmSciTech. 4: 449–454.

Srivastava, V.C. (2012) RSC Advances. 2: 759–783.

Sutterlin, K.A. (2010) Fructophilic yeasts to cure stuck fermentations in alcoholic beverages. PhD Thesis, Stellenbosch University, Institute of Wine Biotechnology, Faculty of AgriSciences, Stellenbosch, South Africa.

Taguchi, G. (1986) Introduction to Quality Engineering: Designing Quality into Products and Processes. Asian Productivity Association, Tokyo, Japan.

Tangaromsuk, J., Borole, A.P., Kruatrachue, M., Pokethitiyook, P. (2008) An integrated biodesulfurization process, including inoculum preparation, desulfurization and sulfate removal in a single step, for removing sulfur from oils. *Journal of Chemical Technology and Biotechnology*. 83: 1375–1380.

Tao, F., Yu, B., Xu, P., Ma, C. (2006) Biodesulfurization in biphasic systems containing organic solvents. *Applied and Environmental Microbiology*. 72: 4604–4609.

Thomas, S. (2000) Production of lactic acid from pulp mill solid waste and xylose using *Lactobacillus delbrueckii* (NRRL B445). *Applied Biochemistry and Biotechnology*. 84–86: 455–468.

Torkamani, S., Shayegan, J., Yaghmaei, S., Alemzadeh, I. (2008) Study of a newly isolated thermophilic bacterium capable of Kuhemond heavy crude oil and dibenzothiophene biodesulfurization following 4S pathway at 60 °C. *Journal of Chemical Technology and Biotechnology*. 83(12): 1689–1693.

Todescato, D., Maass, D., Mayer, D.A., Oliveira, J.V., de Oliveira, D., Guelli Ulson de Souza, S.M.A., Augusto Ulson de Souza, A. (2017) Optimal production of a *Rhodococcus erythropolis* ATCC 4277 biocatalyst for biodesulfurization and biodenitrogenation applications. *Applied Chemistry and Biotechnology*. DOI 10.1007/s12010-017-2505-5.

Selene M. A. Guelli Ulson de Souza1 & Antônio Augusto Ulson de Souza Tsai, S.P., Lee, Y.H. (1990) A model for energy-sufficient culture growth. *Biotechnology and Bioengineering*. 35: 138–145.

Van der Ploeg, J.R., Leisinger, E.E.T. (2001) Sulfonate-sulfur metabolism and its regulation in Escherichia coli. *Archives of Microbiology*. 176: 1–8.

Van Wyk, J.P.H., Mohulatsi, M. (2003) Biodegradation of wastepaper by cellulase from *Trichoderma viride*. *Bioresource Technology*. 86: 21–23.

Verma, N., Sharma, R., Kaur, R. (2016) Microbial desulphurization study of dibenzothiophene and crude oil by a soil isolate. *International Journal of Science and Research Methodology*. 4(4): 133–145.

Wang, J., Davaadelger, B., Salazar, J.K., Butler III, R.R., Pombert, J.-F., Kilbane, J.J., Stark, B.C. (2015) Isolation and characterization of an interactive culture of two *Paenibacillus* species with moderately thermophilic desulfurization ability. *Biotechnology Letters*. 37: 2201–2211.

Wang, M. D., Li, W., Wang, D. H., Shi, Y. (2004) Desulfurization of dibenzothiophene by a newly isolated *Corynebacterium* sp ZD-1 in aqueous phase. *Journal Environmental Science (China)*. 16: 1011–1015.

Wang, M., Li, W., Shi, Y., Wang, D., Feng, H. (2006) Effect of surfactant on biodesulfurization by *Corynebacterium* sp. ZD-1 in the presence of organic phase. *Journal of Zhejiang University Science A*. 7: 371–375.

Wang, M.-d., Li, W., Wang, D.-h., Shi, Y. (2004) Desulfurization of dibenzothiophene by a newly isolated *Corynebacterium* sp. ZD-1 in aqueous phase. *Journal of Environmental Sciences*. 16(6): 1011–1015.

Wang, P., Krawiec, S. (1994) Desulfurization of dibenzothiophene to 2-hydroxybiphenyl by some newly isolated bacterial strains. *Archives of Microbiology*. 161(3): 266–271.

Wang, P., Krawiec, S. (1996) Kinetic analyses of desulfurization of dibenzothiophene by *Rhodococcus erythropolis* in batch and fed-batch cultures. *Applied and Environmental Microbiology*. 61: 1670–1675.

Wang, Z., Wang, D., Li, Q., Li, W., Tang, H., Xing, J. (2011) Enhanced biodesulfurization by expression of dibenzothiophene uptake genes in *Rhodococcus erythropolis*. *World Journal of Microbiology and Biotechnology*. 27: 1965–1970.

Watkins, L.M., Rodriguez, R., Schneider, D., Broderick, R., Cruz, M., Chambers, R., Ruckman, E., Cody, M., Mrachko, G.T. (2003) Purification and characterization of the aromatic desulfinase, 2-(2-hydroxyphenyl)benzenesulfinate desulfinase. *Archives of Biochemistry and Biophysics*. 415: 14–23.

Wen, Z.Y., Liao, W., Chen, S.L., 2004. Hydrolysis of animal manure lignocellulosics for reducing sugar production. *Bioresource Technology*. 91: 31–39.

Wichmann, R, Vasic-Racki, D. (2003) Cofactor regeneration at the lab scale. Advances in Biochemical Engineering / Biotechnology. 92: 225–2260.

Xiaojuan, T., Lingtian, T., Li'e, P., Xinghong, L. (2008) Research on identification and screen of microbial desulfurization strains for petroleum. *Earth Science Frontiers*. 15:192–198.

Xu, P., Li, F., Yu, J., Ma, C., Zhong, J., Qu, Y., Blankespoor, H.D. (2002) Microbial desulfurization of fuel oil. *Chinese Science Bulletin*. 47(5): 365–369.

Yan, H., Kishimoto, M., Omasa, T., Katakura, Y., Suga, K., Okumura, K., Yoshikawa, O. (2000) Increase in desulfurization activity of *Rhodococcus erythropolis* KA2–5-1 using ethanol feeding. *Journal of Bioscience and Bioengineering*. 89(4): 361–366.

Yan, H., Sun, X., Xu, Q., Ma, Z., Xiao, C., Jun, N. (2008) Effects of nicotinamide and riboflavin on the biodesulfurization activity of dibenzothiophene by *Rhodococcus erythropolis* USTB-03. *Journal of Environmental Sciences*. 20: 613–618.

Yang, J., Hu, Y., Zhao, D., Wang, S., Lau, P.C.K., Marison, I.W. (2007) Two-layer continuous-process design for the biodesulfurization of diesel oils under bacterial growth conditions. *Biochemical Engineering Journal*. 37: 212–218.

Yepez, B.O., Maugeri, F. (2005) Agitation, aeration and shear stress as key factors in inulinase production *by Kluyveromyces marxianus. Enzyme and Microbial Technology*. 36: 717–724.

Yu, B., Ma, C., Zhou, W., Wang, Y., Cai, X., Tao, F., Zhang, Q., Tong, M., Qu, J., Xu, P. (2006a) Microbial desulfurization of gasoline by free whole-cells of Rhodococcus erythropolis XP. *FEMS Microbiology Letters*. 258: 284–289.

Yu, B., Xu, P., Shi, Q., Ma, C. (2006b) Deep desulfurization of diesel oil and crude oils by a newly isolated *Rhodococcus erythropolis* strain. *Applied and Environmental Microbiology*. 72: 54–58.

Zahra, E., Giti, E., Sharareh, P. (2006) Removal of dibenzothiophene, biphenyl and phenol from waste by *Trichosporon* sp. *Scientific Research and Essays*. 1(3): 72–76.

Zakharyants, A.A., Murygina, P., Kalyuzhnyi, S. (2004) Screening of *Rhodococcus* species revealing desulfurization activity with regard to dibenzothiophene. In: Biocatalytic technology and nanotechnology. G E. Zaikov. Nova Science Publishers, Inc. Pp. 51–58.

Zeng, A.P., Deckwer, W.D. (1995) A kinetic model for substrate and energy consumption of microbial growth under substrate sufficient conditions. *Biotechnology Progress*. 11: 71–79.

Zhang, C.Y., Dong, Y.S., Li, Y.L., Fu, C.H., Zhao, C.F., Yu, L.J. (2013) Unstructured models for suspension cultures of Taxus media cells in a bioreactor under substrate-sufficient conditions. *Biochemical Engineering Journal*. 71: 62–71.

Zhao, C.H. (2011) Inulin in the application of bio-energy. A*dvanced Materials Research*. (343/344). 556–559.

Zhou, J., Liu, L., Zhongping, S., Guocheng, D., Cheng, J. (2009) ATP in current biotechnology: regulation, applications and perspectives. *Biotechnology Advances* 27: 94–101.

Zubay, G.L. (1998) Biochemistry, 4th edn. William C Brown Pub. Dubuque, Iowa.

9

Kinetics of Batch Biodesulfurization Process

This chapter is co-authored by ***Samiha F. Deriase***, Prof. of Chemical Engineering, Egyptian Petroleum Research Institute, Nasr City, Cairo, Egypt

List of Abbreviations and Nomenclature

C_{DBT_O}	Initial DBT concentration
C_{2HBP}	Concentration of produced 2-HBP
C_x^{max}	Maximum Biomass Concentration (g/L)
R^2_{adj}	Adjusted Correlation Coefficient
q_{O_2}	Specific Oxygen Uptake Rate (h^{-1})
q_p	Specific HBP production rate (h^{-1})
q_s	Specific DBT Consumption rate (h^{-1})
μ	Specific Growth Rate (h^{-1})
μ_a	Apparent viscosity
μ_{max}	Maximum Specific Growth Rate (h^{-1})
2,2'-BHBP	2,2'-Bihydroxybiphenyl

2-HBP	2-Hydroxybiphenyl
2-MBP	2-Methoxybiphenyl
3-MBT	3-Methylbenzothiophene
4,6-DMDBT	4,6-Dimethyl Dibenzothiophene
4-MDBT	4-Methyldibezothiophene
ADS	Adsorptive Desulfurization
ANN	Artificial Neural Network
ATP	Adenosine Triphosphate
BBD	Box-Behnken Design
BDS	Biodesulfurization
BNT	Benzonaphthothiophene
BT	Benzothiophene
C*	Saturation Concentration of Dissolved Oxygen
CCD	Central Composite Design
CFU	Colony Forming Unit
C_o	Initial Concentration of the Dissolved Oxygen
C_x-DBTs	Alkylated dibenzothiophenes
D	Stirrer diameter
DBT	Dibenzothiophene
DBTO	Dibenzothiophene Sulfoxide
$DBTO_2$	Dibenzothiophene Sulfone
DCW	Dry Cell Weight
D_{DBT}	Desulfurizing Capability Index
D_e	Effective diffusivity
D_i	Impeller diameter
DMF	Dimethylformamide
DMSO	Dimethylsulfoxide
DO	Dissolved Oxygen
DOE	Design of Experiment
DPS	Diphenylsulphide
$D_{SZ}A$	Dibenzothiophene monooxygenase enzyme
$D_{SZ}B$	HBPS desulfinase enzyme
$D_{SZ}C$	Flavin reductase enzyme
E	Fractional Approach to Equilibrium
Ea	Activation Energy
ED	Entner–Doudoroff
EDS	Extractive Desulfurization
FWHM	Full Width at Half Maximum
GC-AED	Gas Chromatography-Atomic Emission Detector
Gr	Growth
H/A	Hydrocarbon/Aqueous
HB	Higher is Better
HBPS	2′-Hydroxybiphenyl-2-sulfinic acid
HDS	Hydrodesulfurization

HGO	Heavy Gas Oil
HPLC	High Performance Liquid Chromatography
IC	Ion Chromatography
JA	Jerusalem Artichoke
JAJ	Jerusalem Artichoke Juice
JAJ_t	Treated Jerusalem Artichoke Juice
k_{cat}	turnover number (min^{-1})
k_La	Gas Liquid Mass Transfer Coefficient (s^{-1})
K_m	Michaelis constant (g/L)
LB	Lower is Better
LGO	Light Gas Oil
LSERs	Linear Solvation Energy Relationships
m	Maintenance coefficient ($g_{glucose}/g_{cell}/h$)
MD	Middle Distillate
MDUF	Middle Distillate Unit Feed
MRE	Mean Relative Error
n	Number of substitutions
NB	Nominal is Better
O/W	Oil/Water
OA	Orthogonal Arrays
ODEs	Ordinary Differential Equations
ODS	Oxidative Desulfurization
OFP	Oil Fraction Phase
OSCs	Organosulfur Compounds
OTR	Oxygen Transfer Rate
OUR	Oxygen Uptake Rate
PASHs	Polyaromatic Sulfur Heterocyclic Compounds
$P_{C/O}$	Partition Coefficients between the Biocatalyst and the Oil Phases
$P_{C/W}$	Partition Coefficients between the Biocatalyst and the Aqueous Phases
Q_{max}	Maximum specific desulfurization rate
R	Gas Constant
R^2	Correlation Coefficient
RMSE	Root Mean Square Errors
RPS	Recycled Paper Sludge
RSM	Response Surface Methodology
S	Substrate concentration (g/L)
S/N	Signal to Noise Ratio
SBM	Sugar Beet Molasses
SKIP	Sum Kinetics Interaction Parameters
SSE	Sum of Squares Errors
SSF	Solid State Fermentation
STBR	Stirred Tank Bioreactor
T	Temperature

TCA	Tricarboxylic Acid
t_G	Growth time
Th	Thiophene
U_g	Superficial Gas Velocity (L/min)
V_{tip}	Impeller tip speed
V_{total}	Volumetric desulfurization rate
VVM	Vessel Volume per Minute
WCW	Wet Cell Weight
WOR	Water-to-Oil Ratio
X	Biomass concentration (g/L)
Y_{cal}	Values calculated by the model
YE	Yeast Extract
Y_{exp}	Experimental data
Y_G	Growth yield ($g_{glucose}/g_{cell}/h$)
η	Optimization extent value
ν	Working Volume (L)
ρ	Density

9.1 Introduction

The main drawback of the BDS process is its low rate compared to the other conventional applied methods. Thus, it is important to know more about the kinetics of the biodesulfurization (BDS) mechanism to enhance the process rate. A representative kinetic evaluation would enhance the commercialization of BDS technology since the output assists in the establishment of an upscale process (Calzada *et al.*, 2012). Serval kinetic studies have been performed for BDS throughout the 4S-pathway. This mainly depends on the determination of the DBT-removal rate and 2-HBP production rate and proposes an empiric kinetic model (Folsom *et al.*, 1999; Galán *et al.*, 2000; Kobayashi *et al*, 2001; Luo *et al.*, 2003; Martin *et al.*, 2004; 2005; Rashtchi *et al.*, 2006; Alcon *et al.*, 2008; Calzada *et al.*, 2012; Nassar *et al.*, 2017).

In order to scale-up a BDS process, it is important to study the enzymatic activity of the biodesulfurizing microorganisms with time. Kinetic studies on BDS in literature do not frequently focus on the 4S route as a whole four serial reaction network. However, further analysis on kinetic knowledge about intermediate reactions of this route comes down to in vitro works studying the activities of both monooxygenases, DszC (Lei and Tu, 1996) and DszA (Ohshiro and Izumi, 1999; Konishi *et al.*, 2000). Particularly, the last step of this route presents high interest because it has been proven to act as the controlling step of the overall process (Gray *et al.*, 1996; Santos *et al.*, 2007). That is because the final product of the 4S route,

2-hydroxybiphenyl (2-HBP), is known to cause a competitive inhibition on the desulfinase enzyme, DszB, the enzyme of the last reaction (Nakayama *et al.*, 2002; Watkins *et al.*, 2003).

Moreover, in order to design BDS-reactors, the kinetics of the process must be thoroughly understood. Kinetic equations, which describe the activity of an enzyme or the growth of microorganisms on a particular substrate, are crucial in understanding many phenomena in biotechnological processes.

This chapter gives a simple background about the microbial growth and enzyme kinetics involved in the BDS-process with a summary of the published work concerning the kinetics of the biodesulfurization process. Furthermore, to gain a better understanding and full description of the process governed by the microorganisms, this chapter also covers the published kinetics on the BDS of model compounds and real oil feed.

9.2 General Background

9.2.1 Phases of Microbial Growth

The microbial growth curve in any cultural media means the changes of microbial cell numbers are plotted as a function of incubation time in cultural media. It passes through the main four district phases: lag, log, stationary, and decline phases (Figure 9.1). The graph represents the

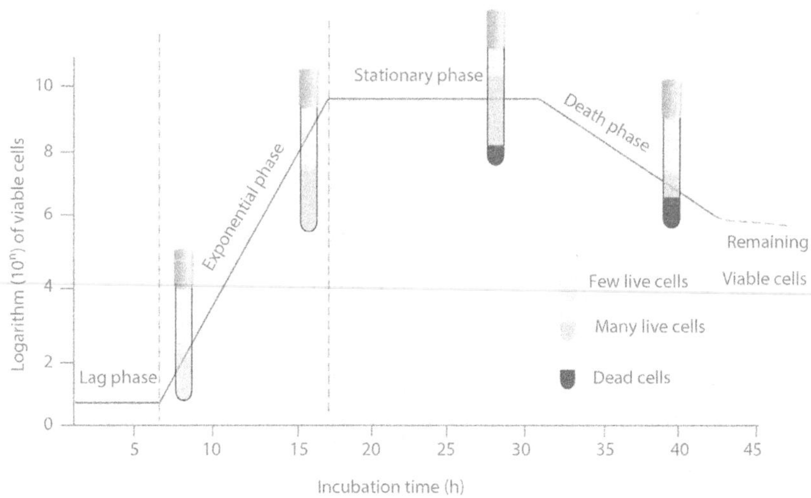

Figure 9.1 Four Main Phases of Microbial Growth.

four main phase patterns of population growth when microorganisms are raised in a batch culture, i.e. closed system, where there is no additional space, no additional nutrients are added, and no waste or dead cells are removed. The mathematical description of the four phases is also summarized in Table 9.1. It should be noted that, usually, there are some physiological changes associated with the shifting from one phase to the next.

9.2.1.1 The Lag Phase

This phase is sometimes known as the adaptation or induction phase. During ***the lag phase***, there is little or no change in the number of cells, but the metabolic activity is high. In a laboratory, microbial cultures are placed in new medium, usually beginning to grow after a "lag phase" due to the physiological adaptation of cells to new physicochemical conditions. The time taken in the lag phase is usually the time to make new enzymes or proteins for the new media.

9.2.1.2 The Log Phase

This phase is also known as the exponential growth or logarithmic phase. During the ***log phase***, the microorganism multiplies at the fastest possible rate (the high rate of cell division and the rate of increase in cells is proportional to the number present at any particular point in time) under the provided conditions (that is a constant supply of fresh media). The log phase represents the period of optimal population growth. Eventually, however, the microorganisms approach the upper limit to their continued growth called the "carrying capacity". Moreover, it should be known that cells in this phase are typically in the healthiest state.

Table 9.1 Mathematical Description of Microbial Growth Phases.

Phase	Mathematical description
Lag Phase	$dx/dt \cong zero$ (1)
Exponential Phase	$dx/dt = \mu x$ (2)
Stationery Phase	$dx/dt = zero$ (3)
Decline Phase	$dx/dt \cong -k_d x$ (4)

where μ is the specific growth rate ($1/\text{time}\ h^{-1}$), t is the incubation time (h), k_d is the specific death rate (h^{-1}), and x is the number of viable microbial cells (CFU/mL).

9.2.1.3 The Stationary Phase

During the ***stationary phase***, there is an equilibrium between cell division and death (i.e. the n ° of new cells = the n ° of dying cells). Thus, no net growth occurs; within this phase, the available nutrients are exhausted, the physical competition for space and nutrients occurs, and, consequently, metabolically active cells stop dividing. Sometimes, production of some toxic substances occurs and waste products are built up. Furthermore, an increase or decrease in pH sometimes occurs; this is because either an essential nutrient is used up or waste products of the organism accumulate in the medium.

9.2.1.4 The Decline Phase

This phase is also known as the death or logarithmic decline phase. During the ***decline phase***, the number of dead cells exceeds the number of produced new cells. In this phase, the cells stop multiplying and die because of nutrient depletion, toxic effects of some products, and/or intermediates and cell aging. This phase is usually a mirror image of the growth phase. During this phase, there is a net loss of viable cells [though some are still growing and dividing.]. If incubation continues after cells reach the stationary phase, the cells will eventually die.

9.2.2 Modeling of Population Growth as a Function of Incubation Time

A number of mathematical models and equations have been developed for the expression of the microbial growth in food and cultural media (Table 9.2). Most of the models are based on some basic mathematical models, such as the exponential, logistic, and Gompertez models. Figure 9.2 illustrates some of the graphical representations of microbial growth model equations.

9.3 Microbial Growth Kinetics

9.3.1 Exponential Growth Model

One of the most famous model for microbial growth is the exponential growth function, where there are plenty available nutrients and growth substrates with reasonable space to grow and no threat from predators. It is applicable only for the subset of growth curve when exponential growth

Table 9.2 Different Microbial Growth Model Equations.

Population growth function	Mathematical expression	
	Differential form	**Integrated form**
Exponential	$\frac{dN}{dt} = rN$ (5)	$N(t) = N_o \exp(rt)$ (6)
Logistic	$\frac{dN}{dt} = rN\left(1-\frac{N}{k}\right)$ (7)	$N(t) = \frac{N_o \exp(rt)}{[1-(N/k)(1-\exp(rt))]}$ (8) $N_{inf} = k/2$ (9)
Ordinary Gompertz	$\frac{dN}{dt} = rN\left[\ln\left(\frac{N}{k}\right)\right]$ (10)	$N(t) = k\exp\left[\ln\left(\frac{N_o}{k}\right)\exp(-rt)\right]$ (11)
Richards	$\frac{dN}{dt} = rN\left[1-\left(\frac{N}{k}\right)^{\beta}\right]$ (12)	$N(t) = k\left[1-\exp(-\beta rt)\left[1-\left(\frac{N_o}{k}\right)^{-\beta}\right]\right]^{-1/\beta}$ (13) $N_{inf} = \left[\left(\frac{1}{1+\beta}\right)^{1/\beta}\right]k$ (14)
Generalized Logistic	$\frac{dN}{dt} = rN^{\alpha}\left[1-\left(\frac{N}{k}\right)^{\beta}\right]^{\gamma}$ (15)	$N_{inf} = \left[\left(1+\frac{\beta\gamma}{\alpha}\right)^{-1/\beta}\right]k$ (16)
Hyper Logistic	$\frac{dN}{dt} = rN^{\alpha}\left[1-\frac{N}{k}\right]^{\gamma}$ (17)	$N_{inf} = \left[\frac{\alpha}{\alpha+\gamma}\right]k$ (18)

where r is the intrinsic growth rate, N_o (i.e. N(0)) and N_t (i.e. N(t)) are the initial population size and the population size at time (t), respectively (dN/dt) is the rate of population growth at time (t), $k=N_{max}$ is the carrying capacity of the environment (i.e. maximum population), $(1/N)(dN/dt)$ is the relative growth rate, N_{inf} is the population size at the inflection point, where growth rate is maximum, and α, β, and γ are positive real numbers.

occurs, where the growth rate tends to increase at a rate that is proportional to the microbial population:

$$\frac{dC_x}{dt} \propto C_x \tag{19}$$

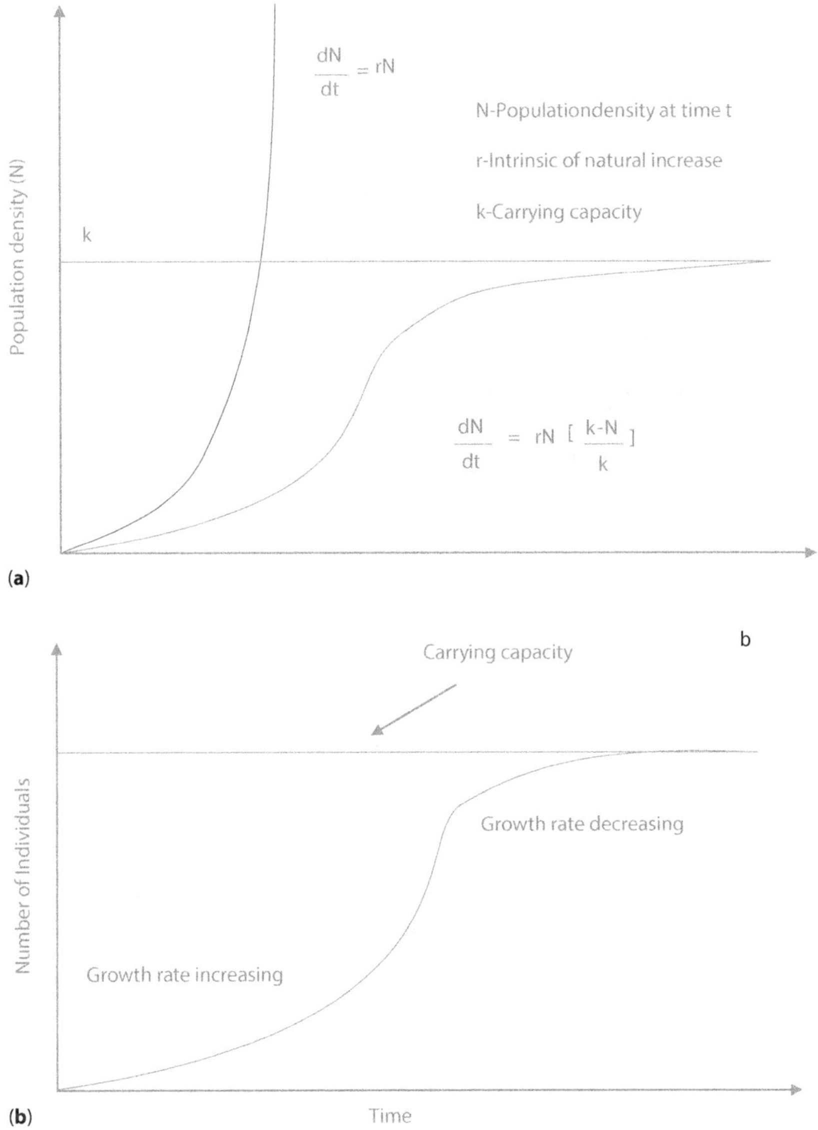

Figure 9.2 Graphical Representation of Microbial Growth (a) Exponential and (b) Logistic Model Equations.

$$\frac{dC_x}{dt} = \mu C_x \tag{20}$$

where dC_x / dt is the growth rate, C_x is the cell concentration (g DCW/L) or change in population, as a function of time t (h), and μ (the proportionality constant) is the specific exponential growth rate (h^{-1}). By integrating equation (20) with initial conditions at $t = 0$ and $C_x = C_{x_o}$:

$$C_x = C_{x_o} \exp(\mu t) \tag{21}$$

Thus, the basic exponential growth model, eq. 21, is good for modeling microbial growth populations that have unlimited resources over a relatively short span of time.

9.3.2 Logistic Growth Model

In the study of population dynamics, one of the most famous models for a growing, but bounded population is the logistic equation.

By including in the model eq. (20), a factor or a term

$$\left\{1 - \frac{C_x}{C_x^{max}}\right\},$$

which suppresses the rate of growth at a high population, the resulting model is the differential form of the logistic growth kinetic model or Verhulst model (reliant to the Belgian statistician, Pierre Francois Verhulst (1804–1849) who worked on population growth in the 19th Century) (Verhulst, 1845; Verhulst, 1847; Blanch and Clark, 1997). In 1847, Pierre F. Verhulst proposed that the population growth depends not only on the population size, but also on the effect of a "carrying capacity" that would limit growth. That formula is called the "logistic model or the Verhulst model":

$$\frac{dC_x}{dt} = \mu C_x \left[1 - \frac{C_x}{C_x^{max}}\right] \tag{22}$$

where C_x^{max} is the maximum biomass concentration reached at the late exponential phase when C_x is much smaller than C_x^{max}, i.e. $C_x <<< C_x^{max}$, and the value of that term is almost 1 and does not affect the growth rate, but as C_x increases to approach C_x^{max}, i.e. $C_x \approx C_x^{max}$, the value of that term approaches zero, thus making the rate of growth, dC_x / dt, almost equal zero, which is the case during the stationary phase. However, if $C_x > C_x^{max}$, so

$$\frac{C_x}{C_x^{max}} > 1$$

and

$$\left[1-\frac{C_x}{C_x^{max}}\right]<0,$$

then $(dC_x / dt) < 0$ and the population will decrease back towards the carrying capacity, i.e. the saturation level forming a numerical upper bound of growth size, in another word, carrying capacity is the amount that when exceeded in value will result in population decline. Thus, the logistic equation is based on exponential growth, but then adds a "braking force" as numbers increase toward the "carrying capacity," C_x^{max}.

Since the obtained experimental data are integral (i.e. biomass concentration varies with time), these data must be differentiated. The integration of the model introduces much less error than data differentiation. Therefore, eq. (22) has been integrated with initial conditions at t = 0 and $C_x = C_{x_o}$, yielding the logistic population growth model (or Verhulst model) equation (eq.23), which is frequently used to model biological populations (Verhulst, 1845; Verhulst, 1847; Blanch and Clark, 1997; Martin *et al.*, 2004) and the growth curves of the biological population are often described well with the logistic model (Nassar *et al.*, 2017).

The logistic function can be obtained as a solution to the logistic differential equation (eq.22):

$$C_x(t) = C_x = \frac{C_{x_o}\exp(\mu t)}{\left[1-\left(C_{x_o}/C_x^{max}\right)(1-exp(\mu t))\right]} \tag{23}$$

where C_{x_o} is the initial biomass concentration (g DCW/L) at time, t = 0, C_x is the biomass concentration (g DCW/L) at time, t, C_x^{max} is the maximum biomass concentation (g DCW/L) (i.e. the carrying capacity), t is the incubation period (h), and μ is the specific growth rate (h^{-1}).

Del Olmo *et al.* (2005) studied the influence of different carbon sources (glucose, glutamic acid, and citrate), nitrogen sources (ammonium and/or glutamic acid), and different DMSO concentrations on the growth of *Rhodococcus erythropolis* IGTS8. The authors employed the logistic kinetic growth model and concluded that the model is able to predict all experiments carried out with a very good fitting of the experimental results and the values of the kinetic parameters (μ and C_x^{max}) were in agreement with the experimental observations. Del Olmo *et al.* (2007) studied the influence of operational conditions (such as temperature, pH, and dissolved oxygen concentration) on the growth of *Rhodococcus erythropolis* IGTS8. A kinetic model based on the logistic equation was also applied to describe biomass concentration during the growth of *Rhodococcus erythropolis* IGTS8 growth. Kinetic model parameters (μ and C_x^{max}) were obtained under several operating conditions and the predicted values of biomass concentration were very close values to those found experimentally. Caro *et al.* (2008) analyzed the values of biomass concentration by applying logistic model equations and determined the kinetic parameters (μ and C_x^{max}) for evaluating the biodesulfurization of dibenzothiophene by growing cells of *Pseudomonas putida* CECT 5279. Chen *et al.* (2008) investigated the effect of 2-hydroxybiphenyl, the end product of biodesulfurization of DBT via 4S pathway during cell growth of *Microbacterium* sp. Strain ZD-M2, using a logistic equation for expressing the cell growth.

9.4 Some of the Classical Kinetic Models Applied in BDS-Studies

For a description of a bioprocess, experimental data can be fitted to several kinetic models to calculate the kinetic parameters for better understanding of the microbial growth, substrate consumption, and product formation.

The relation between the specific growth rate (μ, h^{-1}) of microorganisms and the concentration of substrate (C_s) usually described by the classical empirical growth kinetic models are listed in Table 9.3. Specific growth rates of the culture are fitted to kinetic models, listed above, in order to determine the values of batch growth kinetic parameters like maximum specific growth rate (μ_{max}), half saturation coefficient (K_s), and substrate inhibition constant (K_i). The Monod model (eq.24) has been widely used to describe the growth-linked substrate utilization (Okpokwasili and Nweke, 2005). The Monod kinetic model relates the specific growth rate of microorganisms to substrate concentration via two bio-kinetic parameters (K_s and μ_{max}) (Figure 9.3), where K_s is the half-saturation constant and is defined as the substrate concentration at which μ equals half μ_{max}. The smaller it is, the lower the substrate concentration at which μ approaches $\mu_{max.}$ Also, the value of K_s shows the affinity of the microorganism to the substrate (Nuhoglu and Yalcin, 2005) and K_s is inversely proportional to the affinity of the microbial system for the substrate (Deriase *et al*, 2013).

Table 9.3 Classical Empirical Growth Kinetic Models.

Growth kinetic model	Equation	Reference
Monod	$\mu = \frac{\mu_{max} C_s}{K_s + C_s}$ (24)	Monod (1949)
Haldane	$\mu = \frac{\mu_{max} C_s}{K_s + C_s + \frac{C_s^2}{K_i}}$ (25)	Haldane (1930)
Moser	$\mu = \frac{\mu_{max} C_s^2}{K_s + C_s^2}$ (26)	Layokun *et al.* (1987)
Aiba	$\mu = \mu_{max} \frac{C_s exp\left(-C_s/K_i\right)}{K_s + C_s}$ (27)	Aiba *et al.* (1968)
Tessier	$\mu = \mu_{max} \left[exp\left(-C_s/K_i\right) - \exp\left(-C_s/K_s\right) \right]$ (28)	Edwards (1970)
Webb	$\mu = \mu_{max} \frac{C_s\left(1 + C_s/K_i\right)}{K_s + C_s + C_s^2/K_i}$ (29)	Edwards (1970)
Yano and Koga	$\mu = \mu_{max} \frac{C_s}{K_s + C_s + \left(C_s^3/K_i^2\right)}$ (30)	Yano and Koga (1969)

9.5 Factors Affecting the Rate of Microbial Growth

The main factors affecting the microbial growth are temperature, pH, oxygen requirements, nutrient levels and, finally, radiation and osmotic pressure.

9.5.1 Effect of Temperature

Temperature is the most important environmental factor that determines the rate of growth, multiplication, survival, and death of all living

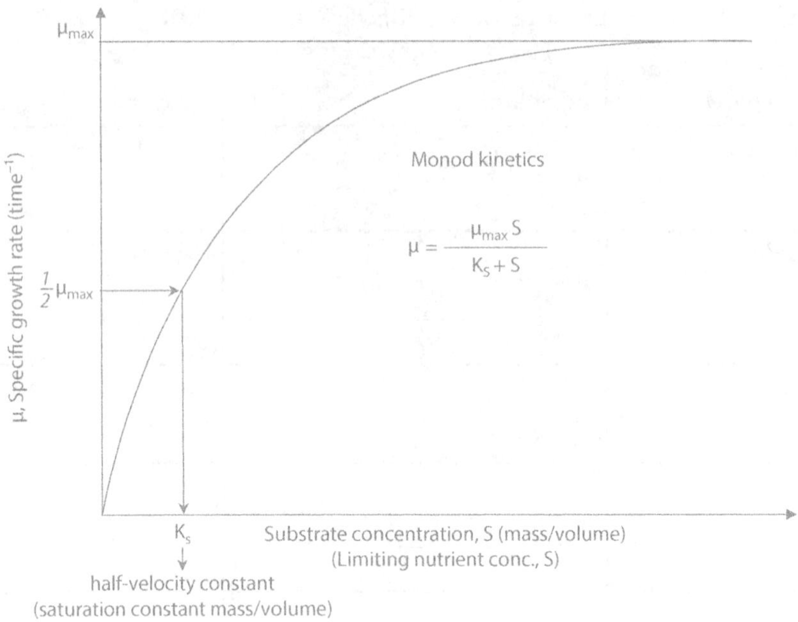

Figure 9.3 Graphical Representation of Monod Kinetic Equation.

organisms. High temperature damages microbes by denaturing enzymes, transport carriers, and other proteins. Microbial membranes are known to be disrupted by temperature extremes. At very low temperatures, membranes also solidify and enzymes do not function properly. Cardinal temperatures are the minimum, optimum, and maximum temperatures at which an organism grows, as illustrated in Figure 9.4. ***Minimum growth temperature*** is the lowest temperature at which organisms grow. ***Optimum growth temperature*** is the temperature at which the most rapid rate of multiplication occurs. ***Maximum growth temperature*** is the highest temperature at which growth occurs. A temperature only slightly above this point frequently kills the microorganisms by inactivating the critical enzymes.

Microorganisms can be classified into groups according to their growth temperature optima, as illustrated in Figure 9.5. A ***psychrophile*** is an organism that can live and thrive at temperatures much lower than normal (< 15 °C). A ***mesophile*** is an organism that can live and thrive at moderate temperatures (25–35 °C). A ***thermophile*** is an organism that can live and thrive at relatively high temperatures (between 45 °C and 80 °C). A ***hyperthermophile*** is an organism that can live and thrive at very high temperatures (i.e. > 80 °C) (Le Borgne and Quintero, 2003).

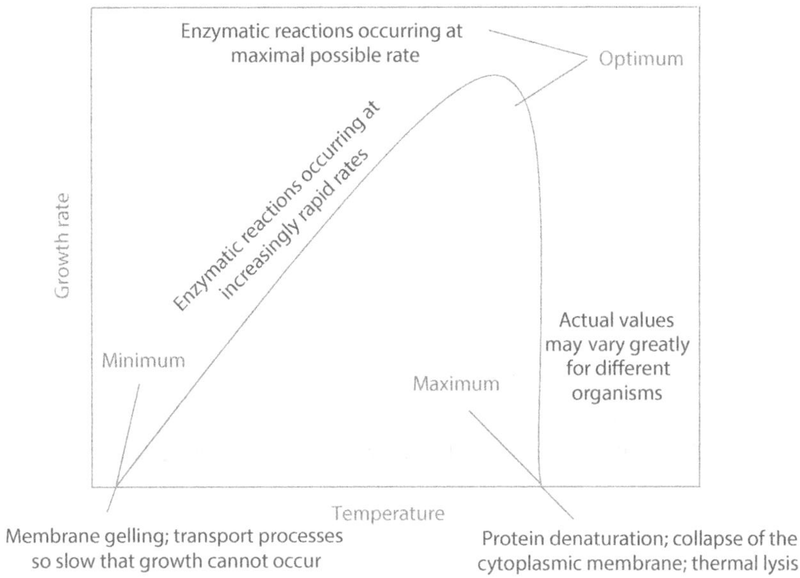

Figure 9.4 The Cardinal Temperatures: Minimum, Optimum, and Maximum.

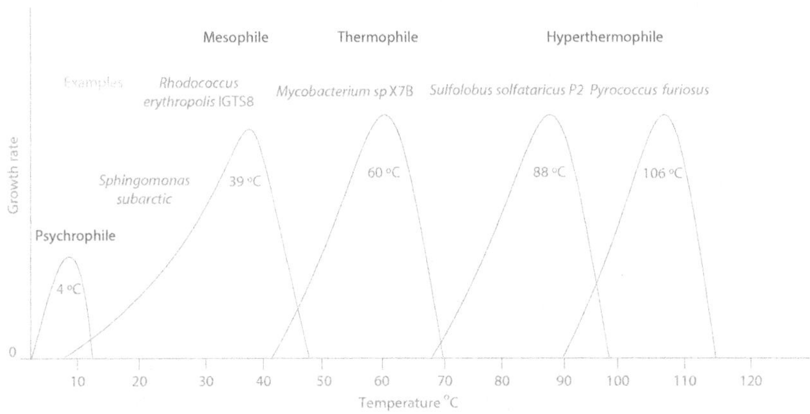

Figure 9.5 Temperature and Growth Rate Response in Different Temperature Classes of Microorganisms.

The effect of temperature on the specific growth rate is represented mathematically by the Arrhenius model equation, as illustrated in Figure 9.6. The Arrhenius equation relates the specific growth rate to the inverse of absolute temperature (1/T), where A is the proportionality factor, Ea is the activation energy for the process, and R is the gas constant.

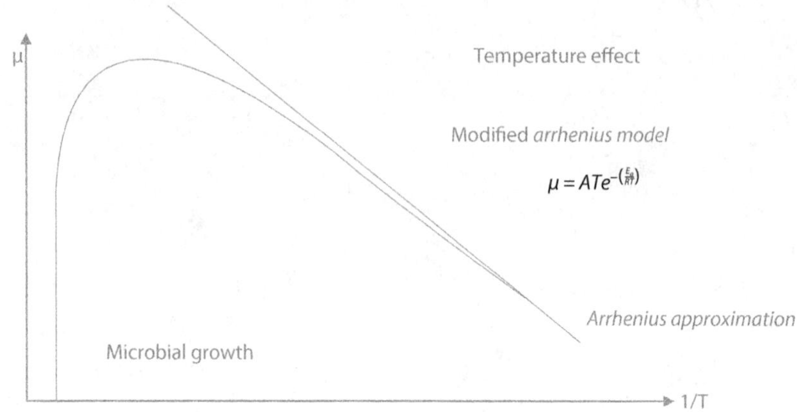

Figure 9.6 Effect of Temperature on Specific Growth Rate of Microorganisms.

9.5.2 Effect of pH

pH refers to the negative logarithm of hydrogen ion concentration, that is pH = - $\log_{10}$ [H^+]. Microbial growth is strongly affected by the pH of the medium. Drastic variations in cytoplasmic pH disrupt the plasma membrane or inhibit the activity of enzymes and membrane transport proteins. Some organisms have evolved to grow best at low or high pH, but most organisms grow best between a pH of 6 and 8 (neutrophiles). Organisms that grow best at a low pH (<6) are known as acidophiles, while those which grow best at a high pH (>9) are known as alkaliphiles, as illustrated in Figure 9.7.

9.5.3 Effect of Oxygen

Oxygen is an important substrate for aerobic organisms. Since metabolic energy production by cells is directly related to the oxygen uptake rate (i.e. the respiration rate), oxygen concentration is very strongly coupled to the growth rate as illustrated in Figure 9.8. The critical dissolved oxygen (DO) concentration refers to the value of DO below which the growth rate is lower than the maximum value. The growth rate sharply rises to its maximum value with DO concentration. The concentration at which the maximum growth rate is attained is often referred to as the critical oxygen concentration $C_{O_2}^{CRIT}$. This value is typically less than 0.5 mg/L for bacteria and yeast.

9.6 Enzyme Kinetics

The living cell is the site of tremendous biochemical activity called metabolism. This is the process of chemical and physical change which goes on

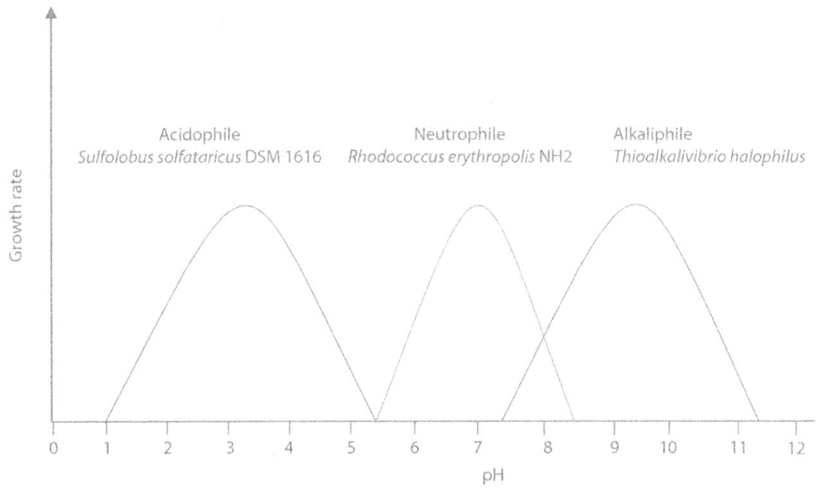

Figure 9.7 Effect of Acidity and Alkalinity on Growth of Microorganism.

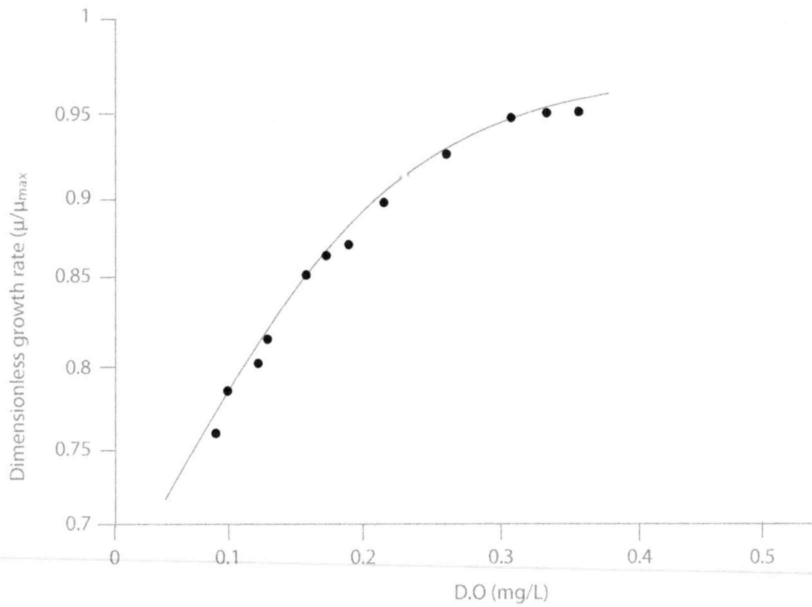

Figure 9.8 Effect of Dissolved Oxygen Concentration on Growth of Microorganisms.

continually in the living organism, including building-up of new tissue, replacement of old tissue, conversion of food to energy, disposal of waste materials, reproduction, etc., briefly, all the activities of "life".

The phenomenon of catalysis makes biochemical reactions necessary for all life processes possible. Catalysis is defined as the acceleration of a chemical reaction by some substance which itself undergoes no permanent change. The catalysts of biochemical reactions are enzymes and are responsible for bringing about almost all of the chemical reactions in living organisms. Without enzymes, these reactions take place at a rate which is too slow for the pace of metabolism. All known enzymes are high molecular weight proteins made up principally of chains of amino acids (Worthington Biochemical Corporation, 2017).

9.6.1 Basic Enzyme Reactions

The basic Enzymatic reaction mechanism can be represented as follows:

$$S + E \rightarrow P + E \tag{9.1}$$

where E represents the enzyme catalyzing the reaction, S represents the substrate (substance being changed), and P represents product of the reaction.

A theory to explain the catalytic action of enzymes was proposed by the Swedish chemist Savante Arrhenius in 1888, where the substrate and enzyme formed some intermediate substance known as the enzyme substrate complex and the reaction can be represented as:

$$S{+}E \longrightarrow ES \tag{9.2}$$

where ES represents enzyme substrate complex.

Combining the aforementioned reactions results in the following reaction:

$$S{+}E \longrightarrow ES \longrightarrow P{+}E \tag{9.3}$$

The study of a large number of chemical reactions revealed that most of these reactions do not go to the true completion, which is likewise true for enzymatically-catalyzed reactions. This is due to the reversibility of most the reactions. Applying this fact to enzymatic reactions results in the following reaction mechanism equation, which is the basic equation upon which most studies related to enzyme activity are based.

$$E{+}S \underset{K_{-1}}{\overset{K_{+1}}{\rightleftharpoons}} ES \xrightarrow{K_{+2}} P{+}E \tag{9.4}$$

where k_{+1} and k_{+2} are the forward reaction rate constants and k_{-1} is the backward reaction rate constant.

Equilibrium (that is the steady state condition) is reached when the forward reaction rates equal the backward rates. Figure 9.9 shows the relatively low and constant concentration of the enzyme – substrate complex (ES).

9.6.2 Factors Affecting the Enzyme Activity

Several factors affect the rate at which enzymatic reactions proceed, such as enzyme and substrate concentrations, temperature, pH, and the presence of inhibitors or activators (Worthington Biochemical Corporation, 2017).

9.6.2.1 Enzyme Concentration

In order to study the effect of increasing the enzyme concentration upon the reaction rate, the substrate must be present in an excess amount, that is, the reaction must be independent of the substrate concentration and any change in the amount of the product formed over a specific period of time will be dependent upon the level of enzyme present (Figure 9.10).

These reactions are said to be "zero order" because the rates are independent of substrate concentration and are equal to the reaction rate constant (k). The formation of products proceeds at a rate which is linear with time. The amount of enzyme present in a reaction is measured by its activity. Table 9.4 represents the reaction order with respect to substrate concentration.

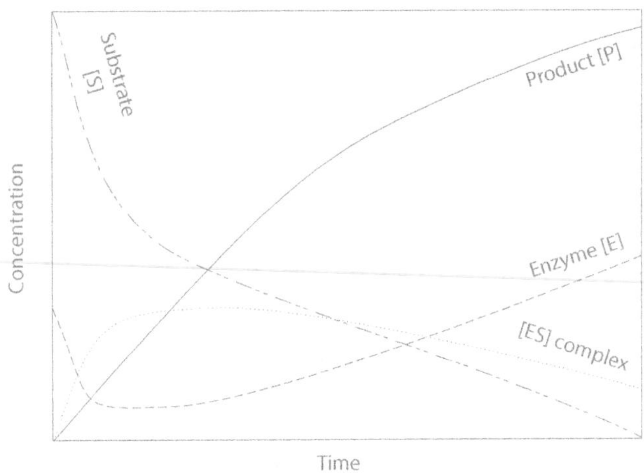

Figure 9.9 Change in Concentrations of substrate (S), product (P), enzyme (E), and enzyme complex (ES).

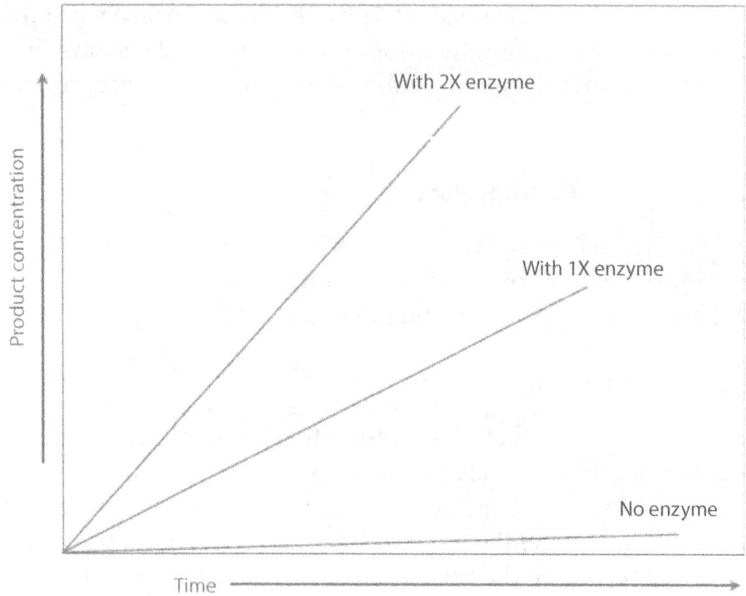

Figure 9.10 Zero Order Reaction Rate is Independent on the Substrate Concentration.

Table 9.4 Reaction Order with Respect to Substrate Concentration.

Order	Rate equation
Zero	Rate = k (31) rate is independent of substrate concentration
First	Rate = $k[C_s]$ (32) rate is proportional to 1st power of concentration
Second	Rate = $k[C_s]^2$ (33) rate is proportional to the square of substrate concentration
Second	Rate = $k[C_{s1}][C_{s2}]$ (34) rate is proportional to 1st power of each substrate concentration

9.6.2.2 Substrate Concentration

Experimentally, it has been shown that if the amount of the enzyme is kept constant and substrate concentration is then gradually increased, the reaction velocity will increase until it reaches a maximum. After this point, the increase in substrate concentration will not increase the velocity (Figure 9.11). When this maximum velocity has been reached, all of the available enzymes would be converted to ES.

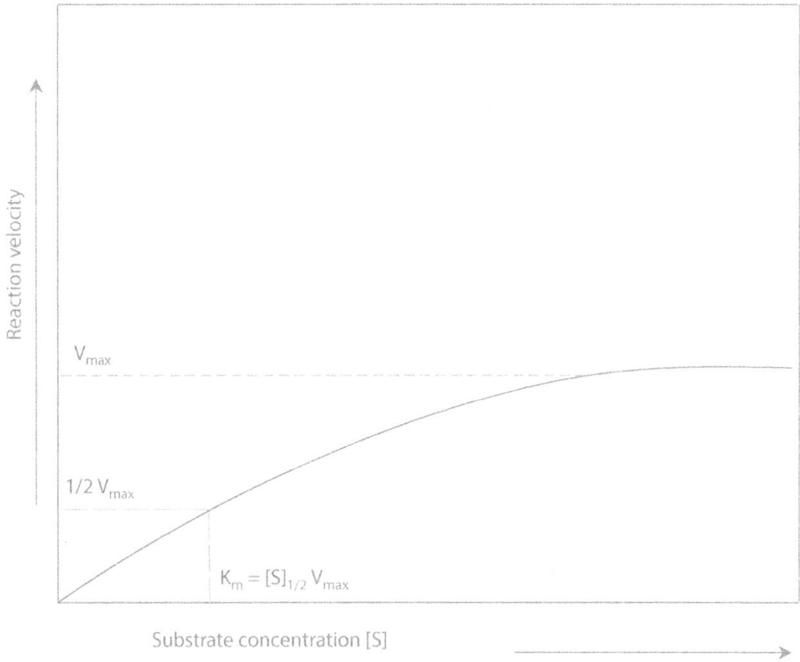

Figure 9.11 Effect of Substrate Concentration on Reaction Velocity.

9.6.2.3 Effect of Inhibitors on Enzyme Activity

Enzyme inhibitors are substances which alter the catalytic action of the enzyme and, consequently, slow down, or in some cases, stop catalysis. There are three types of enzyme inhibition: competitive, non-competitive, and substrate inhibition. Most theories concerning inhibition mechanisms are based on the existence of the enzyme substrate complex (ES). Competitive inhibition occurs when the substrate and a substance resembling the substrate are both added to the enzyme.

A theory called "lock-key theory" of enzyme catalysts can be used to explain why inhibition occurs. This theory utilizes the concept of an "active site". It holds that one particular portion of the enzyme surface has a strong affinity for the substrate. Figure 9.12 represents the "lock-key" theory considering the enzyme as the lock and the substrate as the key. The key is inserted in the lock, turned, the door is opened, and the reaction proceeds. However, when an inhibitor which resembles the substrate is present, it will compete with the substrate for the position in the enzyme lock. When the inhibitor wins, it gains the lock position, but is unable to open the lock.

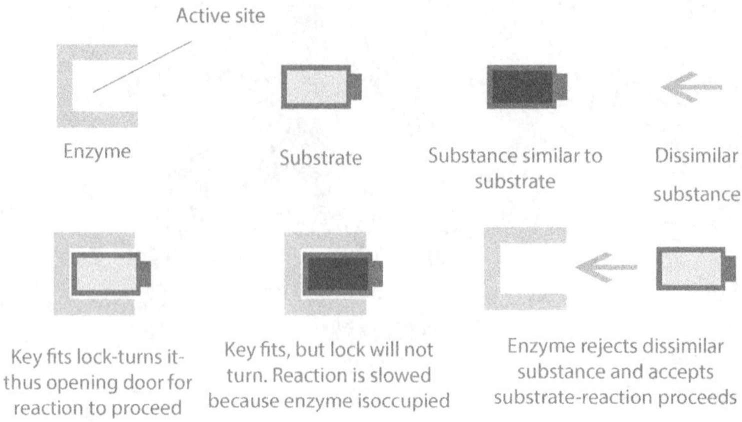

Figure 9.12 Lock – Key Theory.

Hence, the reaction is slowed down because some of the available enzyme sites are occupied by the inhibitor. If a dissimilar substance, which does not fit the site is present, the enzyme rejects it, accepts the substrate, and the reaction proceeds normally.

Non-competitive inhibitors are substances which when added to the enzyme, alter the enzyme in a way where it cannot accept the substrate (Figure 9.13a) When a non-competitive inhibitor is added, the V_{max} is reduced, but the K_m remains the same (Figure 9.13b).

Substrate inhibition will occur when excessive amounts of substrate are present, as according to the Michaelis-Menten equation. Figure 9.14 shows the decrease in reaction velocity after the maximum velocity has been reached, that is excess amounts of substrate added to the reaction mixture decreases the reaction rate. This is thought to be due to the fact that there are many substrate molecules competing for the active sites on the enzyme surfaces that blocked the sites and prevent any other new substrate molecules from occupying them (Figure 9.15).

9.6.2.4 Effect of Temperature

Like most chemical reactions, the rate of an enzyme-catalyzed reaction increases as the temperature is raised, but, in fact, many enzymes are adversely affected by high temperatures and, as a result, the reaction rate increases with temperature to a maximum level, then declines with a further increase of temperature, as shown in Figure 9.16, because most cell enzymes rapidly become denatured at temperatures above 40 °C.

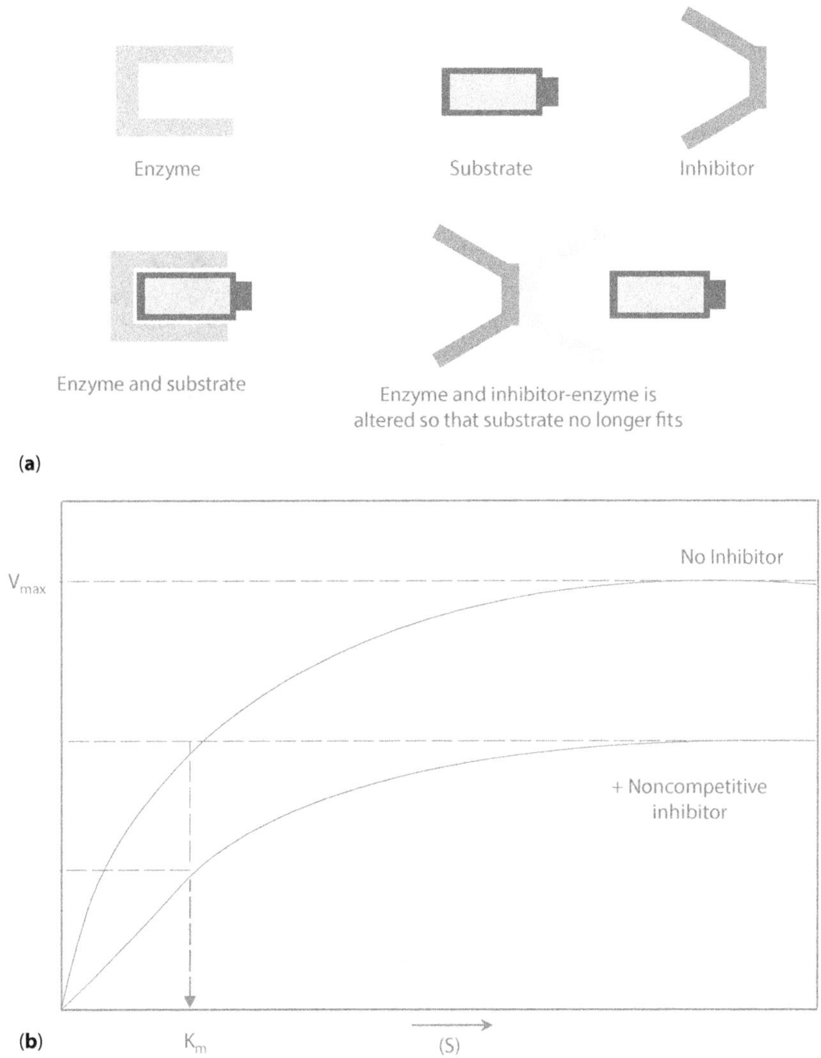

Figure 9.13 Non-Competitive Inhibition.

9.6.2.5 *Effect of pH*

Enzymes are affected by changes in pH. The most favorable pH value, that is, the point where the enzyme is most active, is known as the optimum pH (Figure 9.17). The optimum pH value will vary greatly from one enzyme to another. Extremely high or low pH values generally result in complete loss of activity for most enzymes.

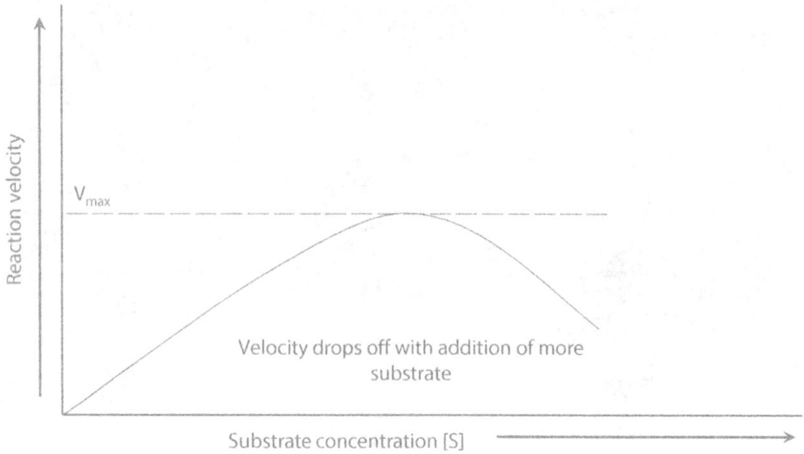

Figure 9.14 Substrate Becomes Rate Inhibition.

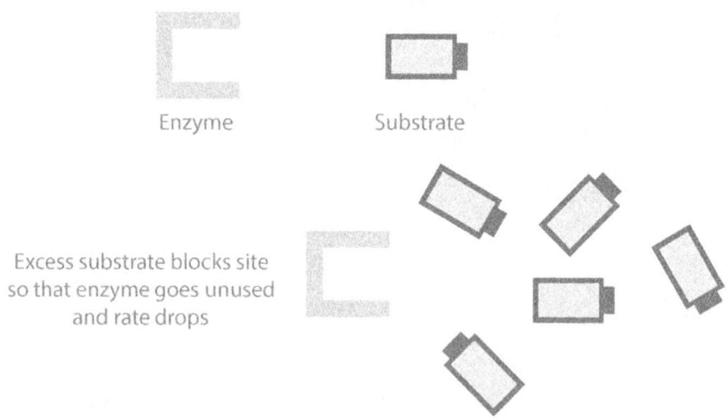

Figure 9.15 Substrate Inhibition.

9.7 Michaelis-Menten Equation

In 1903, French physical chemist Victor Henri found that enzyme reactions were initiated by a bond between the enzyme and the substrate. His work was taken up by German biochemist Leonor Michaelis and Canadian physician Maud Menten, who investigated the kinetics of an enzymatic reaction mechanism. In 1913, they proposed a mathematical model of the reaction. Referring to enzymatic reaction mechanism (**9.4**), it involves the reversible reaction between enzyme [E] and substrate [S], binding together to form the enzyme – substrate complex [ES], which irreversibly yields product [P], regenerating the original enzyme. The enzymatic reaction

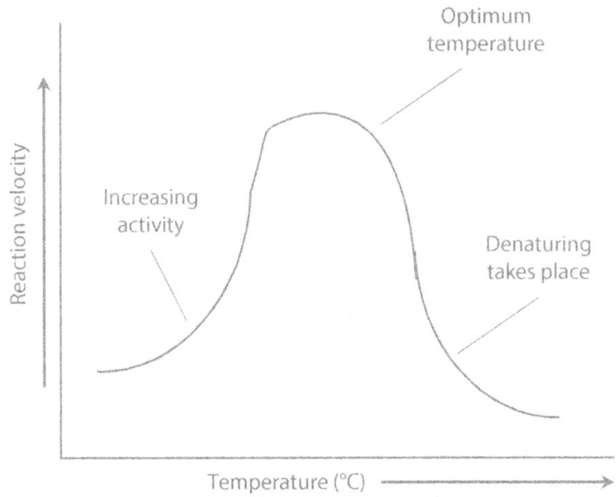

Figure 9.16 Effect of Temperature on Reaction Rate.

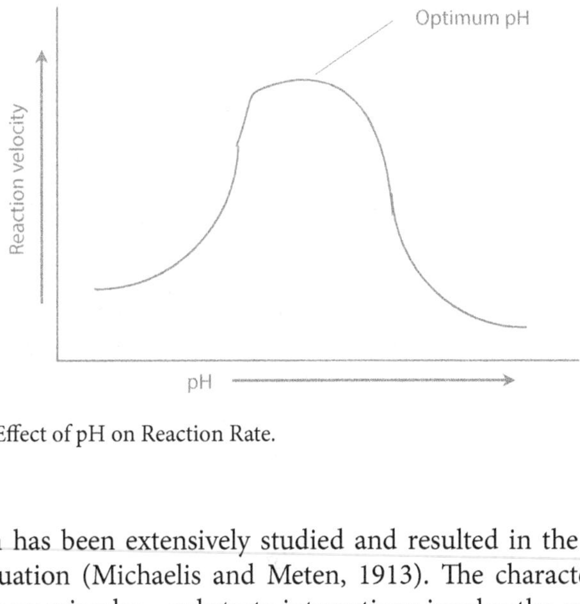

Figure 9.17 Effect of pH on Reaction Rate.

mechanism has been extensively studied and resulted in the Michaelis–Menten equation (Michaelis and Meten, 1913). The characterization of the enzyme or microbe - substrate interactions involve the estimation of several parameters in the kinetic models from experimental data. In order to describe the true behavior of the system, the estimated and measured kinetic parameters used in developing the models are acceptable and can be used in reactor design for biodesulfurization (Kareem *et al.*, 2014). Theoretically, when the maximum velocity, $V_{max,}$ had been reached all of the available enzyme has been converted to ES, the enzyme substrate complex.

Michaelis-Menten developed the following mathematical expression for the reaction velocity in terms of this constant and the substrate concentration-to calculate the enzyme activity in terms of reaction speed from measurable experimental data,

$$V_t = \frac{-dc_s}{d_t} = \frac{V_{max} C_s}{K_m + C_s} \tag{35}$$

where V_t is the velocity at any time t, V_{max} is the limiting maximum velocity (maximum reaction velocity), C_s is the substrate concentration at time, t, and K_m is the Michaelis constant, which is defined as the substrate concentration at half of the maximum velocity, as shown in Figure 9.11_and given by equation (36). A small K_m indicates that the enzyme requires only a small amount of substrate to become saturated to approach V_{max}. Hence, the maximum velocity is reached at a relatively low substrate concentration. A large K_m indicates the need for high substrate concentration to achieve maximum reaction velocity,

$$K_m = \frac{C_s}{\frac{1}{2} V_{max}} = \frac{K_{+1} + K_{+2}}{K_{-1}} \tag{36}$$

where K_{+1}, $K_{+2,}$ and K_{-1} are rate constants.

When resting cells are employed, reaction kinetics could be governed by the Michaelis-Menten mechanism to explain the desulfurization of DBT and alkyl dibenzothiophenes Cx-DBT and calculate the biokinetic constant parameters, K_m and V_{max} (Folsom *et al.*, 1999; Mingfeng *et al.*, 2003a, b; Zhang *et al.* 2003). Nazari *et al.* (2017) reported that for the commercialization of BDS, the removal of sulfur from petroleum compounds needs enzymes with a high specificity and low K_m in a dsz system.

9.7.1 Direct Integration Procedure

To explain the desulfurization pattern, the Michaelis-Menten equation which was originally developed for enzymatic kinetics can be employed.

The direct integration of the Michaelis-Menten equation yielded:

$$-(K_M + C_S)dc_s = V_{max}\ C_s dt \tag{37}$$

$$-\int_{C_{S_o}}^{C_S} \frac{(K_M + C_S)}{C_S} dC_S = \int_{t_o}^{t} V_{max} dt \tag{38}$$

$$-\int_{C_{S_o}}^{C_S} \left(\frac{K_M}{C_S} + 1 \right) dC_S = V_{max} t \tag{39}$$

$$-K_M \ln \frac{C_S}{C_{S_o}} - (C_S - C_{S_o}) = V_{max} t \tag{40}$$

$$K_M \ln \frac{C_{S_o}}{C_s} + (C_{S_o} - C_S) = V_{max} t \tag{41}$$

$$C_{S_o} - C_S = -K_M \ln \frac{C_{S_o}}{C_S} + V_{max} t \tag{42}$$

Further manipulation of equation (42) gives equation:

$$\frac{\left(C_{s_o} - C_s\right)}{ln\left(C_{s_o} / C_s\right)} = -K_m + V_{max} \frac{t}{ln\left(C_{s_o} / C_s\right)} \tag{43}$$

Equation (43) represents the final explicit form of the Michaelis-Menten equation from which the bio-kinetic constants K_m and V_{max} are estimated (Mingfang *et al.*, 2003a).

It is important to note that most kinetic models and their integrated forms are nonlinear. This makes parameter estimation relatively difficult. However, some of these models can be linearized. Various linearized forms of the integrated expressions have been used for parameter estimation. However, the use of a linearized expression is limited because it transforms the error associated with the dependent variable, making it unable to be normally distributed, resulting in inaccurate parameter estimations. Therefore, non-linear least-squares regression is often used to estimate kinetic parameters from nonlinear expressions. However, the application of nonlinear least- squares regression to the integrated forms of the kinetic expressions is complicated. The parameter estimates obtained from the linearized kinetic expressions can be used as initial estimates in the iterative nonlinear least-squares regression (Kareem *et al.*, 2014).

9.7.2 Lineweaver-Burk Plot Method

$$V_t = \frac{V_{max} C_S}{K_m + C_S} \tag{44}$$

$$\frac{1}{V_t} = \frac{K_m + C_S}{V_{max} C_S} \tag{45}$$

$$\frac{1}{V_t} = \frac{K_m}{V_{max} C_S} + \frac{C_S}{V_{max} C_S} \tag{46}$$

$$\frac{1}{V_t} = \left(\frac{K_m}{V_{max}}\right)\left(\frac{1}{C_S}\right) + \left(\frac{1}{V_{max}}\right) \tag{47}$$

Plotting this relationship over X-Y axes (Figure 9.18.a), the result represents a straight line of slop equal to (K_m / V_{max}), where the Y axis intercept is equal to (1 / V_{max}).

Abolfazl *et al.* (2006) applied a Lineweaver-Burk plot method (Lineweaver and Burk, 1934) to calculate the values of maximal velocity, $V_{max,}$ and Michaelis constant, $K_{m,}$ for the biodesulfurization of DBT by a moderately thermophilic bacterium.

9.7.3 Eadie-Hofstee

This is another way to represent the enzyme Michaelis-Menten kinetic model equation from which the bio-kinetic constants K_m and V_{max} can be calculated.

$$V_t(K_m + C_s) = V_{max} C_s \tag{48}$$

$$V_t K_m + V_t C_s = V_{max} C_s \tag{49}$$

$$V_t C_s = V_{max} C_s - V_t K_m \tag{50}$$

$$V_t = V_{max} - K_m (V_t / C_s) \tag{51}$$

Figure (9.18b) represents the plot of the above equation which is the Eadie-Hofstee plot, where the plotting of V_t versus V_t/C_s produces a linear plot, the intercept with the X-axis gives the V_{max}/K_m value, the intercept with the Y- axis gives the value of $V_{max,}$ and the slope of the line gives ($-K_m$).

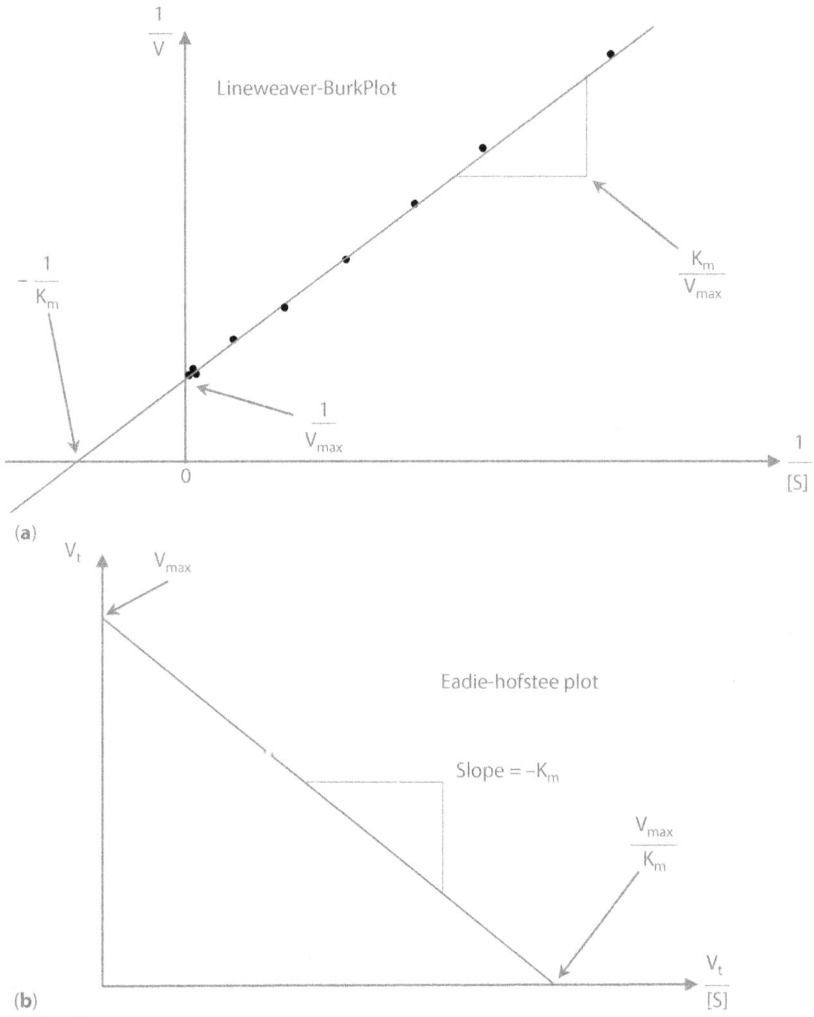

Figure 9.18 Lineweaver-Burk (a) and Eadie-Hofstee (b) Plots Methods.

9.8 Kinetics of a Multi-Substrates System

In a multi–substrates system, the biomass growth is due to the utilization of all available hydrocarbons in the system to the microorganism, accordingly, the specific growth rate will be the summation of all individual ones for each substrate, as represented by the following equation,

$$\mu_T = \sum_{i=1}^{n} \mu_i' \tag{52}$$

where, μ_T is the total specific growth rate (h^{-1}) and μ_i' is the specific growth rate on substrate i, in the multi-substrate system (h^{-1}), which can be calculated from the multi-substrate form of the Monod kinetic equation, represented by eq. (53), where n is the number of substrates in the system,

$$\mu_i' = \frac{\mu_{max,i} C_i}{K_{si} + \sum_{j=1}^{n} \frac{K_{Si}}{K_{sj}} C_j} \tag{53}$$

$$\frac{dC_{si}}{dt} = -\mu_i' C_x / Y_i \tag{54}$$

$$\frac{dC_x}{dt} = \mu_T C_x - bC_x \tag{55}$$

where $\mu_{max,i}$ is the maximum specific growth rate on substrate i, K_{si} and K_{sj} are the half saturation coefficients for substrates i and j, respectively, and C_{si} and C_{sj} are the concentrations of substrates i and j in the multi-substrate system, respectively. The value of the interaction term [K_{si}/K_{sj}], in the multi-substrate Monod kinetic model (eq. 53), represents the effect expressed by substrate j on substrate i (Guha *et al.*, 1999; Knightes and Peters, 2006).

9.9 Traditional 4S-Pathway

In 1990, the specific oxidative desulfurization pathway was elucidated for DBT desulfurization. In this pathway, bacteria are exclusively able to remove sulfur atoms from DBT by breaking the carbon-sulfur bond during the oxidation reactions and producing the final 2-HBP product as a result (Chapter 6). The bacteria are also capable of maintaining the heating value of fuel in this pathway, without decomposition of a carbon skeleton (Caro *et al* 2007; Davoodi-Dehaghani *et al.*, 2010; Nasab *et al.* 2015). The traditional 4S-route is a multi-enzymatic oxidative system (Chapter 7) of four consecutive reactions which convert DBT into dibenzothiophene sulfoxide (DBTO), then into dibenzothiophene sulfone ($DBTO_2$). The sulfone is transformed into 2′-hydroxybiphenyl 2-sulfinic acid (HBPS) and, finally, 2-hydroxybiphenyl (2-HBP) and sulfate are produced as the final end products.

$$\text{DBT} \rightarrow \text{DBTO} \rightarrow \text{DBTO}_2 \rightarrow \text{HBPS} \rightarrow \text{2-HBP} + \text{Sulfate}$$

Current knowledge about the 4S route includes information about both genes and enzymes needed for this transformation (i.e. the transformation of DBT into a sulfur free molecule, 2-HBP). Two monooxygenases, dszC and dszA, and a desulfinase, dszB, participate in the conversion of DBT into 2-HBP, Flavin-dependent monooxygenase, dszC, catalyses the first two steps of DBT oxidation into DBTO and, consecutively, into $DBTO_2$. Monooxygenase dszA catalyzes the transformation of $DBTO_2$ into HBPS. The third enzyme involved in the 4S route, desulfinase dszB, catalyzes the last step of 4S route, which involves the conversion of HBPS into the final product 2-HBP (Calzada *et al.*, 2011). Accordingly, the four genes involved in the multi-enzymatic biodegradation process, which is called 4S metabolic mechanism (Chapter 7) (Mosseini *et al.*, 2006; Calzada *et al.*, 2011), are:

- dszC, encoding a DBT-monooxygenase responsible for the first two oxidations of DBT to sulfoxide and then to sulfone,
- dszA, encoding DBT-sulfone monooxygenase oxidizing DBT-sulfone to 2-HBP-2-sulfinic acid,
- dszB, encoding a 2-HBP- sulfonate desulfinase that converts HBP-sulfinate to 2-HBP and sulfite, and
- dszD, encoding a NADH-flavin mononucleotide oxido-reductase supplying FMNH2 needed for the first three oxidations.

9.9.1 Formulation of a Kinetic Model for DBT Desulfurization According to 4S-Pathway

According to the 4S route, four reactions must be considered as follows (Alcon *et al.*, 2008):

$$DBT + O_2 \xrightarrow{r_1} DBTO + H_2O \tag{9.5}$$

$$DBTO + O_2 \xrightarrow{r_2} DBTO_2 + H_2O \tag{9.6}$$

$$DBTO_2 + O_2 \xrightarrow{r_3} HBPS + H_2O \tag{9.7}$$

$$HBPS + H_2O \xrightarrow{r_4} 2HBP + SO_3^{-2} + H^+ \tag{9.8}$$

The four reactions involved in 4S route have been studied separately. The DBT, DBTO, $DBTO_{2,}$ and HBP have been chosen as key compounds for each of the four reactions of the route.

The first step is the stoichiometric study of the reaction. It is necessary to simplify the reaction scheme as follows:

$$DBT \xrightarrow{r_1} DBT \tag{9.9}$$

$$DBTO \xrightarrow{r_2} DBTO_2 \tag{9.10}$$

$$DBTO_2 \xrightarrow{r_3} HBPS \tag{9.11}$$

$$HBPS \xrightarrow{r_4} 2HBP \tag{9.12}$$

The kinetic equations assumed for each reaction are mainly those given in literature, simple Michaelis-Menten kinetics for reactions (9.9–9.11), and product competitive inhibition for reactions (9.12) according to:

$$r_1 = \frac{k_1 C_{DBT}}{K_1 + C_{DBT}} \tag{56}$$

$$r_2 = \frac{k_2 C_{DBTO}}{K_2 + C_{DBTO}} \tag{57}$$

$$r_3 = \frac{k_3 C_{DBTO_2}}{K_3 + C_{DBTO_2}} \tag{58}$$

$$r_4 = \frac{k_4 C_{HBPS}}{K_4 + \frac{K_4}{K_I} C_{2HBP} + C_{HBPS}} \tag{59}$$

The overall kinetic model for the reaction network is formed by the following set of differential equations, corresponding to the production rates of the four key compounds assumed (Alcon *et al.*, 2008; Calzada *et al.*, 2012; Martinez *et al.*, 2017).

$$R_{DBT} = -r_1 = \frac{k_1 C_{DBT}}{K_1 + C_{DBT}} \tag{60}$$

$$R_{DBT0} = r_1 - r_2 = \frac{k_1 C_{DBT}}{K_1 + C_{DBT}} - \frac{k_2 C_{DBT0}}{K_2 + C_{DBT0}} \tag{61}$$

$$R_{DBTO_2} = r_2 - r_3 = \frac{k_2 C_{DBTO}}{K_2 + C_{DBTO}} - \frac{k_3 C_{DBTO_2}}{K_3 + C_{DBTO_2}} \tag{62}$$

$$R_{2HBP} = r_4 = \frac{k_4 C_{HBPS}}{K_4 + \left(K_4 / K_I \right) C_{2-HBP} + C_{HBPS}} \tag{63}$$

k_i is the kinetic parameter of reaction i (i = 1,2,3,4) μM/min, K_i is the substrate affinity of the reaction substrate i μM, and K_I is the inhibition constant μM.

Employing the production rate method, in order to determine reaction rate in the reaction network, this method is expressed as

$$R' = \gamma' r' \tag{64}$$

R is a vector containing the production rates of the key compounds of the network, γ' is a matrix containing the stoichiometric coefficients involved in the reaction network, andd r' is a vector containing the reaction rates of all reactions in the network.

Accordingly, the above equation is expressed as follows:

$$\begin{bmatrix} R_{DBT} \\ R_{DBTO} \\ R_{DBTO_2} \\ R_{2-HBP} \end{bmatrix} = \begin{bmatrix} -1 & 0 & 0 & 0 \\ 1 & -1 & 0 & 0 \\ 0 & 1 & -1 & 0 \\ 0 & 0 & 0 & 1 \end{bmatrix} \cdot \begin{bmatrix} r_1 \\ r_2 \\ r_3 \\ r_4 \end{bmatrix} \tag{65}$$

Using the inversion matrix of stoichiometric coefficients γ'^{-1}

$$r' = \gamma'^{-1} R' \tag{66}$$

This can be expressed as follows:

$$\begin{bmatrix} r_1 \\ r_2 \\ r_3 \\ r_4 \end{bmatrix} = \begin{bmatrix} -1 & 0 & 0 & 0 \\ -1 & -1 & 0 & 0 \\ -1 & -1 & -1 & 0 \\ 0 & 0 & 0 & 1 \end{bmatrix} \cdot \begin{bmatrix} R_{DBT} \\ R_{DBTO} \\ R_{DBTO_2} \\ R_{2-HBP} \end{bmatrix} \tag{67}$$

According to the above matrix expression, the following set of are equations obtained:

$$r_1 = -R_{DBT} = -\frac{dC_{DBT}}{dt} \tag{68}$$

$$r_2 = -R_{DBT} - R_{DBTO} = -\frac{dC_{DBT}}{dt} - \frac{dC_{DBTO}}{dt} \tag{69}$$

$$r_3 = -R_{DBT} - R_{DBTO} - R_{DBTO_2} = -\frac{dC_{DBT}}{dt} - \frac{dC_{DBTO}}{dt} - \frac{dC_{DBTO_2}}{dt} \tag{70}$$

$$r_4 = R_{2HBP} = \frac{dC_{2HBP}}{dt} \tag{71}$$

The integrated form of the above set of equations takes into account t = 0 and C°_{DBTO} = $C^{\circ}_{DBTO_2}$ = C°_{2-HBP} = 0, thus the concentrations of key compounds involved in the 4S-route can be expressed according to the following equations; these equations are allowed to obtain values of

$$\int_0^t r_i dt,$$

which means calculating values of r_i at every time directly from experimental data.

$$\int_0^t r_1 dt = -\int_{C^{\circ}_{DBT}}^{C_{DBT}} dC_{DBT} = C^{\circ}_{DBT} - C_{DBT}(t) \tag{72}$$

$$\int_0^t r_2 dt = -\int_{C^{\circ}_{DBT}}^{C_{DBT}} dC_{DBT} - \int_{C^{\circ}_{DBTO}}^{C_{DBTO}} dC_{DBTO} = C^{\circ}_{DBT} - C_{DBT}(t) - C_{DBTO}(t) \tag{73}$$

$$\begin{aligned} \int_0^t r_3 dt &= -\int_{C^{\circ}_{DBT}}^{C_{DBT}} dC_{DBT} - \int_{C^{\circ}_{DBTO}}^{C_{DBTO}} dC_{DBTO} - \int_{C^{\circ}_{DBTO_2}}^{C_{DBTO_2}} dC_{DBTO_2} \\ &= C^{\circ}_{DBT} - C_{DBT}(t) - C_{DBTO}(t) - C_{DBTO_2}(t) \end{aligned} \tag{74}$$

$$\int_0^t r_4 dt = \int_{C^{\circ}_{2HBP}}^{C_{2HBP}} dC_{2HBP} = C_{2HBP}(t) \tag{75}$$

According to the above equations, each reaction can be studied separately from integral data and the concentration of each compound versus time.

Calzada *et al.* (2012) proposed a new extended kinetic model by dividing both the numerator and denominator by the affinity of each reaction (Ki) in order to avoid convergence problems during fitting. The concentrations of compounds involved in a 4S-route can be expressed according to the following equations:

$$C_j - C_j^{\circ} = \sum_{i=1}^{NC} \gamma_{ij} \int_{t_o}^{t} r_i dt \tag{76}$$

$$C_{DBT} - C_{DBT}^{\circ} = -\int \frac{k_1' \exp(-S_1 t) C_{DBT} C_X}{1 + K_1' C_{DBT}} dt \tag{77}$$

$$C_{DBTO} = \int_0^t \frac{k_1' \exp(-S_1 t) C_{DBT} C_X}{1 + K_1' C_{DBT}} dt - \int_0^t \frac{k_2' \exp(-S_2 t) C_{DBTO} C_X}{1 + K_2' C_{DBTO}} dt \tag{78}$$

$$C_{DBTO_2} = \int_0^t \frac{k_2' \exp(-S_2 t) C_{DBTO} C_X}{1 + K_2' C_{DBTO}} dt - \int_0^t \frac{k_3' \exp(-S_3 t) C_{DBTO_2} C_X}{1 + K_3' C_{DBTO_2}} dt \tag{79}$$

$$C_{HBPS} = \int_0^t \frac{k_3' \exp(-S_3 t) C_{DBTO_2} C_X}{1 + K_4' C_{DBTO_2}} dt - \int_0^t \frac{k_4' \exp(S_4 t) C_{HBPS} C_X}{1 + K_4' C_{HBPS} + K_5' C_{2HBP}} dt \tag{80}$$

$$C_{2HBP} = \int_0^t \frac{k_4' \exp(S_4 t) C_{HBPS} C_X}{1 + K_4' C_{HBPS} + K_5' C_{2HBP}} dt \tag{81}$$

where $k_i' = k_i / K_i$ and $K_i' = 1/K_i$.

9.10 Different Kinetic Studies on the Parameters Affecting the BDS Process

Wang *et al.* (1996) predicted a model for *R. erythropolis* N1-36, which was based on the Monod growth kinetic equation employing a continuous experimental system. The kinetic parameters (the specific growth rate of cells, the affinity constant for substrate, and the cell yield coefficient) were determined when the sulfur source (dibenzothiophene-sulfone, $DBTO_2$) is growth limiting.

In an attempt to apply biocatalytic oxidation using enzymes as an alternative to BDS using whole cells, Ayala *et al.* (1998) studied the kinetic properties of chloroperoxidase from *Caldariomyces fumago* with different organosulfur compounds and the chemical nature of products. Enzymatic oxidations were performed in media containing pure substrates, such as thiophenes and organosulfides, in the presence of 0.25 mM hydrogen peroxide. The specific activity of the oxidation of pure organosulfur compounds with chloroperoxidase from *Caldariomyces fumago* were ranked in the following decreasing order: ethyl phenyl sulfide (1725 min^{-1}), thianthrene (1310 min^{-1}), bithiophene (840 min^{-1}), phenyl sulfide (831 min^{-1}), benzothiophene (557 min^{-1}), phenyl disulfide (352 h^{-1}), and DBT (126 min^{-1}), and are transformed to their corresponding sulfones. The specific activity is defined as the number of mol of substrate transformed per 1 mol of enzyme per minute (i.e. min^{-1}). The kinetic constants for the oxidation of thianthrene with chloroperoxidase were also estimated in a 1 mL reaction mixture containing 15% acetonitrile and 20 mM KCl in a 60 mM acetate buffer with a pH 3.0 and 1.5 nM of enzyme. The results revealed that the k*cat* for the oxidation reaction was 64 s^{-1} and the K_m for thianthrene was 90-times lower than for hydrogen peroxide, recording 1.45 and 133 μM, respectively. Moreover, the inactivation constants (k_{in} min^{-1}) for chloroperoxidase determined from a first-order equation, in the presence of different concentrations of hydrogen peroxide, were also calculated and recorded as 0.203, 0.248, and 0.287 min^{-1} with 0.25, 0.5, and 1 mM H_2O_2, respectively. So far, the inactivation mechanism has not been clearly elucidated.

DBT biodesulfurization data of two batch cultures inoculated by *Rhodococcus erythropolis* Ac-1514D and *Rhodococcus* ruber Ac-1513D were processed using the Monod equation (Zakharyant *et al.*, 2004). The specific growth rates (μ) were obtained by the analysis of the linear ranges of the growth curves that were measured as cell protein vs. time in the semi-logarithmic coordinates. The results revealed that the two μ were nearly the same, recording 0.08 and 0.086 h^{-1}, respectively. The obtained values for *Rodococcus* sp. are 2.5 times less than those known so far in literature for *R. erythropolis* (Kaufman *et al*, 1998; Chang *et al.*, 2000). The yield coefficients for those cultures were 0.86 and 5.85 g protein/mmol DBT for *R. erythropolis* and *R. ruber*, respectively, while the specific desulfurization activity was measured by the DBT removal and recorded 185 and 30 mmol DBT/kg DCW/h, respectively. 2-HBP production for *R. erythropolis* was also reported and recorded 80 mmol 2-HBP /kg DCW/h. This was in accordance with those determined for the non-mutant *Rhodococcus* sp. (Honda *et al.*, 1998; Kaufman *et al*, 1998; Kobayashi *et al.*, 2000).

Rhodococcus sp. NCIM2891 has been applied for a batch BDS of hydrodesulfurized diesel oil with varied initial S-concentrations of 200–540 ppm at a different O/W phase ratio (0–100% diesel). NCIM2891 was found to follow the classical Monod type of growth kinetics and the maximum specific growth rate, $\mu_{max,}$ was found to be 0.096 h^{-1} and yielded a coefficient of 0.2 g biomass consumed/g of substrate consumed and the half saturation constant K_s was 71 ppm (Mukhopadhyaya *et al.*, 2006).

Del Olmo *et al.* (2005a) studied the effects of media composition on the BDS process using *Rhodococcus erythropolis* IGTS8, which proved great variations in biomass growth and, consequently, the BDS efficiency with the changes in media compositions. First, a kinetic model representing the growth curves, taking into account the composition of the media, was proposed to compare the growth rate in different media compositions throughout the values of the usual parameters employed to describe biomass growth, such as the specific rate (μ) and the maximum biomass concentration reached (C_x^{max}). To evaluate the possibilities of different sources, the biomass yields of the different substrates (that is C, N and S-sources, Y_{CX}, Y_{NX}, and Y_{SX}) have been calculated. Moreover, the desulfurization capability of the cells (X_{BDS}) at different growth times has been evaluated using a desulfurization test of DBT, taking into account only the concentration of the produced 2-HBP. Del Olmo *et al.* (2005a) also proposed a non-structured kinetic model, which is able to describe growth and desulfurizing capability with good statistical parameters and was applied by a non-linear simple response. A parameter considering the maximum desulfurizing capability of cells obtained during growth cycle is defined that involved both the biomass concentration attaining their BDS capability and the time needed to reach this concentration (D_{BDS}). The advantage of this parameter is its ability to show the best medium composition to obtain biodesulfurizing cells of *R. erythropolis* IGTS8 and its ability to be applied to compare the results of different microorganisms produced under different operational media and conditions.

For all the experiments carried out, the yields of substrates (carbon, nitrogen, and sulfur source) into biomass have been calculated, as follows:

$$Y_{jx} = \frac{C_{j_o} - C_j^{X_{max}}}{C_x^{max} - C_{x_o}} \quad (j = C, N, S) \tag{82}$$

Two parameters have been defined in order to calculate the desulfurizing capability of the cells.

The percentage of desulfurization, measured as an HBP conversion, X_{BDS};

$$X_{BDS} = \frac{C_{2HBP}}{C_{DBT,0}} \times 100 \tag{83}$$

To quantify the desulfurizing capability, taking into account all the variables influencing it (desulfurization percentage obtained at each time by resting cells, together with the biomass concentration reached), the desulfurizing development grade during growth (D_{BDS}) was calculated as follows:

$$D_{BDS} = \frac{X_{BDS} C_x}{t_G} \tag{84}$$

where $C_{DBT,0}$ is the initial DBT concentration employed to perform the standard BDS resting cell assay, C_{2HBP} is the concentration of produced 2HBP obtained at the end of the experiment (in that study, 3 h was chosen) of resting cell assay, t_G is the time of growth employed to reach a biomass concentration, C_X, and X_{BDS} is the percentage of biodesulfurization reached by these cells in the standard test of resting cell assay.

The experiments revealed that the parameter $Y_{CX,}$ obtained with citrate was significantly lower than those obtained for the other two C-sources (glucose and glutamic acid). The values obtained for Y_{NX} with glucose and citrate were very similar and lower than those obtained using glutamic acid. The highest value for Y_{SX} was obtained when citrate was employed as a C-source and the values obtained with glucose and glutamic acid were nearly the same. The X_{BDS} was nearly the same with the three studied C-sources, reaching approximately 75%, but the time to reach that maximum value varied, recording 22 h upon the usage of glucose, but 30 h when using glutamic acid or citrate. The D_{BDS}^{max} recorded 5.16, 3.76, and 2.72 using glucose, glutamic acid, and citrate, respectively. The presence of ammonium-nitrogen was found to increase the biomass concentration and expressed higher influence on growth than BDS-activity. However, upon the addition of NH_4Cl, the X_{BDS} reached 80% and decreased to 20% in its absence, recording D_{BDS}^{max} of 3.76 and 0.93, respectively. The growth was faster with DMSO and sulfate, relative to that with DBT. However, the highest value for Y_{CX} was obtained in the experiment carried out using magnesium sulfate while the lowest value was obtained with DMSO as a sulfur source, while the highest values of Y_{NX} and Y_{SX} occurred when employing DBT. Moreover, upon the calculation of X_{BDS}, it was observed that upon the usage of sulfate, the expression of 4S-route genes and desulfurizing capability throughout the growth were suppressed. A maximum value of X_{BDS} (12%) was reached with the cells grown around 25 h when

DMSO was used, but upon the usage of DBT, cells with the same growth time yield had an XBDS value of 9%, but the desulfurizing capability stayed at a constant value for approximately 25 h. The values of Y_{SX} increased with the increase in initial DMSO-concentration. However, the change in initial DMSO concentration expressed no significant effect on Y_{NX}, but the highest Y_{CX}, 9.72 and lowest one occurred at 50 μM and 250 μM, initial DMSO, respectively.

Furthermore, experimental data of biomass concentration with time were fitted to the following kinetic model:

$$\frac{dC_x}{dt} = \mu C_x \left(1 - \frac{C_x}{C_x^{max}}\right) \tag{85}$$

As the integration of the model introduces much less error than the data differentiation (Garcia-Ochoa *et al.*, 1992), thus, upon the integration of the above model, with the initial condition $t = 0$ and $C_x = C_{x_o}$, the following logistic equation was obtained:

$$C_x = \frac{C_{x_o} \exp(\mu t)}{1 - (C_{x_o} / C_x^{max})(1 - \exp(\mu t))} \tag{86}$$

The fitting has been carried out by non-linear regression using the Fischer and Student's tests to evaluate the quality of fitting. Moreover, the sum of square residuals has been used as representative of the fitting:

$$SSR = \frac{\sum_{i=1}^{NE} \left(C_{x_i}^{exp} - C_{x_i}^{cal}\right)^2}{NE} \tag{87}$$

The BDS-model is a modified one for that proposed by Luedeking-Piret (Luedeking and Piret, 1959):

$$\frac{dX_{BDS}}{dt} = \alpha \frac{dC_x}{dt} - \beta C_x \tag{88}$$

with the boundary conditions: t = 10 h and XBDS = 0, which means to take into account a delay time in the development of the desulfurization capability of the cell. Experimental data has been fitted using a fourth order Runge-Kutta algorithm, coupled to a simple-response non-linear algorithm (Marquardt, 1963). The model was able to fit all the runs carried out, showing good results for statistical parameters (Student's t-test and Fischer's F-tests). The value of the fitting parameter "α", considering the production of desulfurizing capability as growth associated, is the highest when glucose was employed. The tendency to decrease this ability during a stationary growth phase (indicated by parameter β) is slightly higher when glucose was employed than that obtained using glutamic acid. The presence of ammonium when glutamic acid was employed also has a clear influence and the value of parameter α presents a value three times higher when ammonium was employed and the decreasing of the desulfurization capability (parameter β) was higher when ammonium was added to the media. When DBT was used as a sulfur source, the value of parameter "α" was lower than that obtained using DMSO. The change in initial DMSO concentration also expressed a clear effect on the values of this parameter; when the higher concentration of DMSO was employed, the higher parameter values were obtained. Thus, the development of the desulfurizing capability of the cells, described as associated to growth (α value), presented the highest value using an initial concentration of 1300 M of DMSO. Although the parameter "β" also had the highest value for an initial concentration of 1300 M of DMSO, the D_{BDS} values always remained higher than that obtained using other concentrations of DMSO.

In a similar study, del Olmo *et al.* (2005b) studied the effect of operational temperature, pH, and dissolved oxygen concentration on the microbial growth of *Rhodococcus erythropolis* IGTS8 and its BDS capability (i.e. the biodesulfurizing degree D_{BDS}). The yields of different substrates into biomass (Y_{iX}), the percentage of desulfurizing capability of the cells (X_{BDS}), and the desulfurization development degree during growth (D_{BDS}) were calculated. The stationary growth phase was attained faster at temperature values from 30 to 32 °C. The highest and lowest cell concentrations occurred at the stationary phase were recorded at 26 °C and 36 °C, respectively. Moreover, the lowest Y_{CX}, $Y_{NX,}$ and Y_{SX} were recorded at 28 °C. Moreover, the highest BDS percentage of 70% of initial DBT concentration 250 M was recorded at a temperature interval from 28 to 32 °C. This was reached within 24 and 30 h from the beginning of the growth. At the late exponential phase, the BDS percentage recorded a maximum of 20% at 26 °C and 36 °C. The maximum biodesulfurization degree (D_{BDS}^{max}) of 5.16 was observed after 24 h of growth at 30 °C. The growth was faster and the biomass concentration was higher when pH was controlled with diluted NaOH at a pH value of 6.5. However, when the pH was not controlled, the lowest Y_{CX}, $Y_{NX,}$ and Y_{SX} were recorded. Also, the highest BDS efficiency was recorded when pH was controlled to 6.5 either by using diluted NaOH or tris-HCl, recording D_{BDS}^{max} 5.15 and 5.16, within 29 or 24 h of growth time, respectively. Both the growth rate and obtained maximum cell concentration increased with the increase of dissolved oxygen concentrations, recording its maximum at 20% of saturation which indicated oxygen limitation during growth. A maximum desulfurization capacity of ≈ 75% occurred within 24 h of growth at 20% dissolved oxygen concentration. Based on a logistic equation, the highest specific growth rate (μ = 0.202 h^{-1}) was obtained at 30 °C and the C_x^{max} parameter decreased with the increase in temperature, recording 2.27 and 0.81 g X/L at 26 °C and 36 °C, respectively.

Both kinetic parameters μ and C_x^{max} are functions of temperature and can be described as follows:

$$\mu(T) = \{C_1(T - T_{min})[1 - exp(C_2(T - T_{max}))]\}^2 \tag{89}$$

$$C_x^{max}(T) = (a - b)T \tag{90}$$

where C_1 and C_2 are parameters of Ratkowsky (Ratkowsky *et al.*, 1983) and a and b are parameters of linear equations. The biomass evolution with time was predicted by the six parameters listed above.

The specific growth rate μ h^{-1} was significantly affected by pH, however the highest C_x^{max} 2.25 g X/L was obtained at pH 6.5 and that was controlled by diluted NaOH, while the lowest 1.43 g X/L was obtained with an uncontrolled pH of 5.5. Furthermore, the specific growth rate of μ h^{-1} and C_x^{max} were found to be increased with the increase of dissolved oxygen concertation, recording their highest values of 0.409 h^{-1} and 4.27 g X/L at a 20% saturation value and its lowest value, 0.202 h^{-1}, was recorded at a 10% saturation value.

The model describing the development of desulfurizing capability during growth that considers X_{BDS} as an associated product to growth and had a production time delay, presented a decreasing tendency as growth finish was used.

$$\frac{dX_{BDS}}{dt} = \alpha \frac{dC_x}{dt} - \beta C_x \tag{91}$$

The experimental data was fitted by using a fourth order Runge–Kutta algorithm to integrate it, with the following boundary condition: $X_{BDS} = 0$ for t = 10 h, coupled to a simple-response non-linear algorithm (Marquardt, 1963). Model parameter values, α and β, that represent rates of biodesulfurization capability development and losing, respectively, were calculated.

It was found also that α recorded its maximum value at 127.79 at 30 °C and parameter β recorded zero at 26 °C, which was attributed to the maximum BDS capability that could not be reached during the growth time employed. The lowest value of α was obtained at an uncontrolled pH of 5.5 and recorded β, since a maximum BDS capability has not been reached during growth time. However, high α values were recorded over the studied range of dissolving oxygen. However, β recorded the lowest value when dissolved oxygen concentration was kept at 10% saturation, which suggested that 10% saturation was a better choice than 20% saturation since a high BDS capability was kept during more growth time.

A similar study was performed by Martin *et al.* (2005) to study the influence of different working conditions on the growth and BDS capacity of the genetically modified *Pseudomonas putida* CECT5979 desulfurizing DBT to 2-HBP. The studied parameters were pH conditions (buffered and non-buffered media) using different carbon sources (glucose, citrate, and glutamic acid), operating temperatures (26–32 °C), and different dissolved oxygen concentrations due to different aeration conditions (different air flows 1 and 3 L/min, using enriched air of 75% in oxygen). For achieving the optimum operating conditions, that is to obtain desulfurizing cells, a parameter (D_{BDS}) that incorporates both biomass concentration and time to reach a particular percentage of desulfurizing capability (X_{BDS}) has been used. This was done by applying the previously mentioned model equations (eqs. 82–84). The effect of different C-sources and changes in pH were performed together and revealed that the growth was better when the microorganism was grown in a non-buffered medium, attaining a value of 4.5 g/L. The maximum biomass concentration reached at the stationary phase was higher when the growth was conducted with glutamic acid as a carbon source in a non-buffered medium. On the other hand, the lowest final concentration of biomass was obtained when citrate was used as a carbon source in a buffered medium, reaching a value of 2.5 g/L. Moreover, the pH increased with time when citrate and glutamic acid were used as carbon sources, but decreased to 4.90 and 5.92 with glucose in non-buffered and buffered BSM, respectively, where the highest Y_{CX}, 3.26, was obtained in the run performed with citrate as the carbon source in a buffered medium. The parameter Y_{NX}, 0.63, obtained with glutamic acid in a buffered medium, was higher than that obtained for other experiments, while the highest X_{BDS}, 64%, was obtained when glutamic acid is used as the carbon source in a non-buffered medium which also recorded the highest D_{BDS} of 21.65 at 30 °C. The lowest value of X_{BDS}, 5%, was obtained with glucose in a non-buffered medium, however it increased to 20% in a buffered medium. Upon studying the effect of temperature, it was depicted that the maximum cell concentration was nearly the same in all temperatures, but lowered at 32 °C. The highest value of the parameter X_{BDS} was obtained when the microorganism grew at 30 °C in the exponential phase of growth (t_G of 9 h) and conserved up to 15 h of growth; after that point, the desulfurizing capability decreased to 40%. The growth and BDS activity increased with the increase of the concentration dissolved oxygen in the growth medium. The highest biomass concentration was obtained when enriched air was used. Nevertheless, although the highest Y_{CX} occurred with enriched air, the lowest Y_{NX} occurred with enriched air. With an air flow of 1 L/min, the X_{BDS} reached approximately 80%. The growth was fitted with a logistic model equation (eq.86) and the highest C_x^{max} of 4.70 g/L was also obtained when glutamic acid was used as the carbon source in a non-buffered medium which gave the highest D_{BDS} of 21.65 at 30 °C. Moreover, the growth parameters, specific growth rate μ (h^{-1}) and maximum biomass concentration C_x^{max} (g/L), increased with the increment of dissolved oxygen concentration recorded at 0.73 h^{-1} and 6.23 g/L, respectively, in an experiment that was performed with enriched air and a flow rate of 3 L/min. Furthermore, upon applying the desulfurization model (eq. 91), the following observations were recorded: α, which considers the production of desulfurizing capability to be growth-associated, was found to be similar when glucose and citrate were used as carbon sources in buffered and non-buffered media, however, when glutamic acid was used, α decreased from 33.72 (in a non-buffered medium) to 18.77 (in a buffered medium). The tendency to decrease the desulfurizing capability during the stationary growth phase (indicated by the parameter β) was found to be higher when glucose was used as the carbon source in a non-buffered medium. The temperature highly influenced the BDS capacity of *P. putida* CECT5279. The α recorded its highest value, 33.72 at 30 °C and its lowest value, 10.52 at 26 °C. The highest value of β, 1.77, was also recorded at 30 °C. This would give an indication that the tendency to decrease the desulfurizing capability of *P. putida* CECT5279 is higher when the cells grow at 30 °C. Further, α and β values were higher at an air flow rate of 1 L/min than those recorded applying other air flow rates or upon applying enriched air.

Upon modeling and simulation of both, the growth of *R. erythropolis* HN2 and consumption of different concentrations of DBT as a function of incubation time, the logistic model equation was found to accurately describe the change in biomass concentration with time using different initial DBT concentrations. Also, the kinetic model equations relating to the rate of growth of microorganisms and that of DBT consumption with the yield coefficient calculated from the Monod equation allowed the prediction of the DBT–time profile along with microbial growth under the predicted optimal operating conditions, although a slight decrease in the specific growth rate, μ h^{-1}, with an increase of initial DBT concentrations was observed. However, the specific desulfurization rate was found to increase with the increase of initial DBT concentration, recording its maximum at 1000 ppm, where regardless of the initial DBT-concentration, the maximum desulfurization rate occurred at a 72 h incubation period. The recorded maximum specific growth and degradation rates were 0.06667 and 0.8818 h^{-1}, respectively. Thus, that study proved the strong influence of S-concentration on the rate of BDS, the well adaptation of *R. erythropolis* HN2 towards high DBT concentrations, and its ability to overcome the feedback inhibition of sulfate and 2-HBP (Nassar *et al.*, 2017). This promotes the application of *R. erythropolis* HN2 in the petroleum desulfurization process.

In another study performed by Dejaloud *et al.* (2017), a logistic growth model (eq.86) combined with a Pirt model were applied to predict the importance of the maintenance energy during a batch BDS of different initial concentrations of DBT (0.03, 0.05, 0.08, and 0.11 mM) at two levels of glucose-concentrations: energy-limited medium 55 mM and energy-sufficient medium 111 mM using growing cells of *Ralstonia eutropha* (PTCC1615). Fitting the experimental data to the logistic model was performed using SigmaPlot 12.3 software considering Levenberg–Marquardt as a nonlinear regression algorithm. The lowest values of the sum squares of the differences between the values calculated by the model (Y_{cal}) and the experimental data (Y_{exp}) were determined $\{SSR = \sum(Y_{exp} - Y_{cal})^2\}$. The standard error of the estimate, $S_{y,x} = \sqrt{SSR/df}$, was described where *df* equals the number of data points minus the number of fitted parameters. The coefficient of determination ($R^2 = 1 - (SSR/SS_{total})$) which gives the ratio of the sum squares of the residuals to sum squares of the total and shows the extent of closeness of the experimental data to the test model showing the goodness of fit was also obtained. The confidence limits for the estimated parameters were expressed in terms of a 95% confidence interval and determined as follows: estimated parameter $\pm t_{0.975,df} \times$ the approximated standard error, where $t_{0.975,df}$ is the 0.975^{th} quantile of *t*-statistic distribution with *df* degrees of freedom (Lapin, 1997). Additionally, the quality of the data fitting process was quantitatively explained using a *t*-test which has been defined as the ratio of the coefficient to the relevant standard error. The *F*-test statistic was also performed to evaluate the contribution of the independent variables in predicting the dependent variable (Lapin, 1997). If *F* is a large number, it is possible to deduce that the independent variables contribute to the prediction of a dependent variable. All the data revealed good fitting of the model. Moreover, it was found that the increase in the initial glucose concentration from 55 to 111 mM had no significant effect on the maximum cell growth, C_x^{max}, and minor variations were observed for the maximum levels of the cell content (2.32 to 2.87 g/L). The maximum carrying capacity (C_x^{max}) of the A and B media appears to be similar, but the specific growth rate for the energy-sufficient media was lower than those for the energy-limited cultures of DBT, while the opposite occurred in the control media prepared with ammonium sulfate. The reported high values of μ (0.32–0.63 h^{-1}) indicated the cells' efforts to be directed toward reproduction and the cells, apparently, are fully capable of using their biotic potential. Moreover, there was an observed difference in the efficiency of substrate consumption and growth and it was described in terms of the maintenance energy by which energy is used for non-growth functions (Pirt, 1965; Russell and Cook, 1995; van Bodegom, 2007). To quantitatively explain the Pirt concept with reference to the overall rate of glucose concentration $(ds/dt)_T$, two terms are important: that for growth $(ds/dt)_G$ and that for cell maintenance $(ds/dt)_M$ (Pirt, 1965):

$$\left(\frac{dS}{dt}\right)_T = \left(\frac{dS}{dt}\right)_G + \left(\frac{dS}{dt}\right)_M \tag{92}$$

$$\left(\frac{dS}{dt}\right)_T = -\frac{1}{Y_G}\frac{dC_x}{dt} - mC_x \tag{93}$$

By dividing eq. (93) by eq. (85), the following expression is obtained:

$$\frac{dS}{dt} = -\frac{1}{Y_G} - \frac{mC_x^{max}}{\mu(C_x^{max} - C_x)} \tag{94}$$

Integration of eq. (94) between C_{x_o} and C_x yields:

$$S = S_o - \frac{1}{Y_G}(C_x - C_{x_o}) + \frac{mC_x^{max}}{\mu} ln\left(\frac{C_x^{max} - C_x}{C_x^{max} - C_{x_o}}\right) \tag{95}$$

The values of the bioenergetic constants, maintenance coefficient (m) and growth yield (Y_G), for different cultures were estimated and could properly describe the bioprocess. For example, in a control system in the presence of glucose and NH_4SO_4, the value of m recorded 0.13 and 0.34 $g_{glucose}/g_{cell}/h$ for case A and B, respectively. This indicated that the maintenance coefficient

(m) is higher for the energy-sufficient culture than for the energy-limited one, but they recorded the same growth yield 0.26 $g_{cell}/g_{glucose}$. The results also indicated that 2-HBP production was higher for energy-sufficient cultures, while the values of the specific growth rate and the maintenance coefficient for these media were lower than those of the energy-limited cultures. m and Y_G values, in limited energy case A, in the presence of DBT were in the range 0.45–0.62 $g_{glucose}/g_{cell}$/h and 0.58–0.88 $g_{cell}/g_{glucose}$, respectively, and were higher than those recorded in the case of sufficient energy, case B, ranged between 0.31 and 0.35 $g_{glucose}/g_{cell}$/h and from 0.21 to 0.28 $g_{cell}/g_{glucose}$, respectively. Thus, the Y_G values for the "A" cultures were higher than those for the "B" cultures ($Y_{G,As} > Y_{G,Bs}$) and the contrary occurred for the "m" coefficient, which was lower for the "B" than for the "A" cultures ($m_{As} > m_{Bs}$). This was explained as the excessive energy in cells controlled by the efficient regulation on substrate (glucose) uptake rather than energy spilling activities (Tsai and Lee, 1990) and the cellular pools of the produced NADH and $FMNH_2$ coenzymes in "B" cultures should be directed toward the biodesulfurization process and glucose uptake regulation could be dominant. Not only this, but it was very obvious from the obtained m and Y_G in sufficient-energy cultures with DBT (case-B) that the increase of initial DBT concentrations did not affect the m or Y_G values. This indicated no inhibition effect within the studied initial DBT-concentrations. Furthermore, the maximum 2-HBP production in "A" and "B" treatments was about 0.05 mM and inhibition effect of products on the cell growth was not noticeable. The Haldane equation (96) was employed to explain and represent the inhibitory effect of high initial concentrations of DBT,

$$Q_{DBT} = \frac{Q_{max} C_{DBT,0}}{K_S + C_{DBT,0}\left(1 + \frac{C_{DBT,0}}{K_I}\right)} \tag{96}$$

where the Q_{max} is the maximum value of the specific desulfurization rate and K_S and K_I are the half-saturation constant and self-inhibition constant, respectively. The results revealed that with increasing the energy-supplying substrate, the Q_{max} and K_I values kept approximately constant, recording 0.002 mmol/g_{cell}/h and 0.076 and 0.076 mM for "A" and "B" cultures, respectively, but the K_S value decreased from 0.1594 mM to 0.0782 mM. Thus, the kinetic studies showed that the half-saturation constant K_S for the energy-limited cultures was 2 times higher than the energy-sufficient ones. The data also indicated that the increase in the carbon and energy source resulted in modifying the affinity of the enzymes for the BDS-process. The maximum specific desulfurization activities of the growing cells in these cultures were predicted at 0.002 mmol/g_{cell}/h. Further tests on resting cells showed that the cells harvested from the mid exponential growth phase had the highest desulfurization activity where the specific desulfurization rate was 20 times higher than the growing cells at the same conditions. The desulfurizing capability index (D_{DBT} %g_{cell}/L/h) was calculated using the following equation:

$$D_{DBT} = \frac{Y_{DBT} \times C_x}{t} \tag{97}$$

where both the biomass concentration (C_x) and time (t) to reach a particular percentage of desulfurizing capability are included in the D_{DBT} parameter and recorded 23% g_{cell}/L/h at an initial DBT concentration of 0.5 mM.

The initial biomass concentration has also had an important effect on the BDS process. Caro *et al.* (2007a) studied the effect of the addition of β-cyclodextrins and different initial biocatalyst concentrations over the BDS process yielded for model oil (50 ppm DBT in –n-hexadecane) in a biphase system (1:1 oil:water), using growing cells of *R. erythropolis* IGTS8 and a 2 L STBR. The BDS-yield (X_{BDS}) was measured as the percentage ratio between the concentration of produced 2-HBP and the initial DBT concentration. This study proved that the addition of 15 ppm β-cyclodextrin in the BDS reaction increased the rate of production of 2-HBP and XBDS, while the DBT removal rate was not affected. The BDS yield increased with the increase of the initial biomass concentration, but up to a certain limit; 2 g DCW/L and the produced 2-HBP decreased at a higher initial biomass concentration.

Caro *et al.* (2007a) applied another form of the Haldane equation to describe the BDS process and the following equations successfully simulated the removal of DBT and production of 2-HBP:

$$\frac{dX}{Xdt} = \mu = \mu_{max} \cdot \left(\frac{S}{K_s + S + \frac{S^2}{K_I}} \right) \tag{98}$$

$$\frac{dS}{Xdt} = -\left(q_{Smax} \cdot \left(\frac{S}{K_s + S + \frac{S^2}{K_I'}} \right) \right) \tag{99}$$

$$\frac{dP}{dt} = -\frac{dS}{dt} \cdot \left(\frac{M_{HBP}}{M_{DBT}} \right) \tag{100}$$

where X is the biomass concentration g/L and S is the substrate concentration g/L and the maximum specific growth rate μ_{max} recorded 70 h^{-1} and the specific maximum desulfurization rate recorded 0.025 gDBT/gDCW/h with an inhibition rate constant of K_I and K_I' of 2 and 2.5 g/L, respectively. The value of the saturation constants, K_s, decreased as the biocatalyst initial concentrations went up, reaching stability from 2 gDCW/L.

Deriase *et al.* (2013) evaluated the biodegradation kinetics of the toxic 2-hydroxybiphenyl (2-HBP) and 2,2′-bihydroxybiphenyl (2,2′-BHBP) with a different initial concentrations range of S_0 (5–50) ppm using suspended cultures of *Corynebacterium variabilis* Sh42 with a constant initial biomass concentration, X_0 315.8 ppm, in a series of batch experiments. The growth kinetics of *C. variabilis* Sh42 does not follow simple Monod's kinetics. The cultures followed substrate inhibition kinetics. By fitting specific growth rates μ (h^{-1}) on suitable substrate inhibition models, evaluating the bio-kinetic constants that are necessary to understand the kinetics of biodegradation process were evaluated, which confirmed good tolerance, growth, and degradation capabilities of Sh42 on the studied concentration ranges of 2-HBP and 2,2′-BHBP. Although Haldane and Yano and Koga bio-kinetic equations for substrate inhibition seemed to be the best adequate expressions for specific growth rates on 2-HBP and 2,2′-BHBP, respectively, with the highest R^2 at 1.0 and 0.997, respectively and the least RMSE values at 6×10^{-5} and 9×10^{-5}, respectively, an evident disagreement was observed between experimental and simulated profiles for bacterial growth, X (mg/L), and substrate concentration, S (mg/L). A new proposed model based on a modified Haldane equation gave better results.

For a batch biodegradation process, the following differential equations (derived from mass balance considerations) are often used for describing both biomass growth and substrate consumption.

$$\frac{dX}{dt} = \mu X \tag{101}$$

The change in substrate concentration can be defined by:

$$\frac{dS}{dt} = -qX \tag{102}$$

where q (h^{-1}) is the specific substrate consumption rate. The relation between q (h^{-1}) and μ (h^{-1}) can be represented by:

$$q = \mu / Y_{x/s} \tag{103}$$

The relationship between biomass formation and substrate consumption can be approximately determined by the yield coefficient, $Y_{x/s}$ (dry weight of biomass / weight of utilized substrate), indicating the maximum conversion of unit substrate to cell mass (Okpokwasili and Nweke, 2005).

$$Y_{x/s} = -\frac{dX/d_t}{dS/d_t} = -dX/dS = -\frac{dX}{dS} = -\frac{X - X_o}{S - S_o} \tag{104}$$

By introducing the decay rate coefficient [b]:

$$\frac{dX}{dt} = \mu X - bX \tag{105}$$

By introducing a constant cell decay rate coefficient b (0.001 h^{-1}), there is still an evident disagreement between measured and simulated profiles observed with an average maximum percentage deviation of 22 and 44% for cell growth and 63 and 67% for substrate degradation, respectively.

However, by dividing the time span of the biodegradation process into subintervals, according to the assumption of the production and accumulation of metabolic intermediates during the process, the decay rate coefficient b′ (h^{-1}) is changed for these subintervals according to logical IF statements included in the implemented computer program.

μ (h^{-1}) was represented by a modified Haldane equation given by:

$$\mu = \frac{\mu'_{\max} S}{K'_s + S + f(i)} \tag{106}$$

where $f(i)$ is analogous to substrate inhibition term (S^2/K_i) in the classical Haldane equation. $f(i)$ represents the functional relationship of effect of metabolite intermediates on the hydroxybiphenyls (HBPs) biodegradation process. It is expressed as follows:

$$f(i) = \frac{(S_o - S)^2}{K'_i} \tag{107}$$

These equations were coupled with equation (102), describing the change in HBPs concentration with time. Since the kinetic model of Yano and Koga (eq.30) has no substrate inhibition term (S^2/K_i), the new proposed model was applied for both 2-HBP and 2,2′-BHBP.

To estimate the degree of toxicity of 2-HBP and 2,2′-BHBP on *Corynebacterium variabilis* Sh42, the Haldane model, which is the most widely used model, was applied. The two bio-kinetic constants are Ki, the inhibition constant which is a measure of sensitivity by inhibitory substances, and Ks, the half saturation constant which is defined as the substrate concentration at which μ equals half of μ_{max}. Since the studied HBPs expressed inhibitory effects on Sh42, according to Nuhoglu and Yalcin, 2005, if the substrate is inhibitory, it is not possible to observe an actual μ_{max}. Thus, Ks could be taken on a hypothetical meaning. It has been shown that the Haldane equation would go through a maximum value of dμ/ds = 0 at substrate concentrations of S* ppm for inhibitory substrates and it is the concentration at which the microorganisms exhibited their maximum utilization rate, $S^* = (Ks.Ki)^{1/2}$ and the corresponding μ value is

$$\mu^* = \frac{\mu_{max}}{\left[2(K_s.K_i)^{1/2} + 1\right]}.$$

This reflects that the degree of inhibition is determined by the [K_s / K_i] ratio and not just by Ki alone. The larger the ratio of [K_s/K_i] is, the smaller the μ* (h^{-1}) relative to μ_{max} and the lower the degree of inhibition. From that work, it can be concluded that the maximum specific growth rate on 2,2-′BHBP (0.053 h^{-1}) was greater than that on 2-HBP (0.045 h^{-1}). Considering the fact that K_s ppm is inversely proportional to the affinity of the microbial system for the substrate, Sh42 showed higher affinity to 2-HBP (Ks =0.894 ppm) than that of 2,2′-BHBP (Ks = 1.88 ppm). The S* for 2-HBP was smaller than that of 2,2′-BHBP (8.21 and 11.45 ppm, respectively). The [K_s/K_i] ratio of 2-HBP is smaller than that of 2,2′-BHBP (0.01 and 0.027, respectively) and the corresponding μ* (2.58 x 10^{-3} and 2.22 x 10^{-3}, respectively). These data indicated that the toxicity and inhibition effects of 2-HBP on Sh42 are higher than those of 2,2′–BHBP (Deriase *et al.*, 2013).

From the quantitative discussion of modeling a relationship between cell growth rate and substrate consumption rate, it was found that the direct coupling of the substrate consumption rate, with the cell growth model is warranted only under certain conditions (e.g. constant cell yield).

However, for HBPs degradation over a wide range of initial substrate concentrations, the cell mass yield was found to vary. Variations of the biomass decay rate coefficient during the time course of batch experiments while adjusting the interactive changeable biomass decay rate coefficient b′, it was found to be in the range 0.0001–0.007 h^{-1} and were found to exert a great influence on the biodegradation process, with the assumption of the existence of some metabolic intermediates that would exert some inhibition on HBPs degradation. Correlation and simulation studies using a new proposed model based on modified the Haldane equation were established and these factors were taken into consideration in the proposed HBPs degradation model. The proposed model is capable of describing growth and HBP degradation profiles very well over the studied initial HBP concentrations (5–50 mg/L). The average maximum percentage deviation reached 6.2 and 7.2% for cell growth and 11 and 10.2% for substrate degradation in the case of 2-HBP and 2,2′-BHBP, respectively (Deriase *et al.*, 2013).

In another study, biodegradation kinetics of different polyaromatic sulfur heterocyclic compounds (PASHs), including thiophene (Th), benzothiophene (BT), dibenzothiophene (DBT), 4- methydibenzothiophene (4-MDBT), and 4,6-dimethyldibenzothiophene (4,6-DMDBT), with different initial concentration ranges of S_o of (100–1000 mg/L), employing suspended cultures of *Bacillus sphaericus* HN1 with initial concentrations of $X_{o,}$ ranging from 291.90–362.01 mg/L dry weight, in a series of batch experiments were investigated (Deriase and El-Gendy, 2010). A mathematical model was predicted to describe the biodegradation kinetics of the studied PASHs using HN1. The predicted model is based on the Haldane bio-kinetic equation (eq. 98) for substrate inhibition which is applied to describe the dependence of the specific growth rate μ (h^{-1}) on the intial substrate concentration, S_o. The Haldane equation seems to be an adequate expression for the cell growth data and the bio-kinetic constants obtained were maximum specific growth rates of μ_{max} = 0.165, 0.231, 2.461, 0.207 h^{-1}, and 0.202 for Th, BT, DBT, 4-MDBT, and 4,6-DMDBT, respectively, a saturation constant of K_s = 3.007, 18.425, 2004.25, 42.25, and 103.43 mg/L for Th, BT, DBT, 4-MDBT, and 4,6-DMDBT, respectively, while the inhibition constant, K_i = 2110.42, 1752.42, 46.849, 2242, and 360.61 mg/L for Th, BT, DBT, 4-MDBT, and 4,6-DMDBT, respectively. This data indicated that the toxicity and inhibition effects of these PASHs on HN1 can be ranked in the following order: Th > BT > DBT. This was attributed to the water solubility of these PASHs, as with the increase of molecular weight and aromatic rings, water solubility decreases. Inhibition phenomenon is often observed when the inhibitory compound is more soluble than other substrates

(Nadalig *et al.*, 2002). Similar observations of a higher toxicity of Th than DBT were reported for *Agrobacterium* MC 501 and the mixed culture XACO (Constanti *et al.*, 1996). Inhibitory effects of BT at higher concentrations were also reported by Van Hamme *et al.* (2004) and Kirkwood *et al.* (2007). Moreover, the toxicity and inhibition effects of these three PASHs on HN1 can be ranked in the following order: 4-MDBT > 4,6-DMDBT > DBT. Kirkwood *et al.* (2007) reported that methylation decreases aqueous solubility, which might explain the lower toxicity effect of 4,6-DMDBT relative to 4-MDBT, but the lowest toxicity effect occurred by DBT, which might be explained by the adaptation of HN1. Regardless of the type of PASHs, growth decreased with an increase in initial substrate concentration. Experimental results have also made it clear that biocatalyst growth was always stopped before the complete removal of the PASHs. This could be attributed to the biodegradation process itself, which might produce toxic metabolites to the cell, but the time profile of each PASH, regardless of the PASHs type and So, expresses the degradation trait at increasing levels during the bacterial growth cycle even after reaching its maximum growth. This indicated that the stationary phase cells have the ability to continue degrading PASHs.

The proposed mathematical model can be described by the following equations:

$$\frac{dS}{dt} = -qX^c \tag{108}$$

$$\frac{dX}{dt} = \mu X^c - bX^c \tag{109}$$

where b (h^{-1}) is the decay rate coefficient found to be in the range 0.0001-0.009 and c is the power of microbial concentration found to be in the range of 0.4–0.6.

The predicted model simulation curves for bacterial growth, X (mg/L), and substrate concentration, S (mg/L), are derived by simultaneously solving the resulting ordinary differential equations (ODEs), 108 and 109 together, with the explicit equations, 103 and 104. These differential equations were numerically solved using POLYMATH 6.10 (professional version). The program used the Felhberg fourth-fifth order Runge-Kutta (FRK45) numerical integration method. The predicted model was adequate in reflecting the PASHs concentration profile. The maximum specific degradation rate (q_{max}, h^{-1}) recorded 0.042, 0.063, 1.53, 0.088, and 0.053 h^{-1} for Th, Bt, DBT, 4-MDBT, and 4,6-DMDBT, respectively. The degradation rates ranked in the following increasing order: Th < BT < DBT and 4,6-DMDBT < 4- MDBT < DBT. The highest degradation efficiency recorded in DBT cultures indicated the adaptation of HN1 biodegrading enzymes towards DBT, as it was previously isolated and enriched on DBT.

Nadalig *et al.* (2002) reported that the presence and position of the methyl group on the PAH molecules govern degradation rates of the compounds. Also, Kirkwood *et al.* (2007) suggested that steric hindrance influences enzymatic activity. This could explain the lower biodegradation that occurred in 4,6-DMDBT cultures compared to those occurring in 4-MDBT cultures, although the toxicity and inhibition effect of 4-MDBT on HN1 was higher than that of 4,6-DMDBT. Fedorak and Westlake (1983 and 1984) reported that, in general, the more the number of alkyl carbons, the more recalcitrant the compounds. Lu *et al.* (1999) reported the preference of *Shingomonas paucimobilis* strain TZS-7 for DBT rather than 4,6-DMDBT. Results from the study performed by Deriase and El-Gendy (2010) are consistent with the results reported by Kropp *et al.* (1997); C1-DBTs are more susceptible to biodegradation than C2-DBTs. Thus, in environments contaminated with crude oil, DBT and methyl-DBTs will be depleted before the isomers of dimethyl-DBTs.

Guchhait *et al.* (2005a and b) also described the evolution of BDS by growing cells of *Rhodococcus* sp. JUBT1 in biphasic media by a typical Haldane equation, as the substrate inhibits both biocatalyst growth and the 4S-pathway. For each OSC and diesel oil with a sulfur content of 500 ppm, having a C1 –DBT of 40 wt.%, C2-DBT of 30 wt.%, C3-DBT of 20 wt.%, and others of 10 wt.%, and an aromatic content of 27.16 wt.%, the value of maximum substrate concentration, $C_{Smax,}$ corresponding to the maximum substrate consumption rate has been determined using the following theoretical concept:

$$\frac{d}{dC_s}(r_S) = 0 \quad \text{or} \quad \frac{d}{dC_s}\left(\frac{\mu_{max} C_s}{K'_S + C_s + \dfrac{C_s^2}{K_{si}}}\right) = 0$$

$$\text{or} \quad \frac{\mu_{max}\left[\left(K'_s + C_s + \dfrac{C_s^2}{K_{si}}\right) - C_s\left(1 + \dfrac{2C_s}{K_{si}}\right)\right]}{\left(K'_s + C_s + \dfrac{C_s^2}{K_{si}}\right)^2} = 0$$

or

$$K'_s + C_s + \frac{C_s^2}{K_{si}} - C_s - \frac{2C_s^2}{K_{si}} = 0 \text{ or } \frac{C_s^2}{K_{si}} = K'_s \text{ or } C_s = \sqrt{K'_s K_{si}} = C_{smax} \tag{110}$$

where r_s (mg/dm^3/h) is the consumption rate of the substrate, t is the time (h), K'_s (mg/dm^3) is the modified half saturation constant for a substrate inhibited system, μ_{max} is the maximum specific growth rate (h^{-1}), and i is any compound.

The values of the half saturation constant (K_S, mg/dm^3), maximum substrate concentration (C_{Smax}, mg/dm^3), and inhibition constant (K_{Si}, mg/dm^3) have been correlated to the number of alkylation in DBTs.

The rate of removal of OSCs was calculated as follows:

$$\frac{dC_{si}}{dt} = -\mu_i C_B \frac{1}{Y_{X/si}} \tag{111}$$

$$\frac{dC_B}{dt} = \mu C_B \tag{112}$$

where the C_B is the biomass concentration mg/dm^3, C_s is the substrate concentration, μ is the specific growth rate (h^{-1}), i is any compound, and $Y_{X/si}$ is the yield coefficient equal to the mass of biomass produced/mass of substrate consumed.

The specific growth rate, μ, has been calculated using Monod type and Haldane type kinetics during simulation using a 4th order Runga Kutta technique. The data analysis revealed that up to 100 mg/dm^3 of each of the studied OSCs', DBT, C1-DBT, C2-DBT, and C3-DBT, Monod type kinetics were able to explain their transient behavior, while Haldane type kinetics were more suitable to predict the higher removal of higher concentrations, up to 1000 mg/dm^3, where the substrate inhibition was validated above 100 mg/dm^3 of initial substrate concentration. The value of C_{Smax} increased with the number of substitution, n, in DBT. The value of μ_{max}, however, showed the reverse trend, while C_{Smax} and K_S have increased with the number of alkylation and K_{Si}, μmax, and $Y_{x/S}$ have decreased with it. The validity of the Haldane model was also established in the case of diesel. Moreover, the applicability of the linear correlations of the number of alkylation, n, with different kinetic parameters in the lower initial concentration range of diesel, up to 200 mg/dm^3, was also confirmed (Guchhait, *et al.*, 2005a).

Caro *et al.* (2008a) studied the effect of organic solvents and initial S-concentrations on the BDS efficiency of *P. putida* CECT 5279, using model oil (DBT/n-hexadecane). CECT 5279 was able to tolerate high concentrations of organic solvent up to 1:1 v/v. A concentration of 400 ppm of DBT was converted at a specific rate of generation of the desulfurized final product, 2-HBP, of 2.3 and 1.5 mg HBP/g DCW/h for 27% and 50% (v/v) of hexadecane, respectively. It was found that upon the usage of a no-selective S-source, such as $MgSO_{4,}$ the growth was not affected even at high concentrations of n-hexadecane (i.e. up to 50% v/v of OFP). Nevertheless, when using a selective S-source, such as DBT, the biomass growth was twice lower than that achieved with $MgSO_4$ and the biomass decreased with increase of the OFP. This was attributed to the hydrophobicity of DBT and, consequently, there would be a lower amount of DBT in an aqueous phase with the increase of OFP, which might also depend on the hydrophobicity of the microbial cell itself (Xu *et al.*, 2006; Caro *et al.*, 2007a; Mohebali *et al.*, 2007). It may be also due to the lower oxygen supply in a bi-phase system, especially with the increase of the oil phase. Strain CECT 5279 was probably not capable of being joined at the oil-to-water interface for the uptake of DBT there and, hence, the lower final optical densities were achieved. It is known that the transfer of OSCs from the oil to the aqueous phase is one of the most determinant parameters in the BDS process, especially when biocatalysts without the capacity to be adhered at the interface are used (Maxwell and Yu, 2000; Monticello, 2000; Abbad-Andaloussi *et al.*, 2003). Usually, in the aqueous phase system, there is a lag phase for the BDS of the organosulfur compounds (e.g. DBT). This is the time during which DBT concentration overcomes its solubility in water (Monticello, 2000) and a nucleation process could be produced, as it has been described before (Jia *et al.*, 2006). This time is not always considered in the posterior kinetic analysis because DBT molecules are only assembled by cells when they are completely dissolved into the broth (Wodzinski and Coyle, 1974).

The values of biomass concentration were analyzed by applying the logistic equation to determine the kinetic parameters, μ and C_x^{max}, while the evolution of the BDS process was analyzed applying the Haldane equation, taking into account that the pathway's final product is stoichiometrically generated. Moreover, the evaluation of the process was done by determining the desulfurization percentage (X_{BDS}). Furthermore, the HBP production rates (r_p) were measured as the average cellular activity into a reaction interval. The saturation and inhibition constants, K_s and $K_{I,}$ were found to be increased with the increment of the OFP. It is normally assumed that the increment of K_S is related to mass transfer limitations as this effect is similar to a decrease of enzymatic affinity. Moreover, the growth inhibition constant was higher than the desulfurization inhibition constant, indicating that the sensitivity of cell growth towards DBT-concentration was higher than that of desulfurization. It is important to note that the calculated amount of 2-HBP was always higher than the real experimental results since the applied model did not consider the accumulation of intermediates, as they were usually detected only at negligible concentrations in the extracellular media. In another study by Caro *et al.* (2008b), it was proven that there was accumulation of several products inside the cells. Not only this, but Abbad-Andaloussi *et al.* (2003) reported that 2-(2′-hydroxyphenyl)benzenesulfinate (HBPSi) was not detected either by gas chromatography (GC) or high performance liquid chromatography (HPLC) due to an analytical selectivity problem since the retention time and spectrum of both HBPSi and HBP are quite similar and the concentrations produced are near the detection limit as well. Upon studying the effect of initial DBT concentration in the model oil (200 to 3200 ppm), the growing cells of *P. putida* CECT 5279 proved good BDS capabilities. However, the biomass growth stopped before the complete conversion of DBT. This was attributed to the feed-back inhibition effects produced by 2-HBP (Setti *et al.*, 1999; Maxwell and Yu, 2000) and sulfate accumulation (Li *et al.*, 1996). However, due to the hydrophobic nature of 2-HBP, it migrates to the oil phase so its inhibition effect on the cell at the aqueous phase will be decreased. Thus, the growth might be stopped for other reasons, such as several other limiting substrates related with 4S-pathway development, such as the co-factors regeneration and oxygen contribution.

Irani *et al.* (2011) determined the batch growth kinetic parameters of the aerobic Gram +ve *Gordonia alkanivorans* RIPI90A for diesel and hydrotreated diesel in biphasic media, with the initial varied range of S-content at 2–5 g/L and 0.007–0.028 g/L, respectively. The experimental results have shown acceptable agreement with the Haldane kinetic for diesel and the Monod kinetic for hyrotreated diesel. The maximum specific growth rate (μ_{max} h^{-1}) on diesel oil and hydrotreated diesel oil recorded 0.459 and 0.095 h^{-1}, with a half saturation coefficient (K_S) of 3.55 and 0.02 g/L, respectively, and a substrate inhibition constant (K_{Si}) of 19.24 g/L for diesel oil. This was attributed to the inhibitory high S-concentration in diesel oil. This study revealed that the strong influence of sulfur concentration on its desulfurization rate. Moreover, the inhibition effect of the produced 2-hydroxybiphenyl throughout the 4S-pathway would decrease the desulfurization rate. However, with a low sulfur concentration range, as the availability of sulfur compound increases, better growth of biomass leads to a higher rate of substrate consumption. However, at a high sulfur concentration range, the substrate inhibition effect occurs and, consequently, the rate of BDS decreases.

To determine the kinetics of desulfurization in a batch BDS of model oil (DBT in dodecane) using resting cells of *Nocardia globerula* R-9, the initial desulfurization rate in a biphasic system (1:1 O/W) was studied with various DBT concentrations in dodecane, at pH7 and 30 °C within 1 h. The obtained results were best represented by the Michaelis–Menten equation:

$$-\frac{dS}{dt} = \frac{(V_{max} S_o)}{(K_m + S_o)} \tag{113}$$

where S_o is the initial DBT-concentration in mmol/L, V_{max} is the maximum reaction velocity in mmol S/kg/h, and K_m is the Michaelis constant.

The values of V_{max} and K_m for the model oil BDS were found to be 11.0 mmol S/kg/h and 0.70 mmol/L, respectively (Mingfang *et al.*, 2003b).

To analyze the desulfurization pattern of model oil (DBT in dodecane) using lyophilized cells of *Pseudomonas delafieldii* R-8, the initial desulfurization rate, under various DBT concentrations in the dodecane with the ratio of oil-to-water and cell density of 1.0 and 20 mg DCW/L of the aqueous phase, was tested. The obtained results were best represented by the Michaelis–Menten equation. The values of the rate constants, the limiting maximal velocity (V_{max}), and Michaelis constant (K_m), for the desulfurization of DBT were fitted as 0.32 mM/h (13.0 mmol/kg DCW/h) and 1.3 mM, respectively. This indicated that the high concentration of DBT, up to 14 mM (corresponding to 448 ppm), did not inhibit the desulfurization activity of R-8 (Luo *et al.*, 2003).

The time course of desulfurization of the mixture of DBT and 4,6-DMDBT by resting cells of *Pseudomonas delafieldii* R-8 in a biphasic system was described by Mingfang *et al.* (2003b). The concentration of each in the dodecane phase was about 0.5 mmol/L. It indicated that the desulfurization process of DBT and 4,6-DMDBT proceeded simultaneously without showing any preference to either one. The desulfurization pattern of DBT was described by the Michaelis–Menten equation. The BDS-initial rate of DBT and 4,6-DMDBT was 1.7 and 0.75 mmol S/kg cell/h, respectively, while the initial desulfurization rate in a single substrate system recorded 6.5 and 2.5 mmol S/kg cell/h for DBT and 4,6-DMDBT.

The initial desulfurization rate in a biphasic system (30 % O/W) of model oil (DBT in hexadecane) was tested under various levels of DBT concentrations, a pH of 6, 30 °C, and 175 rpm, using 11 g DCW/L of resting cells of RIPI-22 bacteria harvested at the end of the logarithmic phase (30 h) (Rashtchi *et al.*, 2006). Considering the fact that BDS is a kind of enzymatic reaction, the Michaelis–Menten equation (eq.113), which is equivalent to Monod, was applied, where dS/dt is the initial specific reaction rate for the production of 2-HBP, S, and the concentration of 2-HBP (mM). The parameter fitting was done using the Curve-Expert 1.3 and the values of V_{max}, K_m, and the correlation coefficient (R^2) were obtained and recorded as 0.21779 mM/h, 0.04099 mM 2-HBP, and 0.9724.

However, in a study performed by Boltes *et al.* (2013), the Michaelis–Menten kinetics used for data simulation in aqueous media were not valid under biphasic conditions due to the low bioavailability of DBT and its alkylated derivatives (C_x-DBTs) in an aqueous phase where microorganisms exist.

Calzada *et al.* (2012) developed a kinetic model that fits the experimental results for resting cell operation with cells of different ages in different concentrations, taking into account the enzyme deactivation. The genetically modified *Pseudomonas putida* CECT 5279 was employed in that study. It is a genetically modified bacterium that successfully performs the 4S-pathway, as it carries the genes dszABC from *Rhodococcus erythtropolis* IGTS8 and a flavin-oxydo-reductase from *Escherichia coli* (hpaC) (Galán *et al.*, 2000).

Aggarwal *et al.* (2013) applied *in silico* modeling to study the BDS efficiency of BT and DBT, using *G. alkanivorans* in batch BDS processes with glucose a carbon source. Based on the assumption of a fixed uptake of glucose (20 mg/g DCW/h), with unlimited supplies of BT and DBT and maximized biomass growth, the model consumed BT with a growth rate of 0.021 h^{-1} and a desulfurization rate of 2.55 mmol/g DCW/h. This corresponds to the minimum *in silico* sulfur requirement of the cell for maximum growth at a glucose uptake of 20 mg/g DCW/h. In subsequent simulations, they gradually reduced the uptake of BT from 2.55 mmol/g DCW/h to zero. As expected, the DBT uptake gradually increased to meet the cellular demand of sulfur. Since the model has no regulatory mechanics and its uptake system has no bias for DBT or BT, Aggarwal *et al.* (2013) ruled out the non-specific uptake factors for explaining the preference of BT over DBT. Instead, energy usage offers a plausible explanation. 1 mole BT requires 2 moles NADH, while 1 mole DBT requires 4 moles NADH (Figure 9.19). Thus, for the cell, BT is a better sulfur source than DBT energetically. For investigating the effect of BT and DBT on growth, two simulations were performed. In the first, BT was provided and DBT was provided in the second. For both cases, their uptake rates were fixed at 20 mg/g DCW/h with an unlimited supply of glucose and other nutrients. The obtained maximum growth rate was 1.24 h^{-1} and 0.90 h^{-1} with BT and DBT, respectively. Thus, BT promotes higher growth than DBT for *G. alkanivorans*. This was also attributed to the lower energy requirements

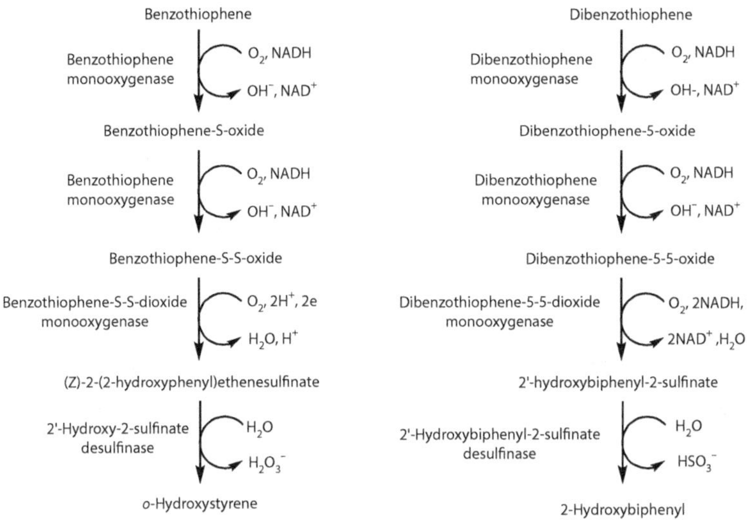

Figure 9.19 Selective Desulfurization of Benzothiophene (BT) and Dibenzothiophene (DBT).

of BT. Also, Aggarwal *et al.* (2013) applied *in silico* modeling to compare between the *R. erythropolis* and *G. alkanivorans* BDS activities. Based on this assumption, maximized cell growth for a fixed glucose uptake of 1 mmol/g DCW/h with unlimited supplies of DBT/BT and other minimal nutrients occurred. Both *R. erythropolis* and *G. alkanivorans* can utilize DBT as the sole sulfur source producing 2-HBP, but *R. erythropolis* cannot utilize BT. Upon simulation for BDS of DBT, based on this assumption, maximized cell growth for a fixed glucose uptake of 1 mmol/g DCW/h, with unlimited supplies of DBT and other minimal nutrients, the growth rate and the corresponding desulfurizing activity were found to be higher for *G. alkanivorans* (0.15 h^{-1}, 18.13 mmol HBP/g DCW/h) than *R. erythropolis* (0.14 h^{-1}, 13.54 mmol HBP/g DCW/h). Furthermore, the desulfurization activity exhibited by the two strains increased with the increase in glucose uptake rates. Based on the predicted two models, the minimum sulfur requirements (in terms of DBT) of the two strains' unit growth rate were calculated to be 120 mmol/g DCW/h and 93.9 mmol/g DCW/h for *G. alkanivorans* and *R. erythropolis*, respectively, based on the assumption of supplying all the nutrients in excess and minimized the DBT uptake for a fixed biomass growth rate of 1 h^{-1}. Thus, these analyses showed that *G. alkanivorans* has higher desulfurization activity than *R. erythropolis* under the same medium conditions, promoting it as better catalyst for biodesulfurization.

Mathematical modeling and evaluation of bio-kinetic constants are mandatory in the designing and scaling up of bioprocesses for industrial scales. Consequently, it is essential to relate the kinetics of BDS in the petroleum feed with that of the culture growth. The most widely used method for the measurement of bacterial growth is optical density, but this is difficult in the presence of a real oil feed. However, it is more applicable to perform this measuring direct cell counts or dry cell weight, based on the experimental conditions. Nassar *et al.* (2016) attempted to isolate and identify a new bacterial strain capable of desulfurization without altering the hydrocarbon skeleton. It also aimed to predict a relationship between different bacterial population units, optical density (OD), total viable count (i.e. colony forming unit (CFU/mL)), and biomass dry cell weight (DCW, mg/mL) for further application in the petroleum industry.

Numerical investigation was performed to find out the relationship between CFU/mL and OD_{600}, DCW mg/mL and OD_{600}, and finally, between CFU/mL and DCW mg/mL. The relationship between the DCW and OD_{600} of DBT-biodesulfurizing bacterium *Brevibacillus brevis* strain HN1 (accession no. KF018281) was found to be represented by a linear polynomial equation:

$$Y = f(X) = P_1 X + P_2 \tag{114}$$

where X represents the OD_{600} and the dependent variable Y represents the corresponding DCW in mg/mL. The coefficients of equation (114) with 95% confidence bounds were $P_1 = 0.1375$ (0.124, 0.151) and $P_2 = 0.01056$ (-0.01481, 0.03593).

The relationship between the total viable count (CFU to cells/mL) of HN1 and OD600 was found to be described by the general Gaussian model for fitting the peak (Giraud, 2008), which can be given by the following equation:

$$CFU = a^{*}exp\left[-\left(\frac{OD-b}{c}\right)^2\right] \tag{115}$$

The converged values of the parameters, a, b, and c (with 95% confidence bounds), were as follows: a = 9.083e+9 (8.6e+9, 9.566e+9), b = 3.066 (2.992, 3.141), and c = 1.272 (1.156, 1.387), where parameter a presents the height of the curve's peak, b represents the position of the center of the peak, and c controls the width of the curve and is related to the full width at half maximum (FWHM) of the peak.

Finally, based on the previously predicted two model equations, a new relationship was derived that correlated CFU (cells/ml) and DCW, which was found to be represented by the following Gaussian equation:

$$CFU = 9.083e+9^{*}\exp\left[-\left(\frac{\left(\left(\frac{DCW-0.01056}{0.1375}\right)-3.066\right)}{1.272}\right)^{2}\right] \quad (116)$$

Thus, depending on how HN1 would be applied, these predicted correlations would facilitate the monitoring of bacterial growth. This would, consequently, facilitate modeling, simulation, and design of different BDS processes using the obtained bacterial isolate *Brevibacillus brevis* strain HN1.

Peng and Zhou (2016) studied the desulfurization activity of *Rhodococcus* sp. MP12 and the specific desulfurization activity in a DBT system was expressed as the amount of 2-HBP produced per gram dry cells weight per hour (μmole HBP/g DCW/h), but in practice, it is difficult to measure 2-HBP production in a biphasic system and a crude oil-aqueous system since 2-HBP production can be divided into oil and water phases. The total HBP buildup in a BDS process of crude oil would include HBP in a buffer phase (i.e. aqueous phase), HBP in a crude oil phase, and HBP absorbed by cells. For the first time, the specific desulfurization activity in a crude oil system was expressed as the amount of sulfate produced per gram dry cells weight per hour (μmole sulfate/g DCW/h) (Peng and Zhou, 2016). This can improve the simplicity and convenience of conventional measurement methods based on the assumption that 1 mol of sulfate and 1 mol of 2-HBP are formed per mole of DBT desulfurized (Gallagher *et al.*, 1993; Oldfield *et al.*, 1997; Rashtchi *et al.*, 2006). Sulfate measurement can be easily fulfilled by a commercially available sulfate kit or ion chromatography (IC). However, Peng and Zhou (2016) used a MOPS-Na buffer instead of a phosphate buffer in crude oil BDS experiments in order to avoid interference of phosphate buffer on sulfate measurement by a sulfate kit. Upon the performance of an experiment on 0.3 (O/W), the sulfur content in desulfurized crude oil, which was measured by an X-ray fluorescent analyzer, was reduced by 8% from 3.42% to 3.13% (w/w) after 168 h of incubation. The analysis of total 2-HBP and total sulfate molar balance showed that the total HBP buildup was found to have a very good match with the total sulfate buildup with no significant difference between total sulfate and 2-HBP buildup by t-test ($p > 0.05$).

Boltes *et al.* (2013) studied the effect of mass transfer on BDS-kinetics of alkylated forms of dibenzothiophene (C_x-DBTs) by *Pseudomonas putida* CECT5279 under aqueous and biphasic resting cell conditions in a 2 L stirred tank reactor. For producing an effective oxygen transfer rate in the broth and volumetric mass transfer coefficient (k_La), the stirring rate was controlled. That study proved that the BDS-process is strongly affected by the mass transfer between liquids, where complete conversion of 10 μmol/L DBT, 4-MDBT, and 4,6-DMDBT was achieved in aqueous reaction media (i.e. one phase system), while the conversions in the presence of an organic liquid (1:1 aqueous:n-hexadecane v:v) were decreased, recording 38%, 19.5%, and 16.5% for DBT, 4MDBT, and 4,6DMDBT, in a mixture with an initial concentration of 271 μmol/L, respectively. The conversion times for DBT, 4-MDBT, and 4,6-DMDBT increased 356, 441, and 498 fold when the oil phase was used. Thus, in a biphasic system, mass transfer between liquids controls the process, increasing reaction time and decreasing the process yield. The 4-MDBT and 4,6-DMDBT conversion times were 4 times slower than DBT in aqueous reactions conducted in a stirred tank reactor, whereas they presented a 5 fold delay compared with DBT conversion times in biphasic reaction media, using the same reactor. This indicates the effect of alkylation and hydrophobicity on the BDS-efficiency.

The desulfurization yield (X_{BDS}) was defined as the ratio of 2HBP accumulated over an established reaction to initial DBT times:

$$X_{BDS} = \left(\frac{C_{2HBP,t}}{C_{DBT,t_o}} \right) \times 100 \tag{117}$$

where $C_{2HBP,t}$ is the concentration of produced 2-HBP at time t (h) and C_{DBT,t_o} is the intial DBT concentration at time zero.

However, the conversion yield Y_{BDS} is calculated as follows:

$$Y_{BDS} = \left(\frac{C_{Cx-DBT,0} - C_{Cx-DBT,t}}{C_{Cx-DBT,0}} \right) \times 100 \tag{118}$$

where $C_{cx\text{-}DBT,t}$ is the remaining DBT concentration at time t (h) and $C_{cx\text{-}DBT,0}$ is the intial DBT concentration at time zero.

The X_{BDS} and Y_{BDS} give an indication about the microbial BDS efficiency, while the degree to which desulfurization occurred during the growing process (D_{BDS}) was used to correlate the BDS capacity of the cells and the biomass obtained over a given growth time provides information about the optimum biocatalyst production process.

$$D_{BDS} = \left(\frac{X_{BDS} \cdot C_x}{t_G} \right) \tag{119}$$

where t_G is the growth time required to attain a biomass concentration, C_x, and X_{BDS} is BDS-yield.

In aqueous media, the kinetic analysis of DBT and C_x-DBT resting cell biodesulfurization under aqueous conditions was based on the Michaelis–Menten equation. Thus, the kinetic equations for 4S-pathway compounds, DBTO and $DBTO_2$, were fitted to a simple Michaelis–Menten equation and to a by-product competitive inhibition for HBP, as follows:

$$v_{DBT} = \frac{1}{S}\frac{ds}{dt} = \frac{v_{max1} S_{DBT}}{K_{S1} + S_{DBT}} \tag{120}$$

$$v_{DBTO} = \frac{1}{S}\frac{ds}{dt} = \frac{v_{max2} S_{DBTO}}{K_{S2} + S_{DBTO}} \tag{121}$$

$$v_{DBTO_2} = \frac{1}{S}\frac{ds}{dt} = \frac{v_{max3} S_{DBTO_2}}{K_{S3} + S_{DBTO_2}} \tag{122}$$

$$v_{HBP} = \frac{1}{S}\frac{ds}{dt} = \frac{v_{max4} S_{HBPS}}{K_{S4}(1 + P_{HBP}/K_1) + S_{HBPS}} \tag{123}$$

where S is the 4S compound concentration (μmol/L), v_{max} is the maximum velocity (μmol/g_x/min), and K_S is the saturation constant (μmol/L).

For Cx-DBT elimination, only the evolution of the initial compounds measured since the intermediates were unknown.

$$v_{4MDBT} = \frac{1}{S}\frac{ds}{dt} = \frac{v_{max5} S_{4MDBT}}{K_{S5} + S_{4MDBT}} \tag{124}$$

$$v_{4MDBT} = \frac{1}{S}\frac{ds}{dt} = \frac{v_{max6} S_{4,6DMDBT}}{K_{S6} + S_{4,6DMDBT}} \tag{125}$$

To minimize miscalculation and to calculate the kinetic parameters and volumetric mass transfer, the experimental data was fitted to a well-known exponential concentration–time function (that is, a first-order kinetic equation) for DBT and Cx-DBTs, as the mass transfer controls the process and the transfer rate depends solely on substrate concentration in the oil phase.

$$C_{Cx-DBT} = C_o + A\exp(-X/t) \tag{126}$$

The kinetic parameters (v_{max}, K_S) were obtained from this data using the Lineweaver–Burk method for DBT and Cx-DBT reactions.

In a double phase system, to determine the effect of mass transfer, Boltes *et al.* (2013) used the concept of characteristic time to compare the BDS process in aqueous media and biphasic liquid reaction media.

$$t_{CW} = \frac{1}{V_{max}} \tag{127}$$

$$t_{CO} = \frac{1}{K_{Cx-DBT}} \tag{128}$$

This study revealed that the maximum specific desulfurization rate of DBT in an aqueous phase was almost four times higher than those of alkylated compound concentrations. Although the saturation constant, Ks (μmol/L), values were similar for all the studied compounds, the calculated values of the maximum velocity (v_{max} μmol/g/min) varied according to the degree of alkylation and the hydrophobicity of the compounds in aqueous reaction media.

In a biphasic system, the volumetric mass coefficient value for HBP (K_{HBP}) was two times higher than those for K_{DBT}, indicating that HBP accumulation in the oil phase is favored over DBT elimination. The initial elimination rate for DBT and initial accumulation rate for HBP were very similar, as expected, demonstrating that no significant amount of 4S intermediates were accumulated either in the organic or aqueous phases.

The volumetric mass transfer coefficient was calculated for DBT and for HBP from the oil to water phase and also for Cx-DBTs, according to the double-film model. The mass transfer coefficients were determined from the decrease in DBT concentration (K_{DBT}) and the accumulation of HBP (K_{HBP}) in the oil phase, using the integral method for kinetic analyses in both cases. If the reaction takes place very close to the interfacial layer under the water side of the double film where the microorganism exists, the change in watery phase concentrations cannot be measured in the presence of the oil layer. However, the net conversion rate of each compound in the model oil phase could be described according to the following first-order equations:

$$v_{DBT,O} = -\frac{dDBT}{dt} = K_{DBT} \times C_{DBT} \tag{129}$$

$$v_{HBP,O} = \frac{dHBP}{dt} = K_{HBP} \times C_{HBP} \tag{130}$$

$$v_{4MDBT,O} = -\frac{d4MDBT}{dt} = K_{4MDBT} \times C_{4MDBT} \tag{131}$$

$$v_{4,6DMDBT,O} = -\frac{d4,6DMDBT}{dt} = K_{4,6DMDBT} \times C_{4,6DMDBT} \tag{132}$$

The initial Cx-DBT-BDS rates in the oil phase decreased with the presence of substituent groups in the aromatic ring, but were unaffected by the number of methyl substituents. The higher first-order constant K = 2.67×10^{-3} min^{-1} was recorded for DBT, which has the lowest degree of alkylation and hydrophobicity. This value was two times higher than those obtained for 4-MDBT and 4,6-DMDBT. Furthermore, the DBT desulfurization yield X_{BDS}% recorded 38.80% and its conversion yield (Y_{BDS}%) was the highest of all the tested compounds, while 4-MDBT and 4,6DMDBT recorded 19.5% and 16.61%, respectively.

In a similar study performed by Caro *et al.* (2007a), *P. putida* CECT5279, using an orbital shaker instead of mechanical stirring, using a biomass cell age of 5.5 h and pH, was controlled at 8.0; the HBP accumulation was favored over DBT elimination and the K_{HBP} value was also two times higher than that for K_{DBT}, but the values of the kinetic coefficients were two orders of magnitude lower than the corresponding results reported in Boltes *et al.* (2013) using mechanical stirring and Tween 80 as a surfactant, which improved the mass transfer process and aeration.

Kobayashi *et al.* (2001) studied the BDS rate of different alkyl-DBTs in model oil and light gas oil (LGO) using *Rhodococcus erythropolis* KA2-5-1 (FERM P-16277) and its genetically improved recombinant, strain rKA2-5-1, as the biocatalysts. The DBS-pattern was represented by the Michaelis-Menten equation and the value of rate contents, the limiting maximum velocity (v_{max}), and the Michaelis rate constant (K_m) were calculated for a biphasic system with 1:1 v/v of different individual DBTs in model oil (DBT, 4-MDBT, 4,6-DMDBT, or 3,4,6-trimethyldibnozothiophene (3,4,6-TMDBT) dissolved in n-tetradecane).

$$\frac{dS}{dt} = \frac{(v_{max} S)}{(K_m + S)} \tag{133}$$

where S is the concentration of Cx-DBTs (mM), v_{max} is the limiting maximum velocity (mmol/kg dry cell weight/h), and K_m is the Michaelis rate constant (mM).

The DBT-desulfurization activity was 196 and 120 mmol/kg dry cell weight/h for rKA2–5-1 and KA2–5-1, respectively, thus the rKA2–5-1 is 63% higher in DBT-BDS efficiency than the original KA2–5-1. Moreover, the DBT-BDS rate is 2 and 2.5 times higher than those for 4-MDBT and 4,6-DMDBT.

This study revealed that as alkylation increases, the desulfurization activities decrease, where alkyl-DBTs with six carbons of alkyl substituent groups were not desulfurized and the hydrophobicity and size increase, which decreases the transfer of the compound from the solvent phase to inside the cell. Since the hydrophobic compounds are expelled from the aqueous phase, it is hard to be in contact with the bacterial cell surface. However, the type or position of alkyl substituent groups has little effect on BDS activity. Moreover, the BDS activity decreases in the presence of mixtures of alky-DBTs due to the competitive inhibition effect. The inhibition effect of phenolic products was also investigated: 2-hydroxt-biphenyl (2-HBP) for DBT, 2-methy-6-(3-methylphenyl)phenol that is 4,6-dimethylhydroxybiphenyl (4,6-DMHBP) for 4,6DMDBT, and 2,3-dimethyl-6-(3-methylphenyl)phenol that is 3,4,6-trimethylhydroxybiphenyl (3,4,6-TMHBP) for 3,4,6-TMDBT, within the concentration range of 0–10 mM. It was surprising that no inhibitory effects occurred except for 2-HBP. However, 80% BDS activity occurred even in the presence of 10 mM 2-HBP. This low inhibition effect was attributed to the enzymatic activity of the whole cell system and solubility of 2-HBP in a bi-phasic system, where, in the case of a whole cell system, 2-HBP is excreted from the reaction environment which is the water phase inside the cell to the organic phase that is n-tetradecane. The distribution coefficient of 2-HBP to n-tetradecane and water was found to be 30.

The apparent competitive inhibition model was applied to study the desulfurization pattern of a system of multiple substrates, but in order to apply this model, the kinetic constants of BDS of individual compounds

(C0, C1, C2, and C3-DBT, that is DBT, 4-MDBT, 4,6-DMDBT, and 3,4,6-TMDBT) calculated from the Michaelis–Menten equation were used. Based on the assumption that in the BDS of a multiple component system the total concentration of cells was the sum of each concentration of substrate-cell complexes and the concentration of free cells, the model of BDS was extended. Scheme 9.1 summarizes the mutual competitive reaction scheme as suggested by Kobayashi *et al.* (2001), where one alkyl-DBT acts as the substrate and the other alkyl-DBTs act as inhibitors. That is, when $k_{-1} > k_{+2}$, $k_m = k_{-1}/k_{+1}$, the dissociation constants for inhibitors are assumed to be $K_I = k_{-3}/k_{+3}$, $K_J = k_{-4}/k_{+4}$, $K_K = k_{-5}/k_{+5}$.

Thus, the reaction velocity of each alkyl-DBT was expressed as follows:

$$r = (v_{max}S)/(K_m(1 + I/K_I + J/K_J + K/K_K) + S) \tag{134}$$

where r is the reaction velocity (mmol/kg drycell weight/h), V_{max} is the limiting maximum velocity (mmol/kg dry cell weight/h), and K_m is the Michaelis constant (mM).

Upon the application of a lumping model to describe the multiple BDS of different alkyl-DBTs in the LGO of initial S-content 360 ppm, first, Kobayashi *et al.* (2001) divided the alkyl-DBTs in LGO to four groups: C1, C2, C3, and Cx, for mono-, di-, tri-, and more than 4- alkyl-substituents. Then, they

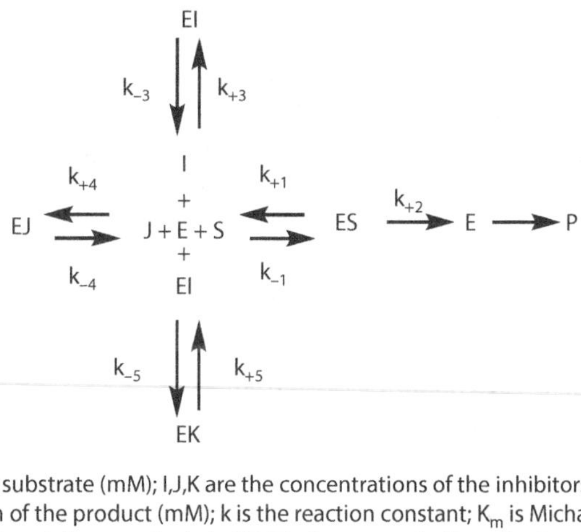

S is the concentration of substrate (mM); I,J,K are the concentrations of the inhibitors (mM); P is the concentration of the product (mM); k is the reaction constant; K_m is Michaelis constant (mM); $K_{I,J,K}$ are dissociation constants (mM); E is the concentartion of active sites; and ES, EI, EJ, EK are the concentrations of intermediates (mM).

Scheme 9.1 Mutual Competitive Reaction as Suggested by Kobayashi *et al.* (2001).

chose 4-MDBT, 4,6-DMDBt, and 3,4,6-TMDBT as representatives for C1-, C2-, and C3- DBTs, respectively. The GC-AED analysis was done and the total S-concentration of each alkyl-group was calculated by summing the concentrations of compounds belonging to the same group, while the concentration of Cx-DBT was calculated by subtraction of the summed amount of C1-, C2-, and C3- DBTs from the total S-content in LGO.

Similarly, Chen *et al.* (2008) described the desulfurization efficiency and specific rate of DBT and 4,6- DMDBT elimination using resting cells of *Mycobacterium* sp. ZD-19 as the biocatalyst, where both values were almost 2 times higher for DBT than for 4,6DMDBT. The rate of BDS of different studied OSCs were reported to be: 12.8, 11.5, 8.58, 3.71, and 1.36 mM/kg DCW/hfor thiophene (Th), benzothiophene (BT), diphenylsulfide (DPS), dibenzothiophene (DBT), and 4,6-dimethyldibenzothiophene (4,6-DMDBT), respectively.

Li *et al.* (2003) reported the specific desulfurization rate of diesel oil by *Mycobacterium* sp. X7B to be 0.14 mg sulfur/g DCW/h, while Li *et al.* (2009) reported the specific desulfurization rates of middle distillate unit feeds (MDUF) and light gas oil (LGO) by resting cells of *Gordonia* sp. CYKS1 to be 0.17 and 0.15 mg sulfur/g DCW/h, respectively. Bordoloi *et al.* (2014) reported the specific desulfurization rate of a hydrodesulfurized diesel oil by *Achromobacter* sp. to be 1.20 mg sulfur/g DCW/h.

According to Soneyink and Lenkins (1980), kinetics of the biodegradation reaction can be described in terms of its order. Goindi *et al.* (2002) reported that the batch BDS in a hydrocarbon aqueous biphasic system of DBT/n-hexadecane by S*taphylococcus* sp. strain S3/C follows the first order kinetic model. The resting cells decreased the sulfur content of the hydrocarbon phase by 57% at 2.2 mg S/L/h in the absence of any additional carbon and sulfur source at an optimum hydrocarbon to aqueous phase ratio (H/A) of 2:1, whereas the specific sulfur removal rate was maximized (0.8 mg S/h/g DCW) at H/A 3:1.

The frequently used first- and second-order kinetic models were employed to investigate the biodegradation rate of different PASHs, including Th, BT, and DBT, in single- or multi-substrate systems using growing cells of *Bacillus sphaericus* HN1 (El-Gendy *et al.*, 2015). It was found that the biodegradation of the three studied PASHs follows the first order kinetic reaction in the single-substrate batch cultures with the regression coefficient $R^2 \geq 0.9497$, with rate constants of $k_{Th} \approx 0.0095\ h^{-1}$, k_{BT} $0.0448\ h^{-1}$, and k_{DBT} $0.0187\ h^{-1}$ and a half life time of $t_{1/2}$ of ≈ 72.96, 15.47, and 37.07 h for Th, BT, and DBT, respectively. Biodegradation of BT ($R^2 \geq 0.9538$), DBT ($R^2 \geq 0.9566$) follows the first order kinetic reaction in the multi-substrate batch systems, while biodegradation of Th follows the second-order kinetic reaction ($R^2 \geq 0.9049$). The biodegradation rate of DBT recorded k_{DBT} values of ≈ 0.0117, 0.0411, and 0.0126 h^{-1} with $t_{1/2}$ of ≈ 59.24, 16.86, and 55.01 h in a binary mixture with Th or BT and tertiary mixture, respectively. This indicates that the biodegradation of DBT was enhanced in the presence of BT, but depleted in the presence of Th and tertiary multi-substrate batch cultures. The same trend was also observed in case of the biodegradation of BT, where k_{BT} recorded ≈ 0.026 and 0.0571 h^{-1} with a $t_{1/2}$ of ≈ 26.66 and 12.14 h in binary mixture with Th or DBT, respectively, while k_{BT} recorded ≈ 0.0405 h^{-1} with a $t_{1/2}$ of ≈ 17.11 h in the case of tertiary mixture. This indicates the synergetic effect between BT and DBT and the antagonistic effect of Th on the biodegradation of BT and DBT. For the biodegradation rate of Th, k_{Th} recorded values of $\approx 2\times10^{-4}$, 1×10^{-4}, and 1×10^{-4} L/mg/h with a $t_{1/2}$ of ≈ 50, 100, and 100 h in a binary mixture with Th or BT and a tertiary mixture, respectively. This indicates that the rate of biodegradation of Th is relatively lower than that of BT and DBT, whether in single- or multi-substrate batch cultures. This also shows that the presence of BT enhanced the rate of biodegradation of Th, but there was a strong antagonistic effect expressed on the Th biodegradation rate in a binary mixture with DBT or tertiary-mixture-substrate batch cultures. Briefly, in binary-substrate-batch cultures, BT enhanced the biodegradation of Th and DBT. Although DBT enhanced the biodegradation of BT, it depleted the biodegradation of Th. However, Th depleted the biodegradation of both BT and DBT, but in tertiary-substrate-batch cultures, the biodegradation efficiency of the three PASHs were depleted. Thus, this study proved the competitive inhibition effects of a multi-substrate system.

In another study, the biodegradation kinetics of different polyaromatic sulfur heterocyclic compounds (PASHs), including thiophene (Th), benzothiophene (BT), and dibenzothiophene (DBT), in a series of batch experiments were investigated as mono-, binary-, and tertiary-substrate systems to study the substrate interaction effects and capabilities of *Bacillus sphaericus* HN1 to utilize PASHs as sole carbon and energy sources in mono-and multi-substrate systems (Deriase *et al.*, 2015). Based on a full factorial design, 2^3, by varying the studied three PASHs concentrations between 0 and 100 mg/L, eight batch experiments were conducted out in duplicate for a period of 7 days in an orbital shaking incubator set at 200 rpm and 30 °C. A multiple comparison test was performed for the growth of microorganism and substrate biodegradation to obtain pairwise comparisons between the three studied systems. Both tests of ANOVA1 and Kruskal-Wallis showed that there was a highly statistically significant difference in the growth of HN1 between the negative control (free of PASHs) and all other treatments (p = 2.2105e-7 and 9.303e-4, respectively). Although there was a statistically significant difference for biodegradation of DBT in mono– and binary-substrate systems within the time interval 96–168 h ($p < 0.05$), there was a non-statistically significant difference in the biodegradation of both Th and BT in the three studied systems ($p > 0.05$). A multi-substrate form of the Monod kinetic model was applied to predict the substrate interactions in binary- and tertiary-substrate systems using the Monod parameters derived from the mono-substrate systems.

Modeling the relationship between cell growth and substrate consumption rates was done by direct coupling of the substrate consumption rate with the cell growth model applying constant cell yield and biomass decay rate coefficients during the time course of batch experiments.

$$\frac{dC_i}{dt} = -\mu_i X / Y_i \tag{135}$$

where the biodegradation rate of a substrate (i) is proportional to both the specific growth rate (μ_i) on that substrate and the biomass concentration (X) in the system.

The change in microorganism concentration with time can be expressed mathematically as follows:

$$\frac{dX}{dt} = \mu_i X - bX \tag{136}$$

where the change in biomass concentration is modeled as if the substrate (i) is the only growth substrate present. The dependent variables were $C_{i,}$ the concentration of the PASHs, that is the substrate, i (mg/L), and X, the biomass concentration (mg/L), while the independent variable was time, t (h). Y_i is the stoichiometric biomass yield coefficient for substrate i (mg biomass/mg substrate) and b is the endogenous decay rate coefficient (h^{-1}) which was estimated using the independent substrate free experiment (–ve control) with a known starting biomass concentration. The endogenous rate of microbial decay, which was characterized by rate coefficient b (h^{-1}), was subtracted from the growth rate of the biomass to account for the energy lost for cell maintenance and b was assumed to be constant for all the studied biodegradation systems. The Monod kinetic model relates growth rate to substrate concentration [$\mu_i = f\ (C_i)$] via the two bio-kinetic Monod parameters K_{Si} and $\mu_{max,i;}$

$$\mu_i = \frac{\mu_{max,i} C_i}{K_{Si} + C_i} \tag{137}$$

where $\mu_{max,i}$ is the maximum specific growth rate on substrate i (h^{-1}) and K_{Si} is the half saturation coefficient for substrate i (mg/L) and the parameter estimation is challenging for the Monod equation. The relationship between biomass formation and substrate consumption can be determined by the yield coefficient Y_i (dry weight of biomass/weight of utilized substrate) indicating the maximal conversion of unit substrate to cell mass.

$$Y_i = \frac{dX/dt}{dC_i/dt} = -\frac{dX}{dC_i} = -\frac{X - X_o}{C_i - C_{io}} \tag{138}$$

where $X>X_o$ and $C_i < C_{io}$.

In a multi-substrate system, the biomass growth is due to the utilization of all available hydrocarbons in the system to the microorganism, accordingly and the specific growth rate will be the summation of all individual ones for each substrate, as represented in eq. (52). In that study, correlation and simulation studies using the Monod kinetic model for a mono-substrate system were established and kinetic model parameters were taken into consideration in the multi-substrate Monod kinetic model (eq.53) used to describe the biodegradation process of the studied PASHs in the binary and tertiary substrate batch cultures. Mathematical expressions for modeling and simulating substrate (C mg/L) and biomass (X mg/L) concentration profiles throughout the time span of the batch experiments were also proposed and successfully applied (model equations 54 and 55). Thus, considering the fact that K_{Si} (mg/L) is inversely proportional to the affinity of the microbial system for the substrate (Agarry *et al.*, 2008; Deriase *et al.*, 2013), according to the obtained data from mono-substrate system, the affinity of HN1 towards the studied PASHs could be ranked in the following decreasing order, BT > DBT > Th, indicating the high toxicity of Th compared to BT and DBT. This was confirmed by the lowest $\mu_{max,i,}$ 0.01434 h^{-1}, obtained with Th.

In the experimental design of that study, the primary concern was to determine the substrate removal kinetics rather than biomass growth kinetics, where the batch experiments conducted in this study provided a deep understanding of the degradation kinetics and the positive and negative interactions of different PASHs, including Th, BT, and DBT in mono-, binary, and tertiary substrate aqueous systems. Experimentally, the biodegradation rates for the individual substrates in the multi-substrate systems may be enhanced or reduced relevant to the mono-substrate system. The effect of one substrate on the degradation of another is given by K_{si}/K_{sj} terms, where K_{Si} and K_{Sj} are the half saturation coefficients for substrates i and j (mg/L) (Guha *et al.*, 1999; Knightes and Peters, 2006). The experimental observations and modeling predictions, i.e. the simulation results for the biodegradation of the three studied PASHs in three studied systems, revealed the following: relative to the biodegradation of the three studied PASHs in mono-substrate-batch cultures, BT enhanced the biodegradation of DBT in the binary substrate-batch cultures of DBT-BT, especially within the time interval of 96–168 h, while, in Th-BT binary substrate-cultures, the BT expressed a negative impact on the biodegradation of Th up to 96 h, but it enhanced the Th-biodegradation within a relatively short time interval of 96–144 h, then the antagonistic effect occurred again with a longer incubation period. Although DBT enhanced the biodegradation of BT in the binary substrate batch cultures of BT-DBT, especially within the time interval 24–168 h, it depleted the biodegradation of Th in the binary substrate batch cultures of Th-DBT. However, Th depleted the biodegradation of both BT and DBT in the binary substrate-batch cultures of BT-Th and DBT-Th, respectively, where the negative impact of Th was very obvious within the time intervals of 12–120 h and 48–168 h, respectively. In the tertiary substrate-batch cultures, Th-BT-DBT, the biodegradation efficiency of the three PASHs were depleted relevant to their biodegradation efficiencies in the mono-substrate batch cultures, where the $\Sigma(K_{si}/K_{sj})$ recorded as 5.785, 0.975, and 1.887 for the biodegradation of Th, BT, and DBT, respectively, which indicated that the antagonistic effect between the three studied PASHs on the rate of their biodegradation can be ranked in the following decreasing order: Th > DBT > BT (Deriase *et al.*, 2015).

Simulation of the applied models to describe and estimate the biodegradation kinetics of the three studied PASHs in the different investigated systems was performed and the validity of the models was confirmed. The simulated effectiveness factor of the kinetic models was depicted by using a mean relative error (MRE), eq. (139), to show the goodness of fit between the obtained experimental and simulated data of PASH depletion with time.

$$MRE = \frac{1}{m}\sum_{i=1}^{m}\frac{|C_{i.cal} - C_{i.exp}|}{C_{i.exp}} \tag{139}$$

where $C_{i,cal}$ and $C_{i,exp}$ are the simulated substrate concentration determined from the model and its observed experimental value at time, t, respectively, while m is the number of experimental points. The results of MRE ranged between 0.096 and 0.282 and percentage MRE ranged from 9.6% to 28.2%, which confirmed the adequacy of the mono- and multi-substrate Monod kinetic models for estimating the biodegradation of the studied PASHs. It was also very obvious in the agreement between the measured and simulated profiles for the substrates' depletion with time. Thus, the approaches of modeling and simulation for the experimental results obtained in that work are considered to be essential in order to facilitate the understanding of the interactive effect of PASH contaminants and their biodegradation processes for better achievement of biotreatment of contaminated sites and industrial effluents.

The total cell-specific desulfurization rate (R_{total}) in a three-component (BDS) experiment, using oil, water, and cells, can be determined according to Abin-Fuentes *et al.* (2014) by the following relationship:

$$R_{total} = \frac{\left(C_{HBP,water}\Big|_{t_f} (1-\varnothing) + C_{HBP,oil}\Big|_{t_f} \right)\varnothing}{X(1-\varnothing)_{t_f}} \tag{140}$$

where t_f is the one-hour length of the experiments, $C_{HBP,water}\big|_{t_f}$ is the HBP concentration in water at t_f, $C_{HBP,oil}\big|_{t_f}$ is the HBP concentration in hexadecane at t_f, Ø is the oil fraction, and X is the cell density. Within the three-component BDS experiments, three different biocatalyst populations exist: free cells in an aqueous phase, cells adhered to oil-droplets, and cells that form aggregates. Thus, the R_{total} could be expressed by the following relationship:

$$R_{total} = f_{free}R_{free} + f_{oil}R_{oil} + f_{agg}R_{agg} \tag{141}$$

where R_{free}, $R_{oil,}$ and R_{agg} are the cell-specific desulfurization rates of free cells in an aqueous phase, cells adhered to oil-droplets, and cells that form aggregates, respectively. The f_{free}, f_{oil}, and f_{agg} are the fractions of free cells in an aqueous phase, cells adhered to oil-droplets, and cells that form aggregates, respectively. The values of R_{oil} and R_{free} were found to be equal to R_{max} as explained in the results obtained by Abin-Fuentes *et al.* (2014). The specific desulfurization rate of cells in aggregates (R_{agg}) was determined by the following relationship:

$$R_{total} = R_{max}\eta \tag{142}$$

where η is the effectiveness factor which is defined as the ratio of the actual rate of desulfurization within the aggregates to the rate of desulfurization if there was no intraaggregate diffusion limitation. Rearranging the previous equation allows one to solve for η:

$$\eta = \frac{R_{total}/R_{max} - f_{free} - f_{oil}}{f_{agg}} \tag{143}$$

The effectiveness factor for spherical aggregates and zero-order kinetics ($C_{DBT,water}$ >> Michaelis constant K_m in all experiments) is determined from the following relationship (Shuler and Kargi, 2002):

$$\eta = 1 - \left(1 - \frac{6D_e C_{DBT,water}}{X/(1-\varnothing) f_{agg} R_{max} (d_{32,agg}/2)^2}\right)^{3/2} \quad (144)$$

where D_e is the effective diffusivity within the aggregates and $d_{32,agg}$ is the Sauter mean aggregate diameter. Three-component BDS experiments provided a data pair of ($d_{32,agg}$, η) values for each mixing speed studied. Then, equation 144 is fit to the ($d_{32,agg}$, η) data set using the non-linear fitting function nlinfit in MATLAB® to obtain the least-square best-fit value of the parameter D_e.

The volumetric desulfurization rate (V_{total}) was determined according to the following relationship:

$$V_{total} = R_{total} X(1-\varnothing) \quad (145)$$

For each mixing speed tested in the small-scale system, the corresponding power input per unit volume (P/V) and impeller tip speed (v_{tip}) were calculated from the following relationships (Garcia-Ochoa and Gomez, 2009):

$$P/V = \rho N_p D_i^2 N^3 \quad (146)$$

$$v_{tip} = ND_i \propto \left(\frac{dV}{dY}\right)_{max} \quad (147)$$

where ρ is the density of the media, N_p is the power number, N is the mixing speed, D_i is the impeller diameter, and v_{tip} is proportional to the maximum shear rate, $(dV/dy)_{max.}$

Arabian *et al.* (2014) studied the effect of process temperature on the kinetics of BDS of a model oil DBT in dodecane using an inoculum size of 3.6×10^7 (cell/mL) *Bacillus cereus* HN, collected at the stationary phase, oil phase fraction of 0.2, initial DBT concentration of 1086 ppmw, and mixing rate of 180 rpm. The general n^{th} order power-law reaction rate expression was chosen to correlate the experimental data. It was found that the best value of n for this process could be 1. Therefore, the rate expression for the BDS of DBT was expressed as follows:

$$\frac{-dC_{DBT}}{dt} = kC_{DBT} \tag{148}$$

$$k = 2.715\exp\left(\frac{-12.79}{RT}\right) \tag{149}$$

where C_{DBT}, t, and k are DBT concentration (ppmw), time (h), and the reaction rate constant (h^{-1}) respectively. R and T represent the universal gas constant (kJ/mol/K) and temperature (K), respectively.

Adlakha *et al.* (2016) studied the kinetics of BDS of hydrodesulfurized diesel oil with an initial S-content of 70 ppm in a shaken flask batch BDS process of a 1/3 O/W phase ratio at 250 rpm at 30 °C using growing cells of *Gordonia* sp. IITR100 over a 12 day incubation period. It was found that after four days almost all of the OSCs, including C_2-DBT and C_x- BTs, in HDS diesel were used up. The BDS-rates of C_2-DBT, C_3-BT, and C_2-BT were found to be 0.34, 0.76, and 0.038 mg/kg/h (i.e. ppm/h), respectively, while the BDS rate for C_{x-} thiophenes was found to be the lowest, recording 0.08 mg/kg/h and a minor fraction of thiophene and its derivatives were remaining, with a total S-decrease of approximately 98% and after 6 days of incubation, the S-compounds were below the detectable levels.

In a study performed by Maass *et al.* (2015) for the evaluation of different kinetic models in the batch BDS of model oil (DBT/n-dodecane) by *Rhodococcus erythropolis* ATCC 4277, several models were applied, including simple ones, such as Monod and Andrews, and more complex ones, such as competitive, non-competitive, uncompetitive inhibition, sum kinetics interaction parameters (SKIP), and kinetic models to inhibition, and/or limitation effects by multiple-substrate concentrations, to predict the one that best describes the BDS process. This was tested using a parameters minimization method in the MATLAB program and their validation was made by the Fisher's test. Monod and competitive inhibition models were found to be the best at describing the studied BDS process and exhibited kinetic parameters with a real physical significance. They were able to describe the microbial growth, consumption of DBT and glucose, and production of 2-HBP. One of the important observations reported by Maass *et al.* (2015) in that study is that the increase in the concentration of dodecane from 20 to 80 % v/v in the studied system did not reflect any change in the kinetic behavior of BDS by *R. erythropolis* ATCC 4277 and the Monod was the model that presented a fit statistically better for both cases. Simulation of kinetic models to thte BDS process using just an organic phase (100 % v/v) has been also performed. Furthermore, that study revealed that *Rhodococcus erythropolis* ATCC 4277 is tolerant to high concentrations of organic phasing with a great affinity towards DBT, which is important to real field applications. The obtained yield coefficient, $Y_{x/s}$ (glucose), in that study indicated that the maintenance coefficient, *R. erythropolis* ATCC4277, is lower than those reported for other *R. erythropolis* IGTS8, which results in lower costs with a fermentation medium (del Olmo *et al.*, 2005a; Guchhait *et al.*, 2005). The higher $Y_{x/s}$ (DBT) indicated higher utilization of DBT as a co-substrate than *R. erythropolis* IGTS8, which means a higher BDS efficiency (del Olmo *et al.*, 2005a; Guchhait *et al.*, 2005).

Martinez *et al.* (2015) studied the effects of the supplementation of BDS-media with some of the metabolites of the Krebs cycle, such as citric acid, succinic acid, and acetic acid, on the BDS efficiency and the intracellular concentrations of both ATP and NADH, using resting cells of *Pseudomonas putida* CECT5279. The study was performed by a mixture of resting cells of two ages: 67% of cells grown for 23 h and 33% of cells grown for 5 h in batch and fed-batch operations.

The effect of the different co-substrates on the BDS capacity was evaluated using the following parameters:

The maximum BDS-capacity:

$$X_{BDS}^{max} = \frac{C_{HBP}^{max}}{C_{DBT}^{0}} \times 100 \tag{150}$$

where C_{DBT}^{0} is the initial DBT concentration (μM) used in the resting cell assay and C_{HBP}^{max} is the maximum concentration of the 2-HBP concentration (μM) produced during the time employed in the reaction.

Variation of the BDS capacity (X_{BDS}):
This parameter is used to compare the BDS capacity when different co-substrates are employed with respect to a control experiment.

$$\Delta X_{BDS} = \left(\frac{X_{BDS}^{max}\Big|_{C_{0-substrate}} - X_{BDS}^{max}\Big|_{control}}{X_{BDS}^{max}\Big|_{control}} \right) \times 100 \tag{151}$$

A positive value means an improvement in the BDS capacity, while a negative value means a decrease, where X_{BDS}^{max} BDS is the maximum percentage of BDS.

Variation of the BDS capacity estimated ($X_{BDS|M}$):
This parameter is employed to estimate the variation of the BDS capacity when the biocatalyst consists of the optimal mixture of cells, taking into account the contribution of each cellular age.

$$\Delta X_{\mathrm{BDS|M}} = [\Delta X_{\mathrm{BDS|5h}} \times 0.33] + [\Delta X_{\mathrm{BDS|23h}} \times 0.67] \qquad (152)$$

HBP productivity (P_{HBP}):
This parameter relates to the HBP production in each batch when the BDS is carried out in a fed-batch process.

$$P_{HBP} = \frac{\Delta C_{HBP}}{\Delta t} \qquad (153)$$

where C_{HBP} is the concentration of the produced 2-HBP (μM) in each batch and Δt is the time duration of each batch (min).
Initial substrate removal rate (R_j^0):
This parameter shows the initial rate of transformation (μmol/L/min) of each sulfur substrate, j (DBT, DBTO, $DBTO_2$ and HBPS), which can be considered as the initial rate of each 4S route reaction (R_i^0).

$$R_j^0 \approx r_i^0 = \left.\frac{dC_j}{dt}\right|_{t \to 0} \qquad (154)$$

where r_i corresponds to each 4S route reaction (i) as follows:

$$DBT \xrightarrow{r_1} DBTO \xrightarrow{r_2} DBTO_2 \xrightarrow{r_3} HBPS \xrightarrow{r_4} HBP \qquad (9.13.)$$

To obtain this parameter, the evolution with time of each 4S route compound was analyzed and derived.
This study showed that it is possible to improve the yield and BDS-rate by adding co-substrates, since the three examined compounds, acetic, citric, and succinic acids, provided positive results; nevertheless, the best improvement in the BDS capacity was achieved using 1.5% acetic acid. The yield of the process increased up to 140% when using single-aged (5h) cells in a batch process and to 122% using an optimized mixture of cells in a fed-batch process. The consumption of both cofactors, ATP and NADH, in a fed-batch process was very similar with and without a co-substrate (0.50 μmol_{NADH}/gX and 30 $nmol_{ATP}$/gX, respectively), using a consortium

of different resting cells' ages, but the initial intracellular concentrations of ATP and NADH also increased to 58 and 42%, respectively. The differences in the concentrations of the cofactors at the end of the process were much higher when acetic acid was supplied (74% for NADH and 181% for ATP). To determine the influence of acetic acid (1.5%) on each reaction of the 4S route, the initial substrate removal rate ($R_j^0 \approx r_i^0$) was also studied using the consortium of different cells' age in a batch process. The results showed that the initial rate of all of the reactions was higher upon the usage of 1.5% acetic acid. The improvements were 42, 30, 62, and 117% for r_1, r_2, r_3, and r_4, respectively. The increases in the rates of the first three reactions were expected due to the higher concentration of NADH. Moreover, a significant improvement has been observed in the fourth reaction rate, which is the rate-determining step, but is not NADH dependent. Briefly, the addition of co-substrate enhanced the reaction rates of the 4S route and, consequently, the overall rates of all of the reactions were enhanced (Martinez *et al.*, 2015).

Gomez *et al.* (2015) applied the logistic kinetic model equation to study the effects of hydrodynamic conditions on bacterial cultures of *R. erythropolis* IGTS8 under different operating conditions in a STBR. Hydrodynamic conditions were changed varying agitation and aeration. Gomez *et al.* (2015) performed a regime analysis based on characteristic time to provide information about which one is the controlling step by comparing the rates of the different steps. The characteristic or critical time constant is defined as the ratio between capacity and flow; therefore, a small time constant represents a fast process and vice versa. From the regime time analysis, the need for the investigation and optimization at a laboratory scale of certain mechanisms and features can be decided.

The characteristic time for oxygen consumption during growth, t_{OUR}, provides a critical oxygen depletion time and can be calculated from the following equation:

$$t_{OUR} = \frac{C^*}{OUR_{max}} \tag{155}$$

where C* is the saturation concentration in the liquid phase or, in another word, concentration in equilibrium (mol/m^3).

Upon the assumption that the growth rate of microbial cells can be modeled according to the logistic model equation, the maximum oxygen uptake rate (OUR_{max}, mol O_2/m^3/s), which is reached at the middle of the exponential growth stage, can be calculated as follows (Garcia-Ochoa *et al.*, 2010):

$$OUR_{max} = (m_{O_2} + Y_{xo}\mu)C_x \tag{156}$$

where μ is the specific growth rate, Y_{xo} presents the overall yield of cells on oxygen, and m_{O_2} is the oxygen consumption coefficient for maintenance (mol O_2 kg/X/s).

The characteristic time for oxygen transfer from gas to liquid, t_{OTR}, can be determined by the reciprocal of the volumetric oxygen transfer coefficient according to:

$$t_{OTR} = \frac{1}{k_L a} \tag{157}$$

where k_La is the volumetric oxygen mass transfer coefficient (s^{-1}). The characteristic time for mixing is the time required for complete mixing of the content of the bioreactor, that is the time necessary to pass from an initial non-homogeneous mixture to a certain degree of homogeneity, while a variable with power input and tank geometry is independent of the impeller type (Ma *et al.*, 2006; Nienow, 1997; 2010). For baffled tanks, the mixing time can be related to the design and operation parameters by the following equation (Godoy-Silva *et al.*, 2010):

$$t_{mix} = 57.4\left(\frac{T}{D}\right)^{1/3} \times \in^{-1/3} \times T^{2/3} \tag{158}$$

This equation indicates that at constant values of vessel diameter (T, m) and Stirrer Diameter (D, m) (fixed for any given equipment), the mixing time decreases only if the energy dissipation rate is increased.

Although the latter equation has been obtained for one single impeller in non-aerated conditions, it is claimed that it can be used for aerated systems by correcting the power input value (Martin *et al.*, 2008). The average energy dissipation rate, ε, i.e. total energy dissipation rate per unit mass (W/kg), can be calculated as:

$$\in = \frac{P}{\rho \times V} \tag{159}$$

where P is power input (W) due to agitation, V is the volume of the tank (m^3), and ρ is the density (kg/m^3).

The power input by the agitation, P, is mainly due to the stirrer speed and, depending on the tank geometry, can be calculated from the power number using the following equation:

$$N_P = \frac{P_0}{\rho \times N^3 \times D^5} \tag{160}$$

where N is stirrer speed (rpm or rps), N_p is the stirrer power number, and P_0 (for an un-aerated tank) is calculated; the equation of Michel and Miller (1962) can be selected to calculate the actual power applied to the gas–liquid dispersion according to:

$$P = \alpha \times \left(\frac{P_0^2 \times N \times D^3}{Q^{0.56}} \right)^{\beta} \tag{161}$$

where the values of the constants α and β depend on the stirrer type and the configuration of the agitation system. For a dual impeller turbine, like the one used in Gomez *et al.*'s (2015) work, values of α and β were of 1.224 and 0.432, respectively (Abrardi *et al.*, 1988).

Gomez *et al.* (2015), upon studying the effect of different stirring rates on the biomass growth and BDS activity of *R. erythropolis* IGTS8, found that the effectiveness factor for growth, η, changed with respect to specific growth rate and the stirrer speed (N) or the impeller Reynolds number:

$$R_e = \frac{\rho \times N \times D^2}{\mu_a} \tag{162}$$

where μ_a is the apparent viscosity (Pa s), D is the stirrer diameter (m), and ρ is the density (kg/m^3).

It was observed that oxygen limitation occurs at agitation conditions from approximately 100 to 250 rpm, with an optimization extent value (η) less than 1. This revealed the strong limitations of mass transfer rates that could occur at lower stirring speeds. However, with the increase of agitation speed from 250 to 450 rpm, the η is independent of the agitation conditions, values close to 1.0 are found (the same value as that is in shaken flasks at 210 rpm), mass transport limitations are overcome, and the growth rate is kinetically limited. This behavior is due to the combined effect of improved mixing and a decreased resistance to the transportation of nutrients resulting in an increment of cell growth rate to a point at which the transfer rate is fast enough and does not affect the growth rate any more. However, at a higher agitation speed (> 450 rpm) a sharp decrease in η occurs as the high stirrer speeds change the turbulence inside the stirred tank reactor, which in turn affects the cells, probably through shear effects. Thus, it was concluded that the cultures carried out runs at stirrer speeds higher than 450 rpm are harmful for growth. The impeller Reynolds number under these conditions is higher than 9375 and the value of the average power drawn (ε) is 0.6 W/kg (Gomez *et al.*, 2015). Garcia-Ochoa *et al.* (2013) reported the same observation for the culture of *Xanthomonas campestris*, where the growth rate was related to the oxygen limitation conditions or damage produced by hydrodynamic stress also using an optimization extent (η).

A bioprocess consists of a series of several successive steps like that of the 4S-pathway BDS process. The overall rate that is generally observed depends on the slower among those. These steps (oxygen transfer, biochemical reaction, etc.) take place at rates dictated by different phenomena (turbulence, heat transfer, and kinetics). The evaluation of each of these particular rates is a key revealing which of them is the one controlling the overall rate. One of the most instinctive methods of doing this is the comparison of the characteristic times of each step. Average values of the calculated characteristic times for oxygen consumption, oxygen transfer, and mixing in runs conducted at different stirrer speeds in batch BDS processes, using *R. erythropolis* IGTS8, have been reported (Gomez *et al.*, 2015), where the average values of m_{O_2} , $Y_{OX,}$ and $C^*$of 2.3×10^{-4} mol O_2 kg/X/s, 18.5 mol O_2 kg/X, and 0.245 mol/m^3, respectively, have been assumed considering the operational conditions at which the runs have been conducted. Mass transfer characteristic times over the range of the operating conditions studied were generally in the interval from 1578.5 to 16.1 s and mixing characteristic times were in the range from 43.3 to 6.2 s. Both characteristic times decrease when the stirrer speed is increased (from 100 to 700 rpm). The values estimated for mixing characteristic periods are within the range reported in the literature (Lu *et al.*, 1997; Jahoda *et al.*, 2007). A characteristic time for mixing much longer than that for oxygen uptake suggests that significant spatial differences in oxygen concentration could occur and the opposite situation indicates that dissolved oxygen (DO) concentration is homogeneous in the liquid phase. A characteristic time for oxygen transfer greater than that for oxygen uptake indicates potential oxygen transfer limitations. According to the obtained results, the characteristic times for oxygen uptake, t_{OUR}, are above 263.5 s. These values suggest that for agitation between 100 and 350 rpm, oxygen transfer limitation will occur in the STBR because the oxygen transfer time is much longer than the oxygen consumption time. However, mixing time is short compared to oxygen transfer time, indicating homogeneous DO concentration.

The regime analysis performed by Gomez *et al.* (2015) showed that the mixing characteristic time is much lower than transfer time and oxygen consumption time, indicating that the mixing time for the different agitation regimes is acceptable and that there are no mixing limitations. The minimum stirrer speed necessary for a satisfactory performance of the bioreactor from the point of view of the IGTS8 cells oxygen demand was found to be 350 rpm, while at stirrer speeds over 450 rpm, growth rate and desulfurization capacity are both negatively affected.

In a similar study, Escobar *et al.* (2016) applied the logistic kinetic model equation to evaluate the effects of hydrodynamic conditions on growth by changing the stirrer speed in a STBR using the genetically engineered *Pseudomonas putida* KTH2. This parameter was calculated as the effectiveness factor for growth and the ratio between the growth rates, in the presence of external limitations (bad mixing conditions, oxygen transfer limitations, or cell damage), was calculated under the best conditions, that is when the growth rate is the controlling step of the overall process rate and is not affected by any other physical phenomena, according to the following equation:

$$\eta = \frac{\mu}{\mu_{max}} \tag{163}$$

In that batch operation, the following mass balance equation for the dissolved oxygen concentration, assuming a well mixed liquid phase has been applied:

$$\frac{dC_{O_2}}{dt} = OTR - OUR \tag{164}$$

where the OTR and OUR are the oxygen- transfer and uptake rates (mol O_2/L/s), respectively, and the OTR is proportional to the mass transfer coefficient, k_L, the specific interfacial area, a, and the difference between the saturation concentration and the dissolved oxygen concentration in the broth.

The OTR was calculated from the following equation:

$$OTR = k_L a(C^*_{O_2} - C_{O_2}) \tag{165}$$

The OUR was calculated using the following equation:

$$OUR = k_L a(C^*_{O_2} - C_{O_2}) - \left(\frac{dC_{O_2}}{dt} \right) \tag{166}$$

where the C_{O_2} and $C^*_{O_2}$ are the dissolved oxygen concentrations in the broth and in equilibrium at the gas–liquid interface, respectively. In that study, $C^*_{O_2}$ was 0.21 mol/m^3.

In another study performed by Martinez *et al.* (2016) to study the effect of the hydrodynamic conditions of the oxygen transfer rate during the scale-up of a BDS process using mixed culture of resting cells with different ages of *Pseudomonas putida* CECT5279, the change of dissolved oxygen concentration with time, (dC_{O_2}/dt), in the broth was expressed as the difference of OTR and OUR (eq. 164). The OTR is described as being proportional to the oxygen concentration gradient using the volumetric mass transfer coefficient, k_{La}, as the proportional constant (eq. 165). The OUR was expressed as the product of the specific oxygen consumption, q_{O_2}, and the biomass concentration:

$$OUR = q_{O_2} \times C_x \tag{167}$$

where q_{O_2} is the specific oxygen consumption rate (mol_{O_2}/g/s) and C_x is the concentration of biomass (g DCW/L).

The k_La in a shaken flask must be determined using empirical equations, such as eq. 168, which was proposed by Liu *et al.* (2006) and modified by Garcia-Ochoa *et al.* (2013):

$$k_L a = 1.44 \times 10^{-3} \times N^{0.88} \left(\frac{V_F}{V_L} \right)^{0.8} \left(\frac{\mu_L}{\mu_W} \right)^{-0.4} \tag{168}$$

where N is the stirrer speed (rpm), V_F and V_L are the volume (mL) of the flask and liquid, respectively, and μ_L and μ_W are the viscosity (Pa s) of the medium and water, respectively.

Martinez *et al.* (2016) evaluated the BDS capacity of X_{BDS} at different stirring speeds of 200–500 rpm, after 180 min, using an airflow of 1 vvm, where the X_{BDS} increased with the stirring speed and recorded its maximum at 400 rpm, where a complete conversion of 25 μmol/L DBT to 2-HBP was obtained at 400 rpm and 1 vvm within 180 min, while upon varying the airflow from 1–3 vvm and decreasing the time to 90 min. The maximum conversion to 2-HBP occurred at 2 vvm within 90 min. Thus, the optimum condition for the BDS of 25 μM DBT in STBR using *P. putida* CECT5279 were 400 rpm, 90 min, and 2 vvm. Upon the study os different stirring speeds, the evolution of the dissolved oxygen (DO) fell to zero and the time needed to deplete the oxygen increased with the k_La value. This suggests that the potential value of OUR was higher than the actual value of OTR. The k_La recorded its highest value of 1.68×10^{-2} s^{-1} at 400 rpm and 3 vvm. It is important to keep in mind the fact that the microorganism consumes all the oxygen and that just because it is transported, does not imply an increase in BDS capacity with OTR. Thus, this study proved that the 4S route is highly sensitive to oxygen availability working under oxygen limiting conditions. Moreover, the hydrodynamic conditions highly affect the 4S route, but not the metabolism of cell growth. Upon comparing the results obtained with shaken flasks and STBR, they were found to give the same BDS-capacity. This occurred because both have the same hydrodynamic conditions since the most important principle for the scale-up of aerobic bioprocesses consists of maintaining a constant OTR defined by k_La at different scales (Garcia-Ochoa and Gomez, 2009), so by applying eq. 168, it was possible to calculate the k_La values for the shaken flask operation at different agitation speeds; when BDS is carried out at 200 rpm, the value of k_La was recorded 1.39×10^{-2} s^{-1} when BDS was conducted in STBR and a predictive method was employed to estimate k_La (Garcia-Ochoa and Gomez, 2009).

The k_La values can also be correlated with the combination of stirrer speed, N, superficial gas velocity, V_S, and liquid effective viscosity, μ_a, obtaining equations such as the following:

$$k_L a = C \times V_s^a \times (P/V)^b \mu_a^c \tag{169}$$

where the constant, C, depends on the geometric parameters of the vessel and the stirrer employed. Other equations proposed substitute the average power input per volume, P/V, by the effect of stirrer speed, N. The exponent (a, b, and c) values, show a wide variation range in the different correlations proposed by different authors: $0.3 \leq a \leq 0.7$, $0.4 \leq b \leq 1$, and $-0.4 \leq c \leq -0.7$. For non-viscous systems, the most frequently used correlation is that from Van't Riet (1979).

Upon the best conditions previously determined (400 rpm and 2 vvm of aeration), the value obtained for k_La was 1.37×10^{-2} s^{-1}, which was almost the same as that calculated at a shaken flask scale. Consequently, Martinez *et al.* (2016) proved with these results the suitability of the constant-k_La criterion for STBR scale-up.

Gomez *et al.* (2015) divided the change of the effectiveness factor for growth, η, with the operating conditions into a bioreactor of aerobic bioprocesses into three stages: the first stage occurs when the fluid dynamic conditions concomitantly occur with the increase of the OTR, which provides an increase on the growth rate (e.g. by changing stirrer speed) because this is the limiting step controlling the overall rate of the process. The second stage is the plateau stage that corresponds to the situation where the oxygen transfer rate (OTR) is equal to or greater than the maximum oxygen uptake rate (OUR), which is the controlling step of the overall rate. The third stage occurs upon further increment in agitation speed; this stage is undesirable because it occurs concomitantly with the occurrence of some cell damage or some change in the metabolism provoked by a hydrodynamic or oxidative stress.

The BDS capacity of the cells accumulated during the growth was monitored through resting cell desulfurization assays. For a quantitative analysis of the desulfurization capacity under different agitation conditions, the dibenzothiophene (DBT) conversion (X_{DBT}) and the yield of 2-hydroxybiphenyl ($X_{2\text{-}HBP}$) were calculated as follows:

$$X_{DBT} = \frac{C_{DBT}^{t,0} - C_{DBT}^{t}}{C_{DBT}^{t,0}} \times 100 \tag{170}$$

$$X_{HBP} = \frac{C_{HBP}^{t}}{C_{DBT}^{t,0}} \times 100 \tag{171}$$

where the C_{DBT}^{t} and C_{HBP}^{t} were DBT and 2-HBP concentrations at 120 min of resting cell assay.

The enzymes activities were determined in resting cell assays from cells of different time courses of growth using DBT, DBTO, $DBTO_2$, and HBPS as sulfur substrates, independently, and then were calculated at the specific time-zero reaction rate according to equation:

$$a_j = \left. \frac{-dC_j}{C_X \cdot dt} \right|_{t=0} \tag{172}$$

where C_j is the concentration of the enzyme substrate, j, and C_X is the concentration of the biomass employed in the resting cell assays.

Gomez *et al.* (2015) found that upon using different *R. erythropolis* biocatalyst ages, the values of X_{DBT} and X_{HBP} change with agitation speed.

For stirrer speeds above 450 rpm, a decrease in the consumption rate of DBT is evident and, therefore, lower HBP concentrations are obtained. Furthermore, an increase of the agitation from 100 to 450 rpm has a positive influence on the development of the desulfurization capacity of the cells during growth, where X_{DBT} recorded 60% and 100% at 100 and 450 rpm, reaching 100%.

Escobar *et al.* (2016) observed results revealing that at stirrer speeds < 700 rpm, the biochemical reactions were limited by mass transfer and η values increased from 0.53 (400 rpm) to 1.0 (700 rpm). These results proved that limitations of the mass transfer rate could occur when agitation < 700 rpm. The increase of η was attributed to the increase of the oxygen transport rate, OUR, and, consequently, the growth rate. Nevertheless, η recorded a constant value of 1 at an agitation speed of 700 to 2000 rpm, which was attributed to the overcome of the mass transport limitations and the process rate was determined by the cell metabolism reaching the maximum growth rate. That is, the OUR increased with the increase of the stirring rate (N) because OUR is taking the highest possible value under these conditions, being equal to the OTR value. When N is increased, k_L and a increase and, therefore, OTR increases and OUR can take higher values until the maximum value is reached. Then, the increase of N will provoke an increase of k_La and OTR, but now OUR value is not affected and the value of the effectiveness factor is constant and equal to 1. The changes of X_{DBT} and X_{HBP} with different agitation speeds and resting cell ages proved that both parameters were constant within the agitation speed of 100 and 300 rpm. However, the cell age at this stirring speed range expressed a significant impact. The X_{DBT} recorded 50% for cells of 5 h and 100% with cells of 15 and 23 h of age, while the X_{HBP} recorded 15% at 5 h and increased to approximately 40 and 60 % for cells of 15 and 23 h, respectively. Then, they were decreased at higher speeds and lost at 800 rpm.

Escobar *et al.* (2016) followed up the kinetics of each enzyme, DszA, DszB, and DszD, where the recombinant protein was expressed in and purified from *E. coli*. The kinetic data obtained were appropriately modeled by the Michaelis-Menten model.

$$v = \frac{k_{cat}[E_0][S]}{K_m + [S]} = \frac{V_{max[S]}}{K_m + [S]} \tag{173}$$

where v is the enzyme activity, k_{cat} is the turnover number, E_0 is the enzyme concentration in each assay, S is the substrate concentration, K_m is the

Michaelis constant, and V_{max} is the maximum enzyme activity. The kinetics of DszC could not be modeled accurately with a simple Michaelis-Menten model because it was found that DszC was inhibited by its substrate, DBT. Thus, Haldane (i.e. the inhibition mode) was applied to fit the data obtained with DszC.

$$v = \frac{V_{max[S]}}{K_m + [S] + \frac{[S]^2}{K_{SI}}} \tag{174}$$

where K_{SI} is the substrate (DBT) inhibition constant. The catalytic efficiency of an enzyme is reported to be best defined by the ratio of the kinetic constants k_{cat} and K_m (Copeland, 2000).

The concentrations of DszA, DszB, and DszC in that study were assumed to be all the same and equal to E_0, which was obtained by allowing its value to vary until the initial desulfurization rate predicted by the model matched the measured desulfurization rate and was found to be 15 mg/mL, so that the total concentration of desulfurization enzymes in the cytoplasm was 45 mg/mL. This value was found to be of the same order as the total cytoplasmic concentration of protein which has been reported to be approximately 200 mg/mL (Ellis, 2001).

The k_{cat} values of DszB, DszA, and DszD were found to be approximately 1.7 ±0.2, 11 ± 2, and 760±10 min^{-1}, respectively. The K_m value of DszB was calculated to be approximately 1.3 ± 0.3 μM. Nevertheless, DszC was inhibited by its substrate (DBT) and its kinetic parameters were found to be approximately 1.6 ± 0.3 min^{-1}, 1.4 ± 0.3 M, and 1.8 ± 0.2 M for k_{cat}, K_m, and K_{SI}, respectively, while the activity of DszC was found to be 31.3 nmol $DBTO_2$/mg DszC/min. The catalytic efficiencies of DszA, DszB, and DszC were calculated to be approximately 3.1, 1.3, and 1.1 $\mu M^{-1} min^{-1}$, respectively. Furthermore, the catalytic efficiency of DszD on NADH and FMN was calculated to be 6.7 and 100 $M^{-1}min^{-1}$, respectively. Consequently, the efficiency of the 4S-enzymes was suggested to be listed in the following decreasing order: DszD > DszA > DszB ≈ DszC.

Copeland (2000) reported the dose-response equation describing the effect of the inhibitor concentration on enzyme activity:

$$\frac{v_i}{v_0} = \frac{1}{1 + \frac{[I]}{IC_{50}}} \tag{175}$$

where v_i is the enzyme's activity at an inhibitor concentration of I, v_0 is the enzyme's activity in the absence of inhibitor, and IC_{50} is the concentration of the inhibitor required to reduce the enzyme's activity by 50% and it is phenomenological and has no mechanistic implications. The value of IC_{50} can be obtained by fitting the data for v_i/v_0 versus I to the dose-response equation using the nonlinear fitting package, nlinfit, in the MATLAB program.

Abin-Fuentes *et al.* (2013) reported the four major inhibitory interactions of the 4S-pathway in an order of decreasing strength, as follows: DszC inhibition by HBPS (IC_{50} = 15 ± 2 μM), DszC inhibition by HBP (IC_{50} 50 ± 5 μM), DszA inhibition by HBP (IC_{50} 60 ± 5 μM), and DszB inhibition by HBP (IC_{50} 110 ± 10 μM). It is important to note that the major inhibitory compounds in the pathway are the last two intermediates, HBPS and HBP. Consequently, the 4S-pathway suffers from the pattern of inhibition from the type of feedback inhibition of linear pathways; for example, in the tricarboxylic acid (TCA) cycle, the first enzyme in the pathway is strongly inhibited by the end product, ATP.

Abin-Fuentes *et al.* (2013) concluded that IC_{50}s of all four inhibitory interactions are significantly less than the estimated cytoplasmic HBP concentration. Even in the best case scenario, when 50 g/L of a highly HBP-selective resin was added to the BDS mixture, the estimated cytoplasmic HBP concentration was still 260 M. This finding suggests that these four inhibitory interactions are likely responsible for the reduction in biocatalyst activity that is observed during a typical BDS process when HBP is generated endogenously from DBT within the biocatalyst.

The model for non-competitive inhibition of an enzyme that obeys Michaelis-Menten kinetics is given by Copeland (2000):

$$v = \frac{V_{max}[S]}{K_m\left(1+\frac{[I]}{K_i}\right)+[S]\left(1+\frac{[I]}{\alpha K_i}\right)} \quad (176)$$

where I is the inhibitor concentration, K_i is the inhibition constant, and α is a parameter that reflects the effect of the inhibitor on the affinity of the enzyme for its substrate and, likewise, the effect of the substrate on the affinity of the enzyme for the inhibitor. Non-competitive inhibition refers to the case in which an inhibitor displays binding affinity for both the free enzyme and the enzyme-substrate binary complex. This form of inhibition is the most general case. In fact, competitive and uncompetitive inhibition can be viewed as special, restricted cases of non-competitive inhibition in which the values of α are infinity and zero, respectively.

Abin-Fuentes *et al.* (2013) reported the occurrence of the telltale sign of non-competitive inhibition, which is a reduced V_{max} without a change in K_m as the inhibitor concentration is increased, in the inhibition of DszC by both HBPS and HBP in *R. erythropolis* IGTS8 and suggested a modified non-competitive model:

$$v = \frac{V_{max[S]}}{K_m\left(1+\frac{[I]}{K_i}\right)+[S]\left(1+\frac{[I]}{\alpha K_i}\right)+\frac{[S]^2}{K_{SI}}} \tag{177}$$

The K_i and α for DszC of *R. erythropolis* IGTS8 were 13.5 μM and 0.13, respectively, for HBPS inhibition and 40 μM and 0.4, respectively, for HBP. These parameters were calculated by graphical plots; the construction of the double-reciprocal Lineweaver-Burk plot and the Dixon plot of 1/ $V_{max,}$ as a function of I, from which the value of $-\alpha K_i$ can be determined as the x intercept, while the slopes of the double-reciprocal lines (from the Lineweaver-Burk plot) are plotted as a function of I. For this plot, the x intercept is equal to Ki. Unfortunately, the mechanism of DszB inhibition by HBP and DszA inhibition by HBPS could not be determined due to the limited resolution of the HPLC detector.

Kareem *et al.* (2013a) carried out the anaerobic biodesulfurization of diesel using an isolated bacterial strain, *Desulfobacterium indolicum*.

The mass balance on the OSCs in diesel resulted in the following model equation:

$$\frac{v_{max,i}C_i}{K_{m,i}+C_i} = -(1+K_i)\frac{dC_i}{dt} \tag{178}$$

where K_i is the distribution coefficient of the OSCs between diesel and water. The direct integration of the Michaelis-Menten equation yielded:

$$K_m ln\left(\frac{C_o}{C}\right)+(C_o - C) = v_{max}t \tag{179}$$

Manipulation of equation 179 gave a more explicit result:

$$\frac{C_o - C}{ln\frac{C_o}{C}} = -K_m + v_{max}\frac{t}{ln\frac{C_o}{C}} \tag{180}$$

Different OSCs in diesel oil were monitored. The kinetic parameters of BDS, like the maximum velocity rate constant, $V_{max,}$ and Michaelis-Menten constant, $K_{m,}$ were determined for BT and DBT were calculated to be 0.540 and 5.992 mg/L/h and 19.837 and 192.782 mg/L, respectively.

The partition coefficient was estimated through the efficacy of combining linear solvation energy relationships (LSERs) developed for pure systems through the application of the linear solvent strength theory of Arey and Gschwend (2005):

$$logK_{i,fw} = log\left[\frac{RT}{V_f P_i^o}\right] - (C_{aw} + r_{aw}R_2 + S_{aw} \sqcap_2^H + a_{aw}\alpha_2^H + b_{aw}\beta_2^H + V_{aw}V_x) \tag{181}$$

where R is the universal gas constant, T is the temperature at which the coefficient is measured, V_f is the molar volume of diesel, P_i^o is the vapor pressure of the components of the fuels parameters, R_2 describes the excess molar refraction of solute i, $\sqcap_2^H$ describes the polarity/polarizability of solute i, α_2^H describes the hydrogen-bonding acidity of solute i, β_2^H describes the hydrogen bonding basicity of solute i, and Vx describes the group-contributable molecular volume of solute i, while c, r, s, a, b, and v are adjusted coefficients specific to the two-phase system. The $K_{i,fw}$ recorded 2.312 and 2.038 for BT and DBT, respectively. A mathematical model showing the pattern of diesel biodesulfurization was developed. The developed model was solved numerically by the Finite Difference method. The model was able to describe all the experimental data with good statistical parameters and the simulated data were compared with the experimental findings to validate the model developed; the simulated results showed very good agreement with the experimental findings.

Kareem *et al.* (2013b) also modelled the BDS of kerosene by taking a mass balance on the substrates, i.e. the OSCs in the kerosene.

The material balance on the solute (substrates in kerosen) over the time period from t to t + Δt over the element of volume of a batch reactor from z to z + Δz was obtained:

(Input at z) - (Output at z + Δz) - (Reaction due to BDS) - (Rate of transfer to the cell surface) = Accumulation over time period (182)

In a batch reactor there is no inflow or outflow of material, thus:

(Accumulation over time period) = (Rate of transfer of the substrate to the cell surface) - (Reaction due to biodesulfurization) (183)

The mass transfer rate to the solid is:

$$r_m = ak_L(C\text{-}CS)A\Delta z \tag{184}$$

The accumulation in the fluid phase is:

$$r_{ac} = A\Delta z\,(dC/dt) \tag{185}$$

There was no reaction in the fluid phase in the reactor since all the reactions were assumed to take place on the cell surface, hence, the reaction term was considered equal to zero. Thus, by substituting eqs. 184 and 185 in eq.182, the following expression is obtained:

$$k_L a(C - C_s)A\Delta z = A\Delta z\left(\frac{dC}{dt}\right) \tag{186}$$

Dividing eq. 186 by $A\Delta z$:

$$k_L a(C - C_s) = \left(\frac{dC}{dt}\right) \tag{187}$$

where C_S is the concentration of the substrate on the surface of the organism which is not easily measurable. This was solved by applying the law of conservation of mass to the absorbable solute contained in the fluid phase and the solid based on the assumption that the adsorption transfers material from the fluid phase and adds to the solid phase, i.e. the microorganism *Desulfobacterium anilini*, taking into consideration that the solid phase loses material by desulfurization and generates none. Thus, the solid phase mass balance for the sulfur specific reductive BDS pathway is expressed as follows:

$$A\Delta z\frac{dq}{dt} - A\Delta z\left(\frac{v_{max}C}{K_{m+C}}\right) = k_L a(C - C_s)A\Delta z \tag{188}$$

Dividing eq. 188 by AΔz:

$$\frac{dq}{dt} - \left(\frac{v_{max}C}{K_{m+C}}\right) = k_L a(C - C_s) \tag{189}$$

Substituting eq. 189 into eq. 187:

$$\frac{dq}{dt} - \left(\frac{v_{max}C}{K_{m+C}}\right) = \frac{dC}{dt} \tag{190}$$

The solutions of eq. 190 were assumed to be simple when q is a linear function of C, that is if the adsorption was assumed to be linear, then dq/dt can be replaced by -KdC/dt, thus:

$$\frac{dq}{dt} = K\frac{dC}{dt} \tag{191}$$

where K is a distribution coefficient:

$$\frac{v_{max}C}{K_m + C} = -(1+K)\frac{dC}{dt} \tag{192}$$

where eq. 192 represents the mass transfer based kinetic rate expression for the specific reductive BDS pathway and it is a first order differential equation which was solved numerically using the Finite Difference Method. The simulated results were found to have a high correlation with known and reliable experimental data. The maximum velocity rate constant and the Michaelis-Menten constant, $K_{m,}$ were obtained by subjecting the obtained parameters from linear plots to the Macquardt's non-linear regression analysis. The V_{max} and K_m were found to be 0.103 mg/L/h and 0.575 mg/L for thiophene (Th) and 1.350 mg/L/h and 54.700 mg/L for 2,5-dimethylthiophene (2,5-DMTh), respectively. The partition (distribution) coefficients ($K_{i,fw}$) of the organic sulfur compounds in the kerosene used in the simulation were estimated through the efficacy of combining linear solvation energy relationships (LSERs) developed for pure systems through the application of the linear solvent strength theory (eq.181). The $K_{i,fw}$ for Th and 2,5-DMTh recorded 1.281 and 1.054, respectively. The quantitative parameters estimated in the course of modelling the anaerobic biodesulfurization of kerosene (Kareem *et al.*, 2013b) would be useful in a bioreactor design of the process that would eventually take the technology to the market.

Moreover, Kareem *et al.* (2014) investigated the kinetics of the anaerobic biodesulfurization of diesel. This was done by simulating the kinetics of the process alone and then with and without the effect of mass transfer. The kinetic parameters V_{max} and K_m of the Michaelis-Menten equation were estimated using linear equations of Hanes, Lineweaver-Buck, and Eadie-Hofstee plots (Eadie, 1942; Hofstee, 1959). However, the values obtained for each parameter from the linear equations were close, but not the same. This necessitated the need to carry out a non-linear regression analysis using Marquardt's algorithm (Marquardt, 1963), where the parameters were estimated and then subjected to Marquardt iterative nonlinear least squares regression and the obtained bio-kinetic parameters were used to model the BDS of diesel by *Desulfobacterium anilini* in a batch reactor. The V_{max} and Michaelis-Menten constant, $K_{m,}$ determined for BT and DBT were calculated to be 0.572 and 6.118 mg/L/h and 18.050 and 182.278 mg/L, respectively. The mass balance was performed by taking the mass balance of the substrates and the sulfur containing hydrocarbons in the fuel and diesel were the same as reported by Kareem *et al.* (2013b). The distribution coefficient, $K_{i,fw}$, was estimated using the model developed by Arey and Gschwend (2005) for sulfur-containing organic substances in the fuel phase (eq. 181). The $K_{i,fw}$ for BT and DBT recorded 2.312 and 2.038, respectively. The Implicit Finite Difference Method was used to obtain a substrate concentration-time data from eq. 192. The results were presented in terms of a percentage biodesulfurization-time profile. There was good agreement between simulated and experimental data proving that the assumptions made in developing the models were correct. The kinetics of the biodesulfurization of diesel by *Desulfobacterium anilini* has been found to be mass transfer driven (Kareem *et al.*, 2014).

Similar work has been published by Kareem (2016), where the kinetics of the anaerobic BDS of kerosene using *Desulfatiglans anilini* comb. nov with and without the influence of mass transfer were investigated. The bio-kinetic parameters were estimated using the various methods of Hanes (eq. 193), Lineweaver-Buck (eq. 194), and Eadie-Hofstee (eq. 195) all from the linear transformation of the Michaelis-Menten equation. The obtained results when compared caused confusion, thus the Michaelis-Menten equation was integrated and the bio-kinetic parameters were estimated. The V_{max} and Michaelis-Menten constant, $K_{m,}$ were determined for Th and 2,5-DMTh were calculated to be 0.104 and 1.426 mg/L/h and 0.548 and 6.7 mg/L, respectively. The partition coefficients of sulfur containing hydrocarbons in kerosene and in water were used instead of the mass transfer coefficient; this was done using the Arey and Gschwend model (eq.181). The developed model equations were solved numerically using the Finite Difference Method. Furthermore, the simulated concentration-time profile of the mass transfer plus kinetics has a better agreement with the known experimental values than those with kinetics alone. The good agreement of simulated data and the experimental ones proved that the assumptions made in developing the models were valid. The result showed that mass transfer played a significant role on the kinetics of the process.

The Hanes-Woolf plot, known as the Hanes plot for short, is obtained by rearranging the Michaelis-Menten equation, such that:

$$\frac{C}{v} = \frac{C}{v_{max}} + \frac{K_m}{v_{max}} \tag{193}$$

The transformation gives the best applying $1 / v_{max}$ instead of v_{max}, but the main drawback of this transformation is dependent of both abscissa and ordinate on the substrate concentration.

The Lineweaver-Burk plot can be obtained by linearly transforming the Michaelis-Menten equation to:

$$\frac{1}{v} = \frac{1}{v_{max}} + \frac{K_m}{v_{max}}\frac{1}{C} \tag{194}$$

The double reciprocal plot is prone to error because it distorts the error structure of the data. In the Eadie-Hofstee plot, the Michaelis-Menten equation can be linearly transformed so that the reaction rate is plotted as a function of the ratio between the rate of reaction and substrate concentration:

$$v = -K_m \frac{v}{C} + v_{max} \tag{195}$$

The abscissa and ordinates are independent variables both dependent on the reaction rate, so, like the Hanes-Woolf plot, any experimental error will be present in both axes. Parameters estimated from this plot are more reliable than those of Eadie-Hofstee and Lineweaver-Burk because they give equal weight to data points in any given range of substrate concentration or reaction rate. All these short comings may be attributed to the nonlinearity of the Michaelis-Menten equation itself, thus a non-linear regression method will give a better estimate of the kinetic parameters.

Zhang *et al.* (2013) studied the interactions among three typical Cx-DBTs, such as dibenzothiophenes (DBT), 4-methyldibenzothiophene (4- MDBT), and 4,6-dimethyldibenzothiophene (4,6-DMDBT), using *Mycobacterium* sp. ZD-19 in an airlift reactor.

The specific desulfurization rate in the airlift reactor, V_i, according to the multi-substrate competitive inhibition kinetics using the Michaelis–Menten pattern was described as follows:

$$V_i = \frac{V_{max,i} C_{si}}{K_{si} + \sum_{J=1} \frac{K_{si}}{K_{sj}} C_{sj}} \tag{196}$$

where V_i is the specific desulfurization rate of substrate i in the multiple substrate system, C_{si} is the concentration of substrate i, K_{si} is the half-saturation constant of substrate i, K_{sj} and C_{sj} are the half-saturation constant and concentration of another co-existing substrate, j, in the multi-substrate system. Additionally, a fractional velocity rate equation (Segel, 1975) was used to estimate the inhibition of a multi-substrate on a sole substrate in the competitive inhibition system.

$$\frac{V_i}{V_o} = \frac{V_{max,i} C_{si}}{K_{si} + \sum_{J=1} \frac{K_{si}}{K_{sj}} C_{sj}} \tag{197}$$

where V_0 is the biodesulfurization rate of substrate i in the sole substrate system and V_i is the biodesulfurization rate of substrate i in the multiple substrates system.

The Michaelis–Menten kinetic parameters were obtained from the sole substrate experiments using a Lineweaver–Burk plot method. The values of V_{max} exhibited a decrease tendency for each additional alkyl substitution. The V_{max} values of DBT, 4-MDBT, and 4,6-DMDBT were 0.328, 0.233, and 0.0.92 mM/h, respectively, indicating that the biodesulfurization rate decreased for each additional alkyl substitution. The K_m values of DBT, 4-MDBT, and 4,6-DMDBT were 0.591, 0.747, and 0.590 mM, respectively. The higher K_m value was considered as an indication of a weaker affinity of the *Mycobacterium* sp. with substrate. Though a weaker affinity of the *Mycobacterium* sp. with 4-MDBT was obtained compared to that with 4,6-DMDBT, a higher desulfurization of 4-MDBT was achieved compared with that of 4,6-DMDBT. These results were attributed to the steric hindrance effect that prevented *Mycobacterium* sp. ZD-19 from attacking C–S for desulfurization of 4,6-DMDBT. Briefly, the experimental results indicated that the desulfurization rates would decrease in the multiple Cx- DBTs system compared to the single Cx-DBT system. The extent of inhibition depended upon the substrate numbers, concentrations, and affinities of the co-existing substrates. For example, compared to the individual desulfurization rate (100 %), the DBT desulfurization rate decreased to 75.2 % (DBT + 4,6-DMDBT), 64.8 % (DBT + 4-MDBT), and 54.7 % (DBT + 4,6-DMDBT + 4-MDBT), respectively. This phenomenon was caused by an apparent competitive inhibition of substrates, which was well predicted by a Michaelis–Menten competitive inhibition model (eq. 196). The DBT desulfurization rate decreased to 0.093 and 0.108 mM/h in the presence of 4-MDBTand 4,6-DMDBT, respectively. The 4-MDBT desulfurization rate decreased from 0.121 to 0.074 and 0.092 mM/h in the presence of DBT and 4,6-DMDBT, respectively, while the 4,6-DMDBT desulfurization rate decreased from 0.023 to 0.013 and 0.014 mM/h in the presence of DBT and 4-MDBT, respectively. The fractional velocity rate equation (eq. 197) estimated the inhibition of multiple substrates on a single substrate. In both binary and ternary systems, the presence of other Cx-DBTs had a significant inhibition effect on the desulfurization of DBT. Similarly, the initial desulfurization rates of 4-MDBT and 4,6-DMDBT also decreased in the binary and ternary mixtures compared with the individual substrate condition. An indication of the competitive effect of the co-existing substrates on a sole substrate were determined by the term

$$\left(\sum_{J=1} \frac{K_{si}}{K_{sj}} C_{sj}\right)$$

in the denominator of the fractional velocity rate equation (eq. 197). The obtained results of the single, binary, and ternary systems indicate that the extent of the competitive inhibition in the multiple substrates system depended upon the numbers, concentrations, and affinities of the co-existing substrates (Zhang *et al.*, 2013).

9.11 Evaluation of the Tested Biocatalysts

The evaluation of the tested biocatalysts can be performed by investigation of:

9.11.1 Kinetics of the Overall Biodesulfurization Reaction

Although BDS the via 4S-pathway takes place in several steps, the overall process can be summarized by BDS of DBT to the dead end product 2-HBP. Moreover, the overall rate kinetics are affected by the concentration and distribution of DBT.

$$\frac{dC_{2HBP}}{dt} = kC_{DBT_O} C_X \tag{198}$$

where k is the rate coefficient (cm³/sec), t is the incubation period (sec), C_{DBT} is the initial DBT concentration (ppm), C_X is the cell concentration (cells/cm) (i.e. the concentration of bacteria), and $C_{2\text{-HBP}}$ is the 2-HBP concentration (ppm).

The integrated form of eq. 198:

$$\ln\left(1 - \frac{C_{2HBP}}{C_{DBT_O}}\right) = -kC_x t \tag{199}$$

$$\left(1 - \frac{C_{2HBP}}{C_{DBT_O}}\right) = \exp(-kC_x t) \tag{200}$$

$$C_{2HBP} = C_{DBT_O}[1 - \exp(-kC_x t)] \tag{201}$$

9.11.2 Maximum Percentage of Desulfurization (X_{BDS}^{MAX} %)

This parameter indicates the maximum desulfurizing capability of cells obtained during the resting cell assay:

$$X_{BDS}^{MAX} = \frac{C_{2HBP}^{MAX}}{C_{DBT_O}} \times 100 \tag{202}$$

where C_{2HBP}^{MAX} is the maximum 2-HBP concentration obtained during the time employed in the resting cell assay and C_{DBT_O} is the initial concentration of DBT used to perform desulfurization with resting cell assay.

9.11.3 Time for Maximum Biodesulfurization $t_{BDS_{max}}$ (min)

This parameter indicates the time at which the maximum percentage of biodesulfurization has been reached in the resting cell assay.

9.11.4 Initial DBT Removal Rate R^{o}_{DBT} (μmol/L/min)

This parameter shows the initial rate of DBT transformation. It is calculated by applying differential method over the evolution of the DBT concentration throughout time in order to estimate the value of the DBT removal rate at time, zero, according to the following expression:

$$R^{O}_{DBT} = \left. \frac{dC_{DBT}}{dt} \right|_{t=0} \tag{203}$$

9.11.5 Maximum Productivity P^{MAX}_{BDS} (%/min)

This parameter relates maximum percentage of desulfurization to the time needed for this conversion:

$$P^{MAX}_{BDS} = \frac{X^{MAX}_{BDS}}{t_{BDS_{max}}} \tag{204}$$

9.11.6 Specific Conversion Rate (SE %L/g/min)

This parameter regards the performance of each tested biocatalyst. It relates X^{MAX}_{BDS}, concentration of biomass, C_X, and time needed for the maximum achieved conversion, according to:

$$SE = \frac{X^{MAX}_{BDS}}{t_{BDS_{max}} C_X} \tag{205}$$

For example, for the genetically modified *Pseudomonas putida* CECT 5279 and the maximum in vivo activities of monooxygenase enzymes (DszA and DszC) are shown when the late exponential growth phase is reached (23 h), while the desulfinase enzyme DszB presents a maximum activity during the early exponential growth phase (5 h) (Calzada *et al.*, 2009b). Also, a consortium of these two cell ages was reported to yield excellent BDS results (Calzada *et al.*, 2009a). Calzada *et al.* (2011) optimized the ratio and total biomass concentration of both 5 h and 23 h growth time in cells in a complex consortium for desulfurization by performing resting cells of biodesulfurization assays using dibenzothiophene (DBT) as a sulfur model compound. A particular cell mixture containing 66.7% of 23 h growth time cells was found to work as the most effective desulfurization biocatalyst. The authors showed that the complex biocatalyst that combines 5 and 23 h cells of $Y^{23} = C_X^{23}/C_X^{5+23}$ achieve better behavior for BDS than the 9 h simple catalyst. That led to a higher DBT conversion into 2-HBP, a higher initial DBT removal rate, and maximum productivity. Selection took into account the highest conversion of DBT into 2-HBP in the shortest operation time with the lowest possible amount of biomass for each tested biocatalyst. The mixture composed by a 1:2 ratio of cells collected at 5 and 23 h of growth time of *Pseudomonas putida* CECT 5279, using a total biomass concentration of 2.1 g DCW/L, showed the best capabilities for desulfurization in resting cells among all the proposed biocatalyst formulations according to the specific conversion rate previously defined. Improving complex biocatalysts can offer better biodesulfurization processes, achieving higher DBT conversions and reducing time needed for biodesulfurization.

References

Abbad-Andaloussi, S., Lagnel, C., Warzywoda, M., Monot, F. (2003) Multicriteria comparison of resting cell activities of bacterial strains selected for biodesulfurization of petroleum compounds. *Enzyme and Microbial Technology*. 32(3/4): 446–454.

Abin Fuentes, A., Leung, J.C., Mohamed, M.E.S., Wang, D.I.C., Prather, K.L.J. (2014) Rate limiting step analysis of the microbial desulfurization of dibenzothiophene in a model oil system. *Biotechnology and Bioengineering*. 111(5): 876–884.

Abin-Fuentes, A., Mohamed, M.E., Wang, D.I.C., Prather, K.L.J. (2013) Exploring the mechanism of biocatalyst inhibition in microbial desulfurization. *Applied and Environmental Microbiology*. 79(24): 7807–7817.

Abolfazl, S.H., Soheila, Y., Mohammad, M.S. (2006) Biodesulfurization of dibenzothiophene by a newly isolated thermophilic bacteria strain. *Iranian Journal of Chemistry and Chemical Engineering*. 25:65–71.

Abrardi, V., Rover, G., Sicardi, S., Baldi, G., Conti, R. (1988) Sparged vessels agitated by multiple turbines. Proceeding of 6th European Conference on Mixing. Pavia, Italy, 24–26 May, 1988. Pp. 329–336.

Adlakha, J., Singh, P., Ram, S. K., Kumar, M., Singh, M. P., Singh, D., Srivastava, P. (2016). Optimization of conditions for deep desulfurization of heavy crude oil and hydrodesulfurized diesel by *Gordonia* sp. IITR100. *Fuel*.184: 761–769.

Agarry, S.E., Solomon, B.O., Layokun, S.K. (2008) Substrate inhibition kinetics of phenol degradation by binary mixed culture of *Pseudomonas aeruginosa* and

Pseudomonas fluorescence from steady state and wash-out data. *African Journal Biotechnology*. 7: 3927–3933.

Aggarwall, S., Karimil I.A., Ivan, G.R. (2013) In silico modeling and evaluation of *Gordonia alkanivorans* for biodesulfurization. *Molecular BioSyststem*. 9: 2530–2540.

Aiba, S., Shoda, M., Nagatami, M. (1968) Kinetics of product inhibition in alcohol fermentation. *Biotechnology and Bioengineering journal*. 10: 845–864.

Alcon, A., Martin, A.B., Santos, V.E., Gomez, E., Garcia-Ochoa, F. (2008) Kinetic model for DBT biodesulphurization by resting whole cells of Pseudomonas putida CECT5279. *Biochemical Engineering Journal*. 39:486–495.

Alcon, A., Santos, V.E., Martin, A.B., Yustos, P., Garcia-Ochoa, F. (2005) Biodesulfurisation of DBT with Pseudomonas putida CECT5279 by resting cells: Influence of cell growth time on reducing equivalent concentration and HpaC activity. *Biochemical Engineering Journal*. 26: 168–175.

Alonso del Aguila, R., Boltes, K., Leton, P., Rodríguez, A., Rosal, R., Perdigón, J., Garcia-Calvo, E. (2007) Biodesulfurization of DBT in model oil by resting cell of Pseudomonas putida CECT5279: Process enhancement. European Congress of Chemical Engineering - 6, (September), 1–6.

Alves, L., Paixao, S.M. (2011) Toxicity evaluation of 2-hydroxybiphenyl and other compounds involved in studies of fossil fuels biodesulphurisation. *Bioresource Technology*, 102(19), 9162–9166.

Ansari, F., Grigoriev, P., Libor, S., Tothill, I. E., Ramsden, J.J. (2009) DBT degradation enhancement by decorating Rhodococcus erythropolis IGST8 with magnetic Fe3O4 nanoparticles. *Biotechnology and Bioengineering*, 102(5): 1505–1512.

Arabian, D., Najafi, H., Farhadi, F., Dehkordi, A.M. (2014) Biodesulfurization of simulated light fuel oil by a native isolated bacteria *Bacillus cereus* HN. *Journal of Petroleum Science and Technology*. 4(1): 31–40.

Arey, J.S., Gschwend, P.M. (2005) Estimating the partition coefficients for fuel – water systems: developing linear salvation energy relationships using linear solvent strength theory to handle mixtures. *Environmental Science and Technology*. 39: 2702 – 2710.

Arnaud, J.P., Lacroix, C., Foussereau, C., Choplin, L. (1993) Shear stress effects on growth and activity of *Lactobacillus delbrueckii* subsp. Bulgaricus. *Journal of Biotechnology*. 29: 157–175.

Ayala, M., Tinoco, R., Hernandez, V., Bremauntz, P., Vazquez-Duhalt, R. (1998) Biocatalytic oxidation of fuel as an alternative to biodesulfurization. *Fuel Processing Technology*. 57: 101–11.

Bahuguna, A., Lily, M.K., Munjal, A., Singh, R.N. Dangwal, K. (2011) Desulfurization of dibenzothiophene (DBT) by a novel strain Lysinibacillus sphaericus DMT-7 isolated from diesel contaminated soil. *Journal of Environmental Sciences*. 23: 975–982.

Bandyopadhyay, S., Chowdhury, R., Bhattacharjee, C. (2013) Steady state performance of a bioreactor for production of near zero sulfur diesel (NZSD) and bio-surfactant. Journal of Clean Energy Technologies. 1(3): 189–193.

Blanch, H.W., Clark, D.S. (1997) Biochemical Engineering. Marcel Dekker Inc., New York.

Boltes, K., Alonso del Aguila, R., García-Calvo, E. (2013) Effect of mass transfer on biodesulfurization kinetics of alkylated forms of dibenzothiophene by *Pseudomonas putida* CECT5279. *Journal of Chemical Technology and Biotechnology*. 88: 422–431.

Bordoloi, N.K., Rai, S.K., Chaudhuri, M.K., Mukherjee, A.K. (2014) Deep-desulfurization of dibenzothiophene and its derivatives present in diesel oil by a newly isolated bacterium Achromobacter sp. to reduce the environmental pollution from fossil fuel combustion. *Fuel Processing Technology*, 119: 236–244.

Bronnenmeier, R., Markl, H. (1982) Hydrodynamic stress capacity of microorganisms. *Biotechnology and Bioengineering*. 24: 553–578.

Calik, P., Yilgör, P., Ayhan, P., Demir, A. (2004) Oxygen transfer effects on recombinant benzaldehyde lyase production. *Chemical Engineering Science*. 59: 5075–5083.

Calzada, J., Alcon, A., Santos, V.E., Garcia-Ochoa, F. (2011) Mixtures of Pseudomonas putida CECT 5279 cells of different ages: optimization as biodesulfurization catalyst. *Process Biochemistry*. 46:1323–1328.

Calzada, J., Alcon, A., Santos, V.E., Garcia-Ochoa, F. (2012) Extended kinetic model for DBT desulfurization using *Pseudomonas Putida* CECT5279 in resting cells. *Biochemical Engineering Journal*. 66: 52–60.

Calzada, J., Heras, S., Alcon, A., Santos, V.E., Garcia-Ochoa, F. (2009a) Biodesulfurization of dibenzothiophene (DBT) using *Pseudomonas putida* CECT 5279: a biocatalyst formulation comparison. *Energy and Fuels*. 23: 5491–5495.

Calzada, J., Zamarro, M.T., Alcon, A., Santos, V.E., Diaz, E., Garcia, J.L., Garcia-Ochoa, F. (2009b) Analysis of dibenzothiophene desulfurization in a recombinant *Pseudomonas putida* strain. *Applied and Environmental Microbiology*. 75: 875–877.

Caro, A,, Boltes, K., Leton, P., Garcia-Calvo, E. (2008b) Description of by-product inhibition effects on biodesulfurization of dibenzothiophene in biphasic media. Biodegradation. 19: 599–611.

Caro, A., Boltes, K., Letón, P., García-Calvo, E. (2007b) Dibenzothiophene biodesulfurization in resting cell conditions by aerobic bacteria. *Biochemical Engineering Journal*. 35:191–197.

Caro, A., Boltes, K., Leton, P., Garcia-Calvo, E. (2008a) Biodesulfurization of dibenzothiophene by growing cells of *Pseudomonas putida* CECT 5279 in biphasic media. *Chemosphere*. 73: 663–669.

Caro, A., Letón, P., García-Calvo, E., Setti, L. (2007a) Enhancement of dibenzothiophene biodesulfurization using β-cyclodextrins in oil-to-water media. Fuel. 86: 2632–2636.

Chalmers, J.J. (1994) Cells and bubbles in sparged bioreactors. *Cytotechnology*. 15: 311–320.

Chang, J.H., Chang, Y.K., Cho, K.-S., Chang, H.N.: (2000) Desulfurization of model and diesel oils by resting cells of *Gordona* sp. *Biotechnology Letters*. 22(3): 193.196

Chen H., Cai Y.B., Zhang W.J., Li W. (2009) Methoxylation pathway in biodesulfurization of model organosulfur compounds with Mycobacterium sp. *Bioresource Technology*. 100:2085–2087.

Chen, H., Cai, Y.B., Zhang, W.J., Li, W. (2009) Methoxylation pathway in biodesulfurization of model organosulfur compounds with Mycobacterium sp. *Bioresource Technology*. 100:2085–2087.

Chen, H., Zhang, W.J., Cai, Y.B., Zhang, Y., Li, W. (2008) Elucidation of 2-hydroxybiphenyl effect on dibenzothiophene desulfurization by Microbacterium sp. strain ZD-M2. *Bioresource Technology*. 99(15): 6928–6933.

Chen, H., Zhang, W.J., Cai, Y.B., Zhang, Y., Li, W. (2008a) Elucidation of 2-hydroxybiphenyl effect on dibenzothiophene desulfurization by Microbacterium sp. strain ZD-M2. *Bioresource Technology*. 99(15): 6928–6933.

Chen, H., Zhang, W.J., Chen, J.M., Cai, Y.B., Li. W. (2008b) Desulfurization of various organic sulfur compounds and the mixture of DBT + 4,6DMDBT by *Micobacterium* sp. ZD-19. *Bioresource Technology*. 99: 3630–3634.

Chisty, Y. (2010) In: M.C. Flickinger (ed) Shear Sensitivity in Encyclopedia of Industrial Biotechnology: Bioprocess, Bioseparation, and Cell Technology. John Wiley & Sons, New York.

Choi, O.K., Cho, K.S., Ryu, H.W., Chang, Y.K. (2003) Enhancement of phase separation by the addition of de-emulsifiers to three phase (diesel oil/biocatalyst/aqueous phase) emulsion in diesel biodesulfurization. *Biotechnology Letters*. 25(1): 73–77.

Constanti, M., Giralt, J., Bordons, A. (1996) Degradation and desulfurization of dibenzothiophene sulfone and other sulfur compounds by *Agrobacterium* NC501 and a mixed culture. *Enzyme and Microbial Technology*. 19: 214–219.

Copeland, R.A. (2000) Enzymes: a practical introduction to structure, mechanism, and data analysis, 2nd ed. Pp. 397, Wiley-VCH, Weinheim, Germany.

Davoodi-Dehaghani, F., Vosoughi, M., Ziaee, A.A. (2010) Biodesulfurization of dibenzothiophene by a newly isolated *Rhodococcus erythropolis* strain. *Journal of Bioresource Technology*. 101: 1102–1105.

Dejaloud, A., Vahabzadeh, F., Habibi, A. (2017) *Ralstonia eutropha* as a biocatalyst for desulfurization of dibenzothiophene. *Bioprocess and Biosystems Engineering*. 40: 969–980.

del Olmo, C.H., Alcon, A., Santos, V.E., Garcia-Ochoa, F. (2005a) Modeling the production of a *Rhodococcus erythropolis* IGTS8 biocatalyst for DBT biodesulfurization: influence of media composition. *Enzyme and Microbial Technology*. 37: 157–166.

del Olmo, C.H., Santos, V.E., Alcon, A., Garcia-Ochoa, F. (2005b) Production of a *Rhodococcus erythropolis* IGTS8 biocatalyst for DBT biodesulfurization: influence of operational conditions. *Biochemical Engineering Journal*. 22: 229–237.

Denome, S.A., Oldfield, C., Nash, L.J., Young, K.D. (1994) Characterization of the desulfurization genes from *Rhodococcus* sp. strain IGTS8. *Journal of Bacteriology*. 176: 6707–6716.

Deriase, S.F., Nassar, H.N., El-Gendy, N.Sh. (2015) Modeling and simulation for biodegradation of mono-, binary-, and tertiary-substrate batch systems of different polyaromatic sulfur heterocyclic compounds. *Petroleum Science and Technology*. 33(10): 1063–1076.

Deriase, S.F., El-Gendy, N.Sh. (2010) Modeling and simulation for degradation of different polyaromatic sulfur heterocyclic compounds by *Bacillus sphaericus* HN1 in a batch reactor. 4(2): 1–24.

Deriase, S.F., Younis, Sh.A., El-Gendy, N.Sh. (2013) Kinetic evaluation and modeling for batch degradation of 2-hydroxybiphenyl and 2,2'-dihdroxybipheny by *Corynebacterium variabilis* Sh42. Desalination and Water Treatment. 51: 4719–4728.

Derikvand p., Etemadifar Z. (2014) Improvement of biodesulfurization rate of alginate immobilized Rhodococcus erythropolis R1. *Jundishapur Journal of Microbiology*. 7:1–7.

Eadie, G. S., (1942) The Inhibition of Cholinesterase by Physostigmine and Prostigmine. *Journal of Biological Chemistry.*146:85–93.

Edwards, V.H. (1970) The influence of high substrate concentrations on microbial kinetics. *Biotechnology and Bioengineering journal.* 12: 679 - 712.

El-Gendy, N.Sh., Nassar, H.N., Younis, S.A. (2015) Main and interactive effects of polyaromatic sulfur heterocyclic compounds on growth and biodegradation efficiencies of *Bacillus sphaericus* HN1: modeling and statistical analysis. *Petroleum Science and Technology*. 33(11): 1167–1181.

Ellis, R.J. (2001) Macromolecular crowding: obvious but under appreciated. *Trends in Biochemical Sciences*. 26:5 97–604.

Escobar, S., Rodriguez, A., Gomez, E., Alcon, A., Santos, V.E., Garcia-Ocho, F. (2016) Influence of oxygen transfer on *Pseudomonas putida* effects on growth rate and biodesulfurization capacity. *Bioprocess and Biosystems Engineering*. 39: 545–554.

Fedorak, P.M., Westlake, D.W.S. (1983) Microbial degradation of organic sulfur compounds in Prudhoe Bay crude oil. *Canadian Journal of Microbiology*. 29: 291–296.

Fedorak, P.M., Westlake, D.W.S. (1984) Degradation of sulfur heterocycles in Prudhoe Bay crude oil by soil enrichments. Water Air Soil Pollution. 21: 225–230.

Folsom B.R., Schieche D.R., Digrazia P.M., Werner J. and Palmer S. (1999) Microbial desulfurization of alkylated dibenzothiophenes from a hydrodesulfurized middle distillate by *Rhodococcus erythropolis* I-19. *Applied Environmental Microbiology*. 65: 4967–4972.

Galán, B., Díaz, E., García, J.L. (2000). Enhancing desulphurization by engineering a flavin reductase encoding gene cassette in recombinant biocatalysts. *Environmental Microbiology*. 2: 687–694.

Gallagher, J.R., Olson, E.S., Stanley, D.C. (1993) Microbial desulfurization of dibenzothiophene: a sulfur-specific pathway. *FEMS Microbiology Letters*. 197: 31–36.

García Camacho, F., Molina Grima, E., Sánchez Mirón, A., González Pascual, V., Chisti, Y. (2001) Carboxymethyl cellulose protects algal cells against hydrodynamic stress. *Enzyme and Microbial Technology*. 29: 602–610.

Garcia-Ochoa, F., Escobar, S., Gomez, E. (2015) Specific oxygen uptake rate as indicator of cell response of *Rhodococcus erythropolis* cultures to shear effects. *Chemical Engineering Science*. 122: 491–499.

Garcia-Ochoa, F., Gomez, E. (2009) Bioreactor scale-up and oxygen transfer rate in microbial processes: An overview. *Biotechnology Advances*. 27: 153–176.

Garcia-Ochoa, F., Gomez, E., Alcon, A., Santos, V.E. (2013) The effect of hydrodynamic stress on the growth of *Xanthomonas campestris* cultures in a stirred and sparged tank bioreactor. *Bioprocess and Biosystems Engineering*. 36: 911–925.

Garcia-Ochoa, F., Gomez, E., Santos. V.E., Merchuk, J.C. (2010) Oxygen uptake rate in microbial processes: an overview. *Biochemical Engineering Journal*. 49: 289–307.

Garcia-Ochoa, F., Romero, A., Santos, V.E. (1992) Comparison of Methods for determining kinetic parameters in complex reactions. I*nternational chemical engineering* 32(3): 538–551.

Giraud, C. (2008). Estimation of Gaussian graphs by model selection. *Electronic Journal of Statistics*. 2:542–563.

Godoy-Silva, C. Berdugo, J.J. Chalmers, Aeration, mixing and hydrodynamics of animal cell bioreactors, in: M.C. Flickinger (Ed.), Encyclopedia of Industrial Biotechnology: Bioprocess, Bioseparation, and Cell Technology, John Wiley & Sons, 2010.

Goindi, H.K., Saini, V.S., Verma, P.S., Adhikari, D.K. (2002) Dibenzothiophene desulfurization in hydrocarbon environment by Staphylococcus sp. resting cells. *Biotechnology Letters*. 24:779–781.

Gomez, E., Alcon, A., Escobar, S., Santos, V.E., Garcia-Ochoa, F. (2015) Effect of fluid dynamic conditions on growth rate and biodesulfurization capacity of *Rhodococcus erythropolis* IGTS8. *Biochemical Engineering Journal*. 99: 138–146.

Gomez, E., Santos, V., Alcon, A., Garcia-Ochoa, F. (2006a) Oxygen transport rate on *Rhodococcus erythropolis* cultures: effect on growth and BDS capability. *Chemical Engineering Science*. 61: 4595–4604.

Gomez, E., Santos, V., Alcon, A., Martin, A.B., Garcia-Ochoa, F. (2006b) Oxygen-uptake and mass-transfer rates on the growth of *Pseudomonas putida* CECT5279: influence on biodesulfurization (BDS) capability. *Energy and Fuels*. 20: 1565–1571.

Gray, K.A., Mrachko, G.T., Squires, C.H. (2003) Biodesulfurization of fossil fuels. *Current Opinion in Microbiology*. 6: 229–235.

Gray, K.A., Pogrebinsky, O.S., Mrachko, G.T., Xi, L., Monticello, D.J., Squires C.H. (1996) Molecular mechanisms of biocatalytic desulfurization of fossil fuels. Nature Biotechnology. 14: 1705–1709.

Guchhait, S., Biswas, D., Bhattacharya, P., Chowdhury, R. (2005a) Biodesulfurization of model organo-sulfur compounds and hydrotreated diesel – experiments and modelling. *Chemical Engineering Journal.* 112: 145–151.

Guha, S., Peters, C.A., Jaffe, P.R. (1999) Multi-substrate biodegradation kinetics of naphthalene, phenanthrene, and pyrene mixtures. *Biotechnology Bioenergy.* 65: 491–499.

Gunam, I.B.W., Yaku, Y., Hirano, M., Yamamura, K., Tomita, F., Sone, T., Asano, K., (2006) Biodesulfurization of alkylated forms of dibenzothiophene and benzothiophenes by *Sphingomonas subarctica* T7b. *Journal of Bioscience and Bioengineering.* 101: 322–327.

Gunam, I.B.W., Yamamura, K., Nengah Sujaya, I., Antara, N.S., Aryanta, W.R., Tanaka, M., Asano, K. (2013) Biodesulfurization of dibenzothiophene and its derivatives using resting and immobilized cells of Sphingomonas subarctica T7b. *Journal of Microbiology and Biotechnology*,.23(4): 473–482.

Guobin S., Huaiying Z., Weiquan C., Jianmin X., Huizhou L. (2005) Improvement of biodesulfurization rate by assembling nanosorbents on the surfaces of microbial cells. *Biophysical Journal.* 89(6): 58–60.

Haldane, J.B.S. (1930) Enzymes. Longmans, Green & Company, London, England, UK.

Hewitt, C.J., Boon, L.A., McFarlane, C.M., Nienow, A.W. (1998) The use of flow cytometry to study the impact of fluid mechanical stress on *Escherichia* coli W3110 during continuous cultivation in an agitated bioreactor. *Biotechnology and Bioengineering.* 59: 612–620.

Hodaifa, G., Martinez, M.E., Orpez, R., Sanchez, S. (2010) Influence of hydrodynamic stress in the growth of *Scenedesmus obliquus* using a culture medium based on olive mill waste water. *Chemical Engineering Process.* 49: 1161–1168.

Hofstee, B.H.J. (1952) Specificity of Esterases: I. Identification of Two Pancreatic Aliesterases, *Journal of Biological Chemistry.* 199:357–364.

Honda, H., Sugiyama, H., Saito, I., Kobayashi, T. (1998) High cell density culture of *Rhodococcus rhodochrous* by pH-stat feeding and dibenzothiophene degradation. *Journal Fermentation and Bioengineering.* 85:334–338. http://chemwiki.ucdavis.edu/Biological_Chemistry/Catalysts/Enzymatic kinetics.

Introduction to enzymes (2017) worthington biochemical corporation. http://www.worthington-biochem.com/introbiochem/introEnzymes.html

Irani, Z.A., Yazdian, F., Mohebali, G., Soheili, M., Mehrnia, M.R. (2011) Determination of growth kinetic parameters of a desulfurizing bacterium, *Gordonia alkanivorans* RIPI90A. *Chemical Engineering Transactions.* 24: 937–942.

Izumi, Y., Oshiro, T., Ogino, H., Hine, Y., Shimao, M. (1994) Selective desulfurization of dibenzothiophene by *Rhodococcus erythropolis* D-1. *Applied and Environmental Microbiology.* 60: 223–236.

Jahoda, M., Mostek, M., Kukukova, A., Machon, V. (2007) CFD modelling of liquid homogenizationin stirred tanks with one and two impellers using large eddy simulation. *Chemical Engineering Research and Design*. 85: 616–625.

Jia, X., Wen, J., Sun, Z., Caiyin, Q., Xie, S. (2006) Modeling of DBT biodegradation behaviors by resting cells of *Gordonia* sp WQ-01 and its mutant in oil–water dispersions. *Chemical Engineering Science*. 61: 1987–2000.

Jorjani, E., Chehreh Chelgani, S., Mesroghli, S. (2007) Prediction of microbial desulfurization of coal using artificial neural networks. *Minerals Engineering*. 20(14): 1285–1292.

Kao, P.-M., Chen, C.-I., Huang, S.-C., Chang, Y.-C., Tsay, P.-Y., Liu, Y.-C. (2007) Effects of shear stress and mass transfer on chitinase production by *Paenibacillus* sp. CHE-N1. *Biochemical Engineering Journal*. 34: 172–178.

Kareem, S.A. (2016) Anaerobic biodesulfurization of kerosene part II: investigating its kinetics. *IOSR Journal of Biotechnology and Biochemistry*. 2(5): 37–45.

Kareem, S.A., Susu, A.A., Aribike, D.S., Nwachukwu, S.C.U. (2013a) Modelling the biodesulfurization of diesel. *International Journal of Engineering Research and Technology*. 2(7): 1010–1015.

Kareem, S.A., Susu, A.A., Aribike, D.S., Nwachukwu, S.C.U. (2013b) Modelling the biodesulfurization of kerosene in a batch reactor. *International Journal of Engineering Science Invention*. 2(4): 61–66.

Kareem, S.A., Susu, A.A., Aribike, D.S., Nwachukwu, S.C.U. (2014) Investigating the kinetics of biodesulfurization of diesel. *Assumption University : AU Journal of Technology*. 18 (1): 27–35.

Kaufman, E.N., Borole, A.P., Shong, R., Sides, J.L., Juengst, C. (1999) Sulfur specificity in the bench-scale biological desulfurization of crude oil by *Rhodococcus* IGTS8. *Journal of Chemical Technology and Biotechnology*. 74: 1000–1004.

Kaufman, E.N., Hankins, J.B., Borole, A.P. (1998) Comparison of batch-stirred and electrospray reactors for biodesulfurization of dibenzothiophene in crude oil and hydrocarbon feedstocks. *Applied Biochemistry and Biotechnology*. 73: 127–143.

Kayser, Y., Bielagajones, B.A., Jackowski, K., Odusan, O., Kilbane, J.J. (1993) Utilization of organusulphur compounds by axenic and mixed cultures of *Rhodococcus rhodochrous* IGTS8. *Journal of General Microbiology*. 139: 3123–3129.

Kilbane, J.J. (1992) Mutant microorganisms useful for cleavage of organic C-S bonds. US patent 5,104,801.

Kilbane, J.J. (2006) Microbial biocatalyst development to upgrade fossil fuels. *Current Opinion in Microbiology*. 17: 305–314.

Kilbane, J.J., Jackowski, K. (1992) Biodesulfurization of water soluble coal derived material by *Rhodococcus rhodochrous* IGTS8. *Biotechnology and Bioengineering*. 40: 1107–1114.

Kim, Y.J., Chang, J.H., Cho, K.-S., Ryu, H.W., Chang, Y.K. (2004) A physiological study on growth and dibenzothiophene (DBT) desulfurization characteristics of *Gordonia* sp. CYKS1. *Korean Journal of Chemical Engineering*. 21(2): 436–441.

Kirkwood, K.M., Andersson, J.T., Fedorak, P.M., Foght, J.M., Gray, M.R. (2007): Sulfur from benzothiophene and alkyl benzothiophenes supports growth of *Rhodococcus* sp. strain J7H1. Biodegradation. 18: 541–549.

Kishimoto, M., Inui, M., Omasa, T., Katakura, Y., Suga, K., Okumura, K. (2000) Efficient production of desulfurizing cells with the aid of expert system. *Biochemical Engineering Journal.* 5: 143–147.

Knightes, C.D., Peters, C.A. (2006) Multi-Substrate biodegradation kinetics for binary and complex mixtures of polycyclic aromatic hydrocarbons. *Environmental Toxicology and Chemistry.* 25:1746–1756.

Kobayashi, M., Horiuchi, K., Yoshikama, O., Hirasawa, K., Ishii, Y., Fujino, K., Sugiyama, H., Maruhashi, K. (2001) Kinetic analysis of microbial desulfurization of model and light gas oils containing multiple alkyl dibenzothiophenes. *Bioscience, Biotechnology, and Biochemistry.* 65: 298–304.

Kobayashi, M., Onaka, T., Ishii, Y., Konishi, J., Takaki, M., Okada, H., Ohta, Y., Koizumi, K., Suzuki, M. (2000) Desulfurization of alkylated forms of both dibenzothiophene and benzothiophene by a single bacterial strain. *FEMS Microbiology Letters.* 187: 123–126.

Konishi, J., Ishii, Y., Osaka, T., Ohta, Y., Suzuki, M., Maruhashi, K. (2000) Purification and characterization of dibenzothiophene sulfone monooxygenase and FMN-dependent NADH oxidoreductase from the thermophilic bacterium *Paenibacillus* sp. strain Al 1-2. *Journal of Bioscience and Bioengineering.* 90 (2000) 607–613.

Konishi, J., Ishii, Y., Osaka, T., Okumura, K., Suzuki, M. (1997) Thermophilic carbon-sulfur-bond-targeted biodesulfurization. *Applied and Environmental Microbiology.* 63(8): 3164–3169.

Kropp, K.G., Andersson, J.T., Fedorak, P.M. (1997) Bacterial transformations of three dimethyl dibenzothiophenes by pure and mixed bacterial cultures. *Environmental Science and Technology.* 31: 1547–1554.

Lapin, L.L. (1997) Modern engineering statistics. 1st ed. Duxbury Press, Belmon.

Layokun, S.K., Umoh, E.F., Solomon, B.O. (1987) A kinetic model for the degradation of dodecane by P. *fluorescens* isolated from the oil polluted area, Warri in Nigeria. *Journal of Nigerian Society of Chemical Engineers.* 16: 48 - 52.

Le Borgne, S., Quintero, R. (2003) Biotechnological processes for the refining of petroleum. *Fuel Processing Technology.* 81: 155–169.

Lei, B., Tu, S.-Ch. (1996) Gene over-expression, purification and identification of a desulfurization enzyme from Rhodococcus sp. Strain IGTS8 as a sulfide/sulfoxide monooxygenase. *Journal of Bacteriology.* 178: 5699–5705.

Li, F.L., Xu, P., Ma, C.Q., Luo, L.L., Wang, X.S. (2003) Deep desulfurization of hydrodesulfurization-treated diesel oil by a facultative thermophilic bacterium *Mycobacterium* sp. X7B, *FEMS Microbiology Letters.* 223: 301–307.

Li, M.Z., Squires, C.H., Monticello, D.J., Chids, J.D. (1996) Genetic analysis of the dsz promoter and associated regulatory regions of *Rhodococcus erythropolis* IGTS8. *Journal of Bacteriology.* 178(22): 6409–6418.

Li, W., Tang, H., Liu, Q., Xing, J., Li, Q., Wang, D., Yang, M., Li, X., Liu, H. (2009) Deep desulfurization of diesel by integrating adsorption and microbial method. *Biochemical Engineering Journal*. 44: 297–301.

Lichtinger, T., Reiss, G., Benz, R. (2000) Biochemical identification and biophysical characterization of a channel-forming protein from *Rhodococcus erythropolis*. *Journal of Bacteriology*. 182: 764–770.

Lineweaver, H., Burk, D. (1934) The Determination of Enzyme Dissociation Constants. *Journal of American Chemical Society*. 56: 658–666.

Liu, Y.S., Wu, J.Y., Ho. K.P. (2006) Characterization of oxygen transfer conditions and their effects on *Phaffia rhodozyma* growth and carotenoid production in shake-flask cultures. *Biochemical Engineering Journal*. 27: 331–335.

Lu, J., Nakajima-Kampe, T., Shigeno, T., Ohbo, A., Nomura, N., Nakahara, T. (1999) Biodegradation of dibenzothiophene and 4,6- dimethyl dibenzothiophene by *Sphingomonas paucimobilis* strain TZS-7. *Bioscience and Bioengineering*. 88(3): 293–299.

Lu, W.-M., Wu, H.-Z., Ju, M.-Y. (1997) Effects of baffle design on the liquid mixing in an aerated stirred tank with standard Rushton turbine impellers. *Chemical Engineering Science*. 21: 3843–3851.

Luedeking, R., Piret, E.L. (1959) A Kinetic study of the lactic acid fermentation batch process at controlled pH. *Biotechnology and Bioengineering*. 1: 393–412.

Luo, M.F., Xing, J.M., Gou, Z.X., Li, S., Liu, H.Z., Chen, J.Y. (2003) Desulfurization of dibenzothiophene by lyophilized of *Pseudomonas delafieldii* R-8 in the presence of dodecane. *Biochemical Engineering Journal*. 13: 1–6.

Ma, N., Mollet, M., Chalmers, J.J. (2006) Aeration, mixing and hydrodynamics in bioreactors, in: W.-S. Hu, S.S. Ozturk (Eds.), Cell Culture Technology for Pharmaceutical and Cell-based Therapies, CRC Press, Boca Raton (FL). Pp. 225–248.

Maass D., Mayer D.A., Moritz D.E., Oliveira D., de Souza A.A.U., Souza S.M.A.G. (2015) An Evaluation of Kinetic Models in the Biodesulfurization of Synthetic Oil by Rhodococcus erythropolis ATCC 4277. *Applied Biochemistry and Biotechnology*. 177(3): 759–770.

Maghsoudi, S., Vossoughi, M., Kheirolomoom, A., Tanaka, E., Katoh, S. (2001) Biodesulfurization of hydrocarbons and diesel fuels by *Rhodococcus* sp. strain P32C1. *Biochemical Engineering Journal*. 8: 151–156.

Marcelis, C.L.M., van Leeuwen, M., Polderman, H.G., Jannsen, A.J.H., Lettinga, G. (2003) Model description of dibenzothiophene mass transfer in oil/water dispersions with respect to biodesulfurization. *Biochemical Engineering Journal*. 16: 253–264.

Märkl, H., Bronnenmeier, R., Wittek, B. (1991) The resistance of microorganisms to hydrodynamic stress. *International Chemical Engineering*. 31: 185–197.

Marquardt, A., W. (1963) An algorithm for least-squares estimation of non-linear parameters. *Journal of the Society for Industrial and Applied Mathematics*. 11: 431–41.

Martin, A.B., Alcon, A., Santos, V.E., Garcia-Ochoa, F. (2004) Production of a biocatalyst of *Pseudomonas putida* CECT5279 for dibenzothiophene (DBT) biodesulfurization for different media compositions. *Energy and Fuel*. 18: 851–857.

Martin, A.B., Alcon, A., Santos, V.E., Garcia-Ochoa, F. (2005) Production of a Biocatalyst of Pseudomonas putida CECT5279 for DBT Biodesulfurization: Influence of the Operational Conditions. *Energy & Fuels.* 19(3): 775–782.

Martin, M., Montes, F.J., Galan, M.A. (2008) On the contribution of the scales of mixing to the oxygen transfer in stirred tanks. *Chemical Engineering Journal.* 145: 232–241.

Martinez, I., Santos, V.E., Alcon, A., Garcia-Ochoa, F. (2015) Enhancement of the biodesulfurization capacity of Pseudomonas putida CECT5279 by co-substrate addition. *Process Biochemistry* 50: 119–124.

Martinez, I., Santos, V.E., Garcia-Ochoa, F. (2017) Metabolic kinetic model for dibenzothiophene desulfurization through 4S pathway using intracellular compound concentrations. *Biochemical Engineering Journal.* 117: 89–96.

Martinez, I., Santos, V.E., Gomez, E., Garcia-Ochoa, F. (2016) Biodesulfurization of dibenzothiophene by resting cells of *Pseudomonas putida* CECT5279: influence of the oxygen transfer rate in the scale-up from shaken flask to stirred tank reactor. *Journal of Chemical Technology and Biotechnology.* 91: 184–189.

Marzona, M., Pessione, E., Di Martino, S., Giunta, C. (1997) Benzothiophene and dibenzothiophene as the sole sulfur source in Acinetobacter: growth kinetics and oxidation products. *Fuel Processing Technology.* 52: 199–205.

Matsui, T., Hirasawa, K., Konishi, J., Tanaka, Y., Maruhashi, K., Kurane, K. (2001) Microbial desulfurisation of alkylated dibenzothiophene and alkylated benzothiophene by recombinant *Rhodococcus* sp. Strain T09. *Applied Microbiology and Biotechnology.* 56: 196–200.

Maxwell, S., Yu, J., (2000) Selective desulphurisation of dibenzothiophene by a soil bacterium: microbial DBT desulphurisation. *Process Biochemistry.* 35: 551–556.

McFarland, B.L. (1999) Biodesulfurization. *Current Opinion in Microbiology.* 2,257–264.

Meijer, J.J., ten Hoopen, H.J.G., van Gameren Y.M., Luyben, K.Ch.A.M., Libbenga, K.R. (1994) Effects of hydrodynamic stress on the growth of plant cells in batch and continuous culture. *Enzyme and Microbial Technology.* 16: 467–477.

Michaelis, L., Menten, M.L. (1913) Die Kinetik der Invertinwirkung. Biochemische Zeitschrift 49:333–369.

Michel, B.J. and Miller, S.A. (1962) Power requirements of gas–liquid agitated systems, *AIChE Journal.* 8: 262–267.

Mingfang, L., Jianmin, X., Zhongxuan, G., Huizhou, L., Jiayong, C. (2003b) Microbial desulfurization of dibenzothiophene and 4,6-dimethyldibenzothiophene in dodecane and straight-run diesel oil. *Korean Journal of Chemical Engineering.* 20(4): 702–704.

Mingfang, L., Zhongxuan, G., Jianmin, X., Huizhou, L., Jiayong, C. (2003a) Microbial desulfurization of model and straight-run diesel oils. *Journal of Chemical Technology and Biotechnology.* 78: 873–876.

Mohebali, G., Ball A.S. (2008) Biocatalytic desulfurization (BDS) of petrodiesel fuels. Microbiology. 154: 2169–2183.

Mohebali, G., Ball, A.S., Rasekh, B., Kaytash, A. (2007) Biodesulfurization potential of a newly isolated bacterium, *Gordonia alkanivorans* RIPI90A. *Enzyme and Microbial Technology*. 40: 578–584.

Monod, J. (1949) The growth of bacterial cultures. *Annual Review of Microbiology*. 3:371–94.

Monticello, D.J. (2000) Biodesulfurization and the upgrading of petroleum distillates. *Current Opinion in Microbiology*. 11: 540–546.

Monticello, D.J., Finnerty, W.R. (1985) Microbial desulfurization of fossil fuels. *Annual Review of Microbiology*. 39. 371–389.

Mukhopadhyaya, M., Chowdhury, R., Bhattacharya, P. (2006) Biodesulfurization of hydrodesulfurized diesel in a trickle bed reactor - Experiments and modeling. *Journal of Scientific and Industrial Research*. 65(5): 432–436.

Nadalig, T., Raymond, N., Nimatuzahroh, M., Budzinsk, H., Bertrand, J.C. (2002) Degradation of phenanthrene, methylphenanthrenes and dibenzothiophene by a *Sphingomonas* strain 2mpII. *Applied Microbiology and Biotechnology*. 59(1): 79–85.

Naito, M., Kawamoto, T., Fujino, K., Kobayashi, M., Maruhashi, K., Tanaka, A. (2001) Long-term repeated desulfurization by immobilized *Rhodococcus erythropolis* KA2–5-1 cells. *Applied Microbiology and Biotechnology*. 55: 374–378.

Nakayama, N., Matsubara, T., Ohshiro, T., Moroto, Y., Kawata, Y., Koizumi, K., Hirakawa, Y., Suzuki, M., Maruhashi, K., Izumi, Y., Kurane, R. (2002) A novel enzyme, 2-hydroxybiphenyl-2-sulfinate desulfinase (DszB), from a dibenzothiphene-desulfurizing bacterium *Rhodococcus erythropolis* KA2–5-1: gene over-expression and enzyme characterization. *Biochimica et Biophysica Acta (BBA)*1598(1/2): 122–130.

Nasab, N.A., Kumleh, H.H., Kazemzad, M., Panjeh, F.G., Davoodi-Dehaghani, F. (2015) Improvement of Desulfurization Performance of Rhodococcus erythropolis IGTS8 by Assembling Spherical Mesoporous Silica Nanosorbents on the Surface of the Bacterial Cells. *Journal of Applied Chemical Research*. 9(2): 81–91.

Nassar, H.N., Deriase, S.F., El-Gendy, N.Sh. (2016) Modeling the relationship between microbial biomass and total viable count for a new bacterial isolate used in biodesulfurization of petroleum and its fractions. *Petroleum Science and Technology*. 34(11–12): 980–985.

Nassar, H.N., Deriase, S.F., El-Gendy, N.Sh. (2017) Kinetic modeling and simulation of a batch biodesulfurization process by *Rhodococcus erythropolis* HN2. *Petroleum Science and Technology*. 35(20): 1995–2002.

Nazari, F., Kefayati, M.E., Raheb, J. (2017) The study of biological technologies for the removal of sulfur compounds. *Journal of Sciences, Islamic Republic of Iran*. 28(3): 205–219.

Nienow, A.W. (1997) On impeller circulation and mixing effectiveness in the turbulent flow regime. *Chemical Engineering Science*. 15(52): 2557–2565.

Nienow, A.W. (2010) Scale-up, stirred tank reactors, in: M.C. Flickinger (Ed.), Encyclopedia of Industrial Biotechnology: Bioprocess, Bioseparation, and Cell Technology, John Wiley & Sons.

Noda, K., Watanabe, K., Maruhashi, K. (2003) Isolation of the *Pseudomonas aeruginosa* gene affecting uptake of dibenzothiophene in n-tetradecane. *Journal of Bioscience and Bioengineering.* 95(5): 504–511.

Nuhoglu, N., Yalcin, B. (2005) Modeling of phenol removal in a batch reactor. *Process Biochemistry.* 40: 1233–1239.

Ohshiro, T., Hine, Y., Izumi, Y. (1994) Enzymatic desulfurization of dibenzothiophene by a cell-free system of *Rhodococcus erythropolis* D-1. *FEMS Microbiology Letters.* 118: 341–344.

Ohshiro, T., Hirata, T., Izumi, Y. (1995a) Microbial desulfurization of dibenzothiophene in the presence of hydrocarbon. *Applied Microbiology and Biotechnology.* 44: 249–252.

Ohshiro, T., Izumi, Y. (1999) Review microbial desulfurization of organic compounds in petroleum. *Bioscience, Biotechnology and Biochemistry.* 63: 1–9.

Ohshiro, T., Kobayashi, Y., Hine, Y., Izumi, Y. (1995b) Involvement of flavin coenzyme in dibenzothiophene degrading enzyme system from *Rhodococcus erythropolis* D-1. *Bioscience, Biotechnology and Biochemistry.* 59(7): 1349–1351.

Ohshiro, T., Suzuki, K., Izumi, Y. (1996) Regulation of dibenzothiophene degrading enzyme activity of *Rhodococcus erythropolis* D-1. *Journal Fermentation and Bioengineering.* 81: 121–124.

Ohshiro, T., Suzuki, K., Izumi, Y. (1997) Dibenzothiophene (DBT) degrading enzyme responsible for the first step of DBT desulfurization by *Rhodococcus erythropolis* D-1: purification and characterization. *Journal of Fermentation and Bioengineering.* 83(2): 233–237.

Ohshiro, T., Suzuki, K., Izumi, Y. (1998) Regulation of dibenzothiophene degrading enzyme activity of *Rhodococcus erythropolis* D-1. *Journal of Fermentation and Bioengineering.* 81: 121–124.

Okpokwasili G.C., Nweke, C.O. (2005) Microbial growth and substrate utilization kinetics. *African Journal of Biotechnology.* 5(4): 305–317.

Oldfield, C., Pogrebinsky, O., Simmonds, J., Olson, E.S., Kulpa, C.H. (1997) Elucidation of the metabolic pathway for dibenzothiophene desulfurization by *Rhodococcus* sp. strain IGTS8 (ATCC 53968). *Microbiology.* 143: 2961–2973.

Olmos, E., Mehmood, N., Haj-Husein, L., Goergen, J.L., Fick, M., Delaunay, S. (2013) Effects of bioreactor hydrodynamics on the physiology of *Streptomyces. Bioprocess and Biosystems Engineering.* 36:259–272.

Omori, T., Saiki, Y., Kasuga, K., Kodama, T. (1995) Desulfurization of alkyl and aromatic sulfides and sulfonates by dibenzothiophene desulfurizing *Rhodococcussp.* strain SY1. *Bioscience, Biotechnology and Biochemistry.* 59(7): 1195–1198.

Onaka, T., Kobayashi, M., Ishii, Y., Okumura, K., Suzuk,i M. (2000) Application of solid-phase extraction to the analysis of the isomers generated in biodesulfurization against methylated dibenzothiophenes. *Journal of Chromatography A* 903: 193–202.

Patel, S.B., Kilbane, J.J., Webster, D.A. (1997) Biodesulphurisation of dibenzothiophene in hydrophobic media by *Rhodococcus erythropolis* sp. strain IGTS8. *Journal of Chemical Technology and Biotechnology.* 69: 100–106.

Peng, B., Zhou, Z. (2016) Study on growth characteristic and microbial desulfurization activity of the bacterial stain MP12. *Biochemical Engineering Journal.* 112: 202–207.

Pirt, S.J. (1965) The maintenance energy of bacteria in growing cultures. Proceedings of the Royal Society of London. Series B. 163: 224–231.

Prokop, A., Bajpai, R.K. (1992) The sensitivity of biocatalysts to hydrodynamic shear stress. *Advances in Applied Microbiology*. 37: 165–232.

Ramirez-Corredores, M.M., Borole, A.P. (2007) Biocatalysis in Oil Refining (Studies in Surface Science and Catalysis). Elsevier Science; 1st edition (September 22, 2011).

Rashtchi, M., Mohebali, G.H., Akbarnejad, M.M., Towfighi, J., Rasekh, B., Keytash, A. (2006) Analysis of biodesulfurization of model oil system by the bacterium, strain RIPI-22. *Biochemical Engineering Journal*, 29(3): 169–173.

Ratkowsky, D.A., Lowry, R.K., Stokes, T.A., Chandler, R.E. (1983) Model for bacterial culture growth rate throughout the entire bio kinetic temperature range. *Journal of Bacteriology*. 154: 1222–1226.

Russell, J.B., Cook, G.M. (1995) Energetics of bacterial growth: balance of anabolic and catabolic reactions. *Microbiological Reviews*. 59: 48–62.

Sahoo, S., Verma R.K., Suresh, A.K., Rao, K.K., Bellare, J., Suraishkumar, G.K. (2003) Macro-level and genetic-level responses of bacillus subtilis to shear stress. *Biotechnology Progress*. 19: 1689–1696.

Santos, V.E., Alcon, A., Martin, A.B., Gomez, E., Garcia-Ochoa, F. (2007) Desulfurization of dibenzothiophene using the 4S enzymatic route: influence of operational conditions on initial reaction rates. *Biocatalysis and Biotransformation*. 25: 286–294.

Schilling, B.M., Alvarez, L.M., Wang, D.I.C., Cooney, C.L. (2002) Continuous desulfurization of dibenzothiophene with *R. rhodochrous* IGTS8 (ATCC 53968). *Biotechnology Progress*. 18: 1207–1213.

Segel, I.H. (1975) Enzyme kinetics. Wiley, New York.

Setti, L., Bonoli, S., Badiali, E., Giuliani, S. (2003) Inverse phase transfer biocatalysis for a biodesulfurization process of middle distillates. Вестник Московского университета. Серия 2. 44(1): 80–83.

Setti, L., Farinelli, P., Di Martino, S., Frassinetti, S., Lanzarini, G., Pifferi, P.G. (1999) Developments in destructive and non-destrucitve pathways for selective desulfurizations in oil biorefining processes. *Applied Microbiology and Biotechnology*. 52: 111–117.

Setti, L., Lanzarini, G., Pifferi, P.G. (1994) Diffusion as a rate controlling step in heavy oil biodesulfurization processes. *Fuel Processing Technology*. 40, 311–317.

Setti, L., Lanzarini, G., Pifferi, P.G. (1995) Dibenzothiophene biodegradation by a *Pseudomonas* sp. in model solutions. *Process Biochemistry*. 30(8): 721–728.

Setti, L., Lanzarini, G., Pifferi, P.G. (1996) Immobilized cells for applications in non-conventional systems. *Progress in Biotechnology.* 11: 777–784.

Setti, L., Lanzarini, G., Pifferi, P.G. (1997) Whole cell biocatalysis for an oil desulfurization process. *Fuel Processing Technology*. 52: 145–153.

Shan, G., Xing, J., Zhang, H., Liu, H. (2005) Biodesulphurization of dibenzothiophene by microbial cells coated with magnetite nanoparticles. *Applied and Environmental Microbiology*. 71: 4497–4502.

Shuler, M.L., Kargi, F. (2002) Bioprocess Engineering: Basic concepts. Second Edition. Prentice Hall, Inc. Pp. 268–273.

Soneyink, V.L., Lenkins, D. (1980) Water Chemistry. John Wiley and Sons, New York, USA.

Sutcliffe, I.C., Brown, A.K., Dover, L.G. (2010) The *rhodococcal* cell envelope: composition, organization and biosynthesis. Biology of *Rhodococcus*. Microbiology monographs 16. Springer-Verlag, Berlin, Germany.

Tao, F., Yu, B., Xu, P., Ma, C. (2006) Biodesulfurization in biphasic systems containing organic solvents. Applied and Environmental Microbiology. 72: 4604–4609.

Tsai, S.P., Lee, Y.H. (1990) A model for energy-sufficient culture growth. *Biotechnology and Bioengineering*. 35: 138–145.

Van Bodegom P (2007) Microbial maintenance: a critical review on its quantification. Microbial Ecology. 53: 513–523.

van Hamme, J.D., Fedorak, P.M., Foght, J.M., Gray, M.R., Dettman, H.D. (2004) Use of novel fluorinated organosulfur compound to isolate bacteria capable of carbon-sulfur bond cleavage. *Applied and Environmental Microbiology*. 70: 1487–1493

Van't Riet, K. (1979) Review of measuring methods and nonviscous gas–liquid mass transfer in stirred vessels. *Industrial and Engineering Chemistry Process Design and Devlopment*. 18: 357–364.

Verhulst, P.F. (1845) Recherches math mematiques sur la loi d'accroissement de la population. Nouv. Mem. Acad. R. Sci. B.-lett. Brux.18: 1–45. gdz.sub.uni-goettingen.de

Verhulst, P.F. (1847) Deuxieme memoire sur la loi d'accroissement de la population. Mem. Acad. R. Sci. Lett. B.-Arts Belg. 20: 142–173. www.digizeit-schriften.de

Wang, J., Butler, R.R., Wu, F., Pombert, J.F., Kilbane, J.J., Stark B.C. (2017) Enhancement of microbial biodesulfurization via genetic engineering and adaptive evolution. PLoS ONE. 12(1): 1–20.

Wang, P., Humphrey, A.E., Krawiec, S. (1996) Kinetic analysis of desulfurization of dibenzothiophene by *Rhodococcus erythropolis* in continuous cultures. *Applied and Environmental Microbiology*. 62: 3066–3068

Wang, P., Krawiec, S. (1996) Kinetic analyses of desulfurization of dibenzothiophene by *Rhodococcus erythropolis* in batch and fed-batch cultures. *Applied and Environmental Microbiology*. 61: 1670–1675.

Watkins, L.M., Rodriguez, R., Schneider, D., Broderick, R., Cruz, M., Chambers, R., Ruckman, E., Cody, M., Mrachko, G.T. (2003) Purification and characterization of the aromatic desulfinase 2-(2′-hydroxyphenyl) benzenesulfinate desulfinase. *Archives of Biochemistry and Biophysics*. 415:14–23.

Wodzinski, R.S., Coyle, J.E. (1974) Physical state of phenanthrene for utilization by bacteria. *Applied Microbiology*. 1091–1094.

Worthington Biochemical Corporation (2017) Introduction to Enzymes http://www.worthington-biochem.com/introbiochem/Enzymes.pdf

Xu X., Cai Y., Song Z., Qiu X., Zhou J., Liu Y., Mu T., Wu D.,Guan Y., Xing, J. (2015). Desulfurization of immobilized sulfur-oxidizing bacteria, Thialkalivibrio versutus, by magnetic nanaoparticles under haloalkaliphilic conditions. *Biotechnology Letters.* 37(8): 1631–1635.

Xu, P., Yu, B., Li, F.L., Cai, X.F., Ma, C.Q. (2006) Microbial degradation of sulfur, nitrogen and sulfur heterocycles. *Trends in Microbiology.* 14: 398–405.

Xu, X., Cai, Y., Song, Z., Qiu, X., Zhou, J., Liu, Y., Mu, T., Wu, D., Guan, Y., Xing, J. (2015) Desulfurization of immobilized sulfur-oxidizing bacteria, Thialkalivibrio versutus, by magnetic nanaoparticles under haloalkaliphilic conditions. *Biotechnology Letters.* 37(8): 1631–1635.

Yan, H., Kishimoto, M., Omasa, T., Katakura, Y., Suga, K., Okumura, K., Yoshikawa, O. (2000) Increase in desulfurization activity of *Rhodococcus erythropolis* KA2-5-1 using ethanol feeding. *Journal of Bioscience and Bioengineering.* 89(4): 361–366.

Yang, J., Hu, Y., Zhao, D., Wang, S., Lau, P.C.K., Marison, I.W. (2007) Two-layer continuous-process design for the biodesulfurization of diesel oils under bacterial growth conditions. *Biochemical Engineering Journal.* 37: 212–218.

Yang, J., Marison I. (2005) Two-stage process design for the biodesulphurisation of a model diesel by a newly isolated *Rhodococcus globerulus* DAQ3. *Biochemical Engineering Journal.* 27: 77–82.

Yano, T., Koga, S. (1969) Dynamic behavior of the chemostat subject to substrate inhibition. *Biotechnology and Bioengineering journal.* 11: 139 - 153.

Yepez, B.O., Maugeri, F. (2005) Agitation, aeration and shear stress as key factors in inulinase production *by Kluyveromyces marxianus. Enzyme and Microbial Technology.* 36: 717–724.

Yu, B., Ma, C., Zhou, W., Wang, Y., Cai, X., Tao, F., Zhang, Q., Tong, M., Qu, J., Xu, P. (2006a) Microbial desulfurization of gasoline by free whole-cells of *Rhodococcus erythropolis* XP. *FEMS Microbiology Letters.* 258: 284–289.

Yu, B., Xu, P., Shi, Q., Ma, C. (2006b) Deep desulfurization of diesel oil and crude oils by a newly isolated *Rhodococcus erythropolis* strain. *Applied and Environmental Microbiology.* 72: 54–58.

Zakharyants, A.A., Murygina, P., Kalyuzhnyi, S. (2004) Screening of *Rhodococcus* species revealing desulfurization activity with regard to dibenzothiophene. In: Biocatalytic technology and nanotechnology. G E. Zaikov. Nova Science Publishers, Inc. Pp. 51–58.

Zeelani, G.G., Ashrafi, A., Dhakad, A., Gupta, G., Pal, S.L. (2016) A Review on Desulfurization Techniques of Liquid Fuels. *International Journal of Science and Research (IJSR).* 5(5): 331–336.

Zhang, S.H., Chen, H., Li, W. (2013) Kinetic analysis of biodesulfurization of model oil containing multiple alkyl dibenzothiophenes. *Applied Microbiology and Biotechnology.* 97(5): 2193–2200.

10

Enhancement of BDS Efficiency

List of Abbreviations and Nomenclature

1-MDBT	1-Methyldibenzothiophene
1-MN	1-Methylnaphthalene
2,2′-DMBP	2,2'-Dimethoxybiphenyl
2-CAT	2-Carboxythiophene
2-ETh	2-Ethyl Thiophene
2-FOT	2-Formylthiophene
2-HBP	2-Hydroxybiphenyl
2-HMT	2-Hydroxymethyl thiophene
2-MBP	2-Methoxybiphenyl
2-MT	2-Methylthiophene
3-MBT	3-Methylbenzothiophene
3-MT	3-Methylthiophene
4,6-DEDBT	4,6-Diethyldibenzothiophene
4,6-DMDBT	4,4-Dimethyldibenzothiophene
4,6-DMHBP	4,6-Dimethyl-Hydroxybiphenyl
4-MDBT	4-Methyldibenzothiophene
AC	Activated Carbon

ADS	Adsorptive Desulfurization
AED	Atomic Emission Detector
Al	Alumina
B.P.N.	Boiling Point Number
BDN	Biodenitrogenation
BDS	Biodesulfurization
BDSM	Biodesulfurizing Microorganisms
BNT	Benzo-Naphtho-Thiophene
BP	Biphenyl
BT	Benzothiophene
BTO	Benzothiophene Sulfoxide
BTO_2	Benzothiophene Sulfone
BTs	Benzothiophenes
CAR	Carbazole
CPI	Carbon Preference Index
CTAB	Cetyl Trimethyl Ammonium Bromide
DBDS	Dibenzyl Disulfide
DBS	Dibenzyl Sulfide
DBT	Dibenzothiophene
DBTO	Dibenzothiophene sulfoxide
$DBTO_2$	Dibenzothiophene Sulfone
DBTs	Dibenzothiophenes
DC	Direct Current
DCW	Dry Cell Weight
DES	Diethyl Sulfide
DMF	Dimethylformamide
DMS	Dimethyl Sulfide
DMSO	Dimethyl Sulfoxide
DPS	Diphenyl Sulfide
EBC	Energy Biosystems Corporation
EDS	Extractive Desulfurization
EK-BDS	Electrokinetic Biodesulfurization
FMN	Flavine Mononucleotide
FT-IR	Fourier Transform Infrared
GC/FID	Gas Chromatography/Flame Ionization Detector
GC/MS	Gas Chromatography/Mass Spectrometry
GC/SCD	Gas Chromatography/Sulfur Chemiluminescence Detector
GMO	Genetically Modified Organism
$HBPSi^-$	2-Hydroxybiphenyl-2′-Sulfinate
HCLN	High Carbon and Low Nitrogen
HDEBP	2-Hydroxy-3,3'-Diethylbiphenyl
HDMBP	Hydroxy Dimethyl Biphenyl
HDS	Hydrodesulfurization
HFBT	3-Hydroxy-2-Formylbenzothiophene

HGO	Heavy Gas Oil
HN	High Nitrogen
HPEal	2-(2′-hydroxyphenyl)ethan-1-al
HPESi$^-$	2-(2′-Hydroxyphenyl)Ethen-1-Sulfinate
HPLC	High Performance Liquid Chromatography
HRP	Horseradish Peroxidase
IBUs	Immobilized Biomass Units
IEP	Isoelectric Points
IGT	Institute of Gas Technology
IPTB	Inverse Phase Transfer Biocatalysis
KEGG	Kyoto Encyclopedia of Genes and Genomes
LGO	Light Gas Oil
LLE	Liquid-Liquid Extraction
LMW	Low Molecular Weight
LN	Low Nitrogen
MAS	Mesoporous Aluminosilicates
MBSM	Modified Basal Salts Medium
MNPs	Magnetic Nano Particles
MSM	Minimal Salts Medium
NADH	Reduced Nicotinamide Adenosine
NPs	Nanoparticles
n-TD	normal-Tetradecane
NTH	Naphtho[2,1-B]Thiophene
O/W	Oil/Water
ODS	Oxidative Desulfurization
OSCs	Organosulfur Compounds
OTR	Oxygen Transfer Rate
PACs	Poly Aromatic Compounds
PASHs	Polyaromatic Sulfur Heterocyclic Compounds
PCR	Polymerase Chain Reaction
PEG	Polyethylene Glycol
Ph	Phytane
Pr	Pristane
PU	Poly Urethane
PVA	Polyvinyl Alcohol
PVAc	Polyvinyl Acetate
RE	Restriction Enzyme
S	Sulfur
SA	Sodium Alginate
Sep	Sepiolite
SFC	Supercritical Fluid Chromatography
Si	Silica
SRB	Sulfate Reducing Bacteria
T2C	Thiophene-2-Carboxylic Acid

TBL	γ-Thiobutyrolactone
TCA	Tricarboxylic Acid
TG-DTA	Thermogravimetric-Differential Thermal Analysis
Th	Thiophene
Ths	Thiophenes
Ti	Titania
TLC	Thin Layer Chromatography
TPO	2(5H)- thiophenone
TPR	Temperature Programmed Reduction
TTC	2,3,5-Triphenyl-2H-Tetrazolium Chloride
UCM	Unresolved Complex Mixtures
ULS	Ultra-Low Sulfur
ULSD	Ultra-Low Sulfur Diesel
UV	Ultraviolet Spectroscopy
VRICR	Vertical Rotating Immobilized Cell Reactor
β-CD	B-Cyclodextrin

10.1 Introduction

The biodesulfurization (BDS) is expected to play an important role in biorefining, however in pre-refining and post-refining of oil products it involves a microbial or enzymatic system that selectively oxidizes the sulfur atoms or removes sulfur without attacking the C-C bond and avoiding the loss of oil value. Historically, between the 1970's and 1980's, advances in BDS yielded low-octane values, that is low quality fuels, because of anaerobic (Kim *et al.*, 1990) or aerobic degradation (Kodama *et al.*, 1973; Van Afferden *et al.*, 1993) of the carbon backbone by microorganisms that obviously did not satisfy industrial expectations, but it has good applicability for the remediation of petroleum hydrocarbon polluted soil and water.

Despite the discovery of specific microorganisms (4S-pathway) that remove sulfur without affecting the octane value of fuels (Malik, 1978), the refining industry was not concerned because of the flexible environmental rules, high light petroleum reserves with low S-content, and profitability of conventional processes (mainly the HDS). However, the resurgence of BDS during the 1990's came with more severe environmental emission regulation, increasing worldwide fuel demand, the depletion of light crude reserves, and the exploitation of new reservoirs of sulfur- and nitrogen rich oils, such as heavy crude, tar sands, and oil shale. Consequently, the oil industry in collaborative efforts with academia aimed to research and develop the applicability of BDS over BDN (Chapter 4) processes.

It has been estimated that the BDS-catalyst must have an S-removal activity within the range of 1–3 mmol DBT/g DCW/h in real petroleum fractions. This means the activity of the current biocatalysts needs to be improved 500 fold to attain a commercially viable process (Kilbane, 2006).

This chapter summarizes the worldwide effort and achievements for enhancement of the BDS efficiency to be applied in the real field industry.

10.2 Isolation of Selective Biodesulfurizing Microorganisms with Broad Versatility on Different S-Compounds

There is increasing research about the isolation of new biodesulfurizing strains with high BDS-capabilities via the 4S-pathway being performed worldwide.

There are many microbial strains capable of degrading DBT to compounds that enter into the ticarboxylic acid (TCA) cycle. These microorganisms are recommended for the bioremediation of petroleum hydrocarbons polluted by environment, but they are not recommended for BDS application as they will affect the calorific value of the biotreated fuels because most of these strains are capable of degrading different polyaromatic compounds (PACs) and other hydrocarbon components of fuels (Arun *et al.*, 2011), for example *Pseudomonas* sp. NCIB 9816-4 (Resnick and Gibson, 1996), *Pseudomonas fluorescens* 17 (Finkelstein *et al.*, 1997), *Janibacter* sp. YY-1 (Yamazoe *et al.*, 2004), *Arthrobacter* sp. P1-1 (Seo *et al.*, 2006), *Terrabcter* sp. DBF63 (Habe *et al.*, 2004), *Sphingomonas* sp. (Gai *et al.*, 2007), *Bacillus sphaericus* HN1 (Nassar *et al.*, 2010), and *Corynebacterium variabilis* sp. Sh42 (El-Gendy *et al.*, 2010).

In 1990, the Institute of Gas Technology (IGT, Chicago, IL, USA) developed new strains through the selection of mutagens (Kilbane and Bielaga 1990). Further, extensive work performed at IGT has led to the identification of the *Rhodococcus* sp. strain IGTS8 as the bacterium of choice in the BDS process and further studies have revealed the ability of this newly recognized bacterium (IGTS8) to remove sulfur from DBT without the carbon–carbon cleavage of the ring structure and save the calorific value of the fuel (Ohshiro *et al.* 1994).

Dibenzothiophene (DBT) serves the archetype condensed thiophene. Many studies have been conducted predicting the fate of DBT and its homologues in petroleum-contaminated environments (Atlas *et al.*, 1981; Fedorak and Westlake, 1984; Hostettler and Kvenvolden, 1994). However, much work has been conducted on the specific desulfurization of DBT without affecting its hydrocarbon skeleton (Krawiec, 1990; Denome *et al.*, 1994; Shennan 1996; Oldfield *et al.*, 1997, 1998; Derikvand *et al.*, 2014; Akhtar *et al.*, 2016). Generally, there are three modes of aerobic attack on DBT: (1) the naphthalene-like attack, which involves the dioxygenase attack at carbons number 1 and 2, followed by meta-cleavage (Grifoll *et al.*, 1995), (2) the five-membered ring attack where a direct attack occurs at the S-atom of DBT, and (3) the angular dioxygenase attack that involves the oxidation of carbons number 4a and 4.

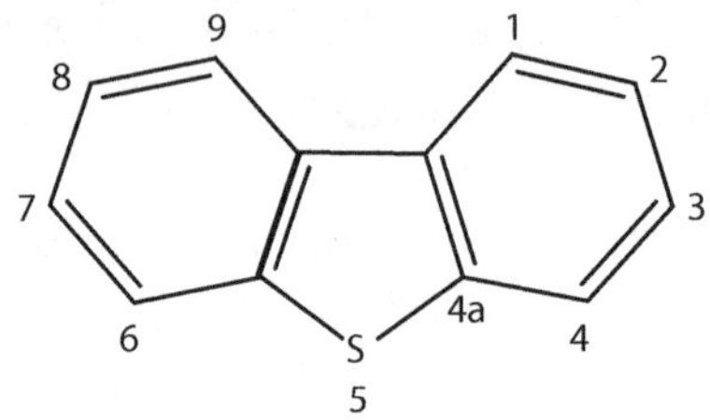

10.2.1 Anaerobic Biodesulfurizing Microorganisms

Anaerobic reactions proceed more slowly than aerobic reactions, but generate the same products as conventional HDS technology, H_2S, and desulfurized oil.

Mechanisms for the desulfurization of crude oil using the hydrogenase enzyme activity of sulfate reducing bacteria (SRB) was proposed in 1953, where a Texaco patent reported the desulfurization of crude oil by hydrogenation in the presence of SRB, such as *Desulfovibrio desulfuricans* (Monticello, 1985).

The microbial metabolisms of thiophene (Th) and benzothiophene (BT) have been reported by Kurita *et al.* (1971). The bacterial cultures in this study were obtained from oily sludge and grown with polypeptone, lactic acid, glucose, meat extract, and yeast extract. Hydrogen sulfide was detected as a product during Th metabolism, but lower C_1-C_4 hydrocarbons could not be found. Benzothiophene could yield hydrogen sulfide as an anaerobic end-product as well as several unknown carbon compounds.

Several anaerobic strains have demonstrated the ability to remove organic sulfur, for example DBT. Köhler *et al.* (1984) reported that mixed cultures containing sulfate reducing bacteria (SRB) desulfurized a variety of model compounds, including DBT, BT, DBS, and dibenzyldisulfide (DBDS).

A mixed methanogenic culture which was derived from a methanogenic sewage digester reductively desulfurized DBDS to toluene with benzylmercaptan as an intermediate in a sulfur limited medium (Miller, 1992). Kim *et al.* (1990a) investigated the specific desulfurization by *Desulfovibrio desulfuricans* M6. This anaerobic strain could degrade 96% and 42% of BT and DBT, respectively, as well as other substrates, including phenylsulfide, benzylsulfide, benzyldisulfide, ethanethiol, butanethiol. Metabolite analyses proved that the strain could convert DBT to biphenyl and H_2S. M6 also removed 21% of the sulfur in Kuwait crude oil and up to 17 wt% from other crude oils and their distillate products (Kim *et al.*, 1990b).

The extreme thermopile *Pyrococcus furiosus*, growing at 98°C, was shown to reduce organic polysulfide to H_2S using the reduction of sulfur as a source of energy (Tilstra *et al.*, 1992).

A *Nocardioform actinomycete* FE9 grown on n-hexadecane with DBT as the sole sulfur source formed biphenyl, traces of 2-HBP, dihydroxybiphenyl, and sulfate under air atmosphere, but biphenyl and H_2S under nitrogen or hydrogen atmospheres where the highest conversion rates were under a hydrogen atmosphere. This strain could remove both the sulfur heteroatoms in thianthrene converting them to benzene and sulfate under an oxygen atmosphere. Terthiophene was transformed to H_2S and an unsaturated product tentatively identified as 1,3,5,7,9,11-dodecahexaene (Finnerty, 1992; 1993).

Three species of Sulfate-Reducing Bacteria (SRB) were reported to be able to grow using DBT as their sole source of sulfur and sole electron acceptor in media containing lactate and citrate. *Desulfotomaculum orientis* and *Desulfovibrio desulfuricans* were grown at 30 °C, while *Thermodesulfobacterium commune* was grown at 60 °C and hydrogen sulfide was the product of dissimilatory sulfur reduction (Lizama *et al.*, 1995).

Bahrami *et al.* (2001) reported the anaerobic DBT-BDS by a thermophilic mixed culture isolated from crude oil. Methyl Viologen is as an electron mediator, which is essential for microbial desulfurization (Kurita *et al.* 1971). The mixed culture recorded 98% BDS of 0.5 mg/mL DBT at 65 °C over 15 days both in the presence and in the absence of Methyl Viologen, but biphenyl (BP) and sulfide were not the only end-products of DBT biodegradation and further investigation was recommended.

The ability of hyper-thermophelic *Sulfolobus solfataricus* P2 to desulfurize dibenzothiophene and its derivatives at 78 °C has also been reported (Gün *et al.*, 2015), but it was not possible to desulfurize benzothiophene.

10.2.2 Bacteria Capable of Aerobic Selective DBT-BDS

Many species were found to be involved in selective desulfurization via the 4S pathway, including *Rhodococcus erythropolis* (Davoodi-Dehaghani *et al.*, 2010; Mohamed *et al.*, 2015), *Klebsiella* (Dudley and Frost, 1994), *Corynebacterium*, *Arthrobacter* and *Xanthomonas* (Constanti *et al.*, 1994), *Paenibacillus* (Konishi *et al.*, 1997), *Nocardia* CYK2 (Chang *et al.*, 1998), *Microbacterium* strain ZD-M2 (Li *et al.*, 2005a), *Mycobacterium goodii* X7B (Li *et al.*, 2005b), *Trichosporon* sp. (Zahra *et al.*, 2006), *Shewanella putrefaciens* NCIMB8768 (Ansari *et al.*, 2007), *Rhizobium* sp. MS4 (Pokethitiyook *et al.*, 2008), *Stachybotrys* sp. WS4 (Torkamani *et al.*, 2008), *Lysinibacillus sphaericus* DMT-7 (Bahuguna *et al.*, 2011), *Stenotrophomonas* sp. NISOC-04 (Papizadeh *et al.*, 2011), *Pantoea agglomerans* D23W3 (Bhatia and Sharma, 2010a), *Bacillus* sp. KS1 (Rath *et al.*, 2012), *Corynebacterium mycetoides* EMB113 (Olabemiwo *et al.*, 2013), and *Chelatococcus* sp. (Bordoloi *et al.*, 2016).

Generally, the bacteria belonging to the genus *Rhodococcus* are the widely applied genera in BDS studies (Izumi *et al.*, 1994; Ohshiro *et al.*, 1995; Omori *et al.*, 1995; Wang and Krawiec, 1996; Kishimoto *et al.*, 2000; Matsui *et al.*, 2001). *R. erythropolis* IGTS8 (previously known as *R. rhodochorus* IGTS8) was the first to be identified (Kilbane and Jackowski, 1992) and has received the most attention. *Rhodococcus* requires the addition of NADH to restore activity. It was found that flavin mononucleotide (FMN) increased the rate of desulfurization. DBT is first converted to DBTO by a monoxygenase, then to $DBTO_2$ again by a monoxygenase both using molecular oxygen and requiring an oxidoreductase enzyme to oxidize NADH. *Rhodococcus erythropolis* is unusual in the sense that it has a separate enzyme (oxidoreductase) from the monooxygenases (Guerinik and Al-Mutawah, 2003). Most oxygenases have a flavin incorporated into their structure. The selective 4S-biodesulfurization (BDS) pathway is a series of three bio-reactions leading to HBP with DBTO and $DBTO_2$ as intermediates. The conversion of DBT to DBTO and $DBTO_2$ depends on the activity of monoxygenase and oxidoreductase. The production of HBP depends on the availability of $DBTO_2$ and activity of a desulfinase. Conceivably, under high activities of monooxygenase DBTO is generated faster and will accumulate as an intermediate and the rates of consumption of DBT will be greater than the HBP rate of production. If, on the other hand, the sulfinase is more active, $DBTO_2$ will be readily converted to 2-HBP as it is produced and, in this case, the whole cycle will be controlled by only the first two steps (Figure 10.1).

However, most *Rhodococci* are unable to show high activity for the alkyl derivatives of DBT, BT, and other thiophenic compounds. Furthermore, there are only limited biochemical and genetic studies of bacteria that exhibit desulfurization of both DBT and BT (Kayser *et al.*, 2002; Alves *et al.*, 2005). Since these compounds exist in significant amounts in fossil fuels, it

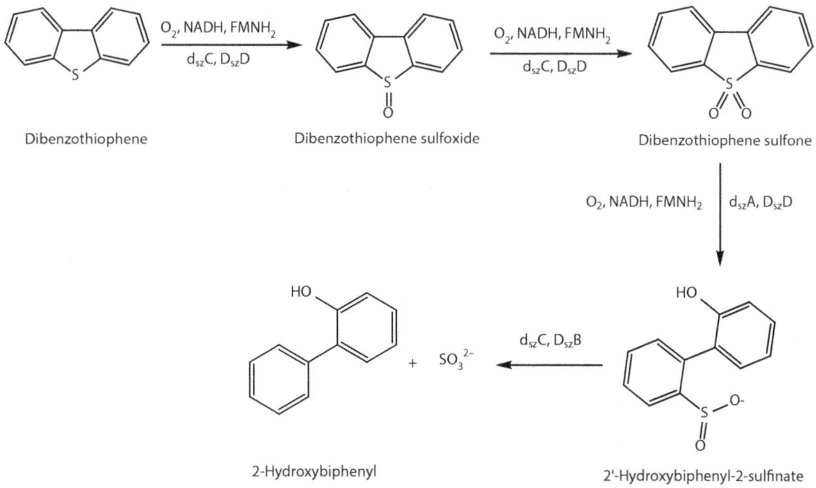

Figure 10.1 The Pathway of Biological Desulfurization of DBT as a Model Compound (Denome *et al.*, 1993 and 1994). NADH is a reduced nicotinamide adenosine dinucleotide, FMN is a flavin mononucleotide, and $D_{SZ}A$, $D_{SZ}B$, $D_{SZ}C$, and $D_{SZ}D$ are the catalytic gene products of *dszA, dszB, dszC,* and *dszD,* respectively.

is important to study microbes that possess activity for compounds other than DBT. Furthermore, because desulfurizing these compounds requires distinct pathways, bacterial strains that possess the associated genes for all these pathways are clearly desirable. *Gordonia* is an attractive genus in this regard because its members exhibit much metabolic versatility (Arenskötter *et al.*, 2004). However, besides DBT, it can also specifically cleave the C–S bond in BT and other thiophenes. One of the most important characteristics of this genus is its ability to perform BDS through the 4S pathway (Kim *et al.*, 1999; Mohebali *et al.*, 2007a). The first representative of the genus *Gordonia* described as being capable of DBT desulfurization was *Gordonia* sp. strain CYKS1 (Rhee *et al.*, 1998). Since then, many others have been described within that genus, such as *Gordonia desulfuricans, Gordonia nitida,* and *Gordonia alkanivorans.* The latter is probably the most studied and the one with greatest number of strains important to BDS, for example RIPI90A (Shavandi *et al.*, 2009, 2010; Mohebali *et al.*, 2007a, 2007b, 2008) and strain 1B. Because of its ability to desulfurize a wider range of PASHs, *G. alkanivorans* appears to offer some advantage over *R. erythropolis* for BDS. Moreover, *G. alkanivorans* strains are reported to show nearly 2–10 times higher desulfurization activities than the desulfurizing *R. erythropolis* strains (Mohebali *et al.*, 2007a). However, the activity levels are still not acceptable for commercial application and BDS studies on *G. alkanivorans* are far more limited than those with *R. erythropolis.* This might be attributed to the adaptability of *R. erythropolis* to high concentrations or pure oil feed, for example heavy gas oil (HGO), due to the fact that these bacteria modulate the viscosity of cell membrane lipids to maintain or increase the fluidity of the same by reducing the degree of saturation, increasing the ratio of cis/trans of fatty acids, and by increasing the relative amount of branched fatty acids (Whyte *et al.*, 1999; Heipieper *et al.*, 2007). Although the hydrophobicity of *R. erythropolis* would increase the contact between cells and the sulfurous compounds, the high viscosity of HGO or heavy crude oil may have decreased access to these compounds, leading the way to a minor removal percentage (Folsom *et al.*, 1999; Kirimura *et al.*, 2001; Carvalho and Da Fonseca, 2005). A solution to this problem has been developed employing surfactants to increase the contact between those two phases (Dinamarca *et al.*, 2014).

The most widely employed bacterium in BDS works is the Gram +ve *Rhodococcus erythropolis* IGTS8, which is able to selectively remove the sulfur atom from DBT throughout the 4S-pathway without affecting its hydrocarbon skeleton and, consequently, the calorific value, producing 2-HBP and sulfate. IGTS8 is the applied biocatalyst in the commercial process of the Energy Biosystems Corporation (Olson *et al.*, 1993; Oldfield *et al.*, 1997; McFarland *et al.*, 1998). Other applied microorganisms belong to other genus, such as *Pseudomonas* (Constanti *et al.*, 1994), *Gordona* (Gilbert *et al.*, 1998; Rhee *et al.*, 1998), and *Brevibacterium* (Van Affender *et al.*, 1990). *Corynebacterium* sp. ZD-1 has been isolated from refinery sludge of the Hanzhou Refinery Co. (Hangzhou, China) for its ability to desulfurize DBT via the 4S-pathaway, producing 2-HBP as a dead end product (Wang *et al.*, 2004). Numerous *Gordonia* strains exhibit higher desulfurization activities (Arenskötter *et al.*, 2004; Alves *et al.*, 2005) than the *rhodococci* for a broader range of PASHs (Rhee *et al.*, 1998; Alves *et al.*, 2005, 2007; Santos *et al.*, 2006; Kilbane and Robbins, 2007; Li *et al.*, 2008a). For example, *G. alkanivorans* (Alves *et al.*, 2005, 2007; Shavandi *et al.*, 2008; Alves and Paixão 2011) desulfurizes DBT via the well-known 4S pathway (Gray *et al.*, 1996). Recently, *Ralstonia eutropha* PTCC1615 was reported for its ability to desulfurize DBT via the 4S-pathway, producing 2-HBP as a dead end product (Dejaloud *et al.*, 2017).

However, Pokethitiyook *et al.* (2008) reported the isolation of *Rhodococcus gordoniae* R3 from oil-contaminated soil, which is able to grow in high concentrations of DBT, but not via the S-selective pathway. It also showed broad versatility over different S-compounds, including DBT, 3,4-benzo-DBT, 4,6-DMDBT, and methyl phenyl disulfide. Pikoli *et al.* (2014) reported the isolation of three bacterial isolates from a coal-mixed soil collected from South Sumarta for DBT-BDS. One was identified as

Bacillus megaterium CSK2 and the other two were identified as *Bacillus subtillus* CSK3 and CSK4. They expressed good growth rates (0.313, 0.27, and 0.194 h^{-1}, respectively) and BDS potential (37.4%, 16.7%, and 25.9%, respectively) in a batch biodesulfurization process of 0.1 mM DBT as an S-source in the presence of 10 g/L glucose as a C-source at 29 °C, 150 rpm, and within 48 h of incubation.

Serval reports about the isolation of biodesulfurizing microorganisms (BDSM) from soil contaminated by petroleum hydrocarbons have been published. Moreover, landfarm soil from tropical regions provides a rich source of alternative microbes and metabolisms for biodesulfurization processes (Buzanello *et al.*, 2014). Jiang *et al.* (2002) reported the isolation of five bacterial isolates from soil and wastewater samples collected from coal mining areas and oil refineries and identified *Bacillus sphaericus* R-6, *Bacillus brevis* R-16, *Nocardia globerula* R-9, and *Pseudomonas delafieldii* R-8 for their ability to desulfurize DBT via the 4S-pathway to 2-HBP. Another *Pseudomonas delafieldii* R-12 was also isolated, but it completely degrades DBT without the production of $DBTO_2$ or 2-HBP as intermediates (Jiang *et al.*, 2002). *Rhodococcus* FMF native bacterium was isolated from soil contaminated with oil in a Tabriz refinery for selective DBT-BDS via the 4S-pathway (Akbarzadeh *et al.*, 2003). Guerinik and Al-Mutawah (2003) reported the isolation of *Rhodococcus erythropolis* and *Rhodococcus globerulus* from soil samples collected from Kuwait for their ability to desulfurize DBT via the 4S-pathway to 2-HBP as a dead end product. Mingfang *et al.* (2003) reported the efficiency of *Nocardia globerula* R-9 for the selective BDS of DBT and 4,6-DMDBT for their corresponding phenolic compounds, 2-HBP and monohydroxy dimethyl biphenyl (HDMBP). *Gordonia alkanivorans* strain 1B was isolated from hydrocarbon-contaminated soil for its ability to desulfurize DBT via the 4S-pathway, producing 2-HBP (Alves *et al.*, 2005). Labana *et al.* (2005) reported the ability of *Rhodococcus* sp. and *Arthrobacter sulfureus* isolated from oil-contaminated soil/sludge to desulfurize DBT throughout the 4S-pathway, producing 2-HBP as a dead end-product. Ardakani *et al.* (2010) reported the isolation of *Stenotrophomonas maltophila* strain Kho1 (NCBI Gene Bank Accession No. HM367710) from oil-contaminated soil in Khuzestan for its ability to desulfurize DBT via the 4S-pathway, producing 2-HBP. *Pantoea agglomerans* D23W3, isolated from contaminated soil collected from a refinery, desulfurizes DBT via the 4S-pathway and has good BDS potential on BT and 4,6-DMDBT, but cannot desulfurize 4,6-DEDBT (Bhatia and Sharma, 2010a). Davoodi-Dehaghani *et al.* (2010) reported the isolation of *Rhodococcus erythropolis* SHT87 from oil-contaminated soil around the Tehran Oil Refining Co., Iran, for its ability to desulfurize

DBT via the 4S-pathway, producing 2-HBP as a dead end product. SHT87 has a broad versatility on different S-compounds; it showed good growth on DBT, dibenzothiophene sulfone, thiophene, 2-methyl thiophene, and dimethylsulfoxide, but the growth on thiophene 2-carboxylic acid and 5,5'-dithio-bis (2-nitrobenzoic acid) was not considerable. *Gordonia* sp. IITR100 (Gene bank accession No. GU084407) has been isolated for the isolated selective enrichment of hydrocarbon contaminated soil collected from the Koyali refinery, Gujarat, using 4,6 dimethyl dibenzothiophene as the only sulfur source (Singh and Srivastava, 2013). *Gordonia* sp. IITR100 reported to desulfurize 4,6-DMDBT to 2-hydroxy-3,3′-dimethylbiphenyl (Adlakha *et al.*, 2016) and benzonaphtothiophene to α-hydroxy-β-phenyl naphthalene (Yu *et al.*, 2006a). Buzanello *et al.* (2014) reported the isolation of *Bacillus pumilus*-like RR-3 from tropical landfarm soil for its ability to rapidly desulfurize high concentrations of DBT via the 4S-pathway, producing 2-HBP. However, the depletion of 2-HBP suggests the occurrence of further degradation or metabolism. Mohamed *et al.* (2015) reported the isolation of *Rhodococcus* sp. strain SA11, *Stenotrophomonas* sp. strain SA21, and *Rhodococcus* sp. strain SA31 from oil contaminated soil for their ability to specifically desulfurize DBT via the 4S-pathway with broad versatility on other alkylated high molecular weight DBTs, such as 1-MDBT, 4-MDBT, 2,3-DMDBT, and 4,6-DMDBT. Not only this, but SA11 also expressed good growth and BDS activity on BT. Derikvand *et al.* (2015a) reported the isolation of *Rhodococcus erythropolis* PD1 (NCBI Gene Bank Accession No. JX625154) from an oil contaminated soil sample for its ability to desulfurize DBT via the 4S-pathway. Derikvand *et al.* (2015b) also reported the isolation of *Paenibacillus validus* (strain PD2) from oil contaminated soil for its ability to desulfurize DBT via the 4S-pathway. Two *Streptomyces* species, namely *Streptomyces* sp. VUR PPR 101 and *Streptomyces* sp. VUR PPR 102, were isolated from oil contaminated sites for their ability to desulfurize DBT via the 4S-pathway (Praveen Reddy and Umamaheshwara Rao, 2015). *Rhodococcus* sp. MP12, which is reported to be used as a desulfurizing strain for DBT via the 4S-pathway, was isolated from oil field soil by repeated cultivation in a selective medium with DBT as the sole sulfur source in Shaanxi, China (Peng and Zhou, 2016). *Bacillus* sp. E1 (Verma *et al.*, 2016) isolated from oil contaminated soil samples was reported to desulfurize DBT throughout the 4S-pathway. A mixed culture AK6 isolated from petroleum hydrocarbons-polluted soil was reported to desulfurize DBT via the 4S-pathway. Furthermore, it is reported to utilize 4-MDBT, 4,6-DMDBT, and BT as sole S-sources, but not as C-sources, but it could not desulfurize dibenzylsulfide (DBS) (Ismail *et al.*, 2016). *Gordonia* sp. JDZX13 was isolated from crude oil-contaminated samples of the Karamay Oil field, Xinjiang, China, for its ability to desulfurize DBT via the 4S-pathway to 2-HBP, while 4,6-DMDBT was desulfurized to 4,6-dimethyl-hydroxybiphenyl (4,6-DMHBP). The sulfur-specific biodesulfurization of thiophene, benzothiophene, dibenzothiophene, 4-methyl dibenzothiophene, and 4,6-dimethyl dibenzothiophene by the Gram-positive bacterial isolate *Brevibacillus invocatus* C19, that was previously isolated from an Egyptian coke sample, has been also reported (Nassar *et al.*, 2013). In another study, Nassar *et al.* (2016) isolated *Brevibacillus brevis* strain HN1 (accession no. KF018281) from an Egyptian coke sample with a sulfur content of 2.74% for its ability to desulfurize DBT via the 4S-pathway. Furthermore, the Gram-negative bacterial isolate *Sphingomonas subarctica* T7b was reported for the BDS of dibenzothiophene, 4,6-dimethyl dibenzothiophene, 4,6-dipropyl dibenzothiophene, 4,6-dibutyl dibenzothiophene, 4,6-dipentyl dibenzothiophene, 4-hexyl dibenzothiophene, 2,3-dihexyl dibenzothiophene, and 3-propyl-4,8-dimethyl dibenzothiophene via the 4S-pathway (Gunam *et al.*, 2013). The search for new desulfurizing psychrophilic strains from the Antarctic continent revealed the presence of important microorganisms isolated from soil and the rhizosphere. These microorganisms show a higher capacity of desulfurization than previously described in the literature (Boniek *et al.* 2010). In earlier studies, one Gram-negative species known as *Sphingomonas* subarctic was isolated in Japan and it was proven that it possesses an efficient ability to desulfurize DBT, considering the industrial perspectives (Gunam *et al.* 2006, 2013).

A Gram –ve *Entreobacter* sp. strain NISOC-03, isolated from petroleum-polluted soil samples, has been reported to show an ability to desulfurize DBT via the 4S-pathway in the presence of benzoate as a C-source, where 64% desulfurization of 0.8 mM dibenzothiophene occurred with the non-stoichiometric production of 0.27 mM phenylphenol (i.e. 2-HBP) (Papizadeh *et al.*, 2017).

Jiang *et al.* (2002) reported the broad versatility of two bacterial isolates, *Pseudomonas delafieldii* R-8 and *Nocardia globerula* R-9, to utilize different S-compounds sole S-sources at 32 °C, pH 7.02, and 170 rpm, where R-8 recorded different efficiencies on different studied S-compounds that can be ranked in the following decreasing order: 4,6-DMDBT>DBT-sulfone>DBT>DBS>BT, recording desulfurization rates of 2.46, 2.35, 2.26, 1.93, and 1.34 mg/g/h, respectively, while R-9 showed a different trend: DBT-sulfone>4,6-DMDBT>DBS>DBT>BT, recording BDS-rates of 2.45, 2.25, 2.06, 1.98, and 1.04 mg/g/h, respectively, which revealed the ability of *Pseudomonas delafieldii* R-8 and *Nocardia globerula* R-9 to desulfurize different OSCs, such as sulfides and thiophenes throughout an S-specific route, promoting their applicability in the BDS process.

Chen *et al.* (2008) reported the complete BDS of Th, BT, DBT, and 4,6-DMDBT by *Mycobacterium goodii* X7B, within 10, 42, 50, and 56 h, respectively, while diphenylsulfide (DPS) expressed the lowest desulfurization at 80% within 90 h.

Akhtar *et al.* (2009) reported the isolation of *Rhodococcus* spp. Eu32, from a soil sample taken from the roots of a eucalyptus tree, for its ability to desulfurize DBT via the 4S-pathway producing 2-HBP and biphenyl as a way to overcome the toxicity of 2-HBP at 30 °C. Eu-32 also showed a broad versatility to utilize different S-compounds as sole S-sources in the presence of glucose as a C-source. However, it expressed better growth on DBT than DBT-sulfone. Eu-32 expressed nearly the same growth capacity on Th and 3-methyl thiophene, which was better than 2-methyl thiophene. Moreover, it expressed good growth efficiency on thiophene-2-acetic acid, thianthrene, thioxenthne-9-one, *p*-tolyle disulfide, sulfanilamide, and 1,3-dithiane, but it expressed lower growth capabilities on thiophene 3-carboxylic acid, *p*-toluenesulphonic acid, and sulfur. Casullo de Araújo *et al.* (2012) also reported the isolation of *Serratia marcescens* UPC1549 from the semi-arid soil of the northeastern region of Brazil (Pernambuco - Brazil) and banana culture which is capable of desulfurizing DBT via the 4S-pathway to 2-HBP at 28 °C and overcome the toxicity of 2-HBP by converting it to biphenyl.

Akhtar *et al.* (2016) reported the isolation of 10 new bacterial isolates of genera *Gordonia, Amycolatopsis, Microbacterium,* and *Mycobacterium* for their ability to desulfurize DBT via the 4S-pathway, producing 2-HBP as a dead end-product, with a maximum 2-HBP formation rate of 3.5 μmol/g DCW/h. Although of the toxic effect of 2-HBP, especially at a high concentration (> 0.6 mM), would inhibit the cell growth and DBT-BDS, the growth of cells would be enhanced in the presence of low concentrations of 2-HBP (≤ 0.1 mM). They also have broad versatility on different OSCs and showed good BDS activity in batch cultures of 0.2 mM of DBT, DBT sulfone, 4-methyl DBT, benzothiophene (BT), 3-methyl BT, thiophene (Th), 2-methyl Th, 3-methyl Th, thioxanthen-9-one, and *p*-tolyl disulfide. Gibb's reagent revealed the formation of phenolic compounds (the formation of blue color) when 4-methyl DBT, benzothiophene (BT), and 3-methyl BT were used as sole sources of organic sulfur.

The Gram –ve, aerobic *Pseudomonas putida* CECT 5279 is a promising biocatalyst for the BDS process (Gallardo *et al.*, 1997; Martin *et al.*, 2004; Caro *et al.*, 2007b, 2008) because it performs the 4S-pathway with increasing process efficiency because the amount of reducing equivalents consumed is reported to be lower (Gallardo *et al.*, 1997). The Gram negative bacteria, such as *Klebsiella* sp. 13T and *Pantoea agglomerans* D23W3, were also reported for thermophilic and mesophilic biodesulfurization of DBT, respectively, from different petroleum oils (Bhatia and Sharma, 2010a, 2012).

Boshagh *et al.* (2014) proposed the combination of electrokinetics and BDS for the first time. It comprises the induction of an electrical field on the acting media of microorganisms for removing sulfur content. The effects of electrokinetics on the BDS of model oil are studied using *Rhodococcus erythropolis* PTCC1767 and *Bacillus subtilis*

DSMZ 3256, capable of BDS via the 4S-pathway. Desulfurization of the model oil was initially evaluated by BDS experiments, then combined electrokinetic-BDS experiments were performed using the same conditions: 1:1 O/W, 10 mM DBT in n-hexadecane, pH 7.4, and 30 °C with an incubation period of 6 days using resting cells harvested at the late exponential phase. The bench-scale model aquifer system packed with glass beads under constant-temperature and pH was employed (Figure 10.2). A peristaltic pump was used to circulate the electrolyte solution from cathode to anode at a rate of 30 mL/min to control ionic concentrations and sudden pH change in the bioreactor, while the temperature was controlled using a circulator which circulated the water in a bypass channel of the reactor. The effects of electric current density on the DBT conversion and desulfurization yield were investigated. Three different current densities of 2.5,

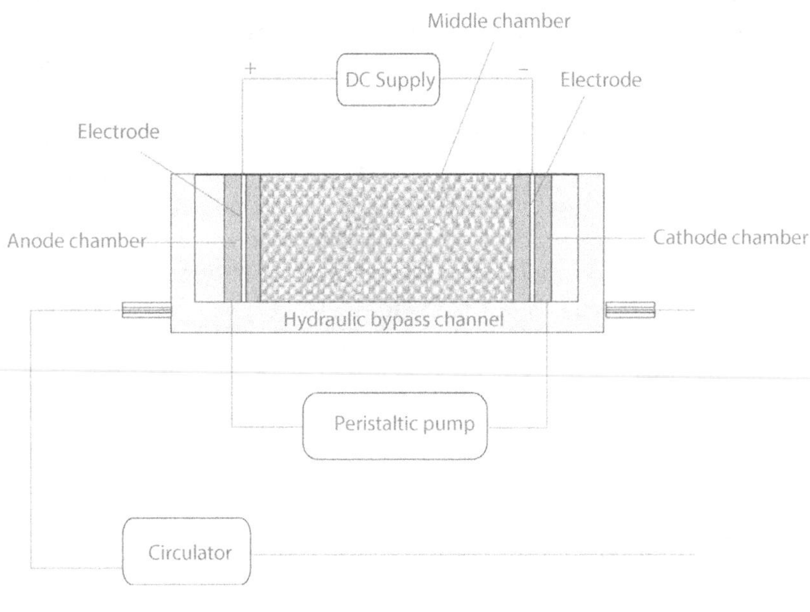

Figure 10.2 Schematic Diagram of Electro-Kinetic Setup for BDS.

5, and 7.5 mA/cm^2 were tested. The samples were collected at each 24-h interval for a 3-day period. The DBT conversion (X_{DBT} %) and biodesulfurization yield (X_{BDS} %) were calculated by eqs. (1) and (2), respectively. The desulfurization yield is defined as the percentage ratio of produced 2-HBP (C_{2HBP}) and initial DBT (C_{DBTo}) concentrations:

$$X_{DBT} = \frac{(C_{DBTo} - C_{DBT})}{C_{DBTo}} \times 100 \quad (1)$$

$$X_{BDS} = \frac{C_{2HBP}}{C_{DBTo}} \times 100 \quad (2)$$

The results indicated that *R. erythropolis* is a more efficient strain than *B. subtilis* for the biodesulfurization and EK-BDS processes. The results represented 34% and 62% DBT conversions after 1 and 6 days by *R. erythropolis* and the biodesulfurization yields were 11% and 36%, respectively. However, the DBT conversions for *B. subtilis* strain after 1 and 6 days were 31% and 55% and the biodesulfurization yields were 9% and 31%, respectively, while the X_{DBT} and X_{BDS} for *R. erythropolis* after 3 days were 76% and 39%, respectively, at the current density of 7.5 mA/cm^2. At the same conditions, the X_{DBT} and X_{BDS} for *B. subtilis* were 71% and 37%, respectively.

Eelectrokinetic enhanced the BDS-activity via one of several transport mechanisms, including electromigration, electroosmosis, and electrophoresis (Kim *et al.*, 2005; Luo *et al.*, 2005; Wick *et al.*, 2007). OSCs are transported by electroosmosis and electromigration to either the cathode or anode. Electroosmosis is the movement of the pore fluid which contains dissolved ionic and non-ionic species toward the cathode. Electromigration is the movement of the dissolved ionic species which are present in the pore fluid toward the opposite electrode. The mechanisms of the bacteria transport are electroosmosis and electrophoresis. When a direct current (DC) is applied through inert electrodes, electrolysis of water molecules occurs in both electrodes as follows:

$$\text{In anode: } 2H_2O - 4e^- \rightarrow 4H^+ + O_2(g)$$

$$\text{In cathode: } 2H_2O + 2e^- \rightarrow 2OH^- + H_2(g)$$

Thus, the microorganisms consume the oxygen released from the electrolysis of water, producing more biomass. Moreover, the microorganisms can be transported in both directions; electroosmosis transports the microorganisms from the anode to the cathode along with the bulk water flow, while the direction of transport by electrophoresis is from the cathode to the anode (Wick *et al.*, 2004; Kim *et al.*, 2005). Boshagh *et al.* (2014) reported that electrophoresis was the key mechanism for microorganism transport in EK-BDS since the microbial population increased in the anode and middle, but decreased in the cathode. This was explained by the direction of electrophoresis flow, which was from the cathode to the anode. Nevertheless, as DBT is a nonionic compound, the removal mechanism by electrokinetics mainly depended on electroosmosis and the highest DBT conversion was observed in the anode; a high residual DBT concentration was observed in the cathode, which indicated DBT was removed mainly by electroosmosis.

10.2.3 Microorganisms with Selective BDS of Benzothiophene and Dibenzothiophene

The condensed thiophenes are the most common form in which sulfur is present in crude oil (Kropp and Gerber, 1998). Dibenzothiophene (DBT), benzothiophene (BT), and their substitutes are the major sulfur-containing aromatic compounds in fuels, accounting for up to 70% of the sulfur content (Kertesz, 2001), because they have higher boiling points (more than 200 °C) and it is difficult to remove them from atmospheric tower outlet streams (e.g. middle distillates) (Shennan 1996; Kawatra and Eisele 2001; El-Gendy and Speight, 2016). Benzothiophene (BT), non-β, single β, and di-β-substituted benzothiophenes (B.P.N. 219 °C) are the typical thiophenic compounds that are found up to 30% in diesel oils (McFarland *et al.* 1998). Thus, it is important to reach for microorganisms capable of selective desulfurization of DBTs and BTs. Polyaromatic sulfur heterocycles are resistant to HDS and form the most abundant organosulfur compounds (OSCs) after HDS (Ma *et al.* 1994; Monticello 1998; El-Gendy and Speight, 2016). Thus, from a practical point of view, a bacterial strain that displays the ability to desulfurize both BT and DBT would be ideal.

The majority of bacteria which are capable of desulfurizing the symmetric heterocyclic S in DBTs such as many *Rhodococcus* spp. (strain IGTS8, Gallagher *et al.*, 1993; strain D-1, Izumi *et al.*, 1994; and strain H-2, Ohshiro *et al.*, 1996), as well as *Agrobacterium* sp. MC501 (Constanti *et al.*, 1996), *Nocardia* strain CYKS2 (Chang *et al.*, 1998), *Mycobacterium* pheli WU-F1 (Furuya *et al.*, 2001), *Bacillus subtilis* WU-S2B (Kirimura *et al.*, 2001), *Stenotrophomonas maltophilia* KHO1 (Ardakani *et al.*, 2010), *Stenotrophomonas* sp. strain SA21 (Mohamed *et al.*, 2015), and *Rhodococcus* sp. strain SA31 (Mohamed *et al.*, 2015) seem to be incapable of desulfurizing the asymmetric heterocyclic sulfur in BT. Similarly, BT desulfurizing bacteria, such as *Gordonia* strain 213 (Gilbert *et al.*, 1998), *Rhodococcus* sp. strain TO9 (Matsui *et al.*, 2000), *Sinorhizobium* sp. strain KT55 (Tanaka *et al.*, 2001), *Rhodococcus* sp. strain WU-K2R (Kirimura *et al.*, 2002), and *Gordonia* terrae strain C-6 (Wang *et al.*, 2013) seem to be lacking the enzymatic capability that expresses the DBT desulfurization phenotype. On the other hand, few bacterial strains, such as *Gordonia* strain CYKS1 (Rhee *et al.*, 1998), *Nocardia* sp. strain CYKS2 (Chang *et al.*, 1998), the thermophilic *Paenibacillus* sp. strainA11-2 (Konishi *et al.*, 2000), mesophilic *Rhodococcus* sp. KT462 (Tanaka *et al.*, 2002), *Mycobacterium phlei* GTIS10 (Kayser *et al.*, 2002), *Gordona rubropertinctus* ICP172 (Acero *et al.*, 2003), the facultative thermophilic *Mycobacterium goodi* X7B (Li *et al.*, 2005b), *Gordonia alkanivorans* strain 1B (Alves *et al.*, 2005), *Candida parapsilosis* NSh45 (El-Gendy *et al.*, 2006), *Sphingomonas subacteria* T7b (Gunam *et al.*, 2006), *Rhodococcus* sp. strain SA11 (Mohamed *et al.*, 2015), and *Brevibacillus invocatus* C19 (Nassar *et al.*, 2013), which exhibit broad desulfurization capability against DBTs and BTs have been reported. In almost all cases, the desulfurization of DBTs and BTs in this group is mediated by different enzymes.

Bacillus sp. KS1, isolated from soil contaminated with crude oil collected at the site of Chennai Petroleum Corporation Limited's refinery in Manali, Tamilnadu, India, has the ability to desulfurize DBT via the 4S-pathway, producing 2-HBP as a dead end product. It also has broad versatility on utilization of different S-coumpounds, including thiophene, benzothiophene, 5,5'-dithio-*bis*-(2-nitrobenzoic acid), and dimethylsulfoxide, but it could not desulfurize thiophene-2-carboxyllic acid (Rath *et al.*, 2012).

Khedkar and Shanker (2015) proved that the actinomycetes present in soil can remove sulfur from different PASHs and have potential as BDS-catalysts. For example, *Oreskovia* sp. R3 has been isolated for its ability to desulfurize DBT via the 4S-pathway. Not only this, but it also has the ability to desulfurize BT and thianthrene (Khedkar and Shanker, 2015).

10.2.4 Microorganisms with Methoxylation Pathway

The conversion of HBP to MBP has been reported in *Mycobacterium* sp. G3 (Okada *et al.* 2002a) and *Microbacterium* sp ZD-M2 (Li *et al.* 2005b). However, it has been indicated that the expression of the methoxylation and desulfurization activities bear no correlation (Okada *et al.* 2002a). The production of 2-methoxybiphenyl and 2,2′-dimethoxy-1,1′-biphenyl as final-end products of biodesulfurization of dibenzothiophene, through the 4S-pathway, using *Mycobacterium* sp. and *Rhodococcus erythropolis* HN2, respectively, has been reported (Chen *et al.*, 2009; El-Gendy *et al.*, 2014). The methoxylation pathway from 2-hydroxybiphenyl to 2-methoxybiphenyl has been examined (Ohshiro *et al.*, 1996a; Monticello, 2000; Kim *et al.*, 2004) to investigate the potential to overcome the toxicity of 2-hydroxybiphenyl since the production of 2-methoxybiphenyl would create a less inhibitory effect on the microbe due to the loss of the hydroxy structure of phenolic compounds; this expresses a lower inhibitory effect on growth and biodesulfurization efficiencies of microorganisms.

Bordoloi *et al.* (2014) reported the isolation of a strain of *Achromobacter* sp., isolated from a contaminated petroleum–oil soil sample of Assam, North-East India, for its ability to desulfurize DBT to 2-HBP, followed by methylation at the hydroxyl group, producing 2-methoxybiphenyl 2-MBP. El-Gendy *et al.* (2014) reported the isolation of Gram +ve *R. erythropolis* HN2 (accession no. KF018282) from an Egyptian coke sample for its ability to desulfurize DBT via the 4S-pathway to 2-HBP, which is further detoxified to 2-methoxybiphenyl (2-MBP), which is further oxidized to form 2-methoxy [1,1′-biphenyl]-2-ol, then, subsequently, methoxylated to yield 2,2′-dimethoxybiphenyl (2,2′-DMBP). Methoxylation of 2-HBP by NH2 was a way for microorganisms to overcome the toxicity of 2-HBP.

Al-Jailawi *et al.* (2015) reported the isolation of three *Pseudomonas aeruginosa* strains M9, M19, and S25 from oil contaminated soil samples that were collected from different sites in Iraq. The three isolates showed good capabilities to desulfurize DBT via the 4S-pathway at 35°C. Nevertheless, S25 showed the best capability to overcome the toxicity of 2-HBP via the methylation of the-OH group producing 2-methoxybiphenyl (2-MBP).

10.2.5 Microorganisms with High Tolerance for Oil/Water Phase Ratio

The *R. globerulus* DAQ3, which was isolated from oil-contaminated soil, demonstrated high specific DBT desulfurization activity and stability in oil/water (O/W) systems. A model diesel with 1452 μM DBT/hexadecane was contacted with the same volume of 10 g DCW/L of aqueous phase. It was reported that resting cells of DAQ3 could transform 92% and 100% of DBT in 9 h and 24 h, respectively. Desulfurization with IGTS8 at identical conditions resulted in 29% and 32% desulfurization. The DBT desulfurization by DAQ3 proved to be significantly higher than the one reported by IGTS8 (Yang and Marison, 2005).

Luo *et al.* (2003) reported that desulfurization of DBT by *Pseudomonas delafieldii* R-8 in dodecane with an optimum O/W ratio of 25% (v/v) phase ratio. Rashtchi *et al.* (2006) reported that biodesulfurization of DBT by strain RIPI-22 was well accomplished in low phase ratio and the highest effect was for the hydrocarbon fraction of 30%.

El-Gendy (2004) reported that yeast, *Candida parapsilosis* NSh45, showed in batch flasks of a 1/3 O/W phase ratio complete desulfurization of 1000 ppm DBT dissolved in n-hexadecane.

Nassar (2015) enhanced the BDS of diesel oil (8,600 ppm initial sulfur content) by using Fe_3O_4 MNPs coated cells, magnetically immobilized cells (alginate + Fe_3O_4 MNPs), and agar immobilized cells of *Brevibacillus invocatus* C19 and *Rhodococcus erythropolis* IGST8. The Fe_3O_4 MNPs coated cells of *R. erythropolis* IGTS8 showed better BDS efficiency (%BDS 78.26 %) in case of a 10% (v/v) O/W phase ratio than that which occurred by free cells (%BDS 74.42%), magnetic immobilized cells (%BDS 57.26%), and agar immobilized cells (%BDS 58.98%). The agar immobilized cells and coated cells of *Brevibacillus invocatus* C19 showed the highest BDS efficiency (%BDS 98.97% and 98.69%, respectively) at 25% (v/v) O/W phase ratio compared to free cells (%BDS 91.31%) and magnetic immobilized cells (%BDS 89.99%). Thus, a 10% and 25% (v/v) phase ratio were considered as the optimum O/W phase ratio of *R. erythropolis* IGTS8 and *Brevibacillus invocatus* C19, respectively.

10.2.6 Thermotolerant Microorganisms with Selective BDS Capability

Mesophilic and thermophilic microorganisms have also been studied (Kargi and Robinson, 1984; Konishi *et al.*, 1997; Kirimura *et al.*, 2001; Li *et al.*, 2003; Furuya *et al.*, 2003). However, the use of mesophilics in fossil fuel desulfurization as an alternative or complement to HDS requires an additional cooling process of the fuel to ambient temperature following HDS. This additional cooling process causes an economical load when used in large scale fossil fuel desulfurization. Thus, moderately thermophilic microorganisms and hyper-thermophilic microbial desulfurization are desirable and make the crude oil BDS process more feasible due to the lower viscosity of the crude at high temperature, as decrease in viscosity, consequently, increases the biodesulfurization rate due to higher mass transfer (Le Borgne and Quintero, 2003; Torkamani *et al.*, 2008; Bhatia and Sharma, 2012).

There are various reported DBT-desulfurizing microorganisms that work via the 4S-pathway that can be categorized according to the optimal working temperature, including mesophilics, such as *Rhodococcus erythropolis* IGTS8 (Patel *et al.*, 1997), *R. erythropolis* H-2 (Ohshiro *et al.*, 1996a), *Corynebacterium* sp. ZD-1 (Wang *et al.*, 2004), *Bacillus subtilis* WUS2B

(Kirimura *et al.*, 2001), and *Rhodococcus erythropolis* NH2 (El-Gendy *et al.*, 2014) that optimally work around 25-30°C, moderately thermophilics, which exhibit a high DBT-desulfurization ability at around 40–50 °C, such as *Mycobacterium pheli* WU-F1 (Furuya *et al.*, 2003), and hyper-thromophilics, which work at high temperatures above 50°C, such as *Sulfolobus acidocaldarius* (Kargi and Robinson, 1984) and *Sulfolobus solfataricus* P2 (Gün *et al.*, 2015).

Highlights should be made on the isolation and characterization of thermophile strains such as *Paenibacillus* sp., capable of desulfurizing benzo- and dibenzothiophenes present in diesel oil (Konishi, 2000; Yoshitaka *et al.*, 2000), as well as the development of several evolution techniques aimed at modifying and obtaining enzymes with new and/or improved characteristics (Arnold *et al.*, 2001).

Kargi and Robinson (1984) reported DBT-BDS to sulfate by the hyperthermophilic, *Sulfolobus acidocaldarius,* at 70 °C, but unfortunately, that study did not include DBT degradation at high temperatures in the absence of the microorganism. Thus, the obtained rate of desulfurization does not represent the real BDS rate. In another study by Constanti *et al.* (1992) by the acidophilic and thermophilic *Sulfolobus solfataricus* DSM 1616 at 68 °C, DBT showed a self-degradation in the absence of microorganisms at high temperatures; no substantial DBT utilization could be observed. This study clearly showed the difficulty of using a DBT model compound at high temperatures in BDS by *S. solfataricus*. Nevertheless, the same study showed the oxidation of thiophene-2-carboxylate by *S. solfataricus*. Therefore, the organic sulfur desulfurization molecular mechanism was shown to be present in that hyper-thermophile. Hyperthermophiles are usually isolated mainly from water-containing volcanic areas, such as solfataric fields and hot springs, where they are unable to grow below 60 °C. The acidophilic *Sulfolobus solfataricus* P2, which belongs to the archaebacteria, grows optimally at temperatures between 75 and 85 °C and at low pHs between 2 and 4, has been reported for DBT-BDS.

Resting cells of *Mycobacterium* sp. G3 expressed broad versatility over alkylated DBTs in a biphasic batch BDS process of model oil (100 mM different alkylated DBTs compounds in n-tetradecane) at 1:1 (O/W), 37 °C, pH7, and 10 g/L glycerol within 2 h. The desulfurization activities against alkyl DBTs decreased with increasing molecular weight of DBT derivatives, where complete removal of DBT was achieved and 85% of 4,6-dimethyl-DBT and 36% of diethylDBT were also removed, but only 4% of 4,6-dipropylDBT was removed (Okada *et al.*, 2002b). It was concluded that G3 was superior to the reported desulfurizing microorganisms in terms of both the permeability of substrates and the substrate specificity of the desulfurizing enzymes. As with a longer incubation period (15 h), it desulfurized 4,6-dipropylDBT three fold as much as within a short incubation period (2 h). Not only this, but in a one phase aqueous system, it desulfurized 4,6-dibuylDBT and 4,6-dipentylDBT. The phenomenon of the difference in reactivity between the oil/water biphasic reaction system and the one aqueous phase reaction system was assumed to be attributed to the efficiency of contact between cells and substrates. In the case of the oil/water two-phase reaction system, the high hydrophobic alkyl-DBTs substrate was dissolved in the entire oil phase at the same concentration. However, in the water phase reaction system, the hydrophobic substrate is dispersed in the water phase and might have the tendency to gather on the cell surface, which is assumed to be a more hydrophobic field in the reaction mixture, and, thus, concentrate around the cells to become more directly bioavailable.

An organism identified as *Mycobacterium pheli* GTIS10 was isolated based on its ability to use DBT as a sole source of sulfur for growth at 30–52 °C to 2-HBP as detected by HPLC. It can also desulfurize BT and thiophene. The temperature at which resting cells of *Rhodococcus erythropolis* IGTS8 reach the highest rate of DBT metabolism is near 30 °C, whereas the temperature that shows the highest activity in resting cell cultures of *M. Pheli* GTIS10 is near 50 °C and activity is detectable at temperatures as high as 57 °C (Kayser *et al.*, 2002).

Furuya *et al.* (2001) reported the thermophilic microorganism *Mycobacterium phlei* strain WU-F1 which is able to desulfurize DBT and its derivatives via the 4S-pathway over a wide temperature range of 20–50 °C with the highest level at 45–50 °C. The maximum BDS rate of DBT to 2-HBP was 0.54 mM/h at 50 °C, while the alkylated-DBTs, 2,8-dimethylDBT and 4,6-dimethylDBT, were converted with a rate of 0.1 mM/h at the same temperature (Furuya *et al.*, 2001). This strain was also used in a study for the conversion of naphthothiophenes (NTH), which can be formed in diesel fuel during severe physicochemical hydrodesulfurization (Furuya *et al.*, 2002). Resting cells of strain WU-F1 converted 67% and 83% of 0.81 mM NTH and 2-ethylNTH, respectively, within 8 h.

Li *et al.* (2005b) reported thermophilic and hydrocarbon tolerant *Mycobacterium goodii* X7B which had been primarily isolated as a bacterial strain capable of desulfurizing dibenzothiophene to produce 2-hydroxybiphenyl via the 4S pathway and was also found to desulfurize benzothiophene to o-hydroxystyrene at 40 °C. Another *Bacillus* sp. was also isolated for the selective desulfurization of DBT to 2-HBP at 45 °C (Abolfaz *et al.*, 2006).

Gordonia sp. strain F.5.25.8, which is able to grow at temperatures up to 42 °C, is reported to be able to use DBT and CAR as sole S and N sources, respectively. F.5.25.8 also has good ability to desulfurize DBT via the 4S-pathway, producing 2-HBP as a dead end product at 30 and 37 °C. Moreover, this strain has broad versatility to grow using various organic sulfur or nitrogen compounds as the sole sulfur or nitrogen sources, including DBT, DBT-sulfone, BT, 2-phenylbenzothiazole, thiazole, 2-methyl-BT, 2-methyl-thiophene, carbazole, quinoline, imidazole, butylpyrrolidine, indole, methylpropylpyrazine, methylnaphtothiazole, and mercaptobenzothiazole. However, when F.5.25.8 was tested for the batch BDS and BDN of both DBT and CAR (3 mM carbazole and 0.5 mM DBT), about 40 and 90% removal occurred within 10 d at 30 °C, respectively (Santos *et al.*, 2006). Abolfazl *et al.* (2006) isolated a new thermophilic strain belonging to the *Bacillus* genus for selective BDS of DBT to 2-HBP at 45 °C, which recorded 87.5% BDS of 0.15 mM DBT within 6 d in the presence of 6 g/L glucose and 4 g/L ammonium chloride at a pH7. Applying the Michaelis-Menten equation, the maximum specific BDS rate (V_{max}) and the Michaelis constant, $K_{m,}$ for desulfurization of DBT were fitted as 0.548 mM/h and 0.458 mM, respectively.

A thermophilic bacterium, *Klebsiella* sp. 13T, was isolated from contaminated soils collected from a petroleum refinery, for its ability to desulfurize DBT via the 4S-pathway to 2-hydroxybiphenyl (2-HBP) at 45 °C (Bhatia and Sharma, 2012).

Lysinibacillus sphaericus DMT-7, isolated form diesel oil contaminated soil and collected from a few oil filling stations situated in Rishikesh, Uttarakhand, India, for its ability to selectively desulfurize DBT to 2-HBP at 37 °C, also possess the ability to utilize a broad range of substrates as a sole source of sulfur, such as benzothiophene, 3,4-benzo-DBT, 4,6-dimethyl-DBT, and 4,6-dibutyl DBT (Bahuguna *et al.*, 2011).

Arabian *et al.* (2014) reported the isolation of Gram +ve *Bacillus cereus* HN from west Paydar oil wells in the south of Iran for its ability to desulfurize DBT via the 4S-pathway over a wide range of temperatures from 30–40 °C. Shahaby and Essam El-din (2017) reported the isolation of 12 biodesulfurizing bacteria isolated belonging to five genera, including *Bacillus*, *Pseudomonas*, *Rodococcus*, *Mycobacterium*, and *Klebsiella* that showed optimum growth at 35 °C.

Wang *et al.* (2015) reported the isolation of thermophilic mixed culture, *Paenibacillus* spp. "32O-W" and "32O-Y", from a soil sample obtained near the Field Museum in Chicago, IL, by repeated passage of a soil sample at up to 55° C in a medium containing dibenzothiophene (DBT) as a sulfur source. They have the ability to desulfurize DBT via the 4S-pathway, producing 2-HBP as a dead end product. This expressed a maximum BDS-activity at 45–50 °C. They can grow on sulfate at temperatures up to 63 °C.

However, they tolerate up to 58 °C in the presence of DBT as sole S-source. Moreover, the amino acid sequences of the DszABC/TdsABC homologs in 32O-Y and 32O-W have 26–48% identity with the corresponding Tds and Dsz proteins from *Paenibacillus* sp. A11-2 and *R. erythropolis* IGTS8, respectively. 320-W enhanced the activity of 320-Y. The optimal temperature for 32O-Y is 40 °C, where activity is 0.13 units and the activity of 32O-Y is 0.1 units at 45 °C (1 unit is equivalent to 1 nmol of 2-HBP produced by 1 mg DCW in 1 min). The activity of a mixed culture selected at 37°C was higher than that of the pure 32O-Y culture by 22, 36, and 74 % at 40, 45, and 50 °C, respectively. Furthermore, the selection of a mixed culture of 32O-Y and 32O-W, at 54 °C, increased DBT metabolism 42 and 36% at 40 and 45 °C, respectively, but was nearly identical at 50 °C compared to that of 32O-Y and 32O-W selected at 37 °C.

Gunam *et al.* (2012) reported the isolation of *Pseudomonas* sp. strain KWN5 from soil samples derived from petroleum-contaminated soil near oil fields in Kawengan, Bojonegoro, East Java, Indonesia, for its ability to desulfurize DBT via the 4S-pathway over a wide range of temperatures between 27–47 °C and is most efficient in the range of 32–37 °C. The S-content of model oil (DBT in n-tetradecane) in a biphasic batch process (1:5 O/W) decreased from 200 ppm to 55.74 ppm DBT within 4 d of incubation at 37 °C, but the BDS activity suddenly decreased at 42 °C. Kirimura *et al.* (2001) also reported that *B. subtilis* WU-S2B exhibited DBT-desulfurizing ability over a wide temperature range from 30 to 55 °C and the activity suddenly decreased at 52 °C. Gunam *et al.* (2016) reported that *Agrobacterium tumefaciens* LSU20, isolated from oil-contaminated soil by enrichment culture for its ability to desulfurize DBT via the 4S-pathway, can tolerate up to 49 °C with a desulfurization capacity of 30.5% in a biphasic batch BDS (1:5 O/W) of model oil at 200 ppm DBT in n-tetradecane using resting cells of LSU20 aged 4 d at 150 rpm, but it showed its highest BDS rate at 37 °C and pH7 in the presence of glucose as C-sources, recording a DBT-removal percentage of 76.9% within 96 h.

10.2.7 BDS Using Yeast and Fungi

Relatively little work has been carried out among eukaryotic organisms; the fungus *Cunninghamella elegans* (Crawford and Gupta, 1990) and the white rot fungus *Pleurotus Ostreatus* (Bezalel *et al.*, 1996) have been found to grow on DBT by forming DBTO and $DBTO_2$, but not biphenyl. The white-rot basidiomycete, *C. versicolor*, and brown-rot basidiomycete, *Tyromyces palustris*, have been reported as capable of cleaving the sulfide bond in thiodiglycol, benzyl sulfide, and bis(2-bromoethyl)sulfide (Itoh *et al.* 1997), where this ability was applied to degrade Yperite, the chemical warfare reagent. *Trametes versicolor* IFO30340 can produce DBTO and $DBTO_2$ from DBT by a cytochrome P-450 mechanism (Ichinose *et al.*, 2002a).

Itoh *et al.* (1997) reported that *T. Versicolor* IFO30340 and *Tyromyces Palustris* IFO30339 metabolize dibenzylsulfide (DBS) to benzylalcohol and benzylmercaptan. *Trametes trogii* UAMH 8156, *Trametes hirsute* UAMH8165, *Phanerochaete chrysosporium* ATCC 24725, *Trametes versicolor* IFO30340 (formerly *Coriolus sp.),* and *Tyromyces Palustris* IFO 303339 all oxidized DBS to dibenzylsulfoxide prior to oxidation to dibenylsulfone (Van Hamme *et al.*, 2003). The fungus *Paecylomyces* sp. carries out sulfur-specific oxidation of DBT by producing 2,2'-dihydoxybiphenyl (Faison *et al.*, 1991). *C. versicolor* has been reported to convert 4-methyl dibenzothiophene (4MDBT) to water-soluble products via S-oxidation to S-oxide and dioxide, hydroxylation of the methyl substituent to a hydroxymethyl group, and xylosylation (Ichinose *et al.* 1999).

The lignin-degrading white basidiomycete *Coriolus versicolor* has a broad versatility over different S-compounds. It is reported to degrade 2-hydroxymethylthiophene (2HMT), 2-formylthiophene (2FOT), 2-carboxythiophene (2CAT), 2(5H)-thiophenone (TPO), and γ-thiobutyrolactone (TBL) under high carbon and nitrogen conditions. However, it only oxidizes DBT, 4-MDBT, 2-MBT, and 7-MBT to their corresponding sulfoxides and sulfones only under the ligninolytic conditions, i.e. high carbon and low nitrogen (HCLN), strongly suggesting that the S-oxidation reactions were catalyzed by ligninolytic enzyme(s). However, no reaction was observed with 4,6-DMDBT in either high nitrogen (HN) or low nitrogen (LN) culture mediums, but upon the exogenous addition of 4,6DMDBT-5-oxide it was effectively oxidized to 4,6-DMDBT-5-dioxide by *C. versicolor* (Ichinose *et al.* 2002b).

Few numbers of yeast strains have been reported to grow desulfurizing DBT and/or related thiophenic compounds that extensively occur in industrial fuel oils.

Rhodosporidium toruloides strain DBVPG6662 was isolated from industrial waste several years ago and grows on thiosulfate as a sulfur source in the presence of glucose as a carbon source (Baldi *et al.*, 1990; Pepi and Baldi, 1992, 1995). *Rhodosporidium toruloides* strain DBVPG6662 was also found to be able to utilize DBT as a sulfur source, producing 2,2'-dihydroxybiphenyl. Moreover, it is able to utilize benzothiophenes and dibenzothiophenes and other organic sulfurs in a large variety of thiophenic compounds that occur extensively in commercial fuel oils by physically adhering to the organic sulfur source (Baldi *et al.*, 2003).

El-Gendy (2004) reported that yeast, *Candida parapsilosis* NSh45, has broad versatility to desulfurize organosulfur compounds, including dimethy sulfide (DMS), dimethylsulfoxide (DMSO), diethylsulfate (DES), dibenzylsulfide (DBS), diphenyldisulfuide (DPDS), thiophene (Th), benzothiophene (BT), DBTO, $DBTO_2$, and 4-methyldibenzothiophene (4-MDBT).

10.3 Genetics and its Role in Improvement of BDS Process

Microorganisms can remove sulfur pollutants from petroleum and will reduce the amount of released sulfur oxides. However, genetic manipulations to improve sulfur-removal efficiencies are necessary before an upgrading process for the removal of harmful sulfur compounds from fossil fuels can be developed. In most times, engineered bacteria are required to remove more sulfur compounds with higher activities. Cultures with improved substrate ranges are also needed to better address the complicated mixture of chemicals present in petroleum.

The commercialization of the biocatalytic desulfurization (BDS) process does not seem to be realistic in the near future because of the low desulfurization rate of the known microorganisms. Hence, future development will depend on either genetically modifying the currently available bacteria or identifying novel biodesulfurizers. Moreover, the known biodesulfurizers have an approximately 500 fold lower desulfurization rate than what is required in industrial processes (Bhatia and Sharma, 2010b).

With the advancement in sequencing technology, the past few years have seen mushrooming of the genomes of various microorganisms. Exploration of this genomic information has accelerated progress in a few fields, for example the areas of natural products and biosynthetic pathways. However, their use has never been exploited in BDS research. The key to increasing the bacterial desulfurization rate is to identify the genes responsible for C–S bond cleavage in organic sulfur compounds and manipulate the system through genetic engineering techniques.

Amino acid sequences of the Dsz proteins of *R. erythropolis* IGTS8 (Gallagher *et al.*, 1993) were used as a query to perform protein-protein BLAST (BLASTp, http://www.ncbi.nlm.nih.gov/blast/index.shtml) in order to search for the homolog protein(s) in the entire finished, non-redundant, and unfinished microbial genome database. Further, microbes reported in the Kyoto Encyclopedia of Genes and Genomes (KEGG, http://

www.genome.jp/kegg) database were also screened for the identification of putative homologs. This approach leads to the identification of protein(s) in microorganisms which have already been shown experimentally to possess dsz genes and their respective proteins.

Bhatia and Sharma (2010b) reported 13 novel putative DBT degraders belonging to 12 genera. The list of these 13 microbes, along with their unique properties and the presence/absence of Dsz proteins are listed in Table 10.1. All the identified microorganisms have DszA, DszC, and DszD proteins except *Methylobacillus flagellatus*, which does not and seems to have DszC and DszD proteins. The rate-limiting step in the "4S pathway" is the final step that leads to the accumulation of 2-hydroxybiphenyl (2-HBP) and sulfate and requires DszB enzyme. Hence, one of the ways to improve the rate of sulfur removal could be to develop/identify the DszB biocatalyst. Further absence of dszB gene would allow the accumulation of hydroxybiphenyl benzene sulfinate (HPBSi), a more valuable product than the sulfate. HPBSi can be recovered from the aqueous phase and used as surfactant. Seven of the identified microbes in that study do not have DszB protein. *T. fusca* and *R. xylanophilus* are thermophilic in nature. These microbes could be advantageous if a BDS reaction could be carried out at a higher temperature and then there would be no need to cool the HDS-treated oil to ambient temperature, which is required in case of mesophiles. Moreover, at higher temperatures, BDS would afford a more practical approach to a large scale industrial process and could result in higher rates and low processing costs. In addition, higher temperature decreases oil viscosity and contamination by undesirable bacteria which affects the BDS process.

FMN reductase is among the four key enzymes of the 4S pathway. When flavin reductase, FMN reductase, or various oxidoreductases were added to the reaction mixture or over-expressed in recombinant constructs, the desulfurization rate increased (Ohshiro *et al.*, 1995; Gray *et al.*, 1996; Squires *et al.*, 1998, 1999; McFarland, 1999; Rambosek *et al.*, 1999). The use of the flavoprotein could result in an approximately 100 fold improvement in the rate of reaction compared with a system where no flavoprotein was added (Squires *et al.*, 1998). Several research groups have focused on over-expression of this enzyme. FMN reductase can be over expressed in desulfurizing microorganisms via mutagenesis. The DNA encoding flavin reductase can be transferred into desulfurizing microorganisms and can be simultaneously or independently transferred into a desired host cell with the DNA, encoding the Dsz enzymes, and can be under the control of the same or a different promoter as the DNA encoding the Dsz enzymes (Squires *et al.*, 1999).

Table 10.1 Potential Dibenzothiophene (DBT) Degrading Microorganisms Discovered on the Basis of dsz Genes and Their Unique Properties.

Organisms	DszA	DszB	DszC	DszD	Unique properties
Burkholderia fungorum[a]	+	+	+	+	Good bio-degraders of polychlorinated biphenyls (PCBs), commercially important for bioremediation, pollutant degrading ability, role in global C-cycle
Burkholderia cepacia	+	+	+	+	Capacity to fix nitrogen, biodegradation of pollutants, biocontrol of root diseases, model for comparative genomic studies
Bradyrhizobium japonicum	+	+	+	+	Nitrogen-fixing ability, excellent model organism for studying respiratory enzymes
Methylobacillus flagellatus	+	+	-	-	Obligate methylotrophic bacterium, ideal producer for biotechnology industry, used for the overproduction of amino acids and vitamins
Magnetospirillum magnetotactium	+	+	+	+	Exhibit magnetotaxis, model for biomineralization, potential geobiological tracer, commercially can be used for magnetic targeting of pharmaceuticals, cell separation, and applications in magnetic resonance imaging
Thermobifida fusca[b]	+	-	+	+	Degrade all major plant cell wall polymers except lignin and pectin, its spores cause the condition farmer's lung disease.

Azotobacter vinelandii	+	+	+	+	Nitrogen-fixing ability, highly amenable to genetic manipulation, produces 5-alkylresorcinols, alginates and poly-β-hydroxybutyric acid, which are of importance in medicine, food and biodegradable plastics industries, respectively
Mesorhizobium loti	+	-	+	+	Nitrogen-fixing ability
Oceanobacillus iheyensis	+	-	+	+	Extremely halotolerant and facultative alkaliphilic capability
Novoshingobium aromaticivorans[a]	+	-	+	+	Glycosphingolipid as a cell wall component, degrades a wide variety of aromatic hydrocarbons including tolune, xylene, naphthalene, fluorine, etc.
Brevibacterium linens	+	-	+	+	Produce volatile sulfur compounds, bacteriocin, self-processing extracellular proteases. Can metabolize heterocyclic and polycyclic ring structures
Rubrobacter xylanophilus[b]	+	-	+	+	Gamma radiation resistant, degrades hemicellulose and xylan
Ralstonia metallidurans	+	-	+	+	Resistant towards various heavy metals like Zn, Cd, Co, Pb, Cu, Hg, Ni and Cr

[a]Unfinished genomes, [b]Thermophile. The amino acid sequences of the enzymes viz. DszA (accession no. AAA99482), DszB (accession no. AAA99483), DszC (accession no. AAA99484), and DszD (accession no. AAC33226) involved in the 4S pathway of DBT desulfurization found in *Rhodococcus erythropolis* IGTS8 was used to BLAST the available genomic databases.

To increase the biodesulfurization activity, Reichmuth *et al.* (1999) studied the desulfurization ability of a cloned *E. coli* DH10B strain that contained the plasmids; pDSR2 contained a Vibrio Harvey NADH:FMN oxidoreductase gene and pDSR3 encoded enzymes, which converted DBT to HBP. In plasmid pDSR3, the native desulfurization control element had been removed. Therefore, *E. coli* DH10B/pDSR3 could show its desulfurization trait even in the presence of a sulfate ion or undefined rich media such as LB. When *E. coli* DH10B/pDSR3 was placed in a medium with DBT as the sole sulfur source, desulfurization activity stopped at 0.2 mM HBP. The repression was attributed to insufficient oxidoreductase supplied by *E. coli* DH10B/pDSR3. As in IGTS8, *E. coli* has an intrinsic NADH:FMN oxidoreductase. However, the oxidoreductase level proved to be insufficient for over-expressed dszABC. Desulfurization activity was tested with a cloned *E. coli* DH10B/pDSR2/pDSR3, however no significant desulfurization activity change was observed under different amounts of inducers.

Genetic engineering has been applied to increase the BDS rate. It has been reported that the desulfurization enzymes are repressed in the presence of inorganic sulfate when the native promoter of the D_{SZ}ABC operon is replaced with a non-repressible promoter for enhancement of biodesulfurization activity (Noda *et al.*, 2002). There is also a high biodesulfurization of dibenzothiophene (50–250 μmol/g DCW/h) using genetically modified *R. erythropolis* KA2-5-1 (Konishi *et al.*, 2005) and there is a 200-fold increase in the biodesulfurization activity of *R. erythropolis* IGTS8 by genetic engineering and the dibenzothiophene desulfurization activity reached 20 μmol/g DCW/min (Le Borgne and Quintero, 2003). The desulfurization rate of a *Mycobacterium* sp. G3 culture was observed to increase from 178 to 211 mmol/g DCW/h as a consequence of genetic manipulation (Takada *et al.*, 2005).

Alves *et al.* (2007) studied the *Gordonia alkanivorans* Strain 1B desulfurization operon (*dsz*), identified its genes, and compared them with previously described bacterial genes from

Rhodococcus erythropolis IGTS8. Three open reading frames were identified: *dsz*A, *dsz*B, and *dsz*C, which showed high similarity to those of IGTS8 (88 % for *dsz*A, 88 % for *dsz*B, and 90 % for *dsz*C). In the same work, Alves *et al.* (2007) proceeded to clone and express *G. alkanivorans dsz*AB genes in *Escherichia coli*. The resulting recombinant strain was able to grow in a culture medium containing dibenzothiophene sulfone ($DBTO_2$) as the only sulfur source desulfurizing the same amount of $DBTO_2$ (0.2 mM) to 2-HBP, but 4.5 fold faster than the original strain 1B (Alves *et al.*, 2007).

Gordonia sp. IITR100 is reported to metabolize benzonapthothiophene (BNT) via the 4S-pathway and utilizes the released sulfur for its growth throughout the formation of BNT-sulfone and BNT-sulfinate, in addition to BNT-hydroxide, as a dead end product. Recombinant *E. coli* cells harboring DszC or DszA are also able to perform the metabolism of BNT to BNT-sulfone or of BNT-sulfone to BNT-sulfinate, respectively. In a batch BDS of 0.3 mM BNT, IITR100 reported a maximum specific growth rate of 0.021 h^{-1} during a specific desulfurization rate of 1.2 mmol BNT/g DCW/h and > 99% BNT-removal within 6 d of incubation period. This increased to 2.6 mmol BNT/g DCW/h with > 99% decrease in the concentration of BNT after eight hours of incubation using the recombinant *E. coli*-DszC. This recorded faster rate, compared to IITR 100, was attributed to the presence of a higher amount of enzyme in the recombinant cells (Chauhan *et al.*, 2015).

In a search for the development of a method to provide the required supply of reduced flavin to a DBT oxygenation system, Galán *et al.* (2000) used hpaC flavin reductase originated from *E. coli* and contacted it, in vitro, with a system of dszABC purified enzymes and an NADH source. They also used catalase in the desulfurization medium to minimize the probability of H_2O_2 formation, which might be produced by the non-enzymatic reoxidation of $FMNH_2$ under high oxygen concentration. Addition of hpaC flavin reductase increased DBT desulfurization 7–10 times within 30 min. The enzyme hpaC flavin reductase and the oxidoreductase originating from IGTS8 were from the same subfamily of flavin: NAD(P)H reductase. To verify if hpaC flavin reductase could increase desulfurization in resting cells, a recombinant Gram-negative strain, *P. putida* KTH2/pESOX3, was designed to contain both dszABC and hpaC genes. To eliminate the effects of sulfate inhibition resulting from DBT desulfurization, the dszABC gene cluster in pESOX3 was designed to express under the control of the Ptac promoter. Results obtained from desulfurization by KTH2/pESOX3 strain showed that the bacteria could keep its desulfurization phenotype even in sulfate containing media. Designing a plasmid that encodes the right amount of FMN:NADH reductase to dszABC enzymes is crucial to reach an optimum desulfurization potential in this bacterium. Insufficient FMN:NADH reductase would make NADH supply the limiting step in DBT oxidation. On the other hand, high concentration of $FMNH_2$ will give rise to H_2O_2 formation which would be lethal to cells (Gaudu *et al.*, 1994; Galán *et al.*, 2000). Matsubara *et al.* (2001) amplified the flavin reductase gene with primers designed by using dszD of *R. erythropolis* IGTS8; the enzyme was over-expressed in *Escherichia coli*. The specific activity in crude extracts of the over-expressing strain was about 275 fold that of wild-type strain, *R. erythropolis* strain D-1.

Rhodococcus erythropolis IGTS8 is the natural bacteria mostly used in DBT desulfurization by the 4S pathway, but the reaction is energetically expensive due to the high reducing equivalents consumed, which supports the three first oxygenation steps. In this sense, the use of a genetically modified organism (GMO), ultimately, has attracted special attention. Accordingly, Gallardo *et al.* (1997) carried out BDS by a GMO, *Pseudomonas putida* CECT 5279, which has the genes dszA, dszB, and dszC cloned from *R. erythropolis* IGTS8 and the flavinoxido reductase, hpaC (to catalyzed $FMNH_2$ production), from *Escherichia* coli (Alcon *et al.*, 2005) because it is effectively capable to develop a similar *Rhodococcus's* pathway. With the cloning and expression of the FMN:NADH hpaC gene, a significant enhancement of the DBT desulfurization efficiency has been observed (Gallardo *et al.*, 1997). The Dsz enzymes are reported to be soluble and presumably found in the cytoplasm of *R. erythropolis* IGTS8 (Gray *et al.*, 1996). Although the intracellular metabolism of DBT by strain IGTS8 has been reported, there is no evidence for DBT being actively transported into the cell (Monticello, 2000) or for the mass transfer limitations in DBT metabolism (Kilbane and Le Borgne, 2004). Nevertheless, it has been reported that in strain IGTS8, desulfurization activity was associated with the external surface of the cells (Kayser *et al.*, 1993; Kilbane, 1991; Patel *et al.*, 1997). A substantial proportion (70%) of the total desulfurization activity expressed by IGTS8 was found in the cell debris fraction, which contains external cell membrane and cell-wall fragments, and the enzyme biocatalyst responsible for desulfurization was reported as being a component of the cell envelope (Monticello, 1996). Moreover, there is no evidence that Dsz enzymes are excreted from IGTS8 cells, but the size of the substrates metabolized and the ability of other bacterial species to successfully compete for sulfur liberated from organosulfur substrates by strain IGTS8 make it likely that desulfurization does not occur intracellularly, but in association with the external surface of cells (Kilbane and Le Borgne, 2004). Therefore, it is believed that the desulfurization pathway may function in association with the cell membrane such that extracellular substrates and intracellular cofactors can both be accessed.

To improve uptake of alkylated sulfur compounds, Matsui *et al.* (2001) introduced dsz genes into *Rhodococcus* sp. T09, a strain capable of desulfurizing benzothiophene (BT). The resulting recombinant strain grew with both DBT and BT as the sole sulfur sources. The recombinant cells desulfurized not only alkylated BTs, but also various Cx -DBTs, producing alkylated hydroxyphenyls as the end products.

An improvement in the uptake of sulfur compounds in oil fractions should be effective in enhancing the biodesulfurization activity. Watanabe *et al.* (2003) transferred the dsz gene cluster from *R. erythropolis* KA2-5-1 into *R. erythropolis* MC1109, which was unable to desulfurize light gas oil (LGO). Resting cells of the resultant recombinant strain, named MC0203, decreased the sulfur concentration of LGO from 120 ppm to 70 ppm in 2 h and the LGO-desulfurization activity of this strain was about twice that of strain KA2-5-1.

To overcome the mass transfer limitation, the microorganism should be able to tolerate a high concentration of hydrocarbons (i.e. oil phase or solvents) so they would function optimally. It is well known that the solvent tolerance of *Rhodococcus* is lower than *Pseudomonas*, although the hydrophobicity of its membrane makes it easily uptake very hydrophobic C_X-dibenzothiophene from the oil phase (McFarland, 1999). Gene cloning of dszABC genes from *R. erythropolis* XP into *P. putida*, constructing a solvent-tolerant, resulted in desulfurization by *P. putida* A4 (Tao *et al.*, 2006). This strain, when contacted with sulfur refractory compounds and dissolved in hydrocarbon solvent, maintained the same substrate desulfurization traits as observed in *R. erythropolis* XP. Resting cells of *P. putida* A4 could desulfurize 86% of dibenzothiophene in 10% (v/v) p-xylene in 6 h. Within the first 2 h, the desulfurization occurred with a rate of 1.29 mM dibenzothiophene/g DCW/h. Further mutagenesis is attempted on $D_{SZ}B$ to increase the biocatalytic activity and thermo-stability of the enzyme (Ohshiro *et al.*, 2007). The design of a recombinant microorganism resulted in the removal of the highest amount of sulfur compounds in fossil fuels (Raheb *et al.*, 2009). The three genes (DszA,B,C) from the desulfurization operon in *Rhodococcus erythropolis* IGTS8 were inserted into the chromosome of a novel indigenous *Pseudomonas putida*. As mentioned before, the reaction catalyzed by products of DszA,B,C genes require $FMNH_2$ supplied by the DszD enzyme. Thus, the pVLT31 vector harboring the dszD gene was transferred into this recombinant strain. This new indigenous bacterium is an ideal biocatalyst for a desulfurizing enzyme system due to the solvent tolerant characteristic and optimum growth temperature at 40 °C, which is suitable for the industrial biodesulfurization process. In addition, this strain produces rhamnolipid biosurfactant which accelerates a two-phase separation step in the biodesulfurization process through increasing emulsification. Moreover, the strain has a high growth rate which causes the removal of sulfur compounds faster than *R. erythropolis* IGTS8 and expresses higher biodesulfurization activity in shorter time.

Introduction of dsz genes into *Rhodococcus* sp. T09, a strain capable of desulfurizing benzothiophene, showed an improved uptake of alkylated sulfur compounds (Matsui *et al.*, 2001). The resulting recombinant strain grew with both dibenzothiophene and BT as the sole sulfur sources and the recombinant cells desulfurized alkylated benzothiophene derivatives with the production of alkylated hydroxyphenyl derivatives as the end products.

It has been reported that the desulfinase $D_{SZ}B$ catalyzes the rate limiting step in the 4S-pathway. Thus, increasing its level, either by removing the overlap between $D_{SZ}A$ and $D_{SZ}B$ or by rearrangement of the genes in the operon, such that $D_{SZ}B$ becomes the first gene in the operon, would result in increased biodesulfurization activity (Li *et al.*, 2008b). Deletion of the $D_{SZ}B$ gene could allow the accumulation of hydroxyphenyl benzene sulfinate (HBBSi) which is a valuable product for the surfactant industry (Yu *et al.*, 2005). Genetic engineering can also be applied on $D_{SZ}C$, which catalyzes the first step in the 4S-pathway to improve the substrate range (Arsendorf *et al.*, 2002).

Noda *et al.* (2003a) reported that *Rhodococcus erythropolis* strain KA2-5-1 is unable to desulfurize 4,6-dipropyl dibenzothiophene in the oil phase. The *dsz* desulfurization gene cluster from *R. erythropolis* strain KA2-5-1 was transferred into 22 *rhodococcal* and *mycobacterial* strains using a transposon-transposase complex. The recombinant strain MR65, from *Mycobacterium* sp. NCIMB10403, was able to grow on a minimal medium supplemented with 1.0 mM 4,6-dipropyl DBT in n-tetradecane (50%, v/v) as the sole sulfur source. The concentration of sulfur in light gas oil (LGO) was reduced by strain MR65 from 126 ppm to 58 ppm and by strain KA2-5-1 from 126 ppm to 80 ppm and within 24 hrs, strain MR65 had about a 1.5 fold higher LGO desulfurization activity than *R. erythropofis* strain KA2-5- 1. The application of a recombinant, which is able to utilize 4,6-dipropyl DBT in the oil phase, was effective in enhancing LGO biodesulfurization and the desulfurization activity for LGO was likely to be higher than that for 4,6-dipropyl DBT in n-TD, since other sulfur are found in LGO, including 4,6-DMDBT or 4,6-DEDBT in addition to 4,6-dipropyl DBT.

Recently, researchers found that bacterial cultures that possess identical dsz gene sequences can have very different Dsz phenotypes. This was clearly illustrated by examining the desulfurization activity of *Mycobacterium phlei* GTIS10, which has dszABC gene sequences identical to *R. erythropolis* IGTS8; the temperature at which maximum desulfurization activity was detected in the cultures was about 50 °C and 30 °C, respectively (Kayser *et al.*, 2002; Kilbane, 2006; Mohebali and Ball, 2008).

Yu *et al.* (2005) designed a novel method for the production of valuable BDS intermediates HPBS/Cx-HPBS from DBT/Cx-DBT. *Rhodococcus erythropolis* XP was isolated from a soil sample as an effective and stable bacterium that can desulfurize DBT, alkyl DBTs, and some of the alkyl benzothiophenes. The *dsz*B gene eliminated the slowest step of the Dsz pathway. Deletion of the *dsz*B gene could allow the accumulation of hydroxyphenyl benzene sulfinate (HPBS), which is a more valuable product than sulfate because it can be recovered easily from the aqueous phase and used as surfactant. Because the feedstock for this material is a low value, high-sulfur refinery stream and the conversion occurs at low temperatures and pressures, the process for producing this material looks quite economical.

The observation that the dsz operon had two apparent temperature maximums in two different bacterial hosts, suggests that if the dsz operon were to be expressed in a thermophilic bacterial host, the desulfurization enzymes could function at even higher temperatures. This possibility was tested in gene expression studies using the extreme thermophile, *Thermus thermophilus* (Park *et al.*, 2004), where the *dszC* gene from *R. erythropolis IGTS8* was successfully expressed in *T. thermophilus*, but the thermolability of the other enzymes prevented the functional expression of the full desulfurization pathway in this host. With the increased availability of desulfurization-competent bacterial cultures, it is likely that the complete desulfurization pathway could eventually be expressed in *T. thermophilus*.

Also, because this host grows at temperatures ranging from 55 °C to 85 °C, it might be possible to employ directed evolution and selective pressure to gradually develop desulfurization enzymes capable of functioning at higher temperatures so that they are suitable for a high temperature bioprocess (Park *et al*, 2004; Nakamura *et al.*, 2005).

To develop an efficient biocatalyst, many investigators have constructed recombinant biocatalysts. For example, the dsz genes from *R. erythropolis* DS-3 were successfully integrated into the chromosomes of *Bacillus subtilis* ATCC 21332 and UV1, yielding two recombinant strains, *B. subtilis* M29 and M28, in which the integrated dsz genes were expressed efficiently under control of the promoter Pspac. The DBT desulfurization efficiency of M29 was significantly higher than that of *R. erythropolis* DS-3, recording 16.2 mg DBT/L/h within 36 h and also showed no product inhibition (Ma *et al.*, 2006a).

Li *et al.* (2007) improved the DBT desulfurization activity of *R. erythropolis* DR-1 by removing the gene overlap in the dsz operon. Desulfurization activity of the redesigned strain was about five fold higher than that of strain DR-1.

Calzada *et al.* (2007) reported that the genetically modified microorganism, *Pseudomonas putida* CECT5279, is employed as a desulfurizing biocatalyst. The combination of two cell age biomasses can be used in order to optimize an effective biodesulfurizing catalyst. Complete transformation of DBT in a minimized reaction time is achieved by mixing different biomass concentrations of both higher DBT removal activity cells and higher HBP production activity ones, where 5 and 23 h growth time cells, respectively, in 0.7 and 1.4 g DCW/L, respectively, are found to be the best combinations. This biocatalyt formulation gets 100% DBT conversion, while reducing BDS time. In addition, this cell combination achieves a higher initial BDS elimination rate than 23 h cells used alone.

Etemadifar *et al.* (2008) reported enhanced desulfurization activity in protoplat transformed *Rhodococcus erythropolis.* Plasmids from *R. erythropolis* strain R1 (HBP-positive), isolated for its rapid capability to desulfurize DBT, were used to produce a stable desulfurizing mutant (mut23) of an HBP-negative strain by polyethylene glycol (PEG)-mediated protoplast transformation. *R. erythropolis* strain R1 and transformed *R. erythropolis* strain (mut23) were able to desulfurize 100% of the original DBT after 72 h. However, mut23 desulfurized DBT and produced HBP more efficiently than wild type R1 strain.

Raheb *et al.* (2009) reported the design of a recombinant microorganism to remove the highest amount of sulfur compounds in fossil fuels. Three genes (*dszA,B,C*) from *dsz* operon are responsible for the 4S pathway (biodesulfurization pathway) in *Rhodococcus erythropolis* IGTS8 were inserted into the chromosome of a novel indigenous *Pseudomonas putida.* The reaction catalyzed by products of *dszA,B,C* genes require $FMNH_2$ supplied by the dszD enzyme. Thus, the *pVLT*31 vector harboring *dszD* gene was transferred into this recombinant strain. This new indigenous bacterium is an ideal biocatalyst for the desulfurizing enzyme system due to the solvent tolerant characteristic and optimum growth temperature at 40 °C which is suitable for the industrial biodesulfurization process. In addition, this strain produces *rhamnolipid biosurfactant* which accelerates the two-phase separation step in the biodesulfurization process through increasing emulsification. Moreover, it has a high growth rate which causes the removal of sulfur compounds faster than *R. erythropolis* IGTS8 and has high biodesulfurization activity in a short time. The comparison of the biodesulfurization activity of recombinant indigenous *P. putida* and *R. erythropolis* IGTS8 was performed and showed that 2-HBP production of recombinant *P. putida* was more than that of *R. erythropolis* IGTS8 in a primary 1-20 hrs 2-HBP production of *R. erythropolis* IGTS8, which increased after a long time of cultivation (approximately 22 h). Therefore, engineered *P. putida* could be a promising candidate for industrial and environmental application in biodesulfurization due to the removal of higher sulfur amounts from oil in the shortest time. In addition to a higher optimal growth temperature, the ability to produce *rhamnolipid* biosurfactant and solvent toleration were the other privileges of this recombinant strain which are applicable in biodesulfurization processes.

Tao *et al.* (2011) combined the solvent tolerance exhibited by *Pseudomonas putida* (DS23) with biodesulfurization capability through inserting gene cluster dszABCD into a solvent-responsive expression vector. About a 56% reduction of 0.5 mM DBT within 12 h in a biphasic system containing 33.3% (v/v) n-hexane was reported while another strain induced by isopropyl β-D-1-thiogalactopyranoside could only degrade 26%. In an attempt to reduce the often high water-to oil ratios (>200) in bi-phasic systems, Kawaguchi *et al.* (2012) metabolically engineered the hydrophobic microorganism *Rhodococcus opacus* so that it was capable of expressing the *dszABC* operon constitutively. A maximum DBT consumption rate of 9.5 mmol/h/g wet cell was observed at a water-to-oil ratio of 4, containing an initial DBT concentration of 0.5 mM. The DBT consumption rate marginally declined with increasing oil content, while in another experiment conducted at an oil/water ratio of 4, an increase in DBT consumption rate was noted with increasing the DBT concentration.

In conclusion, mining of the genomic database to identify novel biodesulfurizing microorganisms helps in screening new isolates, especially those devoid of inhibitory steps in the 4S pathway, which were unknown to possess biodesulfurization capability. These microorganisms might also have the ability to degrade environmental pollutants, a range of industrial waste and aromatic compounds, and would be known to play a role in the production of various valuable industrial products. For the development of an efficient BDS process, there is a need to develop biocatalysts that remove nitrogen (denitrogenation), metals (demetallation), as well as sulfur, simultaneously, resulting in the overall upgrading of crude oil. This will help in the development of the crude oil biorefining to be applied in place of or after the treatment of oils by the HDS process to achieve ultra-deep desulfurization. The bioinformatics approach to discover different microorganisms for the biorefining of fossil fuels would be a cost-effective technique leading to savings of time, labor, and effort.

10.4 Overcoming the Repression Effects of Byproducts

Inhibition of cell growth and desulfurization activity by the end products of DBT desulfurization, including 2-HBP and/or 2,2'-BHBP and sulfate are known to be severe.

Several investigators have reported that the desulfurization activity in various bacteria was completely repressed and the production of DBT desulfurizing enzymes has been shown to be inhibited by sulfate or other readily bioavailable sulfur sources, including methionine, cysteine, taurine, methanesulfonic acid, and Casamino acids (Kayser *et al.*, 1993; Ohshiro *et al.*, 1995, 1996; Rhee *et al.*, 1998; Kertesz, 2000; Chang *et al.*, 2001; Matsui *et al.*, 2002; Noda *et al.*, 2002; Kim *et al.*, 2004; Gunam *et al.*, 2006).

The inhibition of DBT desulfurization activity by sulfate is considered to be a gene-level regulation. The expression of dsz genes that are involved in desulfurization is strongly repressed by sulfate. The removal of this feedback regulation is of a great importance from a process view-point (Kim *et al.*, 2004). Therefore, for efficient desulfurization of fossil fuels, it would be advantageous to develop a new strain that is not susceptible to sulfate repression.

Compared to what was reported by Li *et al.* (1996), *R. erythropolis* IGTS8 shows its highest desulfurization activity on 0.1% DBT at 0.05 mM sulfate. El-Gendy (2004) reported that bacterial isolate *Agrobacterium* sp. NSh10, in batch flask experiments at 30°C and pH7, produced about 210 ppm 2,2'-BHBP from the removal of ≈ 439 ppm of the initially added 1000 ppm DBT when 0.1 mM of $MgSO_4$ was initially added, but the overall BDS activity decreased to about half (%BDS ≈ 44%) that obtained without the external addition of SO_4^{2-} ions (%BDS ≈ 86%). El-Gendy (2004) also reported that the yeast isolate *Candida parapsilosis* NSh45 showed complete removal of 1000 ppm with non-stoichiometric production of 309 ppm 2,2'-BHBP in batch flask experiments at 30°C and pH7 with the initial addition of 0.2 mM $MgSO_4$.

The 4S pathway is a complex enzyme system and its cofactor requirements prohibit the use of purified enzyme systems rather than whole cells for a practical BDS process (Kilbane and Le Borgne, 2004). Therefore, for research into BDS, resting cells (i.e. non-proliferating cells) have been accepted as the best biocatalysts. Implementation of a commercial BDS process consists of several stages including: (i) growing the selected strain in a suitable medium in such a way as to obtain cells (biomass) that exhibit the highest possible level of desulfurizing activity and (ii) harvesting these active cells and using them in the form of resting cells (biocatalysts) (Monot *et al.*, 2002). To make the BDS process economically competitive with the deep hydrodesulfurization process that is currently used, it is necessary to improve several factors including the cost of producing the biocatalyst and its biocatalytic activity. It is well understood that the sulfur source used for growth strongly influences the desulfurization activity of resting cells because desulfurization enzymes are sulfate-starvation-induced proteins (Tanaka *et al.*, 2002). Due to the repression phenomenon, it is difficult to prepare cultures using inexpensive sulfate-containing media. The use of synthetic media with low sulfate content is an expensive option, therefore several authors have reported attempts to overcome sulfate repression (Piddington *et al.*, 1995; Kishimoto *et al.*, 2000; Chang *et al.*, 2001; Matsui *et al.*, 2002; Tanaka *et al.*, 2002; Noda *et al.*, 2003b). The hypothesis is that the repression could be avoided by various microbiological means, including lowering the repressor (sulfate) content of the growth medium and substituting a sulfur source other than DBT for sulfate since mass production of biocatalyst using DBT is impractical because of its high price, low water solubility, and growth inhibition by 2-HBP. Since low-sulfur-containing synthetic media are expensive, the first solution is not commercially attractive. Therefore, using an alternative sulfur source to DBT with subsequent induction using DBT represents a reasonable option.

Ma *et al.* (2006b) investigated DBT, DMSO, sodium sulfate, and mixed sulfur sources to study their influence on cell density, desulfurization activity, and the cost of biocatalyst production. In contrast to that observed from bacteria cultured in DBT, only partial desulfurization activity of strain lawq was induced by DBT after cultivation in a medium containing inorganic sulfur as the sole sulfur source. The biocatalyst, prepared from culture with mixed sulfur sources, was found to possess desulfurization activity. With DMSO as the sole sulfur source, the desulfurization activity was shown to be similar to that of bacteria incubated in a medium with DBT as the sole sulfur source. The biocatalyst prepared by this method with the least cost could remove sulfur from hydrodesulfurization (HDS) treated diesel oil efficiently, providing a total desulfurization percent of 78% and suggesting its cost-effective advantage.

Mohebali *et al.* (2008) used dimethyl sulfoxide (DMSO) as a sulfur source for the production of desulfurizing resting cells of *Gordonia alkanivorans* RIPI90A. DMSO was more efficient as the sulfur source than DBT. This study confirms that active resting cells can be prepared in two steps as follows: (i) production of resting cells using DMSO as the sulfur source for growth and (ii) improvement of their desulfurization activity by using DBT as an inducer.

To alleviate the sulfate repression and enhance the desulfurization activity, several recombinant biocatalysts, including recombinant *Escherichia coli*, *Pseudomonas aeruginosa,* and *Rhodococcus* strains, have been developed (Kilbane, 2006). Recently, *Gordonia alkanivorans* RIPI90A has been reported as a DBT desulfurizing strain with special emulsion stabilization properties which could be of special interest in BDS process design (Mohebali *et al.*, 2007b). Shavandi *et al.* (2009) reported cloning and sequencing of *dszABC* genes of *G. alkanivorans* RIPI90A, identification of a suitable vector system for *Gordonia* genus, and self-cloning of the *dszABC* genes in a *Gordonia* isolate. The recombinant strain was able to desulfurize dibenzothiophene in the presence of inorganic sulfate and sulfur-containing amino acids while the native strain could not. The maximum desulfurization activity by recombinant resting cells (131.8 μM 2-hydroxybiphenyl/g DCW/h) was increased by 2.67-fold in comparison to the highest desulfurization activity of native resting cells.

It is reported that 2-HBP and 2,2'-BHBP repress both growth and desulfurization activity (Ohshiro *et al.*, 1996b; Okada et., 2003; Kim *et al.*, 2004). Compared to what was reported by Setti *et al.* (1999), 2-HBP growth-limiting concentration for IGTS8 is 30-40 ppm. However, El-Gendy (2004) reported that the yeast isolate *Candida parapsilosis* NSh45 can tolerate up to 475 ppm of 2,2'-BHBP with the removal of approximately 71% of the initially added 1000 ppm DBT.

Chen *et al.* (2008) studied the effect of 2-hydroxybiphenyl (2-HBP), the end product of dibenzothiophene (DBT) desulfurization via 4S pathway, on cell growth and desulfurization activity of *Microbacterium* sp. ZD-M2. Experimental results indicated that 2-HBP would inhibit the desulfurization activity. Providing 2-HBP was added in the reaction media, the DBT degradation rate decreased along with the increase of the added 2-HBP. By contrast, cell growth would be promoted in the addition of low concentration 2-HBP (<0.1 mM). On the contrary, at a high concentration of 2-HBP, the inhibition on the cell growth occurred. Meanwhile, the inhibitory effect of 2-HBP on DBT desulfurization activity was tested both in the oil/aqueous two-phase system and the aqueous system. On the other hand, higher DBT degradation activity and lower 2-HBP inhibitory effect in the oil/aqueous two-phase system than that in the sole aqueous system were obtained. There are two reasons for these possibilities. Firstly, the addition of an organic phase would enhance the transfer of DBT to the cell (Okada *et al.*, 2002a; Luo *et al.*, 2003). Secondly, 2-HBP is very soluble in the oil phase and, in practice, finds its way back to the petroleum fraction (Moniticello, 2000). Thus, the inhibition of 2-HBP on DBT desulfurization might be lower in the oil/aqueous two-phase system than that in sole aqueous phase because most 2-HBP would return into the oil (n-hexadecane) phase and minimize the toxicity of 2-HBP to the desulfurization enzyme. By comparing the influence of 2-HBP on biodesulfurization with an aqueous and oil/aqueous system, it is advised that the removal of 2-HBP from the reaction medium is necessary for biodesulfurization to decrease the inhibition of 2-HBP on an enzyme.

Also, overcoming the effect of toxic by-products like phenolic compounds can be done by adding surfactants. Feng *et al.* (2006) reported the enhancement of the BDS of hydrodesulfurized oil by *Rhodococcus erythropolis* lawq by adding surfactant Tween 80. Tween 80 was shown to decrease the product concentration associated with the cells, reducing the product inhibition.

The problems involved in the biodesulfurization of fuels are inhibition of the biocatalyst by the byproducts and slow diffusion between the organic and aqueous phases. Inverse phase transfer biocatalysis (IPTB) uses supramolecular receptors (modified cyclodextrine-like hydroxy propyl-β-cyclodextrine), which selectively pick up the sulfur aromatic compounds in the organic phase and transfers them into the water phase which contains the biocatalyst. The IPTB approach can increase mass transfer of water insoluble substrates between aqueous and organic phases and eliminate or reduce feedback inhibition of the biocatalyst due to accumulation of the byproducts in the water phase. It has been reported that 2-hydroxybiphenyl (HBP) inhibits *Rhodococcus rhodochrous* IGTS8. Setti *et al.* (2003) reported that, in the presence of 10 mM cyclodextrine, the growth of IGTS8 falls from 100 to only 80%. This indicates that cyclodextrine probably picks up the HBP in solution as well as directly from the interphase of the cellular biomembrane, thus protecting the microorganisms from the irreversible inhibition effect of this phenol. It was also observed that hydroxypropyl cyclodextrine can improve the mass transfer of water insoluble substrates such as DBT in the n-hexadecane and aqueous phase. The specific rate of the DBT converted by IGTS8 (i.e., the parts per million of DBT converted per hour per gram of dry cell) at a DBT concentration of 120 ppm in n-hexadecane increased from 2.9 to 4.3 in the absence and presence of 3.14 mM of hydroxypropyl cyclodextrine, respectively.

10.5 Enzymatic Oxidation of Organosulfur Compounds

One of the major problems in the biodesulfurization of petroleum and its fractions using whole cells is the need of an aqueous phase, for the viability of the cells. Thus, a two-phase system reactor is required with the mass transfer limitation problems needed to metabolize the hydrophobic substrate and a microorganism with a broad substrate specificity for the various organosulfur compounds present in oil is also required. This problem can be solved by the application of enzymes with broad specificity instead of whole cells in an anhydrous organic phase or at very low water content, so the mass transfer limitations would be reduced, but as has been mentioned before, the use of enzymes is disadvantageous since extraction and purification of the enzyme is costly and, frequently, enzyme catalyzing reactions require cofactors which must be regenerated (Setti *et al.*, 1997). However, some studies have been carried out on the bio-oxidation of organosulfur compounds using isolated enzymes in the free or immobilized form. It is a technique similar to that of conventional oxidative desulfurization, where firstly, the OSCs are enzymatically oxidized to sulfones, followed by physicochemical processes for the separation of the oxidized organosulfur compounds from the main hydrocarbon mixture, such as column chromatography, precipitation, and complexation with a solid support (Vazquez-Duhalt *et al.*, 2002).

The hemoprotein enzymes like cytochrome P450 (Alvarez and Ortiz de Montellano, 1992), lignin peroxidase from the white rot fungus *Phanerochaete chrysosporium* (Vazquez-Duhalt *et al.*, 1994), lactoproxidase (Doerge *et al.*, 1991), horseradish peroxidase (Kobayashi *et al.*, 1986), and chloroperoxidase from the fungus *Caldariomyces fumago* (Pasta *et al.*, 1994) can be applied for biotransformation of thiophene, dibenzothiophene, and other organosulfur compounds (Ayala *et al.*, 2007; Terres *et al.*, 2008). The non-enzymatic hemoproteins are also able to perform the dibenzothiophene oxidation in vitro, such as hemoglobin (Ortiz-Leon *et al.*, 1995), cytochrome c (Vazquez-Duhalt*et al.*, 1993), and microperoxidase (Colonna *et al.*, 1994).

The bio-oxidation of organosulfur compounds can be performed in batch, semi-continuous, or continuous methods alone or in a combination with one or more additional refining process. The reaction can be carried out in an open or closed vessel. Cytochrome c is a biocatalyst able to oxidize thiophenes and organic sulfides and has several advantages when compared with other hemoenzymes. It is active in a pH range from 2 to 11 and has the hemo-prosthetic group covalently bonded, exhibiting activity at high concentrations of organic solvents and is not expensive (Vazquez-Duhalt *et al.*, 1993). The activity of chloroperoxidase towards the oxidation of 4,6-dimethyl dibenzothiophene (4,6-DMDBT) in a water/acetonitrile mixture, using free-form and immobilized SBA-16 mesoporous material has been investigated and the thermal stability of the immobilized enzyme was found to be three times higher than that of the free enzyme (Terres *et al.*, 2008). Madeira *et al.* (2008) performed a stepwise evaluation of the enzymatic oxidation of DBT by horseradish peroxidase (HRP). The reactions were carried out in monophasic organic media containing 25% (v/v) acetonitrile. The best results were observed in a reaction medium at pH 8.0, presenting HRP 0.06 IU/ml, DBT 0.267 mM, a DBT:H_2O_2 molar ratio of 1:20 (stepwise hydrogen peroxide addition), and incubated at 45 °C for 60 min. Under these conditions, ≈ 60% of DBT was converted into DBT sulfoxide (12%) and DBT sulfone (46%).

10.6 Enhancement of Biodesulfurization via Immobilization

In the last few years, the application of biotechnological processes that involves microorganisms with the objective of solving biodesulfurization (BDS) problems is rapidly growing. The BDS of fuels is still not a commercial technology because some problems have to be solved. Stability and life-time of biocatalyst are two crucial factors. In order to solve these problems, cell immobilization, which has been widely studied and applied in recent years, is considered to be a potential method and is becoming the focus of BDS research.

Immobilization is a natural phenomenon existing in the globe. An immobilized molecule is the one whose movement in space has been restricted either completely or to a small limited region by attachment to a solid structure. In general, the term immobilization refers to the act of the limiting of movement or making incapable of movement (Yu-Qung *et al.*, 2004). Immobilization can be performed for enzymes and/or whole microbial cells.

Most microbial cells, by their nature, tend to attach themselves onto a solid surface in some growth phase of their life cycles. In nature, cell immobilization occurs naturally by adhesion of cells to solid surfaces (biofilm). In some industrial bioprocess flocculation, pellet formation and surface attachment can be considered as immobilization forms (Akin, 1987). Immobilizing the matrix of cells must have properties such as a high biocatalytic activity, long-term stability of cells, possibility of regenerating the biocatalyst, and low loss of activity during immobilization and the BDS process. In addition, it also has low leakage of cells, non-compressible particles, and high resistance to abrasion. It is also very desirable to be an economic support matrix with high resistance to microbial degradation, with low diffusional limitation, high surface area, and appropriate density for the reactor type (Nussinovitch, 1997). Biocatalyst reactions by immobilized microbial cells, compared to immobilized purified enzyme systems, have numerous advantages such as less expenses on separation, isolation, and purification methods, allows multistep reactions, enhances stability of the enzyme in its native state, allows longer enzyme activity in the presence of co-factors, and continues biosynthesis by cell (Jack and Zajic, 1997). Immobilization allows the cells to be used repeatedly and continuously, which helps the maintenance of the high cell density during the process and, consequently, reduces the cost of the bioprocess. Immobilized cells are more resistant to environmental stress since there is an additional support material to provide protection (Akin, 1987). The advantages of immobilized cell systems are summarized in Figure 10.3.

Immobilized enzymes were introduced in the 1950s for laboratory studies once it was realized that enzymes could be in a water-insoluble form without losing their catalytic activity (Dervakos and Webb, 1991; Wandrey, 1996). Attempts to immobilize whole cells began in the 1970s (Dervakos and Webb, 1991). Research on cell immobilization techniques is extensive and numerous immobilization systems have been studied at a laboratory scale. However, few immobilized cell systems have been scaled up (Dervakos and Webb, 1991; Wandrey, 1996).

Cell immobilization has been used in applications such as biotransformation, hazardous material degradation, and material production (Pan *et al.*, 1997; Beshay *et al.*, 2002; Kapoor and Kuhad, 2007; Robatjazi *et al.*, 2012). Microorganisms have been immobilized on a variety of matrices such as agar, silica, calcium alginate, and polyvinyl alcohol (Nigam *et al.*, 1998; Ha, 2005; Huang *et al.*, 2012; Taha *et al.*, 2013).

All the BDS processes reported hitherto are tri-phasic systems composed of cells, water, and oil or, in another, word bi-phasic systems (oil

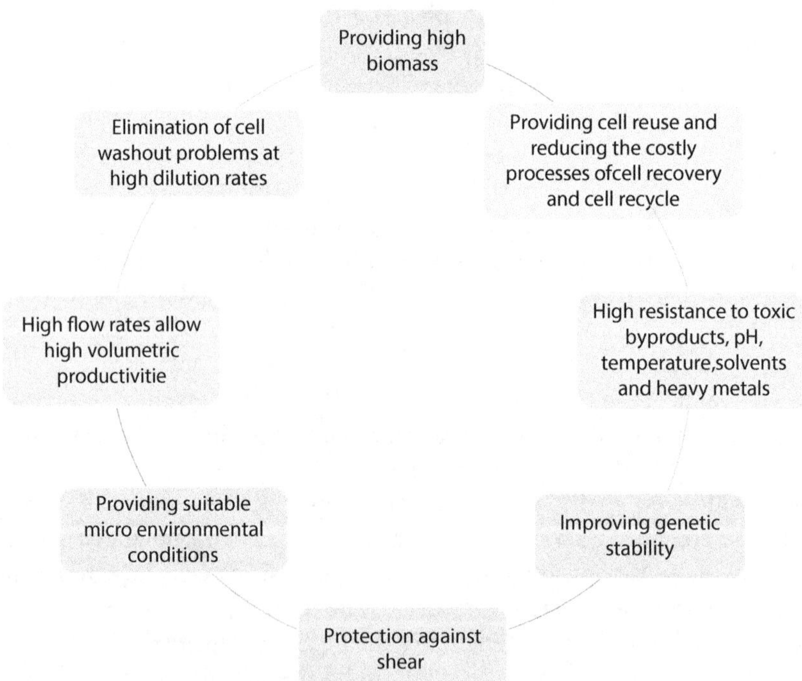

Figure 10.3 Advantages of Immobilized Cell Systems.

and cells in water). There are numerous reports on the treatment of diesel oils or model oil mixtures by using suspensions of growing or resting cells (Chang *et al.*, 2001; Noda *et al.*, 2003a; Labana *et al.*, 2005; Yu *et al.*, 2006b). Treatment of oils using free cells has some limitations such as high cost of the biocatalyst and low volumetric ratio between the organic phase and the aqueous one. In these cases, oil is mixed together with the cells as a suspension, which produces a sort of surfactant, emulsifying the oil. It seems to be very difficult to separate oil and water from the emulsified oil. Furthermore, the recovery of the cells is also difficult. Yu *et al.* (1998) reported an idea to separate oil, water, and cells from the emulsified oil based on a cyclone, but this concept of separation is based on centrifugal force and it seems to be difficult to separate the components completely.

Cell immobilization was considered to be one of the most promising approaches. Compared with cell suspension, BDS with immobilized cells has some advantages: ease of biocatalyst separation from the treated fuels, low risk of contamination, relatively high oil/water volumetric ratios, high stability, and long life-time of the biocatalyst (Chang *et al.*, 2000; Hou *et al.*, 2005). Nevertheless, physical interactions between bacterial cells and sulfured substrates require further studies to resolve problems associated with the limited access of microorganisms to organic substrates to upscale BDS (Tao *et al.*, 2006; Yang *et al.*, 2007). In this context, surfactants and immobilized cells are considered promising solutions to the problem of low solubility. Bio-modification of inorganic supports using cell immobilization increases the interaction between reactants present in two-phase systems, avoiding the need to use expensive surfactants (Feng *et al.*, 2006). So far, very few published papers are available on BDS by immobilized cells.

Generally, entrapment and adsorption are preferred methods for cell immobilization. In entrapment, living cells are enclosed in a polymeric matrix which is porous enough to allow diffusion of substrates to the cells and of products away from the cells. The materials used for entrapment of cells are mainly natural polymers, such as alginate, carrageenan, gelatin, and chitosan. They may also be synthetic polymers, such as polysaccharides, photo-cross-linkable resins, polyurethane, polyvinyl alcohol, polyacrylamide, and so on. Major drawbacks of an entrapment technique are diffusional limitations and steric hindrance, especially when diffusion of macromolecular substrates, such as starch and proteins, is involved. Mass transfer involved in diffusion of a substrate to a reaction site and in the removal of inhibitory or toxic products from the environment may be impeded. Cell immobilization by adsorption is currently gaining considerable importance because of a major advantage, namely, reducing or eliminating the mass transfer problems associated with common entrapment methods. However, the adsorption technique is generally limited by biomass loading, strength of adhesion, biocatalytic activity, and operational stability. This is because immobilization by adsorption involves attachment of cells to the surface of an adsorbent like Celite. Adsorption is a simple physical process in which the forces involved in cell attachment are so weak that cells which are several micrometers across are not strongly adsorbed and are readily lost from the surface of the adsorbent (Shan *et al.*, 2005a).

In the last two decades, numerous studies have been carried out on BDS using whole cells (Maghsoudi *et al.*, 2001; Tao *et al.*, 2006; Yang *et al.*, 2007; Caro *et al.*, 2008) or isolated enzymes (Monticello and Kilbane, 1994) in the free or immobilized form. However, the BDS of DBT occurs via a multi-enzyme system that requires cofactors (e.g. NADH). The use of enzymes is disadvantageous since extraction and purification of the enzyme is costly and, frequently, enzyme catalyzing reactions require cofactors which must be regenerated (Setti *et al.*, 1997). Therefore, BDS can often be designed by using whole cell biotransformation rather than that of the enzyme. Nevertheless, there are still some bottlenecks limiting the commercialization of the BDS process. One of the challenges is improving the current BDS rate by about 500 fold, assuming the target industrial process is 1.2–3 mM/g DCW /h (Kilbane, 2006). When free cells were used for petroleum BDS, deactivation of the biocatalyst and troublesome oil–water–biocatalyst separation were significant barriers (Konishi *et al.*, 2005; Yang *et al.*, 2007). Cell immobilization may give a solution to the problems, providing advantages such as repeated or continuous use, enhanced stability, and easy separation. Suitable models of immobilized cell behavior should include the following basic processes: diffusion of nutrients and products through the support material and the cells, cellular metabolism, and cell proliferation and death.

For BDS, free cells need to be harvested from the culture medium and several separation schemes had been evaluated, including settling tanks (Schilling *et al.*, 2002), hydrocyclones (Yu *et al.*, 1998), and centrifuges (Monticello, 2000), but these procedures are time-consuming and are not cost effective. Magnetic separation technology provides a quick, easy, and convenient alternative over traditional methods in biological systems (Haukanes and Kvam, 1993).

In the biphasic system, the hydrophobic substrates (e.g. DBT and substitute DBTs) can be retained mainly in the organic phase. Therefore, it is necessary to select the biocompatible organic phase for BDS. Clearly, the biocompatible organic phase can act as a reservoir for the toxic substrates to regulate the substrate concentration around biocatalysts (He *et al.*, 2012; Dinamarca *et al.*, 2014). Moslemy *et al.* (2002) reported that immobilizing systems provided the bacterial cells with a protective solid barrier which may have somewhat limited the diffusion of diesel hydrocarbons, reducing the bioavailable concentration in the inner space of the beads with respect to that in the bulk liquid. Thus, the potential substrate inhibition can be decreased. Passos *et al.* (2010) and Mulla (2013) reported that immobilized cell cultures act as a protective cover against the toxicity of phenolic compounds. So, to increase the BDS rate, immobilized bacterial cells can be used as another alternative for BDS (He *et al.*, 2012). The selection of immobilizing material is very important for use in immobilization and immobilizing material must have some important criteria (Figure 10.4).

Immobilizing materials are commonly divided into two main groups, as shown in Figure 10.5: organic and inorganic. Organic carriers include modified celluloses, dextran, and chitosan agarose and inorganic carriers include zeolite, clay, anthracite, porous glass, activated charcoal, and ceramics (Lu *et al.*, 2009). Organic materials are more abundant than inorganic carriers and can be obtained with strictly controlled porosity, but they are usually very sensitive to pressure or pH. Most inorganic supports

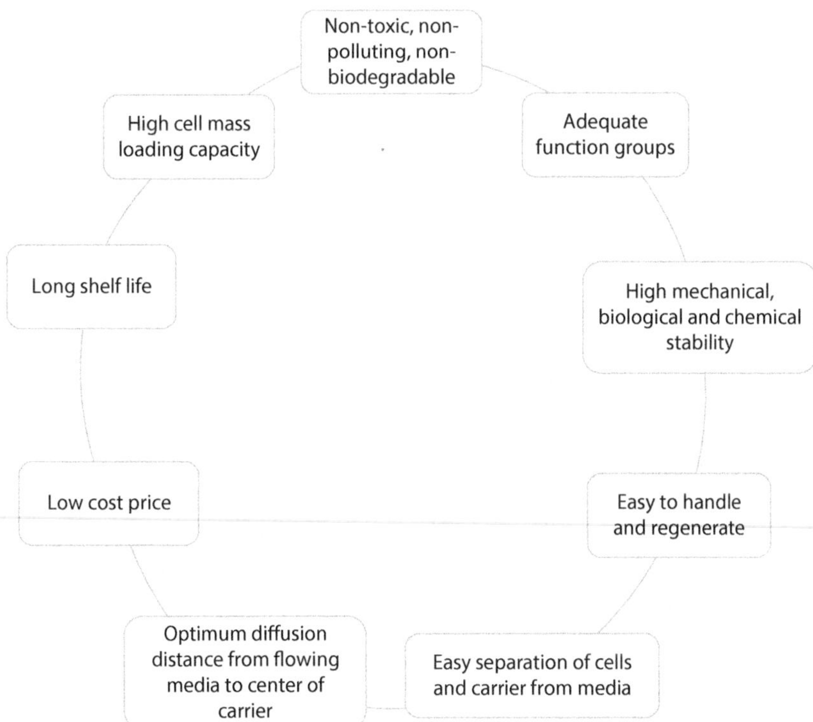

Figure 10.4 Ideal Criteria of Immobilizing Material.

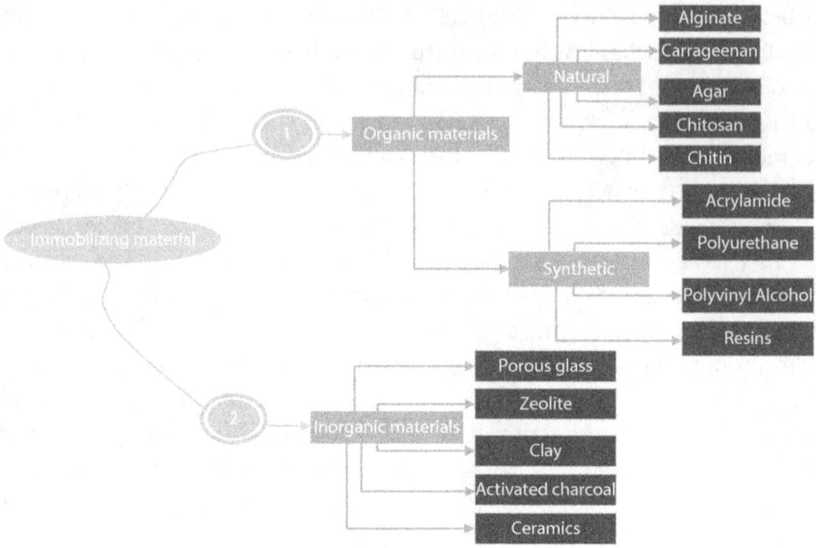

Figure 10.5 Types of Immobilized Material.

are totally inert and resistant to temperature, pH, chemicals, microbial degradation, and crushing or abrasion.

Organic carriers can be also divided into natural and synthetic polymers. Some examples of natural carriers that can be used as supports include alginate, carrageenan, agar, agarose, chitosan, and chitin. A variety of synthetic polymers, such as acrylamide, polyurethane, polyvinyl, and resins are also used for immobilization (Hartman *et al.*, 2005).

10.6.1 Types of Immobilization

Different immobilization techniques are frequently used in the industrial processes and relative to the free cells, immobilization has inherent advantages, including enhanced stability of the system, easy separation of cells, minimizing or eliminating the cell contaminations in the products, convenient recovery, and reuse of cells which enables their frequent use in the process (Zhang *et al.*, 2010). Several immobilization methods have been suggested for industrial applications including adsorption, covalent binding, entrapment, and encapsulation (Figure 10.6) (Martinsen *et al.*, 1989; Biria *et al.*, 2008).

10.6.1.1 Adsorption

This reversible method for the immobilization of cells is based on the physical interaction between the microorganism and surface of water-insoluble

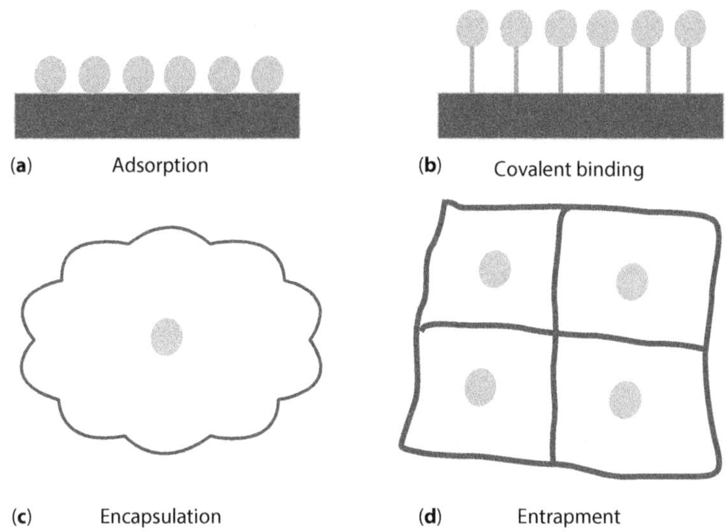

Figure 10.6 Immobilization Techniques.

carriers. Microbial cells can be attached on the porous or non-porous matrix. In the interaction between microorganisms and the surface of the matrix, weak forces include hydrogen bonds, ionic bonds, hydrophobic bonds, and van der Waals forces. Physical adsorption is a simple and less expensive technique, which retains high catalytic activity, therefore, it is used more commonly than other methods. This technique also allows the reuse of expensive support materials. As a result, this reusability of the support material brings economic advantages for industrial productions. On the other hand, this technique does not offer high stability and might cause loss of biomolecules that are immobilized during washing and operation (Salleh *et al.*, 2006). For the immobilization of viable cells, the adsorption process is well-matched when compared with the entrapment technique (Joachim and Holger, 1990). Generally, bacterial immobilization by adsorption is an improvement over the cell entrapment method that reduces mass transference and the steric effect. Bacterial cell adsorption involves the use of inorganic compounds as ideal bio-supports. These materials must have controlled porosity and a high specific area. They must also be inert to biological attack, insoluble in the growth media, and nontoxic to microbial cells. Moreover, adsorbed cells on inorganic supports should be able to maintain the metabolic activity required for the BDS process. Few studies have evaluated the influence of inorganic supports with different physicochemical properties on the BDS activity of DBT or gas oil (Hwan *et al.*, 2000; Zhang *et al.*, 2007); alumina and celite are the most common supports used in these bioprocesses.

Setti *et al.* (1997) reported immobilization by adsorption using hydrophilic natural supports. This technique does not limit the hydrocarbon uptake mechanism. In fact, the ideal conditions for hydrocarbon biodegradation can be restored at the interphase between the oil phase and the hydrophilic support surface. Immobilization also retains water around the catalyst, which is essential for the biocatalytic function. Hydrophilic supports may, however, compete with the biocatalyst for the available water in the reaction system, especially when the water is also a reactant in the DBT-BDS pathway so that complete depletion of water from the reaction system results in no biocatalytic activity. The use of hydrophilic supports permits a process in which a decanter unit downstream of the bioreactor is not required since the water remains entrapped in the support. This factor represents an advantage in terms of finishing costs and times. The activity of the immobilized cells was significantly (up to ten-fold) lower than that of free cells, being in the order of 0.01–0.04 g/g biomass/h, a rate which depended on the kind of support used. The immobilized cells maintained an operational stability for about 100 h, a duration similar to that reported for some cell-free bioconversions in n-alkane.

Celite beads are used as filter aids in pharmaceutical and beverage processing as bulk filters for food and plants and as insulation material for construction. Because of its chemical composition, celite is virtually inert to biological attack, insoluble in culture media, and non-toxic to microbial cells. Celite is composed of silica (SiO_2) (over 90 wt%) and other inorganic oxides including Al_2O_3, Fe_2O_3, and CaO (Hull *et al.*, 1953). Chang *et al.* (2000) reported that BDS efficiency of a model oil (hexadecane containing DBT) and a diesel oil were enhanced by using immobilized DBT-desulfurizing bacterial strains, *Gordona* sp. CYKS1 and *Nocardia* sp. CYKS2. Celite bead (50 to 500 μm bead size and 0.72 porosity) was used as a biosupport for cell immobilization. Celite beads were pretreated by washing three times with distilled water and heated in a furnace overnight at 600 °C to remove volatile materials. The beads were then steam-autoclaved for 1 h at 121 °C and dried for 30 min. For cell immobilization, 3 mL of cell suspension (2.64 g/L) was added to 3.6 g of pretreated celite beads. The total volume of the reaction mixture was 20 mL (celite beads concentration was 50% v/v). 1 mL of model oil hexadecane (10 mM of DBT, sulfur 0.32 g /L_{oil}) or LGO (sulfur 3.0 g /L_{oil}) was added for desulfurization and the volumetric phase ratio of aqueous to oil was 20. Seven to eight cycles of repeated-batch desulfurization were conducted for each strain. Each batch reaction was carried out for 24 h. In the case of model oil treatment with strain CYKS1, about 4.0 mM of DBT in hexadecane (0.13 g sulfur/L_{oil}) was desulfurized during the first batch, while 0.25 g sulfur/L_{oil} was desulfurized during the final, eighth batch. The mean desulfurization rate increased from 0.24 for the first batch to 0.48 mg sulfur/$L_{dispersion}$/h for the final batch. The sulfur content in the light gas oil was decreased from 3 to 2.1 g /L_{oil} by strain CYKS1 in the first batch. The mean desulfurization rate by strain CYKS1 was 1.81 mg sulfur/$L_{dispersion}$/h in the first batch and decreased to 1.43 mg sulfur mg sulfur/$L_{dispersion}$/h in the final batch. However, the desulfurization rate of LGO was about 4-7 times higher than that of model oil. The major reason for the higher desulfurization rate with LGO, presumably, was that the model oil contained only DBT, a recalcitrant compound, while LGO contained various readily desulfurizable compounds, such as thiols and sulfides (Chang *et al.*, 1998). No significant changes in desulfurization rate were observed with strain CYKS2 when the batch reaction was repeated. After the final batch, immobilized cells were collected and washed once with 0.1 M phosphate buffer (pH 7.0). Then, the immobilized cells were stored in a 0.1 M phosphate buffer with a pH 7.0 at 4 °C for 10 days. After 10 days, 20 mL of MSM was added after draining the phosphate buffer and then 1 mL of oil was added to start the desulfurization process. The longevity of immobilized cells was evaluated from the residual desulfurization activity as a fraction of the initial activity. In case of model oil, the residual activities were 70 and 65% for strains CYKS1 and CYKS2, respectively. In the case of LGO, the residual activities were 52 and 65% for strains CYKS1 and CYKS2, respectively.

Li *et al.* (2009) constructed a novel adsorption–bioregeneration system to produce ultra-low sulfur diesel oil by combining adsorption and BDS processes. The adsorption capacity and selectivity towards DBT of the adsorbents were tested under ambient conditions with 8.0 mM/L DBT and 8.0 mM/L naphthalene in n-octane. The ratio of oil to adsorbent was chosen as 100 mL/g. Adsorption properties of Ag-MAS, Ag-MCM-41, and Ag-Y were tested with hydrotreated diesel oil (containing 91µg/g sulfur) at 30 °C. The ratio of adsorbents to oil was 20 mL/g. Adsorption–bioregeneration properties were tested in in situ adsorption bioregeneration systems, as shown in Figure 10.7. The system can be divided into two stages: adsorption and bioregeneration. After the saturation of adsorbents, the adsorption system is shut down and the adsorption reactor was connected with the bioreactor. Then, the desorbed sulfur compounds were converted by P. delafieldii R-8 cells. Finally, the desorbed adsorbents were treated with air at 550 °C to remove water from the adsorption system. The sequence of adsorption capacity of DBT was AC (activated carbon) > NiY > AgY > alumina > 13X. The sequence of selectivity of DBT toward naphthalene

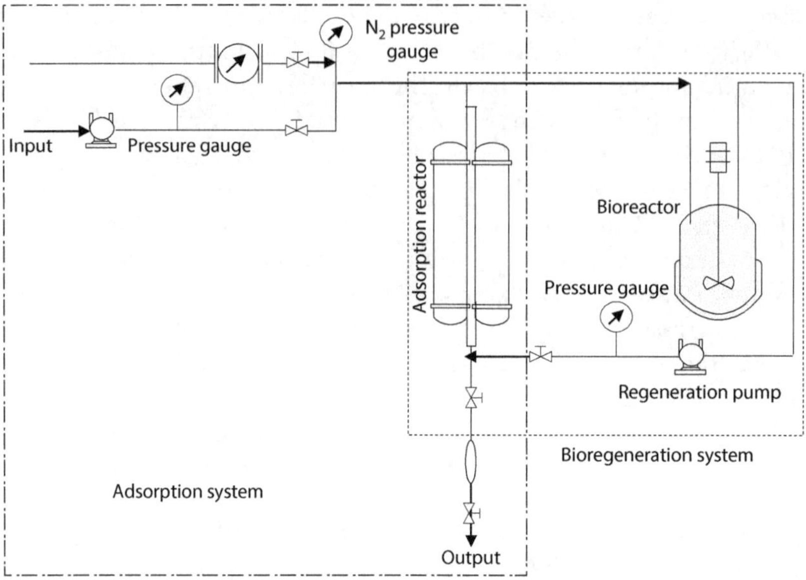

Figure 10.7 In-Situ Adsorption–Bioregeneration System.

was NiY > AgY > alumina ≈ 13X > AC. For hydrotreated diesel oil, MAS (mesoporous aluminosilicates) showed high adsorption capacity, while MCM-41 and NaY showed low adsorption capacity. Adding *P. delafieldii* R-8 cells could improve DBT desorption from adsorbents. The desorption of DBT from adsorbents by bioregeneration followed the sequence: 13X > alumina > AgY > NiY > AC. Ag-MAS could be completely regenerated in an in situ adsorption–bioregeneration system.

Dinamarca *et al.* (2010) studied the efficiency of adsorption of *Pseudomonas stutzeri* on Silica (Si), Alumina (Al), Sepiolite (Sep; a complex magnesium silicate $Mg_4Si_6O_{15}(OH)$), and Titania (Ti) and their influence on the BDS of gas oil, with emphasis on the interaction of inorganic supports on metabolic activity. For the immobilization process, the bacteria cell numbers were adjusted by measuring the turbidity at 600 nm (OD_{600}) (Shan *et al.*, 2005b) in a range from 1×10^9 to 25×10^9. Obtained bacterial cells were contacted with 0.1 g of respective support in 25 mL flasks at 30 °C in a rotary shaker at 200 rpm for 24 h. The number of immobilized cells was measured by losses of turbidity of solution at 600 nm (OD_{600}). The biocatalysts were designated as bacterial cell/solid systems, where the bacterial cells correspond to *P. stutzeri* and the solids correspond to the inorganic supports. For Al and Sep, the maximum of cells adsorbed was established at a value of 9.9 (log adsorbed cells/g supports), which was reached when 14 × 10^8 cells/ mL were used. In the case of Ti, the maximum of adsorbed cells was 9.5 (log adsorbed cells/g supports) when 20 × 10^8 (cells/mL) of initial cells were used in the adsorption process. When the adsorption process of *P. stutzeri* was conducted on Si, saturation of adsorption was not observed. Immobilization results indicated that the lowest interaction between bacterial cells and supports was given between *P. stutzeri* and Ti, whereas adsorption on Al, Sep, and Si was more efficient in relation to the initial number of cells required to reach the saturation. The isoelectric point (IEP) values and metabolic activities indicated that *P. stutzeri* changed the surface of supports and maintained metabolic activity. A direct relation between BDS activity and the adsorption capacity of the bacterial cells was observed at the adsorption/desorption equilibrium level. The bio-modification of inorganic supports by the adsorption process increased the bioavailability of sulfur substrates for bacterial cells, improving the BDS activity. The most active biocatalysts were *P. stutzeri*/Si and *P. stutzeri*/Sep (2.80 x 10^{-13} g S/cell/h and 2.57 x 10^{-13} g S/cell/h, respectively), while *P. stutzeri*/Ti showed the lowest activity (1.03 x 10^{-13} g S/cell/h). The interactive forces produced between the cells and supports, which maintain an appropriate biodynamic equilibrium of adsorbed/desorbed cells, play a major role in the BDS process. The highest interaction was observed in the *P. stutzeri*/Si and *P. stutzeri*/Sep biocatalysts. Thus, the major level of BDS activity was also observed in *P. stutzeri*/Si and *P. stutzeri*/Sep biocatalysts. The bio-modification of inorganic supports generates dynamic biostructures that facilitate the interaction with insoluble organic substrates (Rong *et al.*, 2008), improving the BDS of gas oil in comparison to whole non-adsorbed cells. This study concluded that immobilization by adsorption of bacterial cells is a simple and effective methodology that can be applied in BDS reactions of gas oil.

Li *et al.* (2011) attempted to develop an effective technique for the BDS process by employing chitosan flocculation and integration with cell immobilization onto celite. β-(1–4) linked 2-amino- 2-deoxy-D-glucose (chitosan) is currently one of the most promising biopolymers widely used in separation or stabilization of dispersed systems due to its advantages of biodegradability having less adverse effects on human health and renewable sources. The cationic nature of chitosans makes them especially favorable for flocculation of negatively charged particles. Chitosan was applied as a flocculant of bacteria suspensions. Celite was used as an adsorptive immobilization carrier and functioned as a coagulant aid because of its high adsorption capacity due to the presence of large amounts of micropore space. For cell flocculation/immobilization, a 150 mg/L chitosan solution and 5 g celite beads were mixed and then shaken for 30 min on a rotary shaker at 180 rpm to ensure thorough mixing and allowed to stand for 2 h at 4 °C to complete the sedimentation and adsorptive immobilization process. The amount of 2-HBP formed by chitosan-celite-immobilized *P. delafieldii* R-8 cells was not stoichiometric, which was probably due to intermediate metabolism or adsorption onto the support. The cells could also completely desulfurize 0.5 mM DBT within 2 h and 1 mM DBT within 4 h. High desulfurization activity was also observed with a conversion of 93% in 5 mM DBT model oil (n-dodecane). Compared with the unmodified free cells, the flocculated R-8 cells with chitosan alone exhibited similar time courses and desulfurizing activities during the BDS reaction. The results show that chitosan-celite-immobilized cells were successfully reused for BDS over six batch cycles, retaining at least 85% specific desulfurization activity. By comparison, the same amount of free cells could only be used once and the recycled free cells retained only 15% activity.

Dinamarca *et al.* (2014a) studied the BDS of DBT (0.0623 M) and gas oil (4.7 g S/L) in a bioreactor packed with a catalytic bed of silica (specific surface area of 80 m^2/g) containing immobilized *Rhodococcus rhodochrous* (ATCC 53968). For immobilization, the cells were circulated through the packed bed in a downward direction at 10 mL/min for 72 h. The cell numbers adsorbed were in the range of 74–95 OD_{600}. To remove the non-adsorbed bacteria on the support, saline solution was circulated through the catalytic bed. A glass bioreactor of 125 ml total volume (2 cm diameter, 40 cm length) was constructed. The bioreactor was continuously sparged with air at 70 mL/min in an upward direction. The feed flow rate of sulfured substrates was provided by a peristaltic pump. The temperature inside the bioreactor was maintained at 20 °C by a thermostatically-controlled water bath. The bioreactor was filled with silica particles to evaluate three packed bed heights (5, 10, and 15 cm) for each of two particle sizes (0.71–1.18 mm or 3.35–5.6 mm). DBT or gas oil was fed continuously into the packed bed bioreactor in a downward direction at 0.45 or 0.9 mL/min. The desulfurized substrate was collected after a 2 h trickling operation. The BDS of DBT was dependent on the length of the bed. The BDS efficiency at a flow rate of 0.45 mL/min and bed length of 15 cm was 57 %, whereas lengths of 10 and 5 cm gave 41 and 16 %, respectively, at the same flow rate. With a constant flow, it is clear that a greater length of the bed results in a higher retention time, allowing a sufficient contact time for expression of the absorbed bacterial metabolism. Similar to the BDS of DBT, sulfur removal from gas oil was greater using a larger bed length. However, BDS conversion values were lower for gas oil than for DBT in all the studied systems. It is possible that the presence of organic molecules in gas oil affects the bacterial metabolism and, consequently, the BDS process (Dinamarca *et al.*, 2014a). Dinamarca *et al.* (2014a) reported that the addition of biosurfactant or Tween 80 improves the BDS efficiency of DBT and gas oil in free and immobilized cell systems of *R. rhodochrous* IGTS8 using adsorption on Si, Al, and Sep as an immobilizing technique. The BDS capacity in an immobilized cell system was much higher than a free cell system and BDS with *R. rhodochrous*/Sep, in the presence of biosurfactant, expressed the highest BDS capacity and attributed this to the adsorption efficiency of the cells on the supports where IGTS8 expressed the highest adsorption on Sep (8.9×10^8 cells/g) and more bio-availability of sulfur compounds to the cells is due to the formation and stabilization of S-compound micelles due to the presence of biosurfactants in catalytic media. The mechanism of immobilized cell system action with the studied surfactants was increased through mobility of the solid particles in the reaction medium, increasing the bio-availability of the S-compounds to the cells by reducing the interfacial tension of the aqueous medium (due to the presence of Tween 80) and/or increasing the solubilization of the S-compounds by formation of micelles (i.e. biosurfactant).

Desulfurization of DBT was improved when particles of silica in the range of 3.35–5.6 mm were used to absorb cells. Because the BDS of DBT with *Rhodococcus* is a process that depends on the aeration variable (Del Olmo *et al.*, 2005), the results suggest that higher particle size improves aeration of the bed by offering a greater inter-particle volume that can be occupied by both liquid and air. The effect of a large particle size dominates over the higher specific area for cell immobilization offered by a smaller particle size and shows the importance of the meso- and macro- porosity of the catalytic bed in the configuration of packed bed bioreactors for BDS. The results showed that immobilized *R. rhodochrous* cells maintained BDS activity after three cycles, retaining at least 84% of the initial desulfurization conversion.

Dinamarca *et al.* (2014b) studied the effect of natural and synthetic surfactants on BDS using metabolically active and immobilized bacterial cells. Metabolically active cells of the desulfurizing strain *Rodococcus rhodochrous* IGTS8 were immobilized on Silica (Si), Alumina (Al), and Sepiolite (Sep) by adsorption. For the immobilization process, the bacteria cell numbers were adjusted by measuring turbidity at 600 nm (OD_{600}) in a range from 6.33×10^9 to 1.9×10^{10}. The bacterial cells were placed in contact with 1.0 g of respective support in 25 mL flasks at 30 °C in a rotary shaker at 200 rpm for 24 h. The number of immobilized cells was measured by the turbidity loss of the solution at 600 nm (OD_{600}). The biocatalysts were designated as bacterial cells/solid system, where the bacterial cells were *R. rhodochrous* and the solid system was inorganic supports. The metabolic activity of viable adsorbed cells was determined using the 2,3,5-triphenyl-2H-tetrazolium chloride method (TTC) modified to detect metabolic activity of bacterial cells absorbed or present in solids (Dinamarca *et al.*, 2007). *R. rhodochrous* bacterial cells were adsorbed on each support (ranging from 4.68×10^9 to 8.0×10^9). The BDSM was incubated with 0.01% TTC in the dark for 4 h and then diluted with ethanol to dissolve the insoluble formazan that formed. The amount of TTC formazan produced was determined by measuring absorbance at 490 nm (A_{490}). The reaction mixture of the BDS process contained 10 mL immobilized bacterial cell (1.62×10^{10} to 2.14×10^{10}) on 1 g of each support (Al, Sep, and Si), and Tween 80 or biosurfactant in a range of 0.1–0.6% w/v and 1 mL of DBT (0.0623 M in hexadecane) or gas oil (4700 mg/L of sulfur). The reaction was carried out at 30 °C in a rotary shaker at 200 rpm for 24 h. The desulfurization activity was determined in the presence of biological and synthetic surfactants. The results indicate that adding surfactants to the catalytic system formed by non-immobilized or immobilized cells increases desulfurization of DBT and gas oil. The addition of biosurfactant to non-immobilized cells significantly increased activity to a maximum of 32.6×10^{-13} g_{sulfur} / DCW/ h at a surfactant concentration of 4000 g/L, corresponding to an increase of 54% and a similar trend was observed for Tween 80. There was a significant increase in activity by adding biosurfactant to the immobilized cells with a maximum activity level of 0.75×10^{-13} g_{sulfur} /DCW/ h, representing an increase of 241% in comparison to the non-immobilized cells at a surfactant concentration of 3000 g/L. In the case of Tween 80, there was an observed a lower activity than biosurfactant, with a maximum value of 0.67×10^{-13} g_{sulfur} /DCW/ h, corresponding to an increase of 206% compared to non-immobilized cells. The immobilization by adsorption of *R. Rhodochrous* on the inorganic supports shows an increase in activity compared to non-immobilized cells. BDS activity was evaluated using ICS and N-ICS systems with and without synthetic and biological surfactants. The results indicate that the use of surface-active molecules (synthetic or biological) with immobilized cells improved BDS activity in comparison to free cells. Dinamarca *et al.* (2014b) found a direct relationship between activity and the adsorption capacity of bacterial cells. *R. Rhodochrous* on the supports (*R. Rhodochrous*/Sep: 8.9×10^8 cell/g, *R. Rhodochrous*/Si: 8.3×10^8 cell/g and *R. Rhodochrous*/Al: 8.6×10^8 cell/g) had a similar adsorption capacity to that of the *P. stutzeri* system (Dinamarca *et al.*, 2010), which indicates that the same immobilization mechanism was performed by the bacterial cells, that is electrostatic and non-electrostatic interactions (Yang *et al.*, 2007). In this context, the DBT high desulfurization efficiency observed in the *R. Rhodochrous*/Sep system can be explained by the greater interaction between bacterial cells and the surface groups of Sep. Since the biosurfactant increases BDS activity of non-immobilized systems through the formation of micelles, the effectiveness of Tween 80 in immobilized systems could be related to the greater mobility of the solid particles in the reaction medium, which increases the contact between bacterial cells and DBT molecules. Tween 80 has low inhibitory effect when bacteria are adsorbed on a solid surface.

Carvajal *et al.* (2017) evaluated the environmentally-friendly removal of sulfur-containing organic molecules adsorbed on inorganic materials through the use of BDS. DBT (100 mg/L) and 4,6-dimethyldibenzothiophene (4,6-DMDBT) (100 mg/L) were used as sulfur-containing organic molecules. Silica (Si) (specific area of 80 m^2/g), alumina (Al) (specific area of 200 m^2/g), and sepiolite (Sep) (specific area of 300 m^2/g) were used as inorganic supports. The adsorption of DBT and 4,6-DMDBT was affected by the properties of the supports, such as particle size and the presence of surface acidic groups. The highest adsorption of both sulfur containing organic molecules used particle sizes of 0.43–0.063 mm. The highest percentage removal was with sepiolite (80 % for DBT and 56 % for 4,6-DMDBT) and silica (71% for DBT and 37 % for 4,6- DMDBT). This is attributed to the close interaction between these supports and the bacteria.

10.6.1.2 Covalent Binding

Covalent binding is a reversible immobilization based on covalent bond formation between activated inorganic carriers and cells in the presence of a binding (cross linking) agent. Covalent methods of immobilization are mainly used for enzyme immobilization, but it is rarely applied in whole cell immobilization because the toxicity of the coupling agents often results in loss of cell viability or enzyme activity (Groboillot *et al.*, 1994).

10.6.1.3 Encapsulation

The encapsulation process is based on the entrapment of the cell in a polymeric matrix. The main advantage of this technique is that it allows the transport of low molecular weight (LMW) compounds through the permeable matrix (Guisan, 2006). Therefore, the properties of the gel matrix and the conditions used for encapsulation should be compatible with the cells to be immobilized (Cao, 2006). Among the immobilization methods applicable to BDS, the most common and effective methods are entrapment or encapsulation of cells within a gel or in a solid support (Kuhtreiber *et al.*, 1999).

This method can be achieved by forming a spherical semi-permeable membrane around desulfurizing bacterial cells to be immobilized, denoting the core of the system (Kaila, 2002) in which the inner matrix is protected by means of the outer membrane. Liquid form of the biocatalyst is the core material and the polymeric wall is the outer membrane (Hirech *et al.*, 2003). The ratio of size of pore membrane to size of core material is a significant factor in this phenomenon. This limited availability to the microcapsule inside is one of the main advantages of microencapsulation due to protection of the biocatalyst from the extreme conditions. As in most immobilization methods, it prevents biocatalyst leakage, increasing the efficiency of the process as a result (Park and Chang, 2000).

10.6.1.4 Entrapment

The entrapment method is an irreversible immobilization that is based on capturing particles or cells within a support matrix or inside a hollow fiber. In this technique, a protective barrier is created around the immobilized microbes that prevents cell leakage from the polymers into the surrounding medium while allowing mass transfer of nutrients and metabolites. Entrapment is mostly applied in whole cell immobilization. The advantage of the entrapment of cell immobilization method is that it is fast, cheap, and mild conditions are required for the reaction process. The main disadvantages of this technique are cost of immobilization, injury of support material during usage diffusion limitations, deactivation during immobilization, and low loading capacity as biocatalysts (Trelles and Rivero, 2013).

The high immobilization efficiency of the cells onto the immobilization material and the high affinity between the hydrophobic immobilization material and the substrates caused excellent desulfurization. Increasing availability of the substrates for the cells and a better interaction between the substrates and the immobilized cells, synergistically, results in developing the desulfurization rate (Wilson and Bradley, 1996). Immobilized particle-size to the support material pore-size ratio is probably the most important parameter. When the pores are too big, the material leaks which also decreases loading (Verma *et al.*, 2006).

Various types of support have been used, such as agar, chitosan, alginate, celite, carrageenan, cellulose, and its derivatives, collagen, gelatin, epoxy resin, photo cross-linkable resins, polyester, polystyrene, polyurethane, and acrylic polymers (Lopez *et al.*, 1997). Among them, alginate and Polyvinyl alcohol (PVA) have been widely used most successfully for immobilizing a wide range of BDSM. These matrices have porous structure and, thus, the PASH compounds and various metabolic products can easily diffuse into the matrix.

10.6.1.4.1 Calcium Alginate

Alginic acid is a hetero-polymer of L-guluronic acid and D-mannuronic acid, as shown in Figure 10.8. The alginates are extracted mostly from seaweed like marine algae and become available as water-soluble sodium salts after processing. Depending on the proportions of these two acids, alginate tends to give a different internal pore size to the gel beads (Tampion and Tampion, 1987). Alginate gel is formed by rapid crosslinking of alginate and polyvalent metal ions such as calcium (Ca^{2+}), barium (Ba^{2+}), and strontium (Sr^{2+}). The strength of the complex increases along the series Ca^{2+}, Sr^{2+}, and Ba^{2+} and this affects the physio-chemical properties of the resulting gel beads. Alginates are the polymers of choice in most systems of immobilization because they are easy to handle, non-toxic to humans, the environment, and the entrapped microorganisms, legally safe for human use, available in large quantities, and inexpensive.

Calcium alginate is currently one of the most widely used entrapment carriers for immobilization of whole cells in the field of BDS for its advantages of biocompatibility, cheapness, and simplicity (Houng *et al.*, 1994; Leon *et al.*, 1998). However, the entrapment technique often leads to a decrease of biocatalytic activity, mostly caused by diffusional limitations and steric hindrance (Klibanov, 1983; Gunther and Helmut, 1993; Guobin *et al.*, 2005). The existence of diffusional limitations reduces the catalytic efficiency of immobilized biocatalysts and, therefore, should be minimized.

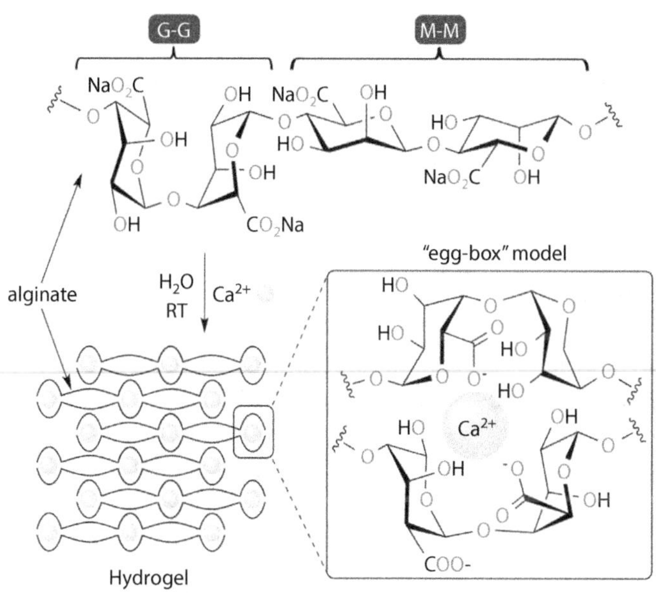

Figure 10.8 Chemical Structures of Calcium Alginate Hydrogel.

The first application in BDS using the entrapment technique for immobilization of free cells is reported by Naito *et al.* (2001), where the BDS process uses immobilized *Rhodococcus erythropolis* KA2-5-1 in n-tetradecane containing 0.54 mM DBT as a model oil (n-tetradecane/immobilized cell biphasic system). The cells were immobilized by entrapment technique with calcium alginate, agar, photo-crosslink-able resin prepolymers (ENT-4000 and ENTP-4000, products of Kansai Paint Company, Osaka, Japan) (Fukui *et al.* 1987), and urethane prepolymers (PU-3 and PU-6) (Fukui *et al.*, 1987). The BDS system consisted of immobilized cells and model oil, but not water and this system offered the advantages of easy separation and possible reactivation of biocatalyst, compared with a system using cell suspension.

The preparation of immobilized BDSM was done as follows:

Calcium alginate gel: 600 mg of sodium alginate (SA) was mixed with 20 mL deionized water and then heated at 100 °C to dissolve the SA completely. After cooling to room temperature, 4 mL of cell suspension (124 mg of dry cells) were added to the solution and mixed well. This mixture was dropped into 0.1 M calcium chloride solution to obtain beads of immobilized cells (about 4 mm diameter).

Agar gel: 600 mg of Agar was dissolved in 200 mM potassium phosphate buffer (pH 7.0, 20 mL) at 100 °C. The solution was cooled to 50 °C and then 4 mL of cell suspension (124 mg of dry cells) were added to the solution and the mixture was rapidly cooled to room temperature. The gel formed was cut into small pieces (5×5×5 mm).

Urethane prepolymers PU-3 and PU-6: 2 g of PU-3 or PU-6 was melted at 120 °C and then cooled to room temperature. Then, 4 mL of cell suspension (124 mg of dry cells) were added to the prepolymer and stirred fully before gelling. The mixture was kept at 4 °C for complete gelation and the gel was cut into small pieces (5×5×5 mm).

Hydrophilic and hydrophobic photo-crosslinkable resin prepolymer (ENT-4000 and ENTP-4000): 2 g of ENT-4000 or ENTP-4000 was mixed with 20 mg of a photosensitizer, benzoin ethyl ether, and 1 mL of 200 mM potassium phosphate buffer (pH 7.0). The mixture was melted at 60 °C and cooled to room temperature. Then, 4 mL of cell suspension (corresponding to 124 mg of dry cells) were added to the molten mixture with stirring. The mixture was gelled by illumination with near UV light (365 nm) for 5 min and the gel sheet formed (about 0.5 mm thick) was cut into small, flat, square pieces (5×5 mm).

Cells immobilized in PU-3, ENTP-4000, and ENT-4000, which are more hydrophobic materials than PU-6, agar, and calcium alginate, exhibited good activities in the model oil. To examine the leakage of cells from each support, the turbidity of n-tetradecane at 660 nm was measured at 68.5 h. Some leakage of cells was observed from calcium alginate gel and agar gel and natural and hydrophilic polysaccharides and these supports gave low activities. These materials were also lacking in mechanical strength. In contrast, the synthetic resins PU-3, PU-6, ENT-4000, and ENTP-4000 showed different results from each other. PU-3-immobilized cells showed a high activity, but massive leakage of cells from the support and PU-6-immobilized cells gave opposite results. However, cells immobilized in ENT-4000 or ENTP-4000 showed high activities and there was almost no leakage of cells from these supports. ENTP-4000 adsorbed 2-HBP (which inhibits desulfurization) more than ENT-4000. Thus, ENTP-4000 expressed more undesirable adsorption characteristics than ENT-4000, so that the ENT-4000-immobilized cells performed the highest DBT desulfurization activity in the model oil system without leakage of cells from the support. Furthermore, ENT-4000-immobilized cells could catalyze BDS repeatedly in this system for more than 900 h with reactivation and the recovery of both the biocatalyst and the desulfurized model oil was easy.

There are a few important factors affecting the rate of desulfurization in an immobilized cells system, such as the concentration of alginate (Zhang *et al.*, 2010), the size of the beads, the presence of surfactants (Li *et al.*, 2008a), and the application of nano γ-Al_2O_3 particles in the system (Zhang *et al.*, 2011). Evidently, maximum BDS efficiency can be achieved by setting these parameters at their optimized values.

Hou *et al.* (2005) reported that the immobilized *Pseudomonas stutzeri* UP-1 cells entrapped in calcium alginate beads can be used to desulfurize DBT efficiently in a model oil and in the aqueous system. The model oil was a mixture of n-dodecane and DBT at 2.7 mM. The optimized operational conditions for immobilization were a temperature of 4 °C, concentration of SA at 3% (w/v), and the ratio of SA (mL) to cells (g) was 20. The BDS efficiency of DBT using immobilized cells reached 65% in the aqueous system and 74% in the model oil system, respectively. The results showed that SA was an appropriate material for immobilization and the stability and life time of immobilized cells (reach 600 h by reactivation) were much better than those of non- immobilized cells, but the maximum desulfurization rate of immobilized cells was lower than that of non-immobilized cells, so it was concluded that the most significant feature of immobilized cells is their long-term lifetime and their ability to be reused for six cycles of reaction.

Li *et al.* (2008c) studied the immobilization of *Pseudomonas delafieldii* R-8 in calcium alginate beads in order to improve BDS activity in oil/water (O/W) biphasic systems. For cell immobilization in calcium alginate beads, 4% w/v of sodium alginate was mixed well with equal volume of cell suspension (5 g DCW/L). Certain amounts of surfactant, such as Tween 20, Tween 80, Span 80, etc., were added, respectively, into the mixture. The target concentration of these surfactants was set at 0.5% w/v. The gas jet extrusion technique, which was modified from Seifert and Phillips (1997), was employed to produce smaller alginate beads (2–5 mm) by extruding the alginate as drops into a calcium salt solution for gelation. The bead formation procedure with a novel instrument is illustrated in Figure 10.9; the resultant slurry was extruded through a cone-shaped needle into a stirred 0.1 M $CaCl_2$ gelling solution. The slurry was intruded as discrete droplets to form calcium-alginate beads with normal sizes (2.5 and 4 mm in

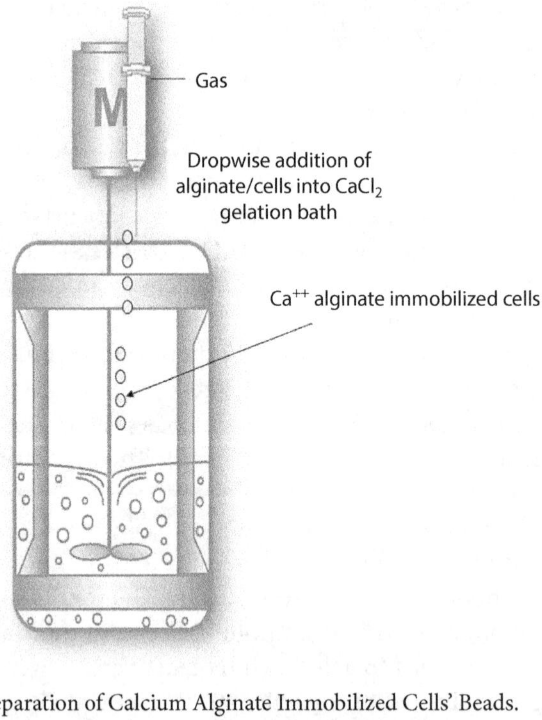

Figure 10.9 Preparation of Calcium Alginate Immobilized Cells' Beads.

diameter). To prepare smaller beads than 2 mm diameter, nitrogen gas was introduced around the tip of the needle to blow off the droplets. By adjusting the gas flow rate to 0.5 L/min, the size of the beads was controlled at 1.5 mm in diameter. The beads were left in a calcium chloride solution for about 2 h for stabilization. Then, the beads were washed with deionized water to remove the residual calcium ions and kept in saline at about 4 °C. Before being used, immobilized cells were activated for 1 day in modified basal salt mediums (MBSM) supplemented with 0.1 mM/L DBT. To avoid the dissolution of alginate- immobilized cells in the BSM, MBSM was used as the aqueous phase for the activation of the immobilized cells. The phosphate components of MBSM were changed to 0.24 g KH_2PO_4 and 1.20 g $Na_2HPO_4.12H_2O$. Other components were the same as those in BSM. The BDS reaction solution contained alginate-immobilized cells (6 g beads, containing 0.15 g of dry cells), 5 mL of model oil (n- dodecane, containing 1 mM/L DBT), and 10 mL of MBSM medium. The results demonstrated that the desulfurization of DBT by alginate-immobilized cells was affected by the size of the immobilized beads. The specific desulfurization rate of 1.5 mm diameter beads was 1.4-fold higher than that of 4.0 mm. Klein *et al.* (1983) reported that a small micro-bead size is important for minimizing the mass-transfer resistance problem normally associated with immobilized cell culture. Cassidy *et al.* (1996) reported that to minimize these effects it is essential to minimize the diffusional distance through a reduction in bead size. Some nonionic surfactants can significantly increase the activity of immobilized cells. Song *et al.* (2005) have employed Tween 20 (polyoxyethylene sorbitan monolaurate) to improve the permeability of the entrapment–encapsulation hybrid membrane. According to Song *et al.* (2005), the immobilized beads without Tween 20 would rupture because of the formation of CO_2 and N_2 as a result of respiration and denitrification. Moreover, the use of Span 80 (sorbitan monooleate) in cell immobilization is beneficial, increasing permeability and mass transfer and, consequently, resulting in higher biocatalytic activity. During cell immobilization in Li *et al.*'s (2008c) process, the addition of non-toxic and non-ionic surfactants, including Span 20 (sorbitan monolaurate), Span 80, Tween 20, and Tween 80 (polyoxyethylene sorbitan monostearate), greatly enhanced the desulfurization rate compared to the control. Span 80 showed the highest effect on desulfurization activity. Within 24 h, the desulfurization rate with the addition of 0.5% Span 80 reported to be 1.8 fold higher than that without Span 80.

Huang *et al.* (2012) found out that using immobilized cells made it easy to recover desulfurized oil and to use the biocatalyst repeatedly for long periods with reactivation. In that work, *Pseudomonas delafieldii* R-8 cells were immobilized in calcium alginate beads and used for BDS. For immobilization, resting cells (20 g DCW/L) were immobilized with conventional methods by extruding the alginate (4% W/V) as drops into a $CaCl_2$ solution (0.1 M/L) for gelation as shown in Figure 10.9. It was found that Th and DBT can be simultaneously metabolized by immobilized R-8 cells. The initial sulfur content in the model oil (n-octane) was 300 mg/kg. After 10 h of treatment, the Th concentration was reduced by 40%, while DBT was reduced by 25%. The utilization rate of Th was faster than that of DBT. Moreover, the oil/water ratio of alginate immobilized cells was studied to reduce the water volume in desulfurization systems. Immobilized cells can be applied for long time recycling for BDS. The oil/water phase ratio was increased from 1/2 to 5/1, which greatly improved the efficiency of the bioreactor. It was demonstrated successfully that immobilized cells could be reused for BDS over fifteen batch cycles, retaining at least 75% specific desulfurization activity. Total desulfurization time of the immobilized beads reaches 450 h. The encouraging results of that work indicated that immobilization of R-8 cells met the demands of an industrial biocatalyst.

Peng and Wen (2010) reported the immobilization of resting cells of *Gordonia* sp. WQ1A, a DBT-desulfurizing strain by calcium alginate, where 4 % (w/v) sodium alginate was added to the resting cell suspension (4.4 g DCW/L) a ratio of 1:1 (v/v) at room temperature and cooled to 4 °C. The alginate/cell mixture was aseptically extruded by an injector through a needle into a stirred solution of sterile 4% calcium chloride (4 °C). The height of the needle and rate of stirring of the calcium chloride solution, about 20 cm and 150 rpm, respectively, were adjusted to obtain uniform spherical gel beads. A batch of DBT BDS experiments using immobilized cells and n-dodecane as the oil phase were conducted in a fermenter under varying operating conditions, such as initial DBT concentration, bead loading, and oil phase volume fraction. When the initial DBT concentrations were 0.5, 1, and 5 mM/L, the DBT concentration dropped almost to zero after 40, 60, and 100 h, respectively. It was concluded from the results of this study that (1) the rate-limiting step in the oil-water-immobilization system is not mass transfer resistance, but bioconversion and (2) oxygen concentration is not an important factor affecting DBT BDS in the immobilized system.

Small beads have a high surface volume ratio which might reduce the mass transfer limitation. Previous studies (Lee and Heo, 2000; Li *et al.*, 2008c) have indicated that reduction in the alginate bead diameter increases the BDS efficiency of the beads. Thus, in a study performed by Derikvand and Etemadifar (2014), the encapsulation of *Rhodococcus erythropolis* R1 in calcium alginate beads enhanced the bioconversion of dibenzothiophene (DBT) to 2-hydroxy biphenyl (2-HBP) as the final product. The 2% (w/v) sodium alginate beads, with 1.5 mm size, were found to be optimum for bead stability and efficient 2-HBP production. The viability of encapsulated cells decreased by 12% after 20 h of desulfurization, compared to free cells. Moreover, adding non-ionic surfactants markedly enhanced the rate of BDS because of increasing mass transfer of DBT to the gel matrix. In addition, Span 80 was more effective than Tween 80. Furthermore, the nano γ-Al_2O_3 particles were found to increase the BDS rate by up to two folds greater than that of the control beads.

Despite the numerous advantages of immobilized cell systems, distinct disadvantages are also apparent. The limited space available for cell growth and proliferation severely restricts their utility for growth-associated products (Lazar, 1991). The most serious disadvantage is the potential limitation in the supply of nutrients and oxygen to the cells. Owing to the presence of the rigid support material, cells receive nutrients by diffusive mechanisms alone; convective flow within the support is negligible. As the cells proliferate, the total nutrient consumption rate increases which sometimes leads to a demand that cannot be met by the prevailing and decaying diffusion rates. These diffusional limitations often lead to uneven distributions of viable cells within the support material (Radovich, 1985). When these limitations become extreme, the amount of nutrients available for cells that are far from the nutrient source is so reduced that the cellular metabolic activity is confined to the vicinity of the interface between the growth media and the cell-containing support (Lazar, 1991). For example, Karel and Robertson (1989) observed that the active cell growth of *Pseudomonas putida* in microporous hollow fibers occurred only within a distance of <25 μm from the oxygen supply. A thorough characterization of these transport limitations is necessary for the proper design of efficient immobilized cell systems.

Aspergillus flavus, previously isolated from an oil contaminated environment, was pregrown on an increasing concentration of sodium metabisulphite $Na_2S_2O_5$ as a surrogate of the sulfur substances in crude-oil at room temperature for 72 h. Then, the harvested spores were immobilized by Ca-alginate. Then, the effects of different incubation periods and temperature and concentration of immobilized spores on batch-BDS of crude oil were observed. Maximum desulfurization efficiency of 94.7% occurred after 3 d at room temperature and 50 g of immobilized spores in of 100 mL crude oil. However, at 35 °C, the BDS efficiency recorded 63.2% and 55.3% at 40°C, but sharply decreased to 10.5% at 45°C. Thus, *A. flavus* can tolerate up to 40 °C and the immobilized microorganism maintained its catalytic properties without any cells being washed out (Adegunlola *et al.*, 2012).

Verma *et al.* (2016) reported that the immobilization of resting cells of *Bacillus* sp. E1 using sodium alginate enhanced the BDS of 2 mM DBT and recorded DBT-BDS of 57.9 ± 1.1% and 2-HBP production of 0.846 ± 0.037 mM with immobilized cells, as compared to the free resting cells that recorded 53.9 ± 2.2% and 0.493 ± 0.084 mM DBT-BDS % and 2-HBP production, respectively, within 24 h of incubation. However, initially, the performance of free resting cells was found to be better than immobilized cells up to 12 h and this was attributed to the low distribution of substrate and products across the matrix (Naito *et al.*, 2001).

10.6.1.4.2 Polyvinyl Alcohol

Polyvinyl alcohol (PVA) is a biodegradable synthetic polymer, odorless and tasteless, translucent, white or cream colored granular powder. PVA was first prepared by Haehnel and Herrmann in 1924 by hydrolyzing polyvinyl acetate in ethanol with potassium hydroxide. The first scientific reports on PVA were published in 1927. PVA is the largest volume synthetic resin produced in the world (Marten, 1927).

PVA is a linear polymer which has a relatively simple chemical structure with a pendant hydroxyl group. PVA is commercially produced through the polymerization of vinyl acetate to polyvinyl acetate (PVAc), followed by partial or complete hydrolysis of PVAc to PVA, removing acetate groups (Figure 10.10) (Hassan and Peppas, 2000).

The polymerization step controls the length (called degree of polymerization) of the vinyl acetate molecular chains and, therefore, the molecular weight. The PVA membrane strength, aqueous solution viscosity, and other properties vary greatly depending on the degree of polymerization.

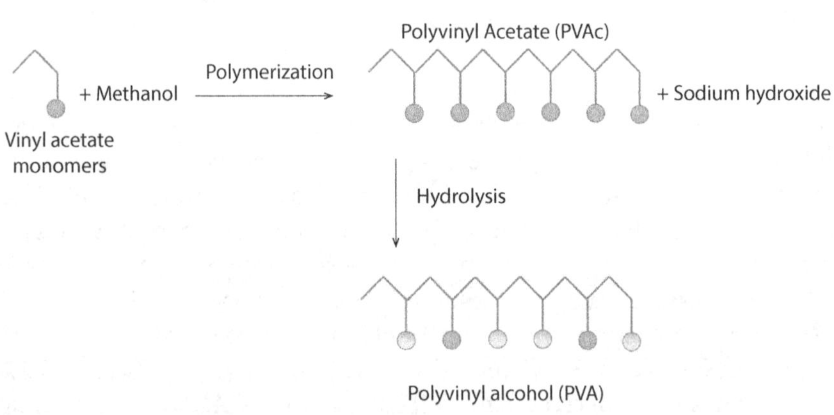

Figure 10.10 The Conversion of PVAc to PVA. In dark gray are the acetate groups, whereas in gray are the hydroxyl groups.

It is also possible to produce PVA with different characteristics by copolymerizing other monomers during this process (Hassan and Peppas, 2000).

The key of the entrapment method is the selection of materials and methods for preparation. The use of polyvinyl alcohol (PVA) for cell entrapment has been investigated (Lozinsky and Plieva 1998). PVA is biologically compatible, readily available, low cost, and nontoxic. It may not have adverse effects upon cells and is becoming one of the most promising materials for entrapment. Ochiai *et al.* (1981) and Hashimoto and Furukawa (1987) used the PVA–boric acid technique to immobilize cells in PVA. Although easy to be operated, the saturated boric acid solution used to crosslink the PVA is highly acidic (pH is approximately 4) and could cause difficulty in maintaining cell viability. In addition, the sphering speed of the PVA is very slow and results in agglomeration of the PVA beads. Moreover, the spherical bead rigidity is weak. A successful attempt to quicken the sphering speed of the PVA, preventing the agglomeration problem of the PVA beads and increasing the beads rigidity, was adding a small amount of CA to the saturated boric acid (Hulst *et al.*, 1985; Wu and Wisecarver, 1992). However, the CA may be easily damaged by a salt of phosphoric acid and tends to be eroded or dissolved when used in reaction systems (Fernandes *et al.*, 2002). In order to overcome the difficulties, a freezing–thawing technique was used to prepare PVA beads (Giuliano *et al.*, 2003). The beads have excellent water resistance, elasticity, and flexibility and show high safety to living bodies because no chemical agent is used for gel formation. In addition, its high water content and porous structure make it ideal for use as a carrier suitable for the culture and propagation of immobilized microorganisms (Freeman and Lilly, 1998).

The process of immobilization could also minimize the toxicity of gasoline to the biocatalyst. Li *et al.* (2005b) reported a comparative study for the immobilization of *Mycobacterium goodie* X7B cells by entrapment with calcium alginate, carrageenan, agar, PVA, polyacrylamide, and gelatin-glutarldehyde. It was found that calcium alginate immobilized cells had the highest desulfurization activity. When immobilized *M. goodii* X7B cells were incubated at 40 °C with Dushanzi straight-run gasoline (DSRG227) in a reaction mixture containing 10% (v/v) oil for 24 h, the sulfur content of the gasoline decreased from 275 to 121 ppm, after a 24 h reaction, and the desulfurization reaction was repeated by exchanging the used immobilized cells for fresh ones. The sulfur content further decreased to 54 ppm, corresponding to a reduction of 81%. Considering that benzothiophenes predominate in gasoline and the toxicity of gasoline to bacterial cells, the BDS of gasoline is considered to be a significant achievement. With this excellent efficiency, strain X7B was considered a good potential candidate for industrial applications for the BDS of gasoline.

In another comparative study, Gunam *et al.* (2013) evaluated the desulfurization ability of *Sphingomonas subarctica* T7b using resting and immobilized cells with DBT and commercial light gas oil (LGO) as the substrates. The resting cells of *S. subarctica* T7b desulfurized 239.2 mg of the initial 250 mg of DBT/L (1.36 mM) within 24 h at 27 °C, while 127.5 mg of 2-hydroxybiphenyl (2-HBP)/L (0.75 mM) was formed representing a 55% conversion of the DBT. The DBT desulfurization activity was significantly affected by the aqueous-to-oil phase ratio. As such, 250 mg of DBT/L was completely desulfurized within 24 h when the water to-oil ratio (W/O) was 4. However, the DBT degraded slowly when the phase ratio (W/O) was 2 and 1. LGO with a total sulfur content of 280 mg/L was desulfurized to 152 mg/l after 24 h of reaction. Immobilized cells were prepared using 4 kinds of support material: polyvinyl alcohol (PVA), poly (ethylene-co-vinyl) alcohol, sodium alginate, and sodium silicate. Only PVA and sodium alginate showed a good performance and were easy to prepare. The cells immobilized in 4% (w/v) sodium alginate, 12% (w/v) PVA, and a mixture of 10% (w/v) PVA and 2% (w/v) sodium alginate exhibited better activity in the model oil and showed a slightly higher activity than the resting cells. In addition, comparative bio-transformations using resting cells or free cells provided similar results, demonstrating that the immobilization procedure did not affect the enzyme activity. The cell-immobilized-PVA beads expressed higher BDS efficiency than free resting cells over a wide range of pH, from 4.5–8.5, with a maximum BDS of 80.9 mg/kg DCW/h at a of pH7 and remained stable at higher pH values up to 8.5. The BDS efficiency of the cell-immobilized-PVA beads at an elevated temperature (37 °C) was higher than that of free resting cells. When the cells were immobilized with sodium alginate, 2.5 mL of n-tetradecane containing 100 mg of DBT/L was completely degraded after 72 h of operation. The cells entrapped in PVA or a mixture of PVA and sodium alginate also exhibited a similar BDS activity, indicating that all the support materials were appropriate for cell immobilization and that cell immobilization is an essential technique for the degradation of sulfur in oil. Cells immobilized by entrapment with PVA exhibited a high DBT desulfurization activity (≈ 72%), including repeated use for more than 8 batch cycles without loss of BDS activity. The desulfurization activity of the cells immobilized in a mixture of PVA and sodium alginate was initially slightly higher than that of the cells immobilized in PVA until the third batch cycle. Thereafter, the desulfurization activity of the cells immobilized in a mixture of PVA and sodium alginate decreased slightly, whereas the desulfurization activity of the cells immobilized in PVA remained more stable, suggesting that the PVA remained stable for a longer period than the mixture of PVA and Na-alginate. Therefore, 12% (w/v) PVA was selected as the appropriate support material for immobilization. The stability of the immobilized cells was better than that of the resting cells at different initial pHs, higher temperatures, and for DBT-BDS in successive desulfurization cycles. The immobilized cells were also easily separated from the oil and water phases, giving this method great potential for oil BDS.

Fatahi and Sadeghi (2017) used *R. erythropolis* supported on polyvinyl alcohol as a biocatalyst for the BDS of thiophenes in gasoline. The biocatalyst desulfurized thiophene to 2-butene-1-ol. The average diameter of supported bacterial cells on PVA was about 68.71 nm. The morphology of the supported cells was maintained for 60 d on the contrary to the free cells which were completely destructed. At optimum conditions, 0.1 g of biocatalyst at thiophene initial concentration of 80 ppm, 97, and 41% of thiophene was degraded after 20 h of contact time at pH and 30 °C. Upon the application of the biocatalyst on gasoline as a real oil feed with an initial S-concentration of 1500 ppm under the optimum operation conditions, BT, Th, 2-MTh, 3-MTh, and 2-ETh recorded desulfurization percentages of 38.89, 26.67, 26.67, 21.03, and 23.62%. respectively. Furthermore, the immobilized cells by adsorption on PVA, i.e. the prepared biocatalyst, were used for three successive cycles without losing its activity.

10.6.1.4.3 Polyurethane Foam

Due to intrinsic slow kinetics of the enzymatic reaction and mass transfer limitation, the overall rate of BDS is much slower than that of oxidative desulfurization (Gray *et al.*, 2003; Bhasarkar *et al.*, 2013, 2014). Amin (2011) reported the optimum growth of *R. erythropolis* ATCC 53968 in minimal salt mediums supplemented with 25 g/L ethanol and 25 g/L glucose, where up to 3.3 g/L of biomass was obtained after 36 h. Then, for enhancement of bacterial cells into polyurethane foams, a vertical rotating immobilized cell reactor (VRICR) was inoculated with the *R. erythropolis*. It was first fed with MSM containing 25 g/L of each of glucose and ethanol at a high feed rate of 320 mL/L. The bioreactor was rotated at 0, 15, and 30 and rpm. The aeration rate was manipulated to ensure an oxygen transfer rate (OTR) of 90.0 mMO_2/L/h. It was found that the biomass concentration in polyurethane foams was highly affected by the rotation speed. A steady state with complete utilization of both carbon sources was achieved only at a rotation speed of 15 rpm. An immobilized cell concentration of 69.8 g/L was maintained. Under such conditions, growth of immobilized *R. erythropolis* cells was apparently balanced by cell leakage into the surrounding medium with a mass ratio of free cells to immobilized cells ($M_{f/I}$) of 21.4%. However, under static conditions, there was a recorded steady increase in immobilized cells, recording 80.2 g/L within 5 d. However, incomplete utilization of carbon source, mainly glucose, was observed. This was attributed to the cell over-growth into polyurethane foams and limitation in mass transfer for both oxygen and nutrients. The $M_{f/I}$ recorded only 2.8% which reflected an extremely slow growth rate of immobilized biomass under static conditions, but at a higher rotation speed (30 rpm), cell leakage from polyurethane foams into the surrounding medium was higher than cell growth in polyurethane foams. Thus, the concentration of immobilized cells was dramatically decreased after reaching a maximum value recording only 12 g/L with a $M_{f/I}$ of 36.9% and a steady state was never reached. Ethanol was completely utilized whereas glucose utilization was only 20%. A similar growth pattern was observed with the bacterium *Z. mobilis* (Amin and Doelle, 1989) and *Corynebacterium glutamicum* (Amin, 1994), but with relatively low immobilized cell concentrations of 25 and 46 g/L, respectively. Thus, *R. erythropolis* appears to attach more successfully to polyurethane foams.

Recently, Bhasarkar *et al.* (2015) reported the application of ultrasonic enhancement sulfur-specific BDS through the physical and chemical effects of ultrasonic enhancement by the generation of intense microturbulence in the medium and the generation of highly reactive oxidizing radicals ($O^{\bullet}$ and $^{\bullet}OH$) from the thermal dissociation of transient collapse that would oxidize DBT to DBTO and $DBTO_2$, respectively. The utilized *R. rhodochrous* MTCC 3552 was immobilized throughout the cross-linking of cells on commercial polyurethane foam; sonication using an ultrasound bath was used in the immobilization process to release all proteins inside the immobilized cells. The BDS reaction mixture was in a liquid-liquid heterogeneous phase, containing organic toluene, containing a different concentration of DBT, aqueous (water), and surfactant β-cyclodextrin (β-CD) as a phase transfer agent to enhance interphase DBT transportation. Bhasarkar *et al.* (2015) attempted to establish a physical mechanism of sonication induced enhancement of BDS by concurrent analysis of Haldane kinetics model parameters and simulations of cavitation bubble dynamics. Both physical and chemical effects of sonication contributed to the enhancement. The cells were immobilized on commercial grade polyurethane foam (cut into cubical pieces of dimensions 1 x 1 x 1 cm). These pieces were autoclaved prior to the addition to the broth at the onset of the log phase of the culture. The polyurethane pieces (or immobilization support) were incubated for 24 h in nutrient broth at 30 °C with shaking at 180 rpm. After incubation, the polyurethane pieces were separated from the broth. The immobilization support was shaken with a phosphate buffer (50 mM, pH 7.0) for 10 min. It was then washed twice with a phosphate buffer and dried for 24 h at room temperature. The immobilized cells on the support were incubated with a 0.1% (1 mM) glutaraldehyde solution for 1 h for cross-linking, followed by washing with a phosphate buffer and, subsequently, drying at room temperature for 24 h. The extent of immobilization of the cells on the support was assessed by sonication of the support (with phosphate buffer as the medium) that would release all proteins inside immobilized cells. Lowry assay was carried out on the extract to determine the presence of proteins. A positive test confirmed the presence of proteins which were released from the cells immobilized on the surface of the support. The optimum amount of immobilized support in the reaction mixture (for both mechanical agitation and sonication) was determined by varying the number of cubes of polyurethane foam added to the reaction mixture. The results revealed that the highest DBT oxidation was achieved for 2 cubes for both mechanical shaking and a sonication system. Lesser cubes reduce the microbial cell population in the reaction medium. Large numbers of cubes in the reaction mixture hinder convection as well as emulsification in the reaction mixture. The β-CD acts as a phase transfer agent for DBT, as it forms complexes with DBT that have high solubility in the aqueous phase. The highest DBT oxidation was observed for 20 ppm concentration of β-CD for both free and immobilized cells. For free cells, sonication of the broth results in a marked increase (57.6%) in extent DBT reduction, as compared to mechanical shaking. Addition of β-CD during mechanical shaking results in a relatively lesser increment of 27.2% in DBT reduction. A combination of sonication and β-CD addition gives an increment of 77% in DBT reduction. For immobilized cells, as compared to mechanical shaking, the increments obtained with sonication and β-CD are relatively smaller than free cells. The percentage increments in DBT oxidation obtained by sonication of the reaction mixture, addition of β-CD, and sonication with β-CD addition are 51%, 23.8%, and 59.8%, respectively. Thus, the immobilized cell system expressed higher BDS than that of the free cell system. For a free cell system, sonication showed better BDS efficiency than mechanical shaking and, additionally, β-CD with sonication improved the BDS capacity of free cells by approximately 77%, but the increment of BDS efficiency in an immobilized cell system by sonication and/or β-CD are relatively smaller than that which occurred in a free cell system; the recoded increment of the addition of β-CD in mechanical shaking was 23.8%, while the application of sonication alone expressed a BDS increment with approximately 51% relative to mechanical shaking. However, application of sonication and β-CD recorded a BDS increment of

59.8%. The intense micro-convection generated by the cavitation bubbles would result in good emulsification of the reaction mixture with a high interfacial area for interphase transport of DBT from the organic phase to the aqueous phase. In addition to the produced micro-turbulence, which enhances the diffusion of DBT and the 2-hydroxy biphenyl sulfinate across the cell membrane, it also enhanced the back diffusion of 2-HBP into the organic phase. Thus, this would have avoided its accumulation in the aqueous phase that, consequently, would decrease its inhibition effect on the cells. The micro-convection enhanced the movement of the microbial cells in the reaction mixture, which would have caused an increase in their inter-collisions and/or their collisions with the walls of the reaction reactor. Thus, kinetic energy gained by the cells through such rapid motion would help in accelerating the enzymatic reaction inside the cells. Sonication produces active $O^{\bullet}$ and $^{\bullet}OH$ that oxidize DBT to its sulfoxide and sulfone, which would diffuse into the cell and undergo further transformation to 2-HBP. Thus, the rate of BDS would be enhanced and the reduction of substrate inhibition would occur. Thus, the strong micro-turbulence generated by sonication enhanced the interfacial mass transport and substrate/product diffusion across the cell. Consequently, the oxidation of DBT to DBT-sulfoxide and DBT-sulfone (intermediates 4S-pathway) by radicals generated by cavitation assisted in faster consumption of DBT by microbial cells. These effects led to an enhancement in reaction velocity and enzyme-substrate affinity, as well as reduction in substrate inhibition.

Agarwal *et al.* (2016) attempted to enhance the kinetics of BDS by identifying links between the physics of ultrasound/cavitation and the chemistry of BDS. The model reaction system is comprised of DBT as a model sulfur compound, toluene as model fuel, and *Rhodococcus rhodochrous* cells (in free and immobilized form) as a microbial culture. DBT was dissolved in toluene at a concentration of 100 ppm and was added to the aqueous fermentation medium to form a biphasic mixture for incubation. The aqueous (fermentation medium) and oil phase (toluene) ratio was maintained at 80:20. For immobilization and cross linking of cells, the PU foam was cut into cubical pieces (dimensions: 1cm x 1cm x 1 cm) for the immobilization of microbial cells. Prior to immobilization, PU foam pieces were autoclaved. Culture broth, at the onset of log phase, was added to PU foam pieces for immobilization. The PU foam pieces were separated from the broth after incubation and were shaken with a phosphate buffer (50 mM, pH 7.0) for 10 min, followed by drying for 24 h at room temperature. Cross-linking of the immobilized cells on the support was carried out by incubation with a 0.1% (1 mM) glutaraldehyde solution for 1 h. This was followed by washing of PU foam pieces with a phosphate buffer and subsequent drying at room temperature for 24 h. The immobilization of microbial cells on PU foam pieces was assessed by sonication of the support (in phosphate buffer as the medium). Sonication released all proteins inside the immobilized cells on the surface of the support. Ultrasound-assisted BDS was carried out in an ultrasound bath (volume: 2 L; Frequency: 35 kHz; Power: 35 W). The medium for ultrasound propagation in a bath was water. In all experiments, the flask was placed in the center of the bath. This same position was maintained in all experiments due to significant spatial variation of ultrasound intensity in the bath. The pressure amplitude of the ultrasound waves generated in the bath was determined as 150 kPa. The temperature of the BDS reaction mixture was monitored throughout the experiment. Water in the ultrasound bath was replaced at regular intervals to maintain the temperature at 30 °C. The total volume of the reaction mixture in all experiments was 100 mL. The sonication of the reaction mixture was carried out with a duty cycle of 20% (i.e. 2 min (on) and 8 min (off) in every 10 min of treatment). 1 mL of reaction mixture was withdrawn at regular intervals to monitor the concentrations profiles of DBT (original sulfur compound) and 2-HBP (the final product of 4S metabolic pathway). The total time of treatment for experiments with mechanical agitation was 28 h while the ultrasound or sonication-assisted experiments were conducted for 6 h. In the case of free cells with mechanical agitation, the microbial cells are distributed in the entire reaction volume, and, thus, the local concentration of the cells was rather low. The concentration of DBT was also low (100 or 200 ppm) due to the probability of interaction between substrate and microbial cells being low, while in the case of immobilization of the cells on porous support, like PU foam, causes confinement of the cells due to their local concentration and probability of interaction with substrate molecule increases. This phenomenon is essentially analogous to increase in degradation of the pollutant in dilute solutions with the addition of an adsorbent. The adsorbent causes a localized concentration of the pollutant molecules due to their probability of interaction with cavitation-generated radicals increasing. For the same reasons, for immobilized cells, the overall order of the kinetics of DBT oxidation increased with increasing substrate concentration. The net manifestation of these observable facts has a higher absolute DBT oxidation for larger initial DBT concentrations. The absolute amount of DBT oxidation in the case of free cells with mechanical agitation remains the same (37 ppm) for both initial concentrations of 100 and 200 ppm. This essentially means that DBT oxidation with free cells under mechanical agitation follows zero-order kinetics. On the other hand, absolute DBT oxidation in immobilized cells shows a more than 2 fold increase (from 50 to 128.7 ppm) as the initial DBT concentration increases from

100 to 200 ppm. Such discrepancy is not seen in absolute DBT oxidation in experimental categories with ultrasound irradiation or sonication. For the free cells, the absolute DBT oxidation shows a 90% increase (from 57.4 to 103.7 ppm) with doubling of an initial DBT concentration from 100 to 200 ppm. For immobilized cells, the corresponding rise in DBT oxidation is approximately 2 fold (or 100%) from 70.4 to 138.9 ppm for initial DBT concentrations of 100 and 200 ppm, respectively. These results clearly indicate that strong micro-convection induced by sonication in reaction mediums enhances the probability of substrate-microbial cell interactions. This is essentially a result of fine emulsification between aqueous and organic phases in reaction mixtures due to sonication. Hence, absolute DBT oxidation, even for free cells, increases with initial DBT concentration indicating higher order reaction kinetics. Higher DBT oxidation in immobilized cells is again attributed to higher localized microbial concentration of microbial cells on PU foam immobilization supports.

10.7 Application of Nano-Technology in BDS Process

In recent years, numerous types of metallic particles of nanometer and micrometer dimensions and composites of these materials have become key components in different areas like catalysis (Shan *et al.*, 2003, 2005a,b; De Windt *et al.*, 2005; Ansari *et al.*, 2009), environmental remediation (Wang and Zhang, 1997; Sharma *et al.*, 2008; Shin and Cha, 2008; Hennebel *et al.*, 2009; Lukhele *et al.*, 2010; Xiu *et al.*, 2010), gene therapy (Ewert *et al.*, 2006; Andreu *et al.*, 2008; Patnaik *et al.*, 2010), drug delivery (Kubo *et al.*, 2000; Yellen *et al.*, 2005; Rawat *et al.*, 2006; Akin *et al.*, 2007), imaging (Lee *et al.*, 2006, 2007; Anceno *et al.*, 2010;), biomarkers (Xie *et al.*, 2010; Ranzoni *et al.*, 2012;), sensors (Li *et al.*, 2009; Fan *et al.*, 2010; Xu *et al.*, 2012), and energy (Che *et al.*, 1998; Ryu *et al.*, 2010).

For BDS, cells are needed to be harvested from the culture medium and several separation schemes have been evaluated, including settling tanks (Schilling *et al.*, 2002), hydrocyclones (Yu *et al.*, 1998), and centrifuges (Monticello, 2000), but these procedures are time-consuming and costly. Magnetic separation technology provides a quick, easy, and convenient alternative over traditional methods in biological systems (Haukanes and Kvam, 1993). Super-paramagnetic nanoparticles are increasingly used to achieve affinity separation of high value cells and biomolecules (Molday *et al.*, 1977). Magnetic supports for cell immobilization offer several advantages such the ease of magnetic collection. The magnetic supports present further options in continuous reactor systems when used in a magnetically stabilized, fluidized bed. In addition, the mass transfer resistance can be reduced by spinning of magnetic beads under a revolving magnetic field (Sada *et al.*, 1981). Among the vast number of unique properties of MNPs (e.g. high surface area and magnetism), additional interest relays on their high chemical activity (Bregar, 2004; He and Zhao, 2005; Moores and Goettmann, 2006; Lee and Sedlak, 2008; Chirita and Grozescu, 2009; Akbarzadeh *et al.*, 2012;). Control of nanoparticle size, shape, and dispersity are the key to selective and enhanced activity of MNPs (He and Zhao, 2007; Tao *et al.*, 2008). With recent advances in nanotechnology and microbiology and due to efforts by different research groups to synthesize, chemically and biologically, various types of metal and metal oxide nanoparticles, novel materials have emerged (Yee *et al.*, 1999; Ma *et al.*, 2004; Chen and Gao, 2007; Shahverdi *et al.*, 2007; Bar *et al.*, 2009; Jain *et al.*, 2010). Easy synthesis in a wide range of sizes and shapes, facile surface conjugation to a variety of chemical and biomolecular ligands, and biocompatibility and high chemical and photostability is possible (Evanoff and Chumanov, 2004; Chen and Gao, 2007; Jain *et al.*, 2008; Koebel *et al.*, 2008). Examples of metal nanoparticles include zerovalent iron, iron oxide, silver, gold, copper, cobalt, platinum, manganese, and nickel nanoparticles.

In all these applications magnetite is typically used in the form of particles. There are three common magnetite iron oxides, FeO, Fe_2O_3, and Fe_3O_4, from these compounds, magnetic Fe_3O_4 is one of the common iron oxides which has many important technological applications.

In order to combine the advantages of immobilization, i.e. ease of separation and microbial longevity with those of free diffusion or good mass transport, another approach is possible, namely to decorate the bacterial cells with magnetic nanoparticles (MNPs). After completion of the reaction, the bacterial cells can be separated from the products using a magnetic field. This is a much milder and more cost-effective process than centrifugation and allows the bacteria to be reused many times and magnetic separation is compatible with any automated platform that can be equipped with a magnet (Ansari, 2008; Li *et al.*, 2009). In recent years, magnetic nanoparticles (MNPs) have been widely used in the field of BDS because of their large surface-to-volume ratios, superparamagnetic properties, and low toxicities (Bardania *et al.*, 2013). MNPs are used for separation, characterization, and purification and are a well-established alternative to centrifugal separation of biological solutions (Ansari *et al.*, 2009; Bardania *et al.*, 2013). At the end of experiments, bacteria coated with MNPs are also easily separated from liquid suspensions by use of an external magnetic field, as shown in Figure 10.11 (Ansari *et al.*, 2009).

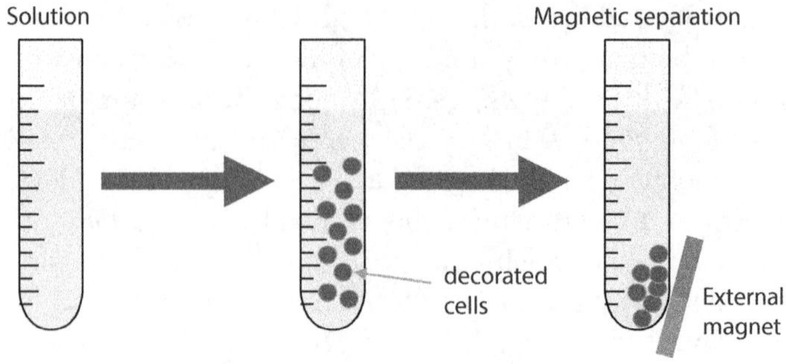

Figure 10.11 Magnetic Separation of Decorated BDSM.

Several expressions, such as decoration and labeling, are used to describe the preparation of magnetically modified cells. The term "decorated cells" usually describes the cells covered with large numbers of nanoparticles and, in some cases, also in several layers; the particles usually do not cross the cell membrane or cell wall. On the contrary, target cells can be magnetically labeled just by a single (or only a few) magnetically responsive particle attached to the cell wall or membrane. However, magnetic labeling also represents the situation when magnetic nanoparticles cross the cell membrane/wall or enter the cytoplasm.

The usage of MNPs to coat the cells not only has an advantage over conventional cell immobilization by adsorption, but can also successfully overcome difficulties of conventional cell immobilization, such as mass transfer problems, cell loss, and separation of carrier with adsorbed cells from the reaction mixture at the end of a desulfurization treatment. Cells coated by MNPs appear not to experience a mass transfer problem because it was found that the desulfurizing activities of coated and uncoated cells exhibited the same time course. This is because DBT transfer from the oil to the water and then from the water to the cells limits the rate of DBT metabolism in both cases. The coating layer of nanoparticles does not change the hydrophilicity of the cell surface. Moreover, the coating layer has a negligible effect on mass transfer because the structure of the layer is looser than that of the cell wall. Thus, the coating layer does not interfere with mass transfer of DBT. The coated cells have good stability and can be reused. This new technique has the advantage of magnetic separation. Therefore, this new technique is rather convenient and easy to perform and it may be promising for large-scale industrial applications (Ansari, 2008; Li *et al.*, 2009).

One of the first applications of nanomaterials in the desulfurization processes was the use of magnetic nanoparticles attached to the surface membrane of *Pseudomonas delafieldii* R-8 (Shan *et al.*, 2005a). In this work, Liu *et al.* (2010) shows that cells coated with Fe_3O_4 nanoparticles have the same desulfurization activity as the uncoated cells. The coating process requires the mixing of both NPs and cells in a saline solution, then the cells can be recovered by applying a magnetic field (Figure 10.11). In addition, coated cells can be reused at least five times, in contrast with free cells that do not show activity if they are used again. Even though the nanoparticles do not directly contribute to the desulfurization process, their use facilitates the recovery of the biomass.

Shan *et al.* (2003) immobilized *Pseudomonas delafieldii* in magnetic polyvinyl alcohol (PVA) beads using a hydrophilic magnetic fluid which was prepared by a co-precipitation method. The beads had distinct super-paramagnetic properties and were compared with immobilized cells in non-magnetic PVA beads. The magnetic fluid prepared with the co-precipitation method is black. The addition of N_2 gas during preparation will prevent the oxidation of ferrous ions in the aqueous solution and also control the particle size. Liu *et al.* (2003) reported that the average particle diameter is about 8 nm and had super-paramagnetic properties when the particle diameter was smaller than 30 nm (Lee *et al.*, 1996). The hydrophilic magnetic fluid was prepared by the conventional co-precipitation method (Liu *et al.*, 2003) with some modifications: 23.5 g $FeCl_3{\cdot}6H_2O$ and 8.6 g $FeCl_2.4H_2O$ were dissolved in 600 ml deionized water under N_2 with mechanical stirring at 85 °C and 30 ml 7.1 M NH_4OH was then quickly added. 16 mL of oleic acid was added dropwise into the suspension over 30 min. After several minutes, the magnetic precipitate was isolated from the solvent by magnetic decantation and washed with deionized water several times. The magnetic precipitate was dissolved in 7.1 M NH_4OH to form a hydrophilic magnetic fluid. Magnetic particles prepared by this method have a large number of hydroxyl groups on their surface in contact with the aqueous phase. The (OH) groups on the surface of Fe_3O_4 particles react readily with carboxylic acid head groups of oleic acid molecules and then the excess oleic acid will be adsorbed to the first layer of oleic acid to form a hydrophobic shell. When this magnetic fluid is put into NH_4OH, the outer layer of oleic acid on the Fe_3O_4 surface will be transformed into ammonium oleate and the hydrophilic magnetic fluid is formed. The hydrophilic magnetic fluid can be extracted directly into hydrophilic support liquids such as PVA and sodium alginate. Immobilization of resting cells in magnetic and non-magnetic supports were prepared as follows: BDSM was harvested in the late growth phase by centrifugation at 5500 g for 5 min. The harvested cells were washed twice with saline and dried for 48 h by vacuum freeze-drying. PVA beads for entrapping cells were prepared by cross-linking in a saturated boric acid solution (Amanda *et al.*, 1992). Magnetic fluid (4% v/v), sodium alginate (1% w/v), and polyvinyl alcohol (9% w/v) were dissolved in deionized water and mixed well with the dried cells (5% w/v). Immobilized beads were formed by extruding the mixture through a syringe into a gelling solution of 0.1 M $CaCl_2$ saturated with boric acid and solidified for 24 h. The immobilized beads formed were washed with saline and then freeze dried for 48 h under a vacuum. The cells in non-magnetic supports were immobilized by similar steps without adding magnetic fluid. A small amount of sodium alginate and Ca^{2+} were added into a PVA aqueous solution and then they were added into a saturated boric acid solution to promote the formation of beads, thus preventing conglomeration of the PVA drops in the preparation process (Hulst *et al.*, 1985). The magnetic fluid is mainly composed of Fe_3O_4 particles, which will not only provide the magnetic property of support, but will also improve the mechanical strength of supports. Therefore, the magnetic immobilized supports could be easily separated and recycled by an external magnetic field and the recovered magnetic supports could be re-dispersed by gentle shaking when the external magnetic field was removed. The desulfurizing activity was increased slightly from 8.7 to 9 mM sulfur / kg DCW/h. The main advantages were that the magnetic immobilized cells maintain a high desulfurization activity and remain in good shape after 7 times of repeated use, while the non-magnetic immobilized cells could only be used 5 times. Furthermore, the magnetic immobilized cells could be easily collected or separated magnetically from the BDS reactor.

Zhang *et al.* (2008) reported that the decrease of DBT concentration is not only caused by the adsorption of Fe_3O_4 to DBT, but also partly by BDS. It is reported that adsorption between Fe_3O_4 and DBT is electrostatic, so the adsorption between them is reversible and DBT can be desorbed easily. The desorbed DBT then transfered to cells for BDS. It is faster for cells to obtain DBT by this way, thus 2-HBP production rate is increased.

Shan *et al.* (2005b) coated the microbial cells of *Pseudomonas delafieldii* with magnetic Fe_3O_4 nanoparticles and then immobilized them by external application of a magnetic field. Magnetic nanoparticles were synthesized and modified by the following method: 23.5 g $FeCl_3 \cdot 6H_2O$ and 8.6 g $FeCl_2 \cdot 4H_2O$ was dissolved in 600 ml deionized water under $N_{2,}$ to prevent the oxidation of ferrous ions in the aqueous solution and control the particle size, with mechanical stirring at 800 rpm and 85 °C and then quickly added 30 ml of 7.1 M NH_4OH. Then, 16 ml of oleic acid was added dropwise to the resulting suspension over a period of 30 min. After several minutes, we separated the magnetic precipitate by magnetic decantation and washed it with deionized water several times. The magnetic precipitate was modified with about 4 mL of 7.1 M NH_4OH to form the hydrophilic magnetic nanoparticles. The surface-modified Fe_3O_4 nanoparticles were monodispersed in an aqueous solution and did not precipitate in over 18 months. Using transmission electron microscopy (TEM), the average size of the magnetic particles was found to be in the range from 10 to 15 nm. Microbial cells were coated by mixing fifteen milliliters of a magnetic suspension (15 g Fe_3O_4 nanoparticles per liter of saline water) with 100 mL of a cell suspension (25 g DCW/ liter saline water). The microbial cells were coated by adsorbing the MNPs. The coated cells could be concentrated on the side of the vessel containing the suspension and separated from the suspension medium by decantation. The TEM cross section analysis of the cells showed further that the Fe_3O_4 nanoparticles were, for the most part, strongly absorbed by the surfaces of the cells and coated the cells. The BDS reaction was divided into three portions and each portion contained the equivalent of 0.255 g (dry weight) of cells, when using free cells (1st portion), Fe_3O_4 nanoparticles coated cells (2nd portion), or immobilized cells by Celite (3rd portion). Then, each portion was suspended in a 15 ml phosphate buffer (0.1 M, pH 7.0) and the suspension was mixed with 8 mL of model oil (2.0 mM DBT in n-dodecane). The coated cells had distinct superparamagnetic properties. The magnetization (δs) was 8.39 emu/g. The results showed that the coated cells and the free cells exhibited similar time courses and the same desulfurizing activities during the first 4 h (16.4 mM/ kg (DCW) cells/h), where DBT, at a concentration of 2.0 mM, was completely desulfurized in 6 h. By contrast, the cells immobilized by Celite desulfurized only 12.6 mM/ kg (DCW) cells/h during the first 4 h. The findings for the cells immobilized by Celite were the result of a small loss of cells from the Celite during washing with the saline solution. By comparison, no MNPs-coated cells appeared to have been lost during washing. Thus, they had greater operational stability than cells immobilized on Celite. The coated cells not only had the same desulfurizing activity as free cells, but could also be reused more than five times. Compared to cells immobilized on Celite, the cells coated with Fe_3O_4 nanoparticles had greater desulfurizing activity and operational stability.

BDS usually occurs in organic phases, therefore, the transference of organosulfur molecules to the cell membrane is limited. The cell membrane needs to be not only resistant, but also permeable enough to capture the sulfur-containing molecules. This issue may be overcome with the use of genetically modified bacteria that are stable in oil phases or by using sorbents of organosulfur molecules. For example, when *P. delafieldii* R-8 are assembled with the nanosorbent γ-Al_2O_3, the desulfurization rate of the cells increased at least two times more than that of the cells alone (Guobin, *et al.*, 2005). The nano γ-Al_2O_3 in the cell membrane adsorbs DBT, which increases the transference rate of the sulfur compound from the media to the cell membrane (Figure 10.12). In a two phase procedure, DBT is first adsorbed in a Cu modified zeolite, then degraded by *P. delafieldii* R-8 to 2-hydroxybiphenyl (HBP), and the nanosorbent is then bioregenerated and may be reused (Li, *et al.* 2006).

Guobin *et al.* (2005) combined the freezing–thawing and magnetic separation technique to immobilize cells for BDS. *Pseudomonas delafieldii* R-8, entrapped in magnetic PVA beads, was used repeatedly in the desulfurization process. Hydrophilic magnetic fluid was prepared by a modified co-precipitation method as follows: 23 g $FeCl_3.6H_2O$ and 8.60 g $FeCl_2.4H_2O$ were dissolved in 600 ml deionized water under N_2 with mechanical stirring at 85 °C and 30 ml 7.1 M/L $NH_3.H_2O$ was then quickly added. A quantity of 16 ml of oleic acid was added dropwise into the suspension within 30 min. After several minutes, the magnetic precipitate was isolated from the solvent by magnetic decantation and washed with deionized water several times. The magnetic precipitate then was dissolved in 7.1 M/L $NH_3.H_2O$ to form hydrophilic magnetic fluid. A mixture of *P. delafieldii* cells suspended in a phosphate buffer (0.1 M, pH 7) with an aqueous solution of PVA (15%, w/v) and hydrophilic magnetic fluid was dropped by an injector into liquid nitrogen for quick freezing to form immobilized beads. Then, the beads were thawed gently with a progressively slow increase of the temperature in a vacuum flask. After being thawed, the beads were put into liquid nitrogen again and were then thawed gently. The freezing–thawing process was repeated for 3–5 times and the immobilized cells (magnetic PVA-R8 beads) were prepared. BDS was carried out using immobilized cells in a model oil of n-dodecane containing 0.5 mM /L DBT. The beads have distinct superparamagnetic properties and their saturation magnetization is 8.02 emu/g.

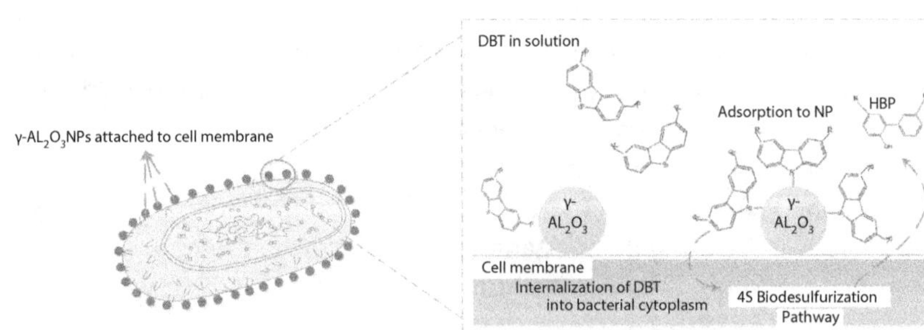

Figure 10.12 Enhancement of BDS Process by Using γ-AL_2O_3 NPs Coated Cells.

The desulfurization rate of the immobilized cells could reach 40.2 mM/kg/h. Desulfurization patterns of DBT in model oil with the immobilized and free cells were represented by the Michaelis–Menten equation. The Michaelis constant for both immobilized and free cells was 1.3 mM/L. The cells immobilized in magnetic PVA beads could be stably stored and repeatedly used 12 times for BDS.

A microbial method has also been used to regenerate desulfurization adsorbents (Li *et al.*, 2006). This occurred throughout a novel integration of a combined physical–biological procedure. During this process, sulfur compounds were removed through adsorption and then biologically removed from the adsorbents in a subsequent stage with a *Pseudomonas delafieldii* R-8 strain in an attempt for regeneration of the adsorbent. To select suitable π-complexation adsorbents for desulfurization, a model oil 15 mmol/L DBT in n-octane was used as a model compound to study the adsorption capacity of these adsorbents where the ratio of the adsorbent to oil was 100 mL/g. Different metal ions, namely Co^{2+}, Ni^{2+}, Ce^{3+}, and Cu^{+}, were used to prepare desulfurization Y zeolite adsorbents using the ion exchange method. The reduction behavior of Cu(II)–Y was studied by using temperature programmed reduction (TPR) with reducing gas (10% H_2 and 90% N_2, v/v). The acidity of the adsorbents was characterized by pyridine adsorption thermogravimetric-differential thermal analysis (TG-DTA) and Fourier transform infrared (FT-IR). The adsorption capacity decreased in the following order: Cu(I)-Y > Co-Y > Ni-Y > H-Y > Ce->Y, which was according to the same sequence as Lewis acidities of the adsorbents. Since the Lewis acid sites provided by transition metal ions and Al^{3+} in the framework of the NaY zeolite acted as adsorption centers for DBT, π-complexation adsorbent Cu(I)–Y was successfully bio-regenerated. Most of the sulfur compounds can be desorbed and removed and heat losses during the bio-regeneration process are markedly reduced. Bioregeneration of the spent adsorbents took place via the interaction between microorganisms and molecules of the adsorbed DBT. One of the important factors for a successful bioregeneration is the porous structure of the adsorbent. As the oil in which DBT was dissolved facilitates contact with biocatalyst, oil, water, and biocatalysts can be fully mixed by agitation and emulsified into microdroplets at the oil/water interface. Consequently, DBT was converted to 2-HBP at the oil/water interface. The adsorption capacity of the regenerated adsorbent is 85% that of the fresh one after being regenerated with *Pseudomonas delafieldii* R-8, washed with n-octane, dried at 100 °C for 24 h, and reduced in H_2 and N_2. After bioregeneration, the adsorbent can be reused. However, the particle size of cells was nearly similar to that of desulfurization adsorbents, which is several microns. Therefore, it was difficult to separate regenerated adsorbents and cells. Super-para-magnetism is an efficient method to separate small particles. Thus, to solve the problem of separation of cells and adsorbents, magnetite nanoparticle modified *P. delafieldii* R-8 cells can be used in the bio-regeneration of adsorbents. Biodesulfurization with *P. delafieldii* R-8 strains coated with magnetite nanoparticles has been reported previously (Shan *et al.*, 2005b), where coated cells with magnetite nanoparticles have been separated and reused several times.

There is also a report (Li *et al.*, 2008c) of the bio-regeneration of desulfurization adsorbents, AgY zeolite, with magnetic cells. Super-paramagnetic magnetite nanoparticles are prepared by the co-precipitation method followed by modification with ammonia oleate. Magnetic *P. delafieldii* R-8 cells can be prepared by mixing the cells with magnetite nanoparticles. The biodesulfurization activity of the magnetic cells is similar to that of free cells. When the magnetic cells were used in the bio-regeneration of desulfurization adsorbents AgY, the concentration of dibenzothiophene and 2-hydroxybiphenyl with free cells is a little higher than that with magnetic cells. The adsorption capacity of the regenerated adsorbent, after being desorbed with magnetic *P. delafieldii* R-8, dried at 100 °C for 24 h, and calcined in the air at 500°C for 4 h, is 93% that of the fresh one. The magnetic cells can be separated from the adsorbent bio-regeneration system after desulfurization with an external magnetic field and, thus, can be reused.

Li *et al.* (2009) developed a simple and effective technique by integrating the advantages of magnetic separation and cell immobilization for the BDS process with *R. erythropolis* LSSE8-1. The Fe_3O_4 NPs were synthesized by co-precipitation followed by modification with ammonium oleate to produce hydrophilic magnetic fluids. The surface-modified NPs were monodispersed and the particle size was about 13 nm with 50.8 emu/g saturation magnetization. After adding the magnetic fluids (30 g Fe_3O_4 NPs/L) to the culture broth, *Rhodococcus erythropolis* LSSE8-1 cells were immobilized by adsorption and then separated with an externally magnetic field. The BDS reaction mixtures contained the magnetic coated cells (0.19 g DCW), 20 mL of BSM, and 5 mL of model oil (2.5 mM/L DBT in n-dodecane). The same amount of free cells served as the control. The maximum amount of cell mass adsorbed was about 530 g (DCW)/g particles to LSSE8-1 cells. Fe_3O_4 NPs were strongly adsorbed on the surfaces of microbial cells and coated the cells because of the large specific surface area and high surface energy of the nanoparticles. The Fe_3O_4 NPs on the cell surface were not washed out by deionized water, ethanol, saline water (0.85 wt.%), or a phosphate buffer (0.1 M, pH 7). Thus, there is little cell loss or decrease in biomass loading when cells are coated with magnetic nanoparticles. This outcome was different from that obtained with cells immobilized by traditional adsorption to a carrier. The coated and free cells exhibited similar time courses and desulfurizing activities during the reaction. The coated LSSE8-1 cells completely degraded 2.5 mM/L DBT in model oil after a 10 h reaction and the desulfurization process was mostly occurring in the first 4 h. The desulfurization rate of the coated LSSE8-1 cells was 14.1 μM DBT/g DCW/ h in the first 4 h. Both the coated cells and free cells had the same desulfurizing activity, suggesting that the coated cells did not experience a mass transfer problem. The magnetic separated/ immobilized cells were successfully reused for BDS over seven batch cycles and they retained at least 80% specific desulfurization activity. By comparison, the uncoated free cells could be used only once and the recycled free cells retained only 15% activity, indicating that it is not economic to reuse free cells. Figure 10.13 shows the postulated adsorption mechanism between the nanoparticles and desulfurizing cells. Firstly, the large specific surface area and the high surface energy of the Fe_3O_4 nanoparticles (i.e. the nano-size effect) make it easy to be strongly adsorbed on the surfaces of microbial cells. Secondly, the hydrophobic interaction between the bacterial cell wall and the hydrophobic tail of oleate-modified Fe_3O_4 NPs may play another important role in cell adsorption. The suspension of oleate-modified magnetite Fe_3O_4 nanoparticles was considered as a bilayer surfactant stabilized aqueous magnetic fluid (Liu *et al.*, 2006; Shen *et al.*, 1999). The iron oxide nanocrystals were first chemically coated with an oleic acid molecule, then the excess oleic was weakly adsorbed on the primary layer through the hydrophobic interaction between the subsequent molecule and the hydrophobic tail of oleate. As we all know, the bacterial cell wall is composed of proteins, carbohydrates, and other substances, for example, peptidoglycan, lipopolysaccharide, mycolic acid, etc. Thus, the

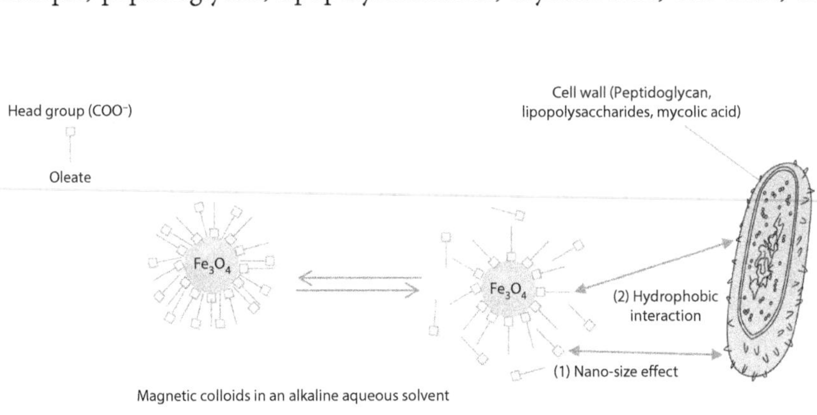

Figure 10.13 Adsorption Mechanism Between Fe_3O_4 NPs and BDSM.

extracellular matrix can form a hydrophobic interaction with the hydrophobic tail of an oleate of NPs.

Magnetic nanoparticles not only allow recovery of bacterial cells, but may also have an important role in cell membrane permeabilization. Ansari *et al.* (2009) clearly demonstrates that Fe_3O_4 nanoparticles actually facilitate transport of DBT from the media to the cytoplasm of BDSM may be due to the self-assembly of the nanoparticles inside the membrane, which may form pore-like structures that increases the surface conductance. In this study, Ansari *et al.* (2009) used a widely studied Gram-positive desulfurizing bacterial strain, *R. erythropolis* IGTS8, in order to investigate the effects of decorating the bacteria with magnetic Fe_3O_4 nanoparticles. The cells were decorated with magnetic nanoparticles as follows: 10 mL of a suspension containing 100 mg/mL Fe_3O_4 nanoparticles per mL of water were mixed with 100 mL of the cell suspension in BSM at a final concentration of 0.5 mM DBT. The ratio of nanoparticle mass to biomass was 1.78 (w/w). It was found that the decorated cells had a 56% higher DBT desulfurization activity in BSM compared to the non-decorated cells. The observation of significantly increased 2-HBP production in the decorated cells suggests that the magnetic nanoparticles might facilitate transport of 2-HBP out of the cells, assuming that it is produced in the cytoplasm. A possible mechanism for enhancement is that the nanoparticles bound to the bacteria make their membranes more permeable. Magnetic nanoparticle-decorated *Rhodococcus* also facilitates its recovery and reuse, hence it offers a number of advantages for industrial applications compared to non-decorated cells.

Also, the size and stability of the nanoparticles in a suspension are important issues that need deep investigation since they could influence the process performance. According to Bardania *et al.* (2013), coating of nanoparticles on the cell membrane of *R. erythropolis* IGTS8 is more effective when smaller nanoparticles of Fe_3O_4 (<5 nm) are used. Additionally, the use of glycine decreases the aggregation tendency of the nanoparticles which may lead to a higher adsorption of DBT by BDSM.

Zhang *et al.* (2011) enhanced BDS activity by assembling nano γ-Al_2O_3 particles on the magnetic immobilized *Rhodococcus erythropolis* LSSE8-1-vgb. The cells can be collected and reused conveniently by an external magnetic field. Firstly, cells were magnetically immobilized by coating with Fe_3O_4 nanoparticles. The optimal ratio of cells to magnetic Fe_3O_4 nano-particles was determined to be 50:1 (g/g). Then, nano γ-Al_2O_3 adsorbents were assembled onto the cells to enhance the desulfurization activity. The nano γ-Al_2O_3 adsorbent had the largest pore volume, as well as specific surface area, and the strongest electrostatic interaction with microbial cells. Both played important roles in enhancing the mass transfer rate in the desulfurization process. Three types of nano γ-Al_2O_3 (A, B, and C), with different characteristics, were used in microbial-adsorptive in-situ integrated desulfurization. The activity of magnetic immobilized cells assembled with nano γ-Al_2O_3 was tested in the desulfurization of model oil (n-octane) 1/2 (oil/water phase ratio) containing 2.0 mM DBT, the highest desulfurization activity obtained when nano γ-Al_2O_3-C was used. DBT in model oil could be completely removed in 9 h. When nano γ-Al_2O_3-A or B was used, the activity was much lower. The total concentration of DBT and its metabolite (2-HBP) in the oil phase kept constant, which meant that all DBT and 2-HBP remained in the oil phase. Adsorbent C had the smallest particle size, largest pore volume, and specific surface area, which can improve the adsorption of DBT in the oil phase most easily. This is one of the reasons that integrated desulfurization activity was enhanced strongest. On the other hand, the interaction between cells (with negative charge) and adsorbent also strongly affects desulfurization activity. Electronic force is an important portion of the interaction force between cells and adsorbent. The three adsorbents carried a positive charge and the charge amount of C is much greater than that of A or B, resulting in the tightest interaction force. The desulfurization rate was raised by nearly 20% when the amount ratio of magnetic particles to nano γ-Al_2O_3 was 1:5 (g/g). The activity of magnetic immobilized cells assembled with adsorbents kept nearly 20% higher than that of magnetic cells and decreased less than 10% throughout three recycles.

Bardania *et al.* (2013) demonstrated the synthesis of magnetite nanoparticles with two average sizes of 5.35 nm (F1 NPs) and 8.74 nm (F2 NPs) (when glycine was added during the synthesis of NPs and when it was absent from the reaction mixture, respectively) and their application to the separation of two desulfurizing bacterial strains, *R. erythropolis* FMF and *R. erythropolis* IGTS8. Magnetite NPs were synthesized via the reverse co-precipitation method. Glycine was added after the synthesis of both F1 and F2 NPs to stabilize the NP dispersion. It was found that the glycine-coated magnetite nanoparticles (F1 NPs) have a strong tendency for immobilization on the surface of bacterial cells and can be efficiently used for the separation of biomass from the suspension using an external magnetic field. Both F1 and F2 MNPs showed superparamagnetic properties with saturation magnetization values of 45.84 and 75.85 emu/g, respectively. BDS efficiency of the F1 MNPs coated with *R. erythropolis* FMF and *R. erythropolis* IGTS8 cells were 70 and 73%, respectively, and were not significantly different from those for free bacterial cells (67 and 69%, respectively). These results indicate that magnetite nanoparticles cannot affect the desulfurization activity of cells examined in this work. Isolation of bacterial cells from the suspension using a magnet and evaluation of desulfurization activity of separated cells showed that Fe_3O_4 nanoparticles can provide a high efficiency recovery of bacterial cells from a suspension with reused MNPs–coated bacterial cells being able to maintain their desulfurization activity efficiently.

Karimi *et al.* (2017) enhanced DBT desulfurization by decorating *Rhodococcus erythropolis* IGTS8 using magnetic Fe_3O_4 nanoparticles in a bioreactor. Magnetite NPs were synthesized via the chemical co-precipitation method. The field emission scanning electron microscope and transmission electron microscopy images showed that the size of the NP is 7–8 nm. Desulfurization activity of the bacteria cells was evaluated in a stirring bioreactor (working volume 5L) containing 2 L BSM with 0.5 mM DBT and 2 mL/L glycerol. At first, the BSM was prepared and a 20 mL aliquot containing 2 mg/mL of Fe_3O_4 NP was added to the BSM in the fermenter. After one day, the fermenters were inoculated with the bacterial cells in the mid-exponential growth phase. The decorated cells showed a longer lag phase (up to 148 h) than the non-coated bacteria (96 h). Production of 2-HBP by decorated cells was significantly different than that by free cells. Gibbs' assay results showed that production of 2-HBP by decorated cells was 0.210 mM at 148 h, while 2-HBP production by non-decorated cells was 0.182 mM at 96 h. Therefore, the decorated cells had a 15.8% higher desulfurization capacity compared to the non-decorated cells. These experimental results can be related to the favorable production of HBP, as the NP action avoids the accumulation of HBP in the cytoplasm and periplasmic space. This study suggests that the addition of NP into a BDS medium does not overcome all of the BDS process limitations, but can avoid HBP inhibition effects.

The observation of significantly increased HBP production in decorated cells suggests that magnetic nanoparticles might facilitate the transport of HBP out of the cells, assuming that they are produced in the cytoplasm. A possible mechanism for enhancement is that the nanoparticles bound to the bacteria make their membranes more permeable. In order to investigate this hypothesis, the possible effect of nanoparticles on membrane permeability was studied by Rahpeyma *et al.* (2017). This study examined the effect of Fe_3O_4, ZnO, and CuO nanoparticles as free factors on BDS. The bacterial strains used as biological catalysts were *Rhodococcus erythropolis* IGTS8 and *Pseudomonas aeroginusa* PTSOX4 to have a comparison between a natural bacteria and a genetically manufactured one in desulfurization activity. Assimilation of ZnO NPs and two tested bacteria showed considerably higher desulfurization of 94% and 58% in the case of using *P. aeroginusa* PTSOX4 and *R. erythropolis* IGTS8, respectively. These results show significant improvement in the BDS process of both bacteria in the absence of NPs, in which 68% and 48% were obtained for PTSOX4 and IGTS8, respectively. The observation of significantly increased HBP production in the presence of ZnO nanoparticles may be attributed to permeabilization of cell membranes, which facilitates the mass transferring of DBT and HBP into or out of bacterial cells. It is also well documented that the first and rate-limiting step in the oxidative BDS of DBT is the transferring of DBT from the oil into the cell (Setti *et al.*, 1999). The system of using Fe_3O_4 nanoparticles alongside bacteria stands as the second by 70% for PTSOX4 and 49% for IGTS8. The obtained results show that Fe_3O_4 NPs have almost no effect on BDS yield compared to the results obtained from the process in the absence of NPs. This is in accordance with the results previously reported by Shan *et al.* (2005a) in which Fe3O4 nanoparticles were utilized as an immobilization matrix for bacterial cells. Their investigation showed that magnetic nanoparticles had no direct effect on the BDS process. The obtained results may also be attributed to the effect of nanoparticles to alter microbial metabolism, which leads to significant changes in the released substances in the surrounding environment. PTSOX4 and IGTS8 also showed decreased BDS activity of DBT in the presence of CuO NPs in comparison with other nanoparticles. Furthermore, the utilization of *Pseudomonas aeruginosa* PTSOX4 as the biocatalyst causes more sulfur removal rather than *Rhodococcus erythropolis* IGTS8. Diminished BDS activities of PTSOX4 and IGTS8 in the presence of CuO may be attributed to its greater particle size, as well as the possible toxicity of these nanoparticles.

Research regarding the synthesis of NPs has been underway for many years, however, its application has been limited due to its easy aggregation during and after synthesis. Thus, the preparation of biocompatible and well-dispersed NPs is key for these applications, especially for biological applications. Many methods have been proposed to solve the problem of agglomeration, however, so far few of them consider the biocompatibility of dispersants. Usually, chemical surfactants were used to control particle sizes.

Kafayati *et al.* (2013) investigated the effect of the MNPs Fe_3O_4 on the growth rate of the genetically engineered *Pseudomonas aeruginosa* PTSOX4 cells (Barzegar *et al.*, 2007) in different media with different MNPs concentrations. The MNPs (10–50 nm) were synthesized using a co-precipitation method. Shan *et al.* (2005b) applied this method to synthesize magnetite nanoparticles and coat bacterial cells and they reported that replacement of air by N_2 has the advantage of preventing the oxidation of ferrous iron during preparation of nanoparticles in the aqueous solution and also has the ability to control size. The surface of nanoparticles has to be modified with a suitable surfactant to use magnetite nanoparticles to coat bacteria. Shan *et al.* (2005b) used oleic acid as a surfactant to functionalize and immobilize magnetite nanoparticles on the surface of bacteria, however, Ansari *et al.* (2009) used glycine to modify the surface of nanoparticles. Fe atoms of magnetite nanoparticles have a strong tendency react to COOH groups, so that the Fe atom of the nanoparticle reacts with COOH of oleic acid or glycine, therefore oleic acid forms a bilayer shell on the surface of nanoparticles (Shi-Yong *et al.*, 2006) and glycine produces an amine layer on the surface of magnetite nanoparticles (Ansari *et al.*, 2009) which leads to the dispersion of magnetic nanoparticles iron oxide in a water phase with hydrophilic characteristics. On the other hand, it is reported that this functionalized magnetite nanoparticle is absorbed on the surface of bacteria simultaneously (Shan *et al.*, 2005b; Ansari *et al.*, 2009). The absorbance of glycine-modified magnetite nanoparticles on the negative-surface of bacterial cells is due to the positive charge of nanoparticles. Previous reports have shown that functionalized magnetite nanoparticles with different surfactants show low toxicity on living eukaryote cells in comparison to free nanoparticles (Mahmoudi *et al.*, 2009). The obtained results from MIC and MBC analysis showed that nanoparticles have a low toxicity on PTSOX4 BDSM. According to this analysis, Pseudomonas bacterial cells could not grow in the presence of more than 5000 ppm MNPs concentration. This may be due to the surface saturating of bacterial cells with magnetite nanoparticles and increasing the contact of nanoparticle to the cell membrane. Thus, cell membrane is injured by them.

Zhang *et al.* (2007) studied the effect of surface modification of γ-Al_2O_3 NPs with gum Arabic (GA) and its applications in adsorption and the BDS process. GA is a natural gum which has good rheological properties and emulsion stability and, thus, has been used as an emulsifier to prevent oil droplet aggregation and coalescence (Islam *et al.*, 1997). The model diesel oil for desulfurization was n-octane containing 2 mM of DBT. The results showed that γ-Al_2O_3 NPs dispersed well in aqueous solutions after GA modification. The optimal modification condition is that 0.5 g of γ-Al_2O_3 to be modified in 100 ml of solution with 1.0 wt% GA. The adsorptive desulfurization capacity of γ-Al_2O_3 NPs was increased from 0.56 mM S/g (Al_2O_3) to 0.81 mM S/g (Al_2O_3) after GA modification. Compared with unmodified γ-Al_2O_3 NPs, the BDS rate by adsorbing GA modified γ-Al2O3 NPs onto the surfaces of *Pseudomonas delafieldii* R-8 cells was increased from 17.8 mM/kg/h to 25.7 mM/kg/h. Therefore, GA modification of γ-Al_2O_3 NPs could be a promising method to obtain biocompatible and well-dispersed γ-Al_2O_3 NPs. The results showed that γ-Al_2O_3 nanoparticles (γ-Al_2O_3 NPs) dispersed well in aqueous solutions after modification with gum Arabic and the adsorptive desulfurization (ADS) capacity of modified γ-Al_2O_3 NPs increased by 1.12 fold than that of the unmodified one. The good dispersion might be attributed to the chemical binding between the negatively charged groups of gum Arabic and the positive sites on the surface of γ-Al_2O_3 NPs, giving rise to the non-DLVO (Derjaguin-Landau-Verwey-Overbeek) surface steric force, which prevents the agglomeration of NPs in the aquoues solution (Leong *et al.*, 2001). The better the dispersion of the adsorbents is, the more the specific surface area the adsorbents would have and, thus, the more the adsorbents can adsorb sulfur, taking into consideration that excess gum Arabic (> 1 wt.%) would take up from the adsorption site on the γ-Al_2O_3 NPs that would decrease the ADS capacity. The adsorption of γ-Al_2O_3 NPs onto *Pseudomonas delafieldii* R-8 increased the BDS efficiency from 14.5 mmol/kg/h to 17.8 mmol/kg/h. Compared with the unmodified γ-Al_2O_3 NPs, the BDS rate by adsorbing the gum Arabic-modified γ-Al_2O_3 NPs onto the surfaces of R-8 cells increased to 25.7 mmol/kg/h (i.e. 1.44 fold), which may be due to the improvement in the dispersion and biocompatibility of γ-alumina nanoparticles after modification with gum Arabic, due to the stronger affinity of gum Arabic-modified γ-Al_2O_3 NPs to the cells than the unmodified ones and its lower toxicity to the cells.

In another study, Zhang *et al.* (2008) applied different kinds of widely used adsorbents (alumina, molecular sieves, and active carbon) throughout in-situ coupling technology of adsorptive desulfurization and biodesulfurization. The procedure was carried out by assembling nano-adsorbents onto surfaces of *Pseudomonas delafieldii* R-8 cells. The data showed that Na-Y molecular sieves restrain the activity of R-8 cells and active carbon cannot desorb the substrate, dibenzothiophene. Thus, they are not applicable to in-situ coupling desulfurization technology. The γ-alumina can adsorb dibenzothiophene from the oil phase quickly and then desorb it and transfer it to R-8 cells for biodegradation, thus, increasing the desulfurization rate. It was also found that nano-sized γ-alumina increases the desulfurization rate more than regular-sized γ-alumina. Therefore, nano-γ-alumina is regarded as a better adsorbent for this in-situ coupling desulfurization technology. However, the eco-friendly MCM-41 mesoporous silica, with a significant number of pores, ordered that porosity and large specific surface area would enhance the adsorption of molecules to microbial cells, relative to zeolites.

Nasab *et al.* (2015) improved the desulfurization performance of *Rhodococcus erythropolis* IGTS8 by assembling spherical mesoporous silica nano-sorbents on the surface of the bacterial cells. MCM-41 mesoporous silica is synthesized based on a self-assembly method, using a quaternary ammonium template and cetyl trimethyl ammonium bromide (CTAB) for the adsorption of sulfur compounds from model oil (1.0 mM/L DBT in dodecane). Then, the adsorption capability of MCM-41 assembled on the surface of bacterium *Rhodococcus erythropolis* IGTS8 is examined regarding the improvement of the BDS process of the oil compound based on the measurement of the DBT consumption rate and 2-HBP production. Study of the model oil desulfurization by the cells assembled with MCM-41 nano-sorbent showed a further improvement in the DBT reduction rate in comparison with the free cells. The results of the investigations showed that the maximum specific desulfurization activity, in terms of DBT consumption rate and 2-HBP production, were 0.34 μM/g DCW/min and 0.126 μM 2-HBP/g DCW/min, which indicated an increase of 19% and 16% compared to the highest specific desulfurization activity of free cells, respectively.

Etemadifar *et al.* (2014) prepared the MNPs by two different methods and their characteristics were determined via transmission electron microscopy (TEM) and X-ray diffraction (XRD). In the first method, 25 mL of ferrous chloride (0.2M) was mixed with 100 mL of ferric chloride solution (0.1 M) under N_2 gas and mechanical stirring and then 3 ml of HCl (2 M) solution was added slowly to make the solution slightly acidic. Then, 1 g of glycine was added and afterward, 11 mL 5 M NaOH solution was added dropwise into the mixture to increase its pH to over 10 to provide an alkaline environment for Fe_3O_4 to precipitate. Next, an additional 3 g of glycine was added and the mixture was stirred for 15 min and then sonicated for 30 min. Finally, 5 mL acetone solution was added and agitated. The Fe_3O_4 NPs were separated with a magnetic field and the supernatant discarded by decantation. The precipitate was washed several times and re-suspended in deionized water. In the second method, 6.76 g of ferric chloride and 2.73 g of ferrous chloride were dissolved in 100 mL deionized water under nitrogen gas with mechanical stirring. The solution temperature was set at 85 °C. Then, 16 mL (25% w) $NH_3.H_2O$ was added and afterward 4 mL of oleic acid was dripped into the suspension by a syringe. The reaction was kept at 85-90 °C for 30 min. The Fe_3O_4 precipitates were separated using a magnetic decantation and washed several times with deionized water. Hydrophilic magnetic NPs were obtained by modification of magnetic precipitate with 7.1M of $NH_3.H_2O$ to pH 8-9, which were mono-disperse in aqueous solution. For preparation of coated cells, 40 ml of the bacterial cell culture at the late exponential phase (5 g·DCW/L) was transferred into a 100 mL Erlenmeyer flask and then 1.5 mL of 30 g/L magnetic suspension was added and mixed thoroughly. After absorption of the magnetic NPs on the cell surface, a permanent magnet was placed at the side of the vessel. The supernatant was decanted and the immobilized cells were washed and suspended in fresh BSM. The model oil was n-tetradecane (containing 6.76mM DBT) in a (2:1 v/v) O/W phase ratio. Both NPs were crystallized and less than 10 nm. Colony count analysis showed that in coated cells produced by method 1, only 78% of the cells had absorbed NPs, while in coated cells produced by method 2, 94% of the cells were decorated by NPs and separated by a magnetic field. The high surface energy and larger specific surface area of the Fe_3O_4 NPs make it strongly adsorbed on the surfaces of microbial cells, but, in oleate-modified NPs, the hydrophobic interaction between the cell membrane and the hydrophobic tail of oleate plays another important role in cell adsorption (Li *et al.* 2009). Obtained results in this study showed that the BDS activity of coated and free cells in the biphasic system were approximately the same and no significant difference was seen between them.

Among the many advantages of immobilization, reusability of cells in successive reaction steps is of great importance. However, BDS activity of the entrapped cells will decrease after each step as a result of the reduction of cofactors such as $NADH_2$ and $FMNH_2$ in the 4S-pathway (Yan *et al.*, 2008). Moreover, the entrapment technique itself often leads to a decrease in biocatalytic activity (Karsten and Simon, 1993).

Dai *et al.* (2014) improved the BDS process by using a combination of magnetic nano Fe_3O_4 particles with calcium alginate-immobilized *Brevibacterium lutescens* CCZU12-1 cells. Entrapment of *Brevibacterium lutescens* CCZU12-1 cells was done as follows: sodium alginate powder was mixed with deionic water at a final concentration of 2% (w/v). This mixture was continuously stirred to prevent the formation of precipitates. Furthermore, equal volume of the Fe_3O_4 MNPs coated cell suspension was added into this sodium alginate solution. The ratio of cell mass to nano-Fe_3O_4 particles was 5:1 (w/w). The obtained alginate-cell mixture was added dropwise to an ice-cold solution of $CaCl_2$ (100 mM) under constant stirring. After stirring for 2 h, the solution was decanted and the resulting beads were stored in a fresh calcium chloride solution until use. Model oil used for the BDS was prepared by 2.5 mM DBT in n-octane (1/9, v/v). The combination of magnetic nano-Fe_3O_4 particles with calcium alginate-immobilized cells appeared not to experience a mass transfer problem because it could be found that the desulfurizing activities of immobilized and free cells exhibited similar time courses. Moreover, 2.5 mM DBT was completely converted into 2-HBP and no 2-HBP accumulation was found in the cells. Furthermore, the combination of Fe_3O_4 MNPs with calcium alginate immobilized cells could be reused for no less than four times. Each batch performed until 2.5 mM DBT was consumed completely. At the end of each batch, the immobilized cells (containing 0.40 g DCW) were collected by application of a magnetic field and then reused in another batch. Compared to the immobilized cells, free cells could be used only once and no desulfurizing activity was found when they were reused. The Fe_3O_4 MNPs have high surface energy and large specific surface area and they can be strongly adsorbed on the surfaces of microbial cells and, thus, coat the cells. Additionally, calcium alginate can prevent the loss of cells and Fe_3O_4 MNPs particles. Thus, the combination of immobilized cells would give a solution to solve some problems in BDS, such as the troublesome process of recovering desulfurized oils and the short life of biocatalysts, etc.

Derikvand *et al.* (2014) optimized the desulfurization process of DBT from the oily phase by a bioprocess employing the immobilized cells. *Rhodococcus erythropolis* R1 cells were encapsulated in calcium alginate beads by considering factors such as the alginate concentration, size of the beads, the concentration of surfactants, and γ-Al_2O_3 NPs for optimizing BDS. The impact of two cofactor precursors (nicotinamide and riboflavin) on long term BDS efficiency was also examined. The biphasic media consisted of MBSM (aqueous phase) and n-tetradecane (organic phase) in a 2:1 ratio and 1 mM/L DBT as the sulfur source. The volumes of nicotinamide and riboflavin were 10 mM/L and 40 mM/L, respectively. The incubation time of DBT utilization and 2-HBP production was 20 h. The results indicated that the highest BDS efficiency ($\approx$ 81%) could be achieved at 20% (w/w) of γ-Al_2O_3 NPs, alginate beads size equal to 1.5 mm, 1% (w/v) of the alginate, and 0.5% (v/v) of span 80. The related statistical analysis showed that the concentration of γ-Al_2O_3 NPs was the most significant factor in the BDS process. Moreover, the addition of nicotinamide and riboflavin significantly decreased the biocatalytic inactivation of the immobilized cells system after successive operational steps, enhancing the BDS efficiency by more than 30% after four steps. So, it can be concluded that a combination of the γ-Al_2O_3 NPs with alginate immobilized cells could be very effective in the BDS process.

The enhancement of anaerobic kerosene biodesulfurization using nanoparticles decorated with *Desulfobacterium indolicum* at 30 °C and atmospheric pressure brought about a decrease in the sulfur content from 48.68 ppm to 13.76 ppm within 72 h (Kareem, 2014).

The immobilization of the haloalkaliphilic S-oxidizing bacteria, *Thialkalivibrio versutus* D301, using super-paramagnetic magnetite nanoparticles under haloalkaliphilic conditions (pH 9.5) to remove sulfide, thiosulfate, and polysulfide from wastewater has been studied (Xu *et al.*, 2015). The magnetite nanoparticle-coated cells expressed similar biodesulfurization activity like that of free cells and could be reused for six batch cycles.

Mesoporous materials in the nano-scale (2 – 50 nm) are characterized by large surface area, ordered pore structures with suitable sizes, high chemical and physical stability, and can be chemically modified with functional groups for the covalent binding of molecules that make them very applicable in adsorption, separation, catalytic reactions, sensors, and immobilization of huge biomolecules like enzymes (Li and Zhao, 2013; Juarez-Moreno *et al.*, 2014). Enzymatic immobilization in nano-materials has some advantages, including increase in catalytic activity, reduction in protein miss-folding, thermal stability, increase in solvent tolerance, and reduction of denaturation caused by the presence of organic solvents and mixtures of water-organic solvent molecules (Carlsson *et al.*, 2014). Also, the immobilization (i.e. conjugation) of the chloroperoxidase (CPO) enzyme with latex nanoparticles after its functionalization with chloromethyl groups can react with the amine groups of the protein (i.e. the enzyme) to form stable covalent bonds. The average size of the conjugated nanoparticles reported to be 180 nm that can be applied in bio-oxidative of fuel streams, followed by physical or chemical removal of the produced sulfoxide and sulfone to produce ultra-low sulfur fuels. Immobilization of chloroperoxidase (as a good candidate for oxidative desulfurization in non-aqueous media) and myoglobin (which, for its availability, represents a good candidate for industrial scale) with La-nanoparticles (40 nm) were modified with chloromethyl groups (Vertegel, 2010). The conjugated and free enzymes were used for dibenzothiophene oxidation in the presence of hydrogen peroxide in a non-aqueous system (dibenzothiophene in hexane) and a two-phase system (dibenzothiophene in acetonitrile/water; 20% v/v). For CPO in a two phase system and non-aqueous one, free and conjugated enzymes expressed nearly the same activity, but the stability of free enzymes was much lower than that of the conjugated ones, where its activity deteriorates much faster. The activity of conjugated chloroperoxidase was approximately 5 times higher than that of free cells after 2 d storage. However, the free myoglobin expressed better performance than the immobilized ones in 2-phase systems, while the immobilized myoglobin expressed twice the activity in a non-aqueous system, which is an encouraging result since this protein, as well as the similar heme-containing protein hemoglobin, can be readily available in large quantities from animal processing industries. The nano-immobilized enzymes are easily separated, stable over time, and can be taken as a good prospect for a recyclable desulfurization agent.

Juarez-Moreno *et al.* (2015) developed the oxidative transformation of DBT by the chloroperoxidase (CPO) enzyme immobilized on (1D)-γ-Al_2O_3 nano-rods. The (1D)-γ-Al_2O_3 nano-rods (Al-nR) were synthesized, thermally treated, and characterized by a variety of techniques. This material was used for CPO enzyme immobilization by physical adsorption and then assayed for the biocatalytic oxidation of DBT. The textural properties of Al-nR nanostructured materials were successfully tuned to perform CPO immobilization. The nanostructures exhibited a good stability throughout the thermal treatments since its crystallinity remained unchanged and the nano-rod arrays were detected in all materials by XRD and HRTEM. A great affinity between the CPO and the nano-rods was observed due to favorable electrostatic interactions during the immobilization process. In a reaction mixture containing 1.0×10^{-7} M of DBT and 1.3 $\times 10^{-12}$ M of CPO in 1 mL, the maximum biotransformation of DBT by free CPO was 42.2% after 120 min. At the same time, CPO immobilized into Al-nR- 700 °C showed a maximum transformation of 89.6% of initial DBT into DBT-sulfoxide, while CPO immobilized in Al-nR-500 °C and Al-nR-900 °C only transforms 65% and 69.8% of initial DBT, respectively. The morphological and textural properties of Al-nR-T material enhanced the enzymatic transformation of DBT to form the sulfoxide derivative, especially when the Al-nR-700 °C material was used.

Nassar (2015) compared the BDS efficiency of Fe_3O_4 MNPs coated cells, immobilized cells prepared from different immobilized material (alginate and the combination of Fe_3O_4 MNPs with calcium alginate and agar), and free cells of Brevibacillus invocatus C19 and Rhodococcus erythropolis IGST8. Immobilized BDSM by sodium alginate was prepared according to Huang *et al.* (2012), Derikvand *et al.* (2014), and Dinamarca *et al.* (2014a). Figure 10.14a represents the beads of Ca-alginate immobilized cells with an average size of 2 mm, as depicted by SEM (Figure 10.14b). The BDSM was immobilized by agar according to Stefanova *et al.* (1998) and Naito *et al.* (2001). Fe_3O_4 MNPs (6–10 nm) were prepared by a reverse (water/oil) microemulsion method (Zaki *et al.*, 2013). Fe_3O_4 MNPs were strongly adsorbed on the surfaces of microbial cells as shown in Figure 10.15.

The immobilized cells of C19 in the agar matrix showed nearly the same BDS activity as that of free cells recording 98.25% and 95.87%, respectively, followed in decreasing order by coated cells (93.56%), magnetically immobilized cells (82.05%), and, finally, alginate immobilized cells (79.53%) after 168 h, while the Fe_3O_4 coated cells of IGTS8 showed higher desulfurization activity (83.65%) than those with magnetically immobilized cells (83.45%),

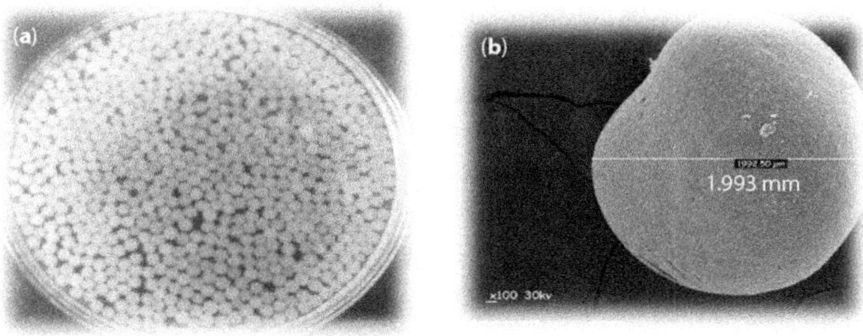

Figure 10.14 Size and Shape of Cells Immobilized Calcium Alginate Beads.

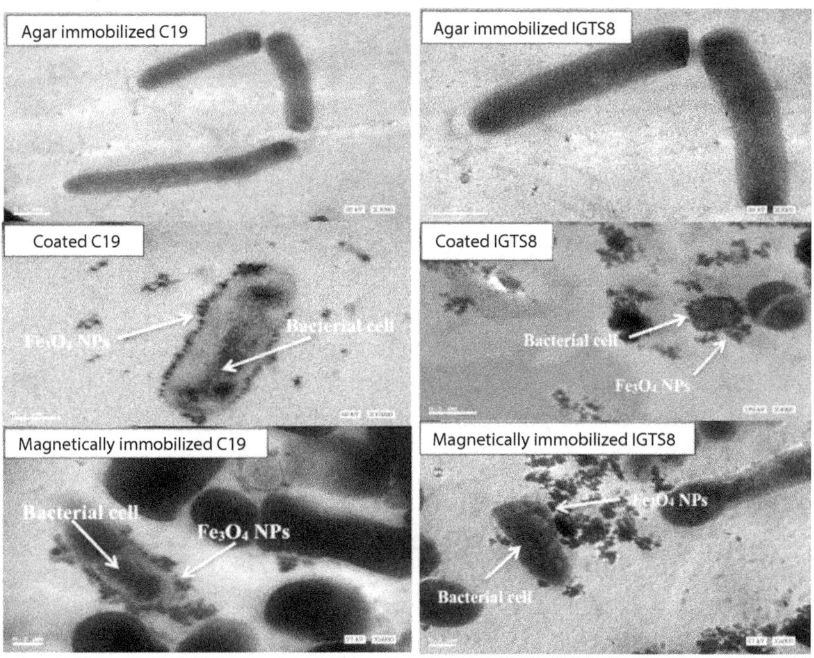

Figure 10.15 TEM Images Showing the Well Adsorption of Fe_3O_4 MNPs on the Surfaces of BDSM.

alginate immobilized cells (72.35%), agar immobilized cells (49.83%), and free cells (71.84%). The agar immobilized cells of C19 successfully retained their BDS efficiency up to five batch cycles, recording DBT removal up to 84.52%. Thus, coated and magnetically immobilized cells sustained up to four and three batch cycles, respectively, recording DBT removal up to 88.17% and 77.09%, respectively, but the uncoated free cells could be used only once and the recycled free cells retained only 52.81% activity, indicating that they is not economically viable to be reused. The Fe_3O_4 coated cells and agar immobilized cells of IGTS8 successfully retained their BDS up to five batch cycles, recording DBT removal up to 81.34% and 43.75%, respectively, but the uncoated free cells could be used only once and the recycled free cells retained only 26.08% activity. The use of IGTS8 and C19, decorated by MNPs, facilitates its recovery (Figure 10.16), reuse, and storage stability, hence it offers a number of advantages for industrial applications compared to free cells. Immobilization and coating of BDSM are important and effective techniques that are usually employed to serve several purposes, including protection of the bacteria from high SO_4^{2-} ions and 2-HBP concentrations, as well as ease of separation and reutilization of the biomass. Free cells of C19 showed high ability to tolerate high concentrations of SO_4^{2-} ions and its efficiency increased after coating and immobilization compared to IGST8, whose BDS efficiency was strongly repressed by the addition of SO_4^{2-} ions either before or after coating and immobilization. The highest BDS efficiency was obtained by agar immobilized C19 at higher 2-HBP concentrations (up to 50 ppm of 2-HBP) compared to coated cell and magnetically immobilized cells which could not tolerate more than 30 ppm, while in the case of IGST8, agar immobilized cells showed less sensitivity to 2-HBP (tolerate up to 30 ppm 2-HBP) compared to free cells.

The results of diesel oil BDS revealed that as the oil phase ratio increases from 10% to 25% (v/v), the sulfur content decreased from 8600 ppm to 343.17 ppm and 88.58 ppm by agar immobilized cells of C19, respectively,

Figure 10.16 Recovery of Magnetically Coated BDSM.

and to 385.31 ppm and 112.67 ppm by coated C19, respectively, which was better than that of IGTS8. Therefore, it was considered that a 25% (v/v) phase ratio is the optimum O/W phase ratio of C19. With this excellent BDS efficiency, agar immobilized cells and coated cells of C19 are considered good potential candidates for industrial applications for the BDS of diesel oil.

10.8 Role of Analytical Techniques in BDS

The organosulfur compounds (OSCs) in crude oil and related materials have been studied because of sulfur sensitivity to oxidation; OSCs are also good markers of the alteration of oils which may result from vaporization, biodegradation, abiotic oxidation, or water washing. OSCs have also a geochemical significance; the distribution of these compounds may be a marker of the origin or maturity of crude oils. It is essential to identify the structures of sulfur compounds in crude oils and petroleum-derived products in order to more effectively choose and optimize their desulfurization processes (Burkow *et al.*, 1990; Damste and De Leeuw, 1990 and Hegazi *et al.*, 2003).

The identification of OSCs started in the nineteenth century with the identification of ten alkyl sulfides in Ohio crude oil (Mabery and Smith, 1891) and thiophene in Gromy crude oil (Charitschkoff, 1899). Ever since, the number and structural complexity of the OSC identified has increased significantly.

The disappearance of sulfur from OSCs in the sample being subjected to biocatalytic treatment can be monitored using X- ray fluorescence (XRF) (IP 336), ultraviolet fluorescence (ASTM D5453), X-ray spectrometry (ASTM D2622), infrared spectrometry (Drushel and Sommers, 1967), atomic emission spectrometry (Rambosek *et al.*, 1994), flame spectrometry, gas chromatography with atomic emission detection (GC-AED) (Mössner and Wise, 1999), and capillary-column gas chromatography/mass spectrometry (GC/MS) (Hegazi *et al.*, 2003). It can also be monitored using molecular emission cavity analysis (MECA) (Burguera and Burguera,1985), gas chromatography with a flame photometric detector (GC-FPD) (Chakhmakhchev *et al.*, 1997; Bradley and Schiller, 1986), gas chromatography with a sulfur chemiluminescence detector (GC-SCD) (ASTM D5623), high performance liquid chromatography with a chemiluminescence detector (HPLC-SCD) (Mishalanie and Birks, 1986), and sulfur K-Edge X-ray absorption spectroscopy (George and Gorbaty, 1989; Sarret *et al.*, 1999).

10.8.1 Gas Chromatography

Gas Chromatography (GC) has a number of advantages over other separation techniques. It is fast and extremely sensitive. The extreme sensitivity is due to the variety of detectors, which are still generally unavailable to other forms of chromatography.

10.8.1.1 Determination of Sulfur Compounds by GC

An attractive feature of the GC method is the availability of a large number of sensitive, universal and selective detectors. The latter detectors are especially useful in the analysis of different contaminants in increasingly complex matrices. Such detectors can reduce analysis time by eliminating laborious and time-consuming procedures of sample preparation, which can also often cause contamination or loss of analytes.

Flame photometric detection (FPD) is still the most widely used sulfur-selective detection method. FPD exhibits a non-linear (exponential) response to sulfur compounds and compound-dependent response factors, but is relatively inexpensive, robust, and adequate for many applications (Bradley *et al.*, 1986). An attractive alternative to FPD is sulfur chemiluminescence detection (SCD). The latest applications of this detection method have shown that SCD proves good performance in terms of detectability, selectivity, linearity, and has uniform sulfur responses. It produces a linear and nearly equi-molar response to sulfur. Because of these advantages, SCD is highly recommended for analysis of extremely complex matrices. The low detection limit, fewer problems with interferences, and noise stability allow SCD more flexibility in a capillary column selection. The combination of fused–silica capillary columns and SCD provides a powerful tool for measurements of trace levels of sulfur-containing compounds in complex matrices (ASTM D5623, Wardencki, 2000; Grossman *et al.*, 1999; Abbad-Andaloussia,b *et al.*, 2003). Atomic emission detection (AED) was also found to have a good combination of specificity and sensitivity for analysis of sulfur-containing compounds (Hegazi *et al.*, 2003). The combination of GC with independent spectroscopic techniques, mainly mass spectrometry (MS), have made the combined techniques extremely versatile sources of qualitative and quantitative information on a variety of environmental samples. The application of the MS or GC–MS systems is still becoming more popular in the analysis of environmental sulfur compounds (Ogata *et al.*, 1983; Rudzinski *et al.*, 2000).

GC analysis coupled with a mass spectrometric detector, MS (Mezcua *et al.* 2008), flame ionization detector, FID (Mezcua *et al.* 2007), or atomic emission detector, AED (Ma *et al.* 2006), remain the most widely used in the BDS process for the determination of volatile and semi-volatile compounds in fossil fuels. By coupling GC with a sulfur chemiluminescene detector (SCD), a specific and very sensitive determination can be made (Garcia *et al.* 2002). The importance of GC–MS stretches beyond ordinary quantitative determination and is also used to establish the structures of organic compounds based on library matches (Nichols *et al.* 1986; Tao *et al.* 2006; Yu *et al.* 2006b).

10.8.1.2 Assessment of Biodegradation

A wide variety of instrumental and non-instrumental techniques are currently used in the analysis of oil hydrocarbons, which include gas chromatography (GC) (Wang *et al.*, 2001), gas chromatography-mass spectrometry (GC/MS) (Anderlini *et al.*, 1981; Wang *et al.*, 1998), high performance liquid chromatography (HPLC) (Krahn *et al.*, 1993; Sink and Hardy, 1994), size exclusion chromatography (Krahn and Stein, 1998), infrared spectroscopy (IR) (ASTM D3414; ASTM D3921), supercritical fluid chromatography (SFC) (ASTM D5186), thin layer chromatography (TLC) (Karlsen and Larter, 1991; Whittaker and Pollard, 1994), ultraviolet spectroscopy (UV) (Burns, 1993), isotope ratio mass spectrometry (Wang *et al.*, 1999), and gravimetric methods (El-Tokhi and Mostafa, 2001). Of all these techniques, GC techniques are the most widely used.

The following groups will be referred to when describing and discussing oil composition changes during biodegradation (Wang *et al.*, 1998b): TPH or total gas chromatographic detectable petroleum hydrocarbons (GC-TPH), which is the sum of all GC-resolved and unresolved hydrocarbons, the unresolvable hydrocarbons, i.e. the unresolved complex mixtures (UCMs) or the 'hump' of hydrocarbons appear as the area between the lower baseline and the curve defining the base of resolvable peaks, total n-alkanes are the sum of all resolved n-alkanes from C8 to C40, total alkanes are the sum of total n-alkanes and branched alkanes and some selected isoprenoids, such as the pristane (Pr) and phytane (Ph), and the carbon preference index (CPI), which is defined as the sum of odd carbon-numbered alkanes to the sum of the even carbon-numbered alkanes (oils characteristically have CPI values around 1.0).

The early effect of microbial degradation is monitored by ratios of biodegradable to less degradable compounds, for example: TRP/UCM, n-C17/Pr, and n-C18/Ph ratios. However, it has been demonstrated that changes in these ratios may substantially underestimate the extent of biodegradation because sometimes isoprenoids also degrade to a significant degree.

10.8.2 Presumptive Screening for Desulfurization and Identification of BDS Pathway

Several interesting developments have occurred in the last few years which are relevant to the microbial removal of organic sulfur from fossil fuel. Among them, two techniques have been widely used for rapid screening, detection, and characterization of microbial cultures that can degrade organosulfur compounds (e.g. DBT). These techniques vary in their rationale, the effort needed to perform the technique, the reliability of obtaining a positive observation, and the likelihood of isolating an organism which only desulfurizes DBT (as opposed to extensively oxidizing the carbon frame of the molecule as well).

The first method, according to Krawiec (1990), can be applied successfully for the detection of organisms which have the ability to degrade DBT to form 3-hydroxy-2-formylbenzothiophene (HFBT) (Kodama pathway). Using DBT-spray plate assay organisms with such capacities was recognized by the presence of rings of bright color around colonies; the color originated from the breakdown products of DBT:

> DBT → 1,2-dihydroxydibenzothiophene → 4-(2-(3-hudroxy)-thionaphthenyl-2-oxo-3-butenoic acid (which is orange-red) → 3-hydroxy-2-formylbenzothiophene (3-HFBT, which is yellow). 3-HFBT is abioticaly condensed to a non-metabolic product, 3-oxo-[3'-hydroxy-thionaphethenyl]-(2)-methylene-dihydrothionaphthene, which is vivid purple.

Krawiec (1988) and Denome *et al.* (1993a,b) used DBT-spray plate assay to identify biodesulfurizing microorganisms (BDSM) following the 4S-pathway. This is based on the observation that some metabolites of DBT that result from sulfur-specific metabolism (the 4S pathway) will fluoresce clearly purple under UV light at 254 nm. This fluorescence was used as a presumptive indication that 2,2'-BHBP or 2-HBP was present. $DBTO_2$ is also fluorescent though the intensity and is much less than with equivalent concentrations of phenolic products. According to Krawiec (1990), Denome *et al.* (1994), and Rhee *et al.* (1998), complete disappearance of the sprayed substrate, beginning as a clear zone around the colony and eventually spreading across the entire plate, indicates +ve BDS strain.

The notable feature of this assay is that it distinguishes bacteria which transforms DBT through the Kodama or 4S-pathway. The assay is quick, easy, positive, and presumptive, but not conclusive and some others don't produce either color or florescence (Krawiec, 1988).

The second method for sulfur bioavailability assay (Kilbane, 1989) for the detection and evaluation of bacteria capable of metabolizing organic sulfur (e.g. DBT) was developed on the basis that all microorganisms require sulfur for growth. The sulfur bioavailability assay is essentially a microbial assay for sulfur; the amount of microbial growth gives a quantifiable index of the utilization of any organic or inorganic compound as a source of sulfur (Matsui *et al.* 2000; Arensdorf *et al.*, 2002). A limitation of the technique should be noted: this assay does not exclude that a sulfur-containing organic compound might also serve as a carbon source. Thus, the carbon frame of a compound of interest may be concomitantly or subsequently oxidized (Krawiec, 1990). To confirm that the fluorescence or the colored product is of the Kodama or 4S pathway metabolites, other presumptive, preliminary, and easy tests can be done: TLC, Gibb's assay, and Phenol assay. Table 10.2 illustrates the results of some published data of TLC and Gibbs assay results with Kodama and 4S-pathways metabolites.

10.8.2.1 *Gibb's Assay*

Determination of accumulated hydroxylated aromatic compounds as desulfurization products can be performed by Gibb's assay. Positive reactions develop blue to purple color (Castorena *et al.*, 2002). For example, Gibbs reagent (2,6-dichloroquinone-4-chlorimide) reacts with aromatic hydroxyl groups, such as 2-HBP or 2,2'-BHBP, at a pH of 8.0 to form a blue-colored complex. To obtain the maximum accuracy using the Gibbs assay, the pH and time of incubation/color development (30 min.) must be precisely controlled (Kayser *et al.*, 1993; Rambosek *et al.* (1994), Kobayashi *et al.* (2000), Tanaka *et al.* (2002), Mohebali *et al.*

(2007), Papizadeh *et al.* (2010)). A reaction between 2,6-dichloroquinone-4-chloroimide (Gibb's reagent) and 2-HBP in the supernatant of microbial isolates produces blue to purple color to signify a positive result, which itself is an indication of the presence and activities of different enzymes (DszABC proteins) involved in the conversion pathway (Bhatia and Sharma (2010b), Davoodi-Dehaghani *et al.* (2010)).

10.8.2.2 *Phenol Assay*

The presence of 2-HBP and/or 2,2'-BHBP or similar phenolic compounds can also be detected with this chromogenic assay, as a presumptive indication for their production in the culture (Wang and Krawiec, 1994). Rapid screening techniques are very often merely presumptive and additional studies are required to confirm that some specific biological activity is indeed present. Other instrumental methods were established to confirm

Table 10.2 TLC and Gibbs Assay Results with Kodama, 4S pathways Metabolites (Monticello *et al.*, 1985; Tanaka *et al.*, 2002)

Compound	TLC		Gibb's Assay
	Daylight	**UV-Illumination (at 245nm)**	
Hemiacetal form of 4-(2-(3-hudroxy)-thionaphthenyl-2-oxo-3-butenoic acid	Orange-red	Purple	-ve
Trans form of 4-(2-(3-hudroxy)-thionaphthenyl-2-oxo-3-butenoic acid	Red	Orange	Purple
3-HFBT	Yellow	Blue	Pink
3-oxo-[3'-hydroxy-thionaphethenyl]-(2)-methylene-dihydrothionaphthene	Purple	Pink	-----
2-HBP	-ve	Bright white-blue	Blue
2,2'-BHBP	-ve	Purple	Blue

the presence of reactant characteristics of the Kodama or 4S-pathway, follow up the utilization of the sulfur compounds, and analyzing and identifying the metabolites and quantifying their amount. UV spectrophotometer, GC/MS (Oldfield *et al.*, 1997; Gilbert *et al.*, 1998; Bressler and Fedorak, 2001a,b; Kirimura *et al.*, 2002; Prince and Grossman, 2003), and HPLC (Baldi *et al.*, 2003) are examples of such instrumental methods.

10.8.3 More Advanced Screening for Desulfurization and Identification of BDS Pathway

10.8.3.1 High Performance Liquid Chromatography

High performance liquid chromatography (HPLC) is used as a separation technique for mixtures of solutes. The stationary phase may be a solid adsorbent, a liquid, an ion exchange resin, or a porous polymer, held in a metal column the liquid mobile phase is forced through under pressure.

The advantages over other forms of liquid chromatography can be summarized as follows: (i) the HPLC column can be used many times without regeneration, (ii) the resolution achieved on such columns far exceeds that of the older methods, (iii) the technique is less dependent on the operator's skill and reproducibility is greatly improved, and (iv) analysis time is much shorter.

It is very useful technique in the field of analysis of (PAHs). The separation and identification of PAHs in crude oils have been reported by several workers (Forster *et al.*, 1989; Zhang *et al.*, 1992). Publications regarding the detection of organosulfur compounds using HPLC are scarce and commonly make use of derivatization techniques that are time-consuming or detection methods that are not highly selective (Pirkle *et al.*, 1979; Reeves *et al.*, 1980; Bossle *et al.*, 1984; Vignier *et al.*, 1985).

There are some HPLC detectors that are designed to be somewhat selective for sulfur-containing species, for example a photoconductivity detector (Popovich *et al.*, 1979). This detector also responds to halogenated compounds, many nitrogen–containing compounds, and sulfur–containing compounds in which the sulfur atom is oxidized. A sulfur chemiluminescence detector (SCD) allows precise measurements of sulfur-containing molecules down to ppb levels. Because SCD ignores the presence of compounds other than those containing sulfur, it generally provides high selectivity and sensitivity (Mishalanie *et al.*, 1986). HPLC with an ultraviolet detector (UV) is frequently employed in the determination of DBT and HBP (Martin *et al.* 2004; Furuya *et al.* 2004).

HPLC in gradient elution mode is commonly used and is vital for separation of compounds with polarity indices that are close and which will otherwise be difficult to separate using isocratic conditions (Rashtchi *et al.*, 2006; Calzada *et al.* 2009). DBT and its metabolites in the BDS process can be detected by HPLC equipped with different types of columns: reverse bonded phase columns (Dahlberg *et al.*, 1993), Synchropak RPC18 (Denome *et al*, 1994), Brownlee columns (Wang *et al.*, 1996), Waters 4μ phenyl Novapak (Patel *et al.*, 1997), Ultremex 5 C18 columns (Serbolisca *et al.*, 1999), Cosmosil 5 C18-AR-300 columns (Maghsoudi *et al.*, 2000), and Zorbax C18 columns (Zhongxuan *et al.*, 2002).

10.8.3.2 X-ray Sulfur Meter and other Techniques for Determining Total Sulfur Content

Sulfur removal from the substrate can be determined by the difference in sulfur content in a sterile control sample and that sample is treated with biodesulfurizing microorganisms (BDSM) using an X-Ray sulfur meter (Forster *et al.*, 1989; Folsom *et al.*, 1999; Okada *et al.*, 2002ab; Abbad-Andaloussi *et al.*, 2003a,b). X-ray fluorescence is suitable for the determination of total sulfur in fossil fuels (Folsom *et al.* 1999; Labana *et al.*, 2005) and an elemental analyzer can also give details about the different elements, including sulfur, in such media (Jorjani *et al.*, 2004). A microcoulomb analyzer was used to determine the total sulfur content by weight

Table 10.3 Analytical Techniques in BDS

Role/determination	Technique/instrument	References
Monitoring the production of HBP from DBT	Gibb's Assay	El-Gendy (2001 ,2004, 2006); Mohebali *et al.* (2007a); Bathia and Sharma (2010); Davoodi-Dehaghani *et al.* (2010); Papizadeh *et al.* (2010); Nassar *et al.* (2013); Nassar (2015)
Monitoring microbial cell growth profile	UV Spectrophotometer	El-Gendy (2001, 2004, 2006); Bouchez-Naitali *et al.* (2004); Li *et al.* (2007); Hai *et al.* (2008); Chen *et al.* (2008a,b); Nassar *et al.* (2013); Nassar, (2015); Nassar *et al.* (2017a,b)
Microbial genotypes	PCR, 16S rRna gene sequencing and computer packages	Ohshiro *et al.* (2007); De Vasconcellos *et al.* (2009); Radwan, (2015)
Microbial phenotypes	Gram stain	El-Gendy, (2001); El-Gendy, (2004); Wubbeler *et al.* (2006); Boniek *et al.* (2010); Nassar, (2015)
Protein purification and quantitation	Bradford's method, etc.	Mirgorodiskaya *et al.* (2000); Kruger (2002); Saxena *et al.* (2009)
Analysis of DszB protein/reaction mechanism	X-ray crystallography/diffraction	Lee *et al.* (2004); Frase *et al.* (2011); Jimenez *et al.* (2011)
Separation and quantitation of DBT, HBP	Reversed phase-HPLC/UV	El-Gendy (2001, 2004); Furuya *et al.* (2004); Martin *et al.* (2004); Rashtchi *et al.* (2006); Calzada *et al.* (2009); Nassar *et al.* (2013); Nassar (2015); Nassar *et al.* (2017a,b)
Separation and quantitation of DBT, HBP and other volatile and semi-volatile organics	GC/MS, GC/FID, GC/AED, GC/SCD	El-Gendy (2001); El-Gendy (2004); Ma *et al.* (2006a,b); Mezuca *et al.* (2007, 2008) ; Nassar *et al.* (2013); Nassar (2015); Nassar *et al.* (2017a,b)
Molecular structure of metabolites	GC/MS	El-Gendy (2004); Nichols *et al.* (1986); Yu *et al.* (2006b); Tao *et al.* (2006); El-Gendy (2014); Nassar (2015); Nassar *et al.* (2016)
TSC	Elemental analyzer/micro-coulomb analyzer, X-ray fluorescence	Labana *et al.* (2005); Folsom *et al.* (1999); Guobin *et al.* (2006)

of diesel oil by measuring the amount of SO_2 released from its combustion (Guobin *et al.*, 2006). This is an especially important determination, keeping in mind the strict regulations that demand controlling the amount of sulfur in fuels due to environmental and health concerns. Table 10.3 summarizes some of the instrumental techniques for monitoring the BDS process.

References

Abbad-Andaloussi S., Warzywoda M., Monot F. (2003a) Microbial desulfurization of diesel oils by selected bacterial strains. Oil and Gas Science and Technology. 58:505–513.

Abbad-Andaloussi, S., Lagnel, C., Warzywoda, M., Monot, F. (2003b) Multicriteria comparison of resting cell activities of bacterial strains selected for biodesulfurization of petroleum compounds. *Enzyme and Microbial Technology*. 32(3/4): 446–454.

Abolfazl, H.S., Soheila, Y., Mohammad, M.S. (2006) Biodesulfurization of dibenzothiophene by a newly isolated thermophilic bacteria strain. *Iranian Journal Of Chemistry & Chemical Engineering*. 25(3): 65–71.

Acero, J., Berdugo, C., Mogollón, L. (2003) Biodesulfurization process evaluation with a *Gordona rubropertinctus* strain. CT&F - Ciencia, Tecnología y Futuro. 2(4): 43–54.

Adegunlola, G.A., Oloke, J.K., Majolagbe, O.N., Adebayo, E.A., Adegunlola, C.O.,

Adewoyin, A.G., Adegunlola, F.O. (2012) Microbial desulphurization of crude oil using *Aspergillus flavus*. *European Journal of Experimental Biology*. 2(2): 400–403.

Adlakha, J., Singh, P., Ram, S.K., Kumar, M., Singh, M.P., Singh, D., Sahai, V., Srivastava, P. (2016) Optimization of conditions for deep desulfurization of heavy crude oil and hydrodesulfurized diesel by *Gordonia* sp. IITR100. 184: 761–769.

Agarwal, M., Dikshit, P. K., Bhasarkar, J. B., Borah, A. J., Moholkar, V. S. (2016) Physical insight into ultrasound-assisted biodesulfurization using free and immobilized cells of *Rhodococcus rhodochrous* MTCC 3552. *Chemical Engineering Journal*. 295: 254–267.

Akbarzadeh, A., Samiei, M., Davaran, S. (2012) Magnetic nanoparticles: preparation, physical properties, and applications in biomedicine. Nanoscale Research Letters. 7: 144–156.

Akbarzadeh, S., Raheb, J., Aghaei, A., Karkhane, A.A. (2003) Study of desulfurization rate in *Rhodococcus* FMF native bacterium. *Iranian Journal of Biotechnology*. 1(1): 36–40

Akhtar, N., Ghauri, M.A., Akhtar, K. (2016) Dibenzothiophene desulfurization capability and evolutionary divergence of newly isolated bacteria. *Archives of Microbiology*. 198: 509–519.

Akhtar, N., Ghauri, M.A., Anwar, M.A., Akhtar, K. (2009) Analysis of the dibenzothiophene metabolic pathway in a newly isolated *Rhodococcus* spp. *FEMS Microbiology Letters*. 301: 95–102.

Akin, C. (1987) In: Russell GE (ed) Biotechnology and Genetic engineering reviews: biocatalysis with immobilized cells. 5(1):319–367.

Akin, D., Sturgis, J., Ragheb, K. (2007) Bacteria-mediated delivery of nanoparticles and cargo into cells. Nature Nanotechnology. 2: 441–449.

Alam K.Y., Worland M.J., Clark D.P. (1990) Analysis and molecular cloning of genes involved in thiophene and furan oxidation by E. coli. *Applied Biochemistry and Biotechnology*. 24:843- 855.

Alcon, A., Santos, V.E., Martin, A.B., Yustos, P., Garcíıa-Ochoa, F. (2005) Biodesulfurisation of DBT with Pseudomonas putida CECT5279 by resting cells: influence of cell growth time on reducing equivalent concentration and HpaC activity. *Biochemical Engineering Journal*. 26 (2005) 168–175.

Al-Jailawi, M.H., Al-Faraas, A.F., Yahia, A.I. (2015) Isolation and identification of dibenzothiophene biodesulfurizing bacteria. *American Journal of Bioscience and Bioengineering*. 3(5): 40–46.

Allen C.C.R., Boyd D.R., Dalton H., Sharma N.D., Haughey S.A., McMordie S., McMurray, B.T., Sheldrake G.N. and Sproule K. (1995) Sulfoxides of high enantiopurity from bacterial dioxygenase-catalyzed oxidation. *Journal of the Chemical Society, Chemical Communications*. 1: 19- 120

Alvarez, J.C., Ortiz de Montellano, P.R. (1992) Thianthrene 5-oxide as a probe of the electrophilicity of hemoprotein oxidizing species. *Biochemistry*. 31: 8315–8322.

Alves, L., Melo, M., Mendonça, D., Simões, F., Matos, F.J. Tenreiro, R., Gírio, F.M. (2007) Sequencing, cloning and expression of the dsz genes required for dibenzothiophene sulfone desulfurization from *Gordonia alkanivorans* strain 1B. *Original Enzyme and Microbial Technology*. 40: 1598–1603.

Alves, L., Paixão (2011) Toxicity evaluation of 2-hydroxybiphenyl and other compounds involved in studies of fossil fuels biodesulphurisation. *Bioresource Technology*. 102: 9162–9166.

Alves, L., Salgueiro, R., Rodrigues, C., Mesquita, E., Matos, J., Gírio, F.M. (2005) Desulfurization of dibenzothiophene, benzothiophene, and other thiophene analogs by a newly isolated bacterium, *Gordonia alkanivorans* Strain 1B. *Applied Biochemistry and Biotechnology*. 120: 199–208.

Alves, L., Salgueiro, R., Rodrigues, C., Mesquita, E., Matos, J., Gírio, F.M. (2005) Desulfurization of dibenzothiophene, benzothiophene, and other thiophene analogs by a newly isolated bacterium, *Gordonia alkanivorans* strain 1B. *Applied Biochemistry and Biotechnology*. 120(3): 199–208.

Amanda, K.Y., Wisecarver, K.D. (1992) Cell immobilization using PVA cross linked with boric acid. *Journal of Biotechnology and Bioengineering*. 39: 447–449.

Amin, G. (1994) Continuous production of glutamic acid by Vertical Rotating Immobilized Cell Reactor of the bacterium *Corynebacterium glutamicum*. *Bioresource Technology* 47: 113–119.

Amin, G. (2011) Integrated two-stage process for biodesulfurization of model oil by vertical rotating immobilized cell reactor with the bacterium *Rhodococcus erythropolis*. Petroleum and Environmental Biotechnology. 2:1 http://dx.doi.org/10.4172/2157-7463.1000107

Amin, G., Doelle, W. (1989) Vertical rotating immobilized cell reactor of the bacterium *Z. mobilis* for stable long-term continuous ethanol production. *Biotech Techniques* 3: 5–100.

Amorelli A., Amos Y.D., Halsig C.P., Kosman J.J., Jonker R.J., de Wind M., Vrieling J. (1992) Estimate feedstock process ability. *Hydrocarbon Processing*. 71: 93–106.

Anceno, A.J., Bonduush, I., Shipin, O.V., Dutta, J. (2010) Nanoparticle self- assembly via facile (bio)chemistry: charge-stabilized metal nanoparticles on microbial cell surfaces. *Journal of Bio nanoscience*. 4: 1–6.

Anderlini, V.C., Al-Harmi, L., DeLappe, B.W., Risebrough, R.W., Walker, W., Simoneit, B.R.T, Newton, A.S. (1981) Distribution of hydrocarbons in the oyster, *Pinctada margaritifera*, along the coast of Kuwait. Marine Pollution Bulletin. 12: 57–62.

Andreu, A., Fairweather, N., Miller, A.D. (2008) Clostridium neurotoxin fragments as potential targeting moieties for liposomal gene delivery to the CNS. Chem Bio Chem, 9, 219–31.

Annweiler E., Michaelis W., Meckenstock R.U. (2001) Anaerobic cometabolic conversion of benzothiophene by a sulfate-reducing enrichment culture and in a tar- oil-contaminated aquifer. *Applied and Environmental Microbiology*. 67:5077–5083.

Ansari, F. (2008) Use of magnetic nanoparticles to enhance biodesulfurization. a Ph.D Thesis. Cranfield university. England.

Ansari, F., Grigoriev, P., Libor, S., Tothill, I. E., Ramsden, J.J. (2009) DBT degradation enhancement by decorating Rhodococcus erythropolis IGST8 with magnetic Fe3O4 nanoparticles. *Biotechnology and Bioengineering*, 102(5): 1505–1512.

Ansari, F., Prayuenyong, P., Tothill, I. (2007) Biodesulfurization of dibenzothiophene by *Shewanella putrefaciens* NCIMB 8768. *Journal of Biological Physics and Chemistry*. 7: 75–78.

Arabian D., Najafi H., Farhadi F., and Dehkordi A.M. (2014) Biodesulfurization of simulated light fuel oil by a native isolated bacteria Bacillus cereus HN. J. Pet. Sci. *Technol.* 4(1): 31–40.

Arafa M.A., Al-Fantoukh S.A., Sano T. (2001) Biodesulfurization activity of dibenzothiophene by mutant strain of *Rhodococcus erythropolis*; 4th Chemistry in Industry Conference and Exhibition. Petroleum Refining and Petrochemicals Session. pp.1–7.

Ardakani, M.R., Aminsefat, A., Rasekh, B., Yazdiyan, F., Zargar, B., Zarei, M., Najafzadeh, H. (2010) Biodesulfurization of dibenzothiophene by a newly isolated *Stenotrophomonas maltophila* strain Kho1. *World Applied Sciences Journal*. 10(3): 272–278.

Arensdorf J.J., Loomis A.K., DiGrazia, P.M., Monticello, D.J., Pienkos P.T. (2002) Chemostat approach for the direct evolution of biodesulfurization gain-of-function mutants. *Applied and Environmental Microbiology*. 68: 691–698.

Arenskötter, M., Bröker, D., Steinbüchel, A. (2004) Biology of the metabolically diverse genus *Gordonia. Applied and Environmental Microbiology*. 70(6): 3195–3204.

Arnold, F.H., Wintrode, P. C., Miyazaki, K. and Gershenson, A. (2001) How enzymes adapt: lessons from directed evoluction. Trends in Bioch. Sci., 26: 100–106.

Arun, K., Ashok, M., Rajesh, S. (2011) Crude oil PAH constitution, degradation pathway and associated bioremediation microflora: an overview. *International Journal of Environmental Sciences*. 1(7): 1420–1439.

Atlas R.M. (1981) Microbial degradation of petroleum hydrocarbons: an environmental perspective. *Microbiological Reviews*. 45(1):180–209.

Ayala, M., Verdin, J., Vazquez-Duhalt, P. (2007) The prospects for peroxidase-based biorefining of petroleum fuels. *Biocatalysis and Biotransformation*. 25(2–4): 114–129.

Bahrami, A., Shojaosadati, S.A., Mohebali, G. (2001) Biodegradation of dibenzothiophene by thermophilic bacteria. *Biotechnology Letters*. 23: 899–901.

Bahuguna, A., Lily, M.K., Munjal, A., Singh, R.N., Dangwal, K. (2011) Desulfurization of dibenzothiophene (DBT) by a novel strain *Lysinibacillus sphaericus* DMT-7 isolated from diesel contaminated soil. *Journal of Environmental Sciences*. 23(6): 975–982.

Baldi, F., Pepi, M., Fava, F. (2003) Growth of *Rhodosporidium toruloides* strain DBVPG 6662 on dibenzothiophene crystals and orimulsion. *Applied and Environmental Microbiology*. 69(8): 4689–4696.

Baldi, F., Vaughan Martini, A.M., Olson, O.J. (1990) Chromium(VI)- resistant yeast isolated from a sewage treatment plant receiving tannery wastes. *Applied Environmental Microbio1ogy*. 56: 913–918.

Bar, H., Bhui, D.K., Sahoo, G.P. (2009) Green synthesis of silver nanoparticles using latex of Jatropha curcas. *Colloids and Surfaces A: Physicochemical and Engineering Aspects*. 339: 134–139.

Bardania, H., Raheb, J., Mohammad-Beigi, H., Rasekh, B., Arpanaei, A. (2013) Desulfurization activity and reusability of magnetite nanoparticle-coated *Rhodococcus erythropolis* FMF and *R. erythropolis* IGTS8 bacterial cells. *Biotechnology and Applied Biochemistry*. 60: 323–329.

Barzegar, F., Raheb, J., Mosavi, M.L. (2007) Cloning and expression of Iran SOX (*dsz A B, C*) Operon in *E. coli* Strain DH5a and comparison its desulfurization activity with *pseudomonas aeroginosa* EGSOX, *Pseudomonas putida* EGSOX and *E. coli* cc118. *Journal of Semnan University of Medical Sciences*. 8(1):1–14.

Beshay, U., Abd-El-Haleem, D., Moawad, H. Zaki, S. (2002) Phenol biodegradation by free and immobilized *Acinetobacter*; *Biotechnology Letters*. 24:1295–1297.

Bezalel, L., Hadar, Y., Fu, P.P., Freeman, J.P., Cerniglia, C.E. (1996) Initial oxidation products in the metabolism of pyrene, anthracene, fluorene, and dibenzothiophene by the white rot fungus *Pleurotus ostreatus*. *Applied and Environmental Microbiology*. 62: 2554–2559.

Bhasarkar, J.B., Chakma, S., Moholkar, V.S. (2014) Investigations in physical mechanism of the oxidative desulfurization process assisted simultaneously by phase transfer agent and ultrasound. Ultson. Sonochem. http://dx.doi.org/10.1016/j.ultsonch.2014.11.008.

Bhasarkar, J.B., Chakma, S., Moholkar, V.S., (2013) Mechanistic features of oxidative desulfurization using sono-Fenton–peracetic acid (ultrasound/Fe^{2+}–CH_3COOH– H_2O_2) system. *Industrial & Engineering Chemistry Research*. 52: 9038–9047.

Bhasarkar, J.B., Dikshit, P.K., Moholkar, V.S. (2015) Ultrasound assisted biodesulfurization of liquid fuel using free and immobilized cells of *Rhodococcus rhodochrous* MTCC 3552: A mechanistic investigation. Bioresource Technology. 187: 369–378.

Bhatia, S., Sharma, D.K. (2010b) Mining of genomic databases to identify novel biodesulfurizing microorganisms. *Journal of Industrial Microbiology and Biotechnology*. 37(4): 425–429.

Bhatia, S., Sharma, D.K. (2012) Thermophilic desulfurization of dibenzothiophene and different petroleum oils by *Klebsiella* sp. 13T. *Environmental Science and Pollution Research*. 19(8): 3491–3497.

Bhatia, S., Sharma, D.K., (2010a). Biodesulfurization of dibenzothiophene, its alkylated derivatives and crude oil by a newly isolated strain *Pantoea agglomerans* D23W3. *Biochemical Engineering Journal*. 50: 104–109.

Biria, D., Zarrabi, A., Khosravi, (2008) The application of corrugated parallel bundle model to immobilized cells in porous microcapsule membranes. J. Membr. Sci. 311, 159e164.

Bohonos N., Chou T.W., Spanggord R.J. (1977) Some observations on biodegradation of pollutants in aquatic systems. Japanese Journal of Antibiotics. 30: 275–285.

Boniek, D., Figueiredo, D., Pylro, V.S., Duarte, G.F. (2010) Characterization of bacterial strains capable of desulphurisation in soil and sediment samples from Antarctica. Extremophiles. 14: 475–481.

Bordoloi, N.K., Bhagowati, P., Chaudhuri, M.K., Mukherjee, A.K. (2016) Proteomics and metabolomics analyses to elucidate the desulfurization pathway of *Chelatococcus* sp. PLoS One 11:e0153547.

Bordoloi, N.K., Rai, S.K., Chaudhuri, M.K., Mukherjee, A.K. (2014) Deep-desulfurization of dibenzothiophene and its derivatives present in diesel oil by a newly isolated bacterium *Achromobacter* sp. to reduce the environmental pollution from fossil fuel combustion. *Fuel Processing Technology*. 119: 236–244.

Boshagh, F., Mokhtarani, B., Mortaheb, H.R. (2014) Effect of electrokinetics on biodesulfurization of the model oil by *Rhodococcus erythropolis* PTCC1767 and *Bacillus subtilis* DSMZ 3256. *Journal of Hazardous Materials*. 280: 781–787.

Bossle, P.C., Martin, J.J., Sarver, E.W., Sommer, H.Z.and 1983. High Performance Liquid Chromatography Analysis of 2-Chloroethyl

Ethvlsulfide and its Decomposition By-Products by Dadvatization. Technical ReportARCSL-TR-83027, U.S. Army Armament, R & D Command, Aberdeen Proving Grounds, Maryland.

Bouchez-Naitali M., Abbad-Andaloussi S., Warzywoda M., Monot F. (2004) Relation between bacterial strain resistance to solvents and biodesulfurization activity in organic medium. *Applied Microbiology and Biotechnology.* 65: 440–445.

Bradley, C. Schiller, J. (1986) Determination of sulfur compound distribution in petroleum by gas chromatography with a flame photometric detector. *Analytical chemistry.* 58:3017–3021.

Bradley, P.B., Engel, G., Feniuk, W., Fozard, J.R., Humphrey, P.P.A., Middlemiss, D.N., Mylecharane, E.J., Richardson, B.P., Saxena, P.R. (1986) Proposals for the classification and nomenclature of functional receptors for 5-hydroxytryptamine. *Neuropharmacology.* 25:563–576.

Bregar, V.B. (2004) Advantages of ferromagnetic nanoparticle compos- ites in microwave absorbers. IEEE Transact. Magnetics. 40: 1679–1684.

Bressler D.C., Fedorak P.M. (2001a) Purification, stability and mineralization of 3-hydroxy-2-formylbenzothiophene, a metabolite of dibenzothiophene. *Applied and Environmental Microbiology.* 67:821–826.

Bressler D.C., Fedorak P.M. (2001b) Identification of disulfides from the biodegradation of dibenzothiophene. *Applied and Environmental Microbiology.* 67:5084– 5093.

Burguera J.L., Burguera M. (1985) Determination of some phosphorus-containing compounds by flow injection with a molecular emission cavity detector. Analytica Chimica Acta. 170: 331–336.

Burkow, I.C., Jørgensen, E., Meyer, T., Rekdal, O., Sydnes, L.K. (1990) Experimental simulation of chemical transformations of aromatic compounds in sediments. *Organic Geochemistry.* 15 (1): 101–108.

Burns, R.G. (1993) Mineralogical applications of crystal field theory. 2nd ed. Cambridge Univ. Press, Cambridge, UK.

Buzanello, E.B., Rezende, R.P., Sousa, F.M.O., Marques, E.L.S., Loguercio, L.L. (2014) A novel *Bacillus pumilus*-related strain from tropical landfarm soil is capable of rapid dibenzothiophene degradation and biodesulfurization. BMC Microbiology. 14:257. http://www.biomedcentral.com/1471–2180/14/257.

Calzada, J., Zamarro, M.T., Alcon, A., Santos, V.E., Diaz, E., Garcia, J.L.,

Garcia-Ochoa, F. (2009) Analysis of dibenzothiophene desulfurization in a recombinant *Pseudomonas putida* strain. *Applied and Environmental Microbiology.* 75: 875–877.

Calzada, J., Zamarro, M.T., Alcon, A., Santos, V.E., Garcia, J.L., Garcia-Ochoa, F. (2007) Intracellular activity of the enzymes involved in 4S route biodesulphurization of DBT by Pseudomonas putida CECET 5279 as function of cell age. *Journal of Biotechnology.* 131(2S): 84–92.

Cao, L. (2006) Carrier-bound immobilized enzymes: principles, applications and design. John Wiley & Sons. pp 131.

Carlsson, N., Gustafsson, H., Thorn, C., Olsson, L., Holmberg, K., Akerman, B. (2014) Enzymes immobilized in mesoporous silica: A physical–chemical perspective. *Advances in Colloid and Interface Science*. 205: 339–360.

Caro, A., Boltes, K., Leton, P., Garcia-Calvo, E. (2007b). Dibenzothiophene biodesulfurization in resting cell conditions by aerobic bacteria. *Biochemical Engineering Journal* 35: 191–197.

Caro, A., Boltes, K., Leton, P., Garcia-Calvo, E. (2008) Description of by-product inhibition effects on biodesulfurization of dibenzothiophene in biphasic media. Biodegradation. 19: 599–611.

Carvajal, P., Dinamarca, A.M., Baeza, P., Camú, E., Ojeda, J. (2017) Removal of sulfur-containing organic molecules adsorbed on inorganic supports by *Rhodococcus Rhodochrous* spp. *Biotechnology Letters*. 39: 241–245.

Carvalho, C.C.C.R., Da Fonseca, M.M.R. (2005) The remarkable *Rhodococcus erythropolis*. analysis. Applied and Environmental Microbiology. 67: 715–726.

Cassidy, M.B., Lee, H., Trevors, J.T. (1996) Environmental applications of immobilized cells: A review. *Journal of Industrial Microbiology and Biotechnology*. 16: 79–101.

Castorena, G., Suarez, C., Valdez, I., Amador, G., Fernandez, L., Le Borgne, S. (2002) Sulfur-selective desulfurization of DBT and diesel oil by newly isolated Rhodococcus sp. Strains, *FEMS Microbiology Letters*. 215:157- 161.

Casullo de Araújo, H.W., Siva, M.C.d.F Lins, C.I.M., Elesbão do Nascimento, A., Alves da Silva, C.A., Campos-Takak, G.M. (2012) Oxidation of dibenzothiophene (DBT) by *Serratia marcescens* UCP 1549 formed biphenyl as final product. Biotechnology for Biofuels. 5:33. http://www.biotechnologyforbiofuels.com/content/5/1/33

Chakhmakhchev, A., Suzuki, M., Takayama, K., (1997) Distribution of alkylated dibenzothiophenes in petroleum as a tool for maturity assessments. *Organic Geochemistry*. 26: 483–489.

Chang J. H., Kim Y.J., Lee B.H., Cho K.S., Rye H.W., Chang Y.K. and Chang H.N. (2001). Production of a desulfurization biocatalyst by two-stage fermentation and its application for the treatment of model and diesel oils. *Biotechnology Progress*. 17: 876–880.

Chang, J.H., Chang, Y.K., Cho, K.S., Chang, H.N. (2000) Desulfurization of model and diesel oils by resting cells of *Gordona* sp. Biotechnology Letters. 22: 193–196.

Chang, J.H., Rhee, S.-K., Chang, Y.K., Chang, H.N. (1998) Desulfurization of diesel oils by a newly isolated dibenzothiophene-degrading *Nocardia* sp. *strain CYKS2. Biotechnology Progress*. 14(6): 851–855.

Charitschkoff, K.V. (1899) Composition of petroleum from Grosny (Caucasus) J. Russ. Phys. Chem. Soc. 31,665–668 (in Russian). J. Soc. Chem. Ind. London 18, 907 (Abstr.).

Chauhan, A.K., Ahmad, A., Singh, S.P., Kumar, A. (2015) Biodesulfurization of benzonaphthothiophene by an isolated *Gordonia* sp. IITR 100. *International Biodeterioration and Biodegradation*. 104: 105–111.

Che, G., Lakshmi, B.B., Fisher, E.R., Martin, C.R. (1998) Carbon nanotubule membranes for electrochemical energy storage and production. *Nature.* 393: 346–348.

Chen T.C., Shen Y.H., Lee W.J., Lin C.C., Wan M.W. (2010) The study of ultrasound-assisted oxidative desulfurization process applied to the utilization of pyrolysis oil from waste tires. *Journal of Cleaner Production.* 18(18): 1850–1858.

Chen, H., Cai, Y., Zhang, W., Li, W. (2009) Methoxylation pathway in biodesulfurization of model organosulfur compounds with Mycobacterium sp. *Bioresource Technology.* 100: 2085–2087.

Chen, H., Zhang, W.J., Cai, Y.B., Zhang, Y., Li, W. (2008a) Elucidation of 2-hydroxybiphenyl effect on dibenzothiophene desulfurization by Microbacterium sp. strain ZD-M2. *Bioresource Technology.* 99(15): 6928–6933.

Chen, H., Zhang, W.J., Chen, J.M., Cai, Y.B., Li, W. (2008b) Desulfurization of various organic sulfur compounds and the mixture of DBT+4,6-DMDBT by *Mycobacterium* sp. ZD-19. *Bioresource Technology.* 99: 3630–3634.

Chen, Z., Gao, L. (2007) A facile and novel way for the synthesis of nearly monodisperse silver nanoparticles. *Materials Research Bulletin.* 42: 1657–1561.

Chirita, M., Grozescu, I. (2009) Fe_2O_3-Nanoparticles, physical properties and their photochemical and photoelectrochemical applications. *Chemical Bulletin.* 54: 1–8.

Colonna, S., Gaggero, N., Carrea, G., Pasta, P. (1994) The microperoxidase-11 catalyzed oxidation of sulfides is enantioselective Tetrahedron Letters. 35: 9103–9104.

Constanti, M., Bordons, A., Giralt, J. (1994) Degradation of dibenzothiophene by *Pseudomonas putida. Letters in Applied Microbiology.* 18: 107–111.

Constanti, M., Giralt, J., Bordons, A. (1996) Degradation and desulfurization of dibenzothiophene sulfone and other sulfur compounds by *Agrobacterium* MC501 and a mixed culture. *Enzyme and Microbial Technology.* 19: 214–219.

Constanti, M.; Giralt, J.; Bordons, A.; Norris, P.R. (1992) Interactions of thiophenes and acidophilic, thermophilic bacteria. *Applied Biochemistry and Biotechnology.* 34: 767–776.

Crawford, D.L., Gupta, R.K. (1990) Oxidation of dibenzothiophene by *Cunninghamella elegans. Current Microbiology.* 21: 229–231.

Cripps R.E. (1973) The microbial metabolism of thiophen-2- carboxylate. *Biochemical Journal.* 134:353–366.

Dahlberg M.D., Rohrer R.L., Fauth D.J., Sprecher R., Olson G.J. (1993) Biodesulfurization of dibenzothiophene sulfone by Arthrobacter sp. and studies with Illinois no. 6 coal. *Fuel.* 72:1645–1649.

Dahlberg, M.D., Rohrer, R.L., Fauth, D.J., Sprecher, R., Olson, G.J. (1993) Biodesulfurization of dibenzothiophene sulfone by *Arthrobacter* sp. and studies with oxidized Illinois No. 6 coal. *Fuel* 72: 1645– 1649.

Dai, Y., Shao, R., Qi, G., Ding, B.B. (2014) Enhanced dibenzothiophene biodesulfurization by immobilized cells of *Brevibacterium lutescens* in n-octane–water biphasic system. *Applied Biochemistry Biotechnology.* 174(6): 2236–2244.

Damste, J.S., de Leeuw, J.W. (1990) Analysis, structure and geochemical significance of organically-bound sulphur in the geosphere: State of the art and future research. *Organic Geochemistry*. 16: 1077–1101.

Davoodi-Dehaghani, F., Vosoughi, M., Ziaee, A.A. (2010) Biodesulfurization of dibenzothiophene by a newly isolated *Rhodococcus erythropolis* strain. *Journal of Bioresource Technology*. 101: 1102–1105.

De Vasconcellos SP, Crespim E, da Cruz GF, Senatore DB, Simioni KCM, Neto EVS, Marsaioli AJ, de Oliveira VM (2009) Isolation, biodegradation ability and molecular detection of hydrocarbon degrading bacteria in petroleum samples from a Brazilian offshore basin. Org Geochem 40:574–588.

De Windt, W., Aelterman, P., Verstraete, W. (2005) Bioreductive deposition of palladium (0) nanoparticles on *Shewanella oneidensis* with catalytic activity towards reductive dechlorination of polychlorinated biphenyls. *Environmental Microbiology*. 7: 314–325.

Dejaloud, A., Vahabzadeh, F., Habibi, A. (2017) *Ralstonia eutropha* as a biocatalyst for desulfurization of dibenzothiophene. *Bioprocess and Biosystems Engineering*. 40: 969–980.

Del Olmo, C.H., Alcon, A., Santos, V.E., Garcia-Ochoa, F. (2005) Modeling the production of a *Rhodococcus erythropolis* IGTS8 biocatalyst for DBT biodesulfurization: influence of media composition. *Enzyme and Microbial Technology*. 37:157–166

Denome S.A., Stanley D.C., Olson E.S., Young K.D. (1993a) Metabolism of dibenzothiophene and naphthalene in Pseudomonas strains: Complete DNA sequence of an upper naphthalene catabolic pathway. *Journal of Bacteriology*. 175:6890–6901.

Denome, S.A., Oldfield, C., Nash, L.J., Young, K.D. (1994) Characterization of the desulfurization genes from *Rhodococcus* sp. strain IGTS8. *Journal of Bacteriology*. 176: 6707–6716.

Denome, S.A., Olson, E.S., Young, K. (1993b) Identification and cloning of genes involved in specific desulfuriza- tion of dibenzothiophene by Rhodococcus sp. strain IGTS8. *Applied and Environmental Microbiology*. 59 (9): 2837–2843.

Derikvand p., Etemadifar Z. (2014) Improvement of biodesulfurization rate of alginate immobilized Rhodococcus erythropolis R1. *Jundishapur Journal of Microbiology*. 7:1–7.

Derikvand, P., Etemadifar, Z., Biria, D. (2014) Taguchi optimization of dibenzothiophene biodesulfurization by *Rhodococcus erythropolis* R1 immobilized cells in a biphasic system. *International Biodeterioration and Biodegradation*. 86: 343–348.

Derikvand, P., Etemadifar, Z., Biria, D. (2015a) RSM optimization of dibenzothiophene biodesulfurization by newly isolated strain of *Rhodococcus erythropolis* PD1 in aqueous and biphasic systems. *Microbiology*. 84(1): 65–72.

Derikvand, P, Etemadifar, Z., Saber, H. (2015b) Sulfur removal from dibenzothiophene by newly isolated *Paenibacillus validus* strain PD2 and process optimization in aqueous and biphasic (model oil) systems. *Polish Journal of Microbiology*. 64: 47–54.

Dervakos, G.A., Webb, C. (1991) On the merits of viable-cell immobilization; *Biotechnology Advances*. 9:559–612.

Dinamarca, M.A., Cereceda, F., Fadic, X., Seeger, M. (2007) Analysis of s-triazine- degrading microbial communities in soil using most-probable-number enumeration and tetrazolium-salt detection. *International Microbiology*. 10: 209–215.

Dinamarca, M.A., Ibacache-Quiroga, C., Baeza, P., Galvez, S., Villarroel, M., Olivero, P., Ojeda, J. (2010) Biodesulfurization of gas oil using inorganic supports biomodified with metabolically active cells immobilized by adsorption. *Bioresource Technology*. 101: 2375–2378.

Dinamarca, M.A., Orellana, L., Aguirre, J., Baeza, P., Espinoza, G., Canales, C., Ojeda, J. (2014b) Biodesulfurization of dibenzothiophene and gas oil using a bioreactor containing a catalytic bed with *Rhodococcus rhodochrous* immobilized on silica. *Biotechnology Letters*, 36(8), 1649–1652.

Dinamarca, M.A., Rojas, A., Baeza, P., Espinoza, G., Ibacache-Quiroga, C., Ojeda, J. (2014a) Optimizing the biodesulfurization of gas oil by adding surfactants to immobilized cell systems. *Fuel*. 116: 237–241.

Doerge, D.R., Cooray, N.M., Brewster M.E. (1991) Peroxidase-catalyzed S-oxygenation: mechanism of oxygen transfer for lactoperoxidase. *Biochemistry*. 30: 8960- 8964.

Drushel, H.V., Sommers, A.L. (1967) Isolation and characterization of sulfur compounds in high-boiling pemleum fraction. Analytical Chemistry. 3:1819- 1829.

Dudley, M.W., Frost, J.W. (1994) Biocatalytic desulurization of aryl sulfonates. *Bioorganic and Medicinal Chemistry*. 2: 681–690.

Eaton R.W., Nitterauer J.D. (1994) Biotransformation of benzothiophene by isopropyl benzene- degrading bacteria. Journal of Bacteriology. 176: 3992–4002.

El-Gendy, N.Sh. (2001) Biodesulfurization of organosulfur compounds in crude oil and its fractions. MSc Thesis, Department of Chemistry, Cairo University, Cairo, Egypt.

El-Gendy, N.Sh. (2004) Biodesulfurization potentials of crude oil by bacteria isolated from hydrocarbon polluted environments in Egypt. PhD Thesis, Faculty of Science, Cairo University, Egypt.

El-Gendy, N.Sh., Farahat, L.A., Mostafa, Y.M., Shaker, N., El-Temtamy, S.A. (2006) Biodesulfurization of crude and diesel oil by *Candida parapsilosis* NSh45 isolated from Egyptian hydrocarbon polluted sea water. *Biosciences Biotechnology Research Asia* 3(1a): 5–16.

El-Gendy, N.Sh., Moustafa, Y.M., Habib, S.A., Ali, Sh. (2010) Evaluation of *Corynebacterium variabilis* Sh42 as a degrader for different poly aromatic compounds. *Journal of American Science*. 6(11): 343–356.

El-Gendy, N.Sh., Nassar, H.N., Abu Amr, S.S. (2014) Factorial design and response surface optimization for enhancing a biodesulfurization process. *Petroleum Science and Technology*. 32(14): 1669–1679.

El-Gendy, N.Sh., Speight J.G. (2016) Handbook of Refinery Desulfurization. CRC Press, Taylor and Francis Group, LLC, 6000 Broken Sound Parkway NW, Suite 300, Boca Raton, FL 33487–2742, USA.

El-Tokhi, M.M., Mostafa, Y.M. (2001) Heavy metals and petroleum hydrocarbons contamination of bottom sediments of El Sukhna area. Gulf of Suez. Egypt. *Petroleum Science and Technology*. 19: 481–494.

Etemadifar, Z., Derikvand, P., Emtiazi, G., Habibi, M.H. (2014) Response surface methodology optimization of dibenzothiophene biodesulfurization in model oil by nanomagnet immobilized *Rhodococcus erythropolis* R1. *Journal of Materials Science and Engineering B* 4(10): 322–330.

Etemadifar, Z., Emtiazi, G., Christofi, N. (2008) Enhanced desulfurization activity in protoplast transformed *Rhodococcus erythropolis. American-Eurasian Journal of Agricultural & Environmental Sciences*. 3: 285–291.

Evanoff, D.D., Chumanov, G. (2004) Size-controlled synthesis of nanoparticles. 1. "Silver-only" aqueous suspensions via hydrogen reduction. *Journal of Physical Chemistry B*. 108: 13948–56.

Evans J., Venables W. (1990) Degradation of thiophene-2- carboxylate, furan-2-carboxylate, pyrrole-2-carboxylate and other thiophene derivatives by the bacterium Vibrio YC1. *Applied Microbiology and Biotechnology*. 32:715–720.

Ewert, K.K., Evans, H.M., Bouxsein, N.F., Safinya, CR. (2006) Dendritic cationic lipids with highly charged headgroups for efficient gene delivery. *Bioconjugate Chemistry*. 17: 877–888.

Faison, B.D., Clark, T.M., Lewis, S.N., Ma, C.Y., Sharkey, D.M., Woodward, C.A. (1991) Degradation of organic sulfur compounds by a coal-solubilizing fungus. *Applied Biochemistry and Biotechnology*. 28/29: 237–251.

Fan, Y., Xu, S., Schaller, R. (2010). Nanoparticle decorated anodes for enhanced current generation in microbial electrochemical cells. Biosens Bioelectron. 26: 1908–1912.

Fasion, B.D., Clark, T.M., Lewis, S.N., Ma, C.Y., Sharkey, D.M., Woodward, C.A. (1991) Degradation of organic sulfur compounds by a coal solubilizing fungus. *Applied Biochemistry and Biotechnology*. 28: 237–251.

Fatahi, A., and Sadeghi, S. (2017) *Biodesulphurization* of gasoline by *Rhodococcus erythropolis* supported on polyvinyl alcohol. *Letters in Applied Microbiology*. 64: 370–378.

Fedorak P.M.; Coy D.L., Peakman T.M. (1996) Microbial metabolism of some 2,5-substituted thiophenes. Biodegradation. 7:313–327.

Fedorak P.N., Grbić-Galić D. (1991) Aerobic microbial cometabolism of benzothiophene and 3-methylbenzothiophene. *Applied and Environmental Microbiology*. 57:932–940.

Fedorak, P.M., Westlake, D.W.S. (1984) Degradation of sulfur heterocycles in Prudhoe Bay crude oil by soil enrichments. *Water Air Soil Pollution*. 21: 225–230.

Feng, J., Zeng, Y., Ma, C., Cai, X., Zhang, Q, Tong, M., Yu, B., and Xu, P. (2006) The Surfactant Tween 80 enhances biodesulfurization. *Applied and Environmental Microbiology*. 72(11): 7390–7393.

Finkelstein, Z.I., Baskunov, B.P., Vavilova, L.N., Golovleva, L.A. (1997) Microbial transformation of dibenzothiophene and 4,6-dimethyldibenzothiophene. *Microbiology*. 66: 402–407.

Finnerty W.R. (1992) Fossil resource biotechnology: Challenges and prospects. *Current Opinion in Biotechnology*. 3: 277–282.

Finnerty, W.R. (1993) Organic sulfur biodesulfurization in non-aqueous media. *Fuel* 72(12): 1631–1634.

Foght J.M., Westlake D.W.S. (1988) Degradation of polycyclic aromatic hydrocarbons and aromatic heterocycles by a *Pseudomonas* species. *Canadian Journal of Microbiology.* 34: 1135–1 141.

Folsom, B.R., Schieche, D.R., Digrazia, P. (1999) microbial desulfurization of alkylated dibenzothiophenes from a hydrodesulfurized middle distillate by *Rhodococcus erythropolis* I-19. *Applied and Environmental Microbiology*. 65: 4967–4972.

Forster, P.G.; Alexander, R.; Kagi, R.I. (1989) Identification and analysis of tetramethylnaphthalenes in petroleum, *Journal of Chromatography.* 483: 384–389.

Frase, H., Smith, C.A., Toth, M., Champion, M.M., Mobashery, S., Vakulenko, S.B. (2011) Identification of products of inhibition of GES-2 b-lactamase by tazobactam by x-ray crystallography and spectrometry. *Journal of Biological Chemistry*. 286: 14396–14409

Freeman, A., Lilly, M.D. (1998) Effect of processing parameters on the feasibility and operational stability of immobilized viable microbial cells. *Enzyme and Microbiology Technology.* 23: 335–345.

Fukui, S., Sonomoto, K., Tanaka, A. (1987) Entrapment of biocatalysts with photocrosslinkable resin prepolymers and urethane resin prepolymers. Methods in Enzymology. 135:230–252

Furuya, T., Ishii, Y., Nada, K., Kino, K., Kirimura, K. (2003) Thermophilic biodesulphurization of hydrodesulfurized light gas oils by *Mycobacterium phlei* WU-F1. *FEMS Microbiology Letters.* 221(1): 137–142.

Furuya, T., Kirimura, K., Kino, K., Usami, S. (2001). Thermophilic biodesulfurization of dibenzothiophene and its derivatives by *Mycobacterium phlei* WU-F1. *FEMS Microbiology Letters*. 204: 129–133.

Furuya, T., Takahashi, S., Ishii, Y., Kino, K., Kirimura, K. (2004) Cloning of a gene encoding flavin reductase coupling with dibenzothiophene monooxygenase through coexpression screening using indigo production as selective indication. *Biochemical and Biophysical Research Communications* 313:570–575.

Gai Z., Yu B., Li L., Wang Y., Ma C., Feng J., Deng Z., Xu P. (2007) Cometabolic degradation of dibenzofuran and dibenzothiophene by a newly isolated carbazole-degradaing *Sphingomonas* sp. strain. *Applied and Environmental Microbiology* 73(9):2832–2838.

Galán, B., Díaz, E., García, J.L. (2000) Enhancing desulphurization by engineering a flavin reductase-encoding gene cassette in recombinant biocatalysts. *Environmental Microbiology*. 2:687–694.

Gallagher, J.R., Olson, E.S., Stanley, D.C. (1993) Microbial desulfurization of dibenzothiophene: a sulfur specific pathway. *FEMS Microbiology Letters*. 107: 31–36.

Gallardo, M.E., Ferrandez, A., de Lorenzo, V., Garcia, J.L., Diaz, E. (1997) Designing recombinant *Pseudomonas* strains to enhance biodesulfurization. *Journal of Bacteriology*. 179: 7156–7160.

Gaudu P., Touati D., Niviere V. and Fontecave M. (1994) The NAD(P)H: flavin oxidoreductase from Escherichia coli as a source of superoxide radicals. *Journal of Biological Chemistry*. 269:8182–8185.

George, G.N., Gorbaty, M.L. (1989) Sulfur K-edge x-ray absorption spectroscopy of petroleum asphaltenes and model compounds. *Journal of the American Chemical Society*, 111(9): 3182–3186.

Gilbert, S.C., Morton, J., Buchanan, S., Oldfield, C., McRoberts, A. (1998) Isolation of a unique benzothiophene-desulfurizing bacterium, *Gordona* sp. strain 213E (NCIMB 40816), and characterization of the desulfurization pathway. *Microbiology*. 144: 2545–2553.

Gray K.A., Machkoyz T., Squires C.H. (2003) Biodesulfurization of fossil fuels. *Current Opinion in Microbiology*. 6: 229–235.

Gray, K.A., Pogrebinsky, O.S., Mrachko, G.T., Xi, L., Monticello, D.J., Squires, C.H. (1996) Molecular mechanisms of biocatalytic desulfurization of fossil fuels. Nature Biotechnology. 14: 1705–1708.

Grbic-Galic, D. (1989) Microbial degradation of homocyclic and heterocyclic hydrocarbons under anaerobic conditions. Developments in industrial microbiology. 30:237–253.

Grifoll, M., Selifonov, S.A., Chapman, P.J. (1995) Transformation of substituted fluorenes and fluorene analogs by *Pseudomonas* sp. strain F274. *Journa of Applied and Environmental Microbiology* 61: 3490–3493.

Grimalt J.O., Grifoll M., Solanas A.M., Albaigés J. (1991) Microbial degradation of marine evaporitic crude oils. Geochimica et Cosmochimica Acta. 55: 1903–1913.

Groboillot, A., Boadi, D.K., Poncelet, D. (1994) Immobilization of cells for application in the food industry. *Critical Reviews in Biotechnology*. 14: 75–107.

Grossman M.J., Lee M.K., Prince R.C., Garrett K.K., George G.N., Pickering I. (1999) Microbial desulfurization of a crude oil middle-distillate fraction: Analysis of the extent of sulfur removal and the effect of removal on remaining sulfur, *Applied and Environmental Microbiology*. 65:181–188.

Guerinik, K., Al-Mutawah, Q. (2003) Isolation and characterization of oil-desulphurizing bacteria. *World Journal of Microbiology and Biotechnology*. 19: 941–945.

Guisan, J.M. (2006) Immobilization of enzymes and cells. Humana Press. Totowa. pp 5–78.

Gün, G., Yürüm, Y., Doğanay, G.D. (2015) Revisiting the bio desulfurization capability of hyperthermophilic archaeon *Sulfolobus solfataricus* P2 revealed DBT consumption by the organism in an oil/water two-phase liquid system at high temperatures. *Turkish Journal of Chemistry*. 39: 255 – 266.

Gunam I.B.W., Yaku Y., Hirano M., Yamamura K., Tomita F., Sone T. and Asano K. (2006) Biodesulfurization of alkylated forms of dibenzothiophene and benzothiophene by *Sphingomonas subarctica* T7b. *Journal of Bioscience and Bioengineering*. 101(4): 322–327.

Gunam, I.B.W., Iqbal, M., Arnata, I.W., Antara, N.S., Anggreni, A.A.M.D., Setiyo, Y., Gunadnya, I.B.P. (2016) Biodesulfurization of dibenzothiophene by a newly isolated *Agrobacterium tumefaciens* LSU20. *Applied Mechanics and Materials*. 855:143–149.

Gunam, I.B.W., Triani, I.G.A.L., Antara, N.S., Duniaji, A.S., Setiyo, Y., Supata, D.A. (2012) Biocatalytic desulfurization of dibenzothiophene by *Pseudomonas* sp. strain KWN5. In: Proceedings of 4th International Conference on Biociences and Biotechnology. Advancing Life Sciences for Health [Anti-Aging Development in Particular] and Food Security. 21st-22nd September, 2012, Udayana University, Denpasar, Bali, Indonesia. Pp. 245–248.

Gunam, I.B.W., Yamamura, K., Nengah Sujaya, I., Antara, N.S., Aryanta, W.R., Tanaka, M., Asano, K. (2013) Biodesulfurization of dibenzothiophene and its derivatives using resting and immobilized cells of Sphingomonas subarctica T7b. *Journal of Microbiology and Biotechnology*,.23(4): 473–482.

Gunther, K., Helmut, S. (1993) Immobilization of Proteus vulgaris for the reduction of 2-oxo acids with hydrogen gas or formate to D-2-hydroxy acids. *Applied Microbiology Biotechnology* 38:441–446.

Guobin S., Huaiying Z., Weiquan C., Jianmin X., Huizhou L. (2005) Improvement of biodesulfurization rate by assembling nanosorbents on the surfaces of microbial cells. *Biophysical Journal*. 89(6): 58–60.

Guobin, S., Huaiying, Z., Jianmin, X., Guo, C., Wangliang, L., Huizhou, L. (2006) Biodesulfurization of hydrodesulfurized diesel oil with *Pseudomonas delafieldii* R-8 from high density culture. *Biochemical Engineering Journal*. 27: 305–309.

Ha, J. (2005) Bioremediation of the organophosphate pesticide, coumaphos, using microorganisms immobilized in calcium- alginate gel beads; A Ph.D. Thesis; Texas A&M University; USA.

Habe, H., Chung, J.S., Kato, H., Ayabe, Y., Kasuga, K., Yoshida, T., Nojiri, H., Yamane, H., Omori, T. (2004) Characterization of the upper pathway genes for fluorene metabolism in *Terrabacter* sp. strain DBF63. *Journal of Bacteriology*. 186: 5938–5944.

Hai, Y., Xudong, S., Qianqian, X., Zhao, M., Chengbin, X., Jun, N. (2008) Effects of nicotinamide and riboflavin on the bio- desulfurization activity of dibenzothiophene by Rhodococcus erythropolis USTB-03. *Journal of Environmental Science*. 20:613–618.

Hashimoto, S., Furukawa, K. (1987) Immobilization of activated sludge by PVA-boric acid method. *Biotechnology and Bioengineering*. 30: 52–59.

Hassan, C.M., Peppas, N.A. (2000) Structure and Applications of Poly (vinyl alcohol) Hydrogels Produced by Conventional Crosslinking or by Freezing-Thawing Methods. Advances in Polymer Science. 153: 37–65.

Haukanes, B.I., Kvam C. (1993). Application of magnetic beads in bioassays. Nature Biotechnology. 11: 60–63.

He, F., Zhao, A. (2005) Preparation and characterization of a new class of starch-stabilized bimetallic nanoparticles for degradation of chlorinated hydrocarbons in water. *Environmental Science and Technology*. 39: 3314–3320.

He, F., Zhao, A. (2007) Manipulating the size and dispensability of zerovalent iron nanoparticles by use of carboxymethyl cellulose stabilizers. *Environmental Science and Technology*. 41: 6216–6221.

He, H., Hong, F.F., Tao, X.X., Li, L., Ma, C.Y., Zhao, Y.D. (2012) Biodesulfurization of coal with Acidithiobacillus caldus and analysis of the interfacial interaction between cells and pyrite; *Fuel Processing and Technology*. 101:73–77.

Hegazi, A.H.; Andersson, J.T. and El-Gayar, M.S. (2003) Application of gas chromatography with atomic emission detection to the geochemical investigation of polycyclic aromatic sulfur heterocycles in Egyptian crude oils. *Fuel Processing and Technology*. 85(1): 1–19.

Heipieper, H.J., Neumann, G., Cornelissen, S., Meinhardt, F. (2007) Solvent-tolerant bacteria for biotransformations in two-phase fermentation systems. *Applied and Environmental Microbiology*. 71: 961–973.

Hennebel, T., Simoen, H., De Windt, W. (2009) Biocatalytic dechlorination of trichloroethylene with bio-palladium in a pilot-scale membrane reactor. *Biotechnology Bioengineering*. 102: 995–1002.

Hirech, K., Payan, S., Carnelle, G., Brujes, L., Legrand, J. (2003) Microencapsulation of an insecticide by interfacial polymerization. Powder Technology. 130: 324–330.

Holland H.L., Khan S.H., Richards D., Riemland E. (1986) Biotransformation of polycyclic aromatic compounds by fungi Xenobiotica.16: 733–741.

Hostettler, F.D., Kvenvolden, K.A. (1994) Geochemical changes in crude oil spilled from the Exxon Valdez super tanker into Prince William Sound Alaska. *Organic Geochemistry*. 21:927–936.

Hou Y., Kong Y., Yang J., Zhang J., Shi D., and Xin W. (2005) Biodesulfurization of dibenzothiophene by immobilized cells of *Pseudomonas stutzeri* UP-1. *Fuel*. 84: 1975–1979.

Houng, J.Y., Chiang, W.P., Chen, K.C. (1994) Hydroxylation of progesterone in biphasic media using alginate entrapped Aspergillus ochraceus gel beads coated with polyuria. *Enzyme Microbiology Technology*. 16:485–491.

Huang, T., Qiang, L., Zelong, W., Daojang Y., Jiamin X. (2012) Simultaneous removal of Thiophene and Dibenzothiophene by immobilized *Pseudomonas delafieldii* R-8 cells. *hinese Journal of Chemical Engineering*. 20:47–51.

Hull, W.Q., Keel, H., Kenny, Jr., K. (1953) Diatomaceous earth. A staff-industry collaborative report. *Journal of Indian Engineering Chemistry*. 54:256–260.

Hulst, H.C., Tramper, J., Rict, K. (1985) Anewtechnique for the production of immobilized biocatalyst in large quantities. *Journal of Biotechnology and Bioengineering*. 27: 870–876.

Hwan, J., Keun, Y., Wook, H., Nam, H. (2000) Desulfurization of light gas oil in immobilized-cell systems of *Gordona* sp. CYKS1 and *Nocardia* sp. CYKS2. *FEMS Microbiology Letters*. 182: 309–312.

Ichinose, H., Nakamizo M, Wariishi, H., Tanaka, H. (2002b) Metabolic response against sulfur-containing heterocyclic compounds by the lignin-degrading basidiomycete *Coriolus versicolor. Applied Microbiology and Biotechnology.* 58(4): 517–26.

Ichinose, H., Wariishi, H., Tanaka, H. (1999) Bioconversion of recalcitrant 4-methyldibenzothiophene to water-extractable products using lignin-degrading basidiomycete *Coriolus versicolor. Biotechnology Progress.* 15: 706–714.

Ichinose, H., Wariishi, H., Tanaka, H. (2002a) Identification and characterization of novel cytochrome P450 genes from white-rot basidiomycete *Coriolus versicolor. Applied Microbiology and Biotechnology.* 58: 97–105.

Ishii, Y., Konishi, J., Suzuki, M., & Maruhashi, K. (2000) Cloning and expression of the gene encoding the thermophilic NAD(P)H-FMN oxidoreductase coupling with the desulfurization enzymes from Paenibacillus sp. A11–2. *Journal of Bioscience and Bioengineering.* 90(6): 591–599.

Islam, A., Phillips, G., Snowden, M., Williams, P. (1997) A review of recent developments on the regulatory, structural and functional aspects of gum Arabic. Food Hydrocolloids. 11:493–505.

Ismail, W., El-Sayed, W.S., Abdul Raheem, A.S., Mohamed, M.E, El Nayal, A.M. (2016) Biocatalytic desulfurization capabilities of a mixed culture during non-destructive utilization of recalcitrant organosulfur compounds. Frontiers in Microbiology. Volume 7: Article 266. doi: 10.3389/fmicb.2016.00266

Itoh, N., Yoshida, M., Miyamoto, T., Ichinose, H., Wariishi, H., Tanaka, H. (1997) Fungal cleavage of thioether bond found in Yperite. FEBS Letters. 412: 281–284.

Izumi, Y., Ohshiro, T., Ogino, H., Hine, Y., Shimao, M. (1994) Selective desulfurization of dibenzothiophene by *Rhodococcus erythropolis* D-1. *Applied and Environmental Microbiology.* 60: 223–236.

Jack, T.R., Zajic, J.E. (1997) Advances in biochemical engineering: the immobilization of whole cells. vol 5. Springer, Berlin, pp 125–145.

Jain, D., Kachhwaha, S., Jain, R. (2010) Novel microbial route to synthesize silver nanoparticles using spore crystal mixture of *Bacillus thuringiensis. Indian Journal of Experimental Biology.* 48: 1152–1156.

Jain, P.K., Huang, X., El-Sayed, I.H., El-Sayed, M.A. (2008) Noble metals on the nanoscale: optical and photothermal properties and some appli- cations in imaging, sensing, biology, and medicine. Accounts of Chemical Research. 41: 1578–86.

Jiang C., Liu, H., Xie, Y., Chen, J. (2002) Isolation of soil bacteria species for degrading dibenzothiophene. *Chinese Journal of Chemical Engineering.* 10(4): 420–426.

Jimenez, V., Bravo, V., Gutierrez, L.G. (2011) integral approach for improving the degradation of recalcitrant petro hydrocarbons in a fixed-film reactor. *Water Air Soil Pollution.* 220: 301–312

Joachim, K., Holger, Ziehr. (1990) Immobilization of Microbial cells by adsorption. *Journal of Biotechnology.* 16:1–16.

Jorjani, E., Rezai, B., Vossoughi, M., Osanloo, M. (2004) Biodesulfurization of the tabas deposit coal by microorganisms. *Journal of Mining Science.* 40(3): 310–320.

Juarez-Moreno K., Diaz de Leon J.N., Zepeda T.A., Vazquez-Duhalt R., Fuentes S. (2015) Oxidative transformation of dibenzothiophene by chloroperoxidase enzyme immobilized on (1D)- γ-Al_2O_3 nanorods. *Journal of Molecular Catalysis B: Enzymatic* . 115: 90–95.

Juarez-Moreno, K., Pestryakov, A., Petranovskii, V. (2014) Engineering of supported nanomaterials. *Procedia Chemistry.* 10: 25–30.

Kafayati, M., Raheb, J., Angazi, M., Alizadeh, S., Bardania, H. (2013) The effect of magnetic Fe3O4 nanoparticles on the growth of genetically manipulated bacterium *Pseudomonas aeruginosa* PTSOX4; Iran Biotech. 11:41–46.

Kaila K. (2002) Microencapsulation of Probiotic Bacteria: Technology and potential Applications. Current Issues in Intestinal Microbiology. 3:39–48.

Kanagawa T. and Kelly D.P. (1987) Degradation of substituted thiophenes by bacteria isolated from activated sludge. Microb. Ecol. 13:47–57.

Kapoor, M., Kuhad, R.C. (2007) Immobilization of xylanase from Bacillus pumilus strain MK001 and its application in production of oligosaccharides; *Applied Biochemistry and Biotechnology.* 142:125–38.

Kareem, S.A., Aribike, D.S., Nwachukwu, S.C., Latinwo, G.K. (2012) Microbial desulfurization of diesel by Desulfobacterium indolicum. *Journals in Environmental Science and Engineering.* 54(1): 98–103.

Karel, S.F., Robertson, C.R. (1989) Cell mass synthesis and degradation by immobilized *Escherichia coli. Biotechnology & Bioengineering.* 34:337–356.

Kargi, F., Robinson, J.H. (1984) Microbial oxidation of dibenzothiophene by the thermophilic organism *Sulfolobus acidocaldarius. Biotechnology and Bioengineering.* 16: 687–690.

Karimi, E., Jeffryes, C., Yazdian, F., Akhavan Sepahi, A., Hatamian, A., Rasekh, B., Ashrafi, S. J. (2017) DBT desulfurization by decorating Rhodococcus erythropolis IGTS8 using magnetic Fe_3O_4 nanoparticles in a bioreactor. *Engineering in Life Sciences*, 17(5): 528–535.

Karlsen, D.A., Larter, S.R. (1991) Analysis of petroleum fractions by TLC-FID: applications to petroleum reservoir description. *Organic Geochemistry.* 17: 603–617.

Karsten, G., Simon, H. (1993) Immobilization of *Proteus vulgaris* for the reduction of

2-oxo acids with hydrogen gas or formate to D-2-hydroxy acids. *Applied Microbiology and Biotechnology.* 38: 441–446.

Kawaguchi, H., Kobayashi, H., Sato, K. (2012) Metabolic engineering of hydrophobic

Rhodococcus opacus for biodesulfurization in oil-water biphasic reaction mixtures. *Journal of Bioscience and Bioengineering.* 113: 360–366.

Kawatra, S.K., Eisele, T.C. (2001) Coal desulfurization, high-efficiency preparation methods. Taylor & Francis, New York.

Kayser, K.J., Cleveland, L., Park, H.S., Kwak, J.H., Kolhatkar, A., Kilbane, J. (2002) Isolation and characterization of moderate thermophile *Mycobacterium phlei* GTIS10, capable of dibenzothiophene desulfurization. *Applied Microbiology and Biotechnology*. 59: 737–746.

Kayser, Y., Bielagajones, B.A., Jackowski, K., Odusan, O., Kilbane, J.J. (1993) Utilization of organusulphur compounds by axenic and mixed cultures of *Rhodococcus rhodochrous* IGTS8. *Journal of General Microbiology*. 139: 3123–3129.

Kertesz M.A. (2000) Riding the sulfur cycle – metabolism of sulfonates and sulfate esters in gram-negative bacteria. FEMS Microbiology Reviews. 24: 135–175.

Kertesz, L. (2001) Building a scientific foundation for prevention. Health Plan. 42: 44–47.

Khedkar, S., Shanker, R. (2015) Isolation and classification of a soil actinomycete capable of sulphur specific biotransformation of dibenzothiophene, benzothiophene and thianthrene. *Journal of Applied Microbiology*. 118(1): 62–74.

Kilbane II, J.J., Jackowski, K. (1992) Biodesulfurization of water soluble coal derived material by *Rhodococcus rhodochrous* IGTS8. *Biotechnology and Bioengineering*. 40: 1107–1114.

Kilbane II, J.J., Robbins, J. (2007) Characterization of the dszABC genes of *Gordonia amicalis* F. 5.25. 8 and identification of conserved protein and DNA sequences. *Applied Microbiology and Biotechnology*. 75(4): 843–851.

Kilbane, J.J. (1989) Desulfurization of coal: the microbial solution. Trends in Biotechnology. 7:97–101.

Kilbane, J.J. (1991) Bacterial produced extracts and enzymes for cleavage of organic C-S bonds. European Patent no. 0 445 896 A2.

Kilbane, J.J. (2006) Microbial biocatalyst development to upgrade fossil fuels. Current

Opinion in Microbiology. 17: 305–314.

Kilbane, J.J., Bielaga, B.A. (1990) Toward sulfur-free fuels. ChemtTch. 20: 747–751.

Kilbane, J.J., Le Borgne, S. (2004) Petroleum biorefining: the selective removal of sulfur, nitrogen, and metals. In Petroleum Biotechnology, Developments and Perspectives, pp. 29–65. Edited by

R. Vazquez-Duhalt & R. Quintero-Ramirez. Amsterdam: Elsevier.

Kim C.G., Chon, B. (2002) Oxygen uptake characteristics of soil inoculum amended with thiophene derivatives. *Korean Journal of Chemical Engineering*. 19:773–779.

Kim, H.Y., Kim, T.S., Kim, B.H. (1990a) Degradation of organic sulfur compounds and the reduction of dibenzothiophene to biphenyl and hydrogen sulfide by *Desulfovibrio desulfuricans* M6. *Biotechnology Letters*. 12: 761–764.

Kim, J.K., Park, J.W. (1999) The biological and chemical desulfurization of crumb rubber for the rubber compounding. *Journal of Applied Polymer Science*. 72(12): 1543–1549.

Kim, S.J., Park, J.Y., Lee, Y.J., Lee, J.Y., Yang, J.W. (2005) Application of a new electrolyte circulation method for the ex situ electrokinetic bioremediation of a laboratory prepared pentadecane contaminated kaolinite. *Journal Hazardous Materials*. 118: 171–176.

Kim, T.S., Kim, H.Y., Kim, B.H. (1990b) Petroleum desulfurization by *Desulfovibrio desulfuricans* M6 using electrochemically supplied reducing equivalent. *Biotechnology Letters*. 12: 757–760.

Kim, Y.J., Chang, J.H., Cho, K.-S., Ryu, H.W., Chang, Y.K. (2004) A physiological study on growth and dibenzothiophene (DBT) desulfurization characteristics of *Gordonia* sp. CYKS1. *Korean Journal of Chemical Engineering*. 21(2): 436–441.

Kirimura, K., Furuya, T., Nishii, Y., Ishii, Y., Kino, K., Usami, S. (2001)

Biodesulfurization of dibenzothiophene and its derivatives through the selective cleavage of carbon-sulfur bonds by a moderately thermophilic bacterium *Bacillus subtilis* WU-S2B. *Journal of Bioscience and. Bioengineering*. 91, 262–266.

Kirimura, K., Furuya, T., Sato, R., Ishii, Y., Kino, K., Usami, S. (2002) Biodesulphurization of naphthothiophene and benzothiophene through selective cleavage of carbon-sulfur bonds by *Rhodococcus* sp. strain WUK2R. *Applied and Environmental Microbiology*. 68: 3867–3872.

Kishimoto, M., Inui, M., Omasa, T., Katakura, Y., Suga, K., Okumura, K. (2000) Efficient production of desulfurizing cells with the aid of expert system. *Biochemical Engineering Journal*. 5: 143–147.

Klein J., Van afforded M., Pfeiffer F., Schacht S. (1994) Microbial desulfurization of coal and oil. Fuel Processing Technology. 40:297–310.

Klein, J., Stock, J., Vorlop, K.D. (1983) Pore size and properties of spherical Ca-alginate biocatalysts. *European Journal of Applied Microbiology and Biotechnology*. 18:86–91

Klibanov, A. M., Tu, T. M., and Scott, K. P. 1983. Peroxidase-catalyzed removal of phenols from coal-conversion wastewaters. Science. 221: 259– 261.

Klyachko N.L., Klibanov A.M. (1992) Oxidation of Dibenzothiophene Catalyzed by Hemoglobin and Other Hemoproteins in Various Aqueous-Organic Media. *Applied Biochemistry and Biotechnology*. 37(1), 53–68.

Kobayashi M., Onaka T., Ishii Y., Konishi J., Takaki M., Okada H., Ohta Y., Koizumi K., Suzuki, M. (2000) Desulfurization of alkylated forms of both dibenzothiophene and benzothiophene by a single bacterial strain. *FEMS Microbiology Letters*. 187:123–126.

Kobayashi, S., Nakano, M., Goto, T., Kimura, T., Schaap, A.P. (1986) An evidence of the peroxidase-dependent oxygen transfer from hydrogen peroxide to sulfides. *Biochemical and Biophysical Research Communications*. 135: 166–171.

Kodama, K., Umehara, K., Shimizu, K., Nakatani, S., Monioda, Y., Yamada, K. (1973) Identification of microbial products from dibenzothiophene and its proposed oxidation pathway. *Agricultural and Biological Chemistry*. 37: 45–50.

Koebel, M.M., Jones, L.C., Somorjai, G.A. (2008) Preparation of size- tunable, highly monodisperse PVP-protected Pt-nanoparticles by seed- mediated growth. *Journal of Nanoparticle Research*. 10:1063–1069.

Köhler, M., Genz, I.-L., Schicht, B., Eckart, V. (1984) Microbial desulfurization of petroleum and heavy petroleum fractions. 4. Anaerobic degradation of organic sulfur compounds of petroleum. Zentralblatt Fur Mikrobiologie. 139: 239–247.

Konishi M., Kishimoto M., Omasa I., Katakura Y., Shioya S., Ohtake H., (2005) Effect of sulfur sources on specific desulfurization activity of Rhodococcus erythropolis KA251 in exponential fed-batch culture. *Journal of Bioscience and Bioengineering*. 24:49–54.

Konishi, J., Ishii, Y., Onaka, T., Okumura, K.,, Suzuki M. (1997) Thermophilic carbon–sulfur bond-targeted biodesulfurization. *Applied and Environmental Microbiology* 63: 3164–3169.

Konishi, J., Onaka, T., Ishii, Y., Suzuki, M. (2000) Demonstration of the carbon-sulfur bond targeted desulfurization of benzothiophene by thermophilic *Paenibacillus* sp. strain A11–2 capable of desulfurizing dibenzothiophene. *FEMS Microbiology Letters*. 187: 151–154.

Krahn, M. M., Stein, J. E. (1998) Peer reviewed: Assessing exposure of marine biota and habitats to petroleum compounds. *Analytical Chemistry*.70: 186–192.

Krahn, M.M., Ylitalo, G.M., Buzitis, J., Chan, S.L., Varanasi, U. (1993) Rapid high-performance liquid chromatographic methods that screen for aromatic compounds in environmental samples. *Journal of Chromatography A*. 642(1): 15–32.

Krawiec, S. (1988): Detection, isolation and initial characterization of bacteria with the ability to desulfurize dibenzothiophene to O,O'-Biphenol; Bioprocessing of Coals Conference; Sheladia Associates; pp. 263–274.

Krawiec, S. (1990) Bacterial desulfurization of thiophenes: screening techniques and some speculations regarding biochemical and genetic bases. *Developments in industrial microbiology*. 31: 103–114.

Kropp K.G., Saftic S., Andersson J.T., Fedorak P.M. (1996) Transformations of six isomers of dimethylbenzothiophenes by three *Pseudomonas* strains. Biodegradation. 7: 203–221

Kropp, P., Gerber, W.D. (1998) Prediction of migraine attacks using a slow cortical potential, the contingent negative variation. Neuroscience Letters. 257:73–76.

Kruger, N.J. (2002) The Bradford method for protein quantitation. In: Walker JM (ed) The protein protocols handbook, 2nd edn. Humana Press Inc., Totowa, pp 15–21.

Kubo, T., Sugita, T., Shimose, S. (2000) Targeted delivery of anticancer drugs with intravenously administered magnetic liposomes in osteosarcoma-bearing hamsters. *International Journal of Oncology*. 17: 309–316.

Kurita S.; Endo T.; Nakamura H.; Yagi T., Tamuja, N. (1971) Decomposition of some organic sulfur compounds in petroleum by anaerobic bacteria. *Journal of General and Applied Microbiology*. 17: 185–198.

Labana, S., Pandey, G., Jain, R.K. (2005) Desulphurization of dibenzothiophene and diesel oils by bacteria. *Letters in Applied Microbiology*. 40: 159–163.

Lazar, A. (1991) Immobilization of animal cells in fixed bed bioreactors. *Biotechnology Advances*. 9:411–424.

Le Borgne S., Ayala M. (2010) Microorganisms utilizing sulfur-containing hydrocarbons. In: Timmis, K.N. (Ed.). Handbook of Hydrocarbon and Lipid Microbiology. Springer-Verlag. Berlin Heidelberg. Germany.

Le Borgne, S., Quintero, R. (2003) Biotechnological processes for the refining of petroleum. *Fuel Process Technology*. 81: 155–169.

Lee M.K., Senius J.D., Grossman M.J. (1995) Sulfur-specific microbial desulfurization of sterically hindered analogs of dibenzothiophene. *Applied and Environmental Microbiology*. 61:4362–4366.

Lee M.Y.; Dordick J.S. (2002) Enzyme activation for non-aqueous media, Current Opinion in Biotechnology. 13(4):376–384.

Lee W.C., Ohshiro T., Matsubara T., Izumi Y. and Tanokura M. (2004) Crystallization and preliminary X-ray analyses of desulfurization enzyme DszB and its C27S mutant complexed with biphenyl-2-sulfinic acid. Acta Crystallographica Section D. 60:1636–1638

Lee, C., Sedlak, D.L. (2008) Enhanced formation of oxidants from bimetallic nickel-iron nanoparticles in the presence of oxygen. *Environmental Science and Technology*, 42: 8528–8533.

Lee, D., Khaja, S., Velasquez-Castano, J.C. (2007) In vivo imaging of hydrogen peroxide with chemiluminescent nanoparticles. Nature Materials. 6: 765–9.

Lee, I.S., Bae, H., Ryu, H.W., Cho, K., Chang, Y.K. (2005) Biocatalytic desulfurization of diesel oil in an airlift reactor with immobilized *Gordonia nitida* CYKS1 cells. Biotechnology Progress. 21(3): 781–785.

Lee, J.H., Huh, Y.M., Jun, Y.W. (2006) Artificially engineered magnetic nanoparticles for ultra-sensitive molecular imaging. Nature Medicine. 13: 95–9.

Lee, K.Y., Heo, T.R. (2000) Survival of *Bifidobacterium longum* immobilized in calcium alginate beads in simulated gastric juices and bile salt solution. *Applied and Environmental Microbiology*. 66(2):869–873.

Leon, S., Quarrell, R., Lowe, G. (1998) Evaluation of resins for on-bead screening: A study of papain and chymotrypsin specificity using PEGA-bound combinatorial peptide libraries. Bioorganic & Medicinal Chemistry Letters. 8: 2997–3002.

Leong, Y.K., Seah, U., Chu, S.Y., Ong, B.C. (2001) Effects of gum Arabic macromolecules on surface forces in oxide dispersions. Colloids and Interfaces.182: 263–268.

Li G-g., Li S-s., Zhang M-j., Wang J., Zhu L., Liang F-l., Liu R-j., and Ma T. (2008b) Genetic rearrangement strategy for optimizing the dibenzothiophene biodesulfurization pathway in *Rhodococcus erythropolis*. *Applied and Environmental Microbiology*. 74(4): 971–976.

Li W., Xing J., Li Y., Xiong X., Li X., Liu H. (2008c) Desulfurization and bioregeneration of adsorbents with magnetic *P. delafieldii* R-8 cells. *Catalysis Communications*. 9: 376–380.

Li, F. L., Xu, P., Feng, J., Meng, L., Zheng, Y., Luo, L., Ma, C. (2005b) Microbial desulfurization of gasoline in a *Mycobacterium goodii X7B* immobilized-cell system. *Applied and Environmental Microbiology*. 71: 276–281.

Li, F.L., Xu, P., Ma, C.Q., Luo, L.L., Wang, X.S. (2003) Deep desulfurization of hydrodesulfurization-treated diesel oil by facultative thermophilic bacterium *Mycobacterium* sp. X7B. *FEMS Microbiology Letters*. 233(2): 301–307.

Li, G.Q., Li, S.S., Qu, S.W., Liu, Q.K., Ma, T., Zhu, L., Liang, F.L., Liu, R.L. (2008a) Improved biodesulfurization of hydrodesulfurized diesel oil using *Rhodococcus erythropolis* and *Gordonia* sp. *Biotechnology Letters*. 30: 1759–1764.

Li, M., Squires, C., Monticello, D., Childs, J. (1996) Genetic analysis of the *dsz* promoter and associated regulatory regions of *Rhodococcus erythropolis* IGTS8. *Journal of Bacteriology*. 178: 6409–6418.

Li, W., Xing, J., Xiong, X., Huang, J., Liu, H. (2006) Feasibility study on the integration of adsorption/bioregeneration of π-complexation adsorbent for desulfurization. *Industrial and Engineering Chemistry Research*. 45(8): 2845–2849.

Li, W., Tang, H., Liu, Q., Xing, J., Li, Q., Wang, D., Yang, M., Li, X., Liu, H. (2009) Deep desulfurization of diesel by integrating adsorption and microbial method. *Biochemical Engineering Journal*. 44: 297–301.

Li, W., Zhang, Y., Wang, D.M., Shi, Y. (2005a) Biodesulfurization of dibenzothiophene and other organic sulfur compounds by a newly isolated *Microbacterium* strain ZD-M2. *FEMS Microbiology Letters*. 247: 45–50.

Li, W., Zhao, D. (2013) An overview of the synthesis of ordered mesoporous materials. *Chemical Communications*. 49: 943–946.

Li, Y., Zhao, S., Wang, Y. (2011) Microbial desulfurization of ground tire rubber by *Thiobacillus ferrooxidans*. *Polymer Degradation and Stability*. 96(9): 1662–1668.

Li, Y.G., Ma, J., Zhang, Q.Q., Wang, C.S., Chen, Q. (2007) Sulfur-selective Desulfurization of dibenzothiophene and diesel oil by newly isolated *Rhodococcus erythropolis* NCC-1. *Chinese Journal of Organic Chemistry*. 25: 400–405.

Liang F., Lu M.M., Birch M.E., Keener M.C., Liu Z.F. (2006) Determination of polycyclic aromatic sulfur heterocycles in diesel particular matter and diesel fuel by gas chromatography with atomic emission detection. *Journal of Chromatography A*. 1114:145–153

Liu, X.Q., Liu, H.Z., Xing, J.M., Guan, Y.P., Ma, Z.Y., Shan, G.B., Yang, C.L. (2003) Preparation and characterization of superparamagnetic functional polymeric microparticles. *China Particuol journal*. 1: 76–79.

Liu, Y., Bisson, T. M., Yang, H., Xu, Z. (2010) Recent developments in novel sorbents for flue gas clean up. *Fuel Processing Technology*. 91(10): 1175–1197.

Liu, Y.S., Wu, J.Y., Ho. K.P. (2006) Characterization of oxygen transfer conditions and their effects on *Phaffia rhodozyma* growth and carotenoid production in shake-flask cultures. *Biochemical Engineering Journal*. 27: 331–335.

Lizama, H.M., Wilkins, L.A., Scott, T.C. (1995) Dibenzothiophene sulfur can serve as the sole electron acceptor during growth by sulfate-reducing bacteria. *Biotechnology Letters*. 17(1): 113–116.

Lopez, A., Lazaro, N., Marques, A.M. (1997) The interphase technique: a simple method of cell immobilization in gel-beads. *Journal of Microbiological Methods*. 30: 231–234

Lukhele, L.P., Bhekie, B.M., Momba, M.N.B, Krause, R.W.M. (2010) Water disinfection using novel cyclodextrin polyurethanes containing silver nanoparticles supported on carbon nanotubes. *Journal of Applied Science.* 10: 65–70.

Luo, M.F., Xing, J.M., Gou, Z.X., Li, S., Liu, H.Z., Chen, J.Y. (2003) Desulfurization of dibenzothiophene by lyophilized cells of *Pseudomonas delafieldii* R-8 in the presence of dodecane. *Biochemical Engineering Journal.* 13: 1–6.

Luo, Q., Zhang, X., Wang, H., Qian, Y. (2005) The use of non-uniform electrokinetics to enhance in situ bioremediation of phenol-contaminated soil. *Journal Hazardous Materials.* 121: 187–194.

Ma C.Q., Feng J.H., Zeng Y.Y., Cai X.F., Sun B.P., Zhang Z.B., Blankespoor H.D. and Xu P. (2006b). Methods for the preparation of a biodesulfurization biocatalyst using *Rhodococcus* sp. *Chemosphere.* 65: 165–169.

Ma T., Li G.Q., Li J., Liang F.L. and Liu R.L. (2006a) Desulfurization of dibenzothiophene by *Bacillus subtilis* recombinants carrying *dszABC* and *dszD* genes. *Biotechnology Letters.* 28:1095–1100.

Ma, H., Yin, B., Wang, S, (2004) Synthesis of silver and gold nanoparticles by a novel electrochemical method. Chem Phys Chem. 5: 68–75.

Ma, T., Tong, M., Zhang, Q., Liang, F., Liu, R. (2006) Screening, identification of the strain Fds1 for microbial desulfurization specially and its use in diesel oil desulfurization. Acta Microbiologica Sinica. 46:104–110.

Mabery, C.F., Smith, A.W. (1891) On the composition of certain petroleum oils and of refining residues. *American Chemistry Journal.*13: 232–243.

MacPherson T., Greer C.W., Zhou E., Jones A.M., Wisse G., Lau, C.K., Sankey B.M., Grossman J., Hawari J. (1998) Application of SPME/GC-MS to characterize metabolites in the biodesulfurization of organosulfurmodel compounds in bitumen. *Environmental Science & Technology.* 32:421–426.

Madeira, L., Santana, V., Pinto, E. (2008) Dibenzothiophene oxidation by horseradish peroxidase in organic media: Effect of the DBT:H_2O_2 molar ratio and H_2O_2 addition mode. *Chemosphere.* 71: 189–194.

Maghsoudi, S., Kheirolomoom, A., Vossoughi, M., Tanaka, E. and Katoh, S. (2000) Selective desulfurization of dibenzothiophene by newly isolated *Corynebacterium* sp. strain P32C1. *Biochemical Engenering Journal.* 5: 11–16.

Maghsoudi, S., Vossoughi, M., Kheirolomoom, A., Tanaka, E., Katoh, S. (2001) Biodesulfurization of hydrocarbons and diesel fuels by Rhodococcus sp. strain P32C1. *Biochemical Engineering Journal.* 8:151–156.

Mahmoudi, M., Simchi, A., Milani, A.S., Stroeve, P. (2009) Cell toxicity of superparamagnetic iron oxide nanoparticles. *Journal of Colloid and Interface Science.* 336(2): 510–518.

Malik, K.A, Claus, D. (1976) Microbial degradation of dibenzothiophene. In: "Abstracts of the Fifth International Fermentation Symposium. (Dellweg H. Ed.)". Berlin. Germany. 23(03): 421.

Malik, K.A. (1978) Microbial removal of organic sulphur from crude oil and the environment: Some new perspectives. *Process Biochemistry.* 13: 10–13.

Mamie Z.L., Charles H.S., Daniel J.M. (1996) Genetic analysis of the *dsz* promoter and associated regulatory regions of *Rhodoccus erythropolis* IGTS8. *Journal of Bacteriology*. 11(178): 6409–6418.

Marten, F.L. (1927) Vinyl alcohol polymers. In Encyclopedia of Chemical Technology, 7: 1–34.

Martin, A.B., Alcón, A., Santos, V.E., García-Ochoa, F. (2004) Production of a biocatalyst of *Pseudomonas putida* CECT5279 for dibenzothiophene (DBT) biodesulfurization for different media composition. *Energy and Fuels*. 18: 851–857.

Martinsen, A., Skjåk-Bræk, G., Smidsrød, O. (1989) Alginate as immobilization ma- terial: I. Correlation between chemical and physical properties of alginate gel beads. *Biotechnology Bioengineering* 33: 79–89.

Matsubara T., Ohshiro T., Nishina Y. and Izumi Y. (2001) Purification, characterization, and over- expression of flavin reduced involved in dibenzothiophene desulfurization by *Rhodococcus erythropolis* D-1. *Applied and Environmental Microbiology*. 67:1179–1184.

Matsui T., Noda K., Tanaka Y., Maruhashi K. and Kurane R. (2002) Recombinant *Rhodococcus* sp. strain T09 can desulfurize DBT in the presence of inorganic sulfate. *Current Microbiology*. 45: 240–244.

Matsui T., Onaka T., Tanaka Y., Tezuka T., Suzuki M., Kurane R. (2000) Alkylated benzothiophene desulfurization by Rhodococcus sp. strain T09. *Bioscience, Biotechnology, and Biochemistry*. 64:596–599.

Matsui, T., Hirasawa, K., Konishi, J., Tanaka, Y., Maruhashi, K., Kurane, K. (2001) Microbial desulfurisation of alkylated dibenzothiophene and alkylated benzothiophene by recombinant *Rhodococcus* sp. Strain T09. *Applied Microbiology and Biotechnology*. 56: 196–200.

McFarland, B.L. (1999). Biodesulfurization. Current Opinion in Microbiology. 2: 257–264.

McFarland, B.L., Boron, D.J., Deever, W., Meyer, J.A., Johnson, A.R., Atlas, RM. (1998) Biocatalytic sulfur removal from fuels: applicability for producing for producing low sulfur gasoline. *Critical Reviews in Microbiology*. 24: 99–147.

Mezcua, M., Fernandez-Alba, A.R., Boltes, K., Del Aguila, R.A., Leton, P., Rodriguez, A., Garcia-Calvo, E. (2008) Determination of PASHs by various analytical techniques based on gas chromatography–mass spectrometry application to a biodesulfurization process. Talanta 75:1158–1166.

Mezcua, M., Fernandez-Alba, A.R., Rodriguez, A., Boltes, K., Leton, P., Garcia-Calvo, E. (2007) Chromatographic methods applied in the monitoring of biodesulfurization processes-state of the art. Talanta 73:103–114

Miller, K.W. (1992) Reductive desulfurization of dibenzyldisulfide. *Applied and Environmental Microbiology*.58(7): 2176–2179.

Mingfang, L., Zhongxuan, G., Jianmin, X., Huizhou, L., Jiayong, C. (2003) Microbial desulfurization of model and straight-run diesel oils. *Journal of Chemical Technology and Biotechnology*. 78: 873–876.

Mirgorodskaya, O.A., Kozmin, Y.P., Titov, M.I., Korner, R., Sonksen, C.P., Roepstorff, P. (2000) Quantitation of peptides and proteins by matrix assisted laser desorption/ionization mass spectrometry using 18O-labeled internal standards. Rapid Commun Mass Spectrom. 14:1226–1232.

Mishalanie, E.A., Birks, J.W. (1986) Selective Detection of Organosulfur Compounds in High Performance Liquid Chromatography. *Analytical Chemistry*. 58: 918–923.

Mishalanie, E.A., Rutkowski, J.C., Hutte, R.S., Birks, J.W. (1986) Ultaviolet absorption spectrum of genus hypochlorous acid. *Journal of physical chemistry*. 90:5578.

Mohamed, M.E., Al-Yacoub, Z.H., Vedakumar, J.V. (2015) Biocatalytic desulfurization of thiophenic compounds and crude oil by newly isolated bacteria. Frontiers in Microbiology. 6: article 112. doi: 10.3389/fmicb.2015.0 0112.

Mohebali, G., Ball, A., Keytash, A., Rasekh, B., (2007b) Stabilization of water/gas oil emulsions by desulfurizing cells of *Gordonia alkanivorans* RIPI90A. *Microbiology*. 153: 1573–1581.

Mohebali, G., Ball, A.S. (2008) Biocatalytic desulphurization (BDS) of petrodiesel fuels. *Microbiology*. 154: 2169–2183.

Mohebali, G., Ball, A.S., Kaytash A., Rasekh, B. (2008) Dimethyl sulfoxide (DMSO) as the sulfur source for the production of desulfurizing resting cells of *Gordonia alkanivorans* RIPI90A. *Microbiology* 154: 878–885.

Mohebali, G., Ball, A.S., Rasekh, B., Kaytash, A. (2007a) Biodesulfurization potential of a newly isolated bacterium, *Gordonia alkanivorans* RIPI90A. *Enzyme and Microbial Technology*. 40: 578–584.

Molday R.S., Yen S.P., Rembaum A. (1977) Application of magnetic microspheres in labelling and separation of cell. Nature. 268: 437–438.

Monot, F., Abbad-Andaloussi, S., Warzywoda, M. (2002) Biological culture containing *Rhodococcus erythropolis erythropolis* and/or *Rhodococcus rhodnii* and process for desulfurization of petroleum fraction. US Patent no. 6337204.

Monticello D.J. (1998) Riding the fossil fuel biodesulfurization wave. Proceedings of the 5th World Petroleum Congress. Published by John Wiley & Sons. pp. 901–906.

Monticello D.J., Bakker D., Finnerty W.R. (1985) Plasmid-mediated degradation of dibenzothiophene by Pseudomonas species. *Applied and Environmental Microbiology*. 49(4): 756–760.

Monticello D.J., Haney III., William M. (1996) Biocatalytic process for reduction of petroleum viscosity. US Pat. 5,529,930.

Monticello, D.J. (2000) Biodesulfurization and the upgrading of petroleum distillates. *Current Opinion in Biotechnology*. 11: 540–546.

Monticello, D.J., Finnerty, W.R. (1985) Microbial desulfurization of fossil fuels. *Annual Review of Microbiology*. 39: 371–389.

Monticello, D.J., Kilbane, J.J. (1994). Microemulsion process for direct biocatalytic desulfurization of organosulfur molecules. US Patent 5,358,870.

Moores, A., Goettmann, F. (2006) The plasmon band in noble metal nanoparticles: an introduction to theory and applications. *New Journal of Chemistry*. 30: 1121–1132.

Moriya K., and Horikoshi K. (1993) A benzene-tolerant bacterium utilizing sulfur compounds isolated from deep sea. Journal of Fermentation and Bioengineering. 76:397–399.

Mormile M.R., Atlas R.M. (1988) Mineralization of the dibenzothiophene biodegradation products 3-hydroxy-2-formyl benzothiophene and dibenzothiophene sulfone. Applied and Environmental Microbiology. 54:3183–3184.

Moslemy P., Guiot S.R., Neufeld, R.J. (2002) Production of size-controlled gellam gum microbeads encapsulating gasoline-degrading bacteria. Enzyme and Microbial Technology. 30:10–18.

Mössner, S.G., Wise, S.A. (1999). Determination of polycyclic aromatic sulfur heterocycles in fossil fuel-related samples. Analytical Chemistry. 71: 58–69.

Mulla, S.I. (2013) Studies on the biodegradation of nitroaromatic pollutants by the isolated bacterial cultures. A Ph.D. Thesis. Karnatak University. India.

Naito, M., Kawamoto, T., Fujino, K., Kobayashi, M., Maruhashi, K., Tanaka, A. (2001) Long-term repeated desulfurization by immobilized *Rhodococcus erythropolis* KA2–5-1 cells. Applied Microbiology and Biotechnology. 55: 374–378.

Nakamura, A., Takakura, Y., Kobayashi, H., Hoshino, T. (2005) In vivo directed evolution for thermostabilization of *Escherichia coli* hygromycin B phosphotransferase and the use of the gene as a selection marker in the host-vector system of *Thermus thermophilus*. Journal of Bioscience and Bioengineering. 100: 158–163.

Nasab, N.A., Kumleh, H.H., Kazemzad, M., Panjeh, F.G., Davoodi-Dehaghani, F. (2015) Improvement of Desulfurization Performance of Rhodococcus erythropolis IGTS8 by Assembling Spherical Mesoporous Silica Nanosorbents on the Surface of the Bacterial Cells. Journal of Applied Chemical Research. 9(2): 81–91.

Nassar, H. N. El-Gendy N. Sh., Farahat, L. A., Hawas A. E., and Mahdy, H. M. (2010) Study of diesel oil biodegradation by *Bacillus sphaericus* HN1. The 13th International Conference in Petroleum, Mineral Wealth and Development, Cairo, Egypt, 7–9 February, 2010.

Nassar, H.N. (2015) Development of biodesulfurization process for petroleum fractions using nano-immobilized catalyst; Ph.D. Degree; Al-Azhar University, Cairo, Egypt.

Nassar, H.N., Deriase, S.F., El-Gendy, N.Sh. (2016) Modeling the relationship between microbial biomass and total viable count for a new bacterial isolate used in biodesulfurization of petroleum and its fractions. Petroleum Science and Technology. 34(11–12):980–985.

Nassar, H.N., Deriase, S.F., El-Gendy, N.Sh. (2017a) Kinetic modeling and simulation of a batch biodesulfurization process by *Rhodococcus erythropolis* HN2. Petroleum Science and Technology. 35(20):1995–2002.

Nassar, H.N., Deriase, S.F., El-Gendy, N.Sh. (2017b) Statistical optimization of biomass production and biodesulfurization activity of *Rhodococcus erythropolis* HN2. Petroleum Science and Technology. 35(20):1951–1959.

Nassar, H.N., El-Gendy, N.Sh., Abo State M., Moustafa, Y.M., Mahdy, H.M., El-Temtamy S.A. (2013) Desulfurization of dibenzothiophene by a novel strain *Brevibacillus invocatus* C19 isolated from Egyptian coke. Biosciences, Biotechnology Research Asia. 10(1): 29–46.

Nichols, P.D., Guckert, J.B., White, D.C. (1986) Determination of monounsaturated fatty acid double-bond position and geometry for microbial monocultures and complex consortia by capillary GC-MS of their dimethyl disulphide adducts. J Microbiological Methods. 5:49–55.

Nigam, J.N.; Gogoi, B.K. and Bezbaruah. R.L. (1998): Alcoholic fermentation by agar immobilized yeast cells; World Microbiology and Biotechnology. 14:457–459.

Noda K., Watanabe K. and Maruhashi K. (2002) Cloning of *Rhodococcus* promoter using a transposon for dibenzothiophene biodesulfurization. Biotechnology Letters. 25: 1875–1888.

Noda K., Watanabe K. and Maruhashi K. (2003b) Recombinant *Pseudomonas putida* carrying both the *dsz* and *hcu* genes can desulfurize dibenzothiophene in n-tetradecane. Biotechnology Letters. 25:1147–1150.

Noda, K.I., Watanabe, K., Maruhashi K., (2003a) Isolation of recombinant desulfurizing 4,6-dipropyl dibenzothiophene n-tetrdecane. Bioscience and Bioengineering 95(4): 354–360.

Nussinovitch, A. (1997) Hydrocolloid applications: gum technology in the food and other industries. Blackie Academic & Prof, London, p 251.

Ochiai, H., Shimizu, S., Tadokoro, Y. and Murakami, I. (1981) Complex formation between poly(vinyl alcohol) and borate ion. Polymer 22, 1456–1458.

Ogata, M., Fujisawa, K. (1983) Capillary GC/MS determination of organic sulfur compounds detected in oyster and mussel caught in the sea as an oil pollution index. Journal of Chromatographic Science. 21:420–424

Ohshiro T, Hine Y, Izumi Y (1994) Enzymatic desulfurization of dibenzothiophene by a cell-free system of *Rhodococcus erythropolis* D-1. FEMS Microbiology Letters. 118: 341–344.

Ohshiro T., Suzuki K., and Izumi Y. (1996b) Regulation of dibenzothiophene degrading enzyme activity of *Rhodococcus erythropolis* D-1. Journal of Fermentation and Bioengineering. 82: 121–124.

Ohshiro, T., Hirata, T., Hashimoto, I., Izumi, Y. (1996a) Characterization of dibenzothiophene desulfurization reaction by whole cells of *Rhodococcus erythropolis* H-2 in the presence of hydrocarbon. Journal of Fermentation and Bioengineering. 82: 610–612.

Ohshiro, T., Hirata, T., Izumi, Y. (1995) Microbial desulfurization of dibenzothiophene in the presence of hydrocarbon. Applied Microbiology and Biotechnology. 44: 249–252.

Ohshiro, T., Ohkita, R., Takikawa, T., Manabe, M., Lee, W.C., Tanokura, M., Izumi, Y. (2007) Improvement of 2'-hydroxybiphenyl-2-sulfinate desulfinase, an enzyme involved in the dibenzothiophene desulfurization pathway from Rhodococcus erythropolis KA2-5-1 by site-directed mutagenesis. Bioscience Biotechnology and Biochemistry. 71(11): 2815–2821.

Okada, H., Nomura, N., Nakahara, T., Maruhashi, K. (2002b) Analyses of substrate specificity of the desulfurizing bacterium *Mycobacterium* sp. G3. Journal of Bioscience and Bioengineering. 93: 228–233.

Okada, H., Nomura, N., Nakahara, T., Saitoh, K., Uchiyama, H., Maruhashi, K. (2003) Analyses of microbial desulfurization reaction of alkylated dibenzothiophenes dissolved in oil phase Biotechnology and Bioengineering. 83(4): 489–497.

Okada, I., Nomura, N., Nakahara, T., Maruhashi K. (2002a) Analysis of dibenzothiophene metabolic pathway in *Mycobacterium* strain G3. *Journal of Bioscience and Bioengineering*. 93(5):491–497.

Olabemiwo, O.M., Adelowo, O.O., Tella, A.C., Bello, I.A. (2013) Preliminary study on biodesulfurization of Nigerian natural bitumen. *International Journal of Basic and Applied Sciences*. 13: 79–86.

Oldfield, C., Pogrebinsky, O., Simmonds, J., Olson, E.S., Kulpa, C.H. (1997) Elucidation of the metabolic pathway for dibenzothiophene desulfurization by *Rhodococcus* sp. strain IGTS8 (ATCC 53968). Microbiology UK. 143: 2961–2973.

Oldfield, C., Wood, N.T., Gilbert, S.C., Murray, F.D., Faure, F.R. (1998) Desulphurisation of benzothiophene and dibenzothiophene by actinomycete organisms belonging to the genus *Rhodococcus*, and related taxa. Antonie van Leeuwenhoek. 74:119–132.

Olson, E.S., Stanley, D.C., Gallaguer, J.R. (1993) Characterization of intermediates in the microbial desulfurization of dibenzothiophene. *Energy and Fuels* 7: 159–164.

Omori, T., Saiki, Y., Kasuga, K., Kodama, T. (1995) Desulfurization of alkyl and aromatic sulfides and sulfonates by Dibenzothiophene desulfurizing *Rhodococcus* sp. Strain SY1. *Bioscience, Biotechnology and Biochemistry*. 59: 1195–1198.

Onaka, T., Kobayashi, M., Ishii, Y., Okumura, K., Suzuki, M. (2000) Application of solid-phase extraction to the analysis of the isomers generated in biodesulfurization against methylated dibenzothiophenes. *Journal of Chromatography A*. 903: 193–202.

Ortiz-Leon M., Velasco L., and Vazquez-Duhalt R. (1995) Biocatalytic oxidation of polycyclic aromatic hydrocarbons by hemoglobin and hydrogen peroxide. *Biochemical and Biophysical Research Communications*. 215: 968–973.

Pan, X.R., Li, G.W., Hu, Y.H., Wang, J.X., Yang, W.Y., An, Z.X. (1997): Effects of diet and exercise in preventing NIDDM in people with impaired glucose tolerance, The Da Qing IGT and Diabetes Study. Diabetes Care. 20:537–544.

Papizadeh, M., Ardakani, M.R., Motamedi, H., Rasouli, I., Zarei, M. (2011) C–S targeted biodegradation of dibenzothiophene by *Stenotrophomonas* sp. NISOC-04. *Applied Biochemistry and Biotechnology*. 165: 938–948.

Papizadeh, M., Roayaei ardakani, M., Ebrahimipour, G., Motamedi, H. (2010) Utilization of dibenzothiophene as sulfur source by *Microbacterium* sp. NISOC-06. *World Journal of Microbiology and Biotechnology*. 26: 1195–1200.

Papizadeh, M., Roayaei Ardakani, M., Motamedi, H. (2017) Growth-phase dependent biodesulfurization of dibenzothiophene by *Enterobacter* sp. strain NISOC-03. *Pollution*. 3(1): 101–111.

Park, H.S., Kayser, K.J., Kwak, J.H., Kilbane, J.J. II (2004) Heterologous gene expression in *Thermus thermophilus*: β-galactosidase, dibenzothiophene monooxygenase, PNB carboxy esterase, 2-aminobiphenyl-2,3-diol dioxygenase, and chloramphenicol acetyl transferase. *Journal of Industrial Microbiology and Biotechnology*. 31 :189–197.

Park, J.K., Chang, H.N. (2000) Microencapsulation of microbial cells. *Biotechnology Advances*. 18: 303–19.

Passos, C.T., Michelon, M., Burkert, J.F., Kalil, S.J., Burkert, C.A. (2010) Biodegradation of phenol by free and encapsulated cells of a new Aspergillus sp. isolated from a contaminated site in southern Brazil; African Biotechnology. 9: 6716–6720.

Pasta P., Carrea G., Colonna S., and Gaggero N. (1994) Effects of chloride on the kinetics and stereochemistry of chloroperoxidase catalyzed oxidation of sulfides. Biochimica et Biophysica Acta. 1209(2): 203–208.

Patel, S.B., Kilbane, J.J., Webster, D.A. (1997) Biodesulphurisation of dibenzothiophene in hydrophobic media by *Rhodococcus* sp. strain IGTS8. *Journal of Chemical Technology and Biotechnology*. 69: 100–106.

Patnaik, S., Arif, M., Pathak, A. (2010). Cross-linked polyethylenimine-hexametaphosphate nanoparticles to deliver nucleic acids therapeutics. Nanomedicine: Nanotechnology, Biology, and Medicine. 6: 344–354.

Peng, B., Zhou, Z. (2016) Study on growth characteristic and microbial desulfurization activity of the bacterial stain MP12. *Biochemical Engineering Journal*. 112: 202–207.

Peng, Y., Wen, J. (2010) Modeling of DBT biodesulfurization by resting cells of *Gordonia* sp. WQ-01A immobilized in alginate gel beads in an oil-water-immobilization system; Chemical and Biochemical Engineering Quarterly. 24:85–94.

Pepi, M., Baldi, F. (1992) Modulation of chromium(VI) toxicity by organic and inorganic sulfur species in yeasts from industrial wastes. BioMetals. 5:179–185.

Pepi, M., Baldi, F. (1995) Chromate tolerance in strains of Rhodosporidium toruloides modulated by thiosulfate and sulfur amino acids. BioMetals. 8:99–104.

Piddington C.S., Kovacevich B.R. and Rambosek J. (1995) Sequence and molecular characterization of a DNA region encoding the dibenzothiophene desulfurization operon of *Rhodococcus* sp. strain IGTS8. *Applied and Environmental Microbiology*. 61:468–475

Pikoli, M.R., Aditiawati, P., Akhmaloka, Indrianiastuti, D., Wijayanti, R. Growth of *Bacillus megaterium* CSK2, *Bacillus subtillus* CSK3 and *Bacillus subtillus* CSK4 from coal mixed soil in dibenzothiophene-containing medium. *Asian Journal of Microbiology, Biotechnology and Environmental Sciences*. 16(2): 455–460.

Pirkle, W.H., Simmons, K.A., and Bolder, C.W. (1979) Dynamic NMR studies of diastereomeric carhamates: Implications toward the determination of relative configuration by NMR. *Journal of Organic Chemistry*. 44:4891–4896.

Pokethitiyook, P., Tangaromsuk, J., Kruatrachue, M., Kalambahcti, C., Borole, A.P. (2008) Biological removal of organic sulphur by bacterial strains isolated in Thailand. *Science Asia*. *34*(4): 361–366.

Popovic, R., Kyle, D., Cohen, A., Zalik, S. (1979) Stabilization of thylakoid membranes by spermine during stress induced senescence of barley leaf disks. Plant Physiology. 64: 721–726.

Praveen Reddy, P., Umamaheshwara Rao, V (2015) Dibenzothiophene desulfurizing *Streptomyces* species isolated from oil contaminated sites. *International Journal of Science and Research*. 5(12): 1671–1674.

Prince, R.C., Grossman, M.J. (2003) Substrate preferences in biodesulfurization of diesel range fuels by Rhodococcus sp. strain ECRD-1. *Applied and Environmental Microbiology*. 69: 5833–5838.

Radovich, J.M. (1985) Mass transfer effects in fermentations using immobilized whole cells. *Enzyme Microbiology Technology*. 7:2–10.

Radwan A. (2015) Genetic Improvement of Desulfurizing Bacteria. MSc Degree, Al-Azhar University.

Raheb, J., Hajipour, M.J., Saadati, M., Rasekh, B., and Memari, B. (2009) The Enhancement of Biodesulfurization Activity in a Novel Indigenous Engineered Pseudomonas putida. *Iran. Biomed. Journal*, 13(4): 141–147.

Rahpeyma, S.S., Mohammadi, M., Rehab, J. (2017) Biodesulfurization of Dibenzothiophene by Two Bacterial Strains in Cooperation with Fe_3O_4, ZnO and CuO Nanoparticles. *Journal of Microbial & Biochemical Technology*. 9(2): 587–591.

Rambosek, J., Piddington, C.S., Kovacevich, B. R., Young., K.D., Piddington, C.S., Kovacevich, B.R., Young, K. D., Denome, S.A. (1994) Recombinant DNA encoding a desulfurization biocatalyst. US Pat. 5,356,801.

Rambosek, J., Piddington, C.S., Kovacevich, B.R., Young, K.D., Denome, S. (1999) Recombinant DNA encoding a desulfurization biocatalyst. US Patent no. 5,879,914.

Ranzoni, A., Sabatte, G., van Ijzendoorn, L.J., Prins, M.W. (2012) One-step homogeneous magnetic nanoparticle immunoassay for biomarker detection directly in blood plasma. ACS Nano, 6: 3134–41.

Rashtchi M., Mohebali G.H., Akbarnejad M.M., Towfighi J., Rasekh B., Keytash, A. (2006) Analysis of biodesulfurization of model oil system by the bacterium strain RIPI-22. *Biochemical Engineering*. 29:169–173.

Rath, K., Mishra, B., Vuppu, S. (2012) Biodegrading ability of organosulfur compound of a newly isolated microbe *Bacillus* sp. KS1 from the oil contaminated soil. *Archives of Applied Science Research*. 4(1): 465–471.

Rawat, M., Singh, D., Saraf, S., Saraf, S. (2006) Nano carriers: promising vehicle for bioactive drugs. Biological & Pharmaceutical Bulletin. 29:1790–8.

Reeves, D. S., Holt, A., Bywater, M. J., Wise, R., Logan, M. N., Andrews, J. M., Broughall, J. M. (1980). Comparison of sensititredried microtitration trays with a standard agar method for determination of minimum inhibitory concentrations of antimicrobial agents. Antimicrob Agents Chemother.18: 844–852.

Reichmuth, D.S., Hittle, J.L., Blanch, H.W., Keasling, J.D. (1999) Biodesulfurization of dibenzothio- phene in Escherichia coli is enhanced by expression of *Vibrio harveyi* oxidoreductase gene. *Biotechnology Bioengineering*. 67:72–79.

Reichmuth, D.S., Hittle, J.L., Blanch, H.W., Keasling, J.D. (2000) Biodesulfurization of dibenzothiophene in Escherichia coli is enhanced by expression of a *Vibrio harveyi* oxidoreductase gene. *Biotechnology and Bioengineering*. 67: 72–79.

Resnick, S.M., Gibson, D.T., (1996) Regio- and stereospecific oxidation of fluorene, dibenzofuran, and dibenzothiophene by naphthalene dioxgenase from *Pseudomonas* sp. strain NCIB 9816-4. *Applied Environmental Microbiology*. 62: 4073–4080.

Rhee, S.K., Chang, J.H., Chang, Y.K., Chang, H.N. (1998) Desulfurization of dibenzothiophene and diesels oils by a newly isolated *Gordona* strain CYKS1. *Applied and Environmental Microbiology*. 64(6): 2327–2331.

Robatjazi, S.M.; Shojaosadati, S.; Khalilzadeh, R.; Farahani, E. and Zeinoddini, M. (2012): Immobilization of magnetic modified *Flavobacterium* ATCC 27551 using magnetic field and evaluation of the enzyme stability of immobilized bacteria; Bioresource Technology. 104:6–11.

Rong, X., Huang, Q., He, X., Chen, H., Cai, P., Liang, W., (2008) Interaction of Pseudomonas putida with kaolinite and montmorillonite: a combination study by equilibrium adsorption, ITC, SEM and FTIR. Colloids and Surfaces B. 64: 49–55.

Ryu, J., Kim, S.W., Kang, K., Park, C.B. (2010) Synthesis of diphenylalanine/ cobalt oxide hybrid nanowires and their application to energy storage. ACS Nano. 4: 159–64.

Sada E., Katon S., and Terashima M. 1981. Enhancement of oxygen absorption by magnetite-containing beads of immobilized glucose oxidase. *Biotechnology and Bioengineering*. 21: 1037–1044.

Sagardia F.; Rigau J.J.; Martinez L.A.; Fuentes F. and Lopez C. (1975) Degradation of benzothiophene and related compounds by a soil Pseudomonas in an oil-aqueous environment. *Journal of Applied Microbiology*. 29: 722–725.

Salleh, A.B., Abdul-Rahman, R.N.Z.R., Basri, M. (2006) New lipases and proteases. Nova Publishers. pp 114.

Sandhya S.; Prabu SK., and Sundari R.B.T. (1995) Microbial degradation of dibenzothiophene by Nocardioides. *Journal of Environmental Science and Health, Part A*. 30:1995–2006.

Santos, S.C.C., Alviano, D.S., Alviano, C.S., Pádula, M., Leitão, A.C., Martins, O.B., Ribeiro, C.M.S., Sassaki, M.Y.M., Matta, C.P.S., Bevilaqua, J., Sebastián, G.V., Seldin, L. (2006) *Applied Microbiology and Biotechnology*. 71: 355–362.

Sarret, G., Connan, J., Kasrai, M., Bancroft, G.M., Charrié-Duhaut, A., Lemoine, S., Adam, P., Albrecht, P., Eybert-Bérard, L., (1999) Chemical forms of sulphur in geological andarcheological asphaltenes from Middle East, France and Spain determined by sulphur K-and L-edge X-ray absorptionnear-edge structure spectroscopy. Geochimica et Cosmochimica Acta. 63: 3767–3779.

Saxena, A, Tripathi, B.P., Kumar, M., Shahi, V.K. (2009) Membrane- based techniques for the separation and purification of proteins: an overview. Advances in Colloid and Interface Science. 145:1–22

Schilling, B.M., Alvarez, L.M., Wang, D.I.C., Cooney, C.L. (2002) Continuous desulfurization of dibenzothiophene with Rhodococcus rhodochrous IGTS8 (ATCC 53968). *Biotechnology Progress*. 18: 1207–1213.

Seifert, D.B., Phillips, J.A. (1997) Production of small, monodispersed alginate beads for cell immobilization. *Biotechnology Progress*. 13: 562 – 568

Selifonov S.A., Grifoll M., Eaton R.W., Chapman P.J. (1996) Oxidation of naphthenoaromatic and methyl-substitued aromatic compounds by naphthalene 1,2-dioxygenase. . *Applied and Environmental Microbiology*. 62: 507–514

Seo J.S., Keum Y.S., Cho I.K., Li Q.X. (2006) Degradation of dibenzothiophene and carbazole by *Arthrobacter* sp. P1-1. *International Biodeterioration and Biodegradation*. 58: 36–43.

Serbolisca L., de Ferra F., Margrit I. (1999) Manipulation of the DNA coding for the desulfurizing activity in a new isolate of Arthrobacter sp. *Applied Microbiology and Biotechnology*. 52: 122–126.

Setti, L., Bonoli, S., Badiali, E., Giuliani, S. (2003) Inverse phase transfer biocatalysis for a biodesulfurization process of middle distillates. 44(1): 80–83.

Setti, L., Farinelli, P., Di Martino, S., Frassinetti, S., Lanzarini, G., Pifferi, P.G. (1999) Developments in destructive and non-destrucitve pathways for selective desulfurizations in oil biorefining processes. *Applied Microbiology and Biotechnology*. 52: 111–117.

Setti, L., Lanzarini, G., Pifferi, P.G. (1997) Whole cell biocatalysis for an oil desulfurization process. *Fuel Processing Technology*. 52: 145–153.

Shahaby, A.F., Essam El-din, K.M. (2017) Desulfurization of crude oil and oil products by local isolated bacterial strains. *International Journal of Current Microbiology and Applied Sciences*. 6(4): 2695–2711.

Shahverdi, A.R., Minaeian, S., Shahverdi, H.R. (2007) Rapid synthesis of silver nanoparticles using culture supernatants of *Enterobacteria*, a novel biological approach. *Process Biochemistry*. 42, 919–23.

Shan G.B., Xing J.M., Guo C., Liu H.Z., Chen J.Y. (2005a) Biodesulfurization using *Pseudomonas delafieldii* in magnetic polyvinyl alcohol beads. *Letters in Applied Microbiology*. 40: 30–36.

Shan G.B., Xing J.M., Zhang H., Liu, H.Z. (2005b) Biodesulfurization of dibenzothiophene by microbial cells coated with magnetite nanoparticles. *Applied and Environmental Microbiology*. 71(8): 4497–4502.

Shan, G.B., Xing, J.M., Luo, M.F., Liu, H.Z., Chen, J.Y. (2003) Immobilization of pseudomonas delafieldii with magnetic polyvinyl alcohol beads and its application in biodesulfurization. *Biotechnology Letters*. 25, 1977–1981.

Sharma, Y.C., Srivastava, V., Upadhyay, S.N., Weng, C.H. (2008) Alumina nanoparticles for the removal of Ni(II) from aqueous solutions. *Industrial & Engineering Chemistry Research*. 47: 8095–100.

Shavandi M., Sadeghizadeh M., Zomorodipour A. Khajeh K. (2009) Biodesulfurization of dibenzothiophene by recombinant Gordonia alkanivorans RIPI90A. *Bioresource Technology*. 100:475–479.

Shavandi, M., Sadeghizadeh, M., Khajeh, K., Mohebali, G., & Zomorodipour, A. (2010) Genomic structure and promoter analysis of the dsz operon for dibenzothiophene biodesulfurization from Gordonia alkanivorans RIPI90A. *Applied Microbiology and Biotechnology*. 87(4): 1455–1461.

Shavandi, M., Sadeghizadeh, M., Zomorodipour, A., Khajeh, K. (2008) Biodesulfurization of dibenzothiophene by recombinant *Gordonia alkanivorans* RIPI90A. *Bioresource Technology*. 100: 475–479.

Shen, L., Laibinis, P.E., Hatton, T.A., (1999) Bilayer surfactant stabilized magnetic fluids: synthesis and interactions at interfaces. Langmuir. 15: 447–453.

Shennan, J.L. (1996) Microbial attack on sulfur-containing hydrocarbons, implications for the biodesulphurization of oils and coals. *Journal of Chemical Technology and Biotechnology*. 67: 109–123.

Shin, K.H., Cha, D.K. (2008) Microbial reduction of nitrate in the presence of nanoscale zero-valent iron. *Chemosphere*. 72: 257–262.

Shi-Yong, Z., Don, K.L., Chang, W.K., Hyun, G.C., Young, H.K., Young, S.K. (2006) Synthesis of Magnetic Nanoparticles of Fe_3O_4 and $CoFe_2O_4$ and Their Surface Modification by Surfactant Adsorption. Bulletin of the Korean Chemical Society. 27(2):237–42.

Singh S.P.; Schwan A.L. (2011) Sulfur metabolism in plants and related biotechnologies. In: Moo-Young, M. (Ed.), Comprehensive Biotechnology, second ed. Elsevier. 4:257–271.

Singh, P., Srivastava, P. (2013) An improved protocol for electroporation in members of the genus *Gordonia*. *Journal of Microbiological Methods*. 95:114–116

Sink, C. W., Hardy, D. R. (1994) Quantification of Compound Classes in Complex Mixtures and Fuels Using HPLC with Differential Refractive Index Detection. *Analytical Chemistry*. 66(8):1334–1338.

Squires, C.H., Ji, W., Xi, L., Ortego, B., Pogrebinsky, O.S., Gray, K.A., Childs, J.D. (1998) Method of desulfurization of fossil fuel with flavoprotein. US Patent no. 5,733,773.

Squires, C.H., Ji, W., Xi, L., Ortego, B.C., Pogrebinsky, O.C., Gray, K.A., Childs, J.D. (1999) Method of desulfurization of fossil fuel with flavoprotein. US Patent no. 5,985,650.

Stefanova, M., Tonkova, A., Dobreva, E., Spasova, D. (1998) Agar gel immobilization of *Bacillus brevis* cells for production of thermostable α-amylase. Folia Microbiologica. 43: 42–46.

Stoner D.L., Wey J.E., Barrett K.B., Jolley J.G. (1990) Modification of water-soluble coal-derived products by dibenzothiophene-degrading microorganisms. *Applied and Environmental Microbiology*. 56:2667–2676.

Taha, T.H., Alamri, S.A., Mahdy, H.M., Hafez, E.E. (2013) The effects of various immobilization matrices on biosurfactant production using hydrocarbon (HC)-degrading marine bacteria via the entrapment technique. *Biological Science*. 13:48–57.

Takada, M., Nomura, N., Okada, H., Nakajima-Kambe, T., Nakahara, T., Uchiyama, H (2005) De-repression and comparison of oil-water separation activity of the dibenzothiophene desulfurizing bacterium, *Mycobacterium* sp. G3. *Biotechnology Letters*. 27: 871–874.

Tampion J, Tampion, M.D. (1987) Immobilized cells: principles and applications. Cambridge Univ Press. pp 91.

Tanaka Y., Matsui T., Konishi J., Maruhashi, K. (2002) Biodesulfurization of benzothiophene and dibenzothiophene by a newly isolated *Rhodococcus* strain. *Applied Microbiology and Biotechnology*. 59:325–328.

Tanaka, Y., Onaka, T., Matsui, T. (2001) Desulfurization of Benzothiophene by the Gram-Negative bacterium, *Sinorhizobium* sp. KT55. *Current Microbiology*. 43: 187–191.

Tao, A.R., Habas, S., Yang, P. (2008) Shape control of colloidal metal nanocrystals. Small. 4: 310–25.

Tao, F., Liu, Y., Luo, Q., Su, F., Xu, Y., Li, F., Xu, P. (2011) Novel organic solvent-responsive expression vectors for biocatalysis: application for development of an organic solvent-tolerant biodesulfurizing strain. *Bioresource Technology*. 102(20): 9380–9387.

Tao, F., Yu, B., Xu, P., Ma, C. (2006) Biodesulfurization in biphasic systems containing organic solvents. *Applied and Environmental Microbiology*. 72: 4604–4609.

Terres, E., Montiel, M., Le-Borge, S., Torres, E. (2008) Immobilization of chloroperoxidase on mesoporous materials for the oxidation of 4,6-dimethyldibenzothiophene, a recalcitrant organic sulfur compound present in petroleum fractions. *Biotechnology Letters*. 30: 173–179.

Tilstra, L., Eng, G., Olson, G.J. and Wang, F.W. (1992) Reduction of sulfur from polysulphidic model compounds by the hyperthermophilic *Archaebacterium Pyrococcus* furiosus. *Fuel*. 71: 779–783.

Torkamani S., Shayegan J., Yaghemaei S., and Alemzadeh I. (2008) Study of a newly isolated thermophilic bacterium capable of kuhemond heavy crude oil and dibenzothiophene biodesulfurization following 4S pathway at 60°C. *Journal Chemical Technology and Biotechnology*. 83: 1689–1693.

Trelles, J.A., Rivero, C.W. (2013) Whole cell entrapment techniques. Methods in Molecular Biology. 1051: 365–374.

Van Affender, M., Schacht, S., Klein J., Truper, H.G. (1990) Degradation of dibenzothiophene by *Brevibacterium* sp. DO. *Archives of Microbiology*. 153: 324–328.

Van Afferden, M., Trappe, D., Beyer, M., Tüpper, H.G., Klein, J. (1993) Biochemical mechanisms for the desulfurization of coal-relevant organic sulfur compounds. *Fuel* 72(12): 1635–1643.

Van Hamme, J.D., Wong, E.T., Dettman, H., Gray, M.R., Pickard, M.A.
(2003) Dibenzyl sulfide metabolism by White Rot Fungi. *Applied and Environmental Microbiology*. 69(2): 1320–1324.

Van Herwijnen R., Wattiau P., Bastiaens L., Daal L., Jonker L., Springael D. (2003) Elucidation of the metabolic pathway of fluorine and cometabolic pathways of phenanthrene, fluoranthene, anthracene and dibenzothiophene by Sphingomonas sp. LB126. Research in Microbiology. 154: 199–206.

Van Herwijnen, R., Wattiau, P., Bastiaens, L., Daal, L. Jonker, L., Springael, D., Govers, H. A. & Parsons, J. R. *Res. Mirobiol.* **3** : p.199–206 2003)

Vazquez-Duhalt R., Westlake D.W.S., and Fedorak P.M. (1994) Lignin peroxidase oxidation of aromatic compounds in systems containing organic solvents. *Applied and Environmental Microbiology*. 60(2): 459–466.

Vazquez-Duhalt, R., Semple, K.M., Westlake, D.W.S., Fedorak, P.M. (1993) Effect of water-miscible organic solvents on the catalytic activity of cytochrome c. *Enzyme and Microbial Technology*. 15: 936–943.

Vazquez-Duhalt, R.V., Bremauntz M.D., Barzana E., and Tinoco R. (2002) Enzymatic oxidation process for desulfurization of fossil fuels.US Patent 6,461,859.

Verma, M., Brar, S.K., Blais, J.F. (2006) Aerobic biofiltration processes- advances in wastewater treatment. Practice Periodical of Hazardous Toxic and Radioactive Waste Management.10: 264–76.

Verma, N., Sharma, R., Kaur, R. (2016) Microbial desulphurization study of dibenzothiophene and crude oil by a soil isolate. *International Journal of Science and Research Methodology*. 4(4): 133–145.

Vertegel A. (2010) 55th Annual Report on Research 2010. Under sponsorship of the American Chemical Society Petroleum Research Fund.https://acswebcontent.acs.org/prfar/2010/reports/P10734.html.

Vignier, V., Berthou, F., Dreano, Y., Floch, H.H. (1985) Dibenzothiophene sulphoxidation: a new and fast high-performance liquid chromatographic assay of mixed-function oxidation. *Xenobiotica*, 15: 991–999.

Wandrey, C. (1996) Immobilized cells: Basics and applications; Proceedings of an International Symposium organized under auspices of the working party on Applied Biocatalysis of the European Federation of Biotechnology; Netherlands.

Wang P., Krawiec S. (1994) Desulfurization of dibenzothiophene to 2 hydroxybiphenyl by some newly isolated bacterial strains. *Archives of Microbiology*. 161:266–271.

Wang, C.B., Zhang, W.X. (1997) Synthesizing nanoscale iron particles for rapid and complete dechlorination TCE and PEBs. *Environmental Science and Technology*. 31: 2154–2156.

Wang, H., Xia, Y., Yang, Z., Natschke, S., Lee K. (1998a) Recent advances in the discovery and development of flavonoids and their analogues as antitumor and anti-HIV agents. Advances in Experimental Medicine and Biology. 439: 191–225.

Wang, J., Davaadelger, B., Salazar, J.K., Butler III, R.R., Pombert, J.-F., Kilbane, J.J., Stark, B.C. (2015) Isolation and characterization of an interactive culture of two *Paenibacillus* species with moderately thermophilic desulfurization ability. *Biotechnology Letters*. 37: 2201–2211.

Wang, L.L., Ando, S., Ishida, Y., Obtain, H., Tsuge, S., Nakayama, T. (2001) Quantitative and discriminative analysis of carnauba waxes by reactive pyrolysis-GC in the presence of organic alkali using a vertical microfurnace pyrolyzer. *Journal of Analytical and Applied Pyrolysis*. 58: 525–537.

Wang, M.-d., Li, W., Wang, D.-h., Shi, Y. (2004) Desulfurization of dibenzothiophene by a newly isolated *Corynebacterium* sp. ZD-1 in aqueous phase. *Journal of Environmental Sciences*. 16(6): 1011–1015.

Wang, P., Krawiec, S. (1996) Kinetic analyses of desulfurization of dibenzothiophene by *Rhodococcus erythropolis* in batch and fed-batch cultures. *Applied and Environmental Microbiology*. 61: 1670–1675.

Wang, W., Ma, T., Lian, K., Zhang, Y., Tian, H., Ji, K., Li, G. (2013). Genetic analysis of benzothiophene biodesulfurization pathway in *Gordonia terrae* strain C-6. PLoS ONE 8:e84386. doi: 10.01371/journal.pone.0084386.

Wang, Z., Fingas, M., Blenkinsopp, S., Sergy, G., Landriault, M., Sigouin, L., Westlake, D.W.S. (1998b) Comparison of oil composition changes due to biodegradation and physical weathering in different oils. *Journal of Chromatography A*, 809(1): 89–107.

Wang, Z., Fingas, M., David, P. (1999) Oil spill identification, *Journal of Chromatography A*. 843: 369–411.

Wardencki, W. (2000) Sulfur compounds: gas chromatography. Technical University of Gdansk, Poland Copyright 2000. Academic Press.P. 4285–4301.

Watanabe K., Noda K., Maruhashi K. (2003) Enhanced desulfurization in a transposon-mutation strain of *Rhodococcus erythropolis Biotechnology Letters*. 25:1299–1304.

Watanapokasin Y., Nuchfoang S., Nilwarangkoon S., Sarangbin S., Kakizono T. (2002) Isolation and characterization of thermophilic benzothiophene degrading Mycobacterium sp. *Applied Biochemistry and Biotechnology*. 98–100:301–309.

Whittaker, M., Pollard, S.J.T (1994) Characterization of refractory wastes at hydrocarbon-contaminated sites. I: Rapid column fractionation and thin layer chromatography of reference oils. *Journal of Planar Chromatography*. 7: 354–361.

Whyte, L.G., Slagman, S.J., Pietrantonio, F., Bourbonniere, L., Koval, S.F., Lawrence, J.R., Inniss, W.E., Greer, C.W. (1999) Physiological adaptations involved in alkane assimilation at a low temperature by *Rhodococcus* sp. strain Q15. *Applied and Environmental Microbiology*. 65: 2961–2968.

Wick, L.Y., Mattlei, P., Wattiau, P., Harms, A. (2004) Electrokinetic transport of PAH degrading bacteria in model aquifers and soil. *Environmental Science and Technology*. 38: 4596–4602.

Wick, L.Y., Shi, L., Harms, H. (2007) Electro-bioremediation of hydrophobic organic soil-contaminants: a review of fundamental interactions. Electrochimica Acta. 52: 3441–3448.

Wilson, N.G., Bradley, G. (1996) Enhanced degradation of petroleum (Slovene diesel) in an aqueous system by immobilized pseudomonas fluorescens. *Journal of Applied Microbiology*. 80: 99–104.

Wu S., Lin J., Chan S.I. (1994) Oxidation of dibenzothiophene catalyzed by heme-containing enzymes encapsulated in sol-gel glass. A new form of biocatalysts, *Applied Biochemistry and Biotechnology*. 47: 11–20.

Wu, K.Y.A. and Wisecarver, K.D. (1992) Cell immobilization using PVA crosslinked with boric acid. Biotechnology and Bioengineering. 39: 447–449.

Wu, S., Lin, j. & Chan, S. I. *Appl. Biochem. Biotechnol.* **47** : p.11–20 (1994)

Xie, C., Xu, F., Huang, X., Dong, C., Ren, J. (2010) Single gold nanoparticles counter: an ultrasensitive detection platform for one-step homogeneous immunoassays and DNA hybridization assays. *Journal of the American Chemical Society*. 131: 12763–70.

Xiu, Z., Jin, Z., Li, T. (2010). Effects of nano-scale zero-valent iron particles on a mixed culture dechlorinating Trichloroethylene. Bioresource Technology. 101: 1141–6.

Xu, S., Liu, H., Fan, Y. (2012) Enhanced performance and mechanism study of microbial electrolysis cells using Fe nanoparticle-decorated anodes. *Applied Microbiology Biotechnology*. 93: 871–80.

Xu, X., Cai, Y., Song, Z., Qiu, X., Zhou, J., Liu, Y., Mu, T., Wu, D., Guan, Y., Xing, J. (2015) Desulfurization of immobilized sulfur-oxidizing bacteria, *Thialkalivibrio versutus*, by magnetic nanoparticles under haloalkaliphilic conditions. *Biotechnology Letters*. 37(8): 1631–1635.

Yamada K.O., Morimoto M., Tani Y. (2001) Degradation of dibenzothiophene by sulfate-reducing bacteria cultured in the presence of only nitrogen gas. *Journal of Bioscience and Bioengineering*. 91:91–93.

Yamazoe, A., Yagi, O., Oyaizu, H. (2004) Degradation of polycyclic aromatic hydrocarbons by a newly isolated dibenzofuran-utilizing *Janibacter* sp. strain YY-1. *Applied Environmental Microbiology*. 65: 211–218.

Yan, H., Sun, X., Xu, Q., Ma, Z., Xiao, C., Jun, N. (2008) Effects of nicotinamide and riboflavin on the biodesulfurization activity of dibenzothiophene by *Rhodococcus erythropolis* USTB-03. *Journal of Environmental Sciences*. 20: 613–618.

Yang B., Hou W., Zhang K., Wang X. (2013) Application of solid- phase micro extraction to the determination of polycyclic aromatic sulfur heterocycles in Bohai Sea crude oils. *Journal of Separation Science*. 16:2646–2655.

Yang, J., Hu, Y., Zhao, D., Wang, S., Lau, P.C.K., Marison, I.W. (2007) Two-layer continuous-process design for the biodesulfurization of diesel oils under bacterial growth conditions. *Biochemical Engineering Journal*. 37: 212–218.

Yang, J., Marison, I.W. (2005) Two-stage process design for the biodesulfurization of a model diesel by a newly isolated *Rhodococcus globerulus* DAQ3. *Biochemical Engineering Journal*, 27: 77–82.

Yee, C.K., Jordan, R., Ulman, A. (1999) Novel one-phase synthesis of thiol-functionalized gold, palladium, and iridium nanoparticles using superhydride. Langmuir. 15: 3486–3491.

Yellen, B.B., Forbes, Z.G., Halverson, D.S. (2005) Targeted drug delivery to magnetic implants for therapeutic applications. *Journal of Magnetism and Magnetic Materials*, 293: 647–654.

Yu B., Xu P., Shi Q., Ma C. (2006a). Microbial desulfurization of gasoline by free whole-cells of *Rhodococcus erythropolis* XP. *Applied and Environmental Microbiology*. 72: 54–78.

Yu, B., Huang, J., Cai, X. F., Liu, X. Y., Ma, C. Q., Li, F. L., Xu, P. (2005) A novel method for conversion of valuable BDS intermediates HPBS/Cx-HPBS from DBT/Cx-DBT. Chinese Chemical Letters. 16(7): 935–938.

Yu, B., Huang, J., Cai, X.F., Liu, X.Y., Ma, C.Q., Li, F.L., Xu, P. (2005) A novel method for conversion of valuable biodesulfurization intermediates HPBS/

Cx-HPBS from dibenzothiophene/Cx-dibenzothiophene. Chinese Chemical Letters. 16(7): 935–938.

Yu, B., Xu, P., Shi, Q., Ma, C. (2006b) Deep desulfurization of diesel oil and crude oils by a newly isolated *Rhodococcus erythropolis* strain. *Applied and Environmental Microbiology*. 72:54–58

Yu, L., Meyer, T., Folsom, B. (1998) Oil/water/biocatalyst three phase separation process. US Patent No. 5,772,901.

Yu-Qung, Z.T., Mei-Lin, S., Wei-De, Z. (2004) Immobilization of L- asparaginase of the microparticles of the natural silk serum protein and its characters. *Biomaterials*. 25: 3751–3759.

Zahra, E., Giti, E., Sharareh, P. (2006) Removal of dibenzothiophene, biphenyl and phenol from waste by *Trichosporon* sp. *Scientific Research and Essays*. 1(3): 72–76.

Zaki T., Saed D., Aman D., Younis S.A., Moustafa Y.M. (2013) Synthesis and characterization of MFe3O4 sulfur nanoadsorbants. *Journal of Sol-Gel Science and Technology*. 65: 269–276.

Zhang T., Li W., Chen V., Tang H., Li Q., Xing J., Liu H. (2011) Enhanced biodesulfurization by magnetic immobilized Rhodococcus erythropolis LSSE8 assembled with nano-γ-AlO. *World Journal of Microbiology and Biotechnology*. 27:299–305.

Zhang, H., Liu, Q.F., Li, Y., Li, W., Xiong, X., Xing, J., Liu, H. (2008) Selection of adsorbents for in-situ coupling technology of adsorptive desulfurization and biodesulfurization. *Science in China Series B-Chemistry*. 51(1): 69–77.

Zhang, H., Shan, G., Liu, H., Xing, J. (2007) Surface Modification of γ-Al_2O_3 Nanoparticles with gum Arabic and its applications in adsorption and biodesulfurization. Surface and Coatings Technology. 201: 6917–6921.

Zhang, J., Huang, W.W., Zhang, J., Wang, Q. (1992), Transport of particulate heavy metals towards the east China Seas: A preliminary study and comparison, Marine Chemistry. 40: 161–178.

Zhang, Y.W., Prabhu, P., Lee, J.K. (2010) Alginate immobilization of recombinant Escherichia coli whole cells harboring L-arabinose isomerase for L-ribulose production. *Bioprocess and Biosystems Engineering*. 33: 741–748.

Zhongxuan G., Huizhou L., Mingfang L. (2002) Isolation and identification of nondestructive desulfurization bacterium. Science in China Series B. 45: 521–531.

11

Biodesulfurization of Real Oil Feed

List of Abbreviations and Nomenclature

4,6-DMBT	4,6-Dimethyldibenzothiophene
4-MDBT	4-Methyldibenzothiophene
ADS	Adsorptive Desulfurization
Al	Alumina
ATF	Aviation Turbine Fuel
BDS	Biodesulfurization
BDSM	Biodesulfurizing Microorganisms
BMS	Benzylmethylsulfide
BNT	Benzonaphthothiophene
BP	Boiling Point
BSA	Bovine Serum Albumin
BTs	Benzothiophenes
BTU	British Thermal Unit
CMC	Critical Micelle Concentration
CPO	Chloroperoxidase
CSTR	Continuous Stirred Tank Reactor
DBDS	Dibenzyldisulfide

DBS	Dibenzyl Sulfide
DBTs	Dibenzothiophenes
DCW	Dry Cell Weight
DGTR	Desulfurated Ground Tire Rubber
DMA	Dynamic Mechanical Analysis
DMS	Dimethyl Sulfide
DMSO	Dimethyl Sulfoxide
DTHT	2–n–Dodecyltetrahydrothiophene
DWLR	Desulfurized Waste Latex Rubber
E_{24}	Emulsification Index
EBC	Energy BioSystems Corporation
EDS	Extractive Desulfurization
FCC	Fluid Catalytic Cracking
FGD	Flue Gas Desulfurization
FTIR-ATR	Fourier-Transform Infrared Spectroscopy-Attenuated Total Reflection
GC/AED	Gas Chromatography/Atomic Emission Detector
GC/FID	Gas Chromatography/Flame Ionization Detector
GC/FPD	Gas Chromatography/Flame photometric Detector
GC/MS	Gas Chromatography/Mass Spectrometry
GC/PFPD	Gas Chromatography/Pulsed Flame Photometric Detector
GC/SCD	Gas Chromatography/Sulfur Chemiluminescence Detector
GTR	Ground Tire Rubber
HCO	Heavy Crude Oil
HDS	Hydrodesulfurization
HDS-diesel	Hydrodesulfurized diesel
HGO	Heavy Gas Oil
HPBSi	2-Hydroxybiphenyl Sulfinate
HPBSo	2-Hydroxybiphenyl Sulfonate
HSFO	High Sulfur Fuel Oil
IPTG	Isopropyl-β-D-1-thiogalctopyranoside
LCO	Light Crude Oil
LGO	Light Gas Oil
MAS	Mesoporous Aluminosilicates
MDFU	Middle-Distillate Unit Feed
MIC	Minimum Inhibitory Concentration
NAPL	Non-Aqueous Phase Layer
NR	Natural Rubber
O/W	Oil/Water
OB	Oregon Basin
ODS	Oxidative Desulfurization
OFP	Oil Fraction Phase
OSCs	Organosulfur Compounds
PASHs	Polyaromatic Sulfur Heterocyclic Compounds

PEG-Cyt	Immobilized Cytochrome c in Poly(Ethylene)Glycol
PFPS	bis-(3-pentaflurophenylpropyl)-sulfide
PT	Phenanthro [4,5-b, c, d] thiophene
PVA	Polyvinyl Alcohol
RSR	Sulfides
RSH	Mercaptans
RSSR	Disulfides
SBR	Styrene Butadiene Rubber
SEM	Scanning Electron Microscopy
Sep	Sepiolite
Si	Silica
SRB	Sulfate Reducing Bacteria
SRG	Straight-Run Gasoline
Ths	Thiophenes
THTA	2-Tetrahydrothipheneacetic acid
THTC	2–Tetrahydrothiphenecarboxylic Acid
Ti	Titania
TLC	Thin Layer Chromatography
TPH	Total Petroleum Hydrocarbons
TRP	Total Resolvable Peaks
TSC	Total Sulfur Compounds
U_g	Superficial Gas Velocity (L/min)
ULS	Ultra-Low Sulfur
ULSD	Ultra-Low Sulfur Diesel
v-IR	Vulcanized Isoprene Rubber
v-SBR	Vulcanized Styrene Butadiene Rubber
WLR	Waste Latex Rubber
XPS	X-Ray Photoelectron Spectroscopy

11.1 Introduction

Sulfur content of residual fuel varies widely according to various local restrictions. The major part of the sulfur contained in fuel oil is in the form of thiophene and its derivatives and are found mainly in the resins (70%) and the remainder is equally distributed in the oils and asphaltenes. It is difficult to dispose high sulfur fuel oil HSFO with an S-content > 3 wt.%. Its primary outlet is the bunkers market, in cement manufacture, or as a liquid fuel for combustion uses in plants with flue gas desulfurization (FGD) or gasification units. The limits of sulfur content of fuel oil for use in most combustion plants without FGD was set to be 1 in 2005, which was then tightened to reach 0.3–0.5%. Although the economic and environmental regulations drove the switch to natural gas, fuel oil sales make up less than 15% of the world petroleum market. This presents a challenge to many refiners, considering the potentially large investments for upgrading the bottom of the barrel.

Heavy gas oil (HGO) is an important fraction of petroleum, as it is an intermediate fraction obtained from vacuum distillation used in the production of diesel and some lubricants. It is a byproduct of oil atmospheric and vacuum distillation that contains a high amount of sulfur-containing organic molecules. Since HGO is used in diesel composition, the application of a BDS process can result in products with low pollutant concentrations and would prevent the poisoning of catalysts (Otsuki *et al.*, 2000; Kilbane and Le Borgne, 2004; Gupta *et al.*, 2016).

There are three major types of transportation fuels, gasoline, diesel, and jet fuels, and there is a growing demand for transportation fuels which are produced from naphtha (BP < 175 °C) and middle distillates (boiling point < 370 °C) (Exxon Mobil 2016). The common types of sulfur compounds in liquid fuels are listed in Table 11.1.

Diesel fuels and domestic heating oils have boiling ranges of bout 400 °-700 °F. The desirable qualities required for distillate fuels include controlled flash and pour points clean burning, no deposit formation in storage tanks, and a proper diesel fuel cetane rating for good starting and combustion. In fractions used to produce diesel oil, most of the sulfur is found in BT, DBT, and their alkylated derivatives (Monticello and Finnerty, 1985). Deep reduction of diesel sulfur (from 500 to < 15ppm sulfur) is dictated largely by 4-methyldibenzothiophene (4-MDBT) and 4,6-dimethydibenzothiophene (4,6-DMDBT), where the sulfur atoms sterically hindered by

Table 11.1 Major Distillates of Crude Oil and Their Main S-Compounds.

Gasoline Range: Naphtha, FCC-naphtha (Selective HDS): • Mercaptans RSH; Sulfides RSR; Disulfides RSSR. • Thiophene and its alkylated derivatives. • Benzothiophene.
Jet Fuel Range: Heavy naphtha, Middle distillate: • Benzothiophene (BT) and its alkylated derivatives.
Diesel Fuel Range: Middle distillate, Light cycle oil: • Alkylated benzothiophenes. • Dibenzothiophene (DBT) and its alkylated derivatives.
Boiler Fuels Feeds: Heavy oils and distillation resides: • ≥3-ring Polycyclic sulfur compounds, including DBT, benzonaphthothiophene (BNT), phenanthro [4,5-b, c, d] thiophene (PT) and their alkylated derivatives.

substitutions in positions 4 and 6 are the most difficult to remove by HDS; 3,6-DMDBT has been shown to be particularly recalcitrant to HDS (Kabe *et al.*, 1997). The deep HDS problem of diesel streams is exacerbated by inhibiting effects of co-existing problems of polyaromatics and nitrogen compounds in the feed, as well as H_2S in the product (Song and Ma, 2003).

According to the Environmental Protection Agency, the standard sulfur level in gasoline fuel is 15 mg/L (Song, 2003). To meet the stringent emission standard postulated by regulatory organics, the petroleum refining industry should reach towards ultra-low sulfur fuels. Thus, ultra-deep removal of sulfur from transportation fuels are mandatory. Nevertheless, to achieve ULS fuels or the recent "no sulfur" specification, the conventional hydrodesulfurization (HDS) method (Chapter 2) needs higher temperature and pressure and expensive catalysts to remove the recalcitrant high molecular weight polyaromatic sulfur heterocyclic compounds (PASHs), such as dibenzothiophene (DBT) and substituted DBTs. Some other technologies, such as oxidative desulfurization (ODS) (Chapter 2), extractive desulfurization (EDS) (Chapter 2), and adsorptive desulfurization (ADS) (Chapter 2) have been proposed.

However, ADS has some drawbacks that need to be solved. If the selectivity of the adsorbent is low, it is easy to be regenerated, but it will lead to heat loss due to the competitive adsorption. Upon the increase of selectivity, the regeneration of spent adsorbent becomes more difficult. Thus, adsorbents should be well designed to achieve suitable selectivity. Solvent extraction and calcination in the air are the two widely used methods to regenerate desulfurization adsorbents. However, they have also some drawbacks that should be also solved. For the solvent extraction method, it is difficult to separate sulfur compounds from organic solvents and reuse these solvents, while for the calcinations method, sulfur compounds and aromatics are burned out causing fuel heat loss.

For reaching ultra-low sulfur fuels, deep HDS processes need large new capital investments, larger reactor volumes, longer processing times, new expensive catalysts, higher temperatures and pressures, and substantial hydrogen and energy inputs. Such extreme conditions to desulfurize high molecular weight recalcitrant polyaromatic sulfur heterocyclic compounds (PASHs) would lead to the deposition of carbonaceous coke on the catalysts. Sometimes, exposure of crude oil fractions to severe conditions, including temperatures above about 360 °C, decreases the fuel value of the treated product. The large amounts of produced H_2S poisons the catalysts and shortens their useful life. Moreover, deep HDS is affected by components in the reaction mixture such as organic hetero-compounds and polyaromatic hydrocarbons. Furthermore, deep HDS has higher operating costs (Mohebali and Ball, 2008, 2016; El-Gendy and Speight, 2016).

The 4S-pathway that removes sulfur without altering the octane value of fuels has been reported (Malik, 1978), but it did not take much concern because of the high light petroleum reserves, flexibility of environmental rules, and profitability of conventional processes at that time. By the 1990s, due to the depletion of light crude oil reserves, increasing worldwide fuel demand, more restrict environmental emission regulations, and the exploration of new reservoirs of sulfur rich oils (e.g. heavy crude oil, tar sands, and oil shale), the petroleum industry shared efforts with academia to research and develop the applicability of biodesulfurization (BDS) throughout specific oxidative BDS. Recently, biodesulfurization (BDS) catches the wave of the applied desulfurization technologies throughout the selective desulfurization of DBT and its derivative without affecting their hydrocarbon skeleton, keeping the heat value of the treated oil. Contrary to HDS, BDS offers mild processing conditions and reduces the need for hydrogen. Both these features would lead to high energy savings in the refinery. Further, significant reductions in greenhouse gas emissions have also been predicted if BDS is used (Linguist and Pacheco, 1999). However, although BDS has the potential benefits of lower operational cost and production of valuable by-products, it suffers from the slow rate of the desulfurization process. Therefore, there is still a need to increase the rate of BDS. Thus, BDS is recommended to be used as a complementary method to the conventional oil refining technologies.

Kilbane (1989) proposed a hypothetical oxidative desulfurization pathway that, if it ever existed in nature, could specifically remove sulfur from DBT. The pathway was named as the '4S-pathway', which implied consecutive oxidation of DBT sulfur to sulfoxide (DBTO), sulfone ($DBTO_2$), sulfinate (HPBSi) and/or sulfonate (HPBSo), and, finally, to the oil soluble product 2-HBP or 2,2'-BHBP, respectively, which finds its way back into the petroleum fractions while retaining the fuel value (Figure 11.1). The use of this pathway has been proposed for the desulfurization of petroleum in production fields as well as refineries (Kim *et al.*, 1996; Soleimani *et al.*, 2007).

There is a particular interest for microorganisms capable of carbon-sulfur (C-S) bond-targeted DBT degradation since they do not alter the calorific value of the fuel. There are 4 genes which are involved in complete C-S bond-targeted DBT degradation (McFarland, 1999; Abbad Andaloussi *et al.*, 2003ab; Gray *et al.*, 2003): dszC, which encodes a DBT-monooxygenase, dszA, which encodes a DBT-sulfone monooxygenase, dszB, which encodes a 2-hydroxybiphenyl sulfinate desulfinase, and dszD,

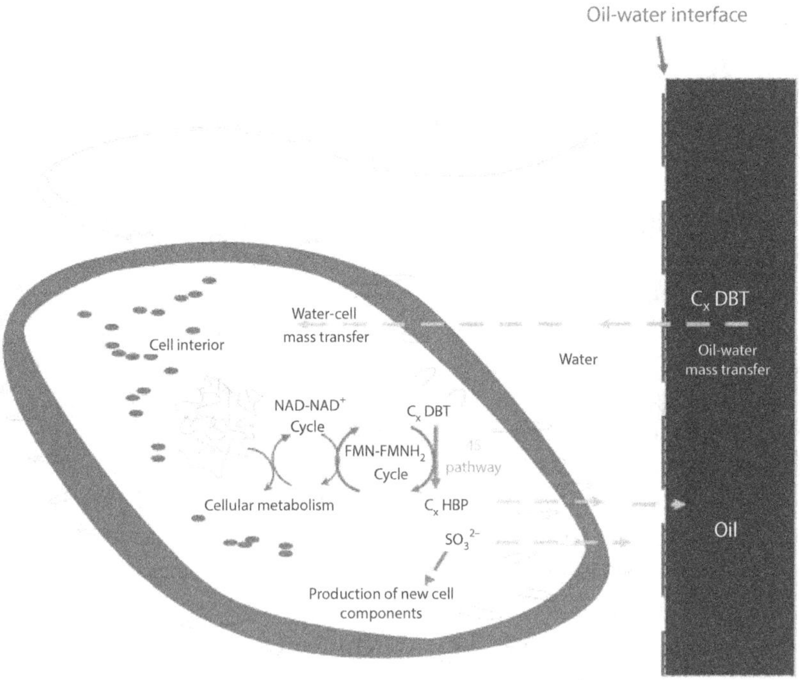

Figure 11.1 A Conceptual Diagram of Some of the Steps in the Biodesulfurization of Oil.

which encodes an NADH-flavin mononucleotide oxidoreductase. The proprietary of a BDS process relies on these four genes.

Although *Rhodococcus erythropolis* IGTS8 is the widely used bacterium in BDS research all over the world and is able to completely desulfurize DBT and some substituted DBTs in model oil system, it is reported to not significantly reduce the sulfur content in crude oil (Kaufman *et al.*, 1999). Generally, work on BDS has focused on model compounds in aqueous systems and, sometimes, in biphasic systems using model solvents, such as tetradecane or hexadecane, little has been reported on the BDS of real refinery feeds. This bears little resemblance to the conditions the biocatalyst would encounter in industrial application and limits the ability to assess the commercial potential of BDS. In fact, the BDS rates of diesel oil were much smaller than those obtained for pure DBT (Rhee *et al.*, 1998). The experimental results indicated that the BDS efficiency would decrease slightly in a diesel oil system compared to a single C_X-DBT system. Folsom *et al.* (1999) and Okada *et al.* (2003) reported that BDS of diesel oil was smaller than that obtained in model oil and attributed that to the existence of other organic sulfur compounds (OSCs), for example 4-MDBT, that might be less desulfurizable than DBT.

The accumulation of toxic metabolites in the medium may cause a decrease in BDS activity (Chang *et al.*, 2001). According to Abbad-Andaloussib *et al.* (2003b), this decrease in BDS might be due either to the inhibitory or toxic effects by certain sulfur compounds on the cells. Zhongxuan *et al.* (2011) and Zhang *et al.* (2011) attributed this phenomenon to an apparent competitive inhibition of substrates. Moreover, according to Reichmuth *et al.* (2000) the sulfur requirement for microbial growth is low compared to the level of sulfur in oil feeds, so microorganisms would cease desulfurization before the bulk of the sulfur is removed from the oil feed. Constanti *et al.* (1996), reported that a Gram-negative *Agrobacterium* sp. MC501 could oxidize alkanes at the terminal or subterminal methyl group and then continue through the β-oxidation pathway and consume more energy and time to obtain sulfur from sulfur compounds for growth which led to a decrease in their BDS efficiency in a real oil feed.

However, considerable research on the BDS of model OSCs, diesel, gasoline, and crude oil by different microorganisms has been done. BDS of diesel has been demonstrated using both growing and resting cells (Mohebali and Ball, 2008). On the contrary, the BDS of real oil feeds has been reported to be higher than model oils with model compounds such as DBT (Chang *et al.*, 2000a, Dinamarca *et al.*, 2010, 2014) since real oil feeds, such as LGO and HGO, contain various easily desulfurizable compounds, such as thiols and sulfides, besides recalcitrant compounds such as DBT (Chang *et al.*, 2000a). Setti *et al.* (1992) reported that the presence of n-alkane is seen to favor the removal of sulfur aromatic compounds, which act as a co-substrate, ensuring the growth of the culture, which also permits the solubilization and the emulsifying of the sulfur aromatic compounds. According to Setti *et al.* (1995), microorganism adsorption to the oil-phase is the most likely mechanism for explaining n-alkane biodegradation. Substrate uptake presumably occurs through diffusion or active transport at the point of contact. It is well known that most of the aerobic microorganisms adhere to the n-alkanes (below n-C_{16}) which are in a liquid form at room temperature, where the n-alkane form a film around the aromatic sulfur compound and, as this film is easily attacked by aerobic microorganisms, the bioavailability of sulfur compounds increases. Studies with a strain of *Candida* in two-phase systems (oil-water mixture) suggested that the rate of biodegradation might be also related to the interfacial area because a large part of the biomass, which characteristically is hydrophobic, adheres to the non-aqueous phase layer (NAPL)-water interface as a biofilm (Ascón-Cabrera and Lebault, 1995), where the smaller the interfacial tension, the larger the uptake of the dissolved compounds in NAPL and the higher the biodegradation of aliphatic hydrocarbons present. According to Setti *et al.* (1997), although aerobic microorganisms can remove high amounts of organic sulfur, the sulfur percentage in the residual heavy oil may increase as a consequence of simultaneous aliphatic compound biodegradation.

The difference in desulfurization efficiency within different microorganisms might be related to differences in cell surface hydrophobicity, mass transport through the cell walls and membranes, as well as metabolic regulations inside the cell (Bustos-Jaimes *et al.*, 2003). Thus, the BDS efficiency in real oil feeds depends mainly on the applied microorganisms and its broad versatility on different OSCs and there is no general trend for the efficiency of microorganisms in model or real feed oils.

BDS of petroleum results in total sulfur removals between 30 and 70% for mid-distillates (Grossman *et al.*, 1999; Pacheco *et al.*, 1999), 24 to 92% for hydrotreated diesel (Rhee *et al.*, 1998; Maghsoudi *et al.*, 2001; Ma *et al.*, 2006), 65–70% for partially HDS-treated mid-distillates (Folsom *et al.*, 1999), ≈ 90% for extensively HDS-treated mid-distillates (Grossman *et al.*, 2001), 20–60% for light gas oils (Chang *et al.*, 1998; Ishii *et al.*, 2005), 75–90% for cracked stocks (Pacheco *et al.*, 1999), and 25–60% for crude oils (Mcfarland, 1999; Premuzic and Lin, 1999b). Although the obtained removals are significant, this level of desulfurization is insufficient to meet the required sulfur level for all oil derivatives (Grossman *et al.*, 1999; Alves, 2007).

The majority of the work on BDS was performed for middle distillate fractions. The US DOE announced that research also started on diesel BDS (Le Borgne and Quintero, 2003). These results may serve as a background for the desulfurization of other streams. The treatment of oils using free cells has encountered some limitations, such as high cost of the biocatalyst and low volumetric ratio between the organic phase and the aqueous one. Also, separation of oil product from oil-water-biocatalyst emulsion is troublesome and the free cells are difficult for recycling (Choi *et al.*, 2003; Konishi *et al.*, 2005). In view of industrial application of BDS, cell immobilization is considered to be one of the most promising approaches (Mukhopadhyaya *et al.*, 2006; Huang *et al.*, 2012).

This chapter summarizes the worldwide achievements in biodesulfurization of crude oil and its distillates, at what point it's reached, and to what point it aims to reach.

11.2 Biodesulfurization of Crude Oil

Oil refineries usually separate crude oil into several fractions and then desulfurize them separately.

Refineries can make substantial cost savings if most of the sulfur is removed from the crude oil before it is fractionated. Also, it has been suggested that due to the high content of water in crude oil, the BDS of crude oil is more practical compared to that of diesel oil and gasoline (Zhou and Zhang, 2004). Nowadays, the petroleum industry is increasingly dependent on heavy crude oil to meet the domestic demand for gasoline and distillate fuels. Heavy oils with high viscosity are expensive to recover, transport, and process and have a lower market value than less viscous oils since asphaltene fraction is a major component of heavy oils. So, to use these as a fuel, they must be upgraded by reducing the average molecular weight of their constituents and remove heteroatoms. For bio-modification of asphaltenes, the reactions with organosulfur moieties could be very significant because sulfur is the third most abundant element in asphaltenes behind carbon and hydrogen and has an important role in the molecular structure of asphaltenes (Sarret *et al.*, 1999). Because the diversity and complexity of asphaltene molecular structures can be attacked, heme proteins were the biocatalysts chosen for investigations on the enzymatic modification of asphaltenes. In a survey of several heme proteins, including horseradish, lignin, manganese chloroperoxidase, and cytochrome c, they were found to be able to modify the greatest number of organosulfur compounds, including sulfur heterocycles and sulfides, and to have superior specific activity (Vazquez-Duhalt *et al.*, 1993).

The high viscosity of many crude oils is a factor contributing to the underutilization of these valuable natural resources. Viscosity greatly complicates and may even defeat the extraction of many types of crude oil from the earth. It remains a concern following extraction, as high viscosity significantly hampers the pumping, transportation, refining, and handling of crude oil. Because of this, the petroleum industry has long recognized the need for a safe, economical, and effective methods for reducing the viscosity of valuable fossil fuel resources.

Under certain circumstances, standard-refining processes such as hydrotreating or hydrodesulfurization (HDS) can favorably affect the viscosity of crude oil during refining. Some reduction in viscosity is also achieved through the breakdown of complex hydrocarbons (e.g. aromatic hydrocarbons) into simpler hydrocarbons of low molecular weight.

A generally accepted and controllable method of modulating crude oil viscosity during refining involves diluting viscous crude oil with low viscosity crude oil refining fractions, usually light-end distillates. Light-end distillates that are used as viscosity lowering diluents are referred to as cutter fractions. Thus, "cutting" it with such a light-end distillate lowers the viscosity of heavy crude oil or bitumen. This technique is useful at some stages of the petroleum refining process, but is not economical for large-scale use or to assist with the extraction of viscous crude oil from the earth.

Microorganisms have been used for the relatively controlled destruction of certain compounds in petroleum with the result that viscosity of the treated product is stabilized. For example, Hitzmann (1962) describes a method for stabilizing the viscosity of jet fuels when stored, as in military installations over seawater. Biocatalytic viscosity reduction uses bacteria to partially transform less valuable crude oil components to surface-active compounds (alcohols and carboxylic acids) that reduce crude oil viscosity is under investigation (Stringfellow, 2001).

A method for reducing the viscosity of viscous petroleum liquids, such as heavy crude oil and bitumen, has been disclosed. The method is appropriate for use with viscous petroleum liquids that contain sulfur-bearing heterocycles, the physicochemical properties of which contribute significantly to the viscosity of the liquid. The method comprises of contacting the viscous petroleum liquid with a biocatalyst that converts sulfur-bearing heterocycles into molecules that make the physicochemical properties conducive to viscosity. The biocatalyst works in a sulfur-specific manner such that the sulfur-bearing heterocycle is altered at the sulfur heteroatom thereof. Via biocatalysis, carbon-sulfur bonds are cleaved and/or polar substituents, such as hydroxyl groups, are joined to the sulfur heteroatom, the hydrocarbon framework of the sulfur-bearing heterocycle, or both. Preferred biocatalysts for viscosity reduction include preparations of *Rhodochrous erythropolis* ATCC 53968 microorganisms and enzymes obtained there from (Monticello *et al.*, 1996; Monticello, 2000). A need remains for a viscosity reducing treatment that can be used to facilitate the handling of viscous petroleum liquids at any desired stage of the extraction and/or refining processes. A suitable viscosity reducing treatment would not require specialized equipment or safety procedures and would not degrade the caloric (fuel) value of the treated petroleum.

In 1982, the Atlantic Richfield Company patented a biodesulfurization process for crude oils using a strain of *Bacillus Sulfasportare* ATCC 39909 (Atlantic Richfield Company, 1986). Referring to the biotreatment of vacuum residual oil, sulfur removal ranged from 17–23 wt.% (originally 4.21 wt.% S) and 45 wt.% for Maya crude oil. The removal of organic sulfur from oil was not combined by a loss of carbon content.

In 1987 it was reported that an aerobic process has been developed by the Unocal Corporation. The process was based on hydroxylation of the sulfur heterocycles to water-soluble products by a *Pseudomonas* strain, which also produced a surfactant, helping oil/water contact. The disadvantages of the process included low efficiency and loss of fuel value (Monticello and Finnerty, 1985).

Treatment of high-sulfur crude oil with the soil isolate *Pseudomonas alcaligenes* (DBT-2) (Finnerty and Robinson, 1986), which efficiently oxidizes DBT to water-soluble products, resulted in the loss of greater than 70% of organic sulfur from the pentane-soluble fraction of crude oil. Major deficiencies exist in this bioprocess technology. First, DBT-2 does not remove organic sulfur from the heavy fractions (asphaltenes and resins) of the crude oil which contain major concentrations of organic sulfur. Secondly, the biocatalyst causes a significant loss of BTU content in the treated oil. Thirdly, a byproduct is produced of sulfur-containing, water-soluble product for which there is no apparent end use.

Removal of organic sulfur from the crude oil or from its different fractions by employing microorganisms such as *Thiobacillus ferroxidans, Thiobacillus thioxidans, Thiobacillus thioparus,* and *Thiophysovolutans* has been attempted, but it is not suitable because catabolic desulfurization of organic molecules mostly involves the utilization of hydrocarbon portions of these molecules as a carbon source, resulting in loss of high calorie petroleum components (Hartdegen *et al.*, 1984; Indian Institute of Petroleum, 1996).

It was reported by Setti *et al.* (1997) that in an experiment of heavy oil desulfurization in which R. *erythropolis* IGTS8 was used, non-destructive desulfurization was followed by the complete degradation of aliphatic fraction which consisted 60% of the oil.

All Egyptian crude oils are characterized by being dominated with mixtures of substituted BTs and DBTs. Specifically, Suez Gulf oils show higher abundance of BTs relative to DBTs, as compared to Western Desert crude oils (Hegazi *et al.*, 2003). Salama *et al.* (2004) reported the BDS of Egyptian crude oil (Balaeem Barry) by *Rhodococcus rhodochrous* ATCC 53968 (which is later identified as *R. erythropolis* IGTS8). The optimized parameters affecting the BDS process were found to be: 24 h, 30 °C, pH 7.2, and 1% oil/water, 10 g/L glucose and 3 g/L ammonium chloride that recorded 45.4% total sulfur compound (TSC) removal. The biodesulfurized crude oil indicated a pronounced drop in sulfur content in asphaltenes, while no difference was monitored for saturates, aromatics, and resins. Microbial desulfurization of Balaeem crude oil was associated with a decrease in paraffins and isoparaffins with C-numbers, C13-C18, with a slight degradation of the biomarker phytane, but it was associated with an increase in those having 21–30 carbon atoms. In another study performed on Balaeem crude oil by Egyptian isolate, *Bacillus subtilis*, the same optimum conditions for maximum BDS-efficiency were found to be the same, but at pH7, recording 38% TSC-removal (Ibrahim *et al.*, 2004). This was accompanied by a pronounced decrease in polynuclear aromatic, phytane, polynuclear naphthene compounds, and asphaltenes, but *B. subtilis* raised paraffins with higher carbon number (C20–30) and isoparaffins (C21 -30). Both ATCC 53968 and isolated *B. subtilis* recorded 36.5% and 22.6% reduction in asphaltene content during the BDS process. IGTS8 was also reported to remove up to 45% of DBT through seven days of incubation from Ras Badran asphaltene fraction. Biodesulfurization of aromatic fractions showed that IGTS8 removed up to 63% of sulfur content from that fraction (Moustafa *et al.*, 2006).

Gunam *et al.* (2006) reported biodesulfurization of the majority of the sulfur (59% w/w) from Liaoning crude oil within 72 hours using *Sphingomonas subarctica* T7B.

Yu *et al.* (2006a) reported the BDS of two crude oils by 16 g DCW/L of resting cells of *Rhodococccus erythropolis* XP in shaken flasks with 1:20 (O/W) and found that the total sulfur content (TSC) of Fushun crude oil was decreased from 0.321 to 0.122 wt.% and that of Sudanese crude oil was decreased from 0.124 to 0.0656 wt.% within 72 h.

The BDS of MFO 380, with adding surfactants, was investigated to explore the potential improvement of BDS efficiency of heavy oil through adding surfactants (Li and Jiang, 2013). According to the hydrophobicity and the solubility of bunker oil, three kinds of surfactants, Tween 20, Triton X-100, and PEG 4000, were used to test the influence of surfactants on the BDS of MFO380. The ratio of surfactant to water was 1:50 (w/w). The ratio of oil phase to aqueous phase was 1:50 (w/w). The biocatalyst was obtained by enrichment with oil sludge as the seed and using DBT (0.5 mM) as the sole sulfur source. The oil sludge was mixed with BSM at a volume ratio of 1:5 at 200 rpm/min and room temperature for 4 days to digest the remaining oil and detach the microbial seed. The results show that the sulfur in MFO 380 has not been utilized during the biotreatment process when received soil or oil sludge was used directly. The possible reason for this is that the specific microorganism for BDS did not exist in the received seed. It is recommended that specific microbial strains should be concentrated by enrichment of a mixed culture system by model compounds first, as this will help in establishing the effectiveness of BDS of heavy oil. On the other hand, the low solubility of MFO 380 in the aqueous phase might be another major limitation to the poor BDS of heavy oil. When MFO 380 was added into the aqueous phase, large particles or irregular balls were formed. A large amount of heavy oil was tightly pasted on the wall of the flasks. Thus, sulfur compounds in heavy oil cannot contact and react with biocatalysts efficiently. Reducing the viscosity of bunker oil and increasing the contact of bunker oil with biocatalysts are the critical steps in the BDS process. The results of bunker oil BDS showed that the removal efficiency of sulfur in MFO380 was only 2.88% after 7 days of incubation, but this could be significantly improved by adding surfactants Triton X-100 or Tween 20. This effect could be attributed to the greatly reduced viscosity of heavy oil and increased mass transfer of sulfur compounds from heavy oil into water. Adding Triton X-100 achieved the highest removal efficiency of sulfur, up to 51.7 % after 7 days of incubation. The optimal amount of Triton X-100 was 0.5 g/50 mL medium.

Chauhan *et al.* (2015) reported the batch BDS of crude oil (1:5 O/W) using *Gordonia* sp. IITR 100, where the GC-MS revealed the production of 2-HBP, 1-methyl hydroxy-biphenyl, and BNT-hydroxide, which were possibly formed from DBT, 1-methyl DBT, and BNT.

Studies on the BDS of heavy crude oil are rather limited. The maximum reduction in sulfur content achieved with different microorganisms using heavy crude oil as the sulfur source are in the range of 47–68% (Setti *et al.*, 1992; EL-Gendy *et al.*, 2006; Torkamani *et al.*, 2008 a,b; Agarwal and Sharma, 2010; Bhatia and Sharma, 2010, 2012; Adlakha *et al.*, 2016).

The BDS of heavy crude oils is considered to be an upgrading process, as it not only decreases the S-content, but also decreases the viscosity of heavy oils. A decrease in viscosity is highly desirable as it can make the heavy crude oil more amenable for the refining process. Viscosity in heavy crude oil is largely due to asphaltenes which contain aromatic heterocyclic rings linked with aliphatic C-S bridges. Thus, a bacterium which will target aliphatic CS bonds will reduce the size of the asphaltene suprastructure and, thereby, viscosity (Kirkwood *et al.*, 2005).

Mohamed *et al.* (2015) reported a 10% reduction in the S-content of heavy crude oil (with initial S-content of 3%) in a batch BDS of 10% O/W, at 30 °C and 250 rpm, within 7 days' incubation period, using resting cells of *Rhodococcus* sp. strain SA11 collected in the exponential phase of DBT-cultures. Moreover, total S and S-speciation analyses of the hexane soluble fraction of the heavy crude oil (i.e. the de-asphaltened oil fraction or the maltenes) revealed 18% total S reduction including a wide range of thiophenic compounds, such as DBT, BT, and their alkylated derivatives.

Adlakha *et al.* (2016) reported a viscosity reduction of 31.28% with 76% BDS of heavy oils in a batch BDS of 1/3 O/W ratio using growing cells of *Gordonia* sp. IITR100, at 30 °C and 250 rpm, within 7 days' incubation period. Adlakha *et al.* (2016) reported approximately 61% BDS of heavy crude oil with an initial S-content of 1.083 wt.% after two successive BDS-rounds, each of 7 days, using *Gordonia* sp. IITR100.

Li and Jiang (2013) and Jiang *et al.* (2014) studied biodesulfurization of model thiophenic compounds and heavy oil by mixed cultures enriched from oil sludge, where the BDS of bunker oil by microbial consortium enriched from oil sludge recorded 2.8 % sulfur removal without de-asphalting within 7 d. After de-asphalting, the BDS efficiency was significantly improved (26.2–36.5 %), which is mainly attributed to fully mixing the oil and water, due to the decreased viscosity of bunker oil. Most recently, Martínez *et al.* (2016) reported enhanced desulfurization of oil sulfur compounds by using engineered synthetic bacterial consortia.

The desulfurization capability of *B. subtilis* Wb600 on light crude oil was examined by Nezammahalleh (2015). This bacterial species grows on a sulfur-free basal salt medium in the presence of light crude oil with a total sulfur content of 1.5%. *B. subtilis* Wb600 utilizing the sulfur-containing compounds of the oil as a source of sulfur without degrading the hydrocarbon skeleton. The microorganism produces biosurfactants which facilitates the transfer of the organic sulfur compounds into the aqueous phase. The quantity of biosurfactants produced in the microbial culture is about two times higher than the ones produced in the control culture with an inorganic sulfur source. This bacterial species decreases the total sulfur content of the light crude oil to about 40% during 35 h with an oil/water phase ratio of 0.2.

11.3 Biodesulfurization of Different Oil Distillates

Biodesulfurization has been studied extensively by researchers at Energy BioSystems Corporation (EBC, named later Enchira) for reducing sulfur content in various hydrocarbon fractions.

Energy Bio Systems Corporation was reported to use a five-barrel/day pilot plant to study the BDS of diesel fuel (Rhodes, 1995) by *R. erythropolis* IGTS8, which was previously known as *R. rhodochrous* IGTS8, before its final identification by 16S rRNA and physiological studies (Monticello, 1994).

Monticello and Kilbane (1994) investigated the formation of a reversible micro-emulsion between the biocatalytic agent and the petroleum liquid in an attempt to facilitate the recovery of the treated product. Firstly, the experiments were performed on model compounds such as DBT and alkylated DBT. The aqueous catalytic agent *Rhodococcus rhodochrous* selectively cleaves organic C-S bonds in the presence of the emulsifying agent Triton. Then, Monticello and Kilbane (1994) presented one example showing the enhanced performance of biocatalytic desulfurization of a residual fuel oil in an emulsion. Six runs were conducted using an emulsifying agent in some of the runs to form an appropriate emulsion between the residual fuel oil and the aqueous phase. BDS was found to be enhanced when an appropriate emulsion between the organic (substrate) and aqueous (biocatalyst) phases is formed. It was concluded that the intimateness of contact between the oil and water phases is the rate-limiting factor of the BDS process.

Applications of growing cells of *Nocardia* sp. CYKS2 in desulfurization of a model oil (10 mM DBT in n-hexadecane), revealed a decrease in S to approximately 1.8 mM within 80 h with a desulfurization rate of 0.279 mg sulfur/L dispersion/h at 1/10 O/W. This was 2.8 times as high as that for a DBT/ethanol system, while in the case of the BDS of diesel oil, the sulfur content decreased from 0.3 to 0.24 and 0.2 wt% within 48 h at 1/10 and 1/20 O/W with desulfurization rates of 0.909 and 0.992 mg sulfur/L dispersion/h, respectively. This was equivalent to 10.0 and 20.8 mg sulfur/L oil/h, respectively. Such results for two different phase ratios implies that the rate of sulfur removal from oil can be increased by decreasing the phase ratio. However, such an advantage is counterbalanced due to an increased reaction dispersion volume requiring almost the same reactor size as for the treatment of a given amount of oil in a given time (Chang *et al.*, 1998).

Paenibacillus sp. strains A11–1 and A11–2 grew well in the presence of LGO (20% O/W) with a recorded decrease in sulfur content from 800 ppm to 720 ppm (Konishi *et al.*, 1997).

Gordona sp. CYKS1, via the S-specific pathway, yielded a 70 and 50% reduction in the S-content of a middle-distillate unit feed (MDUF) and LGO, respectively, in a batch BDS of 190% O/W (Rhee *et al.*, 1998), where the S-content decreased from 0.15% to 0.06% and from 0.3% to 0.25 % after 12 h, respectively. The specific desulfurization rates of MDUF and LGO were 5.3 and 4.7 μM S/g DCW/h, respectively.

The total sulfur in light gas oil which has been hydrodesulfurized decreased from 800 ppm to 310 ppm by resting cells of *R. erythropolis* KA2–5-1. The ratio of light gas oil was 50% in the reaction mixture (Ohshiro and Izumi, 1999).

Studies were carried out at both a shake flask level, as well as in bioreactors, and most of the studies were performed using *Pseudomonas* and *Rhodococcus*. Srivastava (2012) reported that biodesulfurizing bacteria can reduce the S-content of diesel oil from 535 to 75 ppm within only 24 h. McFarland *et al.* (1998) used resting cells of *Rhodococcus* sp. for the BDS of diesel in a continuous stirred tank reactor and observed 50–70% reduction in sulfur.

Rhodococcus sp. strain ECRD-1 was evaluated for its ability to desulfurize a 233–343 °C middle-distillate (diesel range) fraction of Oregon basin (OB) crude oil. Up to 30% of the total sulfur in the straight-run middle distillate cut was removed and oxidized another 35% into oil-soluble products. It has been found that ECRD-1 is able to degrade alkanes, but is unable to attack aromatic hydrocarbons (Grossman *et al.*, 1999).

Chang *et al.* (2000b) reported the decrease of the S-content of non-hydrotreated-middle distillate unit feed from 1500 mg/L to 610 mg/L, with a desulfurization rate of 9.4 mmol/kg DCW/h using resting cells of *Gordona* strain CYKS1. It was worthy to mention that *Gordona* sp. CYK1 expressed a higher desulfurization rate in real oil feed than the model oil (12 mmol DBT in n-hexadecane), recording 0.34 and 0.12 mg sulfur/g cell, respectively.

Jiang *et al.* (2002) found that surfactants could improve the desulfurization rate, where, *Pseudomonas delafieldii* strain R-8 removed 72% of the organic sulfur from low sulfur diesel oil (S<300 mg/L) within 72 h at 250 r/min with the addition of Tween-80.

Mingfang *et al.* (2003) reported that the BDS of straight-run diesel oil was treated by resting cells of *Nocardia globerula* R-9 in a batch process of 1:8 (O/W). The sulfur content of the diesel oil was reduced from 1807 to 741 mg/L after 3 days of reaction. The mean desulfurization rate was found to be 5.1 mmol S/kg DCW/h.

The desulfurization gene cluster from *R. erythropolis* strain KA2–5-1 was transferred into 22 *Rhodococcal* and *Mycobacterial* strains. The resting cells of recombinant strain MR65 from *Mycobacterium* sp. NCIMB could desulfurize 68 ppm of sulfur in light gas oil (LGO) containing 126 ppm sulfur. This strain expressed about 1.5 times higher LGO desulfurization activity as much as of *R. erythropolis* strain KA2–5-1. The application of a recombinant was effective in enhancing LGO BDS (Noda *et al.*, 2003a,b). In another study performed by Watanabe *et al.* (2003b), the dsz gene cluster from *R. erythropolis* KA2–5-1 was transferred into *R. erythropolis* MC1109, which was unable to desulfurize LGO. Resting cells of the resultant recombinant strain, named MC0203, decreased the sulfur concentration of LGO from 120 mg/L to 70 mg/L within 2 h. The LGO desulfurization activity of this strain was about twice that of strain KA2–5-1.

For higher achievement of desulfurization value, sequential BDS of real oil feeds has been reported. Ten cultivation rounds each of 3 days' incubation were done using *Mycobacterium phlei* WU-0103 to achieve 52% reduction in total sulfur in light gas oil (Ishii *et al.*, 2005).

R. erythropolis XP was adopted to desulfurize fluid catalytic cracking (FCC) and straight-run gasoline (SRG) and the first publication on the

BDS of gasoline by free whole cells was conducted. Approximately 29% of the total sulfur of Jilian FCC gasoline (from 495 to 338 ppm) and 32% of that of Qilu FCC gasoline (from 1,200 to 850 ppm) were removed, respectively. About 85% of the sulfur content of SRG (from 50.2 to 7.5 ppm) was also removed (Yu *et al.*, 2006b).

Considering that benzothiophenes predominate in gasoline and the toxicity of gasoline to bacterial cells, the BDS of gasoline is a significant achievement. The alginate immobilization of *M. goodii* X7B is reported to minimize the toxicity of gasoline to the biocatalyst. Li *et al.* (2005a) studied the ability of *Mycobacterium goodie* X7B to desulfurize gasoline in an immobilized-cell system. Cells were immobilized by entrapment with calcium alginate. Cells were harvested in the mid-exponential phase of growth by centrifugation, washed twice with a sodium chloride solution (0.85%), and re-suspended in the same solution containing 0.3% Tween80 and 2% sodium alginate at a concentration of 12.4 mg of dry cells/mL. The mixture was then dropped into a 5% calcium chloride solution containing 0.3% Tween 80 to obtain beads of immobilized cells (about 1.0 mm in diameter). Ten milliliters of gasoline were added for desulfurization and the volumetric phase ratio of the aqueous phase to oil was 9. Dushanzi straight-run gasoline 227 (DSRG227) and DSRG275 were used in the BDS process. The numbers 227 and 275 refer to the concentrations of sulfur in the oil in parts per million. When DSRG227 was treated with immobilized cells of strain X7B for 24 h, the total sulfur content significantly decreased from 227 to 71 ppm at 40 °C, corresponding to a reduction of 69%. In addition, when immobilized cells were incubated at 40 °C with DSRG275, the sulfur content decreased from 275 to 54 ppm in two consecutive reactions corresponding to a reduction of 81%. With this excellent efficiency, strain X7B is considered a good potential candidate for industrial applications for the BDS of gasoline.

A new type of airlift reactor with immobilized *Gordonia nitida* CYKS1 cells on a fibrous support was designed by Lee et al. (2005) and used for the biocatalytic desulfurization (BDS) of diesel oil. Its performance was evaluated at different phase ratios of the oil to aqueous medium (or oil phase fractions) and different sucrose concentrations. A 7-L jar ferment reactor was modified to fabricate a new type of airlift immobilized-cell reactor as shown in Figure 11.2. A glass draft tube (16 and 9 cm in height and diameter, respectively) was placed inside the reactor (32 and 17 cm in height and inner diameter, respectively). Eight strings (5 g for each) of nylon fiber bunches were installed as the cell carrier. When the growth reached an exponential growth phase, the culture was inoculated into the airlift reactor containing a minimal salt medium (MSM) with 0.3 mM DBT and 10g/L

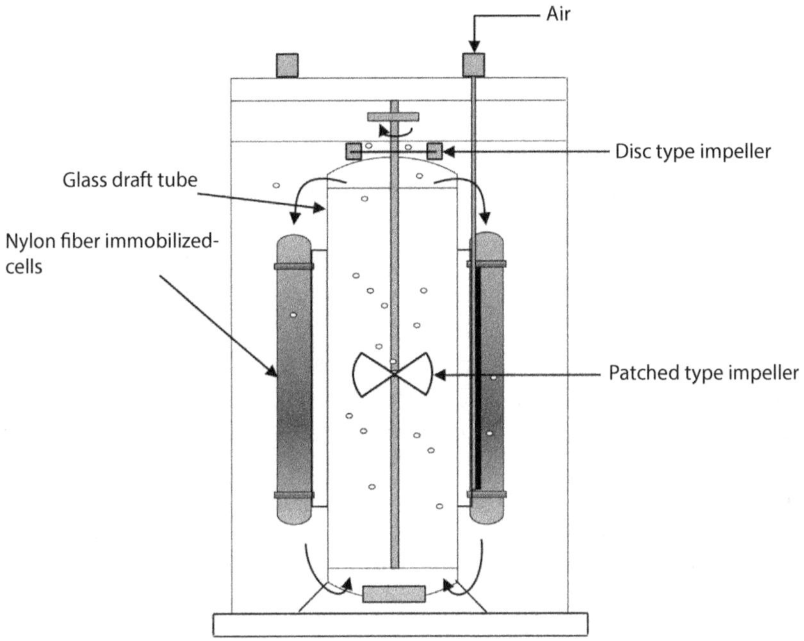

Figure 11.2 Biodesulfurization of Diesel Oil by Airlift Reactor.

sucrose. Cells were successfully immobilized in the nylon fiber bunches by incubating them in MSM with 0.3 mM DBT and 10 g/L sucrose for 3 days without supplying diesel oil. After immobilizing the cells, the medium was drained completely. The reactor was then filled with a mixture of MSM containing 2.9 g/L sucrose and diesel oil. Total biomass immobilized in nylon fibers measured at end of run was 21.6 g and the immobilized cell density was 0.54 g DCW/ g support. When the reaction mixture contained 10% diesel oil (v/v), 61–67% of sulfur was removed and the sulfur content decreased from 202–250 to 76–90 mg/ L within 72 h. The sulfur content did not decrease any further because the remaining sulfur compounds were recalcitrant to BDS. During desulfurization, strain CYKS1 consumed hydrocarbons in the diesel oil, mainly n-alkanes with 10–26 carbons, as a carbon source even though an easily available carbon source, sucrose, was supplied. The results obtained from the airlift reactor operation provided important information on the future development of BDS processes for deep desulfurization of diesel oils and other fuel oils.

Thus, briefly, Lee *et al.* (2005) reported 61–67% S-removal from diesel oil accompanied by a loss of C10–C26 carbons using immobilized *Gordonia nitida* CYKS1 cells in an airlift reactor. Similarly, when using *Pseudomonas delafieldii* R-8 in a 5 L reactor 47% sulfur removal from diesel was achieved (Guobin *et al.*, 2006).

Yang *et al.* (2007) reported 12% BDS of diesel with a sulfur content of 12,600 ppm, where most of the sulfur compounds (>80%) in the diesel were thiophene, benzothiophene, DBT, and its derivatives in a two-layer continuous bioreactor (1/4 O/W) using growing cells of *Rhodococcus globerulus* DAQ3. The total sulfur reduction from 12,600 to 11,100 ppm in 250 mL of non-hydro-treated diesel was performed in a two-layer continuous bioreactor (1:4 O/W), where the organic phase and the cell-containing aqueous phase were kept as two layers and the cells were grown under steady-state conditions in the aqueous phase by carefully controlling the continuous operation. The OSCs diffused from the organic layer into the aqueous phase where BDS occurs in the cells and the produced water-soluble inhibitory compounds from cell metabolism or cell lysis were washed out from the aqueous phase of the bioreactor, consequently, avoiding the accumulation of possible inhibitors.

The total S-content of a middle distillate unit feed (MDUF) decreased from 0.15% (w/w) to 0.06% and that of a light gas oil (LGO) decreased from 0.3% (w/w) to 0.15% by resting cells of *Gordonia* sp. CYKS1 after 12 h incubation (Li *et al.*, 2009).

Bhatia and Sharma (2010) examined the capability of *Pantoea agglomerans* D23W3 to desulfurize different oil feeds: light crude oil (0.38% S), heavy crude oil (2.61% S), diesel oil (0.15% S), hydrodesulfurized diesel (0.07% S), and aviation turbine fuel (ATF, 0.41% S) in a batch BDS of 1:9 (O/W) at 30 °C, 200 rpm, and 120 h. The recorded BDS percentage ranged between 26.38 and 71.42%.

Dinamarca *et al.* (2010) studied the efficiency of adsorption of *Pseudomonas stutzeri* on silica (Si), alumina (Al), sepiolite (Sep), and titania (Ti) and their influence on the BDS of gas oil (4700 mg/L of sulfur). The most active biocatalysts were *P. stutzeri*/Si and *P. stutzeri*/Sep, recording a desulfurization capacity of 2.80×10^{-13} g sulfur/cell/h and 2.57×10^{-13} g sulfur/cell/h, respectively, while *P. stutzeri*/Ti showed the lowest activity, recording 1.03×10^{-13} g sulfur/cell/h.

Nassar (2010) reported that *Bacillus sphaericus* HN1 is considered to have a sufficiently broad substrate specificity to degrade major organic sulfur compounds found in diesel oils. The effect of microbial treatment on the sulfur content and on the hydrocarbon skeleton of the total resolvable components (TRP) of diesel oil at different ratios of oil to water (O/W) (1/9, 1/4, 2/3, 1/1, and 3/2) was studied using GC-FPD and GC-FID analysis, respectively. When the oil phase ratio was increased from 1/9 to 1/4 (O/W) phase ratio, the sulfur content was decreased from 9,594 mg/L to

3,393 mg/L (%BDS 64.63%) and to 1,743 mg/L (%BDS 81.83%), respectively. The efficiency of total sulfur removal was decreased to 78%, 20.94%, and 14.93% at a higher o/w phase ratio of 2/3, 1/1, and 3/2, respectively. The biodegradation capacity of DBT from diesel oil was 62.75%, 78.71%, 74.86%, 29.46%, and 0.53% in 1/9, 1/4, 2/3, 1/1, and 3/2 (O/W) ratio cultures, respectively, and the biodegradation of BT was 60.96%, 41.34%, 35.19%, 32.49%, and 24.75% in 1/9, 1/4, 2/3, 1/1, and 3/2 (O/W) ratio cultures, respectively. HN1 showed an excellent ability to remove high alkylated DBTs (4-MDBT and 4,6-DMDBT) from diesel oil. The biodegradation capacity of 4-MDBT was 66.22%, 78.66%, 75.89%, 14.87%, and 11.07% in 1/9, 1/4, 2/3, 1/1, and 3/2 (O/W) ratio cultures, respectively, and the biodegradation of 4,6-DMDBT was 63.07%, 80.85%, 77.50%, 32.11%, and 32.47% in 1/9, 1/4, 2/3, 1/1, and 3/2 (O/W) ratio cultures, respectively. In this study, the percent of degradation of total petroleum hydrocarbons (TPH) was 63.05%, 86.87%, 57%, 53.90%, and 39.24% in 1/9, 1/4, 2/3, 1/1, and 3/2 (O/W) ratio cultures, respectively, indicating that the maximum TPH degradation occurred in 1/4 (O/W) ratio cultures.

Irani *et al.* (2011) used the growing cells of *Gordonia alkanivorans* RIPI90A for the BDS of diesel oil in an internal airlift bioreactor. The effect of initial sulfur concentration on growth and sulfur reduction during the growth phase were investigated. All the sulfur compounds in diesel oil (30 mg/L) were lumped into a pseudo-compound and then diluted at different dilution ratios using hexadecane to give final sulfur concentrations. Hydrodynamic characterization and BDS assays were evaluated in an internal airlift bioreactor. A reactor vessel was 0.12 m in diameter and its overall height was 0.7 m. The draft-tube was 0.07 m in internal diameter and 0.35 m tall and was located 0.06 m above the bottom of the tank. The vessel was sparged in the concentric zone through a 0.0006 m diameter sparger. The working volume and the overall volume of the reactor were 5 and 8 L, respectively. A dissolved oxygen electrode and a pH-meter were placed in the bioreactor. Air was supplied to the reactor through a filter and a rotameter. Moreover, the effect of superficial gas velocity (U_g) and working volume (v) on volumetric gas liquid mass transfer coefficient was studied in an airlift bioreactor for the BDS of diesel. Under optimum conditions in the airlift bioreactor, the superficial gas velocity (U_g) and working volume (v) was set at 2.5 L/min and 6.6 l, respectively, and an oil/water phase ratio of 30% and initial sulfur concentration of 28 mg/L, about 50% removal of sulfur from diesel occurred within 30 h incubation using *Gordonia alkanivorans* RIPI 90A (Irani *et al.*, 2011).

Tang *et al.* (2013) sonicated a 1 g bunker oil/45 mL BSA solution mixture at different amplitude ratios and times. After 7 d of incubation using a mixed culture from an oil-contaminated soil, approximately 18.4% sulfur reduction in ultrasound pretreated bunker oil compared to 13.8% sulfur reduction in the mechanically stirred and surfactant-supplemented positive control experiment. This experiment proved that the concept of ultrasound pretreatment leads to greater sulfur removal efficiencies in a shorter period of time.

Dinamarca *et al.* (2104a) evaluated the BDS of gas oil in a bioreactor packed with a catalytic bed of silica containing immobilized *Rhodococcus rhodochrous*. For immobilization, the cells were circulated through the packed bed in a downward direction at 10 mL/min for 72 h. Bacterial numbers were adjusted by measuring the OD_{600} and normalized by the mass of support (g) in the catalytic bed. The cell numbers adsorbed were in the range of 74–95 OD_{600}. To remove the non-adsorbed bacteria on the support, saline solution was circulated through the catalytic bed. 0.623 M DBT in n-hexadecane was used as the only sulfur source and gas oil containing 4.7 g sulfur/L. A glass bioreactor of 125 mL total volume (2 cm diameter, 40 cm length) was constructed. The bioreactor was continuously sparged with air at 70 mL/min in an upward direction. A feed flow rate of sulfured substrates was provided by a peristaltic pump. The temperature inside the bioreactor was maintained at 20 °C by a thermostatically-controlled water bath. The bioreactor was filled with silica particles to evaluate three packed bed heights (5, 10, and 15 cm) for each of two particles size (0.71–1.18 mm or 3.35–5.6 mm). DBT or gas oil was fed continuously into the packed bed bioreactor in a downward direction at 0.45 or 0.9 mL/min. The desulfurized substrate was collected after a 2 h trickling operation. Samples were taken over 1 h at intervals of 20 min to confirm steady-state operation. After each BDS run, the bed was washed with a saline solution and then a fresh bacterial suspension was circulated through the packed bed. Results indicated that the BDS of DBT was dependent on the length of the bed. The conversion at a flow rate of 0.45 mL/min and bed length of 15 cm was 57 %, whereas lengths of 10 and 5 cm gave 41 and 16 %, respectively, at the same flow rate. With a constant flow, it is clear that a greater length of the bed results in a higher retention time, allowing a sufficient contact time for expression of the absorbed bacterial metabolism. Similar to the BDS of DBT, sulfur removal from gas oil was greater using a larger bed length. However, BDS conversion values were lower for gas oil than for DBT in all the studied systems. It is possible that the presence of organic molecules in gas oil affects the bacterial metabolism and, therefore, the BDS process. Desulfurization of DBT was improved when particles of silica in the range of 3.35–5.6 mm were used to absorb cells. Because BDS of DBT with *Rhodococcus* is a process that depends on the aeration variable (Del Olmo *et al.* 2005), the results suggest that higher particle size improves aeration of the bed by offering a greater inter-particle volume that can be occupied by both liquid and air. The effect of a large particle size dominates over the higher specific area for cell immobilization offered by a smaller particle size and shows the importance of the meso- and macro-porosity of the catalytic bed in the configuration of packed bed bioreactors for BDS. Immobilized *R. rhodochrous* cells maintained BDS activity after three cycles, retaining at least 84% of the initial desulfurization conversion.

In another study, Dinamarca *et al.* (2014b) studied the effect of adding biological surfactants to immobilized biocatalysts formed by adsorption of *R. rhodochrous* on Si, Al, and Sep and their influence on the BDS activity of gas oil. For the BDS reaction of immobilized cell systems with surfactants, the bacterial cells (4.0×10^9 –1.8×10^{10}) were adsorbed on 1 g of each support (Al, Sep and Si) and were placed in 25-mL flasks containing 10 mL of sulfur-free Medium A and the surfactants (Tween 80 or biosurfactant). One milliliter of DBT or gas oil was added for desulfurization. The reaction was carried out in a 10 mL medium with bacterial cell numbers ranging from 1.62×10^{10} to 2.14×10^{10} with a concentration of biosurfactant and Tween 80 in a range of 0.1–0.6% w/v and 1 mL of gas oil (4700 mg/L of sulfur). The reaction was carried out at 30 °C in a rotary shaker at 200 rpm for 24 h. A greater effect can be noted on the BDS activity in non-immobilized cells with surfactants than in the immobilized cells, essentially in the *Rhodochrous*/Si and *Rhodochrous*/Al systems. In the case of *Rhodochrous*/Sep systems, the greater interaction between bacterial cells and the support optimized the effect of immobilization. The addition of biosurfactant to the immobilized systems increased the desulfurization of the gas oil in the three supports. However, this increase was at its maximum with the *Rhodochrous*/Sep system. The optimal interaction between bacterial cells and Sep and the solubilization of gas oil by the presence of the biosurfactant in the reaction medium explained that behavior. Adding Tween 80 to the immobilized systems increased the BDS of gas oil in the three supports. However, the increase, in the case of the *Rhodochrous*/Sep system, was less than when biosurfactant was added. This effect demonstrated that, in the case of gas oil samples, the formation of micelles is more critical than increasing particle mobility for further solubilization.

Arabian *et al.* (2014) reported the BDS of kerosene with an initial S-content of 2333 ppmw by *Bacillus cereus* HN, at an oil phase fraction of 0.2, using an inoculum size of 3.6×10^7 (cell/mL), 180 rpm at 40 °C. The total S-content decreased to 1557 ppm (≈ 33% BDS) within 72 h.

Maass *et al.* (2014) evaluated the effectiveness of the strain *R. erythropolis* ATCC 4277 for the desulfurization of a synthetic model oil for diesel

(3 mM DBT in n-dodecane) in a batch reactor with different O/W phase ratios of 20, 80, and 100% (v/v). ATCC 4277 was able to degrade 93.3, 98.0, and 95.5 % of the DBT at 20, 80, and 100 % (v/v), respectively. The greatest percentage of DBT desulfurization (98%) and the highest specific rate of 2-HBP production (44 mM DBT/kg DCW/h) were obtained in a batch reactor using 80% (v/v).

Maass *et al.* (2015) reported the biodesulfurization and biodenitrogenation of heavy gas oil (62.5 μg/g S and 37.6 μg/g N) by *Rhodococcus erythropolis* ATCC 4277, in a batch reactor with different initial concentrations of HGO at 28 °C and 200 rpm, within 18 h. The best results were achieved for HGO ratios of 40, 20, and 60% (v/v), where the desulfurization rate of 148, 137, and 123 mg S/kg/h with sulfur percentage removal of 42.7, 39.4, and 35.5% and nitrogen percentage removal of 43.2, 40.2, and 31.2%, were recorded, respectively. Maass *et al.* (2015) examined the culture media of 40 and 100% (O/W) by optical microscopy in an attempt to presumably understand why 40% gave the best result and why it was completely inhibited in 100% (O/W). In a 40% culture, the presence of a considerable amount of refringent cells was observed, which is an indication that the cells of *R. erythropolis* ATCC 4277 were able to adapt well to the reaction medium containing 40% (v/v). Moreover, cells were well dispersed in the medium which is an indication that the medium provides favorable conditions for the microorganism. A good adaptation of the cells to this particular reaction medium was also observed in the result of minimum inhibitory concentration MIC-test by the macro-dilution method, while the cells were aggregated and there was no evidence of refringent cells in the 100% (O/W) culture. This was attributed to the nutritional deprivation growth condition which increases the cell surface hydrophobicity and acts as an ignition power for a cell–cell junction or biogranulation. Adhesion occurs when the electrostatic repulsion is overcome by van der Waals forces and hydrophobic interactions (Carvalho *et al.*, 2009). Furthermore, the cells form an aggregate to protect cell population against the toxicity of hydrophilic compounds since this microorganism resistance decreases with the increase of the system toxicity, restraining the cells to maintain their viability under severe conditions (Carvalho *et al.*, 2004, 2005, 2009; Carvalho and Da Fonseca, 2005; Heipieper *et al.*, 2007). This, consequently, explained the recorded decrease in the BDS and BDN ability of *R. erythropolis* ATCC 4277, at a higher O/W phase ratio.

Nassar (2015) improved the BDS of diesel oil by using Fe_3O_4 MNPs coated cells, magnetically immobilized cells (alginate + Fe_3O_4 MNPs), and agar immobilized cells of *Brevibacillus invocatus* C19 and *Rhodococcus erythropolis* IGST8. The effect of microbial treatment on the BDS of the diesel oil at different ratios of oil to water (O/W) (10%, 25%, 50%, and 75% v/v) was studied using GC-FID and GC-FPD analysis. GC-FPD analysis was used to qualitatively and quantitatively evaluate the effects of free, coated, magnetically immobilized, and agar immobilized cells of C19 and IGTS8 on the sulfur contents of diesel oil as a model of real oil feed with an initial sulfur content of 8,600 ppm. The Fe_3O_4 MNPs coated cells of *R. erythropolis* IGTS8 showed good BDS efficiency (%BDS 78.26 %) in the case of a 10% (v/v) O/W phase ratio than that which occurred by free cells (%BDS 74.42%), magnetic immobilized cells (%BDS 57.26%), and agar immobilized cells (%BDS 58.98%). The desulfurization capability of DBT and BT by coated IGTS8 were 58.63% and 32.98%, respectively, at an optimum O/W phase ratio (10% v/v). The agar immobilized cells and coated cells of *Brevibacillus invocatus* C19 showed the highest BDS efficiency (%BDS 98.97% and 98.69%, respectively) at 25% (v/v) O/W phase ratio, compared to free cells (%BDS 91.31%) and magnetic immobilized cells (%BDS 89.99%). Thus, 10% and 25% (v/v) phase ratios were considered as the optimum O/W phase ratios of *R. erythropolis* IGTS8 and *Brevibacillus invocatus* C19, respectively. From a GC-FPD chromatogram (Figure 11.3), the BDS of DBT from diesel oil by agar immobilized C19 and coated C19 was 93.71% and 92.86%, respectively, and the BDS of BT was 91.42% and 88.76%, respectively, at 25% v/v optimum O/W phase ratio, while the BDS of DBT and BT by coated IGTS8 were 58.63% and 32.98%, respectively, at an optimum 10% v/v O/W phase ratio. The coated cells of IGTS8 showed a low capability to desulfurize 4-MDBT (%BDS 25.34%) and 4,6-DMDBT (%BDS 41.56%) from diesel oil compared to coated cells of C19 (%BDS 83.04%) and agar immobilized cells of C19 (%BDS 87.16%), as shown in Figure 11.3. The effect of microbial treatment on the hydrocarbon skeleton of the total resolvable components (TRP) of the diesel oil at different ratios of oil to water was also studied using GC-FID analysis.

The biodegradation capacity of total petroleum hydrocarbons (TPH) in diesel oil by agar immobilized C19 and coated C19 was 10.05% and 11.73% at an optimum O/W phase ratio (25% v/v), respectively, while the percent of degradation of TPH that occurred by coated IGTS8 was 34.98% at an optimum O/W phase ratio (25% v/v).

Adlakha *et al.* (2016) reported a 70% reduction in the initial S-content (50 ppm) of commercial diesel oil in a 5 L reactor using growing cells of *Gordonia* sp. IITR 100.

Shahaby and Essam El-din (2017) reported the isolation of *Pseudomonas putida* TU-S2, *Bacillus pumilus* TU-S5, and *Rhodococcus erythropolis* TU-S7 from petroleum hydrocarbons polluted soil for their ability to selectively desulfurize DBT to 2-HBP, where the BDS of different oil feeds

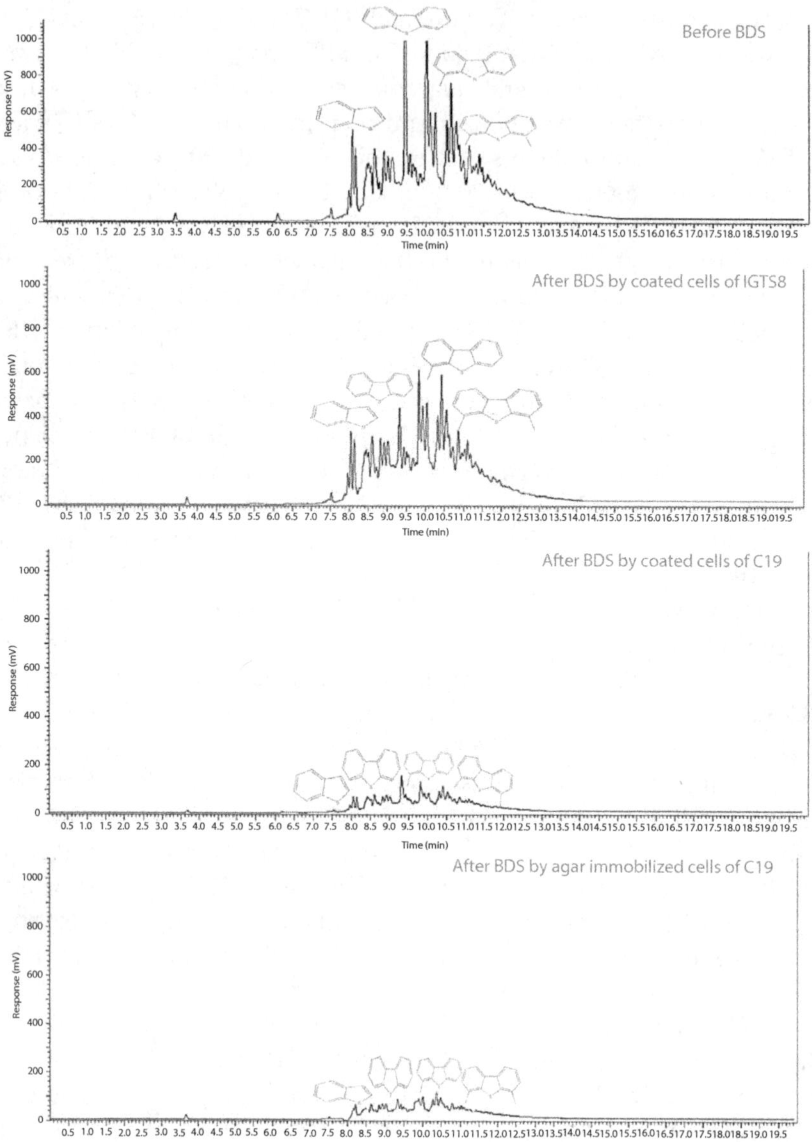

Figure 11.3 GC/FPD Chromatograms of Diesel Oil Before and After BDS Using Different Strains (Nassar, 2015).

(200 g/L) at 35 °C recorded 31, 21.1, 33.2, 21.2, and 21.6 for crude oil, diesel oil, kerosene, gasoline, and motor oil, respectively, using *P. putida* TU-S2, while 25, 19.3, 31, 25, and 29.0, respectively, were recorded using *B. pumilus* TU-S5 and, finally, recorded 26.1, 20.1, 29, 27, and 29%, respectively,

using *R. erythropolis* TU-S7 within 24 h. Moreover, using a consortium of the three strains reported 90% within 72 h.

Fatahi and Sadeghi (2017) reported that the bio-modification of PVA by *R. erythropolis* produced biocatalysts with an active metabolism that facilitates the interaction of the bacterial strain with hydrophobic gasoline thiophenic compounds. According to GC-MS, thiophene and its 2-methyl, 3-methyl, and 2- ethyl derivatives had acceptable BDS efficiencies of about 26.67, 21.03, and 23.62%, respectively. Also, benzothiophene that has been detected in a gasoline sample had 38.89% BDS efficiency. The recovery of biocatalysts has been investigated and after three times of using in BDS activity, its biocatalytic ability had no significant decrement.

The BDS/BDN capability of *R. erythropolis* 4277 cells on different concentrations of HGO (10–90% w/w) was determined by using the best culture conditions: 6.15 g/L of yeast extract, 2.0 g/L of glucose, 5.0 g/L of malt extract, and 1.16 g/L of $CaCO_3$ at 23.7 °C and 180 rpm for 30 h (Todescato *et al.*, 2017). The initial S and N contents of the used HGO were 6944 and 6549 μg/L, respectively. However, the BDN efficiency was 10 times higher than the BDS efficiency, where the presence of nitrogen in the yeast extract allowed the *R. erythropolis* ATCC 4277 cells to be more adapted to the nitrogen consume than to the sulfur. The presence of high amounts of nitrogen sources can be prejudice to BDS capability, as described by Porto *et al.* (2017), possibly because the high concentrations of nitrogen inhibit the formation of oxidoreductase and desulfinase enzymes. Such enzymes are extremely important in the sulfur degrading pathway of recalcitrant heterocyclic compounds (Monticello, 2000). Thus, BDS capability is not associated with optimal growth conditions, however the BDN was quite expressive, where 47% of nitrogen removal was achieved in the assay containing 10% (w/w) of HGO (Todescato *et al.*, 2017).

11.4 BDS of Crude Oil and its Distillates by Thermophilic Microorganisms

Thermotolerant microorganisms are recommended for the BDS of real oil feed, as they will decrease the cost of cooling to ambient temperature if BDS is set down-stream to an HDS unit.

The facultative thermophilic bacterium, *Mycobacterium goodie* X7B, was reported to decrease the total sulfur content (TSC) of Liaoning Crude oil (1:9 O/W phase ratio) from 3,600 to 1,478 mg/L, within 72 h at 40 °C (Li *et al.*, 2007a). In another study, the use of the thermophile *Klebsiella* sp.

13T resulted in 68.08% and 62.43% sulfur removal from heavy and light crude oil, respectively (Agarwal and Sharma, 2010).

Resting cells of *Mycobacterium* strain G3 were reported to desulfurize diesel oil form 116 ppm to 48 ppm within 24 h at 37 °C. When the desulfurization was carried out repeatedly using fresh cells, the sulfur concentration of diesel oil decreased to 44 mg/L (Okada *et al.*, 2002). Upon comparing the initial desulfurization rate of diesel oil (i.e. with the first 2 h) with that of model oil (100 mM DBT in tetratdecane), it recorded 25 and 151 μmol S/g cell /h, respectively. This was attributed to the presence of various kinds of alkyl-DBTs that are assumed to comprise the major sulfur fraction in diesel oil and the desulfurization activity was low due to competitive inhibition.

The total sulfur content of diesel oil was reduced by 86%, using a facultative thermophilic bacterium, *Mycobacterium* sp. X7B, for 24 h at 45 °C (Li *et al.*, 2003).

The thermophilic DBT-desulfurizing bacterium *Mycobacterium pheli* WU-F1 grew in a medium with three hydrodesulfurized light gas oils (LGOs) of different sulfur content (390ppm sulfur, 120 ppm sulfur, or 34 ppm sulfur) as a sole source of sulfur at 45 °C. BDS resulted in around a 60–70% reduction of sulfur content for all the three types of hydrodesulfurized LGOs. In addition, when resting cells incubated at 45 °C with hydrodesulfurized LGOs in the reaction mixtures containing 50% (v/v), BDS reduced the sulfur content from 390 to 100 ppm, from 120 to 42 ppm, and from 34 to 15 ppm. Analysis with GC/AED revealed that the peaks of alkylated DBTs, including 4-MDBT and 4,6-DMDBT, significantly decreased after BDS (Furuya *et al.*, 2003).

Ishii *et al.* (2005) reported that with growing cells of *Mycobacterium phlei* WU-0103, total sulfur content in 12-fold-diluted straight-run LGO was reduced from 1000 to 475 ppm sulfur at 45 °C.

Li *et al.* (2005a) reported thermophilic and hydrocarbon tolerant *Mycobacterium goodii* X7B, which had been primarily isolated as a bacterial strain capable of desulfurizing dibenzothiophene to produce 2-hydroxybiphenyl via the 4S pathway, was also found to desulfurize benzothiophene to *o*-hydroxystyrene at 40 °C. This strain appeared to have the ability to remove organic sulfur from a broad range of sulfur species in gasoline at 40 °C and a 1/9 O/W phase ratio, where desulfurization of gasoline from 227 ppm to 71 ppm sulfur was achieved. The vast majority of biodesulfurization studies target diesel rather than gasoline, so this demonstration of gasoline desulfurization is important. Benzothiophenes are the most abundant organosulfur compounds in gasoline, whereas DBTs predominate in diesel. Gasoline is also considerably more toxic to bacteria than diesel. Thus, that achievement was a great addition to the BDS field.

Bhatia and Sharma (2012) assessed the ability of the thermophilic Klebsiella sp. 13T to desulfurize light (0.35% S) and heavy (2.63%) crude oil along with HDS-diesel (0.05% S) and diesel oil (0.18% S) in a batch BDS process of 10% (O/W) at 45 °C and 120 h. The recorded desulfurization percentage for the studied oil feed ranged between ≈ 22 and 53% in the following increasing order: HDS-diesel < diesel < heavy crude oil < light crude oil.

11.5 Application of Yeast and Fungi in BDS of Real Oil Feed

Few reports have been published about the application of yeast and fungi in the BDS of crude oil and its distillates

Bladi *et al.* (2003) reported the isolation of a yeast strain *Rhodosporidium toruloides* DBVPG 6662 that can grow on a variety of sulfur compounds. When this strain was grown on glucose in the presence of commercial emulsion of bitumen (Orimulsion: a bitumen amended with an emulsifying agent and water), 68% of the benzothiophene derivatives and dibenzothiophene derivatives were removed after 15 days of incubation.

El-Gendy *et al.* (2006) reported a biodesulfurization of crude oil by the halotolerant yeast, *Candida parapsilosis* NSh45, isolated from Egyptian hydrocarbon polluted sea water, in a batch process of 1/3 O/W phase ratio in 7 days at 30 °C with a mixing rate of 200 rpm. The NSh45 reduced the sulfur content of Belayim Mix crude oil (2.74 wt% sulfur) by 75%, with a decrease in the average molecular weight of asphaltenes by approximately 28% and dynamic viscosity by approximately 70%, compared to that of *R. erythropolis* IGTS8 which expressed total sulfur removal of approximately 64% and a decrease in average molecular weight of asphaltenes and crude dynamic viscosity of approximately 24% and 64%, respectively, under the same conditions (El-Gendy *et al.*, 2006).

Torkamani *et al.* (2009) reported the isolation of the native fungus which has been identified as *Stachybotrys* sp. and is able to remove sulfur and nitrogen from heavy crude oil selectively at 30 °C. This fungus is able to desulfurize 76% and 64.8% of the sulfur content of heavy crude oil of the Soroush oil field and Kuhemond oil field in Iran (with the initial sulfur contents of 5 wt % and 7.6 wt %, respectively) in 72 and 144 h, respectively. This fungus strain has been isolated as a part of the heavy crude oil biodesulfurization project initiated by the Petroleum Engineering Development Company (PEDEC), a subsidiary of National Iranian Oil Company.

Adegunlola *et al.* (2010) investigated the ability of the immobilized spores of *Aspergillus flavus* to remove sulfur from crude oil. The spores of A. *flavus* were immobilized by mixing 80 g of the harvested spores in 1% (w/v) sterile alginate solution and then gelled into beads by dropping the suspension into a cold 15 g/L $CaCl_2$ solution. When 50 g of immobilized spores of *A. flavus* was added to crude oil for one, two, three, and seven days, the amounts of sulfur removed were 27.2%, 45.2%, 90.4%, and 91.7%, respectively. When 50 g of immobilized spores of *A. flavus* were introduced into the crude oil for 7 d at 35 °C, 40 °C, and 45 °C, the amount of sulfur removed was 63.2%, 55.3%, and 10.5%, respectively. Lastly, when 10 g, 50 g, and 100 g of immobilized spores of *A. flavus* were added to crude oil for 7 d, the amount of sulfur removed was 49.6%, 94.7%, and 53.9%, respectively.

11.6 Biocatalytic Oxidation

Microbial desulfurization of petroleum derivatives has two main problems: microbial activity is carried out in aqueous phase, thus, a two phase system reactor with intrinsic mass transfer limitations would be needed to metabolize the hydrophobic substrate and the microbial biocatalyst must have a broad substrate specificity for the various organosulfur compounds present in oil. These obstacles can be overcome by the utilization of broad specificity enzymes instead of whole microorganisms. An enzymatic oxidation process for the two-step desulfurization of fossil fuels involves (i) oxidation of diesel oil by chloroperoxidase, lignin peroxidase, manganese peroxidase, or cytochrome c, producing sulfoxide and sulfone, with a higher boiling point, leaving the majority of the hydrocarbons in their original form and (ii) distillation to remove the oxidized organosulfur compounds; other physicochemical processes can be used for the separation of the oxidized organosulfur compounds from the main hydrocarbon mixture, such as column chromatography, precipitation, and complexation with a solid support (Vazquez-Duhalt *et al.*, 2002).

Enzymes are able to perform catalytic reactions in organic solvents (Dordick, 1989) with a reduction in mass transfer limitations. The solvent is the fuel itself. Enzymes can be applied under anhydrous conditions or at very low water activity, enzymes are generally more thermostable, and reactions could be performed at temperatures as high as 100 °C (Mozhaev *et al.*, 1991).

Biocatalytic modification of complex mixtures from petroleum, such as asphaltenes, have been performed in organic solvents (Fedorak *et al.*,

1993). Several enzymes have been reported for the oxidation of thiophenes and organosulfur compounds in vitro, including cytochromes P450 (Nastainzcyk *et al.*, 1975; Fukushima *et al.*, 1978; Takata *et al.*, 1983; Mansuy *et al.*, 1991; Alvarez and Ortiz de Montellano, 1992), lignin peroxidase from the white rot fungus *Phanerochaete chrysosporium* (Vazquez-Duhalt *et al.*, 1994), lactoperoxidase (Doerge, 1986; Doerge *et al.*, 1991), chloroperoxidase from *Caldariomyces fumago* (Alvarez *et al.*, 1992; Kobayashi *et al.*, 1986; Pasta *et al.*, 1994), and horseradish peroxidase (Alvarez *et al.*, 1992; Doerge, 1986; Doerge *et al.*, 1991; Kobayashi *et al.*, 1986). Non-enzymatic hemoproteins are also able to perform DBT oxidation in vitro, such as hemoglobin (Alvarez *et al.*, 1992; Klyachko and Klibanov, 1992; Ortiz-Leon *et al.*, 1995) and cytochrome c (Klyachko and Klibanov, 1992; Vazquez-Duhalt *et al.*, 1993a,b). All the proteins mentioned above are hemoproteins and, in all cases, the products of the biocatalytic oxidations are the respective sulfoxides and/or sulfones.

Ayala *et al.* (1998) reported a biocatalytic oxidation of fuel oil as an alternative to biodesulfurization. The method includes a biocatalytic oxidation of organosulfides and thiophenes with hemoproteins to form sulfoxides and sulfones, followed by a distillation step in which these oxidized compounds are removed from the fuel. The reactions were successfully carried out in aqueous mixtures of diesel fuel (1.6 wt.%S) at room temperature that were biocatalytically oxidized with chloroperoxidase from *Caldariomyces fumago* in the presence of 0.25 mM hydrogen peroxide. The organosulfur compounds (OSCs) were effectively transformed to their respective sulfoxides and sulfones which were then removed by distillation at 50 °C. The resulting fraction after distillation contained only 0.27% sulfur and 71% of the total hydrocarbons were retained. Thus, a biocatalytic treatment of primary diesel fuel with chloroperoxidase from *C. fumago*, followed by a distillation, is able to reduce the sulfur content by 80%.

A simple and effective biochemical method for the desulfurization of bitumen has also been reported by Valentine (1999). A biological agent was used to remove oxidizable sulfur compounds from an emulsion of water and bitumen. Sulfur compounds were oxidized to water-soluble sulfates which could be physically or chemically removed, thus eliminating the SO_x production during combustion.

Vazquez-Duhalt *et al.* (2002) reported that oxidation using chloroperoxidase (CPO) in the presence of H_2O_2, followed by distillation at 50 °C, decreased the sulfur content of straight run diesel fuel from 1.6 to 0.27%, keeping 71% of the original hydrocarbons. A simple estimate of the cost of this technology has been reported; CPO showed a turnover number (i.e. number of substrate molecules that can be converted per molecule of enzyme before inactivation) of 500,000. Thus, 1 g of enzyme could reduce the sulfur content from 500 to 30 ppm of 0.81 tons of fuel (Ayala *et al.*, 2007). Immobilized cytochrome c in poly(ethylene)glycol (PEG-Cyt) has been used to oxidize high sulfur content diesel oil (Zeynalov and Nagiev, 2015). In addition, the use of lipase NOVOZYM™ LC in the presence of H_2O_2 and carboxylic acid in absence of water or any co-factor, followed by furfural extraction of produced sulfoxide and sulfone, has been reported (Singh *et al.*, 2009). The bio-catalyst is active up to 70 °C and preferably in the temperature range from 35 to 60 °C. Three types of diesel oil with different sulfur contents (6400, 1000 and 500 ppm) were used. The sulfur content decreased to 2300, 115, and 29 ppm, respectively.

11.7 Anaerobic BDS of Real Oil Feed

In 1961, ESSO Research and Engineering Co. patented an anaerobic process using *Thiobacillus* strain (ESSO Research and Engineering Company, 1961). Microbial conversion of organic sulfur in a crude petroleum slurry (10–20% water with added surfactant) was described. More than 90% removal of sulfur within a contact time of 10 min was claimed. VEB Petrolchemisches Kombinat Schwedt, in former East Germany, investigated an anaerobic route to crude oil desulfurization in 1974 (Eckart *et al.*, 1980, 1982; Eckart, 1981). Uncharacterized mixed cultures obtained from oil-polluted soils and sludge were contacted with various oils in a 4-dm^3 batch fermenter. Removal of sulfur was achieved, but at the expense of a considerable loss of hydrocarbon, mainly through metabolism of n-alkanes. In 1997, VEB Petrochemisches Kombinat Schwedt patented an anaerobic fermentation process. A 40% reduction in the organic sulfur content of Romaschkino crude oil was claimed after a 2–3 d anaerobic incubation. The sulfur was assumed to be released as H_2S and mercaptans were detected in the waste gases. Some 4–6% of the hydrocarbons were lost, probably through volatilization. Other problems included foaming, increasing viscosity of the crude oil through loss of light molecular weight components, and marked decrease in the activity of the biocatalyst after several days due to the mercaptans dissolved in the aqueous phase, rendering this process non-practical (Kohler *et al.*, 1984).

Eckart *et al.* (1986) examined the anaerobic biodesulfurization of a Romashkino crude oil containing 1.8% sulfur. Mixed cultures grown with lactate and sulfate were employed in these experiments. *Desulfovibrio* spp. were predominant members of the microbial community. When the pH was controlled to maintain circumneutral conditions, desulfurization of 26 to 40% was noted in a few days.

Pifferi *et al.* (1990) performed a continuous anaerobic microbial desulfurization of crude oil, petroleum fractions, and petroleum products using sulfate reducing microorganisms in the presence of H_2 at ambient temperature and atmospheric pressure. The culture medium which was previously isolated from industrial effluents showed the presence of ferrous ions that gave considerable advantages since they enhanced the formation of a colloidal iron sulfide precipitate in the reaction mixture. The desulfurization tests of fuel oils with different sulfur content were conducted in a bench scale laboratory unit with a reactor capacity of 10 L. Desulfurization yield ranged from 25% to 90% for a fuel oil with an S-content 3 wt.%. The SO_4^{2-} ion was kept between 0.3–1.6 g/L. The results showed the effectiveness of the new culture in the desulfurization of the substrate with simultaneous controlled demolition of high molecular weight structure.

Kim *et al.* (1990b) reported that *Desulfovibrio desulfuricans* M6 removed up to 21% of the 3 wt.% sulfur in Kuwait crude oil within 6 days and M6 also reported up to 17 wt.% from other crude oils and their distillate products (Kim *et al.*, 1995).

Some anaerobic microorganisms, such as *Desulfomicrobium scambium* and *Desulfovibrio longreachii*, have been reported to have the ability to desulfurize only about 10% of DBT dissolved in kerosene. The GC analyses of samples showed unknown metabolites, indicating that the bacteria had possibly followed a pathway different from common anaerobic pathways (Yamada *et al.*, 2001).

Aribike *et al.* (2008) reported that *Desulfobacterium indolicum* isolated from oil contaminated soil exhibited very high desulfurizing ability towards kerosene at 30 °C in a 1/9 O/w phase ratio, resulting in reduction of sulfur from 48.68 ppm to 13.76 ppm over a period of 72 h with a significant decrease in thiophene and 2, 5-dimethyl thiophene. The GC/PFPD analysis revealed that kerosene contained 6.955 mg/L of thiophene and 41.724 mg/L of 2, 5-dimethyl thiophene and no benzothiophene or dibenzothiophene is detected in kerosene. The BDS-rate of Th was higher than that of 2, 5-dimethyl thiophene. At the end of 72 hours, 84% of thiophene has been desulfurized, while 70% of 2, 5–dimethyl thiophene was desulfurized. That was explained by the presence of the methyl substituents at positions 2 and 5 which would constitute a steric hindrance to the organism from reaching the sulfur atom in the thiophene ring.

Aribike *et al.* (2009) reported that *Desulfobacterium anilini* isolated from petroleum products-polluted soil showed a significant decrease of benzothiophene and dibenzothiophene in diesel with 82% removal of total sulfur after 72 h at 30 °C with a 1/9 O/W phase ratio.

Agarwal and Sharma (2010) reported the 63.29% and 61.40% biodesulfurization of two crude oil samples: heavy and light crude oils with sulfur contents of 1.88% and 0.378%, respectively, by *Pantoea agglomerans* D23W3. However, the use of *P. agglomerans* D23W3 under anaerobic conditions showed marginally better results than those under aerobic conditions.

Srivastava (2012) mentioned that some anaerobic microorganisms, such as *Desulfomicrobium scambium* and *Desulfovibrio longreachii*, have the ability to desulfurize about 10% of DBT dissolved in kerosene. Kareem *et al.* (2012) reported 81.5% anaerobic biodesulfurization of diesel oil with an initial sulfur concentration of 166.034 ppm using an isolated bacterial strain, *Desulfobacterium indolicum* within 72 h. The strict anaerobe, Gram negative *Desulfatiglans aniline* comb. nov., isolated from petroleum products contaminated soil, is reported for its efficient BDS capacity on kerosene obtained from a retail outlet in Lagos, Nigeria, as it reduced its S-content from 48.68 ppm to 12.32 ppm (representing 75% BDS efficiency) in a batch BDS process of 1:10 (O/W), at 180 rpm, pH7, over a period of 72 h at 30 °C. The gas chromatography analysis with a pulse flow photometric detector (GC/PFPD) analysis revealed that the peaks of thiophene and 2, 5-dimethyl thiophene were significantly decreased after biodesulfurization (Kareem *et al.*, 2016).

The desulfurization of petroleum under anaerobic conditions would be attractive because it avoids costs associated with aeration, it has the advantage of liberating sulfur as a gas, and does not liberate sulfate as a byproduct that must be disposed by some appropriate treatment (Ohshiro and Izumi, 1999). However, due to low reaction rates, safety and cost concerns, and the lack of identification of specific enzymes and genes responsible for anaerobic desulfurization, anaerobic microorganisms effective enough for practical petroleum desulfurization have not been found yet, and an anaerobic BDS process has not been developed. Consequently, aerobic BDS has been the focus of the majority of research in BDS (Le Borgne and Quintero, 2003).

11.8 Deep Desulfurization of Fuel Streams by Integrating Microbial with Non-Microbial Methods

11.8.1 BDS as a Complement to HDS

One of the main drawbacks of HDS is that it is not equally effective in desulfurizing all classes of sulfur compounds present in fossil fuels

(Chapter 2). The BDS process, on the other hand, has broad versatility on different organosulfur compounds (OSCs) (Pacheco, 1999). However, one of the main advantages of HDS (Chapter 2) is that its conditions not only desulfurize sensitive (labile) OSCs, but also (i) remove nitrogen and metals from organic compounds, (ii) induce saturation of at least some carbon–carbon double bonds, (iii) remove substances having an unpleasant smell or color, (iv) clarify the product by drying it, and (v) improve the cracking characteristics of the material (El-Gendy and Speight, 2016).

Therefore, with respect to these advantages, placing the BDS unit downstream of an HDS unit as a complementary technology to achieve ultra-deep desulfurization, rather than as a replacement, should also be considered.

Although some researchers are focusing on implementing the BDS process on a large scale, the BDS rates are still low when compared to HDS. This is due to the limitations that are faced in such processes. The main limitations include the need to enhance the thermal stability of desulfurization, the limited transport of the sulfur compounds from the oil to the membrane of the bacterial cell, and the limited ability to recover the biocatalyst (Kilbane, 2006). Most BDS processes are now focusing on using it as complementary steps for deep desulfurization, where the BDS is integrated with existing HDS units. BDS can be used either before or after the HDS unit. Kleshchev (1989) invented the HDS-BDS process, where HDS of crude oil is carried out as a first step to remove labile organic sulfides, then BDS of recalcitrant sulfur compounds, for example DBT, is performed by *Rhodococcus rhodochrous* (Figure 11.4a). Other researchers believe that employing BDS before HDS (Figure 11.4b) is more efficient for removing a major part of the hydro-treating resistant compounds. This will result in less hydrogen consumption in the HDS unit (Monticello, 2000).

Monticello (1996) suggested a multistage process for desulfurization of fossil fuels. This method was based on subjecting vacuum gas oil to HDS prior to BDS in defined conditions. Pacheco (1999) reported that the Energy BioSystems Corporation (EBC) used BDS downstream of HDS. Fang *et al.* (2006) also showed that combination of HDS and BDS could reduce the sulfur content of catalytic diesel oil from 3358 to <20 mg/g. Thus, with respect to the advantages of both HDS and BDS, integrating BDS with HDS, for example, by placing a BDS unit downstream of an HDS unit as a complementary technology to achieve ultra-deep desulfurization, rather than as a replacement, is very promising and should be considered.

Rhodococcus erythropolis I-19, a genetically engineered bacterium and product of Energy Biosystem Corp, USA, was reported to reduce the S-content of hydrodesulfurized middle distillate from 1850 to 615 μg/mL,

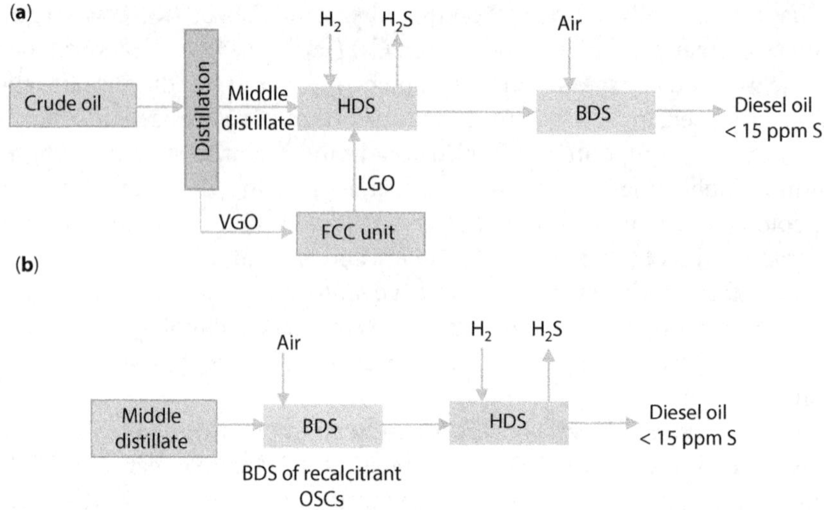

Figure 11.4 Suggested Strategies for Integrating BDS with HDS.

i.e. 67% BDS% (Folsom *et al.*, 1999), where the initial desulfurization rate of the oil fraction was 2.5 μM/g DCW/min based on the change in total sulfur content in the oil phase.

Grossman *et al.* (2001) cultivated *Rhodococcus* sp. strain ECRD-1 in a medium with a hydrocarbon/water ratio of 1:4. The hydrocarbon phase was hydro-treated middle distillate oil with a total sulfur content of 669 ppm. The oil had only 5% of DBT and the majority of the remaining 95% was alkylated DBTs. The final sulfur concentration was detected to be 56 ppm after 7 d of cultivation in a rotary shaker at 25 °C, pH 7 (controlled), and 200 rpm.

The ability of *Rhodococcus* sp. P32C1 for desulfurization of n-hexadecane (n-C_{16}) containing DBT (model oil) was first studied and based on the optimum conditions determined from the BDS of model oil and its capacity for the BDS of real oil feed was studied. Two kinds of diesel oil, hydrodesulfurized diesel oil (HDS-diesel) and another diesel sample, with sulfur contents of 303 and 1000 ppm were used for BDS with *Rhodococcus* sp P32C1 in a batch system of 25% (O/W). About 48.5 and 23.7% of sulfur was removed in 24 h with desulfurization rates of about 0.6 and 1.0 mmol S/kg DCW/h, respectively (Maghsoudi *et al.*, 2001).

A deeply hydrodesulfurized diesel oil containing significant amounts of 4,6-DMDBT was treated with *Rhodococcus* sp. IMP-S02 cells. Up to 60% of the total sulfur was removed at 30 °C and within 7 d and all the 4,6-DMDBT disappeared as a result of this treatment (Castorena *et al.*, 2002).

Xu *et al.* (2002) investigated the BDS efficiency of hydrodesulfurized diesel oil with an initial S-content of 205 μg/mL in a batch BDS of 1:3 (O/W) and pH 7.5 using different bacterial isolates, where *Rhodococcus* sp. 1awq, IG, ZT, ZCR, and a mixture of isolates 5 and 6 recorded a BDS percentage of 80, 69.1, 17.5, 58.5, and 10.6%, respectively, at 30 °C, while isolate X7B expressed 90.3% BDS at 45 °C.

The desulfurization genes (dszABC) were cloned from *Gordonia nitida* (Jae *et al.*, 2003). Nucleotide sequence similarity between the dszABC genes of *G. nitida* and those of *Rhodococcus erythropolis* IGTS8 was 89%. The similarities of deduced amino acids between the two were 86% for DszA, 86% for DszB, and 90% for DszC. The *G. nitida* dszABC genes were expressed in several different *Escherichia coli* strains under an inducible *trc* promoter. The metabolically engineered *E. coli* strains harboring the new desulfurization genes from *G. nitida* were able to efficiently desulfurize DBT and diesel oil. Deregulation of the transcription repression site by employing the inducible *trc* promoter and an artificial *dszABC* operon resulted in a relatively high efficiency of DBT conversion to 2-HBP. The maximum conversion of DBT to 2-HBP was 16% within 60 h. A hydrodesulfurized diesel oil with an S-content of 250 mg/L was examined for desulfurization (SK Corp., Daejeon, Korea). A batch BDS of 5% (v/v) diesel oil of 6 h after the induction with 0.5 mM IPTG at OD_{600} of 0.7 recombinant *E. coli* W3110 (pTrcS1ExABC) removed 15% of sulfur in diesel oil, decreasing sulfur content from 250 to 212.5 mg/L in diesel oil within 60 h (Jae *et al.*, 2003).

Five isolates belonging to the *Rhodococcus*/*Gordonia* cluster isolated by Abbad-Andaloussi *et al.* (2003a) from different soils exhibiting good growth characteristics and high BDS activities in both aqueous and organic media were studied for their abilities to desulfurize different types of diesel oils in order to better assess the potential of BDS, especially as a finishing step complementary to deep HDS. Actually, in spite of their taxonomic similarity, the five strains displayed different activities towards the diesel oil tested. BDS yield was also dependent upon the diesel oil used, especially its sulfur content. Some HDS-recalcitrant compounds, such as 4,6-DMDBT, could be completely removed, but highly alkylated DBTs were resistant to the action of the biocatalysts. Abbad-Andaloussi *et al.* (2003a) observed that BDS was more efficient on diesel oils with a low sulfur content and that biocatalysts were also active on LCO (light cycle oil, a gas-oil type fraction from catalytic cracking units). This result suggests that inhibitory or toxic effects by certain sulfur compounds may occur. A maximum BDS yield of 76% was obtained and it was possible to decrease the sulfur content of a previously deeply-hydrodesulfurized diesel oil down to 14 mg/kg. In addition, the sulfur removal rate was high. Furthermore, some HDS-recalcitrant compounds, such as 4,6-dimethyl DBT, could be totally desulfurized. The desulfurization kinetics were affected by the number and/or lengths of alkyl groups attached to the ring structure of DBT. From a technical point of view, these results showed that BDS could be quite suitable as a finishing step complementary to deep HDS because of its high selectivity. However, the action of the biocatalysts on the highly-alkylated DBTs remained low and these compounds would also have to be removed to decrease the sulfur content of diesel oil to levels as low as 10 mg/kg.

Labana *et al.* (2005) reported 1:3 (O/W) batch-BDS of two different feed diesel from an HDS unit: product diesel from an HDS unit (170 ppm S) and a high-speed diesel (70 ppm S), with resting cells of *Rhodococcus* sp. and *Arthrobacter sulfureus*, collected at mid-log phase. Those recorded BDS percentages of 50 and 53% for hydrodesulfurized diesel oil, respectively, within a 5 d incubation period at 30 °C and 200 rev/min, while *Rhodococcus* sp. recorded only 28% for the high-speed diesel oil sample, but *A. sulfureus* did not express any BDS potential on the sulfur content of that oil sample. The difference in BDS efficiencies was probably because of different chemical forms of sulfur. Nevertheless, it was concluded from that study that HDS and BDS probably have a synergistic effect in removing a broad spectrum of sulfur-containing compounds, as HDS enriches the HDS-refractory sulfur-bearing heterocycles, such as DBT, and depletes the HDS-labile organosulfur compounds (OSCs) (Monticello 1996). Thus, consequently, HDS followed by BDS could help meet the production of ultra-low sulfur fuel in accordance with the regulations.

Mukhopadhyaya *et al.* (2006) used a trickle bed reactor under a continuous mode for BDS of hydrodesulfurized diesel oil with an initial S-content of 200 ppm and pith balls have been used as an immobilizing matrix for *Rhodococcus* sp. NCIM2891, where approximately 99% of sulfur conversion occurred at an inlet diesel flow rate of 0.25 dm^3/h. *R. erythropolis* XP was also reported for the BDS of a hydrodesulfurized diesel oil, where the S-content reduced by 94.5% from 259 to 14 ppm (Yu *et al.*, 2006a). Another study by Nandi (2010) reported an internal loop airlift reactor for the BDS of hydrodesulfurized diesel from 500 ppm to almost zero level within 120 h using *Rhodococcus* sp. NCIM2891.

Ma *et al.* (2006) investigated methods to produce biodesulfurization catalysts and found that cultivation of *Rhodococcus* sp. in 1 mM dimethylsulfoxide was the most cost-effective. The microbial culture was then used in a bi-phasic system (1:9 O/W) and was found to remove 78% of sulfur (200 ppm initial concentration) from an HDS treated diesel oil.

Feng *et al.* (2006) reported the enhancement effect of Tween 80 on the BDS of a hydrodesulfurized FHD200 diesel oil (200 ppm S) with a 1/9

(O/W) in the presence of 2% glucose as a C-source. The BDS increased with increasing the surfactant concentration, reaching 90% at 0.5% Tween 80. Upon its application on another two diesel oils, FHD406 and KHD168 diesel oil, the results revealed 78.1% and 65.0%, respectively, compared to 56.6% and 52.9% without Tween 80.

Li *et al.* (2007b) reported the BDS of a hydrodesulfurized diesel oil with S-contents of 555 μg/g, using cell suspension and resting cells of *R. erythropolis* LSSE8–1 in a batch of 1/2 O/W phase ration at 30 °C and 170 rpm. This recorded BDS-rates of 59.7 and 65.6 μmol sulfur/g DCW/h, respectively, within 4 h of incubation. Upon the direct application of cell suspension obtained after growth for optimized growth for 72 h on another low S-content hydrodesulfurized diesel oil, the sulfur was reduced to 103 μg/g at a 1/2 O/W ratio in the first BDS treatment after 24 h. Then, upon a second treatment it reached 51 μg/g with a total S-removal of 79.4%. Since direct fermentation cell suspension and resting cells expressed high BDS efficiencies, it was concluded from that study that it is feasible that high cell density suspension containing the metabolites of a sulfur source can be directly used for diesel-BDS. Thus, the BDS process was apparently simplified, which had the advantage of saving cost. The simple process is convenient to treat a mass of petroleum fractions and a continuous operation might be applied as a complementary step after HDS to yield an ultra-low sulfur diesel oil (ULSD).

BDS of hydrodesulfurized diesel oils by a newly isolated strain of *Rhodococcus erythropolis* FSD-2 was investigated by Zhang *et al.* (2007). Two representative diesel oils were obtained after distillation and HDS to produce oils with sulfur contents of 666 and 198 μg/mL, respectively. The volumetric phase ratio of the aqueous phase to oil was 5/1. The specific desulfurization rate of diesel oil No. 1 and No. 2 was 0.78 mg S/g DCW/h and 0.26 mg S/g DCW/h, respectively. About 97% of the total sulfur content in the hydrodesulfurized diesel was removed by the two consecutive BDS processes with the majority (~94%) being removed in the first treatment, resulting in diesel with a sulfur content of 5.7 μg/mL.

Several studies have reported 47–94.5% BDS efficiencies of hydrodesulfurized diesel with an initial S-content ranging from 250 ppm to 3000 ppm (Mohebali and Ball, 2008).

Li *et al.* (2008) reported that a microbial consortium (2.2:1 w/w) of *R. erythropolis* DS-3, which is capable of desulfurizing DBT and its derivatives (Ma *et al.*, 2002) and *Gordonia* sp. C-6 w, which is capable of desulfurizing BT and its derivatives (Li *et al.*, 2006) could efficiently desulfurize 86% of the heterocyclic sulfur compounds (from 1260 to 180 ppm) in a hydrodesulfurized diesel oil after 3 cycles of BDS.

Application of chitosan flocculation and integration with cell immobilization onto celite for BDS of hydrotreated diesel oil (1/2 v/v) (O/W phase ratio) was performed. It was observed that the sulfur content of D304 diesel (304 μg/g) reduced by approximately 57% within 24 h by chitosan-celite-immobilized cells. Moreover, the total sulfur of D123 diesel was reduced from 123 to 22 μg/g, corresponding to ≈ 82% BDS. It demonstrated that the flocculated and immobilized cells are capable of removing sulfur from hydrotreated diesel to yield ultra-low sulfur products (Li *et al.*, 2011). BDS was noticed to reach a plateau after 8 h. The initial specific activities (30.6 – 25.6 μmol sulfur/g DCW/h) decreased quickly to between 18.9 – 3.5 and the average activity for the two diesel oils was about 24.8 and 14.6 μmol sulfur/g DCW/h, respectively, within 8 h. Thus, the residence times of the BDS process can be set at 6–8 h.

Bandyopadhyay *et al.* (2013a) reported the production of biosurfactant during the BDS of hydrodesulfurized diesel in a 2 dm^3 B.Braun chemostat with a continuous stirred tank reactor (CSTR) with a working volume of 1.5 dm^3, using *Rhodococcus* sp (NCIM 2891). The stirring rate and aeration rate were maintained at 100 rpm and 25 L/h, respectively. The diesel to aqueous phase ratio was maintained at 80:20 during all runs in the chemostat. The dilution rate was varied in the range of 0.03 to 0.1 and the sulfur concentration in feed diesel was varied from 200 to 540 ppm. Under each operating condition the reactor was operated for 4 days. A chemostat was operated to control/optimize the microbial growth rate, surfactant production rate, and the substrate utilization rate. These were achieved by adjusting the volumetric feed rate or dilution rate. After a certain operating period of 30 hours (>3τ), the concentrations of biomass, substrate, and product become invariant with reaction time. This may be considered to be the onset of a steady state. The concentrations of product and biomass decreased with dilution rate. This was attributed to the fact that as the dilution rate increases, the residence time in the reactor decreases and, consequently, the rates of outlet of both product and biomass outweigh the generation rates of the respective components resulting in an ultimate decreasing trend of concentration with dilution rate. The concentration of both 2-HBP and biomass expressed a decreasing trend with an increase in the dilution rate in the range of 0.03–0.07 h^{-1}. Moreover, both the produced biomass and 2-HBP concentrations at any dilution rate showed increasing trends with an increase of initial S-concentration. Thus, the generation rates increased with the increase of the studied substrate concentrations range. The surface tension of aqueous phase and oil phases of the reaction broth, interfacial tension, as well as the emulsification index of the broth obtained with an initial sulfur concentration of 330 ppm and diesel to aqueous ratio of 80:20 have been plotted with the operating time of the chemostat. It has been observed that the surface tension decreased and the value of E_{24} increased with the increase in reaction time. The surface tension of the aqueous nutrient phase and diesel was observed to decrease from 71 dynes/cm to 30 dynes/cm and to 20 dynes/cm from 30 dynes/cm, respectively. Values of the emulsification index (E_{24}) were found to be varied from 18 to 58 over the growth period of 2 to 48 h in the chemostat. The critical micelle concentration (CMC) was found to be 200 mg/L. This was an indication of the formation of more biosurfactant, namely, 2-HBP and some triglycerides and polar and nonpolar lipids with the propagation of reaction. The increase of values of E_{24} with reaction time also established the effectiveness of the biosurfactants produced as byproducts of the biodesulfurization of diesel. This came with the decrease in the sulfur concentration of the hydrotreated diesel to ≈ 20 ppm with a maximum conversion of 95%. In a similar study, Bandyopadhyay *et al.* (2013b) reported the production of ultra-low sulfur diesel (ULSD) and biosurfactants, namely, 2-hydroxybiphenyl and different lipids. The substituted benzothiophenes (BTs) and dibenzothiophenes (DBTs) in hydrotreated diesel get converted to 2-hydroxybiphenyl, which is a potential bio-surfactant. Kinetics of BDS of deep desulfurized diesel using *Rhodococcus* sp. (NCIM 2891) have been studied with special reference to the removal of OSCs in diesel and the production of 2-hydroxy biphenyl (2-HBP). The sulfur concentration of feed diesel was in the range of 200–540 mg/L. Aqueous phase to diesel ratios have been varied in the range of 9:1 to 1:9. The optimum ratio has been found to be 1:4 (O/W) based on both the ease of separation of the aqueous phase from diesel and the value of the specific microbial growth rate and a maximum conversion of sulfur at 95% has been achieved. A chemostat has been studied using an initial concentration of OSCs in diesel as a parameter. A mathematical model has been developed on the basis of kinetic parameters derived from shake flask analysis. The model has been validated by the comparison with the experimental data. Furthermore, in order to understand the reaction engineering behavior of the biodegradation process, a classical Monod type of a substrate uninhibited kinetic model (eq. 1) has been applied for simulation work. The experimental data generated from the studies in a chemostat have been used to determine the kinetic parameters based on a Monod model:

$$\mu = \frac{\mu_{max} C_s}{K_s + C_s} \tag{1}$$

The rate equation of microbial growth:

$$r_x = \mu C_x \tag{2}$$

The rate equation of S-removal:

$$r_s = -\frac{r_x}{Y_{x/s}} \tag{3}$$

Thus, the rate equation in terms of microbial growth kinetics:

$$\frac{dC_s}{dt} = -\frac{\mu C_x}{Y_{x/s}} \tag{4}$$

where C_s and C_x are the S and biomass concentrations, respectively.

The mathematical model of the system has been developed on the basis of the following assumptions: influent stream of the bioreactor was sterile, there was no external mass transfer resistance present in the system, the OSCs of diesel were the only growth limiting substrates, and, finally, the microbial growth follows Monod Kinetics.

Thus, the system equations for batch mode were as follows:

Substrate (i.e. the S-content):

$$\frac{dS}{dt} = D(S_o - S) - \frac{1}{Y_{x/s}} \frac{\mu_{max} SX}{K_s + S} \tag{5}$$

Biomass:

$$\frac{dX}{dt} = -DX + \frac{\mu_{max} SX}{K_s + S} \tag{6}$$

Product (i.e. the 2-HBP):

$$\frac{dP}{dt} = -DP + Y_{p/x} \frac{\mu_{max} SX}{K_s + S} \tag{7}$$

Taking into account that under a steady state, the mass balance equations for substrate, biomass, and product become:

$$S_s = \frac{K_s D}{\mu_{max} - D} \tag{8}$$

$$X_s = Y_{x/s}\left(S_o - \frac{K_s D}{\mu_{max} - D}\right) \tag{9}$$

$$P_s = \left[\frac{D}{Y_{p/x}}\left(\frac{K_s}{\mu_{max} S_s} + \frac{1}{\mu_{max}}\right)\right]^{-1} X_s \tag{10}$$

Thus, the values of Monod kinetic parameters, the maximum specific growth rate, μ_{max}, half saturation constant, Ks, biomass yield, Yx/s, and 2-HBP yield, Yp/x, were determined by using batch type experimental data obtained with different initial sulfur concentrations (200 to 540 ppm) and the values recorded were 0.096 h^{-1}, 71 mg/L, 0.2, and 17 μmol/g DCW, respectively. Furthermore, the optimum ratio of diesel to aqueous medium was found to be 80:20 (Bandyopadhyay *et al.*, 2013a), while the values of kinetic parameters, namely, μ_{max} and K_s of the microbial growth and yield coefficient of surfactant, with respect to biomass, have been determined to be 0.13 h^{-1}, 68 mg/L, and 18.5 μmol/g DCW, respectively, by conducting batch type experiments at the optimum ratio of oil to aqueous phase of 4:1 (W/O). A maximum conversion of 96% was achieved at a dilution rate of 0.03 h^{-1} when the sulfur concentration in the feed diesel was 430 ppm. Moreover, the aqueous layer also contained extracellular biosurfactants (phospholipid, glycolipid, and rhamnolipid) in the range of 0.03 g/mL, probably resulting from the microbial secretion. The presence of polar and non-polar lipids was confirmed by thin layer chromatography (TLC) (Bandyopadhyay *et al.*, 2013a). Bandyopadhyay *et al.* (2014) also reported the production of biosurfactant during the BDS of spent engine oil, which is a mixture of aliphatic and aromatic compounds, is one of the frequent environmental pollutants. When batch studies were conducted using *Rhodococcus* sp. (NCIM 2891) by varying the oil to aqueous medium ratio of 10:90 to 90:10, the results demonstrated that a maximum desulfurization of 80% was obtained in a batch-BDS at an oil to aqueous phase ratio of 70:30, pH 8.0, 160 rpm, and 28 °C, within a 30 h incubation period. Kinetic parameters of Monod type growth model, μ_{max}, Ks, and Yx/s recorded 0.12 h^{-1}, 73.5 mg/L, and 0.32, respectively, and those were determined by varying the initial sulfur content of spent oil in the range of 0.16 –1.05% (w/v) by diluting with hexadecane. The obtained results indicated a higher growth rate on diesel oil compared to that on spent oil. The biosurfactants, like glycolipids, phospholipids, and rhamnolipids, secreted by the microorganisms are preferentially transferred to the water phase to lower the surface tension. This, consequently, would facilitate the assimilation of nutrients by microorganisms from the aqueous phase (Bandyopadhyay *et al.* 2013a, 2014). Moreover, due to existence of the microorganisms only at the interface can the treated spent oil easily be separated from water phase and the interface containing microorganisms. However, it was observed that the microorganisms growing in the spent engine oil were smaller in size than those grown in the diesel phase (Bandyopadhyay *et al.*, 2013b). This was attributed to the presence of metal contaminants present in the spent engine oil. The surface tensions of the aqueous nutrient phase and spent oil decreased with time from 69 dynes/cm to 24 dynes/cm and from 18 dynes/cm to 10 dynes/cm, respectively, while the E_{24} increased from 14 to 50 and from 24 to 66 within 32 h of incubation period in the aqueous and spent linguine layers, respectively (Bandyopadhyay *et al.*, 2014). This indicated that the production of surface active reagents in the oil phase and other extracellular lipids, like phospholipids, glycolipids, rhamnolipids, etc., occurred in the aqueous phase with the propagation of reaction and was associated with the microbial growth. Moreover, the GC/MS revealed the removal of the PASHs, 1,1-dimethyltetradecyl hydrosulfide, 2-nonadecanone-(2,4-dinitrophenyl-hydrazone), S-[2-((3-[2,4-Dihydroxy-3,3-dimethylbutanoyl)amino] propanyl)amino) ethyl]2tetradecynethioate, etc., which were predominantly present in the untreated spent engine oil, while the mono- and di-aromatic compounds, long, straight, and branched chain hydrocarbons (> C15) were not affected. Bandyopadhyay *et al.* (2014) recommended not only the production of useful biosurfactants during the BDS process, but also the generation of ultra-low sulfur diesel through pyrolysis of the biotreated spent engine oil. In another study, *Rhodococcus* sp (NCIM 2891) was also selected for the production of biosurfactant through biodesulfurization of hydrotreated diesel in a batch and continuous chemostat mode. The best BDS efficiency, cell growth, and biosurfactant production occurred at 80:20 O/W in a batch system. In a batch system, as the diesel proportion is increased, the availability of sulfur compounds, which is the energy source of the microorganism, as well as hydrocarbons, one of the carbon sources for growth, increases resulting in the increased trend in cell concentration up to a ratio of 80:20, where the S-content decreased from 350 ppm to 30 within 48 h at 28 °C. But beyond this OFP, the availability of oxygen might have been decreased at the interfacial zone due to formation of a thick oil layer. Thus causing a decline in the values of biomass concentration, sulfur conversion, and lowering of surface tension (Bandyopadhyay and Chowdhury, 2014).

Bordoloi *et al.* (2014) reported the BDS of hydrodesulfurized diesel oil with an S-content of 420 ppm by *Achromobacter* sp. resting cells with an age of 94 h grown on 0.5 M DBT at 37 °C in a batch process of 1/3 O/W, where the S-content decreased to 19 ppm after 24 h, which corresponds to a reduction of 7.1% of sulfur from the diesel oil /g DCW with a slight effect on the hydrocarbon skeleton of the oil feed.

The BDS of hydrodesulfurized diesel with an initial sulfur content of 70 ppm, using growing cells of *Gordonia* sp. IITR100 in a batch BDS-process of 1/3 O/W, led to a decrease in S-content, reaching to < 2 ppm at 30 °C and 250 rpm within 4 days' incubation period, reached approximately 98%. This occurred with the observed disappearance of most of the peaks corresponding to the aromatic organosulfur compounds, including alkylated dibenzothiophenes and benzothiophenes, as analyzed by GC-SCD. This occurred without affecting the hydrocarbon skeleton of diesel oil as detected by GC-FID analysis. However, after 6 days of incubation, slight changes in the hydrocarbon skeleton occurred. This was explained by the possible depletion of the alternate carbon source in media after four days in the absence of which bacteria started to take up hydrocarbons (Adlakha *et al.*, 2016).

11.8.2 BDS as a Complementary to ADS

The use of a biosorption agent to bind and separate sulfur compounds by forming a biosorption complex was discussed by Johnson *et al.* (1995). The invention included a procedure for the preparation of products resulting from biocatalytic sulfur oxidation such as 2-hydroxybiphenyl (2-HBP). The invention was tested in a two stage adsorption-conversion of 3% DBT and suspended *Rhodococcus rhodochrous* biocatalyst in hexadecane. The results showed rapid DBT adsorption and conversion to 2-HBP.

Li *et al.* (2005b) proposed a method to produce ultra-low sulfur diesel (ULSD) by integration adsorptive desulfurization (ADS) with biodesulfurization (BDS). Briefly, sulfur compounds (for example, dibenzothiophene DBT) are adsorbed on adsorbents and then the adsorbents are regenerated by microbial conversion. The π-complexation adsorbent, Cu(I)-Γ, was obtained by ion exchanging the Γ-type zeolite with Cu^{2+} and then auto-reduction in helium at 450 °C for 3 h was performed. The amounts of DBT desorbed and 2-HBP produced have been increased with addition of *n*-octane. DBT-BDS activity has been improved by increasing cell concentration and the ratio of water-to-adsorbent. The amount of 2-HBP produced was strongly dependent on the volume ratio of oil-to-water, cell concentration, and amount of adsorbent. Moreover, 89% of DBT desorbed from the adsorbents can be converted to 2-HBP within 6 h at almost 100% within 24 h when the volume ratio of oil-to-water was 1/5 mL/mL, the cell concentration was 60 g/L, and the ratio of adsorbent-to-oil was 0.03 g/mL. During the bio-regeneration process, desorption of DBT can be significantly improved by adding an oil phase. The adsorption capacity of the regenerated adsorbent is 95% that of the fresh one after being desorbed with *Pseudomonas delafieldii* R-8, washed with *n*-octane, dried at 100 °C for 24 h, and auto-reduced in He.

Li *et al.* (2009) studied ADS using different types of adsorbents, such as activated carbon (AC), NiY, AgY, alumina, 13X, and bio-regeneration properties of the adsorbents were studied with *P. delafieldii* R-8 (CGMCC 0570). The adsorption capacity of different adsorbents was tested using model oil (8.0 mmol/L DBT and 8.0 mmol/L naphthalene in n-octane) at ambient conditions. The ratio of oil to adsorbent was chosen as 100 mL/g. Hydrotreated diesel oil has also been used in this study. The regeneration system contained n-octane, aqueous phase, lyophilized cells, and spent adsorbents. All reactions were carried out in 100 mL flasks at 30 °C on a rotary shaker operated at 200 rpm. Adsorption–bio-regeneration properties were tested in an in-situ adsorption–bio-regeneration system which can conveniently be divided into two parts: adsorption and bio-regeneration (Figure 11.5). After the saturation of adsorbents, the adsorption system is shut down and the adsorption reactor was connected with the bioreactor. Then, the desorbed sulfur compounds were converted by R-8 cells. Finally, the desorbed adsorbents were treated with air at 550 °C to remove water from the adsorption system. The integrated system is able to efficiently desulfurize DBT and the ADS property of the bio-regenerated adsorbents is similar to fresh ones. The ADS capacity of DBT ranked in the following decreasing order: AC > NiY > AgY > alumina > 13X, while the selectivity of DBT over naphthalene followed the sequence: NiY > AgY > 13X ≈ alumina > AC. For the regeneration of the spent adsorbents, the interaction of the adsorbents with DBT was studied and revealed the following sequence: AC > NiY > AgY > alumina > 13X. The sequence of the DBT desorption ratio was 13X > alumina > AgY > NiY > AC. Since the interaction of DBT with AC was so strong, little DBT has been desorbed. Thus, that revealed that the stronger the interaction of DBT with the adsorbent, the more difficult to be desorbed.

Adsorption properties of NaY, MCM-41, and MAS towards hydrotreated diesel were also carried out at 30 °C in a ratio of 20 mL oil/ g adsorbent. The ADS properties of Ag-MAS were much better than AgMCM-41, while the ADS properties of Ag-MAS and Ag-MCM-41 were much better than those of Ag-Y. This was attributed to the large molecular compounds in diesel which can block the pores of Ag-Y, which, consequently, sharply decreases

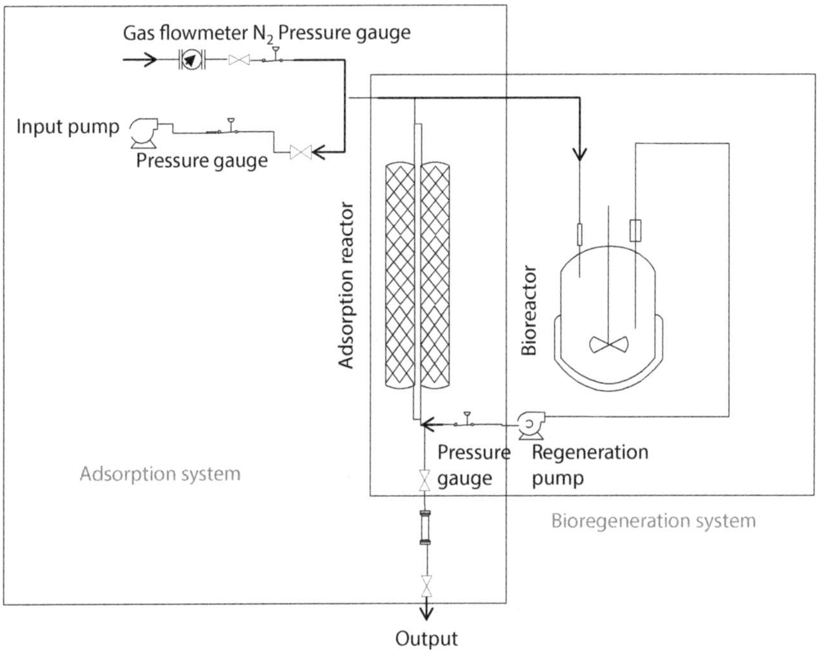

Figure 11.5 The In-Situ Adsorption-Bioregeneration System (Li et al., 2009).

desulfurization efficiency. The bio-regeneration and recycling properties of Ag-MAS were also carried out with 79 g/g hydrotreated diesel in the presence of n-octane as a regeneration solvent (5 mL n-octane/g AgMAS). The spent Ag-MAS can be regenerated in the integrated system and regenerated adsorbents were found to have similar adsorption properties to the fresh ones.

In an attempt to reduce the accumulation of the inhibitory end product of DBT-BDS, 2-HBP and, consequently, increase BDS efficiency, Abin-Fuentes *et al.* (2013) studied the efficiency of different adsorbents: activated charcoal, molecular sieves (pore sizes, 4, 5, and 13 Å), Diaion HP-20, Dowex Optipore L-493, Dowex Optipore SD-2, biobeads, Amberlite XAD4, Amberlite IRC86, and Amberlite IRA958 to adsorb HBP from a hexadecane-water solution (1:1 v/v) that contained 10,000 M DBT and 10,000 M HBP with either 0.1 or 1.0 g of the studied resins. The specific loading of HBP (L_r) was determined by the following equation:

$$L_r = \frac{(C_{HBP}|t_0 - C_{HBP}|_{t_{eq}})MM_{HBP}}{X_r} \tag{11}$$

where C_{HBP} represents the total concentration of HBP (in oil and water), MMHBP is the molecular mass of HBP (170 g/mol), and t_0 and t_{eq} represent the times at initial and equilibrium conditions, respectively. Xr is the concentration of resin employed that was calculated as follows:

$$X_r = \frac{m_r}{V_{total}} \tag{12}$$

Where V_{total} is the total volume of the solution (oil plus water) and m_r is the mass of resin used. Dowex Optipore SD-2 resin was found to express the best results, that is the best HBP loading of 50 mg/g resin with the highest selectivity relative to DBT loading which was 2.5 times greater than that for DBT. The high affinity towards HBP was attributed to the strong hydrophobic interactions between the resins' aromatic side groups and the biphenyl part of HBP. Then, it was applied in a BDS batch process using a resting cell suspension of *R. erythropolis* IGTS8. Ethanol was found to be the best extractant. The cell walls solubilization by treatment with an enzymatic mixture of lysozyme (100 mg/mL) and mutanolysin (5,000 U/mL) did not increase the concentration of DBT or HBP extracted from the cells. Sonication of the cells also did not increase extraction concentrations. Loadings (i.e. the adsorption) of HBP on the biocatalyst (Lc) and resin (Lr) were calculated from the expressions:

$$L_c = \frac{\left(\sum_i C_{extract,i}\right) V_{extract}}{m_{cells}} \tag{13}$$

$$L_r = \frac{\left(\sum_i C_{extract,i}\right) V_{extract}}{m_r} \tag{14}$$

where $C_{extract,i}$ is the concentration of HBP in the ith extract, $V_{extract}$ is the volume of the extract (constant for each fraction), m_{cells} is the mass of cells extracted, and m_r is the mass of resin extracted.

The partition coefficient of HBP between one component (component 1) and another component (component 2) in the four-component (oil, water, cells, and resin) mixtures of BDS experiments ($P_{comp1/comp2}$) was expressed as follows:

$$P_{comp1/comp2} = \frac{C_{HBP,comp1}}{C_{HBP,comp2}} \tag{15}$$

where $C_{HBP,comp1}$ is the HBP concentration measured in component 1 and $C_{HBP,comp2}$ is the HBP concentration measured in component 2. For example, $P_{R/C}$ is the partition coefficient of HBP between the resin and the biocatalyst (cells). Similarly, $P_{O/W}$ is the partition coefficient of HBP between hexadecane (oil) and water.

The resin Dowex Optipore SD-2 was found to have the highest affinity for HBP relative to the other components of the system. The resin's affinity for HBP was about 100 times greater than the affinity of either the hexadecane oil phase or the biocatalyst. Furthermore, its affinity for HBP was about 10,000 times greater than that of the aqueous buffer. The partition coefficients of HBP between oil and the aqueous buffer ($P_{O/W}$) and between the biocatalyst and the aqueous buffer ($P_{C/W}$) were very similar. Also, the partition coefficient of HBP between the biocatalyst and oil ($P_{C/O}$) was calculated to be 1 in the absence of resin. Thus, the oil and biocatalyst components expressed very similar affinities for HBP. Moreover, both the oil and the biocatalyst have an affinity for HBP that is 50 to 60 times greater than that of the aqueous buffer in the absence of resin.

The partition coefficient $P_{C/W}$ was defined as:

$$P_{C/W} = \frac{C_{HBP,intracellular}}{C_{HBP,water}} \tag{16}$$

where C_{HBP},water is the HBP concentration in the water phase and $C_{HBP,intracellular}$ is the HBP concentration within the biocatalyst since the cell is composed of two components, the inner cytoplasmic space and the outer envelope/shell that encompasses the cytoplasm, which is the cell wall. The partition coefficient of HBP between the cytoplasm and water ($P_{cytoplasm/W}$) was estimated from solving its value in the expression:

$$P_{C/W} = P_{cell-wall/W} f_{cell\ wall} + P_{cytoplasm/W} f_{cytoplasm} \tag{17}$$

where $P_{cell-wall/W}$ is the partition coefficient of HBP between the cell wall and the water and it was estimated to be 550 and $f_{cell-wall}$ and $f_{cytoplasm}$ are the fractions of the total volume of a single cell that are occupied by the cell wall and cytoplasm, respectively. The values of $f_{cytoplasm}$ and $f_{cell-wall}$ were calculated as follows:

$$f_{cytoplasm} = \frac{(R_{cell} - W)^3}{(R_{cell})^3} \tag{18}$$

$$f_{cell-wall} = 1 - \frac{(R_{cell} - W)^3}{(R_{cell})^3} \tag{19}$$

where W is the estimated cell wall thickness *Rhodococcus* species, which is approximately 10 nm (Sutcliff *et al.*, 2010) and R_{cell} is its estimated radius, which is approximately 0.5 μm (Kilbane, 1992). The cytoplasmic HBP concentration ($C_{HBP'cytoplasm}$), which decreased with the increase of $X_{r,}$ was calculated as follows:

$$C_{HBP,cytoplasm} = P_{cytoplasm/W} C_{HBP,water} \tag{20}$$

and recorded 1,100, 330, and 260 μM at resin concentrations (X_r) of 0, 10, and 50 g/L, respectively.

Although the resin was effective in reducing HBP retention within the cytoplasm of the biocatalyst, which is where the desulfurization enzymes are present and the corresponding HBP loadings on the biocatalyst (*Lc*) were calculated to be 1.6, 0.5, and 0.2 mg HBP/g DCW, despite the significant decrease in HBP retained within the cytoplasm, the total amount of HBP produced in the system did not increase with increasing resin concentration. Consequently, it was assumed that the biocatalyst and, in particular, the desulfurization enzymes might have been susceptible to cytoplasmic HBP concentrations of less than 260 μM.

In another study, Carvajal *et al.* (2017) studied the ADS efficiency of a model oil of 100 mg/L DBT and 100 mg/L 4,6-DMDBT on different adsorbents: alumina (Al), silica (Si), and sepiolite (Sep), where the highest removal occurred with particle sizes from 0.43 to 0.063 mm and it increased with the decrease of the particles' size, which was attributed to the larger total surface area per volume and the shorter diffusion path for the adsorbate upon the usage of smaller particles. Comparison of the adsorption of both sulfur-containing organic molecules revealed higher adsorption of DBT on Si and Al, while no marked difference was observed in the case of Sep, which was attributed to the different specific areas and density and strength of acid sites of these materials. While higher adsorption of both sulfur-containing organic molecules on the Si support showed that the strength of acid sites is the main factor that affects the adsorption of recalcitrant sulfur-containing molecule, the adsorption capacity values revealed that the strongest interaction of DBT and 4,6-DMDBT was with Sep and Si, respectively. The removal of the adsorbed S-compounds for the recovery of the adsorbents has been done using free cells of *R. erythropolis* IGTS8 and the highest BDS efficiency was performed upon the usage of Sep and Si. The observed differences in BDS were attributed to the interactions between the cells and supports since both physical and biological factors can influence cell–support interactions, such as the size of the bacterial cells, ionic strength, and the support surface structure (Yee *et al.*, 2000; Jeyachandran *et al.*, 2006; Dinamarca *et al.*, 2010). Moreover, the results showed higher activity in systems that had stronger interactions between the bacterial cells and support. There was also a recorded higher removal of DBT compared with that of 4,6-DMDBT. This was attributed to the difficulty of BDS of recalcitrant long-chain alkylated DBT (Bhatia and Sharma, 2010).

11.8.3 Coupling Non-Hydrodesulfurization with BDS

Coupling of a non-hydrodesulfurization process with a BDS process can bring benefits to produce ultra-low sulfur diesel fuels. Under an operating condition of 65 °C and atmospheric pressure, the sulfur content present in diesel fuel decreased from 766 μg/g to 7.14 μg/g by a non-hydrodesulfurization process using persulfate as an oxidizing agent. The non-hydrodesulfurization diesel fuel was followed by BDS using *Rhodococcus* sp. and *Methylomonas methanica*, separately. This occurred in a biphasic batch process with different O/W phase ratios of 15:85 to 100:0. Microbial strains were observed to follow classic Monod type growth kinetics under the present range of substrate, where the maximum growth rate, $\mu_{max,}$ recorded 0.096 and 0.098 h^{-1} for the half saturation constant, $K_{s,}$ of 71 and 77 mg/dm^3 for *Rhodocuccus* sp and *Methylomonas methanica*, respectively. This BDS process was conducted in a trickle bed reactor under continuous mode that was coupled with a non-hydrodesulfurization unit. The sulfur content present after the BDS technique using *Rhodocuccus* sp and *Methylomonas methanica* was 4.22 and 4.06 μg/g, respectively. This was approximately equivalent to 99.93% sulfur removal and was 1.78 times better than Bharat IV norms (Sekar *et al.*, 2016).

11.8.4 Three Step BDS-ODS-RADS

Different desulfurization techniques were conducted on two crude oil samples, the HCO (1.88% S) and the LCO (0.378% S) procured from a local petroleum refinery in India (Agarwal and Sharma, 2010). LCO using *Pantoea agglomerans* D23W3 resulted in 61.40% removal of sulfur, whereas HCO showed 63.29% S removal under similar conditions. The use of *P. agglomerans* D23W3 under anaerobic conditions showed marginally better results than those under aerobic conditions (Table 11.2). Moreover,

Table 11.2 Three-Step Integrated Desulfurization Process of Crude Oil.

Crude Oil	HCO	HCO	LCO	LCO
S-Content	1.88		1.8	
First Step	**Anaerobic BDS**	**Aerobic BDS**	**Anaerobic BDS**	**Aerobic BDS**
S-Content **Desulfurization %**	0.58 69.14	0.69 63.29	0.138 63.4	0.15 61.4
Second Step	**ODS**	**ODS**	**ODS**	**ODS**
S-Content **Desulfurization %**	0.11 94.15	0.23 87.76	0.07 81.48	0.08 78.84
Third Step	**RADS**	**RADS**	**RADS**	**RADS**
S-Content **Desulfurization %**	0.09 95.21	0.1 94.6	0.053 85.9	0.022 94.3

the use of thermophile *Klebsiella* sp. 13T resulted in 62.43% S removal from LCO and 68.08% S removal from HCO. Different adsorbents were also examined for ADS efficiency and the residual coal obtained after the solvent extraction of Samla coal showed the maximum removal of sulfur at 78.90% from LCO and 74.46% from HCO. Finally, Agarwal and Sharma (2010) performed a comparative study on heavy and light crude oil (HCO and LCO) applying a three-step integrated process. Anaerobic or aerobic BDS followed by ODS, then finally reactive adsorptive desulfurization was performed to remove the oxidized S-compounds (Table 11.2), where BDS under anaerobic conditions was followed by oxy-desulfurization and then reactive adsorption integration resulted in a maximum removal of ≈ 95.21% removal of sulfur from HCO and ≈ 94.30% removal of sulfur from LCO.

11.9 BDS of other Petroleum Products

The worldwide increase in the amount of rubber products comes with the rapid industrialization of the modern world. Generally, rubber materials are absolutely necessary for social production and social living. However, the annual waste of rubber is huge and it is hard to be naturally degraded because of the stable cross-linked three-dimensional network structure in vulcanized rubber that causes serious environmental pollution. Recycling rubber is a way to overcome the problem; the cross-links in rubber can be broken by mechanical energy, such as mechanical shearing (Fukumori *et al.*, 2006), ultrasound (Sun and Isayev, 2009), microwave (Vega et al., 2008), and electron beam (Hassan *et al.*, 2007). Chemical desulfurization reagents (De Debapriya et al. 2007; Rajan et al., 2007) can also cut the crosslink bonds or induce active groups on the surface of ground rubber. However, all these methods are expensive, consume a large amount of energy, and release toxic chemicals leading to a secondary pollution. In some developed countries, waste rubber is pulverized into ground rubber to be a replacement of virgin rubber. However, the cross-linked structure restricts the movement of molecular chains and adversely affects reprocessing performance, so the interfacial bonding strength is low when ground rubber is directly blended with raw rubber, resulting in a decrease of the mechanical properties of vulcanized rubber blends. The purpose of vulcanized rubber desulfurization is to break the sulfur crosslinks and allow the rubber to regain mobility for better reprocessability and remoldability. The desulfurization of vulcanized rubber can be obtained by both physical and chemical methods. The desulfurization of vulcanized rubber can significantly improve the coherence of the ground rubber and matrix and the performance of the blend. Microbial desulfurization of rubber gained attention (Holst *et al.*, 1998; Kim and Park, 1999; Fliermans, 2002) for its ability to break sulfur crosslinks with low energy consumption, simple processes, low equipment requirement, and no pollution. However, the main problem in microbial desulfurization is that many microorganisms are sensitive to rubber additives. Moreover, rubber is lipophilic, so it cannot dissolve in an aqueous phase which allows microbes to stay on the rubber surface for a short time. In other words, the BDS effect still needs to improve. Therefore, it is important to increase the affinity between rubber and microorganism, raising the chance of microbial contacting with rubber.

The desulfurization effect of six bacterial strains (*Thiobacillus ferrooxidans, Thiobacillus thiooxidans, Thiobacillus thioparus, Sulfolobus acidocaldarius, Rhodococcus rhodochrous*, and ATCC 39327) on ground tire rubber (GTR) has been studied and *S. acidocaldarius* expressed the best effect (Romine and Snowdon-Swan, 1997; Romine and Romine, 1998). The "4S" pathway was proposed, through which the sulfur crosslinks were metabolized into sulfoxide/sulfone/sulfonate/sulfate. Meanwhile, researchers indicated that treated ground tire rubber (GTR) would achieve a good chemistry reactive if the bioprocess stayed at the first three steps. Thus, the modified GTR could be added into virgin rubber with high loading amounts and perform with good mechanical properties. Bredberg *et al.* (2001a, b) found that the thermophilic *Pyrococcus furiosus* could be used to desulfurize GTR, but the additives in rubber had some adverse effect on this bacterium. Löffler *et al.* (1993) evaluated the effects of *T. thioparus*, *T. ferrooxidanshe*, and *T. thiooxidans* on GTR, where *T. thioparus* expressed the best results and the particle size of GTR was found to have a large influence on the desulfurization effect. The fungal desulfurization of ground rubber, using a white rot basidiomycete, *Ceriporiopsis subvermispora,* to sulfate by selectively breaking the sulfur crosslinks in rubber has been reported (Sato *et al.*, 2004), but microbial desulfurization breaks only the sulfur crosslinks on the surface of ground rubber because ground rubber and culture medium have no affinity and the contacting surfaces on which the desulfurization occurs are different two-phases. In another study, *Ceriporiopsis subvermispora* and *Dichomitus squalens* were cultivated separately in media with natural rubber (NR) for 0–250 days and found that *C. subvermispora* decreased the content of S-C bonds, but *D. squalens* expressed no obvious effect on the content of S-C bonds or S-S bonds (Sato *et al.*, 2009).

Li *et al.* (2011) studied the microbial desulfurization of ground tire rubber (GTR) by *Thiobacillus ferrooxidans,* which was isolated from a soil of an iron mine in Xinlong, Hebei Province, China, and it was characterized by strong sulfur oxidizing capacity. GTR was immersed in 75% ethanol (v/v) for 24 h to kill the microorganisms attached to it and then filtered out and dried for use. *T. ferrooxidans* was cultured in flasks (500 mL with 200 mL medium) on a rotary shaker at 30 °C and the pH was adjusted to 2.5 with HCl. After 24 h, 5% (w/v) detoxified GTR and GTR flakes were added into the medium and desulfurized for 30 days. The biomass of the *T. ferrooxidans* and sulfate content in the medium were measured every 3 days. GTR was desulfurized in the modified Silverman medium during the cultivation of *T. ferrooxidans* for 30 days and *T. ferrooxidans* was able to maintain a high biomass. FTIR-ATR (Fourier-transform infrared spectroscopy-Attenuated total reflection) and XPS (X-ray photoelectron spectroscopy) spectra revealed the occurrence of a rupture of conjugated C=C bonds and a reduction of sulfur content on the surface of GTR during desulfurization. Compared with GTR, the carbon content of DGTR (Desulfurated ground tire rubber) surface slightly decreased from 94.38% to 93.43%, while the oxygen content significantly increased by 30%, from 4.73% to 6.15%, and the sulfur content substantially decreased by 52.8%, from 0.89% to 0.42%. The continuous increase of SO_4^{2-} in the medium indicated that sulfur on the surface of GTR was oxidized. Thus, this study proved that the selected *T. ferrooxidans* had strong effects on the metabolism of the sulfur element, as well as the cross-linked S, on the surface of GTR and expressed a good desulfurization effect on GTR. Microorganisms could cause conjugated C=C to break and changed some chemical structures on the surface of

GTR. After desulfurization, the cross-linked bonds by S on the GTR surface had been partly ruptured and formed a sulfur oxide group. The sol fraction of GTR increased from its original 4.69% to 7.43%. Compared with GTR sheet, desulfurized ground tire rubber (DGTR) sheets had much smoother surfaces, better physical properties, and higher swelling values. As a result of microbial desulfurization and sulfur diffusion, the DGTR filled NR composites had better mechanical properties and lower crosslink density than GTR filled NR composites at the same loading. Compared with NR/GTR composites, NR/DGTR composites had lower internal friction losses of the molecular chain and better interface coherence between the DGTR and matrix.

The surface desulfurization of GTR was carried out via a biological treatment by *Thiobacillus* sp. with a strong sulfur oxidizing capacity (Li *et al.*, 2012a). The bonding states and element content on the surface of GTR and desulfurized GTR (DGTR) were evaluated using an X-ray Photoelectron Spectroscopy (XPS). The contact angle of GTR was 120.5°, which was decreased down to 93.5° after treatment. The surface desulfurization of GTR was realized with biological treatment by *Thiobacillus* sp. for improving interfacial adhesion. After treatment, oxygen content on the surface of waste rubber increased by 30% and cross-linked sulfur bonds were partly cleaved and converted to sulfate or oxygen containing sulfur-based groups. The ration of S-S bonds and S-C bonds were respectively decreased by 18.3% and 42.3%.

The formation of S-O bonds was observed. The cure characteristics, swelling behavior, and crosslink density of natural rubber (NR)/GTR and NR/DGTR were examined. A lower cross-link density was obtained for NR/DGTR vulcanizates, which was attributed to the migration of sulfur being obstructed by GTR with cross-link structure and a lower sulfur content of DGTR. The improvement in mechanical properties was observed for NR/DGTR vulcanizates because of the enhanced interfacial interaction between the DGTR and NR matrix. The dynamic mechanical analysis (DMA) results showed that NR/DGTR vulcanizates had a reduction of molecular chain friction resistance during a glass transition region and scanning electron microscopy (SEM) studies further indicated a good coherency and homogeneity between the DGTR and NR matrix.

In another study by Li *et al.* (2012b), the BDS of GTR was performed by *Sphingomonas* sp. that was selected from coal mine soil in Sichuan Province, China and had sulfur oxidizing capacity. Before desulfurization, GTR was immersed in 75% ethanol (v/v) for 24 h in order to kill the microorganisms attached to it and remove harmful additives. Detoxified GTR was filtered out and dried in a sterile cabinet. After a 3 day incubation,

GTR with glucose was added into the culture medium. The amount of ground rubber was 2.5% (w/v) of the medium. After desulfurization for 20 days, DGTR was filtered out and washed by distilled water several times. The results showed that GTR had low toxicity to *Sphingomonas* sp. and that it was able to maintain a high biomass. After desulfurization, not only a rupture of conjugated C=C bonds, but also a reduction of GTR sulfur content (22.9%) occurred. The sol fraction of GTR increased from its original 4.69% to 8.68% after desulfurization. The carbon content of DGTR (92.16%) was slightly smaller than that of GTR (96.41%), indicating the occurrence of a partial fracture in the carbon chain. The oxygen content of DGTR has obviously increased after desulfurization, indicating the production of some oxidation products on the surface of DGTR. The sulfur content reduced by 22.9% form GTR (0.96%) to DGTR (0.74%). DGTR sheets had better physical properties and higher swelling values than GTR sheets. Similar results were also obtained using other bacterium for rubber desulfurization. *Sulfolobus acidocaldarius* used for rubber desulfurization released 13.4% sulfur of rubber material during 7 d (Romine *et al.* 1997), the fungus could decrease the total sulfur content of the rubber by 29% in 200 d (Sato *et al.* 2004), and the sulfur content of natural rubber was reduced by up to 30% compared to that of the untreated rubber in the case of the 30 d microbial desulfurization by *T. perometabolis* (Jin and Jin, 1999).

The biodesulfurization process consists of three steps. Firstly, microbes grow on GTR surface. Secondly, it produces some desulfurized enzymes. Lastly, these enzymes reacted on the GTR surface. During the bioprocess, microorganisms contact with rubber particles in a bioreactor, for example, a shake flask or fermentation reactor. However, microbes cannot stay on a GTR particle surface for a long time because of incompatibility between these two matters. The effect of three non-ionic surfactants (Tween 20 Polyoxyethylene (20) sorbitan monolaurate $C_{58}H_{114}O_{26}$, Tween 60 Polyoxyethylene (20) sorbitan monostearate $C_{64}H_{128}O_{26}$, and Tween 80 Polyoxyethylene (20) sorbitan monooleate $C_{64}H_{124}O_{26}$) on the BDS of GTR by *Sphingomonas* sp. was examined by Hu *et al.* (2014). Tween 20, among these three surfactants, showed the best effect on the enhancement of BDS, as it has the shortest oil-soluble chain among the three surfactants. The SEM analysis proved that the Tween surfactants could effectively improve affinity between *Sphingomonas* sp. and GTR. After premixing with surfactant, the GTR particle surface was coated by the Tween molecule. Oil-soluble tails contacted with the surface of GTR, while the water-soluble ends extended to the culture medium and contacted with the microbes. The desulfurizing enzymes produced by microbes are water soluble and are needed to pass a surfactant molecule to react with sulfide bonds in GTR particles. Thus, the shorter the length of the hydrophobic part is, the easier for enzymes to reach the rubber surface. Moreover, the results of SEM-EDS (Scanning Electron Microscope-Energy Dispersive X-ray Spectrometer) showed that the amount of sulfur in a rubber surface layer of 0.4 μm was significantly decreased by 67%. XPS (X-ray photoelectron spectroscopy) analysis data showed that the area of S-S bonds and S-C bonds decreased while the S-O bonds obviously increased. The desulfurized effect was evaluated through measuring desulfurized depth, swelling value, and sulfur content of a desulfurized ground tire rubber (DGTR) sheet. The physical properties of desulfurized ground tire rubber/styrene butadiene rubber (DGTR/SBR) composite were measured and they were found to improve. Compared with the tensile strength and elongation of DGTR/SBR composite, those of $DGTR_{20}$/SBR significantly increased by 11.6% and 23.5%, respectively.

A microbe with desulfurizing capability, *Alicyclobacillus* sp., was selected by Yao *et al.* (2013) to recycle waste latex rubber (WLR). The growth characteristics of the microorganism and the technical conditions in the co-culture desulfurization process were studied. The results showed that the toxic effect of WLR on *Alicyclobacillus* sp. was not noticeable and the optimum addition amount of WLR was 5% (w/v) for desulfurization. The surfactant polysorbate 80 (Tween 80) had a toxic effect on *Alicyclobacillus* sp., but the growth of the microbe was vigorous if the proper technique was used; the mixing of WLR with Tween 80 was followed by the addition of the mixture into the culture media. After desulfurization by *Alicyclobacillus* sp., the contents of S and O elements on the surface of DWLR decreased by 62.5% and increased by 34.9%, respectively. The contents of S-C and S-S bonds decreased by 58.0% and 58.2%, respectively, and the content of S-O bonds increased by 79.0% on the surface of DWLR. Briefly, the results showed that the sulfur bonds on the surface of WLR were broken to form sulfones groups and the hydrophilic property of DWLR was improved.

Gordonia amicalisa was used by Hu *et al.* (2016) to treat two kinds of vulcanized synthetic rubbers, vulcanized isoprene rubber (v-IR) and vulcanized styrene butadiene rubber (v-SBR). The effects of microbial treatment on vulcanized rubbers were evaluated by changes of crosslinking density, sol fraction, morphology, element content, and chemical group of treated rubber samples. The results showed that *G. amicalisa* can break the C=C bond on v-IR and S-S bonds on v-SBR. After a microbial treatment of 20 days, the crosslink densities of v-IR and v-SBR decreased by 13.7% and 22.1%, respectively. The carbon content on the v-IR surface decreased by 9.1%. The FTIR analysis showed that C=C bonds transferred into C=O bonds, indicating the main chain scission. In regards to the v-SBR, sulfur content on its surface was decreased by 22.9 %, S-S bonds were broken, and S=O was produced, indicating sulfur cross-link scission. Furthermore, to elucidate the reason that *G. amicalisa* has different decross-linking mechanisms on v-SBR and v-IR, the inducing mechanism of the desulfurizing enzyme produced from *G. amicalisa* was investigated. The v-SBR and v-IR were used as enzyme inducers and DBT was used as a substrate reacting with the desulfurizing-enzyme. Then, the DBT, as a specific substrate, reacted with the cell culture solution, which was induced by different inducers. The concentration of DBT reacted with pure medium did not decrease, indicating that the contents in the medium have no effect on DBT. The reacted concentration of DBT with the control enzyme solution was unchanged, so the biodesulfurizing microorganisms (BDSM) cannot secrete the desulfurizing enzyme from the cell into the solution without the inducer. Regarding the enzyme solution with the inducers v-SBR ($Enzyme_{SBR}$) and v-IR ($Enzyme_{IR}$) respectively, DBT concentration in the $Enzyme_{SBR}$ solution presented a decrease of 28.8%, while DBT in the $Enzyme_{IR}$ solution remained unchanged, which indicated that the $Enzyme_{SBR}$ solution has a desulfurizing enzyme which could metabolize DBT into other products. After a DBT reaction with $Enzyme_{SBR}$ and $Enzyme_{IR}$ solution for 72 h, the metabolic products of $Enzyme_{SBR}$ and $Enzyme_{IR}$, standard solutions of DBT, and 2-HBP were detected by an ultraviolet spectrophotometer. DBT has specific bands at 310 nm and 323 nm and 2-HBP has characterized band at 287 nm; the 2-HBP band was detected and the areas of the DBT-band in $Enzyme_{SBR}$ was smaller than those in standard DBT. This spectrum indicated that v-SBR can induce *G. amicalisa* to produce a desulfurizing enzyme, which can transfer DBT into 2-HBP. Reversely, $Enzyme_{IR}$ solution only had DBT and no 2-HBP product, indicating that v-IR cannot induce a desulfurizing enzyme. The experiment results for v-IR and v-SBR vulcanized used as enzyme inducers showed that v-SBR can induce *G. amicalisa* to secrete a desulfurizing-enzyme from cell inside to outside and this desulfurizing-enzyme can transfer DBT into 2-HBP, while v-IR vulcanized has no such induced effect. This explained why *G. amicalisa* has different effects on v-IR and v-SBR vulcanized rubber.

References

Abbad-Andaloussi S., Warzywoda M., Monot F. (2003a) Microbial desulfurization of diesel oils by selected bacterial strains. *Oil & Gas Science and Technology*. 58:505–513.

Abbad-Andaloussi, S., Lagnel, C., Warzywoda, M., Monot, Frédéric, (2003b). Multicriteria comparison of resting cell activities of bacterial strains selected for biodesulfurization of petroleum compounds. *Enzyme and Microbial Technology*. 32: 446–454.

Abbondanzi F., Bruzzi L., Campisi T., Frezzati A., Guerra R., Iacondini, A. (2006) Biotreatability of polycyclic aromatic hydrocarbon in brackish sediments: preliminary studies of an integrated monitoring. *International Biodeterioration and Biodegradation*. 57(4): 214–221.

Abin-Fuentes, A., Mohamed, M.E., Wang, D.I.C., Prather, K.L.J. (2013) Exploring the mechanism of biocatalyst inhibition in microbial desulfurization. *Applied and Environmental Microbiology*. 79(24): 7807–7817.

Adegunlola, G.A., Oloke, J.K., Majolagbe, O.N., Adebayo, E.A., Oyeyiola, A.G., Adegunlola, F.O. (2010). Microbial desulphurization of crude oil using immobilized spores of *Aspergillus flavus*. Advances in Environmental Biology. 4(2): 155–158.

Adlakha, J., Singh, P., Ram, S.K., Kumar, M., Singh, M.P., Singh, D., Sahai, V., Srivastava, P. (2016) Optimization of conditions for deep desulfurization of heavy crude oil and hydrodesulfurized diesel by *Gordonia* sp. IITR100. 184: 761–769.

Agarwal, P., Sharma, D.K. (2010) Comparative studies on the bio-desulfurization of crude oil with other desulfurization techniques and deep desulfurization through integrated processes. *Energy and Fuels*. 24(1): 518–524.

Aktas, Ö., Cecen, F. (2007) Bioregeneration of activated carbon: a review Int. Biodeterior. Biodegrad. 59: 257–272.

Alvarez, J.C., Ortiz de Montellano, P.R. (1992) Thianthrene 5-oxide as a probe of the electrophilicity of hemoprotein oxidizing species. *Biochemistry*. 31: 8315–8322.

Alves L, Paixão SM, Pacheco R, Ferreira AF, Silva CM (2015) Biodesulphurization of fossil fuels: energy, emissions and cost analysis. RSC Advances. 5: 34047–34057

Alves L. (2007) Dibenzothiophene desulfurization by Gordonia alkanivorans strain 1B. a PhD Thesis. Faculdade de Ciências da Universidade de Lisboa. Lisboa. Portugal.

Arabian, D., Najafi, H., Farhadi, F., Dehkordi, A.M. (2014) Biodesulfurization of simulated light fuel oil by a native isolated bacteria *Bacillus cereus* HN. *Journal of Petroleum Science and Technology*. 4(1): 31–40.

Aribike, D.S., Susu, A.A., Nwachukwu, S.C.U., and Kareem, S.A. (2009) Microbial desulfurization of diesel by *Desulfobacterium* aniline. *Academia Arena*. 1(4): 11–17.

Aribike, D.S., Susu, A.A., Nwachukwu, S.C.U., Kareem, S.A. (2008) Biodesulfurization of kerosene by *Desulfobacterium indolicum*. *Nature and Science*. 6(4): 55–63.

Armstrong, S.M., Sankey, B.M., Voordouw, G. (1995a) Conversion of dibenzothiophene to biphenyl by sulfate-reducing bacteria isolated from oil field production facilities. *Biotechnology Letters*. 17: 1133–1136.

Armstrong, S.M., Sankey, B.M., Voordouw, G. (1995b) Evaluation of sulfate reducing bacteria for desulfurizing bitumen and its fractions. *Fuel.* 76(3): 223–227.

Ascón-Cabrera M.A., Lebeault J.M. (1995) Interfacial area effects of a biphasic aqueous/organic system on growth kinetic of xenobiotic-degrading microorganisms. *Applied Microbiology and Biotechnology*. 43:1136–1141.

Atlantic Richfield Company (1986) Biodesulfurization of carbonaceous materials. United States Patent No. 4632906.

Atlas R.M. (1981) Microbial degradation of petroleum hydrocarbons: an environmental perspective. *Microbiological Reviews*. 45(1):180–209.

Ayala M., Tinoco, R., Hernandez, V., Bremauntz, P., Vazquez-Duhalt, R. (1998) Biocatalytic oxidation of fuel as an alternative to biodesulfurization. *Fuel Processing Technology*. 57: 101–11.

Ayala, M., LeBorgne, S. (2010) Microorganisms utilizing sulfur containing hydrocarbons. In: Timmis KN (ed.) Handbook of hydrocarbon and lipid microbiology. Springer, Berlin.

Ayala, M., Vazquez-Duhalt, R. Morales, M., Le Borgne S. (2016) Application of microorganisms to the processing and upgrading of crude oil and fractions. Lee S.Y. (ed). Consequences of microbial interactions with hydrocarbons, oils, and lipids: Production of fuels and chemicals. Handbook of hydrocarbon and lipid microbiology. Springer International Publishing, AG.

Ayala, M., Verdin, J., Vazquez-Duhalt, P. (2007) The prospects for peroxidase-based biorefining of petroleum fuels. *Biocatalysis and Biotransformation*. 25(2–4): 114–129.

Bahrami, A., Shojaosadati, S.A., Mohebali, G. (2001) Biodegradation of dibenzothiophenes by thermophilic bacteria. *Biotechnology Letters*. 23: 899–901.

Baldi, F., Pepi, M., Fava, F. (2003) Growth of *Rhodosporidium toruloidesStrain* DBVPG 6662 on dibenzothiophene crystals and orimulsion. *Applied and Environmental Microbiology*. 69(8): 4689–4696.

Bandyopadhyay, S., Chowdhury, R. (2014) parameter optimization of surface active properties and quantification of biosurfactant produced in continuous stirred tank bioreactor during biodesulfurization of diesel. *International Journal of Innovative Research in Advanced Engineering*. 1(9): 217–224.

Bandyopadhyay, S., Chowdhury, R., Bhattacharjee, C. (2013b) Steady state performance of a bioreactor for production of near zero sulfur diesel (NZSD) and bio-surfactant. *Journal of Clean Energy Technologies*. 1(3): 189–193.

Bandyopadhyay, S., Chowdhury, R., Bhattacharjee, C., Pan, S. (2013a) Simultaneous production of biosurfactant and ULSD (ultra low sulfur diesel) using *Rhodococcus* sp. in a chemostat. *Fuel*. 113: 107–112.

Bandyopadhyay, S., Chowdhury, R., Bhattacharjee, C. (2014) Production of biosurfactants through biodesulfurization of spent engine oil – an experimental study. *Green Chemistry Letters and Reviews*. 7(3): 288–295.

Becker J.R. (1997) Crude Oil Waxes, Emulsion, and Asphaltenes. Penn Well Books. Tulsa. Oklahoma.

Bhatia, S., Sharma, D.K. (2010) Biodesulfurization of dibenzothiophene, its alkylated derivatives and crude oil by a newly isolated strain *Pantoea agglomerans* D23W3. *Biochemical Engineering Journal.* 50(3): 104–109.

Bhatia, S., Sharma, D.K. (2012) Thermophilic desulfurization of dibenzothiophene and different petroleum oils by *Klebsiella* sp. 13T. *Environmental Science and Pollution Research* 19(8):3491–3497.

Bordoloi, N.K., Rai, S.K., Chaudhuri, M.K., Mukherjee, A.K. (2014) Deep-desulfurization of dibenzothiophene and its derivatives present in diesel oil by a newly isolated bacterium *Achromobacter* sp. to reduce the environmental pollution from fossil fuel combustion. *Fuel Processing Technology.* 119: 236–244.

Borole A.P., Hamilton C.Y., Miller K. (2003) Final Project Report "Improving Enzyme Activity and Broadening Selectivity for Biological Desulfurization and Upgrading of Petroleum Feedstocks". (May 12, 2003).

Bredberg, K., Christiansson, M., Bellander, M., Stenberg, B., Holst, O. (2001a) Properties of rubber materials containing recycled microbially devulcanlzed cryo-ground tire rubber. *Progress in Rubber, Plastics and Recycling Technology.*17(3): 149–161.

Bredberg, K., Persson, J., Christiansson, M. (2001b) Anaerobic desulfurization of ground rubber with the thermophilic archaeon pyrococcus furiosus-a new method for rubber recycling. *Applied Microbiology and Biotechnology.*55(1): 43–48.

Bustos-Jaimes I., Amador G., Castorena G., Le Borgne S. (2003) Genotypic characterization of sulfur-oxidative desulfurizing bacterial strains isolated from Mexican refineries. *Oil and Gas Science. Technology.* 58(4):521–526.

Caro A., Boltes K., Leton P., García-Calvo E. (2007) Dibenzothiophene biodesulfurization in resting cell conditions by aerobic bacteria. *Biochemical Engineering.* 35:191–197.

Caro A., Boltes K., Letón P., García-Calvo E. (2008) Biodesulfurization of DBT by growing cells of *Pseudomonas putida* CECT 5279 in biphasic media. *Chemosphere.* 73:663- 669.

Carvajal, P., Dinamarca, A.M., Baeza, P., Camú, E., Ojeda, J. (2017) Removal of sulfur-containing organic molecules adsorbed on inorganic supports by *Rhodococcus Rhodochrous* spp. *Biotechnology Letters.* 39: 241–245.

Carvalho, C.C.C.R., Da Fonseca, M.M.R. (2005) The remarkable *Rhodococcus erythropolis.* analysis. *Applied and Environmental Microbiology.* 67: 715–726.

Carvalho, C.C.C.R., Parreño-marchante, B., Neumann, G., Da Fonseca, M.M.R. (2005) Adaptation of *Rhodococcus erythropolis* DCL14 to growth on n-alkanes, alcohols and terpenes. *Applied and Environmental Microbiology.* 67: 383–388.

Carvalho, C.C.C.R., Poretti, A., Da Fonseca, M.M.R. (2004) Solvent toxicity in organic-aqueous systems analyzed by multivariate analysis. *Applied and Environmental Microbiology.* 26: 361–375.

Carvalho, C.C.C.R., Wick, L.Y., Heipieper, H.J. (2009) Cell wall adaptations of planktonic and biofilm *Rhodococcus erythropolis* cells to growth on C5 to C16 n-alkane hydrocarbons. *Applied Microbiology and Biotechnology.* 82: 311–320.

Castorena, G., Suarez, C., Valdez, I., Amador, G., Fernandez, L., Le Borgne, S. (2002) Sulfur-selective desulfurization of dibenzothiophene and diesel oil by newly isolated *Rhodococcus* sp. strains *FEMS Microbiology Letters.* 215(1): 157–161.

Chaineau C.H., Rougeux G., Yepremian C., Oudot, J. (2005) Effects of nutrient concentration on the biodegradation of crude oil and associated microbial populations in the soil. *Soil Biology and Biochemistry.* 37:1490–1497.

Chang J.H., Kim Y.J., Lee B.H., Cho K.S., Ryu H.W., Chang Y.K., Chang H.N. (2001) Production of a desulfurization biocatalyst by two-stage fermentation and its application for the treatment of model and diesel oils. *Biotechnology progress.* 17:876–880.

Chang, J.H., Chang, Y.K., Cho, K.S., Chang, H.N. (2000b) Desulfurization of model and diesel oils by resting cells of *Gordona* sp. *Biotechnology Letters.* 22: 193–196.

Chang, J.H., Chang, Y.K., Ryu, H.W., Chang, H.N. (2000a) Desulfurization of light gas oil in immobilized-cell systems of *Gordona* sp. CYKS1 and *Nocardia* sp. CYKS2. *FEMS Microbiology Letters.* 182: 309–312.

Chang, J.H., Rhee, S.K., Chang, Y.K., Chang, H.N. (1998) Desulfurization of diesel oils by a newly isolated dibenzothiophene-degrading *Nocardia* Sp. strain CYKS2. *Biotechnology Progress.* 14:851–855.

Chauhan, A.K., Ahmad, A., Singh, S.P., Kumar, A. (2015) Biodesulfurization of benzonaphthothiophene by an isolated *Gordonia* sp. IITR 100. *International Biodeterioration and Biodegradation.* 104: 105–111.

Choi O., Choi K., Ryu H., Chang Y. (2003) Enhancement of phase separation by the addition of de-emulsifiers to threephase (diesel oil/biocatalyst/aqueous phase) emulsion in diesel biodesulfurization. *Biotechnology Letters.* 25:73–77.

Constanti M., Giralt J., Bordons (1996) Degradation and desulfurization of dibenzothiophene sulfone and other sulfur compounds by *Agrobacterium* NC501 and a mixed culture. *Enzyme and Microbial Technology.* 19: 214–219.

Crawford, D.L., Gupta, R.K. (1990) Oxidation of dibenzothiophene by *Cunninghamella elegans. Current Microbiology.* 21: 229–231.

De Debapriya, De D., Singharoy, G.M. (2007) Reclaiming of ground rubber tire by a novel reclaiming agent. I. virgin natural rubber/reclaimed GRT vulcanizates. *Polymer Engineering and Science.* 47: 1091–1100.

Del Olmo, C.H., Alcon, A., Santos, V.E., Garcia-Ochoa, F. (2005) Modeling the production of a *Rhodococcus erythropolis* IGTS8 biocatalyst for DBT biodesulfurization: influence of media composition. *Enzyme and Microbial Technology.* 37:157–166

Denome S.A., Stanley D.C., Olson E.S., Young K.D. (1993) Metabolism of dibenzothiophene and naphthalene in *Pseudomonas* strains: Complete DNA sequence of an upper naphthalene catabolic pathway. *Journal of Bacteriology.* 175:6890–6901.

Exxon Mobil (2016) Global transportation demand by fuel. In The outlook for energy, a view to 2040, outlook for energy charts. http://corporate.exxonmobil.com/en/energy/energy-outlook/charts-2016/global-transportation-demand-by-fuel?parentId=d7323290-c766-440a-8e68-094d67a30841.

Dinamarca, M.A., Ibacache-Quiroga, C., Baeza, P., Galvez, S., Villarroel, M., Olivero, P., Ojeda, J. (2010) Biodesulfurization of gas oil using inorganic supports biomodified with metabolically active cells immobilized by adsorption. *Bioresource Technology*. 101: 2375–2378.

Dinamarca, M.A., Orellana, L., Aguirre, J., Baeza, P., Espinoza, G., Canales, C., Ojeda, J. (2014b) Biodesulfurization of dibenzothiophene and gas oil using a bioreactor containing a catalytic bed with Rhodococcus rhodochrous immobilized on silica. *Biotechnology Letters*, 36(8), 1649–1652.

Dinamarca, M.A., Rojas, A., Baeza, P., Espinoza, G., Ibacache-Quiroga, C., Ojeda, J. (2014a) Optimizing the biodesulfurization of gas oil by adding surfactants to immobilized cell systems. *Fuel*. 116: 237–241.

Doerge, D.R. (1986) xygenation of organosulfur compounds by peroxidases: evidence of an electron transfer mechanism for lactoperoxidaseArchives of Biochemistry and Biophysics. 244: 678.

Doerge, D.R., Cooray, N.M., Brewster, M.E. (1991) Peroxidase-catalyzed S-oxygenation: mechanism of oxygen transfer for lactoperoxidase. *Biochemistry* 30: 8960–8964.

Dordick, J.S., (1989) Enzymatic catalysis in monophasic organic solvents. *Enzyme and Microbial. Technology*. 11: 194.

Eckart V., Hieke W., Bauch J., Gentzsch H. (1980) Mikrobielle Entschwefelung von Erdöl und schweren Erdölfraktionen: 1. Untersuchungen zur mikrobiellen aeroben Entschwefelung von Romaschkino-Roherdöl. Zentralblatt für Bakteriologie Mikrobiologie . 135: 674–681.

Eckart V., Hieke W., Bauch J., Gentzsch H. (1981) Microbial Desulfurization of Petroleum and Heavy Petroleum Fractions. 1. Studies on Microbial Aerobic Desulfurization of Romashkino Crude Oi. Zentralblatt für Bakteriologie Mikrobiologie . Chem. Abst. 136 152–160.

Eckart V., Hieke W., Bauch J., Gentzsch, H. (1982) Microbial Desulfurization of Petroleum and Heavy Petroleum Fractions. 3. Change in the Chemical Composition of Fuel-D-Oil by Microbial Aerobic Desulfurization. Zentralblatt für Bakteriologie Mikrobiologie . Chem. Abst. 137: 270–279.

Eckart, V., Köhler, M., Hieke, W. (1986) Microbial desulfurization of petroleum and heavy petroleum fractions. 5. Anaerobic desulfurization of Romashkino petroleum. Zentralblatt Fur Mikrobiologie. 141: 291–300.

El-Gendy N.Sh., Speight J.G. (2016) Handbook of Refinery Desulfurization. CRC Press, Taylor and Francis Group, LLC, 6000 Broken Sound Parkway NW, Suite 300, Boca Raton, FL 33487–2742, USA.

El-Gendy, N. Sh., Farahat, L.A., Moustafa, Y.M., Shaker, N., El-Temtamy, S.A. (2006) Biodesulfurization of Crude and Diesel Oil by *Candida parapsilosis* NSh45 Isolated from Egyptian Hydrocarbon Polluted Sea Water. *Biosciences, Biotechnology Research Asia*, 3(1a): 5–16.

El-Gendy, N.Sh. (2001) Biodesulfurization of organosulfur compounds in crude oil and its fractions. MSc Thesis, Department of Chemistry, Cairo University, Cairo, Egypt.

El-Gendy, N.Sh. (2004) Biodesulfurization potentials of crude oil by bacteria isolated from hydrocarbon polluted environments in Egypt. Ph.D. Thesis, Department of Chemistry, Cairo University, Cairo, Egypt.

El-Gendy, N.Sh. (2015) Biodesulfurization of Petroleum and Its Fraction. In: Advances in Petroleum Engineering. In: Chemical Technology Series. Vol. 4 Advances in Petroleum Engineering II: *Petrochemical* Eds. K.K. Pant, Shishir Sinha, S. Bajpai and J.N. Govil, Studium Press LLC, Houston, USA. Chapter 24, Pp. 655–680.

El-Gendy, N.Sh., Nassar, H.N., Abu Amr S.S. (2014) Factorial design and response surface optimization for enhancing a biodesulfurization process. *Petroleum Science and Technology*. 32(14): 1669–1679.

Enchira Biotechnology Corporation. (2001) Gasoline desulfurization program. DE-FC07–97ID13570. Final Report.

Etemadifar, Z., Derikvand, P., Emtiazi, G., Habibi, M.H. (2014) Response surface methodology optimization of dibenzothiophene biodesulfurization in model oil by nanomagnet immobilized *Rhodococcus erythropolis* R1. *Journal of Materials Science and Engineering B* 4(10): 322–330.

Fang, X.X., Zhang, Y.L., Luo, L.L., Xu, P., Chen, Y.L., Zhou, H., Hai, L. (2006) Organic sulfur removal from catalytic diesel oil by hydrodesulfurization combined with biodesulfurization. Mod Chem Ind 26: 234–238 *(Chinese journal; Abstract in English)*.

Fatahi, A., Sadeghi, S. (2017) *Biodesulphurization* of gasoline by *Rhodococcus erythropolis* supported on polyvinyl alcohol. *Letters in Applied Microbiology*. 64: 370–378.

Fedorak, P.M., Semple, K.M., Vazquez-Duhalt, R., Westlake, D.W.S. (1993) Chloroperoxidase-mediated modifications of petroporphyrins and asphaltenes. *Enzyme and Microbial Technology*. 15: 429.

Feng, J., Zeng, Y., Ma, C., Cai, X., Zhang, Q., Tong, M., Yu, B., Xu, P. (2006) The surfactant Tween 80 enhances biodesulfurization. *Applied and Environmental Microbiology*.72: 7390–7393.

Finnerty W.R., Robinson M. (1986) Microbial desulfurization of fossil fuels. A review. *Biotechnology and Bioengineering Symposium Journal*. 16: 207–221.

Finnerty W.R., Singer M.E. (1984) A microbial surfactant: physiology, biochemistry and applications. *Developments in industrial microbiology*. 25:31–36.

Finnerty, W.R. (1992) The biology and genetics of the genus Rhodococcus. *Biodegradation*. 2: 223–226.

Finnerty, W.R., Robinson, M. (1986) Microbial desulfurization of fossil fuels. A review. *Biotechnology and Bioengineering Symposium* 16:205–221

Fliermans, C.B. (2002) Microbial processing of used rubber. US Patent 6479558.

Folsom B.R., Schieche D.R., Digrazia P.M., Werner J., Palmer S. (1999) Microbial desulfurization of alkylated dibenzothiophenes from a hydrodesulfurized middle distillate by *Rhodococcus erythropolis* I-19. *Applied and Environmental Microbiology*. 65:4967–4972.

Fukumori, K., Matsushita, M., Mouri, M., Okamoto, H., Sato, N., Takeuchi, K., Suzuki, Y. (2006) Dynamic devulcanization and dynamic vulcanization for recycling of crosslinked rubber. Kautschuk Gummi Kunststoffe. 59:405–411.

Fukushima, D., Kim, Y.H., Iyanagi, T., Oae, S. (1978) Enzymatic oxidation of disulfides and thiolsulfinates by both rabbit liver microsomes and a reconstituted system with purified cytochrome P-450. *Journal of Biochemistry*. 83: 1019–1027.

Furuya T., Ishii Y., Noda K., Kirimura K. (2003) Thermophilic biodesulfurization of hydrodesulfurized light gas oils by *Mycobacterium phlei* WU-F1. *FEMS Microbiology Letters*. 221(1):137–142.

Gallagher, J.R., Olson, E.S., Stanley, D.C. (1993) Microbial desulfurization of dibenzothiophene: a sulfur-specific pathway. *FEMS Microbiology Letters*. 107(1): 31–35.

Gary, J.H., Handwerk G.E. (1994) Introduction to petroleum refining. Marcel Dekker, New York.

Gilbert S.C., Morton J., Buchanan S., Oldfield C., McRobert A. (1998) Isolation of a unique benzothiophene desulfurizing bacterium, *Gordonia* sp. strain 213E (NCIMB 40816) and characterization of the desulphurization pathway. *Microbiology*. 144:2545–2553.

Goindi H.K., Saini V.S., Verma P.S., Adhikari D.K. (2002) Dibenzothiophene desulfurization in hydrocarbon environment by Staphylococcus sp. resting cells. *Biotechnology Letters*. 24:779–781.

Gray, K.A., Mrachkoyz, G.T., Squiresy, C.H. (2003) Biodesulfurization of fossil fuels. *Current Opinion in Microbiology*. 6: 229–235.

Grossman M.J., Lee M.K., Prince R.C., Garrett K.K., George G.N., Pickering I. (1999) Microbial desulfurization of a crude oil middle-distillate fraction: Analysis of the extent of sulfur removal and the effect of removal on remaining sulfur, *Applied and Environmental Microbiology*. 65:181–188.

Grossman M.J., Lee M.K., Prince R.C., Garrett K.K., Minak-Bernero V., and Pickering I. (2001) Deep desulfurization of extensively hydrodesulfurized middle distillate oil by *Rhodococcus* sp. Strain ECRD-1. *Applied and Environmental Microbiology*. 67: 1949–1952.

Guchhait, S., Biswas, D., Bhattacharya, P., Chowdhury, R. (2005) Biodesulfurization of model organosulfur compounds and hydrotreated diesel – experiments and modeling. *Chemical Engineering Journal*. 112(1): 145–151.

Gunam, I.B., Yaku, Y., Hirano, M., Yamamura, K., Tomita, F., Sone, T., Asano, K. (2006) Biodesulfurization of alkylated forms of dibenzothiophene and benzothiophene by *Sphingomonas subarctica* T7b. *Journal of Bioscience and Bioengineering*. 101(4):322–327.

Guobin, S., Huaiying, Z., Jianmin, X., Guo, C., Wangliang, L., Huizhou, L. (2006) Biodesulfurization of hydrodesulfurized diesel oil with *Pseudomonas delafieldii* R-8 from high density culture. *Biochemical Engineering Journal*. 27: 305–309.

Gupta, A.K., Ibrahim, S., Al Shoaibi, A. (2016) Advances in sulfur chemistry for treatment of acid gases. *Progress in Energy and Combustion Science*. 54: 65–92.

Hartdegen F.J., Coburn J.M., Roberts R.L. (1984) Microbial desulfurization of petroleum. *Chemical Engineering Progress*. 80:63–67.

Hassan, M.M., Mahmoud, G.A., El-Nahas, H.H. (2007) Reinforced material from reclaimed rubber/natural rubber, using electron beam and thermal treatment. *Journal of Applied Polymer Science*. 104: 2569–2578.

Hegazi A.H., Andersson J.T., El-Gayar M.S. (2003) Application of gas chromatography with atomic emission detection to the geochemical investigation of polycyclic aromatic sulfur heterocycles in Egyptian crude oils. *Fuel Processing Technology*. 85: 1–19.

Heipieper, H.J., Neumann, G., Cornelissen, S., Meinhardt, F. (2007) Solvent-tolerant bacteria for biotransformations in two-phase fermentation systems. *Applied and Environmental Microbiology*. 71: 961–973.

Hitzmann D.O. (1962) Treatment of hydrocarbon. US Pat. 3,069,325.

Holst, O., Stenberg, B., Christiansson, M. (1998) Biotechnological possibilities for waste tire-rubber treatment. *Biodegradation* 9(3–4): 301–310.

Hu, M., Zhao, S., Li, C., Wang, B., Fu, Y., Wang, Y. (2016) Biodesulfurization of vulcanized rubber by enzymes induced from *Gordonia amicalisa*. *Polymer Degradation and Stability*. 128: 8–14.

Hu, M., Zhao, S., Li, C., Wang, B., Yao, C., Wang, Y. (2014) The influence of different Tween surfactants on biodesulfurization of ground tire rubber by Sphingomonas sp. *Polymer Degradation and Stability*, *107*: 91–97.

Huang H.P., Larter S.R., Bowler B.F.J. (2004) A dynamic biodegradation model suggested by petroleum compositional gradients within reservoir columns from the Liaohe basin. *Organic Geochemistry*. 35:299–316.

Huang T., Qiang L., Zelong W., Daojang Y., Jiamin X. (2012) Simultaneous removal of Thiophene and Dibenzothiophene by immobilized Pseudomonas delafieldii R-8 cells. *The Chinese Journal of Chemical Engineering*. 20:47–51.

Ibrahim, M.K., Salama, M.F., Abdel-Moghny, T., El-Shahawy, M.R. (2004) A study on biodesulfurization of an Egyptian crude oil. II- valuation of microbial desulfurization capacity of some local bacterial isolates. *Egyptian Journal of Biotechnology*. 17: 329–335.

Indian Institute of Petroleum, Dehradum, "Project proposal on biodesulfurization of petroleum fractions" (Sep.,1996)

Irani, Z.A., Mehrnia, M.R., Yazdian, F., Soheily, M., Mohebali, G., Rasekh, B. (2011) Analysis of petroleum biodesulfurization in an airlift bioreactor using response surface methodology. *Bioresource Technology*. 102: 10585–10591.

Ishii, Y., Kozaki, S., Furuya, T., Kino, K., and Kirimura, K. (2005) Thermophilic biodesulfurization of various heterocyclic sulfur compounds and crude straight-run light gas oil fraction by a newly isolated strain *Mycobacterium phlei* WU-0103. *Current Microbiology*. 50: 63–70.

Jae, P.S., Lee, I-S., Chang, Y.K., Lee, S.Y. (2003) Desulfurization of dibenzothiophene and diesel oil by metabolically engineered *Escherichia coli*. *Journal of Microbiology and Biotechnology*. 13(4): 578–583.

Jeyachandran, Y.L., Narayandassa, S.K., Mangalaraj, D., Bao, C.Y., Li, W., Liao, Y.M., Zhang, C., Xiao, L.Y., Chen, W.C. (2006) A study on bacterial attachment on titanium and hydroxyapatite based films. *Surface and Coatings Technology*. 201: 3462–3474.

Jiang X., Yang S., Li W. (2013) Biodesulfurization of model compounds and deasphalted bunker oil by mixed culture. *Applied Biochemistry and Biotechnology*. 172:62–72.

Jiang, C.Y., Liu, H. Z., Xie, Y.C., Chen, J.Y. (2002) Isolation of Soil Bacteria Species for Degrading Dibenzothiophene. Chinese Journal of Chemical Engineering., 10, 420 (2002).

Jiang, X., Yang, S., Li, W. (2014) Biodesulfurization of model compounds and deasphalted bunker oil by mixed culture. *Applied Biochemistry and Biotechnology*. 172: 62–72.

Jin, K.K., Jin, W.P. (1999) The biological and chemical desulfurization of crumb rubber for the rubber compounding. *Journal of Applied Polymer Science* 72:1543–1549.

Johnson, S., Monticello, D., Kulpa, C. (1995) Biodesulfurization of. Petroleum by *Rhodococcus erythropolis*. WO Patent 95/16762.

Kabe T.A., Kamatsu K., Ishihara A., Otsuki S., Godo M., Zhang Q., Qian W. (1997) Deep hydrodesulfurization of light gas oil. 1. Kinetics and mechanisms of dibenzothiophene hydrodesulfurization. *Industrial and Engineering Chemistry Research*. 36 :5146–5152.

Kareem, S.A., Aribike, D.S., Nwachukwu, S.C., Latinwo, G.K. (2012) Microbial desulfurization of diesel by *Desulfobacterium indolicum*. *Journal of Environmental Science and Engineering*. 54(1): 98–103.

Kareem, S.A., Aribike, D.S., Susu, A.A., Nwachukwu, S.C.U. (2016) Anaerobic biodesulfurization of kerosene part I: Identifying a capable microorganism. *Chemical Engineering and Process Techniques*. 2(2): 1028.

Kaufman E.N., Borole A., Shong R., Sides J.L., Juengst C. (1999) Sulfur specificity in the bench-scale biological desulfurization of crude oil by *Rhodococcus* IGTS8. *Journal of Chemical Technology and Biotechnology*. 74:1000–1004.

Kaufman E.N., Harkins J.B., Barole, A. (1998) Comparison of batch stirred and electrospray reactors for biodesulfurization of dibenzothiophene in crude oil and hydrocarbon feedstocks. *Applied Biochemistry and Biotechnology*. 73:127–143.

Kawaguchi H., Kobayashi H., Sato K. (2012) Metabolic engineering of hydrophobic *Rhodococcus opacus* for biodesulfurization in oil-water biphasic reaction mixtures. *Journal of Bioscience and Bioengineering*. 113:360–366.

Kilbaine, J. J. (1989) Desulfurization of coal: the microbial solution. *Trends in Biotechnology*. 7: 97–101.

Kilbane, J.J. (2006) Microbial biocatalyst developments to upgrade fossil fuels. *Current Opinion in Biotechnology*. 17: 305–314.

Kilbane, J.J., Le Borgne, S. (2004) Chapter 2 Petroleum biorefining: The selective removal of sulfur, nitrogen, and metals. *Studies in Surface Science and Catalysis*. 151: 29–65.

Kim Y.J., Chang J.H., Cho K.S., Ryu H.W., Chang Y.K. (2004) A physiological study on growth and dibenzothiophene desulfurization characteristics of *Gordonia* sp. CYKS1. *Korean Journal of Chemical Engineering*. 21:436–441.

Kim, B.H., Shin, P.K., Na, J.U., Park, D.H., Bang, S.H. (1996) Microbial petroleum desulfurization. *Journal of Microbiology and Biotechnology*. 6(5): 299–308.

Kim, B.Y., Kim, H.Y., Kim, T.S., Park, D.H. (1995) Selectivity of desulfurization activity of *Desulfovibrio desulfuricans* M6 on different petroleum products. *Fuel Processing Technology*. 43: 87–94.

Kim, H.Y., Kim, T.S., Kim, B.H. (1990a) Degradation of organic sulfur compounds and the reduction of dibenzothiophene to biphenyl and hydrogen sulfide by *Desulfovibrio desulfuricans* M6. Biotechnology Letters. 12: 761–764.

Kim, J.K., Park, J.W. (1999) The biological and chemical desulfurization of crumb rubber for the rubber compounding. *Journal of Applied Polymer Science*. 72(12): 1543–1549.

Kim, T.S., Kim, H.Y., Kim, B.H. (1990b) Petroleum desulfurization by *Desulfovibrio desulfuricans* M6 using electrochemically supplied reducing equivalent. *Biotechnology Letters*. 12: 757–760.

Kirkwood, K.M., Ebert, S., Foght, J.M., Fedorak, P.M., Gray, M.R. (2005) Bacterial biodegradation of aliphatic sulfides under aerobic carbon- or sulfur-limited growth conditions. *Journal of Applied Microbiology*. 99: 1444–1454.

Kirshenbaum I. (1961) Bacteriological Desulfurization of Petroleum. Esso Research and Engineering Company.US Pat. 2,975,103.

Klein J., Catcheside D.E.A., Fakoussa R., Gazso L., Fritsche W., Höfer M., Laborda F., Margarit I., Reham H.J., Reich-Walber M., Sand W., Schacht S., Schmiers H., Setti L., Steinbüchel A. (1999) Biological processing of fossil fuels. *Applied Microbiology and Biotechnology*. 52:2–15.

Kleshchev, S.M. (1989) Method of removing organic sulfur compounds from petroleum products. RU-Patent: 1,505,960.

Klyachko, N.L., Klibanov, A.M. (1992) Oxidation of dibenzothiophene catalyzed by hemoglobin and other hemoproteins in various aqueous-organic media. *Applied Biochemistry and Biotechnology*. 37: 53–68.

Kobayashi M., Onaka T., Ishii Y., Konishi J., Takaki M., Okada H., Ohta Y., Koizumi K., Suzuki, M. (2000) Desulfurization of alkylated forms of both dibenzothiophene and benzothiophene by a single bacterial strain. *FEMS Microbiology Letters*. 187:123–126.

Kobayashi, S., Nakano, M., Goto, T., Kimu ra, T., Schaap, A.P. (1986) An evidence of the peroxidase-dependent oxygen transfer from hydrogen peroxide to sulfides. *Biochemical and Biophysical Research Communications*. 135: 166–171.

Köhler, M., Genz, I.-L., Schicht, B., Eckart, V. (1984) Microbial desulfurization of petroleum and heavy petroleum fractions. 4. Anaerobic degradation of organic sulfur compounds of petroleum. Zentralblatt Fur Mikrobiologie. 139: 239–247.

Konishi J., Ishii Y., Onaka T., Okumura K., Suzuki, M. (1997) Thermophilic carbon sulfur bond targeted biodesulfurization. *Applied and Environmental Microbiology*. 63:3164–3169.

Konishi J., Okada H., Hirasawa K., Ishii Y., Maruhashi K. (2002) Purification and characterization of the monooxygenase catalyzing sulfur-atom specific oxidation of dibenzothiophene and benzothiophene from the thermophilic bacterium *Paenibacillus* sp. strain A11-2. *Biotechnology Letters*. 24 :1863–1867.

Konishi, M., Kishimoto, M., Omasa, I., Katakura, Y., Shioya, S., Ohtake, H. (2005) Effect of sulfur sources on specific desulfurization activity of *Rhodococcus erythropolis* KA251 in exponential fed-batch culture. *Journal of Bioscience and Bioengineering*. 24:49–54.

Kurita, S., Endo. T., Nakamura. H., Yagi. T., Tamiya. N. (1971) Decomposition of some organic sulfur compounds in petroleum by anaerobic bacteria. *The Journal of General and Applied Microbiology*. 17: 185–198.

Labana, S., Pandey, G., Jain, R.K. (2005) Desulphurization of dibenzothiophene and diesel oils by bacteria. *Letters in Applied Microbiology*. 40: 159–163.

Le Borgne S., Quintero R. (2003) Biotechnological processes for the refining of petroleum. *Fuel Processing Technology*. 81:155–169.

Leahy J.G., Colwell, R.R. (1990) Microbial degradation of hydrocarbons in the environment. *Microbiological Reviews*. 54(3): 305–315.

Lee, I.S., Bae, H., Ryu, H.W., Cho, K., Chang, Y.K. (2005) Biocatalytic desulfurization of diesel oil in an airlift reactor with immobilized *Gordonia nitida* CYKS1 cells. *Biotechnology Progress*. 21(3): 781–785.

Li , W., Xing, J., Xiong, X., Huang, J., Liu, H. (2006) Feasibility study on the integration of adsorption/bioregeneration of π-complexation adsorbent for desulfurization. *Industrial Engineering Chemistry Research*. 45(8): 2845–2849

Li, F., Xu, P., Feng, J., Meng, L., Zheng, Y., Luo, L., Ma, C. (2005a) Microbial desulfurization of gasoline in a *Mycobacterium goodie* X7B immobilized-cell system. *Applied and Environmental Microbiology*. 71(1):276–281.

Li, F., Xu, P., Ma, C., Luo, L., Wang, X.S. (2003) Deep desulfurization of hydrodesulfurization-tretaed diesel oil by a facultative thermophilic bacterium *Mycobacterium* sp. X7B *FEMS Microbiology Letters*. 223:301–307.

Li, G., Xing, J.M., Xiong, X.C., Li, W., Gao, H., Liu, H.Z. (2008) Improvement of biodesulfurization activity of alginate immobilized cells in biphasic systems. *Industrial Microbiology and Biotechnology*. 35:145–50.

Li, F., Zhang, Z., Feng, J., Cai, X., Xu, P. (2007a) Biodesulfurization of DBT in tetradecane and crude oil by a facultative thermophilic bacterium *Mycobacterium goodii* X7B. *Journal of Biotechnology*. 127:222–228.

Li, G.Q., Lee, K., Ma, T., Li, S.S., Li, H., Liang, F.L., Liu, R.L. (2006). Isolation and characterization of a benzothiophene desulfurizing bacterium. Huan Jing Ke Xue. 27(6): 1181–1185.

Li, L., Hong, L., Jianhua, Q., Jinjuan, X. (2010) Progress in technology for desulfurization of crude oil. *China Petroleum Processing and Petrochemical Technology.* 12(4): 1–6.

Li, W., Jiang, X. (2013) Enhancement of bunker oil biodesulfurization by adding surfactant. World Journal of Microbiology and Biotechnology. 29(1): 103–108.

Li, W., Tang, H., Liu, Q., Xing, J., Li, Q., Wang, D., Yang, M., Li, X., Liu, H. (2009) Deep desulfurization of diesel by integrating adsorption and microbial method. *Biochemical Engineering Journal.* 44: 297–301.

Li, W.L., Xing, J.M., Xiong, X.C., Liu, H.Z. (2005b) Bio-regeneration of π-complexation desulfurization adsorbents. *Science in China Series B: Chemistry* 48(6): 538–544.

Li, Y., Zhao, S., Wang, Y. (2011) Microbial desulfurization of ground tire rubber by *Thiobacillus ferrooxidans.* Polymer Degradation and Stability. 96(9): 1662–1668.

Li, Y., Zhao, S., Wang, Y. (2012a) Improvement of the properties of natural rubber/ground tire rubber composites through biological desulfurization of GTR. *Journal of Polymer Research.*19: 9864. DOI 10.1007/s10965-012-9864-y

Li, Y., Zhao, S., Wang, Y. (2012b) Microbial desulfurization of ground tire rubber by *Sphingomonas* sp.: A novel technology for crumb rubber composites. *Journal of Polymers and the Environment.* 20: 372–380.

Li, Y.G., Ma, J., Zhang, Q.Q., Wang, C.S., Chen, Q. (2007b) Sulfur-selective desulfurization of dibenzothiophene and diesel oil by newly isolated *Rhodococcus erythropolis* NCC-1. *Chinese Journal of Organic Chemistry.* 25: 400–405.

Li, Y.G., Xing, J.M., Xiong, X.C., Li, W.L., Gao, H.S., Liu, H.Z. (2008) Improvement of biodesulfurization activity of alginate immobilized cells in biphasic systems. *Journal of Industrial Microbiology and Biotechnology.* 35: 145–150.

Lin M.S., Premuzic E.T., Yablon J.H., Zhou W.M. (1996) Biochemical processing of heavy oils and residuum. *Applied Biochemistry and Biotechnology.* 57:659–664.

Linguist, L., Pacheco, M. (1999). Enzyme-based diesel desulfurization process offers energy, CO_2 advantages. *Oil and Gas Journal.* 22:45–48.

Lizama, H.M., Wilkins, L.A., Scott, T.C. (1995) Dibenzothiophene sulfur can serve as the sole electron acceptor during growth by sulfate-reducing bacteria. *Biotechnology Letters.* 17(1): 113–116.

Löffler, M., Straube, G., Straube, E. (1993) Desulfurization of rubber by *Thiobacilli. Biohydrometallurgical Technologies.* 2: 673–280.

Luo M.F., Xing J.M., Gou Z.X., Li S., Liu H.Z., Chen J.Y. (2003) Desulfurization of dibenzothiophene by lyophilized cells of *Pseudomonas delafieldii* R-8 in the presence of dodecane. *Biochemical Engineering.* 13:1–6.

Ma, T., Li, J., Tong, M.Y., Zhang, X.P., Liu, R.L. (2002) Desulfurization of dibenzothiophene by *Rhodococcus* sp. DS-3. *Acta Microbiologica Sinica.* 42: 126–131

Ma, T., Tong, M., Zhang, Q., Liang, F., Liu, R. (2006) Screening, identification of the strain Fds-1 for microbial desulfurization specially and its use in diesel oil desulfurization. *Acta Microbiologica Sinica.* 46:104–110.

Maass, D., de Oliveira, D., de Souza, A. A. U., Souza, S.M. (2014) Biodesulfurization of a System Containing Synthetic Fuel Using *Rhodococcus erythropolis* ATCC 4277. *Applied Biochemistry and Biotechnology*. 174(6): 2079–2085.

Maass, D., Todescato, D., Moritz, D.E., Oliveira, J.V., Oliveira, D., Ulson de Souza, A.A., Guelli Souza, S.M. (2015) Desulfurization and denitrogenation of heavy gas oil by *Rhodococcus erythropolis* ATCC 4277. *Bioprocess and Biosystems Engineering*. 38: 1447–1453.

Maghsoudi, S., Vossoughi, M., Kheirolomoom, A., Tanaka, E., Katoh, S. (2001) Biodesulfurization of hydrocarbons and diesel fuels by *Rhodococcus* sp strain P32C1. *Biochemical Engineering Journal*. 8(2): 151–156.

Malik, K.A. (1978) Microbial removal of organic sulfur from crude oil and the environment: Some new perspectives. Process Biochemistry. 13: 10–13.

Mansuy, D., Valadon, P., Erdelmeier, Y., Lopez-Garcia, P., Amar, C., Girault, J.P. Dansette, P.M. *(1991) Journal of American Chemical Society*. 113: 7825.

Martínez, I., Mohamed, M. E., Rozas, D., García, J. L., Díaz, E. (2016) Engineering synthetic bacterial consortia for enhanced desulfurization and revalorization of oil sulfur compounds. Metabolic Engineering. 35: 46–54.

Matsui T., Onaka T., Tanaka Y., Tezuka T., Suzuki M., Kurane R. (2000) Alkylated benzothiophene desulfurization by *Rhodococcus* sp. strain T09. *Bioscience, Biotechnology, and Biochemistry*. 64:596–599.

McFarland B. (1999) Biodesulfurization. Current Opinion in Microbiology. 2:257–264.

McFarland B., Boron D.J., Deever W., Meyer J.A., Johnson A.R., Atlas, R.M. (1998) Biocatalytic sulfur removal from fuels: Applicability for producing low sulfur gasoline. Critical Reviews in Microbiology. 24:99–147.

Miller, K.W. (1992) Reductive desulfurization of dibenzyldisulfide. *Applied and Environmental Microbiology*.58(7): 2176–2179.

Mingfang, L., Zhongxuan, G., Jianmin, X., Huizhou, L., Jiayong, C. (2003) Microbial desulfurization of model and straight-run diesel oils. *Journal of Chemical Technology and Biotechnology*. 78: 873–876.

Mishra S., Sarma P.M., Lal B. (2004) Crude oil degradation efficiency of a recombinant lux tagged *Acinetobacter baumannii* strain and its survival in crude oil contaminated soil microcosm. *FEMS Microbiology Letters*. 235: 323–331.

Mohamed, M.E., Al-Yacoub, Z.H., Vedakumar, J.V. (2015) Biocatalytic desulfurization of thiophenic compounds and crude oil by newly isolated bacteria. Frontiers in Microbiology. 6: article 112. doi: 10.3389/fmicb.2015.0 0112.

Mohebali, G., Ball, A.S. (2008) Biocatalytic desulfurization (BDS) of petrodiesel fuels. *Microbiology*. 154(8): 2169–2183.

Mohebali, G., Ball, A.S. (2016) Biodesulfurization of diesel fuels – past, present and future perspectives. *International Biodeterioration and Biodegradation*. 110:163–180.

Monot, F., Abbad-Andaloussi, S., Warzywoda, M. (2002) Biological culture containing *Rhodococcus erythropolis* and/or *Rhodococcus rhodnii* and process for desulfurization of petroleum fraction. U.S. Patent No. 6,337,204

Monticello D.J. (2000) Biodesulfurization and the upgrading of petroleum distillates. *Current Opinion in Biotechnology*. 11:540–546.
Monticello D.J., Finnerty W.R. (1985) Microbial desulphurization of fossil fuels. *Annual Review of Microbiology* 24: 99–147.
Monticello, D.J., Haney, III., William, M. (1996) Biocatalytic process for reduction of petroleum viscosity. US Pat. 5,529,930.
Monticello, D.J. (1994) Biocatalytic desulphurization, the biorefining of petroleum fractions. *Hydrocarbon Processing*. 73: 39–45.
Monticello, D.J. (1996) Multistage process for deep desulfurization of a fossil fuel. Energy Biosystem Corporation, The Woodlands, TX. US Patent No. 5,510,265.
Monticello, D.J. (2000) Biodesulfurization and the upgrading of petroleum distillates. *Current Opinion in Biotechnology*. 11: 540–546.
Monticello, D.J., Bakker, D., Finnerty, W.R. (1985) Plasmid-mediated degradation of dibenzothiophene by *Pseudomonas* species. *Applied and Environmental Microbiology*. 49(4): 756–760.
Monticello, D.J., Finnerty, W.R. (1985) Microbial desulfurization of fossil fuels. *Annual Review of Microbiology* 39:371–389.
Monticello, D.J., Kilbane, J.J. (1994). Microemulsion process for direct biocatalytic desulfurization of organosulfur molecules. US Patent 5,358,870.
Morales M, Le Borgne S (2014) Protocols for the isolation and preliminary characterization of bacteria for biodesulfurization and biodenitrogenation of petroleum-derived fuels. In: McGenity TJ, Timmis KN, Nogales B (eds.) Hydrocarbon and lipid microbiology protocols. Springer, Berlin.
Moslemy, P., Guiot, S.R., Neufeld, R.J. (2002) Production of size-controlled gellam gum microbeads encapsulating gasoline-degrading bacteria. Enzyme and *Microbial Technology*. 30:10–18.
Moustafa, Y.M.M., El-Gendy, N.Sh. Farahat, L.A. Abo-State, M.A., El-Temtamy S.A. (2006) Biodesulfurization of Ras Badran crude oil and its constituents with special emphasis on its asphaltene fraction. *Egyptian Journal of Petroleum*. 15 (1): 21–30.
Mozhaev, V.V., Poltevsky, K.G., Slepnev, V.I., Badun, G.A., Levashov, A.V. (1991) Homogeneous solutions of hydrophilic enzymes in nonpolar organic solvents. FEBS Letters. 292: 159.
Mukhopadhyaya M., Chowdhury R., Bhattacharya P. (2006) Biodesulfurization of hydrodesulfurized diesel in a trickle bed reactor-experiments and modeling. Journal of Scientific and Industrial Research. 65:432–436.
Nandi, S. (2010) Biodesulfurization of hydro-desulfurized diesel in an airlift reactor. *Journal of Scientific and Industrial Research*. 69: 543–547.
Nassar (2010) Potentials of polycyclic aromatic sulfur heterocycles (PASHs) degrading microorganisms isolated from different Egyptian hydrocarbon polluted sites. MSc Thesis, Al-Azhar University.
Nassar, H.N. (2015) Development of biodesulfurization process for petroleum fractions using nano-immobilized catalyst; Ph.D. Degree; Al-Azhar University, Cairo, Egypt.

Nassar, H.N., Deriase, S.F., El-Gendy, N.Sh. (2016) Modeling the relationship between microbial biomass and total viable count for a new bacterial isolate used in biodesulfurization of petroleum and its fractions. *Petroleum Science and Technology*. 34(11–12): 980–985.

Nastainzcyk, W., Ruf, H.H., Ullrich, V. (1975) Ligand binding of organic sulfides to microsomal cytochrome P-450. *European Journal of Biochemistry*. 60: 615.

Nezammahalleh, H. (2015) Biodesulfurization of light crude oil using *Bacillus subtilis* Wb600. *Journal of Biology and Today's World* 4 (3): 67–73.

Noda K., Watanabe K., Maruhasi K. (2003a) Isolation of a recombinant desulfurizing 4,6-dipropyl dibenzothiophene in n- tetradecane. *Journal of Bioscience and Bioengineering*. 95(4):354–360.

Noda K., Watanabe K., Maruhasi K. (2003b) Isolation of the *Pseudomonas aeruginosa* gene affecting uptake of dibenzothiophene in n-tetradecane. *Journal of Bioscience and Bioengineering*. 95(5):504–511.

Ohshiro T., Hirata T., Izumi Y. (1995) Microbial desulfurization of dibenzothiophene in the presence of hydrocarbon. *Applied Microbiology and Biotechnology*. 44: 249–252.

Ohshiro T., Izumi Y. (1999) Microbial desulfurization of organic sulfur compounds in petroleum. Bioscience Biotechnology and Biochemistry. 63:1–9.

Ohshiro, T. and Izumi, Y. (2000) Purification, characterization and crystallization of enzymes for dibenzothiophene desulfurization. *Bioseparations*. 9: 185–188.

Okada H., Nomura N., Nakahara T., Saitoh K., Uchiyama H., Maruhashi K. (2003) Analyses of microbial desulfurization reaction of alkylated dibenzothiophenes dissolved in oil phase. *Biotechnology and Bioengineering*. 83(1):489–497.

Okada, H., Nomura, N., Nakahara, T., Maruhashi, K. (2002) Analyses of substrate specificity of the desulfurizing bacterium *Mycobacterium* sp. G3. *Journal of Bioscience and Bioengineering*. 93: 228–233.

Ortiz-Leon, M., Velasco, L., Vazquez-Duhalt, R. (1995) Biocatalytic oxidation of polycyclic aromatic hydrocarbons by hemoglobin and hydrogen peroxide. *Biochemical and Biophysical Research Communications*. 215: 968–973.

Otsuki, S., Nonaka, T., Takashima, N., Qian, W., Ishihara, A., Imai, T., Kabe, T. (2000)

Oxidative desulfurization of light gas oil and vacuum gas oil by oxidation and solvent extraction. *Energy and Fuels*. 14: 1232–1239.

Pacheco, M.A., Lange, E.A., Pienkos, P.T., Yu, L.Q., Rouse, M.P., Lin, Q., Linquist, L.K. (1999) Recent advances in desulfurisation of diesel fuel. National petroleum & refiners association- annual meeting, NPRA AM-99–27, San Antonio. Pp. 1–26.

Pasta, P., Carrea, G., Colonna, S., Gaggero, N. (1994) Effects of chloride on the kinetics and stereochemistry of chloroperoxidase catalyzed oxidation of sulfides. Biochimica et Biophysica Acta. 1209: 203–208.

Payzant J.D., Monatgomery D.S., Strausz O.P. (1988) The identification of homologous series of benzo[b]thiophenes, thiophenes, thiolanes and thianes possessing a linear carbon framework in the pyrolysis oil of Athabasca asphaltene. AOSTRA Res. 4:117–131.

Perry J.J. (1979) Microbial cooxidations involving hydrocarbons. *Microbiological Reviews*. 43: 59–72

Pienkos P.T. (1999) Choosing the best platform for the biotransformation of hydrophobic molecules. In Microbial Biosystems: New Frontiers. Proceedings of the 8th International Symposium on Microbial Ecology, [EDs] by C. R. Bell, M. Brylinsky and P. Johnson-Green. Halifax, Canada: Atlantic Canada Society for Microbial Ecology, pp. 875–888.

Pienkos, P. T. (1998) Choosing The Best Platform for The Biotransformation of Hydrophobic Molecules. Proceedings of the 8th International Symposium of Microbial Ecology. Halifax, Nova Scotia, Canada.

Pifferi, P., Lanzarini, G., Malleuzzi, D. (1990) The Desulfurization ofHeavy Oils and Residua. EU Patent 401922.

Porto, B., Maass, D., Oliveira, J.V., de Oliveira, D., Yamamoto, C.I., Ulson de Souzaa, A.A., Ulson de Souza, S.M.A.G. (2017) Heavy gas oil biodesulfurization by *Rhodococcus erythropolis* ATCC 4277: optimized culture medium composition and evaluation of low-cost alternative media. *Journal of Chemical Technology and Biotechnology*. 92: 2376–2382.

Premuzic, E.T., Lin, M.S. (1999a) Biochemical upgrading of oils. US patent No. 5,858,766.

Premuzic, E.T., Lin, M.S. (1999b) Induced biochemical conversions of heavy crude oils. *Journal of Petroleum Science and Engineering*. 22: 171–180.

Rajan, V.V., Dierkes, W.K., Joseph, R. (2007) Effect of diphenyldisulfides with different substituents on the reclamation of NR based latex products. *Journal of Applied Polymer Science*. 104: 3562–3580.

Rashtchi M., Mohebali G.H., Akbarnejad M.M., Towfighi J., Rasekh B., Keytash, A. (2006) Analysis of biodesulfurization of model oil system by the bacterium strain RIPI-22. *Biochemical Engineering*. 29:169–173.

Reichmuth D.S., Hittle J.L., Blanch H.W., Keasling, J.D. (2000) Biodesulfurization of dibenzothiophene in *Escherichia coli* is enhanced by expression of a *Vibrio harveyi* oxidoreductase gene. *Biotechnology and Bioengineering*. 67:72–79.

Rhee S.K., Chang J.H., Chang Y.K., Chang H.N. (1998) Desulfurization of dibenzothiophene and diesel oils by a newly isolated *Gordona* strain CYKSl. *Applied and Environmental Microbiology*. 64: 2327–2331.

Rhodes, A.K. (1995) Enzymes desulphurizing diesel fuel in pilot plant tests. *Oil and Gas Journal*. 93: 39–40.

Romine, R.A, Snowdon-Swan, L. (1997) Enzymatic devulcanization of vulcanized waste rubber to virgin rubber products. US Patent 5597851.

Romine, R.A., Romine, M.F. (1998) Rubbercycle: a bioprocess for surface modification of waste type rubber. *Polymer Degradation and Stability*.59(1–3):353–358.

Rontani J.F., Bosser-Joulak F., Rambelo E. A., Bertrand J.C., Guisti G., Faure R. (1985) Analytical Study of Asthart Crude-Oil Asphaltenes Biodegradation. *Chemosphere*. 14: 1413–1422.

Rosenberg E., Zuckerberg A., Rubinowitz C., Gutnick D.L. (1979) Emulsifier of Arthrobacter RAG-1: isolation and emulsifying properties. *Applied and Environmental Microbiology* 37:402–408.

Sadouk Z., Hacene H., Tazerouti A. (2008) Biosurfactants production from low cost substrate and degradation of diesel oil by a Rhodococcus strain. Oil and Gas Science and Technology. 63:747–753.

Salama, M.F., Ibrahim, M.K., El-Shahawy, M.R. (2004) A study on biodesulfurization of an Egyptian crude oil. I-evaluation of biodesulfurization capacity of *Rhodococcus rhodochrous* ATCC 53968. *Egyptian Journal of Biotechnology.* 17: 321–328.

Sarret G., Connan J., Kasrai M., Bancroft M., Charrié-Duhaut, Lemoine S., Adam P., Albrecht P., Eybert-Bérard L. (1999) Chemical forms of sulfur in geological and archaeological asphaltenes from Middle East, France, and Spain determined by sulfur K- and L-edge X-ray absorption near-edge structure spectroscopy. Geochimica et Cosmochimica Acta 63(22): 3767–3779.

Sato, S., Honda, Y., Kuwahara, M. (2004) Microbial scission of sulfide linkages in vulcanized natural rubber by a white rot basidiomycete, *Ceriporiopsis subvermispora*. Biomacromolecules.5: 511–515.

Sato, S., Ohashi, Y., Kojima, M., Watanabe, T., Honda, Y., Watanabe, T. (2009) Degradation of sulfide linkages between isoprenes by lipid peroxidation catalyzed by manganese peroxidase. *Chemosphere.* 77(6):798–804.

Schulz H.N., Brinkhoff T., Ferdelman T.G., Hernández Mariné M., Teske A., Jørgensen B.B. (1999) Dense populations of a giant sulfur bacterium in Namibian shelf sediments. Science. 284: 493–495.

Sekar, T., Abdullah, M.A., Khuswant, F., Tanizshq, P. (2016) Novel approach to produce ultra low sulfur diesel fuel by non-hydrodesulfurization process coupled with bio-desulfurization technique to meet environmental standards. *Journal of Environmental Biology*. 37.6: 1451–1456.

Setti L., Lanzarini G., Pifferi P.G. (1995) Dibenzothiophene biodegradation by a Pseudomonas sp. in model solutions. *Process Biochemistry.* 30(8): 721–728.

Setti L., Lanzarini G., Pifferi P.G. (1997) Whole cell biocatalysis for an oil desulfurization process. *Fuel Processing Technology.* 52 :145–153.

Setti, L., Rossi, M., Lanzarini, G. and Pifferi, P.G. (1993): Barrier and carrier effects of n-dodecane on the anaerobic degradation of benzothiophene by Desulfovibrio desulfuricans, Biotechnol. Lett. 15:527–530.

Setti, L., Rossi, M., Lanzarini, G., Pifferi, P.G. (1992) The effect of n-alkanes in the degradation of dibenzothiophene and of organic sulfur compounds in heavy oil by a *Pseudomonas* sp. *Biotechnology Letters.* 14(6): 515–520.

Shahaby, A.F., Essam El-din, K.M. (2017) Desulfurization of crude oil and oil products by local isolated bacterial strains. *International Journal of Current Microbiology and Applied Sciences.* 6(4): 2695–2711.

Singh, A. (2012) How specific microbial communities benefit the oil industry: Biorefining and bioprocessing for upgrading petroleum oil. In: Applied microbiology and molecular biology in oilfield systems. Whitby, C., Skovhus, T.L. (eds.). Pp. 121–178.

Singh, M.P., Kumar, M., Kalsi, W.R., Pulikottil, A.C., Sarin, R., Tuli, D.K., Malhutra, R.K., Verma, R.P., Bansal, B.M. (2009) Method for bio-oxidative desulfurization of liquid hydrocarbon fuels and product thereof. US Patent Appl. 2009/0217571 A1.

Soleimani, M., Bassi, A., Margaritis, A. (2007) Biodesulfurization of refractory organic sulfur compounds in fossil fuels. Biotechnology Advances. 25: 570–596.

Song C., Ma X. (2003) New design approaches to ultra-clean diesel fuels by deep desulfurization and deep dearomatization. *Applied Catalysis B: Environmental.* 41:207–238

Song, C. (2003) An overview of new approaches to deep desulphurization for ultra-clean gasoline, diesel fuel and jet fuel. Catalysis Today. 86: 211–263.

Srivastava, V.C. (2012) An evaluation of desulfurization technologies for sulfur removal from liquid fuels. *RSC Advances*. 2: 759–783.

Stringfellow W.T. (2001) Biological Upgrading of Heavy Oil for Viscosity Reduction. Chevron Research and Technology Company, Richmond CA Texaco Exploration and Production Technology Department, Houston, TX.

Sun, X., Isayev, A.I. (2009) Continuous ultrasonic devulcanisation-comparison of carbon black filled synthetic isoprene and natural rubbers. Gummi Fasern Kunststoffe 62: 291–306.

Takata, T., Yamazaki, M., Fujimori, K., Kim, Y.H., Iyanagi, T., Oae, S. (1983) Enzymic oxygenation of sulfides with cytochrome P-450 from rabbit liver. Bulletin of the Chemical Society of Japan. 56: 2300–2310.

Tanaka Y., Matsui T., Konishi J., Maruhashi, K. (2002) Biodesulfurization of benzothiophene and dibenzothiophene by a newly isolated *Rhodococcus* strain. *Applied Microbiology and Biotechnology*. 59:325–328.

Tang, Q., Lin, S., Cheng, Y., Liu, S.-J., Xiong, J.R. (2013) Enhanced biodesulfurization of bunker oil by ultrasound pre-treatment with native microbial seeds. *Biochemical Engineering Journal.* 77: 58–65.

Tilstra, L., Eng, G., Olson, G.J. and Wang, F.W. (1992) Reduction of sulfur from polysulphidic model compounds by the hyperthermophilic *Archaebacterium Pyrococcus* furiosus. Fuel. 71: 779–783.

Todescato, D., Maass, D., Mayer, D.A., Oliveira, J.V., de Oliveira, D., Guelli Ulson de Souza, S.M.A., Augusto Ulson de Souza, A. (2017) Optimal production of a *Rhodococcus erythropolis* ATCC 4277 biocatalyst for biodesulfurization and biodenitrogenation applications. Applied Chemistry and Biotechnology. DOI 10.1007/s12010-017-2505-5.

Tolosa I., De Mora S., Fowler J., Villeneuve J.P., Bartocci J., Cattini, C. (2005) Aliphatic and Aromatic Hydrocarbons in Marine biota and Coastal sediments from the Gulf and the Gulf of Oman. Marine Pollution Bulletin. 50(12): 1619–1633.

Torkamani, S., Shayegan, J., Yaghmaei, S., Alemzadeh, I. (2008a) Study of a newly isolated thermophilic bacterium capable of Kuhemond heavy crude oil and dibenzothiophene biodesulfurization following 4S pathway at 60 °C. *Journal of Chemical Technology and Biotechnology*. 83(12): 1689–1693.

Torkamani, S., Shayegan, J., Yaghmaei, S., Alemzadeh, I. (2008b) Study of the first isolated fungus capable of heavy crude oil biodesulfurization. *Industrial and Engineering Chemistry Research*. 47(19): 7476–7482.

Torkamani, S., Shayegan, J., Yaghmaei, S., Alemzadeh, I. (2009) Annual report of 'heavy crude oil biodesulfurization project' initiated by Petroleum Engineering Development Company (PEDEC), a subsidiary of National Iranian Oil Company.

Valentine, J.M. (1999). Biodesulfurization of fuels. US Patent 5,874,294.

Van Hamme J.D., Wong E.T., Dettman H., Gray M.R., Pickard M.A. (2003) Dibenzylsulfide metabolism by white rot fungi. *Applied and Environmental Microbiology*. 69(2):1320–1324.

Van Houten, R.T., Lettinga, G. (1996) Biological sulfate reduction with synthesis gas: microbiology and technology. p. 793–799. In: Wijffels, R.H., Buitelaar, R.M., Bucke, C., Tramper, J. (Ed.) Progress in Biotechnology. Vol. 11. Elsevier, Amsterdam.

Vazquez-Duhalt, R., Semple, K.M., Westlake, D.W.S, Fedorak, P.M. (1993b) Effect of water-miscible organic solvents on the catalytic activity of cytochrome c. *Enzyme and Microbial Technology*. 15: 936–943.

Vazquez-Duhalt, R., Westlake, D.W.S., Fedorak, P.M. (1993a) Cytochrome c as a biocatalyst for the oxidation of thiophenes and organosulfides. *Enzyme and Microbial Technology*. 15: 494–499.

Vazquez-Duhalt, R., Westlake, D.W.S., Fedorak, P.M. (1994) Lignin peroxidase oxidation of aromatic compounds in systems containing organic solvent. *Applied and Environmental Microbiology*. 60: 459–466.

Vazquez-Duhalt, R.V., Bremauntz M.D., Barzana E., Tinoco R. (2002) Enzymatic oxidation process for desulfurization of fossil fuels. US Patent 6,461,859.

Vega, B., Montero, L., Lincoln, S. (2008) Control of vulcanizing/devulcanizing behavior of diphenyl disulfide with microwaves as the heating source. *Journal of Applied Polymer Science*. 108: 1969–1975.

Walker J.D., Colwell R.R., Petrakis L. (1975) Microbial petroleum degradation: application of computerized mass spectrometry. *Canadian Journal of Microbiology*. 21:1760–1767

Wang M., Li W., Shi Y., Wang D., Feng H. (2006) Effect of surfactant on biodesulfurization by Corynebacterium sp. ZD-1 in the presence of organic phase. *Journal of Zheijang University Science*. 7:371–375.

Wang P., Krawiec S. (1994) Desulfurization of dibenzothiophene to 2-hydroxybiphenyl by some newly isolated bacterial strains. *Archives of Microbiology*. 161:266–271.

Wang Z., Fingas, M. (2003) Development of oil hydrocarbon fingerprinting and identification techniques. Marine Pollution Bulletin. 47: 423–452.

Watanabe K., Noda K., Maruhashi K. (2003a) Selective cleavage of 10-methyl benzo [b]naphtho[2,1-d]thiophene by recombinant *Mycobacterium* sp. strain. *Biotechnology Letters*. 25 :797–803.

Watanabe K., Noda K., Maruhashi K. (2003b) Enhanced desulfurization in a transposon-mutation strain of *Rhodococcus erythropolis Biotechnology Letters.* 25:1299–1304.

Wolf B.P., Summer L.W., Shields S.J., Nielsen K., Gray K.A., Russell D.H. (1998) Characterization of Proteins Utilized in the Desulfurization of Petroleum Products by Matrix-Assisted Laser Desorption Ionization Time-of-Flight Mass Spectrometry. *Analytical Biochemistry.* 260: 117–127.

Xu, P., Li, F., Yu, J., Ma, C., Zhong, J., Qu, Y., Blankespoor, H.D. (2002) Microbial desulfurization of fuel oil. *Chinese Science Bulletin.* 47(5): 365–369.

Yamada, K.O., Morimoto, M., Tani, Y. (2001) Degradation of dibenzothiophene by sulfate-reducing bacteria cultured in the presence of only nitrogen gas. *Journal of Bioscience Bioengineering* 91: 91–93.

Yang, J., Hu, Y., Zhao, D., Wang, S., Lau, P.C.K., Marison, I.W. (2007) Two-layer continuous-process design for the biodesulfurization of diesel oils under bacterial growth conditions. *Biochemical Engineering Journal.* 37(2): 212–218.

Yao, C., Zhao, S., Wang, Y., Wang, B., Wei, M., Hu, M. (2013) Microbial desulfurization of waste latex rubber with *Alicyclobacillus* sp. *Polymer Degradation and Stability*, 98(9), 1724–1730.

Yee, N., Fein, J.B., Daughney, J.C. (2000) Experimental study of the pH, ionic strength, and reversibility behavior of bacteria- mineral adsorption. Geochimica et Cosmochimica Acta. 64(4): 609–617.

Yu, B., Xu, P., Quan, S., Cuiqing, M. (2006a) Deep desulphurization of diesel oil and crude oils by a newly isolated *Rhodococcus erythropolis* strain. *Applied and Environmental Microbiology.* 72:54–58.

Yu, B., Ma, C.Q., Zhou, W.J., Wang, Y., Cai, X.F., Tao, F., Zhang, Q., Tong, M.Y., Qu, J.Y., Xu, P. (2006b) Microbial desulfurization of gasoline by free whole-cells of Rhodococcus erythropolis XP. FEMS Microbiology Letters. 258: 284–289

Yu, L., Meyer, T.A., Folsom, B.R. (1998) Oil/water/biocatalyst three phase separation process. US Patent No. 5,772,901.

Zeynalov, E., Nagiev, T. (2015) Enzymatic catalysis of hydrocarbons oxidation "in vitro" (review) Chemistry and Chemical Technology. 9(2): 157–164.

Zhang Q., Tong M.Y., Li Y.S., Gao H.J., Fang X.C. (2007) Extensive desulfurization of diesel by *Rhodococcus erythropolis. Biotechnology Letters.* 29:123–127.

Zhang T., Li W., Chen V., Tang H., Li Q., Xing J., Liu H. (2011) Enhanced biodesulfurization by magnetic immobilized *Rhodococcus erythropolis* LSSE8 assembled with nano-γ-AlO. *World Journal of Microbiology and Biotechnology.* 27:299–305.

Zhongxuan G., Huizhou L., Mingfang L. (2002) Isolation and identification of nondestructive desulfurization bacterium. Science in China Series B. 45: 521–531.

Zhongxuan, J.; Hongying, L.; Yongna, Z. and Can, L. (2011): Oxidative desulfurization of fuel oils; *Chinese Journal of Catalysis.* 32:707–715.

Zhou, Z.Y., Zhang, K. (2004) Development situation and prospect of oil fields in China. *Petroleum Exploration and Development.* 31:84–87

Zytner R., Salb A., Brook T., Leunissen M., Stiver W. (2011) Bioremediation of diesel fuel contaminated soil. *Canadian Journal of Civil Engineering.* 28: 131–140.

12

Challenges and Opportunities

List of Abbreviations and Nomenclature

2-HBP	2-Hydroxybipheneyl
4,6-DMDBT	4,6-Dimethyldibenzothiophene
ADN	Adsorptive Denitrogenation
ADS	Adsorptive Desulfurization
ALR	Airlift Reactor
Asph	Asphaltenes
b/d	Barrel/Day
BDS	Biodesulfurization
BDSM	Biodesulfurizing Microorganisms
bpd	Barrels per Day
BTs	Benzothiophenes
CA	Calcium Alginate
CD	Cyclodextrin
CED	Conversion Extraction Desulfurization
CMC	Critical Micelle Concentration
CPO	Chloroperoxidase
CSTR	Continuous Stirred Tank Reactor

DBTs	Dibenzothiophenes
DBTS	Dibenzothiophene Sulfone
DCW	Dry Cell Weight
DTABs	Draft-Tube Airlift Bioreactors
EBC	Energy Biosystems Corporation
EDS	Extractive Desulfurization
ENBC	Enchira Biotechnology Corporation
EOR	Enhanced Oil Recovery
FC	Fuel Consumption
FCC	Fluid Catalytic Cracking
GHG	Greenhouse Gas
HBPSi	2-(2′-Hydroxyphenyl) benzene sulfinate
HDS	Hydrodesulfurization
HDS-diesel oil	Hydrodesulfurized-Diesel oil
HLB	Hydrophilic and Lipophilic Balance
HMW	High Molecular Weight
HSFO	High Sulfur Fuel Oil
IBUs	Immobilized Biomass Units
IGT	Institute of Gas Technology
KACST	King Abdulaziz City for Science Technology
La-NPs	Latex Nanoparticles
LCA	Life Cycle Assessment
LCC	Life Cycle Cost
LSFO	Low Sulfur Fuel Oil
MEOR	Microbial Enhanced Oil Recovery
MNPs	Magnetic Nanoparticles
MPI	Mexican Petroleum Institute
MYO	Myoglobin
NPs	Nanoparticles
ODS	Oxidative Desulfurization
OFP	Oil Fraction Phase
ORNL	Oak Ridge National Laboratory
OSCs	Organosulfur Compounds
OTR	Oxygen Transfer Rate
OUR	Oxygen Uptake Rate
P/V	Power Input Per Volume
PASHs	Polyaromatic Sulfur Heterocyclic Compounds
PEC	Petroleum Energy Center
PECJ	Petroleum Energy Center of Japan
PVA	Polyvinyl Alcohol
RSM	Response Surface Methodology
SBM	Sugar Beet Molasses
SBTR	Stirred Tank Bioreactor
SEM	Scanning Electron Microscope

SRB	Sulfate Reducing Bacteria
SRE	Sulfur Removal Efficiency
SSF	Simultaneous Saccharification and Fermentation
STBR	Stirred Tank Bioreactor
STR	Stirred Tank Reactor
TEM	Transmission Electron Microscope
ULSD	Ultra-Low Sulfur Diesel
US DOE	US Department of Energy
VRTCR	Vertical Rotating Immobilized Cell Reactor
v_{tip}	Impeller Tip Speed
vvm	Vessel Volumes per Minute

12.1 Introduction

Petroleum is a naturally occurring gaseous, liquid, or solid mixture which is chiefly composed of hydrocarbons. After carbon and hydrogen, sulfur is characteristically the third most abundant element in petroleum, ranging from 0.05% to 5% in crude oil, but up to 14% in heavier oil (Speight, 1980; van Hamme *et al.*, 2003; El-Gendy and Speight, 2016). Gasoline, diesel, and nonvehicle fuels comprise 75–80% of the total amount of oil distillation products. A continuous rise in the proportion of sulfur and high sulfur heavy crude oils arriving at refineries and the sophisticated stringent environmental requirements towards ultra-low N-and S-product fuels make resolving the problem of deep desulfurization of commercial petroleum products an urgent task.

Stringent regulations on the sulfur content of transportation fuels have been issued in developed countries, but not yet in developing countries. Nonetheless, deep desulfurization of transportation fuels has drawn global research attention. Thiophenic compounds and their derivatives are the sulfur species most difficult to remove and research on deep desulfurization is targeting the removal of this kind of sulfur compounds.

Reducing the sulfur level in diesel to less than 50 mg/kg (ppm) by the conventional hydrodesulfurization (HDS) process is difficult due to the presence of refractory sulfur compounds such as alkyl dibenzothiophenes (DBTs) with one or two alkyl groups at 4 and/or 6-positions. Moreover, the cost of sulfur removal in industrial factories in the HDS process is expensive, although HDS is considered to be a cost-effective method for fossil fuel desulfurization. Aitani *et al.* (2000) reported that there are more than 35 resid HDS units with a total installed capacity of 1.5 million b/d. Atlas *et al.* (2001) estimated the cost of lowering the sulfur content from 500 to 200 mg/kg to be approximately one cent per gallon. To reduce the sulfur content from 200 to 50 mg/kg, the desulfurization cost would be 4 to 5 times or higher. Furthermore, during the HDS process, due to the saturation of double bonds in olefin and/or aromatic components, excessive HDS of gasoline leads to a decrease in octane number and in diesel, an increase of aromatic compounds and nitrides. In addition, deep HDS of diesels is subject to constraints on fuel density, cetane number, and operation cost. Meanwhile, the catalytic activity decreases as a result of catalyst coking in deep HDS.

Aitani *et al.* (2000) reported that HDS is not considered likely to become economically attractive as a "stand-alone" option. This is mainly due to the fact that the price differential between high sulfur fuel oil (HSFO) and low sulfur fuel oil (LSFO) is not projected to reach levels to support this process. If an alternative desulfurization method other than HDS is to be economic, it must require low capital and deal with large volumes of fuels, most likely on a continuous throughput basis. Moreover, the single most important factor in determining the viability of such an option is the price differential between LSFO and HSFO. Currently, the price differential is about $5 in some areas and is expected to widen in the near future.

Different techniques were proposed for the oxidation method (Etemadi and Yen, 2007; Wang *et al.*, 2007; Zhao, 2009). Catalysts used in oxidation and organic acids (Te *et al.*, 2001; Yu, *et al.*, 2005), heteropolyic acids and their salts (Al-Shahrani *et al.*, 2007; Yang *et al.*, 2007a), and Ti-silica molecular sieve-14 was reported. The oxidant used includes 30% aqueous hydrogen peroxide (Te *et al.*, 2001; Kong *et al.*, 2004; Yu *et al.*, 2005; Al-Shahrani *et al.*, 2007), nitrogen dioxide (Tam *et al.*, 1990), oxygen (Ma *et al.*, 2007), and air (Jeyagowry *et al.*, 2006). Polar organic solvents or ionic liquids were used as extractants. The oxidation desulfurization method bears merits in a mild operation condition without the expense of hydrogen. However, this method becomes disabled in fuels containing too many aromatic components and/or dissolved water.

HDS is currently used for the bulk desulfurization of approximately 50 million gallons of gasoline/d (daily gasoline consumption in the U.S.), but the economics of BDS are the subject of much debate (Foght, 1990; McFarland *et al.*, 1998; Shennan, 1996; Coordinated Research Council, 1998), especially since U.S. refineries west of the Rocky Mountains (30% of U.S. capacity) were reported to be capable of achieving sulfur levels in gasoline of <50 ppm using conventional HDS technology (Coordinated Research Council, 1998). Diesel fuels do not enjoy the same degree of facility in the removal of sulfur species because DBT and hexadecane both lie below the boiling cut range and diesel fuels are more difficult to desulfurize in general.

The conventional HDS method allows up to 90% of sulfur to be extracted from petroleum products. However, to reach to ULS-fuels, this requires higher temperature and pressure, more hydrogen, and expensive catalysts, thus, an increase in the capital and operational costs of production occurs, since the increase of the motor fuel production costs is mainly determined by an increase in the hydrogen partial pressure in the system and a decrease of the unit throughput. This makes it mandatory to search for new, non-traditional processes for S-removal during petroleum refining, especially for the refractory S-compounds that are difficult to be removed by the conventional HDS process.

Oxidative desulfurization (ODS), adsorptive desulfurization (ADS), and biodesulfurization (BDS) can be performed under ambient temperature and pressure, thus, the process costs are expected to be substantially decreased. Biodesulfurization (BDS) might have industrial potential, but a lot of fundamental developments are required for better industrial performance. Concerning diesel, significant improvements of HDS catalysts and reactors have been obtained and refiners have been able to revamp existing HDS units and processes that were initially designed in the 1990s to produce 500 ppm sulfur fuels. High hydrogen consumption and the need to add more catalyst, affect process economics, therefore, new non-hydrogenation alternatives, such as oxidative desulfurization (ODS), physical and reactive adsorption, and extractive desulfurization have been proposed. Among these, the most promising seems to be ODS in which sulfur compounds are oxidized to their corresponding sulfones that can be removed by extraction and/or adsorption. Several oil companies are now at this level of ODS process development and five of these processes are now at the commercialization stage (Chapter 2). The integration of ODS to existing conventional HDS units producing 500 ppm sulfur-containing diesel is estimated to be more cost-effective than revamping (Stanislaus *et al.*, 2010). However, one of the drawbacks of ODS is its requirement for a relatively high cost of oxidizing agents, such as hydrogen peroxide, that are not present in a refinery. Nowadays, some approaches take into account the possible integration of a hydrogen peroxide synthesis in the process and/or production of a selective oxidizing catalyst to avoid the usage of H_2O_2 (Campos-Martin *et al.*, 2010). It is also important to improve the reaction system, especially the mass transfer between the oil and polar phase by applying, for example, the micro-structured reactors. Furthermore, it is highly recommended that the catalyst design should be closely integrated with the reactor design taking into consideration the reaction mechanism, catalytic activity, catalyst life time, and exchangeability (Campos-Martin *et al.*, 2010).

In ODS, adsorption, and extraction alternatives, ULSD fuels are produced.

However, these methods imply the removal of organosulfur compounds (OSCs) leading to a decrease in the calorific value of the fuel and to the generation of waste streams concentrated in sulfur compounds that could, subsequently, be treated using BDS to return the desulfurized hydrocarbons into the fuels or to produce valuable compounds (Kilbane and Stark 2016). Few reports of such combined physical-biological procedures can be found in the literature. For example, Li *et al.* (2006, 2008b, 2009b) reported the adsorption of DBT compounds on different adsorbents followed by the action of a *Pseudomonas*-desulfurizing strain to regenerate the adsorbents (Chapter 10).

The presence of nitrogen compounds in some petroleum cuts inhibit deep desulfurization (Chapter 4). Besides sulfur and nitrogen compounds, predominantly H_2S and NH_3, produced during the catalytic reforming of fuels can poison the precious metal electrodes and catalysts in fuel cells. Adsorptive desulfurization and denitrogenation (ADS/ADN) as a method for the selective removal of sulfur and nitrogen compounds from liquid hydrocarbon fuels have attracted much interest (Chapter 2, Chapter 4, Chapter 5). They are cheap, simple, applicable, and environmentally friendly processes. A new proposed option for refinery applications is to use adsorptive desulfurization and a denitrogenation unit after existing HDS units (ADS/ADN/HDS). This combination is able to deeply remove sulfur and nitrogen compounds from hydrocarbon fuels. Fuel streams contain not only nitrogen and sulfur compounds, but also aromatic compounds which have a structure similar to the coexisting refractory nitrogen and sulfur compounds. Therefore, new challenges are present in developing adsorbents for the selective removal of sulfur and nitrogen. Several materials, such as zeolite-based materials, silica-based materials, activated carbons, metal oxides, metal sulfides, and silica gel, have been reported as adsorbents for the selective removal of sulfur and nitrogen containing compounds from liquid hydrocarbon fuels.

The challenge for deep desulfurization of diesel fuels is the difficulty of removing refractory sulfur compounds, particularly 4,6-dimethyldibenzothiophene (4,6-DMDBT), with conventional HDS processes. The problem is exacerbated by the inhibiting effect of polyaromatics and nitrogen compounds which exist in some diesel blend stocks upon deep HDS.

Refineries are in a quest to lower sulfur content in petroleum products. HDS has been widely used in refineries because of its substantial capability to remove sulfur. However, new legislations driven by environmental friendly operations require ultra-low sulfur content in petroleum distillates.

It is difficult to achieve low sulfur contents through HDS due to low reactivity of HDS catalysts towards HMW-S-compounds. An increase in the reactor size and hydrogen consumption, under high operation conditions, is required to achieve high levels of desulfurization. Therefore, HDS may be very costly to employ in order to achieve the ultra-low sulfur fuel.

ADN as a pretreatment step before the HDS-unit would enhance the HDS process (ADN/HDS/BDS). Moreover, ADS as complementary step after the HDS-unit can lower the diesel S-content to < 10 ppm. Since removing refractory sulfur compounds, such as benzothiophenes (BT), dibenzothiophenes (DBT's), and their alkylated derivatives, which are usually, in large amounts, present in diesel fuel is met with high degree of difficulty in the HDS method, the adsorption method is a suitable alternative, gaining the advantages of ADS process, including low-energy consumption, availability of regeneration of spent adsorbent, and ambient operation temperature and pressure.

The above discussed technologies of sulfur removal from refinery streams leads to a wealth of research topics. Not all of them are challenging, of course. Several topics have the character of demonstration of known science. Only an integrated approach (catalyst selection, reactor design, process configuration) will lead to efficient desulfurization processes that will produce fuels with zero sulfur emission. In HDS based technologies, less room for breakthroughs exists. Noble metal-based catalysts with high sulfur tolerance and sufficient kinetics in sulfur removal seem to be the most challenging option for improvement of HDS based technologies. The application of catalytic distillation, in combination with HDS sulfur removal, is also attractive.

Reactive adsorption is applicable at almost all points of refinery where desulfurization is required. Due to very high flexibility in the reactor design and process conditions, reactive adsorption can easily be adapted to streams with different properties and compositions. The ability to remove the last ppm of sulfur also makes reactive adsorption attractive as the finishing step in the fuel production, for instance, for a gasoline stream from the reformer. Development of reactive adsorption-based desulfurization technologies should first be oriented towards the rational selection of sorbent/catalysts formulation, taking into account the nature of the sulfur compounds, hydrocarbon composition, and final product specifications. Sorbent development should be accompanied by evaluation of the kinetics of the sulfur removal processes and design of the optimal reactor configuration.

Selective oxidation of sulfur compounds into hydrocarbons and volatile sulfur products might be also attractive for desulfurization. Substitution of expensive hydrogen, which is normally used in desulfurization, by air will bring about high economic benefits. Thermodynamic feasibility of this process in the presence of different catalytic systems should be evaluated. Based on that, perspective catalyst compositions should be selected and their applicability has to be demonstrated by the experiments.

Other approaches to sulfur removal, such as extractive desulfurization (EDS), look less attractive since the involvement of an additional phase leads to large plants and limited efficiency. The same applies for oxidative extraction in which in addition to the solvent, an oxidant is required, although recycling might reduce the amounts consumed.

BDS has been studied for over 50 years (ZoBell, 1953; Bachmann *et al.*, 2014). Two patents were issued to the Texaco Development Corporation aiming to the BDS of crude oil, where Strawinski (1950) suggested the use of "diverters", such as carbohydrates, for the purpose of sparing the carbon of the petroleum from attack by desulfurizing organisms, which recorded a reduction of about 12.5% of the sulfur in a crude oil sample. The patent also stressed the use of media essentially free of sulfur compounds for isolation procedures and for the treatment of the oil. The second patent claims a method for converting possible oxidized forms of sulfur, resulting from desulfurization, to sulfide by the use of sulfate reducing bacteria (SRB) (Strawinski, 1951).

In 1990, the Institute of Gas Technology (IGT), located in the city of Chicago, USA, isolated indigenous microorganisms from soil samples contaminated with hydrocarbon sulfur. The soil samples were placed in a continuous flow bioreactor where they were exposed to mutagenic agents (NTG -1 methyl-3-nitro-1-nitrosoguanidine) (Kilbane and Bielaga 1990). After the mutagenic action, the remaining species specifically broke the C-S bond (Kilbane and Jackowsky 1992). This new group of microorganisms was called IGTS7. The naming was given due to the fact that seven different types of colonies were obtained from this process. Among these colonies, two were highly capable of selective BDS, but they presented a small population and slow growth. These two new mutant strains were named *Rhodococcus rhodochrous* IGTS8 and *Bacillus sphaericus* IGTS9 (Ohshiro *et al.* 1994). In 1997, an analysis of the nucleotide sequence of the 16S rRNA operon revealed that the strain previously named *R. rhodochrous* IGTS8 presented high similarity with *R. erythropolis* IGTS8. Since then, studies have been conducted for isolating other biodesulfurizing strains and molecular biology with an attempt to have a better understanding on the enzymatic machinery and genetics involving specific sulfur removal. The BDS technology was developed by the USA based Enchira Biotechnology Corporation (ENBC) (formerly Energy Biosystems

Corporation) and patented in 1998. There has been worldwide interest in biodesulfurization as demonstrated by the number of patents and patent holders. The top five patent holders are ENBC (21 patents), the Japanese Petroleum Energy Centre (4), US based Institute of Gas Technology (4), the Korean Advanced Institute of Science and Technology (3), and Exxon Research & Engineering Company (2) (Bachmann *et al.*, 2014).

However, there are still some unsolved aspects, such as stability, lifetime, inhibition effects, and high costs of biocatalyst, reactor design, quantity of water in the media, and separation of aqueous organic phases, low enzyme activity, and low mass transfer rates (Monticello, 2000; Caro *et al.*, 2008; Bhatia and Sharma, 2010a; Srivastava, 2012a; Alves *et al.*, 2015; Kilbane, 2017).

Various reviews on biodesulfurization have been published over the past decade (Grossman, 1996; McFarland, 1999; Ohshiro and Izumi, 1999; Aitani *et al.*, 2000; Monticello, 2000; Gray *et al.*, 2003; Le Borgne and Quintero, 2003; Montiel *et al.*, 2009; Srivastava, 2012b; Nuhu, 2013; Ayala *et al.*, 2016; Kilbane and Stark, 2016) focusing on fundamental and applied aspect of BDS processes.

However, most of the published studies performed for the fundamental understanding of the BDS process have been done using DBT and its alkylated forms both in a one phase system (i.e. aqueous only) and a biphasic system (i.e. aqueous and model oil of DBT dissolved in organic solvent, usually n-tetradecane or n-hexadecane), as representative model compounds for organosulfur compounds (OSCs) in the middle distillate fraction (i.e. diesel fuels) (Monticello and Finnerty, 1985; Chang *et al.*, 2001; Guobin *et al.*, 2005; Aribike *et al.*, 2009; Bandyopadhyay *et al.*, 2013a). The BDS of real oil fractions (middle distillates, gas oil, and diesel) has also been reported and low to ultralow-sulfur levels (less than 50 ppm) were achieved depending on the initial sulfur content and whether the fractions had been previously hydrotreated (Chapter 11).

Biological removal of sulfur has several limitations that prevent it from being applied today. The metabolism of sulfur compounds is typically slow compared to chemical reactions. Also, large amounts of biomass are needed (*typically 2.5 g biomass per g sulfur*) and biological systems must be kept alive to function, which can be difficult under the variable input conditions found in refineries (Nehlsen *et al.*, 2005). The desulfurization rate stops before the complete removal of sulfur compounds. Several authors have reported that the BDS process by growing cells may be deactivated by the accumulation of 2-HBP (Irani *et al.*, 2011b)

For cultures of *R. erythropolis* IGTS8, an excess oxygen transfer rate or operating conditions favoring high hydrodynamic shear may lead to negative effects of oxidative or hydrodynamic stress in a stirred tank bioreactor (STBR). These conditions should not be exceeded in scale-up and operation, in order to obtain a satisfactory bioprocess performance (Gomez *et al.*, 2015).

There are several bottlenecks, such as low rates of microbial metabolism of sulfur heterocycle compounds, requirement of biocatalysts with the ability to remove sulfur from multiple sulfur heterocycle substrates, need for huge amounts of biocatalyst biomass, cost of biocatalyst, reactor design, mass transfer across the oil–water interface, oil–water separation, and biocatalyst recovery, which limit the commercialization of the

BDS process (Mohebali and Ball 2008; Srivastava 2012b). Thus, for future BDS industrial application, it is necessary to overcome some technical issues, such as increasing the process efficiency, reducing the biocatalyst and nutritional environment costs, and the development of a bioreactor and phase separation (Schilling *et al.*, 2002; Yang and Marison, 2005; Yang *et al.*, 2007b).

For commercialization, the desired rate of BDS reported was 20 μmole DBT/min/g DCW (Monticello, 2000; Nazari e al., 2017), 3 mM S/g DCW/h (Singh, 2015), and, in another report, 1.2–3 mM/g DCW/h (Srivastava and Kumar, 2008). The maximum rate achieved to date is 320 μM S/g DCW/h (Kilbane, 2006; Singh, 2015).

Furthermore, some studies also showed that BDS requires approximately two times less capital and 15% less industrial operating costs in comparison with traditional HDS (Kaufman *et al.* 1998; Linguist and Pacheco 1999; Pacheco *et al.* 1999). It was estimated that the investment cost of a biological desulfurization plant is only $2/3^{rd}$ of the traditional hydrodesulfurization and operating costs and carbon dioxide emissions are reduced by 15% and 70–80%, respectively (Aggarwal *et al.*, 2013). However, there are some limitations to this alternative process, such as the cost of the culture medium, representing 30–40% of the total amount, namely the cost of the carbon source (carbohydrate content, which usually is commercial glucose), which makes the process more expensive (Ma *et al.* 2006a; Zhang *et al.* 2007a) and justifies the search for inexpensive and renewable raw materials, such as recycled paper sludge hydrolysate (Alves *et al.* 2008a), carob pulp (Silva *et al.* 2013), or sugar beet molasses (Alves and Paixão 2014).

Numerous attempts to develop improved desulfurization biocatalysts have included the isolation of a plethora of BDS cultures (Kobayashi *et al.*, 2000; Bordoloi *et al.*, 2014; Buzanello *et al.*, 2014; Chauhan *et al.*, 2015; Derikvand *et al.*, 2015; Mohamed *et al.*, 2015; Wang *et al.*, 2015), increasing the copy number of desulfurization genes (Yoshikawa *et al.*, 2001; Konishi *et al.*, 2005a), changing translational sequences (Nakashima and

Tamura, 2004), changing the promoter of the dsz operon (Nakashima and Tamura, 2004; Tao *et al.*, 2011a), altering the gene order in the dsz operon (Reichmuth *et al.*, 2004; Tao *et al.*, 2011a; Chauhan *et al.*, 2015; Martinez *et al.*, 2016a), expressing dsz genes in alternative hosts (Noda *et al.*, 2003a; Park *et al.*, 2003; Aliebrahimi *et al.*, 2015), three-dimensional structure analysis of enzymes combined with site directed mutagenesis (Oshiro *et al.*, 2007), adding surfactants (Dinamarca *et al.*, 2014), immobilization and coating cells with nanoparticles (Guobin *et al.*, 2005; Derikvand *et al.*, 2014), modifying the composition of growth media (Aggarwal *et al.*, 2011; Teixeira *et al.*, 2014; Alves *et al.*, 2015; Martinez *et al.*, 2015), immobilization of cells (Dinamarca *et al.*, 2014; Derikvand *et al.*, 2014), bioreactor design (Wang and Krawiec, 1996; Derikvand *et al.*, 2015), combining ultrasonic oxidative desulfurization with biodesulfurization (Bhasarkar *et al.*, 2015), adding ion exchange resins (Abin-Fuentes *et al.*, 2013), adding cyclodextrins (Bhasarkar *et al.*, 2015), and employing directed evolution (Arensdorf *et al.*, 2002; Pan *et al.*, 2013).

This chapter discusses the challenges that come with withdrawing the commercialization of BDS and the prospective to overcome these drawbacks. It also covers the range of potential applications of BDS as alternative and/or complimentary technology for conventional applied desulfurization methods. Moreover, there are efforts to improve the performance and economics of a BDS process and promote the awareness of alternative applications of BDS enzymes.

12.2 New Strains with Broad Versatility

There are three known pathways that attack the DBT molecule:

Kodama pathway: Analogous to the naphthalene degradation pathway (Kodama *et al.*, 1973), it begins with the dihydroxylation of the peripheral aromatic ring of the DBT molecule, followed by its cleavage. This results in the accumulation of 3-hydroxy-2-formyl-benzothiophene as a water-soluble end product with lower carbon content than DBT, but with the sulfur atom intact.

Van Afferden pathway: This is another ring-destructive pathway that results in the complete mineralization of DBT. Van Afferden *et al.* (1993) isolated *Brevibacterium* sp. DO capable of using DBT for growth as the sole source of carbon, sulfur, and energy. During DBT mineralization, first there is an oxidation into DBT sulfone, then an aromatic dioxygenase causes the rupture of the thiophenic ring to form 2,3-dihydroxybiphenyl-2′-sulfinate DBT. Finally, a second action of the dioxygenase opens the 2,3-dihyroxybenzene nucleus, resulting in production of sulfite, which is oxidized to sulfate, and benzoate, which is mineralized into CO_2 and H_2O.

The ring-destructive pathways may be valuable in biodegradation of DBT in the environment. However, they are not commercially useful for the petroleum industry since the destruction of the carbon skeleton of sulfur compounds by the bacteria reduces the fuels' calorific value.

4S pathway: This is the third and most valuable to the petroleum industry. It owes its name to the formation of four sulfur compounds during the metabolism of DBT. First, a DBT monooxygenase (DszC) catalyzes two oxidation steps and converts DBT into dibenzothiophene sulfone (DBTS). Then, a second mono-oxygenase, DBT sulfone mono-oxegenase (DszA), turns DBTS into 2-(2′-hydroxyphenyl) benzene sulfinate and, finally, a liase, 2′-hydroxybiphenyl-2-sulfinate desulfinase (DszB), promotes the removal of the sulfur, resulting in the formation of 2-hydroxybiphenyl and sulfite. There is a fourth enzyme, FMN-reductase (DszD), with an indirect yet important role since it regulates the activity of the monooxygenases by regulating the levels of reduced flavin (Nomura *et al.*, 2005). This pathway is especially important because with it the microorganisms are able to remove sulfur from DBT and use it as a sulfur source, but they are incapable of destroying the carbon skeleton and maintaining most of the calorific value of the fossil fuel (Wang and Krawiec, 1996). Several studies have been conducted using bacterial genera capable of using 4S pathway for BDS, which include *Arthrobacter*, *Agrobacterium*, *Brevibacterium*, *Klebsiella*, *Mycobacterium*, *Nocardia*, *Paenibacillus*, *Pseudomonas*, *Xanthomonas*, and, the two most important as of now, *Gordonia* and *Rhodococcus* (Clark and Kirk, 1994; Tanaka *et al.*, 2002; Alves *et al.*, 2007; Alves and Paixão, 2011; El-Gendy *et al.*, 2014; Nassar *et al.*, 2016).

DBT has been generally accepted as a model heterocyclic organic sulfur compound in most BDS studies. It has been reported that *Gordonia* sp. CYKS1, *Nocardia* strain CYKS2, *Rhodococcus erythropolis* IGTS8, *Rhodococcus erythropolis* D-1, and *Corynebacterium* sp. SY1 could selectively remove sulfur atoms from DBT under aerobic conditions without destroying the carbon skeleton into low-carbon-number hydrocarbons (Gallagher *et al.*, 1993; Izumi *et al.*, 1994; Kayser *et al.*, 1993; Ohshiro *et al.*, 1995; Piddington *et al.*, 1995; Chang *et al.*, 1998; Rhee *et al.*, 1998). Alkylated DBT can be desulfurized by *Paenibacillus* (Konishi *et al.*, 1997), *Mycobacterium* sp., *Pseudomonas* sp. (Nekodzuka *et al.*, 1997), *Arthrobacter* sp. (Lee *et al.*, 1995), *Pseudomonas* sp. (Kropp *et al.*, 1997), and *Brevibacillus invocatus* (Nassar *et al.*, 2013). The specific desulfurization activity of *R. erythropolis* IGTS8 ranged between 0.58 to 1.2 μmole 2-HBP/min/g DCW depending on the enrichment procedure before its application (Oldifield *et al.*, 1997).

Table 12.1 summarizes most of the isolated bacteria capable of selectively desulfurizing DBT.

The other limitation which might decrease or hinder the overall rate of the BDS process, including substrate acquisition, is the need for reducing equivalents and the preference of the enzymes for specific substrates (Folsom *et al.*, 1999; Gray *et al.*, 2003). Generally, dibenzothiophene and C1-dibenzothiophene are preferentially attacked, followed by the highly-alkylated molecules (Cx-dibenzothiophene). The position of alkylation also influences the BDS rate. Usually, alkylation near the sulfur (e.g. 4-methyl-dibenzothiophene or 4,6-Dimethy-dibenzothiophene) would lower the biodesulfurization rate. However, *Sphingomonas* strain expressed the opposite trend (Monticello *et al.*, 1985). Moreover, the selective BDS of BT and DBT follows two distinctive pathways. Thus, it is desirable to reach for bacterial strains that possess the associated genes for all these pathways.

DBT and its alkylated derivatives, mostly those with substitutions in the 4 and 6 positions, are the most recalcitrant compounds to desulfurize, that is if ULSD fuels are to be produced from 500 ppm sulfur-containing streams obtained from conventional HDS units. A clear advantage of BDS is that most of the microorganisms following the 4S pathway can desulfurize these recalcitrant molecules although the desulfurization rate is affected by the position and degree of alkylation (Mohebali and Ball 2016). Recent studies have shown that mass transfer issues also limit the desulfurization rate of alkylated DBTs by *Pseudomonas putida* in biphasic mediums (Boltes *et al.*, 2013), while a complete conversion of DBT, 4-methyl DBT, and 4,6-dimethyl DBT was obtained in an aqueous medium where conversions of 38%, 19.5%, and 16.5% were obtained for the same compounds in a biphasic medium.

Substrate competitive inhibition has been observed during the desulfurization of mixtures of DBT and 4,6-dimethyl DBT, representing a limiting

Table 12.1 Selective Biodesulfurization of DBT.

Bacterium	DBT-BDS efficiency	References
Achromobacter sp.	90% of 921.30 μg/mL DBT	Bordoloi *et al.* (2014)
Agrobacterium MC50I	19.92% of 0.271 mM DBT	Constanti *et al.* (1994)
Agrobacterium sp. MC501	20% of 0.3 mM DBT	Constanti *et al.* (1994)
Arthrobacter sp. ECRD-1	97.14% of 0.35 mM DBT	Lee *et al.* (1995)
Arthrobacter sulfures	85% of 2 mM DBT	Labana *et al.* (2005)
Brevibacillus invocatus C19 (accession no. KC999852)	66.85% of 1000 ppm DBT	Nassar *et al.* (2013)
Candida parapsilosis NSh45	70% of 1000 ppm DBT	El-Gendy (2004)
Corynebacterium sp. P32C1	Complete removal of 0.25 mM DBT	Maghsoudi *et al.* (2000)
Corynebacterium sp. P32C1	16.7 mmol 2-HBP/kg DCW/h	Maghsoudi *et al.* (2000)
Corynebacterium sp. SY1	87.84% of 0.27 mM DBT	Omori *et al.* (1992)
Cryptococcus laurentii	53.2 % of 250 ppm DBT	Kaewboran (2005)
Entreobacter sp. strain NISOC-03	64% of 0.8 mM DBT	Papizadeh *et al.* (2017)
G. alkanivorans 1B	77% of 200 μM DBT	Alves *et al.* (2008b)
Gordonia alkanivorans 1B	1.03 μmol/g DCW/h	Alves *et al.* (2005)
Gordonia alkanivorans R1P190A	56.34 μM 2-HBP/g DCW/h	Mohebali *et al.* (2007)
Gordonia desulfuricans strain 8N	3.5 μmol/g DCW/h	Akhtar *et al.* (2016)
Gordonia sp. CYKS1	0.917 μmol DBT/g DCW/h	Rhee *et al.* (1998)
Gordonia sp. JDZX5	Complete removal of 0.3 mM DBT	Feng *et al.* (2016)
Gordonia sp. ZD-7	92.85% of 2.8 mM DBT	Li *et al.* (2006)
Lysinibacillus sphaericus DMT-7	60% of 0.2 mM DBT	Bahuguna *et al.* (2011)
Microbacterium sp. NISOC-06	94.8% of 1 mM DBT	Papizadeh *et al.* (2010)
Microbacterium sp. *strain* ZD-M2	11 mM 2-HBP/kg DCW/h	Chen *et al.* (2008a)

Microbacterium strain ZD-M2	Complete removal of 0.2 mM DBT	Li *et al.* (2005)
Mixed culture AK6	0.6 μM 2-HBP/g DCW/h	Ismail *et al.* (2016)
Mycobacterium goodii X7B	90% of 0.5 mM DBT	Li *et al.* (2007a)
Mycobacterium phlei SM120–1	0.17 μmol 2-HBP/g DCW/min	Srinivasaraghavan *et al.* (2006)
Mycobacterium sp. ZD-19	Complete removal of 0.5 mM DBT	Chen *et al.*, (2008b)
Nocardia asteroids	26% of 0.2 mM DBT	Olson (2000)
Nocardia gluberula R-9	95% of 0.2 mM DBT	Jiang *et al.* (2002)
Nocardia sp. strain CYKS2	Complete removal of 0.2 mM DBT	Chang *et al.* (1998)
Pantoea agglomerans D23W3	93% of 100 ppm DBT	Bhatia and Sharma (2010a)
Pedomicrobium sp.	Complete removal of 500 ppm DBT	Ercole *et al.* (1997)
Pseudomonas delafieldii R8	83.33% of 0.18 mM DBT	Jiang *et al.* (2002)
Pseudomonas putida CECT 5279	6.3 mM 2-HBP/kg DCW/h	Caro *et al.* (2007)
R. erythropolis IGTS8	35 μmol DBT/g DCW/h	Oldfield *et al.* (1997)
R. erythropolis R1	45 μmol DBT/g DCW/h	Etemadifar *et al.* (2014)
Rhodococcus erythropolis FSD2	63% of 0.5 mM DBT	Zhang *et al.* (2007)
Rhodococcus erythropolis HN2 (accession no. KF018282)	80% of 1000 ppm DBT	Nassar *et al.* (2017)
Rhodococcus erythropolis IGTS8	72 μmol DBT/g DCW/h	Kilbane and Le Borgne (2004)
Rhodococcus erythropolis IGTS8	50 % of 1000 ppm DBT	Nassar *et al.* (2013)
Rhodococcus erythropolis NCC1	85.7% of 0.2 mM DBT	Li *et al.* (2007b)
Rhodococcus rhodochrous IGTS8 (ATCC 53968)	6.1 mmol 2-HBP /kg DCW/h	Honda *et al.* (1998)
Rhodococcus sp. ECRD1	75% of 0.6 mM	Lee *et al.* (1995)
Stenotrophomonas sp. NISOC-04	96%% of 0.05 M DBT	Papizadeh *et al.* (2011)
Xanthomonas MC701	15.86% of 0.271 mM	Constanti *et al.* (1994)

factor for the desulfurization of real oil fractions (Chen *et al.* 2008). Zhang *et al.* (2013) have shown that the desulfurization rates decreased to 75.2 % in a DBT/4,6-dimethyl DBT mixture, 64.8 % in a DBT/4-methyl DBT mixture, and 54.7 % in a DBT/4- methyl DBT/4,6-dimethyl DBT mixture (considering that DBT desulfurization is 100%). These differences were due to an apparent competitive inhibition of substrates. Most of the studies with alkylated DBT have been performed with axenic cultures.

Despite the significant removal efficiencies achieved, the level of desulfurization is still insufficient to meet the required sulfur levels for all fuels. This is attributed to the highly substrate-specific reactions which are not convenient for a real feed containing species of organosulfur compounds (OSCs). Consequently, the BDS process requires enzymes with a very broad substrate range of the dsz system (Monticello, 2000). This bottleneck can be solved by isolating microbial stains with a broad range of OSCs. Figure 12.1 illustrates the steps for enriching and isolating well adapted

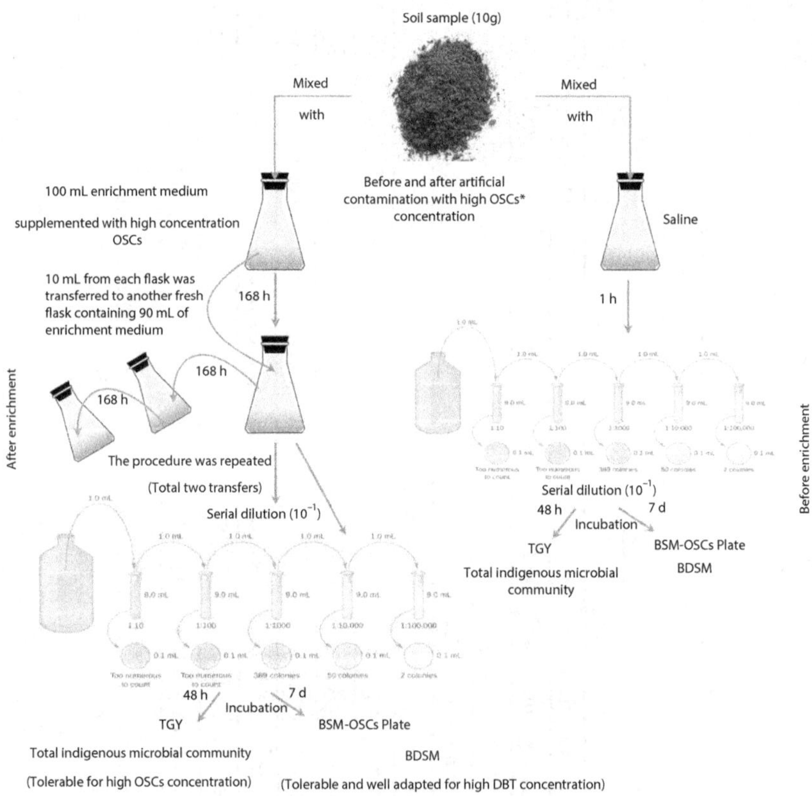

Figure 12.1 Enrichment and Isolation of Tolerable and Efficient BDSM.
*OSCs might be DBT and/or a mixture of different recalcitrant theophoric compounds.

microbial strains with wide versatility to high concentrations of different PASHs. Table 12.2 summarizes most of the isolated bacteria capable of selectively desulfurizing different forms of thiophenic compounds, including thiophenes (Ths), benzothiophenes (BTs), and dibenzothiophenes (DBTs).

The BDS yield is low because it takes a long time to be completed and because of the bioavailability problem. This can be overcome by using electrokinetics, enabling faster OSCs transfer to the bacteria, and/or enhancing the mobility of bacteria. Electrokinetics were reported to reduce the process time and accelerate the BDS of model oil (DBT in n-hexadecane) using resting cells of *Rhodococcus erythropolis* PTCC1767 or *Bacillus subtilis*

Table 12.2 Biodesulfurizing Microorganisms (BDSM) with Broad Versatility.

Bacterium	Reference
Bacillus subtilis WU-S2B	Kirimura *et al.* (2001)
Brevibacillus invocatus C19 (accession no. KC999852)	Nassar *et al.* (2013)
Candida parapsilosis NSh45	El-Gendy (2004)
G. alkanivorans 1B	Alves *et al.* (2008b)
Gordonia alkanivorans 1B	Alves *et al.* (2005)
Gordonia alkanivorans R1P190A	Mohebali *et al.* (2007a)
Gordonia desulfuricans strain 8N	Akhtar *et al.* (2016)
Gordonia sp. CYKS1	Rhee *et al.* (1998)
Lysinibacillus sphaericus DMT-7	Bahuguna *et al.* (2011)
Microbacterium strain ZD-M2	Li *et al.* (2005)
Mycobacterium phlei GT1S10	Kayser *et al.* (2002); Srinivasaraghavan *et al.* (2006)
Mycobacterium sp. G3	Okada *et al.*, 2002
Mycobacterium sp. X7B	Li *et al.*, 2003
Mycobacterium sp. ZD-19	Chen *et al.*, 2008
Nocardia gluberula R-9	Jiang *et al.*, 2002
Paenibacillus sp. A11-2	Konishi *et al.*, 1997
Pantoea agglomerans D23W3	Bhatia and sharma (2010a)
Recombinant Rhodococcus erythropolis KA251	Kobayashi *et al.* (2000)
Rhodococcus sp. ECRD1	Lee *et al.* (1995)

DSMZ 3256, which are capable of BDS via the 4S-pathway (Boshagh et al., 2015). Recently, Bhasarkar et al. (2015) and Agarwal et al. (2016) reported the application of ultrasonic enhancement sulfur-specific BDS through the physical and chemical effects of ultrasonics by the generation of intense micro-turbulence in the medium and the generation of highly reactive oxidizing radicals (O• and •OH) from the thermal dissociation of transient collapse that would oxidize DBT to DBTO and $DBTO_2$, respectively. This enhanced the BDS rate using free and immobilized cells of *Rhodococcus rhodochrous* MTCC 3552 (ATCC 53968).

12.3 New Strains with Higher Hydrocarbon Tolerance

One of the most important rate limiting steps, and sometimes the only rate limiting step, is the transfer of the polycyclic aromatic sulfur heterocyclic derivative compounds from the oil phase into the cell. Access to organic sulfur by microbial cells requires the costly dispersal of heavy oil in an aqueous phase. Since BDS occurs at an oil/water interface, a limiting factor will be the rate of transport of the sulfur compounds from the bulk oil phase to the water interface and the cell membrane of the organisms. The effective limit for industrial application is the critical ratio of an organic/aqueous phase, which affects the costs of reactors and the sizes of the storage tanks. At present, this ratio is not over 1/1 (V/V). Certain technical issues hamper the applicability of BDS, one of them being the biphasic nature of the process. An aqueous phase is needed in order to maintain a viable biocatalyst, thus, significant quantities of water have to be added to the fuel. The values found in the literature generally indicate low oil-to-water volumetric ratios in the desulfurization of real fractions, ranging from 1% to 25% of oil in water. The highest oil-to-water ratios reported have been of 50% and 80% (Yu *et al.* 1998; Monot *et al.* 2002). *Rhodococci* cells are highly hydrophobic and this is advantageous in terms of desulfurization rates because hydrophobic DBTs are efficiently transferred from the oil to the cells. However, the mechanism by which DBT enters into contact with the Dsz enzymes is unknown (Monticello 2000). Nevertheless, these bacteria are difficult to separate due to their strong adherence to oil–water interfaces.

Biodesulfurization with high hydrocarbon phase tolerance is considered an advantage because a lesser amount of water is required for BDS. However, very few reports are available on the BDS of crude oil. Oil refineries usually separate crude oil into several fractions and then desulfurize them separately. Refineries can make lots of cost savings if most of the sulfur can be removed from crude oil before it is fractionated. Further, due to the high content of water in crude oil, the biodesulfurization of crude oil could be more practical compared to that of diesel oil and gasoline.

Despite identification of various DBT desulfurizers to date, commercialization of BDS is still at early stages. A potentially attractive means of implementing a petroleum BDS process could be to treat crude oil on site in the production field prior to the initial separation of petroleum from produced water.

As the O/W phase ratio increased, the CFU/mL decreased, which might be attributed to the mass transfer limitation of O_2 to the cells suspended in the aqueous phase, which, consequently, leads to a decrease in sulfur removal efficiency. Water is also necessary for enzyme activity and too high a concentration of diesel oil resulted in a low desulfurization rate. In addition, as the O/W ratio increased, the quantity of sulfur increased which, consequently, decreased the desulfurization efficiency (Maghsoudi *et al.*, 2001; Goindi *et al.*, 2002; Wang *et al.*, 2006; Bhatia and Sharma, 2012).

Caro *et al.* (2007a) reported that *R. erythropolis* IGTS8 showed a linear decrease of BDS percentages as the oil fraction phase rose because DBT quantity increased and DBT desulfurization was not observed for oil to water ratios higher than one; these results suggest an increase in organic interface surface with the increment of the oil fraction phase (OFP). Similar observations were reported by Klein *et al.* (1999) and explained as: with the increase of oil phase ratio, a decrease of the aqueous phase occurs and, consequently, the trace elements which act as cofactors for the enzymatic activity of the organisms decrease leading to low desulfurization activity.

Caro *et al.* (2008) reported that, whereas the oil phase ratio increased bacterial growth, the BDS process decreased which suggested that higher mass transfer limitations or other negative effects, for instance a lower oxygen supply, were produced when the oil proportion was increased. However, on the contrary, Kawaguchi *et al.* (2012) reported that in biphasic reaction mixtures, 2-HBP accumulation was reduced with an increasing volume ratio of oil-to-water, suggesting that higher concentrations of diesel facilitate 2-HBP diffusion from cells. These findings suggest that the enhanced BDS activity under biphasic conditions results from the combined effects of attenuated feedback inhibition of 2-HBP and reduced mass transfer limitations.

The oil–water ratio of 1:9 i.e. 10% has been used by various investigators to study the desulfurization of different fuel oils (Rhee at al. 1998; Li *et al.*, 2003, 2007a, b). For example, the optimum phase ratio for obtaining the maximum BDS efficiency using *R. erythropolis* IGTS8 was a 10% (v/v) phase ratio (Kaufman *et al.*, 1998, 1999; Folsom *et al.*, 1999; Pienkos,

Table 12.3 BDS of DBT in Biphasic System.

Bacterium	Type of hydrocarbon	Conditions	BDS efficiency	Reference
Agrobacterium tumefaciens strain LSU20	n-tetradecane	200 ppm DBT 1:5 O/W	77%	Gunam *et al.* (2016)
B. subtilis M28*R	n-hexadecane	0.5 mM DBT 1:9 O/W	8.4 mg DBT/L/h	Ma *et al.* (2006b)
B. subtilis M29*R	n-hexadecane	0.5 mM DBT 1:9 O/W	16.2 mg DBT/L/h	Ma *et al.* (2006b)
Corynebacterium sp. ZD-1	n-hexadecane	1 mmol/L DBT 1:2 O/W	64%	Wang *et al.* (2006)
Gordonia sp. JDZX5**	n-hexadecane	0.3 mM DBT 1:2 O/W	0.188 mmol/L	Feng *et al.* (2016)
Mycobacterium goodie X7B*	n-tetradecane	200 ppm S 1:9 O/W	99%	Li *et al.* (2007a)
Nocardia globerula R-9	dodecane	1 mM DBT 1:1 O/W	9.6 mmol sulfur/kg/h	Mingfang *et al.* (2003)
Nocardia sp. strain CYKS2	n-hexadecane	10 mM DBT 1:10 O/W	0.279 mg-sulfur/ (L- dispersion.h)	Chang *et al.* (1998)
Pseudomonas delafieldii R-8	n-octane	300 mg/kg 1:2 O/W	50%	Huang *et al.* (2012)

Pseudomonas delafieldii R-8[R]	n-octane	300 mg/kg 1:5 O/W	75%	Huang *et al.* (2012)
Rhodococcus erythropolis DR1**[R]	n-hexadecane	0.5 mM DBT 1:1 O/W	120 μM DBT/g DCW/h	Li *et al.* (2007b)
Rhodococcus erythropolis DR1**	n-hexadecane	0.5 mM DBT 1:1 O/W	26 μM DBT/g DCW/h	Li *et al.* (2007b)
Rhodococcus erythropolis DS-3	n-hexadecane	0.5 mM DBT 1:9 O/W	13.1 mg DBT/L/h	Ma *et al.* (2006b)
Rhodococcus erythropolis H2	n-tetradecane	3 mM DBT 40% O/W	Complete removal	Ohshiro *et al.* (1995)
Rhodococcus erythropolis LSSE81	n-dodecane	2.0 mM DBT 1:2 O/W	88%	Xiong *et al.* (2007)
Rhodococcus erythropolis NCC1	n-hexadecane	150 ppm DBT 1:9 O/W	87%	Li *et al.* (2007b)
Rhodococcus rhodochrous IGTS8 (ATCC 53968)	n-hexadecane	3.5 g/L DBT 1:1 O/W	1.9 mg HBP/g DCW/ h	Schilling *et al.* (2002)
Sphingomonas subarctica T7B	n-tetradecane	250 ppm DBT 1:5 O/W	91%	Gunam *et al.* (2006)
Sulfolobus solfataricus P2*	xylene	0.1 mM DBT 40% O/W	0.34 μm DBT/g DCW/h	Gün *et al.* (2015)

* Thermophilic; ** Resting; [R] recombinant

1999; Maghsoudi *et al.*, 2001). In other studies, the optimum phase ratio for obtaining the maximum BDS efficiency using IGTS8 was 1/3 (Indian Institute of Petroleum 1996; Wolf *et al.*, 1998).

Table 12.3 summarizes the capabilities of most of the isolated biodesulfurizing strains on model oil biodesulfurization at different oil/water (O/W) ratios.

12.4 Overcoming the Feedback Inhibition of the End-Products

Concerning biocatalyst massive production, renewable carbon sources have been proposed (Alves *et al.* 2008b; Alves and Paixão 2014). However, the main challenges are repression of the dsz operon by sulfate and other easily assimilable sulfur sources, as well as growth inhibition by 2-HBP. The existence of bacteria as *Achromobacter* sp., isolated from an oil-contaminated soil that degrades DBT following the 4S pathway and further transforms 2-HBP to a less toxic compound, 2-methoxybiphenyl, is potentially interesting (Bordoloi *et al.* 2014).

Moreover, a characterization of the nature of the inhibition caused by 2-HBP as allosteric or competitive has not been reported. Only about half of the DBT consumed can be accounted for as 2-HBP in some BDS studies, implying that 2-HBP accumulates intracellularly (McFarland, 1999; Matsui *et al.*, 2002; Guobin *et al.*, 2005; Ismail *et al.*, 2016). 2-HBP was reported to inhibit the desulfurization enzymes DszA, Dsz B, and DszC at concentrations that are significantly lower than the intracellular concentration of 2-HBP that accumulates when cells desulfurize DBT (Abin-Fuentes *et al.*, 2013). Moreover, 2-HBP can be toxic, inhibiting cell growth and metabolism so that it is challenging to distinguish between cell toxicity and enzyme inhibition in studies with bacteria cells (Alves and Paixão, 2011; Abin-Fuentes *et al.*, 2013). The DszB enzyme purified from *R. erythropolis* IGTS8 was reported to retain activity in the presence of 2-HBP (Watkins *et al.*, 2003), while a more recent and more thorough investigation reports that the concentration of 2-HBP needed to decrease the enzymatic activity by 50 % for DszA, DszB, and DszC is 60, 110, and 50 μM, respectively, while the intracellular concentration of 2-HBP in an active desulfurizing cell was determined to be 1100 μM (Abin-Fuentes *et al.*, 2013). DszB is the rate-limiting enzyme in the 4S pathway and it is the enzyme with the highest tolerance to 2-HBP (Reichmuth *et al.*, 2004; Abin-Fuentes *et al.*, 2013).

An examination of the DNA sequences responsible for the translation of DszC has not been reported, but is worthy of investigation. There is a

4-basepair overlap between the dszA and dszB genes of *R. erythropolis* IGTS8 (Piddington *et al.*, 1995) and the bdsA and bdsB genes of *Bacillus subtilis* WU-S2B and *Mycobacterium phlei* WU-F1 (Kirimura *et al.*, 2004), but a 64-bp overlap in the dszA and dszB genes of *Gordonia alkanivorans* strain 1B (Alves *et al.*, 2007). This gene overlap in the desulfurization operon probably exists to allow translational coupling to limit the expression of dszB and minimize the crippling effects of the accumulation of intracellular 2-HBP. The dsz operon has evolved to limit the expression of dszB in response to end-product inhibition by 2-HBP. Altered dsz operons have been constructed that remove the overlap between the dszA and dszB genes (Reichmuth *et al.*, 2004; Pan *et al.*, 2013; Martinez *et al.*, 2016a). This results in some increase in the activity of the 4S-pathway (Martinez *et al.*, 2016a). The desulfurization intermediate HBPSi produced by DszA, as well as 2-HBP, have been shown to be inhibitory to desulfurization enzymes (Martinez *et al.*, 2016a). When a microbial consortium containing *dszCD* and *dszAD* were combined in the right ratio and DszB was supplied extracellularly, they achieved the highest specific activity reported for the conversion of DBT to 2HBP: 23 micromoles 2HBP/gDCW/h (Martinez *et al.*, 2016a). By compartmentalizing reactions, they were able to avoid/minimize the toxic/inhibitory effects of intermediate and end-products.

The effects of the 4S-pathwy end products, 2-HBP and sulfate ions, are considered to be the key rate-limiting factors of the BDS process and have a negative impact on the specific growth rate, desulfurization efficiency, and transcriptional level of dszA/dszB/dszC, where the feedback inhibition effects of 2-HBP and sulfate ions on cell growth, transcription of dszA/dszB/dszC, and BDS efficiency were significant, especially under sulfate ion stress. Taking into consideration the differences in the feedback inhibition effects between 2-HBP and sulfate ions, Feng *et al.* (2016) suggested to overcome such a bottleneck via using the biphasic system with an appropriate O/W phase ratio, where DBT removal efficiency by resting cells of *Gordonia* sp. JDZX13 was improved by 100.7% by better weakening the feedback inhibition effects, using the optimum O/W phase ratio of 1/2.

12.5 Biodesulfurization under Thermophilic Conditions

Biodesulfurization is preferred to occur under thermophilic conditions because they enhance the BDS rate and the operating temperature would be closer to FCC or HDS outlet streams. Higher temperature decreases oil viscosity and makes molecular displacement easier, improves enzymatic rates, and decreases bacterial contamination (Mohebali and Ball, 2008). On the other hand, since distillate fractions are often treated at high temperatures, there may be some cost savings through the use of a moderate thermophile, as it would be unnecessary to cool the fractions to ambient temperature, and, thus, BDS could be more easily integrated to a refining stream after HDS (Campos-Martin *et al.*, 2010; Stanislaus *et al.* 2010). Moreover, desulfurization activity would also be enhanced due to the higher mass transfer rate at high temperatures. For practical BDS, it is useful to obtain microorganisms that exhibit much higher BT and DBT desulfurization activities at high temperatures.

Thermophilic versions of the 4S-pathway have also been discovered and characterized (Konishi *et al.*, 1997; Ishii *et al.*, 2000; Furuya *et al.*, 2001; Kirimura *et al.*, 2001; Kayser *et al.*, 2002; Li *et al.*, 2003) and operons, such as tdsABC (Ishii *et al.*, 2000) and bdsABC (Kirimura *et al.*, 2001), encoding the enzymes involved in these 4S pathways have also been characterized. The tdsABC encoded enzymes are active up to about 60 °C (Konishi *et al.*, 1997).

The extreme thermopile *Pyrococcus furiosus*, growing at 98 °C, was shown to reduce the organic polysulfide to H_2S using the reduction of sulfur as a source of energy (Tilstra *et al.*, 1992). The first thermophilic strain with an aerobic desulfurization trait, *Paenibacillus* sp. A11–2, which desulfurizes DBT at 60 °C, was reported in 1997 (Konishi *et al.*, 1997). The strain proved to have a different desulfurization gene cluster (tdsABC), which was 73%, 61%, and 52% homologous with dszABC genes. It is able to selectively desulfurize DBT without degrading its hydrocarbon matrix. This strain follows the same metabolic pathway of *Rhodococcus erythropolis* IGTS8 (Ishii *et al.*, 2000). *Paenibacillus* enzymes are homologous to *Rhodococcus* enzymes, however, they are active at higher temperatures from 50 to 60 °C. Thus, it is proposed for the development of a BDS process for crude oil at high temperatures where crude oil viscosity is lower and mass transfer limitations are reduced (Konishi *et al.*, 2000). However, the specific desulfurization activity of A11–2, 0.008 mM 2-HBP/h/g DCW (Ishii *et al.*, 2000), was lower relative to the mesophilic cultures, 0.083 to 1.23 μM 2-HBP/h/g DCW, previously reported at that time (McFarland *et al.*, 1998; Rhee *et al.*, 1998; McFarland, 1999; Kobayashi *et al.*, 2000; Hirasawa *et al.*, 2001).

A study financed by the Japan Cooperation Center, Petroleum (JCCP) using resting cells from a thermophile strain of *M. phlei* WU-F1 catalyzed the complete oxidation of 149 ppm DBT (26 ppm sulfur) at 50 °C and 90 min and was also found to possess desulfurizing ability toward naphthaothiophene present in gas oil and hydrodesulfurized light gas oil at 50 and 45 °C, respectively (Furuya *et al.*, 2002). This bacterium also removed 60 to 70% of sulfur content from hydrodesulfurized gas oil (Furuya *et al.*, 2003). In the same way, *M. phlei* WU-0103 was able to reduce 52% of sulfur in a crude straight-run light gas oil fraction not treated with HDS. *B. subtilis* WUS2B is another thermophilic bacterium able to degrade DBT completely at 12 h and its derivatives to approximately 50% at 24 h through selective cleavage of C-S bonds, resulting in the accumulation of 2-HBP (Kirimura *et al.*, 2001). This kind of reaction would be desirable to remove the sulfur atom without C-C bond cleavage, that is, without loss calorific power. The DBT desulfurization gene cluster in WU-S2B was identified and named as bdsABC. The DNA and amino acid sequencing of bds genes with the ones of IGTS8 showed 61% homology (Kirimura *et al.*, 2004). Li *et al.* (2007) reported the BDS of model oil (DBT in tetradecane) by the facultative thermophilic *Mycobacterium goodii* X7B, where the total sulfur level of DBT in tetradecane was reduced by 99% from 200 to 2 ppm within 24 h at 40 °C. Gün *et al.* (2015) reported the ability of hyper-thermophelic *Sulfolobus solfataricus* P2 to desulfurize DBT and its derivatives, but not BT at 78 °C, recording DBT-BDS activity of approximately 1.23 μmol 2-HBP/g DCW/h. It also recorded good DBT-BDS capacity in a biphasic system (DBT dissolved in xylene, 40% O/W), recording 0.34 μmol DBT/g DCW/h.

Table 12.4 summarizes the capabilities of most of the thermotolerant isolated biodesulfurizing strains.

12.6 Anaerobic Biodesulfurization

Reductive or anaerobic biodesulfurization is an attractive option. One approach that has generated a considerable amount of interest uses sulfate reducing bacteria (SRB) for the reductive desulfurization of organic sulfur compounds. These organisms have the unique ability to obtain energy by reducing sulfate to H_2S. However, the driving force of this option is hydrogen, which is more expensive than oxygen (Aitani *et al.*, 2000).

Anaerobic BDS of crude oil and its distillates has been reported using *Desulfovibrio desulfuricans* M6 (Kim *et al.*, 1990a), the extreme thermopile *Pyrococcus furiosus* (Tilstra *et al.*, 1992), *Nocardioform actinomycete* FE9 (Finnerty, 1992 and 1993), *Desulfomicrobium scambium,* and *Desulfovibrio longreachii* (Onodera-Yamada *et al.*, 2001).

The desulfurization of petroleum under anaerobic conditions would be attractive because it avoids costs associated with aeration, it has the advantage of liberating sulfur as a gas, and does not liberate sulfate as a

Table 12.4 BDS of DBT Using Thermotolerant BDSM.

Bacteria	**Temperature**	**BDS efficiency**	**References**
Bacillus subtilis WU-S2B	50 °C	Complete desulfurization of 0.54 mM DBT	Kirimura *et al.* (2001)
Mycobacterium goodii X7B	40 °C	90% of 0.5 mM DBT	Li *et al.* (2007)
Mycobacterium phlei GT1S10	50 °C	1.09 μmol 2-HBP/g DCW/min	Kayser *et al.* (2002)
Mycobacterium phlei SM120–1	45 °C	0.17 μmol 2-HBP/g DCW/min	Srinivasaraghavan *et al.* (2006)
Mycobacterium phlei WU-F1	50 °C	Complete desulfurization of 0.8 mM DBT	Furuya *et al.* (2001)
Mycobacterium sp. X7B	45 °C	82.5% of 0.4 mM DBT	Li *et al.* (2003)
Paenibacillus sp. A11–2	50 °C	Complete desulfurization of 0.065 mM DBT	Konishi *et al.* (1997)
Sulfolobus solfataricus P2	70 °C	1.23μM 2-HBP/g DCW/h	Gün *et al.* (2015)

by-product that must be disposed by some appropriate treatment. It retains the caloric value since C-C bonds are untacked, since the reaction pattern is similar to HDS. Under anaerobic conditions, oxidation of hydrocarbons to undesired compounds, such as colored, acidic, and/or gum forming products, is minimal (McFarland, 1999). Moreover, diluted sulfate is formed as the end-product of the aerobic route that also must be removed, while H_2S that is formed in the anaerobic route can be treated with existing refinery desulfurization plants (e.g. Claus process). Furthermore, the toxic metabolites, such as 2-HBP, which is also involved in the formation of viscous oil sludge ('gum') in the fuel is not produced under anaerobic conditions, but growth under anaerobic conditions proceeds slowly, especially when organic molecules (like thiophenes) are involved in the conversion. Under anaerobic conditions, thiophenes act as the sole electron acceptor and the conversion of thiophenes is coupled with microbial growth. Nevertheless, from a process point of view, the aerobic route has some major drawbacks. Sulfur is used in the assimilatory metabolism of aerobic bacteria. Considering that the sulfur content of biomass is approximately 0.03 wt%, the yield of biomass per mole sulfur removed in the aerobic route is high. In contrast, in a microbial process where dissimilatory sulfur metabolism is involved, accumulation of wasteful biomass will not occur (Lizama *et al.*, 1995). Approximately 50% of the energy produced by aerobic microorganisms will be used for growth, while anaerobic microorganisms use approximately only 10% of their energy for assimilation. However, at high biomass concentrations, down-stream processing is complicated because proteins originating from the biomass emulsify the oil/water mixture. In addition, the mixing efficiency and O_2 availability is less optimal in emulsions with a high biomass concentration.

The requirement of an intimate biomass-substrate contact is the key factor in the preliminary bioreactor design suitable to anaerobically convert non-polar organic sulfur compounds. However, favorable carbon and energy sources are also necessary to convert thiophenes. From a process point of view, H_2 gas is the best option as an electron donor for converting sulfur compounds because of its low costs for application at a relatively large scale (Van Houten *et al.*, 1996) and its availability at refineries (Gary and Handwerk, 1994). Small amounts of acetate and bicarbonate are required for growth and should be supplemented as a carbon source to the aqueous phase. Besides, H_2 and also H_2S is involved as a reaction product. Consequently, the anaerobic process can be considered as a three-phase system, i.e. gas-water-hydrocarbon. To optimize the availability of organic sulfur compounds, it is proposed to disperse the hydrocarbon phase as very fine droplets in the aqueous phase. In order to be able to supply H_2 gas to the dispersion without severe foaming, the hydrocarbon phase is saturated with H_2 gas before introduction in the bioreactor. Then, the hydrocarbon phase is used as a carrier phase for H_2 gas. This approach combines the requirements of a high specific surface area to maximize the availability of both non-polar organic sulfur compounds and H_2 gas. A bioreactor system equipped with a nozzle to disperse the hydrocarbon phase is the most appropriate device to generate fine hydrocarbon droplets, while the mixing energy is only imparted on the hydrocarbon phase.

In order to favor the conversion of organic sulfur compounds, the sulfide concentration must be as low as possible to avoid inhibition. The H_2S produced, in turn, dissolves in the hydrocarbon phase and is stripped off during H_2 saturation. In this way, any sulfide inhibition is avoided.

The aforementioned advantages of anaerobic process can be counted as incentives to continue research on reductive biodesulfurization. However, maintaining an anaerobic process is extremely difficult and the specific activity of most of the isolated strains have been reported to be insignificant for DBTs and oil feed; these include vacuum gas oil, deasphalted oil, or bitumen (Armstrong *et al.*, 1995a, b). Due to low reaction rates, safety and cost concerns, and the lack of identification of specific enzymes and genes responsible for anaerobic desulfurization, anaerobic microorganisms effective enough for practical petroleum desulfurization have not been found yet and an anaerobic BDS process has not been developed. Consequently, aerobic BDS has been the focus of the majority of research in BDS (Le Borgne and Quintero, 2003).

Analogous to reductive desulfurization, oxidative or aerobic biodesulfurization can be divided into dissimilatory and assimilatory systems. The sulfur end products from these systems are oxidized compounds which are water soluble sulfur products (sulfate ions). Dissimilatory oxidation occurs when reduced sulfur compounds are used as electron donors in energy yielding reactions. Assimilatory sulfur oxidation is used to obtain sulfur in a form available for the biosynthesis of organic sulfur compounds (Aitani *et al.*, 2000).

12.7 Biocatalytic Oxidation

The bio-oxidation of organosulfur compounds can be performed in a batch, semi-continuous or continuous methods alone, or in a combination with one or more additional refining processes. The reaction can be carried out in an open or closed vessel. Cytochrome c is a biocatalyst able to oxidize thiophenes and organo-sulfides and has several advantages when compared with other hemoenzymes. It is active in a pH range from 2 to 11, has the hemo prosthetic group covalently bonded, exhibits activity at high concentrations of organic solvents, and is not expensive (Vazquez-Duhalt *et al.*, 1993). The oxidation of chemically diverse OSCs catalyzed by chloroperoxidase (CPO) highlights the potential of this enzyme for applications in the treatment of complex streams (Ayala *et al.* 2000). The largest and unsolved challenge with CPO has been that no heterologous, stable expression of this enzyme in a suitable host has been obtained yet (Conesa *et al.* 2002; de Weert and Lokman 2010). Thus, peroxide stability, substrate partition improvement, and overproduction issues for large-scale applications have been restricted to non-genetic approaches (Ayala *et al.* 2007). CPO showed a turnover number (i.e. number of substrate molecules that can be converted per molecule of enzyme before inactivation) of 500,000. Thus, 1 g of enzyme could reduce the sulfur content from 500 to 30 ppm of 0.81 ton of fuel (Ayala *et al.*, 2007).

However, taking into consideration the current high cost of enzymes, the desulfurization of one barrel of fuel would be too expensive to be implemented on a large scale. It is estimated that the enzyme stability should be increased by two orders of magnitude to be economically attractive. Nevertheless, it should be kept in mind that a biotechnological process would have lower costs in terms of capital investment and energy consumption (Ayala *et al.*, 1998; Linguist and Pacheco, 1999).

12.8 Perspectives for Enhancing the Rate of BDS

Despite all the novel and successful approaches, the achieved BDS rate is still far by two orders of magnitude to develop a commercial BDS for crude oils and refinery streams (Monticillo, 2000; Kilbane, 2006; Xu *et al.*, 2009; Lin *et al.*, 2010; Mohamed *et al.*, 2015). Genetic engineering had to be used to improve the IGTS8 strain, leading to a 200 fold increase in DBT desulfurizing activity allowing shorter residence times in an attempt to be compatible with a commercial application (Pacheco *et al.*, 1999). For this, modifications in media composition and growth conditions, as well as genetic manipulation of the IGTS8 strain, were performed and DBT desulfurization rates around 20 μmol/min/g DCW were obtained (Folsom, 2000).

To improve the BDS process for the removal of sulfur from model compounds such as DBT, different strategies have been adopted by the researchers, including strain improvement through gene manipulation techniques, in silico studies for sulfur metabolism, and deep understanding of the bioprocess development stages (Kilbane 2006b; Abin Fuentes *et al.* 2014).

For higher achievement of desulfurization value, sequential BDS of real oil feeds has been reported. Ten cultivation rounds each of 3 days' incubation were done using *Mycobacterium phlei* WU-0103 to achieve 52% reduction in total sulfur in light gas oil (Ishii *et al.*, 2005). Adlakha *et al.* (2016) reported approximately 61% BDS of heavy crude oil with an initial S-content of 1.083 wt.% after two successive BDS-rounds, each of 7 days using *Gordonia* sp. IITR100.

12.8.1 Application of Genetics in BDS

The bottleneck for the low versatility of some biodesulfurizing microorganisms (BDSM) toward different OSCs can be solved via the application of genetic engineering and recombinant strains. However, to achieve very high levels of expression of the desulfurization pathway, a better understanding of the host factors that contribute to the functioning of the pathway is recommended (Kilbane, 2006). Thus, it has been preferred to use the original strain for technical (e.g. gene expression and codon usage preferences) and regulatory reasons (Monticello, 1998). In addition, self-cloning is generally regarded as more effective than heterologous recombination, as gene expression and DBT permeation are readily achieved (Li *et al.*, 1996; Hirasawa *et al.*, 2001; Matsui *et al.*, 2001a).

The most important development in /enzymatic DBT degradation is the application of directed evolution techniques (gene shuffling) to the dsz system, creating hybrid genes with better rates and extents of BDS (Monticello, 2000; Gray *et al.*, 2003). This technique has been used for the removal of DBT and benzothiophene (BT) using improved biocatalysts (Kilbane, 2006b). Coco *et al.* (2001) used the DNA shuffling method to generate recombined genes and evolved enzymes. An increase in steady-state DBT sulfone concentration of at least 16 fold was observed in pathways containing shuffled dszC genes. Simultaneous improvement of both desulfurization rates and a range of oxidized substrates were obtained. A gain-of-function phenotype allowing the oxidation of 5-methylbenzothiophene was due to the single mutation V261F. The desulfinase DszB is rate limiting for the whole pathway (in terms of quantity of enzyme expressed). The catalytic activity and thermostability of DszB were enhanced by two amino-acids substitutions (Y63F and Q65G) and a double mutant (Y63F, Q65G) presented an improved desulfurizing activity (Ohshiro *et al.*, 2007). Petro Star and Diversa also worked in the development of a host harboring genetically engineered Dsz enzymes (Bonde and Nunn, 2003). The target was monooxygenase DszA that is central in the desulfurization pathway in order to engineer a host for desulfurization of DBT sulfones. These sulfones would be produced from the Petro Star Conversion Extraction Desulfurization (CED) process and the DBT's hydrocarbon skeletons were returned to the fuels. However, no results were published concerning the performance of the obtained DszA variants. Noda *et al.* (2003a) used 22 rhodococcal and mycobacterial strains capable of growth in a hydrophobic medium as hosts for the dsz desulfurization gene cluster from *R. erythropolis* (strain KA 2–5- 1) and found recombinant *Mycobacterium* strain MR 65 was able to desulfurize 68 mg/L of sulfur in light gas oil containing 126 mg/L sulfur in the presence of 4,6-dipropyl DBT. Matsui *et al.* (2001b) introduced dsz genes into *Rhodococcus* sp. T09, a strain capable of desulfurizing benzothiophene (BT). The resulting recombinant strain grew with both DBT and BT as the sole sulfur sources. The recombinant cells desulfurized not only alkylated BTs, but also various Cx-DBTs, producing alkylated hydroxyphenyls as the end products. Pan *et al.* (2013) pursued a novel strategy to improve the performance of the BDS process by increasing the sulfur demand of desulfurization-competent cultures to force increased desulfurization activity via modification of the dsz operon so that it encodes a sulfur-rich polypeptide (Sulpeptide1 S1). This strategy was followed up by directed evolution, i.e. subjecting the transformant to repeated passages in a medium with DBT as the sole sulfur source. After selection for 40 passages, both the dszABC and dszAS1BC expressing *Rhodococcus opacus* strains exhibited a >20 fold increase in specific desulfurization ability and exceeded the specific activity of the control, desulfurization positive strain *Rhodococcus erythropolis* IGTS8. Moreover, Wang *et al.* (2017) reported that a combination of genetic engineering using sulfur sinks and increasing Dsz capability with adaptive selection is a viable strategy to increase the BDS ability, where engineering of desulfurization competent *Rhodococcus qingshengii* to express a peptide was designed to act as a sulfur sink (Sulpeptide 1) combined with adaptive evolution over the course of about 400 generations was successful in substantially increasing both their metabolism of DBT and their growth rate in medium with DBT as the sole sulfur source. The increased growth rate on DBT eventually exceeded (by more than 2-fold) the growth rate of unadapted strains grown with sulfate as the sole sulfur source, while the effect of the expression of Sulpeptide 1 contributed less to these increases than they did to adaptive evolution. However, it was found that the increases in growth rate and DBT metabolism were not correlated to any mutations in genes of obvious connection to DBT metabolism or sulfur metabolism in general, although the eventual increase in DBT metabolism was closely correlated to an increase in the copy number of the genes encoding DBT metabolism. Thus, the important changes that occurred during adaptation may have included epigenetic and stable physiological ones. These results suggested that future research should first optimize growth on sulfate under ideal conditions and only then should the dsz genes be introduced and adaptive evolution experiments be performed.

More recent research work has been concerned with the genetic modification of bacteria to give more efficient and substrate-specific strains (Chapter 7).

R. erythropolis IGTS8 has been the base of the BDS process proposed by Energy BioSystems Corporation (EBC) (named later Enchira). An intensively engineered bacterium was obtained two orders of magnitude more active than the original strain (Monticello, 2000). The copy number of desulfurizing genes was increased, the sulfate repression eliminated by promoter change, and the last gene of the metabolic pathway (dszB) deleted to eliminate this rate-limiting step (in terms of enzyme concentration) and allow the accumulation of HBPSi, a potentially valuable surfactant. The DszC monooxygenase catalyzes the first and limiting step (in terms of oxidation rate and range of oxidized substrates) of the desulfurization pathway using molecular O_2 (Figure 12.2). The activity of DszC was improved through directed evolution using gene shuffling and chemostat enrichment (Arensdorf *et al.*, 2002; Coco *et al.*, 2001). Simultaneous improvement of both desulfurization rates and range of oxidized substrates was obtained. A gain-of-function phenotype allowing the oxidation of 5-methylbenzothiophene was due to the single mutation V261F. The first two steps of the 4S pathway are catalyzed by the cofactor dependent monooxygenases DszC and DszA (Figure 12.2). The complexity of the 4S multi-enzymatic pathway and the requirement for cofactors dimensioned the use of purified enzymes instead of whole cells. As shown in Figure 12.2, the DszC and DszA enzymes use the $FMNH_2$ cofactor that is regenerated by the action of an NADH-FMN oxidoreductase (DszD) (Gray *et al.*, 1996). Other genetic engineering strategies consisted in overexpressing homologous or heterologous reductases for the efficient regeneration of cofactors (Reichmuth *et al.* 2000; Hirasawa *et al.*, 2001; Matsubara *et al.*, 2001) or co-expressing *Vitreoscilla* hemoglobin to increase cell internal oxygenation (Xiong *et al.*, 2007). A recent investigation of the catalytic mechanism of DszD by combined quantum mechanics/molecular mechanics simulation methods opens up the possibility for rational protein engineering of this enzyme (Sousa *et al.*, 2016). The presence of smaller or more positive amino acid residues instead of an active site Thr residue could lower the activation barrier for the rate-limiting step in the reaction of formation of $FMNH_2$ by DszD. Production of the desulfinase DszB (Figure 12.2) could be increased by mutating the untranslated 5′ region of dszB (Reichmuth *et al.* 2004).

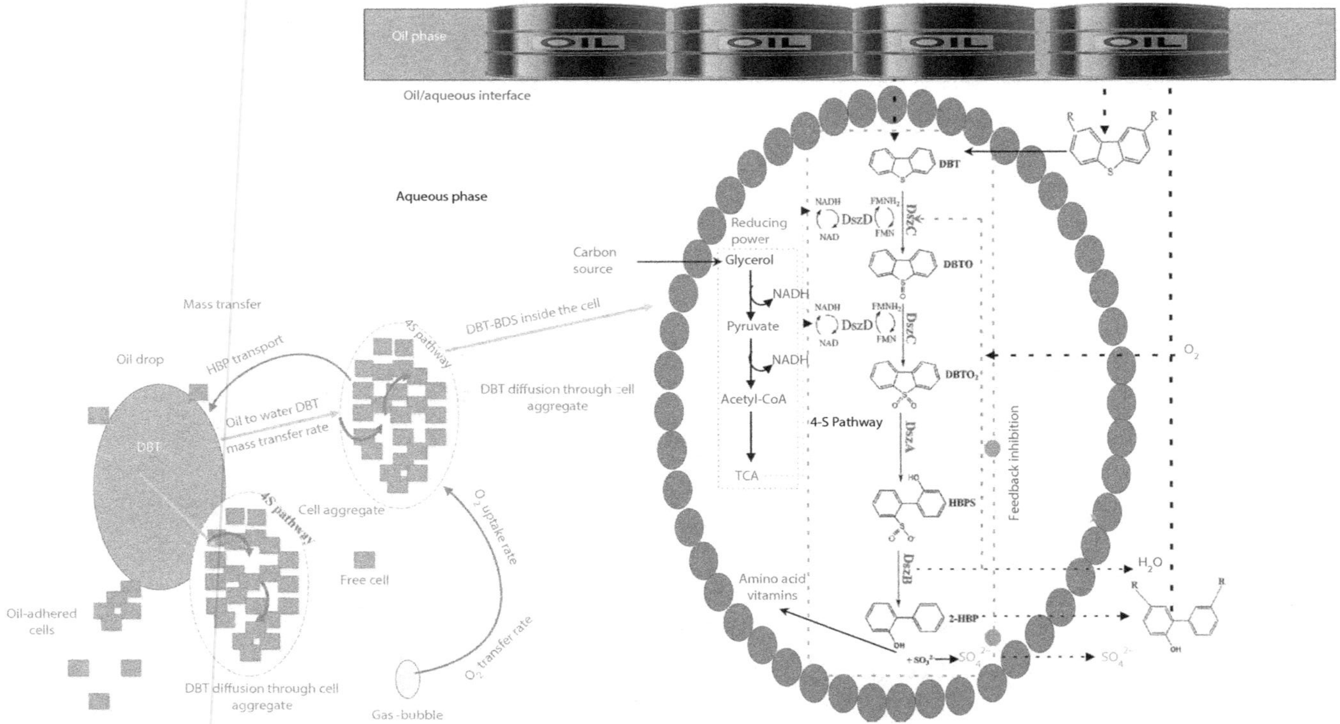

Figure 12.2 Mechanistic Steps in a BDS System at High Cell Density. Biocatalyst may be present in one of three populations: free cells in an aqueous phase, oil-adhered cells, and cells in aggregates. Oxygen transport and uptake is necessary because the 4S pathway is an oxidative pathway that requires 3 moles of O_2 per mole of DBT desulfurized (R may be an alkyl or H).

Removing the gene overlap between *dszA* and *dszB* led to a five-fold increase in desulfurizing activity (Li *et al.* 2007), while cells expressing a rearranged operon (*dszBCA*) presented a 12-fold higher activity (Li *et al.* 2008a). The desulfinase DszB is rate limiting for the whole pathway (in terms of the quantity of the expressed enzyme). The catalytic activity and thermostability of DszB were enhanced by two amino acid substitutions (Y63F and Q65G) and a double mutant (Y63F, Q65G) presented an improved desulfurizing activity (Ohshiro *et al.*, 2007).

Conventional methods for isolating novel microorganisms and improving known biodesulfurizing ones by genetic and protein engineering strategies are time- and labor-consuming. Novel microorganisms were recently identified by using a bioinformatics approach consisting in mining genomic databases to identify DBT-desulfurizing bacteria (Bhatia and Sharma, 2010b). For this, the amino acid sequences of the Dsz enzymes were used to search for homologous proteins in microbial genome databases. Thirteen novel desulfurizing bacteria belonging to 12 genera (*Burkholderia*, *Bradyrhizobium*, *Methylobacillus*, *Magnetospirillum*, *Thermobifida*, *Azotobacter*, *Mesorhizobium*, *Oceanobacillus*, *Novosphingobium*, *Brevibacterium*, *Rubrobacter*, and *Ralstonia*) were identified. Interestingly, seven of these microorganisms lacked the DszB enzyme, which is rate limiting in the 4S pathway and two of them were thermophilic (*Thermobifida* and *Rubrobacter*). Bacteria lacking the DszB enzyme accumulate HPBSi, a potentially valuable surfactant, as mentioned above. The potential advantage of thermophilic microorganisms has also been mentioned above (section 12.5). A novel and interesting approach based on evolutionary engineering to generate strains with an enhanced desulfurization activity has been reported by Pan *et al.* (2013). Bacteria assimilate very small amounts of sulfur for their growth and maintenance. These authors modified the dsz operon to include a synthetic gene encoding a peptide rich in sulfur-containing amino acids (methionine and cysteine). As the Dsz activity is repressed by the presence of sulfate and other readily available sulfur sources as amino acids, desulfurizing strains depend on DBT desulfurization to obtain sulfur for growth. The increased demand of sulfur due to the presence of the sulfur-rich peptide exerts a selective pressure on strains growing fast on DBT as a sulfur source. Evolutionary engineering provides new ways to improve industrially relevant microorganisms based on selectable characteristics (Winkler and Kao. 2014) and does not need previous genetic knowledge; it is particularly suitable when multigene modifications are required as it seems to be the case for desulfurizing microorganisms. Cysteine scanning mutagenesis of desulfurization enzymes, combined with biochemical screening, can identify amino acids that are associated with enzymatic activity, substrate range, and resistance to 2-HBP and can be used to produce derivatives of desulfurization enzymes that are more cysteine rich and disulfide rich. Such enzymes will be better suited for industrial applications because they will have increased resistance to temperature, pH, solvents, and possibly 2-HBP. Cysteine rich desulfurization enzymes will also increase the nutritional demand of cells for sulfur, creating a selective advantage for the evolution of desulfurization enzymes with increased efficiency (Kilbane, 2017).

Petro Star and Diversa also worked in the development of a host harboring genetically engineered Dsz enzymes (Bonde and Nunn 2003). The target was monooxygenase DszA which is central in the desulfurization pathway in order to engineer a host for desulfurization of DBT sulfones. Sulfones would be produced from the Petro Star Conversion Extraction Desulfurization (CED) process, a desulfurization processes involving oxidation and extraction. The CED process first extracts a fraction of the sulfur from diesel, then selectively oxidizes the remaining sulfur compounds into sulfones, and, finally, extracts these oxidized materials based on their increased polarity compared to the parental DBT compounds. The CED process is one of the ODS processes that has reached the commercialization stage. BDS would desulfurize the extracted sulfones and the desulfurized DBT hydrocarbon skeletons returned back to the fuels. No results were published concerning the performance of the obtained DszA variants. Recombinant desulfurizing and solvent-tolerant *Pseudomonas* strains were used to desulfurize DBT sulfones produced by chemical oxidation and solvent separation (ORNL 2000). In that process, 10% of the hydrocarbons were lost after solvent extraction. Thus, BDS of the sulfones concentrated in the solvent was proposed to recover part of the hydrocarbon skeleton. The advantage of this process is that water would be added to the solvent extract and not directly to the fuel. DBT sulfones are more polar than DBT and, hence, less hydrophobic *Pseudomonas* cells were suitable hosts in this case (Tao *et al.* 2006).

Most organic solvents are toxic to microorganisms even at low concentrations (e.g., 0.1 vol.%). Fuel oil is similar to organic solvents regarding toxicity to microorganisms. Microorganisms with a high tolerance to organic solvents are useful and important in many biotechnological fields such as BDS. Considering the difficulty of isolating strains that have both solvent tolerance and the desired catalytic activity from the environment, it might be preferable to combine solvent-tolerance with unique catalytic characteristics using methods of genetic engineering. Tao *et al.* (2006) constructed a solvent-tolerant desulfurizing bacterium, *Pseudomonas putida* A4, by introducing the BDS gene cluster *dszABCD* from *R. erythropolis* XP into the solvent-tolerant strain *P. putida* Idaho. The genes dszA, dszB, dszC, and dszD were amplified from the genome of desulfurizing bacterium *R. erythropolis* XP separately and inserted into the broad host plasmid pMMB66EH, resulting in recombinant plasmid pMMABCD. This plasmid was constructed and introduced into *P. putida* Idaho by the tri-parental mating method. The introduction and the screening method are shown in Figure 12.3. The donor, helper, and receptor strains are *E. coli* HB101 containing plasmid pMMABCD, *E. coli* HB101 containing pRK2013, and *P. putida* Idaho, respectively. They were cultured in liquid LB and harvested and washed separately with a sub-sequential mixture and filtration. Then, the film was detached from the filter and put on an LB plate for an incubation at 37 °C for 10 h, then 30 °C for another 10 h. Finally, the film was washed with a sterile NaCl solution and spread on a selection plate. The resulted strain was named *P.*

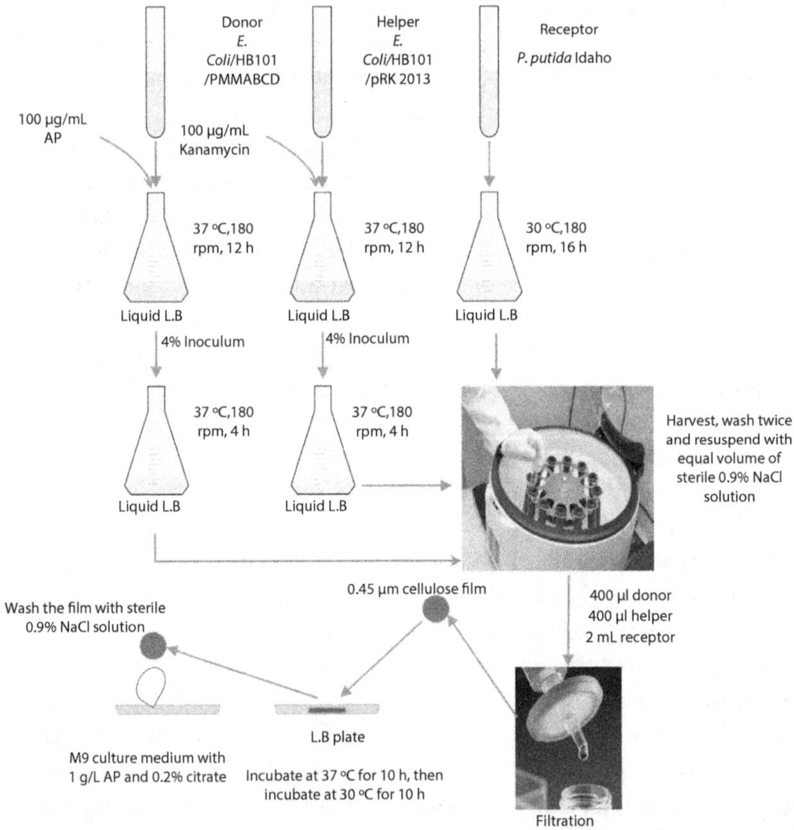

Figure 12.3 Schematic Map of Tri-Parental Mate.

putida A4 which was found to maintain the same substrate desulfurization traits as observed in *R. erythropolis* XP. Resting cells of *P. putida* A4 showed the ability to desulfurize 86% of dibenzothiophene in 10% (v/v) p-xylene within 6 h, where the desulfurization occurred with a rate of 1.29 mM dibenzothiophene/g DCW/h within the first 2 h of incubation.

Table 12.5 summarizes the BDS efficiencies of recombinant or genetically engineered strains.

12.8.2 Implementation of Resting Cells

The implementation of the BDS process via 4S consists of several stages, including (i) growth of the selected strain in a suitable environment in

Table 12.5 DBT-BDS Using Recombinant Strains.

Bacterium	BDS efficiency	Reference
B. subtilis M28	8.4 mg DBT/L/h	Ma *et al.* (2006b)
B. subtilis M29	16.2 mg DBT/L/h	Ma *et al.* (2006b)
E. coli BL21	90%	Rehab *et al.* (2011)
E. coli DH10B	8 mg DBT/g DCW/ h	Reichmuth *et al.* (1999)
Pseudomonas Putida CECT5279	83.6% of 25 μM DBT	Calzada *et al.* (2011)
R. erythropolis DRB	320 μM DBT/g DCW/h	Li *et al.* (2008)
R. erythropolis DS-3	26 mmol DBT/g DCW/h	Li *et al.* (2008)
Rhodococcus erythropolis DR1	120 μM DBT/g DCW/h	Li *et al.* (2007a)
Rhodococcus erythropolis DR2	120 μmol 2-HBP/g DCW/h	Li *et al.* (2007)
Rhodococcus erythropolis I-19	5 μmol DBT/g DCW/min	Folsom *et al.* (1999)
Rhodococcus erythropolis KA2–51	74.2 mM DBT/kg DCW/h	Kobayashi *et al.* (2000)
Rhodococcus sp. B1	------	Denis Larose *et al.* (1997)
Rhodococcus sp. FMF	3.8 μm DBT/g DCW/h	Akbarzadeh *et al.* (2003)
Rhodococcus sp. T09	31 μM DBT/g DCW/h	Matsui *et al.* (2001)
Rhodococcus sp. X309	------	Denis Larose *et al.* (1997)

order to obtain cells that have the highest possible desulfurization activity and (ii) removal of active cells by cells at rest (stationary). The formulation of the growth medium is important for the production of a high density of resting cells, which expressed the highest level of desulfurization activity (Mohebali *et al.*, 2007a).

Moreover, the life of a microorganism in the BDS process is usually short (1–2 days), but this has been extended to 8–16 days. However, new designs would allow the production and regeneration of the biocatalyst with the BDS process, where the biocatalyst life time provides 200–400 h (Babich and Moulijin, 2003; Javadli and de Klerk, 2012; Zeelani and Pal, 2016).

The expression of the desulfurization activity of growing cells is repressed by sulfate and sulfur-containing amino acids and the cell growth and desulfurization activity were inhibited by 2-HBP (Li *et al.*, 1996; Ohshiro *et al.*, 1996; Nekodzuka *et al.*, 1997). However, the sulfate hardly inhibited the activity of *Rhodococcus erythropolis* D-1 even at 1 mM concentration once desulfurization was induced (Ohshiro *et al.*, 1996). The 80% activity of resting cells of *R. erythopolis* KA2-5-1 remained even in the presence of 2-HBP at 10 mM in the model oil system (Morio *et al.*, 2001). The growth and BDS of growing cells of *Pseudomonas delafieldii* R-8 are inhibited at a pH below 5, but the resting cells of R-8 is hardly affected by change in pH. Growing cells are highly affected by media composition and the components of the aqueous phase and need cofactors or reducing agents. However, resting cells of *Pseudomonas delafieldii* R-8 were affected by the aqueous phase and did not need cofactors or reducing agents for achievement of BDS (Luo *et al.*, 2003). With all these factors concerned, resting cells might be preferred to growing cells in practical fuel desulfurization.

Moreover, one of the problems in two-phase BDS systems is the formation of stabilized emulsion which causes difficulty in separation after completion of the reaction. Several suggested solutions for this problem have been reported, including: (i) avoiding the formation of a stable water-in-oil emulsion in order to facilitate oil recovery, (ii) the use of emulsion- destabilizing chemical agents, and (iii) a cell-immobilized BDS process (Chang *et al.*, 2000b; Naito *et al.*, 2001; Lee *et al.*, 2005). Furthermore, to overcome the separation problems a complementary step is needed, which will add to the expense of the process. However, Mohebali *et al.* (2007b) reported the occurrence of a significant decrease in emulsion-stabilizing activity using resting cells of *G. alkanivorans* RIPI90A that were harvested at the time of the transition from the late exponential growth phase to the stationary phase. Thus, Mohebali *et al.* (2007b) suggested that for the enhancement and rationalization of a BDS process the preparation of a stable emulsion, which would have more efficient mass transfer than in the next step, to facilitate oil and cell recovery, it would be required to change the reactor conditions to provide stationary-phase conditions for the breakage of the particle-stabilized emulsions.

A resting cell reaction system is used frequently (Table 12.6). In this approach, cells are pre-grown in the presence of a readily available carbon source (e.g. glucose) with DBT as the sole sulfur source. After harvesting and concentrating the cells, a batch reaction is performed.

12.8.3 Microbial Consortium and BDS

Despite the progress that has been achieved during the last two decades, BDS has not been applied on a commercial scale yet. This is attributed to many issues related to the stability and catalytic efficiency of the microbial biocatalyst, in addition to other technical problems (Monot and Warzywoda, 2008). The majority of the research conducted on microbial desulfurization has adopted axenic cultures of selected microorganisms. However, it is worth investigating BDS capabilities of microbial consortia to benefit from cooperative or synergistic microbe-microbe interactions (McGenity *et al.*, 2012; Mikesková *et al.*, 2012).

Intentional mixtures of the desulfurization competent *R. erythropolis* with *Enterobacter cloacae* that lacks desulfurization genes, were prepared in various ratios spanning six orders of magnitude (Kayser *et al.*, 1993). Then, after growth using DBT as the sole source of sulfur, the ratios of the two cultures were again determined and it was shown that when the specific activity for BDS was calculated, taking into account the relative abundance of the desulfurization competent cells, values that were 200 to 1000 fold higher were obtained for mixed cultures as compared to pure cultures (Kayser *et al.*, 2003). This level of increased activity, if it could be achieved with entire populations of biocatalysts, could enable an economical BDS process for petroleum (Kilbane, 2006).

The effectiveness of a single-species culture can be improved by using resting cells of different ages (Calzada *et al.*, 2008, 2011). Moreover, a mixture of different organisms is also reported to enhance BDS efficiency (Li *et al.*, 2008c; Subashchandrabose *et al.*, 2011). For example, a combination of *R. erythropolis* DS-3 and *Gordonia* sp. C-6 reported 86% reduction in the total sulfur content of a diesel fuel.

Martínez *et al.* (2016) have reported a novel approach utilizing engineered synthetic bacterial consortia for enhanced desulfurization and revalorization of oil sulfur compounds. This new approach was developed to overcome inhibition of the Dsz enzymes by the 4S pathway intermediates and to enable efficient production of value-added intermediates,

Table 12.6 DBT-BDS Using Resting Cells.

Bacteria	BDS efficiency	References
Actinomycete sp. R3	65% of 270 µM DBT	Khedkar and Shanker (2014)
Arthrobacter sulfureus	63% of 27 ppm DBT	Labana *et al.* (2005)
Bacillus pumilus	99.9% of 0.5 mM DBT	Buzanello *et al.* (2014)
Bacillus subtilis WU-S2B	Complete desulfurization of 0.81 mM DBT	Kirimura *et al.* (2001)
Corynebacterium sp. P32C1	37 mmol 2-HBP/kg DCW/ h	Maghsoudi *et al.* (2000)
Corynebacterium sp. P32C1	Complete removal of 0.5 mM DBT	Maghsoudi *et al.* (2000)
Gordonia sp. F.5.25.8	73% of 1 mM DBT	Santos *et al.* (2006)
Mycobacterium phlei WU-*F1*	Complete removal of 0.8 mM DBT	Furuya *et al.* (2001)
Mycobacterium sp. G3	49 µmol DBT/g DCW/h	Okada *et al.* (2003)
Native Rhodococcus erythropolis DS3	26 µmol 2-HBP/g DCW/h	Li *et al.* (2007)
Pseudomonas putida CECT5279	80% of 25 µM DBT	Martin *et al.* (2005)
R. erythropolis DRA	26 µM DBT/g DCW/h	Li *et al.* (2008)

R. globerulus DAQ3	17–18 μmol DBT/g DCW/h	Yang *et al.* (2007)
Recombinant Pseudomonas Putida CECT5279	83.6% of 25 μM DBT	Calzada *et al.* (2011)
Recombinant *R. erythropolis* DRB	320 μM DBT/g DCW/h	Li *et al.* (2008)
Recombinant Rhodococcus erythropolis DR2	120 μmol 2-HBP/g DCW/h	Li *et al.* (2007)
Rhodococcus erythropolis D1	Complete removal of 0.125 mM DBT	Izumi *et al.* (1994)
Rhodococcus erythropolis SHT87	0.36 μmol 2-HBP/min/g DCW	Davoodi-Dehaghani *et al.* (2010)
Rhodococcus globerulus	13 μM 2-HBP/g DCW/h	Guerinik and Al-Mutawah (2003)
Rhodococcus sp.	70% of 27 ppm DBT	Labana *et al.* (2005)
Rhodococcus sp. 1awq	0.26 μmol DBT/g/min	Ma *et al.* (2006a)
Rhodococcus sp. KT462	1.56 mM DBT/mg DCW/h	Tanaka *et al.* (2002)
Rhodococcus sp. P32C1	3.5mM 2-HBP/kg DCW/h	Maghsoudi *et al.* (2001)
Rhodococcus sp. SY1	95% of 0.3 mM DBT	Omori *et al.* (1995)
Shewanella putrefaciens NCIMB	94.22% of 0.3 mM DBT	Ansari *et al.* (2007)

for example 2-(2′-hydroxyphenyl) benzene sulfinate (HBPSi), that are difficult to obtain with monocultures.

Ismail *et al.* (2016) claim that microbial consortia have a better potential for a practical BDS application due to the existence of synergistic or cooperative effects between all the species present. This consortium could desulfurize 4-methyl DBT and 4,6-dimethyl DBT in addition to DBT. The species richness and diversity depended on the substrate used and the 4-methyl DBT culture presented the highest species richness and diversity. Microbial consortia may therefore be more versatile for practical application.

12.8.4 Surfactants and BDS

Surfactants are amphiphilic compounds (containing hydrophobic and hydrophilic portions) that reduce the free energy of the system by replacing the bulk molecules of higher energy at the interface. They have been used industrially for their desirable properties, including solubility enhancement, surface tension reduction, wettability, and foaming capacity. Only a few studies were reported for surfactant's application in BDS. Efficiency of sulfur removal is likely to be related to oil droplet size. This suggests that the heavy oil/water emulsion products currently being developed would be particularly suited as targets for biodesulfurization. One problem to be resolved is whether the chemical surfactants which create such oil/water emulsions would be toxic to the process organisms or against the characteristic adhesion mechanisms of bacteria to oil droplet surfaces (Shenan, 1996). Biosurfactants, produced either intracellularly or extracellularly by stabilize emulsions, show less toxicity and possess higher biodegradability compared to chemical surfactant and exhibit lower Critical Micelle Concentration (CMC) values compared to their chemical counterparts which means less surfactant is required earlier to achieve maximum lowering of surface tension. Biosurfactants show hydrophilic and lipophilic balance (HLB) values ranging from 16 to 20, which is related to the longer hydrocarbon chain length of the molecule (Muthusamy *et al.*, 2008; Banat *et al.*, 2010). Li and Jiang (2013) reported the usage of surfactants, enhanced the BDS of type of heavy oil, are commonly used in marine expeditions. This was attributed to the formation of emulsion, which is one of the most effective methods for reducing the viscosity of heavy oil, thus optimizing the BDS process (Bandyopadhyay *et al.*, 2013a).

The bioprocessing of fuels takes place regularly in oil and water emulsion to retain microbial viability and activity. The oxidation reaction occurs in the cytoplasm requiring mass transfer of DBT from the oil to the cell in the aqueous phase. It is suggested, however, that in the case of *Rhodococcus*, DBTs are assimilated directly from the oil due to their hydrophobic membrane (Monticello, 2000) or by the bio- or exo-surfactant which facilitate the uptake of DBTs from the water phase (Ohshiro and Izumi, 2000; Han *et al.*, 2001).

Indeed, preliminary results from the Oak Ridge National Laboratory (ORNL) showed that *Rhodococcus* sp. forms a better water-hexadecane emulsion (surface tension of 0.35e-3 N/m) than *Escherichia coli* (0.58e-3 N/m), which was attributed to a non-identified biosurfactant and a more hydrophobic cell wall structure (Borole *et al.*, 2002). Because of biosurfactant production and solvent resistance (log P from 3.1 to 3.4) of some *Pseudomonas* strains (Isken *et al.*, 1999), several recombinant *Pseudomonas* were designed to acquire the desulfurization (Dsz) genes (Gallardo *et al.*, 1997; Macfarland *et al.*, 1998). The recombinant bacteria *Pseudomonas aeruginosa* EGSOX was reported to desulfurize 95% of a solution containing 6.4 ppm sulfur (37 ppm DBT) in 24 h more efficiently than the native host *Rhodoccocus erythropolis* with accumulation of a known bactericide, i.e. 2-HBP. This issue was minimized by the expression of Dsz enzymes (C and A) from *R. erythropolis* and a flavin oxidoreductase from *Vibrio harveyi* in a recombinant *E. coli* causing an increase in the DBT removal, but a decrease in the rate of production of 2-HBP (Reichmuth *et al.*, 2000). However, the oxygen mass transfer limitation in the biphasic media should be taken into consideration. The oxygenase, besides using the oxygen as a co-substrate, also uses it in the endogenous metabolism. Therefore, in order to allow the oxygenase to compete for the oxygen in endogenous respiration during biotransformation, it is necessary to increase the oxygen concentration using oxygen-enriched air and/or an increased pressure. The issue of substrate transport in desulfurization was also investigated in *Pseudomonas putida* IFO13696, which is a recombinant strain containing cloned dsz genes (Noda *et al.*, 2003b). This culture can desulfurize DBT in water, but not in tetradecane unless the hbcABC genes (encoding substrate transport proteins) were also provided. Xiong *et al.* (2007) reported Vitreoscilla hemoglobin technology, a strategy to improve the transfer, supply, and store of oxygen in vivo, which was used to enhance BDS activity under low aeration in two liquid phase systems, introducing the vgb gene into *R. erythropolis.* The results revealed 73% of sulfur reduction in a diesel sample oil under hypoxic conditions.

The solubility of DBT in water is very low, about 0.05 mM, although it can be increased with the surfactants produced by the cells. There are many reports concerning aerobic DBT metabolism. DBT-BDS occurs inside the cell with its entrance in the cytoplasm, possibly from the organic phase, after transient adsorption (Monticello, 1998). However, Gallardo *et al.* (1997) reported that DBT is present in the aqueous phase before its entrance into the cell. Marzona *et al.* (1997) found that the binding of benzothiophene (BT) and DBT with cyclodextrins (CD) could strongly enhance their solubility in water. Jiang *et al.* (2002), Feng *et al.* (2006), Wang *et al.* (2006), and Li *et al.* (2008c) reported the enhancement of BDS of diesel oil in the presence of Tween 80. Caro *et al.* (2007b) reported that the addition of β-cyclodextrins enhanced the BDS of DBT in oil–water emulsions by the aerobic *Rhodococcus erythropolis* IGTS8 strain, as it increased the diffusion of DBT into the aqueous phase and avoided the accumulation of 2-HBP, improving the BDS yield. Moreover, it was noticed that higher biocatalyst cell concentrations decreased 2-HBP production rates, which was explained by the combination of both inhibition effects and mass transfer limitations, while Derikvand *et al.* (2014) reported the utilization of Span-80 for enhancement of the BDS process.

In the oil desulfurization process, the production of a large amount of biosurfactants by biocatalysts would enhance the bioavailability of PASHs in fuel oils and at the same time increase the reaction rate by improving the contact between the oil and aqueous phases (Kim *et al.*, 2004).

Production of biosurfactant could promote organic dispersing in the aqueous phase, which would improve the mass transfer of PASHs between organic and aqueous phases. Zytner *et al.* (2006) reported that bacterial numbers and efficiency of biodegradation decreases at higher hydrocarbon concentrations and attributed this to the toxicity of hydrocarbons at these levels. Sadouk *et al.* (2008) reported that *Pseudomonas* rhamnolipid biosurfactant affects the rate of biodegradation in two ways: by increasing solubilization and dispersion of the hydrocarbons and by changing the affinity between microbial cells and hydrocarbons by inducing an increase in cell surface hydrophobicity.

Although in the absence of a chemical surfactant, BDS in *Agrobacterium* sp. NSh10, *Candida parapsilosis* NSh45, and *R. erythropolis* IGTS8 (El-Gendry, 2004) cultures were accompanied by the formation of a strong stable emulsion; this is indirect evidence that there might have been a production of a biosurfactant(s) (Kaufman *et al.*, 1998). Other *Rhodococcus* sp. has been reported to produce glycolipids in the presence of hexadecane, as well as in the presence of some crude and diesel oils (Finnerty and Singer, 1984). Simultaneously with the BDS of petroleum cuts using *Rhodococcus erythropolis, Rhodococcus globerulus, Grodonia alkanivorans, Pseudomonas putida*, and *Bacillus subtilis*, these microorganisms produce some biosurfactants with different chemical nature and molecular size as byproducts of their metabolic pathway (Wang and Krawiec, 1994; Yang and Marison,

2005; Mukhopadhyay *et al.*, 2005, 2007; Ma *et al.*, 2006; Yang *et al.*, 2007b; Mohabali *et al.*, 2007; Alcon *et al.*, 2008). Bandyopadhyay and Chowdhury (2014) also got the advantage of both BDS of hydrodesulfurized diesel oil and production of biosurfactant.

12.8.5 Application of Nanotechnology in the BDS Process

The BDS process consists of a two-phase system in which the whole cell as a biocatalyst in the aqueous phase interacts with the oil phase (Takada *et al.*, 2005). One of the major hurdles facing the application of BDS on a real industrial scale is that, upon using free cells, the cost intensive unit operations, e.g. centrifugation, is downstream of the process. In addition, there is a possibility to have cell contamination in the final products (Guo *et al.*, 2006).

Application of immobilization would solve many free cell drawbacks, including enhanced stability of the system, easy separation of cells, minimizing or eliminating the cell contaminations in the products, convenient recovery, and re-use of cells which enable their frequent use in the process (Zhang *et al.*, 2010).

In the BDS process, the methods for bacterial cell immobilization consider the cells' entrapment (Hou *et al.*, 2005) and adsorption (Hwan *et al.*, 2000; Zhang *et al.*, 2007b).

However, BDS activity of the entrapped cells will decrease after each step as a result of the reduction of cofactors, such as $NADH_2$ and $FMNH_2$, in the 4S-pathway (Yan *et al.*, 2008). Moreover, the entrapment technique itself often leads to a decrease in biocatalytic activity (Karsten and Simon, 1993). This would be solved by the exogenous addition of co-factors.

Bio-modification of chemical supports using cell immobilization via adsorption increases the interaction between reactants present in two-phase systems (Hou *et al.*, 2005; Shan *et al.*, 2005b; Feng *et al.*, 2006). These materials must be inert to biological attack, insoluble in the growth media, and nontoxic to microbial cells, hence, polymeric supports are proper choices for sorption capacity, chemical resistance, and mechanical strength of polymer support (Shao and Huang 2007) and the solubility parameter (Hansen, 1999; Rychlewska *et al.*, 2015) is the main factor in the selection of a polymer support.

However, few studies have evaluated the influence of inorganic supports on BDS activity (Chang *et al.*, 2000a; Zhang *et al.*, 2007b). BDS of dibenzothiophene (DBT) was also carried out by *R. erythropolis* IGST8 decorated with magnetic Fe_3O_4 nanoparticles (Ansari *et al.*, 2009). Moreover, BDS of DBT by Ca-alginate immobilized cells of *R. erythropolis* coated with nano-Al_2O_3 particles and surfactants were studied (Derikvand *et al.*, 2014). Immobilization of recombinant *Pseudomonas* by nano-sorbents and alginate for improvement of the BDS rate in biphasic systems was studied. Cells coated with magnetite (Fe_3O_4) nanoparticles (NPs) were found to express the same biocatalytic activity as free cells, however, they overcame the mass-transfer resistance of traditional entrapment immobilization processes (Guobin *et al.*, 2005; Shan *et al.*, 2005b; Li *et al.*, 2008b).

For a cost effective BDS process, the reaction time and catalyst longevity should be of 1 and 400 h, respectively (Pacheco *et al.*, 1999). Recycling of the biocatalyst is necessary in an industrial process and BDS biocatalysts must have sufficient operational stability. Naito *et al.* (2001) reported the improvement of the BDS process by immobilization of *R. erythropolis* KA2–5-1 by entrapping them with the photocrosslinkable resin prepolymer ENT-4000. The ENT-4000 immobilized cells could catalyze the BDS of model oil (DBT in n-tetradecane) repeatedly in this system for more than 900 h with reactivation. Moreover, the BDS of fuels takes place regularly into a three-phase system oil/water/biocatalyst. The transfer of PASHs from the oil phase to the water phase and then from water to the cells has been reported, where oxidation reaction occurs in the cytoplasm and limits the metabolism of PASHs. The more the hydrophobicity of the microorganism, the higher the rate of BDS as it would assimilate PASHs directly from the oil (Monticello, 2000). The availability of production of bio-surfactant would also help in increasing the rate of BDS due to the formation of oil/water emulsion that retains microbial viability and activity (Han *et al.*, 2001). The hydrophobic nature of *Rhodococcus* microorganisms makes them adhere preferentially to the oil–water interface in water–oil systems. This is advantageous in increasing the BDS rate because substrates are directly transferred from the oil to the cell, avoiding mass transfer limitations at this stage. In the case of *Rhodococcus* biocatalysts, a very stable fine emulsion (droplet sizes between 2 and 50 μm) is formed between the fuel and the cells (Borole *et al.*, 2002). Thus, the separation of the desulfurized fuel from the biocatalyst might not be easy, especially at high cellular densities. The difficulty of breaking such emulsions by centrifugation led to the invention of new separation schemes based on filters and hydrocyclones (Chen and Monticello, 1996; Yu *et al.*, 1998). However, details of these bioprocesses have not been published, so costs are difficult to evaluate. Chen and Monticello (1996) reported that more than one filter would be required for separation of the three phases (fuel, water, and biocatalyst); one filter is required for collecting either the liquid fuel or the aqueous phase as the filtrate, the retentate would then flow to a second filter which would collect the components that were not removed during the first step and, finally, the remaining retantate containing the biocatalyst would be preferably recycled. Other suggested solutions for overcoming the problem of the formation of the stabilized emulsions which cause a difficulty associated with separation include: (i) avoiding the formation of a stable water-in-oil emulsion in order to facilitate oil recovery, (ii) the use of emulsion-destabilizing chemical agents, and (iii) the application of a cell-immobilized BDS process (Chang *et al.*, 2000b; Naito *et al.*, 2001; Lee *et al.*, 2005). Biocatalyst stability, lifetime, and separation are crucial factors for the commercialization of BDS. Immobilized biocatalyst has some advantages: ease of separation, high stability, low risk of contamination, and long lifetime (Hou *et al.*, 2005). Compared to BDS, adsorptive desulfurization (ADS) has a much faster reaction rate (Song, 2003). Adsorbent preparation is the key of ADS. Recently, most adsorbents for desulfurization were based on π-complexation (Shan *et al.*, 2005a) or formations of metal-sulfur bonds (such as Ni-S, La-S) (Tian *et al.*, 2006). Adsorbents based on π-complexation are easy to regenerate, but their selectivity is very low, resulting in a loss of fuel quality. Meanwhile, adsorbents that form metal-sulfur bonds with sulfur have high selectivity, but are difficult to be regenerated. Hence, ADS technology also has a long way to go before being industrialized. If a desulfurization technology has both the high reaction rate of ADS and the high selectivity of BDS, it can increase the desulfurization rate without damaging fuel quality. Adsorbent is an important factor of this coupling technology because different adsorbents have different interactions to organic sulfur compounds and cells, which would affect their assembly onto cell surfaces and the desorption behavior of organic sulfur from them. Because the adsorbents are assembled on the cells' surfaces, the property of the cell surface is another factor which impacts coupling technology. Moreover, desulfurization conditions, such as temperature and volume ratio of oil to water phase, would also affect in-situ coupling technology.

In the last two decades, numerous studies have been carried out on BDS using whole cells (Maghsoudi *et al.*, 2001; Tao *et al.*, 2006; Yang *et al.*, 2007b; Caro *et al.*, 2008) or isolated enzymes (Monticello and Kilbane, 1994) in the free or immobilized form. As has mentioned before, the BDS of DBT occurs via a multi-enzyme system that requires cofactors (e.g. NADH, the reduced form of nicotinamide adenine dinucleotide). The use of enzymes is disadvantageous since extraction and purification of the enzyme is costly and, frequently, enzyme catalyzing reactions require cofactors which must be regenerated (Setti *et al.*, 1997). Therefore, BDS can often be designed by using whole cell biotransformation rather than that of the enzyme. However, there are still some bottlenecks limiting the commercialization of the BDS process.

One of the challenges is to improve the current BDS rate by about 500 fold, assuming the target industrial process is 1.2–3 mmol/g of dry cell weight (DCW)/h (Kilbane, 2006). When free cells are used for petroleum BDS, deactivation of the biocatalyst and troublesome oil–water–biocatalyst separation are significant barriers (Konishi *et al.*, 2005b; Yang *et al.*, 2007b). Cell immobilization may give a solution to the problems, providing advantages such as repeated or continuous use, enhanced stability, and easy separation. However, the BDS rate in an immobilized cell system, especially applying the entrapment technique, would be slightly better or the same and sometimes slightly lower than that in a free cell system due to the diffusional resistance of reactants and products to and from the cells within the support (Shan *et al.*, 2003).

Immobilization has many advantages compared to free cells, including enhanced stability of the system, easy separation of cells, minimizing or eliminating cell contaminations in the products, convenient recovery, and re-use of cells which enable their frequent use in the process (Zhang *et al.*, 2010). Moreover, if magnetic nanoparticles (MNPs) are added to the immobilizing system or used to decorate the applied microbial cells, this will add a more advantageous strategy which is magnetic separation. Magnetic separation is a promising technology in the support systems for immobilization since rapid separation and easy recovery of immobilized cells could be reached in an external magnetic field and the capital and operation costs could also be reduced (Zakaria *et al.*, 2016).

However, Moslemy *et al.* (2002) reported that an immobilizing system provided the bacterial cells with a protective solid barrier, which may have somewhat limited the diffusion of diesel hydrocarbons, reducing the bioavailable concentration in the inner space of the beads with respect to that in the bulk liquid.

Alginate is one of the unique biopolymers that can be employed for the encapsulation of cells and enzymes, which has found extensive applications in industrial processes because of advantages like biocompatibility, simplicity, and low cost (Blandino *et al.*, 1999). Evidently, maximum BDS efficiency can be achieved by setting the parameters at their optimized values, for example optimization of the size of bacterial cells, concentration of alginate, size of the beads, and the application of surfactant, nano γ-Al_2O_3, and magnetic nanoparticles of Fe_2O_3 (Li *et al.*, 2008b; Zhang *et al.*, 2010, 2011).

The efficiency of BDS in immobilized beads is reported to be increased with the decrease of bead diameters, but to a certain limit (Ryu and Wee, 2001; Li *et al.*, 2008d; Derikvand and Etemadifar, 2014; Derikvand *et al.*, 2014). For example, the specific desulfurization rate of 1.5 mm diameter calcium alginate immobilized *P. delafieldii* R-8 beads was found to be 1.4 fold higher than that of 4.0 mm (Li *et al.*, 2008d). This is because at the lower sizes, the surface to volume ratio would increase and more interactions would occur between the cells and DBT (Gautier *et al.*, 2011). Moreover, the stability and re-usability of the beads improved proportionately with an increase in alginate concentration, but to a certain limit. In fact, the beads with higher concentration of alginate are more spherical and could be used in additional cycles in the BDS process. However, the higher mechanical strength of the beads adversely affected BDS efficiency by increasing the mass transfer resistance in the beads' wall. In high alginate concertation, the mass transfer of DBT and 2-HBP through the wall of the beads reduced, which in turn decreased the efficiency of the process. In contrast, beads with low alginate concentration are relatively soft and would favor leakage over mass transfer (Srinivasulu *et al.*, 2003; Derikvand and Etemadifar, 2014). Potumarthi *et al.* (2008) reported that the decreasing alginate concentration leads to an increase in the efficiency of the process. Zhang *et al.* (2010) also reported that the decrease in alginate concentration leads to an increase in efficiency of the beads. The cells of *S. subarctica* T7b immobilized with sodium alginate completely degraded the DBT (0.54 mM) after 72 h (Gunam *et al.*, 2013).

Although 1% alginate concentration was found to be the optimum concentration for maximum BDS efficiency of immobilized *R. erythropolis* R1 (Derikvand *et al.*, 2014), 2-HBP production after 24 h using beads with 1% alginate concentration was 1.75 fold more than that of 3% alginate concentration (Derikvand and Etemadifar, 2014). However, beads with 1% alginate concentration were not rigid enough and had low stability, especially when they were used more than once. Therefore, beads with 2% (w/v) sodium alginate concentration were selected as a better option for multiple applications (Derikvand and Etemadifar, 2014; Derikvand *et al.*, 2014). Span-80 was reported to express a greater impact on the BDS of DBT, as compared to Tween-80, by alginate immobilized *R. erythropolis* R1 which is probably because of a greater reduction in the surface tension of the water medium in presence of Span-80 (Derikvand and Etemadifar, 2014). The desulfurization rate by calcium alginate immobilized *P. delafieldii* R-8 beads was also increased by the addition of 0.5% Span 80 by about 1.8 fold compared with that of the untreated beads (Li *et al.*, 2008d). These results suggested that the rate of mass transfer of substrate to gel matrix could be the limiting step of BDS. Thus, assembling γ-Al_2O_3 nano-sorbent onto the surfaces of cells was developed as a novel BDS technology leading to selective adsorption of DBT from the organic phase. The rate of adsorption was far higher than that of BDS and, thus, the transfer rate of DBT was improved in that system. The desulfurization rate of the cells assembled with nano-sorbents was about twice as high as that of free cells (Guobin *et al.*, 2005).

Other researchers have attempted to increase the efficiency of cells and decrease the cost of operations in a BDS process (Shan *et al.*, 2005b). The magnetic-PVA beads were prepared by a freezing–thawing technique under liquid nitrogen and the beads were found to have distinct super-paramagnetic properties. The desulfurization rate of the immobilized cells has reached 40.2 mmol/kg/h, which is twice that of free cells. The heat resistance of the cells apparently increased when the cells were entrapped in the magnetic-PVA-beads. The cells immobilized in the magnetic polyvinyl alcohol (PVA) beads were found to be stably stored and be repeatedly used over 12 times for BDS. The immobilized cells could be easily separated by magnetic field. In order to understand cell distribution in magnetic-PVA-beads, the sections of the beads after being repeatedly used six times were observed by scanning electron microscopy (SEM). A highly macro-porous structure is found in the beads in favor of diffusion of substrates and dissolved gas. On average, the size of the beads was about 3 mm. It is evident that the R8 cells mainly covered the edges and sub-marginal sections of the bead, with no cells in the center of the bead because of insufficiencies of oxygen and nutrients and gradual autolysis (Shan *et al.*, 2005b).

In-situ coupling of ADS and BDS is a new desulfurization technology for fossil oil. It has the merits of high-selectivity of BDS and a high-rate of ADS. It is carried out by assembling nano-adsorbents onto surfaces of microbial cells. For example, the combination of ADS and BDS and a proposed in-situ coupling technology of them has been described that increases the adsorption of DBT from the oil phase and transfers it quickly into the cell for BDS, thus increasing the overall rate of desulfurization by about 2.5 fold (Shan *et al.*, 2005c).

Generally, bacterial cell surfaces have a negative electrical charge which increases their affinity to adsorb positively charged nano-γ-Al_2O_3 particles. Adsorption of γ-Al_2O_3NPs particles on cell surfaces increases the consumption of DBT and production of 2-HBP in the immobilized system (Derikvand *et al.*, 2014). However, if the accumulation of the NPs on the cell surfaces exceeds a certain limit, the cell activity would decrease drastically. It has been previously reported (Berry *et al.*, 2005; Hayden *et al.*, 2012) that if the adsorbed particles on a cell occupied more than 2/3 of the cell surface area, it would have a negative impact on cell activity. In fact, large specific surface area of γ-Al_2O_3 NPs enables them to absorb DBT and to provide cells with a higher concentration without reducing the cell activity. Moreover, the NPs can create pores in the cell membrane which facilitate transfer of DBT across the cell wall (Zhang *et al.*, 2011). Zhang *et al.* (2008) also showed that both DBT consumption and 2-HBP production rate in the presence of γ-Al_2O_3NPs were higher than those of a free cell system and these particles were more efficient than other adsorbents, such as active carbon particles. The use of gum Arabic as a modifier to the NPs in order to overcome the agglomeration of the inorganic compounds in an aqueous solution, may hinder its optimal performance as a result of decreased dispersion has been reported to increase the ADS capacity of *P. delafieldii* R-8 from 0.56 mmol S/g (Al_2O_3) to 0.81 mmol S/g (Al_2O_3). This was attributed to the enhancing dispersion of Al_2O_3 in the aqueous phase (Zhang *et al.* 2007b). In that study, compared with the unmodified **γ-Al**$_2$O$_3$ NPs, the BDS rate by adsorbing the gum Arabic-modified **γ-Al**$_2$O$_3$ NPs onto the surfaces of *Pseudomonas delafieldii* R-8 cells increased to 25.7 mmol/kg/h (i.e. 1.44 fold) (Zhang *et al.*, 2007b).

Zhang *et al.* (2011) used γ-Al_2O_3NPs on the surface of magnetic immobilized cells and found out that a combination intensified the BDS activity of the immobilized cells. Derikvand *et al.* (2014) also reported the combination of nano-γ-Al_2O_3 with alginate immobilized cells is very effective in BDS. Derikvand and Etemadifar (2014) reported the production of 2-HBP by beads, including *R. erythropolis* R1 cells assembled with nano γ-$Al_2O_{3,}$ after 24 h is two fold more than that of control beads, that is without NPs. This is attributed to the specific surface area of nano-γ-Al_2O_3 and pore creation that causes more adsorption of DBT on the cell surface and easy transfer into the cell. However, for a longer incubation period of 48 h, the concentration of the produced 2-HBP was found to be lower than the control in the absence of γ-Al_2O_3. This was attributed to the adsorption capacity of nano-γ-Al_2O_3 particles and some of the produced 2-HBP would remain at the cell surface and, thus, cannot be released completely in a medium.

The in-situ cell separation and immobilization of bacterial cells for BDS which were developed by using super-paramagnetic magnetite NPs has also been reported (Li *et al.*, 2009a). The magnetite nanoparticles (Fe_3O_4 NPs) were synthesized by co-precipitation followed by modification with ammonium oleate. Surface-modified NPs were mono-dispersed and the particle size was on the order of 13 nm. After adding magnetic fluids to the culture broth, *Rhodococcus erythropolis* LSSE8–1 cells were immobilized by adsorption and then separated with an externally magnetic field. Analysis showed that the NPs were strongly adsorbed to the surface and coated the cells. Compared to free cells, the coated cells not only had the same desulfurizing activity, but could also be easily separated from fermentation broth by magnetic force. It was believed that oleate modified magnetite nanoparticles

(Fe_3O_4 NPs) adsorbed on the bacterial cells mainly because of the nano-size effect and hydrophobic interaction. Ansari *et al.* (2009) reported DBT-BDS using decorated *R. erythropolis* IGT8 with magnetic Fe_3O_4 NPs (the average size of the prepared NPs was 45–50 nm and the ratio of NPs/cells was 1.78 w/w). The BDS efficiency of decorated cells was 56% higher than that of non-decorated cells. About 50.8 emu/g saturation magnetization of *R. erythropolis* LSSE8–1 cells was obtained using supermagnetic Fe_3O_4 NPs modified with oleate (Li *et al.* 2009c). This enhanced the separation after BDS process and also increased the permeability of the organism cells to DBT, consequently, enhancing the performance of the biocatalytic process.

From the point of commercial application, the *Rhodococcus* strains possess several properties favorable for desulfurization over *Pseudomonas* in an oil–water system. First, the hydrophobic nature of *Rhodococcus* makes them access preferential Cx-dibenzothiophenes from the oil, resulting in little mass-transfer limitation (Le Borgne and Quintero, 2003). Therefore, there has been an attempt to develop a simple and effective technique by integrating the advantages of magnetic separation and cell immobilization for the BDS process with Gram +ve *Rhodococcus erythropolis* LSSE8–1 and Gram –ve *P. delafieldii* R-8 (Li *et al.*, 2009b). Cells were grown to the late exponential phase and the culture was transferred into Erlenmeyer flasks. A volume of magnetic fluids was added and mixed thoroughly and the microbial cells were coated by adsorbing magnetic NPs. The ammonium oleate-modified magnetite NPs formed a stable suspension in distilled water and the magnetic fluid did not settle during 8 months of storage at room temperature. The transmission electron microscope (TEM) analysis of Fe_3O_4 NPs showed that the particles have an approximately spherical morphology with an average diameter of about 13 nm. The Fe_3O_4 NPs on the cell surface were not washed out by deionized water, ethanol, saline water (0.85 wt.%), or phosphate buffer (0.1 M, pH 7), thus, there was little cell loss or decrease in biomass loading when cells were coated with Fe_3O_4 NPs. This outcome was different from that obtained with cells immobilized by traditional adsorption to a carrier. The cells coated with Fe_3O_4 NPs were super-paramagnetic. Therefore, the cell-nanoparticle aggregates in aqueous suspension could be easily separated with an externally magnetic field and re-dispersed by gentle shaking after the removal of the magnetic field. For magnetic separation, a permanent magnet can be placed at the side of the vessel. After several minutes (3–5 min), the coated cells can be concentrated and separated from the suspension medium by decantation. This one-step technology, namely in-situ magnetic separation and immobilization of bacteria, efficiently optimized the BDS process flow and much less of Fe_3O_4 NPs was needed.

An adsorption mechanism between the Fe_3O_4 NPs and desulfurizing cells has also been proposed (Li *et al.*, 2009b). The large specific surface area and the high surface energy of the Fe_3O_4 NPs (i.e. the nano-size effect) ensure that the magnetite nanoparticles (MNPs) are strongly adsorbed on the surfaces of microbial cells. Furthermore, the hydrophobic interaction between the bacterial cell wall and the hydrophobic tail of oleate modified Fe_3O_4 NPs may play another important role in cell adsorption. The suspension of oleate-modified Fe_3O_4 NPs was considered a bilayer surfactant stabilized aqueous magnetic fluid (Liu *et al.*, 2006). The iron oxide nano-crystals were, first, chemically coated with oleic acid molecules, after which the excess oleic was weakly adsorbed on the primary layer through the hydrophobic interaction between the subsequent molecule and the hydrophobic tail of oleate. Since the bacterial cell wall is composed of proteins, carbohydrates, and other substances (such as peptidoglycan, lipopolysaccharide, and mycolic acid), the extracellular matrix can form hydrophobic interactions with the hydrophobic tail of oleates of nanoparticles.

A microbial method has also been used to regenerate desulfurization adsorbents (Li *et al.*, 2006). Most of the sulfur compounds can be desorbed and removed and heat losses during the bio-regeneration process are markedly reduced. The particle size of cells is similar to that of desulfurization adsorbents, which is several microns. Therefore, it is difficult to separate regenerated adsorbents and cells. Super-para-magnetism is an efficient method to separate small particles. To solve the problem of separation of cells and adsorbents, magnetite nanoparticle Fe_3O_4 NP modified *P. delafieldii* R-8 cells were used in the bio-regeneration of adsorbents. Biodesulfurization with *P. delafieldii* R-8 strains coated with MNPs has been previously reported (Shan *et al.*, 2005c) and cells coated with MNPs can be separated and reused for several times.

There is also a report (Li *et al.*, 2008c) of bio-regeneration of desulfurization adsorbents AgY zeolite with magnetic cells. Super-paramagnetic magnetite NPs are prepared by the co-precipitation method, followed by modification with ammonia oleate. Magnetic *P. delafieldii* R-8 cells can be prepared by mixing the cells with MNPs. The BDS activity of the magnetic cells is similar to that of free cells. When magnetic cells were used in the bio-regeneration of desulfurization adsorbents AgY, the concentration of DBT and 2-hydroxybiphenyl with free cells was a little higher than that with magnetic cells. The adsorption capacity of the regenerated adsorbent, after being desorbed with magnetic *P. delafieldii* R-8, dried at 100 °C for 24 h and calcined in the air at 500 °C for 4 h, is 93% that of the fresh one. The magnetic cells can be separated from an adsorbent bio-regeneration system after desulfurization with an external magnetic field and, thus, can be reused.

Zhang *et al* (2011) reported the assembling of γ-Al_2O_3 NPs onto magnetic immobilized *R. erythropolis* LSSE8–1-vgb. The optimum ratio of cells/Fe_3O_4 NPS was 50:1 (g/g) for better BDS efficiency and separation of magnetic immobilized cells, taking into consideration that the optimum ratio between Fe_3O_4 NPS/ γ-Al_2O_3 NPs is 1:5 (g/g). Assembling of γ-Al_2O_3 NPs which is characterized by small particle size, large pore volume, large specific surface area, and the strongest electrostatics interaction with desulfurization cells onto magnetic immobilized cells enhanced the BDS efficiency, as it improved the adsorption of DBT from the oil phase. The γ-Al_2O_3 NPs/magnetic immobilized cells have been used in three successive BDS cycles each of 9 h and their BDS efficiency was 20% higher than that of magnetic immobilized cells. The activity decreased by less than 10% throughout the three successive cycles. So, integrating γ-Al_2O_3 NPs with Fe_3O_4 NPS in biocatalyst preparation would enhance the BDS efficiency and the cells can be collected and reused conveniently by an external magnetic field.

Etemadifar *et al.* (2014) reported the decoration of *Rhodococcus erythropolis* R1 with oleate-modified magnetic Fe_3O_4 NPs (average particles size < 10 nm) where response surface methodology (RSM) was used to optimize the BDS conditions of DBT in a biphasic system (DBT in tetradecane/water). The DBT concentration, incubation temperature, and the interaction of these two factors have a significant effect on the BDS efficiency of the decorated cells. The hydrophobic nature of the *Rhodococcus* strains, their resistance to high concentration of DBT and solvents, and the decrease of DBT toxicity as it was dissolved in tetradecane would enhance the adsorption of DBT from the oil phase and, consequently, the BDS efficiency (Monticello, 2000; Bouchez-Naitali *et al.*, 2004). The first and third enzymes of the 4S-pathway ($D_{SZ}C$ and $D_{SZ}B$) are reported to be more sensitive to temperature changes relative to the other enzymes involved in the 4S-pathway and they are BDS rate limiting (Furuya *et al.*, 2001). The BDS efficiency of the free and coated cells was nearly the same and the optimum DBT concentration, temperature, and pH for maximum BDS reported to be 6.67 mM, 29.63 °C, and 6.84, respectively.

Kareem (2014) reported the enhancement anaerobic kerosene BDS using NP decorated *Desulfobacterium indolicum* at 30 °C and atmospheric pressure where sulfur content decreased from 48.68 ppm to 13.76 ppm within 72 h.

Xu *et al.* (2015) reported the immobilization of the haloalkaliphilic S-oxidizing bacteria *Thialkalivibrio versutus* D301 using super-paramagnetic Fe_3O_4 NPs under haloalkaliphilic conditions (pH 9.5) to remove sulfide, thiosulfate, and polysulfide from wastewater. The Fe_3O_4 NP-coated cells expressed similar BDS activity like that of free cells and can be reused for six batch cycles.

Mesoporous materials in the nano-scale (2–50 nm) are characterized by large surface area, ordered pore structures with suitable sizes, high chemical and physical stability, and can be chemically modified with functional groups for the covalent binding of molecules that make them very applicable in adsorption, separation, catalytic reactions, sensors, and immobilization of huge biomolecules like enzymes (Li and Zhao, 2013; Juarez-Moreno *et al.*, 2014). The participation of four enzymes and two cofactors in BDS technology makes it difficult to use the free enzymes. However, recent studies have shown that the use of enzymes immobilized on nanoparticles achieved good levels of efficiency of BDS (Derikvand *et al.* 2014). Enzymatic immobilization in nano-materials has some advantages, such as an increase in catalytic activity, reduction in protein miss-folding, thermal stability, increase in solvent tolerance, reduction of denaturation caused by the presence of organic solvents, and mixtures of water-organic solvent molecules (Carlsson *et al.*, 2014). Carver *et al.* (2009) reported immobilization (i.e. conjugation) of chloroperoxidase (CPO) enzymes with latex nanoparticles (La-NPs) after its functionalization with chloromethyl groups that can react with the amine groups of the protein (i.e. the enzyme) to form a stable covalent bond. The average size of the conjugated NPs was reported to be 180 nm, which can be applied in bio-oxidative fuel streams, followed by physical or chemical removal of the produced sulfoxide and sulfone to produce ultra-low sulfur fuels. Vertegel (2010) reported immobilization of CPO (as a good candidate for ODS in non-aqueous media) and myoglobin (MYO, for its availability, represents a good candidate for industrial scale) with La-NPs (40 nm) that were modified with chloromethyl groups. The conjugated and free enzymes were used for DBT oxidation in the presence of H_2O_2 in a non-aqueous system (DBT in hexane) and two-phase system (DBT in acetonitrile/water, 20% v/v). For CPO in a two-phase system and a non-aqueous one, free and conjugated enzymes expressed nearly the same activity, but the stability of the free enzymes was much lower than conjugated ones because their activity deteriorate much faster. The activity of conjugated CPO was approximately 5 times higher than that of free cells after 2 d of storage. However, the free MYO expressed better performance than the immobilized ones in 2-phase systems, while the immobilized MYO expressed twice the activity in a non-aqueous system, which is an encouraging result, since this protein, as well as the similar heme-containing protein hemoglobin, can be readily available in large quantities from animal processing industries. The nano-immobilized enzymes are easily separated, stable over time, and can be taken as a good prospect for a recyclable desulfurization agent.

Juarez-Moreno *et al.* (2015) reported the immobilization of CPO into (1D) γ-Al_2O_3 nano-rods through physical adsorption. The dimensions of CPO are reported to be 3.1 nm x 5.3 nm x 5.5 nm (Sundaramoorthy *et al.*, 1995), the pore size needed for CPO immobilization is > 5.5 nm, the pore size of the calcined prepared γ-Al_2O_3 nano-rods at 700 °C (> 6 nm) and 900 °C (21.3 nm) are big enough for CPO immobilization. There is a high affinity between negatively charged CPO and positively charged prepared γ-Al_2O_3 nano-rods caused by favorable electrostatic interactions between the enzyme and the support. Immobilized enzymes expressed better oxidation for DBT to DBTO than those of free enzymes and γ-Al_2O_3 nano-rods at 700 °C expressed the highest activity, encouraging its application in bio-oxidative desulfurization of fuel streams.

Although the broad versatility of CPO for the oxidation of chemically diverse organosulfur compounds highlights the potential of this enzyme for applications in the treatment of complex streams (Ayala *et al.*, 2000), the largest unsolved challenge with CPO is that no heterologous, stable expression of this enzyme in a suitable host has been obtained yet (Conesa *et al.*, 2002). Thus, peroxide stability, substrate partition improvement, and overproduction issues for large scale applications have been restricted to nongenetic approaches (Ayala *et al.*, 2007).

Although, of the achieved improvements in BDS, the BDS rate is still low relative to conventional desulfurization techniques. Moreover, other process issues should be taken into consideration, including biocatalyst production and searching for a cheap source for obtaining active biocatalyst with a high selective specific BDS activity, bioreactor design, and configuration, oxygen mass transfer and uptake rate, the substrate (i.e. the fuel) mass transfer, and decrease the cost of the process as much as it could be (Mohebali and Ball, 2016).

12.9 Production of Valuable Products

Development of any process for use in the petroleum industry is extremely challenging because huge volumes of oil must be treated inexpensively. On the other hand, bioprocesses have found the greatest success in applications in the pharmaceutical industry where the production of smaller volumes of higher value products is typical. Thus, it is not surprising that a commercial oil BDS process has not yet been developed, but it may be possible to use currently available biocatalysts for alternative applications, such as the detoxification of chemical warfare agents (Kilbane II and Jackowski, 1996; Prokop *et al.*, 2006) or the production of surfactants, antibiotics, anticancer compounds (Rezanka *et al.*, 2006), diabetes treatments, and polythioesters which can be used to make thermoplastics and biodegradable plastics (Khairy *et al.*, 2015). Oxidative desulfurization and reactive adsorption are two emerging processes for the desulfurization of petroleum, but they produce waste streams that contain high concentrations of organosulfur compounds (Stanislaus *et al.*, 2010; Jiang *et al.*, 2016). These waste streams could be subsequently treated using biodesulfurization to produce sulfoxides, sulfones, sulfonates, phenols, and phenyl styrenes. Additionally, the reaction products of desulfurization enzymes, particularly hydroxylated compounds, can be subsequently chemically reacted to add nitrate, halogen, alkyl, or acyl groups (Li and Chan, 2007), so a variety of higher value byproducts can be made from petroleum by combining oxidative desulfurization and/or reactive adsorption with biodesulfurization. Thus, if desulfurization biocatalysts are considered as tools to selectively modify organosulfur compounds to create chemical derivatives, then the range of applications for desulfurization biocatalysts in industrial processes can be greatly expanded by imaginative individuals (Kilbane, 2016a, b, 2017; Kilbane and Stark, 2016).

Monticello *et al.* (1985) reported that hydroxybiphenyl (which is also known as orthophenyl phenol or 'dowicide number 1') is a potent industrial biocide. Moreover, the 1,3-benzendiol, 5-hexyle, is reported to be produced during DBT-BDS by *Exophiala spinifera* strain FM as a way to overcome the toxicity of 2-HBP (Elmi *et al.*, 2015). Considering the isomer structure of this, the red insoluble 1,3-benzendiol, 4-hexyl, has anthelmintic and topical antiseptic usages and could be an important step for the production of anti-bacterial and anthelmintic drug. On the other hand, the purity of the final product tends to produce high yield drug and easy recovery by an organic phase extraction after fungal treatment (removal by filter).

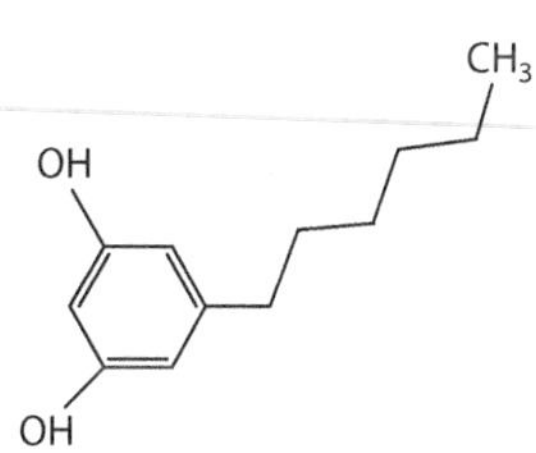

1,3-Benzenediol, 5-hexyl
$C_{12}H_{18}O_2$

Biosurfactants like hydroxybiphenyl, as a hydrotrope, are produced as byproducts (Labana *et al.*, 2005; Ma *et al.*, 2006a). The biosurfactants have many applications in the detergent, food processing, and pharmaceutical industries, as well as in advanced oil recovery from wells. Biosurfactants have similar properties of synthetic surfactants, but are less toxic, biodegradable, and can be produced in-situ at any contaminated petroleum sites. Bandyopadhyay *et al.* (2013a, b, 2014) reported the production of biosurfactant during the BDS of hydrodesulfurized diesel and spent engine oil using a continuous stirred tank reactor (CSTR).

2′-hydroxybiphenyl 2-sulfinic acid (HBPSi) (Figure 12.4) has been shown to have useful hydrotrope properties and to be effective as a

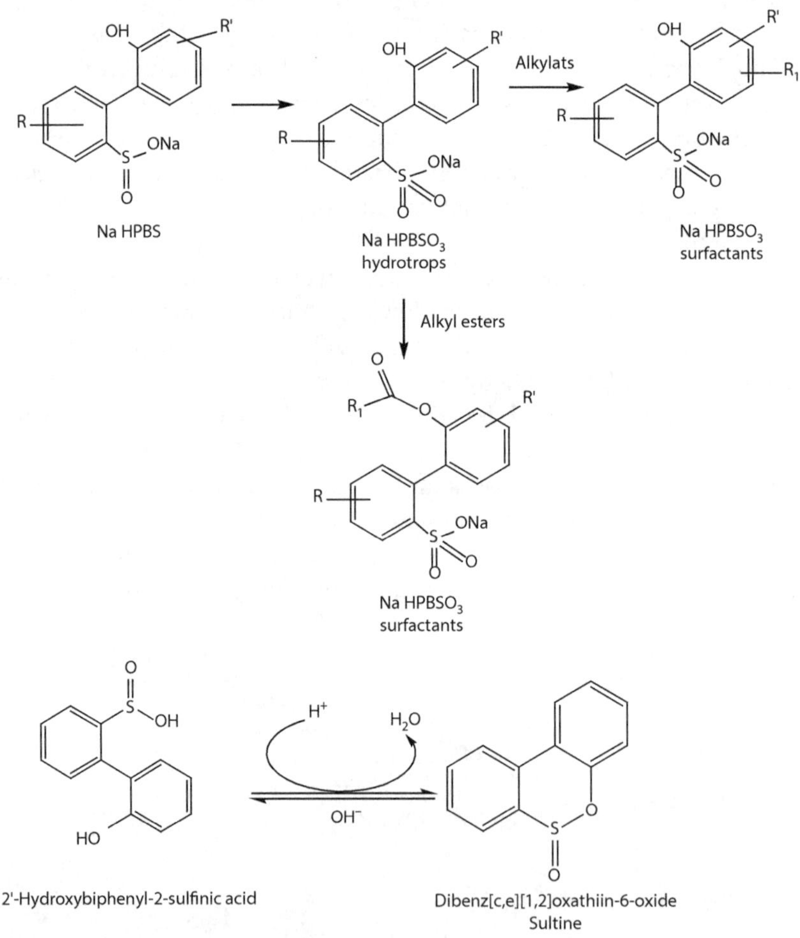

Figure 12.4 Conversion of the BDS Intermediate Cx-HPBSi to Value-Added Surfactants.

detergent when derivatized with a long chain alkyl side chain (Lange and Lin, 1998; Lange *et al.*, 1999; Johnson *et al.*, 2000). Terminating the reaction of 4S-pathway before the last step and production of the phenolic end product (e.g. 2-HBP) to produce hydroxybiphenyl sulfinate (HPBSi) from DBT or Cx-HPBSi from the alkylated DBTs is a recommended technique for the short-term application of biotechnology for producing 'biopetrochemicals' from oil is a spin-off of the biodesulfurization process. These produced biosurfactants can decrease interfacial tension and can be used in enhanced oil recovery (EOR). Thus, using a specially developed biocatalyst in which the dszB gene has been eliminated, Cx-HPBSi can be produced in large quantities as a byproduct of the BDS process. Then, the rings of Cx-HPBSi can be alkylated directly or various esters can be synthesized at the hydroxyl group. The length of the side chain can be varied to produce either water- or oil-soluble detergents (Lange and Lin, 1998; Lange *et al.*, 1999; Johnson *et al.*, 2000) because the feedstock for this material is a low value, high-sulfur refinery stream and the conversion occurs at low temperatures and pressures, thus, the process economics for producing this material look quite favorable. However, the commercialization of this molecule and this process faces the same rate and process issues faced by BDS (Monticello, 2000). Moreover, one of the advantages is that the HBPSi is cyclized under acidic conditions, thus a sultine (Figure 12.4) would be easily recovered during the extraction of the culture supernatant incubation with DBT (Bressler *et al.*, 1998).

12.10 Storage of Fuel and Sulfur

The major microbial problem in the industry is the contamination of the stored products, which can lead to loss in the product quality, formation of sludge, and deterioration of pipe work and storage tanks, both in the refinery and at the end-user (Gaylarde, 1999). The fuel that presents the most serious microbial problems is diesel, mainly due to its hydrocarbon skeleton. The hydrocarbon chains most readily utilized are C10-C18. Shorter chain hydrocarbons may actually inhibit growth of some organisms. For example, SRB is rare to be found as a gasoline contaminant due to the toxic effect of gasoline hydrocarbons. However, *Hormoconis resinae* has been reported to grow in gasoline containing 13% alcohol for only 6 days after inoculation. The low volatility of diesel provides larger time and areas of contact between the microbe and the fuel (Gaylarde, 1999; Yemashova *et al.*, 2007). Long storage of diesel results in microbial growth causing sludge formation, fuel loss, and corrosion in diesel storage tanks

(Hartman, 1991). The use of biocides has proven ineffective due to the resistance developed by the microbes, as well as excessive concentration of biocides proving to be harmful to humans. However, the limiting factor to growth is probably availability of minerals, particularly phosphorus, which is generally present at <1 ppm in the fuel. Nitrogen and iron may also be important limiting nutrients. Moreover, a variety of additives are used to improve the stability of the fuel; these include compounds such as aliphatic amines, chelating agents, detergents, and corrosion inhibitors, some of which can act as a nutrient source for microorganisms. Not only this, but its enhanced affinity to water also contributes to elevated microbial growth. Thus, controlling the water content and addition of biocides would minimize microbial growth (Hill and Hill, 1993). Oxygen is normally present in sufficient quantities in distillate fuels and is continually replenished when tanks are refilled. However, even if the fuel becomes anaerobic, it is not protected from microbial attack since facultative organisms, such as the corroding microorganism SRB, *Hormoconis resinae* fungus, and anaerobes, continue to thrive. The presence of bacteria in diesel fuel is one of the main causes of engine breakdown. Bacteria and fungi will form insoluble particulate matter that can clog fuel filters, resulting in fuel starving and engine stoppage. They can also corrode metal surfaces, including storage tanks and pumps, and will form organic acids that contribute to fuel instability (Rodriguez *et al.*, 2010).

Srivastava and Nandan (2012) tried to overcome such problems through the optimization of sulfur content in diesel so that the losses due to microbial growth and corrosion would be minimal. Samples of 350, 400, 450, and 500 ppm sulfur content were prepared to analyze the corrosion rating and extent of growth of *Hormoconis resinae* in each sample by measuring the loss in the diesel and formation of sludge. Graphs were plotted to analyze the growth and corrosion in terms of loss/year against sulfur content and the sulfur content was optimized to minimize the total loss. It was proven that the increase in S-content increased the corrosion rate and the total corrosion loss/year. Since the increase in S-content increased the acidity of the diesel, which consequently increased the corrosion rate of the storage vessel, maintenance cost and the capital cost of the unit increased. However, microbial growth decreased with the increase of S-concentration and decrease in suspended solids. This was attributed to the increase in acidity which has negative impact on microbial metabolism and growth. Diesel is a food source of *Hormoconis resinae*; it breaks down hydrocarbon and generates sludge, therefore suspended waste is formed which will result in loss of diesel. Thus, upon the decrease in fungal growth, the sludge content formed is decreased. This denotes less loss of diesel and less clogging leading to less maintenance cost and less product loss. The microbial growth losses for diesel samples in refinery capacity per year decreased with the increase of S-content, but the total loss/year increased with the increase of S. As the two ratings showed opposite trends, they check on each other. The ratings obtained from the experiments are integrated to refinery capacity. Total loss was determined in refinery terms from the sum of corrosion loss and microbial growth loss. Thereby, plotting the total cost against S-content to obtain the S-content with lowest total loss, it was termed that C optimum ranged from 380–400 ppm. Thus, in conclusion, if microbial growth is the immediate problem of concern, the S-content can be kept as 400 ppm and if corrosion is given preference, then the S-content can be kept at 380 ppm (Srivastava and Nandan, 2012).

12.11 Process Engineering Research

There is an obvious progress in the field of biocatalysis in organic phase and bioengineering, throughout the design of new bioreactors. Proposals include the use of electro-spray type reactors (Kaufman *et al.*, 1997) or membrane reactors, which favor the contact between the organic and aqueous phases (Setti *et al.*, 1999; Berdugo *et al.*, 2002) and biphasic reactors with and without the reuse of biomass through immobilization of microorganisms. BDS technological issues include good reactor design, product recovery, and oil-water separations. BDS systems consist of three components: oil, aqueous, and cells (Figure 12.2), where the cells are distributed into three populations: free cells in the aqueous phase, oil-drop-adhered cells, and cells in aggregates in the aqueous phase. The number of mechanistic steps involved in the bio-conversion of DBT to HBP depends on the population of cells that is considered. For cells that form aggregates in the aqueous phase, there are three mechanistic steps (Figure 12.2). The first step is the transportation of DBT from the oil to the aqueous phase. The second step is DBT transportation from the external surface of the bacterial aggregate through the aggregate until DBT reaches a single cell's surface. The third and final step is the uptake of DBT by the cells and enzymatic desulfurization of DBT into HBP and sulfate via the 4S pathway. For free cells in the aqueous phase, the second step does not occur. For oil-adhered cells, neither the first nor the second step occurs because cells have access to DBT directly from the oil phase.

In order to make a BDS process competitive with deep HDS, a five-step process is needed: (i) production of active resting cells (i.e. biocatalysts) with a high specific activity, (ii) preparation of a biphasic system containing oil fraction, aqueous phase, and biocatalyst, (iii) BDS of a wide range of organic sulfur compounds at a suitable rate, (iv) separation of desulfurized oil fraction, recovery of the biocatalyst, and its return to the bioreactor, and (v) efficient wastewater treatment that would be more beneficial if it could be recycled for reuse. Each step is affected by a number of factors. Thus, briefly, a typical process consists of charging the biocatalyst, oil, air, and a small amount of water into a batch reactor (Figure 12.5). In the reactor, as the PASHs are oxidized to water-soluble products, the sulfur segregates into the aqueous phase. The oil-water-biocatalyst-sulfur-by-product emulsion from the reactor effluent is separated into two streams, namely, the oil (which is further processed and returned to the refinery) and the water-biocatalyst-sulfur-by-product stream. A second separation is needed to allow most of the water and biocatalyst to return to the reactor for reuse.

The sulfate concentration in a growing culture should be kept limiting for expressing the desulfurization enzymes even in the absence of an inducer (Konishi et at., 2005a). In order to simplify the system and lower the operation cost, Tangaromsuk *et al.* (2008) determined the optimal sulfur to carbon ratio for batch BDS of model oil was 0.6 wt.% DBT in n-hexadecane with a 1:9 O/W using a high cell density culture of *R. erythropolis* IGTS8, which was used directly for desulfurization without harvesting, i.e. in a single-stage reactor (Figure 12.6). That was performed by the addition of a sulfate removal step which was performed by an aqueous bleed stream and the usage of a separation unit to recycle the organic phase to develop

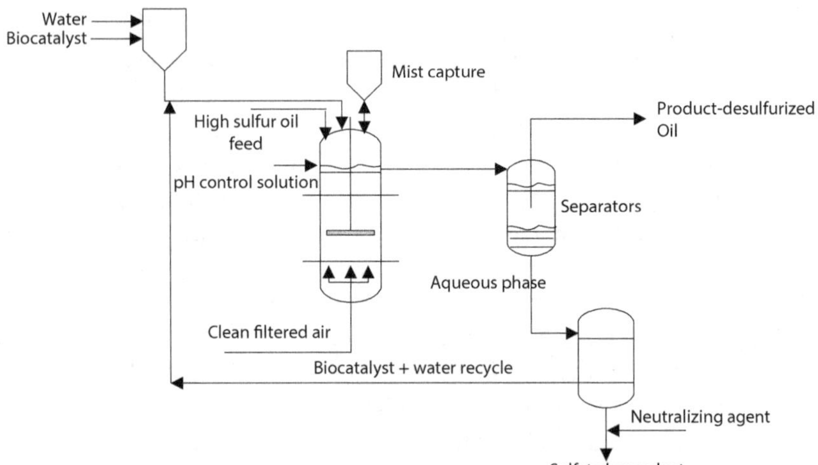

Figure 12.5 Process Flow-Sheet for a BDS Process.

a semi-continuous integrated desulfurization bioprocess. The exponential feed rates were calculated using the following equation:

$$F_{n+1} = F_n \exp(\mu_s \Delta t) \quad (1)$$

where F is the feed rate of ethanol or sulfate (mg/h), μ_s is the exponential coefficient (h^{-1}), Δt is the time interval (h), and n is the number of time points. The exponential coefficient (μ_s) was controlled to obtain a high cell density within a short time. Ethanol was used a s the C-source as it enhanced the growth of IGTS8 and it is the preferred C-source for production of the reduced form of coenzyme flavin mononucleotide ($FMNH^2$). The optimum sulfate to ethanol needed for growth in a sulfate-limiting environment was found to be 1.92 mg/g and 0.96 units OD_{600}/g ethanol/L culture, respectively, giving a cell density of 26.6 within 118 h. The problem of accumulation of acetate during the exponential growth on ethanol was resolved by manipulating the rate of ethanol addition. After the high cell density IGTS8 culture was obtained, in-situ induction was carried out followed by desulfurization of DBT in the same fermenter. The optimal dilution rate to reduce the sulfate is below 0.5 mmol/L since at high cell density the rate of sulfate production was found to be very high due to the high rate of desulfurization per liter of reactor volume, requiring dilution rates much higher than 10%. Significantly higher dilution rates are not desirable from a practical standpoint. Thus, it was found to be necessary to carry out the experiment at a lower cell density (OD_{600} of 2.5). DBT was dissolved in mineral oil instead of n-hexadecane because mineral oil was observed to separate from the aqueous phase faster than n-hexadecane. Furthermore, no sulfate accumulation was observed and the desulfurization continued throughout the period of the experiment (96.5 h). This was attributed to the utilization of sulfate as a source of sulfur for growth of the biocatalyst, and/or it might have been removed throughout the dilution of the aqueous phase. Thus, the continuous replacement of the aqueous phase from the fermenter was to remove sulfate an inactivated biocatalyst. Furthermore, the addition of a carbon source at a controlled rate can maintain desired cell density in the fermenter. Thus, in that study, biocatalyst production, induction, desulfurization reaction, and by-product removal were conducted in the same reactor to develop a semi-continuous process.

There are a few reports that have compared the various mechanistic steps in the BDS process. Abin-Fuentes *et al.* (2014) performed a mechanistic analysis of the various mass transport and kinetic steps in the microbial desulfurization of DBT by *Rhodococcus erythropolis* IGTS8 in a model biphasic (oil-water) in a small-scale system. The biocatalyst was distributed

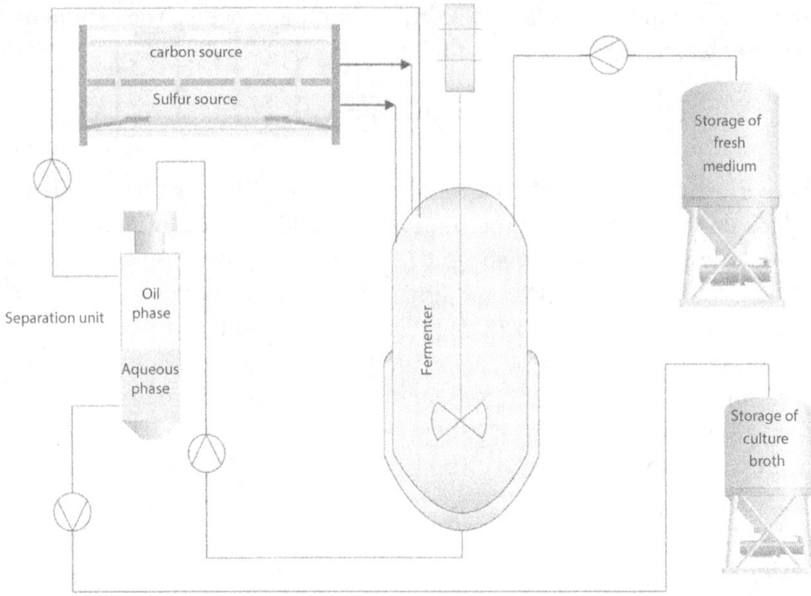

Figure 12.6 An Integrated Biodesulfurization Process. (including inoculum preparation, desulfurization, and sulfate removal in a single step for removing sulfur from oils).

into three populations, free cells in the aqueous phase, cell aggregates, and oil-adhered cells, and the fraction of cells in each population was measured. The power input per volume (P/V) and the impeller tip speed (v_{tip}) were identified as key operating parameters in determining whether the system is mass transport controlled or kinetically controlled. Oil-water DBT mass transport was found to be non-limiting under the conditions tested. Experimental results at both the 100 mL and 4 L (bioreactor) scales suggested that agitation leading to P/V greater than 10,000 W/m^3 and/or v_{tip} greater than 0.67m/s is sufficient to overcome the major mass transport limitation in the system, which was the diffusion of DBT within biocatalyst aggregates.

Jia *et al.* (2004) investigated the BDS of DBT by resting cells of *Gordonia* sp. WQ-01 at cell densities from 10–30 g DCW/L, oil fractions of 0.15–0.25, and DBT concentrations of 1–10 mM in oil. The BDS process experiences a transition in the rate limiting step from bioconversion to mass transfer resistance. However, it was assumed that DBT bioconversion could only take place in the bulk aqueous phase and not at the oil-water interface. This was assumed despite the fact that Gram-positive *Gordonia* sp. WQ-01 has a hydrophobic cell wall and associates well with walls of glass flasks. Similar behavior was performed by *R. erythropolis* and other strains that are able to adhere to an oil-water interface, thus, *Gordonia* may be able to access DBT at such interfaces. Upon the application of *Pseudomonas putida* CECT5279 as the biocatalyst for BDS, the rate limiting step was found to be the oil-to-water DBT mass transport rate in a biphasic system (Boltes *et al.*, 2012). Unlike *R. erythropolis*, *P. putida* is Gram-negative and does not have a hydrophobic cell wall and is not known to adhere to hydrocarbons. Therefore, *P. putida* is not expected to form an oil-water emulsion during biphasic BDS experiments. As a result, the oil droplet size should be significantly larger when *P. putida* is the biocatalyst relative to *R. erythropolis*. In another study, the mass transfer rate of DBT across the oil-water interface was calculated to be 10 to 10^4 times greater than the DBT bioconversion rates of various *R. erythropolis* strains (Marcelis *et al.*, 2003). However, the mathematical model created in that study did not take into account the potential of the cells to form aggregates and/or adhere to oil droplets, which will affect the value of the parameters in the model and the conclusions in the study (Marcelis *et al.*, 2003).

Sulfur containing feedstock and the biocatalyst, usually suspended in the aqueous phase, have to be contacted with each other in a bioreactor. In order to avoid mechanical stirring, which puts stress on the capability of microorganism, different types of bioreactors have been proposed, the majority of which are loop bioreactors (Mehrnia *et al.*, 2004a, b; Yazdian *et al.*, 2009, 2010). Loop bioreactors are characterized by a definitely directed circulation flow, which can be driven in fluid or fluidized systems by a propeller or jet drive and most typically in gas–liquid systems by an airlift drive or liquid pump. They are particularly suitable for fluid systems requiring high dispersion priority. On the other hand, their simple constructions and operation result in low investment and operational costs (Mehrnia *et al.*, 2004a, b; Yazdian *et al.*, 2009, 2010). In the design of the draft tube airlift reactor, higher oxygen utilization is an advantage over mechanically stirred bioreactors. Therefore, if whole cells were used as biocatalysts, less shear damage would occur. The absence of mechanical stirring systems also translates into lower capital and operating costs (Mehrnia *et al.*, 2004b). The draft tube airlift bioreactor was studied using water in kerosene micro emulsions (Mehrnia *et al.*, 2004a). The effect of the draft tube area versus the top section area on various parameters was studied. Also, the effect of gas flow rate on recirculation and gas carry over due to incomplete gas disengagement was considered (Mehrnia *et al.*, 2004b). The oil/water ratio and viscosity effects on gas hold-up and mass transfer coefficient were tested as well (Mehrnia *et al.*, 2005; Yazdian *et al.*, 2009, 2010). In addition, the significance of riser to downcomer cross sectional area, volume of gas–liquid separator value and superficial gas velocity amount on mixing time, gas hold-up, and volumetric gas–liquid mass transfer coefficients in an external airlift loop bioreactor were investigated (Yazdian *et al.*, 2009). BDS of water–kerosene emulsions with resting cells of different strains was studied at 100 ml scale in an airlift bioreactor (Sanchez *et al.*, 2008).

There are very few reports on BDS process designs and cost analysis. In order to ensure that capital and operating costs or BDS will be lower than HDS, it is necessary to design a suitable biocatalytic process (Monticello, 2000). Thus, for successful commercialization of BDS, reactors must be designed to allow sufficient liquid-liquid and gas-liquid mass transfer while simultaneously reducing operating costs. Stirred tank reactors, airlift reactors, and emulsion phase contactors with free cells, as well as fluidized bed reactors with immobilized cells are reported in the BDS review (McFarland *et al.*, 1998; Pacheco *et al.*, 1999). The emulsion phase contactor was designed to create very small water/fuel/biocatalyst droplets to minimize the introduction of water into the fuel. However, very low BDS rates were achieved (McFarland *et al.*, 1998). The fluidized bed configuration could be advantageous because the biocatalyst is immobilized on a matrix allowing continuous operation and easy product separation, however, it was reported that the biocatalyst requires further development for this configuration to be competitive. Generally, batch-stirred tank reactors have been used because of the absence of immobilization technologies and most of the work has been made with it (Pacheco *et al.*, 1999). Maass *et al.* (2014) reported DBT-BDS by the *R. erythropolis* ATCC 4277 in a batch reactor using a two-phase system; DBT dissolved in n-dodecane/water of different ratios 20, 80, and 100 % (v/v), recording BDS efficiencies of 93.3, 98.0, and 95.5 %. The highest value for the specific DBT-BDS rate was 44 mmol DBT/kg DCW/h and was attained in the reactor containing 80% DBT. Yang *et al.* (2007) reported the results for the BDS of model oil (5400 μM DBT in n-hexadecane, 1:4 O/W, pH7, 30 °C) in the fed-batch system, where *Rhodococcus globerulus* DAQ3 was able to remove 5000 μM of sulfur (200 ppm) from the model oil within 72 h, whereas, in the batch system, 3000 μM of sulfur (120 ppm) was removed. This promotes the application of in-situ desulfurization using whole growing cells in the fed-batch system, which is simpler and more convenient than using the resting cells because cell harvest, conservation, and re-suspension are not necessary. However, in order to avoid the apparent deactivation phenomenon observed during the desulfurization process in the fed-batch system, Yang *et al.* (2007b) suggested a structure of a bioreactor (Figure 12.7) with a two-layer (1:4 O/W) and continuous process where the organic and aqueous phases were kept as two separate layers; one mixer (six-blended

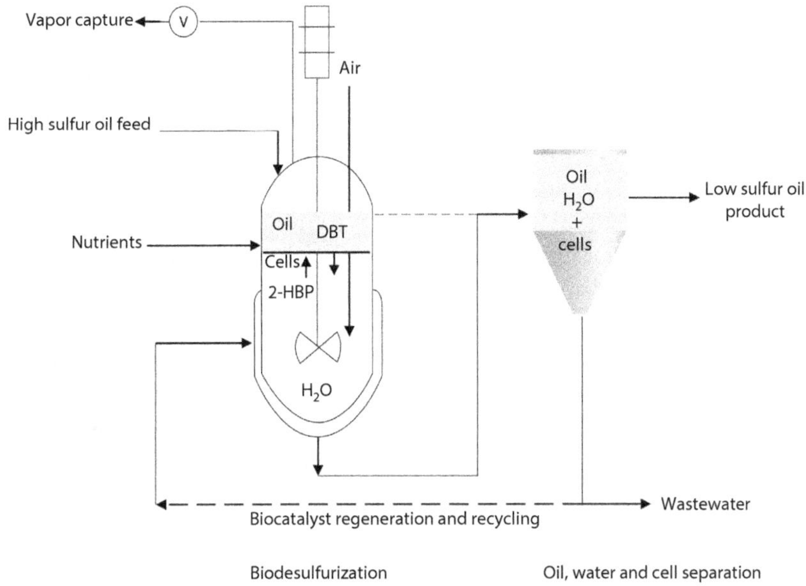

Figure 12.7 Schematic Diagram for a Two-Layer and Continuous Bioreactor.

Rushton turbine) was assembled at the bottom of the bioreactor to mix the aqueous layer. Agitation was controlled at a low speed (600 rpm) and two sintered metal spargers were used to distribute uniform, superfine bubbles to supply oxygen at a low airflow rate (0.5 vvm, based on the aqueous volume). The dilution rate was controlled between 0.06~0.08 h^{-1}. The possible inhibitory compounds produced in the aqueous phase have been washed out of the bioreactor so as not to accumulate to an inhibitory level. The desulfurization activity has been maintained for a longer period (>3 days) and the total sulfur concentration in the model oil was reduced from 1000 ppm (24,000 μM) to 375 ppm within 72 h reaching a 60% conversion.

The microbial desulfurization reaction is usually run in a stirred tank reactor (STR) and requires the creation of an oil/water interface. The formation of the latter depends mainly on the realizable volume-related power input into the reactor, which is scale-dependent. Stirred tank reactors (STR) are frequently used to create fine dispersions since high energy inputs can be achieved (Zhou and Kresta, 1998). Furthermore, general relationships describing the dispersion process in an STR are known. In an agitated medium, the drop size distribution of the dispersed phase depends on droplet breakage and coalescence (Hinze, 1955). Reactor operating conditions, physical properties, and the volume fraction of the dispersed phase are the factors affecting the average droplet size (Calabrese *et al.*, 1986; Wang and Calabrese, 1986).

Alternative reactor systems using an emulsion phase contactor were reported to be able to create more interface per unit power consumption than a conventional STR (Kaufman *et al.*, 1997). The scale-up feasibility of these electrospray bioreactors has yet to be proven. After the microbial desulfurization reaction, the oil/water emulsion has to be separated mechanically. The facility of the phase separation depends largely on the stability of the emulsion that was created in the reactor. The biocatalyst may then be isolated from the aqueous phase and reused. Little is known about the economic feasibility of biocatalyst recycling and possible regeneration or rejuvenation techniques. Whole-cell immobilization reactors offer advantages compared to conventional continuous stirred-tank bioreactors, like biocatalyst recovery, increasing the oil/water volumetric ratio to 90% with respect to the 30–50% of a continuous stirred-tank bioreactor, disposal of a large amount of exhausted medium after treatment, easy separation of oil from biocatalyst, and, lastly, it offers a low risk of microbial pollution (Chang *et al.*, 2000a). Multi-staged airlift reactors can also be used to overcome poor reaction kinetics at low sulfur concentrations and reduce mixing costs. This would enhance the concept of continuous growth and regeneration of the biocatalyst in the reactions system rather than in separate, external tanks (Monticello, 1998). The cost of building a bioreactor can be reduced by changing from a mechanically agitated reactor to airlift designs. An airlift reactor was used at EBC to minimize energy costs (Monticello, 2000; Pacheco, 1999). However, specific details about the EBC process and the results achieved were not published (Kilbane and Le Borgne, 2004). Lee *et al.* (2005) investigated diesel oil desulfurization in a combination of airlift/stirred-tank reactors using immobilized cells of *Gordonia nitida* CYKS1 (Soleimani *et al.*, 2007).

Different configurations of bioreactors have been used by various researchers for biodesulfurization. Kaufman *et al.* (1998) reported the application of an electro-spray bioreactor for BDS due to their reported operational cost savings, relative to mechanically agitated reactors, and their capability of forming emulsions of < 5 μm diameter at a cost of only 3 W/L. However, the power law relationships indicated that mechanically stirred reactors would require 100–1000-fold more energy to create such a fine emulsion. The rates of dibenzothiophene (DBT) oxidation to 2-hydroxybiphenyl (2-HBP) in hexadecane by *Rhodococcus erythropolis* IGTS8 were found to be comparable in the two reactors and ranged from 1–5 mg 2-HBP/g DCW/h, independent of the reactor employed. The batch stirred reactor was capable of forming a very fine emulsion in the presence of the biocatalyst IGTS8, similar to that formed in the electro-spray reactors. This was attributed to the fact that the biocatalyst produces its own surfactant. Thus, Kaufman *et al.* (1998) concluded that electro-spray reactors did not prove to be advantageous for the lGTS8 desulfurization system; it may prove advantageous for systems which do not produce surface-active bio-agents, in addition to being mass transport limited. Further, BDS of actual crude oil with an initial S-content 0.96 wt.% (1:3 O/W) showed that *Rhodococcus* sp. IGTS8 is capable of removing of 90% of DBT and substituted DBT type compounds within 6 d of incubation, but do not affect the remaining portion of the OSCs.

A continuous two phase (organic and aqueous) bioreactor was used for the removal of DBT using *Rhodococcus globerulus* DAQ3, which resulted in 12% sulfur removal (Yang *et al.*, 2007b). An air lift reactor (ALR) can be divided into four main sections (Figure 12.8): (i) the bottom where a gas distributor is placed and the medium is recycled from the downcomer, (ii) the riser where gas liquid flows upwards, (iii) the disengaging section at the top of assembly where some portion of gas is separated from the gas-liquid mixture, and (iv) the downcomer section where gas-liquid flows downwards. Each section has separate mixing characteristics and needs to be modeled separately. Airlift bioreactors were used by Nandi (2010) and Irani

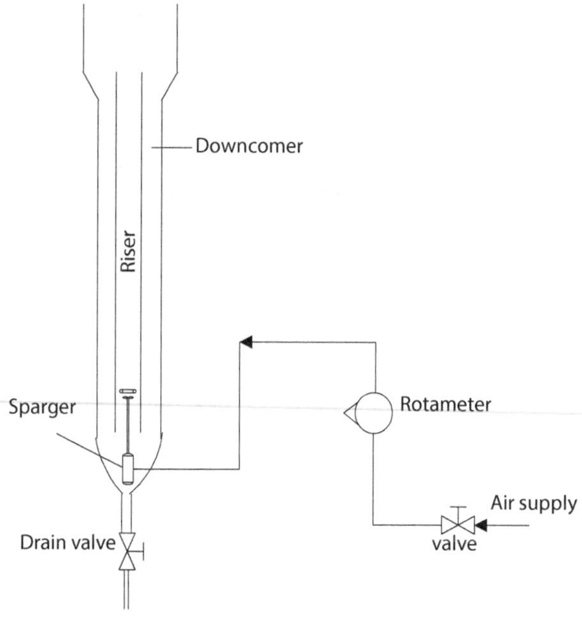

Figure 12.8 Schematic Diagram of Experimental Setup of ALR.

et al. (2011a) with different microorganisms and operating conditions and achieved 100% and 50% sulfur removal, respectively. Other BDS processes of fossil fuels in the airlift bioreactors are reported (Folsom *et al.*, 1999; Guobin *et al.*, 2006; Yang *et al.*, 2007b; Sanchez *et al.*, 2008; Nandi, 2010). The airlift reactor has advantages over a mechanically stirred bioreactor because of its improved use of oxygen, which results from the larger height-to-diameter ratios used, simplified separation of the oil and water phases resulting from the coarser emulsion texture, less shear damage of cells, and lower capital and operational costs attributed to the absence of a mechanical agitation. Both mixing and oxygen transfer rates inside the bioreactor affect the rate of the BDS process. However, most of the published work on the hydrodynamic and oxygen characterization of airlift bioreactors has been limited to a water-based system. However, as explained above, in BDS processes the bioreactor is filled with a water-in-oil emulsion or micro-emulsion. The physicochemical properties of the liquid phase in the bioreactor affects, at least to some extent, the relationship between the design and operating variables of the bioreactor and parameters characterizing hydrodynamics and oxygen transfer inside these bioreactors. Table 12.7 summarizes some examples for the BDS of real oil feed using airlift reactors relevant to other bioreactors.

Throughout the effort of Ecopetrol - Instituto Colombiano del Petróleo (ICP) in Colombia towards reaching the new Colombian requirements set by the Colombian Ministry of the Environment through Resolution 0068 issued in January 18, 2001 of low sulfur content transportation fuels (gasoline and diesel oil sulfur content were 0.03 and 0.05 wt.%, by 2005, respectively), Acero *et al.* (2003) reported the design of a membrane bioreactor prototype and its capability for the application in the BDS process of diesel oil using *Gordona rubropertinctus* ICP172. A normal stirred-tank bioreactor (STR) and a membrane bioreactor were applied for the BDS of model oil DBT in n-hexadecane and diesel oil using resting cells of *Gordona rubropertinctus* ICP172. The latter showed a great potential for the improvement of biodesulfurization reactions and the development of new catalytic/separation systems (Acero *et al.*, 2003).

Among the non-conventional reactors that have been used at industrial levels, membrane reactors have attracted considerable interest due to their possibility of integrating the biocatalytic and the separation processes in one. Membrane bioreactors combine selective mass transportation featuring chemical reactions with the selective removal of inhibitory products from the process, thus increasing reaction conversion (Giorno and Drioli, 2000). Through this system it is possible to immobilize cells or enzymes and to simultaneously reach bioconversion, product separation, and

Table 12.7 BDS of Real Oil Feed Using Different Bioreactors.

Microorganisms	Bioreactor	Substrate	BDS%	Conditions and references
Rhodococcus sp.	Continuous stirred tank	Diesel	50–70	Resting cells, 1:3 O/W, half-life time 20 h (McFarland *et al.*, 1998)
Rhodocccus erythropolis I-19	EBC 2 L reactor	Middle distillate 1850	67	Resting cells, 25:75 O/W (Folsom *et al.*, 1999)
Pseudomonas delafieldii R-8	5-L reactor	Diesel	47	High cell concentration culture 16:84 O/W, 20 h (Guobin *et al.*, 2006)
Rhodocccus globerulus DAQ3	Two-layer Continuous	Diesel	12	Growing cells 20:80 O/W, 120 h (Yang *et al.*, 2007b)
Pseudomonas aeruginosa ATCC 39327 Strain No. 06	Airlift	kerosene	64 53	Resting cells 50:50 O/W, 7 d (Sanchez *et al.*, 2008)
Rhodocccus sp.	Airlift	Diesel	100	Growing cells, 30:70 O/W, 30 h (Nandi, 2010)
Gordonia alkanivorans RIPI90A	Airlift	Diesel	50	Growing cells, 50:50 O/W, 120 h (Irani *et al.*, 2011a)

enrichment in the same operating unit (Cass *et al.*, 2000). The system was also used as a separation mechanism for the emulsion developed during the biocatalytic reaction. The reactor operates as a tube and carcass-type interchanger, where it is possible to separate the organic phase using hydrophobic membranes on the tubing coating through which the emulsion is loaded into the carcass and the organic phase is recovered through the lumen. Two sets of experiments were carried out. In the first one, the emulsion separation was evaluated at different cellular concentrations, keeping the organic/aqueous phase relation of 25/75. In the second set of trials, the separation of phases at different organic/aqueous ratios (60/40, 50/50, 40/60, and 25/75) was evaluated with a cellular concentration of 3 and 7 g/L (Berdugo *et al.*, 2002). Trials carried out with a cellular concentration of 3 g/L and at different organic phase ratios showed that a greater organic/aqueous ratio allows a more effective phase separation. On the other hand, in trials reproduced at 60/40 and 40/60 ratios with a cellular concentration of 7 g/L, the phase separation was similar for 60/40 and 50/50 phase ratios. However, a smaller separation capacity was observed for organic/aqueous phase ratios of 40/60. This preliminary study showed a great potential of applying the membrane bioreactor prototype for the improvement of BDS reactions and the development of new catalytic/separation systems.

One of the least studied aspects of the BDS process is the scale-up from shaken flask to bioreactor. Thus, most of the aspects related to BDS have been studied mainly in resting cells at shaken flask scale (in orbital shakers) and only a few studies have been carried out in different bioreactor configurations, such as the stirred tank bioreactor (STBR) (Wang and Krawiec, 1996; Lin *et al.*, 2012), continuous and semi-continuous reactors (Schilling *et al.*, 2002; Yang *et al.*, 2007b), and airlift bioreactors (Boltes *et al.*, 2008; Zhang *et al.*, 2012). The scale-up in a pilot plant is considered one of the most important challenges of the industrial BDS process.

In order to progress in the development of BDS it is necessary to apply all the knowledge achieved in flask-scale studies, but other aspects have to be taken into account for the scale-up to bioreactor set-up. The influence of variables such as temperature and pH can be studied in shaken flasks, but other variables such as stirrer speed and air flow rate effects must be studied at bioreactor scale. However, more than 30% of the scale-up methods in the industry consist of keeping the volumetric oxygen mass transfer coefficient (k_La) constant, as well as the oxygen transfers rate (OTR) (Garcia-Ochoa and Gomez, 2009).

Process-engineering research can increase the volumetric reaction rate (oil/water O/W ratio). The O/W volume ratio is among the most important technical bottlenecks in the development of petroleum biotechnological processes and, in order to reduce operational costs associated with handling, separation, and disposal of water, ideally, the O/W should be maximized. The application of immobilized biocatalysts is considered to be a potential alternative (Chang *et al.*, 2000a; Naito *et al.*, 2001). Effective oil-cell-water contact and mixing is essential for good mass transfer. Unfortunately, a tight emulsion is usually formed and it must be broken in order to recover the desulfurized oil, recycle the cells, and separate the byproducts. The phases are usually separated by liquid-liquid hydroclones. Another approach is to separate two immiscible liquids of varying densities by using a settling tank, where the liquid mixture is given enough residence time for them to form two layers, which are then drained. Bacteria usually partition to the oil-water interface and move with the discontinuous phase into a two-phase emulsion. In a water-in-oil emulsion, cells associate with water droplets. A small amount of fresh oil can then be added to create an oil-in-water emulsion so that the cells will stick to the oil droplets. Passage of this emulsion through a hydrocyclone will yield a clean water phase and a concentrated cell and oil mixture that can be recycled to the reactor. By manipulating the nature of these emulsions, relatively clean oil and water can be separated from the mixture without resorting to high energy separations. This reduces the capital and operational costs tremendously (US Department of Energy, US DOE, 2003).

There are very few reports on BDS process designs and cost analysis. In order to ensure that capital and operating costs for BDS will be lower than for HDS, it is necessary to design a suitable biocatalytic process (Monticello, 2000). The cost of building a bioreactor can be reduced by changing from a mechanically agitated reactor to airlift designs. An airlift reactor was used at EBC to minimize energy costs (Pacheco, 1999; Monticello, 2000). However, specific details about the EBC process and the results achieved were not published (Kilbane and Le Borgne, 2004). Lee *et al.* (2005) designed a new type of airlift reactor with immobilized *Gordonia nitida* CYKS1 cells on a fibrous support and used it for the biocatalytic desulfurization (BDS) of diesel oil. It was shown that cells immobilized on nylon fibers well sustained their growth accompanied by desulfurization activity during a series of repeated batch runs over an extended period of time. Advantages of easy separation of biocatalyst from the treated fuels, high stability, and long lifetime of the biocatalyst using immobilized cells were concluded. Sanchez *et al.* (2008) reported biodesulfurization of 50:50 water-kerosene emulsions were carried out at a 100 mL scale and in a 0.01 m^3 airlift reactor with resting cells of the reference strain ATCC 39327 and *Pseudomonas* native strains N° 02, 05, and 06. The reactor conditions were 30 °C, pH 8.0, and 0.34 m^3/h air flow. After 7 culture days, the mean sulfur removal for the strains N° 06 and ATCC 39327 were 64 and 53%, respectively, with a mean calorific power loss of 4.5% for both strains. The use of native strain N° 06 and the designed airlift reactor is shown as an alternative for the biodesulfurization process and constitutes a first step for its scale-up to pilot plant.

Mehrnia *et al.* (2004b) studied the effects of important design and operating parameters on the hydrodynamics and oxygen transfer characteristics of draft-tube airlift bioreactors (DTABs). A water-in-kerosene micro-emulsion system was used inside the DTAB (Figure 12.9) to simulate the physicochemical properties of the fermentation broths used in petroleum BDS processes. For this system, the churn-turbulent regime occurred at higher air flow rates compared to those of water-based systems. An increase in draft-tube surface area was found to be a more effective means for enhancement of both mixing and oxygen transfer than an increase in top-section surface area expansion. An increase in the draft-tube height and liquid level had a similar effect on the performance of DTABs containing micro-emulsion or water-based media. Correlations were developed that predicted overall and local gas holdups, liquid velocity, mixing time, and overall volumetric oxygen transfer coefficients as a function of aeration rate, bioreactor geometry, and liquid levels inside the DTAB containing water-in-oil micro-emulsion for all examined experimental conditions. Briefly, the aerobic BDS, the adequate oxygen transfer

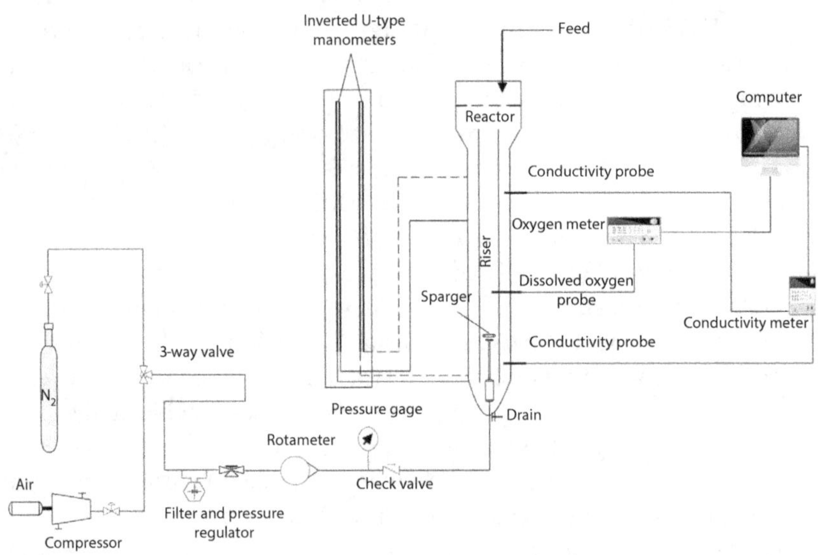

Figure 12.9 Example of a Draft-Tube Airlift Bioreactor Containing Water-In-Oil Micro-Emulsion.

from air bubbles to the water micro-droplets, and also adequate mixing for close and homogeneous contact between the phases are the important considerations in operation, design, and scale-up of DTABs. In contrast to water-based systems where operation of airlift bioreactors at high air flow rates results in the formation of churn or slug flow regimes leading to a reduction of oxygen transfer and mixing efficiency, the use of a micro-emulsion system extends the air rate range, resulting in bubble flow regime. Thus, the bioreactors can be operated profitably at higher rates. An increase in the riser-to-downcomer cross-sectional area ratio was found to be a more effective means for enhancement of both mixing and oxygen transfer than an increase in top-section surface area expansion because of the significant decrease in oxygen transfer coefficient with the latter effect. Similar to water-based media, an increase in the draft-tube height resulted in a significant increase in the mixing time, but a slight increase in oxygen transfer coefficient. Also, similar to water-based media top clearance does not have a significant influence on the hydrodynamics and mass transfer in DTABs (Mehrnia *et al.*, 2004a, b).

The BDS of 2 mM DBT in tetradecane using the UV-mutants of *R. erythropolis* ATCC 53968 was carried out with the swayed (40 strokes/min) agar plate interface bioreactor at 30 °C for 24 h (Oda and Ohta, 2002). Mutant UM-021 did not peel off throughout BDS even after 5 d and expressed 95% BDS efficiency. The bio-toxicity of 2-HBP accumulated in the tetradecane layer was suppressed by using the interface bioreactor, where UM-021 could desulfurize 2 mM DBT to a significant extent even in the presence of 10 mM 2-HBP. In conclusion, the interface bioreactor using a UV-mutant with superior adhering ability, UM-021, is useful for the BDS of DBT present in an organic phase. However, the unfavorable accumulation of water-soluble by-products, such as 2-HBP-sulfinate and sulfate, should be avoided in practical use. Oda and Ohta (2002) suggested the continuous or periodical exchange of the aqueous phase in the carrier to overcome such a problem.

Noda *et al.* (2008) enhanced DBT biodesulfurization in a microchannel reactor. The bacterial cell suspension and n-tetradecane containing 1 mM DBT were introduced separately into the double-Y channel microfluidic device (Figure 12.10) both at 0.2 to 4 μL/min. It was confirmed that a stable n-tetradecane/cell suspension interface was formed in the microchannel and the two liquids were separated almost completely at the exit of the microchannel. An emulsion was generated in the n-tetradecane/cell batch reaction which decreased the recovery rate in the n-tetradecane phase. By contrast, the emulsion was not observed in the microchannel and the n-tetradecane phase and cell suspension separated completely. The rate of

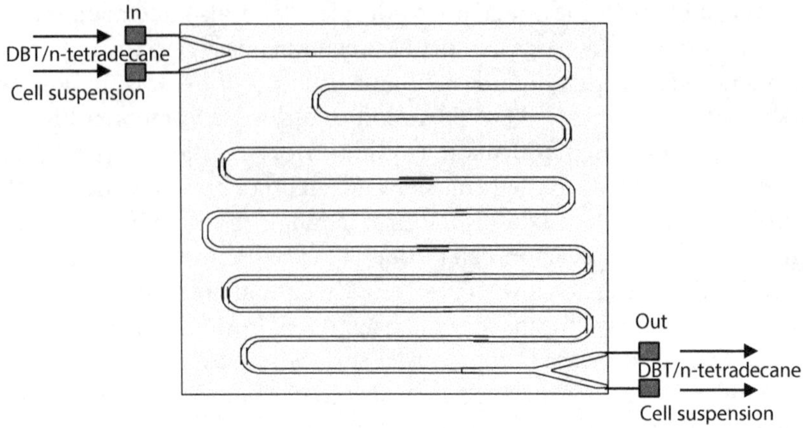

Figure 12.10 Double-Y Channel Design of the Microchannel Reactor.

BDS in the oil/water phase of the microchannel reaction was more than nine-fold that in a batch (control) reaction. In addition, the microchannel reaction system using a bacterial cell suspension degraded alkylated dibenzothiophene that was not degraded by the batch reaction system. This work provided a foundation for the application of a microchannel reactor system consisting of biological catalysts using an oil/water phase reaction.

Mukhopadhyaya *et al.* (2006) reported that *Rhodococcus* sp. NCIM 2891 has shown high activity to reduce sulfur levels in hydrodesulfurized diesel. The initial sulfur concentration was varied in the range of 200–540 ppm. A trickle bed reactor (diameter, 0.066 m; height, 0.6 m), under continuous mode, was studied with the liquid flow rate and inlet sulfur concentration as parameters. Pith balls have been used as the immobilization matrix for the microorganisms with a constant bed porosity of 0.6. The sulfur conversion up to 99% has been achieved from 200 ppm sulfur in feed diesel at the lowest inlet diesel flow rate, 0.25 dm^3/h, while the sulfur removal efficiency corresponding to the highest inlet diesel flow rate of 0.5 dm^3/h recorded ≈ 77.8%, since the increase in flow rate led to a decrease in reactor residence time causing a drop in the ultimate conversion of the reactant. Further, Mukhopadhyaya *et al.* (2006) proposed a mathematical model that describes the BDS of diesel oil in the applied trickle bed bioreactor under the continuous mode of operation (Figure 12.11). The model has been developed under the following assumptions: (1) internal mass transfer resistance is negligible, (2) the effectiveness factor, η, is unity, (3) microbial growth follows Monod kinetics, (4) microbial reaction occurs only at the outer surface of the biofilm, (5) the biofilm thickness, $L_{f,}$ remains

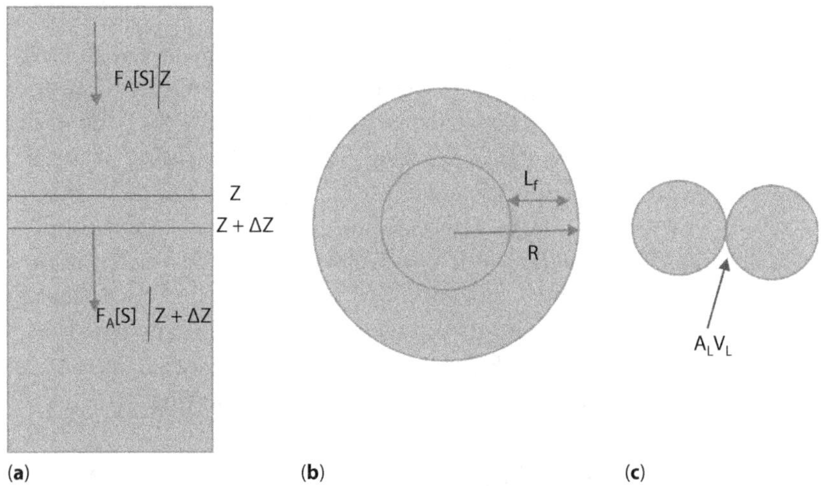

Figure 12.11 Schematic Representation of the Trickle Bed Bioreactor and the Packing Materials (a), Characteristic Sphere (b), and Contacting Pattern of Spheres (c).

constant and it is a stagnant phase, (6) sterile feed, and (7) newly grown microorganisms remain in the liquid phase and are not attached to the packing materials.

The differential mass balance equations for both biomass and organosulfur compounds (OSCs) is within a differential element, ΔZ:

$$\frac{dC_x}{dz} = \frac{1}{F_A}\frac{\mu_{max}[S]C_x}{(K_s + [S])} AL_f a_f (1-\varepsilon) \tag{2}$$

$$\frac{d[S]}{dz} = -\frac{1}{F_A}\frac{\mu_{max}[S]C_x}{Y\left(K_s + [S]\right)} AL_f a_f (1-\varepsilon) \tag{3}$$

$$a_f = \frac{4\pi(R+L_f)^2 - \eta A_L}{4/3\,\pi R^3} \tag{4}$$

$$A_L = 2\pi(R+L_f)L_f \tag{5}$$

where F_A is the volumetric flow rate (dm^3/h), C_x is biomass concentration mg/dm^3 (ppm), [S] is substrate concentration mg/dm^3 (ppm), A is the cross-sectional area of the reactor (m^2), and R is the radius of the immobilization matrixes which are perfect spheres pith balls. Some of the spheres, n, are always in contact with one spherical particle and it leads to loss in biofilm surface area (A_L) and volume (V_L) per unit sphere in contact throughout the operation, while a_f is the specific area considering the loss due to contact area and volume of spheres associated with each packing material, L_f is the biofilm thickness, η is effectiveness factor, Z is the axial position in the reactor (m), and ε is the bed porosity and remains constant throughout the operation. The growth kinetic parameters for desulfurization of diesel were as follows: saturation constant (K_s) 71 mg/dm^3, yield coefficient (Y) 0.2 g biomass consumed/g of substrate consumed, and maximum specific growth rate (μ_{max}),0.096 h^{-1}. Eqs (2) to (5) have been solved by 4th order Runge-Kutta technique with the following initial conditions:

$$Z = 0 \text{ and } \begin{bmatrix} C_x \\ S = S_o \end{bmatrix}.$$

The simulated results of the proposed model were able to satisfactorily explain the experimental data.

Dinamarca *et al.* (2014) immobilized *R. rhodochrous* ATCC 53968 on silica with a specific surface area of 80 m^2/g and investigated its BDS efficiency in a packed trickle bed bioreactor. It was evidenced that the low substrate flow rate, larger bed length, and larger size of immobilization particles enabled a sufficient contact time between the microorganism and the organosulfur compounds for expression of microbial metabolism, thus, enhancing BDS efficiency. Large particle size of the adsorbent improves aeration in a packed bed reactor by offering a greater inter-particle volume that can be occupied by both liquid and air. In a packed bed reactor, large particle size dominates over the higher specific surface for cell immobilization which is offered by smaller particle size. Not only is this, but the importance of the meso- and macro-porosity of the catalytic bed is also significant in the configuration of packed bed bioreactors for BDS. The BDS efficiency of this packed trickle bed reactor was 18% for high sulfur content gas oil (4.7 g sulfur/L). The immobilized cells maintained their BDS activity over three successive cycles and retained approximately 84% of their initial BDS efficiency. In contrast to processes using two phase systems oil/water, the BDS operating with packed trickle bed reactors containing immobilized bacteria on inorganic support continuously fed with S-containing fuel streams is advantageous; it is simple, effective, and ease of in-situ regeneration of the biocatalytic bed for cyclic operation of the BDS of high sulfur content fuel streams.

In most of the published studies on biodesulfurization activity using high cell density of the biocatalyst (Mohebali and Andrew, 2008; Dinamarca *et al.*, 2010), a separate two-stage process was employed: one for cell growth and the second for induction and BDS activity. In an attempt to simplify the process and lower operating costs, Amin (2011) developed a single vertical rotating immobilized cell reactor (VRICR) with the bacterium *R. erythropolis* ATCC 53968, as a biocatalyst, and used it for investigation of the BDS process with its two successive stages of cell growth and desulfurization activity. With a rotation speed of 15 rpm and oxygen transfer rate (OTR) of 90 mM O_2/L/h, immobilized cell concentration of up to 70 g/L in polyurethane foam was achieved during the first stage, with a mass ratio of free cells to immobilized cells ($M_{f/I}$) of 21.4% and further used, in the second, to carry out a stable continuous desulfurization of model oil (11.68 mM dibenzothiophene in hexadecane) at a 1:6 O/W phase ratio. A steady state with a specific desulfurization rate as high as 167 mM 2HBP/kg/h and sulfur removal efficiency of 100% were maintained for more than 120 h with $M_{f/I}$ of 22.5%. The proposed integrated BDS process utilizing VRICR has the potential to lower operating costs and confirm the feasibility of using the VRICR of the bacterium *R. erythropolis* for an efficient long-term continuous BDS of a model oil (Table 12.8). Although the specific BDS rate in a fed-batch, using free suspended cells 200 mM 2-HBP/kg/h (Konishi *et al.*, 2005a), was higher than that reported by Amin (2011), 110.5 mM 2-HBP/kg/h in a continuous system using immobilized cells. However, the VRICR has longevity of operation, final biomass concentration, and a volumetric BDS activity of 1.4, 3.5, and 7.3 times higher, respectively, recoding overall volumetric BDS activities of 5 and 36.5 mM 2-HBP/kg/h, respectively. Moreover, compared to free-suspended cell reactors utilized for BDS processes, VRICR has many advantages. Its scale-up is simple since only IBUs are to be constructed and mounted into the presently existing shaft in conventional bioreactors. Secondly, the open structure of the polyurethane foams, together with the good internal configurations of the IBUs permit simple and direct contact between DBT in the oil phase, dissolved oxygen, and immobilized cells and, hence, a rapid and efficient BDS process. Finally, cell over-growth within polyurethane foams can be effectively eliminated so it is relatively smaller, but enough concentration of actively growing immobilized cells are maintained and used to sustain a long-term BDS process. Thus, those support possibilities of commercial application at the expense of the conventional HDS process.

Martinez *et al.* (2016b) investigated the feasibility of the scale-up of the BDS process from shaken flask to stirred tank bioreactor, using resting cells of *Pseudomonas putida* CECT5279, taking into account the influence

Table 12.8 Comparative Data for BDS of Model Oil by *R. erythropolis.*

Cultivation technique	Carbon source	Biomass g/L	Specific BDS Rate mM 2-HBP/kg/h	Longevity H	References
Fed-batch using free suspended cells	ethanol	37	95.1	48	Yan *et al.* (2000)
Fed-batch using free suspended cells	ethanol	24.3	18.2	10	Matsushita *et al.* (2001)
Fed-batch using free suspended cells	ethanol	20	111.1	89	Yoshikawa *et al.* (2002)
Fed-batch using free suspended cells	ethanol	20	200	89	Konoshi *et al.* (2005)
Continuous using immobilized cells	ethanol + glucose	70	110.5	120	Amin (2011)

of the hydrodynamic conditions of the oxygen transfer rate. The biomass used for inoculation was a mixture of two ages of resting cells, 5 h and 23 h in the ration of 1:2 w/w, with an initial DBT concentration of 25 μmol/L and the time of the experiment was 90 min, pH 8, adding Hepes buffer, 30 °C, and Tween 80 was used to increase the availability of DBT and 200 rpm in a shaken flask. It was found that the 4S route is an oxygen-dependent pathway, but it shows a high sensitivity to the oxygen availability even under oxygen limiting conditions when the rate-limiting step of the overall process is OTR. Therefore, the hydrodynamic conditions (changed by means of the stirrer speed and air flow rate) have a significant effect on the BDS rate. In that study, the OUR was limited by the OTR and the best BDS capacity was obtained under intermediate conditions of OTR. The best results were obtained at a stirrer speed of 400 rpm and an air flow rate of 2 vvm. The oxygen transfer rate has been found to be the rate-limiting step and there is a critical value determined by a value for the volumetric mass transfer coefficient (k_La) of around $1.4 \times 10^{-2}\ s^{-1}$, which cannot be exceeded without affecting the BDS process. It was noticed that under the best conditions it was possible to reproduce the results obtained in a shaken flask and the values of k_La were similar in both situations ($1.37 \times 10^{-2}\ s^{-1}$ in STBR and $1.39 \times 10^{-2}\ s^{-1}$ in a shaken flask), hence, the scale-up criterion of constant k_La was applicable in that process.

12.12 BDS Process of Real Oil Feed

Petroleum contains various organosulfur compounds including DBTs, BTs, and other thiophenic compounds (Matsui *et al.*, 2000). Regulations against sulfur emission are becoming stricter worldwide in light of the need for environmental protection. The removal of organosulfur compounds from crude oil is desirable because low sulfur crude oil will produce low sulfur distillates. The true issue of extent of desulfurization lies in the ability of biocatalysts to remove the more hydrophobic and highly alkylated high molecular weight polyaromatic sulfur heterocyclic compounds (PASHs) from the oil (Pienkos, 1999).

A variety of bacterial cultures has been reported which possess the ability to desulfurize crude oil (McFarland *et al.*, 1998, 1999; Monticello, 2000).

Most of the applications of petroleum biotechnology are at the level of laboratory research except biodesulfurization, for which pilot plants have been established. Biodesulfurization was first developed by ENCHIRA Biotechnology Corporation (ENBC) (formerly; Energy Biosystems Corporation) by the mid-1990s.

Due to the milder and safer process conditions of BDS, the CO_2 emissions and energy requirements of a BDS-based process is estimated to be lower than that of HDS processes (Linguist and Pacheco, 1999; Singh *et al.*, 2012). Moreover, the capital costs to setup a BDS-process is reported to be 50% lower than that of an HDS process (Pacheco *et al.*, 1999; Linguist and Pacheco, 1999; Atlas *et al.*, 1998).

Biodesulfurization has been applied to mid-distillates (Grossman *et al.*, 1999; Pacheco *et al.*, 1999; El-Gendy, 2001; 2004; Li *et al.*, 2005; El-Gendy *et al.*, 2006; Aribike *et al.*, 2009), partially HDS-treated mid-distillates (Folsom *et al.*, 1999), extensively HDS-treated mid-distillates (Grossman *et al.*, 2001), light gas oils (Chang *et al.*, 1998; Pacheco *et al.*, 1999; Furuya, 2003; Noda *et al.*, 2003a; Ishii *et al.*, 2005; Dinamarca *et al.*, 2010), cracked stocks (Pacheco *et al.*, 1999), and crude oils (Premuzic and Lin, 1999; El-Gendy, 2001, 2004; El-Gendy *et al.*, 2006). However, *Pseudomonas* sp. is reported to be an ideal candidate for biodesulfurization in petroleum because they are organic solvent tolerant and have a high growth rate (El-Gendy and Speight, 2016).

BDS of petroleum and its fractions is a promising process. Efforts in microbial screening and development have identified microorganisms capable of BDS of petroleum and its fractions. Table 12.9 summarizes some of the published studies on the BDS of crude oil and its distillates. However, the BDS of crude oil is reported to be lower than that obtained in model oil, which may be due to the accumulation of toxic metabolites in the medium and might be considered to cause such a decrease in desulfurization activity (Chang *et al.*, 2001). The inhibitory or toxic effects by certain sulfur compounds on the cells and/or other different hydrocarbons found in crude oil which might affect the degradation of sulfur aromatic compounds, lead to a decrease in desulfurization activity (Setti *et al.*, 1995; Abbad-Andaloussi *et al.*, 2003). Also, the sulfur requirement for microbial growth is low compared to the level of sulfur in oil feed, so microorganisms would cease desulfurization before the bulk of the sulfur was removed from the crude oil (Reichmuth *et al.*, 2000). According to Setti *et al.* (1997), although aerobic microorganisms can remove high amounts of organic sulfur, the sulfur percentage in the residual heavy oil may increase as a consequence of simultaneous aliphatic compounds biodegradation.

Asphaltene is considered to be the product of complex hetero-atomic aromatic macrocyclic structures polymerized through sulfide linkages (Becker, 1997). The most abundant organosulfur compounds in this fraction are the thiophenes, benzothiophenes, dibenzothiophenes, and sulfide derivatives (Payzant *et al.*, 1988). Moustafa *et al.* (2006) reported that for the biomodification of asphaltenes, reactions with organosulfur moieties

Table 12.9 BDS of Real Oil Feed.

Bacterium	Type of oil	Initial sulfur concentration	BDS efficiency	O/W phase ratio	References
Achromobacter sp.	Light gas oil	0.15% (w/w)	1.2 mg S/g DCW/h	1:3	Bordoloi *et al.* (2014)
Achromobacter sp.	HDS-diesel oil	420 ppm	1.2 mg S/g DCW/ h	1:3	Bordoloi *et al.* (2014)
*Arthrobacter sulfureus***	HDS-diesel	170 ppm	53%	1:3	Labana *et al.* (2005)
Brevibacillus invocatus C19 (accession no. KC999852)	Diesel oil	8600 ppm	91.31%	25%	Nassar (2015)
*Desulfobacterium aniline****	Diesel	1681 ppm	82%	1% (v/v)	Aribike *et al.* (2009)
Escherichia coli W3110[R]	Diesel oil	250 ppm	15%	5% (v/v)	Jae *et al.* (2003)
Gordona sp. CYKS1**	Middle distillate unit feed	1500 ppm	70%	1:9	Rhee *et al.* (1998)
Gordona sp. CYKS1**	Light gas oil	3000 ppm	50%	1:9	Rhee *et al.* (1998)
Gordona sp. CYKS1**	Light gas oil	3000 ppm	35%	1:9	Chang *et al.* (2000b)
Gordona sp. CYKS1**	Middle distillate unit feed	1500 ppm	60%	1:9	Chang *et al.* (2000b)
Gordona sp. CYKS1**	Diesel oil	250 ppm	76%	1:9	Chang *et al.*(2000b)
M. pheli WU-0103*	Light gas oil	1000 ppm	52%	1:9	Ishii *et al.* (2005)
Mycobacterium goodie X7B*	Crude oil	3,600 ppm	59%	1:9	Li *et al.* (2007a)

(*Continued*)

Table 12.9 Cont.

Bacterium	Type of Oil	Initial sulfur concentration	BDS efficiency	O/W phase ratio	References
Mycobacterium pheli WU-F1* &**	B-light gas oil	350 ppm	74%	1:1	Furuya *et al.* (2003)
Mycobacterium pheli WU-F1* &**	F-light gas oil	120 ppm	65%	1:1	Furuya *et al.* (2003)
Mycobacterium pheli WU-F1* &**	X-light gas oil	34 ppm	56%	1:1	Furuya *et al.* (2003)
Mycobacterium phlei SM120–1	Light gas oil	224 ppm	66%	2% (v/v)	Srinivasaraghavan *et al.* (2006)
Mycobacterium sp. X7B*	HDS-diesel	535 ppm	0.14 mg S/g DCW/ h	1:9	Li *et al.* (2003)
Nocardia globerula R-9**	diesel oil	1807 ppm	5.1 mmol S/ kg/h	1:8	Mingfang *et al.* (2003)
Nocardia sp. strain CYKS2	Light gas oil	0.3 wt.%	0.992 mg S/L-dispersion/h	1/20	Chang *et al.* (1998)
Pantoea agglomerans D23W3	Light crude oil	3800 ppm	71%	1:9	Bhatia and Sharma (2010a)
Pantoea agglomerans D23W3	HDS-diesel	70 ppm	26%	1:9	Bhatia and Sharma (2010a)
Pantoea agglomerans D23W3	Diesel	150 ppm	38%	1:9	Bhatia and sharma (2010a)
Pantoea agglomerans D23W3	Heavy crude oil	2.61 wt.%	39%	1:9	Bhatia and sharma (2010a)

Pseudomonas delafieldii R-8**	HDS-diesel	591 ppm	91%	1:1	Guobin *et al.* (2005)
Pseudomonas delafieldii R-8	HDS-diesel	591 ppm	47%	1:1	Guobin *et al.* (2005)
R. erythropolis ATCC 4277	Heavy gas oil	6500 ppm	148 mg S/kg/ h	40% (v/v)	Maass *et al.* (2015)
Rhodococcus erythropolis IGTS8	Diesel oil	8600 ppm	74.42%	10%	Nassar (2015)
R. erythropolis I-19	HDS-middle distillate 1850	1850 ppm	67%	1:9	Folsom *et al.* (1999)
R. erythropolis LSSE81[R]	Diesel oil	251.3 ppm	73%	1:2	Xiong *et al.* (2007)
R. erythropolis NCC1	HDS-diesel oil	554 ppm	51%	1:9	Li *et al.* (2007b)
R. erythropolis XP**	Fushun crude oil	3210 ppm	62%	1:9	Yu *et al.* (2006)
R. erythropolis XP**	Sudanese crude oil	1237 ppm	47%	1:9	Yu *et al.* (2006)
R. erythropolis XP**	Straight run gasoline	50.2 ppm	85%	1:9	Yu *et al.* (2006)
R. erythropolis XP**	Jilian FCC gasoline	1200 ppm	30%	1:9	Yu *et al.* (2006)
R. erythropolis XP**	HDS-diesel oil	259 ppm	95%	1:9	Yu *et al.* (2006)
R. globerulus DAQ3	Diesel oil	12600 ppm	12%	1:4	Yang *et al.* (2007b)
Rhodococcus erythropolis FSD2	HDS-diesel	566 ppm	0.78 mg S/ g DCW/ h	1:5	Zhang *et al.*, 2007c
Rhodococcus erythropolis FSD2	HDS-diesel	198 ppm	0.26 mg S/g DCW) h	1:5	Zhang *et al.*, 2007c
Rhodococcus sp.**	HDS-diesel	170 ppm	50%	1:3	Labana *et al.* (2005)

(*Continued*)

Table 12.9 Cont.

Bacterium	Type of Oil	Initial sulfur concentration	BDS efficiency	O/W phase ratio	References
Rhodococcus sp. (NCIM 2891)	HDS-diesel	430 ppm	96%	1:4	Bandyopadhyay *et al.* (2013)
Rhodococcus sp. ECRD-1	Light cycle oil	669 ppm	92%	2% (v/v)	Grossman *et al.* (2001)
Rhodococcus sp. ECRD-1	Middle distillate fraction of Oregon Basin crude oil	20,000 ppm	8%	1/10	Grossman *et al.* (1999)
Rhodococcus sp. IMP-S02	Diesel	500 ppm	60%	5% (v/v)	Castorena *et al.* (2002)
Rhodococcus sp. P32C1**	HDS-diesel	303 ppm	49%	25% (v/v)	Maghsoudi *et al.* (2001)
Sphingomonas subarctica T7B	LGO	280 ppm	41%	1:5	Gunam *et al.* (2006)
Stachybotrys sp. WS4	Heavy crude oil	5 wt.%	76%	1/20	Torkamani *et al.* (2008)

* Thermophilic ** Resting *** Anaerobic [R] recombinant

could be very significant because sulfur is the most abundant element in Asph besides carbon and hydrogen; sulfur can form up to 8 wt.% of Asph and has an important role in its molecular structure. The organosulfur compounds present in this large molecular-weight fraction of crude oils (asphaltenes) are reported to be converted to lighter fractions with a simultaneous decrease in the total sulfur content (Lin *et al.*, 1996; Le Borgne and Quintero, 2003).

The decline in average molecular weight with the observed decrease in sulfur content of the crude oil might be attributed to the biochemical reactions occurring at the oil-water interface, which are initiated at the heteroatom sites (S-atom) (Rosenberg *et al.*, 1979). These reactions may lead to biodepolymerization within the asphaltene fraction. The microbial degradation of this polar fraction, which has previously been considered relatively recalcitrant to biodegradation (Walker *et al.*, 1975), was reported by Leahy and Colwell (1990) where the microbial degradation of asphaltene up to 74% of its initial concentration occurred. This was ascribed to co-oxidation, in which non-growth hydrocarbons are oxidized in the presence of hydrocarbons which can serve as growth substrates (Perry, 1979 and Atlas, 1981). Evidence for co-oxidation of asphaltenes was proven by Rontani *et al.* (1985), who reported degradation of asphaltenic compounds in mixed bacterial cultures to be dependent upon the presence of n-alkanes 12–18 carbon atoms in length. Asphaltene fraction biodegradation was also reported by Mishra *et al.* (2004).

It was reported by Pineda-Flores and Mesta-Howard (2001) that when microorganisms develop their full metabolic potential, they might have the capacity to use asphaltenes as a source of carbon and energy or of degrading them by co-metabolism mechanisms. This is possible once these compounds contain carbon, hydrogen, sulfur, nitrogen, and oxygen, which are necessary elements for the development of any organisms. Figure 12.12 illustrates asphaltenes structure and regions susceptible for BDS as discussed by Pineda-Flores and Mesta-Howard (2001).

The viscosity of heavy crude oil arises in a significant part from the presence of sulfur-bearing heterocycles, predominantly BTs, DBT, and its derivatives. These sulfur compounds contribute to up to 40% of sulfur compounds in Middle East crude oils. Sulfur-bearing heterocycles such as DBT exist as rigid, planar condensed ring structures, which tend to form highly organized intermolecular interactions (hydrophobic or van der Waals interactions) in liquids. By converting these rigid structures into molecules having physicochemical properties that do not significantly contribute to the residual viscosity of the treated crude oil, the result will be reduction in viscosity. Projects concerning the biocatalytic transformation

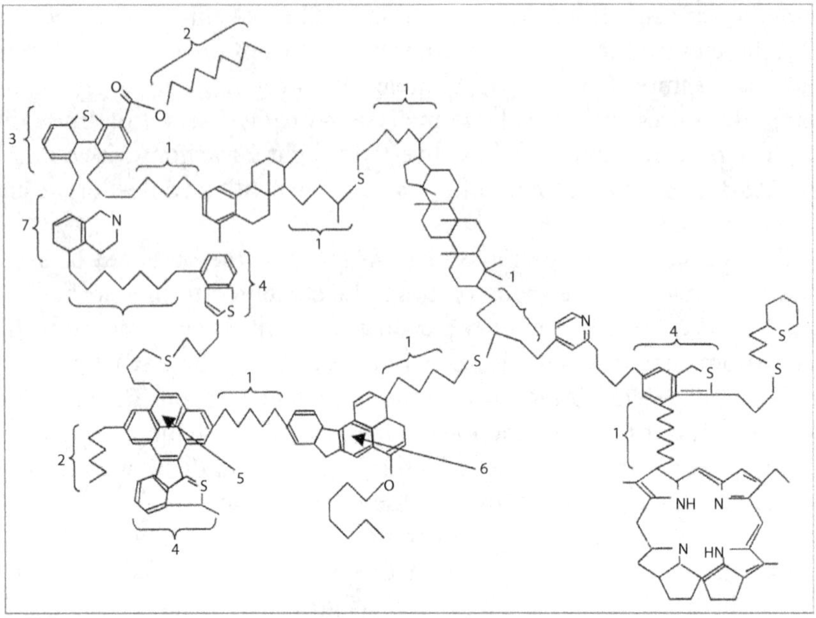

Figure 12.12 Asphaltene Structure and Regions Susceptible of Fragmentation its Molecule. 1: Photooxidation; 2: beta-oxidation; 3: Dibenzothiophene metabolic path; 4: Path similar to dibenzothiophene; 5: Pyrene path; 6: Path similar to benzo(a)pyrene; 7: Similar to carbazoles.

of heavy crude oils into light crudes have also been reported (Borole *et al.*, 2003). The success of the biocatalyst will depend on the ability to enhance biocatalyst activity (i.e. rate and range of substrates) and stability under the conditions found in the petroleum refining industry.

The same reasoning for the decrease in asphaltene content can be applied for the decrease in viscosity because asphaltene is thought to be largely responsible for adverse properties of oil and its high viscosity, so biodepolymerization of asphaltene which occurred as a result of BDS also contributes to the decrease of the viscosity of biodesulfurized crude oil (Monticello, 1996).

According to le Borgne and Quintero (2003), the effect of microorganisms on crude oil depends both on the microorganism and the type of the crude. It is thought that the microorganisms specifically directed their action to the heteroatoms and organometallic structures that serve as the attachments and initial points for microbial activity. The reactions involved probably consisted in oxidations and degradation of higher hydrocarbons. Asphaltenes were probably depolymerized by rapture at active sites containing heteroatoms, allowing the liberation of small molecules previously trapped. The carrier effect of the liquid n-alkanes on the degradation of solid hydrocarbons has been reported for asphaltenes by Rotani *et al.* (1985) and for sulfur aromatic compounds by Setti *et al.* (1992), Konishi *et al.* (2002), and Watanabe *et al.* (2003a-c).

After the performance of selective BDS, the produced products and intermediates, for example, the corresponding sulfoxides, sulfones, sulfonates, and phenols, differ from the originally-present sulfur-bearing heterocycles in that their physicochemical properties do not contribute significantly to the residual viscosity of the biodesulfurized crude oil (Monticello *et al.*, 1996). Not only this, but although in the absence of a chemical surfactant BDS of crude oil is usually accompanied with the formation of a strong and stable emulsion, the formation of that emulsion as a result of the production of biosurfactant (Kaufman *et al.*, 1998) further increases the BDS process and accelerates the rate of viscosity reduction (Folsom *et al.*, 1999).

12.13 BDS as a Complementary Technology

The difficulties in crude oil desulfurization are aggravated by the existence of polycyclic aromatic hydrocarbons (PAHs), nitrogenous compounds, and H_2S. Global crude oil production has nearly reached the limit that technology can break through. By building new devices or optimizing the related technologies, the product yield can be improved and the materials consumption during oil refining can be reduced, which will bring about significant economic returns. In the near future, the key to "zero sulfur" emissions and minimum consumption of oil that is used to produce H_2 applied in the fuel cells is how to produce cleaner fuels by the most effective, environmentally friendly, and economically viable way.

The scenario that has been proposed by EBC for integrating the BDS process with the existing HDS one was that initially crude oil would be distilled to yield products and transportation fuels that include diesel oil, then HDS is used followed by BDS to produce diesel oil with 50 ppm S-content (Figure 12.13). This scenario has the advantage of 70–80% lower CO_2 emissions and energy consumption and safer operating conditions compared to HDS alone (Pacheco *et al.*, 1999).

A comparison study was performed by Linguist and Pacheco (1999) to estimate the energy consumed and the CO_2 generated for the BDS and HDS of diesel oil feedstock. The conducted two cases were BDS standalone and application of BDS downstream of HDS process. Table 12.10 summarizes the obtained results.

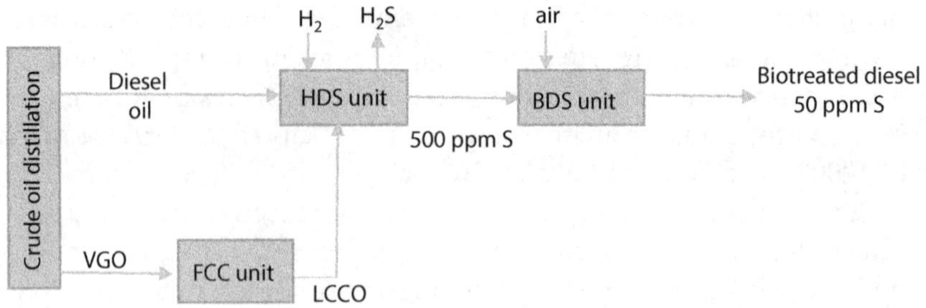

Figure 12.13 Proposed Integration of BDS with HDS.

Table 12.10 Summary of Energy Requirements and CO_2 Generation from BDS vs. HDS of Diesel Oil (Linguist and Pacheco, 1999).

	Desulfurization of diesel oil from 0.2% to 0.005%		**Desulfurization of diesel oil from 0.05% to 0.005%**	
Desulfurization technology	**Energy consumption (MBTU/bbl)**	**CO_2 emission (lb/bbl)**	**Energy consumption (MBTU/bbl)**	**CO_2 emission (lb/bbl)**
HDS	260	29.9	138	15.8
BDS	56	6.1	37	4.3
	Decrease from that of HDS alone			
HDS/BDS	21.5%	20.4%	26.9%	26.9%

The future of biorefining will likely take into consideration mixed, chemical, and/or biological processes that may satisfy the very particular needs of the oil industry, as proposed by Vazquez-Duhalt *et al.* (2002). It was concluded from previous published prospective analysis on the impact biochemical catalysis might have on the oil industry that alternate or complementary biotechnological processes for traditional refining can significantly increase energetic efficiency and reduce the environmental impact of current refining operations (Vazquez-Duhalt *et al.*, 2002; Acero *et al.*, 2003), but this will only occur through the development of the aforementioned areas.

The desulfurization rates obtained with *R. erythropolis* XP (Yu *et al.*, 2006) and *Mycobacterium* sp. X7B (Li *et al.*, 2003) for hydrodesulfurized diesel oil (4 μmol/g DCW/h) had the achievement of a final sulfur concentration of 14 ppm (i.e. ultra-low sulfur content), taking into account that an impressive level of BDS was achieved even though only organosulfur compounds that were relatively recalcitrant to hydrodesulfurization were present. Moreover, *Mycobacterium* sp. X7B was also reported to be capable of desulfurizing gasoline (Li *et al.*, 2005). This is important because the OSCs most commonly present in diesel oil are DBTs, while benzothiophenes predominate in gasoline. Gasoline is also more toxic to bacterial cells than diesel, so the demonstration that gasoline can be successfully biodesulfurized is a significant achievement.

Taking into account the advantages of both HDS and BDS, placing the BDS unit downstream of an HDS unit as a complementary technology to achieve ultra-deep desulfurization, rather than as a replacement, should also be considered. Monticello (1996) suggested a multistage process for desulfurization of fossil fuels. This method was based on subjecting vacuum gas oil to HDS prior to BDS in defined conditions. Pacheco (1999) reported that the Energy BioSystems Corporation (EBC) used BDS downstream of HDS. Fang *et al.* (2006) also showed that combination of HDS and BDS could reduce the sulfur content of catalytic diesel oil from 3358 to < 20 μg/g.

In the ODS process, followed by chemical extraction, approximately 10% of the hydrocarbons are lost after solvent extraction, thus, BDS of the sulfones concentrated in the solvent is proposed to recover the hydrocarbon skeleton. DBT sulfones are more polar than DBT, so less hydrophobic *Pseudomonas* cells are suitable hosts in this case (Tao *et al.*, 2006). Recombinant desulfurizing and solvent-tolerant *Pseudomonas* strains were reported to desulfurize DBT sulfones produced by chemical oxidation and solvent separation (ORNL, 2000). The advantage of that process is that water would be added to the solvent extract and not to the fuel.

It has been reported that combining oxidative desulfurization (ODS) with biodesulfurization (BDS) would achieve 91% sulfur removal from heavy oil (Agarwal and Sharma 2010).

12.14 Future Perspectives

As mentioned before, each of the applied commercial methods for desulfurization of petroleum products have their own deficiencies. It is necessary to improve the desulfurization process to maximize the sulfur removal efficiency and turn the waste into treasure that should be reused. The desulfurization efficiency and economic benefits should be compatible and the existing facilities should be fully utilized to provide quick returns. It is also necessary to support the development of appropriate technologies that can reduce low-value products, increase high-value products, improve economic efficiency, and enhance the competitive edge of the enterprise.

The advanced desulfurization technologies need less investment and are more efficient for treating crude oils, which can solve the problems of pollution, equipment corrosion, equipment overload, and catalyst poisoning during the catalytic cracking process to ensure a long-cycle, full-load, and safe operation of the equipment. Pretreatment of crude oil by means of the desulfurization process is the direction for development of the oil refinery industry to enhance its economic benefits.

A potentially attractive means of implementing a petroleum BDS process could be to treat crude oil on site in the production field prior the initial separation of petroleum from produced water (Shong, 1999).

The understanding of how microorganisms metabolize sulfur heterocyclic compounds in petroleum has increased rapidly. All the studies outlined above are significant steps to explore the biotechnological potential for developing an efficient BDS process. However, these technologies have not yet been valuable for large-scale applications. Any progress that provides the possibility to remove sulfur in crude oil at higher temperature, with higher rate, or longer stability of desulfurization activity is considered to be a significant step toward industry level biodesulfurization. Microorganisms with a wider substrate range and higher substrate affinity in a biphasic reaction containing toxic solvents or higher BDS activities could be engineered if the biocatalysts were to be used for petroleum treatment. In the next few years, the availability of the genome sequences of these biocatalysts will make it possible to regulate metabolism engineering.

For integration of BD with HDS, this would require substantial modification of current operations in a refinery and requires that the BDS process operate at the same speed and reliability as other refinery processes for not disrupting conventional operations. Thus, it is very challenging. BDS could fit well in the petroleum industry if it is performed in conjunction with desalting and dewatering operations. However, this requires thermophilic strains (El-Gendy and Speight, 2016). Bioprocessing of heavy oil can have a beneficial effect on oil properties, especially properties that affect the handling of the bulk oil improvements, such as (1) viscosity reduction, (2) shifts to lower molecular weight distribution in the oil, or (3) lower asphaltene content, that would reduce fluid piping, transportation problems, and costs. Indeed, a higher-grade and cleaner oil (low sulfur content and high API gravity) could well be the outcome of bioprocessing crude oil at the wellhead. Since the biocatalysts which can be applied to crude oil upgrading and desulfurization are always in water the phase and, in an oil field operation, water is coproduced with the oil, the process of water and oil separation is a routine field process and the only added process step would be agitation of the oil and water mixture with the biocatalyst.

The process would generate a wastewater stream which must be handled whether it is generated in the oil field or at the refinery. In a refinery, this new waste stream becomes an added problem. However, in the field, the water containing the formed salts can be diluted and reinjected as part of the field water flood program, which, if allowed under environmental regulations, may have minimal effect on oil field operation. One advantage of such a concept is the potential simplicity of the process; the crude oil is mixed with a water-soluble biocatalyst (either the microorganism or the enzyme) and air. After the reaction, the formed water-oil emulsion is separated to recover the upgraded oil; the biocatalyst remains with the water and is potentially available for reuse. The only added feature (that may not always be available at the wellhead) is a mixing reactor before separation.

Kilbane *et al.* (2006) reported that as petroleum production includes desalting and dewatering steps, then performing BDS/biorefining before or in combination with these oil/water separation steps would allow minimal alterations to existing industry practices. A biorefining process could potentially reduce the sulfur, nitrogen, and metal content of crude oil before the conventional refinery process. This scenario (Figure 12.14) could greatly improve the economics of the BDS process and could be integrated in existing petroleum production procedures with a minor modification, but it would only require a bioreactor (i.e. a BDS-unit) at the production

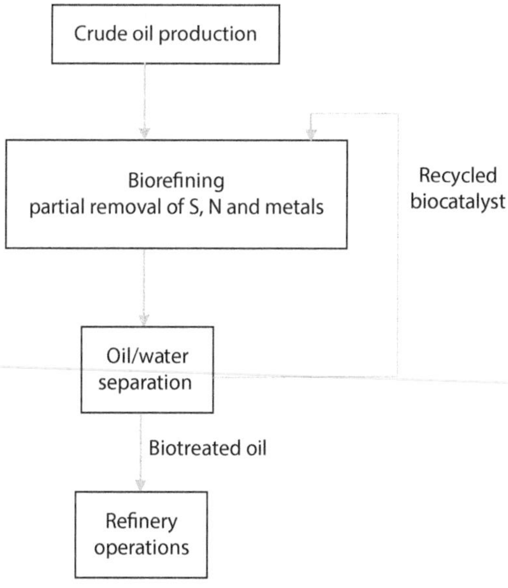

Figure 12.14 Integration of Biorefining Process in Petroleum Industry.

site. It has been suggested that the conventional petroleum storage tanks need to be modified for this BDS approach.

Desalting and dewatering processes for petroleum are normally performed at 60–100 °C and the use of elevated temperatures is very important to deal with the high viscosity of heavy oils. Application of BDS as an upgrading step of heavy oil, decreasing its viscosity and S-content at high temperature prior to its further processing and refining processes, is highly recommended. Thus, it is very practical to reach for solvent tolerant, thermotolerant microbial strains with a broad versatility on different alkylated thiophenic compounds.

12.15 Techno-Economic Studies

Within the years 1972–1985, the US petroleum industry spent approximately $1.4 billion in capital and operating expenditures to control pollution. In 1994, the National Petroleum Refineries Association estimated the cost of meeting Clean Air Act regulations requiring an S-content of 0.05% for diesel fuel has a capital expenditures cost of about $3.3 billion and annual operating cost of $1.2 billion (Pacheco *et al.*, 1999). Accordingly, it can be predicted that the costs associated with upgrading heavy oils and residuum are relatively higher. Kassler (1996) estimated that based on quantity of petroleum consumption in the US and the price differential between high and low S-fuels, the value of an alternative desulfurization technology capable of upgrading the existing production by reducing S-content and the reopening of markets for high S-petroleum reserves is in excess of $10 billion annually in the US. Thus, there is a need for alternative technologies to overcome the problems associated with heavy oils and residuum and there is sufficient economic incentive for the development of appropriate technologies that would overcome the obstacles of the existing ones. Here comes the application of biotechnology in upgrading petroleum, i.e. biorefining.

The key techno-economic challenge of the ability of BDS process is to establish a cost-effective means of implementing the two-phase bioreactor system and de-emulsification steps, as well as the product recovery step (Kaufman *et al.*, 1998; McFarland, 1999; Pacheco *et al.*, 1999). Use of multiple-stage airlift reactors can reduce mixing costs and centrifugation approaches facilitated de-emulsification, desulfurized oil recovery, and recycling of the cells (Ohshiro *et al.*, 1996).

The conventional technologies cannot achieve the target of ULSD (Ultra-low sulfur diesel) cost effectively. Thus, a combination of biodesulfurization and hydrodesulfurization technologies is suggested to achieve the future goals.

Significant progress has been made toward the commercialization of crude oil biodesulfurization. This progress includes the isolation and characterization of crude oil candidates for the BDS process; improved biocatalyst performance directly relates to crude oil biodesulfurization (thermotolerant, solvent tolerant, overcome the toxicity of by-products), development of analytical methodology which led to breakthroughs in the characterization of biodesulfurization-recalcitrant compounds, development of a process concept for crude oil BDS, and construction and testing of a prototype bench unit. Technical hurdles still need to be overcome to achieve commercialization. The major obstacles to the economical biodesulfurization of crude oil include biocatalyst specificity, rate, reusability, and stability for a long time under process conditions. Work continues to modify the biocatalyst to increase its effectiveness and to screen other organisms for additional desulfurization capabilities. In addition, mass transfer and separations hurdles must be overcome in crude oils with increased oil viscosity and density.

The selective removal of the sulfur atom from the hydrocarbon structure (e.g. dibenzothiophene) through the 4S-pathway, retaining its hydrocarbon skeleton and fuel value, is recommended. DBT is not degraded, but transformed into 2-hydroxybiphenyl (2-HBP) or 2,2′-bihydroxybiphenyl (Moniticello, 2000; Nassar *et al.*, 2016), which, in some strains, transformed to the less toxic 2-methoxybipheny and 2,2′-dimethoxy-1,1′-biphenyl (El-Gendy *et al.*, 2014) and both are portioned to the hydrocarbon phase (i.e. the fuel), while the sulfur is eliminated as inorganic sulfate in the aqueous phase containing the biocatalyst (Bordoloi *et al.*, 2014). Several microorganisms are reported retaining the 4S-pathway; the first is *Rhodococcus erythropolis* IGTS8, then different new isolates belonging to the *Rhodococcus*, *Gordonia*, *Nocardia*, *Microbacterium*, and *Mycobacterium* species. *Actinomycetales* have been also reported: *Sphingomonas*, *Pantoea agglomerans*, *Stenotrophomonas*, and *Brevibacillus* species. Moderate thermophiles and thermotolerant species have been also reported which are beneficial in real field operation, especially downstream the HDS process, including *Paenibacillus*, *Bacillus subtilis*, *Mycobacterium*, *Thermobifida*, *Rubrobacter*, and *Klebisella* (Le Borgne and Ayala, 2010; Morales and Le Borgne, 2014; El-Gendy, 2015; Akhtar *et al.*, 2016; Ayala *et al.*, 2016; Mohebali and Ball, 2016; El-Gendy and Speight, 2016). Functional expression of the dszC gene from *R. erythropolis* IGTS8 and the carBaBb genes from *Sphingomonas* sp. GTIN11 was demonstrated in the thermophilic culture *T. thermophiles* (Park *et al.*, 2004), demonstrating the possibility of development of high-temperature biorefining processes for the removal of sulfur and/or nitrogen from petroleum. It has been estimated that the biodesulfurization catalyst must have an S-removal activity within the range of 1–3 mmol DBT/g dry cell weight/h in real petroleum fractions. This means the activity of the current biocatalysts is needed to be improved by 500 fold to attain a commercially viable process (Kilbane, 2006). Alves *et al.* (2015) performed a cost analysis study comparing two BDS process designs: upstream and downstream conventional HDS. The BDS costs and emission estimations were made considering the BDS of DBT as model for S-compounds, while HDS estimations were made on the basis of crude oil HDS. The BDS downstream HDS configuration is found to be the best alternative to be applied in oil refinery, from the point of energy consumption, greenhouse gas (GHG) emissions, and operational costs, to obtain almost S-free fuels with a much lower emission of GHG and CO_2.

Certain technical issues hamper the applicability of BDS. One of them is the biphasic nature of the process. An aqueous phase is needed in order to maintain a viable biocatalyst in the BDS process, thus, significant quantities of water have to be added to the fuel. The values found in the literature generally indicate low oil-to-water volumetric ratios in the desulfurization of real fractions ranging from 1 to 25% of oil in water. The highest oil-to-water ratios have been 50% and 80% as reported by Yu *et al.* (1998) and Monot *et al.* (2002), respectively.

12.16 Economic Feasibility

Although there are many advantages of BDS, the research efforts in BDS methods are at the laboratory to small pilot plant stages and no dates have been set for commercialization. To date, none of the published BDS methods have resulted in a commercially viable process. It is concluded that most of these options are currently uneconomic in the sense that they do not increase the value of the fuel oil enough to justify the capital investment. In all fossil fuel bioprocessing schemes, in order to realize commercial success, there is a need to contact the biocatalyst that is in the aqueous phase with an immiscible or partially miscible organic substrate. Reactors must be designed that allow a sufficient liquid-liquid and/or gas-liquid mass transfer, amenability for continuous operation and high throughput, ability for biocatalyst recovery and reuse, and emulsion breaking; all of this comes with simultaneously reducing operating and capital costs. However, few reports concerning BDS process design and cost analysis have been published (Le Borgne and Quintero 2003; Soleimani *et al.*, 2007; Alves *et al.*, 2015; Ayala *et al.*, 2016). After more than twenty years of research by EBC and Enchira, several process innovations and an intensively engineered bacterium (Monticello, 2000) were obtained. Eventually the project was continued, until 2005, under the auspice of the US Department of Energy, Petro Star and Diversa worked in the development of a host harboring genetically engineered enzymes, but it appears that the biocatalyst activity was insufficient and the project ended without further progress to pilot plant scale (Kilbane 2006). It has been estimated that the BDS catalyst must have a sulfur removal activity within the range of 1–3 mmol DBT/g DCW/h in real petroleum fractions, thus, it may still be necessary to improve 500 fold the activity of current biocatalysts in order to attain a commercially viable process (Kilbane 2006). A recent publication including cost analysis study comparing two BDS process designs, upstream or downstream conventional HDS, in terms of energy consumption, GHG, emissions, and operational costs, showed that the BDS downstream HDS configuration was the best alternative to apply in an oil refinery (Alves *et al.* 2015). According to these authors, BDS integration downstream conventional HDS units may allow obtainment of almost sulfur-free fuels with a much lower emission of GHG and CO_2. However, in that study, BDS costs and emission estimations were made considering the desulfurization of the model compound DBT, while HDS estimations were made on the basis of crude oil desulfurization. The level of desulfurization activity achieved by genetic manipulation appears to be limited by unknown host factors related to the correct supply of cofactors and substrate transport (Kilbane 2006). Complex transcriptional factors involved in the regulation of the utilization of organosulfur compounds as alternative sources of sulfur might also be involved (Tanaka *et al.*, 2002). Diversa initiated the complete sequence determination of the *Rhodococcus* IGTS8 plasmid to gain a greater understanding of the genes required for BDS (Bonde and Nunn 2003), however, the results have not been published. *Rhodococci* appear to have adopted a strategy of hyper-recombination associated with their large genome and, therefore, these factors may be difficult to localize and understand (Larkin *et al.*, 2005). The recent announcements of the genome sequences of two desulfurizing bacteria, *R. erythropolis* XP (Tao *et al.*, 2011b) and *Mycobacterium goodii* X7B (Yu *et al.*, 2015), may provide new elements to elucidate some of the mechanisms limiting desulfurizing activity and propose new strain improvement strategies. In *R. erythropolis* XP, the desulfurizing genes dszABC were found on a plasmid, while the NADH-FMN oxidoreductase gene dszD was located in the chromosome.

From a cost perspective, BDS has favorable features: (1) operation at low temperature and pressure, (2) BDS is estimated to have 70 to 80% lower carbon dioxide emissions and energy consumption (Linguist and Pacheco, 1999; Singh *et al.*, 2012), (3) in the case of reaching adequate BDS efficiency level, the capital cost required for an industrial BDS process is predicted to be two thirds of that for a HDS process, and (4) it is cost effective because the operating cost of BDS is from 10 to 15% less than that of HDS (Kim *et al.*, 1996; Linguist and Pacheco, 1999; Pacheco *et al.*, 1999). Atlas *et al.* (2001) estimated the HDS process cost to decrease sulfur content from 500 to 200 ppm to be approximately 0.26 cent/L, where the desulfurization cost would increase by a factor of four if the sulfur content is further lowered from 200 to 50 ppm. In addition, the capital costs to set up a BDS process were reported to be 50% of that for HDS (Atlas *et al.*, 1998; Pacheco *et al.*, 1999; Linguist and Pacheco, 1999). Caro *et al.* (2007a) reported that BDS requires approximately two times less capital cost and 15% less operating cost as compared with HDS. (5) It has a flexible nature. The process can be applied to many process streams: crude feeds, FCC naphtha/gasoline, and middle distillates.

However, the application of large scale BDS technology is still only at the pilot scale. By the year 2000, the US Department of Energy has awarded a $900,000 grant to Enchira Biotechnology Corp. (EBC) to conduct research and development on the biodesulfurization of gasoline feed stocks through gene-shuffling technology in an attempt to develop a biocatalyst to remove sulfur from gasoline and diesel fuel (Grisham, 2000). A pilot process has been proposed (Monticello, 2000) as a beginning of the industrial process of BDS, where three bioreactors are required to achieve very low sulfur concentration (Figure 12.15). Since the degradation rate and metabolism of currently available microorganisms are still low, three bioreactors are essentially needed to reach low sulfur concentration. Renewing microbial biomass during the BDS process would make the process commercially viable. US-EBC has developed a pilot with a working capacity 5 barrels of

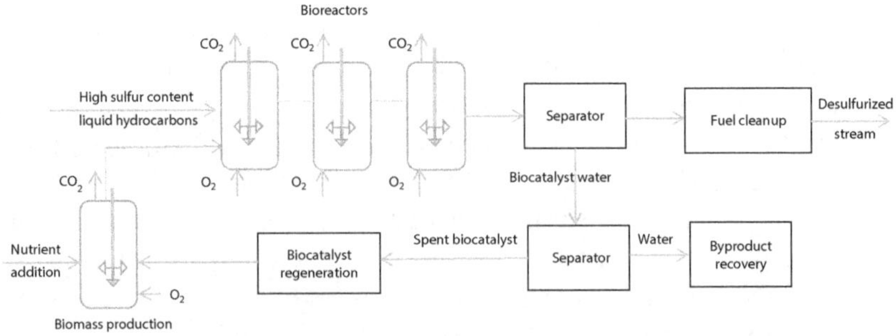

Figure 12.15 Biodesulfurization Pilot Plant (adapted from Monticello, 2000).

oil per day (bpd). In the process, the biocatalyst and fossil fuel are mixed in a bioreactor where BDS takes place, then passes through a series of filters to a container for disposal of sewage. Finally, it is added to a basic aqueous solution for neutralization and removal of sodium sulfate in a wastewater treatment station (Boniek *et al.*, 2015).

Enchira Biotechnology Corporation (2001) reported the process description and the total estimation of capital and operating costs for BDS of gasoline with initial sulfur content of 350 ppmw to reach 35 ppmw, using gasoline resistant *P. putida* PpG1, lead to a prediction of $1.77 per barrel of gasoline. The proposed gasoline BDS process was compared to five existing chemical technologies (the licensors include: UOP (ISAL process), ExxonMobil (Octagain), CD Tech (catalytic distillation), IFP (Prime-G), and Phillips Petroleum (S-Zorb)) and it was found that the BDS process is competitive to the existing technologies, where the capital and operating costs for the CD Tech, IFP, and Phillips processes are very competitive and estimated in the range of $2 per barrel.

The US Department of Energy (DOE) and the Petroleum Energy Center (PEC) of Japan invested about $110 million in research and development on BDS processes (Monticello, 1998). Le Borgne and Quintero (2003) reported the announcement of EBC for the construction and operation of two BDS pilot plants for diesel: one of 5 bpd in 1999 with Total Raffinage and one of 5000 bpd in 2000 with PetroStar in a small Alaskan refinery. Then, the US DOE contracted a 3-year project to PetroStar to develop a BDS biocatalyst using genetic engineering and design a 5000-bpd BDS pilot plant with ENBC. However, scale-up of this process has been reported to be problematic. The PEC announced the operation of a BDS pilot plant of 1 bpd of light crude oil in 2000.

To be economically viable, a BDS process must be competitive with other commercially proven desulfurization routes. The route most chosen and best known to refiners for diesel is HDS. A new BDS facility must be less costly than a comparable new HDS facility or, in the various combination scenarios, using the BDS facility as a pre- or post-treatment facility combined with an existing HDS unit must be less costly than modifications to an existing HDS unit that would be required to achieve the lower sulfur requirement. As part of the study of potential BDS economics, PetroStar Inc. (Nunn *et al.*, 2006) provided process description and total installed cost of a desulfurization process producing 6,000 bpd of highway ULSD at 10 ppmw sulfur using a straight run diesel containing 5,000 ppmw sulfur from PetroStar's Valdez Refinery and *Rhodococcus* sp. The results of the study are summarized in Table 12.11. From this table, the following conclusions can be made regarding the viability of a BDS process to produce

Table 12.11 Estimated Annual Operating Cost of PetroStar Inc. Desulfurization Process (Nunn *et al.*, 2006)

	Annual operating cost	
Desulfurization process	**$**	**Cent/gallon ULSD**
BDS alone	9,232,900	10.7
HDS alone	5,800,000	6.8
BDS followed by HDS	9,400,000	11.1
HDS followed by BDS	8,000,000	9.4

ULSD at PetroStar's Valdez Refinery: an HDS unit has a lower installed cost than a comparable BDS unit, an HDS unit also has substantially lower operating costs than a comparable BDS unit and, finally, the combination of a BDS unit and an HDS unit is not economically viable when compared to either of the standalone units.

In an attempt by the Egyptian Petroleum Research Institute to catch the wave of application of nanotechnology in the field of petroleum biotechnology, Zaki *et al.* (2013) reported the preparation of super-paramagnetic FeO_3O_4 NPs (9 nm) with good pore size, volume, and high specific surface area of 3.2 nm, 0.198 cm ^{3}/g, and 110.47 m^2/g, respectively, using a reverse water/oil micro-emulsion method, which showed a good assembling on the Gram +ve bacterial isolates, *Brevibacillus invocatus* C19, *Micrococcus* lutes RM1, and *Bacillus clausii* BS1 (Figure 12.16), with remarkable adsorption capacity to different polyaromatic compounds, DBT (69 μmole/g), Pyrene (7.66 μmole/g), and carbazole (95 μmol/g), through π complexation bonding. The coated cells are characterized by higher DBT biodesulfurization, Pyrene biodegradation, and carbazole biodenitrogenation rates than the free cells, respectively. Moreover, these Fe_3O_4 NPs-coated cells are characterized by higher storage and operational stabilities, low sensitivity toward toxic by-products, can be reused for four successive cycles without losing their efficiency, and have the advantage of magnetic separation, which would resolve many operational problems in petroleum refinery (Nassar, 2015; Saed *et al.*, 2014; Zakaria *et al.*, 2015).

The application of the Fe_3O_4 NPs-coated *B. clausii* BS1 in the biodesulfurization of diesel oil (1:4 O/W) with initial sulfur concentrations of 8600 ppm showed better efficiency than that of free cells, recording approximately complete removal and 91% within 4 and 7 days of incubation at 35 °C and 200 rpm, respectively. Figure 12.17 illustrates a preliminary bench-scale process design of high-sulfur content diesel oil BDS

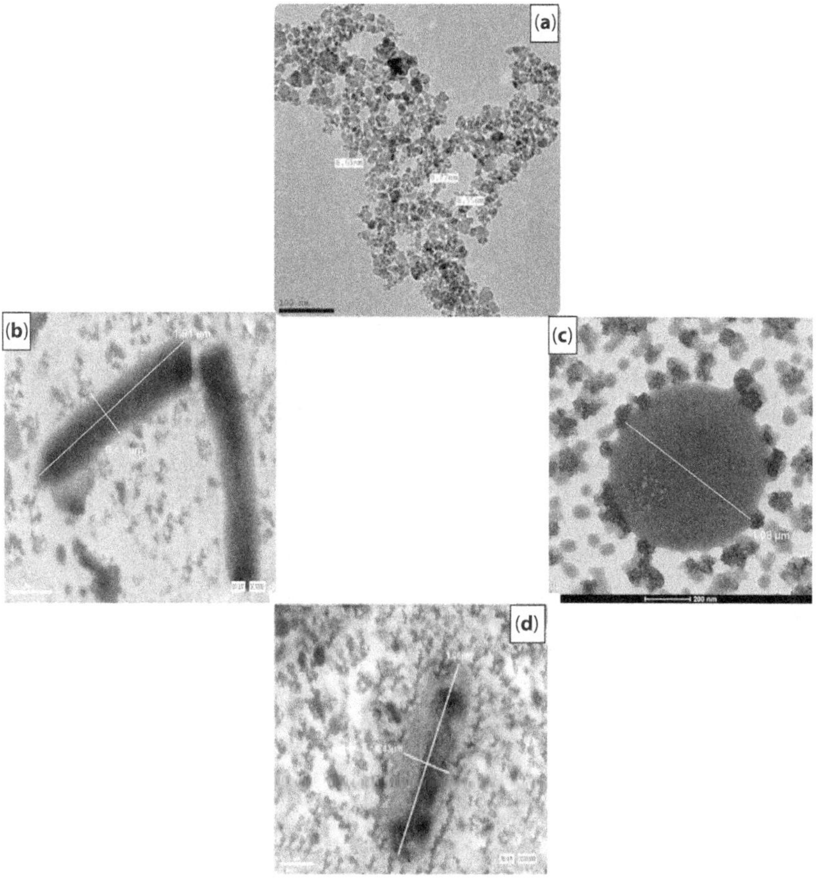

Figure 12.16 TEM Micro-Graphs of Magnetic Fe_3O_4 (a) and Coated Bacterial Isolates, *Brevibacillus invocatus* C19 (b), *Micrococcus* lutes RM1 (c), and *Bacillus clausii* BS1 (d).

using magnetic NPs-coated-BS1. The estimated capital cost for preparation of magnetic NPs is 3.17$/g cell. The summary of the estimated capital and operating cost of free and immobilized cell-systems are listed in Table 12.12. Taking into consideration that the free cells lost their activity after the first BDS cycle of 7 days, for four cycles of BDS the estimated operating and capital cost have been estimated to be 197.64 and 734.08 $/gallon diesel, respectively, with a lower BDS efficiency, while for the magnetic NPs-immobilized cell system, the immobilized cells have been used in four successive cycles, each of 4 days, without losing BDS activity with a higher BDS rate (approximately complete removal of sulfur). Thus, the estimated operating and capital cost were estimated to be 163.42 and 682.87 $/gallon diesel for four successive cycles. The main cost in this process was found

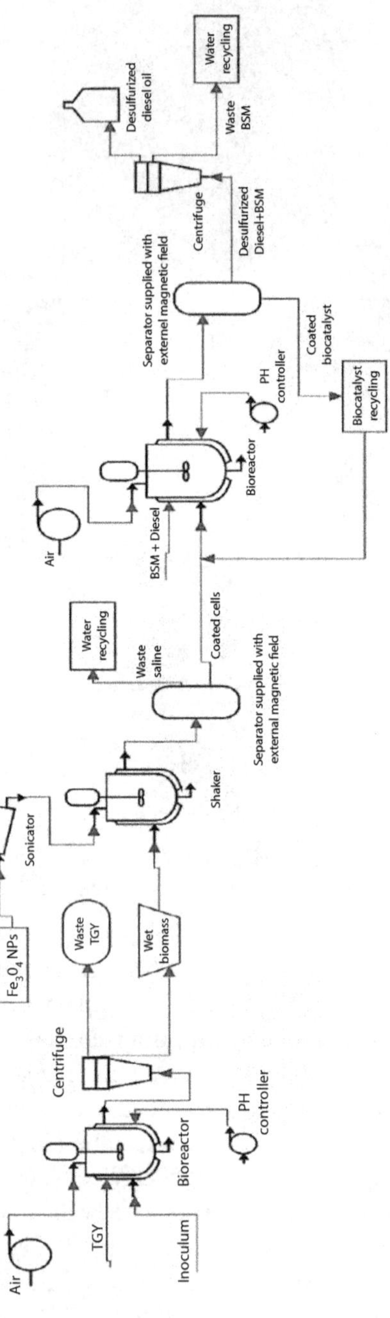

Figure 12.17 Preliminary Bench-Scale Process Design to Test High-Sulfur Content Diesel Oil Biodesulfurization by Magnetic NPs-Coated Bacterial Cells.

Table 12.12 Estimated Operating Cost for BDS of High-Sulfur Content Diesel Oil Using Free and Magnetic NPs-Coated Bacterial Cells

Fe_3O_4 NPs-coated cells decrease the S-content from 8600 to 112.66 ppm	
Operating Cost ($/gallon diesel)	**Capital Cost ($/gallon diesel)**
93.22	376.63
Free-cells decrease the S-content from 8600 to 747.34 ppm	
Operating Cost ($/gallon diesel)	**Capital Cost ($/gallon diesel)**
49.41	183.52

to come mainly from the media cost and preparation of magnetic Fe_3O_4 NPs-coated cells, 4.4 $/g cell and the use of centrifugation for preparation of resting cells and separation of diesel oil from an aqueous phase. Thus, further work is suggested to decrease the cost of media by using low cost and readily available sources of nutrients and carbon as a co-substrate, replace the centrifugation by cyclones, and decrease the cost of magnetic NPs preparation and cell coating processes.

Alves *et al.* (2015) analyzed the energy consumption, greenhouse gas emissions (GHG), and costs of two DBT-BDS process designs by *Gordonia alkanivorans* strain 1B using sugar beet molasses (SBM) as a carbon source either after acid hydrolysis and treatment with 0.255 $BaCl_2$ (to remove excess sulfate) or with treated with $BaCl_2$ and hydrolyzed in a simultaneous saccharification and fermentation (SSF) process with *Zygosaccharomyces bailii* strain Talf1 enzymes (invertase activity) (Alves and Paixão, 2014). In the SSF process, both steps (hydrolysis and sugar consumption) occur simultaneously in the course of BDS at optimal growth and desulfurization conditions for the desulfurizing bacterium (pH 7.5 and 30 °C) (Figure 12.18). This was performed by following a life cycle assessment (LCA) and life cycle cost (LCC) based methodology. The industrial HDS process was used as the reference technology for sulfur removal from fossil fuels. Different theoretical scenarios were considered and the best BDS results were scaled-up to evaluate a case study of providing ultra-low sulfur diesel (ULSD) to an urban taxi fleet. Sulfur (S) mass balance was used to quantify the quantity of its removal from DBT. Each mol of DBT has one mol of S, therefore, each mol of 2-HBP produced is equivalent to one mol of S removed by strain 1B. In Path #1, S removal of 0.0056 g (i.e. 70% SRE – Sulfur Removal Efficiency) within 75 h and 30 °C was observed, while Path #2 recorded an S removal of 0.008 g (99.6% SRE) within 48 h and 30 °C.

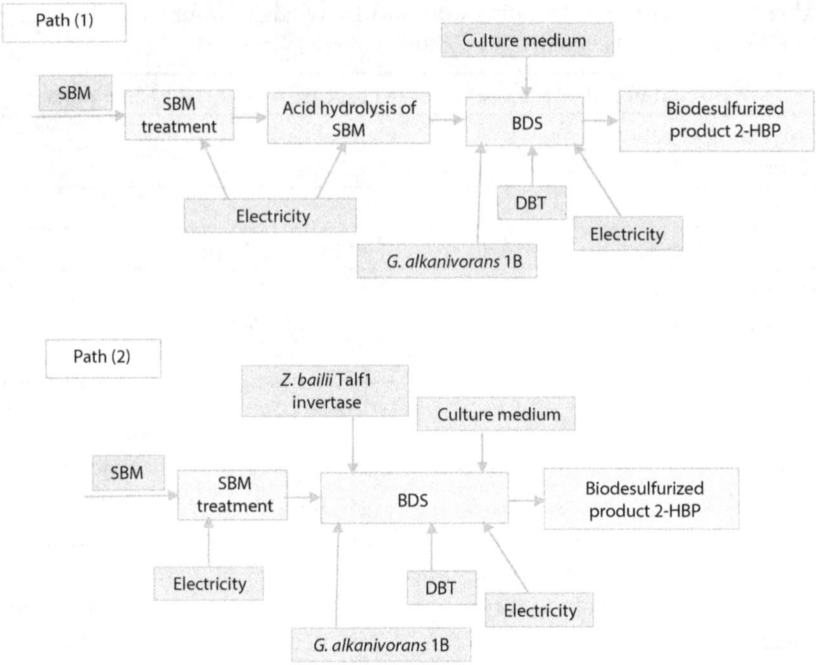

Figure 12.18 The 2 Different Paths Under Study for DBT-BDS by *G. alkanivorans* strain 1B.

The reduction in the sulfur content of refined diesel oil may easily be correlated by direct combustion mass balance with lower SOx emissions at diesel exhaust. For a typical diesel car exhaust, SO_2 emissions are estimated accordingly to the equation, considering a fuel consumption (FC) of 7 L/100 km and a density of 840 g/L (ρ diesel):

$$SO_2(g/km) = [S]ppm \times FC(L/100\text{ km}) \times \rho_{diesel}(g/L) \times \frac{M_{SO_2}}{M_S} \times 10^{-8} \quad (6)$$

This means that a reduction of the 500 ppm limit to the 10 ppm limit implies a reduction of 0.05 g/km to 0.001 g/km. BDS of DBT by strain 1B using SBM, found that the most critical steps in terms of final energy consumption (MJ/g) were centrifugation and incubation. In Path #1, the energy consumption within these steps was 105.57 MJ/g S, while in Path #2 it was 59.88 MJ/g S. However, for both BDS pathways, the energy consumed was found to be >93% of the total energy consumption. In the same way, the incubation and centrifugation steps in both BDS processes were mainly responsible for the overall GHG emissions (kg of CO_2 equivalents per g of

S removed) together with the major nutrients of the culture medium (Path (1): 14.52/15.19 kg_{CO2} eq/g S; Path (2): 8.69/9.14 kg_{CO2} eq/g S). Moreover, the total monetary cost, including consumables and process energy consumption for the BDS process through Path (1) is 8.2 €/g S, whereas for the BDS process through Path (2) is 5.3 €/g S (roughly, 40% due to material consumables and 60% due to electricity consumption). Consequently, Path (2) design is the best BDS concept process in terms of desulfurization ability, cost, energy, and GHG emissions, aiming for a scale-up. Despite the total energy consumption in this BDS process being 3-fold higher than the applied reference value for HDS (20.3 MJ/g S), this value accounted a total desulfurization of the initial product (DBT instead crude oil), while HDS desulfurized the crude oil to a fuel with 500 ppm of sulfur. Moreover, in the BDS process, the microbial biomass produced (5.9 g/L) can be further valorized, for example to single cell protein, biosurfactants, etc., which also contributes to decreasing the overall costs of the process towards its scale-up. It was also found that although SBM is a cheap agroindustrial residue, upon its usage as a C-source for the biocatalysts production during BDS process it was found to be overall more expensive than if commercial sucrose was used as the C-source. In fact, despite sucrose being 3-fold more expensive than SBM, its use for BDS can considerably reduce the overall energy consumption, GHG emissions, and costs associated to the process since the prior C-source pretreatment phase is avoided. The use of sucrose instead SBM led to a decrease in the final energy consumption from 64.1 to 28.1 MJ/g S and, consequently, to a reduction of total cost from 3.8 to 2.9 €/g. This study pointed out that, for the application of BDS, Path (2) downstream HDS is the best cost effective conceptual design to apply into an oil refinery (Table 12.13). Once it is able to desulfurize HDS recalcitrant compounds selectively, BDS integration may led to the accomplishment of the stringent European limit of <10 ppm for S-content on fuels, which otherwise may imply the necessity of more severe conditions within HDS units.

Given the amount of research directed towards biodesulfurization of diesel fuels and the understanding of the fundamental mechanisms gained, further improvements in the cost of biocatalyst production, mass transfer, reactor design, and DBT metabolic rates can be expected in the near future.

12.17 Fields of Developments

- Biological and biotechnology processes are usually regarded to be expensive and tailored for the production of low volume

Table 12.13 Energy Consumption and Cost Analysis of Different Scenarios for Application of BDS Path (2) to Reach ULSD

Energy needs and operation costs	Optimum scenario	BDS upstream HDS[c]	BDS downstream HDS[d]	BDS as alternative to HDS[e]
MJ/L Crude Oil or Hydrodesulfurized Diesel[a]	26% - SRE[b] 99.6% - SRE[b]	17.09 4.44	6.65 1.73	174.99 45.49
€/L Crude Oil or Hydrodesulfurized Diesel	26% - SRE 99.6% - SRE	4.3 1.12	1.63 0.43	44.03 11.45

[a]HDS desulfurized diesel (500 ppm of S). [b]Optimum – 26% sulfur removal efficiency (SRE): 16.15 MJ/g S, 4.06 €/g S; optimum – 99.6% SRE: 4.20 MJ/g S, 1.06 €/g S. [c]Desulfurization of crude oil (1.26% of S) in a BDS unit towards 10% of S removal. [d]Desulfurization of hydrodesulfurized diesel in a BDS unit, from 500 to 10 ppm of S. [e]Desulfurization of crude oil, from 12,600 to 10 ppm of S (the reference value for the energetic needs of HDS process to desulfurize a crude from 12,600 ppm to 500 ppm of S is 20.3 MJ/g S, which corresponds to 210.9 MJ/L crude oil).

high value products. BDS is a slow process and it requires water. In an industrial scale, the size of the two-phase reactor that might meet the required desulfurization is unimaginable. Therefore, a bioprocess that consumes a lower aqueous: hydrocarbon phase ratio is definitely preferred.

- Enzymes, in comparison to microorganisms, require less water to function. However, since several enzymes, NADH, and $FADH_2$ are involved in BDS and their activity is not high, the usage of the whole cells as biocatalysts is more practical. Moreover, cell-free extracts exhibit a lower activity in the order of 0.01 g of DBT removed/g protein/h, as compared with 0.4 g/g DCW/h using aerobic microorganisms and 0.1 g/g DCW/h using anaerobic microorganisms (Setti *et al.* 1997).
- Another point to consider is the hydrocarbon variety. Organic sulfur is found in many different refinery fractions.
- Microbial desulfurization, similar to other microbial processes, is slow and requires much attention to have reproducible results. In this method, either resting cells might be used for desulfurization or microorganisms might be grown in an aqueous/oil two-phase system. Cultivating living

microorganisms in two-phase (aqueous/oil) systems would be impractical because the process would be too slow, with very low desulfurization yield. Production of resting cells by high cell density procedure may be cross contaminated with other microorganisms that can attack C–C bonds or consume nutrients and grow at a higher rate. Furthermore, resting cell production requires huge facilities to prepare cells and to maintain biocatalyst activity while it is shipped to the desulfurization site.

- Biodesulfurization, from a process point of view, consists of feedstock preparation, microorganism/biocatalyst preparation, desulfurization in a bioreactor, and separation and recovery. Among all the stated items, preparation of biocatalysts with long half-life and high and reproducible specific activity is important.
- To date, the most important challenge to approach industry level BDS is the search to isolate a strain with higher BDS activity and wide versatility to different OSCs or to design a recombinant biocatalyst with a stable activity to work in tandem with refining pace.
- Apart from this, more work is required to obtain higher desulfurization specific activity by increasing the driving force from one phase to another and preventing the accumulation of inhibitors. One of the points that might accelerate BDS is the elimination of the cooling time required after HDS. Thus, reaching solvent tolerant, thermophilic biodesulfurizing microorganisms with wide versatility towards different OSCs is one of the important aspects.
- Another important, cost effective technique is applying nanotechnology in preparation of the biocatalyst and its applications in reactors. Alginate immobilized *Pseudomonas stutzeri* pronounced a prolonged cell viability to 600 h by reactivation (Hou *et al.* 2005), while the photo-crosslinkable resin prepolymer (ENT-400 and ENTP-4000), immobilized *R. erythropolis* KA2–51, pronounced a prolonged cell viability of 900 h (Naito *et al.*, 2001). Thus, the application of immobilized bacteria for desulfurization, denitrogenation, and heavy metal removal of heavy crude oil needs more investigation. The schematic diagram of the process is shown in Figure 12.19.

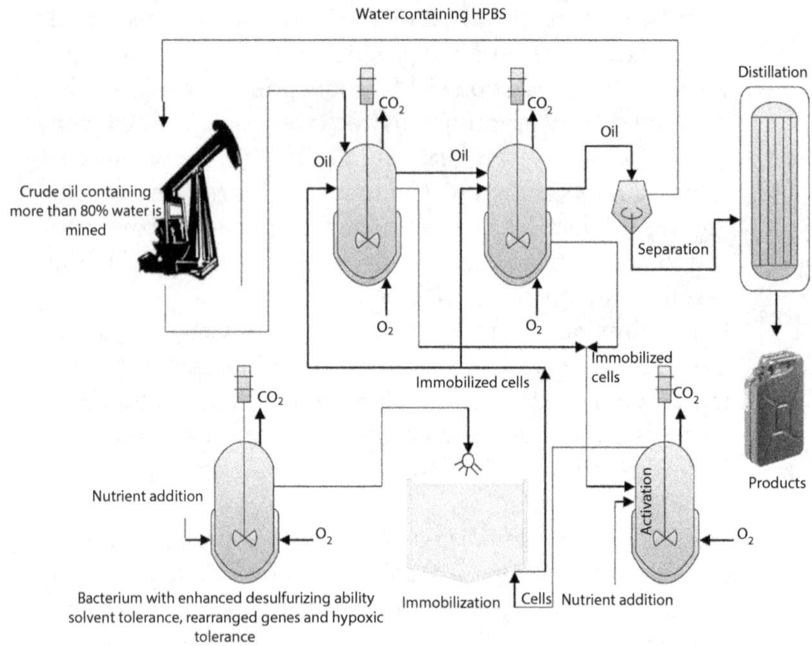

Figure 12.19 Schematic Diagram for an Example of Petroleum Bioprocessing Process - BDS.

12.18 BDS Now and Then

This chapter presented the key issues, advances, and challenges on the path to reach a competitive biodesulfurization process. Microbial desulfurization in nature is different from other, more common biotechnology processes. The delicate enzymes of desulfurization competent cells have to obtain part of their substrates from a different phase in which their survival is not possible. Since the first time that a specific sulfur removal was introduced, research on this field has been steadily continued. Many other strains have been isolated or cloned and different methods have been tried. The most serious problem in the implementation of biodesulfurization as an alternative industrial approach to produce ultra-low sulfur content lies in the isolation or design of a microbial strain with higher efficiency. Any small success that provides the possibility to remove sulfur at a higher temperature, with a higher rate, or longer stability of desulfurization activity is considered a significant step toward industry level biodesulfurization. Moreover, process development and unexpected problems that might occur in big scale operations should be considered.

Upon success, the process may operate in a line after hydrodesulfurization unit. Total desulfurization of fossil fuel by microbial approach is not expected to occur in the early future and more research is needed to design a recombinant strain with a broader range of target sulfur compounds or to use successive desulfurizing microbial systems with high potency. Most studies on desulfurization of refractory compounds have been performed with simple model fuel to understand the nature of desulfurization. Dealing with genuine fossil fuel will open up new challenges to solve.

Comparative study involving BDS and HDS, developed by Enchira Biotechnology Corp. (EBC, formerly Energy BioSystems), demonstrated minor energy consumption and CO_2 generation by the bioprocess to produce a 50 ppm sulfur in diesel from two diesel pools containing 2000 and 500 ppm sulfur. Industries such as Arctic Slope Regional Corp., through its Petro Star subsidiary, installed and operated a 5,000 b/d diesel BDS pilot plant in conjunction with EBC by the end of the 1990s. Other organizations such as the US National Laboratories, the Petroleum Energy Center of Japan (PECJ) together with the King Abdulaziz City for Science and Technology (KACST, Saudi Arabia), and the Mexican Petroleum Institute (MPI) from Mexico are carrying out studies on BDS processes. Scaling, equipment availability, and coupling to conventional infrastructure as well as refiners' mistrust are additional problems to overcome when we look to field applications.

However, there are biorefining processes already applied at the oil industry as shown by the sweetening gas BDS process using *Thiobacillus* bacteria to eliminate the H_2S from gas streams developed by Paques Bio Systems BV and Shell International Oil Products. The gas cleaning manages gases containing up to 80% H_2S or solutions containing up to 10 ppm and produces up to 15 tonnes/day of sulfur (Montiel, 2009).

Currently, it is thought that two major steps should be made soon to make biodesulfurization worthwhile: a rapid progression in understanding and finding optimal ways to implement biotechnology in refineries (Mohebali and Ball, 2008). At the moment, the process is still too slow, too expensive, and not suited for large-scale implementation. It might be used in an additional refinery step (Soleimani *et al.*, 2007), but further development of hydrodesulfurization, adsorptive desulfurization, and oxidative desulfurization might limit its future (Seeberger and Jess, 2010).

Technological challenges still comprise high activity and stability under process conditions (high temperature, low water content) adequate for process configurations to allow high microbial activity, minimization of mass transfer problems, and enhancement of oil-water separation. Microbial processing of fuels clearly differs from other, more common biotechnology processes. Although very intensive research and development has been conducted in BDS and many bacteria have been isolated and intensive genetic engineering improvement realized, the main challenge is still the design of a more active and resistant microbial catalyst. Even with intensive improvement of strains, the desulfurizing activity is still too slow compared to chemical reactions employed in a refinery (El-Gendy and Speight 2016). Any improvement in thermal stability, rate, or overall operational stability could be significant. In this sense, new concepts to be explored include genomes and metagenomes databases mining, the design of microbial consortia to deal with complex hydrocarbon mixtures, and evolutionary engineering of microorganisms to achieve multi-target modifications.

Concerning desulfurization, HDS, ODS, ADS, and BDS technologies should be seen as complementary options where HDS would be the basic technology for obtaining low-sulfur fuels followed by one of these new technologies or a combination of them to remove the most recalcitrant sulfur species for ultra-low sulfur fuel production. Moreover, ADS, BDS, and HDS can be used complementary to each other. In order to develop BDS as a complementary process, the interdisciplinary participation of experts in biotechnology, biochemistry, refining processes, and engineering is essential. Applying nanotechnology to this field is also a point of interest as it leads to economical, cost effective processes.

This is important in the context of increasingly stringent environmental regulations and conventional oil depletion, where both the fuel and the technologies to produce them should be cleaner. BDS and other bioprocessing technologies should also be explored for application at the wellhead to improve crude oil properties and refinery feedstock quality by lowering viscosity of the oil and decreasing the content in high molecular weight hydrocarbons, asphaltenes, sulfur, nitrogen, and/or metals (Speight, 2011). The microbial and enzymatic cracking of PAHs and asphaltene molecules should not be excluded; recent reports on microbial degradation of asphaltenes and the presence of microorganisms with new and previously undetected oxygenases in environmental metagenomes may be a potential source of biocatalysts that justify the exploration of the biotechnological alternatives to upgrading heavy oils. Water is coproduced with the oil and the process of water and oil separation is a routine field process. The biocatalyst could be added at this stage and the only added step would be agitation of the oil and water mixture in the presence of the biocatalyst. The resulting aqueous stream could be reinjected as part of the oil recovery operations or used for the production of valuable chemicals. The concept of valuable chemical production has been recently reintroduced by Kilbane and Stark (2016). This idea was initially proposed by EBC and

Enchira to lower the cost of the entire BDS process by producing HBPSi, a potentially valuable surfactant, instead of HBP as mentioned before. Sulfones are interesting molecules that could be used as building blocks for polymers, plastics, and antibiotic productions (Kilbane and Stark 2016). The main obstacle for the implementation of new technologies is their cost. Not only scientific and technological advances will influence the use of biotechnology in the oil and energy industry (Kilbane 2016a, b). The price of fossil fuels, politics, climate change, land and water use, and resources availability, as well as the increased use of solar and wind energy will also impact the panorama. Biotechnology could be integrated in the refinery of the future that could, in turn, become the base for the development of bio-refineries (Speight 2011; Biernat and Grzelak 2015).

12.19 Conclusion

The two main steps for the successful commercialization of BDS are (i) to continue making rapid technical progress to make it competitive with chemical desulfurization processes and for its implementation in realistic assessments pilot-plant studies and (ii) to find optimum ways to integrate biotechnology into the refineries, which mainly depend on the chemical processes and has also made a lot of progress.

Even when hydrogen and catalyst consumption increases, HDS is still a technically and economically viable unit, especially after the development of non-alumina-based new chemical catalysts. However, the desulfurization chemistry of heavy oils is essentially controlled by strongly inhibiting three and larger ring aromatic hydrocarbon content. BDS for heavily substituted DBTs may, therefore, reclaim some interest.

Integrating a BDS process into a refinery requires a substantial modification of current refinery operations. In addition, the BDS process must operate at the same speed and reliability as other refinery processes so as not to disrupt normal refining operations. Thus, despite the great interest and potential of BDS, it is still challenging to develop it to a stage where it can be practically implemented in refineries. Nevertheless, BDS biocatalysts (whole cells or enzymes) can be used not only for BDS of petroleum and its fractions and products, but it can be alternatively applied for detoxification of chemical warfare agents or for the production of other valuable products, such as surfactants, antibiotics, polythioesters, and various specialty chemicals.

However, in order to develop BDS as a complementary process, the interdisciplinary participation of experts in biotechnology, biochemistry, refining processes, and engineering is essential.

References

Abbad-Andaloussi, S., Lagnel, C., Warzywoda, M., Monot, F. (2003) Multicriteria comparison of resting cell activities of bacterial strains selected for biodesulfurization of petroleum compounds. *Enzyme and Microbial Technology*. 32(3/4): 446–454.

Abin Fuentes, A., Leung, J.C., Mohamed, M.E.S., Wang, D.I.C., Prather, K.L.J. (2014) Rate limiting step analysis of the microbial desulfurization of dibenzothiophene in a model oil system. *Biotechnology and Bioengineering*. 111(5): 876–884.

Abin-Fuentes, A., Mel-S, M., Wang, D.I., Prather, K.L. (2013) Exploring the mechanism of biocatalyst inhibition in microbial desulfurization. *Applied and Environmental Microbiology*. 79(24): 7807–7817.

Acero, J., Berdugo, C., Mogollón, L. (2003) Biodesulfurization process evaluation with a *Gordona rubropertinctus* strain. *CT&F-Ciencia, Tecnología and Futuro*. 2(4): 43–54.

Adegunlola, G.A., Oloke, J.K., Majolagbe, O.N., Adebayo, E.A., Adegunlola, C.O.,

Adewoyin, A.G., Adegunlola, F.O. (2012) Microbial desulphurization of crude oil using *Aspergillus flavus*. *European Journal of Experimental Biology*. 2(2): 400–403.

Adlakha, J., Singh, P., Ram, S.K., Kumar, M., Singh, M.P., Singh, D., Sahai, V., Srivastava, P. (2016) Optimization of conditions for deep desulfurization of heavy crude oil and hydrodesulfurized diesel by Gordonia sp. IITR100. 184: 761–769.

Agarwal, P., Sharma, D.K. (2010) Comparative studies on the biodesulfurization of crude oil with other desulfurization techniques and deep desulfurization through integrated processes. *Energy Fuels*. 24 (1): 518–524.

Aggarwal, S., Karimi, I.A., Ivan, G.R. (2013) In silico modeling of *Gordonia alkanivorans* for biodesulfurization. *Molecular BioSystems*. 9: 2530–2540.

Aggarwal, S., Karimi, I.A., Lee, D.Y. (2011) Reconstruction of a genomescale metabolic network of *Rhodococcus erythropolis* for desulfurization studies. *Molecular BioSystems*. 7(11): 3122–3131.

AitaniCenter for Refining and Petrochemicals, The Research Institute , A.M., Ali, M.F., Al-ADepartment of Chemistry li, H.H. (2000) A review of non-conventional methods for the desulfurization of residual fuel oil. 18(5/6): 537–553.

Akbarzadeh, S., Raheb, J., Aghaei, A., Karkhane, A.A. (2003) Archive of SID Study of desulfurization rate in Rhodococcus FMF native bacterium. *Archive of SID*. 1(1): 36–40.

Akhtar, N., Ghauri, M.A., Akhtar, K. (2016) Dibenzothiophene desulfurization capability and evolutionary divergence of newly isolated bacteria. *Archives of Microbiology*. 198: 509–519.

Alcon, A., Martin, A.B., Santos, V.E., Gomez, E., Garcia-Ochoa, F. (2008) Kinetic model for DBT desulphurization by resting whole cells of *Pseudomonas putida* CECT5279. *Biochemical Engineering Journal*. 39: 486–495.

Alcon, A., Santos, V.E., Martin, A.B., Yustos, P., Garcíıa-Ochoa, F. (2005) Biodesulfurisation of DBT with Pseudomonas putida CECT5279 by resting cells: influence of cell growth time on reducing equivalent concentration and HpaC activity. *Biochemical Engineering Journal.* 26 (2005) 168–175.

Aliebrahimi, S., Raheb, J., Ebrahimipour, G., Bardana, H., Nurollah, M., Aghajani, Z (2015) Designing a new recombinant indigenous *Klebsiella oxytoca* ISA4 by cloning of dsz genes. *Energy Sources, Part A: Recovery, Utilization, and Environmental Effects* . 37(19): 2056–2063.

Al-Shahrani, F., Xiao, T.C., Llewellyn, S.A., Barri, S., Zheng, J., Shi, H.H., Martinie, G., Green, M.L.H. (2007) Desulfurization of diesel via the H2O2 oxidation of aromatic sulfides to sulfones using a tungstate catalyst. *Applied Catalysis B: Environmental.* 73: 311–316.

Alves, L., Marques, S., Matos, J., Tenreiro, R., Gírio, F.M. (2008a) Dibenzothiophene desulfurization by *Gordonia alkanivorans* strain 1B using recycled paper sludge hydrolyzate. *Chemosphere.* 70: 967–973.

Alves, L., Matos, J., Tenreiro, R., Gírio, F. M. (2008b) Evidence for the role of zinc on the performance of dibenzothiophene desulfurization by *Gordonia alkanivorans* strain 1B. *Journal of Industrial Microbiology and Biotechnology.* 35(1): 69–73.

Alves, L., Melo, M., Mendonça, D., Simões, F., Matos, J., Tenreiro, R., Gírio, F.M. (2007) Sequencing, cloning and expression of the dsz genes required for dibenzothiophene sulfone desulfurization from *Gordonia alkanivorans* strain 1B. *Enzyme and Microbial Technology* 40: 1598–1603.

Alves, L., Paixão, S.M. (2011) Toxicity evaluation of 2-hydroxybiphenyl and other compounds involved in studies of fossil fuels biodesulphurisation. *Bioresource Technology.* 102: 9162–9166.

Alves, L., Paixão, S.M. (2014) Enhancement of dibenzothiophene desulfurization by *Gordonia alkanivorans* strain 1B using sugar beet molasses as alternative carbon source. *Applied Biochemistry and Biotechnology.* 172: 3297–3305.

Alves, L., Paixao, S.M., Pacheco, R., Ferreira, A.F., Silva, C.M. (2015) Biodesulfurization of fossil fuels: energy, emissions and cost analysis. *RSC Advances.* 5: 34047–34057.

Alves, L., Salgueiro, R., Rodrigues, C., Mesquita, E., Matos, J., Gírio, F. M. (2005) Desulfurization of dibenzothiophene, benzothiophene, and other thiophene analogs by a newly isolated bacterium, *Gordonia alkanivorans* strain 1B. *Applied Biochemistry and Biotechnology.* 120(3): 199–208.

Amin, G. (2011) Integrated two-stage process for biodesulfurization of model oil by vertical rotating immobilized cell reactor with the bacterium *Rhodococcus erythropolis.* Petroleum and Environmental Biotechnology. 2:1 http://dx.doi.org/10.4172/2157-7463.1000107

Ansari F., Pavel G., Libor S., Tothill I.E., Ramsden J.J. (2009) DBT degradation enhancement by decorating *Rhodococcus erythropolis* IGTS8 with magnetic Fe_3O_4 nanoparticles. *Biotechnology and Bioengineering.* 102(5): 1505–1512.

Ansari, F., Prayuenyong, P., Tothill, I. (2007) Biodesulfurization of dibenzothiophene by *Shewanella putrefaciens* NCIMB 8768. *The Journal of Biological Physics and Chemistry*. 7: 75–78.

Arensdorf, J.J., Loomis, A.K., DiGrazia, P.M., Monticello, D.J., Pienkos, P.T (2002) Chemostat approach for the directed evolution of biodesulfurization gain-of-function mutants. *Applied and Environmental Microbiology*. 68: 691–698.

Aribike, D.S., Susu, A.A., Nwachukwu, S.C.U., Kareem, S.A. (2008) Biodesulfurization of kerosene by *Desulfobacterium indolicum*. *Nature and Science*. 6(4): 55–63.

Aribike, D.S., Susu, A.A., Nwachukwu, S.C.U., Kareem, S.A. (2009) Microbial desulfurization of diesel by Desulfobacterium aniline. Academia Arena 1:11–14.

Armstrong, S.M., Sankey, B.M., Voordouw, G. (1995a) Conversion of dibenzothiophene to biphenyl by sulfate-reducing bacteria isolated from oil field production facilities. *Biotechnology Letters*. 17: 1133–1136.

Armstrong, S.M., Sankey, B.M., Voordouw, G. (1995b) Evaluation of sulfate reducing bacteria for desulfurizing bitumen and its fractions. *Fuel*. 76(3): 223–227.

Atlas, R.M., Boron, D.J., Deever, W.R., Johnson, A.R., McFarland, B.L., Meyer, J.A. (2001) Method for removing organic sulfur from heterocyclic sulfur containing organic compounds. US patent H1, 986.

Atlas, R.M., Boron, D.J., Deever, W.R., Johnson, A.R., McFarland, B.L., Meyer, J.A. (1998) Biodesulfurization of Gasoline: A Technology Roadmap. Coordinating Research Council, Atlanta Georgia, pp. 1e82. CRC Project no. E-7c.

Ayala, M., Robledo, N.R., Lopez-Munguia, A., Vazquez-Duhalt, R. (2000) Substrate specificity and ionization potential in chloroperoxidase-catalyzed oxidation of diesel fuel. *Environmental Science and Technology*. 34: 2804–2809.

Ayala, M., Tinoco, R., Hernandez, V., Bremauntz, P., Vazquez-Duhalt, P. (1998) Biocatalytic oxidation of fuel as an alternative to biodesulfurization. *Fuel Processing Technology*. 57: 101–111.

Ayala, M., Vazquez-Duhalt, R., Morales, M., Le Borgne, S. (2016) Application of microorganisms to the processing and upgrading of crude oil and fractions. In: S.Y. Lee (ed.), Consequences of Microbial Interactions with Hydrocarbons, Oils, and Lipids: Production of Fuels and Chemicals, Handbook of Hydrocarbon and Lipid Microbiology. Springer International Publishing AG. Pp. 1–36. DOI 10.1007/978-3-319-31421-1_205-1.

Ayala, M., Verdin, J., Vazquez-Duhalt, R. (2007) The prospects for peroxidase-based biorefining of petroleum fuels. *Biocatalysis and Biotransformation*. 25: 114–129.

Babich, I.V., Moulijin, J.A. (2003) Science and technology of novel processes of deep desulfurization of oil refinery streams: a review. *Fuel*. 82(6): 607–631.

Bachmann, R.T., Johnson, A.C., Edyvean, R.G.J. (2014) Biotechnology in the petroleum industry: an overview. *International Biodeterioration and Biodegradation*. 86: 225–237.

Bahrami, A., Shojaosadati, S.A., Mohebali, G. (2001) Biodegradation of dibenzothiophenes by thermophilic bacteria. *Biotechnology Letters*. 23: 899–901.

Bahuguna, A., Lily, M.K., Munjal, A., Singh, R.N., Dangwal, K. (2011) Desulfurization of dibenzothiophene (DBT) by a novel strain *Lysinibacillus sphaericus* DMT-7 isolated from diesel contaminated soil. *Journal of Environmental Sciences*. 23: 975–982.

Banat, I.M., Franzetti, A., Gandolfi, I., Bestetti, G., Martinotti, M.G., Fracchia, L., Smyth, T.J., Marchant, R. (2010) Microbial biosurfactants production, applications and future potential. *Applied Microbiology and Biotechnology*. 87:427–444.

Bandyopadhyay, S., Chowdhury, R. (2014) parameter optimization of surface active properties and quantification of biosurfactant produced in continuous stirred tank bioreactor during biodesulfurization of diesel. *International Journal of Innovative Research in Advanced Engineering*. 1(9): 217–224.

Bandyopadhyay, S., Chowdhury, R., Bhattacharjee, C. (2013b) Steady state performance of a bioreactor for production of near zero sulfur diesel (NZSD) and bio-surfactant. *Journal of Clean Energy Technologies*. 1(3): 189–193.

Bandyopadhyay, S., Chowdhury, R., Bhattacharjee, C. (2014) Production of biosurfactants through biodesulfurization of spent engine oil – an experimental study. *Green Chemistry Letters and Reviews*. 7(3): 288–295.

Bandyopadhyay, S., Chowdhury, R., Bhattacharjee, C., Pan, S. (2013a) Simultaneous production of biosurfactant and ULSD (ultra-low sulfur diesel) using *Rhodococcus* sp. in a chemostat. *Fuel*. 113: 107–112.

Becker, J.R. (1997) Crude oil, waxes, emulsions and asphaltenes. Penn Well books. Section III, p.209–250.

Berdugo, C., Caballero, C. and Godoy, R. D. (2002) Aqueous-organic phases separation by membrane reactors in biodesulfurization reactions. *Ciencia, Tecnología y Futuro journal*, 2 (3): 97–112.

Berdugo, C., Mena, J., Acero, J., Mogollón, L. (2001) Increasing the production of desulfurizing biocatalysts by means of fed-batch culture. CT&F-Ciencia, Tecnología and Futuro. 2 (2): 7–15.

Berry, V., Gole, A., Kundu, S., Murphy, C.J., Saraf, R.F. (2005) Deposition of CTAB-terminated nanorods on bacteria to form highly conducting hybrid systems. *Journal of American Chemical Society*. 127(50): 17600–17601.

Bhasarkar, J.B., Dikshit, P.K., Moholkar, V.S. (2015) Ultrasound assisted biodesulfurization of liquid fuel using free and immobilized cells of *Rhodococcus rhodochrous* MTCC 3552: A mechanistic investigation. *Bioresource Technology*. 187: 369–378.

Bhatia S, Sharma DK (2010b) Mining of genomic databases to identify novel biodesulfurizing microorganisms. *Journal of Industrial Microbiology and Biotechnology*. 37: 425–429.

Bhatia, S., Sharma, D.K. (2010a) Biodesulfurization of dibenzothiophene, its alkylated derivatives and crude oil by a newly isolated strain *Pantoea agglomerans* D23W3. *Biochemical Engineering Journal*. 50: 104–109.

Bhatia, S., Sharma, D.K. (2012) Thermophilic desulfurization of dibenzothiophene and different petroleum oils by *Klebsiella* sp. 13T. *Environmental Science and Pollution Research*. 19:3491–3497.

Biernat, K., Grzelak, P.L. (2015) Biorefinery systems as an element of sustainable development. In: Biernat, K. (ed) Biofuels – status and perspective. Croatia, InTech.

Blandino, A., Macías, M., Cantero, D. (1999) Formation of calcium alginate gel capsules: influence of sodium alginate and CaCl2 concentration on gelation kinetics. *Journal of Bioscience and Bioengineering*. 88: 686–689.

Boltes, K., Alonso del Aguila, R., García-Calvo, E. (2013) Effect of mass transfer on biodesulfurization kinetics of alkylated forms of dibenzothiophene by *Pseudomonas putida* CECT5279. *Journal of Chemical Technology and Biotechnology*. 88: 422–431.

Boltes, K., Caro, A., Leton, P., Rodriguez, A., Garcia-Calvo, E. (2008) Gas–liquid mass transfer in oil–water emulsions with an airlift bio-reactor. *Chemical Engineering and Processing*. 47: 2408–2412.

Boltes, K., del Aguila, R.A., Garcia-Calvo, E. (2012) Effect of mass transfer on biodesulfurization kinetics of alkyalted forms of dibenzothiophene by *Pseudomonas putida* CECT5279. *Journal of Chemical Technology and Biotechnology*. 88: 422–431.

Bonde SE, Nunn D (2003) Technical progress report for the biocatalytic desulfurization project. DOE Award Number: DE-FC26-02NT15340 Report Start Date: 9/19/2002 – Report End Date: 12/19/2002.

Boniek, D., Figueiredo, D., dos Santos, A.F.B., de Resende Stoianof, M.A. (2015) Biodesulfurization: a mini review about the immediate search for the future technology. *Clean Technologies and Environmental Policy*. 17: 29–37.

Bordoloi, N.K., Rai, S.K., Chaudhuri, M.K., Mukherjee, A.K. (2014) Deep-desulfurization of dibenzothiophene and its derivatives present in diesel oil by a newly isolated bacterium *Achromobacter* sp. to reduce the environmental pollution from fossil fuel combustion. *Fuel Processing Technology*. 119: 236–244.

Bordoloi, N.K., Rai, S.K., Chaudhuri, M.K., Mukherjee, A.K. (2016) Proteomics and metabolomics analyses to elucidate the desulfurization pathway of *Chelatococcus* sp. PLoS One. 2016 Apr 21;11(4):e0153547. doi: 10.1371/journal.pone.0153547.

Borole, A.P., Kaufman, E.N., Grossman, M.J., Minak-Bernero, V., Lee, M.K. (2002). Comparison of the emulsion characteristics of *Rhodococcus erythropolis* and *Escherichia coli* SOXC-5 cells expressing biodesulfurization genes. *Biotechnology Progress*. 18: 88–93.

Boshagh, F., Mokhtarani, B., Mortaheb, H.R. (2014) Effect of electrokinetics on biodesulfurization of the model oil by Rhodococcus erythropolis PTCC1767 and Bacillus subtilis DSMZ 3256. *Journal of Hazardous Materials*. 280: 781–787.

Bouchez-Naitali M., Abbad-Andaloussi S., Warzywoda M., Monot F. (2004) Relation between bacterial strain resistance to solvents and biodesulfurization activity in organic medium. *Applied Microbiology and Biotechnology*. 65: 440–445.

Bressler, D.C., Norman, J.A., Fedorak, P.M. (1997) Ring cleavage of sulfur heterocycles: how does it happen? *Biodegradation*. 8(5): 297–311.

Bressler, D.C., Norman, J.A., Fedorak, P.M. (1998) Ring cleavage of sulfur heterocycles: how does it happen? Biodegradation. 8: 297–311.

Buzanello, E.B., Rezende, R.P., Sousa, F.M.O., Marques Ede, L.S., Loguercio, L.L. (2014) A novel *Bacillus pumilus*-related strain from tropical landfarm soil is capable of rapid dibenzothiophene degradation and biodesulfurization. BMC Microbiology. 14: 257. http://www.biomedcentral.com/1471-2180/14/257.

Calabrese, R.V., Chang, T.P.K., Dang, P.T. (1986) Drop break-up in turbulent stirred-tank contactors, Part I: Effect of dispersed-phase viscosity. *AIChE Journal*. 32: 657–666.

Calzada, J., Alcon, A., Santos, V.E., Garcia-Ochoa, F. (2011) Mixtures of Pseudomonas putida CECT 5279 cells of different ages: optimization as biodesulfurization catalyst. *Process Biochemistry*. 46:1323–1328.

Calzada, J., Heras, S., Carbajo, J., Alcon, A., Santos, V.E., Garcia, J.L., Garcia-Ochoa, F. (2008) Opitimization of a desulfurizating biocatalyst by combining cells of different age of Pseudomonas putida CECT 5279. *Chemical Engineering Transactions*. 14:259–466.

Campos-Martin, J.M., Capel-Sanchez, M.C., Perez-Presasand, P., Fierro, J.L.G. (2010) oxidative processes of desulfurization of liquid fuels. *Journal of Chemical Technology and Biotechnology*. 85: 879–890.

Carlsson N., Gustafsson H., Thorn C., Olsson L., Holmberg K., Akerman B. (2014) Enzymes immobilized in mesoporous silica: A physical–chemical perspective. *Advances in Colloid and Interface Science*. 205: 339–360.

Caro, A., Boltes, K., Letón, P., García-Calvo, E. (2007a) Dibenzothiophene biodesulfurization in resting cell conditions by aerobic bacteria. *Biochemical Engineering Journal*. 35(2): 191–197.

Caro, A., Boltes, K., Letón, P., García-Calvo, E. (2008). Description of by-product inhibition effects on biodesulfurization of dibenzothiophene in biphasic media. *Biodegradation*. 19(4): 599–611.

Caro, A., Leton, P., Calvo, E.G., Setti, L. (2007b) Enhancement of dibenzothiophene biodesulfurization using β-cyclodextrins in oil-to-water media. *Fuel*. 86: 2632–2636.

Carver J., Barry J., Vertegel A. (2009) Using chloroperoxidase-latex nanoparticles conjugates to desulfurize diesel fuel. Biomaterials Day - Clemson University BBSI/SSBR. http://biomaterials.org/sites/default/files/docs/2014/biodays_2009_clemson_abstracts.pdf

Cass, B.J., Schade, F., Robinson, C.W., Thompson, J.E. Legge, R. L. (2000) Production of tomatoe flavor volatiles from a crude enzyme preparation using a hollow-fiber low-fiber reactor. *Biotechnology and Bioengineering*. 67 (3): 372–377.

Chang, J., Chang, Y.K., Cho, K.-S., Chang, H.N. (2000b) Desulfurization of model and diesel oils by resting cells of *Gordona* sp. *Biotechnology Letters*. 22: 193–196.

Chang, J.H., Chang, Y.K., Ryu, H.W., Chang, H.N. (2000a) Desulfurization of light gas oil in immobilized cell systems of *Gordona* Sp. CYKS1 and *Nocardia* Sp. CYKS2. *FEMS Microbiology Letters*. 182: 309–312.

Chang, J.H., Kim, Y.J., Lee, B.H., Cho, K.-S., Rye, H.W., Chang, Y.K., Chang, H.N. (2001) Production of a desulfurization biocatalyst by two-stage fermentation and its application for the treatment of model and diesel oils. *Biotechnology Progress*. 17: 876–880.

Chang, J.H., Kim, Y.J., Lee, B.H., Cho, K.S., Rye, H.W., Chang, Y.K., Chang, H.N. (2001) Production of a desulfurization biocatalyst by two-stage fermentation and its application for the treatment of model and diesel oils. *Biotechnology Progress*. 17: 876–880.

Chang, J.H., Rhee, S.-K., Chang, Y.K., Chang, H.N. (1998) Desulfurization of diesel oils by a newly isolated dibenzothiophene-degrading *Nocardia* sp. strain CYKS2. *Biotechnology Progress*. 14(6): 851–855.

Chauhan, A.K., Ahmad, A., Singh, S.P., Kumar, A. (2015) Biodesulfurization of benzonaphthothiophene by an isolated *Gordonia* sp. IITR 100. *International Biodeterioration and Biodegradation*. 104: 105–111.

Chen, G.T.C., Monticello, D.J. (1996) Method for separating a petroleum containing emulsion. US Patent No 5,525,235.

Chen, H., Zhang, W.J., Chen, J.M., Cai, Y.B., Li, W. (2008) Desulfurization of various organic sulfur compounds and the mixture of DBT+4,6-DMDBT by *Mycobacterium* sp. ZD-19. *Bioresource Technology*. 99: 3630–3634.

Chen, J.C.T., Monticello, D.J. (1996) Method for separating a petroleum emulsion. US Patent No. 5,525,235.

Clark, P.D., Kirk, M.J. (1994) Studies on the upgrading of bituminousoils with water and transition metal catalysts. *Energy Fuels*. 8:380–387.

Coco, W.M., Levinson, W.E., Crist, M.J., Hektor, H.J., Darzins, A., Pienkos, P.K., Squires, C.H., Monticello, D.J. (2001) DNA shuffling method for generating highly recombined genes and evolved enzymes. *Nature Biotechnology*. 19: 354–359.

Conesa, A., Punt, P.J., van den Hondel, C.A.M.J.J. (2002) Fungal peroxidases: molecular aspects and applications. *Journal of Biotechnology*. 93:143–115.

Constanti, M., Glralt, J., Bordon, A. (1994) Desulfurization of dibenzothiphene by bacteria. *World Journal of Microbial and Biotechnology*. 10: 510–516.

Coordinated Research Council (1998) Report no. 615; CRC project E-7c; Biodesulfurization of gasoline-a technology road map., 82 pages.

Crawford, D.L., Gupta, R.K. (1990) Oxidation of dibenzothiophene by *Cunninghamella elegans*. *Current Microbiology*. 21: 229–231.

Davoodi-Dehaghani, F., Vosoughi, M., Ziaee, A.A. (2010) Biodesulfurization of dibenzothiophene by a newly isolated *Rhodococcus erythropolis* strain. *Bioresource Technology*. 101:1102–1105.

de Weert, S., Lokman, B.C. (2010) Heterologous expression of peroxidases. In: Torres E, Ayala, M. (eds) Biocatalysis based on Heme peroxidases. Springer, Berlin.

Denis-Larose, C., Labbé, D., Bergeron, H., Jones, A. M., Greer, C. W., Al-Hawari, J., Lau, P.C.K. (1997) Conservation of plasmid-encoded dibenzothiophene desulfurization genes in several rhodococci. *Applied and Environmental Microbiology*. 63(7): 2915–2919.

Derikvand, P., Etemadifar, Z. (2014) Improvement of biodesulfurization rate of alginate immobilized *Rhodococcus erythropolis* R1. *Jundishapur Journal of Microbiology*. 7(3): e9123. DOI: 10.5812/jjm.9123.

Derikvand, P., Etemadifar, Z., Biria, D. (2014) Taguchi optimization of dibenzothiophene biodesulfurization by *Rhodococcus erythropolis* R1 immobilized cells in a biphasic system. *International Biodeterioration and Biodegradation*. 86: 343–348.

Derikvand, P., Etemadifar, Z., Saber, H. (2015) Sulfur removal from dibenzothiophene by newly isolated *Paenibacillus validus* strain PD2 and process optimization in aqueous and biphasic (model oil) systems. *Polish Journal of Microbiology*. **64**: 47–54.

Dinamarca, A., Ibacache-Quiroga, C., Baeza, P., Galvez, S., Villarroel, M., Olivero P, Ojeda J. (2010) Biodesulfurization of gas oil using inorganic supports biomodified with metabolically active cells immobilized by adsorption. *Bioresource Technology*. 101(7): 2375–2378.

Dinamarca, M.A., Rojas, A., Baeza, P., Espinoza, G., IbacacheQuiroga, C., Ojeda, J. (2014) Optimizing the biodesulfurization of gas oil by adding surfactants to immobilized cell systems. *Fuel*. 116: 237–241.

Dudley, M.W., Frost, J.W. (1994) Biocatalytic desulurization of aryl sulfonates. *Bioorganic and Medicinal Chemistry*. 2: 681–690.

Eber, J., Wasserscheid, P., Jess, A. (2004) Deep Desulfurization of oil refinery streams by extraction with ionic liquids. *Green Chemistry*. 6: 316–322.

Eckart, V., Köhler, M., Hieke, W. (1986) Microbial desulfurization of petroleum and heavy petroleum fractions. 5. Anaerobic desulfurization of Romashkino petroleum. Zentralblatt Fur Mikrobiologie. 141: 291–300.

El-Gendy N.Sh., Speight J.G. (2016) Handbook of Refinery Desulfurization. CRC Press, Taylor and Francis Group, LLC, 6000 Broken Sound Parkway NW, Suite 300, Boca Raton, FL 33487-2742, USA.

El-Gendy, N. Sh., Farahat, L.A., Moustafa, Y.M., Shaker, N., El-Temtamy, S.A. (2006) Biodesulfurization of Crude and Diesel Oil by Candida parapsilosis NSh45 Isolated from Egyptian Hydrocarbon Polluted Sea Water. *Biosciences, Biotechnology Research Asia*, 3(1a): 5–16.

El-Gendy, N.Sh. (2001) Biodesulfurization of organosulfur compounds in crude oil and its fractions. MSc Thesis, Department of Chemistry, Cairo University, Cairo, Egypt.

El-Gendy, N.Sh. (2004) Biodesulfurization potentials of crude oil by bacteria isolated from hydrocarbon polluted environments in Egypt. A PhD Thesis, Faculty of Science, Cairo University, Cairo, Egypt.

El-Gendy, N.Sh., Nassar, H.N., Abu Amr, S.S. (2014) Factorial design and response surface optimization for enhancing a biodesulfurization process. *Petroleum Science and Technology*. 32(14): 1669–1679.

El-Gendy, N.Sh., Nassar, H.N., Younis, S.A. (2015) Main and interactive effects of polyaromatic sulfur heterocyclic compounds on growth and biodegradation efficiencies of Bacillus sphaericus HN1: modeling and statistical analysis. *Petroleum Science and Technology*. 33(11): 1167–1181.

Elmi, F., Etemadifar, Z., Emtiazi, G. (2015) A novel metabolite (1,3-benzene-diol, 5-hexyl) production by Exophiala spinifera strain FM through dibenzo-thiophene desulfurization. *World Journal of Microbiology and Biotechnology.* 31:813–821.

Enchira Biotechnology Corporation. 2001. Gasoline desulfurization program. DE-FC07-97ID13570. Final Report.

Energy BioSystems advances biodesulfurization research (2000) *Oil and Gas Journal.*

Etemadi, O., Yen, T.F. (2007) Surface characterization of adsorbents in ultrasound-assisted oxidative desulfurization process of fossil fuels. *Journal of Colloid and Interface Science.* 313: 18–25.

Etemadifar, Z., Derikvand, P., Emtiazi, G., Habibi, M.H. (2014) Response surface methodology optimization of dibenzothiophene biodesulfurization in model oil by nanomagnet immobilized *Rhodococcus erythropolis* R1. *Journal of Materials Science and Engineering B* 4(10): 322–330.

Fang, X.X., Zhang, Y.L., Luo, L.L., Xu, P., Chen, Y.L., Zhou, H., Hai, L. (2006) Organic sulfur removal from catalytic diesel oil by hydrodesulfurization combined with biodesulfurization. Mod Chem Ind 26: 234–238 (*Chinese journal*).

Fatahi, A., and Sadeghi, S. (2017) *Biodesulphurization* of gasoline by *Rhodococcus erythropolis* supported on polyvinyl alcohol. *Letters in Applied Microbiology.* 64: 370–378.

Feng, J., Zeng, Y., Ma, C., Cai, X., Zhang, Q., Tong, M., Yu, B., Xu, P. (2006) The surfactant Tween 80 enhances biodesulfurization. *Applied and Environmental Microbiology.*72: 7390–7393.

Feng, S., Yang, H., Zhan, X., Wang, W. (2016) Enhancement of dibenzothiophene biodesulfurization by weakening the feedback inhibition effects based on a systematic understanding of the biodesulfurization mechanism by *Gordonia* sp. through the potential "4S" pathway. RSC Advances. 6: 82872–82881.

Finnerty, W.R. (1992) Biodegradation 2: 223–226.

Finnerty, W.R. (1993) Organic sulfur biodesulfurization in non-aqueous media. *Fuel* 72(12): 1631–1634.

Finnerty, W.R., Singer, M.E.V. (1984) A microbial biosurfactant – physiology, biochemistry and applications. *Developments in Industrial Microbiology.* 25: 31–40.

Foght, J.M., Fedorak, P.M., Gray, M.R., Westlake, D.W.S. (1990) Microbial desulfurization of petroleum. In Microbial Mineral Recovery; Ehrlich, H.L., Brierley, C.L., Eds.; McGraw-Hill: New York. Pp. 379–407.

Folsom, B., Schieche, D., DiGarazia, P., Werner, J., Palmer, S. (1999) Microbial desulfurization of alkylated dibenzothiophenes from a hydrodesulfurized middle distillate by *Rhodococcus erythropolis* I-19. *Applied and Environmental Microbiology.* 65(11): 4967–4972.

Folsom, B.R. (2000) In: Le Borgne, S., Rojas, N., Quintero, R. (Eds.), Proceedings First International Conference on Petroleum Biotechnology: State of the Art and Perspectives, Instituto Mexicano del Petro′ leo, Mexico. Pp. 103.

Furuya, T., Ishii, Y., Noda, K., Kino, K., Kirimura, K. (2003) Thermophilic biodesulfurization of hydrodesulfurized light gas oils by *Mycobacterium phlei* WU-F1. *FEMS Microbiology Letters* 221, 137–142.

Furuya, T., Kirimura, K., Kino, K., & Usami, S. (2001) Thermophilic biodesulfurization of dibenzothiophene and its derivatives by *Mycobacterium phlei* WU-F1. *FEMS Microbiology Letters*. 204(1): 129–133.

Furuya, T., Kirimura, K., Kino, K., Usami, S. (2002) Thermophilic biodesulfurization of naphthothiophene and 2-ethylnaphthothiophene by a dibenzothiophene-desulfurizing bacterium, Mycobacterium phlei WU-F1. *Applied Microbiology and Biotechnology*. 58(2): 237–240.

Gallagher, J.R., Olson, E.S., Stanley, D.C. (1993) Microbial desulfurization of dibenzothiophene: a sulfur specific pathway. *FEMS Microbiology Letters*. 107: 31–36.

Gallardo, M.E., Ferrandez, A., de Lorenzo, V., Garcia, J.L., Diaz, E. (1997) Designing recombinant *Pseudomonas* strains to enhance biodesulfurization. *Journal of Bacteriology*. 179: 7156–7160.

Garcia-Ochoa, F., Gomez, E. (2009) Bioreactor scale-up and oxygen transfer rate in microbial processes: An overview. Biotechnology Advances. 27: 153–176.

Gary, J.H., Handwerk G.E. (1994) Introduction to petroleum refining. Marcel Dekker, New York.

Gautier, A., Carpentier, B., Dufresne, M., Dinh, Q.V., Paullier, P., Legallais, C. (2011)
Impact of alginate type and bead diameter on mass transfers and the metabolic activities of encapsulated C3A cells in bioartificial liver applications. *European cells and materials Journal*. 21: 94–106.

Gaylarde, C.C., Bento, F.M., Kelley, J. (1999) Microbial contamination of stored hydrocarbon fuels and its control. Revista de Microbiologia. 30: 01–10.

Giorno, L., Drioli, E. (2000) Biocatalytic membrane reactors: applications and perspectives. *Trends in Biotechnology*. 18 (8):339–349.

Goindi, H.K., Saini, V.S., Verma, P.S., Adhikari, D.K. (2002) Dibenzothiophene desulfurization in hydrocarbon environment by Staphylococcus sp. resting cells. *Biotechnology Letters*. 24: 779–781.

Gomez, E., Alcon, A., Escobar, S., Santos, V.E., Garcia-Ochoa, F. (2015) Effect of fluid dynamic conditions on growth rate and biodesulfurization capacity of *Rhodococcus erythropolis* IGTS8. *Biochemical Engineering Journal*. 99: 138–146.

Gray K.A., Machkoyz T., Squires C.H. (2003) Biodesulfurization of fossil fuels. Current Opinion in Microbiology. 6: 229–235.

Gray, K.A., Pogrebinsky, O.S., Mrachko, G.T., Xi, L., Monticello, D.J., Squires, C.H. (1996) Molecular mechanisms of biocatalytic desulfurization of fossil fuels. *Nature Biotechnology*. 14: 1705–1709.

Grossman, M.J. (1996) Microbial removal of organic sulfur from fuels: a review of past and present approaches. In: Ocelli, M.L., Chianelli, R. (Eds.), Hydrotreating Technology for Pollution Control. Catalysts, Catalysis, and Processes. Marcel Decker, New York. Pp. 345–359.

Grossman, M.J., Lee, M.K., Prince, R.C., Garrett, K.K., George, G.N., Pickering, I.J. (1999) Microbial desulfurization of a crude oil middle-distillate fraction: analysis of the extent of sulfur removal and the effect of removal on remaining sulfur. *Applied and Environmntal Microbiology*. 65: 181–188.

Grossman, M.J., Lee, M.K., Prince, R.C., Minak-Bernero, V., George, G.N., Pickering, I.J. (2001) Deep desulfurization of extensively hydrodesulfurized middle distillate oil by *Rhodococcus* sp. strain ECRD-1. *Applied and Environmental Microbiology*. 67: 1949–1952.

Guerinik, K., Al-Mutawah, Q. (2003) Isolation and characterization of oil-desulphurizing bacteria. *World Journal of Microbiology and Biotechnology*. 19: 941–945.

Gün G., Yürüm Y., Doğanay G.D. (2015) Revisiting the biodesulfurization capability of hyperthermophilic archaeon *Sulfolobus solfataricus* P2 revealed DBT consumption by the organism in an oil/water two-phase liquid system at high temperatures. *Turkish Journal of Chemistry* 39(2): 255–266.

Gunam, I. B. W., Iqbal, M., Arnata, I. W., Antara, N. S., Dewi Anggreni, A. A. M., Setiyo, Y., Gunadnya, I.B.P. (2016) Biodesulfurization of Dibenzothiophene by a Newly Isolated *Agrobacterium* LSU20. Applied Mechanics and Materials, 855: 143–149.

Gunam, I.B., Yaku, Y., Hirano, M., Yamamura, K., Tomita, F., Sone, T., Asano, K. (2006) Biodesulfurization of alkylated forms of dibenzothiophene and benzothiophene by *Sphingomonas subarctica* T7b. *Journal of Bioscience and Bioengineering*. 101(4):322–327.

Gunam, I.B.W., Kenta, Y., Nengah, I.S., Nyoman, S.A., Wayan, R.A., Michiko, T., Fusao, T., Teruo, S., Kozo, A. (2013) Biodesulfurization of dibenzothiophene and its derivatives using resting and immobilized cells of *Sphingomonas subarctica* T7b. *Journal of Microbiology and Biotechnology*. 23: 473–482.

Guo, X.L., Deng, G., Xu, J., Wang, M.X. (2006) Immobilization of *Rhodococcus* sp. AJ270 in alginate capsules and its application in enantioselective biotransformation of trans-2-methyl-3-phenyl-oxiranecarbonitrile and amide. *Enzyme and Microbial Technology*. 39: 1–5.

Guobin, S., Huaiying, Z., Jianmin, X., Guo, C., Wangliang, L., Huizhou, L. (2006) Biodesulfurization of hydrodesulfurized diesel oil with *Pseudomonas delafieldii* R-8 from high density culture. *Biochemical Engineering Journal*. 27: 305–309.

Guobin, S., Huaiying, Z., Weiquan, C., Jianmin, X., Huizhou, L. (2005) Improvement of biodesulfurization rate by assembling nanosorbents on the surfaces of microbial cells. *Biophysical Journal*. 89: L58–L60.

Han J.W., Park H.S., Kim B.H., Shin P.G., Park S.K., Lim J.C. (2001) Potential use of nonionic surfactants in the biodesulfurization of Bunker-C oil. *Energy and Fuels*. 15(1): 189–196.

Hansen, C.M. (1999) Hansen Solubility Parameters: A User's Handbook: Chapter 1. Hoersholm, Denmark: CRC Press LLC.

Hartman, J., Geva, J., Fass, R.A. (1991) Computerized expert system for diagnosis and control of microbial contamination in jet fuel and diesel fuel storage systems.

In Proceedings of 4th International Conference on Stability and Handling of Liquid Fuels, Orlando, Florida, November 19–22, 1991. Pp. 153–166.

Hayden, S.C., Zhao, G., Saha, K., Phillips, R.L., Li, X., Miranda, O.R., Rotello, V.M., ElSayed, M.A., Schmidt-Krey, I., Bunz, U.H.F. (2012) Aggregation and interaction of cationic nanoparticles on bacterial surfaces. *Journal of American Chemical Society*. 134: 6920–6923.

Hill, E.C., Hill, G.C. (1993). Microbiological problems in distillate fuels. Transactions - Institute of Marine Engineers. 104: 119–130.

Hinze, J.O. (1955) Fundamentals of the hydrodynamic mechanism of splitting in dispersion processes. *AIChE Journal*. 1: 289–295.

Hirasawa, K., Ishii, Y., Kobayashi, M., Koizumi, K., Maruhashi, K. (2001) Improvement of desulfurization activity in *Rhodococcus erythropolis* KA2–5-1 by genetic engineering. *Bioscience, Biotechnology, and Biochemistry*. 65: 239–246.

Honda, H.; Sugiyama, H.; Saito, I. and Kobayashi, T. (1998) High cell density culture of *Rhodococcus rhodochrous* by pH-state feeding and dibenzothiophene degradation. *Fermentation Bioengineering*. 85:334–338.

Hou Y., Kong Y., Yang J., Zhang J., Shi D., Xin W. (2005) Biodesulfurization of dibenzothiophene by immobilized cells of *Pseudomonas stutzeri* UP-1. *Fuel*. 84: 1975–1979.

Huang T., Qiang L., Zelong W., Daojang Y., Jiamin X. (2012) Simultaneous removal of thiophene and dibenzothiophene by immobilized Pseudomonas delafieldii R-8 cells. *The Chinese Journal of Chemical Engineering*. 20:47–51.

Hwan, J., Keun, Y., Wook, H., Nam, H. (2000) Desulfurization of light gasoil in immobilized-cell systems of *Gordona* sp. CYKS1 and *Nocardia* sp. CYKS2. *FEMS Microbiology Letters*. 182: 309–312.

Indian Institute of Petroleum (1996) Project proposal on biodesulfurization of petroleum fractions. (Sep.,1996). Dehradum.

Irani, Z.A., Mehrina, M.R., Yazdian, F., Soheily, M., Mohebali, G., Rasekh, B. (2011a) Analysis of petroleum biodesulfurization in an airlift bioreactor using response surface methodology. *Bioresource Technology*. 102: 10585–10591.

Irani, Z.A., Yazdian, F., Mohebali, G., Soheili, M., Mehrnia, M.R. (2011b) Determination of growth kinetic parameters of a desulfurizing bacterium, *Gordonia alkanivorans* RIPI90A. *Chemical Engineering Transactions*. 24: 937–942.

Ishii, Y., Konishi, J., Okada, H., Hirasawa, K., Onaka, T., Suzuki, M. (2000) Operon structure and functional analysis of the genes encoding thermophilic desulfurizing enzymes of *Paenibacillus* sp. A11–2. *Biochemical and Biophysical Research Communications*. 270: 81–88.

Ishii, Y., Kozaki, S., Furuya, T., Kino, K., Kirimura, K. (2005) Thermophilic biodesulfurization of various heterocyclic sulfur compounds and crude straight-run light gas oil fraction by a newly isolated strain Mycobacterium phlei WU-0103. *Current Microbiology*. 50: 63–70.

Isken, S., Derks, A., Wolfes, P.F.G., de Bont, J.A.M. (1999) Effect of organic solvents on the yield of solvent-tolerant *Pseudomonas putida* S12. *Applied and Environmental Microbiology*. 65(6): 2631–2635.

Ismail, W., El-Sayed, W.S., Abdul Raheem, A.S., Mohamed, M.E., El Nayal, A.M. (2016) Biocatalytic desulfurization capabilities of a mixed culture during non-destructive utilization of recalcitrant organosulfur compounds. Frontiers in Microbiology. 7:266. doi: 10.3389/fmicb.2016.00266.

Izumi, Y., Ohshiro, T., Ogino, H., Hine, Y., Shimao, M. (1994) Selective desulfurization of dibenzothiophene by *Rhodococcus erythropolis* strainD-1. *Applied and Environmental Microbiology*. 60: 223–226.

Jae, P.S., Lee, I-S., Chang, Y.K., Lee, S.Y. (2003) Desulfurization of dibenzothiophene and diesel oil by metabolically engineered *Escherichia coli*. *Journal of Microbiology and Biotechnology*. 13(4): 578–583.

Javadli, A., de Klerk, A. (2012) Desulfurization of heavy oil. Applied Petrochemical Research. 1: 3–19.

Jeyagowry, T.S., Huang, X., Dou, J., Nah, T.Y., Xu, R., Wong, P.K. (2006) A novel oxidative desulfurization process to remove refractory sulfur compounds from diesel fuel. *Applied Catalysis B*. 63: 85–93.

Jia, X., Wen, J., Sun, Z., Caiyin, Q., Xie, S. (2006) Modeling of DBT biodegradation behaviors by resting cells of *Gordonia* sp. WQ-01 and its mutant in oil–water dispersions. *Chemical Engineering Science*. 61: 1987–2000.

Jiang C., Liu, H., Xie, Y., Chen, J. (2002) Isolation of soil bacteria species for degrading dibenzothiophene. *Chinese Journal of Chemical Engineering*. 10(4): 420–426.

Jiang, B., Yang, H., Zhang, L., Zhang, R., Sun, Y., Huang, Y. (2016) Efficient oxidative desulfurization of diesel fuel using amide-based ionic liquids. *Chemical Engineering Journal*. 283: 89–96.

Johnson, S., Monticello, D., Hazan, C., Colin, J.M. (2000) Conversion of organosulfur compounds to oxyorganosulfur compounds for desulfurization of fossil fuels. US Patent No 6,071,738.

Juarez-Moreno K., Diaz de Leon J.N., Zepeda T.A., Vazquez-Duhalt R., Fuentes S. (2015) Oxidative transformation of dibenzothiophene by chloroperoxidase enzyme immobilized on (1D)- γ-Al_2O_3 nanorods. *Journal of Molecular Catalysis B: Enzymatic* . 115: 90–95.

Juarez-Moreno K., Pestryakov A., Petranovskii V. (2014) Engineering of supported nanomaterials. Procedia Chemistry. 10: 25–30.

Kaewboran, W. (2005) Biodesulfurization of Dibenzothiophene by *Cryptococcus laurentii*. A M.Sc. Thesis, Mahidol University, Thailand.

Kanai, T., Uzumaki, T., Kawase., Y. (1996) Simulation of airlift bioreactors: Steady state performance of continuous culture processes. *Computers and Chemical Engineering*. 20: 1089–1099.

Kareem S.A. (2014) Anaerobic microbial desulfurization of kerosene. J. Nanomed. Naotechnol. 5: 5. http://dx.doi.org/10.4172/2157-7439.S1.017

Kareem, S.A., Aribike, D.S., Susu, A.A., Nwachukwu, S.C.U. (2016) Anaerobic biodesulfurization of kerosene part I: Identifying a capable microorganism. *Chemical Engineering and Process Techniques*. 2(2): 1028.

Karsten, G., Simon, H. (1993) Immobilization of *Proteus vulgaris* for the reduction of 2-oxo acids with hydrogen gas or formate to D-2-hydroxy acids. *Applied Microbiology and Biotechnology*. 38: 441–446.

Kassler, P. (1996) World energy demand outlook. In G. Jenkins (ed.) Energy exploration and exploitation, vol. 14. Multi-Science Publishing Co., Berkshire, UK. Pp. 229–242.

Kaufman, E.N., Borole, A.P., Shong, R., Sides, J.L., Juengst, C. (1999) Sulfur specificity in the bench-scale biological desulfurization of crude oil by Rhodococcus IGTS8. *Journal of Chemical Technology and Biotechnology.* 74:1000–1004.

Kaufman, E.N., Harkins, J.B., Barole, A.P. (1998) Comparison of batch-stirred and electrospray reactors for biodesulfurization of dibenzothiophene in crude oil and hydrocarbon feedstocks. *Applied Biochemistry and Biotechnology*. 73: 127–143.

Kaufman, E.N., Harkins, J.B., Rodriguez, M., Jr., Tsouris, C., Selvaraj, P.T., Murphy, S.E. (1997) Development of an electrospray bioreactor for crude oil processing. *Fuel Processing Technology*. 52: 127–144.

Kawaguchi, H., Kobayashi, H., Sato, K. (2012) Metabolic engineering of hydrophobic Rhodococcus opacus for biodesulfurization in oil–water biphasic reaction mixtures. *Journal of Bioscience and Bioengineering*. 113: 360–366.

Kayser, K.J., Cleveland, L., Park, H.-S., Kwak, J.-H., Kolhatkar, A., Kilbane II, J.J. (2002) Isolation and characterization of a moderate thermophile, *Mycobacterium phlei* GTIS10, capable of dibenzothiophene desulfurization. *Applied Microbiology and Biotechnology.* 59: 737–745.

Kayser, Y., Bielagajones, B.A., Jackowski, K., Odusan, O., Kilbane, J.J. (1993) Utilization of organusulphur compounds by axenic and mixed cultures of *Rhodococcus rhodochrous* IGTS8. *Journal of General Microbiology*. 139: 3123–3129.

Khairy, H., Wubbeler, J.H., Steinbuchel, A. (2015) Biodegradation of the organic disulfide 4,4-dithiodibutyric acid by *Rhodococcus* spp. *Applied and Environmental Microbiology*. 81: 8294–8306.

Khedkar, S., Shanker, R. (2015) Isolation and classification of a soil actinomycete capable of sulphur-specific biotransformation of dibenzothiophene, benzothiophene and thianthrene. *Journal of Applied Microbiology*. 118(1): 62–74.

Kilbane II, J.J. (2016a) Future applications of biotechnology to the energy industry. Frontiers Microbiology. 7: 1–4.

Kilbane II, J.J. (2016b) Biodesulfurization: a model system for microbial physiology research. *World Journal of Microbiology and Biotechnology.* 32: 137.

Kilbane II, J.J., Jackowski, K. (1996) Biocatalytic detoxification of 2- chloroethyl ethyl sulfide. *Journal of Chemical Technology Biotechnology*. 65: 370–374.

Kilbane, J.J. (2006) Microbial biocatalyst developments to upgrade fossil fuels. *Current Opinion in Biotechnology*. 17: 305–314.

Kilbane, J.J. II (2017) Biodesulfurization: How to make it work? *Arabian Journal for Science and Engineering*. 42: 1–9.

Kilbane, J.J. II, Stark, B. (2016) Biodesulfurization: a model system for microbial physiology research. *World Journal of Microbiology and Biotechnology*. 32(8): 137. DOI 10.1007/s11274-016-2084-6.

Kilbane, J.J., Bielaga, B.A. (1990) Toward sulfur-free fuels. ChemTech. 20: 747–751.

Kilbane, J.J., Jackowsky, K. (1992) Biodesulfurization of water-soluble coal-derived material by *Rhodococcus rhodochrous* IGTS8. *Biotechnology and Bioengineering*. 40: 1107–1114.

Kilbane, J.J., Le Borgne, S. (2004) Petroleum Biorefining: The Selective Removal of Sulfur, Nitrogen, and Metals. In: Petroleum Biotechnology, Developments and Perspectives, R. Vazquez-Duhalt and R. Quintero-Ramirez (Editors). Elsevier, Amsterdam. Pp. 29–65.

KilbaneII, J.J., Stark, B. (2017) Biodesulfurization: a model system for microbial physiology research. *World Journal of Microbiology and Biotechnology*. 32:137. DOI 10.1007/s11274-016-2084-6.

Kim B.H., Shin P.K., Na J.U., Park D.H., Bang, S.H. (1996) Microbial petroleum desulfurization. *Journal of Microbiology and Biotechnology*. 6(5) 299–308

Kim, H.Y., Kim, T.S., Kim, B.H. (1990a) Degradation of organic sulfur compounds and the reduction of dibenzothiophene to biphenyl and hydrogen sulfide by *Desulfovibrio desulfuricans* M6. *Biotechnology Letters*. 12: 761–764.

Kim, S.B., Brown, R., Oldfield, C., Gilbert, S.C., Iliarionov, S., Goodfellow, M. (2000) *Gordonia amicalis* sp. nov., a novel dibenzothiophene-desulfurizing actinomycete. *International Journal of Systematic and Evolutionary Microbiology*. 50: 2031–2036.

Kim, T.S., Kim, H.Y., Kim, B.H. (1990b) Petroleum desulfurization by *Desulfovibrio desulfuricans* M6 using electrochemically supplied reducing equivalent. *Biotechnology Letters*. 12: 757–760.

Kim, Y.J., Chang, J.H., Cho, K.S., Ryu, H.W., Chang, Y.K. (2004) A physiological study on growth and dibenzothiophene (dibenzothiophene) desulfurization characteristics of *Gordonia* sp. CYKS1. *Korean Journal of Chemical Engineering*. 21: 436–441.

Kirimura, K., Furuya, T., Nishii, Y., Ishii, Y., Kino, K., Usami, S. (2001) Biodesulfurization of dibenzothiophene and its derivatives through the selective cleavage of carbon-sulfur bonds by a moderately thermophilic bacterium *Bacillus subtilis* WU-S2B. *Journal of Bioscience and Bioengineering*. 91: 262–266.

Kirimura, K., Harada K., Iwasawa H., Tanaka, T., Iwasaki, Y., Furuya, T., Ishii, Y., Kino, K. (2004) Identification and functional analysis of the genes encoding dibenzothiophene-desulfurizing enzymes from thermophilic bacteria. *Applied Microbiology and Biotechnology*. 65:703–713.

Kobayashi, M., Onaka, T., Ishii, Y., Konishi, J., Takaki, M., Okada, H., Ohta, Y., Koizumi, K., Suzuki, M. (2000) Desulfurization of alkylated forms of both dibenzothiophene and benzothiophene by a single bacterial strain. *FEMS Microbiology Letters*. 187: 123–126.

Kodama Central Research Institute of Electric Power Industry, Abiko, Chiba Prefecture, K., Umehara Central Research Institute of Electric Power Industry, Abiko, Chiba Prefecture, K., Shimizu Research Laboratories, Fujisawa Pharmaceutical Co., Ltd., Higashiyodogawa-ku, Osaka, K., Nakatani Research Laboratories, Fujisawa Pharmaceutical Co., Ltd., Higashiyodogawa-ku, Osaka,

S., Minoda Department of Agricultural Chemistry, Faculty of Agriculture, The University of Tokyo, Y., Yamada Department of Agricultural Chemistry, Faculty of Agriculture, The University of Tokyo, K. (1993) Identification of Microbial Products from Dibenzothiophene and Its Proposed Oxidation Pathway. *Agricultural and Biological Chemistry*. 37(1): 45–50.

Kodama, K., Umehara, K., Shimizu, K., Nakatani, S., Monioda, Y., Yamada, K. (1973) Identification of microbial products from dibenzothiophene and its proposed oxidation pathway. *Agricultural and Biological Chemistry*. 37: 45–50.

Köhler, M., Genz, I.-L., Schicht, B., Eckart, V. (1984) Microbial desulfurization of petroleum and heavy petroleum fractions. 4. Anaerobic degradation of organic sulfur compounds of petroleum. Zentralblatt Fur Mikrobiologie. 139: 239–247.

Kong, L.Y., Li, G., Wang, X.S. (2004) Mild oxidation of thiophene over TS-1/H_2O_2. *Catalysis Today*. 341, 93–95.

Konishi M., Kishimoto M., Tamesui N., Omasa I., Shioya S., Ohtake H. (2005b) The separation of oil from an oil–water–bacteria mixture using a hydrophobic tubular membrane. *Biochemical Engineering Journal*. 24: 49–54.

Konishi, J., Ishii, Y., Onaka, T., Okumura, K., Suzuki, M. (1997) Thermophilic carbon–sulfur-targeted biodesulfurization. *Applied and Environmental Microbiology*. 63: 3164–3169.

Konishi, J., Okada, H., Hirasawa, K., Ishii, Y., Maruhashi, K. (2002) Comparison of the substrate specificity of the two bacterial desulfurization systems. *Biotechnology Letters*. 24: 1863–1867.

Konishi, J., Onaka, T., Ishii, Y., Susuki, M. (2000) Demonstration of the carbon-sulfur bond targeted desulfurization of benzothiophene by thermophilic *Paenibacillus* sp. strain A11–2 capable of desulfurizing dibenzothiophene. *FEMS Microbiology Letters*. 187: 151–154.

Konishi, M., Kishimoto, M., Omasa, T., Katakura, Y., Shioya, S., Ohtake, H. (2005a) Effect of sulfur sources on specific desulfurization activity of *Rhodococcus erythropolis* KA2–5-1 in exponential fed-batch culture. *Journal of Bioscience and Bioengineering*. 99: 259–263.

Kropp, K.G., Andersson, J.T., Fedorak, P.M. (1997b) Bacterial transformations of 1,2,3,4-tetrahydrodibenzothiophene and dibenzothiophene. *Applied and Environmental Microbiology*. 63(8): 3032–3042.

Kurita, S., Endo. T., Nakamura. H., Yagi. T., Tamiya. N. (1971) Decomposition of some organic sulfur compounds in petroleum by anaerobic bacteria. *The Journal of General and Applied Microbiology*. 17: 185–198.

Labana, S., Pandey, G., Jain, R.K. (2005) Desulfurization of dibenzothiophene and diesel oils by bacteria. *Letters in Applied Microbiology*. 40(3): 159 - 163.

Lange, E., Lin, Q. (1998) Preparation of surfactants from a byproduct of fossil fuel biodesulfurization. Symposium on Chemistry of Diesel Fuels. 21 6th National Meeting: 1998 Aug 23–27; Boston, Massachusetts. Washington, DC: Division of Petroleum Chemistry, Inc., American Chemical Society. 43: 550–552.

Lange, E., Lin, Q., Nielsen, K., Dooyema, C. (1999) Surfactants derived from 2-(2-hydroxyphenyl) benzenesulfinate and alkyl-substituted derivatives. US Patent No 5,973,195.

Larkin, M.J., Kulakov, L.A., Allen, C.C. (2005) Biodegradation and *Rhodococcus* – masters of catabolic versatility. *Current Opinion in Biotechnology*. 16: 282–290.

Le Borgne, S., Ayala, M. (2010) Microorganisms utilizing sulfur-containing hydrocarbons. In: Timmis, K.N. (Ed.), Handbook of Hydrocarbon and Lipid Microbiology. Springer-Verlag, Berlin Heidelberg, Germany.

Le Borgne, S., Quintero, R. (2003) Biotechnological processes for the refining of petroleum. *Fuel Processing Technology*. 81: 155–169.

Le Borgne, S., Quintero, R., (2003) Biotechnological processes for the refining of petroleum. *Fuel Processing Technology*. 81: 155–169.

Lee, I.S., Bae, H., Ryu, H.W., Cho, K., Chang, Y.K. (2005) Biocatalytic Desulfurization of Diesel Oil in an Airlift Reactor with Immobilized *Gordonia nitida* CYKS1 Cells. *Biotechnology Progress*. 21: 781–785.

Lee, M.K., Senius J.D., Grossman, M.J. (1995) Sulfur-specific microbial desulfurization of sterically hindered, analogs of dibenzothiophene. *Journal of Applied and Environmental Microbiology*. 61:4362–4366.

Lee, M.K., Senius, J.D., Grossman, M.J. (1995) Sulphur-specific microbial desulphurization of sterically hindered analogs of dibenzothiophene. *Applied and Environmental Microbiology*. 61: 4362–4366.

Li W., Tang H., Liu Q., Xing J., Li Q., Wang, D., Yang M., Li, X., Liu, H. (2009b) Deep desulfurization of diesel by integrating adsorption and microbial method. *Biochemical Engineering Journal*. 44: 297–301.

Li W., Xing J., Li Y., Xiong X., Li X., Liu H. (2008c) Desulfurization and bioregeneration of adsorbents with magnetic *P. delafieldii* R-8 cells. *Catalysis Communications*. 9: 376–380.

Li W., Xing J., Li Y., Xiong X., Li, X., Liu, H. (2006) Feasibility study on the integration of adsorption/bioregeneration of π-complexation adsorbent for desulfurization. *Industrial and Engineering Chemistry Research*. 45(8): 2845–2849.

Li W., Zhao D. (2013) An overview of the synthesis of ordered mesoporous materials. *Chemical Communications*. 49: 943–946.

Li Y.G., Gao H.S., Li W.L., Xing J.M., Liu H.Z. (2009a) In situ magnetic separation and immobilization of dibenzothiophene-desulfurizing bacteria. *Bioresource Technology*. 100: 5092–5096.

Li, C.-H.; Chan, T.-H. (2007) Comprehensive Organic Reactions in Aqueous Media. Wiley-Interscience (Wiley), Hoboken. ISBN 978–0-471–76129-7.

Li, F., Xu, P., Feng, J., Meng, L., Zheng, Y., Luo, L., Ma, C. (2005) Microbial desulfurization of gasoline in a *Mycobacterium goodii* X7B immobilized-cell system. *Applied and Environmental Microbiology*. 71: 276–281.

Li, F., Zhang, Z., Feng, J., Cai, X., Ping, X. (2007a) Biodesulfurization of DBT in tetradecane and crude oil by a facultative thermophilic bacterium *Mycobacterium goodii* X7B. *Journal of Biotechnology*. 127: 222–228.

Li, F.L., Xu, P., Ma, C.Q., Luo, L.L., Wang, X.S. (2003) Deep desulfurization of hydrodesulfurization-treated diesel oil by a facultative thermophilic bacterium *Mycobacterium* sp. X7B. *FEMS Microbiology Letters*. 223: 301–307.

Li, G., Li, S., Qu, S., Liu, Q., Ma, T., Zhu, L., Liang, F., Liu, R. (2008c) Improved biodesulfurization of hydrodesulfurized diesel oil using *Rhodococcus erythropolis* and *Gordonia* sp. *Biotechnology Letters*. 30(10): 1759–1764.

Li, G., Ma, T., Zhang, Q., Wang, C., Chen, Q. (2007b) Sulfur-selective desulfurization of dibenzothiophene and diesel oil by newly isolated *Rhodococcus erythropolis* NCC1. *Chinese Organic Chemistry*. 25:400–405.

Li, G.Q, Ma, T., Li, S.S., Li, H., Liang, F.L., Liu, R.L. (2007c) Improvement of dibenzothiophene desulfurization activity by removing the gene overlap in the dsz operon. *Bioscience Biotechnology Biochemistry*. 71:849–854.

Li, G.Q., Li, S.S., Zhang, M.L., Wang. J., Zhu, L., Liang, F.L., Liu, R.L., Ma, T. (2008a) Genetic rearrangement strategy for optimizing the dibenzothiophene biodesulfurization pathway in *Rhodococcus erythropolis*. *Journal of Applied and Environmental Microbiology*. 74: 971–976.

Li, M.Z., Squires, C.H., Monticello, D.J. Childs, J.D. (1996) Genetic analysis of the dsz promotor and associated regulatory region of *Rhodococcus erythropolis* IGTS8. *Journal of Bacteriology*. 178: 6409–6418.

Li, W., Jiang, X. (2013) Enhancement of bunker oil biodesulfurization by adding surfactant. World Journal of Microbiology and Biotechnology. 29: 103–108.

Li, W., Xing, J., Li, Y., Xiong, X., Li, X., Liu, H. (2008b) Desulfurization and bioregeneration of adsorbents with magnetic *P. delafieldii* R-8 cells. *Catalysis Communications*. 9: 376–380.

Li, W., Zhang, Y., Wang, M.D., Shi, Y. (2005a) Biodesulfurization of dibenzothiophene and other organic sulfur compounds by a newly isolated *Microbacterium* strain ZD-M2. *FEMS Microbiology Letters*. 247: 45–50.

Li, Y., Xing, J., Xiong, X., Li, W., Gao, H., Liu, H. (2008d) Improvement of biodesulfurization activity of alginate immobilized cells in biphasic systems. *Journal of Industrial Microbiology and Biotechnology*. 35: 145–150.

Li, Y.-G., Gao, H.-S., Li, W.-L., Xing, J.-M., Liu, H.-Z. (2009c) In situ magnetic separation and immobilization of dibenzothiophene-desulfurizing bacteria. *Bioresource Technology*. 100: 5092–5096.

Lin, L., Hong, L., Jianhua, Q., Jinjuan, X. (2010). Progress in the technology for desulfurization of crude oil. *China Petroleum Processing and Petrochemical Technology*. 12: 1–6.

Lin, M.S., Chaffin, M., Liu, J., Glover, J., Davison, R., Bullin, J. (1996) the effect of asphalt composition on the formation of asphaltenes and their contribution to asphalt viscosity. *Petroleum Science and Technology*. 14:139–162.

Lin, X., Liu, H., Zhu, F., Wei, X., Li, Q., Luo, M. (2012) Enhancement of biodesulfurization by *Pseudomonas delafieldii* in a ceramic microsparging aeration system. *Biotechnology Letters*. 34: 1029–1032.

Linguist, L., Pacheco, M. (1999) Enzyme-based diesel desulfurization process offers energy, CO_2 advantages. *Oil and Gas Journal*. 45–48.

Liu X., Kaminski M.D., Guan Y., Chen H., Liu H., Rosengart A.J. (2006) Preparation and characterization of hydrophobic superparamagnetic gel. *Journal of Magnetism and Magnetic Materials*. 306: 248–253

Lizama, H.M., Wilkins, L.A., Scott, T.C. (1995) Dibenzothiophene sulfur can serve as the sole electron acceptor during growth by sulfate-reducing bacteria. *Biotechnology Letters*. 17(1): 113–116.

Luo, M.F., Xing, J.M., Gou, Z.X., Li, S., Liu, H.Z., Chen, J.Y. (2003) Desulfurization of dibenzothiophene by lyophilized of *Pseudomonas delafieldii* R-8 in the presence of dodecane. *Biochemical Engineering Journal*. 13: 1–6.

Ma, C.Q., Feng, J.H., Zeng, Y.Y., Cai, X.F., Sun, B.P., Zhang, Z.B., Blankespoor, H.D., Xu, P. (2006a) Methods for the preparation of a biodesulfurization biocatalyst using *Rhodococcus* sp. *Chemosphere*. 65: 165–169.

Ma, T., Li, G., Li, J., Liang, F., Liu, R. (2006b) Desulfurization of dibenzothiothene by *Bacillus subtilis* recombinants carrying dsz ABC and dszD genes. *Biotechnology Letters*. 28: 1095–1100.

Ma, X.L., Zhou, A.N., Song, C.S. (2007) A novel method for oxidative desulfurization of liquid hydrocarbon fuels based on catalytic oxidation using molecular oxygen coupled with selective adsorption. *Catalysis Today*. 123: 276–284.

Maass, D., de Oliveira, D., de Souza, A. A. U., Souza, S.M. (2014) Biodesulfurization of a System Containing Synthetic Fuel Using *Rhodococcus erythropolis* ATCC 4277. *Applied Biochemistry and Biotechnology*. 174(6): 2079–2085.

Maass, D., Todescato, D., Moritz, D.E., Oliveira, J.V., Oliveira, D., Ulson de Souza, A.A., Guelli Souza, S.M. (2015) Desulfurization and denitrogenation of heavy gas oil by *Rhodococcus erythropolis* ATCC 4277. *Bioprocess and Biosystems Engineering*. 38: 1447–1453.

Madeira, L., Santana, V., Pinto, E. (2008) Dibenzothiophene oxidation by horseradish peroxidase in organic media: Effect of the DBT:H_2O_2 molar ratio and H_2O_2 addition mode. *Chemosphere*. 71: 189–194.

Maghsoudi, S., Kheirolomoom, A., Vossoughi, M., Tanaka, E., Katoh, S. (2000) Selective desulfurization of dibenzothiophene by newly isolated *Corynebacterium* sp. strain P32C1. *Biochemical Engineering Journal*. 5(1):11–16.

Maghsoudi, S., Vossoughi, M., Kheirolomoom, A., Tanaka, E., Katoh, S. (2001) Biodesulfurization of hydrocarbons and diesel fuels by Rhodococcus sp. strain P32C1. *Biochemical Engineering Journal*. 8:151–156.

Marcelis, C.L.M., van Leeuwen, M., Polderman, H.G., Janssen, A.J.H., Lettinga, G. (2003) Model description of dibenzothiophene mass transfer in oil/water dispersions with respect to biodesulfurization. *Biochemical Engineering Journal*. 16: 253–264.

Martin, A. B., Alcon, A., Santos, V. E., Garcia-Ochoa, F. (2005) Production of a biocatalyst of *Pseudomonas putida* CECT5279 for DBT biodesulfurization: Influence of the operational conditions. *Energy and Fuels*. 19(3), 775–782.

Martínez, I., Mohamed, M. E., Rozas, D., García, J. L., Díaz, E. (2016a) Engineering synthetic bacterial consortia for enhanced desulfurization and revalorization of oil sulfur compounds. Metabolic Engineering. 35: 46–54.

Martinez, I., Santos, V.E., Alcon, A., Garcia-Ochoa, F. (2015) Enhancement of the biodesulfurization capacity of *Pseudomonas putida* CETCT5279 by co-substrate addition. *Process Biochemistry*. 50: 119–124.

Martinez, I., Santos, V.E., Gomez, E., Garcia-Ochoa, F. (2016b) Biodesulfurization of dibenzothiophene by resting cells of *Pseudomonas putida* CECT5279: influence of the oxygen transfer rate in the scale-up from shaken flask to stirred tank reactor. *Journal of Chemical Technology and Biotechnology*. 91: 184–189.

Marzona, M., Pessione, E., Di Martino, S., Giunta, C. (1997) Benzothiophene and dibenzothiophene as the sole sulfur source in *Acinetobacter*: Growth kinetics and oxidation products. *Fuel Processing Technology*. 52: 199–205.

Matsubara, T., Ohshiro, T., Nishina, Y., Izumi, Y. (2001) Purification, characterization, and overexpression of flavin reductase involved in dibenzothiophene desulfurization by *Rhodococcus erythropolis* D-1. *Applied and Environmental Microbiology*. 67: 1179–1184.

Matsui, T., Hirasawa, K., Koizumi, K., Maruhashi, K., Kurane, R. (2001a). Optimization of the copy number of dibenzothiophene desulfurizing genes to increase the desulfurization activity of recombinant *Rhodococcus* sp. *Biotechnology Letters*. 23: 1715–1718.

Matsui, T., Hirasawa, K., Konishi, J., Tanaka, Y., Maruhashi, K., Kurane, R. (2001b) Microbial desulfurization of alkylated dibenzothiophene and alkylated benzothiophene by recombinant *Rhodococcus* sp. strain T09. *Applied Microbiology and Biotechnology*. 56: 196–200.

Matsui, T., Noda, K.I., Tanaka, Y., Maruhashi, K., Kurane, R. (2002) Recombinant *Rhodococcus* sp. strain T09 can desulfurize DBT in the presence of inorganic sulfate. *Current Microbiology*. 45: 240–244

Matsushita, H., Sugiyama, T., Hond.a, H. Kobayashi, T. (2001) High cell density culture of Rhodococcus erythropolis KA2–5-1 for in situ induction of dibenzothiophene degrading activity. *Journal of the Chinese Institute of Chemical Engineers*. 32:219–225.

McFarland, B.L. (1999) Biodesulfurization. Current Opinion in Microbiology. 2: 257–264.

McFarland, B.L., Boron, D.J., Deever, W., Meyer, J.A., Johnson, A.R., Atlas, R.M. (1998) Biocatalytic sulfur removal from fuels: Applicability for producing low sulfur gasoline. *Critical Reviews in Microbiology*. 24: 99–147.

McGenity, T.J., Folwell, B.D., McKew, B.A., Sanni, G.O. (2012) Marine crude-oil biodegradation: a central role for interspecies interactions. *Aquatic Biosystems*. 8: 10–28.

Mehrnia, M.R., Bonakdarpour, B., Towfighi, J., Akbarneja, M.M. (2004a) Design and operational aspects of airlift bioreactors for petroleum biodesulfurization. *Environmental Progress*. 23(3): 2016–214.

Mehrnia, M.R., Towfighi, J., Bonakdarpour, B., Akamejad, M.M. (2005) Gas holdup and oxygen transfer in a draft-tube airlift bioreactor with petroleum-based liquids. *Biochemical Engineering*. 22, 105–110.

Mehrnia, M.R., Towfighi, J., Bonakdarpour, B., Akbarnegad, M.M. (2004b). Influence of top-section design and draft-tube height on the performance of airlift bioreactors containing water-in-oil microemulsion, *Journal of Chemical Technology and Biotechnology*. 79(3): 260–267.

Mikesková, H., Novotný, C., Svobodova, K. (2012) Interspecific interactions in mixed microbial cultures in a biodegradation perspective. *Applied Microbiology and Biotechnology*. 95: 861–870.

Miller, K.W. (1992) Reductive desulfurization of dibenzyldisulfide. *Applied and Environmental Microbiology*.58(7): 2176–2179.

Mingfang, L., Zhongxuan, G., Jianmin, X., Huizhou, L., Jiayong, C. (2003) Microbial desulfurization of model and straight-run diesel oils. *Journal of Chemical Technology and Biotechnology*. 78: 873–876.

Mishra, S., Sarma, P.M., Lal, B. (2004) Crude oil degradation efficiency of a recombinant lux tagged Acinetobacter baumannii strain and its survival in crude oil contaminated soil microcosm; *FEMS Microbiology Letters*. 235: 323–331.

Mohabali, G., Ball, A., Kayatash, A. (2007) Stabilization of water/gas oil emulsions by desulfurizing cells of *Gordonia alkanivorans* RIPI90A. *Microbiology*. 153: 1573–1581.

Mohamed, M.E., Al-Yacoub, Z.H., Vedakumar, J.V. (2015) Biocatalytic desulfurization of thiophenic compounds and crude oil by newly isolated bacteria. Frontiers in Microbiology. 6: article 112. doi: 10.3389/fmicb.2015.0 0112.

Mohebali, G., Andrew, S. (2008) Biocatalytic desulfurization (BDS) of petrodiesel fuels. *Microbiology*. 154: 2169–2183.

Mohebali, G., Ball, A.S. (2016) Biodesulfurization of diesel fuels – past, present and future perspectives. *International Journal of Biodeterioration and Biodegradation*. 110: 163–180.

Mohebali, G., Ball, A.S., Kaytash, A., Rasekh, B. (2007b) Stabilization of water/gas oil emulsions by desulfurizing cells of *Gordonia alkanivorans* RIPI90A. *Microbiology*. 153: 1573–1581.

Mohebali, G., Ball, A.S., Rasekh, B., Kaytash, A. (2007a) Biodesulfurization potential of a newly isolated bacterium *Gordonia alkanivorans* RIPI90A. *Enzyme and Microbial Technology*. 40: 578–584.

Monot, F., Abbad-Andaloussi, S., Warzywoda, M. (2002) Biological culture containing *Rhodococcus erythropolis* and/or *Rhodococcus rhodnii* and process for desulfurization of petroleum fraction. U.S. Patent No. 6,337,204.

Monot, F., Warzywoda, M. (2008) Microbial desulfurization of petroleum products. In: Petroleum Microbiology: Concepts, Enviromental Implications. Industrial Applications. Vol. 2 ed J.-P. Vandecasteele (Paris: Editions Technip) Pp.757–775.

Monticello D.J., Bakker D., Finnerty W.R. (1985) Plasmid-mediated degradation of dibenzothiophene by Pseudomonas species. *Applied and Environmental Microbiology*. 49(4): 756–760.

Monticello, D. J. (2000). Biodesulfurization and the upgrading of petroleum distillates. *Biotechnolology*. 11: 540–546.

Monticello, D.J. (1994) Biocatalytic desulphurization, the biorefining of petroleum fractions. *Hydrocarbon Processing*. 73: 39–45.

Monticello, D.J. (1996) Multistage process for deep desulfurization of a fossil fuel. US Patent no. 5,510,265.

Monticello, D.J. (1998) Riding the fossil fuel biodesulfurization wave. Chemtech. 28(7): 38–45.

Monticello, D.J. (2000) Biodesulfurization and the upgrading of petroleum distillates. *Current Opinion Biotechnology*. 11:540–546.

Monticello, D.J., Finnerty, W.R. (1985) Microbial desulfurization of fossil fuels. *Annual Review of Microbiology*. 39: 371–389.

Monticello, D.J., Haney, I.I.I., William, M. (1996) Biocatalytic process for reduction of petroleum viscosity. US patent. 5529930 A.

Montiel, C., Quintero, R., Aburto, J. (2009) Petroleum biotechnology: Technology trends for the future. *African Journal of Biotechnology*. 8(12): 2653–2666.

Morales, M, Le Borgne, S. (2014) Protocols for the isolation and preliminary characterization of bacteria for biodesulfurization and biodenitrogenation of petroleum-derived fuels. In: McGenity T., Timmis K., Nogales B. (eds) Hydrocarbon and Lipid Microbiology Protocols. Springer Protocols Handbooks. Springer, Berlin, Heidelberg. Pp. pp 201–218.

Morio, K., Keizo, H., Yoshikawa, O., Hirasawa, K., Ishii, Y., Fujino, K., Sugiyama, H., Maruhashi, K. (2001) Kinetic analysis of microbial desulfurization of model and light gas oils containing multiple alkyl dibenzothiophenes. *Bioscience, Biotechnology, and Biochemistry*. 65: 298–304.

Moslemy, P.; Guiot, S.R. and Neufeld, R.J. (2002) Production of size-controlled gellam gum microbeads encapsulating gasoline-degrading bacteria; *Enzyme and Microbial Technology*. 30:10- 18.

Moustafa, Y.M., El-Gendy, N.Sh., Farahat, L.A., Abo-State, M.A., El-Temtamy, S.A. (2006) Biodesulfurization of Ras Badran crude oil and its constituents with special emphasis on its asphaltene fraction. *Egyptian Journal of Petroleum*. 15:21–30.

Mukhopadhyay, M., Chowdhury, R., Bhattacharya, P. (2005) Two phase mathematical model for a bio-trickling reactor for the production of ultra low sulfur diesel (ULSD) from deeply hydrodesulfurized diesel. *Catalysis Today*. 106: 180–15.

Mukhopadhyay, M., Chowdhury, R., Bhattacharya, P. (2007) Trickle bed biodesulfurizer of diesel with backwash and recycle. *AICHE Journal*. 53(8): 2188–2197.

Mukhopadhyaya, M., Chowdhury, R., Bhattacharya, P. (2006) Biodesulfurization of hydrodesulfurized diesel in a trickle bed reactor-experiments and modelling. *Journal of Scientific and Industrial Research*. 65: 432–436.

Muthusamy, K., Gopalakrishnan, S., Ravi, T.K., Sivachidambaram, P. (2008) Biosurfactants: Properties, commercial production and application. *Current Science*. 94(6): pp 736–47.

Naito, M., Kawamoto, T., Fujino, K., Kobayashi, M., Maruhashi, K., Tanaka, A. (2001) Long-term repeated desulfurization by immobilized *Rhodococcus erythropolis* KA2–5-1 cells. *Applied Microbiology and Biotechnology*. 55: 374–378.

Nakashima, N., Tamura, T. (2004) A novel system for expressing recombinant proteins over a wide temperature range from 4 to 35 °C. *Biotechnology and Bioengineering*. 86: 136–148.

Nandi, S. (2010) Biodesulfurization of hydro-desulfurized diesel in an airlift reactor. Journal of Scientific and Industrial Research. 69: 543–547.

Nassar, H.N. (2015) Development of biodesulfurization process for petroleum fractions using nano-immobilized catalyst; Ph.D. Degree; Al-Azhar University, Cairo, Egypt.

Nassar, H.N., Deriase, S.F., El-Gendy, N.Sh. (2016) Modeling the relationship between microbial biomass and total viable count for a new bacterial isolate used in biodesulfurization of petroleum and its fractions. *Petroleum Science and Technology*. 34(11/12): 980–985.

Nassar, H.N., Deriase, S.F., El-Gendy, N.Sh. (2017) Statistical optimization of biomass production and biodesulfurization activity of Rhodococcus erythropolis HN2. *Petroleum Science and Technology*. 35(20): 1951–1959.

Nassar, H.N., El-Gendy, N.Sh., Abo State M., Moustafa, Y.M., Mahdy, H.M., El-Temtamy S.A. (2013) Desulfurization of dibenzothiophene by a novel strain *Brevibacillus invocatus* C19 isolated from Egyptian coke. *Biosciences, Biotechnology Research Asia*. 10(1): 29–46.

Nazari, F., Kefayati, M.E., Raheb, J. (2017) The study of biological technologies for the removal of sulfur compounds. *Journal of Sciences, Islamic Republic of Iran*. 28(3): 205–219.

Nehlsen, P.J. (2005) Developing Clean Fuels: Novel Techniques for Desulfurization. Ph.D. Thesis, Princeton University.

Nekodzuka, S., Toshiaki, N., Nakajima-Kambe, T., Nobura, N., Lu, J., Nakahara, Y. (1997) Specific desulfurization of dibenzothiophene by *Mycobacterium* strain G3. *Biocatalysis and Biotransformation*. 15: 21–27.

Noda, K., Kogure, T., Irisa, S., Murakami, Y., Sakata, M., Kuroda, A. (2008) Enhanced dibenzothiophene biodesulfurization in a microchannel reactor. *Biotechnology Letters*. 30: 451–454.

Noda, K., Watanabe, K., Maruhashi, K. (2003b) Recombinant *Pseudomonas putida* carrying both the dsz and hcu genes can desulfurize dibenzothiophene in n-tetradecane. *Biotechnology Letters*. 25: 1147–1150.

Noda, K.I., Watanabe, K., Maruhashi, K. (2003a) Isolation of a recombinant desulfurizing 4,6-dipropyldibenzothiophene in n-tetradecane. J*ournal of Bioscience and Bioengineering*. 95: 354–360.

Nomura, N., Takada, M., Okada, H., Shinohara, Y., Nakajima-Kambe, T., Nakahara, T., Uchiyama, H. (2005) Identification and functional analysis of genes required for desulfurization of alkyl dibenzothiophenes of *Mycobacterium* sp. G3. *Journal of Bioscience and Bioengineering*. 100: 398–402.

Nuhu, A.A. (2013) Bio-catalytic desulfurization of fossil fuels: a mini review. Reviews in Environmental Science and Bio/Technology. 12: 9–23.

Nunn, D., Boltz, J., DiGrazia, P.M., Nace L. (2006) The biocatalytic desulfurization project. Final technical progress report. Prepared for National Energy Technology Laboratory, U.S. Department of Energy.

Oda, S., Ohta, H. (2002) Biodesulfurization of dibenzothiophene with *Rhodococcus erythropolis* ATCC 53968 and its mutant in an interface bioreactor. *Journal of Bioscience and Bioengineering.* 94(5): 474–477.

Ohshiro T, Hine Y, Izumi Y (1994) Enzymatic desulfurization of dibenzothiophene by a cell free system of *Rhodococcus erythropolis* D-1. *FEMS Microbiology Letters.* 118: 341–344.

Ohshiro T, Ohkita R, Takikawa T, Manabe M, Lee WC, Tanokura M, Izumi Y (2007) Improvement of 2′-hydroxybiphenyl-2-sulfinate desulfinase, an enzyme involved in the dibenzothiophene desulfurization pathway, from Rhodococcus erythropolis KA2-5-1 by site-directed mutagenesis. *Bioscience, Biotechnology, and Biochemistry.* 71: 2815–2821.

Ohshiro, T., Hirata, T., Izumi, Y. (1995) Microbial desulfurization of dibenzothiophene in the presence of hydrocarbon. *Applied Microbiology and Biotechnology.* 44: 249–252.

Ohshiro, T., Izumi, Y. (1999) Microbial desulfurization of organic sulfur compounds in petroleum. *Bioscience, Biotechnology, and Biochemistry.* 63: 1–9.

Ohshiro, T., Suzuki, K., Izumi, Y. (1996) Regulation of dibenzothiophene degrading enzyme activity of *Rhodococcus erythropolis* D-1. *Journal of Fermentation and Bioengineering.* 81: 121–124.

Okada H., Nomura N., Nakahara T., Saitoh K., Uchiyama H., Maruhashi K. (2003) Analyses of microbial desulfurization reaction of alkylated dibenzothiophenes dissolved in oil phase. *Biotechnology and Bioengineering.* 83(4):489–497.

Okada, H., Nomura, N., Nakahara, T., Maruhashi, K. (2002) Analyses of substrate specificity of the desulfurizing bacterium *Mycobacterium* sp. G3. *Journal of Bioscience and Bioengineering.* 93: 228–233.

Olabemiwo, O.M., Adelowo, O.O., Tella, A.C., Bello, I.A. (2013) Preliminary study on biodesulfurization of Nigerian natural bitumen. *International Journal of Basic and Applied Sciences.* 13: 79–86.

Oldfield, C., Pogrebinsky, O., Simmonds, J., Olson, E.S., Kulpa, C.F. (1997) Elucidation of the metabolic pathway for dibenzothiophene desulfurization by *Rhodococcus* sp. strain IGTS8 (ATCC 53968). Microbiology UK. 143: 2961–2973.

Olson, G.J. (2000) Microbial catalyst for desulfurization of fossil fuels. US Patent no. 6,124,130.

Omori, T., Monna, L., Saiki, Y. Kodama, T. (1992) Desulfurization of dibenzothiophene by *Corynebacterium* sp. strain SY1. *Applied and Environmental Microbiology.* 58(3):911–5.

Omori, T., Saiki, Y., Kasuga, K., Kodama, T. (1995) Desulfurization of Alkyl and Aromatic Sulfides and Sulfonates by Dibenzothiophene-desulfurizing *Rhodococcus* sp. Strain SY1. *Bioscience, Biotechnology, and Biochemistry. 59*(7), 1195–1198.

Onodera-Yamada, K., Morimoto, M., and Tani, Y. (2001). Degradation of dibenzothiophene by sulfate-reducing bacteria cultured in the presence of only nitrogen gas. *Journal of Bioscience and Bioengineering* 91(1): 91–93.

ORNL Oak Ridge National Laboratory (2000) An emissions mission: solving the sulfur problem. Oak Ridge National Laboratory Review. 33: 6–8.

Oshiro, T., Ohkita, R., Takikawa, T., Manabe, M., Lee, W.C., Tanokura, M., Izumi, Y. (2007) Improvement of 2-hydroxybiphenyl-2-sulfinate desulfinase, an enzyme involved in the dibenzothiophene desulfurization pathway, from *Rhodococcus erythropolis* KA2-5-1 by site-directed mutagenesis. *Bioscience Biotechnology and Biochemistry*. 71(11): 2815–2821.

Pacheco, M.A., Lange, E.A., Pienkos, P.T., Yu, L.Q., Rouse, M.P., Lin, Q., Linguist, L.K. (1999) Recent advances in biodesulfurization of diesel fuel. NPRA Annual Meeting, March 21–23, San Antonio, Texas. National Petrochemical and Refiners Association, paper AM-99-27. Pp. 1–26.

Pan, J., Wu, F., Wang, J., Xu, L., Khayyat, N.H., Stark, B.C., Kilbane II, J.J. (2013) Enhancement of desulfurization activity by enzymes of the *Rhodococcus* dsz operon through coexpression of a high sulfur peptide and directed evolution. *Fuel*. 112: 385–390.

Papizadeh, M., Ardakani, M. R., Ebrahimipour, G., Motamedi, H. (2010) Utilization of dibenzothiophene as sulfur source by *Microbacterium* sp. NISOC-06. *World Journal of Microbiology and Biotechnology*. *26*(7), 1195–1200.

Papizadeh, M., Ardakani, M. R., Motamedi, H., Rasouli, I., Zarei, M. (2011) C-S targeted biodegradation of dibenzothiophene by *stenotrophomonas* sp. NISOC-04. *Applied Biochemistry and Biotechnology*. *165*(3–4): 938–948

Papizadeh, M., Roayaei Ardakani, M., Motamedi, H. (2017) Growth-phase dependent biodesulfurization of Dibenzothiophene by *Enterobacter* sp. strain NISOC-03. *Pollution*. 3(1): 101–111.

Park, H.S., Kayser, K.J., Kwak, J.H., Kilbane, J.J. II (2004) Heterologous gene expression in *Thermus thermophilus*: β-galactosidase, dibenzothiophene monooxygenase, PNB carboxy esterase, 2-aminobiphenyl-2,3-diol dioxygenase, and chloramphenicol acetyl transferase. *Journal of Industrial Microbiology and Biotechnology*. 31 :189–197.

Park, S.J., Lee, I.S., Chang, Y.K., Lee, S.Y. (2003) Desulfurization of dibenzothiophene and diesel oil by metabolically engineered Escherichia coli. *Journal of microbiology and biotechnology*.13 (4): 578–583.

Payzant, J.D., Montgomery, D.S., Strausz O.P. (1988) The identification of homologous series of benzo[b]thiophenes, thiophenes, thiolanes and thianes possessing a linear carbon framework in the pyrolysis oil of Athabasca asphaltene. *AOSTRA Journal of Research*. 4:117–131.

Perry, J.J. (1979) Microbial cooxidations involving hydrocarbons. *Microbiology Reviews*. 43(1): 59–72.

Petro Star to develop biological sulfur removal process (2000) Oil and Gas Journal.

Piddington C.S., Kovacevich B.R., Rambosek J. (1995) Sequence and molecular characterization of a DNA region encoding the dibenzothiophene desulfurization operon of *Rhodococcus* sp. strain IGTS8. *Applied and Environmental Microbiology*. 61:468–475

Pienkos, P.T. (1999) Choosing the best platform for the biotransformation of hydrophobic molecules. In Microbial Biosystems: New Frontiers. Proceedings of the 8th International Symposium on Microbial Ecology; [EDs] by C. R. Bell,

M. Brylinsky & P. Johnson-Green. Halifax, Canada: Atlantic Canada Society for Microbial Ecology; pp. 875–888.

Pineda-Flores, G., Mesta-Howard, A.M. (2001) Petroleum asphaltenes: generated problematic and possible biodegradation mechanisms. Rev Latinoam Microbiol. 43: 143–150.

Pokethitiyook, P., Tangaromsuk, J., Kruatrachue, M., Kalambaheti, C., Borole, A.P. (2008) Biological removal of organic sulfur by bacterial strains isolated in Thailand. Science Asia 34: 361–366.

Potumarthi, R., Subhakar, C., Pavani, A., Jetty, A. (2008) Evaluation of various parameters of calcium-alginate immobilization method for enhanced alkaline protease production by Bacillus licheniformis NCIM-2042 using statistical methods. *Bioresource Technology*. 99: 1776–1786.

Premuzic, E., Lin, M.S., Bohenek, M., Zhou, W.M. (1999) Bioconversion reactions in asphaltenes and heavy crude oils. *Energy Fuels*. 13:297–304.

Prokop, Z., Oplustil, F., DeFrank, J., Damborsky, J. (2006) Enzymes fight chemical weapons. *Biotechnology Journal*. 1: 1370–1380.

Raheb, J., Rasekh, B., Irani, S., Hajipour, M. J., Mozaffari Tabatabaei, M., Kefayati, M. E., Memari, B. (2012) The study of biodesulfurization activity in recombinant E. coli strain by cloning the dsz genes involve in 4S pathway. *Journal of Sciences, Islamic Republic of Iran*,.22(3): 213–219.

Rath, K., Mishra, B., Vuppu, S. (2012) Biodegrading ability of organosulfur compound of a newly isolated microbe *Bacillus* sp. KS1 from the oil contaminated soil. *Archives of Applied Science Research*. 4(1): 465–471.

Reichmuth, D.S., Blanch, H.W., Keasling, J.D. (2004) Dibenzothiophene biodesulfurization pathway improvement using diagnostic GFP fusions. *Biotechnology and Bioengineering*. 88:94–99.

Reichmuth, D.S., Hittle, J.L., Blanch, H.W., Keasling, J.D. (1999) Biodesulfurization of dibenzothiophene in *Escherichia coli* is enhanced by expression of *Vibrio harveyi* oxidoreductase gene. *Biotechnology Bioengineering*. 67:72–79.

Reichmuth, D.S., Hittle, J.L., Blanch, H.W., Keasling, J.D. (2000) Biodesulfurization of dibenzothiophene in Escherichia coli is enhanced by expression of a *Vibrio harveyi* oxidoreductase gene. *Biotechnology and Bioengineering*. 67: 72–79.

Rezanka, T., Sobotka, M., Spizek, J., Sigler, K. (2006) Pharmacologically active sulfur-containing compounds. *Anti-Infective Agents in Medicinal Chemistry*. 5(2): 187–224.

Rhee, S.K., Chang, J.H., Chang, Y.K., Chang, H.N. (1998) Desulfurization of dibenzothiophene and diesel oil by a newly isolated *Gordona* strain, CYKS1. *Applied and Environmental Microbiology*. 64(6): 2327–2331.

Rhodes, A.K. (1995) Enzymes desulphurizing diesel fuel in pilot plant tests. *Oil and Gas Journal*. 93: 39–40.

Rodriguez, R.C.E., Rodriguez, E., Blanco, R., Cordero, I., Segura, D. (2010). Fungal contamination of stored automobile-fuels in a tropical environment. *Journal of Environmental Sciences*. 22: 1595–1601.

Rontani, J. F., Bosser-Jonlak, F., Rambeloarisoa, E., Bertrand, J. C., Giusti, G. (1985) Analytical study of Asthart crude oil. *Chemosphere*, 14 (9):1413–22.

Rosenberg, E., Zuckerberg, A., Rubinovitz, C., Gutinck, D.L. (1979) Emulsifier ArthrobacterRAG-1: Isolation and emulsifying properties. *Applied Environmental Microbiology*, 37: 402–408.

Rotani, J.F., Bosser-Joulak, E., Rambeloarisoa, J.E., Bertrand, G., Fiusti, G., Faure, F. (1985) Analytical study of Ashart crude oil asphaltenes biodegradation. *Chemosphere*. 14:1413 - 1422.

Rychlewska, K., Konieczny, K., Bodzek, M. (2015) Pervaporative desulfurization of gasoline–separation of thiophene/n-heptane mixture. *Archives of Environmental Protection*. 41: 3–11.

Ryu, H.W., Wee, Y.J. (2001) Characterization of bioconversion of fumarate to succinate by alginate immobilized *Enterococcus faecalis* RKY1. *Applied Biochemistry and Biotechnology*. 91: 525–535.

Saed, D., Nassar H.N., El-Gendy N.Sh., Zaki T., Moustafa Y.M., Badr I.H.A. (2014) The enhancement of pyrene biodegradation by assembling MFe_3O_4 nano-sorbents on the surface of microbial cells. *Energy Sources, Part A: Recovery, Utilization, and Environmental Effects*. 36(17): 1931–1937.

Sanchez, O.F., Almeciga-Diaz, C.J., Silva, E., Cruz, J.C., Valderrama, J.D., Caicedo, L.A. (2008) Reduction of sulfur levels in kerosene by *Pseudomonas* sp. strain in an airlift reactor. *Latin American Applied Research*. 38: 329–335.

Santos, S.C.C., Alviano, D.S., Alviano, C. S., Pádula, M., Leitão, A.C., Martins, O.B., Seldin, L. (2006) Characterization of *Gordonia* sp. strain F.5.25.8 capable of dibenzothiophene desulfurization and carbazole utilization. *Applied Microbiology and Biotechnology*. 71(3), 355–362.

Schilling, B.M., Alvarez, L.M., Wang, D.I.C., Cooney, C.L. (2002) Continuous desulfurization of dibenzothiophene with *Rhodococcus rhodochrous* IGTS8 (ATCC 53968). *Biotechnology Progress*. 18: 1207–1213.

Seeberger A, Jess A. (2010) Desulfurization of diesel oil by selective oxidation and extraction of sulfur compounds by ionic liquids—a contribution to a competitive process design. *Green Chemistry*.12(4): 602–608

Setti, L., Farinelli, P., Di Martino, S. (1999) Development in destructive and non-destructive pathways for selective desulfurizations in oil-biorefining processes. *Applied Microbiology and Biotechnology*. 52: 111–117.

Setti, L., Lanzarini, G., Pifferi, P.G. (1995) Dibenzothiophene biodegradation by a Pseudomonas sp. in model solutions. *Process Biochemistry*. 30(8): 721–728.

Setti, L., Lanzarini, G., Pifferi, P.G. (1997) Whole cell biocatalysis for an oil desulfurization process. *Fuel Processing Technology*. 52: 145–153.

Setti, L., Rossi, M., Lanzarini, G., Pifferi, P.G. (1992) The effect of dibenzothiophene and organic sulfur compounds in heavy oil by a Pseudomonas sp. *Biotechnology Letters*. 14: 515–520.

Shan G.B., Xing J.M., Guo C., Liu H.Z., Chen J.Y. (2005b) Biodesulfurization using Pseudomonas delafieldii in magnetic polyvinyl alcohol beads. *Letters in Applied Microbiology*. 40: 30–36.

Shan G.B., Xing J.M., Zhang H., Liu, H.Z. (2005c) Biodesulfurization of dibenzothiophene by microbial cells coated with magnetite nanoparticles. *Applied and Environmental Microbiology*. 71(8): 4497–4502.

Shan G.B., Zhang H., Liu H., Xing, J.M. (2005a) π-Complexation studied by fluorescence technique: application in desulfurization of petroleum product using magnetic π-complexation sorbents. *Separation Science and Technology*. 40(14): 2987–2999.

Shan, G.B., Xing, J.M., Luo, M.F., Liu, H.Z., Chen, J.Y. (2003) Immobilization of pseudomonas delafieldii with magnetic polyvinyl alcohol beads and its application in biodesulfurization. *Biotechnology Letters*. 25, 1977–1981.

Shao, P., Huang, R.Y.M. (2007) Polymeric membrane pervaporation. *Journal of Membrane Science*. 287: 162–179.

Shennan, J. (1996) Microbial attack on sulfur-containing hydrocarbons: Implications for the biodesulfurization of oils and coals. *Journal of Chemical Technology and Biotechnology*. 67: 109.

Shong, R. (1999) Bioprocessing of crude oil. Preprints of Papers- American Chemical Society, Division of Fuel Chemistry. 44: 1–4.

Silva, T.P., Paixão, S.M., Teixeira, A.V., Roseiro, J.C., Alves, L. (2013) Optimization of low sulfur carob pulp liquor as carbon source for fossil fuels biodesulfurization. *Journal of Chemical Technology and Biotechnology*. 88: 919–923.

Singh, A., Singh, B., Ward, O. (2012) Potential applications of bioprocess technology in petroleum industry. Biodegradation. 23: 865–880.

Singh, P (2015) Improved biodesulfurization of persistent organosulfur compounds. PhD Thesis, Department of Biochemical Engineering and Biotechnology, Indian Institute of Technology Delhi, New Delhi, India.

Soleimani, M., Bassi, A., Margaritis, A. (2007) Biodesulfurization of refractory organic sulfur compounds in fossil fuels. *Biotechnology Advances*. 25: 570–596.

Song C.S. (2003) An overview of new approaches to deep desulfurization for ultra-clean gasoline, diesel fuel, and jet fuel. *Catalysis Today*. 86: 211–263.

Sousa, S.F., Sousa, J.F.M., Barbosa, A.C.C., Ferreira, C.E., Neves, R.P.P, Rubeiro, A.J.M., Fernandes, P.A., Ramos, M.J. (2016) Improving the biodesulfurization of crude oil and derivatives: a QM/MM investigation of the catalytic mechanism of NADH-FMN oxidoreductase (DszD). *Journal of Physical Chemistry*. 120:5300–5306.

Speight, J.G. (1980) The Chemistry and Technology of Petroleum. Marcel Dekker. New York.

Speight, J.G. (2011) The refinery of the future. 1st ed., Elsevier, Oxford UK.

Srinivasaraghavan, K., Sarma, P. M., & Lal, B. (2006) Comparative analysis of phenotypic and genotypic characteristics of two desulfurizing bacterial strains, *Mycobacterium phlei* SM120-1 and *Mycobacterium phlei* GTIS10. *Letters in Applied Microbiology*, 42(5), 483–489.

Srinivasulu, B., Adinarayana, K., Ellaiah, P. (2003) Investigations on neomycin production with immobilized cells of *Streptomyces marinensis* Nuv-5 in calcium alginate matrix. AAPS PharmSciTech. 4: 449–454.

Srivastava, N.K., Nandan, N.K. (2012) Microbial growth control in diesel by optimization of

Sulphur. *International Journal of Environmental Pollution and Remediation*. 1(1): 119–125.

Srivastava, P., Kumar, A. (2008) Biodesulfurization of petroleum fractions for a cleaner environment. 28(3): 8–12.

Srivastava, V.C. (2012) An evaluation of desulfurization technologies for sulfur removal from liquid fuels. *RSC Advances*. 2: 759–783.

Stanislaus, A., Marafi, A., Rana, M.S. (2010) Recent advances in the science and technology of ultra low sulfur diesel (ULSD) production. *Catalysis Today*. 153: 1–68.

Strawinski, R.J. (1950) Method of desulfurizing crude oil. US Patent No. 2,521,761. Assigned to Texaco Development Corporation.

Strawinski, R.J. (1951) Purification of substances by microbial action. US Patent No. 2,574,070. Assigned to Texaco Development Corporation.

Subashchandrabose, S.R., Ramakrishnan, B., Megharaj, M., Ven- kateswarlu, K., Naidu, R. (2011) Consortia of cyanobacteria/ microalgae and bacteria: Biotechnological potential. *Biotechnology Advances*. 29:896–907.

Sundaramoorthy M., Terner J., Poulos T.L. (1995) The crystal structure of chloroperoxidase: a heme peroxidase-cytochrome P450 functional hybrid. Structure 3(12): 1367–1377.

Suzuki.

Takada, M., Nomura, N., Okada, H., Nakajima-Kambe, T., Nakahara, T., Uchiyama, H. (2005) De-repression and comparison of oil–water separation activity of the dibenzothiophene desulfurizing bacterium, *Mycobacterium* sp. G3. *Biotechnology Letters* 27: 871–874.

Tam, P.S., Kittrell, J.R., Eldridge, J.W. (1990) Desulfurization of fuel oil by oxidation and extraction. I. Enhancement of extraction oil yield. *Industrial and Engineering Chemistry Research.* . 29: 321–324.

Tanaka, Y., Matsui, T., Konishi, J., Maruhashi, K., Kurane, R. (2002) Biodesulfurization of benzothiophene and dibenzothiophene by a newly isolated *Rhodococcus* strain. *Applied Microbiology and Biotechnology*. 59: 325–328.

Tangaromsuk, J., Borole, A.., Kruatrachue, M., Pokethitiyook, P. (2008) An integrated biodesulfurization process, including inoculum preparation, desulfurization and sulfate removal in a single step, for removing sulfur from oils. *Journal of Chemical Technology and Biotechnology*. 83:1375–1380.

Tao, F., Liu, Y., Luo, Q., Su, F., Xu, Y., Li, F., Yu, B., Ma, C., Xu, P. (2011a) Novel organic solvent-responsive expression vectors for biocatalysis: application for development of an organic solvent tolerant biodesulfurizing strain. *Bioresource Technology*. 102: 9380–9387.

Tao, F., Yu, B., Xu, P., Ma, C.Q. (2006) Biodesulfurization in biphasic systems containing organic solvents. *Applied and Environmental Microbiology*. 72: 4604–4609.

Tao, F., Zhao, P., Li, Q., Su, F., Yu, B., Ma, C., Tang, H., Tai, C., Wu, G., Xu, P. (2011b) Genome sequence of *Rhodococcus erythropolis* XP, a biodesulfurizing bacterium with industrial potential. *Journal of Bacteriology*. 193: 6422–6423.

Te, M., Fairbridge, C., Ring, Z. (2001) Oxidation reactivities of dibenzothiophenes in polyoxometalate/H_2O_2 and formic acid/H_2O_2 systems. *Applied Catalysis A: General*. 219: 267–280.

Teixeira, A.V., Paixao, S.M., DaSilva, T.L., Alves, L. (2014) Influence of the carbon source on *Gordonia alkanivorans* strain 1B resistance to 2-hydroxybiphenyl toxicity. *Applied Biochemistry and Biotechnology*. 173(4): 870–882.

Tian, F., Wu, W., Jiang, Z., Liang, C., Yang, Y., Ying, P. Li, C. (2006) The study of thiophene adsorption onto La(III)-exchanged zeolite NaY by FT-IR spectroscopy. *Journal of Colloid and Interface Science*. 301(2): 395–401.

Tilstra, L., Eng, G., Olson, G.J. and Wang, F.W. (1992) Reduction of sulphur from polysulphidic model compounds by the hyperthermophilic *Archaebacterium Pyrococcus* furiosus. *Fuel*. 71: 779–783.

Torkamani, S., Shayegan, J., Yaghmaei, S., Alemzadeh, I. (2008) Study of the first isolated fungus capable of heavy crude oil biodesulfurization. *Industrial and Engineering Chemistry Research*. 47: 7476–82.

US Department of Energy. (2003) Gasoline biodesulfurization. Office of Energy Efficiency and Renewable Energy, May. United States Department of Energy, Washington, DC.

Van Afferden, M., Trappe, D., Beyer, M., Tüpper, H.G., Klein, J. (1993) Biochemical mechanisms for the desulfurization of coal-relevant organic sulfur compounds. *Fuel* 72(12): 1635–1643.

van Hamme, J.D., Wong, E.T., Dettman, H., Gray, M.R., Pickard, M.A. (2003) Dibenzyl sulfide metabolism by white-rot fungi. *Applied Environmental Microbiology*. 69: 1320–1324.

Van Houten, R.T., Lettinga, G. (1996) Biological sulfate reduction with synthesis gas: microbiology and technology, p. 793–799 In: Wijffels R.H., Buitelaar, R.M., Bucke C., Tramper J. (Ed.) Progress in Biotechnology. Vol. 11. Elsevier, Amsterdam.

Vazquez-Duhalt, R., Semple, K.M., Westlake, D.W.S., Fedorak, P.M. (1993) Effect of water-miscible organic solvents on the catalytic activity of cytochrome c. *Enzyme and Microbial Technology*. 15(11): 936–943.

Vazquez-Duhalt, R., Torres, E., Valderrama, B., Le Borgne, S. (2002) Will biochemical catalysis impact the petroleum refining industry?". *Energy and Fuels*. 16: 1239–1250.

Verma, N., Sharma, R., Kaur, R. (2016) Microbial desulfurization study of dibenzothiophene and crude oil by a soil isolate. *International Journal of Science and Research Methodology*. 4(4): 133–145.

Vertegel A. (2010) Under sponsorship of the American Chemical Society Petroleum Research Fund. 55th Annual Report on Research 2010. https://acswebcontent.acs.org/prfar/2010/reports/P10734.html.

Walker, J.D., Colwell, R.R., Petrakis, L. (1975) Microbial petroleum degradation: application of computerized mass spectrometry. *Canadian Journal of Microbiology*. 21: 1760–1767.

Wang, C.Y., Calabres, e R.V. (1986) Drop breakup in turbulent stirred-tank contactors, Part II: Relative influence of viscosity and interfacial tension *AIChE Journal*. 32: 667–681.

Wang, J., Butler, III, R.R., Wu, F., Pombert, J.F., Kilbane, II, J.J., Stark, B.C. (2017) Enhancement of microbial biodesulfurization via genetic engineering and adaptive evolution. PLOS ONE DOI:10.1371/journal.pone.016883.

Wang, J., Davaadelger, B., Salazar, J.K., Butler III, R.R., Pombert, J.-F., Kilbane, J.J., Stark, B.C. (2015) Isolation and characterization of an interactive culture of two *Paenibacillus* species with moderately thermophilic desulfurization ability. *Biotechnology Letters*. 37(11): 2201–2211.

Wang, M., W., Li, W., Shi, Y., Wang, D.-h., W., Feng, H. (2006) Effects of surfactant on biodesulfurization by *Corynebacterium* sp. ZD-1 in the presence of organic phase. *Journal of Zhejiang University Science A*. 7(Suppl. II): 371–375.

Wang, P., Krawiec, S. (1994) Desulfurization of dibenzothiophene to 2-hydroxybiphenyl by some newly isolated bacterial strains. *Archives of Microbiology*. 161(3): 266–271.

Wang, P., Krawiec, S. (1996) Kinetic analyses of desulfurization of dibenzothiophene by *Rhodococcus erythropolis* in batch and fed-batch cultures. *Applied and Environmental Microbiology*. 62: 1670–1675.

Wang, W.B., Wang, S.J., Liu, H.Y., Wang, Z.X. (2007) Desulfurization of gasoline by a new method of electrochemical catalytic oxidation. *Fuel*. 86: 2747–2753.

Watanabe K., Noda K., Maruhashi K. (2003a) Selective cleavage of 10-methyl benzo [b]naphtho[2,1-d] thiophene by recombinant *Mycobacterium* sp. strain. *Biotechnology Letters*. 25 :797–803.

Watanabe K., Noda K., Maruhashi K. (2003b) Enhanced desulfurization in a transposon-mutation strain of *Rhodococcus erythropolis Biotechnology Letters*. 25:1299–1304.

Watanabe, K., Noda, K., Konishi, J., Maruhashi, K. (2003c). Desulfurization of 2,4,6,8-tetraethyl dibenzothiophene by recombinant *Mycobacterium* sp. strain MR65. 25:1451–1456.

Watkins, L.M., Rodriguez, R., Schneider, D., Broderick, R., Cruz, M., Chambers, R., Ruckman, E., Cody, M., Mrachko, G.T. (2003) Purification and characterization of the aromatic desulfinase 2-(2-hydroxybiphenyl) benzenesulfinate desulfinase. *Archives of Biochemistry and Biophysics*. 415: 14–23.

Winkler, J.D., Kao, K.C. (2014) Recent advances in the evolutionary engineering of industrial biocatalysts. *Genomics*. 104: 406–411.

Wolf, B. P., Sumner, L.W., Shields, S.J., Nielsen, K., Gray, K.A., Russell, D.H. (1998) Characterization of proteins utilized in the desulfurization of petroleum products by matrix assisted laser desorption ionization time-of-flight mass spectrometry. *Analytical Biochemistry*. 260:117–127.

Xiong, X, Xing, J., Li, X., Bai, X., Li, W., Li, Y., Liu, H. (2007) Enhancement of biodesulfurization in two-liquid systems by heterogeneous expression of *Vitreoscilla* hemoglobin. *Applied Environmental Microbiology*. 73: 2394–2397.

Xu X., Cai Y., Song Z., Qiu X., Zhou J., Liu Y., Mu T., Wu D., Guan Y., Xing J. (2015) Desulfurization of immobilized sulfur-oxidizing bacteria, *Thialkalivibrio versutus*, by magnetic nanoparticles under haloalkaliphilic conditions. *Biotechnology Letters*. 37(8): 1631–1635.

Xu, P., Feng, J., Yu, B., Li, F., Ma, C. (2009). Recent developments in biodesulfurization of fossil fuels. *Advances in Biochemical Engineering/Biotechnology*. 113: 255–274.

Yan, H., Kirshimoto, M., Omasa, T., Katakura, Y., Suga, K.I., *et al.* (2000) Increase in desulfurization activity of *R. erythropolis* KA2–5-1 using ethanol feeding. *Journal of Bioscience and Bioengineering*. 89: 361–366.

Yan, H., Sun, X., Xu, Q., Ma, Z., Xiao, C., Jun, N. (2008) Effects of nicotinamide and riboflavin on the biodesulfurization activity of dibenzothiophene by *Rhodococcus erythropolis* USTB-03. *Journal of Environmental Sciences*. 20: 613–618.

Yang J., Hu Y., Zhao D., Wang S., Lau P.C.K., Marison I.W. (2007b) Two-layer continuous-process design for the biodesulfurization of diesel oils under bacterial growth conditions. *Biochemical Engineering Journal*. 37(2): 212–218.

Yang, J., Marison, I.W (2005) Two-stage process design for the biodesulfurisation of a model diesel by a newly isolated *Rhodococcus globerulus* DAQ3. *Biochemical Engineering Journal*. 27: 77–82.

Yang, L., Li, J., Yuan, X.D., Shen, J., Qi, Y.T. (2007a) One step non-hydrodesulfurization of fuel oil: Catalyzed oxidation adsorption desulfurization over HPWA-SBA-15. *Journal of Molecular Catalysis A: Chemical*. 262: 114–118.

Yazdian, F., Pesaran Hajiabbas, M., Shojaosadati, S.A., Nosrati, M., Vashegani-Farahani, E., Mehrnia, M.R. (2010) Study of hydrodynamics, mass transfer, energy consumption and biomass production from natural gas in a forced-liquid vertical tubular loop bioreactor. *Biochemical Engineering Journal*. 49, 192–200.

Yazdian, F., Shojaosadati, S.A., Nosrati, M., Pesaran Hajiabbas, M., Vashegani-Farahani, E. (2009) Investigation of gas properties, design and operational parameters on hydrodynamic characteristics, mass transfer and biomass production from natural gas in an external airlift loop bioreactor. *Chemical Engineering Science*. 64, 2455–2465.

Yemashova, N.A., Murygina, V.P., Zhukov, D.V., Zakharyantz, A.A., Gladchenko, M.A., Vasu Appanna, V., Kalyuzhnyi, S.V. (2007) Biodeterioration of crude oil and oil derived products: A review. *Reviews in Environmental Science and Bio/Technology* 6(4):315–337.

Yoshikawa, O., Ishii, Y., Koizumi, K.I., Ohshiro, T., Izumi, Y. Maruhashi, K. (2002) Enhancement and stabilization of desulfurization activity of *Rhodococcus erythropolis* KA2–5-1 by feeding ethanol and sulfur components. *Journal of Bio science and Bioengineering*. 94:447–452 (2002).

Yoshikawa, O., Maruhashi, K., Kobayashi, M., Ishii, Y., Hirasawa, K. (2001) Improvement of desulfurization activity in *Rhodococcus erythropolis* KA2-5-1 by genetic engineering. *Bioscience, Biotechnology, and Biochemistry.* 65(2): 239–246.

Yu, B., Tao, F., Li, F., Hou, J., Tang, H., Ma, C., Xu, P. (2015) Complete genome sequence of Mycobacterium goodii X7B, a facultative thermophilic biodesulfurizing bacterium with industrial potential. *Journal of Biotechnology.* 212:56–57.

Yu, B., Xu, P., Shi, Q., Ma, C. (2006) Deep desulfurization of diesel oil and crude oils by a newly isolated *Rhodococcus erythropolis* strain. *Applied and Environmental Microbiology.* 72: 54–58.

Yu, G., Lu, S., Chen H., Zhu, Z. (2005) Oxidative desulfurization of diesel fuels with hydrogen peroxide in the presence of activated carbon and formic acid. *Energy and Fuels.* 19: 447–452.

Yu, L., Meyer, T., Folsom, B. (1998) Oil/water/biocatalyst three phase separation process. US Patent No. 5,772,901.

Zahra, E., Giti, E., Sharareh, P. (2006) Removal of dibenzothiophene, biphenyl and phenol from waste by *Trichosporon* sp. Scientific Research and Essays. 1:72–76.

Zakaria B.S., Nassar H.N., Saed D., El-Gendy N.Sh. (2015) Enhancement of carbazole denitrogenation rate using magnetically decorated *Bacillus clausii* BS1. *Petroleum Science and Technology.* 33(7): 802–811.

Zakaria, B.S., Nassar, H.N., Saed, D., EL-Gendy, N.Sh. (2016) Denitrogenation of carbazole by a novel strain Bacillus clausii BS1 isolated from Egyptian coke. Energy Sources, Part A: *Recovery Utilization and Environmental Effects* 38(13): 1840–1851.

Zaki, T., Saed, D., Aman, D., Younis, S.A. Moustafa, Y.M. (2013) Synthesis and characterization of MFe2O4 sulfur nanoadsorbents. Sol-Gel Sci. Technol. 65:269–276.

Zeelani, G.G., Pal, S.L. (2016) A review on desulfurization technique of liquid fuels. *International Journal of Science and Research.* 5(5): 2413–2419.

Zhang, H., Shan, G., Liu, H., Xing, J. (2007b) Surface modification of γ-Al_2O_3 nanoparticles with gum Arabic and its applications in adsorption and biodesulfurization. Surface and Coatings Technology. 201: 6917–6921.

Zhang, H.Y., Liu, Q.F., Li, Y.G., Li, W.L., Xiong, X.C., Xing, J.M., Liu, H.Z. (2008) Selection of adsorbents for in-situ coupling technology of adsorptive desulfurization and biodesulfurization. *Science in China Series B-Chemistry.* 51: 69–77.

Zhang, Q.; Tong, M.Y.; Li, Y.S.; Gao, H.J. and Fang, X.C. (2007c) Extensive desulfurization of diesel by *Rhodococcus erythropolis. Biotechnology Letters* 29:123–127.

Zhang, S.H., Chen, H., Li, W. (2012) Kinetic analysis of biodesulfurization of model oil containing multiple alkyl dibenzothiophenes. *Applied Microbiology and Biotechnology.* 97: 2193–2200.

Zhang, S.H., Chen, H., Li, W. (2013) Kinetic analysis of biodesulfurization of model oil containing multiple alkyl dibenzothiophenes. *Applied Microbiology Biotechnology* 97: 2193–2200.

Zhang, T., Li, W.L., Chen, X.X., Tang, H., Li, Q., Xing, J.M., Liu, H.Z. (2011) Enhanced biodesulfurization by magnetic immobilized *Rhodococcus erythropolis* LSSE8-1-vgb assembled with nano- γ-Al_2O_3. *World Journal of Microbiology and Biotechnology*. 27: 299–305.

Zhang, Y.W., Prabhu, P., Lee, J.K. (2010) Alginate immobilization of recombinant Escherichia coli whole cells harboring L-arabinose isomerase for L-ribulose production. *Bioprocess and Biosystems Engineering*. 33: 741–748.

Zhang, Z.Y., Jin, B., Kelly, J.M. (2007a) Production of lactic acid and byproducts from waste potato starch by *Rhizopus arrhizus:* role of nitrogen sources. *World Journal of Microbiology and Biotechnology*. 23: 229–236.

Zhao, D.S., Li, F.T., Zhou, E.P., Sun, Z.M. (2008) Kinetics and mechanism of the photo-oxidation of thiophene by o2 adsorbed on molecular sieves. *Chemical Research in Chinese Universities*. 24: 96–100.

Zhao, J. (2009) Turning to Nanotechnology for Pollution Control: Applications of Nanoparticles. *Dartmouth undergraduate journal of science*.

Zhou, G., Kresta, S.M. (1998) Correlation of mean drop size and minimum drop size with the turbulence energy dissipation and the flow in an agitated tank. *Chemical Engineering Science*. 53: 2063–2079.

ZoBell, C.E. (1953) Process for removing sulfur from petroleum hydrocarbons and apparatus. USA Patent 2,641,564.

Zytner, R.G., Salb, A.C., Stiver, W.H. (2006) Bioremediation of diesel fuel contaminated soil: comparison of individual compounds to complex mixtures. *Soil and Sediment Contamination*. 15:277–297.

Glossary

Acid Catalyst: a catalyst having acidic character, for example aluminas

Acid Deposition: acid rain; a form of pollution depletion in which pollutants, such as nitrogen oxides and sulfur oxides, are transferred from the atmosphere to soil or water; often referred to as atmospheric self-cleaning. The pollutants usually arise from the use of fossil fuels.

Acidity: the capacity of an acid to neutralize a base, such as a hydroxyl ion (OH−)

Acidizing: a technique for improving the permeability (q.v.) of a reservoir by injecting acid

Acid Number: a measure of the reactivity of petroleum with a caustic solution given in terms of milligrams of potassium hydroxide that are neutralized by 1 g of petroleum

Acid Rain: the precipitation phenomenon that incorporates anthropogenic acids and other acidic chemicals from the atmosphere to the land and water (see acid deposition)

Activation Energy (E_a): the energy that is needed by a molecule or molecular complex to encourage reactivity to form products

Additive: a material added to another (usually in small amounts) to enhance desirable properties or to suppress undesirable properties

Adsorption: transfer of a substance from a solution to the surface of a solid resulting in relatively high concentration of the substance at the place of contact; see also chromatographic adsorption

Air Pollution: the discharge of toxic gases and particulate matter introduced into the atmosphere, principally as a result of human activity

Air Sweetening: a process in which air or oxygen is used to oxidize lead mercaptide to disulfides instead of using elemental sulfur

Alcohol: the family name of a group of organic chemical compounds composed of carbon, hydrogen, and oxygen. The series of molecules vary in chain length and are composed of a hydrocarbon plus a hydroxyl group, $CH_3(CH_2)_nOH$ (e.g., methanol, ethanol, and tertiary butyl alcohol)

Alicyclic Hydrocarbon: a compound containing carbon and hydrogen only, which has a cyclic structure (e.g., cyclohexane); also, collectively called naphthenes

Aliphatic Hydrocarbon: a compound containing carbon and hydrogen only, which has an open-chain structure (e.g., as ethane, butane, octane, or butene) or a cyclic structure (e.g., cyclohexane)

Alkalinity: the capacity of a base to neutralize the hydrogen ion (H^+)

Alkali Treatment: see caustic wash

Alkali Wash: see caustic wash

Alkanes: hydrocarbons that contain only single carbon–hydrogen bonds. The chemical name indicates the number of carbon atoms and ends with the suffix "-ane."

Alkenes: hydrocarbons that contain carbon–carbon double bonds. The chemical name indicates the number of carbon atoms and ends with the suffix "-ene."

Alkylate: the product of an alkylation (q.v.) process

Alkylate Bottoms: residual from fractionation of alkylate; the alkylate product that boils higher than the aviation gasoline range; sometimes called heavy alkylate or alkylate polymer

Alkylation: in the petroleum industry, a process by which an olefin (e.g., ethylene) is combined with a branched-chain hydrocarbon (e.g., isobutane); alkylation may be accomplished as a thermal or as a catalytic reaction

Alkyl Groups: a group of carbon and hydrogen atoms that branch from the main carbon chain or ring in a hydrocarbon molecule. The simplest alkyl group, a methyl group, is a carbon atom attached to three hydrogen atoms.

Alumina (Al_2O_3): used in separation methods as an adsorbent and in refining as a catalyst

ASTM International (formerly American Society for Testing and Materials): the official organization in the United States for designing standard tests for petroleum and other industrial products

Analysis: determining the properties of a feedstock before refining; inspection (q.v.) of feedstock properties

Aniline Point: the temperature, usually expressed in °F, above which equal volumes of a petroleum product are completely miscible; a qualitative indication of the relative proportions of paraffins in a petroleum product that are miscible with aniline only at higher temperatures; a high aniline point indicates low aromatics

API (American Petroleum Institute) Gravity: a measure of the lightness or heaviness of petroleum that is related to density and specific gravity

$$API = (141.5/\text{sp gr at } 60°F) - 131.5$$

Apparent Bulk Density: the density of a catalyst as measured; usually loosely compacted in a container

Apparent Viscosity: the viscosity of a fluid or several fluids flowing simultaneously measured in a porous medium (rock) and subject to both viscosity and permeability effects; also called effective viscosity

Aromatics: a group of hydrocarbons of which benzene is the parent; named because many of their derivatives have sweet or aromatic odors

Aromatization: the conversion of nonaromatic hydrocarbons to aromatic hydrocarbons by (1) rearrangement of aliphatic (noncyclic) hydrocarbons (q.v.) into aromatic ring structures and (2) dehydrogenation of alicyclic hydrocarbons (naphthenes)

Asphalt: highly viscous liquid or semisolid composed of bitumen and present in most crudes; can be separated from other crude components by fractional distillation; used primarily for road paving and roofing shingles

Asphaltene Constituents: molecular species that occur within the asphaltene fraction and that vary in polarity (functional group content) and molecular weight

Asphaltene Fraction: the fraction of petroleum, heavy oil, or bitumen that is precipitated when a large excess (40 vol) of a low-boiling liquid hydrocarbon (e.g., pentane or heptane) is added to (1 vol of) the feedstock; usually a dark brown to black amorphous solid that does not melt before decomposition and is soluble in benzene or aromatic naphtha or other chlorinated hydrocarbon solvents

Asphaltic Constituents: a general term usually meaning the asphaltene fraction plus the resin fraction

Associated Gas: natural gas that is in contact with and/or dissolved in the crude oil of the reservoir. It may be classified as gas cap (free gas) or gas in solution (dissolved gas)

Associated Gas in Solution (or dissolved gas): natural gas dissolved in the crude oil of the reservoir, under prevailing pressure and temperature conditions

Associated Molecular Weight: the molecular weight of asphaltenes in an associating (nonpolar) solvent, such as toluene

Atmospheric Distillation: distillation at atmospheric pressure; the refining process of separating crude oil components at atmospheric pressure by heating to temperatures of about 600-750°F (depending on the nature of the crude oil and desired products) and subsequent condensing of the fractions by cooling

Barrel: the unit of measurement of liquids in the petroleum industry; the traditional measurement for crude oil volume: 1 barrel is equivalent to 42 US gallons (159 L) and 6.29 barrels is equivalent to 1 m^3 of oil

Barrel of Oil Equivalent: a unit of energy based on the approximate energy released by burning 1 barrel of crude oil

Base Number: the quantity of acid expressed in milligrams of potassium hydroxide per gram of sample that is required to titrate a sample to a specified end point

Basic Nitrogen: nitrogen (in petroleum) that occurs in pyridine form

Bentonite: montmorillonite (a magnesium–aluminum silicate); used as a treating agent

Benzene: a colorless aromatic liquid hydrocarbon (C_6H_6); present in small proportion in some crude oils and made commercially from petroleum by the catalytic reforming of naphthenes in petroleum naphtha; also made from coal in the manufacture of coke; used as a solvent in the manufacture of detergents, synthetic fibers, petrochemicals, and as a component of high-octane gasoline

Billion: 1×10^9.

Bioaccumulation (or bioconcentration): the tendency of substances to accumulate in the body of organisms; the net uptake from their diet, respiration, or transfer across skin and loss due to excretion or metabolism. The bioaccumulation factor (BAF) or bioconcentration factor (BCF) is the ratio of concentrations in tissue to concentrations in a source, i.e., water or diet.

Bioavailability (or biological availability): compound that is in a physical or chemical form that can be assimilated by a living organism; also the proportion of a chemical in an environmental compartment (e.g., water) that can be taken up by an organism

Biocide: any chemical capable of killing bacteria and bioorganisms

Biodegradation: a natural process of microbial transformation of chemicals, such as oil, under aerobic or anaerobic conditions; oil biodegradation usually requires nutrients, such as nitrogen and phosphorus; transformation may be complete producing water, carbon dioxide, and/or methane or incomplete, producing partially-oxidized chemicals

Biogenic: material derived from bacterial or vegetation sources

Biological Lipid: any biological fluid that is miscible with a nonpolar solvent. These materials include waxes, essential oils, chlorophyll, etc.

Biological Oxidation: the oxidative consumption of organic matter by bacteria by which organic matter is converted into gases

Biomass: biological organic matter; materials produced from the processing of wood, corn, sugar, and other agricultural waste or municipal waste. It can be converted to syngas via a gasification process.

Bioremediation: an intervention strategy to enhance biodegradation of spilled oil (or other contaminants) ranging from no remedial action other than monitoring (natural attenuation) to nutrient addition (biostimulation) to inoculation with competent microbial communities (bioaugmentation)

Bitumen: a semisolid to solid organic material found filling pores and crevices of sandstone, limestone, or argillaceous sediments; contains organic carbon, hydrogen, nitrogen, oxygen, sulfur, and metallic constituents; usually has an API gravity (<10°), but other properties are necessary for inclusion in a more complete definition; in its natural state, tar sand (oil sand) bitumen is not recoverable at a commercial rate through a well because it is too viscous to flow; bitumen typically makes up approximately 10% w/w of tar sand (oil sand), but saturation varies

Boiling Point: a characteristic physical property of a liquid at which the vapor pressure is equal to that of the atmosphere and the liquid is converted to a gas

Boiling Range: the range of temperature, usually determined at atmospheric pressure in a standard laboratory apparatus, over which the distillation of oil commences, proceeds, and finishes

Bottom-of-the-Barrel: residuum (q.v.)Brønsted Acid: a chemical species that can act as a source of protons

Brønsted Base: a chemical species that can accept protons

BTEX: benzene, toluene, ethylbenzene, and the xylene isomers

BTU (British Thermal Unit): the energy required to increase the temperature of 1 lb of water to 1°FBunker C Oil: see No. 6 fuel oil

Burner Fuel Oil: any petroleum liquid suitable for combustion

Burning Oil: illuminating oil, such as kerosene suitable for burning in a wick lamp

Burning Point: see fire point

Carbon Dioxide-Augmented Waterflooding: injection of carbonated water or water and carbon dioxide to increase water flood efficiency; see immiscible carbon dioxide displacement

Carbon Residue: the amount of carbonaceous residue remaining after thermal decomposition of petroleum, a petroleum fraction, or a petroleum product in a limited amount of air; also called the coke- or carbon-forming propensity; often prefixed by the terms Conradson or Ramsbottom in reference to the inventor of the respective tests

Catalyst: a chemical agent that, when added to a reaction (process), will enhance the conversion of a feedstock without being consumed in the process; used in upgrading processes to assist cracking and other upgrading reactions

Catalyst Plugging: the deposition of carbon (coke) or metal contaminants that decreases the flow through the catalyst bed

Catalyst Poisoning: the deposition of carbon (coke) or metal contaminants that causes the catalyst to become nonfunctional

Catalyst Selectivity: the relative activity of a catalyst with respect to a particular compound in a mixture or the relative rate in competing reactions of a single reactant

Catalyst Stripping: the introduction of steam at a point where spent catalyst leaves the reactor in order to strip, i.e. remove, deposits retained on the catalyst

Catalytic Activity: the ratio of the space velocity of the catalyst under test to the space velocity required for the standard catalyst to give the same conversion as the catalyst being tested; usually multiplied by 100 before being reported

Catalytic Cracking: the conversion of high-boiling feedstocks into lower-boiling products by means of a catalyst that may be used in a fixed bed (q.v.) or fluid bed (q.v.)

Catalytic Distillation: a process that combines reaction and distillation in a single vessel resulting in lower investment and operating costs, as well as process benefits

Catalytic Hydrocracking: a refining process that uses hydrogen and catalysts with relatively low temperatures and high pressures for converting middle-boiling or residual material to high-octane gasoline, reformer charge stock, jet fuel, and/or high-grade fuel oil. The process uses one or more catalysts, depending on product output, and can handle high-sulfur feedstocks without prior desulfurization.

Catalytic Hydrotreating: a refining process for treating petroleum fractions from atmospheric or vacuum distillation units (e.g. naphtha, middle distillates, reformer feeds, residual fuel oil, and heavy gas oil) and other petroleum (e.g., cat cracked naphtha, coker naphtha, gas oil, etc.) in the presence of catalysts and substantial quantities of hydrogen. Hydrotreating includes desulfurization, removal of substances (e.g. nitrogen compounds) that deactivate catalysts, conversion of olefins to paraffins to reduce gum formation in gasoline, and other processes to upgrade the quality of the fractions.

Catalytic Reforming: rearranging hydrocarbon molecules in a gasoline-boiling-range feedstock to produce other hydrocarbons having a higher antiknock quality; isomerization of paraffins, cyclization of paraffins to naphthenes (q.v.), dehydrocyclization of paraffins to aromatics (q.v.)

Caustic Wash: the process of treating a product with a solution of caustic soda to remove minor impurities; often used in reference to the solution itself

Cetane Number: a number indicating the ignition quality of diesel fuel; a high cetane number represents a short ignition delay time; the ignition quality of diesel fuel can also be estimated from the following formula:

Diesel index = [aniline point (°F) × API gravity]100

Chelating Agents: complex-forming agents having the ability to solubilize heavy metals

Chemical Composition: the makeup of petroleum in terms of distinct chemical types such as paraffins, isoparaffins, naphthenes (cycloparaffins), benzenes, diaromatics, triaromatics, polynuclear aromatics; other chemical types can also be specified

Chemical Octane Number: the octane number added to gasoline by refinery processes or by the use of octane number (q.v.) improvers such as tetraethyl lead

Chemical Waste: any solid, liquid, or gaseous material discharged from a process and that may pose substantial hazards to human health and the environment

Chevron Deasphalted Oil Hydrotreating Process: a process designed to desulfurize heavy feedstocks that have had the asphaltene fraction (q.v.) removed by prior application of a deasphalting process (q.v.)

Chevron RDS and VRDS Processes: processes designed to remove sulfur, nitrogen, asphaltene, and metal contaminants from heavy feedstocks consisting of a once-through operation of the feedstock coming into contact with hydrogen and the catalyst in a downflow reactor (q.v.)

Chromatographic Adsorption: selective adsorption on materials such as activated carbon, alumina, or silica gel; liquid or gaseous mixtures of hydrocarbons are passed through the adsorbent in a stream of diluent and certain components are preferentially adsorbed

Chromatography: a method of separation based on selective adsorption; see also chromatographic adsorption.

Clay: silicate minerals that also usually contain aluminum and have particle sizes <0.002 micron; used in separation methods as an adsorbent and in refining as a catalyst

Cloud Point: the temperature at which paraffin wax or other solid substances begin to crystallize or separate from the solution imparting a cloudy appearance to the oil when the oil is chilled under prescribed conditions

Coal: an organic rock; a readily combustible black or brownish-black rock whose composition, including inherent moisture, consists of >50% w/w and >70% v/v of carbonaceous material; formed from plant remains that have been compacted, hardened, chemically altered, and metamorphosed by heat and pressure over geologic time

Coal Tar: the specific name for the tar (q.v.) produced from coal

Coal Tar Pitch: the specific name for the pitch (q.v.) produced from coal

Cogeneration: the simultaneous production of electricity and steam

Coke: a gray to black solid carbonaceous material produced from petroleum during thermal processing; characterized by having a high carbon content (95%+ by weight) and a honeycomb type of appearance and is insoluble in organic solvents

Coking: a process for the thermal conversion of petroleum in which gaseous, liquid, and solid (coke) products are formed; e.g. delayed coking (q.v.) or fluid coking (q.v.)

Condensate: a mixture of light hydrocarbon liquids obtained by condensation of hydrocarbon vapors, predominately butane, propane, and pentane with some heavier hydrocarbons and relatively little methane or ethane; see also natural gas liquids

Conradson Carbon Residue: see carbon residue

Contaminant: a substance that causes deviation from the normal composition of an environment

Conventional Crude Oil: a mixture mainly of pentane and heavier hydrocarbons recoverable at a well from an underground reservoir and liquid at atmospheric pressure and temperature; unlike some heavy oils and tar sand bitumen, conventional crude oil flows through a well without stimulation and through a pipeline without processing or dilution; generally, conventional crude oil includes light- and medium-gravity crude oils; crude oil containing >0.5% w/w sulfur is considered to be sour crude oil, while crude oil with <0.5% w/w sulfur is to be sweet crude oil

Conventional Gasoline: finished automotive gasoline not included in the oxygenated or reformulated gasoline categories; excludes reformulated gasoline blendstock for oxygenate blending (RBOB) as well as other blendstocks

Cracking: the thermal processes by which the constituents of petroleum are converted to lower molecular weight products

Cracking Temperature: the temperature (350°C, 660°F) at which the rate of thermal decomposition of petroleum constituents becomes significant

Crude assay: a procedure for determining the general distillation characteristics (e.g. distillation profile, q.v.) and other quality information of crude oil

Crude Oil: see petroleum

Cumene: a colorless liquid [$C_6H_5CH(CH_3)_2$] used as an aviation gasoline blending component and as an intermediate in the manufacture of chemicals

Cyclization: the process by which an open-chain hydrocarbon structure is converted to a ring structure, e.g. hexane to benzene

Cyclone: a device for extracting dust from industrial waste gases. It is in the form of an inverted cone into which the contaminated gas enters tangentially from the top; the gas is propelled down a helical pathway and the dust particles are deposited by means of centrifugal force onto the wall of the scrubber.

Deactivation: reduction in catalyst activity by the deposition of contaminants (e.g. coke, metals) during a process.

Dealkylation: the removal of an alkyl group from aromatic compounds

Deasphaltened Oil: the fraction of petroleum after the asphaltenes have been removed using liquid hydrocarbons, such as n-pentane and n-heptane

Deasphaltening: removal of a solid powdery asphaltene fraction from petroleum by the addition of the low-boiling liquid hydrocarbons, such as n-pentane or n-heptane, under ambient conditions

Deasphalting: the removal of asphalt (tacky, semisolid, higher molecular weight) constituents from petroleum (as occurs in a refinery asphalt plant) by the addition of liquid propane or liquid butane under pressure; also, the removal of the asphaltene fraction from petroleum by the addition of a low-boiling hydrocarbon liquid, such as n-pentane or n-heptane

Dehydrogenation: the removal of hydrogen from a chemical compound; for example, the removal of two hydrogen atoms from butane to make butene(s) as well as the removal of additional hydrogen to produce butadiene

Density: the mass (or weight) of a unit volume of any substance at a specified temperature; also, the heaviness of crude oil indicating the proportion of large, carbon-rich molecules, generally measured in kilograms per cubic meter (kg/m3) or degrees on the API gravity scale; in some countries, oil up to 900 kg/m3 is considered light to medium crude; see also specific gravity

Desalting: removal of mineral salts (mostly chlorides) from crude oils

Desulfurization: the removal of sulfur or sulfur compounds from a feedstock; a process that removes sulfur and its compounds from various streams during the refining process; desulfurization processes include catalytic hydrotreating and other chemical/physical processes, such as absorption; the desulfurization processes vary based on the type of stream treated (e.g. naphtha, distillate, heavy gas oil, etc.) and the amount of sulfur removed (e.g. sulfur reduction to 10 ppm)

Dewaxing: see solvent dewaxing

Diesel Fuel: fuel used for internal combustion in diesel engines; usually the fraction that distills after kerosene

Diesel Hydrotreater: a refinery process unit for production of clean (low-sulfur) diesel fuel

Diesel Index: an approximation of the cetane number (q.v.) of diesel fuel (q.v.) calculated from the density (q.v.) and aniline point (q.v.)

Diesel Knock: the result of a delayed period of ignition and the accumulation of diesel fuel in the engine

Diluted Crude: heavy crude oil to which a diluent (thinner) has been added to reduce viscosity and facilitate pipeline flow

Dispersant: a chemical or mixture of chemicals applied, for example to an oil spill to disperse the oil phase into small droplets in the water phase

Dispersion: suspension of oil droplets in water accomplished by natural wind and wave action, production of biological materials (biosurfactants) and/or chemical dispersant formulations

Distillate: the products of distillation formed by condensing vapors

Distillate Fuel Oil: a general classification for one of the petroleum fractions produced in conventional distillation operations. It includes diesel fuels and fuel oils. Products known as No. 1, No. 2, and No. 4 diesel fuel are used in on-highway diesel engines, such as those in trucks and automobiles, as well as off-highway engines, such as those in railroad locomotives and agricultural machinery. Products known as No. 1, No. 2, and No. 4 fuel oils are used primarily for space heating and electric power generation.

Distillation: a process for separating liquids with different boiling points without thermal decomposition of the constituents (see destructive distillation)

Distillation Curve: see distillation profile

Distillation Loss: the difference, in a laboratory distillation, between the volume of liquid originally introduced into the distilling flask and the sum of the residue and the condensate recovered

Distillation Profile: the distillation characteristics of petroleum or a petroleum product showing the temperature and the percent distilled

Distillation Range: the difference between the temperature at the initial boiling point and at the end point, as obtained by the distillation test

Domestic Heating Oil: see No. 2 fuel oil

Downcomer: a means of conveying liquid from one tray to the next below in a bubble tray column (q.v.)

Downstream: a sector of the petroleum industry that refers to the refining of crude oil and the products derived from crude oil

Dry Gas: a gas that does not contain fractions that may easily condense under normal atmospheric conditions

Drying: removal of a solvent or water from a chemical substance; also referred to as the removal of solvent from a liquid or suspension

Dry Point: the temperature at which the last drop of petroleum fluid evaporates in a distillation test

Effective Viscosity: see apparent viscosity

Effluent: any contaminating substance, usually a liquid, that enters the environment via a domestic industrial, agricultural, or sewage plant outlet

Electric Desalting: a continuous process to remove inorganic salts and other impurities from crude oil by settling out in an electrostatic field

Electrical Precipitation: a process using an electrical field to improve the separation of hydrocarbon reagent dispersions; may be used in chemical treating processes on a wide variety of refinery stocks

Electrofining: a process for contacting a light hydrocarbon stream with a treating agent (acid, caustic, doctor, etc.), then assisting the action of separation of the chemical phase from the hydrocarbon phase by an electrostatic field

Electrolytic Mercaptan Process: a process in which aqueous caustic solution is used to extract mercaptans from refinery streams

Electrostatic Precipitators: devices used to trap fine dust particles (usually in the size range 30–60 microns) that operate on the principle of imparting an electric charge to particles in an incoming air stream and which are then collected on an oppositely charged plate across a high-voltage field

Emission Control: the use of gas cleaning processes to reduce emissions

Emission Standard: the maximum amount of a specific pollutant permitted to be discharged from a particular source in a given environment

Emulsification: formation of water droplets in an oil matrix (water-in-oil) or, conversely, oil droplets in a water matrix (oil-in-water) achieved by the action of agitation, such as wind and wave activity; can be unstable, separating into oil and water phases again soon after formation or stable for months or years (e.g. 'chocolate mousse', a water-in-oil emulsion).

Emulsion Breaking: the settling or aggregation of colloidal-sized emulsions from suspension in a liquid medium

End-of-Pipe Emission Control: the use of specific emission control processes to clean gases after production of the gases

Energy: the capacity of a body or system to do work, measured in joules (SI units); also, the output of fuel sources

Energy from Biomass: the production of energy from biomass (q.v.)

Enhanced Oil Recovery: petroleum recovery following recovery by conventional (i.e. primary and/or secondary) methods; the third stage of production during which sophisticated techniques that alter the original properties of the oil are used. Enhanced oil recovery can begin after a secondary recovery process or at any time during the productive life of an oil reservoir; the purpose is not only to restore formation pressure, but also to improve oil displacement or fluid flow in the reservoir. The three major types of enhanced oil recovery operations are chemical flooding (alkaline flooding or micellar-polymer flooding), miscible displacement (carbon dioxide injection or hydrocarbon injection), and thermal recovery (steam flood). The optimal application of each method depends on reservoir temperature, pressure, depth, net pay, permeability, residual oil and water saturations, porosity, and fluid properties, such as oil API gravity and viscosity.

ETBE [Ethyl Tertiary Butyl Ether, $(CH_3)_3COC_2H$]: an oxygenated blend stock formed by the catalytic etherification of isobutylene with ethanol

Ethane (C_2H_6): a straight-chain saturated (paraffinic) hydrocarbon extracted predominantly from the natural gas stream, which is gaseous at standard temperature and pressure; a colorless gas that boils at a temperature of −88°C (−127°F)

Ethanol: see ethyl alcohol

Ether: a generic term applied to a group of organic chemical compounds composed of carbon, hydrogen, and oxygen characterized by an oxygen atom attached to two carbon atoms (e.g. methyl tertiary butyl ether)

Ethyl Alcohol (ethanol or grain alcohol): an inflammable organic compound (C_2H_5OH) formed during fermentation of sugars; used as an intoxicant and as a fuel

Evaporation: the physical loss of low molecular weight components of an oil to the atmosphere by volatilization

Ethylene (C_2H_4): an olefin hydrocarbon recovered from refinery or petrochemical processes that is gaseous at standard temperature and pressure. Ethylene is used as a petrochemical feedstock for many chemical applications and the production of consumer goods.

Extractive Distillation: the separation of different components of mixtures that have similar vapor pressures by flowing a relatively high-boiling solvent, which is selective for one of the components in the feed, down a distillation column as the distillation proceeds; the selective solvent scrubs the soluble component from the vapor

FCC: fluid catalytic cracking

FCCU: fluidized catalytic cracking unit

Feedstock: petroleum as it is fed to the refinery; a refinery product that is used as the raw material for another process; the term is also generally applied to raw materials used in other refinery processes or industrial processes

Filtration: the use of an impassable barrier to collect solids, but which allows liquids to pass

Fischer-Tropsch Process: a process for synthesizing hydrocarbons and oxygenated chemicals from a mixture of hydrogen and carbon monoxide

Fixed Bed: a stationary bed (of catalyst) to accomplish a process (see fluid bed)

Flame Ionization Detector (FID): used with analytical instruments like gas chromatographs to detect components of petroleum by combustion ionization, hence GC-FID

Flash Point: the lowest temperature to which the product must be heated under specified conditions to give off sufficient vapor to form a mixture with air that can be ignited momentarily by a flame

Flue Gas: gas from the combustion of fuel, the heating value of which has been substantially spent and which is, therefore, discarded to the flue or stack; gas that is emitted to the atmosphere via a flue, which is a pipe for transporting exhaust fumes

Fluid Bed: a bed (of catalyst) that is agitated by an upward passing gas in such a manner that the particles of the bed simulates the movement of a fluid and has the characteristics associated with a true liquid; cf. fixed bed

Fluid Catalytic Cracking: cracking in the presence of a fluidized bed of catalyst

Fluid Coking: a continuous fluidized solids process that cracks feed thermally over heated coke particles in a reactor vessel to gas, liquid products, and coke

Fluidized Catalytic Cracking: a refinery process used to convert the heavy portion of crude oil into lighter products, including liquefied petroleum gas and gasoline

Fly Ash: particulate matter produced from mineral matter in coal that is converted during combustion to finely divided inorganic material and which emerges from the combustor in the gases

Fossil Fuel Resources: a gaseous, liquid, or solid fuel material formed in the ground by chemical and physical changes (diagenesis, q.v.) in plant and animal residues over geological time; natural gas, petroleum, coal, and oil shale

Fraction: a group of hydrocarbons that have similar boiling points; a portion of crude oil defined by boiling range; naphtha, kerosene, gas oil, and residuum are fractions of crude oil

Fractional Composition: the composition of petroleum as determined by fractionation (separation) methods

Fractional Distillation: the separation of the components of a liquid mixture by vaporizing and collecting the fractions or cuts, which condense in different temperature ranges; a common form of separation technology in hydrocarbon-processing plants wherein a mixture (e.g. crude oil) is heated in a large, vertical cylindrical column to separate compounds (fractions) according to their boiling points

Fractionating Column: a column arranged to separate various fractions of petroleum by a single distillation which may be tapped at different points along its length to separate various fractions in the order of their boiling points

Fractionation: the separation of petroleum into the constituent fractions using solvent or adsorbent methods; chemical agents such as sulfuric acid may also be used

Free Sulfur: sulfur that exists in the elemental state associated with petroleum; sulfur that is not bound organically within the petroleum constituents

Fuel Oil: also called heating oil, is a distillate product that covers a wide range of properties; see also No. 1 to No. 4 fuel oils

Functional Group: the portion of a molecule that is characteristic of a family of compounds and determines the properties of these compounds

Furfural Extraction: a single-solvent process in which furfural is used to remove aromatic components, naphthene components, olefins, and unstable hydrocarbons from a lubricating oil charge stock

Furnace Oil: a distillate fuel primarily intended for use in domestic heating equipment.

Gaseous Pollutants: gases released into the atmosphere that act as primary or secondary pollutants

Gasification: a process to partially oxidize any hydrocarbon, typically heavy residues, to a mixture of hydrogen and carbon monoxide; the process can be used to produce hydrogen and various energy by-products

Gasohol: a term for motor vehicle fuel comprising between 80% and 90% unleaded gasoline and 10–20% ethanol (see also ethyl alcohol)

Gas Chromatography (GC): an analytical method used to characterize petroleum components; GC is combined with different detection methods, hence GC-FID, GC-MS, etc.

Gas Oil: a petroleum distillate with a viscosity and boiling range between those of kerosene and lubricating oil; a middle-distillate petroleum fraction; usually includes diesel, kerosene, heating oil, and light fuel oil; a liquid petroleum distillate having a viscosity intermediate between that of kerosene and lubricating oil. It derives its name from having originally been used in the manufacture of illuminating gas. It is now used to produce distillate fuel oils and gasoline; heavy gas oil is petroleum distillates with an approximate boiling range from 345°C to 540°C (650–1000°F)

Gas-Oil Ratio: ratio of the number of cubic feet of gas measured at atmospheric (standard) conditions to barrels of produced oil measured at stock tank conditions

Gasoline: fuel for the internal combustion engine that is commonly, but improperly, referred to simply as gas

Gasoline Blending Components: naphtha fractions that will be used for blending or compounding into finished aviation or automotive gasoline (e.g. straight-run gasoline, alkylate, reformate, benzene, toluene, and xylene); excludes oxygenates (alcohols, ethers), butane, and pentanes plus

Gas-to-Liquids (GTL): a process used to convert natural gas into longer-chain hydrocarbons such as diesel and jet fuel. Methane-rich gases are converted into liquid syngas (a mix of carbon monoxide and hydrogen) produced using steam methane reforming or autothermal reforming, followed by the Fischer–Tropsch process. Hydrocracking is then used to produce finished fuels.

Greenhouse Effect: warming of the earth due to entrapment of the energy of the sun by the atmosphere

Greenhouse Gases: gases that contribute to the greenhouse effect (q.v.)

Gum: an insoluble tacky semisolid material formed as a result of the storage instability and/or the thermal instability of petroleum and petroleum products

Heat Exchanger: a device used to transfer heat from a fluid on one side of a barrier to a fluid on the other side without bringing the fluids into direct contact

Heating Oil: see fuel oil

Heavy Feedstock: any feedstock of the type heavy oil (q.v.), bitumen (q.v.), atmospheric residuum (q.v.), vacuum residuum (q.v.), and solvent deasphalter bottoms (q.v.)

Heavy Fuel Oil: fuel oil having a high density and viscosity; generally residual fuel oil such as No. 5 and No 6. fuel oil (q.v.)

Heavy Gas Oil: a petroleum distillate with an approximate boiling range from 345°C to 540°C (650–1000°F)

Heavy Oil: petroleum having an API gravity of <20°; other properties are necessary for inclusion in a more complete definition

Heavy Petroleum: see heavy oil.

Heavy Residue Gasification and Combined Cycle Power Generation: a process for producing hydrogen from residuals

Heteroatom: in petroleum, an atom such as nitrogen, sulfur, and/or oxygen that is part of a hydrocarbon skeleton, such as found in the resins fraction of crude oils

Heteroatom Compounds: chemical compounds that contain nitrogen and/or oxygen and/or sulfur and/or metals bound within their molecular structure(s)

High-Boiling Distillates: fractions of petroleum that cannot be distilled at atmospheric pressure without decomposition, e.g. gas oils

High Molecular Weight (HMW): relative term referring to the molecular mass of chemicals; in oil, asphaltenes would be typical of HMW compounds

High performance liquid chromatography (HPLC): an analytical method for separating chemicals in solution

High-Sulfur Diesel (HSD): diesel fuel containing >500 ppm sulfur

High-Sulfur Petroleum: a general expression for petroleum having >1% wt. sulfur; this is a very approximate definition and should not be construed as having a high degree of accuracy because it does not take into consideration the molecular locale of the sulfur. All else being equal, there is little difference between petroleum having 0.99% wt. sulfur and petroleum having 1.01% wt. sulfur.

Hydrocarbon: a chemical that is composed of only carbon and hydrogen; chemicals containing heteroatoms, such as nitrogen, sulfur and/or oxygen, are not hydrocarbons, even though they may contain petroleum constituents

Hydrocarbon Gasification Process: a continuous, noncatalytic process in which hydrocarbons are gasified to produce hydrogen by air or oxygen

Hydrocarbon Gas Liquids (HGL): a group of hydrocarbons including ethane, propane, n-butane, isobutane, natural gasoline, and their associated olefins, including ethylene, propylene, butylene, and isobutylene; excludes liquefied natural gas

Hydrocarbon Resource: resources such as petroleum and natural gas that can produce naturally occurring hydrocarbons without the application of conversion processes

Hydrocarbon-Producing Resource: a resource such as coal and oil shale (kerogen) that produces derived hydrocarbons by the application of conversion processes; the hydrocarbons produced are not naturally occurring materials

Hydroconversion: a term often applied to hydrocracking (q.v.)

Hydrocracker: a refinery process unit in which hydrocracking occurs

Hydrocracking: a catalytic high-pressure, high-temperature process for the conversion of petroleum feedstocks in the presence of fresh and recycled hydrogen; carbon–carbon bonds are cleaved in addition to the removal of heteroatomic species

Hydrocracking Catalyst: a catalyst used for hydrocracking that typically contains separate hydrogenation and cracking functions

Hydrodenitrogenation: the removal of nitrogen by hydrotreating (q.v.)

Hydrodesulfurization: the removal of sulfur by hydrotreating (q.v.)

Hydrodemetallization: the removal of metallic constituents by hydrotreating (q.v.)

Hydrodesulfurization: a refining process that removes sulfur from liquid and gaseous hydrocarbons

Hydrofining: a fixed-bed catalytic process to desulfurize and hydrogenate a wide range of charge stocks from gases through waxes

Hydroforming: a process in which naphtha is passed over a catalyst at elevated temperatures and moderate pressures, in the presence of added hydrogen or hydrogen-containing gases to form high-octane motor fuel or aromatics

Hydrogen: the lightest of all gases occurring chiefly in combination with oxygen in water; exists also in acids, bases, alcohols, petroleum, and other hydrocarbons

Hydrogenation: the chemical addition of hydrogen to a material. In nondestructive hydrogenation, hydrogen is added to a molecule only if, and where, unsaturation with respect to hydrogen exists; classed as destructive (hydrocracking) or nondestructive (hydrotreating)

Hydroprocessing: a term often equally applied to hydrotreating (q.v.) and to hydrocracking (q.v.); also, often collectively applied to both

Hydrotreater: a refinery process unit that removes sulfur and other contaminants from hydrocarbon streams

Hydrotreating: the removal of heteroatomic (nitrogen, oxygen, and sulfur) species by treatment of a feedstock or product at relatively low temperatures in the presence of hydrogen

Hydrovisbreaking: a noncatalytic process conducted under similar conditions to visbreaking, which involves treatment with hydrogen to reduce the viscosity of the feedstock and produce more stable products than is possible with visbreaking

Hydrogen Sulfide (H_2S): a toxic, flammable, and corrosive gas sometimes associated with petroleum

Inhibitor: a substance, the presence of which, in small amounts, in a petroleum product prevents or retards undesirable chemical changes from taking place in the product or in the condition of the equipment in which the product is used

Iodine Number: a measure of the iodine absorption by an oil under standard conditions; used to indicate the quantity of unsaturated compounds present; also called iodine value

Ion Exchange: a means of removing cations or anions from solution onto a solid resin

Isomerate Process: a fixed-bed isomerization process to convert pentane, hexane, and heptane to high-octane blending stocks

Isomerization: the conversion of a normal (straight-chain) paraffin hydrocarbon into an iso- (branched-chain) paraffin hydrocarbon having the same atomic composition; a refining process that alters the fundamental arrangement of atoms in the molecule without adding or removing anything from the original material; used to convert normal butane into isobutane (C4), an alkylation process feedstock, and normal pentane and hexane into isopentane (C5) and isohexane (C6), high-octane gasoline components

Isomers: chemicals that have the same molecular formula (i.e. elemental composition), but different structures; may also have different properties, including water solubility, biodegradability, and toxicity

Jet Fuel: fuel meeting the required properties for use in jet engines and aircraft turbine engines

Kaolinite: a clay mineral formed by hydrothermal activity at the time of rock formation or by chemical weathering of rocks with high feldspar content; usually associated with intrusive granite rocks with high feldspar content

Kerogen: a complex carbonaceous (organic) material that occurs in sedimentary rocks and shale formations; generally insoluble in common organic solvents

Kerosene (kerosine): a fraction of petroleum that was initially sought as an illuminant in lamps; a precursor to diesel fuel; a light petroleum distillate that is used in space heaters, cook stoves, and water heaters and is suitable for use as a light source when burned in wick-fed lamps. Kerosene has a maximum distillation temperature of 400°F at the 10% recovery point, a final boiling point of 572°F, and a minimum flash point of 100°F. Included are No. 1-K and No. 2-K, the two grades recognized by ASTM Specification D 3699, as well as all other grades of kerosene called range or stove oil, which have properties similar to those of No. 1 fuel oil.

Kerosene-Type Jet Fuel: a kerosene-based product having a maximum distillation temperature of 400°F at the 10% recovery point and a final maximum boiling point of 572°F meeting ASTM Specification D 1655 and Military Specifications MIL-T-5624P and MIL-T-83133D (Grades JP-5 and JP-8). It is used for commercial and military turbojet and turboprop aircraft engines.

Kinematic Viscosity: the ratio of viscosity (q.v.) to density, both measured at the same temperature

Knock: the noise associated with self-ignition of a portion of the fuel–air mixture ahead of the advancing flames front

K_{ow}: the partition coefficient describing the equilibrium concentration ratio of a dissolved chemical in octanol versus in water, in a two-phase system at a specific temperature; used in prediction of toxicity

Lewis Acid: a chemical species that can accept an electron pair from a base

Lewis Base: a chemical species that can donate an electron pair

Light Crude Oil: crude oil with a high proportion of light hydrocarbon fractions and low metallic compounds; sometimes defined as crude oil, with a gravity of 28° API or higher; a high-quality light crude oil might have a gravity of approaching 40° API, such as light Arabian crude oil (32–34° API) and West Texas Intermediate crude oil (37–40° API)

Light Ends: the lower-boiling components of a mixture of hydrocarbons; see also heavy ends, light hydrocarbons

Light Gas Oil: liquid petroleum distillates that are higher boiling than naphtha

Light Hydrocarbons: hydrocarbons with molecular weights less than that of heptane (C7H16)

Light Oil: the products distilled or processed from crude oil up to, but not including, the first lubricating oil distillate

Light Petroleum: petroleum having an API gravity >20°

Liquefied Natural Gas (LNG): natural gas cooled to a liquid state

Liquefied Petroleum Gas: propane, butane, or mixtures thereof, gaseous at atmospheric temperature and pressure, held in the liquid state by pressure to facilitate storage, transport, and handling

Liquefied Refinery Gases (LRG): hydrocarbon gas liquids produced in refineries from processing of crude oil and unfinished oils. They are retained in the liquid state through pressurization and/or refrigeration;

includes ethane, propane, n-butane, isobutane, and refinery olefins (ethylene, propylene, butylene, and isobutylene).

Liquid Fuels: products of petroleum refining, natural gas liquids, biofuels, and liquids derived from other sources (including coal-to-liquids and gas-to-liquids); liquefied natural gas and liquid hydrogen are not included

Low-Boiling Distillates: fractions of petroleum that can be distilled at atmospheric pressure without decomposition

Low Molecular Weight: relative terms referring to the molecular mass of chemicals; in oil,mono-aromatics and aliphatics up to C_{10} would be typical of these compounds

Low-Sulfur Petroleum: a general expression for petroleum having <1% wt. sulfur; this is a very approximate definition and should not be construed as having a high degree of accuracy because it does not take into consideration the molecular locale of the sulfur. All else being equal, there is little difference between petroleum having 0.99% wt. sulfur and petroleum having 1.01% wt. sulfur

Lube: see lubricating oil

Lubricants: substances used to reduce friction between bearing surfaces or incorporated into other materials used as processing aids in the manufacture of other products or used as carriers of other materials. Petroleum lubricants may be produced either from distillates or residues; includes all grades of lubricating oils, from spindle oil to cylinder oil to those used in grease.

Lubricating Oil: a fluid lubricant used to reduce friction between bearing surfaces

Maltenes: the fraction of petroleum that is soluble in, for example, pentane or heptane; deasphaltened oil (q.v.); also, the term arbitrarily assigned to the pentane-soluble portion of petroleum that is relatively high boiling (>300°C, 760 mm) (see also petrolenes)

Mass Spectrometry: an analytical method used for detailed characterization of petroleum components, often in combination with GC, hence GC-MS

Marine Engine Oil: oil used as a crankcase oil in marine engines

Marine Gasoline: fuel for motors in marine service

Medium Crude Oil: crude oil with gravity between (approximately) 20° and 28° API

Medium Molecular Weight: relative terms referring to the molecular mass of chemicals; in oil, 3 to 6-ringed PAH and aliphatics up to C_{20} would be typical of MMW compounds

Methanol: see methyl alcohol

MEOR: microbial enhanced oil recovery

Mercaptans: odiferous organic sulfur compounds with the general formula R-SH

Metagenesis: the alteration of organic matter during the formation of petroleum that may involve temperatures above 200°C (390°F); see also catagenesis and diagenesis

Methanol (methyl alcohol, CH_3OH): a low-boiling alcohol eligible for gasoline blending

Methyl Alcohol (methanol; wood alcohol): a colorless, volatile, inflammable, and poisonous alcohol (CH3OH) traditionally formed by destructive distillation (q.v.) of wood or, more recently, as a result of synthetic distillation in chemical plants

Methyl T-Butyl Ether: an ether added to gasoline to improve its octane rating and to decrease gaseous emissions; see oxygenate

Microemulsion (micellar/emulsion) Flooding: an augmented water flooding technique in which a surfactant system is injected in order to enhance oil displacement toward producing wells

Mid-Boiling Point: the temperature at which approximately 50% of a material has distilled under specific conditions

Middle Distillate: distillate boiling between the kerosene and lubricating oil fractions; a general classification of refined petroleum products that includes distillate fuel oil and kerosene

Mineralization: complete oxidation of a compound (e.g., hydrocarbon) to carbon dioxide and water; may be accomplished by a single species of organism or by a community of microbes

Mitigation: identification, evaluation, and cessation of potential impacts of a process product or by-product

Mode of Action (MoA): describes a functional or anatomical change at the cellular level, resulting from the exposure of a living organism to a substance

Molecular Weight: the mass of one molecule

Mono-Aromatics: aromatic hydrocarbons having only a single benzene ring; may also have one or more alkyl side chains

MTBE [Methyl Tertiary Butyl Ether, $(CH_3)_3COCH_3$]: an ether intended for gasoline blending; see methyl t-butyl ether, oxygenates

Naphtha: a generic term applied to refined, partly refined, or unrefined petroleum products and liquid products of natural gas, the majority of which distills below 240°C (464°F); the volatile fraction of petroleum that is used as a solvent or as a precursor to gasoline

Naphtha-Type Jet Fuel: a fuel in the heavy naphtha boiling range having an average gravity of 52.8° API, 20% to 90% distillation temperatures of 290–470°F, and meeting Military Specification MIL-T-5624L (Grade JP-4); primarily used for military turbojet and turboprop aircraft engines because it has a lower freeze point than other aviation fuels and meets engine requirements at high altitudes and speeds

Naphthenes: cycloparaffins; one of three basic hydrocarbon classifications found naturally in crude oil; used widely as petrochemical feedstock

Natural Gas: the naturally occurring gaseous constituents that are found in many petroleum reservoirs; also, there are reservoirs in which natural gas may be the sole occupant

Natural Gas Liquids (NGL): the hydrocarbon liquids that condense during the processing of hydrocarbon gases that are produced from an oil or gas reservoir; see also natural gasoline

Natural Gasoline: a mixture of liquid hydrocarbons extracted from natural gas (q.v.) suitable for blending with refinery gasoline

Neutralization: a process for reducing the acidity or alkalinity of a waste stream by mixing acids and bases to produce a neutral solution; also known as pH adjustment

No. 1 Fuel Oil: very similar to kerosene (q.v.) and is used in burners where vaporization before burning is usually required and a clean flame is specified

No. 2 Diesel Fuel: a distillate fuel oil that has a distillation temperature of 640°F at the 90% recovery point and meets the specifications defined in ASTM Specification D 975. It is used in highspeed diesel engines that are generally operated under uniform speed and load conditions, such as those in railroad locomotives, trucks, and automobiles.

No. 2 Fuel Oil: also called domestic heating oil; has properties similar to diesel fuel and heavy jet fuel; used in burners where complete vaporization is not required before burning

No. 4 Fuel Oil: a light industrial heating oil; is used where preheating is not required for handling or burning; there are two grades of No. 4 fuel oil, differing in safety (flash point) and flow (viscosity) properties

No. 5 Fuel Oil: a heavy industrial fuel oil that requires preheating before burning

No. 6 Fuel Oil: a heavy fuel oil; more commonly known as Bunker C oil when it is used to fuel ocean-going vessels; preheating is always required for burning this oil

Octane Number: a number indicating the antiknock characteristics of gasoline

Octane Rating: a number used to indicate gasoline's antiknock performance in motor vehicle engines. The two recognized laboratory engine test methods for determining the antiknock rating, i.e. octane rating, of gasolines are the research method and the motor method. To provide a single number as guidance to the consumer, the antiknock index (R + M)/2 is the average of the research and motor octane numbers; see octane number

Oil Shale: a fine-grained impervious sedimentary rock that contains an organic material called kerogen; the term oil shale describes the rock in lithological terms, but also refers to the ability of the rock to yield oil upon heating which causes the kerogen to decompose; also called black shale, bituminous shale, carbonaceous shale, coaly shale, kerosene shale, coorongite, maharahu, kukersite, kerogen shale, and algal shale.

Olefins: a class of unsaturated double-bond linear hydrocarbons recovered from petroleum; examples include ethylene, propylene, and butene. Olefins are used to produce a variety of products, including plastics, fibers, and rubber.

Organic Sedimentary Rocks: rocks containing organic material such as residues of plant and animal remains/decay

Paraffins: a group of generally saturated single-bond linear hydrocarbons; also celled alkanes

Partitioning: the diffusion of compounds between two immiscible liquid phases, including water and oil droplets and water and lipid membranes

Petrol: a term commonly used in some countries for gasoline

Petroleum (crude oil): a naturally occurring mixture of gaseous, liquid, and solid hydrocarbon compounds usually found trapped deep underground beneath impermeable cap rock and above a lower dome of sedimentary rock such as shale; most petroleum reservoirs occur in sedimentary rocks of marine, deltaic, or estuarine origin

Petroleum Coke: a solid carbon fuel derived from oil refinery cracking processes such as delayed coking; also called pet coke

Petroleum Products: products obtained from the processing of crude oil (including lease condensate), natural gas, and other hydrocarbon compounds. Petroleum products include unfinished oils, liquefied petroleum gases, pentanes plus, aviation gasoline, motor gasoline, naphtha-type jet fuel, kerosene-type jet fuel, kerosene, distillate fuel oil, residual fuel oil, petrochemical feedstocks, special naphtha, lubricants, waxes, petroleum coke, asphalt, road oil, still gas, and miscellaneous products.

Petroleum Refining: a complex sequence of events that result in the production of a variety of products

Phase Separation: the formation of a separate phase that is usually the prelude to coke formation during a thermal process; the formation of a separate phase as a result of the instability/incompatibility of petroleum and petroleum products

pH Adjustment: neutralization

Photo-Enhanced Toxicity: increased toxicity due to photo-oxidation in vivo

Photo-Oxidation: oxidation due to the influence of photic energy, usually from UV light

Pollution: the introduction into the land, water, and air systems of a chemical or chemicals that are not indigenous to these systems or the introduction into the land, water, and air systems of indigenous chemicals in greater-than-natural amounts

Polycyclic Aromatic Hydrocarbons: (PAHs) a subclass of aromatic hydrocarbons having two or more fused benzene rings; may also have one or more alkyl side chains, generating large suites of isomers; some are considered 'priority pollutants' because of their toxicity and/or potential carcinogenicity

Pore Diameter: the average pore size of a solid material, e.g. catalyst

Pore Space: a small hole in reservoir rock that contains fluid or fluids; a 4-in cube of reservoir rock may contain millions of interconnected pore spaces

Pore Volume: total volume of all pores and fractures in a reservoir or part of a reservoir; also applied to catalyst samples

Porosity: the percentage of rock volume available to contain water or other fluid

Porphyrins: organometallic constituents of petroleum that contain vanadium or nickel; the degradation products of chlorophyll derivatives that became included in the protopetroleum

Pour Point: the lowest temperature at which oil will pour or flow when it is chilled without disturbance under definite conditions

Propane (C_3H_8): a straight-chain saturated (paraffinic) hydrocarbon extracted from natural gas or refinery gas streams, which is gaseous at standard temperature and pressure; a colorless gas that boils at a temperature of −42°C (−44°F) and includes all products designated in ASTM D1835 and Gas Processors Association specifications for commercial (HD-5) propane

Propylene (C_3H_6): an olefin hydrocarbon recovered from refinery or petrochemical processes which is gaseous at standard temperature and pressure; an important petrochemical feedstock

Protopetroleum: a generic term used to indicate the initial product formed as a result of chemical and physical changes that have occurred to the precursors of petroleum; a paleobotanical soup

Pyrolysis: exposure of a feedstock to high temperatures in an oxygen-poor environment

Ramsbottom Carbon Residue: see carbon residue

Reactor: a vessel in which a reaction occurs during processing; usually defined by the nature of the catalyst bed, e.g. fixed-bed reactor, fluid-bed reactor, and by the direction of the flow of feedstock, e.g. upflow, downflow

Recycling: the use or reuse of chemical waste as an effective substitute for commercial products or as an ingredient or feedstock in an industrial process

Refinery: a series of integrated unit processes by which petroleum can be converted to a slate of useful (salable) products

Refining: the process(es) by which petroleum is distilled and/or converted by application of a physical and chemical process to form a variety of products

Reforming: the conversion of hydrocarbons with low octane numbers (q.v.) into hydrocarbons having higher octane numbers, for example, the conversion of a n-paraffin into an iso-paraffin

Regeneration: the reactivation of a catalyst by burning off the coke deposits

Regenerator: a reactor for catalyst reactivation

Renewable Energy Sources: solar, wind, and other non-fossil fuel energy sources

Reserves: well-identified resources that can be profitably extracted and utilized with existing technology

Reservoir: a domain where a pollutant may reside for an indeterminate time resid: the heaviest boiling fraction remaining after initial processing (distillation) of crude oil; see residuum.

Residual Fuel Oil: obtained by blending the residual product(s) from various refining processes with suitable diluent(s) (usually middle distillates) to obtain the required fuel oil grades; a general classification for the heavier oils, known as No. 5 and No. 6 fuel oils, that remain after the distillate fuel oils and lighter hydrocarbons are distilled away in refinery operations. It conforms to ASTM Specifications D396 and D975 and Federal Specification VV-F-815C. No. 5, a residual fuel oil of medium viscosity, is also known as Navy Special and is defined in Military Specification MIL-F-859E, including Amendment 2 (NATO Symbol F-770). It is used in steam-powered vessels in government service and inshore power plants. No. 6 fuel oil includes Bunker C fuel oil and is used for the production of electric power, space heating, vessel bunkering, and various industrial purposes.

Residual Oil: see residuum

Residuals: heavy fuel oils produced from the nonvolatile residue from the fractional distillation process; also called resids

Residuum (resid; pl. residua): the residue obtained from petroleum after nondestructive distillation (q.v.) has removed all the volatile materials from crude oil, e.g. an atmospheric (345°C, 650°F+) residuum

Resins: the portion of the maltenes (q.v.) that is adsorbed by a surface-active material such as clay or alumina; the fraction of deasphaltened oil that is insoluble in liquid propane but, soluble in n-heptane

Resource: the total amount of a commodity (usually a mineral, but can include nonminerals such as water and petroleum) that has been estimated to be ultimately available

Riser: the part of the bubble-plate assembly that channels the vapor and causes it to flow downward to escape through the liquid; also, the vertical pipe where fluid catalytic cracking reactions occur

Rock Asphalt: bitumen that occurs in formations that have a limiting ratio of bitumen-to-rock matrix

SARA Separation: a method of fractionation by which petroleum is separated into saturates, aromatics, resins, and asphaltene fractions

Saturates: paraffins and cycloparaffins (naphthenes). They are a class of hydrocarbons that may be straight-chain, branched-chain, or cyclic, in which all carbon atoms have single bonds to either carbon or hydrogen.

Sediment: an insoluble solid formed as a result of the storage instability and/or the thermal instability of petroleum and petroleum products

Separation Process: an upgrading process in which the constituents of petroleum are separated, usually without thermal decomposition, e.g. distillation and deasphalting

Shale Oil: also known as 'tight oil' (not to be confused with 'oil shale'); liquid petroleum that is produced from shale oil reservoirs, typically by hydraulic fracturing methodsSludge: a semisolid to solid product that results from the storage instability and/or the thermal instability of petroleum and petroleum products.

Slurry Hydroconversion Process: a process in which the feedstock is contacted with hydrogen under pressure in the presence of a catalytic coke-inhibiting additive

Solvent Extraction: a process for separating liquids by mixing the stream with a solvent that is immiscible with part of the waste, but that will extract certain components of the waste stream

Sour Crude Oil: crude oil containing an abnormally large amount of sulfur compounds (a >1% total sulfur content); see also sweet crude oil; crude oil containing free sulfur, hydrogen sulfide, or other sulfur compounds

Specific Gravity: the mass (or weight) of a unit volume of any substance at a specified temperature compared with the mass of an equal volume of pure water at a standard temperature; see also density

Spent Catalyst: catalyst that has lost much of its activity due to the deposition of coke and metals

Steam Cracking: a conversion process in which the feedstock is treated with superheated steam; a petrochemical process sometimes used in refineries to produce olefins (e.g. ethylene) from various feedstock for petrochemicals manufacture; the feedstock ranges from ethane to vacuum gas oil, with heavier feeds giving higher yields of by-products, such as naphtha. The most common feedstocks are ethane, butane, and naphtha and the process is carried out at temperatures of 815–870°C (1500–1600°F) and at pressures slightly above atmospheric pressure. Naphtha produced from steam cracking contains benzene which is extracted before hydrotreating and high-boiling products (residua) from steam cracking are sometimes used as blend stock for heavy fuel oil.

Steam Distillation: distillation in which vaporization of the volatile constituents is affected at a lower temperature by introduction of steam (open steam) directly into the charge

Storage Stability (storage instability): the ability (inability) of a liquid to remain in storage over extended periods of time without appreciable deterioration as measured by gum formation and the depositions of insoluble material (sediment)

Sulfonic Acids: acids obtained by the reaction of petroleum or a petroleum product with strong sulfuric acid

Sulfur: a yellowish nonmetallic element sometimes known by the Biblical name of brimstone; present at various levels of concentration in many fossil fuels whose combustion releases sulfur compounds that are considered harmful to the environment. Some of the most commonly used fossil fuels are categorized according to their sulfur content, with lower sulfur fuels usually selling at a higher price.

Sulfur Recovery Unit (SRU): a refinery process unit used to convert hydrogen sulfide to elemental sulfur using the Claus process

Sweet Crude: petroleum with low total sulfur content, variously defined as <0.5% or <1% sulfur

Sweetening: the process by which petroleum products are improved in odor and color by oxidizing or removing the sulfur-containing and unsaturated compounds

Synthetic Crude Oil (syncrude): a hydrocarbon product produced by the conversion of coal, oil shale, or tar sand bitumen that resembles conventional crude oil; can be refined in a petroleum refinery (q.v.)

Tail Gas: the lightest hydrocarbon gas released from a refining process

Tail Gas Treating Unit (TGTU): a refinery process unit used to control emissions of sulfur compounds; generally integrated with a sulfur recovery unit

Tar: the volatile, brown to black, oily, viscous product from the destructive distillation (q.v.) of many bituminous or other organic materials, especially coal; a name used for petroleum in ancient texts

Tar Sand: see bituminous sand

Thermal Cracking: a process that decomposes, rearranges, or combines hydrocarbon molecules by the application of heat without the aid of catalysts

Thermal Process: any refining process that utilizes heat without the aid of a catalyst

Thermal Stability (thermal instability): the ability (inability) of a liquid to withstand relatively high temperatures for short periods of time without the formation of carbonaceous deposits (sediment or coke)

Toluene ($C_6H_5CH_3$): a colorless liquid of the aromatic group of petroleum hydrocarbons made by the catalytic reforming of petroleum naphtha containing methyl cyclohexane; a high octane, gasoline-blending agent, solvent, and chemical intermediate, and a base for TNT (explosive)

Topped Crude: petroleum that has had volatile constituents removed up to a certain temperature, e.g. 250°C+ (480°F+) topped crude; not always the same as a residuum (q.v.)

Total Acid Number (TAN): a measure of the acidity determined by the amount of potassium hydroxide in milligrams that is needed to neutralize the acids (typically naphthenic acids) in one gram of oil; used by refineries as an indicator of potential corrosion and scale production

Total Petroleum Hydrocarbons (TPHs): the total mass of all hydrocarbons in an oil or environmental sample, including the volatile and extractable (non-volatile) hydrocarbons; may be further defined by stating the analytical method used, e.g. GC-detectable TPH or TPH-F (TPH measured by fluorescence), which vary in their rigor

Total Polycyclic Aromatic Hydrocarbons (TPAHs): including alkyl-PAHs and parent (unsubstituted) PAHs; the sum of all concentrations of PAHs measured by GC-MS

Tower: equipment for increasing the degree of separation obtained during the distillation of oil in a still

Trace Element: those elements that occur at very low levels in a given system

Treatment: any method, technique, or process that changes the physical and/or chemical character of petroleum

Trickle Hydrodesulfurization: a fixed-bed process for desulfurizing middle distillates

Trillion: 1×10^{12}

True Boiling Point (true boiling range): the boiling point (boiling range) of a crude oil fraction or a crude oil product under standard conditions of temperature and pressure

Unconventional Crude Oils: petroleum that does not flow readily in the reservoir and/or must be produced by using unconventional methods, such as surface mining of shallow bitumen deposits, steam-assisted gravity drainage (SAGD) for in situ extraction of deep bitumen deposits, cyclic steam injection for heavy oils, or horizontal drilling with hydraulic fracturing for recovery of light

shale oilsUnfinished Oils: all oils requiring further processing, except those requiring only mechanical blending. Unfinished oils are produced by partial refining of crude oil and include naphtha, kerosene, light gas oils, heavy gas oils, and residuum.

Upgrading: the conversion of petroleum to value-added salable products; includes hydroprocessing, hydrocracking, fractionation, and any other catalytic or noncatalytic processes that improve the value of the products. During upgrading, the products of the Fischer-Tropsch process are converted to diesel, jet fuel, naphtha, or bases for synthetic lubricants and wax.

Upstream: a sector of the petroleum industry referring to the searching for, recovery, and production of crude oil and natural gas; also known as the exploration and production sector

Unresolved Complex Mixture (UCM): petroleum constituents that are not resolved by conventional GC and appear as a 'hump' in the gas chromatogram; comprises many hundreds or thousands of unresolved isomers

Vacuum Distillation: distillation (q.v.) under reduced pressure; distillation under reduced pressure (less than atmospheric) which lowers the boiling temperature of the liquid being distilled. This technique, with its relatively low temperatures, prevents cracking or decomposition of the charge stock.

Vacuum Gas Oil: a product of vacuum distillation; a preferred feedstock for cracking units to produce gasoline; abbreviated as VGO

Vacuum Residuum: a residuum (q.v.) obtained by distillation of a crude oil under a vacuum (reduced pressure); that portion of petroleum that boils above a selected temperature, such as 510°C (950°F) or 565°C (1050°F)

VI: see viscosity index

Visbreaking: a (relatively) mild process for reducing the viscosity of heavy feedstocks by controlled thermal decomposition; a process designed to reduce residue viscosity by thermal means, but without appreciable coke formation

Viscosity: a measure of the ability of a liquid to flow or a measure of its resistance to flow; the force required to move a plane surface of 1 m^2 area over another parallel plane surface 1 m away at a rate of 1 m/s when both surfaces are immersed in the fluid

Viscosity Index (VI): an arbitrary scale used to show the magnitude of viscosity changes in lubricating oils with changes in temperature

Volatile Organic Compounds (VOCs): chemicals having high vapor pressure at room temperature (and corresponding low boiling point) that tend to evaporate or sublimate into the air; for example, BTEX

Water-Accommodated Fraction of Oil (WAF): hydrocarbons that will partition from oil to water during gentle stirring or mixing; may contain droplets, in contrast to water-soluble fractions (WSF)

Water-Soluble Fraction of Oil (WSF): aqueous solution of hydrocarbons that partition from oil; does not include droplet or particulate oil; see also CEWAF and HEWAF.

Weathering: a suite of changes in spilled oil composition and properties brought about by a variety of environmental processes including spreading, evaporation, photo-oxidation, dissolution, emulsification and biodegradation, among others

Wet Gas: gas containing a relatively high proportion of hydrocarbons that are recoverable as liquids; see also lean gas

Xylene [$C_6H_4(CH_3)_2$]: a colorless liquid of the aromatic group of hydrocarbons made from the catalytic reforming of certain naphthenic petroleum fractions; used as high-octane motor and aviation gasoline blending agents, solvents, and chemical intermediates; isomers are ortho-xylene (o-xylene), meta-xylene (m-xylene), and para-xylene (p-xylene)

Zeolite: a crystalline aluminosilicate used as a catalyst and having a particular chemical and physical structure

Index

CPSIA information can be obtained
at www.ICGtesting.com
Printed in the USA
BVHW04*0430101018
529002BV00039BA/138/P

9 781119 223580